The Wiley Blackwell Companion to Tourism

Wiley Blackwell Companions to Geography

Wiley Blackwell Companions to Geography is a blue-chip, comprehensive series covering each major subdiscipline of human geography in detail. Edited and contributed by the disciplines' leading authorities each book provides the most up to date and authoritative syntheses available in its field. The overviews provided in each *Companion* will be an indispensable introduction to the field for students of all levels, while the cutting-edge, critical direction will engage students, teachers, and practitioners alike.

Published

The Wiley Blackwell Companion to Tourism
Edited by C. Michael Hall

A Companion to Feminist Geography
Edited by Lise Nelson and Joni Seager

A Companion to Environmental Geography
Edited by Noel Castree, David Demeritt, Diana Liverman, and Bruce Rhoads

A Companion to Health and Medical Geography
Edited by Tim Brown, Sara McLafferty, and Graham Moon

A Companion to Social Geography
Edited by Vincent J. Del Casino Jr., Mary Thomas, Ruth Panelli, and Paul Cloke

The Wiley-Blackwell Companion to Human Geography
Edited by John A. Agnew and James S. Duncan

The Wiley-Blackwell Companion to Economic Geography
Edited by Eric Sheppard, Trevor J. Barnes, and Jamie Peck

The Wiley-Blackwell Companion to Cultural Geography
Edited by Nuala C. Johnson, Richard H. Schein, and Jamie Winders

The Wiley Blackwell Companion to Tourism
Edited by Alan A. Lew, C. Michael Hall, and Allan M. Williams

The Wiley Blackwell Companion to Political Geography
Edited by John Agnew, Virginie Mamadouh, Anna J. Secor, and Joanne Sharp

Also available:

The New Blackwell Companion to the City
Edited by Gary Bridge and Sophie Watson

The Blackwell Companion to Globalization
Edited by George Ritzer

The Handbook of Geographic Information Science
Edited by John Wilson and Stewart Fotheringham

The Wiley Blackwell Companion to Tourism

Second Edition

Edited by

C. Michael Hall

WILEY Blackwell

This edition first published 2024

Edition History
[John Wiley & Sons, Ltd., 1e, 2014]

Registered Offices
John Wiley & Sons, Inc., 111 River Street, Hoboken, NJ 07030, USA
John Wiley & Sons Ltd, The Atrium, Southern Gate, Chichester, West Sussex, PO19 8SQ, UK

For details of our global editorial offices, customer services, and more information about Wiley products visit us at www.wiley.com.

Wiley also publishes its books in a variety of electronic formats and by print-on-demand. Some content that appears in standard print versions of this book may not be available in other formats.

Library of Congress Cataloging-in-Publication Data Applied for:
Hardback ISBN: 9781119753742
ePDF ISBN: 9781119753766
epub ISBN: 9781119753780
oBook ISBN: 9781119753797

Cover Design: Wiley
Cover Image: © Victor Porof/wikimedia

Set in 9/12.5pt SabonLTStd by Straive, Pondicherry, India
Printed and bound by CPI Group (UK) Ltd, Croydon, CR0 4YY

C9781119753742_030724

To Alan, Allan and good companions

Contents

List of Figures

List of Tables

Notes on Contributors

Bailey Ashton Adie is a Postdoctoral Researcher in Spatial Resilience (part of the FRONT research programme) in the Geography Research Unit at the University of Oulu, Finland. Her research interests include community resilience, second homes, natural hazards and tourism, community-based tourism, World Heritage tourism, tourism and development, dark tourism, film tourism, and heritage tourism. She is the author of *World Heritage and Tourism: Marketing and Management* (Routledge, 2019) and co-editor of *Second Homes and Climate Change* (Routledge, 2023).

Alberto Amore is Assistant Professor in Tourism Geography at the University of Oulu, Finland. His research interests include urban tourism, destination governance, tourism and post-disaster governance, destination resilience, regenerative tourism and tourism and biodiversity. He is author of *Tourism and Urban Regeneration* (Routledge, 2019) and co-author of *Tourism and Resilience* (Channel view, 2018). He is Associate Editor of *Tourism Management Perspectives* and Social Media Editor for *Current Issues in Tourism*.

Richard S. Aquino is a Senior Lecturer in the Business School, University of Canterbury Christchurch, New Zealand. His research interests include tourism for sustainable development, social entrepreneurship, community development, tourism impact assessment, tourist behaviour, and decolonising tourism knowledge production.

Inoormaziah Azman is a senior lecturer at the Universiti Teknologi MARA, Malaysia, serving the Tourism Department in the Faculty of Hotel and Tourism Management. She teaches undergraduate courses in the Tourism Management area, including Principles of Tourism, Fundamentals of Hospitality and Tourism Industry, Tourism Planning and Development, Tourist Behaviour, Event Management, Park and Recreation; and supervises undergraduate and postgraduate student dissertations.

Bernhard F. Bichler is a Lecturer FHWien der WKW (University of Applied Sciences for Management & Communication), Austria, and Referent Forschung, Entwicklung und Unternehmensservice, Bundesministerium für Land- und Forstwirtschaft, Regionen und Wasserwirtschaft. He focus on current issues in tourism research, ranging from customers to companies and destinations.

Gareth Butler is a Senior Lecturer in Tourism at Flinders University, Australia. He is also a Senior Research Affiliate at the University of Johannesburg, South Africa.

Ning (Chris) Chen is a Senior Lecturer in the Department of Management, Marketing and Tourism at the University of Canterbury, Christchurch, New Zealand. His primary areas of interest include identity and attachment theory in different context including place, branding, and sports, and the motivations and impact of word-of-mouth behaviours especially in social media. His research focus include place attachment, brand attachment, fandom, word-of-mouth (mouse), and personal connection.

Prem Chhetri is Professor at RMIT University in Australia. Prem obtained a Ph.D. in Geospatial Science from RMIT University in 2003. He is known internationally for the research in spatially-integrated supply chain analytics and urban logistics. His recent research focused on port logistics, climate change, urban modelling, tourism potential mapping, emergency response, skills and training, and the application of GIS and GPS in transport, infrastructure and logistics planning.

Tim Coles is a Professor in the Department of Management at the University of Exeter Business School in the United Kingdom. His research interests are in the sustainable development of tourism, and he has a longstanding interest in the social responsibility of organisations and how this is studied.

Edward Commons is a former graduate student of the Business School, University of Canterbury, Christchurch, New Zealand.

Cecilia de Bernardi is a lecturer in sustainable destination development at Uppsala University, Sweden, and a postdoctoral student at Mid Sweden University. She has a PhD in social sciences with a focus on tourism and previously studied communication science, political science, Nordic culture and history and cultural as well as natural heritage in tourism. She also focuses on qualitative methods.

B. Bynum Boley is Associate Professor, Parks, Recreation and Tourism Management, University of Georgia, United States, and Director of the university's Tourism Research Lab. His research focuses on sustainable tourism development and how the unique natural and cultural resources of communities can be protected, packaged and marketed to jointly increase sustainability, resident quality of life, and a community's competitiveness as a tourism destination.

Hervé Corvellec is a Professor of Business Adminustration in the Department of Service Stucies, Lund University, Campus Helsingborg, Sweden. His fields of research and teaching are sustainability and organization theory, and he has been working for a long time with infrastructure issues (e.g., public library development, railroad planning, wind power siting, risk in public transportation, and waste management).

Larry Dwyer is a Professor in the Business School, University of Technology, Sydney, New South Wales, Australia. He is a former President of the International Academy for the Study of Tourism and former President of the International Association for Tourism Economics. His research interests include economics of tourism, tourism management and policy.

David Dyason is a Senior Lecturer in Property Studies at Lincoln University, New Zealand, with a focus on land economics and property market analysis. He is an economist with an interest in property, economics, and impact assessment. His research interest and publications are in property, economics, geography, and planning and he continues to be involved as an advisor to the private and public sectors in economic and land-use planning. Through his research, he aims to reveal the causal relationship between property and economics and how it influences economic development in the urban and rural landscape.

Eke Eijgelaar is a Researcher, Academy for Tourism, Centre for Sustainability, Tourism and Transport (CSTT), Breda University of Applied Sciences in the Netherlands. He specializes in the environmental impacts of tourism, with a focus on tourism energy use and emissions. This takes him to many related topics, such as high and low carbon forms of tourism (e.g., cruise tourism), and mitigation strategies and tools such as carbon management and carbon calculators, and more recently also to topics such as overtourism and aviation politics.

Peter Fieger is a Senior Lecturer in the School of Education at Federation University, Australia and Adjunct Associate Professor at the University of New England. Previously he was a Senior Research Fellow at the National Centre for Vocational Education Research (NCVER) in Adelaide, where his main areas of research were in the quantitative analysis of surveys and administrative data collections in vocational education and its social, economic and labour market outcomes; and the Senior Economist at ChristchurchNZ, in Christchurch, New Zealand. His research focus also includes the service, hospitality and tourism industries and their role in the economy.

David A. Fennell is a Professor, Geography and Tourism Studies, Faculty of Social Sciences, Brock University, St. Catherines, Ontario, Canada, and is also the Editor-in-chief of the *Journal of Ecotourism*. His research is focused on five main themes: ecotourism, tourism ethics, moral issues tied to the use of animals in tourism, sustainable tourism, and fly-fishing.

Viachaslau Filimonau is Reader in Hospitality Innovation, Centre for Sustainability and Wellbeing in the Visitor Economy, School of Hospitality and Tourism Management, University of Surrey, UK and the Hotelschool The Hague, Den Haag, the Netherlands. His research interests include sustainable mobilities, water and carbon footprint management in tourism and hospitality operations, and environmental management in tourism and hospitality enterprises.

Jamie Gillen is Director and Associate Professor of the Global Studies Programme at the Faculty of Arts, Cultures, Languages and Linguistics, University of Auckland, New Zealand. His research interests centre on societies and everyday life in mainland Southeast Asia, particularly Vietnam. He has written extensively about tourism, with a concentration on the politics of tourism, and maintains active research on urbanization, rural-urban relations, agrarian change, and culture.

Sandhiya Goolaup is an Assistant Professor at the Department of Marketing and Tourism, School of Business and Economics, University of Linnaeus, Sweden. Her research interests include phenomenology, hermeneutic phenomenology, hermeneutics hospitality and tourism, food consumption, consumer experience, consumer behavior, tourist experience, value creation, tourism research, extraordinary experiences, and destination marketing.

Martin G. Gren is a Professor in Human Geography in the Department of Cultural Sciences, Linnaeus University, Kalmar, Sweden, and is a member of the Center for Climate Emergency Studies (Centre CES) at Linnaeus University, which is a transdisciplinary platform for research, education and action related to the climate emergency. He teaches human geography and his research interest is focused on the new earthly climate regime and the Anthropocene.

Ulrike Gretzel is senior research fellow at the Annenberg Center for Public Relations, University of Southern California and the Director of Research at Netnografica. Her research focuses on persuasion in human-technology interactions with an emphasis on credibility perceptions, information search and processing, electronic decision aids, smart technologies, online and social media marketing, adoption and use of technologies, as well as non-adoption and digital detox. Her research has explored diverse topics such as the detection of fake TripAdvisor reviews, the creation of persuasive interfaces for mobile apps, the phenomenon of selfie-taking, and humor co-creation in Chinese social media conversations.

Dikte Grønvold is the producer and host of On The Green Track, a think tank currently based in Copenhagen, exploring the intersectionality of travel by diving deeper into its complexities with the help of podcasts, deep destination analysis, and nclusive digital events. She is also a Project Manager for the German National Tourist Board in Copenhagen, Denmark.

C. Michael Hall is Ahurei Professor at University of Canterbury, New Zealand; Eminent Scholar, Kyung Hee University; Guest Professor, Department of Service Studies, Lund University Campus Helsingborg, Sweden; Visiting Professor, Linnaeus University, Kalmar, Sweden; Visiting Professor, Taylor's University, Malaysia; and Visiting Professor and Docent in Geography, University of Oulu, Finland. Co-editor of *Current Issues in Tourism* and Field Editor of *Frontiers in Sustainable Tourism,* his research interests include tourism, regional development, global environmental change, food, sustainability, and World Heritage.

Kevin Hannam is a Visiting Senior Research Fellow at the University of Johannesburg, South Africa. He is a founding co-editor of the journals Mobilities and Applied Mobilities with Routledge. He is the co-author of the books *Understanding Tourism* and *Tourism and India.*

Fahimeh Hateftabar holds a PhD in Tourism Marketing Management from Pantheon-Sorbonne University in France and is currently an Assistant Professor in Marketing and Tourism at FERRANDI Paris, France. With a strong mathematical background, Fahimeh is an expert in quantitative research approaches and possesses extensive experience in consumer behavior, tourism economics, and sustainable development.

Edward H. Huijbens is professor and chair of Wageningen University cultural geography research group in the Netherlands. He works on tourism theory, issues of regional development, landscape perceptions, and the role of transport in tourism and polar tourism. He co-edited *Technology in Society/Society in Technology* (2005, University of Iceland Press), *Sensi/able Spaces: Space, Art and the Environment* (2007, Cambridge Scholars Press), *The Illuminating Traveller* (2008, University of Jyväskylä), and *Tourism and the Anthropocene* (2016, Routledge); he co-authored *Icelandic Tourism* (Forlagið, 2013), and authored *Developing Earthly Attachments in the Anthropocene* (Routledge, 2021).

Johan Hultman is Professor and Research coordinator, Department of Service Studies, Lund University, Campus Helsingborg. He investigates issues of sustainable development in service societies and writes about how social and economic aspects of sustainability relate to each other, how services for sustainability can be conceptualised, how local economies can be developed and how research and political change can benefit from each other. Working with colleagues, he moves through different worlds whose common denominator is that they are embedded in a political discourse of sustainable development through economic growth: small-scale fisheries, waste management and rural development.

Dimitri Ioannides is Chaired Professor of Human Geography in the Institution of Economics, Geography, Law and Tourism and a researcher in the European Tourism Research Institute at Mid-Sweden University. Previously, he taught at Missouri State University and held a part-time position at the Centre for Regional and Tourism Research in Bornholm, Denmark. He has published extensively on issues such as the economic geography of tourism and sustainable development. He is especially interested in matters relating to the social-equity dimension of sustainability, including the geographies of tourism workers and work.

Michal Isaacson is a Lecturer, Faculty of Social Welfare and Health Sciences, Department of Gerontology, University of Haifa, Israel. She has extensive experience in research and in using data to design, develop and promote products that address health and wellbeing. She began her career analyzing data collected using GPS receivers to understand the association between mobility and health. Over the years she has broadened the types of sensors and data she has analyzed. She is an expert at combining tools that measure emotional aspects of wellbeing with natural setting data collected using sensors, wearables, video, and other digital data flows.

Jonas Larsen is a Professor of mobility and urban studies at Roskilde University. He has a long-standing interest in tourist photography, tourism, cities, and mobility more broadly. More recently, he has written extensively about urban cycling, running, sport and walking, including a book on urban marathons in 2022 (Routledge). He is currently an academic expert in a research project on urban walking concerning mobility and planning (with COWI). As part of this project, he is writing about sensuous walking practices, design interventions and the social life of pavements and public squares.

Kerryn Lyes is a doctoral student in the Business School at the University of Canterbury, New Zealand. In addition to her marketing practice, her research focuses on non-Western marketing thought, behavioural change and the history of marketing.

Junshuo Li is a Postgraduate Student at the Institute of Geographic Sciences and Natural Resources Research, Chinese Academy of Science, Beijing, China.

Aapo Lundén is a PhD Researcher at the Geography Research Unit at the University of Oulu, Finland, and a member of Tourism Geographies and Sustainable Mobilities (TOGSUM) and Transformation and Transnationalization of state spaces: Geo-politics of borders, mobilities and planning in the Arctic (SPARCTIC) research groups and an affiliate of the Biodiverse Anthropocenes (ANTS) research program. His research interests include nature-based tourism, tourism governance and regulation, wilderness studies, nature conservation, tourism development in Arctic areas, environmental policy, political ecology, and institutional theory.

M. Mau is affiliated with Health, Social Work and Welfare Research at UCL University College, Odense, Denmark. He was formerly affiliated with the Department of Psychology and the Health Sciences Research Centre at the same university. His research focuses on wellbeing and the benefits of walking.

Bob McKercher is a Professor in Tourism in the Business School at the University of Queensland, in St. Lucia, Queensland, Australia. He received his PhD from the University of Melbourne in Australia, a Master's degree from Carleton University in Ottawa, Canada and his undergraduate degree from York University in Toronto, Canada. Prof McKercher has wide ranging research interests. He is the author of *The Business of Nature-based Tourism* (1998) and co-authored *Tourism Theories, Concepts and Models*, *Cultural Tourism,* now in its third edition. He has also co-edited four other books.

Claudio Minca is Professor, Dipartimento di Storia Culture Civiltà, University of Bologna, Italy, and Honorary Professor, Macquarie School of Social Sciences, Macquarie University, Australia. He is a human geographer with strong interest in cultural and political theory. His main research projects have focused on the geographies of tourism and travel and on the spatial theories of modernity, including conceptualizations of landscape, place and power. He has also written extensively on the relationship between space and biopolitics, with a particular focus on the work of philosopher Giorgio Agamben and legal theorist Carl Schmitt. In empirical terms, his work has been on Trieste (cosmopolitanism, border thinking, geographies of absence), Morocco (postcolonial geographies of travel) and the Mediterranean more broadly.

Jan Mosedale is Professor and Head of Research at FHGR Institute for Tourism and Leisure ITF, University of Applied Sciences, Chur, Graubünden, Switzerland. Recent research projects include increasing innovation capabilities in tourism, positioning the Alpine region as globally attractive health promoting place (HEALPS), developing innovations in agritourism, and developing an online platform to support the Swiss hotel industry on the path towards digital transformation.

Dieter K. Müller is a Professor of Human Geography and Deputy Vice-chancellor of Umeå University, Sweden. His research addresses second homes and tourism in northern peripheral areas. Among his books are the co-edited *Tourism, Mobility and Second Homes* (Channel View, 2004) and the *Routledge Handbook of Second Home Tourism and Mobilities* (2018). Müller is one of the editors of the Springer book series *Geographies of Tourism and Global Change* and a resource editor for the *Scandinavian Journal of Hospitality and Tourism*. From 2012 to 2020 he was the Chairperson of the IGU Commission on the Geography of Tourism, Leisure and Global Change.

Jan-Henrik Nilsson is Assistant head of department and Associate Professor, Department of Service Studies, Lund University, Campus Helsingborg, Sweden. His main research interests include tourism geography; mobility, transport and tourism; urban tourism and sustainability; and the history of tourism and hospitality.

Robin Nunkoo is a Professor in the Department of Management, Faculty of Law and Management, University of Mauritius. His research draws widely from political science and investigates the interplay among citizens' trust (distrust) in government (and institutions), power issues in development, and political support for development and government policies. His research interests include: politics, tourism, and economic development; voting behavior; good governance; political support for development; community-based resource management; political trust and good governance; citizens' trust in government institutions and participative planning; statistical modeling; role of theory in social science research; conceptual/structural model development; structural equation modeling; multiple regression; grounded theory; and meta analysis.

Tim Oakes is Professor of Geography at University of Colorado, Boulder. His work focuses on social and cultural transformation in contemporary China and, in particular, the uses and reinventions of local culture as a resource for economic development and governance objectives. He has explored this theme in the contexts of ethnic tourism and craft commodity production, cultural heritage development, and urban redevelopment and planning. His most recent research explores the development and use of leisure and consumption spaces in China's urban areas, as well as in urbanizing areas of rural China.

Paul Peeters is a Professor, Academy for Tourism, Centre for Sustainability, Tourism and Transport (CSTT), Breda University of Applied Sciences, the Netherlands. He is responsible for the CSTT and specialises in the environmental impacts of tourism and tourism transport in general and climate change mitigation in particular. He works on diverse projects like airship reintroduction, slow transport modes, cost and benefits of speed reduction, technological development of aircraft fuel efficiency, and transport accessibility for disabled people; common to almost all projects are the environment and the future.

Nhat Tan Pham is affiliated with the IInternational University, Ho Chi Minh City, Vietnam, and Vietnam National University, also in Ho Chi Minh City. His research focuses on sustainable human resource management, digital human resource management, sustainable tourism, and hospitality management.

Birgit Pikkemaat is an Assistant Professor at the Department of Strategic Management, Marketing and Tourism, University of Innsbruck, Austria. Her research interests include innovation and offer development in tourism, trends in tourist behaviour, and regional and rural destination development.

Girish Prayag is a professor in the Business School at the University of Canterbury, New Zealand. He is the author of numerous books and journal articles. The overarching focus of his research is the consumption experiences in the services industry with a particular focus on the airline, tourism and hospitality industries. His past and current research themes include: consumption emotions, disaster management and recovery, place/brand attachment, anthropomorphism, resilience, service design, destination marketing and tourism/event impacts.

Guiqiang Qiao is an Assistant Professor at Zhejiang International Studies University, China. His research interests include event management, climate change perception and local-migrant interaction.

Yael Ram is an Associate Professor and Director of Academic International Relations at Ashkelon College in Israel. Her main research interests are in the fields of the psychological aspects of tourism and sustainability and tourist behavior. She has co-authored and edited books on public transport and tourism and walking studies, and is the lead author of the cultural services chapter of the Israeli assessment of ecosystem services.

Haywantee Ramkissoon is a Professor of Tourism at UniSA Business at the University of South Australia and a Fellow of the Peer Review College at the British Academy of Management. She has worked in domains of place attachment, pro-environmental behaviour, sustainable community tourism, wellbeing, public health and quality-of-life. She engages in collaborative research with national and international academic and industry partners for the benefits of individuals and society. She serves on several editorial advisory boards of top-tier journals and as a reviewer across disciplines.

K.K. Roessler is a Professor in the Department of Psychology, University of Southern Denmark, Odense, Denmark. Her research includes environmental psychology; how the physical environment and psychological processes are linked together; Health psychology and medical psychology; the psychological determinants of health and illness through life course, and physical activity research; and how physical activity affect health promotion, prevention and treatment of sickness.

Jarkko Saarinen is Professor of Human Geography at the University of Oulu, Finland; Distinguished Visiting Professor at the School of Tourism and Hospitality, University of

Johannesburg, South Africa; and Visiting Professor at the Division of Quality Science, Uppsala University, Sweden. His research interests include tourism development, impacts and sustainability, tourism and climate change adaptation, community and community-based natural resource management studies. He is currently an Associate Editor for *Tourism Geographies* and the *Journal of Ecotourism.*

Anna Dóra Sæþórsdóttir is a Professor in the Faculty of Life and Environmental Sciences at the University of Iceland. She has long conducted research on tourist use of the Icelandic wilderness. Her research interests include nature based tourism, wilderness tourism, tourism carrying capacity, tourism and power production, power plant location and impact, land use planning, and tourism management and planning.

Daniel Scott is a Professor and Research Chair in the Department of Geography and Environmental Management at the University of Waterloo, Canada. He is also an International Research Fellow at the School of Hospitality and Tourism at the University of Surrey. He works extensively on sustainable tourism, with a focus on the transition to a low carbon tourism economy and adaptation to the complex impacts of a changing climate. He has advised and led projects for a wide range of government agencies and tourism organizations, including the UNWTO, UNEP, World Bank, European Tourism Commission, WTTC, International Olympic Committee, OECD, amd the Caribbean Tourism Organization. He has also been a contributor to the UN Intergovernmental Panel on Climate Change Third, Fourth, Fifth, and Sixth Assessments and their 1.5°C special report.

Siamak Seyfi is an Assistant Professor in the Geography Research Unit at the University of Oulu, Finland. Using a multi-/interdisciplinary approach and informed by diverse disciplinary perspectives, his research stems from an interest in geography of tourism and, more particularly, critical tourism geographies, impacts of tourism, resilience, sustainability and politics of tourism and peace through tourism. Guided by generational and lifestyle theories, his research focuses on political and ethical consumerism, boycotts and buycotts with a special focus on Gen Z. His research is informed theoretically by critical theories and draws on diverse qualitative methodologies and sociological/ethnographic research.

Noam Shoval is a Professor and Director of the European Forum and the Director of the Center for Urban Innovation at the Hebrew University of Jerusalem; from 2017 to 2022 he served as the Provost of the Rothberg International School. He completed his PhD at The Hebrew University (2000) and conducted post-doctoral research at the Department of Geography, King's College, University of London (2000–2001). His main research interests are urban geography and planning, urban tourism and the implementation of advanced tracking technologies in various areas of spatial research, such as tourism and urban studies and medicine.

Uglješa Stankov is Professor at the Department of Geography, Tourism and Hotel Management, Faculty of Sciences, University of Novi Sad, Novi Sad, Serbia His main research areas are the strategic role of information technology in tourism experiences and mindful tourist experiences.

Marianna Strzelecka is Docent and Associate Professor in the Department of Marketing and Tourism Studies, Linnaeus University, Kalmar, Sweden. Her work draws from the fields of political science, sociology, psychology, and political ecology to shed light on the socio-cultural aspects of the relationships between communities of place and local 'natures', and how tourism allows to re-negotiate these relationships. She is also an affiliated researcher working with the Environmental Social Science Research Team at Jagiellonian University in Krakow, Poland, where she is examining the role of tourism and outdoors recreation in shaping human-nature relationships.

Theano S. Terkenli is Professor of Geography (Landscape, Cultural Geography, and Tourism), at University of the Aegean, Greece. Her research interests include landscape geographies, geographies of everyday life; critical and cultural geographical perspectives on tourism and recreation; cultural change in contemporary societies; the concepts of cultural identity, cultural sustainability, place, home and attachment in geographic thought and the lifeworld; the making and spatio-temporal change of the Aegean cultural landscape; tourism geographies of "development", modernity and postmodernism; communication technologies and new spatialities of everyday life; qualitative methodological approaches in the social sciences; and humanistic epistemologies in geography.

Dallen J. Timothy is a Professor in the School of Community Resources and Development at Arizona State University, USA. He holds a BSc in geography from Brigham Young University, an Master's degree in geography from the University of Western Ontario, and a PhD in geography from the University of Waterloo. His research interests in tourism include heritage, community-based planning, political boundaries and supranationalism, shopping and consumption, peripheral and rural regions, less-developed countries, and religiously-motivated travel.

Tran Hoang Tuan is at the University of Labor and Social Affairs (CSII), Ho Chi Minh City Vietnam, focusing on socially responsible human resource management, common good human resource management, employee behaviour, and leadership.

Gustav Visser is Professor of Geography, Department of Geography and Environmental Studies, Stellenbosch University, South Africa. An urban geographer by training, he holds a NRF B rating and has received generous research funding support from that institution. Currently funded research deals with queer sexualities, as well as the tourism and development nexus. He is past treasurer and president and now fellow of the Society of South African Geographers. He serves on the editorial boards of several academic journals and is a book series editor for Springer.

Tan Vo-Thanh is a Full Professor at the Department of Marketing, Excelia, La Rochelle, France. He is a member of the Board of Directors of the AFMAT (French Association of Tourism Management), and a member of the CEntre de REcherche en GEstion of University of Poitiers, France (CEREGE). His current research mainly focuses on digitalization process, smart technology adoption, sustainable tourism, customer experience, consumer behavior, customer engagement, employee engagement, and organizational behavior.

Ling-en Wang is an Associate Professor in the Institute of Geographic Sciences and Natural Resources Research, Chinese Academy of Science, Beijing, China. His research interests include food waste and protected area management.

Adi Weidenfeld is an Assistant Professor at the Faculty of Economic Sciences, University of Warsaw, Poland. His research interests include visitor attractions, tourism innovation, cross border tourism, knowledge networks, regional innovation systems and evolutionary economic geography.

Margrét Wendt Is a research assistant in Tourism Studies at the University of Iceland. Her research has focused on overtourism and wilderness tourism in Iceland, and migrant workers in Icelandic hotels.

Lim Xin Jean is a research fellow in Business School, Sun Yat-Sen University, China and Senior Lecturer in the School of Business and Economics, Universiti Putra Malaysia. Her research focuses on consumer behavior in online or offline retail environments, social media marketing, digital marketing, technology management, tourism, and customer relationship management.

Qi Zhang is a Research Assistant at the Institute of Geographic Sciences and Natural Resources Research, Chinese Academy of Science, Beijing, China.

Xiaoyu Zhang is an Associate Professor in the Tourism College, Beijing Union University, Beijing, China. Her research interests include tourist behaviour, as well as film and literary tourism. This includes research relating to the Chinese film tourism industry and theme parks based on film in Beijing, specifically Universal Studios, and in China generally. She is also an associate editor for *Tourism Critiques: Practice and Theory*.

Malin Zillinger is a Senior Lecturer in the Department of Service Studies, Lund University, Campus Helsingborg, Sweden. She is interested in mobilities and experiences. In her research projects, she tries to find out how people move through time and space, and which kinds of experiences they have along the way. During the last years, her focus has been on people involved in tourism and leisure industries, including both tourists and entrepreneurs. The study of mobility includes a special interest on method development.

Judit Zoltan is a Junior Assistant Professor in the Department of Economics, at the University of Bologna. She obtained a PhD in Economics and an MA with major in International Tourism at the University of Lugano, Switzerland, and her BS in Economics of Catering and Hotel Management from the Budapest Business School, Hungary. Her research interests lie in the fields of tourist behavior, destination planning and sustainable tourism.

Acknowledgements

This second edition of the Companion has been completed in very difficult times. It would not have been possible without my co-editors of the two previous editions Alan Lew and Allan Williams, to whose immense contribution to scholarship on tourism and geography this book is a testimony. I would also like to gratefully acknowledge the proofreading of Jody Cowper-James of many of the chapters in the book along with the support of the staff at Wiley for their patience, editing and production. Finally, I would like to thank all the contributors for their support for the book, while on a personal front I would like to thank Sara for all her support through difficult times. Never forget.

List of Abbreviations

ABCD	asset-based community development
ADS	approved destination system
AI	artificial intelligence
AIEST	Association Internationale d'Experts Scientifiques du Tourisme
ANP	analytical network process
AR	augmented reality
ATLAS	Association for Tourism and Leisure Education
BRICS	Brazil, Russia, India, China, and South Africa
CAUTHE	Council for Australian University Tourism and Hospitality Education
CB	consumer behaviour
CBA	cost benefit analysis
CBD	Convention on Biological Diversity
CBT	community-based tourism
CES	cultural ecosystem services
CGE	computable general equilibrium
CHRIE	Council of Hotel, Restaurant and Institutional Education
CO_2e	CO_2 equivalent
CoE	Council of Europe
CSER	corporate social environmental responsibility
CSR	corporate social responsibility
DDW	Dutch Design Week
DEMATEL	decision-making, trial and evaluation laboratory
DMO	destination management organisation/destination marketing organisation
EEG	electroencephalography
EIA	economic impact analysis
EKB	Engel-Kollat-Blackwell model
ELC	European Landscape Convention
ENoLL	European Network of Living Labs
EP	evolutionary psychology

EPA	United States Environmental Protection Agency
ESG	environment, social and governance
ETC	European Travel Commission
ETEZ	effective tourism exclusion zone
EU	European Union
FAO	Food and Agriculture Organization of the United Nations
FCC	Federal Communications Commission (USA)
FCO	Foreign and Commonwealth Office (UK)
FDI	foreign direct investment
FFMQ	five facet mindfulness questionnaire
FOMO	fear of missing out
GBF	global biodiversity framework
GDP	gross domestic product
GEC	global environmental change
GHRM	green human resource management
GIS	geographic information system
GNP	gross national product
GoT	*Game of Thrones*
GPS	global positioning system
GRI	global reporting initiative
GST	goods and services tax
GVCs	global value chains
HOE	hierarchy-of-effects
HR	human resources
HRM	human resource management
ICT	information and communications technology
IFRS	International Financial Reporting Standards Foundation
ILO	International Labour Organization
IO	input-output
IOC	International Olympic Committee
IoT	internet of things
IPBES	Intergovernmental Science-Policy Platform on Biodiversity and Ecosystem Services
IPCC	Intergovernmental Panel on Climate Change
IS	innovation system
IUCN	International Union for Conservation of Nature
KDE	kernel density estimation
LGBTQI+	lesbian, gay, bisexual, transgender, queer, intersex
LOTR	*Lord of the Rings*
LSTM	long short-term memory
MAAS	mindful attention and awareness scale
MBCT	mindfulness based cognitive therapy
MBSR	mindfulness based stress reduction
MENA	Middle Eastern and North African
MGB	model of goal directed behaviour
MICE	meetings, incentives, conferences and exhibitions
MLT	migration led tourism

MNE	multinational enterprises
MRT	mass rapid transit
MSMEs	micro-, small, and medium-sized enterprises
MSPs	multi-stakeholder platforms
NCPs	national contact points
NDC	nationally determined contributions
NEP	new ecological paradigm
NUTS	nomenclature of territorial units for statistics
NVB	nonverbal behaviour
OECD	Organisation for Economic Co-operation and Development
OTAs	online travel agencies
P2P	peer-to-peer
PAT	product potential and attractiveness tool
PDA	personal digital assistant
PERMA	positive emotion, engagement (which overlaps with flow), relationships, meaning, and accomplishment
PPGIS	participation geographical information systems
QoL	quality of life
QuaNGO	quasi non-governmental organization
R&D	research and development
RBSim2	recreational-behaviour simulator
RETS	resident empowerment from tourism scale
RIS	regional innovation system
ROI	return on investment
RV	related variety
SAF	sustainable aviation fuels
SAM	social accounting matrix
SAMs	sequence-alignment methods
SCR	scale Change and Resilience
SDGs	sustainable development goals
SEPA	Swedish Environmental Protection Agency
SET	social exchange theory
SF-MST	statistical framework for measuring the sustainability of tourism
SHIFT	social influence, habit, individual self, feelings and cognition, and tangibility
SIDS	small island developing States
SIS	sectoral innovation system
SmartCulTour	Smart Cultural Tourism as a Driver of Sustainable Development of European Regions
SME	small to medium enterprises
SSC	social sciences citation index
SSP	social signal processing
SSS	sea, sun and sand vacations
TALC	tourism area life-cycle
TSIS	tourism sectoral innovation system
TLC	travel ladder career
TNC	transnational corporation
TOURPAT	tourism and culinary atlas linked to a geospatial database

TPB	theories of planned-behaviour
TRA	theories of reasoned actions
TSA	tourism satellite account
TSV	tourist shopping village
TPCC	Tourism Panel on Climate Change
TTRA	The Tourism and Travel Research Association
UAE	United Arab Emirates
UN	United Nations
UNEP	United Nations Environment Programme
UNESCO	United Nations Educational, Scientific and Cultural Organization
UNFCCC	United Nations Framework Convention on Climate Change
UNWTO	United Nations World Tourism Organization
VAT	value-added tax
VFR	visiting friends and family
VR	virtual reality
WBCSD	World Business Council for Sustainable Development
WCMC	World Conservation Monitoring Centre
WCPSA	Wireless Communications and Public Safety Act
WEF	World Economic Forum
WFSR	Weber's theory of formal and substantive rationality
WFTA	World Food Travel Association
WMO	World Meteorological Organization
WRAP	Waste and Resources Action Programme
WTO	United Nations World Tourism Organization
WTTC	World Travel and Tourism Council
WWAP	World Water Assessment Programme
WWF	World Wide Fund for Nature

Part 1 Introduction

Chapter 1

Tourism
Conceptualisations, Disciplinarity, Institutions and Issues

C. Michael Hall

Introduction

Although any contemporary global account of tourism has to acknowledge at the outset the continued growth of international and domestic tourism, at the same time there is an even greater awareness of the effects of crises. Nevertheless, despite temporary setbacks since the turn of the twenty-first century, tourism, as an industry, has probably achieved a higher profile in the public consciousness of the developed world than ever before. Ironically, this is as much because of too little tourism, e.g. because of the impacts of COVID-19, than it is because of too much, e.g. debates about overtourism (see & Wendt, this volume).

Every edition of this companion has seemingly been written in the shadow of crises that have affected the tourism sector. The first edition was developed after the impacts on international tourism of the terrorist attacks of September 11 2001, and the American-led invasion of Iraq. The second after the 2008–2012 international economic and financial crisis, and the 'Arab Spring'. This third edition is written following the COVID-19 pandemic, which had the proportional largest effect on the growth of international tourism since the second world war (although it did also boost domestic tourism in many countries when borders were closed), as well as the Russian invasion of Ukraine. A future edition will undoubtedly face future political and soco-economic crises and, as several chapters in this and previous editions point out, climate and environmental crises, as markers of the Anthropocene (Huijbens & Gren, this volume).

Although such crises have all proven transitory with respect to tourism growth at the global scale, their impacts at the destination, community and firm levels remain significant. Regional and local changes in consumer confidence and travel behaviour and corresponding shifts in travel patterns, such as a reported growth in 'staycations' and more local travel, have in the main been short-term as economic conditions improve in generating regions. Indeed, economic problems in many destinations, such as Cyprus, Egypt, Greece, Italy, Portugal, Spain, Thailand and Tunisia, only serve to make tourism more important than ever. This is not

The Wiley Blackwell Companion to Tourism, Second Edition. Edited by C. Michael Hall.

only with respect to business and the economy, but also its socio-cultural, environmental and political context. An argument continually repeated by international tourism bodies, such as the UN World Tourism Organisation (UNWTO) and the World Travel and Tourism Council (WTTC), as well as in the introductions to many undergraduate assignments. Furthermore, the continued significance of crises has arguably changed the discourse of the tourism academy, whereby not only is sustainability a major concern, especially with respect to climate and environmental change (Scott, this volume; Eijgelaar & Peeters, this volume), but there is also increased attention to issues of resilience (Amore, this volume).

It is also important to acknowledge that despite the severe impacts of COVID-19 on international tourism, international arrivals have returned to pre-COVID levels in most countries as of late 2023, and it is likely that the UNWTO's forecasts for 2030 will still be met as international tourism was substantially exceeding those forecasts when COVID-19 hit. The UNWTO predicts the number of international tourist arrivals will increase by an average 3.3 percent per year between 2010 and 2030 (an average annual increase of 43 million arrivals), reaching an estimated 1.8 billion arrivals by 2030 (UNWTO, 2011, 2012). Upper and lower forecasts for global tourism in 2030 are between approximately two billion arrivals ('real transport costs continue to fall' scenario) and 1.4 billion arrivals ('slower than expected economic recovery and future growth' scenario) respectively (UNWTO, 2011). Most growth was forecast to come from the emerging economies and the Asia-Pacific, and by 2030 it was estimated that 57 percent of international arrivals will be in what are currently classified as emerging economies (UNWTO, 2011, 2012), although many of the emerging economies have been disproportionately affected by COVID-19 potentially affecting growth in this area. In addition, and as COVID-19 has demonstrated, it is also important to stress that although international tourism is often the focus of government policy because of its trade and foreign exchange implications, international tourism only accounts for around 16 percent of all tourist trips, with domestic tourism being fundamental to tourism development (Mensah & Boakye, 2023).

Tourism research does not exist in a vacuum. It is affected not only by changes in tourism consumption and production, but also wider global events and the shifting priorities and policies of higher education institutions and research funders which have placed a stronger emphasis on assessments research output quality, public-private partnerships and relevance (however defined). The constant fluidity and change in their environment has led many researchers to question many of their assumptions about tourism and reassess the relevance of their work, not only in terms of policy and other applications, but also, more fundamentally, to revisit the ways in which the subject is theorised and conceptualised (Gibson, 2008, 2021; Williams et al., 2014). A history of the sociology of tourism knowledge, unlike a history of tourist activity, has not yet been completed. As previous editions of this book have noted, while this was not explicitly the aim of this volume, the range and depth of the chapters in this book do provide an opportunity to assess many of the key themes and issues in tourism studies, as well as the intellectual context within which they were prepared (Williams et al., 2014). Indeed, several chapters, such as Larsen on the tourist gaze and Minca and Oakes on postmodernism, provide extremely valuable retrospective understandings on previous moments in tourism theory and implications for the future.

This introductory chapter is divided into two further sections. First, an account is presented of some of the issues surrounding the conceptualisation of tourism and, hence, its study. Second, a discussion of some of the key themes and issues which have emerged in the institutionalisation of tourism as a field of social scientific endeavour.

Conceptualising Tourism

Although many students of tourism have long sympathised with the sentiments of Williams and Shaw's observation that 'the definition of tourism is a particularly arid pursuit' (1988, p. 2) it is, as they also acknowledged, 'crucially important'. This is in part because of the continuing need to determine tourism's economic impacts and employment effects, and its broader effects and policy ramifications. Based on generally accepted international agreements for collecting and comparing tourism statistics, the term tourism trip has come to refer to a trip of not more than 12 months and for a main purpose other than being employed at the destination (UN and UNWTO 2007). However, despite UN and UNWTO recommendations, substantial differences continue to remain between countries with respect to the length of time that they use to define a tourist, as well as how employment is defined.

Undoubtedly, a substantial amount of government, industry body, and academic research effort has gone into the determination of 'supply side' or industry approaches to the definition of tourism, such as the development of Tourism Satellite Accounts (TSAs), which have become significant policy tools for organisations such as the WTTC to promote the economic and employment value of tourism. From a production or supply-side perspective, which TSAs seek to measure, the tourism industry may be defined as 'the aggregate of all businesses that directly provide goods or services to facilitate business, pleasure, and leisure activities away from the home environment' (Smith, 1988, p. 183). However, such production oriented approaches, while potentially useful for economic studies, fail to convey the manner in which the production and consumption of tourism are interwoven, the complexity of tourism labour market structure, and even business innovation and strategy (Aquino, this volume). They also do not address the implications that this has for understanding the broader social, environmental and political dimensions of tourism (Ioannides, this volume), as well as fundamental economic issues of commodification, distribution, tourism labour and the appropriate role of the state in tourism (Mosedale, this volume).

An adequate conceptualisation of tourism therefore clearly requires that we go beyond the narrowly economic. Most obviously, there is a need to appreciate the relationships between leisure, recreation and tourism and other social practices, and behaviour (Figure 1.1). As Featherstone (1987, p. 115) argues, tourism phenomenon need to be socially situated: 'The significance and meaning of a particular set of leisure choices ... can only be made intelligible by inscribing them on a map of the class-defined social field of leisure and lifestyle practices in which their meaning and significance is relationally defined with reference to structured oppositions and differences'.

In Figure 1.1 broken lines are used to illustrate how the boundaries between concepts are 'blurred'. Work is differentiated from leisure, with there being two main realms of overlap: first, business travel, which is seen as a work oriented form of tourism in order to differentiate it from leisure based travel; and, second, 'serious leisure', which refers to the breakdown between leisure and work pursuits and the development of leisure career paths with respect to hobbies and interests (Stebbins 1979). Although the blending or even convergence of work and leisure in a travel context has been recognised in the academic literature since the early 1980s it has now become a more common-place and recognised with the growth of working holidays, business travel, tele-commuting from second homes, and volunteer tourism and gap years (Wu & Pearce, 2018; Hannonen, 2020; Lee & Zhang, 2020; Pinho & Marques, 2021; Zhu & Gao, 2022). In fact, government policies have often been set to enable such travel because of its perceived economic benefits to destination areas, including remedying seasonal labour deficiencies, for example through the provision of working visas.

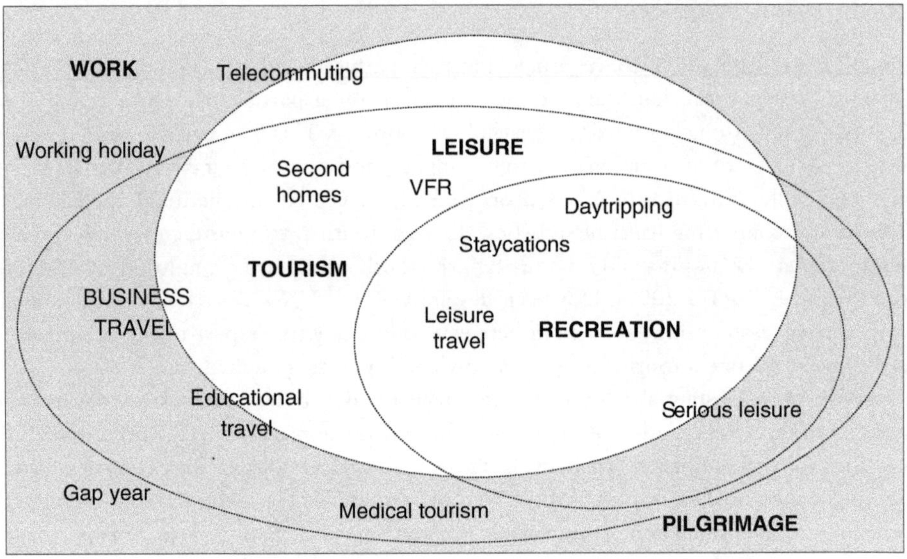

Figure 1.1. Relationships between leisure, recreation and tourism (after Hall 2005).

In addition to being defined in relation to its production and consumption, tourism is also interpreted as but one, albeit highly significant, dimension of temporary mobility and circulation (Hannam & Butler, this volume; Müller, this volume) (Figure 1.2). The merging of aspects of leisure, recreation and tourism research, along with the study of migration, circulation and mobility has had a substantial influence on how tourism is perceived as an area of research interest in relation not only to the movement of people, including the mundane mobilities of the everyday, but also the movement of ideas. Indeed, it is only really since the turn of the century that temporary movements away from home, such as travel for work or education, or travel for health reasons, or even going overseas after finishing university have begun to substantially catch the awareness of migration researchers (e.g., Bell & Ward, 2000). Yet it is increasingly evident to those seeking wider perspectives on tourism, that supposedly 'different' forms of mobility are highly inter-related. Thus, the inclusion of same-day travel 'excursionists' within technical definitions of tourism makes the division between recreation and tourism even more arbitrary (UN and UNWTO 2007). Given innovations in transport technology, same-day travel is becoming increasingly important at widening spatial scales, while communications technology advances are also blurring the boundaries between 'virtual' and 'real' travel and tourism (Lu et al., 2022). This emphasises the need for those interested in tourism to address the arbitrary boundaries between tourism and leisure, and tourism and migration. Tourism constitutes just one form of leisure oriented temporary mobility, and it constitutes part of that mobility, being shaped by and shaping it.

While stressing the need to conceptualise tourism in terms of mobility, Flavell (2001, pp. 391–392) reminds us that there is also a need 'to assess really the extent or nature of movement, or indeed even see it sometimes, you have in fact to spend a lot of the time studying things that stand still: the borders, institutions and territories of nation states; the sedimented 'home' cultures of people that do not move'. This directs our attention to the non-mobile and to constraints on becoming a tourist. Indeed, in much 'mainstream' business research on tourism, issues of equity and rights of access are little considered and there often appears to be an

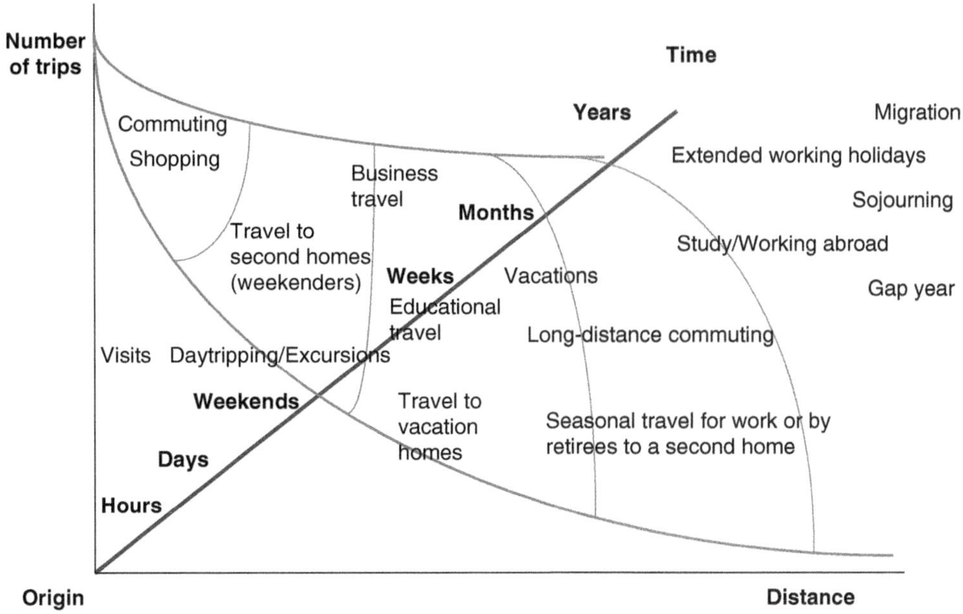

Figure 1.2. Extent of temporary mobility in space and time (after Hall, 2005).

implicit assumption that everyone can travel, has a car and/or a mobile phone (Seow & Brown, 2018; Seyfi & Hall, 2020; Hall et al., 2022). Nevertheless, geographers have long recognised that a basic precondition for tourism mobility is that absences from the stations of the daily world are for certain periods socially and institutionally sanctioned. The opportunity to travel has always depended on the right to be absent from home and work, with such rights having historically been reserved for very few groups in the (usually male) population (Frändberg, 1998; Hall et al., 2022). Indeed, Hägerstrand (1984) describing the breakaway from the time-space prism of everyday life that tourism represents, refers to this as an 'escape from the cage of routines.' Similarly, the growing recognition of the role of spatial settings by sociologists has direct implications for understanding tourism as a social practice with Giddens (1984, p. xxv) observing, 'Time-space "fixity" also means social fixity; the substantially "given" character of the physical milieux of day-to-day life interlaces with routine and is deeply influential in the contours of institutional reproduction.'

However, the embeddedness of tourism in modern social and economic practices has created a significant space for social science research which may not only be of relevance for tourism itself but for a deeper understanding of the everyday as well as wider patterns of mobility. Nevertheless, the notion of tourism is open to multiple conceptualisations which rest on the ontological, epistemological and paradigmatic assumptions of the viewer (Goolaup & Nunkoo, this volume; Lees, this volume). This means that the conceptualisation of tourism remains open to substantial contestation that may almost seem at odds with a lay understanding of what tourism represents. However, it should be noted that some commentators even questioned the utility of tourism as a concept.

> We will begin by interrogating the very category of 'tourism'. Is there such an entity? Does the term serve to demarcate a usefully distinct sphere of social practice? Where does tourism end and leisure or hobbying and strolling begin? πtourism is a term that is waiting to be deconstructed. Or as

> Marx might have said it is a chaotic conception, including within it too wide a range of disparate phenomenon ... It embraces so many different notions that it is hardly useful as a term of social science, although this is paradoxical since Tourism Studies is currently being rapidly institutionalised within much of the academy (Rojek & Urry, 1997, p. 1).

Yet concerns over whether the term 'tourism' is useful is not an issue for the vast majority of the tourism academy, nor industry and policy makers, as the next section of the introduction that takes up the theme of the institutionalisation of tourism illustrates.

The Institutionalisation of Tourism Studies

Tourism studies is well institutionalised in academic terms. Arguably, one of the reasons for conceptual confusion is because of the multiplicity of disciplinary and paradigmatic approaches that have been brought to bear on tourism phenomena, as indeed is true of many of the phenomena which are studied in the social sciences. Tribe (1997, p. 638), described tourism analysis as interdisciplinary, multidisciplinary, and 'conscious of its youthfulness'. Yet while such statements about the state of tourism studies are widespread, they fail to appreciate that the study of tourism within the social sciences has a far longer history than is often imagined, and is less 'youthful' than often implied. For example, with respect to the geography of tourism, Hall and Page (2014) chart Anglo-American and European traditions of social scientific scholarship on tourism that date to the 1920s and 1930s.

The predominant attitude long held among tourism researchers, perhaps best summed up by Bodewes, who argued that 'tourism is usually viewed as an application of established disciplines, because it does not possess sufficient doctrine to be classified as a full-fledged academic discipline' (1981, p. 37). Tribe (1997) even suggests that the search for tourism as a discipline should be abandoned, and that the diversity of the field should be celebrated. Nevertheless, this has to be set against the increasing recognition that tourism is now recognised as a legitimate area of study in its own right, and that it has many of the characteristics of a discipline (Hall, 2005). For example, Johnston (1991), in his landmark review of Anglo-American Geography, identified three key characteristics of a discipline:

- a well established presence in universities and colleges, including the appointment of professorial positions;
- formal institutional structures of academic associations and university departments; and
- avenues for academic publication, in terms of books and journals. Indeed, 'It is the advancement of knowledge – through the conduct of fundamental research and the publication of its original findings – which identifies an academic discipline; the nature of its teaching follows from the nature of its research' (Johnston, 1991, p. 2).

These characteristics clearly apply to the field of tourism studies. There are departments and/or degree programmes established throughout the world. The first undergraduate degree programme in tourism in the United Kingdom was established at the University of Surrey in 1973. The first programmes in Australia were established at Gatton College (now a part of the University of Queensland) and Footscray CAE (now a part of the Victoria University) in the late 1970s. There are also a number of institutional structures for tourism both within universities and colleges of higher learning, that is, departments and schools of tourism, as well as on a national or international basis. For example, at a national level institutions such as the

Council for Australian University Tourism and Hospitality Education (CAUTHE) and the Tourism Society in the United Kingdom run annual research conferences and provide fora for discussion on tourism education. Specialty tourism research groups also operate within national academic associations, such as the Association of American Geographers, the Canadian Association of Geographers and the Institute of British Geographers.

At the international level social scientific unions in the fields of anthropology and ethnology, economic history, geography, history and sociology have tourism commissions or working groups. For example, various International Geographical Union commissions or study groups on tourism have existed since 1972. A number of other international tourism research and education organizations have also made substantial contributions to tourism studies. For example, the first refereed academic journal on tourism *Revue de Tourisme/The Tourist Review* now the *Tourism Review* was established as early as 1946, as the official organ of de l'Association Internationale d'Experts Scientifiques du Tourisme (AIEST) based in Switzerland. The Council of Hotel, Restaurant and Institutional Education (CHRIE), which has a strong tourism component, was also established in 1946 in the United States. The Tourism and Travel Research Association (TTRA) had its beginnings in the merger of the Western Council of Travel Research and the Eastern Travel Research Association in 1970. Although it retains a strong North American base, TTRA is now a substantial international network with European and Asia-Pacific chapters. In Europe, the institutionally-based Association for Tourism and Leisure Education (ATLAS) was established in 1991 to develop transnational educational initiatives in tourism and leisure. Since that time ATLAS has expanded rapidly to include chapters from around the world.

In terms of the advancement of knowledge, there is now a substantial body of tourism literature as evidenced in journals, books, conference proceedings and electronic publications. Although the exact figures are hard to identify given name changes of publications and the growth of on-line publishing, there are over 180 journals, published in English either in full or in part, that can be identified as having a substantial academic component devoted to tourism research (McKercher, 2024). This compares to the 77 journals identified in the first edition of this volume (Hall et al., 2004). According to McKercher (2024) there were more than 12,500 papers published in 316 tourism, hospitality and events journal in 2021, of which some 188 journals were published in whole or in part in English; 39 in whole or in part in Portuguese, mostly from Brazil; 27 in Spanish, and the rest in other languages.

Nevertheless, the market of ideas in tourism publishing is clearly affected by gatekeepers (Cooper & Hall, 2024) and 'the uneven geographies of international journal publishing spaces' (Paasi, 2005, p. 769) that are shaped by different national and institutional research agendas as well as language. No matter how important local and national knowledge is, unless it is conveyed in English it has little chance to enter the global marketplace of ideas and be reproduced and recirculated. This does not mean that tourism research in other languages does not have significance, rather it just means that its international reach is not as great. Somewhat ironically, given the express desire in tourism literatures to give voice to local and indigenous perspectives, unless that voice can be spoken in English it is likely not to be heard, at least in the academic literature (Hall, 2013b). Furthermore, it is also important to point out that much tourism related knowledge is not published in tourism journals. For example, substantial contributions to the tourism literature such as Butler's (1980) life cycle model was published in the *Canadian Geographer*.

A further questions arise with respect to the institutional nature of tourism. First, does tourism studies constitute a discipline? This is a difficult question, especially as there are many disciplinary approaches to tourism, e.g. the geography of tourism, while there is also substantial interest in tourism in the concept of postdisciplinarity (Coles et al., 2009, 2016) where there is a focus on learning rather than with disciplines. According to Coles et al. (2009, p. 81) 'there is the danger that we may be attempting to address contemporary subjects (such as tourism) through outdated and ageing frameworks for scholarly activity and academic administration'. Coles et al. (2009, 2016) also argue that a postdisciplinary approach is especially well-suited to the study of various forms of tourist mobilities as well as being central to progress in understanding and responding to climate and environmental change, noting that human behaviour and its impact on the environment is complex, messy and uncertain and that postdisciplinarity is needed to capture the complexity that may be overlooked by more disciplinary-focussed studies. From this perspective it is interesting to note Johnston's (1991, p. 9) reflections that:

> there is no fixed set of disciplines, nor any one correct division of academic according to subject matter. Those disciplines currently in existence are contained within boundaries established by earlier communities of scholars. The boundaries are porous so that disciplines interact. Occasionally the boundaries are changed, usually through the establishment of a new discipline that occupies an enclave within the pre-existing division of academic space.

The growth of tourism studies of course, as well as geographies of tourism, has helped to reshape such boundaries, as well as being influenced by them. Nevertheless, the field has long been criticised for its theoretical base. As Meethan (2001, p. 2) commented, 'for all the evident expansion of journals, books and conferences specifically devoted to tourism, at a general analytical level it remains under-theorised, eclectic and disparate'. The comments of Franklin and Crang (2001, p. 5) were similarly astringent:

> The first trouble with tourism studies, and paradoxically also one of its sources of interest, is that its research object, 'tourism' has grown very dramatically and quickly and that the tourism research community is relatively new. Indeed at times it has been unclear which was growing more rapidly – tourism or tourism research. Part of this trouble is that tourist studies has simply tried to track and record this staggering expansion, producing an enormous record of instances, case studies and variations. One reason for this is that tourist studies has been dominated by policy led and industry sponsored work so the analysis tends to internalize industry led priorities and perspectives ... Part of this trouble is also that this effort has been made by people whose disciplinary origins do not include the tools necessary to analyse and theorize the complex cultural and social processes that have unfolded.

Williams (2014), in the previous edition of this volume, had a more positive perspective of the different disciplinary dimensions of tourism:

> Tourism studies is a meeting ground, if not a melting pot, for competing theoretical and methodological perspectives. This contributes to its diversity, compared to many individual disciplines, but is also a source of discord, which at its best, is a source of creativity and conversation, and at worse, of posturing whereby different perspectives talk not with each other, but at each other. This meeting ground of competing, and contested perspectives constitutes a shifting set of overlapping borders which fluctuate over time in response to broader scientific paradigmatic shifts, exogenous economic, political and social agenda changes, and the dynamics of academic communities (Williams, 2014, p. 27).

Academic communities argue on empirical, theoretical and territorial grounds and what constitutes evidence 'becomes a way of challenging the very meaningfulness of a particular concept from those affiliated to ... some competing concept or theory' (Agnew &

Livingstone, 2011, p. 13). Claiming that tourism is atheoretical or poorly theorised (Franklin & Crang, 2001) becomes a way of challenging or demeaning the value of research not undertaken within a particular theoretical context (Hall, 2013a). Nevertheless, their assessment does point at one of the persistent tensions in tourism research, between the often contradictory requirements of critical social science and the extent to which industry and policy-makers influence the research agenda, particularly through funding and commercialisation strategies. There are similar contradictions in several of the social sciences, but they are particularly sharp in tourism (Bianchi, 2009; Gibson, 2021). Nevertheless, as the various chapters in this volume illustrate. the understanding of a field as complex and multi-scalar as tourism is unlikely to be the sole domain of either a single paradigm or approach, while there are also a wide range of theoretical approaches that are employed.

The boundaries of tourism studies are constantly changing and will continue to change in light of internal discourses, engagement with debates across boundaries, and exogenous factors. For good or bad, it is also almost inevitable that, given how academic institutions function in capitalist societies, that industry and government agencies (including research funding) will continue to shape the agenda of tourism research, alongside the tradition of critical social and theoretical social scientific enquiry. These permeable boundaries and the space within them lies at the heart of this work. The present volume is therefore a snapshot of some of the dominant discourses in the social science of tourism: where it has come from, where it is now and some thoughts on where it might go in the future. The outcome, inevitably, is that the collection of essays in this book illustrate that tourism is a diverse field, in terms of its concerns, theories and methodologies. But they also demonstrate that it is characterised by substantive debate and continuing innovation, and that it is also engaged in some of the major debates that characterise social science and, in the case of research on climate and environmental change, are reaching across to natural science. Indeed, the continued growth of tourism as well as recognition of its role in contemporary society, the environment and economic development can only serve to reinforce the importance of tourism as a field of study.

Acknowledgements

This chapter extends and revises the previous introductions to the companion (Hall et al., 2004, 2014) with thanks to Alan Lew and Allan Williams.

References

Agnew, J., & Livingstone, D. (2011). Introduction. In J. Agnew & D. Livingstone (Eds.), *The SAGE handbook of geographical knowledge* (pp. 1–18). Sage.

Bell, M., & Ward, G. (2000). Comparing temporary mobility with permanent migration. *Tourism Geographies*, 2, 87–107. https://doi.org/10.1080/146166800363466

Bianchi, R.2009. The 'critical turn' in tourism studies: A radical critique. *Tourism Geographies*, *11*, 484–504. https://doi.org/10.1080/14616680903262653.

Bodewes, T. (1981). Development of advanced tourism studies in Holland. *Annals of Tourism Research*, *8*, 35–51. https://doi.org/10.1016/0160-7383(81)90066-9

Butler, R.1980). The concept of a tourist area cycle of evolution, implications for management of resources. *Canadian Geographer*, *24*, 5–12. https://doi.org/10.1111/j.1541-0064.1980.tb00970.x

Coles, T., Hall, C. M., & Duval, D. T. (2009). Post-disciplinary tourism. In J. Tribe (Ed.), *Philosophical issues in tourism* (pp. 80–100). Channel View.

Coles, T., Hall, C. M., & Duval, D. T. (2016). Tourism and postdisciplinarity: Back to the future? *Tourism Analysis*, *21*(4), 373–387. https://doi.org/10.3727/108354216X14679788636113

Cooper, C. & Hall, C. M. (Eds.) (2024). *How to publish in the best tourism journals*. Elgar.

Featherstone, M. (1987). Leisure, symbolic power and the life course. In J. Horne, D. Jary, & A. Tomlinson (Eds.), *Sport, leisure and social relations* (pp. 113–138). Routledge and Kegan Paul.

Flavell, A. (2001). Migration, mobility and globaloney: metaphors and rhetoric in the sociology of globalization. *Global Networks*, 1, 389–398. https://doi.org/10.1111/1471-0374.00022

Frändberg, L. (1998). *Distance matters: An inquiry into the relation between transport and environmental sustainability in tourism*, Humanekologiska skrifter 15. Göteborg Universitet Department for Interdisciplinary Studies of the Human Condition.

Franklin, A., & Crang, M. (2001). The trouble with tourism and travel theory? *Tourist Studies*, 1, 5–22. https://doi.org/10.1177/146879760100100101

Gibson, C. (2008). Locating geographies of tourism. *Progress in Human Geography*, *32*, 407–422. https://doi.org/10.1177/0309132507086877

Gibson, C. (2021). Critical tourism studies: New directions for volatile times. *Tourism Geographies*, *23*(4), 659–677. https://doi.org/10.1080/14616688.2019.1647453

Giddens, A. (1984). *The constitution of society: Outline of the theory of structuration*. University of California Press.

Hall, C. M. (2005). *Tourism: Rethinking the social science of mobility*. Prentice-Hall.

Hall, C. M. (2013a). Framing tourism geography: Notes from the underground. *Annals of Tourism Research*, *43*, 601–623. https://doi.org/10.1016/j.annals.2013.06.007

Hall, C. M. (2013b). Framing behavioural approaches to understanding and governing sustainable tourism consumption: Beyond neoliberalism,"nudging" and "green growth"?*Journal of Sustainable Tourism*, *21*(7), 1091–1109. https://doi.org/10.1080/09669582.2013.815764

Hall, C. M., & Page, S. J. (2014). *The geography of tourism and recreation: Space, place and environment* (4 th ed.). Routledge.

Hall, C. M., Seyfi, S., & Rasoolimanesh, S. M. (Eds.). (2022). *Contemporary Muslim travel cultures: Practices, complexities and emerging issues*. Routledge.

Hall, C. M., Williams, A. M., Lew, A. (2004). Tourism: Conceptualisations, institutions and issues. In A. A. Lew, C. M. Hall, & A. M. Williams (Eds.), *A companion to tourism* (pp. 3–21). Blackwell Publishing.

Hall, C. M., Williams, A. M., Lew, A. (2014). Tourism: Conceptualisations, disciplinarity, institutions and issues. In A. A. Lew, C. M. Hall, & A. M. Williams (Eds.), *The Wiley Blackwell companion to tourism* (pp. 3–24). Wiley-Blackwell Publishing.

Hannonen, O. (2020). In search of a digital nomad: defining the phenomenon. *Information Technology & Tourism*, *22*, 335–353. https://doi.org/10.1007/s40558-020-00177-z

Hägerstrand, T. (1984). Escapes from the cage of routines: observations of human paths, projects and personal scripts. In J. Long, & R. Hecock (Eds.), *Leisure, tourism and social change* (pp. 7–19). Dunfermline College of Physical Education.

Johnston, R. (1991). *Geography and geographers: Anglo-American human geography since 1945*, 4th ed. Edward Arnold.

Lee, H. Y., & Zhang, J. J. (2020). Rethinking sustainability in volunteer tourism. *Current Issues in Tourism*, *23*(14), 1820–1832. https://doi.org/10.1080/13683500.2019.1653267

Lu, J., Xiao, X., Xu, Z., Wang, C., Zhang, M., & Zhou, Y. (2022). The potential of virtual tourism in the recovery of tourism industry during the COVID-19 pandemic. *Current Issues in Tourism*, *25*(3), 441–457. https://doi.org/10.1080/13683500.2021.1959526

McKercher, B. (2024). The journal publishing landscape. In C. Cooper & C. M. Hall (Eds.), *How to publish in the best tourism journals*. Elgar.

Meethan, K. (2001). *Tourism in global society: Place, culture and consumption*. Palgrave.

Mensah, E. A., & Boakye, K. A. (2023). Conceptualizing post-COVID 19 tourism recovery: A three-step framework. *Tourism Planning & Development*, *20*(1), 37–61. https://doi.org/10.1080/21568316.2021.1945674

Paasi, A. (2005). Globalisation, academic capitalism, and the uneven geographies of international journal publishing spaces. *Environment and Planning A*, 37, 769–789. https://doi.org/10.1068/a3769

Pinho, M., & Marques, J. (2021). The bleisure tourism trend and the potential for this business-leisure symbiosis in Porto. *Journal of Convention & Event Tourism*, 22(4), 346–362. https://doi.org/10.1080/15470148.2021.1905575

Rojek, C., & Urry, J. (1997). Transformations of travel and theory, In C. Rojek & J. Urry (Eds.), *Touring cultures: Transformations of travel and theory* (pp. 1–19). Routledge.

Seow, D., & Brown, L. (2018). The solo female Asian tourist. *Current Issues in Tourism*, *21*(10), 1187–1206. https://doi.org/10.1080/13683500.2017.1423283

Seyfi, S., & Hall, C. M. (2020). Sanctions and tourism: Effects, complexities and research. *Tourism Geographies*, *22*(4–5), 749–767. https://doi.org/10.1080/14616688.2019.1663911

Smith, S. L. J. (1988). Defining tourism: A supply–side view. *Annals of Tourism Research*, 15, 179–90. https://doi.org/10.1016/0160-7383(88)90081-3

Stebbins, R. A. (1979). *Amateurs: On the margin between work and leisure*. Sage.

Tribe, J. (1997). The indiscipline of tourism. *Annals of Tourism Research*, *24*, 638–57. https://doi.org/10.1016/S0160-7383(97)00020-0

United Nations and United Nations World Tourism Organization (UN and UNWTO). (2007). *International Recommendations on Tourism Statistics (IRTS) Provisional Draft*. UN and UNWTO.

United Nations World Tourism Organization (UNWTO). (2011). *Tourism Towards 2030: Global Overview*. UNWTO General Assembly, 19th Session, Gyeongju, Republic of Korea, 10 October 2011. UNWTO.

United Nations World Tourism Organization (UNWTO). (2012). *UNWTO Tourism Highlights. 2012 Edition*. UNWTO.

Williams, A. M. (2014). Introduction: Perspectives on tourism. In A.A Lew, C M. Hall & A.M. Williams (Eds.), *Companion to tourism* (pp. 27–32). Blackwell-Wiley.

Williams, A. M., Hall, C.M., & Lew, A. (2014). Theoretical and methodological challenges for tourism – just out of sight, or just within our reach? In A.A Lew, C M. Hall & A. M. Williams (Eds.), *Companion to tourism* (pp. 625–634). Blackwell-Wiley.

Williams, A. M., & Shaw, G. (1988). Tourism and development: Introduction. In A. M. Williams & G. Shaw (Eds.), *Tourism and economic development: Western European experiences* (pp. 1–11). Belhaven Press.

Wu, M., & Pearce, P. L. (2018). Gap time and Chinese tourists: exploring constraints. *Current Issues in Tourism*, *21*(10), 1171–1186. https://doi.org/10.1080/13683500.2017.1415868

Zhu, P., & Guo, Y. (2022). Telecommuting and trip chaining: Pre-pandemic patterns and implications for the post-pandemic world. Transportation Research Part D: Transport and Environment, 113, Art. 103524. https://doi.org/10.1016/j.trd.2022.103524

Part 2 Disciplinary and Methodological Perspectives on Tourism

Introduction
Disciplinary and Methodological Perspectives on Tourism

C. Michael Hall

As Chapter 1 suggested, there are a wide range of disciplinary perspectives on tourism. As a result, there are also a large number of different qualitative and quantitative methods that are also applied to tourism related problems (Dwyer et al., 2012; Hall & Page, 2014; Khoo-Lattimore et al., 2019). This section provides an overview of some of the main approaches to tourism. Chapter 2 by McKercher and Zoltan provides what may be regarded as a more traditional approach to geography which is a strong emphasis on spatial analysis and flows. They argue that geography has had a major influence in the development of tourism as a field of study. The chapter reflects ways of understanding the movement of tourists in space and time between their home environments and the destination as well as between destinations. McKercher and Zoltan note the importance of intervening factors in tourist's spatial movements including distance, time and monetary budgets, as well as trip and personal characteristics.

Chapter 3 by Prayag also discusses the trip and personal characteristics of tourists but from the perspective of consumer behaviour. As Prayag notes, consumer behaviour is arguably the most studied aspect of tourism. He notes that much tourist behaviour research has been influenced by economics and psychology, and he pays particular attention to the emergence of evolutionary psychology. The chapter reviews tourist decision-making models as well as the tourist experience and its components. However, he also notes how such research is influential in studies of the sociology of tourism and are important for understanding behavioural interventions.

Chapter 4 by Ram reviews a major area of tourist behavioural research which is that of tourist motivations and decision making and how they contribute to tourist demand. She uses understanding of decision-making to identify different capacities for behavioural interventions considering different assumptions of decision-making. Ram notes how these approaches help better understand the constraints facing sustainable behavioural interventions as well as actual behaviours. Importantly, she highlights that intrinsic motivations are insufficient for adoption of sustainable behaviours. As motivations have hedonic components and the decision-making process has irrational parts, Ram argues that only a combination of

The Wiley Blackwell Companion to Tourism, Second Edition. Edited by C. Michael Hall.

socio-psychological influences (including social media channels) and pro-environmental policies can shape future tourism demand in sustainable and responsible directions.

In Chapter 5, Feiger and Dyason examine economic perspectives on tourism. Their chapter provides an overview of some of the major economic issues in tourism at a global scale, and detail the effects that COVID-19 has had on tourism demand and the costs and benefits that arise from tourism. In addition, they also detail some of the issues arising in relation to tourism forecasting, productivity and overtourism, providing a valuable connection to several other chapters in this volume. Nevertheless, as they note, 'rising competitiveness between tourism destinations and greater public and political demands for increased returns from tourism will increase efforts to achieve greater efficiency in all aspects of tourism.'

Chapters 6 and 7 take a different tangent an explore some of the political dimensions of tourism. In Chapter 6 Seyfi and Hall discuss the significance of international relations approaches to tourism, highlighting that despite the economic importance of international tourism it's role within the bi- and multi-lateral relations between countries has been given nowhere near the attention it deserves. However, they note that approaches to international relations have been revitalised in recent years by the development of new understandings of geopolitical analysis, suggesting, 'in tourism geopolitics is recognized as being significant not just in terms of state and sub-state actors and territories, but also with respect to the activities of transnational business interests and the micro-geopolitics of everyday tourist encounters' (Seyfi & Hall, this volume).

In Chapter 7 Mosedale also examines a relatively small, but nevertheless significant field, that of political economy. Mosedale observes that political economy is primarily concerned with the commodification of resources (see also Chapter 20 by Hultman et al. on resourcification and tourism) and the resulting accumulation and distribution of economic, social, and natural capital. However, in an integrated manner he notes that the historical, political, sociocultural, and geographical contexts of the economy influence the structures that organize capital accumulation and its distribution. Importantly, the emphasis placed on each of these factors has resulted in different understandings of political economy including regulation theory, radical political economy, cultural political economy, poststructural political economy, and the political economy of nature/political ecology. With the latter increasingly becoming a distinct tourism subfield, combining radical and poststructural political economy and utilising social and environmental justice approaches.

Chapter 8 by Hannam and Butler provides a meta-theoretical perspective on the application of a mobilities paradigm to the subject of tourism mobilities. According to Hannam and Butler the concept of mobilities is concerned with simultaneously understanding the movements of people, objects, capital, ideas, and information at different scales in space and time, from the global to the local processes. For them, mobilities is important for tourism because it places tourism at the centre of social and economic life. Significantly, they also argue that COVID-19 provided a renewed understanding of mobilities and immobilities that require the use of innovative, experimental, and increasingly sophisticated technologies (see Chapter 11 by Shoval et al.).

The final three chapters in this section deal with potentially such innovative methods to examine tourist mobilities. Chapter 9 by Goolaup and Nunkoo discuss the state of qualitative methods in tourism and hospitality research. They provide a short historical overview of qualitative methods in tourism before focusing in more detail on phenomenology and grounded theory. Chapter 10 by Lees examines a surprisingly little used approach in tourism which is that of historiography, Lee's uses tourism and tourism marketing as a context to argue the value of historiographical approaches and their embeddedness within wider marketing thought.

The final chapter in the section is that Shoval et al. with the respect to the use of GPS and smartphones in tourism research (Chapter 11). As they note, the rapid adoption of these technologies in contemporary society, has revealed how the social interactions, economic engagement, and individual existence of people's lives, including the personal activities of people as tourists, are now not only being shaped by using these devices, but are also trackable and recordable. They provide a valuable overview of the development of the field in a tourism context and the potential for complex spatiotemporal modelling and data mining. However, they also note the potential for misuse and the substantial ethical issues that arise in the use of personal data. Importantly, they also highlight that these technologies do not replace 'traditional' research approaches, such as surveys, diaries, or interviews, which will remain an important source of information on the motives underlying tourist activity. Instead, they stress that new technologies will complement and enrich the findings of traditional research practices.

References

Dwyer, L., Gill, A., & Seetaram, N. (Eds.). (2012). *Handbook of research methods in tourism: Quantitative and qualitative approaches*. Edward Elgar Publishing.

Hall, C.M., & Page, S. J. (2014). *Geography of Tourism and recreation* (4th ed.). Routledge.

Khoo-Lattimore, C., Mura, P., & Yung, R. (2019). The time has come: A systematic literature review of mixed methods research in tourism. *Current Issues in Tourism*, 22(13), 1531–1550. https://doi.org/10.1080/13683500.2017.1406900

Chapter 2

Tourist Flows and Spatial Behaviour

Judit Zoltan & Bob McKercher

Introduction

Tourism involves the movement of people through time and space, either between their home and destinations, or within destination areas. Understanding tourist movements and the factors that influence the time/space relationships tourists have with destinations have profound implications for policy, infrastructure, transport and product development, as well as the management of tourism's impacts (Edwards & Griffin, 2013). Early research focused on inter-destination travel patterns and relied largely on paper-based data collection tools such as surveys, trip diaries and small-scale maps. The emphasis has shifted recently to studies of tourist movements within destination areas, aided by passive, electronic data collection tools, such as GPS (Global Positioning System) tracking devices, mobile phones, PDA (Personal Digital Assistant) instruments, and geospatial metadata online. Technology now permits more sophisticated analysis including network analysis, time budget allocation and market segmentation based on different observed behavioural patterns.

This chapter examines tourism flows and spatial behaviour. It begins with a review of pioneering research conducted in the 1990s that examined inter-destination movements. An overview of intra-destination movements follows before concluding with a discussion of moderating factors that can affect spatial behaviour.

Modeling Inter-destination Tourist Movements

The study of movements between the tourist's home and destination or between destination areas dominated research conducted in the 1990s. This research established the field of study and identified dominant themes that are still prevalent today. The following section reviews briefly key works that laid the foundation for this study area.

Mings and McHugh (1992) identified four types of touring routes taken by domestic American tourists who visited Yellowstone National Park. Three involved automobile travel

The Wiley Blackwell Companion to Tourism, Second Edition. Edited by C. Michael Hall.

exclusively, while the fourth involved a combination of air and automobile transport. Respondents who displayed a 'direct route' itinerary took the most direct path to and from Yellowstone National Park and followed exactly the same route in both directions. The 'partial orbit' itinerary consisted of taking the most direct route to a large destination area, such as the Rocky Mountains, then embarking on a touring loop in the area. The return trip follows the original outward bound transit route. These types of itineraries are typified by a significant transit journey followed by an extensive tour visiting the key attractions and staying in different destinations in an area some distance from home. By contrast, the 'full orbit' tour itinerary involves visiting a number of destinations with no overlap in the tour route. The 'fly-drive' itinerary is similar to the partial orbit itinerary except that the transit leg is made by air.

Lue et al. (1993) focused their research on multi destination trips, but also recognised the existence of single destination, direct-route trips. Four types of itineraries were described. Tourists make a number of short stops on their way to or from the main destination in the 'en-route' model. The 'base camp' model represents a further elaboration of the single destination model. Conceptually, it resembles a hub and spoke. Tourists base themselves in one main destination and then venture out from that destination in a series of short, day tours to nearby attractions and destinations. In the 'regional tour', tourists travel to a destination region, but rather than basing themselves in one locale, they stop overnight in a number of places in a sequential pattern before returning home. The 'trip-changing' pattern involves a multi-foci touring trip visiting a number of destinations without overlapping any leg of the trip.

Oppermann (1995), focusing on international travel, identified seven possible itinerary types. In addition to the five previously mentioned by Mings and McHugh (1992), and Lue et al. (1993), he added two other itineraries that are relevant to long-haul air travel. The 'open jaw loop' model applies to tourists who enter a country through one gateway and leave through another. In between, they embark on a linear tour connecting the two gateways. For example, a European visiting the United States may arrive in New York, travel overland to San Francisco and then return home from there. The 'multiple destination areas loop' itinerary model is the most complex, for it recognises some tourists will visit many countries and tour extensively through these different destinations. The person may engage in different travel patterns at any given stop. Thus, it may be possible for someone on an extended trip to participate in any or all of the 'single destination stopover', 'base camp tours', 'full tours' or 'open jaw' itineraries.

Finally, Flognfeldt (1999), identified four modes of recreation and vacation travel. The 'resort trip' (direct travel, single destination), 'base holiday' and 'round trip' are similar to other itinerary models discussed previously. In addition, he identified 'recreational day trips' from the individual's home community as a fourth travel type. While technically not a tourism trip because no overnight stay is involved, day trips must certainly be considered when examining the full of range of touring options.

These studies identified 26 different itinerary types, but closer inspection reveals the distinction between them is rather forced and arbitrary. Mode of transport, distance and domestic versus international travel delineate different models, when the overall patterns described are largely similar. Indeed, as Figure 2.1 illustrates, the 26 models proposed can be classified into four broad themes. The simplest itinerary type involves a single destination, there-and-back trip that may or may not include side trips. A second type involves a transit leg to the destination area, followed by a circle tour, stopping overnight at different places. A third type involves a circle tour with or without multiple access and egress points. Lastly, hub and

 Single destination, with or without side trips	Lue et al. (1993) × 2 Mings and McHugh (1992) × 1 Oppermann (1995) × 2 Flognfeldt (1999) × 1 Lew and McKercher (2006) × 1
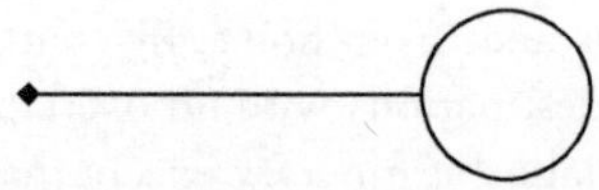 Transit leg and circle tour at a destination (transport modes may vary)	Mings and McHugh (1992) × 2 Lue et al. (1993) × 1 Oppermann (1995) ×1 Lew and McKercher (2006) × 2
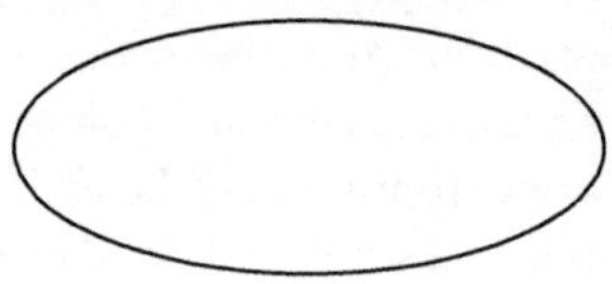 Circle tour with or without multiple access, egress points; different itinerary styles possible at different destination areas (transport mode may vary)	Mings and McHugh (1992) × 1 Lue et al. (1993) × 1 Oppermann (1995) × 3 Flognfeldt (1999) × 1 Lew and McKercher (2006) × 2
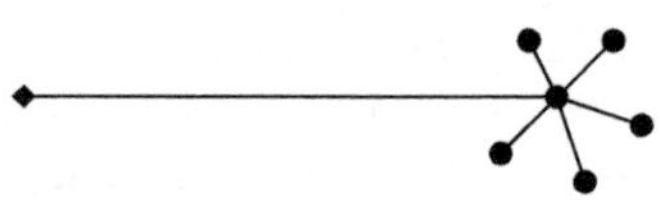 Hub and Spoke Style (From home community or destination area)	Lue et al. (1993) × 1 Oppermann (1995) × 1 Flognfeldt (1999) × 2 Lew and McKercher (2006) × 1

Figure 2.1. Itinerary types.

spoke itineraries may be evident where tourists base themselves in a destination area and take side trips to other destinations. These types of itineraries have not changed over time, related current research streams focus more on the combination of different transport means to follow these spatial patterns (Liu et al., 2023).

This research illustrates that tourist movement patterns involve both a transit and a destination touring component. The different types are shown in Figure 2.2. The various tour combinations identified result from different mixing and matching of transit and touring elements. An outbound and inbound transit leg following the same route is implied in the single destination, base camp, stopover/en-route and the regional tour/partial orbit or destination area loop models. Multiple transit legs are needed for the various loop tours identified. In rare instances, tourists may embark on a single transit leg and then have an extended return tour home.

Research in multi destination travel evolved in terms of data collection and precision of details. Cross-border (Kádár & Gede, 2021) and cross-regional (Peng et al., 2016) movements

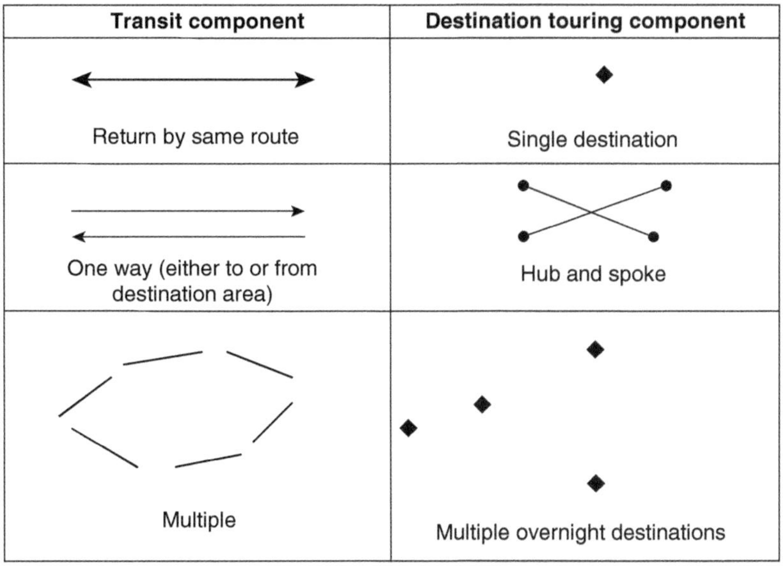

Figure 2.2. Different transit and destination touring component of itineraries.

are tracked with geotagged metadata or mobile applications. With the possibility to analyse tourist flows across nations through 'big data', the implications open up to more detailed cross-border tourism management and strategic cooperation.

Modeling Intra-destination Movements

The examination of intra-destination movements is a much younger area of study, due to three factors. The first issue relates to fineness of data. Unlike inter-destination studies where the destination represents the unit of analysis, intra-destination studies require a level of precision in the order of metres or tens of metres. This first issue leads directly to the second issue, that of the reliability of the tourist as researcher. Fineness of detail depends on how meticulous tourists are in reporting their movements, which, experience shows is rarely finer than the level of attraction node (McKercher & Lau, 2008). The emergence of passive electronic tracking devises has resolved both issues as precise movements can be tracked without any need for tourists to record their activities. However, without the integration of individual data of personal and trip characteristic variables, the explanatory power of the results are limited.

A lack of theoretical framework represented the third limiting factor. While inter-destination movement studies can inform intra-destination research, the spatial organisation of destinations adds a layer of complexity that limits their utility. Traditional urban transport models are of limited value for they are predicated on the assumption the majority of people have perfect knowledge of the road system and will take the shortest or otherwise most-efficient route possible: something that cannot be assumed for tourists who may be destination unaware and who want to explore it widely (Lew & McKercher, 2006).

Caldeira and Kastenholz (2020) categorised intra-destination spatiotemporal tourist behaviour into movement and multi-attraction dimensions that are influenced by tourist- and destination-related factors. Within the movement dimension are the elements of territoriality, linearity, locomotion and wayfinding. The multi-attraction dimension contains intensity and specificity. Territoriality relates to the distance travelled from the accommodation locus (Lew & McKercher, 2006). The path dimension builds on inter-destination studies and suggests tourists could exhibit one of seven types of movement patterns from a simple point to point pattern to a complex, rather random pattern. The types of patterns outlined by Lew and McKercher (2006) have not changed over time (Park et al., 2020). The element of wayfinding, in the conceptual framework, refers to how tourists orient themselves, while locomotion is related to the ability to move around, mostly expressed in transport usage. Intensity and to a lesser degree specificity appear to influence spatial patterns. Intensity relates to the number of stops made during the day and specificity refers to whether tourists tend to confine their movements to a specific node or wander widely throughout the destination.

Studies using network analysis support this latter finding by showing the strength of the relationship between individual attractions, attraction nodes and paths followed. These studies reveal that while movements of individual tourists may be stochastic, there is an underlying order to them. Orellana et al. (2012), for example, noted that while many places are visited often, they are hardly visited in the same order. Indeed, Shoval et al. (2015) applied the sequence alignment method and defined 15 spatio-temporal typologies of tourist activity in Hong Kong. Their study focused on the order in which the attractions are visited, captured by GPS data. Zoltan and McKercher (2015) relied on data retrieved from destination card usage, for clustering tourist experience according to movements and consumption of different attractions. Leung et al. (2012) identified the cornerstone or icon attractions that formed the foundation for these networks.

Park et al. (2020) relied on mobile network data. They used spatial clustering analysis and sequential pattern mining to reveal the popular areas of destinations (polycentricity) and spatial interactions across these places. In their analysis of three South Korean cities, they found that only 8–30 percent of those cities were visited by international tourists. Similarly, Zheng et al. (2022) combined adaptive spatial clustering with frequent pattern mining by analysing tourists mobile positioning data. Their results revealed hierarchical clustering of the destination from core to peripheral areas with temporal dimensions. Shoval et al. (2018) combined four data collection techniques, including spatio-temporal data, physiological objective and subjective measures of emotion and semantic contextual information. Using a single application that collects two types of data eases the work of the researcher and allows real time visualisation of the data. Undoubtedly, the integration of different data improves the understanding of the tourists, explains segments according to movements and aids the managerial implications that can be derived from the analysis (McKercher, 2008a). Yet, it is more likely to be achieved only with smaller sample size.

Interestingly, Zillinger (2007) raised the idea of being able to identify a rhythm which is noticeable in tourist mobility over time and space. Rhythmic movements are defined by a combination of mobility and stationariness. She found that mobility and stationariness alternate in a rhythmic and occurring way, but within each day of the journey and over the entire holiday. Travel rhythms are highly individual, influenced by the tourists' own interests, past visitation history and the spatial organisation of that destination.

Intra-destination studies also highlight the hitherto unrecognised impact that the stage of visit and hotel location have on movements (Zoltan & McKercher, 2015). Movements are heavily restricted during the arrival and departure days, while more unfettered during middle days of a visit (McKercher & Lau, 2008; Shoval et al., 2011). The tourists' effective unrestricted timed budget is, in reality, shortened by two days, limiting how widely they can explore the destination area. For example, a three day, two night stay produces only one full day where tourists can be expected to travel widely. This finding also explains why tourists on limited time budgets often pre-plan their itinerary prior to the visit and are reluctant to change once in the destination (Shoval & Raveh, 2004). In a similar manner, hotel location has been discovered to exert a pronounced effect on movements, influencing when certain places are likely to be visited during the day, sequence of visitation and probability of visiting lower order attractions (Shoval et al. 2011).

Intervening Factors

A series of intervening factors can affect tourist movements. This last section discusses the impact of distance decay and its cousin market access, time, trip and personal characteristics on movements.

Distance Decay and Market Access

Tobler (1970) identified distance decay as the *first law of geography*, noting that "everything is related to everything else, but near things are more related than distant things" (Tobler, 1970, p. 234). Distance decay states demand for activities varies inversely with the distance travelled or with increased time, money or effort required to reach a place. Rationally, tourists aim to maximise the time spent at the destination and at the same time minimise the cost of travel. The associated concept of market access builds on this idea, but argues that the number of intervening opportunities offering similar experiences affects the rate of decay (Pearce, 1989). Thus, while in aggregate tourist flows will decline with distance, the pattern will vary depending on the product type. For example, residents of the sub-tropical city of Brisbane, Australia, have the choice of literally dozens of beaches within a 150 km radius of the city, but must travel more than 2,000 km to access Australia's nearest downhill ski resort, Falls Creek. As a result, a beach located 100 km away may be deemed to have poor market access, while a ski resort located 2,000 km away might enjoy strong market access. Distance decay has been studied empirically in various tourism settings since the late 1970s (Greer & Wall, 1979) and found to be broadly applicable.

Three broad types of decay curve have been identified (McKercher & Lew, 2003). The standard curve shows demand peaking at some distance close to the tourists' home and then declining exponentially. The shape of the curve recognises tourists must travel a certain distance before they feel sufficiently removed from their home environment to make the journey worthwhile (Greer & Wall, 1979). Alternately, demand can plateau for some distance before declining, as a result of a finite number of destination options and accommodation supply along a linear touring route (McKercher, 1998a). Here, market access influences the shape of the curve. The third type has a secondary peak some distance away from the source market, in recognition of the fact that some distant destinations may have such great market appeal that their pulling power supersedes the normal frictional effect of distance.

A global study evaluating 92 source markets and 4,700 discrete destinations pairs (Figure 2.3) confirmed the ubiquity of the impact of distance on demand (McKercher & Mak, 2019). This study concluded that 53 percent of all travel departures are to land neighboured countries, while only 3 percent of travel involves a distance greater 5,000 km. Gateway cities within 2,000 km distance from source markets receive almost 80 percent of all arrivals. Decay curve patterns from most source markets adhered to the one of the three distance decay models. The origin of

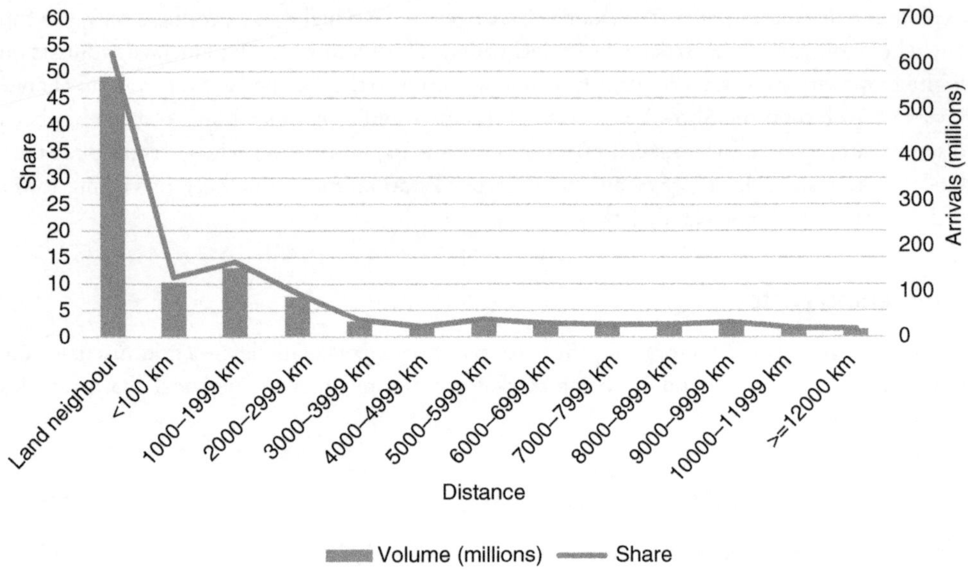

Figure 2.3. Cumulative volume of arrivals.
Source: McKercher and Mak (2019)/with permission of Elsevier.

the decay curve for island economies was shifted, but otherwise one of the three patterns was observed. Japan, New Zealand and Australia represented the only exceptions, where multiple decay curves were noted. Outbound tourism from Japan and New Zealand displayed two distinct short-haul and long-haul distance decay patterns, while three distinct peaks were noted among Australians, coinciding with travel to the South Pacific (peaking in New Zealand), South East Asia (peaking in Thailand) and Europe (peaking in the UK).

Interestingly, a similar decay pattern has also been observed within destinations (Shoval et al., 2011), where tourists are likely to spend up to 40 percent of their total daily time budget within close proximity to their hotel. Remote attractions are unlikely to be visited, unless they have attained the status of icon sites that help brand the destination. In fact, Zoltan and McKercher (2015) conceptualised the decision process in an intra-destination level, by finding that a destination's spatial structure tends to affect more tourists behaviour than its product structure.

Market access has been found to influence the type of visitor attracted to different destinations (McKercher, 1998b). Areas with strong market access did not necessarily attract more visitors or more visitor nights. But they do attract short break vacationers, through travellers and international tourists seeking a short escape from gateway cities. Destinations with poor market access tend to attract repeat visitors and those who stay for long periods. Families travelling with children seek places with strong market access for short break vacations. On the other hand, families who have more time to travel seek destinations with poor market access. Couples with no children choose to vacation at destinations with modest market access, bypassing the most proximate destinations. Market access is not fixed and, therefore, can be altered by a range of factors, including the development of new destinations in closer proximity to source markets, technological innovations, the emergence of low-cost carriers or the elimination of political barriers to travel.

Two other relationships between distance and tourist flows are worthy of discussion. First, an Effective Tourism Exclusion Zone (ETEZ) has been observed in much outbound travel (McKercher & Lew, 2003). The ETEZ represents a geographic zone where little or no tourism activity occurs that is relevant to the market under consideration. The ETEZ is generated by physical or political geography factors. Physical factors refer to natural barriers, such as oceans and mountains, while political factors relate to access issues and/or the lack of services and facilities to cater tourism. The size and proximity of the zone to the market can distort the rate of decay. An ETEZ close to the source market essentially shifts the demand curve outward. A similar zone a moderate distance from the source market accentuates the demand peak leading up to its inner edge, and can produce a secondary demand peak beyond its outer edge. This pattern exists for both inter- and intra-destination travel. More importantly, the zone represents a psychological threshold that distinguishes long- from short-haul travel. Trips taken prior to encountering the ETEZ are typically short break, short duration, single destination trips. Those that occur beyond its outer edge tend to be longer duration, multiple destination, touring-oriented trips.

Second, distance exerts a significant filtering effect on the type of person willing or able to travel long distances, which in turn affects behaviour and consumption patterns within destinations. Essentially, anyone who can travel short distances, but the extra time, cost, effort and willingness to enter culturally dissimilar environments tends to act as a 'filter' that effectively excludes certain segments (McKercher, 2008b, 2018). As a result, short haul visitors tend to be younger, have lower levels of education, lower incomes and have less travel experience. Families, young couples and groups of friends are more common. They travel for escape, fun and relaxation and seek activities that satisfy those needs, such as shopping, theme parks and beach activities. Conversely, the long-haul tourist tends to be older, more affluent and more experienced. Exceptions are backpackers, who have no time constraints. Long-haul travellers are likely to be travelling for aspirational and self-development reasons and will choose activities that reflect those goals. Exploring unique nature, and cultural heritage of faraway places would fit the description of their activities. Related to the effect of geographical distance, also cultural distance has a negative effect on international flows, according to Yang et al. (2019) analysing data on international tourist arrivals from 94 countries.

Time and Financial Budgets

Time is one of the few absolutes tourists face when they travel. Vacation times are usually fixed, with little opportunity to extend holidays. However, tourists do have control over how they choose to spend their time budgets, and this depends on whether they value time as a scarce resource to be rationed or as commodity that can be spent in many places. It is for this reason that McKean et al. (1995) argue time rationing may be the most important factor in the travel cost equation. Those who adopt a resource valuation approach see travel time as something that cannot be saved, stored or accumulated for future use. Transit time is seen as a cost, where time spent travelling must be traded-off for a shorter stay at a destination (Truong & Henscher, 1985). They can be considered as outcome-oriented individuals who seek to maximise time spent at the end point. Alternately, some people see the transit component as having a positive commodity value in itself, where the journey can be as important as the goal (Chaves et al. 1989; Walsh et al., 1990). These people can be described as process-oriented tourists. People with limited time budgets tend to adopt a more resource-oriented approach to travel and want to get to the destination or attraction as quickly as possible. Those with larger time budgets tend to adopt more of a commodity-oriented approach and will engage in touring, sightseeing and exploration.

Money also plays a role, but it is more subtle for it allows people to effectively 'buy' time or distance. For example, a self-drive tourist can likely travel 300 km or less in four hours, limiting the choice of destination. Yet, a flying time of four hours enables the tourist to travel 3,000 km or more. Conversely, someone wishing to drive 3,000 km may have to invest five or six days for each transit leg, meaning only those people with large time budgets can afford to do so. Money can thus buy time, in the sense that it can shorten transit times or for the same amount of time, it can buy distance, opening up a much wider array of destination opportunities for the tourist. With the introduction of low-cost carriers, in the early 2000s, travel distance in short time by air travel became more affordable. Indeed, this invention that went against the idea of spending more for distance when saving time, managed to create new demand for the destinations involved (Dileep & Kurien, 2021).

Trip and Personal Characteristics

Length of stay influences both multi-destination travel and the number of activities tourists can pursue within a destination area (Oppermann, 1997). The greater distance travelled, and the more difference in climate and culture perceived at the destination compared to the origin, likely to lead to longer length of stay (Jackman et al., 2020). Substantial differences have also been noted in the movements of first time and repeat visitors (Caldeira & Kastenholz, 2018). First-timers usually follow more generalist itinerary patterns, and visit iconic attractions, while repeat visitors move unrestrictedly at the destination with higher attraction and activity specificity. Moreover, McKercher et al. (2012) using Geographic Information System (GIS) analysis revealed differences in the amount of time spent at attractions, the times of day when attractions were likely to be visited and overall daily movement patterns, where first-time visitors are more likely to make one long extended trip from the hotel, while repeaters tend to make a number of short forays, returning to the hotel intermittently during the day. The role of the place visited as a main or stopover destination also affects behaviour (McKercher, 2001), due in part to the length of stay and also in part to the psychological investment made in the destination. Stopover destinations are in fact airline-dependent secondary visits, that are affected by price sensitivity, frequent-flyer membership, trip duration, and sensation seeking (Masiero et al., 2020). Furthermore, transport mode selection and movement patterns are two correlated choices of tourists (Masiero & Zoltan, 2013), where car users more likely engage in wider intra-destination or multi-destination visits.

Motivation is another key factor of tourist movement patterns. Pleasure travellers are far more likely to explore a destination than business travellers. People visiting friends and relatives tend to do less, while spending more time with family. When they travel, they may go to areas not predominantly identified as tourism nodes (Koo et al., 2012). Masiero and Zoltan (2013) looked at the effect of motivation on the number of regions visited within a destination, and found tourists who travel to visit historical places and try new foods were more likely to engage extensively in intra-destination travel, while those who were interested in experiencing landscape/nature or to be physically active demonstrate more confined, but more intense spatial movements. Special interest tourists will tend to confine their actions to activities that relate to the specialised reason for visiting, while the generalist sightseeing tourist will tend to travel more widely with no clearly evident pattern (Zakrisson & Zillinger, 2012).

Finally, regarding travel party composition, Zoltan and McKercher (2015) analysed visitation clusters in a Swiss canton, and found the presence of children restricted the movements and type of activities consumed, compared to only-adult groups. Le-Klähn et al. (2015) and others have observed substantial differences between domestic and international tourists; in attraction selection, public transport use and spatial movements at the destination.

Conclusion

An intricate relationship exists among time, space and tourism movements. Over the years, a variety of models has been developed to portray the movement of tourists from their homes to destination areas or between destination areas. These models recognise that tourism movements involve two components, a destination component, and a transit component. Movements of tourists are, in turn, moderated by a number of factors, including the frictional effect of distance on demand, the number of intervening opportunities available, tourists' total time budget and how they choose to spend that time, and trip variables.

The discipline of geography has played a central role in the evolution of tourism as a field of study. The desire to understand the spatial interactions of tourists with a destination and the movement of tourists between destinations has played a critical role in developing the phenomenon of tourism. The geography of tourism seems to have become relatively less important over the last decades, however, as other disciplines have discovered tourism. Yet, an appreciation of spatial relationships forms one of the foundations of tourism by which any study, regardless of the discipline, is based. Many exciting research opportunities exist to build on the existing knowledge base or to reexamine other tourism concepts from a temporal/spatial perspective.

References

Caldeira, A. M., & Kastenholz, E. (2018). Tourists' spatial behaviour in urban destinations: The effect of prior destination experience. *Journal of Vacation Marketing*, *24*(3), 247–260. https://doi.org/10.1177/1356766717706102

Caldeira, A. M., & Kastenholz, E. (2020). Spatiotemporal tourist behaviour in urban destinations: a framework of analysis. *Tourism Geographies*, *22*(1), 22–50. https://doi.org/10.1080/14616688.2019.1611909

Chavas, J. P., Stoll, J., & Sellar, C. (1989). On the commodity value of travel time in recreational activities. *Applied Economics*, *21*(6), 711–722. https://doi.org/10.1080/758520269

Dileep, M., & Kurien, A. (2021). *Air transport and tourism: Interrelationship, operations and strategies*. Routledge.

Edwards, D., & Griffin, T. (2013). Understanding tourists' spatial behaviour: GPS tracking as an aid to sustainable destination management. *Journal of Sustainable Tourism*, *21*(4), 580–595. https://doi.org/10.1080/09669582.2013.776063

Flognfeldt, T. (1999). Traveler geographic origin and market segmentation: The multi trips destination case. *Journal of Travel and Tourism Marketing*, *8*(1), 111–124. https://doi.org/10.1300/J073v08n01_07

Greer, T., & Wall, G. (1979). Recreational hinterlands: A theoretical and empirical analysis. In G. Wall (Ed.), *Recreational land use on Southern Ontario: Department of Geography publication series #14*, (pp. 227–246). Waterloo University.

Jackman, M., Lorde, T., Naitram, S., & Greenaway, T. (2020). Distance matters: the impact of physical and relative distance on pleasure tourists' length of stay in Barbados. *Annals of Tourism Research*, *80*, 102794. https://doi.org/10.1016/j.annals.2019.102794

Kádár, B., & Gede, M. (2021). Tourism flows in large-scale destination systems. *Annals of Tourism Research*, *87*, 103113. https://doi.org/10.1016/j.annals.2020.103113

Koo, T. T., Wu, C. L., & Dwyer, L. (2012). Dispersal of visitors within destinations: Descriptive measures and underlying drivers. *Tourism Management*, *33*(5), 1209–1219. https://doi.org/10.1016/j.tourman.2011.11.010

Le-Klähn, D. T., Roosen, J., Gerike, R., & Hall, C. M. (2015). Factors affecting tourists' public transport use and areas visited at destinations. *Tourism Geographies*, *17*(5), 738–757. https://doi.org/10.1080/14616688.2015.1084527

Leung, X. Y., Wang, F., Wu, B., Bai, B., Stahura, K. A., & Xie, Z. (2012). A social network analysis of overseas tourist movement patterns in Beijing: The impact of the Olympic Games. *International Journal of Tourism Research*, *14*(5), 469–484. https://doi.org/10.1002/jtr.876

Lew, A., & McKercher, B. (2006). Modeling tourist movements: A local destination analysis. *Annals of Tourism Research*, *33*(2), 403–423. https://doi.org/10.1016/j.annals.2005.12.002

Liu, S., Zhang, J., Liu, P., Xu, Y., Xu, L., & Zhang, H. (2023). Discovering spatial patterns of tourist flow with multi-layer transport networks. *Tourism Geographies*, *25*(1), 113–135. https://doi.org/10.1080/14616688.2020.1850849

Lue C., Crompton, J., & Fessenmaier, D. (1993) Conceptualization of multi-destination pleasure trips. *Annals of Tourism Research*, *20*(2), 289–301. https://doi.org/10.1016/0160-7383(93)90056-9

Masiero, L., & Zoltan, J. (2013). Tourists intra-destination visits and transport mode: A bivariate probit model. *Annals of Tourism Research*, *43*, 529–546. https://doi.org/10.1016/j.annals.2013.05.014

Masiero, L., Qiu, R. T., & Zoltan, J. (2020). Long-haul tourist preferences for stopover destination visits. *Journal of Travel Research*, *59*(5), 811–827. https://doi.org/10.1177/0047287519867143

McKean, J. R., Johnson, D. M., & Walsh, R. G. (1995). Valuing time in travel cost demand analysis: An empirical investigation. *Land Economics*, *71*(1), 96–105. https://doi.org/10.2307/3146761

McKercher, B. (1998a). The effect of market access on destination choice. *Journal of Travel Research*, *37*(1), 39–47. https://doi.org/10.1177/004728759803700105

McKercher, B. (1998b). The effect of distance decay on visitor mix at coastal destinations. *Pacific Tourism Review*, *2*(3/4), 215–224. https://www.ingentaconnect.com/content/cog/ptr/1998/00000002/f0020003/art00003

McKercher, B. (2001). A comparison of main destination and through travelers at a dual-purpose destination. *Journal of Travel Research*, *39*(4): 433–441. https://doi.org/10.1177/004728750103900410

McKercher, B. (2008a). Segment transformation in urban tourism. *Tourism Management*,*29*(6), 1215–1225. https://doi.org/10.1016/j.tourman.2008.03.005

McKercher, B. (2008b). The implicit effect of distance on tourist behaviour: A comparison of short and long haul pleasure tourists to Hong Kong. *Journal of Travel and Tourism Marketing*, *25*(3/4), 367–381. https://doi.org/10.1080/10548400802508473

McKercher, B. (2018). The impact of distance on tourism: a tourism geography law. *Tourism Geographies*, *20*(5), 905–909. https://doi.org/10.1080/14616688.2018.1434813

McKercher, B., & Lau, G. (2008). Movement patterns of tourists within a destination. *Tourism Geographies*, *10*(3), 355–374. https://doi.org/10.1080/14616680802236352

McKercher, B., & Lew, A. (2003). Distance decay and the impact of effective tourism exclusion zones on international travel flows. *Journal of Travel Research*, *42*(2), 159–165. https://doi.org/10.1177/0047287503254812

McKercher, B., & Mak, B. (2019). The impact of distance of international tourism demand. *Tourism Management Perspectives*, *31*, 340–347. https://doi.org/10.1016/j.tmp.2019.07.004

McKercher, B., Shoval, N., Ng, E., & Birenboim, A. (2012). First and repeat visitor behaviour: GPS tracking and GIS analysis in Hong Kong. *Tourism Geographies*, *14*(1), 147–161. https://doi.org/10.1080/14616688.2011.598542

Mings, R. C., & McHugh, K. (1992). The spatial configuration of travel to Yellowstone National Park. *Journal of Travel Research*, *30*(4), 38–46. https://doi.org/10.1177/004728759203000406

Oppermann, M. (1995). A model of travel itineraries. *Journal of Travel Research*, *33*(4), 57–61. https://doi.org/10.1177/004728759503300409

Oppermann, M. (1997). Length of stay and travel patterns. In R. Bushell (Ed.), *Tourism research: Building a better industry*, CAUTHE, BTR, Canberra, pp 471–480.

Orellana, D., Bregt, A., Ligtenberg, A., & Wachowicz, M. (2012). Exploring visitor movement patterns in natural recreational areas. *Tourism Management*, *33*(3), 672–682. https://doi.org/10.1016/j.tourman.2011.07.010

Park, S., Xu, Y., Jiang, L., Chen, Z., & Huang, S. (2020). Spatial structures of tourism destinations: A trajectory data mining approach leveraging mobile big data. *Annals of Tourism Research*, *84*, 102973. https://doi.org/10.1016/j.annals.2020.102973

Pearce, D. (1989). *Tourist development* (2nd ed.). Longman.

Peng, H., Zhang, J., Liu, Z., Lu, L., & Yang, L. (2016). Network analysis of tourist flows: a cross-provincial boundary perspective. *Tourism Geographies*, *18*(5), 561–586. https://doi.org/10.1080/14616688.2016.1221443

Shoval, N., & Raveh, A. (2004). Categorization of tourist attractions and the modeling of tourist cities: based on the co-plot method of multivariate analysis. *Tourism Management*, *25*(6), 741–750. https://doi.org/10.1016/j.tourman.2003.09.005

Shoval, N., McKercher, B., Birenboim, A., & Ng, E. (2015). The application of a sequence alignment method to the creation of typologies of tourist activity in time and space. *Environment and Planning B: Planning and Design*, *42*(1), 76–94. https://doi.org/10.1068/b38065

Shoval, N., McKercher, B., Ng, E., & Birenboim, A. (2011). Hotel location and tourist activity in cities. *Annals of Tourism Research*, *38*(4), 1594–1612. https://doi.org/10.1016/j.annals.2011.02.007

Shoval, N., Schvimer, Y., & Tamir, M. (2018). Real-time measurement of tourists' objective and subjective emotions in time and space. *Journal of Travel Research*, *57*(1), 3–16. https://doi.org/10.1177/0047287517691155

Tobler, W. R. (1970). A computer movie simulating urban growth in the Detroit region. *Economic Geography*, *46*, 234–240. http://www.jstor.org/stable/143141

Truong, T. P., & Hensher, D. A. (1985). Measurement of travel time values and opportunity cost from a discrete-choice model. *The Economic Journal*, *95*(378), 438–451. https://doi.org/10.2307/2233219

Walsh, R. G., Sanders, L. D., & McKean, J. R. (1990). The consumptive value of travel time on recreation trips. *Journal of Travel Research*, *29*(1), 17–24. https://doi.org/10.1177/004728759002900105

Yang, Y., Liu, H., & Li, X. (2019). The world is flatter? Examining the relationship between cultural distance and international tourist flows. *Journal of Travel Research*, *58*(2), 224–240. https://doi.org/10.1177/0047287517748780

Zakrisson, I., & Zillinger, M. (2012). Emotions in motion: tourist experiences in time and space. *Current Issues in Tourism*, *15*(6), 505–523. https://doi.org/10.1080/13683500.2011.615391

Zheng, W., Li, M., Lin, Z., & Zhang, Y. (2022). Leveraging tourist trajectory data for effective destination planning and management: A new heuristic approach. *Tourism Management*, *89*, 104437. https://doi.org/10.1016/j.tourman.2021.104437

Zillinger, M. (2007). Tourist routes: A time-geographical approach on German car-tourists in Sweden. *Tourism Geographies*, *9*(1), 64–83. https://doi.org/10.1080/14616680601092915

Zoltan, J., & McKercher, B. (2015). Analysing intra-destination movements and activity participation of tourists through destination card consumption. *Tourism Geographies*, *17*(1), 19–35. https://doi.org/10.1080/14616688.2014.927523

Chapter 3

Consumer Behaviour in Tourism

Girish Prayag

Introduction

Consumer behaviour (CB) remains one of the most studied aspects in the tourism field (Cohen et al., 2014). The diversity and depth of topics that encompass tourist behaviour cannot be reviewed comprehensively here, but this chapter summarises some of the key issues that surround the topic and identify those that require further attention in the tourism literature. The chapter starts with an overview of the evolving nature of CB, commonly referred to as tourist behaviour, using the lens of evolutionary psychology. From the outset, much of what has been written on CB has been drawn primarily from the psychology and economics literatures, with an increasing number of other disciplines such as sociology, geography and anthropology continuing to infuse the discipline of marketing in general and by implication the study of CB. From there on, the chapter reviews decision-making models in tourism, the tourist experience and its components, and further hones on sociological and political factors that inform the study of CB. The chapter embeds several areas of further research on each topic throughout.

Evolutionary Psychology and Consumer Behaviour

Economists have traditionally understood CB as stemming from private preferences and rational decision-making when faced with several choices, which are simply revealed through actual choice behaviours (Faber & Frenken, 2009). For example, consumers are thought to apply decision rules such as conjunctive or disjunctive to produce a set of acceptable alternatives, but they can also use compensatory approaches to arrive at a single best alternative (Gilbride & Allenby, 2004). In a travel and tourism context, these rules can facilitate destination and activity choices. As a result, the dominant approach to analysing CB has been to view consumer choice as an act of constrained optimisation. Consequently, for a change in behaviour to occur, it is assumed that a change in relative prices or in access to information that

The Wiley Blackwell Companion to Tourism, Second Edition. Edited by C. Michael Hall.

results in the consumer judging their satisfaction or wellbeing will improve is required (Earl, 2016). This has led to a strong focus on consumer information processing models of CB in the early days and this will be discussed later in the chapter. In these early days to change behaviour, policies were aimed at changing consumption patterns through the use of price instruments such as taxes and subsidies (Faber & Frenken, 2009). A good example is the smoking ban in indoor places, including restaurants and hotels that was introduced in the mid 2000s in several countries in Europe (European Network for Smoking Prevention, 2011). There was also the use of subsidies by governments to national airlines to make prices more accessible for tourists for travel.

An evolutionary approach views CB as 'architecture', that is, the design of complex structures or systems that consumers construct as a means for coping with life. The two approaches, private preferences vs. evolutionary, are not necessarily in competition for policy-making purposes aimed at changing behaviour (Earl, 2016). The evolutionary approach focuses on the dynamics of changing preferences through examining the particular interaction of demand agents with each other or with other players on the market (Faber & Frenken, 2009). This is critical to support sustainable tourism initiatives, of which changing tourist behaviour is an important one. For example, New Zealand has introduced the Tiaki Promise to educate international visitors on responsible behaviours when travelling. Tiaki means to care for people, place and culture. The evolutionary approach questions the use of pricing based policies for promoting change in behaviour and emphasises the role of non-price factors and complementarities in choice (Earl, 2016). For example, the SHIFT framework has recently been proposed as a way to shift CB to address climate change. The SHIFT factors are Social influence, Habit, Individual self, Feelings and cognition, and Tangibility (Habib et al., 2021). Hence, the evolutionary view suggests that change may be resisted, not because incentives are weak, but because making the change would compromise the architecture of a system that the consumer has created (Earl, 2016). As highlighted in several tourism studies, attempts to change tourist travel behaviour to more sustainable ones have been largely ineffective (Buckley, 2012). Improvements are driven principally by regulatory changes, often against political resistance and with poor implementation. Market measures are largely ineffective, with little direct public demand for sustainability in tourism (Buckley, 2012). Tourists are often the least engaged stakeholder compared to government and businesses in driving and sustaining the implementation of sustainable tourism (Rasoolimanesh et al., 2020).

Consumer motivation and learning have been central to understanding CB and driving change in CB over the years. Consumers learn to consume and develop routines for consumption activities (Faber & Frenken, 2009). Several people go on holidays to the same destination every year as they have developed routines and coping mechanisms that are repetitive, which causes less stress and anxiety throughout the entire holiday experience. There are many forms of learning, including classical conditioning, operant conditioning and ecology (Nord & Peter, 1980). It has been argued that the notion of operant conditioning in CB is most closely aligned to an evolutionary explanation of behaviour. Operant conditioning is the procedure in which a positive or negative response is determined by the prior consequences of similar behaviour, whereby reinforcers augment the chance of a positive response, and punishers reduce the occurrence of a negative response (Foxall, 1993). As an example, early booking discounts encourage consumers to book their flights early, while on the day tickets tend to be very expensive. Budget airlines such as Easyjet, Ryanair and Jetstar also encourage consumers to pre-purchase luggage, seats and meals, offering online discounts instead of on the spot buying at the airport check-in desk, which often charges exorbitant prices. As a result, consumers select choices based on avoiding negative consequences (e.g., early booking) and the contingencies of reinforcement (e.g., positive rewards such as airline points and speedy boarding) that shape and maintain operant behaviour (Foxall, 1993). For change to occur in consumption behaviour, it is necessary to disrupt the cumulating effects of evolutionary processes so that alternative sets of stored neural connections get tried for fit or new sets of neural connections are constructed and found to fit (Earl, 2016). This approach has been recommended to encourage more carbon-offsetting by consumers when they travel rather than relying on their voluntary off-set behaviour.

Socialisation based theorists argue that all behaviours are learnt, without providing the genesis and reasons for such learning (Saad & Peng, 2006). However, learnt behaviours that have a cultural component are not devoid of evolution. The vast majority of behaviours can be explained using three components of evolution (adaptations, by-products and noise) (Durante & Griskevicius, 2016). Products themselves are typically systems of interconnected subsystems (Earl, 2016), of which culture is often an important component. For example, the destination product reflects a complex set of systems, processes and stakeholder involvement that have to work together to deliver the destination experience. Therefore, the evolutionary perspective recognises that different adaptive opportunities and threats (e.g., launching a new product or service in a culturally different market) demand qualitatively different solutions (Griskevicius et al., 2010). Global tour operators such as Flight Centre, Expedia, and American Express Global Business Travel have to tailor packages to suit different markets such as Asia, Europe and the Middle East.

Evolutionary psychology (EP), in particular, can be used to explain why a particular behaviour is manifested in the market place rather than what the behaviours manifested in the market place are (Saad & Gill, 2000). It has been argued that despite EP being a scientific discipline with no social agenda, greater knowledge of evolved physiological mechanisms and the environmental cues that activate them can, in principle, aid the effectiveness of pre-existing widely desired social policies (Confer et al., 2010). Thus, EP may have an important role to play in understanding behavioural change that aligns with increasing social expectations from individuals and communities to combat poverty, climate change and prevent social exclusion.

Several harmful and dangerous behavioural patterns (e.g., gambling and drunk driving) seem also to cluster within very specific demographic groups (Saad & Peng, 2006), exacerbated by poverty and social exclusion that often have their roots in colonisation, and intervention strategies to curb such behaviours do not always work. Often, these strategies fail not because of incomplete information but a lack of understanding that many of the targeted behaviours may have a Darwinian-etiology (Saad & Peng, 2006) and systemic issues within communities and societies in general. Darwinian theorising (e.g., EP, life-history theory, gene-culture co-evolution, and memetic theory) can be used to develop more efficacious social intervention strategies (Saad & Peng, 2006) and for social marketing strategies. Evolutionary approaches can also explain the corporate social responsibility (CSR) consumer paradox (Janssen & Vanhamme, 2015). Consumers are more likely to buy CSR products that involve a public display of their altruistic behaviour, of a conspicuous category or that are available in purchase settings that enhance the public visibility of the purchase act (Janssen & Vanhamme, 2015). When status motive is activated, people tend to choose greener products over conventional products if the purchase occurs in public and if the green product costs more than the conventional product (Griskevicius et al., 2010). Knowing such subconscious influences can prove useful for encouraging pro-social alternatives and thus societal wellbeing at large (Janssen & Vanhamme, 2015).

In the field of tourism, Crouch (2013) argues that factors such as motivation, culture, decision-making related to consumption, perception, attitude formation and change, emotions, personality, shopping behaviour, gift giving, conspicuous consumption, and risk related behaviour are the enablers of tourism consumption underpinned by Darwinian's drivers such as the search for improved resources, display of health, status and capability, and fleeing danger. The need for escape as a tourist motive is a classic example of the 'fleeing danger' etiology. Similarly, EP provides insights into material possessions (van Tilburg et al., 2015) and thus can be used to understand why tourists buy second homes, souvenirs, and yachts. Holidays are typical examples of conspicuous consumption and destinations such as Venice, Capri, and Taormina in Italy attract a segment of tourists that are status conscious. It has been argued that tourism is an exemplar of conspicuous consumption, and this reflects sexual signalling that caters to the Darwinian mating module (Saad & Peng, 2006). Likewise, the desire to conform to arbitrary fashion trends or the need to travel to certain destinations that reinforces one's status have at their roots the Darwinian need to belong (Saad & Peng, 2006). Much of tourist behaviour can be explained by evolutionary approaches, but often these approaches are not considered or explicitly identified in tourism studies. A few exceptions have emerged in recent years. For example, Kim and Seo (2019) establish the causal relationship between the activation of mating motives and gender-specific preferences for risk-taking in tourism. Through four experimental studies, they illustrate that the activation of mating motives in men increases their propensity for risk-taking in tourism activities compared to women. Individuals have demonstrated an attraction toward foreignness, which, arguably, constitutes a central reason to travel (Nørfelt et al., 2020). Drawing on social and EP, Nørfelt et al. (2020) demonstrate that tourist xenophilia is linked to willingness to engage with locals, willingness to stay at a bed-and-breakfast, intention to try local food, resident hospitality, support for immigration policies, and travel to foreign destinations. Kock et al. (2018) introduce five fundamental motives that are considered to be different to the proximal motives of tourists for travel, the latter being the main focus of tourism studies so far. They argue that an EP lens enhances our understanding of tourist behaviour: evading harm, exploration, affiliation, status, and mating. The COVID-19 pandemic can also be considered as a paradigm shift in research on tourists' behaviour and decision-making (Kock et al., 2020). This is because the pandemic unsheathed and amplified many deep-rooted human anxieties of contagious diseases, physical harm, and social isolation (Kock et al., 2020). The Evolutionary Tourism Paradigm as proposed by Kock et al. (2020) can be used to understand tourist xenophobia, overcrowding perceptions and tourism ethnocentrism. Next, the chapter identifies some of the major models that have been used to understand tourist behaviour.

Decision-making Models of Tourist Behaviour

While traditional CB models are drawn mainly from rational decision-making processes (Swarbrooke & Horner, 2007), explaining the strong influence of the discipline of economics on studying tourist behaviour, the influence of psychology and marketing on CB cannot be ignored. One of the early models of CB proposed by Andreason (1965) highlights the role of various types of information on attitudes, beliefs, norms, feelings, personality, values and demographic and household characteristics on the buying decision process. This information processing approach underpins many popular theories used to explain CB such as Theory of Planned Behaviour (TPB) (Ajzen, 1985) and Model of Goal Directed Behaviour (MGB) (Perugini & Bagozzi, 2001). The TPB emphasises that human behaviours are governed not only by personal attitudes and social pressures (subjective norms) but also by a sense of behavioural control (Ajzen, 1985). The MGB was proposed as an alternative to TPB in an endeavour to enhance the capacity and address the limitations of the TPB by incorporating motivational and affective processes and past behaviour (Perugini & Bagozzi, 2001). Also, the Howard and Seth (1969) Theory of Buyer Behaviour examining consumption from the lens of extensive, limited and habitual problem solving has been influential on the CB discipline.

One particular aspect of CB that has galvanised research attention in the tourism literature is that of understanding and explicating the motives of tourists to visit places and understanding the factors that drive tourist demand. For example, models that attempt explain motivation include Crompton's (1979) push-pull factors, where push factors have been described as internal drivers such as the need for escape, relaxation, novelty, socialisation, and prestige/status, among others, push individuals to go on holiday. The pull factors are thought of as tourist activities, amenities and infrastructure available at a destination to fulfill the travel needs of tourists (Prayag & Ryan, 2011). An interaction of the push and pull factors determines destination choice (Prayag & Ryan, 2011). The push-pull model of tourist motivation can be linked also to the tourism system, where the push factors arise in the generating region and the pull factors are either available or not in the destination region. Pearce and Lee's (2005) travel career ladder suggested that host-site-involvement motivation (e.g., experiencing different cultures) and nature related motivation (e.g., being close to nature) were more important factors to more experienced travellers, whereas motivations such as stimulation, personal development, relationship (security), self-actualisation, nostalgia, romance, and recognition had a higher priority for the less experienced ones. Motivation studies have shifted over the years from examining pleasure travellers to business travellers and niche segments travel motivation (Cohen et al., 2014). More recently, motivation of travellers through the lens of online reviews has become a popular topic of study. For example, Kwok et al. (2017) identify several streams of research in understanding not only motivation but other aspects of online behaviours of tourists such as quantitative evaluation features, verbal evaluation features, reputation features and social features of online reviews that can have important outcomes for consumer decision-making and business performance.

While Howard and Seth's (1969) model proposed different decision-making processes for different types of problems, there is also an assumption that a staged approach is involved in decision-making embedded in notions of pre, during and post-consumption (Cohen et al., 2014). Not surprisingly, there are several studies examining pre-consumption decision-making in relation to destination choice (see Wong et al., 2017) An implicit premise of this view corroborates the persistency of tourists in fulfilling these needs by developing a travel career (Pearce & Lee, 2005). The information processing approach of decision-making is embedded in the assumption that markets are stable and that visitors have limited constraints. In fact, traditional models of consumer behaviour and those pertaining specifically to tourist behaviour have taken for granted the availability of time, budget, and good health as the building blocks for explaining tourists' decisions to travel or not. An alternative perspective of travel demand and choice rests on travel resistance and constraints. This view focuses on those factors that inhibit tourists from travelling (Nyaupane & Andereck, 2008) and their travel risk perceptions (Rittichainuwat & Chakraborty, 2009). During the COVID-19 pandemic, despite travellers potentially having the time and budget, fear for their health and stronger border controls seem to have dampened the desire to travel for many. For example, Shin et al. (2022) highlight the importance of different factors such as tourist trust (political, destination, and interactional trust), travel constraint (intrapersonal, interpersonal, and 'social distancing' structural constraint), and elements of TPB (travel attitude, perceived behavioural control, subjective norm, perceived health risk, past travel experience), as impacting travel decision-making during and after the pandemic. However, others factors such as mortality salience may be at play, but has yet to be researched in relation to the pandemic and travel behaviours.

Given that most grand models of consumer behaviour in tourism view tourists as rational decision makers (Swarbrooke & Horner, 2007), is not only a limitation to understand tourist behaviour but may also be unable to explain irrational behaviours that require further research as well. Deviant tourist behaviour is an important issue in tourism management (Li & Chen, 2017). For example, as reported in media reports in several countries, at the beginning of the pandemic younger tourists such as backpackers and independent travellers were willing to compromise their own safety in relation to COVID-19 to still travel as they perceived they were at lower risk compared to others. The drivers of this particular behaviour must be studied as well as other types of misbehaviour by tourists that have emerged as a result of this pandemic (e.g., not following isolation protocols and physical distancing and endangering the health of local residents through non-compliance of government advice and rules). Within the decision-making process and destination choice behaviours, to what extent individual and group travel decision making have been impacted by COVID-19 is starting to attract research attention (Shamshiripour et al., 2020), with tourists employing self-protection, coping and resilience strategies to better navigate the new travel challenges (Zheng et al., 2021).

Decision-making can also be understood from other perspectives such as the cognition-emotion school of thought or the emotion-cognition perspective as well as behavioural learning perspectives. The cognition-emotion school of thought posits the casual role of cognition as a necessary but not sufficient condition in order to elicit emotions (Lazarus, 1991). From this perspective, psychological factors such as motivation, values, and beliefs are cognitive concepts that have an influence on attitudes and behaviors. However, hedonic consumption such as tourism, second home and holiday yacht purchase, are good examples of experiences that are goal directed and evoke emotions (Hosany, 2012; Otto & Ritchie, 1996). When consumers' motives are intrinsically rewarding rather than as a means to an end, affective considerations should be more important determinants of behaviour (Pham, 1998). If a situation lacks significance for the person's wellbeing, little motivational force will be generated and no emotions will occur (Lazarus, 1991).

In the marketing and tourism literature, need fulfillment is positively related to positive outcome emotions while failure to fulfill needs is negatively related to negative outcome emotions (Prayag et al., 2022). In this way, motivation determines the valence of the emotional response and differentiates between positive and negative emotions (Roseman et al., 1990). Positive emotions such as happiness, elation and joy are usually associated with goal attainment while negative emotions such as frustration, disappointment and anxiety result from failure to achieve desired goals (Bagozzi et al., 1999). Thus, in a destination setting the emotion of joy would indicate the fulfillment of holiday motives, whereas sadness would indicate failure of needs fulfillment (Hosany, 2012). This illustrates the link between motivation and emotion. Emotion has a central role to play in the formation of other psychological factors such as attitudes and behaviours (Cohen et al., 2014). The tourism literature has taken a critical turn in recognising that tourism experiences are laden with emotions, with often both positive and negative consequences on post-consumption behaviours such as satisfaction and loyalty (Hosany et al., 2021), and more recently, the wellbeing of tourists (Wang et al., 2021).

An alternative view to the cognition-emotion school of thought is that emotion has a strong influence on cognitive processes (Clore et al., 1994). Individuals are more likely to recall information from memory that is congruent with their current moods and feelings (Isen et al., 1978). Both mood-congruent recall and the use of one's feelings as a basis of judgement, may influence decision-making by affecting the accessibility and evaluation of valenced features of the decision situation (Schwarz, 2000). In addition, affective states influence which strategy of information processing individuals are likely to adopt (Schwarz, 2000). Nonetheless, it must be acknowledged that the relationship between emotions and decision-making is bidirectional, and the positive or negative outcome of a decision can profoundly affect the decider's feelings (Schwarz, 2000). While the tourism literature has primarily researched emotions from the cognition-emotion school of thought through theories such as cognitive appraisal theories (Hosany, 2012) and attribution theories (Zhang et al., 2021), there has been several calls for emotions to be studied differently (Hosany et al., 2021; Volo, 2021). Thus, there is an opportunity to expand the breadth and depth of emotions research to better understand tourists' decision-making processes. Likewise, from a learning theory perspective, most studies consider 'think' or 'feel' concepts as driving a certain behaviour but the affect-driven, hierarchy-of-effects (HOE) model (Vaughn, 1986), support the premise that 'do' will ultimately lead to an outcome of 'feel'. The HOE model seems to explain behavioural change in consumers in the area of social marketing (Russell-Bennett et al., 2016).

Irrespective of which school of thought one adopts, psychological factors such as motivation, values, beliefs, perceptions, attitudes, satisfaction and behavioural intentions remain the core facets of tourist behaviour (Cohen et al., 2014). The way in which these concepts affect each other can shape the tourist experience. Despite these factors being well researched in tourism, their collective influence and inter-relationships lack empirical evidence. For example, capturing how motivation to travel, for example, is formed and/or dampened in times of global tourism industry instability and its influence on attitudes, perceptions and risks could shed further light on tourist behaviour. Recent studies (e.g., Aebli et al., 2022) have started to provide empirical evidence of some of the ways in which these factors are related (e.g., motivation, travel risk and wellbeing). Visitors travel for motives such as escape, personal growth, socialisation, enjoyment, health and wellbeing (Cohen et al., 2014) but how health and wellbeing concerns have and will change tourist motives requires further research (Aebli et al., 2022). Given that COVID-19 disproportionately impacted older people (Miller, 2020), the motives, attitudes and behaviours of the senior tourism market will require further scrutiny. Existing studies on motivations and behaviours of this market abound in the literature (Otoo & Kim, 2020; Prayag, 2012) but as the review of the literature by Otoo and Kim (2020) showed, personal safety and mental/physical wellbeing motives (e.g, travel for peace of mind, sense of connectedness) appear to have been given lower importance in relation to other travel motives but are acutely important in relation to the COVID-19 pandemic (Aebli et al., 2022). Next, once the tourist travels to the destination, managing the in-situ experience becomes important for the destination to ensure that positive behavioural outcomes emerge from the tourist experience. Outcomes beyond satisfaction and loyalty such as tourist wellbeing and resilience should be devoted more attention in CB models.

Tourist Experience

Both the marketing and tourism literatures acknowledge the need to broaden and deepen knowledge on the concept of experience and its antecedent factors and outcomes. The concept requires the integration of not only the three original elements (cognition, affect and conation) of attitude but also other relatively new and emerging elements (Prayag et al., 2020).

For example, recent definitions of consumer experience suggest that it consists of elements such as sensorial, social and spiritual (Gentile et al., 2007; Prayag et al., 2020; Taheri et al., 2021). However, terms such as 'consumer experience', 'tourist experience', 'customer experience' and 'tourism experience' are often used interchangeably in the tourism literature without recognising the subtleties within those terms. For example, tourism experience can be considered as the supply side perspective of the offer while the term tourist experience is actually the demand side perspective representing the consumption of that tourism experience. Undoubtedly the concept of tourist experience has some parallels to the key marketing concept of customer experience in its theorisation (Homburg et al., 2017), yet the growing number of studies on this topic within both the marketing and tourism literatures add to the fragmentation and theoretical confusion (Becker & Jaakkola, 2020; Taheri et al., 2021). The lack of a unified view creates not only considerable challenges for theory development but also for operationalisation of the concept (Taheri et al., 2021).

While tourist experience research implicitly assumes that experiences from tourism are generally positive and therefore worth designing and managing, COVID-19 raises the question of whether expectations of travellers with respect to tourist experiences locally and globally have really changed and in what ways (Aebli et al., 2022). For example, recent studies (e.g., Yang et al., 2021) suggest that tourists' real-time on-site emotional experience after the outbreak of COVID-19 is significantly lower than that of the period before the COVID-19 outbreak, suggesting that tourism destinations should not only focus on the recovery of tourist arrivals, but also pay attention to the tourist experience recovery during the tourism recovery stage. The mantra of value co-creation (Assiouras et al., 2019) so prevalent in tourism studies has changed as tourists have to consider health and risk factors in their travel decision-making and providers have incorporated such risks in their pricing strategies. Thus, value creation becomes more challenging. Therefore, how tourists perceive value, their engagement in co-creating value and the relative importance of emotional versus functional value will need to be teased out by researchers. The relationship between perceived value and risk perceptions is not new in tourism research with lower risks often associated with higher perceived value (Wang et al., 2019). However, considerations have to be given to potential moderating factors such as personality, sensation-seeking and resilience of visitors that can have a crucial role to play in determining the wellbeing of tourists. More importance must be given to the application of positive psychology principles to further enhance the potential contribution of tourism experiences to tourists' wellbeing, or lack of (Chen & Yoon, 2019).

Related to the tourist experience, negative experiences with operators, airlines, travel agents/online travel agencies (OTAs) and communities need to be researched to fully understand the impacts on wellbeing and resilience of tourists. Satisfaction and dissatisfaction are well researched concepts in tourism (Prayag et al., 2019) and satisfaction can be studied either as an outcome or as a process (Cohen et al., 2014), and the interplay between these two in the face of COVID-19 should provide additional insights into how satisfaction remains high or morphs into dissatisfaction during a crisis. Tourists globally have experienced lock-downs and have had to self-isolate in hotels, apartments, Airbnb, and youth hostels among other forms of accommodation, their satisfaction or lack of, will provide valuable insights into how the accommodation sector has coped with the pandemic and the toll on both tourists' psychological health and their experience of destinations. Along with satisfaction, other manifestations of customer citizenship behaviours such as customer feedback, advocacy, customer-to-customer assistance and tolerance in less satisfactory future services can be investigated (Assiouras et al., 2019).

Already, the media have reported that the mere presence of tourists in some communities have created conflicts as tourism restarts after the pandemic. Tourists are perceived as a contributing factor to the spread of the virus. Hostility and discrimination against tourists has gained research attention and tourist harassment by locals has substantial effects on tourists' cognitive assessment, emotional response, coping strategy, destination image, and future intentions (Otoo et al., 2019; Tse & Tung, 2022). Tourists are subjected to both aggressive and passive forms of harassment (Otoo et al., 2019) from residents and other tourists, and their coping strategies are important for understanding tourist wellbeing and resilience. Therefore, COVID-19 offers opportunities to study these phenomena in more depth, while incorporating the resident view in such research could expand our understanding of host-guest interactions (Tse & Tung, 2022).

Sociological Factors of CB

In relation to sociological factors, the influence of societal norms and values on decision-making and the tourist experience still remains an area that requires much research. For example, Gitelson and Kerstetter (1995) showed how friends and family influenced travel decision-making processes. In Chinese culture, Song et al. (2017) identify the conflict resolution mechanisms used by tourists in group-travel. Peer-pressure as another example, has been shown to change the mode of transport used by travellers (Feygin & Pozdnoukhov, 2018) enabling more socially responsible and environmentally conscious mobility habits. The influence of significant others is part of the broader reference group influence on travel. At the same time, it must be recognised that cultural factors such as religious beliefs, values and ideologies have a strong bearing on psychological factors of CB. Subcultures based on religion, gender, sexuality, and status are also important sociological factors to consider in understanding CB. Each of these factors have their own stream of tourism literature. As an example, cultural factors may influence the perceived cultural distance in the tourist-host relationship. Generally, six dimensions have been evaluated in understanding this relationship: purposes, determinants, activities, intensity, impacts and attitudes (Fan et al., 2017), and these dimensions help understand some of the animosity of communities toward tourists, especially during COVID-19 (see also Hateftabar, this volume).

Related to this, substantial research has examined emotional solidarity between residents and tourists with this line of research arguing that emotional closeness, sympathetic understanding and making tourists feel welcomed by residents create emotional solidarity (Woosnam, 2011). Shared beliefs and behaviours as well as interactions can explain emotional solidarity (Woosnam, 2011). Some have argued that domestic tourists contribute more to the happiness of locals than foreign tourists (Okulicz-Kozaryn & Strzelecka, 2017) and this will need to be researched further in light of the negative impacts of COVID-19 on communities and the closure of borders to international tourists. The issue of tourists volunteering during a crisis can also be explored building on the idea that tourists who volunteer contribute to the building of social capital in a community (Thompson & Taheri, 2020). Another concept that is worth exploring further to better understand tourist decision-making is the extent to which tourists feel any sense of responsibility toward hosts given that previous studies have examined tourists' attribution of destination social responsibility (Su et al., 2020). In essence, tourists' ethical beliefs and senses of self and how these materialises in group decisions may need to be explored further.

Political Factors of CB

Through legislation and social marketing, political actors such as governments and NGOs can cooperate to curb or eliminate undesirable tourist behaviours and promote sustainable behaviours through positive reinforcement. While having much influence on the supply side of tourism through tourism policy and tourism business specific legislations, there is little understanding of how political players achieve and sustain behavioural change among tourists. While pro-environmental attitudes and behaviours of tourists are well understood in specific contexts such as natural protected areas (Esfandiar et al., 2022), how health and biosecurity regulations can affect tourist behaviour are not fully understood. For example, governments can also impose taxes on tourist behaviour or offer incentives to, for example, improve sustainability, reduce carbon emissions and travel more in rural than urban areas. The way tourists perceive these initiatives must be researched further. Transnational organisations such as the UNWTO, World Bank, and World Economic Forum (WEF) through their directives and involvement in global tourism can shape pathways for tourism development that can then affect tourist behaviour. For example, some of the latest reports from these organisations suggest that destinations should focus on developing rural tourism, wellness tourism and community-based tourism. From their perspective, destinations should also build resilience by considering the types of products, tourists and markets they attract. Such directives can then prompt governments to invest in developing the supporting infrastructure for these segments, which then attract tourists that are interested in such activities. Thus, indirectly and directly political actors can shape behaviour of tourists and tourism providers.

Conclusion

The chapter sought to elucidate some of the key concepts and perspectives on CB in tourism. With a strong influence from disciplines such as economics, marketing and psychology on CB models, it is not surprising that the concept of tourist behaviour can be understood from several perspectives. While the area of CB is rapidly growing with access to 'big data' where it is possible to understand CB at the aggregate level in a way that scholars and political actors could not before, there are emergent issues such as consumer privacy and information-sharing by third-parties that need to be further understood so that policy makers can extend consumer protection laws to the digital sphere. The rise of consumer misbehaviour has also been noted in several studies and in tourism but the topic remains sparsely studied. The choices and behaviours that unfold in individualistic versus collectivistic cultures need to be further understood in the context of adversity, whether disaster or pandemic related. At the same time, theoretical developments in the sphere of online behaviours are much needed. To conclude, the chapter offers a brief overview of psychological, sociological and political factors affecting CB and reviews tourist decision-making models. Each sub-area has a significant body of literature that must be considered when attempting to research CB and behavioural change.

References

Aebli, A., Volgger, M., & Taplin, R. (2022). A two-dimensional approach to travel motivation in the context of the COVID-19 pandemic. *Current Issues in Tourism*, *25*(1), 60–75. https://doi.org/10.1080/13683500.2021.1906631

Ajzen, I. (1985). From intentions to actions: A theory of planned behavior. In J. Kuhl & J. Beckman (Eds), *Action-control: From cognition to behavior* (pp. 11–39). Springer.

Andreason, A. R. (1965). Attitudes and consumer behaviour: A decision model in new research in marketing. In L. Preston (Ed.), *Institute of business and economic research* (pp. 1–16). University of California.

Assiouras, I., Skourtis, G., Giannopoulos, A., Buhalis, D., & Koniordos, M. (2019). Value co-creation and customer citizenship behavior. *Annals of Tourism Research*, *78*, 102742. https://doi.org/10.1016/j.annals.2019.102742

Bagozzi, R. P., Gopinath, M., & Nyer, P. U. (1999). The role of emotions in marketing. *Journal of the Academy of Marketing Science*, *27*(2), 184–206. https://doi.org/10.1177%2F0092070399272005

Becker, L., & Jaakkola, E. (2020). Customer experience: fundamental premises and implications for research. *Journal of the Academy of Marketing Science*, *48*(4), 630–648. https://doi.org/10.1007/s11747-019-00718-x

Buckley, R. (2012). Sustainable tourism: Research and reality. *Annals of Tourism Research*, *39*(2), 528–546. https://doi.org/10.1016/j.annals.2012.02.003

Chen, C. C., & Yoon, S. (2019). Tourism as a pathway to the good life: Comparing the top–down and bottom–up effects. *Journal of Travel Research*, *58*(5), 866–876. https://doi.org/10.1177%2F0047287518775282

Clore, G. L., Schwarz, N., & Conway, M. (1994). Affective causes and consequences of social information processing. In R. S. Wyer & T. K. Srull (Eds), *Handbook of social cognition* (2nd ed., Vol. 1, pp. 323–418). Erlbaum.

Cohen, S. A., Prayag, G., & Moital, M. (2014). Consumer behaviour in tourism: Concepts, influences and opportunities. *Current Issues in Tourism*, *17*(10), 872–909. https://doi.org/10.1080/13683500.2013.850064

Confer, J. C., Easton, J. A., Fleischman, D. S., Goetz, C. D., Lewis, D. M., Perilloux, C., & Buss, D. M. (2010). Evolutionary psychology: Controversies, questions, prospects, and limitations. *American Psychologist*, *65*(2), 110–126. https://psycnet.apa.org/doi/10.1037/a0018413

Crompton, J. L. (1979). Motivations for pleasure vacation. *Annals of Tourism Research*, *6*(4), 408–424. https://doi.org/10.1016/0160-7383(79)90004-5

Crouch, G. I. (2013). Homo sapiens on vacation: What can we learn from Darwin? *Journal of Travel Research*, *52*(5), 575–590. https://doi.org/10.1177%2F0047287512475219

Durante, K. M., & Griskevicius, V. (2016). Evolution and consumer behavior. *Current Opinion in Psychology*, *10*, 27–32. https://doi.org/10.1016/j.copsyc.2015.10.025

Earl, P. E. (2016). The evolution of behavioural economics. In R. Frantz, S. H. Chen, K. Dopfer, F. Heukelom & S. Mousavi (Eds), *Routledge handbook of behavioral economics* (pp. 5–17). Routledge.

Esfandiar, K., Pearce, J., Dowling, R., & Goh, E. (2022). Pro-environmental behaviours in protected areas: A systematic literature review and future research directions. *Tourism Management Perspectives*, *41*, 100943. https://doi.org/10.1016/j.tmp.2022.100943

European Network for Smoking Prevention (2011). **UPDATED* European smoking bans – Evolution of the legislation*. https://epha.org/updated-european-smoking-bans-evolution-of-the-legislation/#:~:text=From%20now%20on%2C%20smoking%20is,their%20rooms%20open%20to%20smokers

Faber, A., & Frenken, K. (2009). Models in evolutionary economics and environmental policy: Towards an evolutionary environmental economics. *Technological Forecasting and Social Change*, *76*(4), 462–470. https://doi.org/10.1016/j.techfore.2008.04.009

Fan, D. X., Zhang, H. Q., Jenkins, C. L., & Lin, P. M. (2017). Does tourist–host social contact reduce perceived cultural distance? *Journal of Travel Research*, *56*(8), 998–1010. https://doi.org/10.1177%2F0047287517696979

Feygin, S., & Pozdnoukhov, A. (2018). Peer pressure enables actuation of mobility lifestyles. *Transportation Research Part C: Emerging Technologies*, *87*, 26–45. https://doi.org/10.1016/j.trc.2017.12.008

Foxall, G. R. (1993). Consumer behaviour as an evolutionary process. *European Journal of Marketing*, *27*(8), 46–57. https://doi.org/10.1108/03090569310042936

Gentile, C., Spiller, N., & Noci, G. (2007). How to sustain the customer experience: An overview of experience components that co-create value with the customer. *European Management Journal*, *25*(5), 395–410. https://doi.org/10.1016/j.emj.2007.08.005

Gilbride, T. J., & Allenby, G. M. (2004). A choice model with conjunctive, disjunctive, and compensatory screening rules. *Marketing Science*, *23*(3), 391–406. https://doi.org/10.1287/mksc.1030.0032

Gitelson, R., & Kerstetter, D. (1995). The influence of friends and relatives in travel decision-making. *Journal of Travel & Tourism Marketing*, *3*(3), 59–68. https://doi.org/10.1300/J073v03n03_04

Griskevicius, V., Shiota, M. N., & Neufeld, S. L. (2010). Influence of different positive emotions on persuasion processing: a functional evolutionary approach. *Emotion*, *10*(2), 190–206. https://psycnet.apa.org/doi/10.1037/a0018421

Habib, R., White, K., Hardisty, D. J., & Zhao, J. (2021). Shifting consumer behavior to address climate change. *Current Opinion in Psychology*, *42*, 108–113. https://doi.org/10.1016/j.copsyc.2021.04.007

Homburg, C., Jozić, D., & Kuehnl, C. (2017). Customer experience management: toward implementing an evolving marketing concept. *Journal of the Academy of Marketing Science*, *45*(3), 377–401. https://doi.org/10.1007/s11747-015-0460-7

Hosany, S. (2012). Appraisal determinants of tourist emotional responses. *Journal of Travel Research*, *51*(3), 303–314. https://doi.org/10.1177%2F0047287511410320

Hosany, S., Martin, D., & Woodside, A. G. (2021). Emotions in tourism: Theoretical designs, measurements, analytics, and interpretations. *Journal of Travel Research*, *60*(7), 1391–1407. https://doi.org/10.1177%2F0047287520937079

Howard, J. A., & Sheth, J. N. (1969). *The theory of buyer behavior*. John Wiley and Sons.

Isen, A. M., Shalker, T. E., Clark, M., & Karp, L. (1978). Affect, accessibility of material in memory, and behavior: A cognitive loop? *Journal of Personality and Social Psychology*, *36*(1), 1–12. https://psycnet.apa.org/doi/10.1037/0022-3514.36.1.1

Janssen, C., & Vanhamme, J. (2015). Theoretical lenses for understanding the CSR–consumer paradox. *Journal of Business Ethics*, *130*(4), 775–787. https://doi.org/10.1007/s10551-014-2111-1

Kim, J., & Seo, Y. (2019). An evolutionary perspective on risk taking in tourism. *Journal of Travel Research*, *58*(8), 1235–1248. https://doi.org/10.1177%2F0047287518807579

Kock, F., Josiassen, A., & Assaf, A. G. (2018). On the origin of tourist behavior. *Annals of Tourism Research*, *73*, 180–183. https://doi.org/10.1016/j.annals.2018.04.002

Kock, F., Nørfelt, A., Josiassen, A., Assaf, A. G., & Tsionas, M. G. (2020). Understanding the COVID-19 tourist psyche: The evolutionary tourism paradigm. *Annals of Tourism Research*, *85*, 103053. https://doi.org/10.1016/j.annals.2020.103053

Kwok, L., Xie, K. L., & Richards, T. (2017). Thematic framework of online review research: A systematic analysis of contemporary literature on seven major hospitality and tourism journals. *International Journal of Contemporary Hospitality Management*, *29*(1), 307–354. https://doi.org/10.1108/IJCHM-11-2015-0664

Lazarus, R. S. (1991). *Emotion and adaptation*. Oxford University Press.

Li, T., & Chen, Y. (2017). The destructive power of money and vanity in deviant tourist behavior. *Tourism Management*, *61*, 152–160. https://doi.org/10.1016/j.tourman.2017.02.001

Miller, E. A. (2020). Protecting and improving the lives of older adults in the COVID-19 era. *Journal of Aging & Social Policy*, *32*(4–5), 297–309. https://doi.org/10.1080/08959420.2020.1780104

Nord, W. R., & Peter, J. P. (1980). A behavior modification perspective on marketing. *Journal of Marketing*, *44*(2), 36–47. https://doi.org/10.1177%2F002224298004400205

Nørfelt, A. W., Kock, F., & Josiassen, A. (2020). Tourism xenophilia: Examining attraction to foreignness. *Journal of Travel Research*, *59*(8), 1386–1401. https://doi.org/10.1177%2F0047287519883037

Nyaupane, G. P., & Andereck, K. L. (2008). Understanding travel constraints: Application and extension of a leisure constraints model. *Journal of Travel Research*, *46*(4), 433–439. https://doi.org/10.1177%2F0047287507308325

Okulicz-Kozaryn, A., & Strzelecka, M. (2017). Happy tourists, unhappy locals. *Social Indicators Research*, *134*(2), 789–804. https://doi.org/10.1007/s11205-016-1436-9

Otoo, F. E., & Kim, S. (2020). Analysis of studies on the travel motivations of senior tourists from 1980 to 2017: Progress and future directions. *Current Issues in Tourism*, *23*(4), 393–417. https://doi.org/10.1080/13683500.2018.1540560

Otoo, F. E., Badu-Baiden, F., & Kim, S. (2019). A qualitative cognitive appraisal of tourist harassment. *International Journal of Tourism Research*, *21*(5), 575–589. https://doi.org/10.1002/jtr.2274

Otto, J. E., & Ritchie, J. B. (1996). The service experience in tourism. *Tourism Management*, *17*(3), 165–174. https://doi.org/10.1016/0261-5177(96)00003-9

Pearce, P. L., & Lee, U. I. (2005). Developing the travel career approach to tourist motivation. *Journal of Travel Research*, *43*(3), 226–237. https://doi.org/10.1177%2F0047287504272020

Perugini, M., & Bagozzi, R. P. (2001). The role of desires and anticipated emotions in goal-directed behaviours: Broadening and deepening the theory of planned behaviour. *British Journal of Social Psychology*, *40*(1), 79–98. https://doi.org/10.1348/014466601164704

Pham, M. T. (1998). Representativeness, relevance, and the use of feelings in decision making. *Journal of Consumer Research*, *25*(2), 144–159. https://doi.org/10.1086/209532

Prayag, G. (2012). Senior travelers' motivations and future behavioral intentions: The case of nice. *Journal of Travel & Tourism Marketing*, *29*(7), 665–681. https://doi.org/10.1080/10548408.2012.720153

Prayag, G., & Ryan, C. (2011). The relationship between the 'push' and 'pull' factors of a tourist destination: The role of nationality–an analytical qualitative research approach. *Current Issues in Tourism*, *14*(2), 121–143. https://doi.org/10.1080/13683501003623802

Prayag, G., Hassibi, S., & Nunkoo, R. (2019). A systematic review of consumer satisfaction studies in hospitality journals: Conceptual development, research approaches and future prospects. *Journal of Hospitality Marketing & Management*, *28*(1), 51–80. https://doi.org/10.1080/19368623.2018.1504367

Prayag, G., Pung, J. M., Lee, C., & Del Chiappa, G. (2022). The self-concept and psychological antecedents of intention to recommend a heritage site: The moderating effects of visitor type. *Tourism Management Perspectives*, *42*, 100962. https://doi.org/10.1016/j.tmp.2022.100962

Prayag, G., Spector, S., & Finsterwalder, J. (2020). Customer experience in tourism: An overview. In S. Dixit (Ed.), *The Routledge handbook of tourism experience management and marketing* (pp. 67–76). Routledge.

Rasoolimanesh, S. M., Ramakrishna, S., Hall, C. M., Esfandiar, K., & Seyfi, S. (2020). A systematic scoping review of sustainable tourism indicators in relation to the sustainable development goals. *Journal of Sustainable Tourism*. https://doi.org/10.1080/09669582.2020.1775621

Rittichainuwat, B. N., & Chakraborty, G. (2009). Perceived travel risks regarding terrorism and disease: The case of Thailand. *Tourism Management*, *30*(3), 410–418. https://doi.org/10.1016/j.tourman.2008.08.001

Roseman, I. J., Spindel, M. S., & Jose, P. E. (1990). Appraisals of emotion-eliciting events: Testing a theory of discrete emotions. *Journal of Personality and Social Psychology*, *59*(5), 899–915.

Russell-Bennett, R., Leo, C., Rundle-Thiele, S., & Drennan, J. (2016). A hierarchy-of-effects approach to designing a social marketing game. *Journal of Nonprofit & Public Sector Marketing*, *28*(2), 105–128. https://doi.org/10.1080/10495142.2014.988081

Saad, G., & Gill, T. (2000). Applications of evolutionary psychology in marketing. *Psychology & Marketing*, *17*(12), 1005–1034. https://doi.org/10.1002/1520-6793(200012)17:12<1005::AID-MAR1>3.0.CO;2-H

Saad, G., & Peng, A. (2006). Applying Darwinian principles in designing effective intervention strategies: The case of sun tanning. *Psychology & Marketing*, *23*(7), 617–638. https://doi.org/10.1002/mar.20149

Schwarz, N. (2000). Emotion, cognition, and decision making. *Cognition & Emotion*, *14*(4), 433–440. https://doi.org/10.1080/026999300402745

Shamshiripour, A., Rahimi, E., Shabanpour, R., & Mohammadian, A. K. (2020). How is COVID-19 reshaping activity-travel behavior? Evidence from a comprehensive survey in Chicago. *Transportation Research Interdisciplinary Perspectives*, *7*, 100216. https://doi.org/10.1016/j.trip.2020.100216

Shin, H., Nicolau, J. L., Kang, J., Sharma, A., & Lee, H. (2022). Travel decision determinants during and after COVID-19: The role of tourist trust, travel constraints, and attitudinal factors. *Tourism Management*, *88*, 104428. https://doi.org/10.1016/j.tourman.2021.104428

Song, H., Sparks, B. A., & Wang, Y. (2017). Exploring disagreement prevention and resolution in travel decision-making of young Chinese travelers. *Journal of Travel & Tourism Marketing*, *34*(2), 257–273. https://doi.org/10.1080/10548408.2016.1156615

Su, L., Lian, Q., & Huang, Y. (2020). How do tourists' attribution of destination social responsibility motives impact trust and intention to visit? The moderating role of destination reputation. *Tourism Management*, *77*, 103970. https://doi.org/10.1016/j.tourman.2019.103970

Swarbrooke, J., & Horner, S. (2007). *Consumer behaviour in tourism*. Routledge.

Taheri, B., Prayag, G., & Muskat, B. (2021). Introduction to the special issue: Consumer experience management and customer journeys in tourism, hospitality and events. *Tourism Management Perspectives*, *40*, 100877. https://doi.org/10.1016/j.tmp.2021.100877

Thompson, J., & Taheri, B. (2020). Capital deployment and exchange in volunteer tourism. *Annals of Tourism Research*, *81*, 102848. https://doi.org/10.1016/j.annals.2019.102848

Tse, S., & Tung, V. W. S. (2022). Understanding residents' attitudes towards tourists: Connecting stereotypes, emotions and behaviours. *Tourism Management*, *89*, 104435. https://doi.org/10.1016/j.tourman.2021.104435

van Tilburg, M., Lieven, T., Herrmann, A., & Townsend, C. (2015). Beyond "pink it and shrink it" perceived product gender, aesthetics, and product evaluation. *Psychology & Marketing*, *32*(4), 422–437. https://doi.org/10.1002/mar.20789

Vaughn, R. (1986). How advertising works: A planning model revisited. *Journal of Advertising Research*, *26*, 57–66.

Volo, S. (2021). The experience of emotion: Directions for tourism design. *Annals of Tourism Research*, *86*, 103097. https://doi.org/10.1016/j.annals.2020.103097

Wang, L., Hou, Y., & Chen, Z. (2021). Are rich and diverse emotions beneficial? The impact of emodiversity on tourists' experiences. *Journal of Travel Research*, *60*(5), 1085–1103. https://doi.org/10.1177%2F0047287520919521

Wang, Y., Gu, J., Wang, S., & Wang, J. (2019). Understanding consumers' willingness to use ride-sharing services: The roles of perceived value and perceived risk. *Transportation Research Part C: Emerging Technologies*, *105*, 504–519. https://doi.org/10.1016/j.trc.2019.05.044

Wong, I. A., Law, R., & Zhao, X. (2017). When and where to travel? A longitudinal multilevel investigation on destination choice and demand. *Journal of Travel Research*, *56*(7), 868–880. https://doi.org/10.1177%2F0047287516670269

Woosnam, K. M. (2011). Testing a model of Durkheim's theory of emotional solidarity among residents of a tourism community. *Journal of Travel Research*, *50*(5), 546–558. https://doi.org/10.1177%2F0047287510379163

Yang, Y., Ruan, Q., Huang, S. S., Lan, T., & Wang, Y. (2021). Impact of the COVID-19 outbreak on tourists' real-time on-site emotional experience in reopened tourism destinations. *Journal of Hospitality and Tourism Management*, *48*, 390–394. https://doi.org/10.1016/j.jhtm.2021.07.014

Zhang, Y., Prayag, G., & Song, H. (2021). Attribution theory and negative emotions in tourism experiences. *Tourism Management Perspectives*, *40*, 100904. https://doi.org/10.1016/j.tmp.2021.100904

Zheng, D., Luo, Q., & Ritchie, B. W. (2021). Afraid to travel after COVID-19? Self-protection, coping and resilience against pandemic 'travel fear'. Tourism Management, 83, 104261. https://doi.org/10.1016/j.tourman.2020.104261

Chapter 4

Tourism Motivations and Decision-making

Yael Ram

Introduction – Free Will and Choice in Leisure Tourism

Motivations and decision-making are the building blocks of tourism demand. Accordingly, they have been studied extensively over the years by tourism scholars. As a social and economic activity, tourism demand involves three components: voluntary, temporary, and mobility (Hall, 2005). The main segment within tourism is leisure travels, recreation and holidays, which refers to more than fifty percent of all tourism activities. Additionally, tourism includes other segments such as business and professional travels, visiting friends and relatives, health tourism, and religious travels (UNWTO, 2019).

Leisure tourism demand is different from other types of tourism because it is characterised by two additional core concepts, apart from the threesome of voluntary, temporary and mobility; it is based on intrinsic motivation and free choice (Mannell & Iso-Ahola, 1987). When a business traveller participates in a business conference, they have almost no choice about if, where, when, and what to do. On the other hand, a leisure tourist goes on vacation because they want to visit a specific place and choose a particular destination and activities. Both travellers are not forced to travel abroad temporarily, but only the latter freely satisfies their intrinsic motivation and choices.

Since the 1970s, tourism scholars (such as Cohen, 1979; Crompton, 1979; Dann, 1977; Mannell & Iso-Ahola, 1987) have tried to explain why people travel and how they make their choices. Dann (1977) argued that these are two separate but dependent questions. The question 'why travel' is linked to personal predispositions and thus is more substantial than questions of how and where to travel. However, Dann (1977) also noted that the former question, which he labeled as 'push' factors, gains less attention than the latter question he coined as 'pull' factors. In later years, the push and pull factors became highly popular in tourism studies and were widely used for simplicity and clarity (Cohen et al., 2014).

In this chapter, intrinsic motivation is linked to push factors, while the element of free choice corresponds to decision-making and pull factors. The chapter tries to answer two big

The Wiley Blackwell Companion to Tourism, Second Edition. Edited by C. Michael Hall.

questions: why people travel and how they decide where, when and what to do while travelling. It also discusses the segment of non-traveller and their constraints. The conclusion presents current developments in the research of motivation and decision-making in the presence of the environmental crisis.

Motivations and Push Factors

Everyone knows what it means to be motivated and refers to it by using words such as wants, needs, drives, or desires (Petri & Govern, 2012). In tourism, motivation refers to the question 'why people travel?' (Pearce & Packer, 2013). Over the years, four central answers have been given to this question: escape/novelty (Iso-Ahola, 1983; Crompton, 1979; Fridgin, 1991), quest (Cohen, 1979, MacCannell, 1973; Wang, 2000), social factors (Sharpely, 2018; Uriely et al., 2011), and wellbeing (Smith & Diekmann, 2017). A fifth popular answer, interpersonal differences, was also considered. However, personality traits such as sensation seeking (Lepp & Gibson, 2008) and the big five traits (Šagovnović & Kovačić, 2022) were used to explain differences in tourist behavior, influences and preferences and not to answer the fundamental question of why travel at all. The four answers will be described below. Figure 4.1 presents the main theme of each perspective, including possible overlapping between them.

Escape/Novelty

Earlier researchers such as Iso-Ahola (1983), Crompton (1979) and Fridgin (1991) argued that the intrinsic motivation for travel is combined by two main forces: avoiding routine and seeking novelty. Although the two forces may seem similar, they produce different and even competing arousal levels. The escaping force aims to decrease everyday pressure on interpersonal and intrapersonal levels, while the seeking force is directed to gain excitement and arousal at these two levels. The tension between the two forces on inter/intrapersonal levels, together with the intent to achieve an equilibrium, activate the motivation for leisure travel.

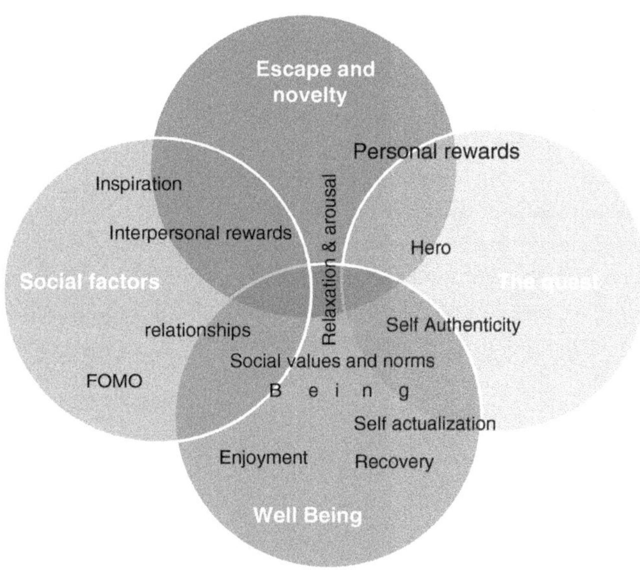

Figure 4.1. Motivations for tourism - main perspectives and themes.

Mannell and Iso-Ahola (1987) noted that the split between escaping everyday life and seeking novelty is not an either/or case. However, they assumed the optimum level to be in the direction of the escape factor, as leisure tourism, and especially get-away vacations, satisfies more escape and relaxation motivations.

When Pearce and Lee (2005) tried to support their Travel Ladder Career (TLC) approach, they found that all tourists reported high novelty and high escape motivations, whether highly experienced or less experienced travellers. These findings imply that the two forces, escape and novelty, are both critical for travellers and have a strong positive connection, rather than a dialectical association and tension. Both novelty and escape were found as leading motivations for leisure travels, although not the only motivations (Assiouras et al., 2015; Chen et al., 2020; Liu et al., 2021; Özdemir & Çelebi, 2018; Voigt et al., 2011, Wong & Rosenbaum; 2012). In summary, escape and novelty are important motivations for leisure tourism, and they probably reinforce each other. The empirical findings do not support any dialectical association between these two motivations and do not reflect tension or equilibrium.

The Quest

The idea of tourism as a quest was made by several scholars, mainly from sociology (Cohen, 1979; MacCannell, 1973; Wang, 2000). But even before this, Campbell (1969) linked modern tourism and Carl Jung's notions of the collective unconscious and the idea of archetypes. Campbell (1969) defined a hero as one who departs from routine on a challenging journey and returns home to share their gained rewards with friends and family. Berger (2009) had suggested that the folktales of heroes are congruent with various stages of the tourist experience, including leaving home, transferring and then coming back home. Here, the motivation for tourism is activated by an unconscious archetype and a quest for adventure.

Authenticity is probably the most popular subject for tourist quests. According to MacCannel (1973), the quest for authenticity is doomed and produces only disappointment due to over commodification. Cohen (1979) argued that aiming for a 'generic' quest for authenticity misses interpersonal differences. Instead, he suggested that the quest is directed to find a personal 'center'. His typology of 'quests for centers' includes five types of tourist motivations. At one end of this typology, there is a quest for pleasure and recreation. A quest for self-meaning is located on the other end of the typology (Cohen, 1979).

Later, Wang (2000) referred to this quest as the longing for existential authenticity. It relates to MacCannel (1973) by pointing to authenticity as the main driver. However, Wang (2000) argues that authenticity is subjective and varies between people, similarly to Cohen's (1979) viewpoint. His main argument is that tourists seek their authentic selves and intersubjective authenticity. In other words, their quest is to find a 'Being', a state in which they are authentic to themselves and can feel their true selves. The separation between interpersonal and intrapersonal existential authenticity means that both can be the targets of the quest of tourists (Kim & Jamal, 2007; Yi et al., 2017). However, Yi et al. (2017) suggested that the quest focuses on intrapersonal domains.

Social Motivations

"No man is an island", wrote the English poet John Donne (1624) about 400 years ago. The same is true regarding tourist motivations, as tourists are always influenced by others. In the famous hierarchy of needs by Maslow (1954), only the last category of needs – self-actualisation – can be achieved without the involvement of others. The fulfillment of all other needs – physiological, safety and security, belongingness, and esteem – includes other people and is also derived from their presence. Therefore, it is not surprising that when Pearce and Lee (2005) tried to support their TLC approach, which was loosely based on the hierarchy of needs, all their participants reported strengthening relationships as one of the main motivations for tourism, alongside escape and novelty.

Culture and society shape tourist's motivations, attitudes, and consumption. Sharpley (2018) argued that extrinsic societal motivations are internalised to become tourism needs. Mehmetoglu (2012) demonstrated how the cultural milieu (family, upbringing, friends, and peers) strongly affects people's interest in holidaying. Yet, the societal influence on tourism goes beyond holiday-making because it also enables individuals to connect with social values and rules (Uriely et al., 2011). In this context, the internalised social norms or religious imperatives motivate tourism, including pilgrimage (Collins-Kreiner & Kliot, 2000) and heritage tourism (Poria et al., 2004). Even social transitions are being reflected in tourism as rites of passage (White & White, 2004). These include tourism forms such as honeymoons (Chen et al., 2020), backpacking (Noy & Cohen, 2012), and post-retirement tourism (Nimrod, 2008; White & White, 2004).

Social media has multiple functions in motivating tourism. Potential tourists consider social media before their planned trips and are influenced by social media during their experiences (Leung et al., 2019). Social media can affect motivation by exposing users to reviews, choices, or others' lives and lifestyles (Stephen, 2016). Aldao and Mihalic (2020) added that social media provokes travel inspiration, raises familiarity with remote destinations, and spreads the adverse feeling of 'FOMO' (fear of missing out), which is the anxiety not to have experiences that others have.

Wellbeing

Positive tourism is the study of "what makes life worth living" (Seligman & Csikszentmihalyi, 2000, p. 5). In their seminal paper, Seligman and Csikszentmihalyi (2000) explained that positive psychology addresses concepts such as wellbeing, flow, happiness, and satisfaction. All these concepts are also relevant to the subjective experience of leisure tourists before, during, and after their travels. In particular, the two perspectives of wellbeing, hedonic and eudaimonic, have vital functions regarding motivations for travel (Smith & Diekmann, 2017).

The hedonic perspective relates to immediate sensory pleasure, happiness and enjoyment (Rahmani et al., 2018) and experience as much pleasure as possible (Smith & Diekmann, 2017). Smith and Diekemann (2017) linked this view of wellbeing with motivations to gain short-term benefits from Sea, Sun and Sand vacations (SSS), including parties, shopping, and nightlife. More recently, Kim and Hall (2019) found that the motivation for enjoyment can be associated with virtual reality tourism. Ram et al. (2021) showed that people plan to return to travelling abroad after Covid-19 because it represents a return to normality and a source of recovery.

The eudaimonic perspective is a meaning-related aspect of self-growth and self-actualisation (Rahmani et al., 2018). It corresponds to the quest for authenticity (Cohen, 1979), pilgrimage and heritage tourism (Collins-Kreiner & Kliot, 2000; Poria et al., 2004). Smith and Diekemann (2017) also added motivations for ethical and altruistic tourism, self-development and transformation retreats, adventure tourism and extreme sports, where tourists are not enjoying the experience but push their body and mind to the limit.

Decision-making

Tourism decision-making is generally analysed using a descriptive approach, e.g., how tourists decide from available choices (Wattanacharoensil & La-ornual, 2019). Within the descriptive approach, two main models have been adopted: the first is the Engel-Kollat-Blackwell (EKB) model (Han et al., 2020a, b), a consequential cognitive process of sub-decisions (Sirakaya & Woodside, 2005). The second is social psychology Theories of Reasoned Actions (TRA), (Ajzen & Fishbein, 1980) and Planned-Behaviour (TPB), (Ajzen, 1991). These two theories explained how multiple and reciprocal beliefs, attitudes and norms shape intentions to behaviour and actual behaviours. The latter (TPB) also includes a component of behavioural control (Ajzen, 1991; Ulker-Demirel & Ciftci, 2020). Table 4.1 presents the similarities and differences between the two approaches of decision-making – EKB and TRA/TPB.

The Engel-Koolat-Blackwell (EKB) Model of Decision-making

The EKB model of decision-making generally describes the process of tourism-related decisions (Han et al., 2020a, b). This model follows a funnel-like procedure of narrowing down choices among options (Sirakaya & Woodside, 2005), and is divided into five consecutive stages: (1) problem recognition; (2) information search; (3) evaluation of alternatives; (4) making a choice and; (5) evaluation of post-purchase (Han et al., 2020a, b). In tourism, this model involves many elements, sub-decisions, considerations, and co-deciders and is taken before, during and even after the trip (Han et al., 2020a, b; Sirakaya & Woodside, 2005).

Problem Recognition – 1st Stage

Klenosky (2002) referred to decision-making as a means-end investigation, where the 'end' is the push motivations, and the 'means' are the pull attributes of destinations. Hence, the 'end' is not the final stage but the trigger of the whole process of decision-making, i.e., problem

Table 4.1. Similarities and differences between EKB and TRA/TPB.

Features/elements	*EKB*	*TRA/TPB*
Psychology branch	Cognitive psychology	Social psychology
Rationality	Assumed	Assumed
Can elements of the process be analysed separately?	Yes, and often done	Rarely done
Constraints	Can be included in the 'making-choice' stage	Not included/have to be added to the model as external factors
Push/pull factors	Push factors are included in the 'problem recognition' stage; Pull factors are evaluated in the third stage of the decision-making process	Not included/have to be added to the model as external factors
Motivations	Can be included in the first stage, 'problem recognition'	Have to be added to the model as an external factor
Feedback	The last stage	Not included
Personalattitudes/norms and beliefs	Not included; same process for all people	Included

recognition. Dai et al. (2022) added inspiring sources to the first stage of push motivation. Among these sources, they referred to stories and movies, social media and virtual reality.

Information Search – 2nd Stage

Information can be delivered by different sources (travel agents or tourism providers, for example) and various channels (e.g., internet and mess media) (Grønflaten, 2009), and it also can be targeted by active or passive searches. Coromina and Camprubí (2016) found that active information search uses online channels and includes sources such as search engines, official websites and blogs. On the other hand, passive information search relied on personal resources and media.

The growth in online information resources brought an overload of information. Confused tourists use more shortcuts today than in the past and rely more on recommendations of friends, relatives and online reviews (Gorsuy, 2019). These shortcuts in the decision-making process were also noted by Kahnemen (2011), who argued that most of our decisions are based on intuitive and emotional routes rather than rational and logical reasoning. Hence, potential tourists may search for information, but many will not undertake a systematic search due to cognitive overload and interpersonal differences (McCabe et al., 2016).

Evaluation of Alternative Pull Factors – 3rd Stage

The outputs of the information search are the attributes of tourism destinations and activities. Um and Crompton (1990) analysed the decision-making process of leisure destinations and defined two main concepts: the awareness set, which contains all the possible optional attributes of destinations, and evoked set, which is the limited and chosen attributes after considerations of constraints and preferences.

The chosen attributes include the pull features that attract people to a specific place or activity (Chen & Chen, 2015; Klenosky, 2002). The tourism literature identified five leading groups of pull factors:

- cultural and natural attractions (Klenosky, 2002; Michael et al., 2020; Nikjoo & Ketabi, 2015; Prayag & Ryan, 2011);
- facilitators include infrastructures, safety and value for money (Nikjoo & Ketabi, 2015; Prayag, 2010; Prayag & Ryan, 2011);
- appealing augmented tourism product, which frequently embraces friendly people, good food and shopping (Nikjoo & Ketabi, 2015; Prayag, 2010);
- climate and scenery (Klenosky, 2002; Prayag & Ryan, 2011); and
- the positive image of a place or activity (Michael et al., 2020; Prayag & Ryan, 2011).

Making a Choice – 4th Stage

The choice of tourism activity is complicated and probably not entirely rational (Kahnemen, 2011; McCabe et al., 2016). It also does not take place in a vacuum. Sirakaya and Woodside (2005) emphasised that tourism-related decisions mainly occur in a social context with family, relatives, friends, and the travelling group. Cohen et al. (2014) pointed to interdependencies between tourism and other consumption decisions, and Decrop (2010) included constraints' consideration in the choice making process. Another aspect of choice is timing: if to plan and purchase in advance or at the site and time of the trip (Han et al., 2020a, b).

Some choices have transformed since the Covid-19 pandemic. Factors such as risk perception (Gallego et al., 2022; Neuburger & Egger, 2021), safety-seeking (Kim et al., 2022), third-party communication (Neuburger & Egger, 2021; Villacé-Molinero et al., 2021), and health and hygiene (Pappas & Glyptou, 2021) not only influence the decisions if, and where to travel, but also amplify the tendency to select compromising and satisfying options over optimum solutions (Kim et al., 2022; Pappas & Glyptou, 2021).

Post-purchase Evaluations

Evaluating tourism services and products combines consumption, purchase and cognitive and emotional dimensions (Sanchez et al., 2006). This evaluation contributes to following purchases decisions (Han et al., 2020a, b), satisfaction, recommendations and loyalty (Lee et al., 2007; Sanchez et al., 2006; Yoon & Uysal, 2005) or otherwise to regret and disappointment (Jang et al., 2013). According to Sanchez et al. (2006) the overall value of tourism products and services is composed of four functional values (physical, service, quality and money) and two emotional and social values. They concluded that "cognitive elements such as quality and price play a fundamental role, especially the latter, but it would not be possible to understand the behaviour of tourists without incorporating affective components, especially social value. . ." (Sanchez et al., 2006, p. 404). Although Lee al al. (2007) did not study the social value, they pointed out how the overall value directly influences satisfaction and recommendations.

Theories of Reasoned Actions (TRA) and Planned Behaviour (TPB)

In their systematic literature review on TRA and TPB in tourism, leisure and hospitality management, Ulker-Demirel and Ciftci, (2020) found 259 relevant publications. Most of them addressed the theories as to the main frameworks for the study. In the case of TPB, the studies tested how behavioural, normative and control beliefs influence corresponding attitudes and norms and how the attitudes and norms influence behaviour intention and hence actual behaviour. The TPB studies focused mainly on two main issues: environmental and responsibility concerns and using internet technologies and social media. These two subjects, in particular, evoke growing interests, involve norms and perception of control, and thus were found as good candidates for decision-making analysis using the TPB conceptualisation.

Extensions to the TPB for the tourism context include externalities to the decision-making process, such as visa exemption (Han et al., 2011) environmental education (Leelapattana et al., 2019), and internal effects such as Self-Congruity Theory (Chiu et al., 2015), persuasive information behaviour (Meng & Choi, 2019), place attachment (Han et al., 2019), age and risk perceptions (Wang et al., 2022), and perceived knowledge of Covid-19 (Han et al., 2020a, b). These personal, cognitive and emotional factors were used as additional factors to explain intentions and behaviours in the tourism context, alongside the TPB components.

The Missing Segment – People Who Don't Travel

Hall (2010) argued that tourism research tends to ignore the largest segment of non-travellers. Indeed, the number of non-travellers is larger than the number of travellers. Regarding motivations and decision-making, the questions are more focused: don't the non-travellers have motivations? Or – do they decide not to travel? Both travellers and non-travellers share travel constraints that act as barriers and prevent them from travelling in general or travelling to the extent they would like (Hall & Page, 2014). However, they differ in the form and magnitude of constraints (Ram & Hall, 2020).

Crawford et al. (1991) suggested a three-dimensional leisure constraints model consisting of intra-personal, interpersonal, and structural constraints. The three types of constraints were also found in various tourism contexts (Hudson & Gilbert, 2000; Hung & Petrick, 2010; Ram & Hall, 2020). Bozic et al. (2017) proposed a different model of constraints, including lack of information and recommendation, lack of time, and a unifying factor of inter/intrapersonal constraints.

Intra-personal constraints can correspond to the issue of motivations, suggesting that people may not have a desire to travel. Therefore the first stage of decision-making ('problem recognition') is not activated. Li et al. (2016) indicated how Chinese non-domestic tourists were influenced by personal factors, such as motivation and attitude, and also by the economic situation (travel expenses) and leisure time. Covid-19 time evoked health and risk-related constraints (Neuburger & Egger, 2021). In a small-scale explorative study, de Bloom et al. (2017) found that people who stayed a weekend at home did not have a different sense of wellbeing compared to people who went for a weekend away. This finding implies that not just personal motivations, time and money budget reduce travel motivations, but also wellbeing may not serve as a motivation for travelling away.

Inter-personal constraints are linked to both motivations and decision-making. In terms of motivations, interpersonal constraints for vacations can hinder optional social rewards. For example, Tan and Chen (2021) found that work-related FOMO decreased social motivations for first-time vacationers to Hualien, Taiwan. Zhu et al. (2020) noted a list of social-related stressors and mentioned conflicts as the major obstacle for decision-making regarding activities and a source of dissatisfaction. Dissatisfaction from social aspects and travel accompanying persons was also mentioned by Sirgy et al. (2011), decreasing overall satisfaction with life. Given that the post-purchase evaluation leads to subsequent purchases (Han et al., 2020a, b), negative social experiences possibly reduce following tourism motivation.

Structural constraints reduce the motivation for travel by influencing the decision-making to reach a 'No-Go' decision. Possible structural constraints can be lack of information and equipment (Hudson & Gilbert, 2000; Ram & Hall, 2020), crime, terror and accidents concerns, reservations concerns (Larsen et al., 2009), money and time budgets, other commitments (Bozic et al., 2017; Hudson & Gilbert, 2000; Nyaupane et al., 2004), and more recently – Covid-19 induced regulations (Ram et al., 2021).

The Need for Behavioural Change

Leisure tourism is about free choice and intrinsic motivations. Models and theories that describe travel motivations and decision-making processes assumed almost unlimited options for travel. However, real life is full of constraints. First, most people do not travel (Hall, 2010), suggesting that external and internal barriers are more potent than the motivations for travel. Second, the extent of the legal restrictions on travelling has increased since the outbreak of Covid-19 (Neuburger & Egger, 2021; Ram et al., 2021), preventing people from travelling as much as they want. Third, environmental concerns and the climate change crisis require a substantial behavioural change (Hall, 2013; Ram et al., 2013).

There is a need for behavioural change to mitigate tourism's negative environmental consequences. Hall (2013) linked the required behavioural change to tourism motivations and decision-making. The first approach to behavioural change is utilitarian, involving constraints on the demand, such as taxation, incentives for green consumption, green labelling, and carbon trading. The second approach is social-psychological, and it includes three main tools for managing the demand: education, nudging, which is 'a choice architecture that alerts people's behavior in a predictable way without forbidding any options or significantly changing their economic incentives' (Thaler & Sunstein, 2008, p. 6), and social marketing. The third approach, systems of provision, focuses on the supply chain rather than the demand side and refers to a permanent change in the supply system through short supply chains and local production.

The utilitarian approach depends on the information search stage and is effective only if consumers are aware of taxing, incentives, labelling and trading. The problem is that tourism consumption is often emotional and impulsive (Cohen et al., 2014; McCabe et al., 2016) and may not include a systematic information search. However, if the tourism decision-making is rational, the utilitarian factors (such as taxation and labelling) will be considered through the stages of evaluations of alternatives and making choices in the EKB model. From the perspective of the TPB model, the utilitarian approach is associated with control beliefs, perceptions and actual behaviour.

The utilitarian approach does not affect the push factors and the post-purchase stages in the EKB model. In addition, it does not shape norms and attitudes towards tourism as part of the TPB model. According to these two models of decision-making, the 'fuel' for tourism is still active as the motivations for tourism are not changed. If the policy is altered with the removal of taxation and labelling, there is a risk of a rebound effect: frustrated consumers will hyper-consume tourism as compensation after scarcity.

The social-psychological approach addresses motivations for tourism. With education, nudging, and social marketing, this approach tries to reshape motivations for tourism so that escape, novelty, quest, wellbeing, and social recognition can be achieved through local and sustainable forms of tourism. This approach influences the TPB's attitudes and norms of tourism and reforms the problem recognition and the post-purchase evaluation of the EKB model. However, this kind of behavioural change is supposed to take generations and is not suited to immediate action (Hall, 2013).

The third approach, systems of provision, concentrates on the supply chain; it bypasses the demand side and does not change or affect the decision-making process or motivation but creates a new reality. It shapes the behaviour without changing the cognitive and emotional processes. Instead of putting constraints, it changes the infrastructure and conditions. In terms of impact, its behavioural outputs will be noticed instantly. However, its feasibility is weak since it depends on policy paradigm shift and social-psychology change. Figure 4.2 describes the interrelationships between the three behavioural change approaches according to Hall (2013) and the two models of decision-making, EKB and TPB.

Conclusions – Towards Sustainable and Responsible Tourism

Studying motivations and decision-making for leisure tourism assumes intrinsic motivations and free choice. However, this chapter shows that these assumptions are not always supported. First, motivations are frequently external and influenced by others. Second, the choice is not free from constraints and external influences. We can identify four overlapping motivations for leisure travel: novelty and escape; quest; social factors; and wellbeing. However, most

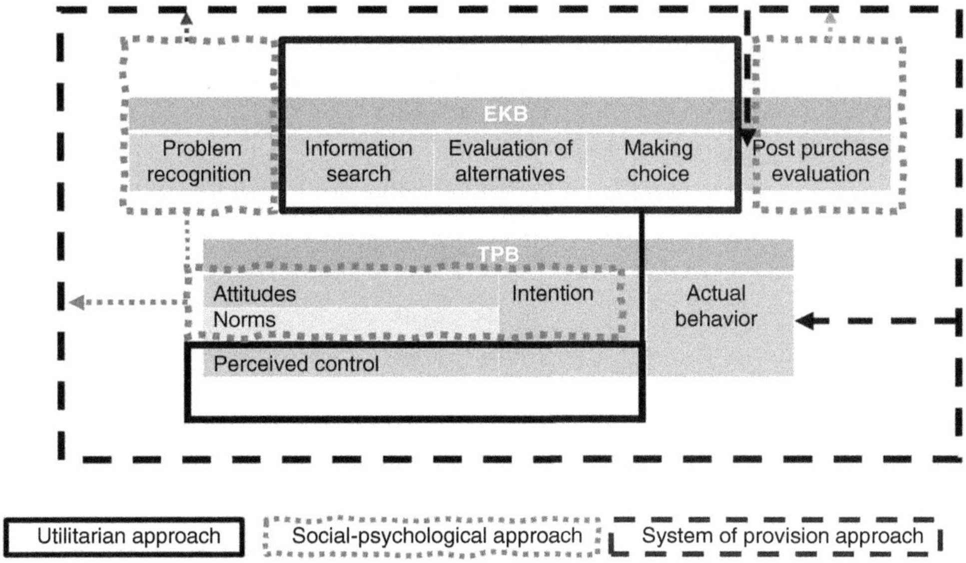

Figure 4.2. How behavioural change approaches shape tourism decision-making (after Hall, 2013).

people do not travel. Motivated leisure tourists make decisions regarding tourism. Even if their decision-making process is not entirely rational, some parts of the process are more systematic than others.

Social media has reshaped motivations for tourism and its decision-making process and can challenge the known push-pull conceptualisation (Dann, 1977). The separation between pull and push factors gets blurred with social media, as it pushes people and pulls them simultaneously. Future studies may focus on these transformations, especially among young generations and prospective tourists.

Constraints are an inseparable part of decision-making. People that want to travel have to compromise and adapt to reality requirements. The Covid-19 crisis set travel restrictions that changed motivations and decision-making. Whether these behavioural changes are prominent depends on the nature of the transformations: if they are utilitarian, socio-psychological or system-provisioning related. Utilitarian steps have short-term impacts, and the travel behaviour, alongside its motivations and decision-making patterns, will return to the business-as-usual mode when restrictions are lifted. Socio-psychological influences have long-term consequences and can lead to behavioural change through motivational transformation. However, only system-provisioning change can ensure that these changes are continued. In addition, system-provisioning change will not be possible without corresponding socio-psychological change.

Future environmental challenges require a change in travel behaviour. This chapter shows that it will not happen by relying solely on free choice and intrinsic motivations. As motivations have hedonic components and the decision-making process has irrational parts, only a combined effort of socio-psychological influences (including social media channels) and pro-environmental policies can shape future tourism demand in sustainable and responsible directions.

References

Ajzen, I. (1991). The theory of planned behavior. *Organizational Behavior and Human Decision Processes*, *50*(2), 179–211. https://doi.org/10.1016/0749-5978(91)90020-T

Ajzen, I., & Fishbein, M. (1980). *Understanding attitudes and predicting social behavior*. Englewood Cliffs.

Aldao, C., & Mihalic, T. A. (2020). New frontiers in travel motivation and social media: The case of Longyearbyen, the High Arctic. *Sustainability*, *12*(15), 5905. https://doi.org/10.3390/su12155905

Assiouras, I., Skourtis, G., Koniordos, M., & Giannopoulos, A. A. (2015). Segmenting East Asian tourists to Greece by travel motivation. *Asia Pacific Journal of Tourism Research*, *20*(12), 1389–1410. https://doi.org/10.1080/10941665.2014.982140

Berger, A. A. (2009). The hero as traveler, the traveler as hero. *Journal of Travel & Tourism Marketing*, *26*(1), 98–104. https://doi.org/10.1080/10548400802660258

Bozic, S., Jovanovic, T., Tomic, N., & Vasiljevic, D. A. (2017). An analytical scale for domestic tourism motivation and constraints at multi-attraction destinations: the case study of Serbia's lower and middle Danube region. *Tourism Management Perspectives*, *23*, 97–111. https://doi.org/10.1016/j.tmp.2017.05.002

Campbell, J. (1969). *The hero with a thousand faces*. Princeton University Press.

Chen, G., So, K. K. F., Poomchaisuwan, M., & Hu, X. (2020). Examining affection-based travel: Development and validation of a measurement scale for honeymooners' motivation. *Journal of Destination Marketing & Management*, *17*, 100452. https://doi.org/10.1016/j.jdmm.2020.100452

Chen, L. J., & Chen, W. P. (2015). Push-pull factors in international birders' travel. *Tourism Management*, *48*, 416–425. https://doi.org/10.1016/j.tourman.2014.12.011

Chiu, P., Yen, T., & Chu, K. (2015). Tourists behavioral intentions in rural area: an integrated aspect of SCT and TPB. *Journal of Food, Agriculture & Environment*, *13*(3/4), 168–174. https://www.researchgate.net/publication/284748005_Tourists_behavioral_intentions_in_rural_area_An_integrated_aspect_of_sct_and_tpb

Cohen, E. (1979). A phenomenology of tourist experiences. *Sociology*, *13*(2), 179–201. https://doi.org/10.1177/003803857901300203

Cohen, S. A., Prayag G., & Moital, M. (2014) Consumer behaviour in tourism: Concepts, influences and opportunities. *Current Issues in Tourism*, *17*(10), 872–909. https://doi.org/10.1080/13683500.2013.850064

Collins-Kreiner, N., & Kliot, N. (2000). Pilgrimage tourism in the Holy Land: The behavioural characteristics of Christian pilgrims. *GeoJournal*, *50*(1), 55–67. https://doi.org/10.1023/A:1007154929681

Coromina, L., & Camprubí, R. (2016). Analysis of tourism information sources using a Mokken Scale perspective. *Tourism Management*, *56*, 75–84. https://doi.org/10.1016/j.tourman.2016.03.025

Crawford, D. W., Jackson, E. L., & Godbey, G. (1991). A hierarchical model of leisure constraints. *Leisure Sciences, 13*(4), 309–320. https://doi.org/10.1080/01490409109513147

Crompton, J. L. (1979). Motivations for pleasure vacation. *Annals of Tourism Research*, *6*(4), 408–424. https://doi.org/10.1016/0160-7383(79)90004-5

Dai, F., Wang, D., & Kirillova, K. (2022). Travel inspiration in tourist decision making. *Tourism Management*, *90*, 104484. https://doi.org/10.1016/j.tourman.2021.104484

Dann, G. M. (1977). Anomie, ego-enhancement and tourism. *Annals of Tourism Research*, *4*(4), 184–194. https://doi.org/10.1016/0160-7383(77)90037-8

de Bloom, J., Nawijn, J., Geurts, S., Kinnunen, U., & Korpela, K. (2017). Holiday travel, staycations, and subjective well-being. *Journal of Sustainable Tourism*, *25*(4), 573–588. https://doi.org/10.1080/09669582.2016.1229323

Decrop, A. (2010). Destination choice sets: an inductive longitudinal approach. *Annals of Tourism Research, 37*(1), 93–115. https://doi.org/10.1016/j.annals.2009.08.002

Donne, J. (1624). No man is an island. *PoemHunter.com*. https://www.poemhunter.com/poem/no-man-is-an-island/

Fridgin, J. D. (1991) *Dimensions of tourism*. Educational Institute of the American Hotel and Motel Association.

Gallego, I., Font, X., & González-Rodríguez, M. R. (2022). The impact of COVID-19 on European tourists' attitudes to air travel and the consequences for tourist destination evoked set formation. *Tourism Management Perspectives, 41*, 100945. https://doi.org/10.1016/j.tmp.2022.100945

Grønflaten, Ø. (2009). Predicting travelers' choice of information sources and information channels. *Journal of Travel Research, 48*(2), 230–244. https://doi.org/10.1177/0047287509332333

Gursoy, D. (2019). A critical review of determinants of information search behavior and utilization of online reviews in decision making process. *International Journal of Hospitality Management, 76*, 53–60. https://doi.org/10.1016/j.ijhm.2018.06.003

Hall, C. M. (2005). *Tourism: Rethinking the social science of mobility*. Prentice-Hall.

Hall, C. M. (2010). Equal access for all? Regulative mechanisms, inequality and tourism mobility. In S. Cole & N. Morgan (Eds), *Tourism and inequality: Problems and prospects* (pp. 34–48). CABI.

Hall, C. M. (2013). Framing behavioural approaches to understanding and governing sustainable tourism consumption: beyond neoliberalism, "nudging" and "green growth"? *Journal of Sustainable Tourism, 21*(7), 1091–1109. https://doi.org/10.1080/09669582.2013.815764

Hall, C. M., & Page, S. J. (2014). *The geography of tourism and recreation: Environment, place and space* (4th ed.). Routledge.

Han, H., Al-Ansi, A., Chua, B. L., Tariq, B., Radic, A., & Park, S. H. (2020a). The post-coronavirus world in the international tourism industry: Application of the theory of planned behavior to safer destination choices in the case of US outbound tourism. *International Journal of Environmental Research and Public Health, 17*(18), 6485. https://doi.org/10.3390/ijerph17186485

Han, H., Lee, S., & Lee, C. K. (2011). Extending the theory of planned behavior: Visa exemptions and the traveller decision-making process. *Tourism Geographies, 13*(1), 45–74. https://doi.org/10.1080/14616688.2010.529930

Han, J. H., Kim, J. S., Lee, C. K., & Kim, N. (2019). Role of place attachment dimensions in tourists' decision-making process in Cittáslow. *Journal of Destination Marketing & Management, 11*, 108–119. https://doi.org/10.1016/j.jdmm.2018.12.008

Han, Y., Zhang, T., & Wang, M. (2020b). Holiday travel behavior analysis and empirical study with Integrated Travel Reservation Information usage. *Transportation Research Part A: Policy and Practice, 134*, 130–151. https://doi.org/10.1016/j.tra.2020.02.005

Hudson, S., & Gilbert, D. (2000). Tourism constraints: The neglected dimension in consumer behaviour research. *Journal of Travel & Tourism Marketing, 8*(4), 69–78. https://doi.org/10.1300/J073v08n04_05

Hung, K., & Petrick, J. F. (2010). Developing a measurement scale for constraints to cruising. *Annals of Tourism Research, 37*(1), 206–228. https://doi.org/10.1016/j.annals.2009.09.002

Iso-Ahola, S. E. (1983). Towards a social psychology of recreational travel. *Leisure Studies, 2*(1), 45–56. https://doi.org/10.1080/02614368300390041

Jang, Y. J., Cho, S. B., & Kim, W. G. (2013). Effect of restaurant patrons' regret and disappointment on dissatisfaction and behavioral intention. *Journal of Travel & Tourism Marketing, 30*(5), 431–444. https://doi.org/10.1080/10548408.2013.803388

Kahneman, D. (2011). *Thinking, fast and slow*. Macmillan.

Kim, H., & Jamal, T. (2007). Touristic quest for existential authenticity. *Annals of Tourism Research, 34*(1), 181–201. https://doi.org/10.1016/j.annals.2006.07.009

Kim, J., Park, J., Lee, J., Kim, S., Gonzalez-Jimenez, H., Lee, J., Choi, Y.K., Lee, J.C., Jang, S., Franklin, D., Spence, M.T., & Marshall, R. (2022). COVID-19 and Extremeness Aversion: The role of safety seeking in travel decision-making. *Journal of Travel Research, 61*(4), 837–854. https://doi.org/10.1177/00472875211008252

Kim, M. J., & Hall, C. M. (2019). A hedonic motivation model in virtual reality tourism: Comparing visitors and non-visitors. *International Journal of Information Management, 46*, 236–249. https://doi.org/10.1016/j.ijinfomgt.2018.11.016

Klenosky, D. B. (2002). The "pull" of tourism destinations: A means-end investigation. *Journal of Travel Research*, *40*(4), 396–403. https://doi.org/10.1177/004728750204000405

Larsen, S., Brun, W., & Øgaard, T. (2009). What tourists worry about–Construction of a scale measuring tourist worries. *Tourism Management*, *30*(2), 260–265. https://doi.org/10.1016/j.tourman.2008.06.004

Lee, C. K., Yoon, Y. S., & Lee, S. K. (2007). Investigating the relationships among perceived value, satisfaction, and recommendations: The case of the Korean DMZ. *Tourism Management*, *28*(1), 204–214. https://doi.org/10.1016/j.tourman.2005.12.017

Leelapattana, W., Hsu, S. Y., Thongma, W., Chen, C., & Chiang, F. M. (2019). Understanding the impact of environmental education on tourists' future visit intentions to leisure farms in mountain regions. *Sustainability*, *11*(6), 1567. https://doi.org/10.3390/su11061567

Lepp, A., & Gibson, H. (2008). Sensation seeking and tourism: Tourist role, perception of risk and destination choice. *Tourism Management*, *29*(4), 740–750. https://doi.org/10.1016/j.tourman.2007.08.002

Leung, X. Y., Sun, J., & Bai, B. (2019). Thematic framework of social media research: state of the art. *Tourism Review*, *74*(3), 517–531. https://doi.org/10.1108/TR-05-2018-0058

Li, H., Meng, F., & Zhang, Z. (2016). Non-participation of domestic tourism: Analyzing the influence of discouraging factors. *International Journal of Tourism Research*, *18*(6), 567–578. https://doi.org/10.1002/jtr.2074

Liu, S., Lai, D., Huang, S., & Li, Z. (2021). Scale development and validation of anime tourism motivations. *Current Issues in Tourism*, *24*(20), 2939–2954. https://doi.org/10.1080/13683500.2020.1854198

MacCannell, D. (1973). Staged authenticity: Arrangements of social space in tourist settings. *American Journal of Sociology*, *79*(3), 589–603. https://doi.org/10.1086/225585

Mannell, R. C., & Iso-Ahola, S. E. (1987). Psychological nature of leisure and tourism experience. *Annals of Tourism Research*, *14*(3), 314–331. https://doi.org/10.1016/0160-7383(87)90105-8

Maslow, A. H. (1954). *Motivation and personality*. Harper & Row.

McCabe, S., Li, C., & Chen, Z. (2016). Time for a radical reappraisal of tourist decision making? Toward a new conceptual model. *Journal of Travel Research*, *55*(1), 3–15. https://doi.org/10.1177/0047287515592973

Mehmetoglu, M. (2012). What determines holidaying interest? Extrinsic versus intrinsic motivations. *Journal of Social, Evolutionary, and Cultural Psychology*, *6*(1), 93–110. https://doi.org/10.1037/h0099222

Meng, B., & Choi, K. (2019). Tourists' intention to use location-based services (LBS): Converging the theory of planned behavior (TPB) and the elaboration likelihood model (ELM). *International Journal of Contemporary Hospitality Management*, *31*(8), 3097–3115. https://doi.org/10.1108/IJCHM-09-2018-0734

Michael, N., Nyadzayo, M. W., Michael, I., & Balasubramanian, S. (2020). Differential roles of push and pull factors on escape for travel: Personal and social identity perspectives. *International Journal of Tourism Research*, *22*(4), 464–478. https://doi.org/10.1002/jtr.2349

Neuburger, L., & Egger, R. (2021). Travel risk perception and travel behaviour during the COVID-19 pandemic 2020: A case study of the DACH region. *Current Issues in Tourism*, *24*(7), 1003–1016. https://doi.org/10.1080/13683500.2020.1803807

Nikjoo, A. H., & Ketabi, M. (2015). The role of push and pull factors in the way tourists choose their destination. *Anatolia*, *26*(4), 588–597. https://doi.org/10.1080/13032917.2015.1041145

Nimrod, G. (2008). Retirement and tourism themes in retirees' narratives. *Annals of Tourism Research*, *35*(4), 859–878. https://doi.org/10.1016/j.annals.2008.06.001

Noy, C., & Cohen, E. (Eds). (2012). *Israeli backpackers: From tourism to rite of passage*. SUNY Press.

Nyaupane, G. P., Morais, D. B., & Graefe, A. R. (2004). Nature tourism constraints: A cross-activity comparison. *Annals of Tourism Research*, *31*(3), 540–555. https://doi.org/10.1016/j.annals.2004.01.006

Özdemir, G., & Çelebi, D. (2018). Exploring dimensions of slow tourism motivation. *Anatolia*, *29*(4), 540–552. https://doi.org/10.1080/13032917.2018.1460854

Pappas, N., & Glyptou, K. (2021). Accommodation decision-making during the COVID-19 pandemic: Complexity insights from Greece. *International Journal of Hospitality Management, 93*, 102767. https://doi.org/10.1016/j.ijhm.2020.102767

Pearce, P. L., & Lee, U. (2005). Developing the travel career approach to tourist motivation. *Journal of Travel Research, 43*(3), 226–237. https://doi.org/10.1177/0047287504272020

Pearce, P. L., & Packer, J. (2013). Minds on the move: New links from psychology to tourism. *Annals of Tourism Research, 40*, 386–411. https://doi.org/10.1016/j.annals.2012.10.002

Petri, H. L., & Govern, J. M. (2012). *Motivation: Theory, research, and application.* Wadsworth, Cengage Learning.

Poria, Y., Butler, R., & Airey, D. (2004). Links between tourists, heritage, and reasons for visiting heritage sites. *Journal of Travel Research, 43*(1), 19–28. https://doi.org/10.1177/0047287504265508

Prayag, G. (2010). Images as pull factors of a tourist destination: A factor-cluster segmentation analysis. *Tourism Analysis, 15*(2), 213–226. https://doi.org/10.3727/108354210X12724863327768

Prayag, G., & Ryan, C. (2011). The relationship between the 'push'and 'pull'factors of a tourist destination: The role of nationality–an analytical qualitative research approach. *Current Issues in Tourism, 14*(2), 121–143. https://doi.org/10.1080/13683501003623802

Rahmani, K., Gnoth, J., & Mather, D. (2018). Hedonic and eudaimonic wellbeing: A psycholinguistic view. *Tourism Management, 69*, 155–166. https://doi.org/10.1016/j.tourman.2018.06.008

Ram, Y., & Hall, C. M. (2020). The camp not taken: analysis of preferences and barriers among frequent, occasional and non-campers. *Leisure Sciences.* https://doi.org/10.1080/01490400.2020.1731885

Ram, Y., Collins-Kreiner, N., Gozansky, E., Moscona, G., & Okon-Singer, H. (2021). Is there a COVID-19 vaccination effect? A three-wave cross-sectional study. *Current Issues in Tourism, 25*(3), 379–386. https://doi.org/10.1080/13683500.2021.1960285

Ram, Y., Nawjin, J., & Peeters, P. (2013). Happiness and limits to sustainable tourism mobility: a new conceptual model. *Journal of Sustainable Tourism, 21*(7), 1017–1035.

Šagovnović, I., & Kovačić, S. (2022). Predicting travel motivation with personality and personal values–The roles of big five plus honesty/humility personality traits and Kahle's values. *International Journal of Hospitality & Tourism Administration.* https://doi.org/10.1080/15256480.2021.2025189

Sanchez, J., Callarisa, L., Rodriguez, R. M., & Moliner, M. A. (2006). Perceived value of the purchase of a tourism product. *Tourism Management, 27*(3), 394–409. https://doi.org/10.1016/j.tourman.2004.11.007

Seligman, M., & Csikszentmihalyi, M. (2000). Positive psychology: An introduction. *American Psychologist, 55*(1), 5–14. https://doi.org/10.1037/0003-066X.55.1.5

Sharpley, R. (2018). *Tourism, tourists and society.* Routledge.

Sirakaya, E., & Woodside, A. G. (2005). Building and testing theories of decision-making by travellers. *Tourism Management, 26*(6), 815–832. https://doi.org/10.1016/j.tourman.2004.05.004

Sirgy, M. J., Kruger, P. S., Lee, D. J., & Yu, G. B. (2011). How does a travel trip affect tourists' life satisfaction? *Journal of Travel Research, 50*(3), 261–275. https://doi.org/10.1177/0047287510362784

Smith, M. K., & Diekmann, A. (2017). Tourism and wellbeing. *Annals of Tourism Research, 66*, 1–13. https://doi.org/10.1016/j.annals.2017.05.006

Stephen, A. T. (2016). The role of digital and social media marketing in consumer behavior. *Current Opinion in Psychology, 10*, 17–21. https://doi.org/10.1016/j.copsyc.2015.10.016

Tan, W. K., & Chen, Y. C. (2021). Tourists' work-related smartphone use at the tourist destination: making an otherwise impossible trip possible. *Current Issues in Tourism, 24*(11), 1526–1541. https://doi.org/10.1080/13683500.2020.1814706

Thaler, R., & Sunstein, C. R. (2008). *Nudge: Improving decisions about health, wealth, and happiness.* Yale University Press.

Ulker-Demirel, E., & Ciftci, G. (2020). A systematic literature review of the theory of planned behavior in tourism, leisure and hospitality management research. *Journal of Hospitality and Tourism Management, 43*, 209–219. https://doi.org/10.1016/j.jhtm.2020.04.003

Um, S., & Crompton, J. L. (1990). Attitude determinants in tourism destination choice. *Annals of Tourism Research, 17*(3), 432–448. https://doi.org/10.1016/0160-7383(90)90008-F

UNWTO. (2019). *International tourism highlights, 2019 edition*. https://doi.org/10.18111/9789284421152

Uriely, N., Ram, Y., & Malach-Pines, A. (2011). Psychoanalytic sociology of deviant tourist behavior. *Annals of Tourism Research*, *38*(3), 1051–1069. https://doi.org/10.1016/j.annals.2011.01.014

Villacé-Molinero, T., Fernández-Muñoz, J. J., Orea-Giner, A., & Fuentes-Moraleda, L. (2021). Understanding the new post-COVID-19 risk scenario: Outlooks and challenges for a new era of tourism. *Tourism Management*, *86*, 104324. https://doi.org/10.1016/j.tourman.2021.104324

Voigt, C., Brown, G., & Howat, G. (2011). Wellness tourists: in search of transformation. *Tourism Review*, 66(1/2), 16–30. https://doi.org/10.1108/16605371111127206

Wang, L. H., Yeh, S. S., Chen, K. Y., & Huan, T. C. (2022). Tourists' travel intention: revisiting the TPB model with age and perceived risk as moderator and attitude as mediator. *Tourism Review*. *77*(3), 877–896. https://doi.org/10.1108/TR-07-2021-0334

Wang, N. (2000). *Tourism and modernity: A sociological analysis*. Sage.

Wattanacharoensil, W., & La-ornual, D. (2019). A systematic review of cognitive biases in tourist decisions. *Tourism Management*, *75*, 353–369. https://doi.org/10.1016/j.tourman.2019.06.006

White, N. R., & White, P. B. (2004). Travel as transition: Identity and place. *Annals of Tourism Research*, *31*(1), 200–218. https://doi.org/10.1016/j.annals.2003.10.005

Wong, I. A., & Rosenbaum, M. S. (2012). Beyond hardcore gambling: Understanding why mainland Chinese visit casinos in Macau. *Journal of Hospitality & Tourism Research*, *36*(1), 32–51. https://doi.org/10.1177/1096348010380600

Yi, X., Lin, V. S., Jin, W., & Luo, Q. (2017). The authenticity of heritage sites, tourists' quest for existential authenticity, and destination loyalty. *Journal of Travel Research*, *56*(8), 1032–1048. https://doi.org/10.1177/0047287516675061

Yoon, Y., & Uysal, M. (2005). An examination of the effects of motivation and satisfaction on destination loyalty: a structural model. *Tourism Management*, *26*(1), 45–56. https://doi.org/10.1016/j.tourman.2003.08.016

Zhu, M., Gao, J., Zhang, L., & Jin, S. (2020). Exploring tourists' stress and coping strategies in leisure travel. *Tourism Management*, *81*, 104167. https://doi.org/10.1016/j.tourman.2020.104167

Chapter 5

Economic Perspectives on Tourism

Peter Fieger & David Dyason

Introduction

Worldwide tourism has experienced an astonishing growth trajectory over the past 30 years, driven by an increase in disposable incomes, reduced cost of travel, economic ascendancy of several nations and the general liberalisation of domestic and international travel conditions. More recently, the emergence of the sharing economy has further reduced the cost of travel and thus made travelling more accessible. This growth has been so pronounced that in some tourism hot spots such as Venice or Iceland, tourism's impact on infrastructure and the environment has reached critical limits. Acceptance of further growth came into question in host populations, and calls for a more sustainable tourism framework have become louder. Nonetheless, policymakers in most tourism destinations have been striving to stimulate further tourism growth.

There are multiple obvious economic benefits from such endeavours, such as the creation of employment, the prospect of levying and raising taxes, the stimulation of economic growth via the consumption of goods by tourists and people employed by tourism, the construction and maintenance of infrastructure, and the supply of seasonal labour. In addition, international tourism provides the added benefit of the ability for the host country to earn foreign exchange, and thus international tourism can be seen as an export commodity. Some other non-economic benefits of tourism are more challenging to measure, such as the promotion of a region, the experience of different cultures by tourists and host populations alike, and the promotion of specialised interests, such as the increase of the environmental appeal of a region.

Several disadvantages partially offset these benefits of tourism to the host location. Like other business sectors, tourism consumes resources and contributes to environmental degradation via pollution, emissions and erosion. Tourists may be allocated resources at the expense of local populations. Many destinations' tourism is seasonal, which has negative implications

The Wiley Blackwell Companion to Tourism, Second Edition. Edited by C. Michael Hall.

with respect to employment and the viability of investment. In tourism dependent destinations, the presence of tourism may discourage the development of economic diversity. Tourism jobs are often lowly paid and do not exert a strong requirement of a highly educated resident workforce, which lessens the attractiveness of a location to employers requiring a more sophisticated workforce.

Tourism is entrenched within society as an active contributor to the world's economies through its benefits and costs, which requires us to study it. Tourism economics, as an applied branch of micro-economics, is the study of tourism in the context of the domestic and international economy. At the most basic level this includes the factors impacting demand and supply. In addition, tourism economics is concerned with optimising the trade-offs between the benefits and disadvantages. This heightens the importance of tourism economics in the contemporary destination management environment.

In this chapter, we will examine some of the issues in the economics of tourism that have gained the attention of scholars, industry representatives and policymakers in recent times. This includes tourism consumption, seasonality, optimising the economic returns from tourism, minimising the undesired economic aspects of tourism, over-tourism, and the use of econometric techniques required to analyse strategies aimed to achieve these goals.

In 2020, the emergent COVID-19 pandemic disrupted the world economy to an extent not seen since the end of World War II (Gössling et al., 2020). Border closures, lock-downs, supply chain disruptions and other events had a destructive impact on many sectors of the world's economies, but the impact on tourism has been especially significant. The decades long growth in this industry came to an abrupt halt, with many stakeholders in the tourism sector unsure how the industry will recover. While many governments have instituted various measures designed to minimise business failures and unemployment in the sector, the true long-term extent of the social and economic damage incurred is only likely to be seen once the pandemic is completely over and normalcy in economic life has resumed. We will conclude this chapter with an overview of the impact of the COVID-19 pandemic on tourism and some thoughts on the way forward for the time after the pandemic. This chapter will largely focus on international aspects of tourism, but reference to domestic tourism will be made where necessary.

Tourism Consumption and the Economy

In economics, the consumption of goods and services involves a value-chain process that draws on inputs to deliver a product suitable for consumption. This process is mainly unseen to the general public, who consume the goods and services, but requires coordination and inputs from various role-players to succeed. In other words, serving a coffee to a consumer requires various ingredients to be sourced, ultimately leading to benefits and costs for those involved. Consumption primarily occurs to satisfy a need, which could change depending on the individual's tastes and preferences. Changes to tastes and preferences do occur, which require a reaction from those involved in the value-chain process. The interaction between the role-players leads to transactions, which are measurable and comparable over time and illustrate how tastes and preferences change.

Tourism consumption within the economy is based on the same principles that drive consumer consumption. It has evolved from the need to connect, explore, learn and gain knowledge outside of one's current place of residence. People value the activity associated with tourism, and as a result, we see increasing tourism consumption over time. The continued growth in tourism consumption reveals how valued the consumption of tourism goods and services has become within the world economy. Between 2010 and 2019, international tourism spending increased by 51.8 percent, outpacing global economic activity growth, measured as gross domestic product (GDP), of 31.7 percent over the same period. This is significant, especially given what GDP represents within economics.

GDP represents the market value of the total output of goods and services produced in the economy within a year. The value represents the total income of everyone within the world economy. With this in mind, the growth in international tourism spending reveals the growing importance of tourism consumption from consumers. It also shows that consumers are increasing their spending share in tourism relative to other goods and services.

Figure 5.1 reveals the international tourism spending and worldwide GDP between 2010 and 2019. The comparison highlights the positive association between tourism consumption and total worldwide income.

The consumption of tourism goods and services is strongly and positively associated with the prevailing economic climate (Anagnostou et al., 2021; Seghir et al., 2015). Considering the results in Figure 5.1, it is evident that periods of GDP growth correspond with similar levels of growth in tourism spending. Over the past two decades, the growth in tourism consumption has followed GDP growth closely, a relationship well documented in the literature. Figure 5.2 provides a similar view of this relationship for individual markets. The results reveal that countries that experience high growth in GDP experienced strong growth in outbound tourism spending.

While a robust economic climate is beneficial for increasing tourism consumption, consumption value is not spread evenly worldwide. Figure 5.3 shows the outbound international tourism spending for various countries between 1995 and 2019. The long-term growth of tourism spending is evident in almost all the nations illustrated, while an apparent change in outbound spending is evident from 2007 onwards. There is a significant rise in international tourism spending from countries in Asia, including China, the Republic of Korea, and India.

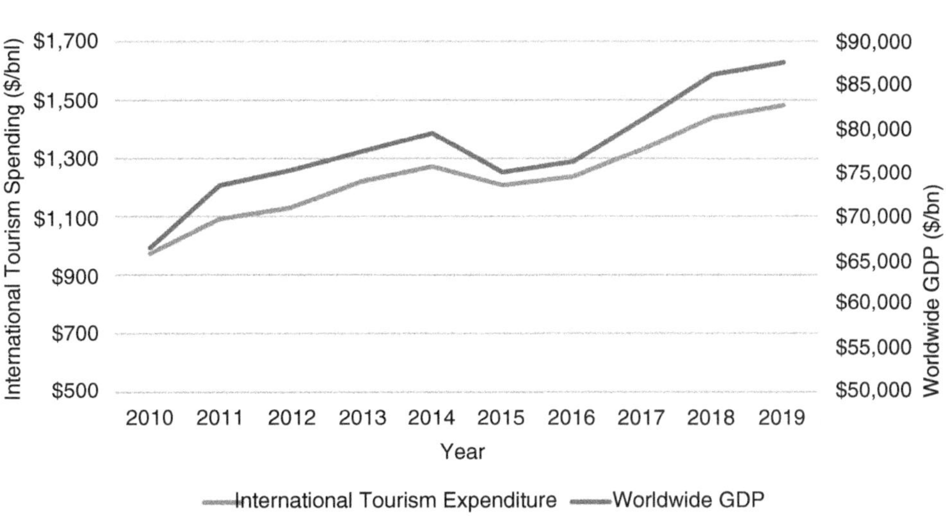

Figure 5.1. International tourism expenditure and worldwide GDP.
Source: World Tourism Organization (UNWTO) & International Labour Organisation (ILO) (2022); UNWTO, 2022.

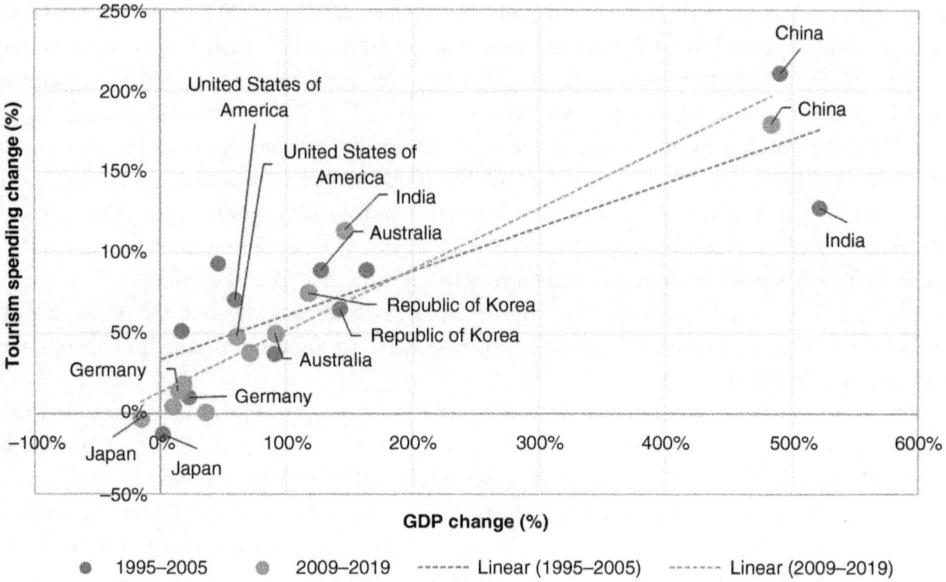

Figure 5.2. GDP and tourism spending association.
Source: Author calculations based on UNWTO & ILO, 2022; UNWTO, 2022.

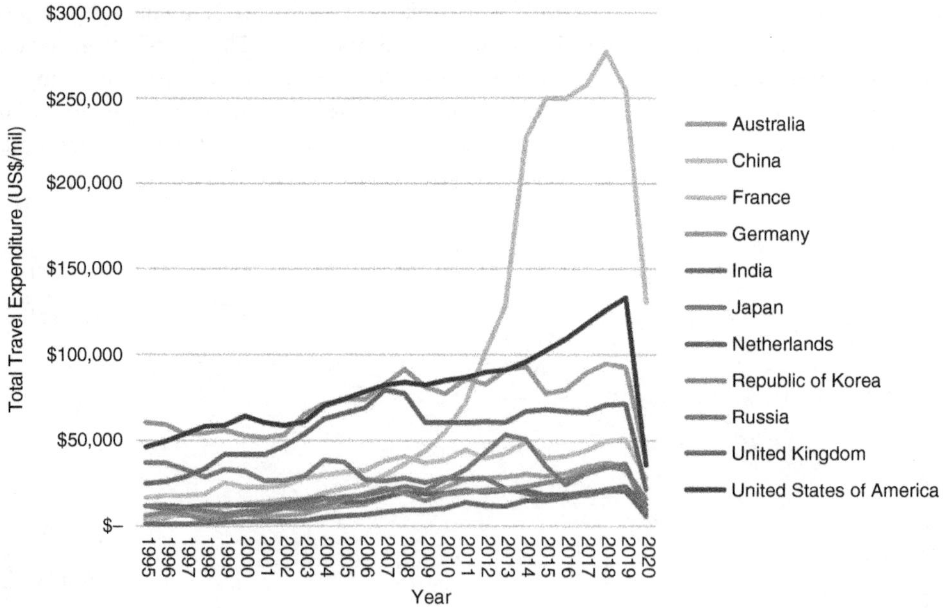

Figure 5.3. Outbound tourism expenditure per country.
Source: Data taken from UNWTO, 2022.

These countries have experienced significant economic growth between 2009 and 2019, with China growing at an average annual rate of 10.5 percent, India at 7.9 percent and the Republic of Korea at 5.5 percent. The underlying economic environment and the income that countries generate, expressed in GDP, are essential for tourism consumption.

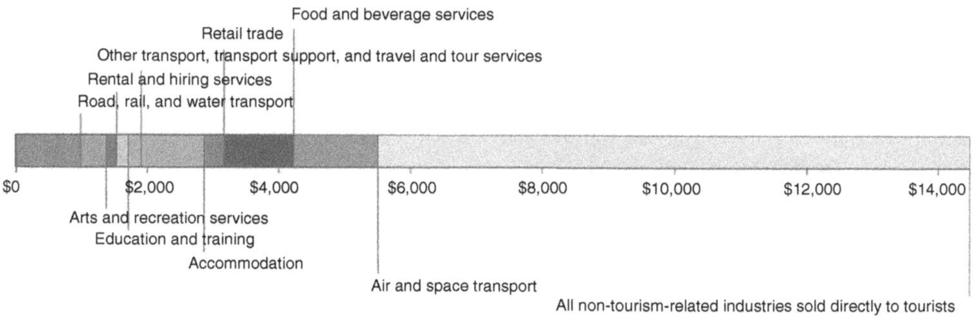

Figure 5.4. Tourism expenditure, by type of product, the year ended March 2020 (NZ$/Mil). **Source:** Stats NZ, 2022.

Tourism consumption and its economic benefits are measured with spending data on tourism goods and services (Li et al., 2018). As tourism has various inter-industry relationships, it implies that growing tourism spending benefits various industries through these relationships within the economy. Using tourism spending data from New Zealand, the industries that benefit from tourism spending are illustrated in Figure 5.4.

The growth of economies worldwide has led to increases in disposable incomes, the ascendancy of several economies that represent new tourism source markets and relative improved access to travel to previously hard to reach places. This, amongst other things, is driving tourism consumption and benefitting various industries that form part of the tourism value chain. Measuring tourism consumption through transactional spending data has provided insight into the benefits and costs that are part of tourism.

Tourism's Economic Benefits and Costs

The promotion of tourism is instinctively perceived as beneficial to the economic development of regions. However, there are often overlooked costs related to attracting visitors with these benefits. A discussion on various benefits and sometimes overlooked costs follow.

The unique attribute of the tourism industry is that it incorporates the use of a variety of activities across multiple sectors. As a result, assessing its significance within the economy often involves various measuring approaches. Since tourism is not regarded as a conventional sector within the economy, many countries have established a national account that measures the industry's activity as a standalone sector. The Tourism Satellite Account (TSA) is a popular method to measure the contribution of tourism to the national economy (Frechtling, 2010). The TSA allows for a yearly comparison of the industry's contribution to the national economy on aspects such as spending, taxing and employment, and comparing the contribution to the economy of domestic and international visitors. Although the TSA is widely used on a national level, efforts are underway to develop it for sub-national economies to measure the benefits to employment and spending at the regional level (Frenţ & Frechtling, 2020).

Another approach that reveals the benefits of tourism consumption is through economic impact analysis, which uses the sale and purchase relationship between producers, households, and consumers in an economy. The most popular of these models include the computable general equilibrium (CGE) model, social accounting matrix (SAM) model and Input-Output (IO) model (Baggio, 2019). In all of the abovementioned approaches, the key measures of benefit to the economy from tourism consumption are illustrated through employment and economic value add (GDP). At the same time, benefits to household income and taxes are often referenced.

Economic Benefits

Employment creation is a significant benefit of increasing tourism activity, predominantly due to the labour-intensive nature of the industry (UNWTO, 2022). The benefit of tourism consumption extends across various industries and, in the process, leads to employment creation. For example, on a sub-national level, a study of Central Finland (Tohmo, 2018) found that visitor spending by domestic and international tourists on hotels, restaurants and transport activity generates 30 percent of the region's 8,411 industry employment. The spending injection through visitors to the region supports a significant share of employment, and without it, employment in these industries is likely to be lower. Beyond the spending injection from visitors, evidence from Dogru, et al. (2020) reveal how investment in tourism supports overall employment within the economy. Their study examined the effects of hotel investment on employment within the USA and found that a 1 percent increase in hotel investment increases total employees by 0.136 percent for the country and a 0.48 percent increase in tourism employment. On a global scale, tourism creates 1 in every 11 jobs (UNWTO, 2017).

Beyond employment, the gains in providing an income for the government through taxation are also evident. Most countries have a tax system that provides an avenue for the government to collect income for public spending. In most countries, spending on goods and services incurs a value-added tax (VAT) or a goods and services tax (GST). The purchases of tourists provide an additional source of revenue to the government's fiscal budget. Apart from taxes collected from domestic visitor spending, the injection from international tourist spending brings an additional benefit to government tax income. This money can be spent on public services to benefit the broader community (Gooroochurn & Sinclair, 2005).

Benefits to the economy through infrastructure provision, which is foundational to tourism development and tourism commerce (Dwyer, 2007), represent a significant benefit to society. Not only are infrastructure developments beneficial for economic activity through improved connectivity of economic nodes, but are also perceived as beneficial in supporting tourism activity by the resident population (Kanwal et al., 2020). Investment in infrastructure further allows for opportunities to support the economy by reducing transport costs, allowing economic growth driven by tourism activity (Adeola & Evans, 2020). The infrastructure that comes to mind within the tourism context relates to transportation, such as airports, cruise berths, railways and other public transport services. At the same time, public infrastructure such as convention centres for conferences and sports facilities supports event tourism.

A further benefit of tourism activity is the provision of seasonal labour. For instance, locations where tourism destinations are near horticultural hot spots often benefit from visiting backpacker populations, of whom many engage in short-term employment during peak times. This provides a source of seasonal labour to benefit the local economy.

Tourism as a standalone industry is a vital contributor to the economy through its benefits in job creation, infrastructure development, taxation and seasonal employment. Beyond this, it allows areas that might have nothing other than cultural heritage or aesthetic value to partake and benefit in economic activity. The industry's ability to bring people together from different parts of the world makes a remarkable contribution to the overall economy (Ahmad et al., 2019). Finally, it helps to improve the image of countries.

Economic Costs

While the benefits of tourism are well presented and often the main reasons to support tourism development in regions, these benefits play a part and contribute to environmental degradation through pollution, erosion, and emissions. There seems to be a causal relationship between tourism development and environmental degradation (Katircioglu et al., 2014). This requires governments to pay attention to protection measures and environmental protection policies to encourage sustainable tourism. While these protective measures are required, the transition of the global economy into a knowledge-led economy, replacing industrial activity as the main economic driver, is helping. Recent studies suggest that this shift is beneficial for the global economy since a move away from industrialisation to services, such as tourism, improves the environmental quality by reducing CO_2 emissions and promoting cleaner technologies (Ahmad et al., 2019; Balsalobre-Lorente et al., 2020).

In today's global economy, international migration continues to be an essential contributor to labour supply. However, filling low-skilled tourism jobs remains problematic due to the pay and work conditions. At the same time, those employed in the industry seldom continue with further education, as few career progression options exist (Booyens, 2020). As a result, the industry continues to experience a fluid labour market, with Migration Led Tourism (MLT) evident in areas dependent on tourism. With growing labour and skills shortages, employers continue to look for workers with lower pay expectations since jobs with lower-skill requirements struggle to attract native-born workers (Joppe, 2012).

Finally, the restrictions placed by governments worldwide in response to COVID-19 have once again highlighted how connected the world economy has become. The limitations have not only impacted mobility on a national level but the interconnectedness of the global economy was affected. As the global economy adjusts to a post-pandemic environment, tourism-related labour mobility is expected to remain an essential aspect in support of the industry (Salazar, 2022).

Current Trends in Tourism

International tourism has, in recent decades, grown at a rapid pace. While there were 930 million departures in 1997, in the year just before the COVID-19 pandemic departures had grown to 2 billion, representing a 215 percent increase (UNWTO, 2022). This growth was not evenly distributed. China's outbound tourism increased from 5.3 million in 1997 to 154.6 million in 2019 (a 1,917 percent increase). This growth in China was especially fast after 2008. More mature outbound markets grew at a slower rate, albeit still exhibiting impressive growth rates. Outbound tourism originating from the European Union increased from 271.5 million to 468.7 million (73 percent growth) and North America from 103.30 to 208.8 million (102.2 percent growth) over the 1997 to 2019 period.

The traditional source markets for international tourism have seen a significant adjustment from 2008 with the rise of Asia, and China in particular, as a new emergent source market for tourism. Figure 5.5 illustrates the rapid growth experienced in the East-Asian and Pacific market since 2008, mainly driven by the growing Chinese market. The period since 2008 is seen as a transitional period with mixed growth in the traditional international source markets such as Europe and North America and high growth in the Asian market.

The growing Chinese tourism market has not gone unnoticed and has been labelled the most influential social and economic phenomenon in recent years (Johnson et al., 2020).

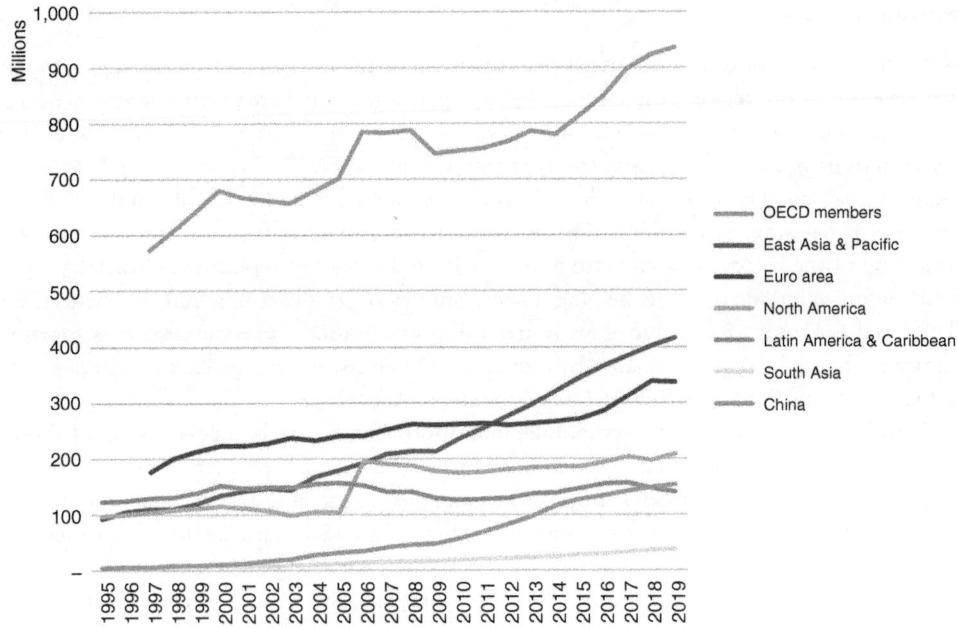

Figure 5.5. International tourism departures from the source country, 1995–2019.
Source: Data taken from UNWTO, 2022.

This has created new opportunities within the tourism industry, for example, through integrating and using technology for payment, such as WeChat and Alipay (Dichter et al., 2018). Furthermore, as Chinese outbound tourists travel to new destinations, the popularity of semi-guided tours and high-end tour packages is on the rise (Dichter et al., 2018). The expanding non-traditional outbound tourism markets continue to create new opportunities that complement the existing traditional outbound tourism market. This is especially relevant to tourism destinations that rely on inbound international tourism. Understanding the opportunities created from new markets, their motivation for travel and preferences could benefit destinations. In this instance, national and regional tourism promotion organisations' role in identifying such opportunities and responding is vital. It requires a concerted effort within the organisation to collaborate across disciplines through marketing, destination development, business attraction, and insights to identify market changes. The ability to make informed decisions largely depends on the organisation's ability to use insights to forecast tourism demand.

Tourism Forecasting

From an economic perspective, recent operational trends in tourism have seen the effort to understand better visitor characteristics in relation to their destination and how this knowledge can be leveraged to increase efficiencies and marketing efforts. Specific opportunities have emerged from a contemporary abundance of big data and survey data. The interrogation of such data offers the opportunity to apply a multitude of analytical techniques to use the insights gained from such analysis to increase visitor numbers, improve marketing campaigns' efficiency, or forecast visitor numbers.

Survey data and big data, originating from sources as diverse as mobile phone network providers, search engines, rental vehicle providers, airlines, and transport operators, are increasingly being used to model tourist behaviours, destinations, activities, itinerary patterns and to connect visitor characteristics to preferences (Mutinda & Mayaka, 2012; Fieger et al., 2019). These analytical endeavours are aimed to optimise the match between visitor and destination and unlock more significant economic benefits to the host location. Other analytical targets are the ability of destinations to forecast visitor numbers over the short-, mid-term and long horizon.

Analytical techniques aimed at providing insights on visitor behavioural patterns have gained popularity and become more diversified (Jiao & Chen, 2019). Traditionally, forecasting models frequently employed time-series models where historical data in auto-regressive moving average models, such as Chu (2009). More recently, auxiliary data from search engines and the use of econometric techniques such as multivariate times series analysis (for instance, Fieger & Rice, 2018) have become more popular. Econometric models have gained prominence in the search for cause-and-effect relationships between tourism demand and the economy (Song et al., 2019). The recent popularity of big data is driven by improved access to technology, news and information, making analysable data more freely available. As a result, data-driven artificial intelligence (AI) models continue to be developed to examine non-linear data to forecast tourism demand (Song et al., 2019).

Although there is substantial heterogeneity of econometric techniques applied in the tourism demand forecasting context, recent reviews of demand forecasting by Jiao and Chen (2019) and Song et al. (2019) suggest that no one methodological technique is preferred above another. Combining models and developing new techniques continues to take place. Nonetheless, the increasing availability of primary and secondary data and the emergence of new data sources and computational evolution will continue to inspire the development of new econometric methods to optimise economic benefits from tourism.

Over-tourism

Rising tourist numbers over the last two decades, combined with an increasing awareness of the ecological implications of mass tourism, have in several more mature tourist destinations (for instance, Venice, see Bertocchi et al. (2020); or Iceland (OECD, 2017), see Sæþórsdóttir et al. (2020)) led to the consideration of reducing negative externalities of tourism and the emphasising of 'quality over quantity', e.g. a refocusing of marketing efforts toward higher spending visitors while limiting the growth in overall tourist numbers. This increased accentuation of sustainability stresses both the efficiency and sufficiency in terms of the natural capital and ecological resources on which economic benefit is based. It therefore aims at a tourism system that encourages qualitative development but not aggregate quantitative growth that unsustainably reduces natural capital (Hall, 2009).

The increase in visitation following the emergence of low-cost transport, low-cost shared accommodation and rising disposable incomes have led to what is perceived as over-tourism in some destinations. While rising visitor arrivals can lead to increased economic returns for the host location, the issue of over-tourism arises from considerations of sustainability, erosion, environmental emissions and degradation, overcrowding, and its adverse effects on the resident host population. These include the increasing cost of living, reduced access to affordable housing and resulting difficulty for employers in attracting required staff.

For example, the popularity of home-sharing as an alternative source of accommodation has been studied intensively since its emergence, with researchers trying to understand how this new trend could have a spill-over effect on the commercial accommodation and residential market. Rodriguez-Perez de Arenaza et al. (2019) suggest that home-sharing shifts residential supply from the resident population to cater to tourism demand. At the same time, Franco and Santos (2021), and Elíasson and Ragnarsson (2018) find that this leads to increasing rental rates in the residential market.

This has prompted scholars and policymakers alike to assess whether these costs of over-tourism can be mediated via modified approaches to destination management. Measures proposed include utilising mechanisms that modify price, taxes and levies, length of stay, and marketing strategies (Oklevik et al., 2019; Fieger et al., 2021; Seraphin & Ivanov, 2020). The breadth of these counter measures applied in different locations does indicate that there is no universal solution to over-tourism that applies across all destinations, and that not only a broad range of stakeholders should be involved in the development of a destination, but also that a long-term framework is required to be implemented by political bodies which goes beyond a short-term visitor destination strategy (Innerhofer et al., 2019).

Pursuing the quest for more sustainable tourism may entice destination management organisations to focus on more affluent market segments of customers while limiting overall tourist numbers. Pursuing this qualitative rather than quantitative growth would require tourist destinations to consider which forms of tourism to encourage and which to discourage and to set incentives and disincentives accordingly (Benner, 2019).

Tourism, Economic Development and Productivity

The question whether tourism represents a viable tool to shape economic development and stimulate growth has been considered for some time, although early attempts at tourism development in the 1950s and 1960s were mostly concerned with simple tourism planning without consideration for the economic impact (Baud-Bovy, 1982). Increasing awareness of the economic impacts of tourism development led to more elaborate planning which considered economic factors such as the maximisation of foreign currency receipts, employment, or the welfare of the local population. The costs and benefits of these alternative policies and their economic impacts were evaluated often in comparison with other economic sectors (Baud-Bovy, 1982).

The subject of tourism as part of an economic development strategy began to receive greater academic and policy maker attention with the advent of mass tourism from the 1970s. Hunt et al. (1972) recognised that tourism development, while not representing a panacea for solving economic problems, can nonetheless play a major role in the economy with significant potential. Obstacles for tourism development were identified as not solely lying in a lack of capital and infrastructure, but also a lack of understanding of how tourism functions as an economic development tool. While tourism's potential for economic development became more recognised, Cohen (1984) observed that the positive effects of tourism often fell significantly short of expectations and predictions. The more contemporary view of tourism's role in the economic development realm recognises its potential for high growth, increasing income per capita, foreign currency and government revenue (Sinclair, 1998) and also argues that tourism causes significant local economic gains, including spillover effects into other industries (Faber & Gaubert, 2019) and regions (Andraz et al., 2016).

The exports of a nation represent a constituent part of economic activity and contribute to economic growth by improving the efficiency in the allocation of the factors of production and by expanding their volume (Balassa, 1978). As a quasi or invisible export, tourism contributes thus to a nation's export capability via the earning of foreign exchange. Exports have long been recognised as a driver of economic growth (Feder, 1983; Kwan & Cotsomotis, 1991), and in 2002, Balaguer and Cantavella-Jorda examined the hypothesised causality between tourism and economic growth, which became known as the *tourism led growth hypothesis*. Subsequently, this hypothesis has been shown to hold in many empirical studies, either in the context of tourism-dependent developing economies or mature diversified economies (Neuts, 2020; Prayag et al., 2019).

The value that economies extract from tourism is not uniform across countries. Obviously, the larger the share of tourism within an economy, the larger the extent to which an economy can be considered dependent on tourism. Therefore, the tourism industry is making a crucial contribution to the economy of many smaller countries where it accounts for a substantial share of national GDP. Nonetheless, most developed economies consider tourism a fundamental industry for their economic growth because it depends on other productive sectors making the infrastructure of a zone and cycling its economic wheel (Zurub et al., 2015).

Tourism is generally seen as exerting positive impulses to economies via promoting economic growth and the provision of employment. Given tourism's potential role as an engine of economic growth, it is evident that productivity is instrumental in realising desired economic growth ambitions. Productivity is a critical economic measure describing the relationship between inputs and outputs. Increasing productivity is necessary for remaining competitive, and can be achieved by increasing outputs at a greater rate than inputs or reducing inputs at a greater rate than outputs. The interpretation of productivity in tourism is nuanced differently from other industries (Liu & Wu, 2019), but is generally defined by the amount of labour (and/or capital) required to produce a dollar of GDP. It is recognised that tourism is plagued with low productivity compared with other industries. This is due to tourism's high reliance on manual labour and the low degree of automatability in many sub-domains of tourism (specifically the accommodation and hospitality sectors). Nonetheless, while in some mature tourist destinations productivity is fairly stagnant (see, for instance Pham, 2020), other research indicates that innovation in specific areas such as technology and marketing (Nguyen et al., 2021) can still lead to increased tourism productivity. Efforts by destinations who heavily rely on tourism as an engine for growth will thus benefit from striving for improved productivity in tourism, specifically in domains where it is likely that such gains can be made. This is a particularly pertinent strategy in developing destinations where tourism accounts for a substantial part of export earnings.

Seasonality

Seasonality in tourism is a phenomenon that is present in most tourist locations. It refers mostly to recurring events, such as public and school holidays in the source country of visitors, festivals, sporting events or the seasons themselves, resulting in varying attractiveness to potential visitors of a destination during a given year. Resulting from this, seasonality manifests itself through large variability of visitor numbers and visitor spending throughout the year. While seasonality generates some positive aspects for a destination, for instance, hosts can conduct maintenance during less busy times of the year, the overwhelming economic impact is generally negative.

Adverse effects of seasonality in tourism include the difficulty of obtaining and retaining reliable staff. While labour sourcing is a function of seasonality itself, it further affects labour productivity, motivation, and long-term career commitment. The ability of tourism operators to source skilled labour is limited to the availability of work-seekers in the market at that particular time, potentially affecting the training and retention of staff. This represents an additional cost to the business owner.

Furthermore, seasonality impedes the efficient allocation of resources, leads to lower returns on equity and may thus impede investment and access to capital. Existing facilities may be prone to overuse and increased wear and tear during peak times. Individual operators and destination management organisations do generally attempt to mitigate this effect, with varying success.

Mitigation measures include creating events aimed at attracting visitors during less busy times of the year. Extending the main season via promotions, scheduling events, conferences, adding additional attractions, and the shifting school holidays are some of the approaches to mitigation that have been employed. Some locations, especially cities, have successfully implemented some or all of these measures. Nonetheless, seasonality continues to lead to a sub-optimal allocation of resources, particularly in destinations with little diversity in their attractions to appeal to potential visitors.

The seasonal nature of tourism means that swings between the upper- and lower end of employment are possible within weeks or even days of each other. For example, in Queenstown, New Zealand the difference in employment between the low season in June and the high season in September endure a swing of nearly 2,000 jobs. This change in employment represents a 10 percent increase in total employment for the local economy within a short space of time. This affects sectors such as short-stay accommodation and rental-housing markets as changes in demand increase the short-term availability of supply within the economy. Not only is there demand from visitors, but added short-term need for short-stay accommodation to accommodate additional workers during the peak times and vice versa. Seasonality within tourism is evident beyond the annual seasons and is further apparent between weekends and weekdays, between school holidays and school terms. While seasonality is generally seen as a problem for tourism destinations (Duro & Turrión-Prats, 2019), it is a part of the industry that is likely to remain.

COVID and Beyond: Outlook for Tourism

Since its beginning in late 2019, the COVID-19 pandemic has taken several turns and setbacks concerning its severity and outlook. Given travel restrictions, lock-downs and other measures designed to counter COVID-19's impact on populations and economies, tourism has been an industry that was particularly affected by the COVID pandemic.

Responding to and transitioning through this period remained a challenge for businesses in the industry, especially since tourism operators recognise their reliance on visitor spending and the spending injection as a key contributor to a successful operation. The expectation of failure and significant negative effects on the industry were evident at the onset of the pandemic (Dyason & Fieger, 2021).

The preponderance of owner-operators in this industry has led to hardship on such entrepreneurs, and while many governments have made efforts to support this sector, mostly via the provision of support payments, liquidity, tax breaks and low interest rates (Şengel et al., 2022), the prolonged downturn in tourism has led to business closures and consolidation. Larger corporate operators such as airlines and hotel chains had additional support mechanisms. In this regard, national airlines were supported as deemed 'too big to fail', and many larger hotels were repurposed as quarantine facilities during the height of the pandemic. Nonetheless, as one of the industries experiencing the heaviest impact of the pandemic, furloughing and laying off of workforce has been widespread. The economic impact of the pandemic on tourism varied significantly from country to country, being predominantly a function of the restrictiveness of COVID countermeasures, the degree of economic dependence on domestic and international tourism, and the course that the pandemic took in respective countries.

Two and a half years into the pandemic (at time of writing), mid-2022, there is cautious optimism concerning a prospective return to normality as far as tourism is concerned. Several European countries have started to lift Covid related border restrictions. Some scholars have argued that this point in time should give cause to pause and reflect on the prospect of resuming pre-COVID growth patterns with the benefits and dangers that such return to established economic policies entails. Rather, they argue, the immediate post-COVID environment should mark a turning point, a return to pre-COVID exponential growth is undesirable from a number of environmental and sustainability concerns (Hall & Seyfi, 2021; Gössling & Schweiggart, 2022). While this appears to be a desirable course of action, the possibility can however, not be discounted that governments will feel compelled to follow economic necessities and view tourism as one potential revenue generator to fill holes in public finances. Policies adopted during the height of the COVID-19 crisis have included quantitative easing as a means to finance measures such as lock-downs, border closures, mandates, etc designed to counter the economic fallout from the pandemic. As a result, many governments have accumulated hitherto unseen amounts of debt which will require increasing shares of GDP to service payments on this debt. Moreover, resulting from the unprecedented injection of money into the economies of many developed nations, inflationary pressures have started to build, which central banks will attempt to counter via the raising of interest rates. This in turn will force governments of smaller and tourism-dependent countries even more to plug budgetary holes by any means possible, including the stimulating of tourism as a means to raise funds. Furthermore, government funding is required to support and maintain tourism-related infrastructure such as roads, airports and accommodation establishments (Brien, et al., 2022). Without sufficient tourism, the burden of maintaining these investments could likely fall on the resident populations.

It is thus possible that themes and policy considerations that immediately preceded the COVID-19 pandemic, such as addressing the issues of sustainability, over-tourism, minimisation of carbon emissions and so on, may take a backseat for the foreseeable future with the view of policy makers to get their respective economies back on track first. It will be of interest if a balance is struck with respect to competing interests such as adherence to committed carbon reduction goals in light of the need to return to a path of economic growth again in the post-pandemic environment.

The effect of lockdowns and the uncertainty and economic conditions the COVID-19 pandemic has created within the tourism industry continues to receive attention from scholars and policy makers. Therefore, one current focus area of research is geared at understanding the possible changes or displacement in spending and its effect on tourism moving forward.

Conclusion

Irrespective of the impact of the COVID-19 pandemic, the relevance of tourism in the modern global economy is likely to become more pronounced. What makes it difficult to assess the exact trajectory in the immediate future is the confluence of several secondary economic developments that manifest themselves to varying degrees and depend on a number of factors present in respective host locations. Many developed economies with significant tourism industries are already jousting with finding the right balance between extracting better monetary value from a growing economic share of tourism, while at the same time aiming to satisfy increasingly stronger demands to consider the ecological and sustainability aspects of mass tourism. On the other hand, less developed nations who depend on tourism will have an incentive to further stimulate tourism development with less regard for environmental issues. Independent from environmental considerations, rising competitiveness between tourism destinations and greater public and political demands for increased returns from tourism will increase efforts to achieve greater efficiency in all aspects of tourism. This means that some of the contemporary tourism facets discussed in this chapter will receive increased attention from operators, environmentalists, policymakers and industry representatives in the near-term future. These efforts should particularly focus on addressing the challenges related to seasonality, over-tourism, better forecasting and infrastructure planning, as potential for the most substantial gains in economic returns lies here. It remains to be seen to what extent the COVID-19 pandemic and its aftermath will shape these developments, but it will undoubtedly exert significant impact. One possible scenario is that low GDP nations that depend on income from tourism will aim to quickly recover 'lost ground' without focusing overly on these challenges, while public and political pressure in high GDP nations will ensure that these challenges do not go unaddressed. Either way, tourism will continue to play an important economic role in many locations, though its fortunes will depend on how its challenges are addressed.

References

Adeola, O., & Evans, O. (2020). ICT, infrastructure, and tourism development in Africa. *Tourism Economics*, *26*(1), 97–114. https://doi.org/10.1177/1354816619827712

Ahmad, F., Draz, M. U., Su, L., & Rauf, A. (2019). Taking the bad with the good: The nexus between tourism and environmental degradation in the lower middle-income Southeast Asian economies. *Journal of Cleaner Production*, *233*, 1240–1249. https://doi.org/10.1016/j.jclepro.2019.06.138

Anagnostou, A., Ekonomou, G., & Kallioras, D. (2021). The nexus of tourism spending with economic performance: A panel data analysis for the Eurozone area. *Geographica Pannonica*, *25*(1), 53–62. https://doi.org/10.5937/gp25-29541

Andraz, J. M., Norte, N. M., & Goncalves, H. S. (2016). Do tourism spillovers matter in regional economic analysis? An application to Portugal. *Tourism Economics*, *22*(5), 939–963. https://doi.org/10.5367/te.2015.0469

Baggio, R. (2019). Measuring tourism: methods, indicators, and needs. In E. Fayos-Solà & C. Cooper (Eds), *The future of tourism: Innovation and sustainability* (pp. 255–269), Springer.

Balaguer, J., & Cantavella-Jorda, M. (2002). Tourism as a long-run economic growth factor: the Spanish case. *Applied Economics*, *34*(7), 877–884. https://doi.org/10.1080/00036840110058923

Balassa, B. (1978). Exports and economic growth: Further evidence. *Journal of Development Economics*, *5*(2), 181–189. https://doi.org/10.1016/0304-3878(78)90006-8

Balsalobre-Lorente, D., Driha, O. M., Shahbaz, M., & Sinha, A. (2020). The effects of tourism and globalization over environmental degradation in developed countries. *Environmental Science and Pollution Research*, *27*(7), 7130–7144. https://doi.org/10.1007/s11356-019-07372-4

Baud-Bovy, M. (1982). New concepts in planning for tourism and recreation. *Tourism Management*, *3*(4), 308–313. https://doi.org/10.1016/0261-5177(82)90056-5

Benner, M. (2019). From overtourism to sustainability: A research agenda for qualitative tourism development in the Adriatic. *German Journal of Economic Geography*, 2(64), 74–87. https://core.ac.uk/download/pdf/214011208.pdf

Bertocchi, D., Camatti, N., Giove, S., & van der Borg, J. (2020). Venice and overtourism: Simulating sustainable development scenarios through a tourism carrying capacity model. *Sustainability*, *12*(2), 512. https://doi.org/10.3390/su12020512

Booyens, I. (2020). Education and skills in tourism: Implications for youth employment in South Africa. *Development Southern Africa*, *37*(5), 825–839. https://doi.org/10.1080/0376835X.2020.1725447

Brien, A., Ahmed, H., & Dyason, D. (2022). Perspectives: NZ needs an ambitious target for its tourism recovery. *Tourism Ticker.* https://www.tourismticker.com/2022/02/11/perspectives-nz-needs-an-ambitious-target-for-its-tourism-recovery/

Chu, F. L. (2009). Forecasting tourism demand with ARMA-based methods. *Tourism Management*, *30*(5), 740–751. https://doi.org/10.1016/j.tourman.2008.10.016

Cohen, E. (1984). The sociology of tourism: approaches, issues, and findings. *Annual Review of Sociology*, *10*(1), 373–392. https://www.jstor.org/stable/2083181

Dichter, A., Chen, G., Saxon, S., Yu, J., & Suo, P. (2018). *Chinese tourists: Dispelling the myths. An in-depth look at China's outbound tourist market.* McKinsey.

Dogru, T., McGinley, S., & Kim, W. G. (2020). The effect of hotel investments on employment in the tourism, leisure and hospitality industries. *International Journal of Contemporary Hospitality Management*, *32*(5), 1941–1965. https://doi.org/10.1108/IJCHM-11-2019-0913

Duro, J. A., & Turrión-Prats, J. (2019). Tourism seasonality worldwide. *Tourism Management Perspectives*, *31*, 38–53. https://doi.org/10.1016/j.tmp.2019.03.010

Dwyer, L. (2007). *International handbook on the economics of tourism.* Edward Elgar Publishing.

Dyason, D., & Fieger, P. (2021, September 13–14). *Successful visitor market transition for the hospitality industry. A case study of Christchurch, New Zealand.* [Paper presentation]. In Proceedings of the 32nd Annual Conference of the South African Institute of Management Scientists, Online: North-West University. https://researcharchive.lincoln.ac.nz/items/fc773444-b5b4-4e0e-9b40-ed3d8cb069f6.

Elíasson, L., & Ragnarsson, Ö. P. (2018). Short-term renting of residential apartments. Effects of Airbnb in the Icelandic housing market. *Economics, wp76*, Department of Economics, Central Bank of Iceland.

Faber, B., & Gaubert, C. (2019). Tourism and economic development: Evidence from Mexico's coastline. *American Economic Review*, *109*(6), 2245–93. https://doi.org/10.1257/aer.20161434

Feder, G. (1983). On exports and economic growth. *Journal of Development Economics*, *12*(1–2), 59–73. https://doi.org/10.1016/0304-3878(83)90031-7

Fieger, P., & Rice, J. (2018) Modeling Chinese inbound tourism arrivals into Christchurch. *e-Review of Tourism Research*, *15*(4/5). https://ertr-ojs-tamu.tdl.org/ertr/index.php/ertr/article/view/234

Fieger, P., Prayag, G., & Bruwer, J. (2019). 'Pull' motivation: an activity-based typology of international visitors to New Zealand. *Current Issues in Tourism*, *22*(2), 173–196. https://doi.org/10.1080/13683500.2017.1383369

Fieger, P., Prayag, G., Hall, C. M., Rice, J., & Gössling, S. (2021). When staying long enough is enough? *Annals of Tourism Research*, *87*, 102974. https://doi.org/10.1016/j.annals.2020.102974

Franco, S. F., & Santos, C. D. (2021). The impact of Airbnb on residential property values and rents: Evidence from Portugal. *Regional Science and Urban Economics*, *88*, 103667. https://doi.org/10.1016/j.regsciurbeco.2021.103667

Frechtling, D. C. (2010). The tourism satellite account: A primer. *Annals of Tourism Research*, *37*(1), 136–153. https://doi.org/10.1016/j.annals.2009.08.003

Frenţ, C., & Frechtling, D. C. (2020). Regional tourism satellite accounts: towards international comparability. *Tourism Review*, *77*(1), 97–112. https://doi.org/10.1108/TR-07-2020-0315

Gooroochurn, N., & Sinclair, M. T. (2005). Economics of tourism taxation: Evidence from Mauritius. *Annals of Tourism Research*, *32*(2), 478–498. https://doi.org/10.1016/j.annals.2004.10.003

Gössling, S., & Schweiggart, N. (2022). Two years of COVID-19 and tourism: what we learned, and what we should have learned. *Journal of Sustainable Tourism*, *30*(4), 915–931. https://doi.org/10.1080/09669582.2022.2029872

Gössling, S., Scott, D., & Hall, C. M. (2020). Pandemics, tourism and global change: a rapid assessment of COVID-19. *Journal of Sustainable Tourism*, *29*(1), 1–20. https://doi.org/10.1080/09669582.2020.1758708

Hall, C. M. (2009). Degrowing tourism: Décroissance, sustainable consumption and steady-state tourism. *Anatolia*, *20*(1), 46–61. https://doi.org/10.1080/13032917.2009.10518894

Hall, C. M., & Seyfi, S. (2021). COVID-19 pandemic, tourism and degrowth. In C. M. Hall, L. Lundmark & J. J. Zhang (Eds), *Degrowth and tourism: New perspectives on tourism entrepreneurship, destinations and policy* (pp. 220–238). Routledge.

Hunt, J. D., Brown, P. J., & Schomaker, J. H. (1972). Answering questions about tourism: A growing economic development tool. *Utah Science*, *33*(1), 7–9. https://scholarworks.umt.edu/cgi/viewcontent.cgi?article=1024&context=forest_pubs

Innerhofer, E., Erschbamer, G., Pechlaner, H. (2019). Overtourism: The challenge of managing the limits. In H. Pechlaner, E. Innerhofer & G. Erschbamer (Eds), *Overtourism: Tourism management and solutions* (pp. 3–13). Routledge.

Jiao, E. X., & Chen, J. L. (2019). Tourism forecasting: A review of methodological developments over the last decade. *Tourism Economics*, *25*(3), 469–492. https://doi.org/10.1177/1354816618812588

Johnson, P. C., Xu, H., & Arlt, W. G. (2020). Outbound Chinese tourism: looking back and looking forward. *Journal of Policy Research in Tourism, Leisure and Events*, *12*(1), 1–7. https://doi.org/10.1080/19407963.2018.1505098

Joppe, M. (2012). Migrant workers: Challenges and opportunities in addressing tourism labour shortages. *Tourism Management*, *33*(3), 662–671. https://doi.org/10.1016/j.tourman.2011.07.009

Kanwal, S., Rasheed, M. I., Pitafi, A. H., Pitafi, A., & Ren, M. (2020). Road and transport infrastructure development and community support for tourism: The role of perceived benefits, and community satisfaction. *Tourism Management*, *77*, 104014. https://doi.org/10.1016/j.tourman.2019.104014

Katircioglu, S. T., Feridun, M., & Kilinc, C. (2014). Estimating tourism-induced energy consumption and CO2 emissions: The case of Cyprus. *Renewable and Sustainable Energy Reviews*, *29*, 634–640. https://doi.org/10.1016/j.rser.2013.09.004

Kwan, A. C., & Cotsomitis, J. A. (1991). Economic growth and the expanding export sector: China 1952–1985. *International Economic Journal*, *5*(1), 105–116.

Li, K. X., Jin, M., & Shi, W. (2018). Tourism as an important impetus to promoting economic growth: A critical review. *Tourism Management Perspectives*, *26*, 135–142. https://doi.org/10.1016/j.tmp.2017.10.002

Liu, A., & Wu, D. C. (2019). Tourism productivity and economic growth. *Annals of Tourism Research*, *76*, 253–265. https://doi.org/10.1016/j.annals.2019.04.005

Mutinda, R., & Mayaka, M. (2012). Application of destination choice model: Factors influencing domestic tourists destination choice among residents of Nairobi, Kenya. *Tourism Management*, *33*(6), 1593–1597. https://doi.org/10.1016/j.tourman.2011.12.008

Neuts, B. (2020). Tourism and urban economic growth: A panel analysis of German cities. *Tourism Economics*, *26*(3), 519–527. https://doi.org/10.1177/1354816619833553

Nguyen, V. K., Natoli, R., & Divisekera, S. (2021). Innovation and productivity in tourism small and medium enterprises: a longitudinal study. *Tourism Management Perspectives*, *38*, 100804. https://doi.org/10.1016/j.tmp.2021.100804

OECD. (2017). *2017 OECD economic survey of Iceland. Preserving sustainable and inclusive growth.* https://web-archive.oecd.org/2017-06-26/444002-Iceland-2017-OECD-economic-survey-preserving-sustainable-and-inclusive-growth.pdf

Oklevik, O., Gössling, S., Hall, C. M., Steen Jacobsen, J. K., Grøtte, I. P., & McCabe, S. (2019). Overtourism, optimisation, and destination performance indicators: A case study of activities in Fjord Norway. *Journal of Sustainable Tourism*, *27*(12), 1804–1824. https://doi.org/10.1080/09669582.2018.1533020

Pham, T. D. (2020). Tourism productivity theory and measurement for policy implications: The case of Australia. *Journal of Travel Research*, *59*(2), 247–266. https://doi.org/10.1177/0047287519835972

Prayag, G., Fieger, P., & Rice, J. (2019). Tourism expenditure in post-earthquake Christchurch, New Zealand. *Anatolia*, *30*(1), 47–60. https://doi.org/10.1080/13032917.2018.1496941

Rodriguez-Perez de Arenaza, D., Hierro, L. Á., & Patiño, D. (2019). Airbnb, sun-and-beach tourism and residential rental prices. The case of the coast of Andalusia (Spain). *Current Issues in Tourism*. https://doi.org/10.1080/13683500.2019.1705768

Sæþórsdóttir, A. D., Hall, C. M., & Wendt, M. (2020). Overtourism in Iceland: Fantasy or reality? *Sustainability*, *12*(18), 7375. https://doi.org/10.3390/su12187375

Salazar, N. B. (2022). Labour migration and tourism mobilities: Time to bring sustainability into the debate. *Tourism Geographies*, *24*(1), 141–151. https://doi.org/10.1080/14616688.2020.1801827

Seghir, G. M., Mostéfa, B., Abbes, S. M., & Zakarya, G. Y. (2015). Tourism spending-economic growth causality in 49 countries: A dynamic panel data approach. *Procedia Economics and Finance*, *23*, 1613–1623. https://doi.org/10.1016/S2212-5671(15)00402-5

Şengel, Ü., Işkın, M., Çevrimkaya, M., & Genç, G. (2022). Fiscal and monetary policies supporting the tourism industry during COVID-19. *Journal of Hospitality and Tourism Insights*. https://doi.org/10.1108/JHTI-08-2021-0209

Seraphin, H., & Ivanov, S. (2020). Overtourism: A revenue management perspective. *Journal of Revenue and Pricing Management*, *19*(3), 146–150. https://doi.org/10.1057/s41272-020-00241-7

Sinclair, M. T. (1998). Tourism and economic development: A survey. *The Journal of Development Studies*, *34*(5), 1–51. https://doi.org/10.1080/00220389808422535

Song, H., Qiu, R. T., & Park, J. (2019). A review of research on tourism demand forecasting: Launching the Annals of Tourism Research Curated Collection on tourism demand forecasting. *Annals of Tourism Research*, *75*, 338–362. https://doi.org/10.1016/j.annals.2018.12.001

Stats NZ. (2020). *Tourism satellite account: Year ended March 2020*. https://www.stats.govt.nz/assets/Uploads/Tourism-satellite-account/Tourism-satellite-account-2020/Download-data/Tourism-satellite-account-Year-ended-March-2020.pdf

Tohmo, T. (2018). The economic impact of tourism in Central Finland: a regional input–output study. *Tourism Review*, *73*(4), 521–547. https://doi.org/10.1108/TR-04-2017-0080

World Tourism Organization (UNWTO) and International Labour Organisation (ILO). (2022). *Employment and decent work in tourism — ILO-UNWTO joint project*. https://www.unwto.org/project/employment-and-decent-work-tourism-ilo-unwto-joint-project

World Tourism Organization (UNWTO). (2017). *The travel & tourism competitiveness report 2017*. https://reports.weforum.org/travel-and-tourism-competitiveness-report-2017/tourism-and-job-creation-advancing-the-2030-development-agenda/

World Tourism Organization (UNWTO). (2022). *Yearbook of tourism statistics, compendium of tourism statistics and data files*. https://data.worldbank.org/indicator/ST.INT.DPRT?view=chart

Zurub, H. H., Ionescu, A., & Constantin, V. D. (2015). Measuring the economic impact of tourism in European emerging markets. *Procedia Economics and Finance*, *32*, 95–102. https://doi.org/10.1016/S2212-5671(15)01369-6

Chapter 6

International Relations and Geo-political Approaches to Tourism

Siamak Seyfi & C. Michael Hall

Introduction

With more people traveling abroad, tourism has become a vital source of international relations. Thus, as Lanfant and Graburn (1992, p. 94) noted, 'tourism is not just a matter of national growth but must be conceptualized as part of international relations'. Likewise, Hall (1994) asserts that tourism's increasingly international nature makes it inextricably linked to the field of international relations. According to the World Tourism Organization (2020), worldwide cross-border tourism reached a record number of 1.5 billion travelers in 2019. Global tourism is expected to reach 1.8 billion by 2030 at a growth rate of 4–5% annually, with pre-COVID-19 growth returning in 2023 to many countries and regions. This large mobility of individuals of different nationalities requires considerable cooperation on political, economic, and cultural levels (Mosedale, 2011). Much of the tourism industry, particularly tour operations and air transport, is dominated by multinational corporations, which have belatedly gained recognition as international actors (Zapata Campos et al., 2018). Such massive movement of individuals is not only shaped by the economic and political relations between different states, but it has affected the course of those relations.

The increasingly growing number of tourists crossing national borders annually, as well as globalization processes and the growing internationalization of tourism businesses and their supply chain, has made tourism as one of the most far-reaching geopolitical practices of the twenty-first century (Hall, 2017; Mostafanezhad, 2018, Gillen & Mostafanezhad, 2019). Thus, it has long been recognized that a stable, peaceful economy is conducive to tourism's flourishing, and political or economic change can have a serious impact on tourism flows and destination perceptions (Richter, 1989; Hall, 1994). Although, democracy is not a prerequisite for tourism. In many cases authoritarian states may even be conducive for the development of tourism, as in the case of Portugal and Spain in the 1960s and 1970s, as it can allow for the rapid development of resort and transport infrastructure (Hall, 1994). International tourism development and regulations are therefore directly and indirectly connected with politics and

The Wiley Blackwell Companion to Tourism, Second Edition. Edited by C. Michael Hall.

international relations. This could be marked by a number of areas (political, economic, and sociocultural), as well as issues (intercultural understanding, travel security, democracy development, and environmental protection), and stakeholder groups (tourists, destinations, hosts, employees, businesses, governments, developing countries, non-governmental organizations, multinational companies).

Research into tourism development has until recently remained largely disconnected from questions of international relations, political economy and critical geopolitics. However, interest in this area continues to grow. The chapter is structured as follows. The key theories in international relations are first reviewed and then discussed in relations to international tourism development. The next sections then discuss tourism within a wider globalisation, political economy, and geopolitical context before conclusions and a future research agenda are noted.

Key Theories in International Relations

As a multidisciplinary subject, international relations refer to the study of the relations of states with each other and with international organizations and certain sub-national entities (e.g., bureaucracies, political parties, and interest groups) (McClelland, 1960, 1971; Tanter & Ullman, 2015). International relations encompass a number of theories that attempt to explain how states operate within the international systems. Table 6.1 summarizes some of the key theories of international relations in the context of international tourism.

In order to better understand the motivations and goals driving policy decisions, it is important to understand key theories in international relations. International relations can be classified into three major strands, namely realism, liberalism, and constructivism (Table 6.2). The theories explain how international systems work as well as how nations view the world and engage with one another. Ranging from straightforward realist concepts to liberal, equality-centric strategies, these theories explain the directions that governments take in regard to an international political issue or concern.

Table 6.1. Key theories in international relations.

Theory	*Definition and description*
Realism	Focuses on power and state-to-state interaction, viewing international politics as a struggle for power among states. According to this theory, states are motivated by national interests and governments act primarily to maintain their security and influence over other countries.
Liberalism	Focuses on how states can cooperate with each other to create an international society that promotes peace and prosperity for all individuals regardless of their nationality or status. This theory emphasizes the importance of individual rights and freedoms as well as democracy.
Constructivism	International relations are regarded as being constructed through reciprocal interactions between states. This theory purports that international relations were about power, not about collective security or economic development.
Marxism	Marxism is a philosophy of thought that advocates class struggle as the motor force of history. Marxists believe that capitalism creates the conditions for conflict among countries by creating economic inequalities among countries and by forcing some countries to compete for resources with others.
Feminism	Focuses on gender inequality and how it affects nations and their relationship with each other.

Table 6.2. Three main perspectives on international relations.

Theories	*Realism*	*Liberalism*	*Constructivism*
Key contributors	Thucydides (460-395 BC) Niccolò Machiavelli (1469–1527) Thomas Hobbes (1588–1679)	John Locke (1632–1704) Immanuel Kant (1724–1804) Adam Smith (1723–1790) Woodrow Wilson (1856–1924)	Alexander Wendt (1958-) Friedrich Kratochwil (1944-) Nicholas Onuf (1941-) John Ruggie (1944–2021)
Main theoretical proposition	Self-interested states compete for power and security, anarchy, self-help and national interest	Concern for power overridden by economic/political considerations (desire for prosperity, commitment to liberal views), collective security, international regimes, complex interdependence and transnational relations	State behavior shaped by elite beliefs, collective norms and values, and social and cultural identities.
Major variations	Neo-realism: Distribution of power decides outcome	Neo-liberal institutionalism: international system anarchic, but centric institutions created by states in their self-interest do constrain anarchy	Critical constructivism: challenges the state-centric constructivism of Wendt
Main actors	States	States, international institutions, global corporations, and commercial interest	Individuals (especially elites), NGOs and transnational networks
Main instruments	Military power and state diplomacy	Varies (international institutions, economic exchange, promotion of democracy)	Ideas and discourse
Core concerns	War and security; how vulnerable, self-interested states survive in an environment where they are uncertain about the intentions and capabilities of other states	Institutionalized peace and prosperity: how self-serving actors learn to see the advantages of coordinating behavior through rules / organizations in order to achieve collective gains	Social groups' shared meanings and images: how ideas, images, and identities develop, change, and shape world politics
Approach to peace	Protect sovereign autonomy and deter rivals through military preparedness and alliances	Institutional reform through democratization, open markets, and international law and organization	Activists who promote progressive ideas and encourage states to adhere to norms for appropriate behavior

(*Continued*)

Table 6.2. (*Continued*)

Theories	*Realism*	*Liberalism*	*Constructivism*
Global outlook	Pessimistic; great powers are locked in a relentless security competition	Optimistic: cooperative view of human nature and a belief in progress	Agnostic: global prospect hinges on the content of prevailing ideas and values
Main limitation	Does not account for progress and change in international relations	Tends to ignore the role of power	Does not explain which power structures and social conditions allow for changes in values, better at describing the past than anticipating the future.
Current significance in policymaking	High	High	Low

Source: Adapted from Walt (1988).

These theories offer fundamental arguments such as the description of the international political structure, the key actors of that structure, main instruments adopted by these actors, core concerns, as well as the way in which they approach peace and global outlook. The next sections positions tourism within the international relations theoretical framework.

International Relations and Tourism

Following the theoretical discussion on key theories in international relations, we focus on how international relations and tourism are intertwined. The tourism industry and international relations encompass a wide range of topics, and their relationships develop in relation to globalization, borders, tourism mobilities, and their potential contribution to world peace. As discussed earlier, while tour operators and airline companies are among the tourism actors operating internationally in tourism, tourism mobilities can also be affected by political and economic turmoil in one part of the world, altering tourism flows and affecting the economy in other parts (Hall, 2010). For example, the Russian invasion of Ukraine not only affected the flow of tourists in and out of the two countries, but also access to their airspace, increasing flight costs and times to destinations reached via their airspace and therefore the flow of tourists.

International tourism has always been subject to international politics. The mobility of people across national boundaries has political implications (Baerenholdt, 2013), which sometimes can be manipulated by government for geopolitical purposes as an instrument of political leverage between nations. This reflects the observation of Molz (2010) that views tourism is as much a political terrain as a cultural practice. Azcárate et al. (2021, p. 18) concur, framing this as 'mobility geopolitics', and they argue that tourism is increasingly used by states as a 'geopolitical and political economic weapon of choice that often fuels discriminatory social imaginaries of domestic and international destinations' (p. 18). One notable example of mobility restrictions was the trade embargo against the apartheid regime in South Africa which included restrictions on travel, notably on overflying and landing rights for South African airlines (Pirie, 1990). This also had an interesting indirect effect on tourism in the region as some international connections, for example from Australia to South Africa, shifted to Zimbabwe as a result of government sanctions and restrictions of trade with South Africa. The longstanding US sanctions and embargo against Cuba is another noteworthy example of mobility restrictionsm which significantly affects US citizens and residents wishing to visit Cuba as well as sanctions against cruise lines (Gordon, 2016). Furthermore, tourism has always been an effective policy tool for politicians to use due to its international and high-profile nature and its consequent economic and social benefits, e.g. visiting friends and family (VFR). For instance, the relaxations of travel restrictions by the US toward Cuba are one of the examples of using tourism as a political instrument and tool to achieve foreign policy objectives while also responding to domestic pressure groups.

Tourism has also been used as a tool by governments in their wider sanction regime. For instance, following the shooting down of a Russian military aircraft over Turkey, Russia immediately reacted and imposed a range of sanctions on travel to Turkey, including the sale ban of charter vacations to Turkey for Russian citizens. The sanctions against Russia following the Russian military intervention in Ukraine and the annexation of Crimea 2014 by many Western countries also include travel bans which affected the overall mobility of Russian citizen wishing to visit EU and other sanctioning countries. As a result, tourist flows from Russia to Europe were significantly reduced and sanctions contributed to the collapse of the Russian ruble, impacted the image of Russia in tourism and investment terms, and contributed to the Russian financial crisis which led to the decline of leading Russian players in the tourist market (Ovcharov et al., 2015). In another case, Popesku and Hall (2004) noted that Serbia became isolated from the international tourism market throughout the 1990s due to economic sanctions that were then in place, along with regional and political change. The resulting currency devaluation in sanctioned countries is also likely to restrict outbound travel, which could lead to the overall reduction of tourists and tourist expenditures even without sanctions on nationals (Seyfi & Hall, 2020a).

In addition, governments have deliberately used tourism to further their purposes, influencing where and how it occurs. In some cases, governments block their citizens from traveling either by outright banning them (as was the case in Taiwan and the two Koreas previously), often for economic reasons; by restricting the countries they allow their citizens to visit (as in the former Soviet bloc); by issuing exit visas at the government's discretion (again, the former Soviet bloc); by imposing prohibitive travel taxes (as in Thailand); or restricting the export of domestic currency and by issuing travel advisories (Toubes & Araújo-Vila, 2021). In addition, the Japanese and Taiwanese governments, for example, have historically actively promoted overseas travel among their citizens in order to redress their enormous trade surplus with the USA (Hall, 1994).

Tourism has increasingly become a significant sector for economic and social development of many developing countries although this role has often been subject to criticism with respect to the outflow of capital and uneven distribution of wealth. The tourism industry is an important contributor to closer international relations, cultural, civilization, and economic relations between countries of the world, particularly between neighboring countries which, it has long been argued potentially leads to greater peace, security, and stability (Ap & Var, 1990), although more recent research questions this view (Becken & Carmignani, 2016). Brown (1998) also viewed tourism as a 'first cover of international relations' and argues that tourism contributes to showing the reality of the countries and societies throughout the world in order to obtain gain and uphold political positions. Table 6.3 summarizes some of the key international relations concerns and their impacts of tourism.

Table 6.3. The relationship between international relations and tourism.

International relations concerns	*Tourism-related effects*
War/conflict	Discourages visitors to within a wide radius; knock-on economic impact; tourism infrastructure damaged
Economic competition	Tourism chosen as 'easy' to implement
Currency movements/ devaluations/ inflation	Tour operators and tourists switch to cheaper countries
Global integration	Tourism draws 'traditional' or isolationist societies into global mainstream
Growth/development/ restructuring	Tourism supplants agriculture in LDCs; replaces manufacturing in DCs
Neocolonialism, core-periphery relationships	Tourism as agent of perpetuation of colonially imposed structural dependency
Secessionist/independence/ radical change movements	Target tourists to hurt government financially or draw attention to cause
Promotion of ideology/way of life	Tourism as image enhancer, notably via mega-events
Discouragement of others' ideology or policies	Travel embargoes; extra-territorial legislation
International co-operation	Regional marketing strategies
Deregulation	Higher/lower fares; better/worse service; travel safety issues
Sovereignty	Country may facilitate tourism to disputed territory to strengthen its claim
Flows of people across borders	Possible regional integration; may foreshadow or predict aid flows
Currency flows across borders	International balance of trade affected
Foreign/outside investment in destinations	New political power arrangements; rise of new Interests; capital flows
Imposition or removal of visa requirements	Barometer of countries' relations and alliances

Source: After Hall (1994)

Globalization and International Tourism

With globalization, the world is becoming increasingly interconnected, demonstrated by the expanded flow of information, technology, capital, goods, services, and people (Dwyer, 2015). Globalization is the process of integrating national and regional economies, societies and cultures through the global network of trade, communication, finance, immigration, and transportation (Hall, 2005). In his study, Dwyer (2015) identified five main types of forces that drive globalization: economic, technology, demographic, social and political drivers. However, he concluded that: 'No single driver or trend dominates. Each influences the others to a greater and lesser degree and will have varying impacts on tourism destination different regions, countries, industry sectors and tourist behaviour' (2015, pp. 329–330).

Tourism development is shaped by many contemporary factors due to globalization's complex and multifaceted structure. By expanding markets and integrating societies, globalization affects tourism development (Song et al., 2018). As Brown (1998) noted:

> ... tourism exemplifies many aspects of globalization. The spread of tourism to the farthest reaches of the planet, and people's willingness to travel to places their parents had never even heard of ... demonstrate the extent to which tourists themselves conceptualize the world as a single place (Brown, 1998, p. 12).

Tourism and globalization contribute to economic growth and connectedness (Fahimi et al., 2018). Tourism, for instance, stimulates economic growth by increasing revenue, employment opportunities, infrastructure development, and currency balance (Ehigiamusoe, 2021). As a result of globalization, increased trade openness can increase market access, service access, and commerce, accelerating tourism growth and improving infrastructure. Despite many crisis events affecting globalization (Witt, 2019a; Witt 2019b), such as economic and financial crises, no force had effectively resisted or stopped it before COVID-19 (Niewiadomski, 2020). According to Dwyer (2015) globalization and tourism are inextricably linked; that is, while tourist development contributes to the economic, social, and political dimensions of globalization, it has also become a major beneficiary of this phenomenon. In addition, the globalization of the tourism industry has contributed to the fragmentation of tourism manufacturing systems and the transnationalization of ownership structures, marketing arrangements, the outsourcing of services, and the transfer of expertise (McGrew, 2011).

Cohen (2012) contends that the growing number of political, economic, and environmental crises poses a threat to both globalization and the international travel and tourism industry. Furthermore, economic integration because of globalization potentially limits national policy capability in some areas and has served to reinforce the privatization and market liberalization of national political-economic frameworks by facilitating the transfer of international capital (Hall, 2022). Sabir and Gorus (2019) criticize globalisation and argue that it has weakened environmental protection while promoting economic expansion. Bridge (2002) also argued that environmental challenges such as climate change, cross-boundary water, air pollution and over-fishing of the ocean are linked to globalization. A point also picked up in some of the tourism and global environmental change literature (Hall, 2022). Moore (2017, p. 3) argued, 'While there is no question that environmental change accelerated sharply after 1850, and especially after 1945, it seems equally fruitless to explain these transformations without identifying how they fit into patterns of power, capital and nature'. As a result, Hall (2022) asked whether the question should be that is it a case that tourism is not just part of the Anthropocene (the age of humans) geological epoch, but is instead a part of the age of capital—the capitalocene— 'an ugly word for an ugly system' (Moore, 2017, p. 15), which is shaped by the seemingly endless accumulation of capital and its corresponding effects on nature and society. As Harvey (2018b, p. 424) describes it: 'compound growth (endless accumulation of capital) at three percent forever, which becomes more and more stressful as the exponential growth curve leaps upwards.' For example, Cohen (2012) views the threat of a profound structural crisis in the global economic system as the most significant emerging counter trend to globalization which would have significant implications for tourism. Nevertheless, Hall (2022) suggested that tourism is often regarded as a way to respond to or fix overaccumulation but, that the nature of capitalism is such that the 'fix' only serves to reinforce the system that created the problem in the first place because it remains predicated on expansion and growth.

Political Economy of International Tourism

The term political economy refers to the study of the relationship between what is called the economy and its non-economic (political, social, psychological and geographical) contexts (Mosedale, 2011). Bianchi (2018, p. 88) defines political economy as a term that 'comprises the study of the socio-economic forces and power relations that are constituted in the process of the production of commodities for the market and the divisions, conflicts and inequalities that arise from this'. Mosedale (2011) identified four main approaches to political economy which are widely used in social sciences. These are *Marxian political economics* which rely on a close interpretation of Marx's writings and historical materialism (e.g., Harvey, 2006, 2010, 2018a, b); *regulationists* who are interested in the regulatory framework (structure) of capitalism (e.g., Boyer, 1990); *comparative and international political economists* who examine the regulatory structures and the trade relations between nations-states, and finally, and more recently, *poststructural political economists* or *post-Marxists* who are influenced by poststructural concepts and focus on alternatives to capitalism (e.g., Gibson-Graham, 2006; Brouder, 2018).

Bianchi (2002, p. 265) applies a historic-geographical political economic approach to analyse the structures governing tourism development. He posits that capital restructuring and economic globalization have resulted in a changing configuration within the tourism production system, thus requiring a detailed examination of relationships between all actors in the system in order to unveil the impacts of the 'transnationalization' of the tourism system (cited in Mosedale, 2011, p. 4). Drawing on a Marxian approach to the political economy of tourism, Bianchi (2018) analyzed the global tourism political economy by discussing labour relations in tourism, access to travel, and equitable distribution in the destination. Guided by a comparative political economy approach, Webster and Ivanov (2015) explored and compared the institutional organization of tourism in different political systems and conclude that tourism and its institutional organization are influenced by the paradigms of political systems. Nevertheless, Marxian perspectives are still significant. Hall (2011) argues that tourism studies should continue to engage with the concept of class. He argues that stratification (e.g., people, capital, knowledge) is caused by unequal access to time and money and class is a representation of the stratification of capitalist, socio-economic organizations.

According to Bevir (2009), political economy holds that governments, particularly the state, frequently intervene in social production in order to facilitate capital accumulation and economic growth. In tourism, political economy is an important approach for understanding government involvement in tourism development and political trust (Bramwell & Lane, 2011; Nunkoo & Smith, 2013; Nunkoo, 2015). Political economic theory allows us to understand how economic and political variables shape power dynamics (Wan & Bramwell, 2015). According to a political economic perspective, tourism institutions, particularly the central government, play a crucial role in the development and planning of tourism (Nunkoo, 2015; Wang & Bramwell, 2012). Hence, residents tend to view local government as responsible for tourism policy and for enhancing coordination over tourism issues (Bramwell & Lane, 2011). In fact, tourism sectors are characterized by 'various labor market conflicts, multiple subsistence tactics, and complicated power interactions' (Madsen Camacho, 1996, p. 33). Thus, national political economies and the economic structures of tourism capitalism play a role in the determination of these parameters (Gibson, 2009). Tourism and political economy have been the subject of research since the early 1980s, although there is substantial need for additional study in this field (Bianchi, 2018).

Despite tourism's unquestionable importance in many nations' national accounts and international trade, tourism development research has, until recently, been mostly detached from political economy issues (Steiner, 2006). This stands in stark contrast to the industry of tourism in terms of economics (Bianchi, 2018). In fact, this may be partly due to the fact that anthropology, sociology, and tourism tourist policy, planning, and sustainability have been the subjects of a lot of political economy-related topics (Bianchi, 2018). Consequently, this has led to a significant deal of conceptual inconsistencies and theoretical fuzziness (Steiner, 2006). The notion of political economy is fundamental to tourism development and trust in tourism institutions (see Mosedale, 2011; and Mosedale, this volume). Thus, Bramwell (2011) suggests that a political economic approach could be useful for analysing tourism's complex challenges, such as power relations and network structures. Nonetheless, tourism literature is yet lacking with the dynamics of political economy and power.

Geopolitics and International Tourism

Although tourism is considered subject to the outcome of geopolitical activity, the link between geopolitics and tourism is an emerging theme in tourism-related studies (Hall, 2017). Tourism and geopolitics are open to numerous interpretations, methodologies, and theories, some of which are in direct antagonism to one another. Based on geopolitical perspectives, it is possible to analyze how tourism builds and represents particular political discourses, and, in turn, spatializes international politics (Mostafanezhad & Norum, 2016). Indeed, tourism can help to reimagine geopolitics as a mode of discourse that depicts international politics as a social landscape populated by a diverse range of people and places (Ó Tuathail & Agnew, 1992).

Geopolitics is a multiple layered concept (Hepple, 1986). According to Seyfi et al. (2023) geopolitics refers to the impact of issues such as geography, economics, demographics, and culture on the governance of a state, which ultimately frames and influences how enterprises and destinations act and the capacity of tourists to visit a destination. In a comprehensive approach to the geopolitical dimensions of tourism, Mostafanezhad and Norum (2016) suggest it can be defined as the competition of powers between attracting tourists at the regional and global levels in line with economic policies for the development of the territorial space and obtaining national interests. Hence, such research (e.g., Seyfi & Hall, 2020b) has shown how international tourism is essentially geopolitical, as it is dependent on cooperative linkages and competition between administrations of various blocs, countries, and regions.

The literature on tourist geopolitics has often focused on investigating the relationship between tourism, space, and power, as well as the geopolitical significance of tourism (Pfoser & Yusupova, 2022). Indeed, tourism geopolitics is a component of critical geopolitics (Ó Tuathail & Dalby, 1998; Dodds, 2007). The focus of this body of literature is the presuppositions about space that underpin many geopolitical discourses and practices. A recent trend in popular and feminist geopolitics play an important role in shaping the spaces of international politics, demonstrating the importance of popular representations (Dittmer & Gray, 2010) and highlighting the link between international politics and interpersonal relations (Hyndman & Amarasingam, 2014). Thus, the literature on tourism geopolitics has been influenced by these techniques, studying a variety of players and media to shed light on the geopolitical significance of tourism (Pfoser & Yusupova, 2022). Recently, the post-structural critical turn in geopolitics has spurred the development of new methods for analyzing the role of speech, literature, and meaning in the production of geopolitical conceptions and their tangible consequences (Agnew, 2001, 2004, 2016; Dalby, 2008; Ó Tuathail & Agnew, 1992). Hall (2017), Gillen and Mostafanezhad, (2019) and Mostafanezhad et al. (2020), have also highlighted the geopolitical significance of personal, every-day, micro-political contacts. Indeed, the realization of geopolitics as a daily experience and tourist views that occurs outside of policymaking and academic discourse was a significant component of the development of critical geopolitics (Dittmer & Gray, 2010, p. 1667). Therefore, as long recognized, tourist encounters perpetuate or challenge conceptions of other people and places and play an important part in larger geopolitical discourses, which in turn mediate geopolitical actions.

As an arena in which people and businesses from different countries come into contact, tourism also impacts on and is impacted by international relations, and can be a significant instrument of geopolitics. Although there is often a tendency to marginalize tourist activities within mainstream geopolitical research, there is an emerging geopolitics of tourism. In the past decade, tourism scholars have started to address the role of geopolitical discourse and practice in tourism, ushering in what could be described as a geopolitical turn in tourism geographies (Mostafanezhad, 2018; Devine & Ojeda, 2017; Rowen, 2016; Hall, D., 2017, Hall, C. M., 2017). Nevertheless, the powerful connections between tourism and geopolitics as well as various geopolitical approaches have yet to be fully conceptualized in tourism studies and empirically grounded. Furthermore, new approaches of geopolitics such as the geopolitics of the platform economy and digital governance (including topics such as economic globalization and the liberalization of telecoms/internet policy, the role of US hegemony on the global platform economy, and the emerging role of BRICS countries in challenging US hegemony on the digitalized world) have not yet been substantially studied in relation to tourism even though the field is incredibly important for so-called disruptive companies such as Uber and Airbnb. The geopolitics of the environment including in relation to biodiversity and climate change is also under-researched in tourism studies. Similarly, the geopolitics of tourism mobilities and diaspora and associated geographies of identity are also areas of increasing interest to students of tourism.

Concluding Remarks

This chapter has provided an overview of a significant subfield of the politics of tourism, international relations and geopolitics. By its nature international tourism is inherently bound up with issues of crossing national borders as well as, because of the mobility of tourists, the transport and other infrastructure that support such mobility, as well as the businesses and governments that supply such services. Clearly, this means that tourism is inherently political as the granting of a passport or visa is bound up within a range of concerns over citizenship, rights, and the relationships between state actors. Indeed, tourist mobility is often caught up in international diplomacy and policy because it is a relatively easy target in which states can show their largesse in enabling or disabling the capacity of individuals to move, and in some cases prevent mobility and access altogether.

The chapter has indicated something of the interplay of different theories of international relations with tourism although, as noted, there is a real need for further examination of the extent to which such theoretical linkages are recognized in a research context as well as in the assumptions made by state and non-state actors. Such theoretical approaches are also connected to the three main perspectives on international relations: realism, liberalism, and constructivism that, in turn, frame instruments, actors and how policy making is treated, including with respect to tourism. The latter part of the chapter deal with the emergence of geopolitics as a concern in tourism studies, and therefore as a specific branch of international relations research. However, significantly, in tourism geopolitics is recognized as being significant not just in terms of state and sub-state actors and territories, but also with respect to the activities of transnational business interests and the micro-geopolitics of everyday tourist encounters.

The limitations on mobility implemented because of the COVID-19 pandemic as well as the implications for tourism of Russia's invasion of Ukraine highlights how tourism is bound up in the actions of state and other actors within the international system of nations. Surprisingly, the international relations aspects of these events have been little understudied in a tourism context, although the broader role of political consumption by tourists and state sanctions has become a significant research theme. Nevertheless, more than 40 years on from Richter's (1983, 1989) path breaking research on tourism, politics, and the actions of state governments with respect to tourism, the subject of tourism in an international relations context remains extremely limited within the broader political dimensions of tourism. Global events may change this situation, given the importance of the international institutional structure for tourism and tourist mobility. However, what is more likely is that the field will continue to be the domain of a small number of interested researchers, particularly those with connections to political geography and political science, and with only a small but possibly significant impact on tourism studies as a whole.

References

Agnew, J. (2001). The new global economy: time-space compression, geopolitics, and global uneven development. *Journal of World-Systems Research*, *7*(2), 133–154. https://doi.org/10.5195/jwsr.2001.167

Agnew, J. (2004). *Geopolitics: Re-visioning world politics*. Routledge.

Agnew, J. (2016). The origins of critical geopolitics. In Dodds, K., Kuus, M., & Sharp, J. (Eds.) *The Ashgate research companion to critical geopolitics* (pp. 19–32). Routledge.

Ap, J., & Var, T. (1990). Does tourism promote world peace? *Tourism Management*, *11*(3), 267–273. https://doi.org/10.1016/0261-5177(90)90054-D

Azcárate, M. C., Mostafanezhad, M., & Norum, R. (2021). Introduction: Tourism geopolitics. In M. Mostafanezhad. M. C. Azcárate, & R. Norum (Eds.), *Tourism geopolitics: Assemblages of infrastructure, affect, and imagination* (pp. 3–28). The University of Arizona Press.

Baerenholdt, J. O. (2013). Governmobility: The powers of mobility. *Mobilities*, *8*(1), 20–34. https://doi.org/10.1080/17450101.2012.747754

Becken, S., & Carmignani, F. (2016). Does tourism lead to peace? *Annals of Tourism Research*, *61*, 63–79. https://doi.org/10.1016/j.annals.2016.09.002

Bevir, M. (2009). *Key concepts in governance*. Sage.

Bianchi, R. V. (2002). Towards a new political economy of global tourism. In Sharpley, R. & Telfer, D. (Eds.), *Tourism and development: Concepts and issues* (pp. 265–299). Channel View Publications.

Bianchi, R. (2018). The political economy of tourism development: A critical review. *Annals of Tourism Research*, *70*, 88–102. https://doi.org/10.1016/j.annals.2017.08.005

Boyer, R. (1990). *The regulation school: a critical introduction*. Columbia University Press.

Bramwell, B. (2011). Governance, the state and sustainable tourism: a political economy approach. *Journal of Sustainable Tourism*, *19*(4–5), 459–477. https://doi.org/10.1080/09669582.2011.576765

Bramwell, B., & Lane, B. (2011). Critical research on the governance of tourism and sustainability. *Journal of Sustainable Tourism*, *19*(4–5), 411–421. https://doi.org/10.1080/09669582.2011.580586

Bridge, G. (2002). Grounding globalization: The prospects and perils of linking economic processes of globalization to environmental outcomes. *Economic Geography*, 78(3), 361–386. https://doi.org/10.1111/j.1944-8287.2002.tb00191.x

Brouder, P. (2018). The end of tourism? A Gibson-Graham inspired reflection on the tourism economy. *Tourism Geographies*, *20*(5), 916–918. https://doi.org/10.1080/14616688.2018.1519721

Brown, F. (1998). *Tourism reassessed: Blight or blessing*? Butterworth Heinemann.

Cohen, E. (2012). Globalization, global crises and tourism. *Tourism Recreation Research*, *37*(2), 103–111. https://doi.org/10.1080/02508281.2012.11081695

Dalby, S. (2008) Imperialism, domination, culture: the continued relevance of critical geopolitics. Geopolitics, 13, 413–436.

Devine, J., & Ojeda, D. (2017). Violence and dispossession in tourism development: A critical geographical approach. *Journal of Sustainable Tourism*, *25*(5), 605–617. https://doi.org/10.1080/09669582.2017.1293401

Dittmer, J., & Gray, N. (2010). Popular geopolitics 2.0: Towards new methodologies of the everyday. *Geography Compass*, *4*(11), 1664–1677. https://doi.org/10.1111/j.1749-8198.2010.00399.x

Dodds, K. (2007). *Geopolitics: A very short introduction*. Oxford University Press.

Dwyer, L. (2015). Globalization of tourism: Drivers and outcomes. *Tourism Recreation Research*, *40*(3), 326–339. https://doi.org/10.1080/02508281.2015.1075723

Ehigiamusoe, K. U. (2021). The nexus between tourism, financial development, and economic growth: Evidence from African countries. *African Development Review*, *33*(2), 382–396. https://doi.org/10.1111/1467-8268.12579

Fahimi, A., Saint Akadiri, S., Seraj, M., & Akadiri, A. C. (2018). Testing the role of tourism and human capital development in economic growth. A panel causality study of micro states. *Tourism Management Perspectives*, *28*, 62–70. https://doi.org/10.1016/j.tmp.2018.08.004

Gibson, C. (2009). Geographies of tourism: Critical research on capitalism and local livelihoods. *Progress in Human Geography*, *33*(4), 527–534. https://doi.org/10.1177/0309132508099797

Gibson-Graham, J. K. (2006). Imagining and enacting a postcapitalist feminist economic politics. *Women's Studies Quarterly*, *34*(1/2), 72–78. https://www.jstor.org/stable/40004733

Gillen, J., & Mostafanezhad, M. (2019). Geopolitical encounters of tourism: A conceptual approach. *Annals of Tourism Research*, *75*, 70–78. https://doi.org/10.1016/j.annals.2018.12.015

Gordon, J. (2016). Economic sanctions as 'negative development': The case of Cuba. *Journal of International Development*, *28*(4), 473–484. https://doi.org/10.1002/jid.3061

Hall, C. M. (1994). *Tourism and politics: policy, power and place*. John Wiley & Sons.

Hall, C. M. (2005). *Tourism: Rethinking the social science of mobility*. Pearson.

Hall, C. M. (2010). Crisis events in tourism: Subjects of crisis in tourism. *Current issues in Tourism* 13(5), 401–417. https://doi.org/10.1080/13683500.2010.491900

Hall, C. M. (2011). Yes, Virginia, there is a tourism class. Why class still matters in tourism analysis. In Mosedale, J. (Ed.), *Political economy and tourism: A critical perspective* (pp. 111–125). Routledge.

Hall, C. M. (2017). Tourism and geopolitics: The political imaginary of territory, tourism and space. In D. Hall (Ed.) *Tourism and geopolitics: Issues from central and eastern Europe* (pp. 15–24). CABI.

Hall, C. M. (2022). Tourism and the capitalocene: From green growth to ecocide. *Tourism Planning & Development*, *19*(1), 61–74. https://doi.org/10.1080/21568316.2021.2021474

Hall, D. (Ed.) (2017). *Tourism and geopolitics: Issues from central and eastern Europe*. CABI.

Harvey, D. (2006). Spaces of global capitalism. Verso.

Harvey, D. (2010). *The enigma of capital: The crises of capitalism*. Oxford University Press.

Harvey, D. (2018a). Universal alienation. *Journal for Cultural Research*, *22*(2), 137–150. https://doi.org/10.1080/14797585.2018.1461350

Harvey, D. (2018b). Universal alienation and the real subsumption of daily life under capital: A response to Hardt and Negri. *tripleC: Communication, Capitalism & Critique. Open Access Journal for a Global Sustainable Information Society*, *16*(2), 424–439. DOI: https://doi.org/10.31269/triplec.v16i2.1027

Hepple, L. (1986). The revival of geopolitics. *Political Geography Quarterly*, *5*(4, S1), S21–S36. https://doi.org/10.1016/0260-9827(86)90055-8

Hyndman, J., & Amarasingam, A. (2014). Touring 'terrorism': landscapes of memory in post-war Sri Lanka. *Geography Compass*, *8*(8), 560–575. https://doi.org/10.1111/gec3.12149

Lanfant, M. F., & Graburn, N. H. (1992). International tourism reconsidered: The principle of the alternative. In V. Smith and W. Eadington (eds), *Tourism alternatives: Potentials and problems in the development of tourism* (pp. 88–112). University of Pennsylvania Press

Madsen Camacho, M. (1996). Dissenting workers and social control: A case study of the hotel industry in Huatulco, Oaxaca. *Human Organization*, *55*(1), 33–40. https://doi.org/10.17730/humo.55.1.76086h6l775457g1

McClelland, C. A. (1960). The function of theory in international relations. *Journal of Conflict Resolution*, *4*(3), 303–336. https://doi.org/10.1177/002200276000400304

McClelland, C. A. (1971). Power and influence. In Champlin, J.R. (Ed.), *Paradigms of political power* (pp. 35–66). Routledge.

McGrew, L. (2011). Pathways to reconciliation in Cambodia. *Peace Review*, *23*(4), 514–521. https://doi.org/10.1080/10402659.2011.625851

Molz, J. G. (2010). Performing global geographies: Time, space, place and pace in narratives of round-the-world travel. *Tourism Geographies*, *12*(3), 329–348. https://doi.org/10.1080/14616688.2010.494684

Moore, J. W. (2017). The capitalocene, part I: On the nature and origins of our ecological crisis. *The Journal of Peasant Studies*, *44*(3), 594–630. https://doi.org/10.1080/03066150.2016.1235036

Mosedale, J. (Ed.). (2011). *Political economy of tourism: A critical perspective*. Routledge.

Mostafanezhad, M. (2018). The geopolitical turn in tourism geographies. *Tourism Geographies*, *20*(2), 343–346. https://doi.org/10.1080/14616688.2018.1434820

Mostafanezhad, M., Cheer, J. M., & Sin, H. L. (2020). Geopolitical anxieties of tourism: (Im) mobilities of the COVID-19 pandemic. *Dialogues in Human Geography*, *10*(2), 182–186. https://doi.org/10.1177/2043820620934206

Mostafanezhad, M., & Norum, R. (2016). Towards a geopolitics of tourism. *Annals of Tourism Research*, *61*(C), 226–228. http://dx.doi.org/10.1016/j.annals.2016.08.003

Niewiadomski, P. (2020). COVID-19: from temporary de-globalisation to a re-discovery of tourism? *Tourism Geographies*, *22*(3), 651–656. https://doi.org/10.1080/14616688.2020.1757749

Nunkoo, R. (2015). Tourism development and trust in local government. *Tourism Management*, *46*, 623–634. https://doi.org/10.1016/j.tourman.2014.08.016

Nunkoo, R., & Smith, S. L. (2013). Political economy of tourism: Trust in government actors, political support, and their determinants. *Tourism Management*, *36*, 120–132. https://doi.org/10.1016/j.tourman.2012.11.018

Ó Tuathail, G., & Agnew, J. (1992). Geopolitics and discourse: practical geopolitical reasoning in American foreign policy. *Political Geography*, *11*(2), 190–204. https://doi.org/10.1016/0962-6298(92)90048-X

Ó Tuathail, G., & Dalby, S. (Eds.) (1998). *Rethinking geopolitics*. Routledge.

Ovcharov, A. O., G. Ismagilova, I. Ziganshin, & M. Rysayeva (2015). The influence of Western countries' sanctions on the development of the Russian exit tourism. *Asian Social Sciences*, 11, 289–296.

Pfoser, A., & Yusupova, G. (2022). Memory and the everyday geopolitics of tourism: Reworking post-imperial relations in Russian tourism to the 'near abroad'. *Annals of Tourism Research*, *95*, 103437. https://doi.org/10.1016/j.annals.2022.103437

Pirie, G. H. (1990). Aviation, apartheid and sanctions: air transport to and from South Africa, 1945–1989. *GeoJournal*, 22, 231–240. https://doi.org/10.1007/BF00192821

Popesku, J., & Hall, D. (2004). Sustainability as the basis for future tourism development in Serbia. In Hall, D.R. (Ed.), *Tourism and transition: Governance, transformation and development* (pp. 95–103). CABI.

Richter, L. K. (1983). Tourism politics and political science: A case of not so benign neglect. *Annals of Tourism Research*, *10*(3), 313–335. https://doi.org/10.1016/0160-7383(83)90060-9

Richter, L. K. (1989). *The politics of tourism in Asia*. University of Hawaii Press.

Rowen, I. (2016). The geopolitics of tourism: Mobilities, territory, and protest in China, Taiwan, and Hong Kong. *Annals of the American Association of Geographers*, 106(2), 385–393. https://doi.org/10.1080/00045608.2015.1113115

Sabir, S., & Gorus, M. S. (2019). The impact of globalization on ecological footprint: empirical evidence from the South Asian countries. *Environmental Science and Pollution Research*, *26*, 33387–33398. https://doi.org/10.1007/s11356-019-06458-3

Seyfi, S., & Hall, C. M. (2020a). Sanctions and tourism: Effects, complexities and research. *Tourism Geographies*, *22*(4–5), 749–767. https://doi.org/10.1080/14616688.2019.1663911

Seyfi, S., & Hall, C. M. (2020b). *Tourism, sanctions and boycotts*. Routledge.

Seyfi, S., Hall, C. M., & Shabani, B. (2023). COVID-19 and international travel restrictions: the geopolitics of health and tourism. *Tourism Geographies*, 25(1), 357–373. https://doi.org/10.1080/14616688.2020.1833972

Song, H., Li, G., & Cao, Z. (2018). Tourism and economic Globalization: An emerging research agenda. *Journal of Travel Research*, 57(8), 999–1011. https://doi.org/10.1177/0047287517734943

Steiner, C. (2006). Tourism, poverty reduction and the political economy: Egyptian perspectives on tourism's economic benefits in a semi-rentier state. *Tourism and Hospitality Planning & Development*, *3*(3), 161–177. https://doi.org/10.1080/14790530601132286

Tanter, R., & Ullman, R. (2015). *Theory and policy in international relations*. Princeton University Press.

Toubes, D. R., & Araújo-Vila, N. (2021). The treatment of language in travel advisories as a covert tool of political sanction. *Tourism Management Perspectives*, *40*, 100866. https://doi.org/10.1016/j.tmp.2021.100866

United Nations World Tourism Organization (UNWTO). (2020). *World tourism barometer*, UNWTO.

Walt, S. M. (1998). International relations: one world, many theories. *Foreign Policy* (110): 29–46. https://doi.org/10.2307/1149275

Wang, Y., & Bramwell, B. (2012). Heritage protection and tourism development priorities in Hangzhou, China: A political economy and governance perspective. *Tourism Management*, *33*(4), 988–998. https://doi.org/10.1016/j.tourman.2011.10.010

Wan, Y. K. P., & Bramwell, B. (2015). Political economy and the emergence of a hybrid mode of governance of tourism planning. *Tourism Management*, *50*, 316–327. https://doi.org/10.1016/j.tourman.2015.03.010

Webster, C., & Ivanov, S. (2015). Geopolitical drivers of future tourist flows. *Journal of Tourism Futures*, *1*(1), 58–68. https://doi.org/10.1108/JTF-12-2014-0022

Witt, M. A. (2019a). De-globalization: Theories, predictions, and opportunities for international business research. *Journal of International Business Studies*, 50(7), 1053–1077. https://doi.org/10.1057/s41267-019-00219-7

Witt, M. A. (2019b). China's challenge: Geopolitics, de-globalization, and the future of Chinese business. *Management and Organization Review*,15(4), 687–704. https://doi.org/10.1017/mor.2019.49

Zapata Campos, M. J., Hall, C. M., & Backlund, S. (2018). Can MNCs promote more inclusive tourism? Apollo tour operator's sustainability work. *Tourism Geographies*, *20*(4), 630–652. https://doi.org/10.1080/14616688.2018.1457074

Chapter 7

Political Economy Approaches in Tourism

From Radical Political Economy via Cultural Political Economy to Political Ecology

Jan Mosedale

Introduction

The challenge of writing an overview of political economy research as applied to tourism lies in a number of issues. Firstly, political economy means different things to different people; it is far from being a unified concept. It spans various disciplines, ranging from economics to anthropology, each entertaining their own interpretations of political economy (economics focuses on individual behavior and rational choice theory, whereas anthropology is interested in culturally determined inequalities), and each discipline has made distinct theoretical contributions to it, leading to an array of approaches labeled as political economy. Secondly, although the critical academic discussions of tourism in the 1970s (e.g., Young, 1973; de Kadt, 1979) and applications of dependency theory (Bryden, 1973; Turner, 1976; Britton, 1989) were influenced by political economy, there has not been a particularly strong tradition of applying political economy approaches to tourism (see Mosedale, 2011a), leading Williams (2004, p. 62) to conclude that '. . .fruitful theoretical developments in political economy have largely bypassed tourism.' While I stated in the previous edition of the *Wiley Blackwell Companion to Tourism* that not many tourism scholars undertaking research that could be considered as following a political economy approach would not consider themselves to *be* 'political economists' or *do* political economy, political economy has received more attention in tourism research since. With the term 'political economy' having become more prominent. This makes a truly comprehensive overview of the political economy literature in tourism difficult.

At a rudimentary level, political economy is concerned with the commodification of resources (material or symbolic) and the resulting accumulation and distribution of capital (not restricted merely to monetary gains but including social and natural capital). The economy is embedded in historical, political, sociocultural, and geographical contexts as these influence the structures that organize capital accumulation and its distribution across society. The emphasis placed on each of these factors differs depending on the disciplinary lens of the

The Wiley Blackwell Companion to Tourism, Second Edition. Edited by C. Michael Hall.

researcher and has resulted in different understandings of political economy. In this chapter I want to provide an overview of different political economy approaches and how they have been adopted in tourism research.

Of particular interest to political economy are the roles of structures (social, cultural, political, and economic) that guarantee the reproduction of the organization of production, commodification, distribution, exchange, and consumption and how these structures sustain specific power constellations. The concept of power is thus central to political economy as its location determines who gains from capitalist production and who is disadvantaged (Bramwell & Meyer, 2007; Coles & Church, 2007; Carrier & MacLeod, 2010). Power is not distributed equally, even within stakeholder collaborations (Saito & Ruhanen, 2017) and varies over time as demonstrated by a temporal analysis of tourism development in the Maldives (Bowen et al., 2017). Structural forces in our societies determine the level and type of power that particular individuals, social groups, businesses, organisations and governments can wield.

Tourism and Regulation Theory

Regulation theory is one of several approaches of political economy (see Mosedale, 2011a) and examines the relationships between states, governance, and society.

Regulation theory is concerned with analyzing the combination of *regime of accumulation* – the organization and process of production and consumption – and the *mode of regulation* – the structures that stabilize the regime of accumulation and guarantee its reproduction (see Cornelissen, 2005, 2011). Much of the tourism literature on regulation to date has focused on the regime of accumulation and on the discussion whether the production and consumption of tourism has experienced a shift from pre-Fordism (small-scale, artisanal production) to Fordism (mass production of an identical product and mass consumption via package tours; economies of scale), post-Fordist (small-scale firms offering individually tailored packages focusing on niche products; economies of scope), and neo-Fordist (adaptation of Fordist production to become more flexible and appear to be more individual; economies of both scale and scope) regimes of accumulation (Ioannides & Debbage, 1998). An analysis of the mode of production and consumption in the mass tourism resort of Cancun, Mexico, by Torres (2002) demonstrates that the destination comprises features from various modes of production: new individualized tourism products (e.g., ecotourism and heritage tourism) are offered alongside mass tourism products which have themselves been adapted to cater to different consumer tastes and niche markets. A shift in demand has led to greater diversity in tourism products and further niche market segmentation, therefore resulting in more consumer choice and flexible production (in terms of specialized packaging).

The mode of regulation consists of a collection of formal and informal norms, rules, codes, social practices, and institutions which sustain the survival of the regime of accumulation by providing ground rules for economic behavior and for the resolution of conflict. Institutional arrangement, in contrast, represents the specific configuration of organizations that are involved in regulation and have been the primary focus of tourism research to date, in particular the role of government institutions and public policy on tourism development (e.g., Hall, 1994, 1997; Hall & Jenkins, 1995; Church, 2004). It is indeed important to analyze the organization of institutions and the allocation of roles and responsibilities is important as it influences '. . .the process which the policy agenda for tourism is shaped, the way in which tourism problems are defined and alternatives are considered and how choices are made and decisions and actions taken' (Hall & Jenkins, 1995, p. 19). Institutions play a key role in the configuration of social processes and are thus responsible for the resulting structural and spatial characteristics of tourism (Britton, 1991). However, it is important to analyze not just the institutional arrangement but the entire institutional regime or mode of regulation including more implicit norms, values, and discourses (Earl & Hall, 2021).

Institutional political economy recognizes that the organization of capitalism is influenced by the relationship between the sociopolitical and cultural structures and institutions to create the economic landscape: '. . .economic activity is *socially and institutionally situated*: it cannot be explained by reference to atomistic individual motives alone, but has to be understood as enmeshed in wider *structures of social, economic and political rules, procedures and conventions*' (Martin, 2003, p. 79; original emphasis). Most of the institutional contributions that focus on tourism have concentrated on analyses of the institutional arrangement with very few studies incorporating the explicit institutional environment (e.g., Vail & Heldt, 2000) and a dearth of literature on implicit rules and norms.

The most obvious contributions to institutional political economy have been place-specific analyses of political and economic contexts of tourism development mostly approached from a political science perspective. Such studies have been undertaken for Africa (Dieke, 2000), Cambodia (Chheang, 2008), Egypt (Richter & Steiner, 2008), the Gambia (Dieke, 1994), Iran (Seyfi & Hall, 2019, 2020), Jordan (Scherle & Pillmayer 2016), the Middle East (Hazbun, 2004, 2008), Morocco (Scherle, 2011), Syria (Gray, 1997), Tunisia (Poirier & Wright, 1993; Hazbun, 2007/2008), and Western Europe (Williams & Shaw, 1998). These studies offer broad political-historical descriptions of tourism in a national – and sometimes regional – context, often with a focus on economies, institutions, infrastructure, policies, and plans. They demonstrate that political economies are path-dependent; they are dependent on prior political and economic decisions, institutional structures, and social relations: '[tourism] attitudes are heavily influenced by ideas, values, perceptions and beliefs developed over time, and which are deeply embedded in the institutions of local government and are manifested through its structures and practices' (Dredge, 2001, p. 357). Place-specific analyses provide contextual insights into particular local political economies and the importance of tourism to them, explain varieties in both tourism development and national political economies, and provide a basis for more detailed, theoretical reflections, for instance on neoliberalism (Scherle, 2011) or on the changing role of the state (Dredge, 2001; Coles et al., 2012).

Institutional regimes are not static but adapt to internal and external changes. Of particular interest to regulationists are the temporal variations of relationships between the institutional environment and the institutional arrangements and their effects on tourism. By its nature, institutions are designed to provide the structural framework that guarantees the stability of economic exchange and thus enables social economic processes and activity over time and space.

How modes of regulation either adapt to environmental changes (internal or external) or how societies create new institutional regimes is of particular interest to regulationists. The role of the state in market economies has also been challenged and debated during the last decades (Dicken, 1994; Goodwin & Painter, 1996; MacLeod, 1999). The shift to post-Fordism production and consumption (although contested as a comprehensive transformation) has resulted in an adaptation of the mode of regulation as the Keynesian welfare state was not deemed fit to stabilize the relations between capital, government, and labor. Due to the failure of Fordism to adequately address the increasing internationalization and interconnectedness of economies, which led to the internationalization of economic crises and the collapse of the welfare state, neoliberalism has widely been adopted as a mode of regulation (Jessop, 2002).

This political theory argues for the primacy of the market to take advantage of comparative advantages without social or political interventions. The application of neoliberal theory has resulted first in a 'roll-back' of government as the economy was deregulated – freed of government intervention – and then a 'roll-out' of new institutional arrangements, so-called reregulation (Peck & Tickell, 2002).

Desforges (2000, p. 184) offers an analysis of the 'rolling back' of the state in Peru as the newly elected Fujimori government in 1990 implemented their version of the neoliberal project in response to an economic crisis in the 1980s and the ensuing '. . .maelstrom of political and economic chaos.' Cuts in state spending resulted in the withdrawal of the state from entrepreneurial activities in tourism, the national hotel chain was privatized, and the roles and responsibilities for tourism planning and development were reorganization within a streamlined institutional landscape. A similar situation is described by Clancy (1999) for Mexico, suggesting the contestation of certain roles and responsibilities of nation states and the increasing decentralization of government (Yüksel et al., 205). This decentralization resulted in local and regional governments needing to increase their involvement in tourism promotion and development (Shone, 2016; Shone et al., 2016).

Not just the role of 'the state' but its composition has changed as new actors have emerged, creating new and complex and dynamic relations between capital, institutions, and society (see for example Fitchett et al., 2021, for a case study of the Lofoten Islands). The multitude of actors engaged in the tourism policy process at various scales makes for a complex institutional landscape of national, regional, and local state institutions, quangos, nongovernmental organizations, public–private partnerships, and elites: 'Tourism is a diverse and at times chaotic policy arena. ...[with] numerous policy strands at a variety of geographical scales but without an overall strategic underpinning' (Church et al., 2000, p. 330). The role of the nation state may no longer be to actively form policy and thus to take an active part in the development of tourism but may instead retreat further to coordinate efforts among various actors, to '...play a central role in mediating the mechanics, ethos and outcomes of this [collaborative governance] process' (Church et al., 2000, p. 332). Such a retreat of the state from policymaking raises questions about democratic legitimacy within the collaborative process. Reregulation and decentralization may lead to increased democratization when decisions are made at more local scales and including a large variety of interest groups but may at the same time lead to decisions being made outside of democratic processes when special interest groups seize the initiative in public–private partnerships (Yüksel et al., 2005; Geoffrey Deladem et al., 2021).

Radical Political Economy and Tourism

Radical Political Economy analyses the relational processes involved in the production and consumption of tourism to reveal underlying power structures which not only allow the system to function but also allow its continuous reproduction despite instability and crises (Mosedale, 2011a). Recently, a select few have taken up Gibson's (2009) call for critical research on tourism and capitalism and addressed the role of tourism within capitalism and vice versa (Fletcher, 2011), in respect to profits and land rents (Liodakis, 2019; Yrigoy 2021), capital accumulation and its inherent structural violence (Büscher & Fletcher, 2017; Devine & Ojeda, 2017), the role of tourism within a neoliberal project (Lapointe et al., 2018) and post-capitalist futures (Fletcher et al., 2023). These studies position tourism within a broader political economic context of capitalism.

Another focus of radical political economy are labor relations (Bianchi, 2011, 2018) and the precarious nature of tourism labor (Lee et al., 2015, Robinson et al., 2019; Shircliff, 2020). While the exploitation of labor was historically linked to class struggles (Hall, 2011; Mansilla López, 2019), these have resurfaced in many destinations experiencing overtourism (Milano et al., 2019). The strained relations between capital accumulation and labor as well as different classes 'are both a symptom and constitutive of processes of tourism capitalism in a particular historical-geographic context' (Bianchi, 2018, p. 97).

While power is at the forefront of radical political economy, the topics of race and ethnicity have only recently received some attention in the tourism literature (Chio et al., 2020). As otherness, race and ethnicity are commodified in tourism and thus central to the representation of some tourism destinations. Yet '[t]ourism ultimately conceals the realities of institutional racism, and touristic objectifications of the racial Other behind the veneer of, respectively, mutually beneficial economic practice and cultural uniqueness' (Jamerson, 2016, p. 1039). While race becomes part of the accumulation of capital, its integration into the production and consumption of tourism may obscure institutionalised racism within the tourism industry, not only as representation of otherness to be experienced by tourists but also as tourists who experience racism when travelling (Benjamin & Dillette, 2021).

Gender perspectives are another important research field for political economists as the accumulation of capital via the exploitation of various kinds of labour and therefore the wider political economy '... cannot be understood without examining how gender is embedded in all social relations' (Dickinson & Schaeffer, 2001, p. 23). According to Peterson (2006) there are two ways gender can be understood and analysed within political economy: on the one hand, how the effects of political economy differ depending on gender and on the other hand, how the meaning of gender is socially constructed, produced and reproduced - and hence how gender becomes a social structure (Figueroa-Domecq & Segovia-Perez, 2020). For instance, Mkhize and Cele (2017) highlight these gendered relations of power in tourism in a case study of KwaZulu-Natal.

The factors of inequality, such as class, race, ethnicity, gender identity, sexual orientation, age and disability do not only occur in isolation but may often intersect and combine to create unique dynamics, effects and exploitation (see for example Chambers, 2022, in the context of tourism or Sumerau et al., 2021, in general). Also, they are not only restricted to the phenotype but these identities are socially constructed and also have cultural elements which are not appropriately addressed via Radical Political Economy.

Cultural Political Economy

Cultural Political Economy has emerged as an addition to Radical Political Economy to counter its perceived economic determinism (while attempting to avoid cultural determinism). The shift of focus from conservative interpretations of an asocial economy towards an economy that both influences and is influenced by social and cultural contexts leads to more nuanced understandings of tourism as a set of practices - influenced by culture - are key to the enactment, reproduction of dominant structures in particular places but also across geographical scales (Bramwell & Lane, 2014). At the same time, a focus on practices allows researchers to take human agency into account: 'Both structure and agency meet and become visible in practices' (Mosedale, 2012, p. 166). A focus on social practices within certain cultural contexts allow a more nuanced analysis of how political economy is enacted and how agency and structures coalesce to form a particular type of political economy (Mosedale, 2016; Su et al., 2018; Stoffelen, 2022b).

Despite earlier contributions (Ateljevic, 2000; Ateljevic & Doorne, 2002, 2003), cultural political economy has only recently re-emerged explicitly (Stoffelen 2022b). Inspired by developments in the field of urban studies, Su et al. (2018) have re-theorised urban heritage tourism research by adopting a cultural political economy perspective and thus combining cultural and discursive elements with political economic factors, and have applied this concept to two case studies of heritage tourism in Nanjing, China with differing relevance of the economic respectively the political sphere. Also, in the sphere of heritage tourism/production, Stoffelen (2022a) analyses the concept of heritage (dissonance) in the context of a particular historic production process in beer-making in context of changing consumption patterns. Different actors interpret the concept of tradition/heritage differently, often in relation to their commercialisation strategies.

Like the neglect by tourism regulation theory of the implicit institutional structures that affect tourism policymaking, planning, and development, regulationist research to date has mainly focused on production to the detriment of consumption. Changing demand has clear repercussions on production systems and the organization of production, as shown by Torres (2002) in the case of Cancun, but the consequences of consumption practices and choices on the organization of production are underrepresented in contrast to studies of production processes. Market societies provide consumers with certain 'freedoms' of consumption: 'The simple fact that demand is required before profit can be made, places enormous economic power in consumers' hands' (Hartwick, 2000, p. 1184). Invoking this agency to choose ethical consumption influences production processes and their organization, as ethical consumption practices such as alternative forms of tourism are increasing (Miller, 2003; Butcher, 2008).

Yet consumption in itself is of interest as social differentiations no longer seem to be based on production and social class but on cultural status (Mavroudeas, 2002) with its emphasis on what is consumed, and how it is consumed. Tourism consumption plays a key role in creating social differences (Hall, 2011) and political economists should further engage with the divisive aspect of tourism consumption (in particular in times of crisis).

As discussed, a sociocultural understanding of regulation that includes consumption is necessary to appreciate the complexities of tourism regulation. Research on ethnic food and fashion (Cook & Crang, 1996; Jackson, 1999), for instance, have challenged globalization theories of cultural homogenization, global culture theory, or McDonaldization theory (Ritzer, 2000) and demonstrated that consumers actively shape their personal relationship and interaction with goods and thus create their own meanings for commodities (Ateljevic & Doorne, 2003).

To integrate consumption and production processes it is therefore necessary to combine a political economy approach (to unravel the complex system of economic actions and regulation) and a cultural approach (to understand the meaning attached to goods and services at different times, places, and phases of consumption and production) in order to connect '... two utterly unequal and contrasting real spaces: the world of consumption and the world of production' (Hartwick, 2000: 1,182).

Poststructural Political Economy

While Radical and Cultural Political Economy are concerned with identifying the structural conditions of inequality in processes of capital accumulation, poststructural political economy analyses how structures are formed and institutionalised via discursive power which is embedded in the current, dominant system (Gibson-Graham, 2006; Mosedale, 2011b; Brouder, 2018).

This is succinctly demonstrated by the following quote by Margaret Thatcher (then Prime Minister of the UK) in 1980 in support of free market economics: 'There is no alternative'. This statement, which has since regularly been used in the political arena, is an attempt at ignoring and flatly rejecting alternative discourses. Despite these attempts at silencing alternatives, multiple discourses coexist at any one time and there may be shifts in the acceptance or dominance of one particular discourse over others, as regulating discourse is a manifestation of power (Foucault, 1980). Sudden crises are occasions where drastic shifts in discourse occur, as shown in the example of tourism discourse during the 2008 economic crisis in Iceland (Jóhannesson & Huijbens, 2010) or the recent COVID-19 pandemic (Tremblay-Huet & Lapointe, 2021). Language is an important aspect of representing regimes of accumulation and modes of regulation and thus equally serves to reproduce dominant projects and the inequalities ingrained in these projects. Critical discourse analysis would greatly contribute to the political economy of tourism by enhancing our understanding of the discursive strategies used to reproduce political projects and would provide opportunities for discussing alternative projects. Deconstructing dominant discourses, which limit alternative discourses ('There is no alternative', Margaret Thatcher) and thus reproduce their dominance, allows the highlighting of alternatives ('Another world is possible', World Social Forum) thus making them imaginable and (maybe) attainable. Poststructuralists also challenge binary constructions (economy/culture, man/woman, natural/artificial, reason/emotion, homosexual/heterosexual) to open up space for non-binary differences. While the Critical Tourism Studies movement has managed to galvanise tourism research around critical research and the deconstruction of binaries and reconstruction of non-binary alternatives, thus far, little research has specifically been undertaken on poststructural political economy of tourism, leading Brouder (2018, p. 916) to comment that this is a topic '… research on the tourism economy is most likely to gravitate towards in the years ahead'.

From the Political Economy of Nature to Political Ecology

Nature is of paramount importance to tourism and the power relations associated with the representation, commodification and consumption of nature for tourism purposes are central to a political economy of tourism. Analysing the political, economic, social and ecological relations in society-nature relationships and the unequal distribution of power with its implications for ecological health and local livelihoods are thus necessary (Douglas, 2014; Mosedale, 2015). Political economy of nature is concerned with the increasing neoliberalisation of nature (Rytteri & Puhakka, 2012, Duffy, 2015, Cive & Avar, 2019), while a strand more aligned with cultural political economy analyse the relations between nature and society. Fletcher and Toncheva (2021), for instance, have analysed conflict and coexistence of humans and wildlife and how these relations differ across space and time.

With a focus on addressing inequality, political ecology combines Radical with Poststructural Political Economy (Fletcher. 2016; Rainer, 2018) and applies social & environmental justice approaches to environmental issues (see for example Cole, 2017, or Mkono et al., 2021). With increasing scarce resources, climate change, increasing pressure from tourism development and resulting questions of social and environmental justice, research in the political ecology of tourism will be crucial to uncovering and (hopefully) addressing social and environmental injustices.

The recent COVID-19 pandemic, its regulation and its impact on the tourism industry has led to market interventions from governments across the world. National and regional states have actively supported tourism and hospitality businesses. The question is whether this crisis will change the balance between the state, society and the market for the long-term. Furthermore, the pandemic has affected marginal communities disproportionately (Kantamneni, 2020; Cohen & van der Meulen Rodgers, 2021). Various political economy approaches have a role to play in analysing how the pandemic is reconfiguring the relationships between the market, the state and society, how different set of actors are renegotiating these relationships in the long-term, how discourses about tourism are changing and being employed to drive particular agendas and how to ensure the social and environmental justice.

References

Ateljevic, I. (2000). Circuits of tourism: Stepping beyond the "production/consumption" dichotomy, *Tourism Geographies*, 2: 369–388.

Ateljevic, I. and Doorne, S. (2002). 'Representing New Zealand: Tourism imagery and ideology', *Annals of Tourism Research*, 29: 648–667.

Ateljevic, I. and Doorne, S. (2003). 'Culture, economy and tourism commodities: Social relations of production and consumption', *Tourist Studies*, 3: 123–141

Benjamin, S., & Dillette, A. K. (2021). Black travel movement: Systemic racism informing tourism. *Annals of Tourism Research*, 88: 103169.

Bianchi, R. (2018). The political economy of tourism development: A critical review. *Annals of Tourism Research*, 70: 88–102.

Bianchi, R. V. (2011). Tourism, capitalism and Marxist political economy. In *Political Economy of Tourism: A Critical Perspective*, J. Mosedale (ed.), pp. 17–37. Abingdon: Routledge.

Bowen, D., Zubair, S., & Altinay, L. (2017). Politics and tourism destination development: The evolution of power. *Journal of Travel Research*, 56(6), 725–743.

Bramwell, B. and Meyer, D. (2007). Power and tourism policy relations in transition. *Annals of Tourism Research*, 34(3): 766–788.

Bramwell, B., & Lane, B. (2014). The "critical turn" and its implications for sustainable tourism research. *Journal of Sustainable Tourism*, 22(1), 1–8.

Britton, S. (1989). Tourism, dependency and development: a mode of analysis. In *Towards Appropriate Tourism: The Case of Developing Countries*, T.V. Singh, H.L. Theuns, and F.M. Go (eds), pp. 93–116. Frankfurt am Main: Peter Lang.

Britton, S. (1991). Tourism, capital, and place: towards a critical geography of tourism. *Environment and Planning D: Society and Space*, 9(4): 451–478.

Brouder, P. (2018). The end of tourism? A Gibson-Graham inspired reflection on the tourism economy. *Tourism Geographies*, 20(5), 916–918. https://doi.org/10.1080/14616688.2018.1519721

Bryden, J. (1973). *Tourism and Development: A Case Study of the Commonwealth Caribbean*. Cambridge: Cambridge University Press.

Büscher, B., & Fletcher, R. (2017). Destructive creation: Capital accumulation and the structural violence of tourism. *Journal of Sustainable Tourism*, 25(5), 651–667.

Butcher, J. (2008). Ecotourism as life politics. *Journal of Sustainable Tourism*, 16(3): 315–326.

Carrier, J. G. and Macleod, D. V. L. (2010). *Tourism, Power and Culture: Anthropological Insights*. Bristol: Channel View.

Chambers, D. (2022). Are we all in this together? Gender intersectionality and sustainable tourism. *Journal of Sustainable Tourism*, 30(7), 1586–1601.

Chheang, V. (2008). The political economy of tourism in Cambodia. *Asia Pacific Journal of Tourism Research*, 13(3): 281–297.

Chio, J., Gill, T., Gonzalez, V. V., Harp, S. L., McDonald, K., Rosenbaum, A. T., Rugh, S. S. & Thomas, L. L. (2020). Discussion: tourism and race. *Journal of Tourism History*, 12(2), 173–197.

Church, A. (2004). Local and Regional Tourism Policy and Power. In *A Companion to Tourism*, A. A. Lew, C. M. Hall and A. M. Williams (eds), pp. 555–568. Oxford: Blackwell Publishing.

Church, A., Ball, R., Bull, C., and Tyler, D. (2000). Public policy engagement with British tourism: the national, local and the European Union. *Tourism Geographies*, 2(3): 312–336.

Cive, Y. Ö., & Avar, A. A. (2019). Neoliberal Governance and Accumulation by Dispossession in Karaburun Peninsula, Izmir, Turkey. *Proceedings of the AESOP*.

Clancy, M. J. (1999). Tourism and development: evidence from Mexico. *Annals of Tourism Research*, 26(1): 1–20.

Cohen, J., & van der Meulen Rodgers, Y. (2021). The feminist political economy of Covid-19: Capitalism, women, and work. *Global Public Health*, *16*(8–9), 1381–1395.

Cole, S. (2017). Water worries: An intersectional feminist political ecology of tourism and water in Labuan Bajo, Indonesia. *Annals of Tourism Research*, *67*, 14–24.

Coles, T. and Church, A. (2007). *Tourism, Power and Space*. Abingdon: Routledge.

Coles, T., Dinan, C., and Hutchison, F. (2012). May we live in less interesting times? Changing public sector support for tourism in England during the sovereign debt crisis. *Journal of Destination Marketing & Management*, 1(1–2): 4–7.

Cook, I. and Crang, P. (1996). The World On a Plate: Culinary Culture, Displacement and Geographical Knowledges. *Journal of Material Culture*, 1(2): 131–153.

Cornelissen, S. (2005). *The Global Tourism System: Governance, Development And Lessons from South Africa*. Aldershot: Ashgate.

Cornelissen, S. (2011). Regulation theory and its evolution and limitations in tourism studies. In *Political Economy of Tourism: A Critical Perspective*, J. Mosedale (ed.), pp. 39–54. Abingdon: Routledge.

de Kadt, E. J. (1979). *Tourism: Passport to Development? Perspectives on the Social and Cultural Effects of Tourism in Developing Countries*. New York: Oxford University Press.

Desforges, L. (2000). State tourism institutions and neo-liberal development: a case study of Peru. *Tourism Geographies*, 2(2): 177–192.

Devine, J., & Ojeda, D. (2017). Violence and dispossession in tourism development: A critical geographical approach. *Journal of Sustainable Tourism*, *25*(5), 605–617.

Dicken, P. (1994). Global-local tensions: firms and states in the global space economy. *Economic Geography*, 70(2): 101–128.

Dickinson, T. D., & Schaeffer, R. K. (2001). *Fast Forward: Work, Gender, and Protest in a Changing World*. Rowman & Littlefield.

Dieke, P. U. C. (1994). The political economy of tourism in the Gambia. *Review of African Political Economy*, 21(62): 611–626.

Dieke, P. U. C. (2000). *The Political Economy of Tourism Development in Africa*. Elmsford: Cognizant Communication Corp.

Douglas, J. A. (2014). What's political ecology got to do with tourism? *Tourism Geographies*, *16*(1), 8–13.

Dredge, D. (2001). Local government tourism planning and policy-making in New South Wales: institutional development and historical legacies. *Current Issues in Tourism*, 4(2–4): 355–380.

Duffy, R. (2015). Nature-based tourism and neoliberalism: Concealing contradictions. *Tourism Geographies*, *17*(4), 529–543.

Earl, A., & Hall, C. M. (2021). *Institutional theory in tourism and hospitality*. Routledge.

Figueroa-Domecq, C., & Segovia-Perez, M. (2020). Application of a gender perspective in tourism research: a theoretical and practical approach. *Journal of Tourism Analysis: Revista de Análisis Turístico*, 27(2), 251–270.

Fitchett, J., Lindberg, F., & Martin, D. M. (2021). Accumulation by symbolic dispossession: Tourism development in advanced capitalism. *Annals of Tourism Research*, *86*, 103072.

Fletcher, R. (2011). Sustaining tourism, sustaining capitalism? The tourism industry's role in global capitalist expansion. *Tourism Geographies*, *13*(3), 443–461.

Fletcher, R. (2016). Cannibal tours revisited: The political ecology of tourism. *Turismo y crisis, turismo colaborativo y ecoturismo. XV Coloquio de Geografía del Turismo, el Ocio y la Recreación de la AGE. Monografies de la Societat d'Història Natural de les Balears*, *18*(23), 19–29.

Fletcher, R., & Toncheva, S. (2021). The political economy of human-wildlife conflict and coexistence. *Biological Conservation*, *260*, 109216.

Fletcher, R., Blanco-Romero, A., Blázquez-Salom, M., Cañada, E., Murray Mas, I., & Sekulova, F. (203). Pathways to post-capitalist tourism. *Tourism Geographies*, 25, 707–728.

Foucault, M. (1980). *Power/knowledge: Selected interviews and other writings, 1972–1977*. Vintage.

Geoffrey Deladem, T., Xiao, Z., Siueia, T. T., Doku, S., & Tettey, I. (2021). Developing sustainable tourism through public-private partnership to alleviate poverty in Ghana. *Tourist Studies*, *21*(2), 317–343.

Gibson, C. (2009). Geographies of tourism: Critical research on capitalism and local livelihoods. *Progress in Human Geography*, *33*(4), 527–534.

Gibson-Graham, J.-K. (2006). *The End of Capitalism (as we knew it): A Feminist Critique of Political Economy*. Minneapolis: University of Minnesota Press.

Goodwin, M. and Painter, J. (1996). Local governance, the crises of Fordism and the changing geographies of regulation. *Transactions of the Institute of British Geographers*, 21(4): 635–648.

Gray, M. (1997). The political economy of tourism in Syria: State, society, and economic liberalization. *Arab Studies Quarterly*, 19(2): 57–74.

Hall, C. M. (1994). *Tourism and Politics: Policy, Power and Place*. London: John Wiley.

Hall, C. M. (1997). The institutional setting: tourism and the state. In *The Economic Geography of the Tourist Industry: A Supply-Side Analysis*, D. Ioannides and K.G. Debbage (eds), pp. 199–219. London: Routledge.

Hall, C. M. (2011). Yes, Virginia, there is a tourism class: why class still matters in tourism analysis. In *Political Economy of Tourism: A Critical Perspective*, J. Mosedale (ed.), pp. 111–125. Abingdon: Routledge.

Hall, C. M. and Jenkins, J. M. (1995). *Tourism and Public Policy*. London: Routledge.

Hartwick, E. R. (2000). Towards a geographical politics of consumption. *Environment and Planning A*, 32(7): 1177–1192.

Hazbun, W. (2004). Globalisation, reterritorialisation and the political economy of tourism development in the Middle East. *Geopolitics*, 9(2): 310–341.

Hazbun, W. (2007/2008). Images of openness, spaces of control: the politics of tourism development in Tunisia. *The Arab Studies Journal*, 15/16(2/1): 10–35.

Hazbun, W. (2008). *Beaches, Ruins, Resorts: The Politics of Tourism in the Arab World*. Minneapolis: University of Minnesota Press.

Ioannides, D. and Debbage, K. G. (1998). Neo-Fordism and flexible specialization in the travel industry: dissecting the polyglot. In *The Economic Geography of the Tourist Industry: A Supply-Side Analysis*, D. Ioannides and K. G. Debbage (eds), pp. 99–121. London: Routledge.

Jackson, P. (1999). Commodity Cultures: The Traffic in Things. *Transactions of the Institute of British Geographers*, 24(1): 95–108.

Jamerson, W. T. (2016). Race and critical tourism studies: An analytical literature review. *Sociology Compass*, *10*(11), 1038–1045.

Jessop, B. (2002). Liberalism, neoliberalism, and urban governance: a state-theoretical perspective. *Antipode*, 34(3): 452–472.

Jóhannesson, G. T. and Huijbens, E. H. (2010). Tourism in times of crisis: exploring the discourse of tourism development in Iceland. *Current Issues in Tourism*, 13(5): 419–434.

Kantamneni, N. (2020). The impact of the COVID-19 pandemic on marginalized populations in the United States: A research agenda. *Journal of vocational behavior*, *119*, 103439.

Lapointe, D., Sarrasin, B., & Benjamin, C. (2018). Tourism in the sustained hegemonic neoliberal order. *Revista Latino-Americana de Turismologia*, *4*(1), 16–33.

Lee, D., Hampton, M., & Jeyacheya, J. (2015). The political economy of precarious work in the tourism industry in small island developing states. *Review of International Political Economy*, *22*(1), 194–223.

Liodakis, G. (2019). Transnational political economy and the development of tourism: A critical approach. *Social Sciences*, *8*(4), 108.

MacLeod, G. (1999). Place, politics and 'scale dependence': exploring the structuration of euro-regionalism. *European Urban and Regional Studies*, 6(3): 231–253.

Mansilla López, J. A. (2019). No es turismofobia, es lucha de clases. Políticas urbanas, malestar social y turismo en un barrio de Barcelona. *Revista Nodo, 13*(26), 42–60.

Martin, R. (2003). Institutional approaches in economic geography. In *A Companion to Economic Geography*, E. Sheppard and T. J. Barnes (eds), pp. 77–94. Oxford: Blackwell Publishing.

Mavroudeas, S. (2002). The French regulation approach and its theory of consumption. *Storia del Pensiero Economico*, 43: 233–255.

Milano, C., Novelli, M., & Cheer, J. M. (2019). Overtourism and tourismphobia: A journey through four decades of tourism development, planning and local concerns. *Tourism Planning & Development, 16*(4), 353–357.

Miller, G. A. (2003). Consumerism in sustainable tourism: a survey of UK consumers. *Journal of Sustainable Tourism*, 11(1): 17–39.

Mkhize, G., & Cele, N. (2017). The role of women in tourism in KwaZulu-Natal: Case studies from the South Coast of KwaZulu-Natal. *Agenda, 31*(1), 128–139.

Mkono, M., Rastegar, R., & Ruhanen, L. (2021). Empowering women to protect wildlife in former hunting tourism zones: a political ecology of Akashinga, Zimbabwe. *Journal of Sustainable Tourism*, 1–21.

Mosedale, J. (2011a). Re-introducing tourism to political economy. In *Political Economy of Tourism: A Critical Perspective*, J. Mosedale (ed.), pp. 1–13. Abingdon: Routledge.

Mosedale, J. (2011b). Thinking outside the box: alternative political economies in tourism. In *Political Economy of Tourism: A Critical Perspective*, J. Mosedale (ed.), pp. 93–108. Abingdon: Routledge.

Mosedale, J. (2012). Diverse economies and alternative economic practices in tourism. In I. Atelvevic, N. Morgan, & A. Pritchard (Eds.), *The critical turn in tourism studies: Creating an academy of hope* (pp. 194–207). London: Routledge.

Mosedale, J. (2015). Critical engagements with nature: Tourism, political economy of nature and political ecology. *Tourism Geographies, 17*(4), 505–510.

Mosedale, J. (2016). Neoliberalism and the political economy of tourism: Projects, discourses and practices. In J. Mosedale (Ed.). *Neoliberalism and the political economy of tourism* (pp. 13–32). Routledge.

Peck, J. and Tickell, A. (2002). Neoliberalizing space. *Antipode*, 34(3): 380–404.

Peterson, V. S. (2006). How (the meaning of) gender matters in political economy. In A. Payne (Ed.) *Key debates in new political economy* (pp. 87–113). Routledge.

Poirier, R. A., and Wright, S. (1993). The political economy of tourism in Tunisia. *The Journal of Modern African Studies*, 31(1): 149–162.

Rainer, G. (2018). Producing nature for tourism: A political ecology angle. *Annals of Tourism Research, 71*(C), 62–63.

Richter, T. and Steiner, C. (2008). Politics, Economics and Tourism Development in Egypt: insights into the sectoral transformations of a neo-patrimonial rentier state. *Third World Quarterly*, 29(5): 939–959.

Ritzer, G., & Malone, E. L. (2000). Globalization theory: Lessons from the exportation of McDonaldization and the new means of consumption. *American studies, 41*(2/3), 97–118.

Robinson, R. N., Martins, A., Solnet, D., & Baum, T. (2019). Sustaining precarity: Critically examining tourism and employment. *Journal of Sustainable Tourism*, 27(7), 1008–1025.

Rytteri, T. and Puhakka, R. (2012). The art of neoliberalizing park management: commodification, politics and hotel construction in Pallasyllästunturi National Park, Finland. *Geografiska Annaler B* 94(3), 255–268.

Saito, H., & Ruhanen, L. (2017). Power in tourism stakeholder collaborations: Power types and power holders. *Journal of Hospitality and Tourism Management, 31*, 189–196.

Scherle, N. (2011). Tourism, neoliberal policy and competitiveness in the developing world: the case of the master plan of Marrakech. In *Political Economy of Tourism: A Critical Perspective*, J. Mosedale (ed.), pp. 207–224. Abingdon: Routledge.

Scherle, N., & Pillmayer, M. (2016). Developing markets–the neoliberalization of tourism structures in Jordan: the example of Aqaba Special Economic Zone. In J. Mosedale (Ed.) *Neoliberalism and the Political Economy of Tourism* (pp. 71–87). Routledge.

Seyfi, S., & Hall, C. M. (2019). International sanctions, tourism destinations and resistive economy. *Journal of Policy Research in Tourism, Leisure and Events, 11*(1), 159–169.

Seyfi, S., & Hall, C. M. (2020). Political transitions and transition events in a tourism destination. *International Journal of Tourism Research, 22*(4), 493–506.

Shircliff, J. E. (2020). Nomadic by nature? Contradictions and precarious work in Mongolian tourism. *Central Asian Survey, 39*(3), 361–377.

Shone, M. C. (2016). The changing role of the state: Neoliberalism and regional tourism development in New Zealand. In J. Mosedale (Ed.) *Neoliberalism and the political economy of tourism*, 55–70. Routledge.

Shone, M. C., Simmons, D. G., & Dalziel, P. (2016). Evolving roles for local government in tourism development: a political economy perspective. *Journal of Sustainable Tourism, 24*(12), 1674–1690.

Stoffelen, A. (2022a). Revitalising place-based commercial heritage: A Cultural Political Economy approach to the renaissance of lambic beers in Belgium. *International Journal of Heritage Studies, 28*(1), 16–29.

Stoffelen, A. (2022b). Theorizing tourism meaning creation: An exploration of the cultural political economy framework. *Annals of Tourism Research, 95*, 103431.

Su, R., Bramwell, B., & Whalley, P. A. (2018). Cultural political economy and urban heritage tourism. *Annals of Tourism Research, 68*, 30–40.

Sumerau, J. E., Forbes, T. D., Denise, E. J., & Mathers, L. A. (2021). Constructing allyship and the persistence of inequality. *Social Problems, 68*(2), 358–373. https://doi.org/10.1093/socpro/spaa003

Torres, R. (2002). Cancun's development from a Fordist spectrum of analysis. *Tourist Studies*, 2(1): 87–116.

Tremblay-Huet, S., & Lapointe, D. (2021). The new responsible tourism paradigm: the UNWTO's discourse following the spread of COVID-19. *Tourism and Hospitality, 2*(2), 248–260.

Turner, L. (1976). The international division of leisure tourism and the Third World. *Annals of Tourism Research*, 4(1): 12–24.

Vail, D. and Heldt, T. (2000). Institutional factors influencing the size and structure of tourism: comparing Dalarna (Sweden) and Maine (USA). *Current Issues in Tourism*, 3(4): 283–324.

Williams, A. M. (2004). Toward a political economy of tourism. In *A Companion to Tourism*, A. A. Lew, C. M. Hall, and A. M. Williams (eds), pp. 61–73. Malden: Blackwell.

Williams, A. M. and Shaw, G. (1998). *Tourism and Economic Development: European Experiences.* Chichester: John Wiley.

Young, G. (1973). *Tourism: Blessing or Blight?* Harmondsworth: Penguin.

Yrigoy, I. (2021). Strengthening the political economy of tourism: Profits, rents and finance. *Tourism Geographies*, ahead-of-print, 1–20.

Yüksel, F., Bramwell, B., and Yüksel, A. (2005). Centralized and decentralized tourism governance in Turkey. *Annals of Tourism Research*, 32(4): 859–886.

Chapter 8

Tourism Mobilities

Kevin Hannam & Gareth Butler

Introduction

Many commentators argued that the world would change due to the coronavirus pandemic heralding a new normal in terms of societal health, politics, and sustainability. The 'virus mutates as it moves through our bodies, awakening our own antibodies against it: both host and virus are changed as a result, genetically' (Adey et al., 2021, p. 3). However, in many ways it has stayed the same as old patterns and practices of tourism mobilities have simply returned to pre-covid levels or even exceeded them after lockdown with little thought for a change in practices to make life more sustainable. As Adey et al. (2021, p. 1) write: 'Viral mobilities have unleashed not just a disruption of human mobilities, but also a vast intensification of existing uneven relations of (im)mobilities'.

This chapter reviews recent work from what has been termed the 'new mobilities paradigm' (Sheller & Urry, 2004) and what has become known as the study of 'tourism mobilities' in terms of the pandemic and the future of tourism. The study of tourism has often been seen as on the periphery of the social sciences, but the mobilities paradigm arguably allows us to place travel and tourism at the core of social and cultural life rather than at the margins (Coles & Hall, 2006; Hannam, 2009). Thus, the concept of tourism mobilities is not just a fad or fashionable turn as some researchers have argued (Hall, 2011). Rather, the impact of the Covid pandemic highlighted the importance of tourism mobilities to everyday life as travel systems shut down, affecting travel and tourism, but also migration and working in the tourism, hospitality, and events industries (Zuev & Hannam, 2021).

Tourism mobilities are viewed as being bound up with both every day and mundane journeys, as well as with the more exotic encounters that have been the mainstay of much analysis in contemporary tourism. Tourism is then analysed not as an ephemeral aspect of social life that is practised outside normal, everyday life. Rather it is seen as integral to wider processes of economic and political development processes and even constitutive of everyday life (Franklin & Crang, 2001; Franklin, 2003; Coles & Hall, 2006; Edensor, 2007; Hannam & Knox, 2010).

The Wiley Blackwell Companion to Tourism, Second Edition. Edited by C. Michael Hall.

It is not *just* that tourism is a form of mobility like other forms of mobility such as commuting or migration, but that different mobilities inform and are informed by tourism. Thus, we need to continually examine the multiple mobilities in any situation: mobilities involve the movement of people but also the movement of a whole range of material things as well as the movement of imaginative thoughts and fantasies. In short, proponents of the mobilities paradigm argue that the concept of mobilities is concerned with mapping and understanding both the large-scale movements of people, objects, capital, and information across the world, as well as the more local processes of daily transportation, movement through public space, and the travel of material things within everyday life simultaneously (Hannam et al., 2006).

Zuev and Hannam (2021) document how, given our reliance on being mobile, the pandemic led to mental health anxieties or 'anxious immobilities' during the long period of lockdown when people were prevented from travelling overseas to see loved ones: 'a total disruption of everyday rhythms and specifically anxious waiting for the normalization of activity while being the subject of biosocial narratives of quarantine and social responsibility of remaining immobile' (Zuev & Hannam, 2021, p. 35). They base their research on Macau, a city state of China, 'a significant example of a global highly mobile place with a high concentration of tourist and expat flows, high-end luxury travel and low-end migrant workers mobility, mass cross-border travel and large mobility-facilitating infrastructure projects, significant mobilities of money through remittances and offshoring, as well as specific cargo and ferry mobilities' (Zuev & Hannam, 2021). However, while many studies focused on the impacts and consequences of (predominantly Western) tourist immobilities, Rogerson and Baum (2020) noted that in the context of the Global South, the rapid spread of the virus presented not only economic anxieties amongst marginalised or remote communities but amplified the precarious nature of being socially and geographically immobile. Such immobilities are always relationally co-constitutive of mobilities (Hannam et al., 2006). Mobility dependency often becomes only visible when it is disabled, impaired or ruptured. Immobility enforcing stillness complies mobile subjects to appreciate their relationship to mobility through remobilisation and the reclaiming of mobility as a resource (Cook & Butz, 2016).

A great deal of mobilities research has formerly analysed forms and experiences of so called 'corporeal' travel – the body in motion. This involves the travel of people for work, leisure, family life, pleasure, migration and escape, organised in terms of contrasting time-space modalities (ranging from daily commuting to a once-in-a-lifetime exile). Such rhythms of everyday life are negotiated through the relational dynamics of affect, care and familial bonding, such as a family holiday (Jensen et al., 2015) where sparingly clothed bodies play and relax in a dense choreography (Edensor, 2007).

Embodying Tourism Mobilities

Tourism mobilities research has examined aspects of such corporeal travel conceptualised as embodied experiences (Crouch, 2000; Edensor, 2007) beyond the visual gaze. Covid has brought to the fore the importance of touch in terms of bodily encounters, so called 'skin studies' (Howes, 2018) of how we interact with various environments picking up viruses as we travel on planes, trains or automobiles. The wearing of gloves or masks, forsaking bodily interactions such as the handshake altogether and measured social distancing became significant for fear of contagion during Covid.

Walking is a significant embodied tourism mobilities experience (Hannam et al., 2021). In terms of tourism mobilities, rhythm analysis has offered a fruitful avenue of investigation for understanding relationships between recreational walkers, walking experiences and places (e.g., Küpers & Wee, 2018; Sarmento, 2016). As Edensor (2010) outlined, walking practices and values are informed by particular socio-cultural discourses and historic contexts. For example, Sarmento (2016) investigated how tourists move through unknown places using rhythm analysis, finding improvisation to exist alongside patterns of unease and mirroring local rhythms, contributing to place-making. Johinke (2018) discusses the importance of sound in terms of the embodied experience of walking tours. In terms of China, Witte and Hannam (2017) argue that the development of online communities has allowed the formation of new hybrid worlds enabling spatially and socially more stretched social networks. In a country with rapid urban development, nature and walking in nature in particular, is viewed increasingly by the Chinese middle classes as a means of slowing down. In terms of South Africa, Rogerson and Rogerson (2021), discuss how tourism mobilities found 'less crowded places' with the growing pursuit of physiological and psychological wellbeing providing important opportunities to support rural economies that promote tourism structured around 'tranquillity' and 'seclusion', especially during the pandemic.

Like walking, cycling tourism can contribute to health and wellbeing, but also has environmental benefits and positive impacts facilitating an intimate contact with nature and contact with local populations, as well as access to places off the beaten track (Gazzola et al., 2018). Until recently, there has been relatively little research on touring cultures and the embodied experiences of cycling tourists (Liu et al., 2018). A characteristic feature of independent cycling tourism is that cyclists use their own bicycles and other gear for (or when) transiting to the destination and moving through the destination. However, it is also possible for many mass tourists to fit an autonomous cycling trip into their vacations and thus shift gears in their vacationing rhythm (Hannam et al., 2021).

Larsen (2001, p. 80), meanwhile, notes how historically 'in their early years, the train and the car were perceived as shocking speed machines that radically changed people's experiences of distance, movement, time and landscape and became discursively associated with various bodily pleasures and pains among different "movements" of tourists'. Recent tourism mobilities research has thus examined how different modes of transport are quite literally felt by tourists of different genders and ethnic identities. Adey's (2007) early research into the immobilities of airports is somewhat prescient, particularly as airports began to re-open after Covid but with increased bodily regulation and 'sorting'. Adey's (2007, p. 515) research was 'concerned with the immobilities which airports induce and force to occur by positioning passengers through various and perhaps unconscious technological arrangements and performances of spectatorship'. Here we might also consider that embodied aspects of air sickness on tourism mobilities as a felt experience.

In terms of train use, Jensen et al. (2015) focus on interrailing as embodied tourism mobilities in terms of soundscapes and thermalscapes; the latter analysis how temperatures on trains affect the tourism experience. Heritage railways provide a good example of heritage tourism mobilities, of literally heritage on the move (Schivelbusch, 1986; Löfgren, 2008; Bissell, 2009; Aguiar, 2011). Several critiques have explored different aspects of railway mobility, such as the role of the body combined with the visual nature of being on the move (Bissell, 2010; Johnson, 2010). The visual experience of railway mobility potentially offers a mobile travel glance which is more of a 'cinematic' experience of moving landscape images for the tourist who is corporeally, largely an immobile 'armchair' spectator (Larsen, 2001). Moreover, in a study of the Darjeeling Himalayan Railway (DHR), Roy and Hannam (2013, p. 587) note that:

> One of the most significant performances that the DHR induces because of its track and its relatively slow speeds is that one can get on and off the train while it is still on the move, and it is in this process that we argue that the DHR negotiates between place and people, not by being simply a stage for performance but by being itself an actor and taking initiative into this process.

Driving, or automobility, is a significant aspect of many tourists embodied mobilities. Automobility has been identified as the simultaneous achievement of autonomy and mobility, which in essence, involves 'modes of autonomous self-directed movement' (Featherstone, 2004, p. 1). Through the notion of automobility, personal vehicles may effectively offer the tourist freedom and flexibility other modes of travel do not. Cars enable tourists to tailor their own itineraries and allow them to control the routes they travel between destinations. They also permit the driver (and their passengers) to decide when and where to stop, and consequently allow direction and velocity to be controlled. As Urry (2000, p. 61) noted, the road has the power to 'set people free' as drivers are offered a range of choices, inaccessible to those using public methods of transport. Cars are therefore identified as important tools that may permit adventure (Sheller & Urry, 2000, 2003; Featherstone, 2004; Huijbens & Benediktsson, 2007; Collin-Lange & Benediktsson, 2011; Butler & Hannam, 2012) and have consequently fostered a 'dynamic culture of individualism' (Jacobsen, 2004, p. 7) but with significant consequences for the global environment (Vanderheiden, 2006). These promises are now increasingly sought after and utilised by emerging 'car cultures', including indigenous communities like the Vedda people of Sri Lanka (see Ranasinghe & Cheng, 2018). Here, increasing motorcycle ownership amongst younger cohorts was attributed to exposures to inbound tourism-induced mobilities. However, these 'modern' forms of mobility (as well as other items including televisions and radio sets) were quickly hidden once non-indigenous tour groups arrived so that traditional lifestyles could be reenacted. Such behaviours reveal the tensions and conflicts exposures to automobilities present, especially as the 'car system' remains deeply entrenched in many urban and rural societies across the globe. Indeed, as revealed by Butler et al. (2022), the rapid growth of domestic tourist automobilities during China's 'Golden Week' public holidays now trigger significant (im)mobility paradoxes, while Henderson (2018) argues that increasing traffic flows in many Asian cities' present greater levels of risk to pedestrians and tourists alike.

From the perspective of the car driver, Butler and Szili (2020) question how emergent technologies increasingly impede quests for challenge and adventure to be realised through automobility. Car technologies, including GPS, inbuilt WiFi connectivity, 'lane assist' systems, autonomous emergency breaking (AEB), and parking sensors, as well as connections to ancillary devices such as smartphones and tablets, continue to transform and sterilise driving experiences and, with growing frequency, disempower the car user. From an alternative perspective, the proliferation of automobilities shaped by the shared economy are an important direction for future inquiry. For example, Giddy (2019) reveals the conflicting issues presented by the emergence of Uber in South Africa, and more broadly, how 'Mobility as a service' (MaaS) providers now challenge existing transport systems in the Global South. Here, even though control is ceded to the Uber driver, 'freedom' and 'independence' are realised as those typically deterred from using conventional taxis or public transport in South African cities, experience what they perceive to be safer and more trustworthy modes of mobility.

Nonetheless, Pirie's (2009, p. 22) assertion that Africa remains synonymous with the 'kinetic underclass' persists as new forms of mobility and, by extension, economic opportunity, remain beyond reach (especially if phone credit and WiFi connectivity are required to access them). Indeed, MaaS providers merely amplify opportunities amongst those already in positions of privilege to move safely and more efficiently, while those on the margins remain susceptible to the risks and inefficiencies of inadequate public transportation systems and informal or paratransit ride-share providers (see Rogerson & Rogerson, 2021).

Additional mobility inequalities are framed by Rink (2017), who noted that while scenic flights in Cape Town offer the 'aeromobile elite' alternative 'ways of seeing, sensing and selling the landscape below' (Rink, 2017, p. 884) they remain highly exclusive and largely inaccessible to domestic or African tourists. Moreover, referring back to tourists' embodied experiences of particular modes of transport, Rink (2017) considers these flights to be 'detached' mobilities that by avoiding the risks and complexities of life in Cape Town reduce the city to 'simplistic and narrow frames of reference' (Rink, 2017, p. 888) such as landmarks or landscapes.

In terms of tourism mobilities, Yang et al. (2022) further discuss how Chinese tourists inhabit the hire car through an embodied habitation, interacting with strangers and feeling the discomfort of a long-distance drive. Wilson and Hannam (2017), meanwhile, focus on the embodied frictions involved in slow tourism such as campervan travel where gradients, weather, surfaces and other vehicles affect the tourism experience. In terms of water-based tourism, Kaaristo et al. (2020) examine how in canal-based tourism, 'associated tempos, are governed through creative interplays of freedom and control, and hierarchy and etiquette' (Kaaristo et al., 2020, p. 844). Like air travel, we might also note the significance of sea sickness on many tourists' bodies. In terms of Covid, however, cruise ships were notable early sites of viral transmission and quarantine for the tourist body (Iaquinto, 2020).

Health tourism and its research has a long history as a significant embodied experience beyond the scope of this chapter (Smith & Puczko, 2014). Such research focused on the management on spas and wellness treatments as well as travel for medical purposes such as hospital operations. Hall (2011) expands this to include dental tourism, stem-cell tourism, transplant tourism, abortion tourism, and xeno-tourism. Hall (2011, p. 4) emphasises the 'extent to which regulation encourages individuals to engage in cross-border purchase of health services and products'. This regulation has been a key factor in the governance of Covid related tourism mobilities. However, recent work in health tourism mobilities has emphasised other aspects of health tourism – 'the health risks and benefits as encountered 'on the move' rather than focusing on the risks and benefits incurred at fixed locations' (Gatrell, 2016, p. 1). Kaspar et al. (2019, p. 1) put forward the idea of therapeutic mobilities in this context:

> which consist of multiple movements of health-related things and beings, including, though not limited to, nurses, doctors, patients, narratives, information, gifts and pharmaceuticals. The therapeutic emerges from the encounters of mobile human and non-human, animate and inanimate subjects with places and environments and the individual components they are made of.

They see such mobilities as not just about transnational health care, but also about how human bodies become transformed by mobilities. For instance, due to Covid much needed health tourism mobilities have been delayed leading to mental health issues and even death as ageing populations have not been able to access the care needed.

Health tourism mobilities has frequently been associated with ageing populations and second homes. In the case of China, Wu et al. (2018) note how for Chinese 'snowbirds' seasonal migration to Sanya involves interactions between the multiple affordances of home fostered through continual corporeal, material and imaginative mobilities, sustaining a continued sense of belonging. Second homes have mainly been conceptualised as personal/family projects, but also places and spaces of new lifestyle mobilities and identities (Hall & Müller, 2018). Second homes have begun to take on different forms: they are increasingly owned or used by multiple dwellers, on the basis of time or home sharing as the case of Airbnb indicates. Second home ownership has also been linked to the development of new spaces of hospitality such as hotels, clubs, cafes, spas, restaurants, and digital homepages. Such dwelling in motion involves the use of technological, digital platforms to build trust (Molz, 2018; Knaus, 2020; Christensen, 2022). For many cities the rise of Airbnb has prompted significant urban change with whole districts being transformed as the platform became more commercialised. In Barcelona, for example, (Sans & Quaglieri-Domínguez, 2016) local residents have been pushed out, leading to anti-tourism protests due to real and perceived over-tourism in some districts. During Covid, meanwhile, second home owners in Belgium successfully lobbied to be able to travel to their second homes, highlighting how different mobility regimes continued to be practiced for elites (Salazar, 2021).

(Re)Bordering

Political, technological, financial and transportational changes have been critical in significantly lowering the mobility barriers for many but not for all. Indeed, immobilities become particularly apparent in the so-called tourism 'contact zones' at the borders of different countries where notions of citizenship can become highly contested and where multiple identities can become increasingly fluid. Travel writer Kassabova (2017) rightly identifies the seductive nature of borders which hold an enduring fascination for tourists (Timothy, 2000), while Rogerson (2015) argues that they represent critical economic frontiers for informal 'business tourists', such as borderland traders and vendors in West and Southern Africa. In the context of the Global South, cross-border, business tourist mobilities are also heavily influenced by VFR travel, especially in regions where ethnic connections transcend political borders and produce 'geographically stretched or multi-locational households' (Rogerson, 2017a, p. 4). These seasonal or circular non-permanent migration flows represent a significant component of the 'new geographies of migration' in many parts of Africa and Asia (see Dick & Reuschke, 2012; Rogerson, 2017b) and reveal new opportunities for scholars to explore the 'economic mutual dependence between migrants and their home areas' (Rogerson, 2017b, p. 473). Indeed, Rogerson (2015, p. 49) challenges us to question the 'the hegemonic Western treatment of the 'business traveller'' that continues to exclude significant domestic and international flows across the Global South.

The right and ability of humans to travel freely and without hindrance across international borders have thus been increasingly subject to intense political scrutiny and even wider public debate (Bianchi et al., 2020). Such mobility rights raise serious questions when set against the political imperative to ensure the continued growth and expansion of tourism. Indeed, the increase in numbers of refugees has led many western countries to increasingly regulate immigration, even so far as the UK government's decision to send certain asylum seekers to Rwanda. Moreover, the border between the United States and Mexico together with Europe's militarised maritime frontier spanning the Mediterranean, are amongst the world's deadliest (Jones, 2016). This has been conceptualised as in terms of mobility justice, whereby the 'mobility crisis' is constituted out of inter-connected crises of the environment/climate, urbanisation and mass migration/refugee crises (Sheller, 2018). As Andreas (2003, p. 3) has argued:

> The celebrated debordering of the state ... is far more selective than the inflated rhetoric of globalization would suggest. Debordering is being accompanied in many places by a partial rebordering in the form of enhanced policing. Even as many borders have been demilitarized in the traditional realm of national security, as well as economically liberalized to facilitate commercial exchange, they are also now more criminalized to deter those who are perceived as trespassers. Thus, it may be more accurate to say that the importance of territoriality is shifting rather than simply diminishing.

For example, if we take the example of the island of Penang in Malaysia, western tourists have frequently used the island in order to renew their visas to extend their stay in Thailand, extending their mobility. On the other hand, the island is home to many locals and migrant workers who may have multiple affiliations and who do not have the same mobility privileges that western tourists demonstrate (see Bianchi, 2000). From an alternative perspective, differences in mobility privilege are also identified. Taylor et al. (2015) noted that while the freedom to move is typically deemed to be a positive process when undertaken by non-indigenous Australians, the mobilities of indigenous Australians produce derogatory and stereotypical assumptions relating to anti-social behaviours and alcohol consumption.

Macau, meanwhile, has been developed as a destination primarily for rich Chinese gambling tourists but relies heavily upon both elite western expatriate expertise and Filipino migrant labour to maintain the tourism product. Both groups of migrants make use of multiple tourism mobilities to connect with their countries of origin for family reasons, to send money, to communicate online and to travel back and forth from work to home and back again.

Places have thus become important for tourism mobilities in this context, particularly during Covid. Often a clear distinction is made between places and those travelling to places; pushing or pulling people to visit. The mobilities paradigm argues against the ontology of distinct 'places' and 'people'. Rather there is a complex relationality of places and persons. Places are thus not so much fixed but are implicated within complex networks through which 'hosts, guests, buildings, objects and machines' (Urry, 2006, p. vii) are contingently brought together to produce certain performances. Moreover, places are also 'about proximities, about the bodily co-presence of people who happen to be in that place at that time, doing activities together, moments of physical proximity between people that make travel desirable or even obligatory for some' (Hannam et al., 2006, p. 13). In their discussion of Singapore as the archetypal 'mobile city', Oswin and Yeoh (2010, p. 170) argue that the:

> notion of the 'mobile city' thus makes us think about flows and movements in and through the global city in specific ways. It of course facilitates the study of migration as more than a movement from one end point to another. Attention is called to the lines that connect points, to journeys and their continuous negotiations.... The interrelationships between mobile experiences also come into view. Further, the study of the mobile city extends our interest to the study of the movement of ideas and material things that may or may not coincide with the movement of people. ... A process-orientation enables examination of interrelationships of movements of people, objects, capital and ideas in and through the overlapping scales of the local, the bodily, the national, and the global.

Mobilities research emphasises the inter-relation of different scales as discussed above. In terms of mapping the larger-scale movements of people, objects, capital, and information across the world a mobilities perspective allows us to analyse the connections between tourism and geopolitics critically rather than from just a managerial perspective. Geopolitics has been defined in terms of the ways in which we view the world; how the global landscape is structured into various nation-states, and this informs and is informed by various foreign policy agendas of different governments (Dodds, 2007). Such discourses make their way into everyday use in the media, and, indeed, tourism. In terms of tourism, foreign policy discourses can have profound effects on when, who and for what reason people are able to move freely across international borders. Geopolitical discourses or 'scripts' as shown in a variety of institutional and popular media, are thus powerful, and as they divide up the world, can lead to conflicts over space and resources (O'Tuathail, 2002). Moreover, websites have become increasingly powerful actors on the geopolitical scene as destination marketing has become more complex.

Waleed Hazbun (2004) has been at the forefront of linking the analysis of tourism with geopolitics in his analysis of the contemporary Middle East. Hazbun (2004) goes on to demonstrate changes in the tourism development strategies in the context of wider geopolitical agendas of many Middle Eastern and North African (MENA) countries in the wake of 9/11. Tourism in some MENA countries, such as Tunisia, that had been wholly dependent on Western tourists initially collapsed for all practical purposes. Egypt and Jordan with a good share of Arab tourists were quickly able to re-direct their marketing strategies towards Arab countries to compensate the reduction in Western tourists. The UAE, meanwhile, was able to increase the numbers of in-coming Arab tourists especially those Arabs who, due to visa restrictions and xenophobic attitudes, preferred not to spend holidays in North America and Europe but rather to visit an Arab country (Al-Hamarneh & Steiner, 2004).

Bianchi (2006) analysed the relationships between tourism, the freedom to travel and the geopolitics of security. He argues that implicit in much of contemporary geopolitics is a western liberal idealised discourse of tourism as freedom (for some but not for others). Tourism destinations can become drawn into political conflicts when accumulated local grievances and wider geopolitical issues coincide. Moreover, '[w]here perhaps tourism becomes even more closely intertwined with global geopolitics is in the mapping of global risk and threats to security through the mechanism of state travel advisories' (Bianchi, 2006, p. 70). Advisories such as the Foreign and Commonwealth Office (FCO) in the UK are extremely powerful in portraying a dominant Western worldview and became more important during Covid in terms of providing advice. The mobilities of global tourism, then, remain intimately entwined with broader geopolitical issues.

Imagining Tourism Mobilities

Moreover, tourism mobilities have become intimately connected with issues of climate change as more and more people travel both by air and by car (Urry, 2008). The freedoms promised by air and car travel are now recognised as being unsustainable given current technologies but, despite the Covid crisis, are still expected to grow. In terms of aeromobilities, Urry (2003) noted that there were four million air passengers each day and at any one point in time over 300,000 air passengers are in the skies above the USA. Forecasts suggested that as many as 400 million passengers would use the UK's airports by 2030 (Budd, 2008). The perceived freedom of air travel and its democratisation through the expansion of low-cost airlines is still heavily regulated, leading to the return of queues at airports for security post-pandemic. At airports, tourists continue to have their bodies biometrically scanned or 'read' and are continually monitored on CCTV. That symbol of national allegiance and identity - the passport - is now 'chipped' and used by governments as a tool to monitor the population's movements. The objects tourists carry, too, are also subject to processes of control through x-rays and searches. Moreover, the very material and immaterial architecture of the modern airport is calculated to control the corporeal movements of the tourist (Adey, 2008). The socio-technical system of air travel has been turned into a form of mass mobility requiring an exceptionally extensive and immobile place, the airport-city with tens of thousands of workers orchestrating the millions of air journeys taking place each day (Hannam et al., 2006). Since the pandemic, a key problem for air systems has been the return of these workers from furlough and the recruitment and training of new staff.

When air systems break down due to an unexpected environmental event such as the Icelandic ash cloud eruption the consequences for tourism mobilities and immobilities are extreme (Birtchnell & Urry, 2015; Adey & Anderson, 2011). Similarly, the Covid pandemic has also caused significant systemic disruptions leading to greater global awareness (Jensen, 2021) as well as increased public awareness of scientific statements as being the truth. In this case we need to ask the question: 'How will this pandemic disruption change our practices, cultures, and interaction norms for the future?' (Jensen, 2021, p. 68). Covid might herald various imaginaries and reactions, from creative community activism and utopian visions to more destructive theories of conspiracy and nationalist protectionism (Jensen, 2021). One forecast was for the end of face-to-face societies where work shifts online to programmes such as Zoom and Teams and outside play is replaced by increased online gaming, social media apps and Netflix. During COVID there were calls for tourism to move beyond 'business as usual' and to find a pathway to more regenerative tourism, particularly for the small island states of the Caribbean who had been reliant on the mass cruise ship industry (Sheller, 2021). Such regenerative tourism would embrace alternative non-capitalist forms of ownership, non-monetary exchange and beneficial community-based development. Moreover, the shift to 'non-tourism', argue Gössling et al. (2020), indicates an unprecedented crisis for tourism, which 'holds important messages regarding the resilience of the tourism system, also in regard to other ongoing crises that are not as immediate, but potentially more devastating than COVID-19, such as climate change' (Gössling et al., 2020, p. 3). In this context, Sheller (2021, p. 1445) argues that:

> Rather than yearning for the return of cruise ships, the rebuilding of bankrupt hotels, or the arrival of time-share tourists and second-home buyers as the panacea to the region's current woes, we might do better to work with Caribbean-based communities to reimagine regenerative economies and resilient ecologies that are grounded in more just relations of mobility and connection.

Tourism in this context also shifts further online into more immersive and imaginative experiences.

Studies of imaginative travel have been a mainstay of recent tourism research particularly as destinations have sought to explicitly brand themselves as cultural or creative places (Cronin, 2008) or as mobile cities as in the case of Singapore (Oswin & Yeoh, 2010) using ever more complex destination management systems (Ek & Hultman, 2008). Salazar (2012, p. 865), defined the tourism imaginary as 'socially transmitted representational assemblages that interact with people's personal imaginings and are used as meaning-making and world-shaping devices'.

In her paper 'Gifting Mecca', Kenny (2007) analyses pilgrimage and its gifts, linking heritage and imaginative tourism mobilities. She further argues that in the 'extended family households of Kankan [Guinea], participation in pilgrimage creates a new kind of globally implicated person and also may influence the relative status of other members of the household' (Kenny, 2007, p. 363). Imaginative travel thus involves the movement of spiritual capital for some African societies, or in the case of Pacific societies like those in Samoa, a diverse range of tourism practices including *malaga* that 'conventional Western understandings cannot readily account for' (Gibson et al., 2020, p. 2). Indeed, there have been increasing calls to apply the mobilities paradigm as an alternative to rigid, Eurocentric notions of travel (see Cohen & Cohen, 2015; Gibson et al., 2020) that fail to adequately distil the myriad mobilities of emerging tourists, including indigenous peoples (Richards, 2018). Understanding the mobilities of indigenous tourists is pertinent due to persistent assumptions that they remain 'immobile' and 'fixed' to their host community (Wei et al., 2021) and because studies exploring tourism systems involving indigenous people often remain focused on their role as suppliers and objects for non-indigenous tourists (see Richards, 2018).

The use of mobile communications while travelling has become central to twenty-first century life and a great deal of research has examined the connections between mobile phone use and forms of tourism. In this context, Smith (2009, p. 220) reminds us that: '[w]ork and leisure are scarcely differentiated as tourists increasingly check e-mails, carry around laptops, and are glued to the ubiquitous mobile phone whilst on holiday!' In the Somali context, Greg Collins (2009) has examined the extraordinary demand for telecoms services that lies behind the emergence of Somalia's telecoms industry in the wake of state collapse. He argues that this demand is not only driven by the size of the Somali diaspora and the astounding amount of money they remit each year, 'but is a reflection of an ethno-national logic that has shaped the ways in which Somalis at home and abroad have dealt with unpredictability in the wake of state collapse – by staying mobile and staying connected' (Collins, 2009, p. 203). He concludes that this has 'created an extraordinary demand for telecoms services in post-collapse Somalia for the very reason that telecoms were necessary to adapt and use these culturally and historically informed strategies on a global scale' (Collins, 2009 p. 203).

Dawood et al.'s (2010) research, meanwhile, has explored the business practices of mobile entrepreneurs, specifically mobile food vendors in Indonesia, and the potential use of location aware and mobile phone-based applications to support their information needs. Mobile food vendors are a ubiquitous phenomenon in the developing world and can be seen selling their wares in carts, bicycles, or motorcycles. Dawood et al. (2010) discuss the key characteristics and features of a mobile phone-based application that will enable these vendors to advertise their current location, accept orders from customers, enable customers' recommendation of vendors, and inform fellow vendors of various special events thus transforming the tourist experience.

All sorts of imaginative and communicative tourism mobilities – from film to food thus go into our understandings of mobile places and have seen increased demand under Covid.

Conclusions: Theoretical and Methodological Reflections

Law and Urry (2004, p. 403) have argued that existing social science methodologies:

> ...deal, for instance, poorly with the fleeting – that which is here today and gone tomorrow, only to re-appear again the day after tomorrow. They deal poorly with the distributed – that is to be found here and there but not in between – or that which slips and slides between one place and another. They deal poorly with the multiple – that which takes different shapes in different places. They deal poorly with the non-causal, the chaotic, the complex.

If we are to adequately understand the ontology of contemporary mobilities then we also need to have mobile methodologies not necessarily to 'capture' but to keep pace with the fluid (dis) order and (dis)embeddedness of (de)territorialised social life (D'Andrea, 2006).

Such research may provide a new critical window on the mobilities, immobilities and moorings of contemporary tourism by utilising innovative, experimental and increasingly sophisticated technologies (Hannam et al., 2006; Fincham et al., 2010; Büscher et al., 2011; see Shoval et al., this volume). However, the impacts of Covid also emphasise that we need to make sure that we continue to pay attention to both established and innovative methods beyond the social sciences which allow us to examine other histories, artistic and scientific practices (Merriman, 2014). It is important to recognise that, as in lockdown and quarantine, '[s]tillness, waiting, slowness and boredom may be just as important... as sensations and experiences of speed, movement, excitement and exhilaration' Merriman, 2014, p. 182) when doing tourism research. Hence a great deal of research under lockdown has had to have been desk top based or through online interviews rather than face to face. This chapter has sought to give some insights into the multiple mobilities that may be involved in the study of tourism mobilities post-covid. The 'new mobilities paradigm' arguably allows us to place travel and tourism at the centre of social and cultural life rather than at the margins and reminds us of the importance of selective immobilities. Everyday embodied practices of tourism such as walking, cycling or driving have become more significant post lockdown. However, issues of sustainable mobilities remain to be adequately addressed as governments have sought to support unsustainable travel practices through subsidies (Ioannides & Gyimóthy, 2020). As soon as mobility restrictions were lifted old mobility patterns were quickly restored as people preferred a return to their known tourism practices instead of making changes for the better (Salazar, 2021). The transition to a post-carbon future will potentially open up many new configurations for tourism mobilities but at present the demand for further aeromobilities and automobilities seem continued to grow and places and politics will remain paramount. Finally, returning to the need for greater attention to be afforded to indigenous tourist mobilities, Wei et al. (2021, p. 2) argue that there is an important opportunity to 'theorise indigenous people as tourists'.

References

Adey, P. (2007). 'May I have your attention': airport geographies of spectatorship, position, and (im) mobility. *Environment and Planning D: Society and Space*, *25*(3), 515–536. https://doi.org/10.1068/d69j

Adey, P. (2008). Airports, mobility and the calculative architecture of affective control. *Geoforum 39*(1), 438–451. https://doi.org/10.1016/j.geoforum.2007.09.001

Adey, P., & Anderson, B. (2011). Anticipation, materiality, event: The Icelandic ash cloud disruption and the security of mobility. *Mobilities*, *6*(1), 11–20. https://doi.org/10.1080/17450101.2011.532919

Adey, P., Hannam, K., Sheller, M., & Tyfield, D. (2021). Pandemic (im) mobilities. *Mobilities*, *16*(1), 1–19. https://doi.org/10.1080/17450101.2021.1872871

Aguiar, M. (2011). *Tracking modernity: India's railway and the culture of mobility*. University of Minnesota Press.

Al-Hamarneh, A., & Steiner, C. (2004). Islamic tourism: Rethinking the strategies of tourism development in the Arab world after September 11, 2001. *Comparative Studies of South Asia, Africa and the Middle East*, *24*(1), 173–182.

Andreas, P. (2003). Redrawing the line: Borders and security in the twenty-first century. *International Security*, *28*(2), 78–111. https://www.jstor.org/stable/4137469

Bianchi, R. (2006). Tourism and the globalisation of fear: Analysing the politics of risk and (in) security in global travel. *Tourism and Hospitality Research*, *7*(1), 64–74. https://doi.org/10.1057/palgrave.thr.6050028

Bianchi, R. V. (2000). Migrant tourist-workers: Exploring the 'contact zones' of post-industrial tourism. *Current Issues in Tourism*, *3*(2), 107–137. https://doi.org/10.1080/13683500008667869

Bianchi, R. V., Stephenson, M. L., & Hannam, K. (2020). The contradictory politics of the right to travel: mobilities, borders & tourism. *Mobilities*, *15*(2), 290–306. https://doi.org/10.1080/17450101.2020.1723251

Birtchnell, T., & Urry, J. (2015). The mobilities and post-mobilities of cargo. *Consumption Markets & Culture*, *18*(1), 25–38. https://doi.org/10.1080/10253866.2014.899214

Bissell, D. (2009). Conceptualising differently-mobile passengers: geographies of everyday encumbrance in the railway station. *Social & Cultural Geography*, *10*(2), 173–195. https://doi.org/10.1080/14649360802652137

Bissell, D. (2010). Passenger mobilities: affective atmospheres and the sociality of public transport. *Environment and Planning D: Society and Space*, *28*(2), 270–289. https://doi.org/10.1068/d3909

Budd, L. (2008). Air craft: Producing UK airspace. In S. Cwerner, S. Kesselring & J. Urry (Eds), *Aeromobilities*, (pp. 115–134). Routledge.

Büscher, M., Urry, J., & Witchger, K. (2011). Introduction: mobile methods. In M. Büscher, J. Urry & K. Witchger (Eds), *Mobile methods* (pp. 1–19). Routledge.

Butler, G., & Hannam, K. (2012). Independent tourist's automobilities in Norway. *Journal of Tourism and Cultural Change*, *10*(4), 285–300. https://doi.org/10.1080/14766825.2012.718441

Butler, G., & Szili, G. (2020). Mediating technologies on the move: tourists' automobilities through the Limestone Coast, South Australia. *Asia Pacific Journal of Tourism Research*, *25*(4), 369–381. https://doi.org/10.1080/10941665.2019.1708756

Butler, G., Szili, G., & Huang, H. (2022). Cultural heritage tourism development in Panyu District, Guangzhou: community perspectives on pride and preservation, and concerns for the future. *Journal of Heritage Tourism*, *17*(1), 56–73. https://doi.org/10.1080/1743873X.2021.1881524

Christensen, M. D. (2022). Doing digital discipline: how Airbnb hosts engage with the digital platform. *Mobilities*. https://doi.org/10.1080/17450101.2022.2060756

Cohen, E., & Cohen, S. A. (2015). Beyond Eurocentrism in tourism: A paradigm shift to mobilities. *Tourism Recreation Research*, *40*(2), 157–168. https://doi.org/10.1080/02508281.2015.1039331

Coles, T., & Hall, C. M. (2006). The geography of tourism is dead. Long live geographies of tourism and mobility. *Current Issues in Tourism*, *9(4–5)*, 289–292. https://doi.org/10.1080/13683500608668250

Collin-Lange, V., & Benediktsson, K. (2011). Entering the regime of automobility: car ownership and use by novice drivers in Iceland. *Journal of Transport Geography*, *19*(4), 851–858. https://doi.org/10.1016/j.jtrangeo.2010.10.012

Collins, G. (2009). Connected: exploring the extraordinary demand for telecoms services in post-collapse Somalia. *Mobilities*, *4*(2), 203–223. https://doi.org/10.1080/17450100902905139

Cook, N., & Butz, D. (2016). Mobility justice in the context of disaster. *Mobilities*, *11*(3), 400–419. https://doi.org/10.1080/17450101.2015.1047613

Cronin, A. M. (2008). Mobility and market research: outdoor advertising and the commercial ontology of the city. *Mobilities*, *3*(1), 95–115. https://doi.org/10.1080/17450100701797349

Crouch, D. (2000). Places around us: Embodied lay geographies in leisure and tourism. *Leisure Studies*, *19*(2), 63–76. https://doi.org/10.1080/026143600374752

D'Andrea, A. (2006). Neo-nomadism: A theory of post-identitarian mobility in the global age. *Mobilities*, *1*(1), 95–119. https://doi.org/10.1080/17450100500489148

Dawood, R., Jackson, S. J., & Yew, J. (2010, December). Supporting the information needs of mobile microentrepreneurs in the developing world: The case of Indonesian food cart vendors. In *Proceedings of the 4th ACM/IEEE International Conference on Information and Communication Technologies and Development* (pp. 1–10). https://doi.org/10.1145/2369220.2369226

Dick, E., & Reuschke, D. (2012). Multilocational households in the Global South and North: Relevance, features and spatial implications. *DIE ERDE–Journal of the Geographical Society of Berlin*, *143*(3), 177–194. https://www.die-erde.org/index.php/die-erde/article/view/13

Dodds, K. (2007). *Geopolitics: A very short introduction*. Oxford University Press.

Edensor, T. (2007). Mundane mobilities, performances and spaces of tourism. *Social & Cultural Geography*, *8*(2), 199–215. https://doi.org/10.1080/14649360701360089

Edensor, T. (2010). Walking in rhythms: Place, regulation, style and the flow of experience. *Visual Studies*, *25*(1), 69–79. https://doi.org/10.1080/14725861003606902

Ek, R., & Hultman, J. (2008). Sticky landscapes and smooth experiences: The biopower of tourism mobilities in the Öresund region. *Mobilities*, *3*(2), 223–242. https://doi.org/10.1080/17450100802095312

Featherstone, M. (2004). Automobilities: an introduction. *Theory, Culture & Society*, *21*(4–5), 1–24. https://doi.org/10.1177/0263276404046058

Fincham, B., McGuinness, M., & Murray, L. (2010). Introduction. In B. Fincham, M. McGuinness & L. Murray (Eds), *Mobile methodologies* (pp. 1–10). Palgrave Macmillan.

Franklin, A. (2003). *Tourism: An introduction*. Sage.

Franklin, A., & Crang, M. (2001). The trouble with tourism and travel theory? *Tourist Studies*, *1*(1), 5–22. https://doi.org/10.1177/146879760100100101

Gatrell, A. (2016). *Mobilities and health*. Routledge.

Gazzola, P., Pavione, E., Grechi, D., & Ossola, P. (2018). Cycle tourism as a driver for the sustainable development of little-known or remote territories: The experience of the Apennine regions of Northern Italy. *Sustainability*, *10*(6), 1863. https://doi.org/10.3390/su10061863

Gibson, D., Pratt, S., & Iaquinto, B. L. (2020). Samoan perceptions of travel and tourism mobilities–the concept of Malaga. *Tourism Geographies*. https://doi.org/10.1080/14616688.2020.1780632

Giddy, J. K. (2019). The influence of e-hailing apps on urban mobilities in South Africa. *African Geographical Review*, *38*(3), 227–239. https://doi.org/10.1080/19376812.2019.1589732

Gössling, S., Scott, D., & Hall, C. M. (2020). Pandemics, tourism and global change: a rapid assessment of COVID-19. *Journal of Sustainable Tourism*, *29*(1), 1–20. https://doi.org/10.1080/09669582.2020.1758708

Hall, C. M. (2011). Health and medical tourism: a kill or cure for global public health? *Tourism Review*, *66*(1/2), 4–15. https://doi.org/10.1108/16605371111127198

Hall, C. M., & Müller, D. (Eds.) (2018). *The Routledge handbook of second home tourism*. Routledge.

Hannam, K. (2009). The end of tourism? Nomadology and the mobilities paradigm. In J. Tribe (Ed.), *Philosophical issues in tourism* (pp. 101–113). Channel View.

Hannam, K., & Knox, D. (2010). *Understanding tourism*. Sage.

Hannam, K., Butler, G., Witte, A., & Zuev, D. (2021). Tourist's mobilities: Walking, cycling, driving and waiting. *Tourist Studies*, *21*(1), 57–69. https://doi.org/10.1177/1468797621992931

Hannam, K., Sheller, M., & Urry, J. (2006). Editorial: mobilities, immobilities and moorings. *Mobilities*, *1*(1), 1–22. https://doi.org/10.1080/17450100500489189

Hazbun, W. (2004). Globalisation, reterritorialisation and the political economy of tourism development in the Middle East. *Geopolitics*, *9*(2), 310–341. https://doi.org/10.1080/14650040490442881

Henderson, J. (2018). Making cities more walkable for tourists: A view from Singapore's streets. *International Journal of Tourism Cities*, *4*(3), 285–297. https://doi.org/10.1108/IJTC-11-2017-0059

Howes, D. (2018). The skinscape: Reflections on the dermalogical turn. *Body & Society*, *24*(1–2), 225–239. https://doi.org/10.1177/1357034X18766285

Huijbens, E. H., & Benediktsson, K. (2007). Practising highland heterotopias: Automobility in the interior of Iceland. *Mobilities*, *2*(1), 143–165. https://doi.org/10.1080/17450100601106518

Iaquinto, B. (2020). Tourist as vector: Viral mobilities of COVID-19. *Dialogues in Human Geography* *10*(2), 174–177. https://doi.org/10.1177/2043820620934250

Ioannides, D., & Gyimóthy, S. (2020). The COVID-19 crisis as an opportunity for escaping the unsustainable global tourism path. *Tourism Geographies*, *22*(3), 624–632. https://doi.org/10.1080/14616688.2020.1763445

Jacobsen, J. K. S. (2004). Roaming romantics: Solitude-seeking and self-centredness in scenic sightseeing. *Scandinavian Journal of Hospitality and Tourism*, *4*(1), 5–23. https://doi.org/10.1080/15022250410003475

Jensen, M. T., Scarles, C., & Cohen, S. A. (2015). A multisensory phenomenology of interrail mobilities. *Annals of Tourism Research*, *53*, 61–76. https://doi.org/10.1016/j.annals.2015.04.002

Jensen, O. (2021). Pandemic disruption, extended bodies, and elastic situations - Reflections on COVID-19 and Mobilities. *Mobilities, 16*(1), 66–80. https://doi.org/10.1080/17450101.2021.1867296

Johinke, R. (2018). Take a walk on the wild side: Punk music walking tours in New York City. *Tourist Studies, 18*(3), 315–331. https://doi.org/10.1177/1468797618771694

Johnson, J. (2010). Euro-railing: A mobile ethnography of backpacker train travel. In K. Hannam & A. Diekmann (Eds), *Beyond backpacker tourism: Mobilities and experiences* (pp. 102–113). Channel View.

Jones, R. (2016). *Violent borders: Refugees and the right to move.* Verso.

Kaaristo, M., Medway, D., Burton, J., Rhoden, S., & Bruce, H. L. (2020). Governing mobilities on the UK canal network. *Mobilities, 15*(6), 844–861. https://doi.org/10.1080/17450101.2020.1806507

Kaspar, H., Walton-Roberts, M., & Bochaton, A. (2019). Therapeutic mobilities. *Mobilities, 14*(1), 1–19. https://doi.org/10.1080/17450101.2019.1565305

Kassabova, K. (2017). *Border: A journey to the edge of Europe.* Granta.

Kenny, E. (2007). Gifting Mecca: importing spiritual capital to West Africa. *Mobilities, 2*(3), 363–381. https://doi.org/10.1080/17450100701597376

Knaus, K. (2020). At home with guests–discussing hosting on airbnb through the lens of labour. *Applied Mobilities, 5*(1), 68–85. https://doi.org/10.1080/23800127.2018.1504600

Küpers, W., & Wee, D. (2018). Tourist cities as embodied places of learning: Walking in the "feelds" of Shanghai and Lisbon. *International Journal of Tourism Cities, 4*(3), 376–390. https://doi.org/10.1108/IJTC-12-2017-0076

Larsen, J. (2001). Tourism mobilities and the travel glance: Experiences of being on the move. *Scandinavian Journal of Hospitality and Tourism, 1*(2), 80–98. https://doi.org/10.1080/150222501317244010

Law, J., & Urry, J. (2004). Enacting the social. *Economy and Society, 33*(3), 390–410. https://doi.org/10.1080/0308514042000225716

Liu, X., Huang, D., & Li, Z. (2018). Examining relationships among perceived benefit, tourist experience and satisfaction: the context of intelligent sharing bicycle. *Asia Pacific Journal of Tourism Research, 23*(5), 437–449. https://doi.org/10.1080/10941665.2018.1466814

Löfgren, O. (2008). Motion and emotion: learning to be a railway traveller. *Mobilities, 3*(3), 331–351. https://doi.org/10.1080/17450100802376696

Merriman, P. (2014). Rethinking mobile methods. *Mobilities, 9*(2), 167–187. https://doi.org/10.1080/17450101.2013.784540

Molz, J. G. (2018). Discourses of scale in network hospitality: from the Airbnb home to the global imaginary of 'belong anywhere'. *Hospitality & Society, 8*(3), 229–251. https://doi.org/10.1386/hosp.8.3.229_1

O'Tuathail, G. (2002). Post-cold war geopolitics: Contrasting superpowers in a world of global dangers. In R. J. Johnson, P. Taylor & M Watts (Eds), *Geographies of global change* (2nd ed., pp. 174–190). Blackwell.

Oswin, N., & Yeoh, B. S. (2010). Introduction: mobile city Singapore. *Mobilities, 5*(2), 167–175. https://doi.org/10.1080/17450101003665002

Pirie, G. H. (2009). Virtuous mobility: moralising vs measuring geographical mobility in Africa. *Afrika Focus, 22*(1), 21–35. https://brill.com/view/journals/afoc/22/1/article-p21_4.xml

Ranasinghe, R., & Cheng, L. (2018). Tourism-induced mobilities and transformation of indigenous cultures: where is the Vedda community in Sri Lanka heading to? *Journal of Tourism and Cultural Change, 16*(5), 521–538. https://doi.org/10.1080/14766825.2017.1393081

Richards, G. (2018). Cultural tourism: A review of recent research and trends. *Journal of Hospitality and Tourism Management, 36*, 12–21. https://doi.org/10.1016/j.jhtm.2018.03.005

Rink, B. (2017). The aeromobile tourist gaze: Understanding tourism 'from above'. *Tourism Geographies, 19*(5), 878–896. https://doi.org/10.1080/14616688.2017.1354391

Rogerson, C. M. (2015). Unpacking business tourism mobilities in sub-Saharan Africa. *Current Issues in Tourism, 18*(1), 44–56. https://doi.org/10.1080/13683500.2014.898619

Rogerson, C. M. (2017a). Visiting friends and relatives travel matters for sub-Saharan Africa. *African Journal of Hospitality, Tourism and Leisure, 6*(3). http://www.ajhtl.com/uploads/7/1/6/3/7163688/article-38-vol-6-3-2017.pdf

Rogerson, C. M. (2017b). Unpacking directions and spatial patterns of VFR travel mobilities in the global South: Insights from South Africa. *International Journal of Tourism Research*, *19*(4), 466–475. https://doi.org/10.1002/jtr.2101

Rogerson, C. M., & Baum, T. (2020). COVID-19 and African tourism research agendas. *Development Southern Africa*, *37*(5), 727–741. https://doi.org/10.1080/0376835X.2020.1818551

Rogerson, C. M., & Rogerson, J. M. (2021). COVID-19 and changing tourism demand: Research review and policy implications for South Africa. *African Journal of Hospitality, Tourism and Leisure*, *10*(1), 1–21. -21. https://doaj.org/article/56d78984428f406f994957649cc5b280

Roy, S., & Hannam, K. (2013). Embodying the mobilities of the Darjeeling Himalayan Railway. *Mobilities*, *8*(4), 580–594. https://doi.org/10.1080/17450101.2012.745695

Salazar, N. (2012). Tourism imaginaries: A conceptual approach. *Annals of Tourism Research*, *39*(2), 863–882. https://doi.org/10.1016/j.annals.2011.10.004

Salazar, N. (2021) Existential vs. essential mobilities: insights from before, during and after a crisis. *Mobilities*, *16*(1), 20–34. https://doi.org/10.1080/17450101.2020.1866320

Sans, A. A., & Quaglieri-Domínguez, A. (2016). Unravelling airbnb: Urban perspectives from Barcelona. In A. P. Russo & G. Richards (Eds), *Reinventing the local in tourism: Producing, consuming and negotiating place* (pp. 209–228). Channel View.

Sarmento, J. (2017). Tourists' walking rhythms: 'doing' the Tunis Medina, Tunisia. *Social & Cultural Geography*, *18*(3), 295–314.

Schivelbusch, W. (1986). *The railway journey: Trains and travel in the nineteenth century*. Blackwell.

Sheller, M. (2018). *Mobility justice: The politics of movement in an age of extremes*. Verso.

Sheller, M. (2021). Reconstructing tourism in the Caribbean: connecting pandemic recovery, climate resilience and sustainable tourism through mobility justice. *Journal of Sustainable Tourism*, *29*(9), 1436–1449. https://doi.org/10.1080/09669582.2020.1791141

Sheller, M., & Urry, J. (2000). The city and the car. *International Journal of Urban and Regional Research*, *24*(4), 737–757. https://doi.org/10.1111/1468-2427.00276

Sheller, M., & Urry, J. (2003). Mobile transformations of 'public' and 'private life. *Theory Culture & Society*, *20*(3), 107–125. https://doi.org/10.1177/02632764030203007

Sheller, M., & Urry, J. (Eds). (2004). *Tourism mobilities: Places to play, places in play*. Routledge

Smith, M. (2009). *Issues in cultural tourism studies* (2nd ed.) Routledge.

Smith, M., & Puczkó, L. (2014). *Health, tourism and hospitality: Spas, wellness and medical travel*. Routledge.

Taylor, A., Carson, D. B., Carson, D. A., & Brokensha, H. (2015). 'Walkabout' tourism: The indigenous tourism market for outback Australia. *Journal of Hospitality and Tourism Management*, *24*, 9–17. https://doi.org/10.1016/j.jhtm.2015.04.002

Timothy, D. J. (2000). Borderlands: An unlikely tourist destination. *IBRU Boundary and Security Bulletin*, *8*(1),57–65.https://www.durham.ac.uk/media/durham-university/research-/research-centres/ibru-centre-for-borders-research/maps-and-databases/publications-database/boundary-amp-security-bulletins/bsb8-1_timothy.pdf

Urry, J. (2000). *Sociology beyond societies*. Routledge.

Urry, J. (2003). *Global complexity*. Polity.

Urry, J. (2006). Preface. In Minca, C & Oakes, T. (Eds). *Travels in paradox: Remapping tourism* (p. vii – xi). Rowan & Littlefield.

Urry, J. (2008). Climate change, travel and complex futures. *British Journal of Sociology*, *59*(2), 261–279. https://doi.org/10.1111/j.1468-4446.2008.00193.x

Vanderheiden, S. (2006). Assessing the case against the SUV. *Environmental Politics*, *15*(1), 23–40. https://doi.org/10.1080/09644010500418688

Wei, L., Qian, J., & Zhu, H. (2021). Rethinking indigenous people as tourists: Modernity, cosmopolitanism, and the re-invention of indigeneity. *Annals of Tourism Research*, *89*, 103200. https://doi.org/10.1016/j.annals.2021.103200

Wilson, S., & Hannam, K. (2017). The frictions of slow tourism mobilities: Conceptualising campervan travel. *Annals of Tourism Research*, *67*, 25–36. https://doi.org/10.1016/j.annals.2017.07.021

Witte, A., & Hannam, K. (2017). Walking online: A netnography of China's emerging hiking communities. In C. M. Hall, Y. Ram, & N. Shoval (Eds), *The Routledge international handbook of walking* (pp. 147–156). Routledge.

Wu, Y. F., Hannam, K., & Xu, H. G. (2018). Reconceptualising home in seasonal Chinese tourism mobilities. *Annals of Tourism Research*, *73*, 71–80. https://doi.org/10.1016/j.annals.2018.09.005

Yang, P., Xu, H., & Hannam, K. (2022). A touristic habitation: automobilities of Chinese driving tourists. *Tourism Geographies*. https://doi.org/10.1080/14616688.2022.2046148

Zuev, D., & Hannam, K. (2021). Anxious immobilities: an ethnography of coping with contagion (Covid-19) in Macau. *Mobilities*, *16*(1), 35–50. https://doi.org/10.1080/17450101.2020.1827361

Chapter 9

The State of Qualitative Research Methods in Tourism and Hospitality

Research Design Issues and The Way Forward

Sandhiya Goolaup & Robin Nunkoo

Introduction: Qualitative Research Method

It is tempting to start this chapter by defining what is exactly qualitative research to indicate its breadth and depth. Yet, given the range of definitions that various scholars have provided, it is hard to find one specific definition. Denzin and Lincoln, whose work have been seminal in legitimatizing qualitative research in social sciences, defined it as 'studying things in their natural setting by attempting to make sense of or interpret phenomena in terms of the meanings people bring to them' (1994, p. 2). Essentially, qualitative research is a method that relates to the idea that meanings are socially constructed by individuals. It is based on using a range of tools to inductively explore, interpret and understand a specific phenomenon. As a method it is often used when there is a lack of theory or existing theory that can explain a phenomenon.

The purpose of qualitative research is mainly to generate what Geertz (1973) describes as 'thick description' of a phenomenon. In general, there are various tools of qualitative research such as interviews, focus group, photo elicitation techniques, text and discourse analysis, observation (participant or non-participant), ethnography. These empirical materials are used to describe the routine and problematic moments and meanings in an individual's life. The advantages of qualitative research have been described by Tracy (2013) as consisting of:

> focusing on lived experience, keeping participants' meanings; preserving the chronological flow, documenting which events lead to which consequences, and useful in explaining why a specific chronology may have occurred, and illustrating how a multitude of interpretations are possible through the interpretation of participant viewpoints and stories.

Within the field of tourism and hospitality, qualitative research as a method gained recognition only in the early 1990s. Qualitative research in tourism and hospitality research has primarily taken inspiration from the anthropology, sociology, and geography disciplines (Ren, 2016). Prior to the 1990, the field was largely dominated by research based on positivistic philosophies.

The Wiley Blackwell Companion to Tourism, Second Edition. Edited by C. Michael Hall.

Since then, researchers have started to question the fundamental nature of quantitative research, as it couldn't address the questions of meanings and understandings (Riley & Love, 2000). Qualitative research was, therefore, considered as an alternative method offering richer interpretations and understanding of lived realities. While there has been a surge in qualitative methods over the years, quantification is still a dominant paradigm in much of the tourism and hospitality scholarship. Qualitative research is still portrayed as less methodologically robust, lacking rigor and credibility. Hence various attempts have been made to overcome the weaknesses. Decrop (1999), for instance, suggested triangulation as a tool to enhance the robustness of a qualitative research. Triangulation relates to the act of looking at a particular phenomenon through the combination of various data sources, methods, theories, and investigators to facilitate the emergence of richer and valid interpretation. Dimitriadis (2016) suggested using the word 'qualitative inquiry' instead of qualitative research to emphasize the open-endedness and ambiguity of the method. While other researchers have related the problem emanating from authors' failure to explain and justify the appropriateness of their qualitative approaches (e.g., Decrop, 1999). Despite the controversies surrounding qualitative research, the acceptance of qualitative studies has increased over the last decade (Wilson et al., 2020) and the trend is likely to continue in the future.

To understand how qualitative studies can be further developed to generate robust theoretical advancement, it is important to first understand the historical development of qualitative research within the field. We also provide insight on some of the most common types of qualitative inquiries used in tourism and hospitality research as well as their inherent challenges.

Historical Development of Qualitative Research in Tourism and Hospitality

Qualitative research as a method has evolved over the years. Scholars from various disciplines with an interest in qualitative research has been instrumental in legitimatizing this form of research approach. Of particular interest is the work of Denzin and Lincoln (2011, 2018) that have been instrumental in legitimatizing the qualitative approach to research within the field of social science. Much of their work concerns the historical analysis of qualitative methodologies. They identified eight period through which qualitative inquiry has transitioned; (1) traditional; (2) modernist; (3) blurred genres; (4) crisis in representation (5) postmodern; (6) post experimental inquiry; (7) methodologically contested present and, finally, (8) the fractured future. Each of these periods are guided by specific ontological, epistemological, and methodological assumptions. The different periods also reflect distinct worldviews that inform the practice of research. While Denzin and Lincoln present them in a chronological order, in practice these different periods tend to overlap over each other (Mura & Khoo-Lattimore, 2018) and the prevailing methodological approach largely depend on the disciplines and phenomenon being studied.

The traditional period occurred between 1900 and 1950, and relates to the time when social scientists held positivist belief. Early works related to tourism ethnographies and anthropologies were conducted by 'quantifying' qualitative data. The modernist (1950–1970) and blurred genres (1970–1986) are both connected to the appearance of post-positivist paradigm. The modernist marked the period when researchers started to question the objectivity in social reality and the world was seen as consisting of multiple social realities. However, during this period, qualitative research was mainly used as a support for quantitative research. Triangulation was implemented to establish 'rigor'. The blurred genres were the period when researchers started to engage with multiple and multi-disciplinary methodologies (Denzin & Lincoln, 2018). More reflexive and innovative methods like in-depth interviews and personal diaries were used.

Unlike the first three periods, the crisis in representation (1986 – mid-1990s) period forges the way to new approaches that perceived realities as being socially, culturally, and politically situated. There was also more acceptance towards scholars turning towards intensive self-reflection and introspection to establish the authenticity of their works (Mura & Pahlevan Sharif, 2015). During this phase the researchers also became more aware of their emotions during the inquiry process. Subsequently, the post-modern (1990 - 1995), post experimental inquiry (1995- 2000), methodologically contested present (2000- 2004) and the (8) fractured future (2005 – onwards) all share a form of inquiry that encourages participatory approach and are grounded on local realities. These periods also gave rise to researchers who read widely and have knowledge about the many interpretive paradigms (e.g., feminism, Marxism, cultural studies) and the researchers became what Denzin and Lincoln (2018) called the 'theoretical bricoleur'.

The progress in qualitative research in tourism studies have been documented by a few scholars (e.g., Mura & Pahlevan Sharif, 2015; Riley & Love 2000; Wilson et al. 2020; Phillimore & Goodsone, 2004). These studies have found that much of the qualitative studies are confined within the first three period: traditional, modernist, and blurred genres. For instance, Wilson et al. (2020), in an analysis of 1,541 articles found most of the qualitative studies were anchored in the traditional period and these studies have not moved beyond the first three period. Arguing in a similar line, Mura and Pahlevan Sharif (2015) also found that much of the qualitative research conducted within an Asian context (e.g., Singapore, Thailand, Indonesia, Malaysia) are confined within the first three period (traditional, modernist, and blurred genres). Indeed, this is reflected in a vast majority of tourism and hospitality studies where qualitative studies are used mainly as supportive data or as tool to collect variable data that will be further tested using survey. The construction of knowledge in these papers is centered towards scientific rationality, replicability, and generalization. In many of these papers, there is a lack of reflexibility and introspection and with observation and interviews being the most common method. As further argued by these scholars there is a lack of innovative and alternative strategies such as ethnographic, textual, and visual analysis, which are situated within the post-modern and later period of qualitative inquiry.

Hence, within the field of tourism, much of the studies are still anchored towards the traditional positivist paradigm. This provides further support to the fact that qualitative research is still perceived as soft and non-scientific in nature. The lack of support and acceptance for qualitative studies can be explained by the fact that in comparison to other fields in the social sciences, tourism and hospitality is relatively new, inter-disciplinary and immature. Phillimore and Goodsone (2004) argued that it can be a reflection that much of the analysis conducted provide an oversimplification that occurs in research. Nevertheless, it is also important to note that there is an increasing number of researchers who are calling for the development of more robust and co-creative method within qualitative research (e.g., Nunkoo et al. 2020; Ivanova et al., 2021; Pritchard et al., 2011). To describe this trend towards embracing transdisciplinary, hybridity and criticality in qualitative research studies, Ivanova et al. (2021) used the termed 'creative and disruptive' methodologies, while Pritchard et al. (2011) provided a 'hopeful tourism perspective'.

Indeed, there are a few studies that have attempted to introduce innovative approaches in areas that have traditionally been studied using quantitative approach. For instance, in a study of how residents respond to the economic and socio-cultural impacts of mega-events in Rio de Janeiro, Duignan, and McGillivray (2021), used a creative amalgam of physical and digital methods. Physical data in the form of observation data, geo tapped photo and video, and interviews were used. Further various forms of digital data such as vlogging, blogging on twitter and media were combined to understand 'what' and 'how' spatial arrangêment are produced in mega events. Likewise, Ogle (2018) introduced Sensory Quasi-Q-Sort (SQQS) as an alternative method where the participants had to use all their human senses to gauge the perception of service products. Hindley and Font (2018) introduced a collection of five types of projective techniques: collage, list of values, word association, photo elicitation and scenarios as useful tool in studying travelers' behavior. Through these qualitative methodologies, researchers have been able to reveal the subconscious feelings and thoughts of the participants and circumvent socially desirable responses and develop theoretically robust studies.

While there is progress in terms of the use of qualitative research method, it is also clear that it will take some time before it is legitimatized in tourism and hospitality research. The transition to the phase of post-modernism is more likely to occur when researchers are reflexively vigilant and situationally open to the contesting cultural research settings (Wilson & Hollinshead, 2015). This will allow the researcher to interrogate previously assumed realities and misinterpreted understandings can be revealed. Since, qualitative research method in tourism and hospitality is dominated mainly by the traditional; modernist; and blurred genres approach, the next section will provide an overview of the most common qualitative strategies inquiries used and the challenges involve in designing the study.

Types of Qualitative Strategies for Inquiry

Qualitative research strategies for inquiry have evolved over the years and various classifications have been made by different scholars based on their disciplines. Lancy (1993), using an educational perspective, organized qualitative research into five categories: anthropology, sociology, biology, cognitive psychology, and history. Denzin and Lincoln (2005) identified several qualitative research approaches that are used in social sciences: ethnography, case studies, grounded theory, life history, narrative, participatory action research and clinical research. Creswell (2007) using a generic perspective, identified five approaches: narrative research, phenomenology, grounded theory, ethnography, and case studies. Within the field of tourism and hospitality research, some of the most common types of qualitative research method that has been used include, but are not limited to, phenomenology, and grounded theory. The choice of these qualitative research strategies for elaboration are also a reflection of what previous researchers have found (e.g., Mura & Pahlevan Sharif, 2015; Wilson et al., 2020) as dominating qualitative studies within tourism and hospitality. These two qualitative strategies for inquiry are also selected since they have been poorly executed in most of the tourism and hospitality studies. Hence, the section below provides an overview of the two qualitative research approach, namely phenomenology and grounded theory. The section below briefly explores the types of studies conducted and the challenges involve in designing such type of research inquiry.

Phenomenology

Phenomenology is concerned with understanding the lived experiences of individuals. The goal of phenomenology in tourism research is to study tourists being in the world. Phenomenology is both a philosophy and methodological approach that draws heavily on the work of Edmund Husserl (1859–1938), a German mathematician. His views were further expanded by Heidegger, Sartre and Merleau- Ponty. As a result, this has led to different phenomenological approach, commonly known as 'Husserlian', 'Heideggerian' as well as a merger between these two, termed as 'existential phenomenology'. Within the field of tourism and hospitality research, phenomenology is developed in two different strands: *descriptive phenomenology* building on the works of Husserl and *hermeneutic phenomenology* following Heidegger (Pernecky & Jamal, 2010). Recently, there has also been an increase application of *existential phenomenology*, a merger between Husserlian and Heideggerian phenomenology (Goulding, 2005; Thompson et al., 1989).

The Husserlian approach is based on describing 'pure' consciousness by isolating the empirical world and describing the structure of the conscious experience. Husserl's approach was often described as being positivist in nature since it focused on rigid and descriptive exercise. Descriptive phenomenology is based on suspending all beliefs (i.e., bracketing) as a means to acquire knowledge (Spurling, 2013). In contrast, Heidegger approach is based on the notion that our understanding of the world is situated, temporal 'being in the world' and historical. The hermeneutic phenomenology is based an individual experience as being formed through interpretation of the world. Existential phenomenology, on the other hand, is based on understanding human experiences as it is lived and how individuals ascribe meaning to and perceive lived experience (Thompson et al., 1989). As a method, it is widely used in consumer research and marketing and recently used in tourism studies (e.g., Ziakas & Boukas, 2013; Goolaup & Nunkoo, 2021). In an analysis of qualitative studies using phenomenology from the period 2005 – 2016 in some of the top tourism journals, such as *Annals of Tourism Research*, *Tourism Management*, *Journal of Travel Research*, *Current Issues in Tourism*, and *International Journal of Contemporary Hospitality Management*, Goolaup and Soler (2019) found there was a dominance of Heideggerian over the Husserlian approach. In contrast, Pernecky and Jamal's (2010) analysis of research articles pre-2010 found that Husserl's positivist approach was favoured over Heidegger approach.

While phenomenology as a method has been popularized in the field of tourism and hospitality, Pernecky and Jamal, (2010) and Kirillova (2018) found that it has also been largely misunderstood and applied. For instance, in most of the pre-2010 research articles reviewed by Pernecky and Jamal (2010), it was found that researchers do not address the ontological and epistemological assumptions of phenomenology. As a result, these have translated to inappropriate approaches to data collection, analysis, and interpretation. Likewise, a lack of theoretical justification in the use of phenomenology was observed. Similarly, Kirillova (2018) analyzing articles post 2010, pointed that most of the studies using phenomenological approach can be classified as ambiguous in nature, since they do not distinguish between the methodologies (e.g., Marschall 2015; Podoshen et al. 2018) and this resulted in confusing design of the study. As an example, Marschall (2015) in a study related to 'memory trips' using a combination of phenomenology and auto-ethnographic approach failed to demonstrate an understanding of the underlying philosophical foundation and theoretical justification for using phenomenology. Likewise, the study applied a 're-designed thematic strands' with the use of deductive thematic analysis, which is in direct contradiction with the epistemological assumption of phenomenology.

Hence, to design appropriate phenomenological studies, it is important for researchers to understand that phenomenology is not only a method but a philosophical approach that shapes how knowledge is constructed as well as how research needs to be approached. It is also vital for researchers to have knowledge of the philosophical underpinning, given that different approaches (Husserlian, Heidegger or a combined approach) will shape the study differently. Indeed, as stated by Pernecky and Jamal (2010), staying true to the philosophical origins of phenomenology is vital. There is need to not treat phenomenology as a prescribed method but rather to recognize the diversity within the phenomenological tradition. Kirillova (2018) suggested that authors consider tracing the intellectual roots of phenomenology and then proceed towards assessing which phenomenological approach will best suit the study at hand. Indeed, being aware of the diversity in approaches will help in designing robust research studies that can have significant theoretical contribution.

Grounded Theory

Grounded theory is traditionally rooted in sociology (Glaser & Strauss, 1967; Glaser, 1978; Strauss & Corbin, 1990), an area of enquiry focused on the study of individuals and society. The methodology is influenced by the broader sociological schools of symbolic interactionism, in which an individual behavior is seen as being goal driven and evolving from symbolic social interaction (Goulding, 2005). Charmaz (2006) defined it as an inductive method related to theory building from the empirical field through a systematic process of data collection and analysis. As the name suggest, grounded theory is grounded in the words and actions of the individuals under study. In grounded theory, a wide range of data can be used; in-depth interviews, observations and memos are usually used to keep track of feelings, events and describe situation.

Grounded theory as a method has evolved over the years. There are various authors whose works have been seminal in developing the method (e.g., Glaser, 1978; Strauss & Corbin, 1990; Charmaz, 2006; Clarke, 2005). There are basically three main approaches that reflect the evolution of grounded theory: objectivist, post-positivist, and constructivist. The objectivist approach is linked to Glaser's classic grounded theory, based on producing objective data and a unidimensional reality. The post-positivist approach is related to the work of Strauss and Corbin recognizing the fluid nature of reality. Finally, there is the constructivist nature of grounded theory which is representative of the work by Charmaz (2006). This perspective posits that the research process emerges from interactions, and that the production of data and analysis are socially constructed. However, there are two core approaches of grounded theory that has been widely applied and popularized: (1) Glaserian and (2). Straussian schools. The Glaserian school of thoughts considers that the field should be the main place where data emerge. The researcher should minimize their impact by not bringing their knowledge or experience to the research study. In contrast, the Straussian school emphasize the researcher as being an interventionist who influences the data through his/her acquired knowledge. These two schools of thoughts also differ in terms of the development of theory. Glaser believes that theory development should be based on an abstract of time, place, and people and be conceptually dense and integrated. While Strauss and Corbin expand Glaser perspective and argued that it can also be used to developing non-theory and descriptive studies. However, these two approaches have similar ground when it comes to the notion of theoretical sampling (i.e., generating theory by jointly collecting and analyzing), constant comparison of data by exploring differences and similarities and the process of abstraction by moving to higher order theoretical categories. Refer to Jones and Noble (2007) and Charmaz (2006) for a full comparison between these two approaches to grounded theory.

Grounded theory has also found its way in tourism and hospitality research. In the last two decades, there has been an upsurge of studies using grounded theory, albeit it seems to be more common in tourism rather than hospitality related research. In a review undertaken by Decrop and Masset (2018), it was found that grounded theory was mainly used to study tourist travel behaviours and experiences (e.g., Goolaup & Mossberg, 2017; Jiménez-Barreto et al., 2022; Matteucci & Filep, 2017; Tung & Ritchie, 2011). However, there are major issues in terms of how grounded theory has been applied in tourism and hospitality research. In a review of grounded theory of 48 tourism studies, Matteucci and Gnoth (2017) found 10 percent of the papers used the Glaserian approach, 19 percent Straussian, another 19 percent following Charmaz constructivist approach and with the vast majority not specifying the approach. Rather, these studies use a mixture of different approaches. This shows that there is a lack of clarity and confusion when it comes to the epistemological and ontological underpinnings that shape grounded theory. For instance, in a study related to slow travel, Lumsdon and McGrath (2011) position their works as constructivist in nature but Glaser's grounded theory method was adopted, and they used an inductive approach aligning with the Straussian approach. In addition, the authors neither justify the choice of grounded theory nor addresses the issue of reflexivity. As a result, this has significant consequences in terms of the theoretical contribution and make the overall research design questionable. Indeed, Stumpf et al. (2016), in an analysis of 31 articles in 11 tourism journals, found that 68 percent of these studies published provide minor to no theoretical advancement. This further support the fact that grounded theory as a theory development tool in tourism research is under-utilized.

To design robust grounded theory tourism and hospitality studies, Stumpf et al. (2016) urged researchers to use a consistent variant of the method. This is important to legitimize the analysis and foster the theory development process. Since much of the current grounded theory studies have failed in developing theories, Urquhart et al. (2010) urged researchers to apply the notion of 'theoretical integration'. This involves the process of integrating existing theoretical understanding of the phenomenon under study to the emerging grounded theory to highlight new revelations and contrasting characteristics. Furthermore, as urged by Matteucci and Gnoth (2017), to develop robust methodological applications, researchers should ensure that there is greater integration of the core element of constant comparison by exploring differences and similarities, and ensuring it is an iterative process. Researchers should also ensure that grounded theory is not a pick and mix box of method but rather being aware that there are various variants that are underpinned by different epistemological and ontological philosophies.

Designing Robust Qualitative Research Methods

Qualitative research has tremendous potential for theoretical advancement. Theoretical advancement occurs when there is theoretical elaboration beyond the current research context. However, as demonstrated poor execution of a qualitative study can have significant impact in terms of the contribution it can make. Hence, there is a need for researcher to consider more robust methodological applications. At the core, developing robust qualitative research methods, require the need to firstly being aware of the diversity of approaches present, the philosophical underpinning of each approach as well as the ontological and epistemological considerations of the qualitative inquiry under question. As demonstrated, using different variants of grounded theory or phenomenology in one research have severe impact in the overall research design of the study and in terms of the theoretical contribution of the study. Indeed, arguing in a similar line, Stumpf et al. (2016) urged researchers to be transparent in terms of the version that are being used. Therefore, using an established grounded theory and phenomenology variant is important for providing structure to the theory development process and legitimizing the study. In other words, the chosen research inquiry should give the researcher the best opportunity to develop theory considering individual preference (i.e., epistemological, and ontological philosophy) and the research question being addressed. Hence, this will necessitate that researchers have appropriate knowledge. It is also advisable that authors consider including a couple of sentences on how and why the selected qualitative inquiry was utilized in the research was chosen, and how it supports the research question(s).

Last but not the least, there is a need to use alternative and innovative approach. Moving beyond the first three period postulated by Denzin and Lincoln (2011, 2018) will require the need to engage and have more acceptance towards innovative forms of collecting, analyzing, and writing of qualitative studies. There is a need to have a variety of textual, visual, and haptic method that can be useful in studying for instance, topic related to tourist experience. Further, there is a need to embrace the vast methodological possibilities that virtual, online environments present.

Conclusion

Tourism and hospitality researchers now recognize the legitimacy of qualitative research for progressing the field. This chapter provides a historical overview of how qualitative research has evolved as well as the challenges faced in qualitative inquiry method such as phenomenology and grounded theory. As highlighted, there is a need for researchers to be competent analysts of the multiple ways available for analyzing qualitative data, as this is likely to improve the rigor of the study (Lincoln & Guba, 1985). Likewise, being aware of the different variants available and the epistemological and ontological underpinning the method is crucial. While recognizing the fact that qualitative research in the field of tourism and hospitality is still developing, this chapter calls for qualitative researchers to move freely between disciplinary boundaries and assumptive frameworks as it is at the core to advance qualitative method. There is indeed much that can be learnt from the field of social sciences such as anthropology, sociology and psychology where qualitative inquiries are well established.

References

Charmaz, K. (2006). *Constructing grounded theory: A practical guide through qualitative analysis*. Sage.

Clarke, A. E. (2005). *Situational analysis: Grounded theory after the postmodern turn*. Sage.

Creswell, J. W. (2007). *Qualitative inquiry and research design: Choosing among five approaches*. Sage.

Decrop, A. (1999). Triangulation in qualitative tourism research. *Tourism Management*, 20(1), 157–161.

Decrop, A., & Masset, J. (2018). Building knowledge from the tourism field: the grounded theory approach. In R. Nunkoo (Ed.), *Handbook of research methods for tourism and hospitality management* (pp. 136–156). Elgar.

Denzin, N. K., and Lincoln Y. S. (1994). Introduction: Entering the field of qualitative research. In N. K. Denzin & Y. S. Lincoln (Eds.), *Handbook of qualitative research* (pp. 1–7). Sage.

Denzin, N. K., & Lincoln, Y. S. (2005). *The Sage handbook of qualitative research*(3rd ed). Sage.

Denzin, N. K., & Lincoln, Y. S. (Eds.). (2011). *The Sage handbook of qualitative research*. Sage.

Denzin, N. K., & Lincoln, Y. S. (2018). Introduction. In N. K. Denzin & Y. S. Lincoln (Eds.), *The Sage Handbook of Qualitative Research* (5th ed., pp. 1–26) Sage.

Dimitriadis, G. (2016). Reading qualitative inquiry through critical pedagogy: some reflections. *International Review of Qualitative Research*, 9(2), 140–146.

Duignan, M., Pappalepore, I., Smith, A., & Ivanescu, Y. (2022). Tourists' experiences of mega-event cities: Rio's Olympic 'double bubbles'. *Annals of Leisure Research*, 25(1), 71–92. https://doi.org/10.1080/11745398.2021.1880945

Geertz, C. (Eds.) (1973). *The Interpretation of Cultures: Selected Essays*, Basic Books.

Glaser, B. (1978). *Theoretical sensitivity*. The Sociology Press.

Glaser, B., & Strauss, A. (1967). *The discovery of grounded theory: Strategies for qualitative research*. Aldine.

Goolaup, S., & Mossberg, L. (2017). Exploring the concept of extraordinary related to food tourists' nature-based experience. *Scandinavian Journal of Hospitality and Tourism*, 17(1), 27–43.

Goolaup, S., & Nunkoo, R. (2021). Reconceptualizing tourists' extraordinary experiences. *Journal of Travel Research*, 00472875211064632.

Goolaup, S., & Solér, C. (2018). Existential phenomenology: Understanding tourists' experiences. In R. Nunkoo (Eds.), *Handbook of research methods for tourism and hospitality management*. Edward Elgar Publishing.

Goulding, C. (2005). Grounded Theory, ethnography and phenomenology. *European Journal of Marketing*, 39(3/4), 294–308.

Hindley, A., & Font, X. (2018). The use of projective techniques to circumvent socially desirable responses or reveal the subconscious. In R. Nunkoo (Ed.), *Handbook of Research Methods for Tourism and Hospitality Management* (pp. 202–211). Edward Elgar Publishing.

Ivanova, M., Buda, D. M., & Burrai, E. (2021). Creative and disruptive methodologies in tourism studies. *Tourism Geographies*, 23(1–2), 1–10.

Jiménez-Barreto, J., Rubio, N., Mura, P., Sthapit, E., & Campo, S. (2022). "Ask Google Assistant where to travel" tourists' interactive experiences with smart speakers: An assemblage theory approach. *Journal of Travel Research*, 00472875221094073.

Jones, R., & Noble, G. (2007). Grounded theory and management research: a lack of integrity? *Qualitative Research in Organizations and Management: An International Journal*, 2(2), 84–103.

Kirillova, K. (2018). Phenomenology for hospitality: Theoretical premises and practical applications. *International Journal of Contemporary Hospitality Management*, 30(11), 3326–3345.

Lancy, D. F. (1993). *Qualitative research in education: An introduction to the major traditions*. Longman.

Lincoln, Y. S., & Guba, E. G. (1985). *Naturalistic inquiry*. Sage.

Lumsdon, L. M., & McGrath, P. (2011). Developing a conceptual framework for slow travel: A grounded theory approach. *Journal of Sustainable Tourism*, 19(3), 265–279.

Marschall, S. (2015). 'Travelling down memory lane': Personal memory as a generator of tourism. *Tourism Geographies*, 17(1), 36–53.

Matteucci, X., & Filep, S. (2017). Eudaimonic tourist experiences: The case of flamenco. *Leisure Studies*, 36(1), 39–52.

Matteucci, X., & Gnoth, J. (2017). Elaborating on grounded theory in tourism research. *Annals of tourism research*, 65, 49–59.

Mura, P., & Khoo-Lattimore, C. (Eds.). (2018). *Asian qualitative research in tourism: Ontologies, epistemologies, methodologies and methods*. Springer.

Mura, P., & Pahlevan Sharif, S. (2015). The crisis of the 'crisis of representation'–Mapping qualitative tourism research in Southeast Asia. *Current Issues in Tourism*, 18(9), 828–844.

Nunkoo, R., Thelwall, M., Ladsawut, J., & Goolaup, S. (2020). Three decades of tourism scholarship: Gender, collaboration, and research methods. *Tourism Management*, 78, 104056.

Ogle, A. (2018). Sensual Quasi-Q-Sort (SQQS): enriching qualitative hospitality and tourism research via the human senses. In R. Nunkoo (Ed.), *Handbook of research methods for tourism and hospitality management*. Edward Elgar Publishing.

Pernecky, T., & Jamal, T. (2010). (Hermeneutic) phenomenology in tourism studies. *Annals of Tourism Research*, 37(4), 1055–1075.

Phillimore, J., & Goodson, L. (2004). Progress in qualitative research in tourism. In J. Phillimore & L. Goodson (Eds.), *Qualitative research in tourism: Ontologies, epistemologies, methodologies* (pp. 3–29). Routledge.

Podoshen, J. S., Yan, G., Andrzejewski, S. A., Wallin, J., & Venkatesh, V. (2018). Dark tourism, abjection and blood: A festival context. *Tourism Management*, 64, 346–356.

Pritchard, A., Morgan, N., & Ateljevic, I. (2011). Hopeful tourism: A new transformative perspective. *Annals of Tourism Research*, 38(3), 941–963.

Ren, C. (2016) Qualitative research, tourism. In Jafari, J., & Xiao, H. (Eds.), *Encyclopedia of Tourism* (s. 1–4). Springer Publishing Company.

Riley, R. W., & Love, L. L. (2000). The state of qualitative tourism research. *Annals of Tourism Research*, 27(1), 164–187. https://doi.org/10.1016/S0160-7383(99)00068-7

Spurling, L. (2013). *Phenomenology and the social world: The philosophy of Merleau-Ponty and its relation to the social sciences*. Routledge.

Corbin, J. M., & Strauss, A. (1990). Grounded theory research: Procedures, canons, and evaluative criteria. *Qualitative Sociology*, 13(1), 3–21. https://doi.org/10.1007/BF00988593

Stumpf, T. S., Sandstrom, J., & Swanger, N. (2016). Bridging the gap: Grounded theory method, theory development, and sustainable tourism research. *Journal of Sustainable Tourism*, 24(12), 1691–1708.

Thompson, C. J., Locander, W. B., & Pollio, H. R. (1989). Putting consumer experience back into consumer research: The philosophy and method of existential-phenomenology. *Journal of Consumer Research*, 16(2), 133–146.

Tracy, S. J. (2013). *Qualitative research methods: Collecting evidence, crafting analysis, communicating impact*. John Wiley & Sons.

Tung, V. W. S., & Ritchie, J. B. (2011). Exploring the essence of memorable tourism experiences. *Annals of Tourism Research*, 38(4), 1367–1386.

Urquhart, C., Lehmann, H., & Meyers, M.D. (2010). Putting the 'theory' back into grounded theory: Guidelines for grounded theory studies in information systems. *Information Systems Journal*, 20(4), 357–381.

Wilson, E., & Hollinshead, K. (2015). Qualitative tourism research: Opportunities in the emergent soft sciences. *Annals of Tourism Research*, 54, 30–47.

Wilson, E., Mura, P., Sharif, S. P., & Wijesinghe, S. N. (2020). Beyond the third moment? Mapping the state of qualitative tourism research. *Current Issues in Tourism*, 23(7), 795–810.

Ziakas, V., & Boukas, N. (2013). Extracting meanings of event tourist experiences: A phenomenological exploration of Limassol Carnival. *Journal of Destination Marketing and Management*, 2(2), 94–107.

Chapter 10

Historiography, Tourism and Tourism Marketing

Kerryn Lyes

Doing marketing history requires a mind opening, allowing awareness of change over time to take a large role (Fullerton, 2011, p. 446).

Introduction

Etymologically, the word history is rooted in 'story' as well as the Greek verb *historie*, meaning 'to see' (Tucker, 2004). Significantly, historical enquiry has been adopted across many of the social sciences with varying degrees of focus on content, context, or methodology (Bailyn, 1982). Considering the proliferation of historical writing, the sheer scope of possible interpretations brings with it a risk of the obscuration of historical movements, events, and motives, rather than the illumination historical writers intend to uncover when digging through archival materials and artifacts (Bailyn, 1982). Conversely, even fictional works offer some level of historical value with regards to how people work, buy, sell, or travel. Take an historical novel, like Steinbeck (1998) *The Grapes of Wrath*, which weaves a familial narrative within the collective themes of the Great Depression. While it is historical fiction, it is characterised by a quality of hyper-reality – it is viscerally painful and honest. And it is dynamically representative of a particular time and place.

The topic of tourism history is vast and mutable (Gordon, 2015), however the direct application of historiography in tourism is relatively limited. With tourism histories emerging across disciplines ranging from archeology to architecture (Gordon, 2015), historiography holds relevance to the cultural and cross-disciplinary aspects of tourism research. Existing tourism histories are relatively Euro-centric, while the *Journal of Tourism History*, founded in 2009, represents a growing interest in global tourism histories (Gordon, 2021). To date, historical methods have been employed to explore the relational dynamics between tourism activities and other areas of cultural influence such as sports (Huggins, 2013), migrant food districts (Innis-Jiménez, 2018), or political influences on the development of tourism markets

The Wiley Blackwell Companion to Tourism, Second Edition. Edited by C. Michael Hall.

(Afinoguénova, 2014; Gordon, 2015). One such case explores how the establishment of a tourism board may influence regional cuisine (Afinoguénova, 2014). Even if in an allegorical sense rather than a direct fact-based approach, there is some form of knowledge or understanding imbued simply by virtue of the written word. Like many readily assimilated cultural influences or artifacts (Marwick, 1989), the institutions, systems, and physical premises in which tourism activities take place were commonly built in the past with the needs of the people of the past in mind. In a tourism context, there are further considerations at play. Take the touristic use of an historic site (Austin, 2002), the use, management and marketing of which will be inherently embroiled with local and international politics, legislation, and historical events relating to special interest or landmark status.

While there are formal tourism histories, there are many other instances in which historical methods are implied in a tourism context, sometimes unknowingly. For example, the focus on understanding the role of tourism in economic, environmental, and social change often necessitates the development of micro-histories of tourism development while demands for more longitudinal research in tourism also require greater historical appreciation (Sæþórsdóttir & Hall, 2020). Similarly, archival research and fieldwork will also necessitate historical research skills (Hall, 2010) as will accounts of the development of tourism as a field of study or its subfields (Smith, 2010; Wilson, 2012; Hall, 2013; Müller, 2019, 2021). While studies on historical methodology in tourism are very limited, tourism studies are anything but ahistorical. Beyond simply seeing how 'where we were' informs 'where we are' in terms of our understanding of tourism, what other findings, pitfalls, and challenges await those seeking to understand tourism histories?

While researchers may employ the methods discussed here to interrogate events of the past, the present discussion is primarily concerned with the philosophies involved in the use of historical method in tourism, and tourism marketing research in particular, given its relative dominance over much of the tourism studies field in both approach and citation. Furthermore, and as is widely acknowledged, the concerns of tourism scholars are often shared across sister disciplines, with many studies on tourism being published in 'non-tourism' journals. Therefore, while this chapter will likely be of most interest to those engaged in tourism marketing, the implications of historical method in marketing should be relevant to most tourism scholars. As a relatively forward-facing discipline with a tradition of prioritising recency (Tadajewski & Saren, 2008), the status quo in broader marketing discourse is often centered around cross-sectional findings (Rindfleisch et al., 2008), or problem-solving for a 'brighter' future (Sheth, 2021), an approach shared with much of tourism studies. This brings to light the need for deeper considerations of what the temporal, contextual, and historical contents of tourism marketing phenomenon might contribute to the discipline. Therefore, adopting the principles of a problematising review, as opposed to an integrative one, this chapter is an attempt to undertake a close reading of some existing historical literature to 'open up' rather than 'build' a picture of history in marketing (Alvesson & Sandberg, 2020) as a frame for history enquiries in tourism.

Marwick (1989, p. 398), presents historiography as 'the systematic study of historians' interpretations of (or writings about) the past'. In this vein, an historiographical approach is often backed by the intention of assessing the value of historical works for present day research (Becker, 1938). As such, historiography is not necessarily a direct history of an event, object, or school of thought, and may also include adjacent areas of influence, confluence, or impact:

> Historiographic interpretations include knowledge generated by scientific historiography, but also ethical, aesthetic, political and other value judgements (Tucker, 2004, p. 2).

Considering this, historiography may be appropriate for answering research questions like: What are the enduring characteristics of the writing of tourism history? Or what are the strengths and blind spots of historical method(s)? In this chapter, historiography is employed as a vehicle to consider the myriad ways in which historical contribution has been both conducted and considered in the broader canons of marketing, tourism, and social scientific thought, as well as covering examples and potential applications in tourism and tourism marketing. The contrasting approaches to history in marketing discussed here are in many cases informed by the same historical texts, artifacts, and contexts in ways that lead to often conflicting conclusions regarding the meaning of said artefacts, destinations, and associated events (Tucker, 2004). By both virtue of the historical methods and marketing and tourism thought's emphasis on written publications, as well as consideration of the parameters of an historiography, this chapter also includes a discussion on the aesthetics of historical writing within marketing and tourism. Alternate methods of framing, collecting, and disseminating historical findings are also discussed. With these elements in mind, along with a consideration of how historical writing fits in to the wider marketing literature, resistant curiosity is employed through iterative questioning, an intellectual stance that is decidedly pliable (Tadajewski, 2022a) and with the ideal of reflexivity positioned as central to the review process (Alvesson & Sandberg, 2020).

Methodological Show or Tell?

Fullerton (2011), a trained-historian-turned-marketer, iterates that his historical training did not include a symposium in methodology – rather, 'methodology should be used, not discussed' (Fullerton, 2011, p. 437). This approach puts greater emphasis on smooth writing, careful transitions, and the use of primary sources – original documents, ephemera, artifacts, as the main input into historical scholarship. Schwarzkopf (2015) argues that historical method privileges the perspectives of managers and academics, with limited scope to consider the perspectives of buyers. There are also calls to incorporate a wider breadth of sources such as archival data and oral histories (Crawford & Bailey, 2018) or photographical histories (White, 1988). While oral history may be further adopted academically (Crawford & Bailey, 2018; Trapp-Fallon, 2003), there is also some use of oral history in tour commentaries (Roberts, 2014). There is further scope for comparison between existing tourism research with richer consideration of political, social, or environmental complexities (Walton, 2005).

Methodological Perspectives

> Evaluating scientific work is no easy task. Those who claim they can provide a strict system are misled. They can't; science as a reflection of 'reality' and 'truth' is too complex and varied. (Fuat Fırat, 2010, p. 438).

There are at least three markedly differentiated *general* methodological approaches for historical research in marketing as presented by Savitt (1980), Smith and Lux (1993), and Golder (2000). These streams of historical thought are summarised in the following section with papers outlining method discussed in chronological order, ending with more recent approaches such as the instructions for critical historical method being presented by Yan and Hyman (2018).

In 1980, Savitt put forward a case that historical marketing research was being predominantly undertaken outside of marketing scholarship, partially due to a lack of well-delineated method for applying an historical lens to marketing problems (Savitt, 1980). According to Savitt (1980), historical method entails a process of both description and comparison. In this case, history in marketing thought accounts for the adoption of theories from disciplines such as psychology, sociology, or even law, as well as opening the possibility for further cross-disciplinary enquiry. Savitt (1980) demonstrates this aspect in the following excerpt drawing on a legal approach to marketing.

> The legal approach to the study of marketing, even though it is not well developed, provides insights into what practices are acceptable and insights into the evolving legal structure which surrounds marketing decision making. The failure of a contract to meet the test of legality through litigation provides much understanding about the ways in which marketing practices evolve, though it would be incorrect to accept the test of legality as a measure of the success or failure of a specific practice (Savitt, 1980, p. 56).

In *Historical Research in Marketing,* Savitt (1980) posits that historical analysis provides the discipline with a level of robustness that is otherwise lacking. In Savitt's (1980) approach, the use of historical analysis may be historical predominantly in the sense that it employs chronology while also extending to analysis and comparison. In this view, historical examination provides a contextual basis for understanding the roots, qualities, and patterns of change within the discipline as a whole (Savitt, 1980). However, Savitt (1980) does offer the caveat that undertaking historical research requires a certain level of aptitude in dealing with the treatment of time:

> What is required of the historian is recognition of the complexities in the meaning of time. This important topic is beyond the scope of the article; however, anyone undertaking historical studies should become acquainted with the problems (Whitrow, 1975) (Savitt, 1980, p. 53).

In the passage above, Savitt (1980) cites Whitrow (1975) as recommended reading for marketing historians to introduce the conceptual complexities of dealing with time. Remarkably, a citation search for papers within marketing which cite Whitrow's (1975) *The Nature of Time* returns no more than 10 results, while Savitt's (1980) paper returns more than 300 citations (Google Scholar retrieved, 12 June 2022). While by no means the definitive authority on time, Whitrow (1975) does provide a relatively accessible discussion of time that includes both physical and philosophical intimations as to the 'nature' and significance of time with consideration of the value in both logical and speculative approaches.

Savitt's (1980) prescribed approach to historical methods in marketing rests on an explicit assumption that marketing is the discipline which describes pre-purchase and post-purchase activities in relation to transactions of ownership, usage, or use rights (Narver & Savitt, 1971). Savitt (1980, p. 53) elaborates 'A historical study in marketing would focus on the elements of want satisfaction, rather than on resource allocation as in economics'. Given the distending of marketing thought following the broadening movement (Kotler, 2005), this warrants further consideration. Historical analysis may look inherently more complicated when marketing concepts and value exchange are explicitly (not just incidentally) prescribed in political, moralistic, and even cultural spheres. This also points to the potential for investigating and compiling marketing histories in a more direct sense, drawing on archival materials or even internal documents that may still be in use. However, there is also the consideration of legal factors, commercial sensitivity, and the protection of 'trade secrets' that in real time may prevent inquiry into marketing practice, but after a considerable passing of time these documents may be made available, through the release by an estate, or even post-mortem publishing.

In the following passage, Savitt (1980) places emphasis on historical research as a means to enable 'problem solvers' with better tools for decision making through establishing theories alongside methods for application:

> As an applied discipline, marketing must cater to its client market of decision makers. Their concerns are directed toward, understandably, making better decisions in today's market. Looking backward to them is a luxury. Moreover, the discipline itself has had to absorb an ever-increasing set of new concepts and methods and then establish appropriate ways to apply them to the problems at hand (Savitt, 1980, p. 54).

According to Savitt (1980), most of the historical research in marketing has been centered around marketing thought with less attention being directed at marketing practice. This is particularly significant considering the role of primary sources in historical research, as well as the opportunity to create greater synergy between marketing thought and practice. A theme that runs though the historical marketing literature is that there is a dynamic interplay between marketing history, thought, and practice rather than a clean division in cause and effect (Domegan, 2010; Savitt, 1980; Shapiro, 2011; Shaw & Tamilia, 2001; Wooliscroft, 2011).

While experimentation may not be applied in history as it is applied in the biological and physical sciences, or even some social sciences, Savitt (1980) outlines a process for assessment and mediation of the researcher's levels of subjectivity before perceptions are compared with the real world. Of course, those with a more post-modern view may challenge that the 'real-world' comparison takes place through the same perception as the review process. In one stroke, the controlled 'lab' environment becomes both further removed from the 'marketing reality', while also avoiding a layer of seemingly 'un-scientific' researcher interpretation:

> Greater reliance is placed on archival materials, corporate records, and public and legal documents as well as business, economic, social, and urban histories. Artifact evidence including buildings and relics also is used. Attention must be given to the distinction between primary and secondary sources as a means of *minimizing the gap between the actor and the observer* (emphasis added). Drawing a precise line between primary and secondary sources is difficult in historical research. Simply because a document is old does not mean that it is primary. On a spectrum from 'purely primary data' to 'purely secondary data' there is a greater overlap for historical data than for contemporary data. The marketing scholar can in part minimize this problem by attempting to view historical data as if he or she were a contemporary with them. This is no mean task and it requires that the researcher have some knowledge about the period of the study (Savitt, 1980, p. 55).

While Savitt (1980) presents what appears to be a more systematic, less interpretive approach – the above passage essentially advocates for a sort of imaginative-time-travel-roleplay in which the researcher is advised to attempt to view an archival document as if they were a contemporary of the original author. This points to both the care required for consulting written, visual, or verbal records, as well as the integral part played by researcher understanding. Just as in a legal hearing, the verbal testimony differs from the written record (Savitt, 1980) – the fidelity of information is changed at a subtle level, through dimensions of tone, expression, or even timing.

By contrast, Smith and Lux (1993) describe historical research as a 'qualitative interpretive' method that can be applied to theory development, theory testing, or the explanation of anomalies in existing literature or historical artefacts. In this approach, rather than devising research questions and collecting data, the research question must be successively adapted and molded to fit the evolving understanding of the historical data being reviewed (Smith & Lux, 1993). They list four 'nominal' tools of historical method including 'chronicles, analyses of continuity and trends, analyses of change, and methodology' (Smith & Lux, 1993, p. 596) and put forward that many historians refuse to come to a lawlike generalization. In this approach, history is presented as placing emphasis on complexity of events rather than simplicity (Smith & Lux, 1993). Smith and Lux (1993) also list use-cases in which historical method has been employed to debunk overstated findings from previous research. Smith and Lux (1993) present a dichotomy of historical research consisting of change and continuity. On assessing six methodological papers that existed during the time of their review, Smith and Lux (1993) lament a restriction in application of historical method based on the positions of historical researchers at the time, particularly in terms of knowledge value or epistemology.

> Savitt's (1980) article (published before the 1983 starting date of our literature review) remains the most complete and general methodological exposition available for consumer researchers. But, as will become clear in the following section, his equation of historical analysis and scientific method severely limits history's usefulness while also straining the limits of its ontological and epistemological capabilities. Methodological articles published since Savitt reveal a narrower focus in that they discuss specific philosophies of history (Fullerton 1987) or describe how generalized historical methods might be applied to a specific consumer or marketing question (see Kumcu 1987; Pollay 1987; Rassuli and Hollander 1987). To date, however, no authors have undertaken a comprehensive, general explication of the potential usefulness of historical method for consumer research (Smith & Lux, 1993, p. 597).

Conversely, having boundaries for enquiry can be extraordinarily useful in distilling a piece of work into a usable, comprehendible contribution to knowledge. However, there is a level of discernment required in selecting constructive rather than constrictive boundaries for enquiry, which again points to the value in researcher reflexivity and the deep work associated with refining ones' point of view (Tadajewski, 2022a).

Thirdly, Golder's (2000) paper on historical method begins with a discussion over the 'scientific grey areas' of both marketing and historical methods – with some advocates presenting both disciplines as scientific as being representative of an empirical, objective, materialist approach. Golder's (2000) seminal paper on historical method (Thompson, 2010) indirectly demonstrates how the history of marketing thought appears to be entangled with efforts towards legitimising marketing as a science, or a social science. The paper outlines a five-step process which presents historical methods for assessing historical evidence as well as the validity (or as it turns out invalidity) of a particular finding. In this sense, the steps of historical method are designed to draw conclusions from the data, not preconceived notions. In Golder's (2000) method, emphasis is placed on prioritising relevancy over what is interesting, due to both the proliferation of available data as well as the risk of researcher-overwhelm (Golder, 2000). Golder (2000) also lists the limitations of his historical method as being centralised around the ability to assess and generate causal theories:

> First, because the historical method evaluates phenomena in their complex real-world environment, developing and confirming theory precisely can be difficult. However, even when a single study does not generate theory, it can contribute to new theory. . . A second limitation of historical method is that it is more capable of descriptions than of explanations or predictions. . . A final limitation of the historical method, which prevents its use by most researchers, is that it is very time consuming (Golder, 2000, p. 167).

The passage above is reflective of prior debates in marketing research (Arndt, 1985; Carman, 1980; Deshpande, 1983; Hunt & Speck, 1985; Lemel, 1993; Lutz, 1989; Reidenbach & Robin, 1991) and more broadly in the social sciences (Heyl, 1975; Willis, 2007) in which there appears to be some underlying rhetoric that an emphasis on narrative aligns one with an interpretivist approach. However, historical research at least in part serves as a reminder that the astounding contributions to knowledge attained through scientific methods and materialistic worldviews do not entirely negate the social nature of academic endeavors – and neither is the inverse necessarily true. One purported strength of the historical method is the ability to triangulate findings in primary sources by incorporating sources longitudinally. As such, critically revisiting historical research may also serve to pressure-test the foundation of existing theories — a task which seems particularly relevant in a period of perceived market, economic, emotional, and climate volatility (Brown et al., 2010) there is a growing collective requirement for anti-fragile systems and robust ideation (Taleb, 2013).

While Golder (2000) proposed historical method may be used to both confirm established theory or construct new theory from new analyses of preexisting data, the tools of historical research may be equally useful for interrogating received theories. However, Golder (2000), also cautions against extrapolating second-order meaning from historical sources stating, 'The process of interpretation becomes complicated when historians try to provide a higher level of meaning for a collection of events' (Golder, 2000, p. 161). Fullerton (2011) also stresses the importance of careful reading with awareness of potential distortions:

> When reading primary sources it was important to read them critically since much writing – whenever it is done and by whomever it is done – is self-serving and distorted, or incomplete, or it advocates a definite point of view. By reading critically, I mean challenging the authors' viewpoints, trying to determine what viewpoint they held and why, and assessing just how believable their explanations are. One should expect differences, but one has to try to reconcile them – or accept them as presenting a complex contradiction-filled reality – or completely discard one or another. A distorted viewpoint can still be usable as long as the historian knows that it is distorted' (Fullerton, 2011, p. 439).

This stressing of criticality is further discussed in Yan and Hyman's (2018) recent paper on critical historical method which posits critical historical research should be adopted to probe longitudinal change (or lack of change) within the discipline. Their proposed method attempts to build a case for contexualising research using methods that treat marketing phenomenon as part of dynamic and open systems (Yan & Hyman, 2018).

The Tourism Marketing Historians Toolbelt

> For researchers in hospitality and tourism seeking to expand their range of research methodologies, historical approaches offer the dual benefit of both transparency in data and methodological rigour allowing for a more complete research approach (MacKenzie et al., 2020, p. 1482).

Historical Case Study or 'History of. . .'

As demonstrated by Fagan (2018), deeply contextual case studies can draw on primary sources and historical fact, while also incorporating elements of narrative or storytelling. In a history of rock climbing in Kentucky's Red River Gorge, Maples (2021) incorporates elements of natural and social history in the development of rock climbing tourism in a particular location. Using a broader social context, Clark and Lopez (2014) detail the emergence of tourism in colonial Victoria, Australia. Both these cases exemplify the resource heaviness of historical method (Fullerton, 2011), with a thorough historical case study perhaps being more suited to book-length pieces than journal articles.

Periodization

A conceptual linkage of distance through time and the qualification of history is mediated by proximity. Last week does not feel like history, and yet through other framings it is as equally in the past as the 1800s — this leads to surprisingly controversial use of periodisation, both around the periods named and the knowledge value provided by the technique. Periodisation may be used incidentally (Hollander et al., 2005), but it can also provide structural parameters for a particular piece of historical writing such as in Jones and Shaw's (2002) 'A History of Marketing Thought', which periodises from Ancient and medieval marketing through to post-Alderson, modern marketing thought. Periodisation segments marketing history into digestible pieces, which in turn may facilitate a deeper contextualisation of changing and enduring aspects of marketing theory as anchored in the time, events, locations, and inflection points in the development of ideas (Domegan, 2011).

Time-series and Citation Analysis

Through Google Scholar and other tools like Scopus and Web of Science, citation analysis is a relatively straightforward way to see how many times a particular article has been referenced, but from a critical perspective would need to be accompanied by an understanding, or at least investigation of the structural and social influences on the adoption, or rejection, of the ideas in question (Hall & Page, 2010; Yan & Hyman, 2018).

Biography

Biographies are often undertaken after an author's death (e.g. see Hall & Page (2010), on Neil Leiper), or when they are no longer active 'in the field', which, in turn, influences how conclusions are drawn in relation to the proximal distance from the life and events being explored. However, a number of biographical profiles have been developed in tourism as both a way of 'celebrating' their work, e.g., the profiles regularly occurring in the journal *Anatolia*, or as part of an examination of contribution to a field. Savitt (2011) begins his instructions for biographical method with a treatise on the importance of weaving biographical writing with elements of brevity, style, and intrigue, to draw the researcher and the prospective reader into the lives of the subjects in a meaningful and compelling way.

Autobiography and Self-reflection

While autobiography or self-reflection may not have been explicitly called out as an historical method, it still presents a method to contextualise discussion in time and place, while maintaining close, or even absolute, proximity to the source of information. In a detailed and relatively unconventional paper Gould (1991), provides a 'thick' description of how they manage personal sexual drive and energy using changes in habits of consumption. Both the level of 'gory' detail and visceral response elicited by reading translate the qualities of a particular consumption context in a way that would be difficult to achieve via experiment, survey, or interview participants. This example also draws on a significant level of diversion from the methodological norm, an approach that has not been widely replicated or applied on the other side of the marketer/consumer dyad. In the tourism context, this raises further questions regarding the variations in positionality between researchers, visitors, and residents — consider how the approach to studying a tourist attraction might differ when written by a visitor or resident to the locality.

Pitfalls of Historical Analysis

The degree of attachment or detachment of the scholar to the field appears to influence their approach – an historian studying tourism marketing presents a different view to a tourism researcher employing historical method. Brown (2011), laments a stylistic shortfall in marketing literature, presenting it as a high output, low insight research domain with the target benefactors being represented by amorphous stakeholders. Brown (2011) goes on to advocate for historical method(s) to uphold a certain level of stylistic readability and enjoyment in the literature. On a more technical level, Fullerton (2011) warns against the 'sin of anachronism' – or reading the present into the past. This also highlights how researchers are socialised into a particular field of thought, and the importance of incorporating methods that encourage a productive line of questioning beyond simply reinforcing existing understandings (Tadajewski, 2022a). The historian, as demonstrated by Fullerton (2011) is aiming to hear the voices of the past, while the marketing historian is often, not always, aiming to demonstrate findings through analysis of historical information (Golder, 2000; Savitt, 1980). Fullerton (2011) speaks to the dual process of both collating facts and abstracting an interpretation from historical materials, with the abstraction of meaning from materials providing greater scope for conceptual wondering, or overdrawn conclusions.

One issue with the hard separation between fact driven and interpretive approaches is a limited understanding of one's own point of view, a limitation on the clarity, and consequential impact of an individual piece of historically informed writing. Overall, whether with a critical or celebratory stance, historical method in marketing research seems at least in part focused on the requirement to look a little deeper, a little closer, and a little longer at the perspectives of our scholarly predecessors and the traditions of business scholarship. Historical methods also provide a sober and nuanced approach for addressing commonly subjugated or disregarded concepts that go against the dominant narrative, as in Tadajewski's (2022b) enquiry into the role of telepathy in marketing as influenced by new thought.

As mentioned in the prior discussion, a rhetoric for scientising marketing and tourism phenomenon is embedded in historical marketing research methods (Golder, 2000; Savitt, 1980; Yan & Hyman, 2018) and in a sense creates a, perhaps unintentional, disregard for non-material principles, or less tangible forms of meaning creation and dissemination. Other voices in historical marketing research are building a case for why interesting research (to conduct and to read) is important — because it is enriching, inspiring and brings a humanitarian quality to the work for both students and academics (Domegan, 2010; Shaw, 2015). At a basic level, marketing, and indeed tourism marketing, is partially about resolving practical matters of human existence – what, where, and how to eat, live, celebrate, explore, travel, and individuate. In the following passage, Yan and Hyman (2018) warn against treating historical research objects as if they exist separately from social structures.

> Any procrustean methodology that presumes all congeneric research objects share identical meanings and implications regardless of cultural, structural, or mechanical context is problematic. Instead, critical HRM can clarify marketing's multidimensional, holistic, and dynamic characteristics by substantiating market evolutions ..., social institutions that mould and instigate consumers' behaviours ..., and revised social parameter estimates ... (Yan & Hyman, 2018, p. 857).

However, this does present somewhat of a quandary in distinguishing marketing as a social science as operating under different parameters from the physical sciences, while also co-opting bio-centric vocabulary like 'congeneric' and 'evolution'. The epistemic preference towards scientising also comes through indirectly, for example, Zaltman (1982) employs the use of a David Bohm quote at the beginning of a book on theory construction in marketing. This may seem relatively arbitrary but could also hint at a sign of respect for a physicist's approach to knowledge as an ideal for marketing scholars to strive for.

Lawson and Wooliscroft (2004) point to the risk in oversimplifying our models and disregarding the dynamism that is inherent in any facet of human activity. In this sense, historical analysis provides an openly-cross-disciplinary approach to engaging with marketing thought and the history of its practice in tandem, rather than in isolation. Like history in a wider sense (Eley, 2008), despite every attempt at being objective, scientific, or even clinical – both marketing and marketing research are to some degree entwined with matters of culture, politics, and even public opinion. This is not a critique on the value of at least some degree of epistemic objectivity, nor it is a statement as to whether an absolute truth does or doesn't exist. While scientific annotations, calculations and descriptions are accessible to highly trained specialists, the associated meaning, outcome, or actionable insight may be translated into popular science through the written word (Gregory, 1988). The same could be said for history. In some sense, formal training, reading, experimentation, and life experience may influence our reading comprehension of a particular piece of work.

An academic who is also running a business will have a different view of the 'marketing reality' or 'tourism reality' than one who is firmly positioned on only one side of the theory/practice dichotomy. However, the supplied methodologies for historical research in marketing reflect the paradigm debates that emphasise the philosophies of science within social science (Carman, 1980; Heyl, 1975; Lowe, 2001). This 'pick your side' approach is most broadly represented by the dichotomy of positivist/interpretivist or discussed with high granularity regarding the subtleties of each point of view. However, scientific endeavours as typified by a systematic and empirical approach, are still commonly *communicated* largely via the written word (Gregory, 1988). Both practically and intellectually, historical methods are both labour and time intensive (Golder, 2000; Tadajewski & Jones, 2014), without necessarily producing a replicable framework or conceptual model of the kind that is reified in tourism marketing management research. Historical research in marketing may be messy, difficult, time-consuming (to both read and write), but nonetheless holds significant promise for tourism marketing scholars to create scholarly works that are much more illustrative, contextually nuanced, and in turn contemporarily applicable. While cited with the intention of maintaining methodological rigour, existing approaches may disregard some of the nebulous, imaginative elements involved in creating brands, marketing products, or even initiating movements.

Uses and Contributions of Historical Method(s)

Historical analysis may also provide a capacity for interrogating our domain's capacity for self-deception or 'unreliable narration' regarding:

a. what significant contribution is;
b. what contributed to the current state of marketing research, practice, and pedagogy; and
c. what other options are available.

By drawing on temporal influence, historical methods have a capacity to provide contrast to teleological descriptions in marketing textbooks in which marketing phenomenon are treated as a function of the ends they serve, rather than the cause by which they arise (Tadajewski & Saren, 2008). As demonstrated by Lawson and Wooliscroft (2004), one contribution of the historical method is the use of intellectual histories to demonstrate the foundational '-isms' involved in areas of marketing thought such as determinism and functionalism. In this sense, historical methods may serve to illuminate foundational ideas on which tourism marketing theories have been built, as well as how tourism marketing sits within marketing thought and within broader philosophical movements and intellectual fashions the social sciences. Incorporating an example from his own historical research, Fullerton (2011) describes how descriptions of the consumer misconduct attributed to kleptomania remained fairly consistent through time, the attributed cause behind the phenomena changed significantly according to what psychological theories were au fait in a particular given time and place. Historical elements may also be employed incidentally (not directly called out as method or methodology) to add depth to analysis, or to illustrate oppositional approach. For example, Lawson and Wooliscroft (2004) disclose biographical elements, such as the parental relationships, of Hobbes and Rousseau to illustrate their contrasting positions on the underlying characteristics or qualities of human nature. This approach, in line with Fullerton (2011), makes for a much smoother and more engaging read than a standalone review of their respective bodies of work.

Both in theory and practice, tourism and marketing history appears to be informed by the supposition that public and private perception and decision making is malleable, that there is a cumulative effect of instilling a particular belief or behaviour on a selected public, or target audience. This can be seen through the influence of the work of Edward Bernays (Tadajewski, 2022b), regarding the increased attention paid to public opinion and the degree to which both personal and public opinions are influenced by outside sources 'the most obvious of these forces are parental influence, the school room, the press, motion pictures, advertising, magazines, lectures, the church, the radio' (Bernays, 1923, p. 42).

Like Fullerton's (2011) reference to the changing attribution of kleptomania, the connections between theory and practice may be more moveable, and more integrated than currently understood. Proponents of historical methods present it as holding a considerable opportunity for greater contextualisation of research. In this sense, a more personal take or embedding research in the experiential lifeworld of subjects need not necessarily be at odds with but can act as complimentary to quantitative theory building methods. Historical detail may also be used in tandem with fieldwork (Hall, 2010), such as Babb's (2011) ethnographic account of tourism in Cuba through the 1990s and 2000s. This approach gives primacy to one topical focal point, while not entirely disregarding adjacent contributing factors. In this sense, historiography offers a method for greater dynamism and contextualisation in tourism studies. With scope to engage with the cultural politics of tourism and consider both commercial and non-commercial forces at play within a given tourism market.

Where To From Then?

An understanding of history surely must be tackled in conjunction with a developing understanding of time. While time is generally accepted as the linear through-line of life experience, present-day technology in some sense contributes to a fragmented experience of time, or an atemporal engagement with information. Outside of academia, through group platforms like Dischord or Telegram, sharing information is more fragmented, more searchable, and with greater layers of anonymity and much less time-bound at a microlevel. In this way, the tracking of movements, ideas, and protection of intellectual property all become somewhat less bound by traceable, linear movements through written text. The existing methods present a very forward arrow approach to time; 'historical thought uses concepts to create a descriptive narrative of the flow of ideas, whereas theory uses constructs to create an explanation of phenomena that can be empirically tested' (Shaw, 2011, p. 492). As such, historic methods may need to be adjusted as more decentralised and informal marketing and tourism systems are created. A consideration of what history was like prior to written documentation, and it is changing given the proliferation of increasingly 'virtual' forms of media and information transfer, could provide some grounds for further discussion. While the earlier methods may appropriate for studying a time of predominantly print media, or at least largely 'fixed' media like PDFs, the influence of editable and network based communication in digital media warrants consideration. The existing complexity around source or fact checking, and editable documentation is certainly a valid concern for historical research, and for the wider body of research moving forward. The sheer volume of information available somewhat negates an articles existence as proof of its legitimacy. There are more than three million tourism articles on Google scholar, conversations conducted over Zoom are recordable, and the advent of AI-video production and 'deep fakes' means digital documents can be readily hacked, deleted, edited, or replicated. By 2080, there's a very substantial possibility that there would be enough data available to build a compelling, even 'data-driven' argument from largely unsubstantiated sources or paper mills (Campbell et al., 2000), the existence of which raises further concern as to why students feel compelled to cheat via outsourcing their reading and writing work rather than engaging with the literature directly.

The above dilemma may be reflective of an underlying need for better writers, and that may come through teaching students how to write, and, how to *read* via historical methods while also providing a vehicle for a deeper consideration of the role of invention or imagination in the development of new business structures, technology, and forms of marketing. To equip students and academics with close reading, critical thinking, and writing skills, and to allow space alongside serious research for topics and discussions that are interesting, imaginative, and fun. As such, historiography may assist in forming a braided understanding of the interrelation across tourism studies ranging from regional human interest to global sociopolitical issues. In this way, it proffers an invitation for tourism scholars to lean in to, and perhaps even celebrate, the complexities and nuances of their chosen subject matter.

References

Afinoguénova, E. (2014). An organic nation: State-run tourism, regionalism, and food in Spain, 1905–1931. *The Journal of Modern History*, *86*(4), 743–779. https://doi.org/10.1086/678951

Alvesson, M., & Sandberg, J. (2020). The problematizing review: A counterpoint to Elsbach and Van Knippenberg's argument for integrative reviews. *Journal of Management Studies*, *57*(6), 1290–1304. https://doi.org/10.1111/joms.12582

Arndt, J. (1985). On making marketing science more scientific: Role of orientations, paradigms, metaphors, and puzzle solving. *Journal of Marketing*, *49*(3), 11–23. https://doi.org/10.1177/002224298504900302

Austin, N. K. (2002). Managing heritage attractions: Marketing challenges at sensitive historical sites. *International Journal of Tourism Research*, *4*(6), 447–457. https://doi.org/10.1002/jtr.403

Babb, F. E. (2011). Che, Chevys, and Hemingway's daiquiris: Cuban tourism in a time of globalisation. *Bulletin of Latin American Research*, *30*(1), 50–63. https://doi.org/10.1111/j.1470-9856.2010.00450.x

Bailyn, B. (1982). The challenge of modern historiography. *The American Historical Review*, *87*(1), 1–24. https://doi.org/10.2307/1863306

Becker, C. (1938). What is historiography? *The American Historical Review*, *44*(1), 20–28. https://doi.org/10.2307/1840848

Bernays, E. L. (1923). *Crystallizing public opinion*. Project Gutenberg.

Brown, S. (2011). Yo, Clio! Can historiography save marketing scholarship? *Journal of Historical Research in Marketing*, *3*(4), 449–468. https://doi.org/10.1108/17557501111183617

Brown, V. A., Harris, J. A., & Russell, J. Y. (2010). *Tackling wicked problems through the transdisciplinary imagination*. Earthscan. https://doi.org/10.4324/9781849776530

Campbell, C. R., Swift, C. O., & Denton, L. (2000). Cheating goes hi-tech: Online term paper mills. *Journal of Management Education*, *24*(6), 726–740. https://doi.org/10.1177/105256290002400605

Carman, J. M. (1980). Paradigms for marketing theory. *Research in Marketing*, *3*(1), 1–36.

Clark, I., & Lopez, L. (2014). *An historical geography of tourism in Victoria,* Australia*: Case studies*. De Gruyter Open Poland. https://doi.org/10.2478/9783110370119

Crawford, R., & Bailey, M. (2018). Speaking of research: Oral history and marketing history. *Journal of Historical Research in Marketing*, *10*(1), 107–128. https://doi.org/10.1108/JHRM-02-2017-0007

Deshpande, R. (1983). "Paradigms lost": On theory and method in research in marketing. *Journal of Marketing*, *47*(4), 101–110. https://doi.org/10.1177/002224298304700411

Domegan, C. (2010). The history of marketing thought: A teaching reflection. *Journal of Historical Research in Marketing*, *2*(4), 457–466. https://doi.org/10.1108/17557501011092493

Domegan, C. (2011). Co-authoring tomorrow: On how past marketing knowledge enlightens future marketing theory. *Marketing Theory*, *11*(4), 495–498. https://doi.org/10.1177/1470593111418801

Eley, G. (2008). The profane and imperfect world of historiography. *The American Historical Review*, *113*(2), 425–437. https://doi.org/10.1086/ahr.113.2.425

Fagan, K. S. (2018). *A historical perspective on LP marketing and payola in 1962: The case of Robby and The Troubadours*. [Master's thesis, Florida Atlantic University]. ProQuest LLC. https://www.proquest.com/openview/79745628ef47b9f7ff2ab1e4f11fcfa4/1?pq-origsite=gscholar&cbl=18750

Fuat Fırat, A. (2010). Commentaries on the state of journals in marketing. *Marketing Theory*, *10*(4), 437–455. https://doi.org/10.1177/1470593110382827

Fullerton, R. A. (1987). Historicism: what it is, and what it means for consumer research. In M. Wallendorf & P. F. Anderson (Eds.), *ACR North American Advances* (Vol. 14, pp. 431–434). Association for Consumer Research.

Fullerton, R. A. (2011). Historical methodology: The perspective of a professionally trained historian turned marketer. *Journal of Historical Research in Marketing*, *3*(4), 436–448. https://doi.org/10.1108/17557501111183608

Golder, P. N. (2000). Historical method in marketing research with new evidence on long-term market share stability. *Journal of Marketing Research*, *37*(2), 156–172. https://doi.org/10.1509/jmkr.37.2.156.18732

Gordon, B. M. (2015). Touring the field: The infrastructure of tourism history scholarship. *Journal of Tourism History*, *7*(1–2), 135–156. https://doi.org/10.1080/1755182X.2015.1059493

Gordon, B. M. (2021). Introduction to the Journal of Tourism History – 13/3. *Journal of Tourism History*, *13*(3), 225–228. https://doi.org/10.1080/1755182X.2022.2075606

Gould, S. J. (1991). The self-manipulation of my pervasive, perceived vital energy through product use: An introspective-praxis perspective. *The Journal of Consumer Research*, *18*(2), 194–207. https://doi.org/10.1086/209252

Gregory, B. (1988). *Inventing reality: Physics as language*. Wiley.

Hall, C. M. (2013). Framing tourism geography: Notes from the underground. *Annals of Tourism Research*, *43*, 601–623. https://doi.org/10.1016/j.annals.2013.06.007

Hall, C. M. (Ed.). (2010). *Fieldwork in tourism: Methods, issues and reflections*. Routledge.

Hall, C. M., & Page, S. (2010). The contribution of Neil Leiper to tourism studies. *Current Issues in Tourism*, *13*(4), 299–309. https://doi.org/10.1080/13683500.2010.482652

Heyl, J. D. (1975). Paradigms in social science. *Society*, *12*(5), 61–67. https://doi.org/10.1007/BF02699921

Hollander, S. C., Rassuli, K. M., Jones, D. G. B., & Dix, L. F. (2005). Periodization in marketing history. *Journal of Macromarketing*, *25*(1), 32–41. https://doi.org/10.1177/0276146705274982

Huggins, M. (2013). Sport, tourism and history: Current historiography and future prospects. *Journal of Tourism History*, *5*(2), 107–130. https://doi.org/10.1080/1755182X.2013.828782

Hunt, S. D., & Speck, P. S. (1985). Does logical empiricism imprison marketing? *Research in Marketing* (2), 27–35.

Innis-Jiménez, M. D. (2018). Mexican Chicago's Colonia Hull House: Food, tourism, and belonging in the 1930s. *Global Food History*, *4*(1), 22–39. https://doi.org/10.1080/20549547.2018.1413915

Jones, D. G. B., & Shaw, E. (2002). A history of marketing thought. In B. A. Weitz & R. Wensley (Eds), *Handbook of marketing* (pp. 39–65). Sage.

Kotler, P. (2005). The role played by the broadening of marketing movement in the history of marketing thought. *Journal of Public Policy & Marketing*, *24*(1), 114–116. https://doi.org/10.1509/jppm.24.1.114.63903

Lawson, R., & Wooliscroft, B. (2004). Human nature and the marketing concept. *Marketing Theory*, *4*(4), 311–326. https://doi.org/10.1177/1470593104047641

Lemel, R. (1993). Scientific inquiry in marketing: An empirical investigation. *American Business Review*, *11*(2), 6–36.

Lowe, S. (2001). In the kingdom of the blind, the one-eyed man is king. *International Journal of Cross Cultural Management*, *1*(3), 313–332. https://doi.org/10.1177/147059580113004

Lutz, R. J. (1989). Presential address positivism, naturalism and pluralism in consumer research: Paradigms in paradise. *Advances in Consumer Research*, *16*, 1–8.

MacKenzie, N. G., Pittaki, Z., & Wong, N. (2020). Historical approaches for hospitality and tourism research. *International Journal of Contemporary Hospitality Management*, *32*(4), 1469–1485. https://doi.org/10.1108/IJCHM-03-2019-0273

Maples, J. N. (2021). *Rock climbing in Kentucky's Red River Gorge: An oral history of community, resources, and tourism*. West Virginia University Press.

Marwick, A. (1989). *The nature of history* (3rd ed.). Macmillan Education.

Müller, D. K. (Ed.). (2019). *A research agenda for tourism geographies*. Elgar.

Müller, D. K. (2021). 20 years of Nordic second-home tourism research: a review and future research agenda. *Scandinavian Journal of Hospitality and Tourism*, *21*(1), 91–101. https://doi.org/10.1080/15022250.2020.1823244

Narver, J. C., & Savitt, R. (1971). *Instructor's manual for the marketing economy: An analytical approach*. Holt, Rinehart and Winston.

Reidenbach, R. E., & Robin, D. P. (1991). Epistemological structures in marketing: Paradigms, metaphors, and marketing ethics. *Business Ethics Quarterly*, *1*(2), 185–200. https://doi.org/10.2307/3857262

Rindfleisch, A., Malter, A. J., Ganesan, S., & Moorman, C. (2008). Cross-sectional versus longitudinal survey research: Concepts, findings, and guidelines. *Journal of Marketing Research*, *45*(3), 261–279. https://doi.org/10.1509/jmkr.45.3.261

Roberts, C. (2014). From the ground up: Exploring the use of oral history in tourism. *Oral History Australia Journal*, (36), 23–30. https:/search.informit.org/doi/10.3316/informit.835046423076437

Sæþórsdóttir, A. D., & Hall, C. M. (2020). Visitor satisfaction in wilderness in times of overtourism: A longitudinal study. *Journal of Sustainable Tourism*, *29*(1), 123–141. https://doi.org/10.1080/09669582.2020.1817050

Savitt, R. (1980). Historical research in marketing. *Journal of Marketing*, *44*(4), 52–58. https://doi.org/10.1177/002224298004400408

Savitt, R. (2011). On biography in marketing. *Journal of Historical Research in Marketing*, *3*(4), 486–506. https://doi.org/10.1108/17557501111183635

Schwarzkopf, S. (2015). Marketing history from below: Towards a paradigm shift in marketing historical research. *Journal of Historical Research in Marketing*, *7*(3), 295–309. https://doi.org/10.1108/JHRM-06-2015-0021

Shapiro, S. J. (2011). Marketing thought and the further development of marketing theory. *Marketing Theory*, *11*(4), 479–481. https://doi.org/10.1177/1470593111418803

Shaw, E. H. (2011). A comment on the relationship between the history of marketing thought and the development of marketing theory. *Marketing Theory*, *11*(4), 491–494. https://doi.org/10.1177/1470593111418807

Shaw, E. H. (2015). Teaching the history of marketing thought: An approach. *Journal of Historical Research in Marketing*, *7*(2), 256–271. https://doi.org/10.1108/JHRM-08-2014-0025

Shaw, E. H., & Tamilia, R. D. (2001). Robert Bartels and the history of marketing thought. *Journal of Macromarketing*, *21*(2), 156–163. https://doi.org/10.1177/0276146701212006

Sheth, J. (2021). New areas of research in marketing strategy, consumer behavior, and marketing analytics: The future is bright. *Journal of Marketing Theory and Practice*, *29*(1), 3–12. https://doi.org/10.1080/10696679.2020.1860679

Smith, R. A., & Lux, D. S. (1993). Historical method in consumer research: Developing causal explanations of change. *The Journal of Consumer Research*, *19*(4), 595–610. https://doi.org/10.1086/209325

Smith, S. L. (Ed.). (2010). *Discovery of tourism*. Emerald Group Publishing.

Steinbeck, J. (1998). *The grapes of wrath*. Folio Society.

Tadajewski, M. (2022a). On being critically oriented in precarious times: For resistant curiosity. *Journal of Marketing Management*. https://doi.org/10.1080/0267257X.2021.2016892

Tadajewski, M. (2022b). Writing telepathy back into marketing theory. *Marketing Theory*. https://doi.org/10.1177/14705931221095611

Tadajewski, M., & Jones, D. G. B. (2014). Historical research in marketing theory and practice: A review essay. *Journal of Marketing Management*, *30*(11–12), 1239–1291. https://doi.org/10.1080/0267257X.2014.929166

Tadajewski, M., & Saren, M. (2008). The past is a foreign country: Amnesia and marketing theory. *Marketing Theory*, *8*(4), 323–338. https://doi.org/10.1177/1470593108096539

Taleb, N. N. (2013). 'Antifragility' as a mathematical idea. *Nature*, *494*(7438), 430–430. https://doi.org/10.1038/494430e

Thompson, A.-M. K. (2010). Golder's historical method in research in marketing. *Journal of Business Research*, *63*(12), 1269–1272. https://doi.org/10.1016/j.jbusres.2009.09.005

Trapp-Fallon, J. M. (2003). Searching for rich narratives of tourism and leisure experience: How oral history could provide an answer. *Tourism and Hospitality Research*, *4*(4), 297–305. https://doi.org/10.1177/146735840300400403

Tucker, A. (2004). *Our knowledge of the past: A philosophy of historiography*. Cambridge University Press.

Walton, J. K. (2005). *Histories of tourism: Representation, identity and conflict*. Channel View Publications Ltd.

White, H. (1988). Historiography and historiophoty. *The American Historical Review*, *93*(5), 1193–1199. https://doi.org/10.2307/1873534

Whitrow, G. J. (1975). *The nature of time*. Penguin.

Willis, J. W. (2007). *Foundations of qualitative research: Interpretive and critical approaches*. Sage Publications.

Wilson, J. (Ed.). (2012). *The Routledge handbook of tourism geographies*. Routledge.

Wooliscroft, B. (2011). Marketing theory as history. *Marketing Theory*, *11*(4), 499–501. https://doi.org/10.1177/1470593111418800

Yan, T., & Hyman, M. R. (2018). Critical historical research method and marketing scholarship. *Journal of Marketing Management*, *34*(9–10), 841–864. https://doi.org/10.1080/0267257X.2018.1515783

Zaltman, G. (1982). *Theory construction in marketing: Some thoughts on thinking*. Wiley.

Chapter 11

GPS, Smartphones, and the Future of Tourism Research

Noam Shoval, Michal Isaacson, & Prem Chhetri

The development of Global Positioning System (GPS) receivers, cell phones, and recently smartphones is a very dynamic domain, with the potential to capture ubiquitous and real-time spatial human movement and behaviour. The rapid adoption of these devices shows that our social interactions, economic engagement, and individual existence including personalised life activities are now being shaped by the use of these devices, which have recently become easily accessible and more affordable. Although research employing these technologies in tourism is not yet mature, in our view enough has been accomplished at this stage to be documented. Furthermore, the possibilities for implementation of these technologies for the purposes of research are increasing and have the power to change tourism mobility research as it stands today.

Until recently, the most common method for gathering information on human time–space patterns was the time–space diary. This method provides a systematic record of the way in which individuals occupy their time in space over a limited period, be it a few hours, a day, or a week (Anderson, 1971). While time–space diaries have been used to great effect (see, for example, Goodchild & Janelle, 1984; Janelle et al., 1988) they do have several disadvantages as research tools. In particular, time–space diaries require that the subjects are actively involved in the process of data collection by recording, in detail and at length, their activities throughout the entire experiment (Thornton et al., 1997). Since participants often fail to record their actions faithfully, the data obtained are often of questionable credibility. Nonetheless, the temporal dimension is what makes geography different from geometry (Parks & Thrift, 1980). The expansion of navigation technology, the increased ease of gathering digital spatial data, and the growth in computing capacity to manage great volumes of data have led to a situation where it is time to consider both the spatial and temporal behaviour of tourist movement.

The Wiley Blackwell Companion to Tourism, Second Edition. Edited by C. Michael Hall.

The Introduction of GPS to Social Science Research and to Tourism Studies

GPS devices offer researchers the opportunity of continuous and intensive high-resolution data collection in time (seconds) and space (meters) for long periods of time; this was never possible before in spatial science research. GPS allows the spatial and temporal positioning of tourists to be captured through a network of satellites orbiting in space around the Earth. Each tourist visit involves some form of movement and a series of social interactions with other visitors. The movement of a tourist across space and over time can be ascertained from recorded *x* and *y* coordinate pairs as well as a recorded time collected through a GPS receiver. Acquiring tourist movements in time and space has the potential in providing arrival and departure times, attractions visited, the sequence of attractions visited, as well as walking speed and orientation. Spatiotemporal characteristics of tourist movements are recorded in GPS devices to capture parameters such as sites visited, time of day (when), duration (how long), and sequence (what order). Changes in these characteristics indicate changes in tourist positioning that in turn reflect exhibited tourist behaviour. Scheduling of recreational activities that a tourist plans to undertake at a destination is therefore inextricably embedded within the constraints of space and time. Understanding of mobility of tourist within the limits imposed by space and time will help answering queries such as: What are the typical patterns exhibited by day visitors compared to those visitors on a longer stay?; and Where and when are peak capacities occurring?

GPS and other tracking technologies are now used in a wide variety of fields aside from tourism, such as environmental health (Phillips et al., 2001; Elgethun et al., 2003); the medical field, in such subjects as Alzheimer's disease (Miskelly, 2004, 2005; Shoval et al., 2008), physiology (Terrier & Schutz, 2005), and cardiology (Le Faucheur et al., 2008); and as a tool to assist in navigation for visually impaired and blind pedestrians (Golledge et al., 1991, 1998; Maeda, 2002). However, to date, most of the research conducted based on material gathered by advanced technologies has been in the field of transportation studies, usually in regard to tracking the spatial paths of motor vehicles (see, for example, Zito et al., 1995; Quiroga & Bullock, 1998; Wolf et al., 2001; Bohte & Maat, 2009).

Emerging technologies have resolved both data-collection and analysis problems, and in doing so have potentially revolutionised research into tourist behaviour in urban destinations and in open areas. GPS loggers with an accuracy of at least 10 m can track a tourist's location as frequently as every second at every point on Earth. These instruments have resolved the data-collection challenge by providing a level of fineness of detail that was previously unavailable. Geographic Information System (GIS) software can analyse these high-resolution data quickly and efficiently, and importantly provide a range of analytical options including tracking tourist movements, time analysis, space analysis, and length of stay (Shoval & Isaacson, 2007a).

One concern in GIS is the storage, representation, and querying of mobile objects such as humans, animals, or vehicles that move within geographical space. Mobile objects are entities that are able to perform actions and have properties. Time geography is one such framework developed in geography, during the 1950s and 1960s, through which data logged in GPS devices can be data mined to analyse mobile objects in space and time (Hägerstrand, 1970; Raper, 2000; Peuquet, 2002). In recent years a growing body of work has demonstrated the efficacy and the limitations of using tracking technologies to explore leisure and tourist activities (for a review see Shoval & Isaacson, 2010).

As with any emerging technology, tourism researchers are still experimenting to determine the limits of its application. To date, much of the research implementing tracking technologies has tended to be rather descriptive and small scale. Some more sophisticated studies have been conducted, but they have mostly been tightly spatially bound, for example, focusing on small historic cities (Modsching, 2008; Shoval, 2008; van der Spek, 2008; Tchetchik et al., 2009), confined attractions like theme parks and zoos (Russo et al., 2010; Pettersson & Zillinger, 2011), natural parks (Arrowsmith & Chhetri, 2003; Harder et al., 2008; Hallo et al., 2012), and small islands (Xia et al., 2009; Nielsen, 2010). Each of these locales has a clearly defined entry and exit point, making both the selection of potential participants and the modelling of their movements easier tasks. Only few large, complex, and multifunctional urban settings like Rome (Calabrese & Ratti, 2006) and Hong Kong have also been investigated successfully using GPS (McKercher et al., 2012; Shoval et al., 2011). Recently, we are aware of wide scale GPS-based surveys on a national scale.

An Estonian group of researchers (Ahas et al., 2007, 2008) used cellular-network information on a national scale to gain insight into the activity of tourists in Estonia. In the first paper, published in 2007, cellular phone data was used to establish spatial patterns of tourists in different seasons. The second paper focused on the flows of tourist movement as can be observed by the records describing the roaming phones in the cell-phone network's database. We now have strong reasons to believe that this field of research has matured and evolved from an exploratory to an integrative phase and begin to show features that characterise a predictive phase (see Figure 11.1).

The exploratory phase denotes the initial research conducted to merely map location behaviour of tourists using GPS and mobile devices. They rely on few studies and GPS-logged data to build new theories and behavioural models. Qian et al. (2021) used mobile phone data to analyse spatial interactions of visitors between destinations, resting places and transportation hubs in Shanghai. Xu et al. (2021) capture the mobility patterns and behavioural characteristics of tourists. They analysed mobile phone trajectories of international travellers in South Korea using nine mobility indicators (e.g., duration of stay in a city, spatial extent of activities, location visited, trips conducted, and mobility diversity). The study by Dane et al. (2020) also tracked event visitors at the Dutch Design Week (DDW) 2017 event in Eindhoven, the Netherlands to explore the area of interest locations, visitor flows, visitor clusters and area of interest choices. Location-based data are also captured through devices attached to micro-mobility services such as docked bikes, docked e-bikes, dockless e-bikes, and dockless e-scooters. Li et al. (2020) investigated the variations of micro-mobility behaviour before and during the COVID-19 pandemic in Zurich, Switzerland. Similarly, a study by Buning and Lulla (2020) examined the spatial-temporal usage patterns of bikeshare by visitors using GPS based big data. This augmented data enabled generating measures of trip behaviour such as route selection, time of rental, checkout/check-in locations, distance, speed, duration, and physical activity intensity.

The integrative phase involves studies that integrates location data with other sources of data, such as expenditure data, photos, and social media, to provide comprehensive understanding of tourist behaviour and thought process. Studies in this phase have created theoretical models and advanced the inclusion of diverse methodologies. Walden-Schreiner et al. (2018) integrated geotagged photos on Flickr, a popular photo-sharing social-media site, with location data to link visuals with temporal and spatial patterns of use in Aconcagua Provincial Park (Argentina) and Kosciuszko National Park (Australia). Sinclair et al. (2020) use the metadata of geotagged photographs, sourced through passive crowdsourced data from social media, and estimate state-wide visitation to individual wetlands to improve the provision of recreational services and management practices in German national parks.

RMIT Classification: Trusted

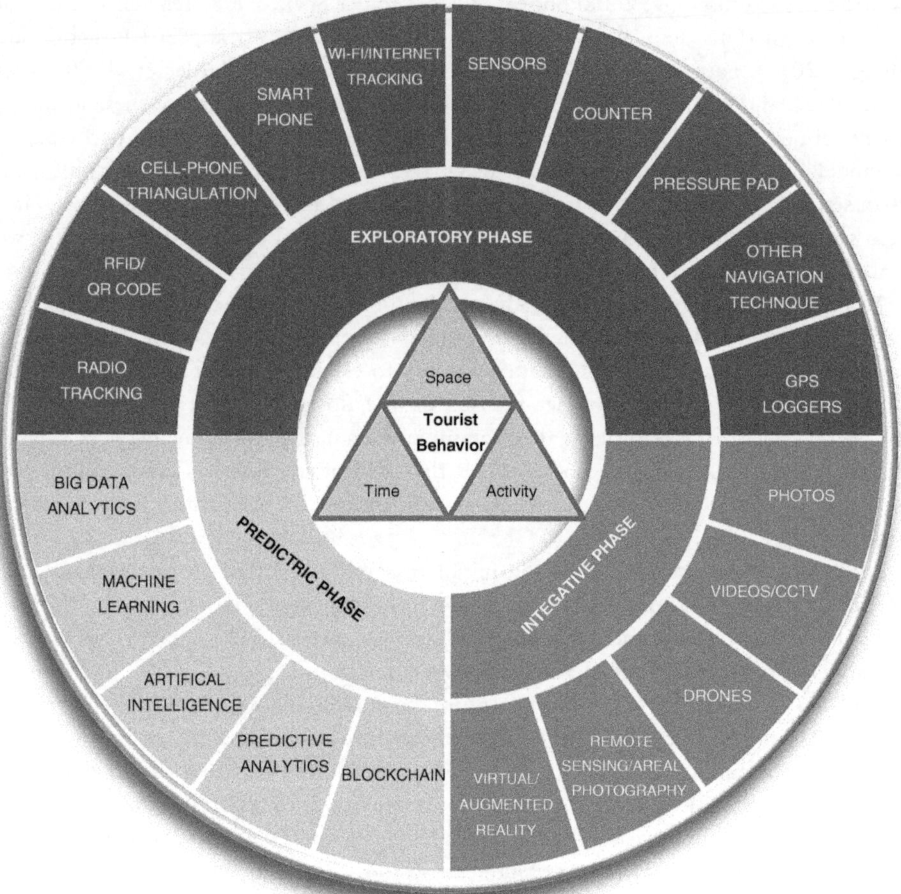

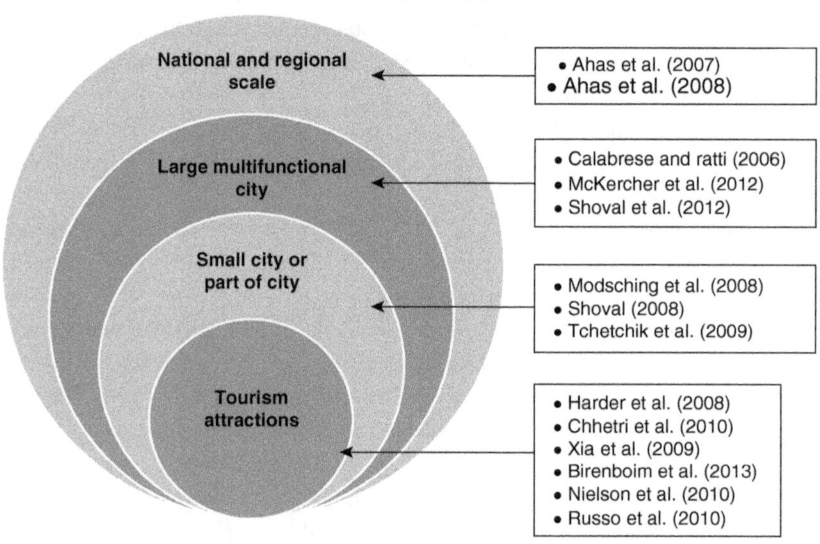

Figure 11.1. Different scales of studies using tracking technologies in the recent decade.

Navarro-Ruiz et al. (2020) adopted a multi-method approach to triangulate multiple datasets (reflecting spatial movement, time consumption, and participation patterns with attractions) collected through questionnaires, trip diaries and GPS tracking technologies to capture intra-destination behaviour of cruise visitors in Valencia. Munro et al. (2019) added public participation geographical information systems (PPGIS) to integrate spatially explicit data on place values, activities and development preferences along the coastline in Western Australia. Similarly, Soe and Sein (2017) built a tourist guide information system using location-based service of the mobile phone and Google Map services. User's current location, reachable attractions and amenities and the routes to reach nearest place to visits were embedded in the systems.

The study of nonverbal behaviour (NVB), and in particular kinesics (i.e., face and body motions), is typically seen as cost-intensive. However, the development of new technologies (e.g., ubiquitous sensing, computer vision, and algorithms) and approaches to study social behaviour (i.e., social signal processing (SSP)) makes it possible to train algorithms to automatically code NVB, from action/motion units to inferences. Online social media, such as Twitter and Instagram, democratised information broadcast, allowing anyone to share information about themselves and their surroundings at an unprecedented scale. Domènech et al. (2020) integrated expenditure data with mobility patterns of cruise passengers in the city of Tarragona (Catalonia) to establish that tourists with a higher per-capita expenditure are likely to spend more time onshore, visit fewer tourist sites and spend more time in areas which provide tourist-oriented, mixed commercial and recreational functions. Visitor behaviour in crowded places such as heritage sites was analysed using zenith images using drones and time-lapse. Location data were integrated with drone captured data to enhance the analytical capability to model patterns of visitor use (Donaire et al., 2020).

The predictive phase includes studies that intend to predict outcomes, consequences, or effects. Researchers have applied advanced analytics to make predictions about unknown future events such as the likely location of tourists in time and space as well as their behaviour manifested in the form of consumption, destination choice or group-behaviour dynamics. Widely applied techniques include data mining, analytics, modelling, machine learning, and artificial intelligence to analyse big and integrated datasets to generate predictions about tourist behaviour patterns. Zheng et al. (2017) attempted to accurately predict the tourist's next location within a given destination using a heuristic method based on data mining of past visitor movements at the Summer Palace in Beijing, China (see also Huang et al., 2017). A deep learning-based approach is also adopted by Crivellari and Beinat (2020) who data mine the collective mobility of tourists. Their study also predicted the next visited site in the trajectory based on a long short-term memory (LSTM) neural network model trained on pre-processed location sequences. Hardy and Aryal (2020) built an automated spatio-temporal dependence model via machine-based learning to operate in real-time to enable automated tracking of tourists' behaviour.

Volchek et al. (2019) have developed an application that uses higher frequency search query data to generate weekly predictions of visits to five London museums to improve attraction- and destination-level planning. Dietz et al. (2020) used a data-driven method to mine trips from location-based social networks to analyse tourists travel patterns globally. They used Twitter and Foursquare to integrate social aspects with location data and generate metrics to reflect mobility patterns.

The Smartphone Revolution and its Implications for Tourists and Researchers

The implication of the fast and wide introduction of smartphones since about 2005 has created immense availability of information for their users at any time and any place; this also has implications for the ability of tourists to receive information while touring destinations and changes the whole paradigm about tourists' knowledge of destinations (Tussyadiah & Zach, 2012). However, until recently, due to the high roaming prices for phones crossing international borders, the use of such data by tourists was relatively limited. Nonetheless, recently use of local SIM cards and attractive international roaming programmes, make smartphones relevant for tourists. The other implication is for tourism research, as will be discussed below.

Smartphones are highly programmable cell-phone devices which integrate powerful computing abilities, advanced communication, and geolocation technologies, as well as several other useful functionalities that may be practical for researchers. The 'always on always worn' (unlike GPS loggers used in research so far) status of smartphones turns them into an accessible medium through which surveys can be delivered and self-completed by respondents in their immediate, natural environment. Other than self-report questionnaires, smartphones can supply researchers with physiological, behavioural, and locomotion information which can be *generated* automatically based on built-in and external sensors. Several researchers outside the tourism research realm have already taken advantage of smartphone capabilities and employed them in their studies. The majority of these studies come from the fields of psychology, psychiatry, and medicine, but several examples from social science can be found as well.

Raento et al. (2009) were one of the first groups to discern the potential of smartphones as a research tool in general and for social science in particular. While their initial motivation comes from the field of human–computer interaction, which explores how people use and interact with computers, they do note that smartphones could be useful tools for social science researchers.

Gaggioli et al. (2011) developed Psychlog, an open source cell-phone platform that allows recording and analysis of individuals' psychological, physiological, and activity information based on self-report questionnaires and sensor information. They conducted a small pilot study with six students for a period of 1 week, during which they recorded heart rate samples (via electrocardiography) and collected self-assessment digital forms for the purpose of monitoring participants' concurrent stress and psychological arousal during their daily lives. Their results 'suggest the feasibility and accuracy of using [their] system to monitor the dynamic psycho-physiological profile of mental health constructs such as stress in ecological situations' (Gaggioli et al., 2011, p. 9). However, their system has two prominent disadvantages. First, it works offline and therefore requires manual downloading of the information collected. Second, the system runs on the Windows Mobile operating system, which today is not among the more popular mobile operating systems.

The use of smartphones in general is fraught with a number of disadvantages. The relatively short battery life of the phones, at the time of writing this chapter, may limit the use of the devices for surveying purposes; however, new battery types introduced recently may be able to tackle this shortcoming. Another drawback has to do with the validity of the samples: to date, fewer than 50 percent of people in developed countries own a smartphone, which might lead to sample biases. However, this situation is expected to change in the near future, with the increase of smartphone penetration rates. A third disadvantage that might hinder the smooth implementation of smartphones as a research tool is that most of the systems and software, such as those reviewed above, require some knowledge of programming or at least familiarity with the structure of XML files and how to write them. Finally, smartphones run on several different operating systems; this requires researchers to develop several different applications to meet the specifications of each operating system.

GPS Data and the Potential for Complex Spatiotemporal Modelling and Data Mining

Due to the availability of high-resolution spatial and temporal data there are new horizons for complex spatiotemporal modelling and data mining. For example, Arrowsmith and Chhetri (2003) adopted GPS technology to identify typical spatial behaviour sequences within a park site including where visitors were going and how long they spent at each location. Using a range of network matrices, visiting sequences were quantified to reveal the complexity of movement patterns of tourists. The information obtained through GPS devices, which captured the behaviour sequence patterns, were then integrated in the Recreational-Behaviour Simulator (RBSim2), a computer programme that simulates the behaviour of human visitors to a recreational park (Itami, 2003). The probabilistic matrix generated to map tourists' movement patterns was used to set behavioural rules for autonomous agents. The simulated data generated for different growth scenarios were then utilised to evaluate the carrying capacity of various lookouts and allowed infrastructure planning such as upgrading of car-parking facilities.

GPS-logged data opens up further opportunities for conducting complex spatiotemporal modelling and data mining. For example, Xia et al. (2008, 2009) and Chhetri et al. (2010) have explored the richness of GPS-logged data using methods such as Markov chains and circular statistics to ascertain predictability of tourist visits. Chhetri et al. (2010) calculated measures of circular mean and circular dispersion to capture the mobility patterns of tourists in Port Campbell national park in Melbourne, Australia. Their study also visualised temporal data collected by handheld GPS receivers that tourists carried while visiting the park using KDE (Kernel Density Estimation) plots over a 24-hour period (see Figure 11.2). Statistically significant differences in the overall trip duration – that is, the total time that tourists spent at the site – were found between different groups and trip types.

One form of analysis that has tremendous potential for creating typologies of tourists based on their spatial behaviour while taking into account the sequence of locations can be seen in sequence-alignment methods (SAMs). These methods, which have been used since the 1980s, were introduced to the social sciences by Abbot (1995) and Wilson (1998) and to the spatial sciences by Shoval and Isaacson (2007b) and Wilson (2008). The methods have been developed and refined to improve the accuracy with which they compare sequences.

Ethical Questions and the Tracking of Tourists

The question of privacy in relation to the increasing use in daily life of such devices as smartphones, which are actually, inherently, and among other things, tracking devices, has become more pronounced (Curry, 1997; Fisher & Dobson, 2003; Hall & Ram, 2020). This concern has been widely debated by the popular media, academia, courts, and lawmakers. However, in most countries, legal systems have failed to tackle the question fully and, as such, the question is of considerable and global importance (Renenger, 2002).

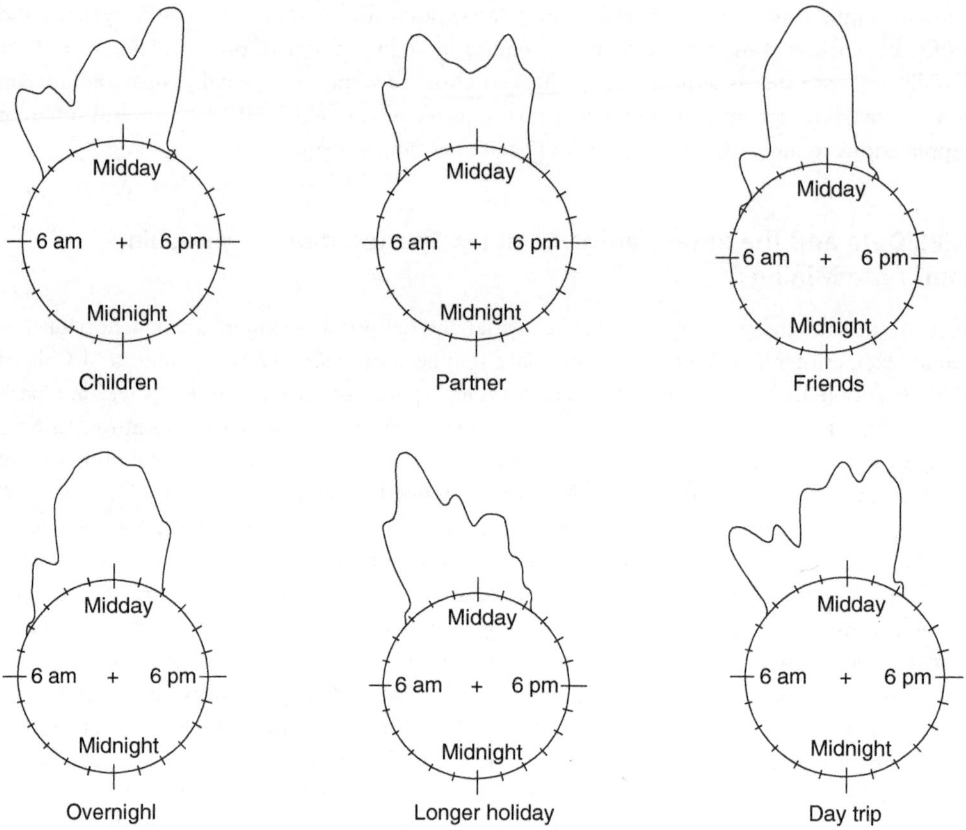

Figure 11.2. Circular plots by region and by group and trip type.

Advanced tracking technologies have the ability and potential for implementation in tourism research. These methods can provide a richer understanding of the mobility patterns of tourists in time and space and their impact on destinations. However, the use of such technologies, which are able to obtain the exact locations of research participants at any given moment, can infringe the privacy of the participants and adds a geographical dimension to the *surveillance society* (Lyon, 2001) and the ability to better track the *digital individual* (Curry, 1997). This raises different ethical and moral issues arising when conducting research that involves such technologies. One of the most prominent among these issues is whether, and to what extent, such research projects impinge on their participants' right to privacy.

The need to undertake research in a manner that will protect the privacy, dignity, and well-being of its subjects is obviously a concern to anyone engaging in any kind of study. The need to receive the approval of an ethics committee for research that involves humans is clear, accepted, and part of any research design in the social sciences. In the specific case of studies of the spatiotemporal activities of tourists, however, we believe that the potential abuse of privacy from the point of view of the participant is less of a problem due to the fact that people being tracked are monitored for a limited and well-defined time period and are not in their regular habitat. In our experience with various tracking projects in tourism and other areas, tourists are less reluctant to participate in such surveys than other types of subjects.

Due to the relative novelty of using tracking technologies in tourism research, its implications have probably not yet received the attention it deserves in academic literature (Fennell, 2017). Unsurprisingly, then, little within the current law pertains directly to the use of land-based or satellite tracking systems. Nonetheless, one particular statute (and its implementing regulations) merits discussion for its applicability to tracking technologies: the US *Wireless Communications and Public Safety Act* (WCPSA). Although the WCPSA is not overtly concerned with privacy issues, it has nonetheless shaped the way the government limits the sharing of GPS data. The WCPSA provided the Federal Communications Commission (FCC) with congressional encouragement to continue to develop and implement the then nascent 'enhanced 911' (or E911) infrastructure.

The enhanced 911 scheme mandated that wireless providers develop their network capabilities to pinpoint the location of 911 calls (calls made to the US emergency services) made from cell phones so that they may be relayed to the emergency services. Even more pertinently, the WCPSA expanded the definition of 'customer proprietary network information' under the *Telecommunications Act* to include information pertaining to the location of mobile network use. The *Telecommunications Act* places restrictions on how customers' proprietary network information can be shared. Telecommunication carriers are only permitted to disclose such information either with the consent of the customer or as related to its provision of services. Therefore, in most cases mobile providers may not distribute locational information about their consumers.

This protective scheme does contain a few notable loopholes. Firstly, the restriction on disclosure is limited to 'individually identifiable' information, so it does not apply to information that has been anonymised or where the personal identity of the customer is not discernable. Secondly, mobile carriers have been afforded leeway to obtain customer consent through fingerprint service agreements.

In our view conducting tourism research with tracking technologies is less problematic in terms of potential privacy abuse for participants, because they are monitored for a limited and well-defined time in a destination away from where they live. However, the privacy issue is not the only ethical issue in research and, in any case, the obligation of the researcher is to conform to certain moral and ethical principles.

Privacy and ethical issues differ between two types of research using tracking technologies in terms of the participants. The first is direct participation, in which participants are asked individually to take part in the study and give their consent. As will be explained, there is an additional subdivision within this category between cases in which designated tracking devices are given to participants and cases in which subjects are tracked using their own tracking devices (i.e., their smartphones). The second is indirect participation, when participants are tracked through the operator of a cellular-network provider.

Privacy issues, as we have seen, have evolved over time, leading to a state in which the protection of the privacy of the individual is an uppermost value in most Western countries. This must be taken into account when designing studies which examine human behaviour. In the field of tourism research, the issue of maintaining the privacy of the participants is important and an essential part of any study. However, we are confident that, by following strict guidelines and avoiding the pitfalls outlined in this chapter, researchers can and should use tracking technologies of all kinds to learn more about the time–space activities of tourists.

Theoretical and Methodological Reflections

Providing extremely accurate data in time and space, these technologies are opening up new, previously unfeasible, lines of inquiry. In practical terms, the high-resolution data thus obtained could be used, among other things, to monitor and regulate the carrying capacity of tourist sites more rationally, improve the allocation of accommodation services, reduce friction between tourists and local populations, and assist in planning tourism/transport infrastructures. All in all, an important point to be clarified is that these technologies do not replace questionnaires, diaries, or interviews, which will remain an important source of information on the activity – and especially the motives underlying the activity – of tourists. Instead, the new technologies will complement, add to, and enrich the findings of these more traditional research tools.

The examples presented in this chapter were mostly carried out on a limited scale both in terms of the numbers of participants and the duration of sampling. In the future, we will no doubt see research projects that are executed on larger scales covering wider geographic regions, including bigger samples over longer periods of time. Cities and countries that flourish on tourism will be able to gain knowledge about the flow of tourists throughout their locales by carrying out ongoing surveys of the locations of tourists within their boundaries. Visitor satisfaction surveys and questionnaires about tourist spending patterns will be considered incomplete without information on where the visitor has spent his or her day and spatial patterns of satisfied visitors will become an industry standard. Tourism planners will be able to consider the impacts of different development scenarios more accurately and objectively. This will enable a much higher degree of efficiency in the management of tourism destinations of different scales.

Key Questions for Future Research

One open question that still must be investigated is whether visitors, once they know they are being followed, change their activity. And, if so, how? This question should be further explored in empirical studies. However, even the more traditional methods for gathering information about the time–space activities of tourists raise concerns as to whether reports are inaccurate (in questionnaires and time budgets) and whether tourists change their patterns of activity due to participation in a study, feeling that they must behave in a certain way.

An additional direction of inquiry that has yet to be explored is the combination of tracking data with additional digital sources of data. Items that can be added to a tracking kit and carried by the participant can include, for example, sensors that measure excitement and physical effort. Sensors can also be placed throughout a destination to measure, for instance, noise levels, pollution, and other qualities of the environment.

Another remaining question relates to the large amounts of data that are accumulated when using these methods, which will require the development of more algorithms to enable automatic scripts to analyse the data in a fast and practical way. It would be beneficial if at some point the software being developed today on an ad-hoc basis by research teams around the globe could be standardised so that common measures are developed. More importantly, this could lead to an increase in the number of researchers in the field, because the current challenge of data analysis no doubt limits the number of prospective users of these methods.

The extant literature on the use of location-based technology in modelling spatial mobility and behaviour of tourists highlights significant growth in recent years (Hardy & Aryal, 2020;

and Shoval & Isaacson, 2010). The data collected by GPS receivers, mobile cell phones, and recently smartphones have added enormous value to augmenting capabilities of big data, machine learning and AI on predicting consumer behaviour and individual-level profiling of tourists. Padrón-Ávila and Hernández-Martín (2020) have identified 31 different tracking techniques deployed to track movement patterns of tourists. Lee et al. (2021) reviewed the key studies that implemented (Internet of things) IoT related technologies to explore smart tourism. The use of smartphones, linked to location-based services and navigation capability, can change the perception of tourism spaces and may potentially alter the behaviour of tourists (Calvignac & Smolinski, 2020). The large amounts of accurate data collected using these methods will make possible the development of new theories regarding the spatial activity of tourists, new understandings of tourists and their influence on destinations, and more accurate ways to calculate both physical and social carrying capacity in order to ensure the sustainability of tourism within a destination. These developments were not possible using the data sources that were available up until the introduction of these technologies. This technological transformation has opened new frontiers for research in building complex spatial decision systems and predictive analytics of tourists' behaviour.

Further research is necessary to develop an understanding of what patterns of spatial activity are beneficial to both the destination and the tourists visiting the destination. Recognising patterns of spatial activity throughout a destination that are not beneficial, that may lead to damage of delicate attractions (such as historical locations) or infrastructure, and that may cause frustration and displeasure to the tourists can be of great assistance in the smooth and efficient management of tourist destinations. Being able to read into the spatial activity, to understand the different meanings that these patterns of behaviour have for tourist destinations, and to project the possible results of various interventions are essential skills and can be developed through research. We have the ability to recognise areas of high and low activity and even to classify destinations into different types of space according to spatial activity, but we do not yet have the knowledge to identify over-usage, when activity levels have reached a point that is no longer beneficial for the destination.

The potential improvements in the implementation of tracking technologies for tourism research and practice mentioned above could enhance the benefits of using those methods for researchers and practitioners alike. The rapid advances in tracking technologies and the growing possibilities in implementing them for use in different areas in general, and in the realm of tourism research in particular, as elaborated in this chapter, leave no doubt that the future of tracking technologies in tourism will be exciting and dynamic, providing researchers with invaluable insight and information.

References

Abbott, A. (1995). Sequence analysis: new methods for old ideas. *Annual Review of Sociology*, *21*, 93–113. https://www.jstor.org/stable/2083405

Ahas, R., Aasa, A., Mark, U., Pae, T., & Kull, T. (2007). Seasonal tourism spaces in Estonia: case study with mobile positioning data. *Tourism Management*, *28*(3), 898–910. https://doi.org/10.1016/j.tourman.2006.05.010

Ahas, R., Aasa, A., Roose, A., Mark, U., & Silm, S. (2008). Evaluating passive mobile positioning data for tourism surveys: an Estonian case study. *Tourism Management*, *29*(3), 469–486. https://doi.org/10.1016/j.tourman.2007.05.014

Anderson, J. (1971). Space-time budgets and activity studies in urban geography and planning. *Environment and Planning*, *3*(4), 353–368. https://doi.org/10.1068/a030353

Arrowsmith, C., & Chhetri, P. (2003). *Port Campbell National Park: Patterns of use: A report for the development of visitor typology as input to a generic model of visitor movements and patterns of use*. Parks Victoria.

Bohte, W., & Maat, K. (2009). Deriving and validating trip purposes and travel modes for multi-day GPS-based travel surveys: a large-scale application in The Netherlands. *Transportation Research Part C*, *17*(3), 285–297. https://doi.org/10.1016/j.trc.2008.11.004

Buning, R. J., & Lulla, V. (2020). Visitor bikeshare usage: tracking visitor spatiotemporal behavior using big data. *Journal of Sustainable Tourism*, *29*(4), 711–731. https://doi.org/10.1080/09669582.2020.1825456

Calabrese, F., & Ratti, C. (2006). Real time Rome. *Networks and Communications Studies, NETCOM*, *20*(3–4), 247–258.

Calvignac, C., & Smolinski, J. (2020). How can the use of a mobile application change the course of a sightseeing tour? A question of pace, gaze and information processing. *Tourist Studies*, *20*(1), 49–74. https://doi.org/10.1177/1468797619885955

Chhetri, P., Corcoran, J. C., & Arrowsmith, C. (2010). Investigating the temporal dynamics of tourist movements: an application of circular statistics. *Tourism Analysis*, *15*(1), 71–88. https://doi.org/10.3727/108354210X12724734223676

Crivellari, A., & Beinat, E. (2020). LSTM-based deep learning model for predicting individual mobility traces of short-term foreign tourists. *Sustainability*, *12*(1), 349. https://doi.org/10.3390/su12010349

Curry, M. R. (1997). The digital individual and the private realm. *Annals of the Association of American Geographers*, *87*(4), 681–699. https://doi.org/10.1111/1467-8306.00073

Dane, G., Borgers, A., Kaya, D. I., & Feng, T. (2020). Visitor flows at a large-scale cultural event: GPS tracking at Dutch Design Week. *ISPRS International Journal of Geo-Information*, *9*(11), 661. https://doi.org/10.3390/ijgi9110661

Dietz, L. W., Sen, A., Roy, R., & Wörndl, W. (2020). Mining trips from location-based social networks for clustering travelers and destinations. *Information Technology & Tourism*, *22*, 131–166. https://doi.org/10.1007/s40558-020-00170-6

Domènech, A., Gutiérrez, A., & Anton Clavé, S. (2020). Cruise passengers' spatial behaviour and expenditure levels at destination. *Tourism Planning & Development*, *17*(1), 17–36. https://doi.org/10.1080/21568316.2019.1566169

Donaire, J. A., Galí, N., & Gulisova, B. (2020). Tracking visitors in crowded spaces using zenith images: Drones and time-lapse. *Tourism Management Perspectives*, *35*, 100680. https://doi.org/10.1016/j.tmp.2020.100680

Elgethun, K., Fenske, R. A., Yost, M. G., & Palcisko, G. J. (2003). Time-location analysis for exposure assessment studies of children using a novel global positioning system instrument. *Environmental Health Perspectives*, *111*(1), 115–122. https://doi.org/10.1289/ehp.5350

Fennell, D. A. (2017). *Tourism ethics* (2nd ed.). Channel View Publications.

Fisher, P., & Dobson, J. (2003). Who knows where you are, and who should, in the era of mobile geography? *Geography*, *88*(4), 331–337. https://www.jstor.org/stable/40573887

Gaggioli, A., Pioggia, G., Tartarisco, G., Baldus, G., Corda, D., Cipresso, P., & Riva, G. (2011). A mobile data collection platform for mental health research. *Personal and Ubiquitous Computing*, *15*, 241–251. https://doi.org/10.1007/s00779-011-0465-2

Golledge, R. G., Klatzky, R. L., Loomis, J. M., Speigle, J., & Tietz, J. (1998). A geographical information system for a GPS based personal guidance system. *International Journal of Geographical Information Science*, *12*(7), 727–749. https://doi.org/10.1080/136588198241635

Golledge, R. G., Loomis, J. M., Klatzky, R. L., Flury, A., & Yang, X. L. (1991). Designing a personal guidance system to aid navigation without sight: progress on the GIS component. *International Journal of Geographical Information Systems*, *5*(4), 373–395. https://doi.org/10.1080/02693799108927864

Goodchild, M. F., & Janelle, D. G. (1984). The city around the clock: Space—time patterns of urban ecological structure. *Environment and Planning A*, *16*(6), 807–820. https://doi.org/10.1068/a160807

Hägerstrand, T. (1970). What about people in regional science? *Papers of the Regional Science Association*, *24*, 7–21.

Hall, C. M., & Ram, Y. (2020). Protecting privacy in tourism–A perspective article. *Tourism Review*, *75*(1), 76–80. https://doi.org/10.1108/TR-09-2019-0398

Hallo, J. C., Beeco, A., Goetcheus, C., McGee, J., McGehee, N. G., & Norman, W. C. (2012). GPS as a method for assessing spatial and temporal use distributions of nature-based tourists. *Journal of Travel Research*, *51*(5), 591–606. https://doi.org/10.1177/0047287511431325

Harder, H., Bro, P., Tradisauskas, N., Alexander, T., & Nielsen, S. (2008). Tracking visitors in public parks – experiences with GPS in Denmark. In J. van Schaick & S. C. van der Spek (Eds), *Urbanism on track: Application of tracking technologies in Urbanism*, (pp. 65–77). IOS Press BV.

Hardy, A., & Aryal, J. (2020). Using innovations to understand tourist mobility in national parks. *Journal of Sustainable Tourism*, *28*(2), 263–283. https://doi.org/10.1080/09669582.2019.1670186

Huang, X., Zhang, L., & Ding, Y. (2017). The Baidu Index: Uses in predicting tourism flows–A case study of the Forbidden City. *Tourism Management*, *58*, 301–306. https://doi.org/10.1016/j.tourman.2016.03.015

Itami, R. M. (2003). *RBSim3: Agent-based simulations of human behaviour in GIS Environments using hierarchical spatial reasoning*. [Paper presentation]. ModSim 2003 International Congress on Modelling and Simulation, Townsille, Australia.

Janelle, D. G., Goodchild, M. F., & Klinkenberg, B. (1988). Space-time diaries and travel characteristics for different levels of respondent aggregation. *Environment and Planning A: Economy and Space*, *20*(7), 891–906. https://doi.org/10.1068/a200891

Le Faucheur, A., Abraham, P., Jaquinandi, V., Bouyé, P., Saumet, J. L., & Noury-Desvaux, B. (2008). Measurement of walking distance and speed in patients with peripheral arterial disease: a novel method using a global positioning system. *Circulation*, *117*(7), 897–904. https://doi.org/10.1161/CIRCULATIONAHA.107.725994

Lee, P., Zach, F. J., & Chung, N. (2021). Progress in smart tourism 2010–2017: A systematic literature review. *Journal of Smart Tourism*, *1*(1), 19–30. https://doi.org/10.52255/smarttourism.2021.1.1.4

Li, A., Zhao, P., He, H., & Axhausen, K. W. (2020). Understanding the variations of micro-mobility behavior before and during COVID-19 pandemic period. *Arbeitsberichte Verkehrs-und Raumplanung*, *1547*. https://doi.org/10.3929/ethz-b-000430395

Lyon, D. (2001). *Surveillance society: Monitoring everyday life*. Open University Press.

Maeda, Y. (2002). *Evaluation of a GPS-based guidance system for visually impaired pedestrians*. [Paper presentation]. Proceedings of the Technology and Persons with Disabilities Conference, Los Angeles, CA.

McKercher, B., Shoval, N., Ng, E., & Birenboim, A. (2012). First and repeat visitor behaviour: GPS tracking and GIS analysis in Hong Kong. *Tourism Geographies*, *14*(1), 147–161. https://doi.org/10.1080/14616688.2011.598542

Miskelly, F. (2004). A novel system of electronic tagging in patients with dementia and wandering. *Age and Ageing*, *33*(3), 304–306. https://doi.org/10.1093/ageing/afh084

Miskelly, F. (2005). Electronic tracking of patients with dementia and wandering using mobile phone technology. *Age and Ageing*, *34*(5), 497–499. https://doi.org/10.1093/ageing/afi145

Modsching, M. (2008). Using location-based tracking data to analyze the movements of city tourists. *Information Technology & Tourism*, *10*(1), 31–42. https://doi.org/10.3727/109830508785059011

Munro, J., Kobryn, H., Palmer, D., Bayley, S., & Moore, S. A. (2019). Charting the coast: Spatial planning for tourism using public participation GIS. *Current Issues in Tourism*, *22*(4), 486–504. https://doi.org/10.1080/13683500.2017.1346589

Navarro-Ruiz, S., Casado-Díaz, A. B., & Ivars-Baidal, J. (2020). Modelling the intra-destination behaviour of cruise visitors based on a three-dimensional approach. *Journal of Destination Marketing & Management*, *18*, 100470. https://doi.org/10.1016/j.jdmm.2020.100470

Nielsen, N. (2010). *Approaches to GPS-survey of tourist movements within a North Sea Island destination*. [Paper presentation]. ENTER 2010 Conference, Lugano, Switzerland.

Padrón-Ávila, H., & Hernández-Martín, R. (2020). How can researchers track tourists? A bibliometric content analysis of tourist tracking techniques. *European Journal of Tourism Research*, *26*, 2601. https://doi.org/10.54055/ejtr.v26i.1932

Parks, D. N., & Thrift, N. (1980). *Times, spaces and places: A chronogeographic perspective*. John Wiley & Sons.

Pettersson, R., & Zillinger, M. (2011). Time and space in event behaviour: tracking visitors by GPS. *Tourism Geographies*, *13*(1), 1–20. https://doi.org/10.1080/14616688.2010.529932

Peuquet, D. (2002). *Representations of space and time*. The Guilford Press.

Phillips, M. L., Hall, T. A., Esmen, N. A., Lynch, R., & Johnson, D. L. (2001). Use of global positioning system technology to track subject's location during environmental exposure sampling. *Journal of Exposure Analysis and Environmental Epidemiology*, *11*, 207–215. https://doi.org/10.1038/sj.jea.7500161

Qian, C., Li, W., Duan, Z., Yang, D., & Ran, B. (2021). Using mobile phone data to determine spatial correlations between tourism facilities. *Journal of Transport Geography*, *92*, 103018. https://doi.org/10.1016/j.jtrangeo.2021.103018

Quiroga, C. A., & Bullock, D. (1998). Travel time studies with global positioning and geographic information systems: an integrated methodology. *Transportation Research C: Emerging Technologies*, *6*(1–2), 101–127. https://doi.org/10.1016/S0968-090X(98)00010-2

Raento, M., Oulasvirta, A., & Eagle, N. (2009). Smartphones: An emerging tool for social scientists. *Sociological Methods & Research*, *37*(3), 426–454. https://doi.org/10.1177/0049124108330005

Raper, J. (2000). *Multidimensional geographic information science*. Taylor and Francis.

Renenger, A. (2002). Satellite tracking and the right to privacy. *Hastings Law Journal*, *53*, 549–565.

Russo, A. P., Clave, S. A., & Shoval, N. (2010). Advanced visitor tracking analysis in practice: explorations in the PortAventura theme park and insights for a future research agenda. In U. Gretzel, R. Law, & M. Fuchs (Eds), *Information and communication technologies in tourism*, (pp. 159–170). Springer.

Shoval, N. (2008). Tracking technologies and urban analysis. *Cities*, *25*(1), 21–28. https://doi.org/10.1016/j.cities.2007.07.005

Shoval, N., Auslander, G. K., Freytag, T., Landau, R., Oswald, F., Seidl, U., Wahl, H.-W., Werner, S., & Heinik, J. (2008). The use of advanced tracking technologies for the analysis of mobility in Alzheimer's disease and related cognitive diseases. *BMC Geriatrics*, *8*, 1–12. https://doi.org/10.1186/1471-2318-8-7

Shoval, N., & Isaacson, M. (2007a). Tracking tourists in the digital age. *Annals of Tourism Research*, *34*(1), 141–159. https://doi.org/10.1016/j.annals.2006.07.007

Shoval, N., & Isaacson, M. (2007b). Sequence alignment as a method for human activity analysis in space and time. *Annals of the Association of American Geographers*, *97*(2), 282–297. https://doi.org/10.1111/j.1467-8306.2007.00536.x

Shoval, N., & Isaacson, M. (2010). *Tourist mobility and advanced tracking technologies*. Routledge.

Shoval, N., McKercher, B., Ng, E., & Birenboim, A. (2011). Hotel location and tourist activity in cities. *Annals of Tourism Research*, *38*(4), 1594–1612. https://doi.org/10.1016/j.annals.2011.02.007

Sinclair, M., Mayer, M., Woltering, M., & Ghermandi, A. (2020). Using social media to estimate visitor provenance and patterns of recreation in Germany's national parks. *Journal of Environmental Management*, *263*, 110418. https://doi.org/10.1016/j.jenvman.2020.110418

Soe, H., & Sein, M. M. (2017). Tourist guide information system using google map and GPS. *International Journal of Advanced Engineering Research and Science*, *4*(3), 237097. https://dx.doi.org/10.22161/ijaers.4.3.32

Tchetchik, A., Fleischer, A., & Shoval, N. (2009). Segmentation of visitors to a heritage site using high resolution time-space data. *Journal of Travel Research*, *48*(2), 216–229. https://doi.org/10.1177/0047287509332307

Terrier, P., & Schutz, Y. (2005). How useful is a satellite positioning system (GPS) to track gait parameters? A review. *Journal of NeuroEngineering and Rehabilitation*, 2, 28. https://doi.org/10.1186/1743-0003-2-28

Thornton, P. R., Williams, A. M., & Shaw, G. (1997). Revisiting time-space diaries: An exploratory case study of tourist behaviour in Cornwall, England. *Environment and Planning A: Economy and Space*, *29*(10), 1847–1867. https://doi.org/10.1068/a291847

Tussyadiah, L. P., & Zach, F. J. (2012). The role of geo-based technology in place experiences. *Annals of Tourism Research*, *39*(2), 780–800. https://doi.org/10.1016/j.annals.2011.10.003

van der Spek, S. C. (2008). Spatial metro: tracking pedestrians in historic city centres. In J. van Schaick & S. C. van der Spek (Eds), *Urbanism on track: Application of tracking technologies in urbanism* (pp. 79–102). IOS Press.

Volchek, K., Liu, A., Song, H., & Buhalis, D. (2019). Forecasting tourist arrivals at attractions: Search engine empowered methodologies. *Tourism Economics*, *25*(3), 425–447. https://doi.org/10.1177/1354816618811558

Walden-Schreiner, C., Rossi, S. D., Barros, A., Pickering, C., & Leung, Y. F. (2018). Using crowd-sourced photos to assess seasonal patterns of visitor use in mountain-protected areas. *Ambio*, *47*(7), 781–793. https://doi.org/10.1007/s13280-018-1020-4

Wilson, C. (1998). Activity pattern analysis by means of sequence-alignment methods. *Environment and Planning A: Economy and Space*, *30*(6), 1017–1038. https://doi.org/10.1068/a301017

Wilson, C. (2008). Activity patterns in space and time: calculating representative Hagerstrand trajectories. *Transportation*, 35, 485–499. https://doi.org/10.1007/s11116-008-9162-z

Wolf, J., Guensler, R., & Bachman, W. (2001). *Elimination of the travel diary: Experiment to derive trip purpose from GPS travel data.* [Paper presentation]. 80th annual meeting of the Transportation Research Board, Washington, DC.

Xia, J., Arrowsmith, C., Jackson, M., & Cartwright, W. (2008). The wayfinding process relationships between decision-making and landmark utility. *Tourism Management*, *29(3)*, 445–457. https://doi.org/10.1016/j.tourman.2007.05.010

Xia, J., Zeephongsekul, P., & Arrowsmith, C. (2009). Modelling spatio-temporal movement of tourists using finite Markov chains. *Mathematics and Computers in Simulation*, *79*(5), 1544–1553. https://doi.org/10.1016/j.matcom.2008.06.007

Xu, Y., Xue, J., Park, S., & Yue, Y. (2021). Towards a multidimensional view of tourist mobility patterns in cities: A mobile phone data perspective. *Computers, Environment and Urban Systems*, *86*, 101593. https://doi.org/10.1016/j.compenvurbsys.2020.101593

Zheng, W., Huang, X., & Li, Y. (2017). Understanding the tourist mobility using GPS: Where is the next place? *Tourism Management*, *59*, 267–280. https://doi.org/10.1016/j.tourman.2016.08.009

Zito, R., D'este, G. M., & Taylor, M. A. P. (1995). Global positioning in the time domain: how useful a tool for intelligent vehicle-highway systems? *Transportation Research C: Emerging Technologies*, *3*(4), 193–209. https://doi.org/10.1016/0968-090X(95)00006-5

Part 3 Theories of Tourist Practice, Experience, and Encounter

Introduction
Theories of Tourist Practice, Experience, and Encounter

C. Michael Hall

The third section continues the examination of theoretical ideas, but in a more specific tourism context that reflects the diversity of the tourism experience. Landscapes are strongly promoted and even shaped to attract tourists and are an important part of destination encounters. Terkenli (Chapter 12) describes the relationship between tourism and landscape as 'essential'. The chapter positions tourism landscapes within the wider body of landscape studies before examining the place of landscape in tourism geographies. Terkenli notes the visual and relational aspects of landscape in tourism and how tourist landscapes enter the tourist consciousness. Indeed, commoditised images of tourist destinations are now conveyed by a range of visual information and communications technologies (ICTs), social media (e.g., Facebook, Instagram, Whatsapp), virtual reality tourism, and mass media that blur geographical differentiation and the distinction between 'authentic' and staged and the familiar and the 'exotic', in the landscape images conveyed. Arguably, such theorisations are part of moving 'tourism theory forward from the conception of Urry's gaze, by bringing landscape theory into consideration' (Knudsen et al., 2008, p. 132) (see also Larsen, Chapter 18; Knudsen et al., 2012).

Chapter 13 by de Bernadi builds upon some of the same territory as Terkenli in her discussion of the development of authenticity in tourism scholarship. As she notes, authenticity is probably one of the most used concepts in tourism but at the same time is also a widely debated and contested. She observes that issues of authenticity are particularly important to understanding the tangible and intangible aspects of heritage as well as representation in marketing, while the concept is becoming reassessed in the context of fictional and virtual realities. Shehade and Stylianou-Lambert (2020) argued that there should be a balance between different ways to conceptualise authenticity to accommodate the different values of several stakeholders. De Bernadi accommodates this perspective in stressing the importance of different paradigmatic assumptions in approaching research on authenticity and of researcher reflexivity.

Similarly, Ramkissoon highlights the multi-dimensional nature of place attachment as a construct (Chapter 14). She highlights how the concept intersects with social, psychological,

The Wiley Blackwell Companion to Tourism, Second Edition. Edited by C. Michael Hall.

environmental, and behavioural sciences, this chapter discusses the multi-dimensional nature of place attachment and its relevance to a range of tourism research areas with respect to tourists, businesses, and residents. It notes how place attachment is central to place memorable experiences in tourism but also more recent relevance with respect to resilience and COVID-19. In addition, Ramkissoon suggests that an understanding of place attachment will contribute to community-based sustainable tourism and wellbeing.

An area of applied social psychology that is also concerned with wellbeing is that of mindfulness (Moscardo, 2009). In Chapter 15 Stankov et al. examine the diverse relationships between mindfulness and tourism, while also noting its connectivity to place and setting. However, they note that more informational material, industry case studies, and education and training opportunities are needed to help the tourism sector benefit from the mindfulness trend. A similar focus on wellbeing is also noted in Chapter 16 by Mau and Roessler which focuses on long-distance walking. As they argue the meaning of long-distance walking more broadly has shifted from pilgrimage to concern with the demands of modern life and contact with nature. However, they note that nearly all forms of long-distance walking show how long-distance walking is often part of a search for deeper and more rewarding experiences rather than hedonistic pleasure. As a result, they position such behaviours with the slow tourism movement, an observation that may also hold for much of the interest in mindfulness and tourism.

The last three chapters of the section move away from the focus on more individual tourist practices and experiences to provide a wider perspective on the influences on tourism encounters. In Chapter 17 Gillen examines the relations between tourism and development theory and the extent to which the development and tourism industries reproduce or challenge 'developed/developing' and centre/periphery relations and understandings through their commodities and promotion is subject to historical and ongoing concern. Importantly, he argues that the late twentieth and early twenty-first centuries have not seen separation from earlier colonial encounters between host and guest that are epitomized through tourism, with tourism and development shaping relations between the Global North and Global South over time. As a result, Gillen concludes by challenging the notion that tourism and development are tenets of globalization in a borderless world (see also Dwyer, Chapter 21) and that the language and tools tourist and development researchers and practitioners use to achieve greater equity and justice has not been adequate.

As Gillen noted in Chapter 17, the goal of modernizing places and peoples has been fuelled through the process of raising distinctions between 'us' and 'them' that are foundational to both development and tourism (Said, 1978). Arguably, one of the tourism concepts that most engages in such notions is that of the tourist gaze (Urry, 1990). Larsen (Chapter 18) reflects on the different stages of the tourist gaze concept. He suggests that the 1.0 version explored the discursive social and cultural ordering of the tourist gaze, the 2.0 version described its global ordering, and the 3.0 version detailed the embodied and resource ordering of the gaze. Thus, reflecting the intellectual movement of Urry and other gaze scholars as the concept became further developed as well as new external realities and paradigmatic and theoretical lenses. One of these lenses is that of postmodernism and, in the final chapter of the section (Chapter 19) Minca and Oakes present a fascinating reflection on their previous work on postmodernism and tourism (Minca & Oakes, 2014) and its wider development and engagement with critical social theory since the last edition of this book. As they note, they did not consider that their 2014 argument that the biopolitical should be a key element in understanding the regimes of spatio-temporal control of the tourist in some specific spaces. Yet, post COVID-19 and the growth and implementation of biometric technologies and the associated forms of bordering, body scanning and profiling have now become a travel norm. Far more than what they could have expected. Yet, significantly, they still suggest 'that tourism be understood as a comprehensive analytic capable of reassembling our understandings of contemporary life, an analytic through which new insights into the 'social,' 'cultural,' and 'political,' along with their related conceptualizations, are gained' (Minca & Oakes, 2014, p. 301).

Many of the various chapters in this section reflect on the changes that have occurred in the past decade. Importantly, they note that changes, such as COVID-19 and the growth of virtual and communication technologies, affect not only tourist practices and experiences but also the theoretical lenses by which we come to understand them.

References

Knudsen, D. C., Metro-Roland, M. M., Soper, A. K., & Greer, C. E. (Eds). (2008). *Landscape, tourism, and meaning*. Ashgate.

Knudsen, D. C., Rickly-Boyd, J. M., & Metro-Roland, M. M. (2012). Landscape perspectives on tourism geographies. In J. Wilson (Ed.), *The Routledge handbook of tourism geographies* (pp. 201–206). Routledge.

Minca, C., & Oakes, T. (2014). Tourism after the postmodern turn. In A. A. Lew, C. M. Hall & A. M. Williams (Eds), *Wiley Blackwell companion to tourism* (2nd ed., pp. 294–303). Wiley-Blackwell.

Moscardo, G. (2009). Understanding tourist experience through mindfulness theory. In M. Kozak & A. Decrop (Eds), *Handbook of tourist behaviour: Theory and practice* (pp. 99–115). Routledge.

Said, E. (1978). *Orientalism*. Routledge.

Shehade, M., & Stylianou-Lambert, T. (2020). Revisiting authenticity in the age of the digital transformation of cultural tourism. In V. Katsoni & T. Spyriadis (Eds), *Cultural and tourism innovation in the digital era* (pp. 3–16). Springer, Cham.

Urry, J. (1990). *The tourist gaze: Leisure and travel in contemporary societies*. Sage.

Chapter 12

Tourism and Landscape

Theano S. Terkenli

Context and Scope: An Introduction

It has been previously argued (Cartier & Lew, 2005; Knudsen et al., 2012; Terkenli, 2014) that the relationship between tourism and landscape is irrevocable, uncontested — even essential to tourism. During the past four decades, the socio-cultural constitution of the landscape has been steadily gaining ground in all scientific fields pertaining to the landscape (Zube & Pitt, 1981; Nassauer, 1988; Cosgrove, 1998; Tress & Tress, 2001), thus opening the way for the study of the implications of landscape for tourism and vice versa. Likewise, the continuous explosion of tourism studies has been paving the ground for the development of this interdisciplinary field. Accordingly, scientific interest in the landscape-tourism interrelationship and in landscapes of/and tourism has been growing in the twenty-first century, as reflected in the increase of relevant debates, research and publications.

The inter-/multi-disciplinary field of landscape studies is ever evolving, consolidating gains and breaking ground, in slow but steady steps. This development has been very significantly based, at least in the Western world, on the landscape concept as promulgated by the European Landscape Convention (ELC) (Council of Europe (CoE), 2000). The ELC landscape definition encapsulates the evolution of approaches, knowledge and understanding of the landscape historically gained by Geography and the Social Sciences: 'landscape is an area, as perceived by people, whose character is the result of the action and interaction of natural and/or human factors' (CoE, 2000, p. 9). Nonetheless, landscape remains a highly complex, synthetic and controversial concept and a challenging field of scientific research, policy and practice. It has been conceptualised as a literal and/or metaphorical image of and stage set for the humanised environment (Sauer, 1925/2007; Daniels & Cosgrove, 1988; Cosgrove, 1998) and viewed as representing both a medium and an outcome of human perception, experience, and action, where vision predominates (Lothian, 1999), whereby people's landscape notions and preferences for certain types and characteristics of landscape are largely culturally driven, socially

The Wiley Blackwell Companion to Tourism, Second Edition. Edited by C. Michael Hall.

contested and politically/institutionally articulated (Meinig, 1979; Jackson, 1984; Cosgrove, 1998; Burton, 2012; Roe, 2014; Burley & Machemer, 2016; Olwig, 2019).

Reference to landscape in connection with tourism has come in various forms and guises: either explicitly or implicitly; as a background to other analyses or as a key issue (Terkenli, et al., 2021; Land, 2021). It has been assigned varying degrees of weight and centrality, in various related subfields - all areas with remarkable growth and potential for future research expansion (Knudsen et al., 2012; Terkenli, in press). Tourism advances, at the same time, have embraced a multitude of cultural perspectives to the study of the human environment, such as corporeality, materiality, sustainability, identity, locality, consumption, natural and cultural heritage, justice and participatory practices — all in close connection to landscape. As will be shown further down, progress in the study of the relationship between tourism and landscape(s), through various ontological and epistemological approaches in various related scientific areas, has recently taken off, although still few studies clearly address this relationship and/or dissect it in depth.

During the 1980s and 1990s, critical approaches to landscape conceptualisation and analysis were firmly established in Geography and other Social Sciences, with an emphasis on addressing the representational qualities of landscape, the contestation within landscape, and landscape as a text to be read (Knudsen et al., 2012). Since the 1990s, new cultural geography has focused on variable and contested identity dimensions in landscape formation, as well as their representational and ethical qualities, including perspectives on race, ethnicity, class, gender, sex and sexuality. Twenty-first century approaches consciously seek to complicate established positions by signaling multiple and shifting points of view, in the context of leisure economy production and consumption (Cartier & Lew, 2005). More specifically, a fairly new paradigm in geography, *more-than-representational geographies*, brings out and delves into the relationship of the tourist with the visited landscape, through lived, interactive experience (Lorimer, 2005; Wylie, 2007), with an emphasis on 'sensescapes' and emotions/affect, tied to place and landscape. These ontological and epistemological advances have laid the ground for the growth and development of an interdisciplinary scientific area of study of the interrelationship between tourism and landscape.

The consequent proliferation of critical, post-humanist, post-structuralist and postmodern perspectives and insights into landscapes of tourism has opened up their study beyond traditional scientific boundaries, and rendered this broader field distinctly more inter-disciplinary, multi-disciplinary and trans-disciplinary. However, as this chapter continues to argue, the engagement with inter-subjective, embodied and multisensorial tourism performance in the landscape needs to be played out in more complexity and the tourist's encounter with space, generally speaking, developed further. This review builds on, updates and expands on its previous edition (Terkenli, 2014).

The Place of Landscape in Tourism Geography: A Critical Theorisation

The Relationship of Tourism and Landscape

The tourism industry has been both denounced as a defiler of landscapes and hailed as a quintessentially modern medium of globalising standards of identity and development for contemporary landscapes (Jackson, 1984; Sack, 1992; Urry, 1995). Relevant literature increasingly describes and cautions against the despoliation of landscape by tourism — often destroying the very basis of its development — as well as both positive and negative challenges posed by trends of globalisation to landscapes of tourism, such as practices of branding destination landscapes or risks induced by 'overtourism' (Terkenli, in press). On the other hand, the landscape is opened up through tourism; the tourist, like a modern pilgrim, seeks regeneration in the realms of pleasure, dream, tradition, arts and sports, prompted to an on-going quest for novel tourism landscape destinations. Landscape is key to the development, marketing/promotion, and consumption of tourism destinations, to triggering and sustaining tourism markets, and to enticing tourist dreams, fantasies, and behaviours. Landscapes of tourism increasingly stage and reflect recreational trends, multifunctional livelihood systems, conflicts and opportunities for employment and income generation, as well as human, cultural, and natural resource management and use.

Thus, on the basis of its both tangible and intangible character, landscape constitutes a most significant geographical medium in the analysis of relationships that develop between tourist and visited location. 'As a geographical medium conceived and appropriated through the senses and the power of cognition and symbolism, landscape represents the first and most enduring medium of contact between tourist and prospective or consumed place of travel; through acquired photographs, it becomes a traveler's lasting memoir' (Terkenli, 2014, p. 283). Accordingly, Knudsen et al. (2012) stress the co-constructional attribute of tourism that a 'landscape approach' to tourism captures: 'Landscapes, besides being material objects, are the intentional or unintentional symbols of power relations and ideology from which locals and tourists seek to make meaning wherein in making meaning those very same locals and tourists alter the underlying power relations and ideologies' (Knudsen et al., 2012, p. 6).

Through promotion, sustenance and transformation of their specific functions, tourist/tourism landscapes are among the most significant and conflictual cultural grounds in the formation of identity and the negotiation of local development (Terkenli, 2000). The importance of the role of tourism for identity formation in the landscape comes through the multiplicity of contextualised tourist knowledges, imaginaries and pursuits, in dialectical relationship with place and the landscape. Due to its relational character, landscape becomes a medium of cultural and personal identity construction both for the local side and for the side that engages in the gazing and the consuming (Figure 12.1). 'It is in its immediacy, intertextuality

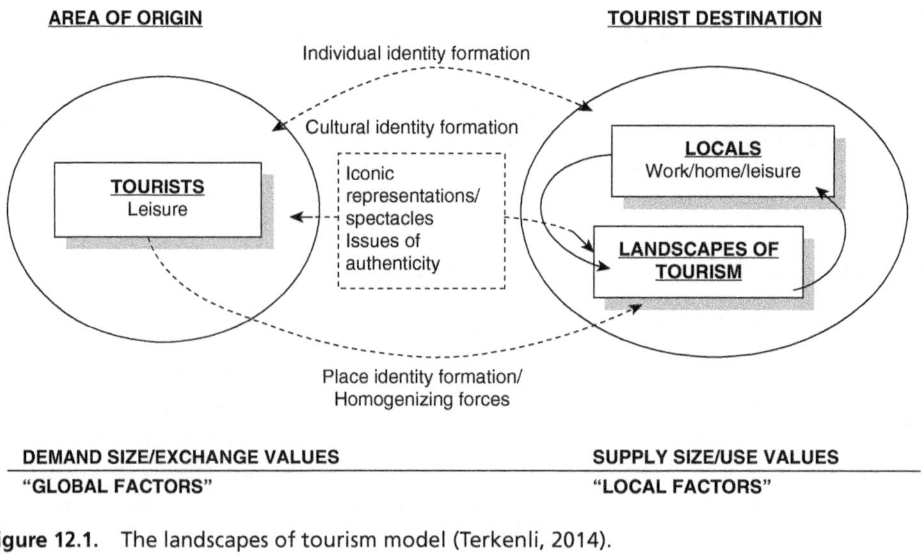

Figure 12.1. The landscapes of tourism model (Terkenli, 2014).

and experiential character that the...landscape becomes a social interface where local and global perspectives and other dimensions of tourism studies come together in the ready construction and consumption of place identity' (Terkenli, 2000, p 185–186). Furthermore, through intricately interwoven demand and supply tourism processes, place and landscape identities are under processes of constant construction and transformation.

As a focus of research, landscape requires contextual interpretation and cannot be detached from questions of positionality and from its historical and socio-cultural context. Objectivity and subjectivity also intertwine in the creation of meaning, through landscape, necessitating variable readings of the contact between the subject and the object (here the observer and the landscape, respectively). This contact takes place through variable geographical, social, ethical and historical lenses of perceiving, understanding and communicating individual or cultural reflections of the world. This exercise finds direct application in the construction and study of landscapes of tourism, where such issues became especially relevant, in recent decades, through the proliferation of critical scientific perspectives into historical, anthropological, sociological and psychological dimensions of the tourist experience. It finds application in processes of social construction of space and identity, while elucidating the articulation of new power relations, which are emerging out of contemporary tourism landscape processes and practices. It also finds application in the establishment of common grounds and understandings in cultural readings and representations of landscape. Knudsen et al. (2008, p. 5) concede that 'the locus of study for tourism is and should be the landscape. Tourism, by definition, takes place in a "tourism landscape"'.

This brings us to the confusion that exists in scholarly tourism literature around the term's 'tourism' vs. 'tourist' in conjunction with the concept of landscape (Skowronek et al., 2018; Terkenli et al., 2021). The term 'tourism landscape' is intended to refer to the processes through which a landscape, activity, development and so forth is shaped to serve purposes of tourism (also, i.e., 'landscapes of tourism'). The term 'tourist landscape' (or 'touristed' landscapes, in Knudsen et al. (2012)) indicates the ways/reasons/processes in/through which such landscapes are substantiated and/or appropriated via the phenomenon of tourism. In other words, the term 'tourism landscapes' implies the ways in which these landscapes are produced, whereas 'tourist landscapes' rather implies the ways in which these landscapes are consumed (Terkenli et al., 2021). The term 'touristed landscapes' (Cartier & Lew, 2005; Knudsen et al., 2012) carries the same latter connotation and tends to be employed in similar ways as the term 'tourist landscapes', although heavier or lighter emphasis may be placed on the resultant state of such landscapes stemming from tourism development (Terkenli, in press).

The Visual and Relational Aspects of Landscape in Tourism

Characteristics of modern European cultural landscapes were already established by the seventeenth century, imbuing the definition of the landscape with notions of vistas, prospects, or views of scenery of the land (Daniels & Cosgrove, 1998). These notions have accompanied landscape development and appropriation, up to our times, i.e., in panoramic views, in landscapes of characteristic forms, in the ephemeral, in the picturesque, in the perspective, and in other landscape principles and models, ubiquitous to the tourism industry and the 'tourist gaze' (Urry, 1990): photographic mementos, postcards, advertisements and, generally speaking, tourism marketing iconography. In all of these cases, the landscape has been *staged* by tourism planning and development initiatives for purposes of tourist consumption. Through tourism, often, the imaginary, the artificial, the staged and the make-believe replace the original; the sight is turned into spectacle, reality into performance.

At the most basic level of tourist landscape analysis, it has been recognised that tourist landscapes, through their images, are advertised, replicated, signified and communicated around the world electronically, or through travel literature, prompting tourists to become, in MacCannell's (1976) words, amateur semioticians and collectors of signs. 'Imagery is one of the most researched aspects of tourism marketing' (Pritchard & Morgan, 2001, p. 167). Tourism marketing reproduces images and discourses about landscapes, through representations of cultural signs, on the basis of which the tourist, through processes of experiential re-interpretation of the sign, may assess the sight and validate the meanings of the visited landscape within the predominant discourse (Terkenli, 2014). Iconographical methods of construction, signification, decodification, and deconstruction are central to the making of tourist landscapes (Norton, 1996; Stefanou, 2000). Images of tourist destinations, mass-produced by the tourist industry with the purpose of establishing and promoting tourism landscape identities, accordingly, serve to shape 'distinctive' characteristics of the tourism destination. Such processes, supported and supplemented by other visual means of contemporary information and communications technologies (ICTs), social media (e.g., Facebook, Instagram), and mass media (TV, video) blur geographical differentiation and the distinctions 'authentic'- staged and familiar - 'exotic', in the landscape images conveyed.

The observer-landscape relationship, as staged and played out in tourist landscapes, has been increasingly explored in the context of tourism studies, during the past couple of decades, e.g., in the analysis of aspects of tourism destinations as cultural landscapes (Ringer et al. cited in Pritchard & Morgan, 2001; Daugstad, 2008; Norton, 1996). Through its relationship with a viewer or tourist, the landscape essentially becomes a mediational 'stage' where historical, socio-economic and cultural processes articulating and articulated in space are mirrored on the land. Decidedly, neither is it desirable to generalise the multiplicity of shifting positions of viewing or gazing, nor, however, to dismiss or diminish the importance of historically predominant landscape interpretations (Terkenli, 2014). In other words, what is upheld here is not an essentialist notion of the tourist landscape, but rather a culturally ambivalent, socially constructed and historically specific notion of the landscape that invites multiple and fluid interpretations, among which certain particular interpretations have historically prevailed.

Finally, the connection between landscape and tourism is not restricted either to the representational/performative, or to the essentially geographical/physical nature of the travel experience (Terkenli, 2018). It extends to the *pleasure* sought in the experience, a component of tourism that has become much more central and predominant in the historical evolution of twentieth-century forms of tourism (Löfgren, 1999) and acknowledged through theories of emotion and affect, as well as *more-than-representational geographies* of human-landscape interaction (Lorimer, 2005). The readiness and amenability of landscape to human relatedness and intervention transform it into a veritable stage for play and recreation (Carmichael, 1998). Indeed, the enduring intensity of pleasure sought and found in the landscape since the Renaissance, in the context of an emerging European bourgeoisie, expresses something profound and constant about the human condition (Rose, 1996), something that links landscape and pleasure or attraction inextricably together, thus highlighting the great significance of the emotional/affective dimension to the relationship of the visitor with the visited landscape.

Contemporary Trends of Landscape Transformation

In the context of a fast-changing postwar world, processes of geographical change have been finding in the landscape their most natural and expressive medium and outlet. Such forces, functioning at various geographical scales and subject to complex political, economic and ideological formulations and repercussions, have been creating new schemata of geographical movement, differentiation, connectivity, consumption, and pleasure. Their imprint and reflection on the landscape is invariably substantiated and articulated through landscape forms, practices, functions and meanings indicative of *a new cultural economy of space* (Terkenli & d'Hauteserre, 2006). Some of the characteristics of this new cultural economy of space - especially intensified through globalising patterns of tourism and recreation — are: fusion of boundaries; merging of disparate worlds; dissolution of geographical particularity and identity; spatio-temporal deconstruction, redefinition and decentering, as well as the dissemination and communication of all above changes through new ICT's. They are characterised by a growing disassociation of these new schemata from geographical location and distinct place characteristics. Thus, for instance, the omnipresence and flexibilisation of contemporary forms of recreation, leisure and tourism, as fairly new occurrences, have far-reaching implications for the diversity and rich variety of European landscapes, while the historical heritage of European landscapes is coming under increasing pressure (Hazendonk et al., 2008).

These transformative processes have been taking on increasingly globalised dimensions, although their manifestations vary over space, time, and social context (Pritchard & Morgan 2001). For instance, post-national forms of government usher into tourism development notions of the magic, of tropicalisation and of exoticised gardening and landscaping, whereby 'entire inhabited spaces—islands, regions, deserts, city centers, mountains—are gardened as living artworks within a global modernity, invoking and thus bringing to life the grand modernist paradigms of beauty, nature and time. . . cultivated as 'human gardens' within a tourism specific form of modern world system; a fair ground in which a global middle class meets, parties, resources itself and recreates the symbolic order of their worlds' (Picard, 2011, p. 5).

In the meantime, crises brought about by — or exacerbated in — the twenty-first century (climate, energy, hygienic, economic/financial etc.) have been challenging the grand theme and outlook of globalisation and curbing its momentum, with the 'greening' of social-cultural practices (including tourism), the 4th industrial/technological revolution and the current geopolitical re-organisation of state allegiances. Global tourism has been experiencing various significant quantitative and qualitative shifts, which recent crises are contributing to or reconfiguring, with a strong bearing on the landscape: a shift from 'mass' to 'mild' tourism, from special-interest tourism to special-purpose tourism, and from tangible to intangible destination experiences. Through these shifts in tourism, the landscape seems to be regaining its place, role and purpose in spatial/land stewardship and decision-making. Such developments are presently central and conducive to the transformation of landscape for purposes of tourism, and clearly pose problems for hegemonic and totalising discourses, so far accommodated by tourism studies, while calling for new theoretical frameworks and epistemologies in tourism-landscape research.

Key Issues and Research Advances in Tourism Landscapes

Setting the General Context

The cultural turn of the social sciences, in the 1990s, signaled an acceleration in the breakdown and blurring of disciplinary boundaries within the academy. A more flexible approach to engagement with space has thus allowed insights from a range of disciplines and subject fields addressing leisure-related issues into the relationship between tourism and the landscape to infiltrate into the geography of tourism (Aitchison et al., 2000), indeed into a series of new 'geographies of tourism'. Simultaneously, the opening-up to matters of space and geographical analysis of other tourism-related disciplines has encouraged the latter to turn their attention to the landscape in the context of leisure and tourism (Terkenli, 2014), resulting into an effort to restructure this interdisciplinary field into a more concerted theoretical and empirically endowed scientific area.

The growth of interest in the relationship of tourism with the landscape rests on the convergence of certain distinct and on-going larger tendencies. Firstly, there is an overarching realisation of the growing degree of modification that tourism development has been imparting on many landscapes around the world, highlighting the shift from landscapes of production to landscapes of consumption, surpassing landscape carrying capacity, and leading to variable change in the pre-existing landscape. 'Sometimes the changes are so profound that it is possible to talk of tourism landscapes in which tourism dominates the uses of the land and the appearance of the area' (Wall, 2000, cited in Jafari, 2000, p. 347). Secondly, international, and especially European, interest in the landscape, landscape policy, planning, assessment, values and analytical methods and tools has skyrocketed in the recent decades, calling for organised action to evaluate, protect and enhance the quality and multifunctionality of landscapes at the local, regional, national and international levels (Klijn et al., 1999; Tress & Tress, 2001). Thirdly, the dominance of structuralist (and more recently postmodern, posthumanist and poststructuralist) critical perspectives to landscape in social and cultural geography(ies) of tourism has been increasingly highlighting the complex interrelationships between the phenomenon of tourism and the construction, representation and consumption of landscape in and out of the context of everyday life.

Two larger periods in the recent history of geographical inquiry, corresponding to two distinct areas of geographical ontology and epistemology, may be identified here regarding the analysis and interpretation of space/landscape in relation to leisure and tourism. The first overarching period — and epistemological approach — to landscape study gathers momentum through the quantitative revolution of the 1950s and the 1960s. It finds application in a large body of work on tourism landscape analysis that is largely apolitical, informed more by economic concerns for landscape development. This body of work on the interrelationships of landscape and tourism constitutes an extensive domain of landscape research in the field of tourism to date and continues to grow. The second period and corresponding epistemological approaches are formulated around more critical perspectives revolving around the historical and material processes of space and landscape production, reproduction, and consumption. At the outset of this chronological period, humanistic methodologies, also prevalent in geographical inquiry at the time, were not much explicitly employed in the study of the relationship tourism-landscape at that time (Relph, 1976; Meinig, 1979; Jackson, 1984), but have been re-instated in geography and the social sciences since then, through more than representational forays into the tourism-landscape interrelationship. More recently, the dialectical relationships society-individual and structure-culture built into the landscape, as illustrated by structuralist interpretations of tourism landscapes in the 1970s and 1980s, have acquired a prominent position in tourism-landscape research, through postmodern, poststructuralist, post-humanist and other epistemological approaches. They have been gaining substantial ground over the previously dominant ontology and epistemology of tourism-landscape scientific inquiry of the first period, shifting from perspectives stressing the physical, material, and absolute nature of space to an emphasis on the socio-cultural, symbolic and relative nature of space.

In the remainder of this section, an epistemological overview of the infusion of tourism-related issues into landscape geography and vice-versa will trace a series of themes in landscape geographies of tourism or tourism landscape analysis. Each of these themes represents an important position in geographical discourses of leisure/tourism, landscape, and culture. A more exhaustive presentation of even the most significant work in these fields would be ineffable, within the constraints of this review.

The First Overarching Period

Admittedly, the earliest geographical research on leisure and tourism landscapes, during the first half of the twentieth century, was purely concerned with the classification of scenic quality and land use (Aitchison et al., 2000). This preoccupation with landscape classification, assessment and evaluation, with the aid of land use mapping and other quantitative methodologies adhering to positivist epistemologies, continues to inform a large portion of research in the area, i.e. through spatial analysis (Hall & Page, 1999; Roovers et al., 2002; García Pérez, 2002). In the late 1960s and early 1970s, landscape assessment was mainly applied to the evaluation of recreation landscapes, i.e. aesthetic studies measuring human preferences and responses to landscape alterations. In the 1980s, landscape assessment came to be used as a tool for description/classification of landscape character. These trends have since been superseded by more holistic approaches to landscape, based on the concept of Landscape Character Assessment, with a clearer link to sustainable development and environmental protection (Hall & Page, 2006; Coccossis, 2004; Cassar, 2010), and, more recently, to the impacts of European economic recession on tourism and place/landscape.

Aitchison et al. (2000, p. 12) point out that 'while leisure studies was embracing sociological concepts from Marxist, structuralist and humanist theory in the 1960s and 1970s, tourism studies and recreation research appeared to hold on to their more positivist paradigm'. They add that analyses of recreation and tourism continued to rely on quantitative methods to devise theories which mapped, modeled, coded and classified recreation and tourism development, provision participation, and impacts. Löfgren (1999) places the period of popularity of such landscape evaluation and measurement techniques in environmental studies and planning during the span of several decades and especially in the 1970s, when scores of studies appeared, with titles like 'Eye pupillary measurement of aesthetic response to forest scenes' (Wenger & Videbeck, 1969) or 'Modeling and predicting human response to the visual recreation environment' (Peterson & Neumann, 1969). Such attempts to generate rich bodies of empirical data and to create tools for natural resource management and planning, from wilderness aesthetics to scenic roads design, 'reflect a period in tourist and leisure management when hard facts, models, and clear taxonomies were in demand' (Löfgren, 1999, p. 83). These ongoing attempts have also aimed towards landscape impact analysis and physical resource assessment for tourism and recreation. They also continue to illustrate the importance of human perception and recognition of what is attractive and valued by different people in relation to recreational time (Hall & Page, 2006). Thus, application of such techniques in a wider recreational context, in the 1990s and 2000s, moved away from land use zoning methods to ones where the impacts of specific activities are evaluated (Hall & Page 2006).

Meanwhile, by the late 1930s and 1940s, another related quantitative approach to tourism landscape analysis had already gained ground, in support of tourism development and planning, and focusing on systematic geographical approaches to tourism landscape morphology, through the employment of spatial analytical techniques. In this context, resort morphology was recognised by tourism geographers and others as early as the 1930s, generated a considerable body of urban and economic analysis in the 1970s, and remains today 'a small but important research thread among planners and tourism geographers' (Meyer-Arendt, 2000, cited in Jafari 2000, pp. 504–505). Thus, the widespread employment of descriptive and predictive models in early recreation and tourism research within Geography continues to be a prevalent and ongoing trend in the geography of tourism to date (Pearce, 1995), but has yet to produce either a strong conceptual or a theoretical base for tourism research. Altogether, a number of models dealing with various aspects of the spatial structures of tourism emerged in the late 1960s, the 1970s and the 1980s, most of which were developed independently of one another, with little or no recognition of, or attempt to build on previous ones (Pearce, 1995). Themes under this umbrella of positivist modeling approaches to tourism and the landscape vary enormously: land use change, planning and management; landscape as a resource for urban, rural, regional, and national development; travel motives and tourist types; natural and cultural resource assessment for recreation and tourism; landscape impact analyses; tourism carrying capacity, etc. Within this positivist paradigm, perhaps the most significant theme in tourism analysis, during the late 1990s-early 2000s, has been a series of perspectives revolving around sustainable development issues, with applications in sustainable tourism development per se — extending to the forms, functions and symbolisms of landscapes of tourism.

A third strand of tourism-landscape research that spans both chronological periods in the development of relevant ontological and epistemological research perspectives, during the second half of the twentieth century, has been the historical perspective into landscape change through tourism. Although the historical perspective has always been present in tourism studies as an introductory overview of current issues, complete historical geographical perspectives to tourism emerged only in the 1990's (Towner, 1996). These have tended to examine the interrelationships between tourism and the landscape historiographically, stressing the impact of changing motives and socio-economic constraints of travel on the landscape (Towner, 1996). By the turn of the century, their tendency towards an analysis of the social history and cultural underpinnings of the relationship tourism-landscape assumed more critical perspectives (Löfgren, 1999). Although analysis of the spatial patterns arising from locational attributes of supply and demand predominated as an organisational concept in this area of research, the particular role of place has also been explored (Towner, 1996). In the past two or three decades, it has been supplemented by anthropological and sociological inputs and lately also by concerns about public participation in landscape governance (Jones & Stenseke, 2011). Within this third strand of tourism-landscape research of the first chronological period, landscape experience and relevance have been negotiated from an increasingly critical standpoint than in the other two research strands.

The Second Overarching Period

This leads us to the second overarching chronological period, which traces the development of perspectives into the relationship of tourism and landscape with the aid of humanistic and stucturalist (to be followed by post-humanist and post-structuralist) approaches. Although a more critical shift to geographical research was already apparent in the 1960s, a Marxist and structuralist reorientation of landscape geography gained momentum only in the 1970s and 1980s. With the post-modern 'cultural turn' of the social sciences in the 1990s, cultural meanings pertaining to landscapes of tourism were built into landscapes of both built and rural milieux (Aitchison et al., 2000), illustrating dialectical relationships between social structures and cultures, on the one hand, and the role of the individual, on the other. Research within the structuralist paradigm on the relationship between tourism and the landscape was initially scant but substantial (Shaw & Williams, 1994; Britton, 1982, 1991; Porter & Sheppard, 1998). However, following contemporary trends in geography and the social sciences more generally, such research was soon superseded by postmodern, postcolonial and poststructuralist approaches, whereby this type of research has since been flourishing.

Postmodern and post-structural theoretical perspectives, together with the growing impact of the field of cultural economy, have significantly contributed to the study of 'the social-cultural nexus of the material and the symbolic in the production, representation and consumption of leisure and tourism landscapes. . .as regimes of signification. . .mediated by sites and processes of leisure and tourism' (Aitchison et al., 2000, pp. 4, 6). Contemporary social and cultural geographies of landscapes of tourism demonstrate how the latter serve as sites and sights of social and cultural inclusion/exclusion, which are not fixed, but are in a constant state of transition (Urry, 1995). Such approaches to landscapes of tourism acknowledge specific landscape contexts of transformation, rather than placing landscape in the context of an overall system of global change and re-organisation — which comes a bit later on. They include a wide variety of themes, such as landscape and place experiences and practices (Crang, 1999; Daugstad, 2008), gender and identity formation (Kinnaird & Hall, 1994; Pritchard & Morgan, 2000), spectacle and performance (Hollinshead, 1999), the gaze and surveillance (Crang, 1999; Hollinshead, 1999; Urry, 1990), attraction and desire (Cartier & Lew, 2005), mobility and materiality (della Dora, 2009), the experiential and non-representational (Prince, 2018, 2019), and other. They call for the development of socio-cultural *geographies* of leisure and tourism, rather than of a more systematic science of a geography of tourism.

Four monographs provide excellent illustrations of work within this paradigm. In *Leisure and tourism landscapes: Social and cultural geographies,* Aitchison et al. (2000), identify a series of poststructuralist discourses on the spatiality of leisure and tourism landscapes, such as the role of gender in constructing and contesting leisure and tourism spaces, places, and landscapes; the place of ethnicity in the production and reproduction of urban landscape; and the interrelation between sexuality and spatiality in city landscapes. *Leisure/tourism geographies: Practices and geographical knowledge* (Crouch, 1999) is a collection of essays on the 'fractured, multiple geographical knowledge... [that is] closer to our lived practice' (Crouch, 1999, p. 13); it considers leisure/tourism as an encounter that exists in place and landscape between people, between people and space, and among people as socialised and embodied subjects with expectations, experience and desire. Cartier and Lew's (2005) edited volume, *Seductions of place. Geographical perspectives on globalization and touristed landscapes*, examines experiences and meanings of the *touristed* landscape, dependent on 'understanding subject positions and subject formations of the tourism, the toured, and those who would work at being both or neither, and from one moment or place to the next' (Cartier & Lew, 2005, p. 4). In the spirit of this past period under consideration, aspects of desire, as well as multiple positions of sensory engagement, attraction, and legibility — ways in which landscapes can be read, imagined, and experienced, from diverse points of view and positions of orientation—are more generally explored. Finally, *Landscape, tourism and meaning*, (Kundsen et al., 2008) offers different views of the intersection between landscape, tourism and identity, and the ways in which these interweave with meaning, in an attempt to 'move tourism theory forward from the conception of Urry's gaze, by bringing landscape theory into consideration' (Knudsen et al., 2008, p. 132) (see also Chapter 18, this volume). However, more in-depth and incisive probes into the subject matter have taken place in the past decade.

First, three substantial efforts to organise and review theory and epistemology in this area of study have been undertaken in recent years, aiming at highlighting key issues and empirical advances, as regards the state of the art and future horizons. The first such undertaking was the chapter 'Landscape perspectives on tourism geographies' by Knudsen et al. (2012), in *The Routledge handbook of tourism geographies,* edited by Wilson (2012), which points to the multi-methodological context of such study, and describes the main research trends in this field in terms of:

- the materiality of landscapes as cultural artifacts;
- the representation of these landscapes through various media; and
- the interpretation of landscapes as meaningful places.

This review was very recently updated by Terkenli (in press) in the chapter 'Landscape perspectives in/and tourism geographies' in the new *Routledge handbook of tourism geographies*, edited by Wilson and Muller (in press). The third undertaking was the chapter 'Landscapes of tourism', also by Terkenli (2014) in *The Wiley-Blackwell companion to tourism*, edited by Lew et al. (2014), to which the present chapter serves as a sequel, building on, updating and complimenting the 2014 review, by summarising predominant themes and laying out emergent prospects.

Secondly, significant advances in the scientific area at the interface of landscapes and/of tourism are compiled in the special issue 'Landscape and tourism, landscapes of tourism' recently published in *Land* (2021), aiming 'to enhance the interdisciplinary scientific dialogue on these issues and challenges, while highlighting their range and significance for tourism and the landscape, in terms of theory, empirical practice, approach, policy, ethics, and future prospects'. The issue contains 16 research articles which contribute to the latter objective by covering a broad spectrum of relevant scholarly advances, in terms of which three articles stand out. Meneghello (2021) delves into the theorisation of the tourism-landscape interrelationship by performing a bibliographic analysis of 'tourist landscapes' and other relevant terms and arrives at three main research topics predominating in this interdisciplinary field:

- planning and governance,
- situated spatial-social-symbolic interrelations; and
- impact evaluation.

Terkenli et al. (2021) conduct a comprehensive, in-depth investigation of the relationship between landscape and tourism, with the aid of a survey of relevant European experts, revealing the experts' socio-environmental concerns regarding the tourism-landscape interrelationship, alongside their support for the principles of sustainability, locality, and participatory governance, for purposes of appropriate future planning. Finally, Jiménez-García et al. (2020) map this field of knowledge with bibliometric techniques, identifying research themes which have been basic to the construction of the field but are currently in decline (e.g., cultural heritage, national parks), but also point to sustainability — as a term and as related issues — as a transversal presence and value throughout this scientific research area.

Theoretical and Methodological Reflections

Landscape and tourism are irrevocably linked together and co-constitutive, in a variety of ways. We concede that 'landscape as an approach to tourism is not a mutually exclusive perspective but is quite compatible with numerous other theories, frameworks, and perspectives used to understand tourism, thus making a landscape perspective a robust tool of inquiry' (Knudsen et al., 2012). The interface between these two broad and complex areas of scientific study, tourism and landscape, has accordingly elicited a variegated body of research in recent years, in terms of its nature, focus and approach, e.g. analysis of aspects of tourism destinations as cultural landscapes, or in the context of sustainable development. Interest in the study of landscapes of/and tourism has been growing, especially in the last decade, as reflected in the increase of publications and research questions addressed in this area of study (Land, 2021). Nonetheless, there is, as of yet, no comprehensive and cohesive conceptual/theoretical framework to support this increasingly interdisciplinary body of work — consisting mostly of disparate efforts and perspectives — while few attempts have been made towards the construction of tourism landscape typologies (Gkoltsiou et al., 2013; Skowronek et al., 2018). This is even more so the case as regards frameworks or typologies that sufficiently addresses the spatial and social imprint of a rapidly changing world geography of tourism.

The prevailing discourses and other forms of interdisciplinary contribution presented in this chapter are reflective and indicative of wider contemporaneous perspectives on landscape and on tourism, in all disciplinary fields related to the study of landscapes of/and tourism. They indicate the need to study the ways in which space and landscape literally and metaphorically shape the experience of leisure and tourism, not only through the gaze, but also through multi-sensorial sensitivity, materiality, embodiment and expressive practice (della Dora, 2009). Furthermore, they elicit future research in the ways in which contextualised, as well as overarching, leisure and tourism experiences inform and substantiate landscapes of tourism. Many challenges are involved in this task. The need for re-articulation and renewal of an awareness of the place of landscape in (the) geography(ies) of tourism continues to be always relevant. However, this endeavor towards theory building must be carried out while embracing the complexity of interrelationships between tourism and the landscape, in the context of a newly-emerging cultural economy of space (Terkenli & d'Hauteserre, 2006), informing new geographies of tourism and signaling tourism 'as continual practice' (Rojek & Urry, 1997). In this context, practices that we understand by leisure/tourism today have been merging with other areas of life (Crouch, 1999; Harvey, 1989), through processes of spatial de-differentiation of leisure and tourism, shopping, work, culture, satisfaction of basic needs, comfort, play, familiarity etc.

Finally, the crisis context in which we are presently called to analyse and assess the tourism-landscape interrelationship, is one of new global challenges and emergent local opportunities. As a result of the latter challenges, tourism patterns and practices have been diversifying in space-time-cultural terms, seeking ever new thrills and attractions in response to society's changing concepts and demands (i.e. the local, the 'authentic', the 'green'), providing the incentive and the grounds for more sustainable place-making and development. The landscape seems poised to figure very prominently in these new and developing forms of travel (i.e. return to locally-sourced products or quest for outdoors alternative activities and individualised tourism experiences, such as gastronomic tourism and glamping). Such forms of tourism may contribute actively to landscape management (i.e., through revitalisation of traditions and resource protection), while fostering bottom-up involvement in landscape stewardship, both for inhabitants and for tourists, strengthening identity formation, participatory practices and social cohesion. In order for landscape to become such a catalyst for an emerging new model for human life, however, further and more in-depth scientific investigation into the role of landscape in visitor/tourist experiences is certainly necessary and called for.

References

Aitchison, C., MacLeod, N. E., & Shaw, S. J. (2000). *Leisure and tourism landscapes: Social and cultural geographies*. Routledge.

Britton, S. G. (1982). The political economy of tourism in the third world. *Annals of Tourism Research*, *9*(3), 331–358. https://doi.org/10.1016/0160-7383(82)90018-4

Britton, S. G. (1991). Tourism, capital and place: towards a critical geography of tourism. *Environment and Planning D: Society and Space*, *9*(4), 451–478. https://doi.org/10.1068/d090451

Burley, J. B., & Machemer, T. (2016). *From eye to heart: Exterior spaces explored and explained*. Cognella Academic Publishing.

Burton, R. J. F. (2012). Understanding farmers' aesthetic preference for tidy agricultural landscape: A Bourdieusian perspective. *Landscape Research*, *37*(1), 51–71. https://doi.org/10.1080/01426397.2011.559311

Carmichael, B. (1998). *Destinations: Cultural landscapes of tourism*. Routledge.

Cartier, C., & Lew A. A. (Eds). (2005). *Seductions of place*: Geographical perspectives on globalization and touristed landscapes. Routledge.

Cassar, L. F. (2010). *A landscape approach to conservation: Integrating ecological sciences & participatory methods*. UNESCO and International Environment Institute.

Coccossis, H. (2004). Sustainable development, landscape conservation and tourism in the small islands of Greece: The Nature Conservation-Society Interface. In D. Martin & J. van der Straaten (Eds), *Cultural landscapes and land use* (pp. 111–124). Kluwer Academic Publishers.

Cosgrove, D. (1998). *Social formation and symbolic landscape*. University of Wisconsin Press.

Council of Europe (CoE). (2000). *Council of Europe landscape convention*. https://www.coe.int/en/web/landscape

Crang, M. (1999). Knowing, tourism and practices of vision. In D. Crouch (Ed.), *Leisure/tourism geographies: Practices and geographical knowledge* (pp. 238–256). Routledge.

Crouch, D. (Ed). (1999). *Leisure/tourism geographies: Practices and geographical knowledge*. Routledge.

Daniels, S., & Cosgrove, D. (1988). Introduction: iconography and landscape. In D. Cosgrove & S. Daniels (Eds), *The iconography of landscape: Essays on the symbolic representation, design and use of past environments* (pp. 1–10). Cambridge University Press.

Daugstad, K. (2008). Negotiating landscape in rural tourism. *Annals of Tourism Research*, *35*(2), 402–426. https://doi.org/10.1016/j.annals.2007.10.001

della Dora, V. (2009). Travelling landscape-objects. *Progress in Human Geography*, *33*(3), 334–354. https://doi.org/10.1177/0309132508096348

García Pérez, J. (2002). Ascertaining landscape perceptions and preferences with pair-wise photographs: planning rural tourism in Extremadura, Spain. *Landscape Research*, *27*(3), 297–308. https://doi.org/10.1080/01426390220149539

Gkoltsiou, A., Terkenli, T. S., & Koukoulas, S. (2013). Landscape indicators for the evaluation of tourist landscape structure. *International Journal of Sustainable Development and World Ecology*, *20*(5), 461–475. https://doi.org/10.1080/13504509.2013.827594

Hall, C. M., & Page S. J. (2006). *The geography of tourism and recreation. Environment, place and space* (3rd ed.). Routledge.

Hall, C. M., & Page, S.J. (1999). *The geography of tourism and recreation: Environment, place and space*. Routledge.

Harvey, D. (1989). *The condition of postmodernity*. Blackwell.

Hazendonk, N., Hendriks, M., & Venema, M. (Eds). (2008). Greetings from Europe: *Landscape and leisure*. OIO Publishers.

Hollinshead, K. (1999). Surveillance of the worlds of tourism: Foucault and the eye-of-power. *Tourism Management*, *20*(1), 7–23. https://doi.org/10.1016/S0261-5177(98)00090-9

Jackson, J. B. (1984). *Discovering the vernacular landscape*. Yale University Press.

Jafari, J. (Ed.). (2000). *Encyclopedia of tourism*. Routledge.

Jiménez-García, M., Ruiz-Chico, J., & Peña-Sánchez, A. R. (2020). Landscape and tourism: Evolution of research topics. *Land*, *9*(12), 488–505. https://doi.org/10.3390/land9120488

Jones, M., & Stenseke, M. (Eds). (2011). *The European landscape convention*: Challenges of participation. Springer.

Kinnaird, V., & Hall, D. (Eds). (1994). *Tourism: a gender analysis*. John Wiley.

Klijn, J. A., Bethe, F., Wijermans, M., & Ypma, K. W. (1999). *Landscape assessment method at a European level: a case study of polder landscapes. Report 173*. Winand Staring Centre.

Knudsen, D. C., Metro-Roland, M. M., Soper, A. K., & Greer, C. E. (Eds). (2008). *Landscape, tourism, and meaning*. Ashgate.

Knudsen, D. C., Rickly-Boyd, J. M., & Metro-Roland, M. M. (2012). Landscape perspectives on tourism geographies. In J. Wilson (Ed.), *The Routledge handbook of tourism geographies* (pp. 201–206). Routledge.

Land. (2021). Special Issue "Landscape and tourism, landscapes of tourism". *Land* (ISSN2073-445X). https://www.mdpi.com/journal/land/special_issues/landscape_tourism

Löfgren, O. (1999). *On holiday: a history of vacationing*. University of California Press.

Lorimer, H. (2005). Cultural geography: the busyness of being more-than-representational. *Progress in Human Geography*, *29*(1), 83–94. https://doi.org/10.1191/0309132505ph531pr

Lothian, A. (1999). Landscape and the philosophy of aesthetics: Is landscape quality inherent in the landscape or in the eye of the beholder? *Landscape and Urban Planning*, *44*(4), 177–198. https://doi.org/10.1016/S0169-2046(99)00019-5

MacCannell, D. (1976). The past and future of 'symbolic interactionism'. *Semiotica*, *16*(2), 99–114. https://doi.org/10.1515/semi.1976.16.2.99

Meinig, D. (1979). The beholding eye: ten versions of the same scene. In D. Meinig (Ed.), *The interpretation of ordinary landscapes: Geographical essays* (pp. 33–48). Oxford University Press.

Meneghello, S. (2021). The tourism–landscape nexus: Assessment and insights from a bibliographic analysis. *Land*, *10*(4), 417. https://doi.org/10.3390/land10040417

Nassauer, J. I. (1988). The aesthetics of horticulture: neatness as a form of care. *HortScience*, *23*(6), 973–977. http://hdl.handle.net/2027.42/49345

Norton, A. (1996). Experiencing nature: the reproduction of environmental discourse through safari tourism in East Africa. *Geoforum*, *27*(3), 355–73. https://doi.org/10.1016/S0016-7185(96)00021-8

Olwig, K. R. (2019). *The meanings of landscape: Essays on place, space, environment and justice*. Routledge.

Pearce, D. (1995). *Tourism today: A geographical analysis* (2nd ed.). Longman.

Peterson, G. L., & Neumann, E. S. (1969). Modeling and predicting human response to the visual recreation environment. *Journal of Leisure Research*, *1*(3), 219–237. https://doi.org/10.1080/00222216.1969.11969734

Picard, D. (2011). *Tourism, magic and modernity: Cultivating the human garden*. Berghahn Books.

Porter, P. W., & Sheppard, E. S. (1998). *A world of difference: Society, nature, development*. The Guilford Press.

Prince, S. (2018). Dwelling in the tourist landscape: Embodiment and everyday life among the craft-artists of Bornholm. *Tourist Studies*, *18*(1), 63–82. https://doi.org/10.1177/1468797617710598

Prince, S. (2019). Dwelling and tourism: Embracing the non-representational in the tourist landscape. *Landscape Research*, *44*(6), 731–742. https://doi.org/10.1080/01426397.2018.1518520

Pritchard, A., & Morgan, N. J. (2000). Privileging the male gaze: gendered tourism landscapes. *Annals of Tourism Research*, *27*(4), 884–905. https://doi.org/10.1016/S0160-7383(99)00113-9

Pritchard, A., & Morgan, N. J. (2001). Culture, identity and tourism representation: marketing Cymru or Wales? *Tourism Management*, *22*(2), 167–179. https://doi.org/10.1016/S0261-5177(00)00047-9

Relph, E. (1976). *Place and placelessness*. Pion.

Roe, M. (2014). Exploring future cultural landscapes. In M. Roe & K. Taylor (Eds), *New cultural landscapes* (pp. 241–269). Routledge.

Rojek, C., & Urry J. (Eds). (1997). *Touring cultures*: Transformations of travel and theory. Routledge.

Roovers, P., Hermy, M., & Gulinck, H. (2002). Visitor profile, perceptions and expectations in forests from a gradient of increasing unbanisation in central Belgium. *Landscape and Urban Planning*, *59*(3), 129–145. https://doi.org/10.1016/S0169-2046(02)00011-7

Rose, G. (1996). Geography and the science of observation: the landscape, the gaze and masculinity. In J. Agnew, D. N. Livingstone & A. Rogers (Eds), *Human geography: An essential anthology* (pp. 341–350). Blackwell.

Sack, R. D. (1992). *Place, modernity and the consumer's world*. The Johns Hopkins University Press.

Sauer, C. O. (1925/2007). *The morphology of landscape*. University of California Publications in Geography, 2(2), pp. 19–54.

Shaw, G., & Williams, A. M. (1994). *Critical issues in tourism: a geographical oerspective*. Blackwell.

Skowronek, E., Tucki, A., Huijbens, E., & Jóźwik, M. (2018). What is the tourist landscape? Aspects and features of the concept. *Acta Geographica Slovenica*, *58*(2), 73–85. https://doi.org/10.3986/AGS.3311

Stefanou, J. (2000). The contribution of the analysis of the image of a place to the formulation of tourism policy. In H. Briassoulis & J. van der Straaten (Eds), *Tourism and the environment: Regional, economic, cultural and policy issues* (revised 2nd ed., pp. 229–237). Kluwer Academic Publishers.

Terkenli, T. S. (2000). Landscapes of tourism: a cultural geographical perspective. In H. Briassoulis & J. van der Straaten (Eds), *Tourism and the environment: Regional, economic, cultural and policy issues* (revised 2nd ed., pp. 179–202). Kluwer Academic Publishers.

Terkenli, T. S. (2014). Landscapes of tourism. In A. A. Lew, C. M. Hall & A. M. Williams (Eds), *The Wiley-Blackwell companion to tourism* (pp. 282–293). John Wiley & Sons Ltd.

Terkenli, T. S. (2018). Tourism: beyond the pleasure of practicing Geography. *Tourism Geographies.*, *20*(1), 170–171. https://doi.org/10.1080/14616688.2017.1400094

Terkenli, T. S. (in press). Landscape perspectives in/and tourism geographies. In J. Wilson & D. Müller (Eds), *The Routledge handbook of tourism geographies*. Routledge.

Terkenli, T. S., & d'Hauteserre, A.-M. (Eds). (2006). *Landscapes of a new cultural economy of space*. Springer.

Terkenli, T. S., Skowronek, E., & Georgoula, V. (2021). Landscape and tourism: European expert views on an intricate relationship. *Land*, *10*(3), 327. https://doi.org/10.3390/land10030327

Towner, J. (1996). *An historical geography of recreation and tourism in the Western World, 1540–1940*. John Wiley and Sons.

Tress, B., & Tress, G. (2001). Capitalising on multiplicity: a transdisciplinary systems approach to landscape research. *Landscape and Urban Planning*, *57*(3–4), 143–157. https://doi.org/10.1016/S0169-2046(01)00200-6

Urry, J. (1990). *The tourist gaze: Leisure and travel in contemporary societies*. Sage.

Urry, J. (1995). *Consuming places*. Routledge.

Wenger Jr, W. D., & Videbeck, R. (1969). Eye pupillary measurement of aesthetic responses to forest scenes. *Journal of Leisure Research*, *1*(2), 149–161. https://doi.org/10.1080/00222216.1969.11969721

Wilson, J. (Ed.). (2012). *The Routledge handbook of tourism geographies*. Routledge.

Wylie, J. (2007). *Landscape*. Routledge.

Zube, E. H., & Pitt, D. G. (1981). Cross-cultural perceptions of scenic and heritage landscapes. *Landscape and Urban Planning*, *8*(1), 69–87. https://doi.org/10.1016/0304-3924(81)90041-1

Chapter 13

From Pseudo-Events to Virtual Reality
The Development of Authenticity in Tourism Scholarship

Cecilia de Bernardi

Introduction

Authenticity can probably be seen as one of the most commonly used concepts in tourism. The layman talk of authenticity is usually tinged by phrases such as that things should be 'kept authentic' or this and that 'did not feel authentic'. From an academic viewpoint, the initial development of the concept was a dichotomy, either something is authentic or not. The following phase is to actually conceptualise authenticity as constructed, depending on 'the eyes of the beholder'. Later conceptualisations of authenticity describe it as existential and mostly connected to the experience of the tourist (or the practitioner) no matter where or how this experience takes place. In between there has also been a discussion of staged authenticity and hyperreality, in which the authentic does not really exist. It is either something that is organised for the tourists or simply fake. These different ways to theorise authenticity are not mutually exclusive and not entirely in chronological order. The studies that have adopted these concepts have used them in several situations and even combined them. Since there are a lot of parallels and overlapping between how the terms have been used, this chapter will be centered around two main questions: how do the tourists experience authenticity? And how do the practitioners and the locals experience it?

These questions will be answered through an overview of the authenticity scholarship by discussing a series of thematic areas. With the help of examples from authenticity research and from my research, I illustrate how this concept is treated empirically and theoretically to provide an overview of its development from the beginning, up to the challenges we face nowadays. Some examples of areas in which authenticity has been studied, both theoretically and empirically, are philosophy, heritage, marketing, souvenirs, virtual reality and fiction connecting places to literature, series and movies. Authenticity is used to appraise places that should become part of UNESCO World Heritage, both tangible and intangible. Authenticity is also a concept pertaining to marketing communication and the representations places. When the tourists go home some of them will bring home things and this raises the question of the

The Wiley Blackwell Companion to Tourism, Second Edition. Edited by C. Michael Hall.

authenticity of souvenirs. When a movie, a series or a book takes place somewhere, this may result in tourists travelling there. The fact that this is the place where this particular story took place is what makes the place authentic even though what is described in the story does not actually 'exist'. As technology advances, we are also able to experience destinations in a virtual setting, which is something that has developed even faster due to the COVID-19 pandemic. These are some of the examples that will be discussed through practical examples during the course of the chapter.

The Origins

When people ask about my view on authenticity, I usually start by saying that authenticity is the equivalent of a conceptual black hole, since it is a big body of research with a long history. It has been discussed in the 19th and 20th centuries by philosophers such as Heidegger, Adorno and Kierkegaard (Varga & Guignon, 2020). This concept has developed in different directions as these philosophers have very different views on what authenticity is and how it should be achieved. One of the common themes is usually the self and how people can become the most authentic version of themselves vis-á-vis with society and the barriers that it may pose. If we go back to our underlying questions about achieving an authentic experience, in this case it is about different kinds of tourism experiences leading to 'a sense of desired self' (Kim & Jamal, 2007, p. 181).

However, existential authenticity in tourism research has developed later. In the beginning the focus was on people visiting places, their experiences and the authenticity of what they see. In the book *The Image: A Guide to Pseudo-events in America*, Boorstin (1961) described the attractions that tourists can experience and how these are 'pseudo-events' and are not actually authentic. The reason is that what the tourists see is the result of careful planning and while being out travelling to see something that is exotic, the tourists do not want be away from the comforts of home (Boorstin, 1961).

In a development of this same argument, MacCannell (1973) talks about authenticity as something that is 'staged' and to which there is a backstage. Sightseeing tours and attractions are set up for the tourists and they rarely get to experience the real life of the locals. In fact, in response to Boorstin's work, Umberto Eco (2014) conceptualised attractions and travelling in the United States as hyperreality. According to Eco (2014), a society which lacks history and is highly consumeristic focuses on the spectacular and on the unusual. Tourism becomes a prime example of this hyperreality, even though not all places display the same degree of it (Buchmann et al., 2010; Eco, 2014). This has also been connected to Baudrillard's (1994) concept of 'simulacra', which is a sign for the real thing, so reality becomes populated by signs, replacing actual objects and events (Buchmann et al., 2010). According to these approaches, there is no possibility for the tourists to experience the authentic.

Despite the fact that these approaches describe authenticity as something unattainable, tourists travel searching for it and, in this process, they can be a trigger for commoditisation. Products change for both locals and tourists due to this process. Cohen (1988) argues that authenticity can still remain even in a product created for tourism, both for those who produced it and those who 'consume' it. In this case, authenticity can be found in a constructed experience, which is the result of a negotiation between the host and the guest. Cohen (1979) also developed different tourism typologies that are based on their motivations. From mere leisure, tourists may also look for more meaning and a more authentic experience. This takes us to Wang (1999), who described the development of the concept of authenticity as it was presented here and then discussed how authenticity depends on whom experiences it and how it is communicated. 'In this sense, the authenticity of toured objects is in fact symbolic authenticity' (Wang, 1999, p. 352). Wang (1999) proposes to connect authenticity to activities, which activate an existential experience. This is what constitutes the authenticity in the experience, the existential state (Wang, 1999). And here we have come full circle. From Heidegger through different ways to see authenticity and back to Heidegger again. Reisinger and Steiner (2006) argued that object authenticity (meaning the dichotomic view of authentic versus inauthentic) should be abandoned in favour of an approach to authenticity based on Heidegger's philosophy. From a theoretical viewpoint existential authenticity could be linked to Heidegger's insights on the topic to provide the concept with a solid theoretical foundation to be tested empirically (Steiner & Reisinger, 2006). As we will see in the coming section, existential authenticity has been studied empirically in a multitude of contexts.

Authenticity and Heritage: A Multifaceted Relationship

Now that I have presented a summary of how authenticity scholarship has developed in very short terms, I will now present some examples to illustrate how these different conceptualisations are used in empirical research. In the beginning the discussion focused on objects so heritage is a good starting point. The first thing that comes to mind could be an old building, which may signal to the tourists some kind of historical authenticity. However, this is usually just one piece of the puzzle. The tourist experience of heritage is usually made up of many different aspects. So how do people experience authenticity at heritage sites?

McIntosh and Prentice (1999) describe how the experience of heritage is closely related to how the tourists themselves interpret the sites. Some examples are events organised to stage something that took place in the past, which the visitors felt was authentic (Chhabra et al., 2003). In fact, the experience is more relevant than the actual place in heritage tourism (Park et al., 2019) and it can generate meaningful self-reflection (Crang, 1996). However, the experiences that are generated by the places that the tourists visit is not the only important aspect. Tourists also look for authentication by experts, especially in the case of heritage reconstruction (Holtorf, 2014). As we will see in more detail in the section dealing with intangible heritage, authentication is also important for the practitioners when working with heritage reconstruction (Halewood & Hannam, 2001). One aspect that is important to remember regarding how history is presented is that the tourists perceive reconstructions and the information given as authentic and sometimes the viewpoint showcased at heritage sites may be specific, ignoring other perspectives on important historical facts (Waitt, 2000).

Another important aspect of heritage, food, is also often part of discussions of authenticity. Food tourism can be very diversified. It can be connected to people looking for something different than the homogenised fast food, but also, especially in the context of tourism enclaves, people may instead look for the comfort of the food they know. The authentic experience of food is usually tied to taste. However, what constitutes 'authentic food' is adapted to tourism audiences and is highly impacted by trends. Therefore, food offered at certain destinations may become very homogenous (Cohen & Avieli, 2004).

Furthermore, tourists usually also look for more than just food, they are after an exposure to authentic local culture, which transcends the food, making the experience more existential (Sims, 2009). This means that food tourists are on a quest for authenticity as mentioned by MacCannell (1973) (Ellis et al., 2018). Similar preferences are manifested for restaurants, where the place itself should feel 'authentic'. This is exemplified by the San Angel Inns.

One can be found in Mexico City and the other at Disney World. Both are supposed to be authentically Mexican (Salamone, 1997). While people usually claim that the authentic restaurant is the one in Mexico, Salamone (1997) argues that both are authentic because both reflect different dimensions of Mexican culture. Tourists may expect culture to be frozen, but both these restaurants try to fight this assumption (Salamone, 1997).

Intangible Heritage

Authenticity is also used as a criterion for appraisal and inclusion in the UNESCO World Heritage list. This includes a set of criteria for both intangible and tangible heritage.

The first criteria for authenticity were mainly based on built heritage, which classify authenticity based on 'design, material, workmanship and setting' (Jokilehto, 2006, p. 8). The revisions add that authenticity can be met also in relation to traditions, language and feeling, to name a few (Jokilehto, 2006). This shows an evolution in the World Heritage criteria to include also aspects of culture that evolve and change with time. Furthermore, even built heritage can be included in the list based on the virtue of its meaning rather than its historical authenticity (Jokilehto, 2006).

In the case of intangible heritage, we will try to answer our second question, which deals with how practitioners achieve an authentic experience through different kinds of intangible heritage practice. Practitioners think that authenticity is a very important element of intangible cultural heritage. This is meant as 'a holistic notion integrating the transmitted custom, inherited meanings and the practitioner identity' (Kim et al., 2019, p. 432). Intangible heritage is not fixed, the same can be said about this 'authentic self', which is also constantly changing and evolving. Therefore, authenticity in the context of intangible heritage should also include an aspect dealing with the authentic self and subjectivity since it can be a way to construct the practitioners' identity (Su, 2018). When a ritual is performed for the tourists for instance, it could be argued that it is staged and therefore not authentic, but in some cases, this is a very positive occasion for both parties and it can lead to self-awareness (Seraphin et al., 2018). However, tourism development can bring negative impacts on intangible heritage and should therefore be monitored (Kim et al., 2019).

Souvenirs

It may be not be intuitively logic to include souvenirs in the section about heritage, but there is a reason for this structure. Souvenirs have been treated as part of the process of commoditisation connected to tourism. Part of the problem with souvenirs is inherent in their name, as traditional heritage crafts are sold and taken home by the tourists, thus becoming souvenirs. The word souvenir is tied to the word 'remember' in French and in fact people buy souvenirs as a way to remember their trip (Edelheim, 2015).

Tourists achieve an authentic experience through souvenirs. They may bestow emotional value to objects that are not authentic in terms of criteria used to appraise souvenirs (Swanson & Timothy, 2012). These can be connected to aesthetics, workmanship, historical ties or the things that are used to make the souvenir as well as the person who made it (Littrell et al., 1993). Furthermore, tourists are interested in 'traditional cultural motifs' and even regarding decorative items, the element of traditional is considered very important (Xie et al., 2012). These criteria can be related to the previous discussion of object authenticity and the souvenirs' essential characteristics (Edelheim, 2015). Even though we know that the tourists assign different meaning to souvenirs (Edelheim, 2015), they still would be happier to purchase when there is a label certifying that the handicraft they would like to buy as a souvenir is authentic (Asplet & Cooper, 2000). One example of this kind of 'authenticity' label is the Sámi[1] Duodji, which certify that Sámi handicraft has been produced in a certain way and with certain materials (De Bernardi, 2020).

As mentioned in the beginning of this section, souvenirs may be considered problematic (especially by locals) because cultural objects are turned into products for the tourists to purchase and bring home. In the case of the indigenous population Sámi, the Duodji are not designed, created and sold to be taken home by the tourists, one criterion to be given the certification is that Sámi Duodji are made for everyday use (De Bernardi, 2020). Furthermore, questions of labelling and authenticity are not homogenous in the Sámi community (De Bernardi, 2020). The same issues also apply to other indigenous populations; designs and meanings change and evolve (Xie et al., 2012). However, the criteria for authenticity that have been mentioned for the tourists, also apply to the artists making them. The fact that the souvenirs are handmade, integrating local culture and requiring local skills are some examples of what makes them authentic according to the artists (Soukhathammavong & Park, 2019).

It has been argued that different kinds of tourists will buy different kinds of souvenirs (Edelheim, 2015). Therefore, we can imagine that the tourists wanting to buy handicraft such as the Duodji will probably have a specific interest in this kind of 'souvenir'. Trends regarding crafts that are locally produced and which contribute to the local economy are becoming increasingly popular. One example is a study on travel blogs showing that people do not feel that mass-produced souvenirs mean something to the ones who made them and that purchasing locally sourced and made souvenirs may be of help to the local economy and local businesses (Torabian & Arai, 2016).

Authenticity, Marketing and Representation

No matter if we are talking about objects or experiences, there is a great deal of communication in tourism, mostly for marketing purposes. In fact, there is a strong focus in the tourism scholarship on how certain aspects of a destination may be marketed as authentic. One aspect of achieving an authentic experience for the tourists is related to what is communicated to them before they begin their trip; 'Authenticity is an influential factor altering tourists' evaluations and perceptions [. . .], but also develops the underlying rationale that tourists' perceived authenticity positively influences satisfaction via forming a favorable image' (Lu et al., 2015, p. 93). Some examples are the connections that have been established in research between destination image, place attachment and authenticity. Place attachment influences perceived authenticity while heritage attractions play a big role in mediating this relationship (Ram et al., 2016). Attractions such as architecture, scenery and food influence the perception the tourists have of the destination (Lu et al., 2015). Other studies have shown that authenticity is an antecedent that influences place attachment (Jiang et al., 2017). This can be seen as a question of what came first, the chicken or the egg. Future research should look more into how authenticity, place attachment and destination image are connected to one another (Jiang et al., 2017).

As there are different aspects that influence how tourists experience a destination, how local populations are presented is a particularly thorny issue. Tourism marketing communication has a strongly essentialist character, which means that the representation of a whole destination, a whole country has been squeezed into a few pictures and short messages.

Furthermore, what is being communicated in the messages also needs to be attractive to the tourists, which has created even more problems. One consequence of this is that indigenous and other local populations are presented in a way that is exoticising, essentialist and in many cases also stuck in certain historical era (Silver, 1993). 'Local residents, when pictured, are configured as genericised icons of exoticism that serve to imbue the tourist's experience with authenticity' (Smith, 2018, p. 173). This kind of representation of indigenous populations, minorities and other locals is a way to evoke authenticity in the mind of the tourists (Yang, 2011). In the case treated in my PhD thesis, the Sámi have been presented in the same way for a long time (De Bernardi, 2020).

As mentioned by Silver (1993), often there is no possibility for the represented populations to influence how they are portrayed, which is another aspect of this complex problem. This was the same issue for the Sámi. In fact, one of the recommendations made both in my PhD thesis and in the included articles is to let the Sámi participate in the creation of marketing materials (De Bernardi, 2020). This is a way to balance the power relations that have been around for a very long time and which will not be solved short-term (Koot, 2016). These power relations are usually highly situational and contextual, but it is good to always keep in mind that indigenous and other local populations are not just powerless victims of the marketing communication producers, they have agency and are capable of resistance (De Bernardi, 2020; Koot, 2016). This issue of the representation of people and culture is where the answer to our two questions meet. How does the authentic experience of the tourists impact how the people working with tourism achieve authenticity?

Things have changed over time. Destination Management Organisations (DMOs) do not 'control' the images with the output of brochures or a website that can influenced people. One example is social media. The results of my PhD thesis showed that brochures are not a medium in which it is easy to show how multifaceted indigenous and other local are. If we take another example, Smith (2018) has studied Instagram. This is a space where previously challenged colonial imagery receives a new space in which the tourists can describe places as empty paradises to which they belong (Smith, 2018). On the other side of the coin there is the fact that social media becomes a platform even for the people working with tourism, which can spread information to potential visitors about the everyday life of the indigenous and local populations. This information is more well-rounded and multifaceted (De Bernardi, 2022).

The Fictional and Virtual Reality

Gilmore and Pine (2007) drafted a practical guide for companies to be able to achieve as well as sell authenticity. The core argument is that authenticity is something that is man-made, artificial and therefore there is no connection between reality and authenticity without human action (Gilmore & Pine, 2007). Indeed, we have discussed many examples throughout the chapter that show how authenticity is something that is highly subjective and dependent on a number of contextual factors. To some extent, Gilmore and Pine's (2007) book is a practical guide to answer both of the questions guiding this chapter.

I have discussed how an authentic experience can be achieved in a number of contexts, both regarding objects but also intangible phenomena such as performances. We now delve a little bit deeper into the grey areas of authenticity research. When tourists visit places that are tied to fiction, fantasy, books, and TV, how do the tourists experience something authentic? If the tourists travel to places from their living room using technological means to 'travel', how is this experience authentic?

Disney, Literature and the Moving Pictures

One of the most notable examples of film tourism is people travelling to New Zealand to visit the places where the *Lord of the Rings* (LOTR) movies have been filmed. The authentic experience for these visitors is related to being where the picture was filmed as well as being there with others who are also fans of the movies (Buchmann et al., 2010). Buchmann et al. (2010) compare this kind of trip to a pilgrimage and to a religious experience. In other cases, LOTR tourism has been classified as an example of staged authenticity and as a way to profile New Zealand to attract tourists (Tzanelli, 2004) as well as a way to construct a national identity in which authenticity is fabricated (Jones & Smith, 2005). The tourists achieve authenticity by visiting the places where their favorite movies have been filmed, while for the destination it is a way to construct an identity and attract tourists. As in the cases presented before in relation to heritage, it is important to keep track of tourism development in order to avoid negative effects on local culture (Tzanelli, 2004).

If movie sets are places in which authenticity is created and recreated in different ways, the *New York TV Tour* is described in terms of a constant movement between reality and fiction. This is not seen as something negative, but as a way to counteract views of authenticity that strongly rely on reproduction or simulacra as a sign of avoidance (e.g., Eco, 2014). This 'restless movement' is what attracts tourists in the first place (Torchin, 2002, p. 252). In addition to the TV show and how the places are connected to it, authenticity in TV-series tourism is not only dependent on the moving picture context but also on other services and amenities offered and present at the destination (Rittichainuwat et al., 2018).

Film and TV-series tourism often develops in places where there is already a local identity and connected heritage. In the case of New Zealand there was a worry that LOTR tourism may take over the local identity. This kind of tourism is dependent on the local characteristics and this includes local culture and heritage. In the case of *Game of Thrones* (GoT) and Dubrovnik, the issue has been to reconcile the authenticity tied to the designation as a World Heritage Site and moving picture tourism (Šegota, 2018). For what it concerns the tourists' own experience, both *King's Landing* and the city of Dubrovnik are seen as authentic. Visitors who may travel to Dubrovnik because of GoT are then reported to discover the city. In this case, both authenticities related to objects and to experiences are seen as important for the tourists to achieve an authentic experience (Li et al., 2021). The preservation of a place's identity is very important when this kind of phenomena are associated to a place. This has been the strategy of the two Italian towns connected to *The Twilight Saga* both from a literature and film perspectives (Larson et al., 2013).

In the case of literary places, authenticity may be connected to something real, such as the place where the author was born, but then the experience develops into something subjective (Herbert, 2001). Literature gives the tourists the possibility to live out fantasies and to redefine their own heritage experiences. Authenticity is seen as fluid and the authentic experience is subjective (Squire, 1994). Literary places may also concentrate on authenticity because of rising competition from other sites related to the same literary piece (Fawcett & Cormack, 2001). When it comes to places that can be found in literature, but do not actually exist, the source of authenticity can come from the author. The theme park dedicated to *Harry Potter* (Wizarding World of Harry Potter in the United States) is said to derive its authenticity from the approval of author J. K. Rowling and from the multisensory experiences of the visitors tied to other aspects such as shops selling food and drinks with a British style (Waysdorf & Reijnders, 2018).

While J. K. Rowling is a guarantee of authenticity for tourists visiting the *Harry Potter* theme park in the United States, *Harry Potter* tourism taking place in the UK is different. Authenticity here is also connected to the places that actually exist and have a long history, such as the dining hall of Oxford University. The authentic experience is then derived from a mix of the interest in the books and the places that are connected to them (Lee, 2012). Other literary places have so-called 'site guardians' who have their own idea of what authenticity is in relation to the site they work at. In this case, authenticity is achieved through a mix of personal commitments and contextual factors (Fawcett & Cormack, 2001).

When it comes to attractions that are connected to fantasy, such as Disney, authenticity is also said to have different meanings and changes over time (Houston & Meamber, 2011). We have previously discussed the role of aesthetics in the choice of handicraft and souvenirs, in the case of *Disney Future World* and *World Showcase* aesthetics are expressed through buildings, staff in costume and performances. This is an idealised version of the history of the countries represented, which is strongly sanitised (Houston & Meamber, 2011). As previously mentioned, what is communicated to the tourists about history is considered authentic and therefore the perspective presented is very important (Waitt, 2000).

The authenticity provided by the setting and the performances is a common theme in studies on Disney. The same has been argued for Buffalo Bill's Wild West (Lovell & Hitchmough, 2020). Furthermore, Disney has been described as evolving into part of American culture, which means that it is considered more and more authentic (Lovell & Hitchmough, 2020).

Virtual Reality

Experiences in a virtual reality (VR), augmented reality (AR) and virtual trips are described in the same terms as other kinds of tourism experiences. Authenticity cannot be assessed from an objective viewpoint and how authentic the experience will depend on the person, their interests and how VR is applied to tourism (Guttentag, 2010). Furthermore, both object as well as existential approaches to authenticity are relevant for VR and virtual trips (Gao et al., 2022). VR and virtual trips also fulfil the purpose of providing the tourists with satisfying sensory experiences that are unique, exhilarating (Gao et al., 2022), immersive, escapist and a 'replenishment of psychological resources' (Wong et al., 2022, p. 216). Empirical findings on tourism students show that none of them would like to replace their in-person experience of tourism with a virtual one (Mura et al., 2017), while virtual tourism has been proposed as an alternative because of the COVID-19 pandemic (Schiopu et al., 2021). Tourists who think that VR may be a viable substitute are motivated to use it and the more they perceive VR as authentic, the more they will want to utilise it (Schiopu et al., 2021).

VR and AR have not been used to only 'substitute' the tourism experience, but also to enhance it. Shehade and Stylianou-Lambert (2020) argue that in order to appraise authenticity in these contexts we need to stop applying either object or existential authenticity. Instead 'we should rather acknowledge that a fair balance should be found so that the different needs and values of involved stakeholders are accommodated' (Shehade & Stylianou-Lambert, 2020, p. 14). Furthermore, cultural tourism should also include a discussion of both physical elements as well as emotional ones since both are part of the experience (Shehade & Stylianou-Lambert, 2020).

Finding a Compromise

The remarks made by Shehade and Stylianou-Lambert (2020) regarding cultural tourism and VR/AR highlight a direction that acknowledges the necessity to find some middle-way that does not rely solely on subjective or objective views on authenticity. This has also been highlighted in different parts of previous sections. This is why one additional direction in the authenticity discussion has opted for finding a compromise between the different ways to conceptualise authenticity.

If we go back to our underlying questions and we focus on the experience of the tourists, Belhassen et al. (2008) have argued that postmodern and existential authenticity still need to be combined with objective authenticity to not separate the experience from the objects. This is necessary for the tourists to achieve an authentic experience. Cohen and Cohen (2012) proposed to include in authenticity two dimensions named 'hot' and 'cool'. The latter is comparable to the authentication provided in heritage and museum contexts. The former is related to constructed parts of the tourist experience and to the subjective feelings involved in both its production and consumption (Cohen & Cohen, 2012). In fact, there is a negotiation between essentialist (object) and constructivist views on authenticity even from the viewpoint of museum curators (Chhabra, 2008). This answers the second question guiding the chapter.

Authenticity as a compromise was also the topic of my PhD dissertation. In my research, the focus was on indigenous Sámi tourism. One crucial aspect related to authenticity was that 'fake' was mentioned by the respondents in relation to companies that pretend to be Sámi or that showcase Sámi products, but in fact have no connection to the Sámi (De Bernardi, 2020). Both from the perspective of the tourists and of the practitioners, there is a manifested thought of authentic versus inauthentic, which is usually explained in terms of social construction (e.g., Cohen & Avieli, 2004; Sims, 2009). If we would answer both of our questions for this section, we would then say that a feeling of authenticity for both the tourists and the practitioners is tied to both experiences and some essential characteristics possessed by places and objects that they encounter in their travels. This question was especially relevant regarding the Sámi. In order for tourism to be a positive force for Sámi culture, a realist view on authenticity needed to be re-established. This is done through the philosophical approach critical realism and the concept of concrete universals. Authenticity has then a universal component that is shared by different phenomena as well as many different contextual and individual manifestations that do not necessarily look alike and that cannot be reduced to the universal characteristic (De Bernardi, 2020).

Conclusions

This chapter has presented an overview of authenticity research from the origins up until the modern developments in line with technological advancement. Two main points that can be made about the development of authenticity scholarship in tourism are, on the one hand, the concept has become overly subjective and, on the other hand, some connection between authenticity and objects has been reestablished. Authenticity is indeed strongly tied to experiences and these can be very subjective. However, to exclusively tie authenticity to individual experiences bears the risk of eliminating from the equation the locals, the built heritage, the tourism workers and everything else the tourists come in contact with.

Shehade and Stylianou-Lambert (2020) argued that there should be a balance between different ways to conceptualise authenticity to accommodate the different values of several stakeholders. Regarding both people and objects that the tourists encounter at the destination, some guidance is sometimes needed. As previously mentioned in the case of the Sámi, while it is often considered important that the tourists come in touch with Sámi culture, an establishment of some kind of 'authenticity standards' supports the tourists in their knowledge about their experiences. This is similar to what has been proposed by Jamal and Hill (2004) in the form of indicators in the context of heritage tourism.

Regarding authenticity and the representation of history, it has been argued that there is a need for truth in the account on the history of the First Australians (Smallwood, 2015), which are used here as an example regarding many other indigenous and other local populations. As argued across the chapter, the presentation of history is important in relation to which viewpoint is presented (Waitt, 2000). A measure of authentication in the process of historical and heritage display could make sure that the account of events is as balanced as possible since subjective experiences can always lean too much towards one side or another. The same can be argued for marketing representations. As mentioned in the chapter, in my PhD thesis I dealt with the representation of the Sámi in marketing communication. When representing the Sámi and other local populations for tourism purposes, it is important to remember what their viewpoint on authenticity might be. Throughout the chapter I have presented many examples of how practitioners experience authenticity. The indicators developed by Jamal and Hill (2004) are an example of the conceptualization of a compromise between different dimensions of authenticity that also takes context (such as sociopolitical dimensions) into consideration.

As technology advances, the line between reality and fiction will increasingly be blurred. While we should consider approaches based on authentication and the different perspectives of the actors at the destination and working with tourism, we will also face a multiplicity of enhanced experiences that will be increasingly closer to the 'real thing'. Therefore, future research should not only focus on the development of how authenticity is conceptualized in different contexts, especially with regard to how authenticity dimensions are balanced, but the focus should also be on locating authenticity in the virtual and fantastic phenomena, preferably with the support of other disciplines to have a well-rounded grip on the issue.

In conclusion, it is important to follow the paradigmatic assumptions that you have already adopted (or to reflect on them before starting) when considering how to approach authenticity in a research project. As shown in this chapter, there are plenty of different perspectives that can be embraced to bring the discussion further. However, it is also important to keep in mind the many diversified viewpoints that are inherently part of tourism and of authenticity; when studying authenticity there is almost always 'another side of the coin'. Therefore, it is important that we, as researchers, pause and take some extra time to reflect on the different aspect of our topic so that we can embark in our research journey on authenticity in tourism with the best premises.

Note

1. The Sámi are an indigenous population in Europe who live mostly in Sweden, Norway, Finland and Russia. The total population is estimated to be between 80 000 and 100 000 individuals. The Sámi are a minority and have self-governing institutions such as a national parliament, at least in Europe. Even though sharing a common identity, there are many different expressions of Sámi culture, which is modern and constantly evolving. More information about the Sami can be found here: https://sanningskommissionensamer.se/en/about-the-indigenous-sami/

References

Asplet, M., & Cooper, M. (2000). Cultural designs in New Zealand souvenir clothing: The question of authenticity. *Tourism Management*, *21*(3), 307–312. https://doi.org/10.1016/S0261-5177(99)00061-8

Baudrillard, J. (1994). *Simulacra and simulation*. University of Michigan Press.

Belhassen, Y., Caton, K., & Stewart, W. P. (2008). The search for authenticity in the pilgrim experience. *Annals of Tourism Research*, *35*(3), 668–689. https://doi.org/10.1016/j.annals.2008.03.007

Boorstin, D. J. (1961). *The image: A guide to pseudo-events in America*. Vintage Books.

Buchmann, A., Moore, K., & Fisher, D. (2010). Experiencing film tourism: Authenticity & fellowship. *Annals of Tourism Research*, *37*(1), 229–248. https://doi.org/10.1016/j.annals.2009.09.005

Chhabra, D. (2008). Positioning museums on an authenticity continuum. *Annals of Tourism Research*, *35*(2), 427–447. https://doi.org/10.1016/j.annals.2007.12.001

Chhabra, D., Healy, R., & Sills, E. (2003). Staged authenticity and heritage tourism. *Annals of Tourism Research*, *30*(3), 702–719. https://doi.org/10.1016/S0160-7383(03)00044-6

Cohen, E. (1979). A phenomenology of tourist experiences. *Sociology*, *13*(2), 179–201. https://doi.org/10.1177/003803857901300203

Cohen, E. (1988). Authenticity and commoditization in tourism. *Annals of Tourism Research*, *15*(3), 371–386. https://doi.org/10.1016/0160-7383(88)90028-X

Cohen, E., & Avieli, N. (2004). Food in tourism: Attraction and impediment. *Annals of Tourism Research*, *31*(4), 755–778. https://doi.org/10.1016/j.annals.2004.02.003

Cohen, E., & Cohen, S. A. (2012). Current sociological theories and issues in tourism. *Annals of Tourism Research*, 39(4), 2177–2202. https://doi.org/10.1016/j.annals.2012.07.009

Crang, M. (1996). Magic kingdom or a quixotic quest for authenticity? *Annals of Tourism Research*, *23*(2), 415–431. https://doi.org/10.1016/0160-7383(95)00070-4

De Bernardi, C. (2020). *Authenticity as a compromise: a critical realist perspective on Sámi tourism labels* [Doctoral dissertation]. University of Lapland. https://lauda.ulapland.fi/handle/10024/64414

De Bernardi, C. (2022). Sámi tourism in marketing material: a multimodal discourse analysis. *Acta Borealia*, 39(2), 115–137. https://doi.org/10.1080/08003831.2022.2104544

Eco, U. (2014). *Faith in fakes: Travels in hyperreality*. Random House.

Edelheim, J. R. (2015). *Tourist attractions: From object to narrative*. Channel View Publications.

Ellis, A., Park, E., Kim, S., & Yeoman, I. (2018). What is food tourism? *Tourism Management*, *68*, 250–263. https://doi.org/10.1016/j.tourman.2018.03.025

Fawcett, C., & Cormack, P. (2001). Guarding authenticity at literary tourism sites. *Annals of Tourism Research*, *28*(3), 686–704. https://doi.org/10.1016/S0160-7383(00)00062-1

Gao, B. W., Zhu, C., Song, H., & Dempsey, I. M. B. (2022). Interpreting the perceptions of authenticity in virtual reality tourism through postmodernist approach. *Information Technology & Tourism*, *24*(1), 31–55. https://doi.org/10.1007/s40558-022-00221-0

Gilmore, J. H., & Pine, B. J. (2007). *Authenticity: What consumers really want*. Harvard Business Press.

Guttentag, D. A. (2010). Virtual reality: Applications and implications for tourism. *Tourism Management*, *31*(5), 637–651. https://doi.org/10.1016/j.tourman.2009.07.003

Halewood, C., & Hannam, K. (2001). Viking heritage tourism: Authenticity and commodification. *Annals of Tourism Research*, *28*(3), 565–580. https://doi.org/10.1016/S0160-7383(00)00076-1

Herbert, D. (2001). Literary places, tourism and the heritage experience. *Annals of Tourism Research*, *28*(2), 312–333. https://doi.org/10.1016/S0160-7383(00)00048-7

Holtorf, C. (2014). The time travellers' tools of the trade: some trends at Lejre. *International Journal of Heritage Studies*, *20*(7–8), 782–797. https://doi.org/10.1080/13527258.2013.818568

Houston, H. R., & Meamber, L. A. (2011). Consuming the "world": reflexivity, aesthetics, and authenticity at Disney World's EPCOT Center. *Consumption Markets & Culture*, *14*(2), 177–191. https://doi.org/10.1080/10253866.2011.562019

Jamal, T., & Hill, S. (2004). Developing a framework for indicators of authenticity: The place and space of cultural and heritage tourism. *Asia Pacific Journal of Tourism Research*, 9(4), 353–372. https://doi.org/10.1080/1094166042000311246

Jiang, Y., Ramkissoon, H., Mavondo, F. T., & Feng, S. (2017). Authenticity: The link between destination image and place attachment. *Journal of Hospitality Marketing & Management*, *26*(2), 105–124. https://doi.org/10.1080/19368623.2016.1185988

Jokilehto, J. (2006). Considerations on authenticity and integrity in world heritage context. *City & Time*, *2*(1), 1. http://www.ceci-br.org/novo/revista/docs2006/CT-2006-44.pdf

Jones, D., & Smith, K. (2005). Middle-earth meets New Zealand: Authenticity and location in the making of *The Lord of the Rings*. *Journal of Management Studies*, *42*(5), 923–945. https://doi.org/10.1111/j.1467-6486.2005.00527.x

Kim, H., & Jamal, T. (2007). Touristic quest for existential authenticity. *Annals of Tourism Research*, *34*(1), 181–201. https://doi.org/10.1016/j.annals.2006.07.009

Kim, S., Whitford, M., & Arcodia, C. (2019). Development of intangible cultural heritage as a sustainable tourism resource: The intangible cultural heritage practitioners' perspectives. *Journal of Heritage Tourism*, *14*(5–6), 422–435. https://doi.org/10.1080/1743873X.2018.1561703

Koot, S. P. (2016). Contradictions of capitalism in the South African Kalahari: Indigenous Bushmen, their brand and *baasskap* in tourism. *Journal of Sustainable Tourism*, *24*(8–9), 1211–1226. https://doi.org/10.1080/09669582.2016.1158825

Larson, M., Lundberg, C., & Lexhagen, M. (2013). Thirsting for vampire tourism: Developing pop culture destinations. *Journal of Destination Marketing & Management*, *2*(2), 74–84. https://doi.org/10.1016/j.jdmm.2013.03.004

Lee, C. (2012). "Have magic, will travel": Tourism and Harry Potter's United (magical) Kingdom. *Tourist Studies*, *12*(1), 52–69. https://doi.org/10.1177/1468797612438438

Li, S., Tian, W., Lundberg, C., Gkritzali, A., & Sundström, M. (2021). Two tales of one city: Fantasy proneness, authenticity, and loyalty of on-screen tourism destinations. *Journal of Travel Research*, *60*(8), 1802–1820. https://doi.org/10.1177/0047287520961179

Littrell, M. A., Anderson, L. F., & Brown, P. J. (1993). What makes a craft souvenir authentic? *Annals of Tourism Research*, *20*(1), 197–215. https://doi.org/10.1016/0160-7383(93)90118-M

Lovell, J., & Hitchmough, S. (2020). Simulated authenticity: Storytelling and mythic space on the hyperfrontier in Buffalo Bill's Wild West and Westworld. *Tourist Studies*, *20*(4), 409–428. https://doi.org/10.1177/1468797620937912

Lu, L., Chi, C. G., & Liu, Y. (2015). Authenticity, involvement, and image: Evaluating tourist experiences at historic districts. *Tourism Management*, *50*, 85–96. https://doi.org/10.1016/j.tourman.2015.01.026

MacCannell, D. (1973). Staged authenticity: Arrangements of social space in tourist settings. *American Journal of Sociology*, 79(3), 589–603. https://doi.org/10.1086/225585

McIntosh, A. J., & Prentice, R. C. (1999). Affirming authenticity: Consuming cultural heritage. *Annals of Tourism Research*, *26*(3), 589–612. https://doi.org/10.1016/S0160-7383(99)00010-9

Mura, P., Tavakoli, R., & Pahlevan Sharif, S. (2017). 'Authentic but not too much': exploring perceptions of authenticity of virtual tourism. *Information Technology & Tourism*, *17*(2), 145–159. https://doi.org/10.1007/s40558-016-0059-y

Park, E., Choi, B. K., & Lee, T. J. (2019). The role and dimensions of authenticity in heritage tourism. *Tourism Management*, *74*, 99–109. https://doi.org/10.1016/j.tourman.2019.03.001

Ram, Y., Björk, P., & Weidenfeld, A. (2016). Authenticity and place attachment of major visitor attractions. *Tourism Management*, *52*, 110–122. https://doi.org/10.1016/j.tourman.2015.06.010

Reisinger, Y., & Steiner, C. J. (2006). Reconceptualizing object authenticity. *Annals of Tourism Research*, *33*(1), 65–86. https://doi.org/10.1016/j.annals.2005.04.003

Rittichainuwat, B., Laws, E., Scott, N., & Rattanaphinanchai, S. (2018). Authenticity in screen tourism: Significance of real and substituted screen locations. *Journal of Hospitality & Tourism Research*, *42*(8), 1274–1294. https://doi.org/10.1177/1096348017736568

Salamone, F. A. (1997). Authenticity in tourism: The San Angel Inns. *Annals of Tourism Research*, *24*(2), 305–321. https://doi.org/10.1016/S0160-7383(97)80003-5

Schiopu, A. F., Hornoiu, R. I., Padurean, M. A., & Nica, A. M. (2021). Virus tinged? Exploring the facets of virtual reality use in tourism as a result of the COVID-19 pandemic. *Telematics and Informatics*, *60*, 101575. https://doi.org/10.1016/j.tele.2021.101575

Šegota, T. (2018). Creating (extra) ordinary heritage through film-induced tourism: The case of Dubrovnik and the Game of Thrones. In C. Palmer & J. Tivers (Eds), *Creating heritage for tourism* (pp. 115–126). Routledge.

Seraphin, H., Yallop, A. C., Capatîna, A., & Gowreesunkar, V. G. (2018). Heritage in tourism organisations' branding strategy: The case of a post-colonial, post-conflict and post-disaster destination. *International Journal of Culture, Tourism and Hospitality Research*, *12*(1), 89–105. https://doi.org/10.1108/IJCTHR-05-2017-0057

Shehade, M., & Stylianou-Lambert, T. (2020). Revisiting authenticity in the age of the digital transformation of cultural tourism. In V. Katsoni & T. Spyriadis (Eds), *Cultural and tourism innovation in the digital era* (pp. 3–16). Springer, Cham.

Silver, I. (1993). Marketing authenticity in third world countries. *Annals of Tourism Research*, *20*(2), 302–318. https://doi.org/10.1016/0160-7383(93)90057-A

Sims, R. (2009). Food, place and authenticity: local food and the sustainable tourism experience. *Journal of Sustainable Tourism*, *17*(3), 321–336. https://doi.org/10.1080/09669580802359293

Smallwood, G. (2015). *Indigenist critical realism: Human rights and first Australians' well-being*. Routledge.

Smith, S. P. (2018). Instagram abroad: performance, consumption and colonial narrative in tourism. *Postcolonial Studies*, *21*(2), 172–191. https://doi.org/10.1080/13688790.2018.1461173

Soukhathammavong, B., & Park, E. (2019). The authentic souvenir: What does it mean to souvenir suppliers in the heritage destination? *Tourism Management*, *72*, 105–116. https://doi.org/10.1016/j.tourman.2018.11.015

Squire, S. J. (1994). The cultural values of literary tourism. *Annals of Tourism Research* *21*(1), 103–120. https://doi.org/10.1016/0160-7383(94)90007-8

Steiner, C. J., & Reisinger, Y. (2006). Understanding existential authenticity. *Annals of Tourism Research*, *33*(2), 299–318. https://doi.org/10.1016/j.annals.2005.08.002

Su, J. (2018). Conceptualising the subjective authenticity of intangible cultural heritage. *International Journal of Heritage Studies*, *24*(9), 919–937. https://doi.org/10.1080/13527258.2018.1428662

Swanson, K. K., & Timothy, D. J. (2012). Souvenirs: Icons of meaning, commercialization and commoditization. *Tourism Management*, *33*(3), 489–499. https://doi.org/10.1016/j.tourman.2011.10.007

Torabian, P., & Arai, S. M. (2016). Tourist perceptions of souvenir authenticity: An exploration of selective tourist blogs. *Current Issues in Tourism*, *19*(7), 697–712. https://doi.org/10.1080/13683500.2013.820259

Torchin, L. (2002). Location, location, location: The destination of the Manhattan TV Tour. *Tourist Studies*, *2*(3), 247–266. https://doi.org/10.1177/14687976020023002

Tzanelli, R. (2004). Constructing the 'cinematic tourist': The 'sign industry' of The Lord of the Rings. *Tourist Studies*, *4*(1), 21–42. https://doi.org/10.1177/1468797604053077

Varga, S., & Guignon, C. (2020). Authenticity. In E. N. Zalta (Ed.), *The Stanford encyclopedia of philosophy* (Spring 2020 Edition). https://plato.stanford.edu/entries/authenticity/

Waitt, G. (2000). Consuming heritage: Perceived historical authenticity. *Annals of Tourism Research*, *27*(4), 835–862. https://doi.org/10.1016/S0160-7383(99)00115-2

Wang, N. (1999). Rethinking authenticity in tourism experience. *Annals of Tourism Research*, *26*(2), 349–370. https://doi.org/10.1016/S0160-7383(98)00103-0

Waysdorf, A., & Reijnders, S. (2018). Immersion, authenticity and the theme park as social space: Experiencing the Wizarding World of Harry Potter. *International Journal of Cultural Studies*, *21*(2), 173–188. https://doi.org/10.1177/1367877916674751

Wong, I. A., Lin, S. K., Lin, Z. C., & Xiong, X. (2022). Welcome to stay-at-home travel and virtual attention restoration. *Journal of Hospitality and Tourism Management*, *51*, 207–217. https://doi.org/10.1016/j.jhtm.2022.03.016

Xie, P. F., Wu, T. C., & Hsieh, H. W. (2012). Tourists' perception of authenticity in indigenous souvenirs in Taiwan. *Journal of Travel & Tourism Marketing* *29*(5): 485–500. https://doi.org/10.1080/10548408.2012.691400

Yang, L. (2011). Ethnic tourism and cultural representation. *Annals of Tourism Research*, *38*(2), 561–585. https://doi.org/10.1016/j.annals.2010.10.009

Chapter 14

Place Attachment in Tourism

Haywantee Ramkissoon

Introduction

Place attachment research in tourism has important implications for sustainable tourism development. This is increasingly recognised by scholars and practitioners as we continue to battle with the COVID-19 pandemic. Place attachment in tourism studies has been commonly defined as the emotional bonding between people and environmental place settings (Ramkissoon, 2023a, 2023b). Place settings often provide venues for leisure and recreation for both residents and tourists. These venues allow interactions and involvement between people and place which can develop into an immediate to a long-lasting sense of deep affect. Place attachment is about how these attractions get immersed in values and feelings (Kyle et al., 2004; Jiang et al., 2017), and how people build social bonds and friendships (Ramkissoon, 2022b). Some examples of place settings providing interactions influencing people's experiences include famous beaches, national parks, peri-urban parks, cultural heritage attractions (e.g., museums, industrial heritage sites, wineries), local cafes and restaurants, and shopping malls in the neighbourhood among others. Place attachment scholars have been struggling with the multiple terminologies used in place attachment research. Some examples of terms coined in addition to the term place attachment are 'sense of place', 'neighbourhood attachment', 'community attachment' and 'connectedness to nature' (e.g., Chen et al. 2021; Eslami et al., 2019; Pearce et al. 2022). These terms have also been used to describe the relationship that people share with spatial settings. The term 'place attachment' however, remains the most commonly used in tourism research.

Defining Place Attachment

The term 'place attachment' has been borrowed from environmental psychology and is now largely applied in tourism research. Place attachment draws on attachment theory which was developed by Bowlby (1988). Scholars argue that it is one of the most popular explanations

The Wiley Blackwell Companion to Tourism, Second Edition. Edited by C. Michael Hall.

of how early relationships influence other social relationships (Bosmans et al., 2022; Ramkissoon, 2022c). Attachment theory suggests that we first develop a relationship with our carer and this relationship influences other social relations we develop over the years. Bowlby (1988) explains the infant-mother relationship as a template for love and trust that the infant follows, often without realising it. These childhood experiences lay the foundations for the attachment theory.

Attachment theory has been applied to a wide range of contexts in tourism. Drawing on Bowlby's (1988) attachment theory, tourism researchers study hosts' and tourists' affective and cognitive aspects of human attachment in tourism. Some areas include virtual reality (Wang et al. 2022), customer engagement and social media (Shoukat & Ramkissoon, 2022; Hinson et al. 2019), human behaviour in nature (Akhshik et al., 2022b; Ramkissoon et al., 2013a, 2013b) and cultural heritage (Buonincontri et al., 2017).

Place Attachment: A Multi-dimensional Construct in Tourism

Place attachment is an important component of the human-place interaction (Kyle et al., 2004; Scannell & Gifford, 2010, 2017) and deserves significant attention. With operationalisations of the place attachment construct being very diverse across disciplines, there has been several calls to investigate how the term is operationalised. A number of scholars have argued that the construct deserves further investigation and concluded that it needs to be treated as a multi-dimensional construct (e.g., Kyle et al., 2004; Ramkissoon et al., 2012).

The quest for evolving research in place attachment in tourism also stemmed from researchers questioning place attachment as a unidimensional construct (or measured with only the two most widely used dimensions of the construct) (Ramkissoon et al., 2012; Scannell & Gifford, 2010). Most often, place attachment in literature has been conceptualised as a two-dimensional construct with place dependence (Stokols & Shumacker, 1982) and place identity (Vaske & Kobrin, 2001; Walker & Chapman, 2003). Ramkissoon et al. (2012) proposed place attachment as a multi-dimensional construct with four sub-dimensions namely place dependence, place identity, place affect, and place social bonding in a national park context. By testing place attachment as an overarching construct with these four sub-dimensions, Ramkissoon et al. (2012) brought theoretical, methodological and practical insights to place attachment scholars in tourism and across other disciplines. In addition, Ramkissoon et al's (2013a) overall measurement model for place attachment showed a good fit to the data which required no re-specification. Their measurement model imposed for the place attachment construct was consistent with the place attachment observations in the extant literature. Further, from a methodological perspective, the confirmatory factor analytic method applied did support the validity of the place attachment scale as a second-order factor. It is not surprising that place attachment as a second-order construct continues to attract a lot of interest from tourism scholars and researchers across a number of disciplines wishing to study place attachment in other contexts (e.g., Buonincontri et al., 2017; Dwyer et al., 2019).

Place Dependence Behaviours in Tourism

Place attachment research in tourism has also focussed on the functional purpose of place. It is described in recreational settings as visitors' functional attachment to a specific place (Stokols & Shumacker, 1982). People may become place dependent on a place's setting as their awareness of the uniqueness of a setting grows and meets their visitation goals (Williams et al., 1992).

The attachment to place here is demonstrated in the physical settings of place e.g., hiking trails, camping sites, rock climbing routes which essentially provide the required services for desired recreational activities. Place dependence hence is useful in understanding how the current setting compares with other similar ones (Williams et al., 1992). This highlights its importance to destination managers and marketers in promoting their attractions' functional purpose to attract visitation.

Place Identity Behaviours in Tourism

Consumers' place identity in tourism refers to the connection between a place and one's personal identity. It contains both cognitive and affective elements (Proshansky, 1978). Consumers of nature, culture and heritage can also experience a sense of place identity with the cultural heritage offerings (Nekmahmud et al., 2022). Often, the latter allows them to reflect on the meanings and significance they attach to the cultural attractions which could potentially lead to revisit intentions (Marschall, 2012; Rasoolimanesh et al., 2021; Seyfi et al., 2023). This is particularly important as meaningful and positive visitor experiences have potential to promote value co-creation and contribute to tourism product development.

One example could be a special cultural interest in food and wine/gastronomic tourism (Baird et al., 2020; Hall, 2013, 2020; Carvalho et al., 2021). Food and wine lovers can strongly identify with wineries and restaurants that give rise to extraordinary and memorable gastronomic experiences (Goolaup & Mossberg, 2017). The place's uniqueness needs to be recognised and promoted. Referring to the culinary experiences, co-creation of new food recipes with locals can create more positive memorable experiences (Chen & Rahman, 2018; Lee, 2015). Gastronomic experiences remain a defining marker of culture and heritage and have important relevance for promoting place identity and hence revisits. Taste and smell of food and wine can contribute to preservation of gastronomic identity and give rise to a sense of gastronomic uniqueness. This is an important aspect to consider in developing regional food and wine businesses. Emphasising their distinctiveness, tourism settings can offer visitors the opportunity to develop a sense of identity with a place (Ramkissoon et al., 2013b).

Place Affective Behaviours in Tourism

Researchers of place note that people develop affective connections (Rollero & De Picolli, 2010) by building their sentiments about the place (Chanchaichujit et al., 2018; Tuan, 1977). In a tourism/leisure context, emphasising the affective connection people develop with cultural and heritage environments is another important area deserving attention in place attachment research. A thorough understanding of the affective connection with place will further contribute to the place attachment literature and bring forward important recommendations for managers/marketers of place.

Some evidence suggests that natural and cultural tourist settings do tend to further increase positive affect in individuals (Hartig et al., 1996) and generates a sense of psychological wellbeing for visitors (Kaplan & Talbot, 1983; Korpela et al., 2010; Townsend et al., 2018). As visitors develop and share emotional bonds, an increase in place affect with places can help to combat stress and foster more positive emotions (Ramkissoon, 2022e) and attachment to place. The latter can further contribute to visitors' wellbeing (Correia et al., 2017). Visitors experiencing greater affect with those natural and cultural settings may express a stronger sense of emotional attachment with those environments than those with lesser experience (Akhshiq et al., 2022a, 2022b; Hinds & Sparks, 2008, 2011).

Places can be restorative and provide the stimuli to restore a person's self-evaluation and contribute to their emotional, mental and physical wellbeing (Ramkissoon, 2021). Places can give rise to a 'good feel' factor and influence other people's experiences positively (Akshiq et al., 2020; Rezapouraghdam et al., 2021). Research in tourism also show how places can contribute to psychological restorativeness and help one attain their spiritual goals (Ramkissoon, 2021). Atsiz et al. (2023) emphasise the benefits of affective connection to place in Yoga tourism. They argue the affective connection to place can give rise to a sense of awe for yoga travellers. Yoga travellers report restorative effects, experiencing inner peace, and unique spiritual experiences to place (Tian et al., 2015).

Place Social Bonding and Pro-social Behaviours in Tourism

As tourists seek new, exciting and meaningful experiences at tourist attractions, they are often also thrilled to make new connections with others. They develop social ties in their neighbourhoods, and when encountering others in places with natural beauty, and cultural and heritage significance (Ramkissoon, 2022b). As such, people in places co-create meaningful memorable visitor experiences (Chen & Rahman, 2018; Ramkissoon et al., 2018) often with an excitement to fulfil higher-order goals. The latter can include a range of positive cultural exchanges including being kind and offering their help to others, showing respect and compassion for others, which can also extend to supporting others to overcome stressful situations (Ramkissoon, 2022e; Stigsdotter et al., 2010).

Place social bonding can foster pro-social behaviours, with people helping one another. People's behaviours can often be influenced by what others approve or disapprove of. Hence, the fear of losing one's distinctive place can encourage collective actions such as a desire to learn from others on how to protect their place. Social engagements and affiliations can re-affirm ones' attachment to his or her place and lend support to promote sustainable tourism development.

Place Attachment and Sustainable Tourism in Nature-based and Cultural Settings

The multivalent nature of place as discussed above demonstrates the cultivation of place attachment through subdimensions of place dependence, place identity, place affect, and place social bonding. The multidimensional constructs of place attachment are conceptually different from one another. Several studies used the four sub-constructs of place attachment in a single theoretical model (e.g., Akhshik et al., 2022a, 2022b; Ramkissoon & Mavondo, 2015) and tested their associations with other constructs across tourism settings and beyond. There continues to be an increased focus on place attachment being an overarching construct with other sub-dimensions in tourism studies (e.g., Dwyer et al., 2019).

Place Attachment and Memorable Experiences in Tourism

A deeper focus on understanding visitors' place attachment in tourism and how to create memorable experiences through promoting their levels of place attachment with the place settings and social connections is required in tourism studies. The social and emotional ties formed and shared can play a crucial role in influencing the sustainability of the natural and cultural heritage offerings and hence contribute in developing memorable visitor experiences. Chen et al. (2021) used two terms, namely place memory and place expectations, suggesting that they are important dimensions of place attachment to be considered when people are interacting with a place. They define place memory as attachment reflected through memories of one's interaction with a place and place expectations as one's reflected expectations of their future experiences in this place. They conclude that the explanatory power of place attachment was further enhanced with place memory and place expectations.

Meanings that consumers of place create for collective purposes can further trigger sustainable behaviours at the attractions (Ramkissoon, 2022a) and has potential to promote active participation. Further, the engagement can also bring co-creation (Zhang et al., 2019; Suntikul & Jachna, 2016) of newer memorable visitor experiences and in turn encourage their revisits to these cultural heritage attractions (Megeirhi et al., 2020). Ramkissoon (2022c, 2022d) further argues that people's attachment to place plays a significant role in spirituality as people attach meanings to places. Hassan and Ramkissoon (in press) argue that virtual reality (VR) and attachment to spiritual places can create memorable experiences. VR in tourism has been attracting much attention in the current global health pandemic (e.g., Lu et al., 2022; Manchanda & Deb, 2022). Place attachment researchers would benefit from further investigation of VR, spirituality and connection to place.

Place Attachment in Therapeutic Tourist Settings

As the tourism industry depends heavily on our natural and cultural resources, promoting place attachment and place protective behaviours for both human and planetary health has substantial relevance. The COVID-19 pandemic fuelled more research on attachment to place and place meanings. Behavioural scientists and clinical practitioners now argue that attachment to nature needs to be promoted among people. This can have important health and wellbeing benefits especially as we continue to struggle with a decline in mental and physical wellbeing (NHS Foundation Trust, 2021). An investigation of how people can become further attached to natural settings and culture demands that we explore further how place meanings are created.

Place attachment has been used in therapeutic settings (Bretherton, 1992). Scholars have argued that place settings play an important role in physical, emotional and mental health (Ramkissoon, 2022a, 2022e) and hence promote quality-of-life (Gatrell, 2013). Individuals' health and wellbeing can be determined by their psychological expectations of green and blue spaces (Völker & Kistemann, 2011) (e.g., land, forest, cities, sea). Their emotions can influence their travel decisions (Majeed & Ramkissoon, 2020). The association of place settings with health and wellbeing is referred to as psychological rootedness (Gesler, 1992). This finds support in the theory of therapeutic landscape suggesting that human health and wellbeing is a place-based concept (Gesler, 1992) with healing benefits. Some examples of touristic therapeutic sites are the hot springs of South Dakota, Epidaurus in Greece, and spiritual retreats in Lourdes attracting repeat visitations (English et al., 2008). These tourist attractions present the stimulus that restores visitors' positive evaluations of themselves (Stigsdotter et al., 2010) which can promote place attachment and re-visit intentions (Majeed & Ramkissoon, 2020). With the COVID-19 pandemic leaving many people dealing with loss and separation and coping with grief, policy makers and destination managers and marketers need to invest further efforts in promoting place attachment in therapeutic landscapes for health and wellbeing benefits.

Place Attachment and Place Protective Behaviours

Place attachment in sustainable tourism studies has focussed largely on nature-based attractions and continue to attract interest from tourism scholars interested in culture and heritage (e.g., Ramkissoon et al., 2014; Xu et al., 2022; Zhang et al., 2019). Ramkissoon et al's (2013a, 2013b) findings suggest that place attachment is associated with visitor satisfaction with a spatial setting, and pro-environmental behaviours. Recognising the negative impacts of tourism and severe environmental consequences at a number of tourist attractions, tourism scholars continue to explore the associations between tourists' attachment to place and their pro-environmental behaviour (e.g., Ramkissoon et al, 2013a, 2013b; Qu et al., 2019; Wu et al., 2021).

While most studies concentrate on tourists' place attachment and their engagement in efforts to reduce their environmental footprint, place attachment has also attracted interest from quality of life researchers (e.g., Mihaylov & Perkins, 2014; Rollero & De Piccoli, 2010). Some scholars argue place attachment plays a significant role in promoting quality of life in public places (Firouzmakan & Daneshpour, 2015; Ramkissoon et al., 2018). Ramkissoon (2023a) explored how residents' place attachment could lead to engagement in pro-social and pro-environmental place protective behaviours and support for tourism development. She argues residents are important stakeholders and as consumers of place settings, they can play an important role in safeguarding their place. The residents' interaction with other social connections developed in place settings can create meanings. These assigned meanings have potential to lead to collective efforts among residents to protect their place (Ramkissoon et al., 2018). Her proposed framework posits that residents' engagement in pro-social and pro-environmental behaviour can influence support for tourism development and contribute to their overall quality of life.

There has been important scholarly works on residents' attitudes to tourism demonstrating that happy residents can generate more support for tourism development (Nunkoo & Ramkissoon, 2012; Nunkoo et al., 2013). The role of affective connections to place can address some important aspects of further support for sustainable tourism development. Connecting people to place and their engagement in the protection of the planet's resources aligns with the United Nations *2030 Agenda for Sustainable Development* and planetary health research calls (Khetrapal & Bhatia, 2020; Ramkissoon, 2020, 2023a; Rasoolimanesh et al., 2020). Ramkissoon (2022a) further argues that spending time in nature and with social connections (with social distancing in the current pandemic) has potential to bring physical and mental health benefits to both residents and tourists. This aligns closely with strategic directions of healthcare providers (e.g., NHS Foundation Trust, 2021).

In the current dire times, place attachment can help deal with the restarting of tourism dependent economies in a much more equitable way as compared to the COVID-19 pandemic times. A multi-stakeholder engagement approach will however, need to be in place when working on fostering place attachment. This demands that several actors, including residents and tourists play an important and active role in revitalising local tourism businesses (Hall et al., 2020). With the pressing desire from local stakeholders to keep their tourism sector going especially in tourism dependent economies with limited socio-economic alternatives, fostering place attachment in tourism has emerged as a fundamental issue that needs to be addressed.

Place Attachment and Resilience

Drawing on the current context of the COVID-19 pandemic, a discussion on place attachment and resilience has been brought to the table. Drawing on multi-disciplinary theories and existing place attachment research in tourism, this chapter calls for more interdisciplinary and multidisciplinary research to inform place attachment and resilience in tourism research. The aim is to promote place attachment to revive and promote a sustainable tourism industry. The COVID-19 pandemic can be a transformative opportunity for tourism-dependent economies (and others) (Hall et al., 2020; Ramkissoon, 2022b, 2023a) to reflect, re-visit and re-design their business strategies aligning with the sustainable development goals (Gössling et al., 2020; Ramkissoon, 2022a, 2022b).

Scholars argue that the increased ticket prices, economic instability, and COVID-19 travel anxiety (fear of contracting the virus), virtual travel and the growing interest in metaverse and its potential still remain a possibility for many (Gössling & Schweiggart, 2022; Ramkissoon, 2022a, 2022b). However, while the year 2020 was seen as being the worst in tourism history in international tourism arrivals because of COVID-19, the subsequent easing of border restrictions and summer holidays in many parts of the globe has fuelled travel. While this could have important implications with opportunities to promote the economic recovery of several nations, some tourism scholars have started questioning whether we are simply not going back to the pre-pandemic times and the heated debate on 'over-tourism'? Several airports are now flooded with travellers suggesting that attractions are getting massive tourist numbers (Assaf et al., 2022). This has consequently resulted in tourism induced pollution emissions (Bamidele et al., 2022) leaving staff struggling to manage the visitor flow at tourist attractions. Place attachment could play an even bigger role as we address tourist/residents' behaviours and behaviour change for tourism sustainability (Ramkissoon, 2023b).

Conclusion and Future Directions

Drawing on social, psychological, environmental, and behavioural sciences, this chapter discusses the multi-dimensional nature of place attachment and its relevance in sustainable tourism research. It explores the role of place attachment in building pride in place and collective place protective behaviours. It provides a number of important insights into how place attachment can contribute to place memorable experiences and finally discusses place attachment and resilience using the current COVID-19 pandemic as a context.

It further draws on the extant literature and provides implications for place attachment tourism researchers, tourism policy makers and destination marketers, and managers to consider. Place attachment research in tourism is timely and demands that we explore the missing mechanisms to promote sustainable consumption for a sustainable future of the tourism industry and hence minimise our environmental footprint.

Place attachment among both tourists and residents needs to be promoted with the aim of achieving a balance between consumption and visitation. This chapter stresses the importance of adopting a multistakeholder engagement approach in place attachment research in tourism to align to the United Nations Sustainable Development Goals. This will require that residents, visitors, policy makers, employees, business enterprises and entrepreneurs, and other co-actors play an important role in the place making planning and development process. This chapter discussed important practical implications for sustainable community tourism and also potential contributions to achieving the broader sustainable tourism and sustainable development goals e.g., stakeholder engagement in place attachment (tourism) research and people wellbeing and planetary health. The author recognises that she has made just a selection of important observations requiring attention in place attachment research in tourism. Readers are encouraged to further explore this important and timely research area in tourism studies.

References

Akhshik, A., Rezapouraghdam, H., & Ramkissoon, H. (2022). Industrialization of nature in the time of complexity unawareness: The case of Chitgar Lake, Iran. *Journal of Hospitality & Tourism Research*, *46*(3), 583–606. https://doi.org/10.1177/1096348020964949

Akhshik, A., Rezapouraghdam, H., & Ramkissoon, H. (2022a). Industrialization of nature in the time of complexity unawareness: The case of Chitgar Lake, Iran. *Journal of Hospitality & Tourism Research*, *46*(3), 583–606. https://doi.org/10.1177/1096348020964949

Akhshik, A., Rezapouraghdam, H., Otzuren, A., & Ramkissoon, H. (2022b). Memorable experiences and critical outcomes among nature-based visitors: a fuzzy set qualitative comparative analysis approach. *Current Issues in Tourism*. https://doi.org/10.1080/13683500.2022.2106196

Assaf, A. G., Kock, F., & Tsionas, M. (2022). Tourism during and after COVID-19: An expert-informed agenda for future research. *Journal of Travel Research*, *61*(2), 454–457. https://doi.org/10.1177/00472875211017237

Atsız, O., Ramkissoon, H., & Öğretmenoğlu, M. (2023). Yoga travellers' experiences in guided tours: a multiple case study approach. *Current Issues in Tourism*, *26*(2), 177–183.

Baird, T., Hall, C. M., Castka, P., & Ramkissoon, H. (2020). Migrant workers' rights, social justice and sustainability in Australian and New Zealand wineries: A comparative context. In S. L. Forbes, T.-A. De Silva & A. Gilinsky (Eds), *Social sustainability in the global wine industry: Concepts and cases* (pp. 107–118). Palgrave MacMillan.

Bamidele, R., Ozturk, I., Gyamfi, B. A., & Bekun, F. V. (2022). Tourism-induced pollution emission amidst energy mix: evidence from Nigeria. *Environmental Science and Pollution Research*, *29*, 19752–19761. https://doi.org/10.1007/s11356-021-17233-8

Bosmans, G., Van Vlierberghe, L., Bakermans-Kranenburg, M. J., Kobak, R., Hermans, D., & van IJzendoorn, M. H. (2022). A learning theory approach to attachment theory: Exploring clinical applications. *Clinical Child and Family Psychology Review*. https://doi.org/10.1007/s10567-021-00377-x

Bowlby, J. (1988). Developmental psychiatry comes of age. *The American Journal of Psychiatry*, *145*(1), 1–10. https://doi.org/10.1176/ajp.145.1.1

Bretherton, I. (1992). The origins of attachment theory: John Bowlby and Mary Ainsworth. *Developmental Psychology*, *28*(5), 759–775. https://doi.org/10.1037/0012-1649.28.5.759

Buonincontri, P., Marasco, A., & Ramkissoon, H. (2017). Visitors' experience, place attachment and sustainable behaviour at cultural heritage sites: A conceptual framework. *Sustainability*, *9*(7), 1112. https://doi.org/10.3390/su9071112

Carvalho, M., Kastenholz, E., & Carneiro, M. J. (2021). Co-creative tourism experiences–a conceptual framework and its application to food & wine tourism. *Tourism Recreation Research*. https://doi.org/10.1080/02508281.2021.1948719

Chanchaichujit, K., Holmes, K., Dickinson, S., & Ramkissoon, H. (2018). *An investigation of how user generated content influences place affect towards an unvisited destination* [Paper presentation]. 8th Advances in Hospitality and Tourism Marketing and Management (AHTMM) Conference, Bangkok, Thailand.

Chen, H., & Rahman, I. (2018). Cultural tourism: An analysis of engagement, cultural contact, memorable tourism experience and destination loyalty. *Tourism Management Perspectives*, *26*, 153–163. https://doi.org/10.1016/j.tmp.2017.10.006

Chen, N. C., Hall, C. M., & Prayag, G. (2021). *Sense of place and place attachment in tourism*. Routledge.

Correia, A., Oliveira, C., & Pereira, R. (2017). From emotions to place attachment. In A. Correia, M. Kozak, J. Gnoth & A. Fyall (Eds.), *Co-creation and well-being in tourism* (pp. 163–177). Springer.

Dwyer, L., Chen, N., & Lee, J. (2019). The role of place attachment in tourism research. *Journal of Travel & Tourism Marketing*, *36*(5), 645–652. https://doi.org/10.1080/10548408.2019.1612824

English, J., Wilson, K., & Keller-Olaman, S. (2008). Health, healing and recovery: therapeutic landscapes and the everyday lives of breast cancer survivors. *Social Science and. Medicine*, 67, 68–78. https://doi.org/10.1016/j.socscimed.2008.03.043

Eslami, S., Khalifah, Z., Mardani, A., Streimikiene, D., & Han, H. (2019). Community attachment, tourism impacts, quality of life and residents' support for sustainable tourism development. *Journal of Travel & Tourism Marketing*, *36*(9), 1061–1079. https://doi.org/10.1080/10548408.2019.1689224

Firouzmakan, S., & Daneshpour, S. A. (2015). Promotion quality of life by increasing place attachment in public places. *Procedia-Social and Behavioral Sciences*, *201*, 418–428. https://doi.org/10.1016/j.sbspro.2015.08.197

Gatrell, A. C. (2013). Therapeutic mobilities: walking and 'steps' to wellbeing and health. *Health & Place*, *22*, 98–106. https://doi.org/10.1016/j.healthplace.2013.04.002

Gesler, W. M. (1992). Therapeutic landscapes: medical issues in light of the new cultural geography. *Social Science & Medicine*, *34*(7), 735–746. https://doi.org/10.1016/0277-9536(92)90360-3

Goolaup, S., & Mossberg, L. (2017). Exploring the concept of extraordinary related to food tourists' nature-based experience. *Scandinavian Journal of Hospitality and Tourism*, *17*(1), 27–43. https://doi.org/10.1080/15022250.2016.1218150

Gössling, S., & Schweiggart, N. (2022). Two years of COVID-19 and tourism: what we learned, and what we should have learned. *Journal of Sustainable Tourism*, *30*(4), 915–931. https://doi.org/10.1080/09669582.2022.2029872

Gössling, S., Scott, D., & Hall, C. M. (2020). Pandemics, tourism and global change: a rapid assessment of COVID-19. *Journal of Sustainable Tourism*, *29*(1), 1–20. https://doi.org/10.1080/09669582.2020.1758708

Hall, C. M. (Ed.). (2013). *Wine, food, and tourism marketing*. Routledge.

Hall, C. M. (2020). Improving the recipe for culinary and food tourism? The need for a new menu. *Tourism Recreation Research*, *45*(2), 284–287. https://doi.org/10.1080/02508281.2019.1694243

Hall, C. M., Scott, D., & Gössling, S. (2020). Pandemics, transformations and tourism: Be careful what you wish for. *Tourism Geographies*, 22(3), 577–598. https://doi.org/10.1080/14616688.2020.1759131

Hartig, T., Böök, A., Garvill, J., Olsson, T., & Gärling, T. (1996). Environmental influences on psychological restoration. *Scandinavian Journal of Psychology*, *37*(4), 378–393. https://doi.org/10.1111/j.1467-9450.1996.tb00670.x

Hassan, A., & Ramkissoon, H. (in press). Virtual space, spirituality and health in tourism. In V. Counted, H. Ramkissoon, L. Captari & R. G. Cowden (Eds), *Place, spirituality, and wellbeing: A global and multidisciplinary approach*. Springer.

Hinds, J., & Sparks, P. (2008). Engaging with the natural environment: The role of affective connection and identity. *Journal of Environmental Psychology*, *28*(2), 109–120. https://doi.org/10.1016/j.jenvp.2007.11.001

Hinds, J., & Sparks, P. (2011). The affective quality of human-natural environment relationships. *Evolutionary Psychology*, *9*(3). https://doi.org/10.1177/147470491100900314

Hinson, R., Boateng, H., Renner, A., & Kosiba, J. P. B. (2019). Antecedents and consequences of customer engagement on Facebook: An attachment theory perspective. *Journal of Research in Interactive Marketing*, *13*(2), 204–226. https://doi.org/10.1108/JRIM-04-2018-0059

Jiang, Y., Ramkissoon, H., Mavondo, F. T., & Feng, S. (2017). Authenticity: The link between destination image and place attachment. *Journal of Hospitality Marketing & Management*, *26*(2), 105–124. https://doi.org/10.1080/19368623.2016.1185988

Kaplan, S., & Talbot, J. F. (1983). Psychological benefits of a wilderness experience. In I. Altman & J. F. Wohlwill (Eds), *Behavior and the natural environment* (pp. 163–203). Springer.

Khetrapal, S., & Bhatia, R. (2020). Impact of COVID-19 pandemic on health system & Sustainable Development Goal 3. *The Indian Journal of Medical Research*, *151*(5), 395–399. https://doi.org/10.4103/ijmr.IJMR_1920_20

Korpela, K. M., Ylén, M., Tyrväinen, L., & Silvennoinen, H. (2010). Favorite green, waterside and urban environments, restorative experiences and perceived health in Finland. *Health Promotion International*, *25*(2), 200–209. https://doi.org/10.1093/heapro/daq007

Kyle, G. T., Mowen, A. J., & Tarrant, M. (2004). Linking place preferences with place meaning: An examination of the relationship between place motivation and place attachment. *Journal of Environmental Psychology*, *24*(4), 439–454. https://doi.org/10.1016/j.jenvp.2004.11.001

Lee, Y. J. (2015). Creating memorable experiences in a reuse heritage site. *Annals of Tourism Research*, *55*, 155–170. https://doi.org/10.1016/j.annals.2015.09.009

Lu, J., Xiao, X., Xu, Z., Wang, C., Zhang, M., & Zhou, Y. (2022). The potential of virtual tourism in the recovery of tourism industry during the COVID-19 pandemic. *Current Issues in Tourism*, *25*(3), 441–457. https://doi.org/10.1080/13683500.2021.1959526

Majeed, S., & Ramkissoon, H. (2020). Health, wellness, and place attachment during and post health pandemics. *Frontiers in Psychology*, *11*, 573220. https://doi.org/10.3389/fpsyg.2020.573220

Manchanda, M., & Deb, M. (2022). Effects of multisensory virtual reality on virtual and physical tourism during the COVID-19 pandemic. *Current Issues in Tourism*, *25*(11), 1748–1766. https://doi.org/10.1080/13683500.2021.1978953

Marschall, S. (2012). 'Personal memory tourism'and a wider exploration of the tourism– memory nexus. *Journal of Tourism and Cultural Change*, *10*(4), 321–335. https://doi.org/10.1080/14766825.2012.742094

Megeirhi, H. A., Woosnam, K. M., Ribeiro, M. A., Ramkissoon, H., & Denley, T. J. (2020). Employing a value-belief-norm framework to gauge Carthage residents' intentions to support sustainable cultural heritage tourism. *Journal of Sustainable Tourism*, *28*(9), 1351–1370. https://doi.org/10.1080/09669582.2020.1738444

Mihaylov, N., & Perkins, D. D. (2014). Community place attachment and its role in social capital development. In L. C. Manzo & P. Devine-Wright (Eds), *Place attachment: Advances in theory, methods and applications*, (pp. 61–74). Routledge.

Nekmahmud, M., Ramkissoon, H., & Fekete-Farkas, M. (2022). Green purchase and sustainable consumption: A comparative study between European and Non-European tourists. *Tourism Management Perspectives*, *43*, 100980. https://doi.org/10.1016/j.tmp.2022.100980

NHS Foundation Trust (2021). *How spending time in nature can benefit your mental health*. https://www.humber.nhs.uk/news/?postid=367439

Nunkoo, R., & Ramkissoon, H. (2012). Power, trust, social exchange and community support. *Annals of Tourism Research*, *39*(2), 997–1023. https://doi.org/10.1016/j.annals.2011.11.017

Nunkoo, R., Smith, S. L., & Ramkissoon, H. (2013). Residents' attitudes to tourism: A longitudinal study of 140 articles from 1984 to 2010. *Journal of Sustainable Tourism*, *21*(1), 5–25. https://doi.org/10.1080/09669582.2012.673621

Pearce, J., Huang, S. S., Dowling, R. K., & Smith, A. J. (2022). Effects of social and personal norms, and connectedness to nature, on pro-environmental behavior: A study of Western Australian protted area visitors. *Tourism Management Perspectives*, *42*, 100966. https://doi.org/10.1016/j.tmp.2022.100966

Proshansky, H. M. (1978). The city and self-identity. *Environment and Behavior*, *10*(2), 147–169. https://doi.org/10.1177/0013916578102002

Qu, Y., Xu, F., & Lyu, X. (2019). Motivational place attachment dimensions and the pro-environmental behaviour intention of mass tourists: A moderated mediation model. *Current Issues in Tourism*, *22*(2), 197–217. https://doi.org/10.1080/13683500.2017.1399988

Ramkissoon, H. (2020). COVID-19 Place confinement, pro-social, pro-environmental behaviors, and residents' wellbeing: A new conceptual framework. *Frontiers in Psychology*, *11*, 2248. https://doi.org/10.3389/fpsyg.2020.02248

Ramkissoon, H. (2021). Place affect interventions during and after the COVID-19 pandemic. *Frontiers in Psychology*, 3864. https://doi.org/10.3389/fpsyg.2021.726685

Ramkissoon, H. (2022a). Behaviour change in tourism. In D. Buhalis (Ed.), *Encyclopedia of tourism management and marketing*. Edward Elgar Publishing.

Ramkissoon, H. (2022b). Cultural tourism impacts and place meanings: Focusing on the value of domestic tourism. In A. Stoffelen & D, Ioannides (Eds), *Handbook of tourism impacts: Social and environmental perspectives* (pp. 75–87). Edward Elgar Publishing.

Ramkissoon, H. (2022c). Perceived visitor impacts of cultural heritage tourism: The role of place attachment in memorable visitor experiences. In D. Agapito, M. A. Ribeiro, & K. M. Woosnam (Eds), *Handbook on the tourist experience: Design, marketing and management* (pp. 166–175). Edward Elgar Publishing.

Ramkissoon, H. (2022d). COVID-19 Adaptive interventions: Implications for wellbeing and quality-of-life. *Frontiers in Psychology, 13*. https://doi.org/10.3389/fpsyg.2022.810951

Ramkissoon, H. (2022e). Pro-sociality in times of separation and loss. *Current Opinion in Psychology, 45*, 101290. https://doi.org/10.1016/j.copsyc.2021.11.008

Ramkissoon, H. (2023a). Perceived social impacts of tourism and quality-of-life: A new conceptual model. *Journal of Sustainable Tourism, 31*(2), 442–459.

Ramkissoon, H. (2023b). Introduction to the Handbook on Tourism and Behaviour Change. In *Handbook on tourism and behaviour change* (pp. 1–19). Edward Elgar Publishing.

Ramkissoon, H., & Mavondo, F. T. (2015). The satisfaction–place attachment relationship: Potential mediators and moderators. *Journal of Business Research, 68*(12), 2593–2602. https://doi.org/10.1016/j.jbusres.2015.05.002

Ramkissoon, H., Mavondo, F., & Uysal, M. (2018). Social involvement and park citizenship as moderators for quality-of-life in a national park. *Journal of Sustainable Tourism, 26*(3), 341–361. https://doi.org/10.1080/09669582.2017.1354866

Ramkissoon, H., Smith, L. D. G., & Kneebone, S. (2014). Visitor satisfaction and place attachment in national parks. *Tourism Analysis, 19*(3), 287–300. https://doi.org/10.3727/108354214X14029467968402

Ramkissoon, H., Smith, L. D. G., & Weiler, B. (2013a). Testing the dimensionality of place attachment and its relationships with place satisfaction and pro-environmental behaviours: A structural equation modelling approach. *Tourism Management, 36*, 552–566. https://doi.org/10.1016/j.tourman.2012.09.003

Ramkissoon, H., Smith, L. D. G., & Weiler, B. (2013b). Relationships between place attachment, place satisfaction and pro-environmental behaviour in an Australian national park. *Journal of Sustainable Tourism, 21*(3), 434–457. https://doi.org/10.1080/09669582.2012.708042

Ramkissoon, H., Weiler, B., & Smith, L. D. G. (2012). Place attachment and pro-environmental behaviour in national parks: The development of a conceptual framework. *Journal of Sustainable Tourism, 20*(2), 257–276. https://doi.org/10.1080/09669582.2011.602194

Rasoolimanesh, S. M., Seyfi, S., Hall, C. M., & Hatamifar, P. (2021). Understanding memorable tourism experiences and behavioural intentions of heritage tourists. *Journal of Destination Marketing & Management, 21*, Art. 100621 https://doi.org/10.1016/j.jdmm.2021.100621

Rasoolimanesh, S. M., Ramakrishna, S., Hall, C. M., Esfandiar, K., & Seyfi, S. (2020). A systematic scoping review of sustainable tourism indicators in relation to the sustainable development goals. *Journal of Sustainable Tourism*. https://doi.org/10.1080/09669582.2020.1775621

Rezapouraghdam, H., Akhshik, A., & Ramkissoon, H. (2021). Application of machine learning to predict visitors' green behavior in marine protected areas: Evidence from Cyprus. *Journal of Sustainable Tourism*. https://doi.org/10.1080/09669582.2021.1887878

Rollero, C., & De Piccoli, N. (2010). Does place attachment affect social well-being? *European Review of Applied Psychology, 60*(4), 233–238. https://doi.org/10.1016/j.erap.2010.05.001

Scannell, L., & Gifford, R. (2010). Defining place attachment: A tripartite organizing framework. *Journal of Environmental Psychology, 30*(1), 1–10. https://doi.org/10.1016/j.jenvp.2009.09.006

Scannell, L., & Gifford, R. (2017). The experienced psychological benefits of place attachment. *Journal of Environmental Psychology, 51*, 256–269. https://doi.org/10.1016/j.jenvp.2017.04.001

Seyfi, S., Rastegar, R., Rasoolimanesh, S. M., & Hall, C. M. 2023, A framework for understanding media exposure and post-COVID-19 travel intentions. *Tourism Recreation Research*, 48(2), 305–310. https://doi.org/10.1080/02508281.2021.1949545

Shoukat, M. H., & Ramkissoon, H. (2022). Customer delight, engagement, experience, value co-creation, place identity, and revisit intention: a new conceptual framework. *Journal of Hospitality Marketing & Management, 31*(6), 757–775. https://doi.org/10.1080/19368623.2022.2062692

Stigsdotter, U. K., Ekholm, O., Schipperijn, J., Toftager, M., Kamper-Jørgensen, F., & Randrup, T. B. (2010). Health promoting outdoor environments-Associations between green space, and health, health-related quality of life and stress based on a Danish national representative survey. *Scandinavian Journal of Public Health, 38*(4), 411–417. https://doi.org/10.1177/1403494810367468

Stokols, D., & Shumaker, S. A. (1982). The psychological context of residential mobility and well-being. *Journal of Social Issues, 38*(3), 149–171. https://doi.org/10.1111/j.1540-4560.1982.tb01776.x

Suntikul, W., & Jachna, T. (2016). The co-creation/place attachment nexus. *Tourism Management, 52*, 276–286. https://doi.org/10.1016/j.tourman.2015.06.026

Tian, Y., Lu, D., & Powpaka, S. (2015). Tourist's awe and loyalty: An explanation based on the appraisal theory. *Tourism Tribune, 30*(10), 80–88.

Townsend, M., Henderson-Wilson, C., Ramkissoon, H., & Werasuriya, R. (2018). Therapeutic landscapes, restorative environments, place attachment, and well-being. In M. van den Bosch & W. Bird (Eds), *Oxford textbook of nature and public health: The role of nature in improving the health of a population* (pp. 57–62)

Tuan, Y. F. (1977). *Space and place: The perspective of experience*. University of Minnesota Press.

Vaske, J. J., & Kobrin, K. C. (2001). Place attachment and environmentally responsible behavior. *The Journal of Environmental Education, 32*(4), 16–21. https://doi.org/10.1080/00958960109598658

Völker, S., & Kistemann, T. (2011). The impact of blue space on human health and well-being – Salutogenetic health effects of inland surface waters: A review. *International Journal of Hygiene and Environmental Health, 214*(6), 449–460. https://doi.org/10.1016/j.ijheh.2011.05.001

Walker, G. J., & Chapman, R. (2003). Thinking like a park: The effects of sense of place, perspective-taking, and empathy on pro-environmental intentions. *Journal of Park & Recreation Administration, 21*(4).

Wang, F., Huang, S., Morrison, A. M., & Wu, B. (2022). The effects of virtual reality tourism involvement on place attachment and behavioral intentions: virtual reality tourism of the Yellow Crane Tower in Wuhan. *Asia Pacific Journal of Tourism Research, 27*(3), 274–289. https://doi.org/10.1080/10941665.2022.2061363

Williams, D. R., Patterson, M. E., Roggenbuck, J. W., & Watson, A. E. (1992). Beyond the commodity metaphor: Examining emotional and symbolic attachment to place. *Leisure Sciences, 14*(1), 29–46. https://doi.org/10.1080/01490409209513155

Wu, J., Font, X., & Liu, J. (2021). Tourists' pro-environmental behaviors: Moral obligation or disengagement? *Journal of Travel Research, 60*(4), 735–748. https://doi.org/10.1177/0047287520910787

Xu, Y., Wu, D., & Chen, N. C. (2022). Here I belong!: Understanding immigrant descendants' place attachment and its impact on their community citizenship behaviors in China. *Journal of Environmental Psychology, 79*, 101743. https://doi.org/10.1016/j.jenvp.2021.101743

Zhang, C. X., Fong, L. H. N., & Li, S. (2019). Co-creation experience and place attachment: Festival evaluation. *International Journal of Hospitality Management, 81*, 193–204. https://doi.org/10.1016/j.ijhm.2019.04.013

Chapter 15

Mindfulness and Tourism

Uglješa Stankov, Ulrike Gretzel, & Viachaslau Filimonau

Introduction

Tourism research has acknowledged the importance of mindfulness as a theoretical concept with important implications for tourism practice, but it has done so from a very specific point of view. The most common way to think about mindfulness in tourism research is through the lens of applied social psychology (Moscardo, 2009). This way of thinking focuses on ideas like heightened awareness, focused attention, and flow. From a cognitive point of view, mindfulness is linked to processing information and learning, but sometimes it is also linked to the possibility of having more meaningful experiences (Moscardo & Ballantyne, 2008). However, the broad phrase 'mindfulness' encompasses a variety of behaviours, pursuits, and character attributes. The word 'mindfulness' has the following definition in standard English: 'a mental state achieved by concentrating on the present moment, while calmly accepting the feelings and thoughts that come to you, used as a technique to help you relax ' (Oxford Advanced Learner's Dictionary, 2022). As such, mindfulness was originally developed from an Eastern concept but then modified for use in Western therapeutic procedures, and more recently the definition of mindfulness has been altered to reflect the variety of contemporary uses (Bishop et al., 2004; Kabat-Zinn, 2013; Shonin et al., 2015). Thus, mindfulness refers to the capacity [of an individual] to be present in the moment and pay attention to what is going on around and within an individual (Kabat-Zinn, 2013). Drawing upon this notion of meditative mindfulness, this chapter seeks to provide an overview of its relevance in/for tourism.

An increasing number of recent publications have set to promote a meditative perspective on mindfulness as it applies to tourism. Besides inner experiences and habits affected by mindfulness (Chen et al., 2017; Sheldon, 2020), the practice itself comes in many forms (Lea et al., 2015; Williams & Kabat-Zinn, 2011). Different perspectives talk about the social and cultural effects of mindfulness (Farkic et al., 2021; Joiner, 2017), economic influences (Andrews, 2007; Govia et al., 2020), technological (Diamond et al., 2017; Stankov et al., 2020a), or even political ones (Leggett, 2021; Lengyel, 2015a).

The Wiley Blackwell Companion to Tourism, Second Edition. Edited by C. Michael Hall.

Interest in mindfulness within tourism contexts spreads beyond the academic realm. Mindfulness is becoming more and more popular as a product and service in the tourism industry (Habeeb, 2021). This is reflected in the growth of consumer groups that are interested in and actively looking for mindfulness themed products and services. The trend is also seen in how mindful states change people's buying habits and how the tourism industry is putting more emphasis on mindfulness related experiences (Stankov et al., 2022).

The market's growing demand for promoting and enabling mindfulness practices was first addressed by travel industry operators. At the moment, the majority of mindfulness offerings in tourism contexts are connected to spa and wellness services, nature-based travel, meditation courses, and other similar, wellbeing focused, goods and services (Chen, 2015; Chen et al., 2017; McGroarty et al., 2019; Stankov et al., 2022). For example, the mindfulness based offers of holiday retreat operators typically combine a central wellness theme with the inclusion of specialised services, sports and relaxation activities, and specific diets to help visitors feel calm and reach spiritual or lifestyle transcendence (Chen et al., 2017).

Characteristics of Meditative Mindfulness

Mindfulness has been defined as both a *state* attained through mindfulness meditation and a *personality trait*. When conscious attention to experience is fostered through an active process of responding in a non-judgmental manner to the present thoughts, feelings, and sensations, mindfulness can be conceived of as a mode, or state-like quality, that can only be sustained under certain conditions (Lau et al., 2006). As a personality trait, mindfulness refers to an individual's inherent ability to pay attention to and maintain awareness of present moment experiences with an open and non-judgmental attitude. Even in the absence of mindfulness interventions, this capacity remains rather steady over time (Brown & Ryan, 2003).

Self-report questionnaires are usually employed to assess trait mindfulness (for example, the Mindful Attention and Awareness Scale (MAAS), or the Five Facet Mindfulness Questionnaire (FFMQ) (Chowdhury, 2022; Van Dam et al., 2018). New neurofeedback approaches are now being developed that employ electroencephalography (EEG) data to objectively measure and validate mindfulness states (Hunkin et al., 2021).

As an *intervention*, mindfulness can be defined as either formal or informal. Formal mindfulness is practiced over a specified period of time and in specific situations, whereas informal mindfulness can be performed at any time and in nearly any context (Birtwell et al., 2018). Mindfulness based interventions are typically composed of both formal and informal activities. Mindfulness Based Stress Reduction (MBSR) (Kabat-Zinn, 2013) and Mindfulness Based Cognitive Therapy (MBCT) are the two most common standardised mindfulness based interventions (Segal et al., 2012).

The importance of focused daily activities, such as mindful walking or eating, is emphasised by several mindfulness-based treatment techniques. Within the practice of 'everyday mindfulness' people focus on one item, such as their breath or physical sensations, for a predetermined amount of time (Thompson & Waltz, 2007).

From a medical point of view, the original goal of putting mindfulness based therapies into clinical practice is to reduce patient stress (Kabat-Zinn, 2003). However, mindfulness has developed into a lucrative commodity for many industries, including publishing, leisure, travel, internet services, the mobile application market, and technology gadgets, to name a few. In light of growing public interest in mindfulness at a worldwide level, it is projected that mindfulness will continue to increase in popularity (European Travel Commission (ETC), 2016).

Again, meditative mindfulness presented in this chapter should not be confused with the socio-cognitive idea of mindfulness or non-meditative mindfulness, based on Ellen Langer's (1989) work. Both meditative and socio-cognitive mindfulness are intimately related since they are concerned with awareness and attention, but the ways of interaction with present moment experiences differ (Chen et al., 2017). Internal in nature, present moment awareness is the foundation of meditative mindfulness. It is also built on non-judgment. The socio-cognitive idea of mindfulness, by contrast, relates to a person's cognitive flexibility, generally in connection to external and social situations that support the active construction of new categories and/or differences, and with a learning objective in mind (Chen et al., 2017; Pirson et al., 2018).

Factors Promoting Mindfulness

There are many factors that determine why mindfulness is popular. All factors that contribute to the mindfulness trend can be divided into two categories: *external factors* influenced by forces in the economy, politics, environment, technology, and society; and *internal factors* generated by the needs and wants of consumers (Table 15.1).

E-mindfulness: Technology-assisted Mindfulness

A wave of technological solutions aimed at helping users learn, achieve, and maintain a mindful state has resulted from the widespread and practical understanding of mindful meditation and its advantages (Diamond et al., 2017; Taylor et al., 2021). This technology enhanced mindfulness idea is referred to as 'e-mindfulness' and it is primarily supported by the use of smartphone applications, web and streaming services, and body sensing equipment (Stankov et al., 2020a; Tedder et al., 2015). Most applications and technologies for mindfulness are not meant to take the place of meditative awareness in everyday life. Most of the time, they are meant to motivate and teach people who are just starting out to engage with mindfulness or who want to keep their mindfulness practice going well.

Critique of Mindfulness

The contemporary applications of mindfulness are sometimes criticised. This critique focuses on the efficacy of contemporary mindfulness applications, the potential effects and outcomes of mindfulness on individuals, and the social, political, and economic ramifications of the wider applications of mindfulness. This is exacerbated by the constant pressure of religious or spiritual leaders concerned with preserving the integrity of mindfulness as a religious or cultural practice.

The mindfulness movement has reached the mainstream of culture in Europe and North America and some other Westernised countries. It brings with it all of the possible benefits of contemplative practice for individuals and all of the possible risks of contemplative practice being misused (Gotojones, 2013). Common descriptions of mindfulness at the individual level portray it as a 'miracle cure' for suffering, as opposed to a long-term practice of reorienting, reframing, and interacting with everyday experience (Choi et al., 2021). Mindfulness is seen as a seduction that leads people away from facing life's obstacles, encouraging them to avoid problems. Laypeople may misinterpret acceptance as passive acceptance of experience which impedes problem solving (Choi et al., 2021). In this context, focusing on individuals as the

Table 15.1. Factors promoting mindfulness.

External factors	*Internal factors*
Support from multiple domains and science Meditative mindfulness has been used for a growing number of purposes in applied sciences and different areas of economic activity, such as education, organisational behaviour, the military, law, and the arts (Brown et al., 2015; Williams & Kabat-Zinn, 2011).	***Health issues*** Increasing numbers of individuals who want to lessen anxiety, stress, and depression, or who want to boost their immune systems, use mindfulness as a form of informal practice. (Lengyel, 2015a; Ludwig & Kabat-Zinn, 2008).
Societal acceptance, socio-economic and environmental pressure Mindfulness acceptance in Western cultures has resulted from societal changes and the need to alter one's fundamental approach to coping with the stresses of modern life (Schmidt, 2011). Meanwhile, mindfulness is gaining popularity as a potential social change agent that could foster pro-social behaviour and general wellbeing (Panno et al., 2018; Steidle, 2017). By enabling a more meaningful way of living that is more aligned with human nature, mindfulness is promoted as 'a way out' of the rising stress and burnout in modern societies and workplaces (Axelrod, 2006; Steidle, 2017).	***Wellbeing and self-development aspirations*** Since mindfulness practice is linked to better monitoring of somatic and emotional experiences, less impulsivity, and other good things, it is often seen as an important tool for people who want to improve themselves (Segall, 2005) and with people's general pursuit of subjective wellbeing (Zheng et al., 2020). In addition, self-help and mental health services are becoming more mainstream and less stigmatised in society as a whole (Bernstein et al., 2019).
Institutional acknowledgement Many institutions are now adding mindfulness as a new and different agenda to their official programmes and domains (Vogus & Sutcliffe, 2012). Many official development policies and plans openly promote the idea of mindfulness, either as a way to help people feel less stressed and save money on health care costs or as an investment in mindfulness themed public spaces like mindfulness trails and parks (Stankov et al., 2022).	***Spiritual and religious motivation*** Despite stripping mindfulness of its canonical and sociocultural traditions (Compson, 2017; Joiner, 2017), some practitioners view nonsecular motivations as valid internal motivations for contemporary meditative mindfulness practice (Kristeller, 2007; Miller, 1999).
Technological progress and acceptance Technology has made it possible to offer tools and resources to help with meditative and mindful practices (Diamond et al., 2017). Software applications make it easy and scalable to deliver many mindfulness training programmes and interventions that are based on mindfulness (Bernstein et al., 2019). Many people believe that technology and meditation are compatible, and the increasing popularity of mindfulness smartphone apps backs this belief up (Stankov et al., 2020a).	

primary unit of attention obscures the larger picture. Some critics label aspects of the mindfulness movement in popular culture as 'McMindfulness ' (Purser & Loy, 2013), a product of the consumer society in the form of a wellness brand that highlights short-term relief of personal suffering over the involved, in-depth investigation of the sources of distress that may contribute to real organisational or sociocultural change (Purser, 2019).

These criticisms seem to have little impact on the popularity of mindfulness and its successful commercial exploitation. Especially in tourism, mindfulness is increasingly seen as an enabler of positive experiential outcomes.

Characteristics of Mindful Tourists

In essence, tourists who are mindful accept their current feelings and thoughts and pay attention to the here and now, attending to the somatic sensations they experience at a destination in an open, non-reactive, and non-judgmental way (Loureiro et al., 2020). However, it must be noted, there are no operational definitions of mindful tourists in the literature that are currently globally available and accepted, nor descriptions of their profiles (Chen et al., 2017).

The broad definition of a mindful tourist provided above, characterises mindfulness as a transient state. However, trait mindfulness can help understand more stable mindful tourist characteristics. For example, older adults may have higher trait mindfulness levels than younger adults (Kim et al., 2018). This means that older people are more likely to be in the group of mindful tourists. Thus, while state mindfulness can facilitate learning how mindful tourists experience their trips and how these experiences may differ or how they may affect the tourism product or service offering, trait mindfulness can aid in learning about how mindful tourists have more stable qualities and how these qualities can be developed and built upon during a trip (Stankov et al., 2022).

Mindful tourists may be more likely to choose environmentally friendly transportation and lodging options (Errmann et al., 2021), visit ethical attractions, buy local products, consume less and thus leave less of an impact on the environment (Stankov & Filimonau, 2021), enjoy just being in nature (Farkic et al., 2021), look for meaningful interactions with hotel staff and locals, and generally slow down their travel and consumption habits (Stankov et al., 2020b).

Consumers who practice mindfulness may be more aware of the indirect effects of travel, such as reminiscing about holidays, collecting impressions, deriving new meanings, and social enrichment (Nawijn, 2011; Stankov et al., 2021). Thus, traditional tourism advertising methods that are often too simplistic and refer to mass consumers can be perceived as patronising (Szmigin & Carrigan, 2000) by more mindful tourists. Mindful tourists may prefer tourism marketing that promotes vacations as a way to live a more diverse life, be better involved in the outer world, and stay socially and environmentally responsible while travelling (Stankov et al., 2021).

Mindful tourists may have a different idea of what a destination should offer. Thus, mindful tourists who are open-hearted, open-minded, focused, aware, accept the moment, and do not judge their thoughts and perceptions may have a more positive attitude towards certain destinations or tourism product and service offerings and develop different levels of attachment (Kang & Gretzel, 2012; Loureiro et al., 2020).

Applications of Mindfulness in the Tourism Industry

The phrase 'mindful travel' has a variety of market specific meanings, and it is frequently used in specialised online travel forums to describe current trends, both among consumers and in relation to the products and services offered by travel companies and their host destinations. It reflects consumers' rising interest in meditation, health, and broader sociological and environmental challenges brought on by contemporary culture and it is mostly associated with wellness tourism establishments that gladly provide mindfulness based product and service offerings (Barrie, 2019).

More and more hospitality businesses are looking into ways to give their visitors access to meditation or mindfulness. Visitors are frequently encouraged to practice mindfulness on their own, in designated meditation locations on hotel premises, or by enrolling in specially designed mindfulness programmes led by hotel staff (JW Marriott Hotel Nara, 2022). In addition, numerous popular mindfulness programmes being offered in hotels are linked to the idea of digital wellbeing (Stankov & Gretzel, 2021) wherein mindfulness is promoted as a remedy or a means of coping with information overload.

Transportation businesses have begun to offer products and services with a mindfulness theme, primarily on full-service flights of large, legacy airlines. Partnerships between airlines and developers of mindfulness based mobile applications typically aim to aid clients who experience boredom or anxiety related to flying.

Tourist Experience and Mindfulness

The underlying mechanisms of mindfulness become particularly relevant for understanding the effect of mindfulness on tourist experiences and subsequent evaluations (Chen et al., 2017; Kang & Gretzel, 2012). The underlying mechanisms of mindfulness include paying attention to the experience, living in the present, and cultivating non-judgmental awareness.

Due to tourists' inherent willingness to try new things, the tourism sector offers an environment that is conducive to the adoption of mindfulness practices (Lengyel, 2015a), that is combined with a change from tourists' usual surroundings. However, many common mindfulness based practices in daily life are equally applicable to experiences in the context of tourism, such as practicing mindfulness while moving or eating or when focusing attention on a specific stimulus, such as body sensations or breath (Thompson & Waltz, 2007).

When in a state of mindfulness, tourists pay attention to a specific emotion, object, or the entirety of the present moment and surroundings (Chittaro & Vianello, 2014). This enables present focused tourists to engage more fully with all that is taking place at that very moment and to understand what it means to live in the now. Research shows that people who are more mindful are less likely to be defensive in social situations (Deci et al., 2015), potentially making interactions easier in unfamiliar contexts.

Mindfulness is frequently coupled with a misunderstanding of non-judgmental awareness (Salvati et al., 2019). Tourists, contrary to a popular belief, will not scrutinise unpleasant aspects of tourism or will silently tolerate their own bad habits. Non-judgment in the context of mindfulness refers to the elimination of automatic judgments, that is, inner judgments based on one's experiences, ideas, and taught stereotypes, as well as a habit of being critical of oneself in general (Antonova et al., 2021). In other words, non-judgmental awareness is non-elaborative because it comprises only the recognition of an item rather than its conceptualisation or appraisal. It is an open embrace of anything (such as a thought, a sentiment, or an experience) that has not been thoroughly investigated or comprehended (Chen et al., 2017). Mindfulness is linked to lower negative interpretation bias via lowering bad mood (Gibb et al., 2022). As a result, it can be argued that cautious travellers will be more tolerant of mistakes occurring in the tourism and hospitality service supply chain (Stankov et al., 2020a).

The Co-creation of Mindful Tourism Experiences

To capitalise on the business potential presented by the mindfulness movement, it is necessary to begin with the idea that co-creation of tourism experiences is a relevant managerial approach to embracing mindfulness. The reason for this is that it is impossible for tourism providers to create mindfulness experiences that will be simply consumed by tourists (Stankov et al., 2020a). A wellbeing service, mindfulness in tourism connects individual tourist demands for enhanced experiences and wellbeing with the business goals of tourism companies. The mindful tourist experience can only be realised when both parties offer key resources (or 'contributions') and integrate these resources together (Lusch & Vargo, 2014) (Table 15.2).

The key benefits of co-creating mindful tourist experiences at the tourist level may include the following: increased subjective wellbeing; more meaningful and transformational experiences; and the development of new skills and behaviours. At the level of

Table 15.2. Contributions to co-creation of mindful tourism experiences.

Tourist contributions	*Tourism provider contributions*
Travel motivations Even though it seems logical that tourists who are already open to mindfulness should enjoy it the most to improve their travel, many tourists are willing to try new things (Stankov et al., 2020a). Most of the time, tourism is a good way to introduce mindfulness to people who are new to it or who have trouble practicing it in their everyday lives (Lengyel, 2015b). Even when travelling for business, there is evidence that a short course on mindfulness can help people deal with stress, anxiety, and depression (Rundle et al., 2018).	**Physical environments that are supportive** Certain locations are better than others suited for promoting mindfulness (Porter et al., 2017). Typically, mindfulness practices are thought to be carried out in specially designed meditation spaces (Porter et al., 2017). By default, these spaces ought to be created to foster awareness of the present moment or inspire wonder (Leddy, 2015). This includes the design of music, smells, biophilic elements of the surroundings, etc. (Stankov et al., 2022).
The development stage of mindfulness The general development stages of mindfulness meditation are significant as factors that may affect the co-creation process. Generally speaking, there are three significant milestones: the initial phase, which occurs after relatively little mindfulness practice; the intermediate phase, which begins after more extensive mindfulness practice, but has a great deal of individual variation in its extent; and the advanced phase, which is characterised by profound personal transformations (Kristeller, 2007). In relation to these landmarks, mindfulness practices can range from a brief exercise to a regular informal practice that can be integrated into virtually any tourist activity (Birtwell et al., 2018; Chittaro & Vianello, 2014).	**Supportive frontline personnel and organisational settings** Workers who are directly involved in a mindful tourist experience are critical here, as they must be sensitive enough to understand, support, and manage mindful visitors (Hanley et al., 2015; Jang et al., 2020; Stankov & Filimonau, 2020). This is especially crucial because tourists may have a different viewpoint on the mindful experience or be at different stages of mindfulness growth, requiring varying amounts of support (Stankov et al., 2022). In addition, the supporting organisational setting, which enables the general conditions for mindfulness in a tourism company or destination, must also be considered.
Technological contributions Additionally, the relationship to technology and its use should be considered as additional inputs for co-creation if technology assisted mindfulness mediates tourism experience (Stankov & Filimonau, 2020; Stankov et al., 2020a). Because of greater digital elasticity (Pearce & Gretzel, 2012), technology utilised in daily life can also be used as a tool in the e-mindful tourist experience. The novelty effect of technology applied in the case of mindfulness, on the other hand, can be an extra magnet, notably during the early stages of mindfulness development (Stankov & Filimonau, 2019).	

tourism employees, a reduction in work-related stress can increase job performance and boost job satisfaction (Stankov et al., 2022).

In addition, mindfulness is increasingly being recognised as a major driver of favourable behavioural change (Dharmesti et al., 2020). Mindfulness has been linked to long-term sustainability goals because of its effects on halting mindless patterns, establishing greater alignment of one's thoughts and behaviours, enhancing non-materialistic wellbeing, and encouraging pro-social behaviour (Fischer et al., 2017).

Barriers to Broader Adoption of Mindfulness in Tourism

In spite of the tourism sector's increased interest in mindfulness, there remain barriers to the adoption of mindfulness. These barriers stem from the features of mindfulness itself: mindfulness is an ethereal psychological quality; its expressions can only be felt directly by tourists; in most cases mindfulness takes practice; and its results are frequently indirect and gradual. These features affect both the demand for, and the supply of tourism related to mindfulness.

The difficulty of implementing mindfulness in tourism and hospitality environments is additionally increased by the widespread public perception — or misperception — of mindfulness. When considering the introduction of mindfulness into mainstream tourism services, uncritical advertising of mindfulness as a solution to all problems is not always desired and has a negative connotation. This is sometimes linked to anti-mindfulness sentiment that criticises mindfulness methods for their excessive focus on the present moment (Purser, 2015).

Although mindfulness has been standardised, de-sacralised, and mainstreamed in the West, it is still associated with Eastern cultural symbols in branding (Borup, 2016). Thus, the practice of mindfulness remains considered overly mystical or having roots in other religions (Lewis, 2016). Additionally, religious groups who place a strong emphasis on mindfulness based techniques occasionally oppose the commercialisation of mindfulness.

Conclusions and Further Research

Mindfulness, as a socio-spatial phenomenon, has significant implications for tourism in general because people all over the world are incorporating mindfulness practices into their recreational and leisure activities (Whitehead et al., 2016). It also has specific implications for tourism providers and destinations. This chapter argues that there are numerous pragmatic benefits to incorporating mindfulness into the travel and tourism sector, especially the industry's need to become more sustainable (Errmann et al., 2021; Lengyel, 2018). It also shows that tourism providers and destinations can make significant contributions in support of mindfulness practices. Beyond internal processes or behaviours (Chen et al., 2017; Sheldon, 2020), the practice is influenced by or reflects a certain place or setting (Lea et al., 2015). Fostering tourists' mindfulness through co-creation is therefore an important and potentially very lucrative opportunity for the tourism sector.

While the chapter outlines ways in which tourism providers and destinations can promote mindfulness, little knowledge currently exists in terms of the breadth of available alternatives and their relative effectiveness. More informational material, industry case studies, and training opportunities are needed to help the tourism sector benefit from the mindfulness trend. At the same time, given its popularity and potential contribution to wellbeing and sustainability in and through tourism, mindfulness should become a more central object of tourism research. Several research avenues can be identified.

Exploratory Evidence of Mindful Tourist Experiences

Following a larger trend of research that has set out to assess the usefulness of mindfulness applications in other disciplines, mindfulness is gaining ground as a viable new study field in tourism (Baminiwatta & Solangaarachchi, 2021). Additionally, research is needed to give tourism businesses proof of what and how best to operate when promoting the use of mindfulness techniques.

Multi-disciplinary and Multi-method Investigations of Mindful Tourism

It is important to note that mindful tourism research needs to be aware of and should remain open to the many different aspects, philosophical roots, and different ways of looking at mindfulness and its possible effects. Despite the fact that mindfulness has been around for a long time, its methods continue to evolve. This needs to be looked into from different angles and with the help of different methods.

Mindfulness and Technology

The world may see a shift in how people think about mindfulness as a result of technology's increasing influence on everyday life and tourism. This should be reflected in the definition of mindfulness. Researchers have to be aware of the changing connotation of mindfulness in order to properly examine the development of mindful tourist experiences.

It is expected that more mindfulness research in tourism will further propel interest in the practice by consumers and practitioners and will also be of relevance to other sectors and academic disciplines (e.g., health sciences and engineering).

References

Andrews, S. M. (2007). *Is mindfulness a luxury? Examining the role of socioeconomic status in the relationship between dispositional mindfulness and psychological distress* [Master's thesis, Southern Illinois University at Carbondale]. https://www.proquest.com/openview/d04176c2d15450e4b6ec26dff60149a0/1?cbl=18750&pq-origsite=gscholar

Antonova, E., Schlosser, K., Pandey, R., & Kumari, V. (2021). Coping with COVID-19: Mindfulness-based approaches for mitigating mental health crisis. *Frontiers in Psychiatry*, *12*, 563417. https://doi.org/10.3389/fpsyt.2021.563417

Axelrod, R. (2006). *The evolution of cooperation*. Basic Books.

Baminiwatta, A., & Solangaarachchi, I. (2021). Trends and developments in mindfulness research over 55 years: A bibliometric analysis of publications indexed in Web of Science. *Mindfulness*, *12*, 2099–2116. https://doi.org/10.1007/s12671-021-01681-x

Barrie, L. (2019). Can wellness find its way into the cramped confines of economy class? *Skift*. https://skift.com/2019/08/22/can-wellness-find-its-way-into-the-cramped-confines-of-economy-class/

Bernstein, A., Vago, D. R., & Barnhofer, T. (2019). Understanding mindfulness, one moment at a time: an introduction to the special issue. *Current Opinion in Psychology*, *28*, vi–x. https://doi.org/10.1016/j.copsyc.2019.08.001

Birtwell, K., Williams, K., van Marwijk, H., Armitage, C. J., & Sheffield, D. (2018). An exploration of formal and informal mindfulness practice and associations with wellbeing. *Mindfulness*, *10*, 89–99. https://doi.org/10.1007/s12671-018-0951-y

Bishop, S. R., Lau, M., Shapiro, S., Carlson, L., Anderson, N. D., Carmody, J., Segal, Z. V., Abbey, S., Speca, M., Velting, D., & Devins, G. (2004). Mindfulness: A proposed operational definition. *Clinical Psychology: Science and Practice*, *11*(3), 230–241. https://doi.org/10.1093/clipsy/bph077

Borup, J. (2016). Branding Buddha - mediatized and commodified Buddhism as cultural narrative. *Journal of Global Buddhism*, *17*, 41–56. https://doi.org/10.5281/zenodo.1305859

Brown, K. W., Creswell, J. D., & Ryan, R. M. (2015). Introduction: The evolution of mindfulness science. In K. W. Brown, J. D. Creswell R. M. Ryan (Eds), *Handbook of mindfulness theory, research, and practice* (pp. 1–6). Guilford Press.

Brown, K. W., & Ryan, R. M. (2003). The benefits of being present: Mindfulness and its role in psychological well-being. *Journal of Personality and Social Psychology*, *84*(4), 822–848. https://doi.org/10.1037/0022-3514.84.4.822

Chen, I.-L. (2015). *An exploration of mindfulness and its experiential benefits: Taiwanese backpackers in Australia* [Doctoral dissertation]. The University of Queensland. https://core.ac.uk/download/pdf/43380777.pdf

Chen, I.-L., Scott, N., & Benckendorff, P. (2017). Mindful tourist experiences: A Buddhist perspective. *Annals of Tourism Research*, *64*, 1–12. https://doi.org/10.1016/j.annals.2017.01.013

Chittaro, L., & Vianello, A. (2014). Computer-supported mindfulness: Evaluation of a mobile thought distancing application on naive meditators. *International Journal of Human-Computer Studies*, *72*(3), 337–348. https://doi.org/10.1016/J.IJHCS.2013.11.001

Choi, E., Farb, N., Pogrebtsova, E., Gruman, J., & Grossmann, I. (2021). What do people mean when they talk about mindfulness? *Clinical Psychology Review*, *89*, 102085. https://doi.org/10.1016/j.cpr.2021.102085

Chowdhury, M. R. (2022). The five facet mindfulness questionnaire (FFMQ). *PositivePsychology.com.* https://positivepsychology.com/five-facet-mindfulness-questionnaire-ffmq/

Compson, J. F. (2017). Is mindfulness secular or religious, and does it matter? In L. M. Monteiro, J. F. Compson & F. Musten (Eds), *Practitioner's guide to ethics and mindfulness-based interventions* (pp. 23–44). Springer.

Deci, E. L., Ryan, R. M., Schultz, P. P., & Niemiec, C. P. (2015). Being aware and functioning fully: Mindfulness and interest taking within self-determination theory. In K. W. Brown, J. D. Creswell & R. M. Ryan (Eds), *Handbook of mindfulness: Theory, research, and practice* (pp. 112–129). The Guilford Press.

Dharmesti, M., Merrilees, B., & Winata, L. (2020). "I'm mindfully green": examining the determinants of guest pro-environmental behaviors (PEB) in hotels. *Journal of Hospitality Marketing & Management*, *29*(7), 830–847. https://doi.org/10.1080/19368623.2020.1710317

Diamond, M., Zheng, P., & Zoogman, S. (2017). Mindfulness and technology. In E. Zerbo, A. Schlechter, S. Desai & P. Levounis (Eds), *Becoming mindful: Integrating mindfulness into your psychiatric practice*, (pp. 163–182). American Psychiatric Association Publishing.

Errmann, A., Kim, J., Lee, D. C., Seo, Y., Lee, J., & Kim, S. S. (2021). Mindfulness and pro-environmental hotel preference. *Annals of Tourism Research*, *90*, 103263. https://doi.org/10.1016/j.annals.2021.103263

European Travel Commission (ETC). (2016). *Lifestyle trends & tourism: How changing consumer behaviour impact travel to Europe.* https://etc-corporate.org/uploads/reports/Trend1-LifestyleTrends%26Tourism-TheEndofAdventure(2016).pdf

Farkic, J., Isailovic, G., & Taylor, S. (2021). Forest bathing as a mindful tourism practice. *Annals of Tourism Research Empirical Insights*, *2*(2), 100028. https://doi.org/10.1016/j.annale.2021.100028

Fischer, D., Stanszus, L., Geiger, S., Grossman, P., & Schrader, U. (2017). Mindfulness and sustainable consumption: A systematic literature review of research approaches and findings. *Journal of Cleaner Production*, *162*, 544–558. https://doi.org/10.1016/j.jclepro.2017.06.007

Gibb, A., Wilson, J. M., Ford, C., & Shook, N. J. (2022). Does mindfulness reduce negative interpretation bias? *Cognition and Emotion*, *36*(2), 284–299. https://doi.org/10.1080/02699931.2021.2008322

Gotojones, C. (2013). Zombie apocalypse as mindfulness manifesto (after Žižek). *Postmodern Culture*, *24*(1). https://doi.org/10.1353/pmc.2013.0062

Govia, I., Robinson, J., Amour, R., Palmer, T., & Stubbs, M. (2020). Practicing mindfulness in low- and middle-income countries: Young workers in hospitality and tourism. *Wellcome*, 1–36. https://wellcome.org/sites/default/files/2021-05/mindfulness-hospitality-tourism-low-middle-income-countries-wellcome-workplace-mental-health.pdf

Habeeb, Z. (2021). Mindful travel. *International Journal of Spa and Wellness*, 4(2–3), 204–212. https://doi.org/10.1080/24721735.2021.1986972

Hanley, A. W., Warner, A. R., Dehili, V. M., Canto, A. I., & Garland, E. L. (2015). Washing dishes to wash the dishes: Brief instruction in an informal mindfulness practice. *Mindfulness*, 6(5), 1095–1103. https://doi.org/10.1007/s12671-014-0360-9

Hunkin, H., King, D. L., & Zajac, I. T. (2021). EEG neurofeedback during focused attention meditation: effects on state mindfulness and meditation experiences. *Mindfulness*, *12*(4), 841–851. https://doi.org/10.1007/s12671-020-01541-0

Jang, J., Jo, W., & Kim, J. S. (2020). Can employee workplace mindfulness counteract the indirect effects of customer incivility on proactive service performance through work engagement? A moderated mediation model. *Journal of Hospitality Marketing & Management*, 29(7), 812–829. https://doi.org/10.1080/19368623.2020.1725954

Joiner, T. (2017). *Mindlessness: The corruption of mindfulness in a culture of narcissism*. Oxford University Press.

JW Marriott Hotel Nara. (2022). Sophisticated tranquil escape. *American Express*. https://www.americanexpress.com/en-py/travel/discover/property/Jw-Marriott-Hotel-Nara

Kabat-Zinn, J. (2003). Mindfulness-based interventions in context: Past, present, and future. *Clinical Psychology: Science and Practice*, *10*(2), 144–156. https://doi.org/10.1093/clipsy/bpg016

Kabat-Zinn, J. (2013). *Full catastrophe living: Using the wisdom of your body and mind to face stress, pain, and illness* (revised edition). Bantam Books.

Kang, M. & Gretzel, U. (2012). Effects of podcast tours on tourist experiences in a national park. *Tourism Management*, *33*(2), 440–455. https://doi.org/10.1016/J.TOURMAN.2011.05.005

Kim, K., Milne, G. R., & Bahl, S. (2018). Smart phone addiction and mindfulness: an intergenerational comparison. *International Journal of Pharmaceutical and Healthcare Marketing*, *12*(1), 25–43. https://doi.org/10.1108/IJPHM-08-2016-0044

Kristeller, J. L. (2007). Mindfulness meditation. In P. M. Lehrer, R. L. Woolfolk & W. E. Sime (Eds), *Principles and practice of stress management* (pp. 393–427). The Guilford Press.

Langer, E. J. (1989). *Mindfulness*. Pireus Books.

Lau, M. A., Bishop, S. R., Segal, Z. V., Buis, T., Anderson, N. D., Carlson, L., Shapiro, S., Carmody, J., Abbey, S., & Devins, G. (2006). The Toronto mindfulness scale: Development and validation. *Journal of Clinical Psychology*, *62*(12), 1445–1467. https://doi.org/10.1002/jclp.20326

Lea, J., Cadman, L., & Philo, C. (2015). Changing the habits of a lifetime? Mindfulness meditation and habitual geographies. *Cultural Geographies*, *22*(1), 49–65. https://doi.org/10.1177/1474474014536519

Leddy, T. (2015). Experience of awe: An expansive approach to everyday aesthetics. *Contemporary Aesthetics (Journal Archive)*, *13*(8). https://digitalcommons.risd.edu/liberalarts_contempaesthetics/vol13/iss1/8/

Leggett, W. (2021). Can mindfulness really change the world? The political character of meditative practices. *Critical Policy Studies*, *16*(3), 261–278. https://doi.org/10.1080/19460171.2021.1932541

Lengyel, A. (2018). *Spatial aspects of sustainablity mindfulness and tourism* [Unpublished doctoral dissertation]. Szent Istvan University.

Lengyel, A. (2015a). Mindfulness and sustainability: Utilizing the tourism context. *Journal of Sustainable Development*, *8*(9), 35–51. https://doi.org/10.5539/jsd.v8n9p35

Lengyel, A. (2015b). *Extending tourism's role in the tourism - meditation - creativity-innovation - sustainability trajectory* [Paper presentation]. 5th International Conference on Management 2015. Management, Leadership and Strategy for SMEs' Competitiveness. http://real.mtak.hu/24797/1/ICoM_2015-paper100.pdf

Lewis, B. (2016). Mindfulness, mysticism, and narrative medicine. *Journal of Medical Humanities*, *37*(4), 401–417. https://doi.org/10.1007/s10912-016-9387-3

Loureiro, S. M. C., Stylos, N., & Miranda, F. J. (2020). Exploring how mindfulness may enhance perceived value of travel experience. *The Service Industries Journal*, *40*(11–12), 800–824. https://doi.org/10.1080/02642069.2019.1600672

Ludwig, D. S., & Kabat-Zinn, J. (2008). Mindfulness in medicine. *JAMA*, *300*(11), 1350–1352. https://doi.org/10.1001/jama.300.11.1350

Lusch, R., & Vargo, S. (2014). *Service-dominant logic: Premises, perspectives, possibilities*. Cambridge University Press.

McGroarty, B., Cavanah, C., Powell, L., Lo, J., Walsh, J., & Ellis, S. (2019). 2019 global wellness trends report. *Global Wellness Summit*. https://www.globalwellnesssummit.com/wp-content/uploads/2019/01/2019GlobalWellnessTrendsReport_GWS.pdf?inf_contact_key=911c57787517a178b672fb09b1687ce2d18a532c4142cb79caf2b269de1401fa

Miller, W. R. (1999). *Integrating spirituality into treatment: Resources for practitioners*. American Psychological Association.

Moscardo, G. (2009). Understanding tourist experience through mindfulness theory. In M. Kozak & A. Decrop (Eds), *Handbook of tourist behaviour: Theory and practice* (pp. 99–115). Routledge.

Moscardo, G. & Ballantyne, R. (2008). Interpretation and attractions. In A. Fyall, B. Garrod, A. Leask & S. Wanhill (Eds), *Managing visitor attractions* (pp. 237–252). Elsevier.

Nawijn, J. (2011). Happiness through vacationing: Just a temporary boost or long-term benefits? *Journal of Happiness Studies*, *12*(4), 651–665. https://doi.org/10.1007/s10902-010-9221-y

Oxford Advanced Learner's Dictionary. (2022). *Mindfulness*. https://www.oxfordlearnersdictionaries.com/definition/english/mindfulness

Panno, A., Giacomantonio, M., Carrus, G., Maricchiolo, F., Pirchio, S., & Mannetti, L. (2018). Mindfulness, pro-environmental behavior, and belief in climate change: The mediating role of social dominance. *Environment and Behavior*, *50*(8), 864–888. https://doi.org/10.1177/0013916517718887

Pearce, P., & Gretzel, U. (2012). Tourism in technology dead zones: documenting experiential dimensions. *International Journal of Tourism Sciences*, *12*(2) 1–20. https://doi.org/10.1080/15980634.2012.11434656

Pirson, M. A., Langer, E., & Zilcha, S. (2018). Enabling a socio-cognitive perspective of mindfulness: The development and validation of the langer mindfulness scale. *Journal of Adult Development*, *25*(3), 168–185. https://doi.org/10.1007/s10804-018-9282-4

Porter, N., Bramham, J., & Thomas, M. (2017). *Mindfulness and design: Creating spaces for well being* [Paper presentation]. Creative Practices for Improving Health and Social Inclusion. 5th International Health Humanities Conference, Sevilla 2016. Universidad de Sevilla, Vicerrectorado de Investigación, Sevilla, pp. 199–209. https://idus.us.es/handle/11441/65446

Purser, R. (2015). The myth of the present moment. *Mindfulness*, *6*(3), 680–686. https://doi.org/10.1007/s12671-014-0333-z

Purser, R., & Loy, D. (2013, July 1). Beyond McMindfulness. Huffington Post. https://www.huffpost.com/entry/beyond-mcmindfulness_b_3519289

Purser, R. E. (2019). *McMindfulness: How mindfulness became the new capitalist spirituality*. Repeater Books.

Rundle, A. G., Revenson, T. A., & Friedman, M. (2018). Business travel and behavioral and mental health. *Journal of Occupational and Environmental Medicine*, *60*(7), 612–616. https://doi.org/10.1097/JOM.0000000000001262

Salvati, M., Chiorri, C., & Baiocco, R. (2019). The relationships of dispositional mindfulness with sexual prejudice and internalized sexual stigma among heterosexual and gay/bisexual men. *Mindfulness*, *10*(11), 2375–2384. https://doi.org/10.1007/s12671-019-01215-6

Schmidt, S. (2011). Mindfulness in East and West – is it the same? In H. Walach, S. Schmidt & W. B. Jonas (Eds), *Neuroscience, consciousness and spirituality* (pp. 23–38). Springer.

Segal, Z. V., Williams, J. M. G., & Teasdale, J. D. (2012). *Mindfulness-based cognitive therapy for depression*. Guilford Press.

Segall, S. (2005). Mindfulness and self-development in psychotherapy. *Journal of Transpersonal Psychology*,*37*(2),143–163.https://www.sethzuihosegall.com/wp-content/uploads/2019/08/Mindfulness-and-Self-JTP-2005.pdf

Sheldon, P. J. (2020). Designing tourism experiences for inner transformation. *Annals of Tourism Research*, *83*, 102935. https://doi.org/10.1016/j.annals.2020.102935

Shonin, D. E., Van Gordon, W., & Singh, N. (2015). Mindfulness and Buddhist practice. In D. E. Shonin, W. Van Gordon & N. Singh (Eds), *Buddhist foundations of mindfulness* (pp. 1–6). Springer.

Stankov, U., Ćikić, J., & Armenski, T. (2021). New age tourism - challenges and positioning in a changing society. *Tourism Culture & Communication*, *21*(2), 69–80. https://doi.org/10.3727/109830421X16191799471944

Stankov, U., & Filimonau, V. (2019). Reviving calm technology in the e-tourism context. *The Service Industries Journal*, *39*(5–6), 343–360. https://doi.org/10.1080/02642069.2018.1544619

Stankov, U., & Filimonau, V. (2020). Technology-assisted mindfulness in the co-creation of tourist experiences. In Z. Xian, M. Fuchs, U. Gretzel & W. Höpken (Eds), *Handbook of e-tourism* (pp. 1–26). Springer.

Stankov, U., & Filimonau, V. (2021). Here and now – the role of mindfulness in post-pandemic tourism. *Tourism Geographies*. https://doi.org/10.1080/14616688.2021.2021978

Stankov, U., Filimonau, V., Gretzel, U., & Vujičić, M. D. (2020a). E-mindfulness – the growing importance of facilitating tourists' connections to the present moment. *Journal of Tourism Futures*, *6*(3), 239–245. https://doi.org/10.1108/JTF-11-2019-0135

Stankov, U., Filimonau, V., & Vujičić, M. D. (2020b). A mindful shift: an opportunity for mindfulness-driven tourism in a post-pandemic world. *Tourism Geographies*, *22*(3), 703–712. https://doi.org/10.1080/14616688.2020.1768432

Stankov, U., & Gretzel, U. (2021). Digital well-being in the tourism domain: mapping new roles and responsibilities. *Information Technology & Tourism*, *23*(1), 5–17. https://doi.org/10.1007/s40558-021-00197-3

Stankov, U., Gretzel, U., & Filimonau, V. (2022). *The mindful tourist: The power of presence in tourism*. Emerald Publishing Limited.

Steidle, G. K. (2017). *Leading from within: Conscious social change and mindfulness for social innovation*. MIT Press.

Szmigin, I., & Carrigan, M. (2000). Does advertising in the UK need older models? *Journal of Product & Brand Management*, *9*(2), 128–143 https://doi.org/10.1108/10610420010322170

Taylor, G., Alpert, J., Waddell, T. F., & Bylund, C. L. (2021). Streaming mindfulness: Well-being and mindfulness among subscribers to a video streaming service. *Internet Interventions*, *25*, 100419. https://doi.org/10.1016/j.invent.2021.100419

Tedder, M., Shi, L., Si, M., Franco, R., & Chen, L. (2015). eMindfulness therapy—A study on efficacy of blood pressure and stress control using mindful meditation and eating apps among people with high blood pressure. *Medicines*, *2*(4), 298–309. https://doi.org/10.3390/medicines2040298

Thompson, B. L., & Waltz, J. (2007). Everyday mindfulness and mindfulness meditation: Overlapping constructs or not? *Personality and Individual differences*, *43*(7), 1875–1885. https://doi.org/10.1016/j.paid.2007.06.017

Van Dam, N. T., van Vugt, M. K., Vago, D. R., Schmalzl, L., Saron, C. D., Olendzki, A., Meissner, T., Lazar, S. W., Kerr, C. E., Gorchov, J., Fox, K. C. R., Field, B. A., Britton, W. B., Brefczynski-Lewis, J. A., & Meyer, D. E. (2018). Mind the hype: a critical evaluation and prescriptive agenda for research on mindfulness and meditation. *Perspectives on Psychological Science*, *13*(1), 36–61. https://doi.org/10.1177/1745691617709589

Vogus, T. J., & Sutcliffe, K. M. (2012). Organizational mindfulness and mindful organizing: A reconciliation and path forward. *Academy of Management Learning & Education*, *11*(4), 722–735. https://doi.org/10.5465/amle.2011.0002C

Whitehead, M., Lilley, R., Howell, R., Jones, R., & Pykett, J. (2016). (Re) inhabiting awareness: geography and mindfulness. *Social & Cultural Geography*, *17*(4), 553–573. https://doi.org/10.1080/14649365.2015.1089590

Williams, J. M. G., & Kabat-Zinn, J. (2011). Mindfulness: diverse perspectives on its meaning, origins, and multiple applications at the intersection of science and dharma. *Contemporary Buddhism*, *12*(1), 1–18. https://doi.org/10.1080/14639947.2011.564811

Zheng, S., Sun, S., Huang, C., & Zou, Z. (2020). Authenticity and subjective well-being: The mediating role of mindfulness. *Journal of Research in Personality*, *84*, 103900. https://doi.org/10.1016/j.jrp.2019.103900

Chapter 16

When the Process is the Destination

Long-distance Walking as Slow Tourism

M. Mau & K.K. Roessler

Introduction

Along with broader societal changes, peoples' motives for going on long-distance walks, also known as *hikes*, or more traditionally, *pilgrimages*, have changed (Reader, 2007). What once was a risky activity reserved for the few who were willing to take the necessary chances in search of religious goals (Edensor, 2000), has now become a modern, tourist activity (den Breejen, 2007), appealing broadly, to both young, middle-aged, and older people (see for example Yang et al., 2018; Mau et al., 2021a).

The increasing popularity of long-distance walking (Crust et al., 2011) reflects individual needs, hopes, and aspirations (Olsen, 2013), societal *advances*, including the ability to travel, practically and economically (Reader, 2007), but also societal *demands*, which for some necessitate a long walk (Mau et al., 2021a). One popular story, which reflects some of the psychological topics this chapter will delve into, is the book and movie *Into the Wild*, written by Jon Krakauer (2018), on the authentic story of a young man, Christopher McCandless, who ventured out, first by car, but soon thereafter, by foot, to explore and develop as a person.

> Christopher McCandless initiates his walk to break free from the expectations and values that the people around him, the society, and his family, are trying to instill in him. He leaves, abruptly, freeing himself from the estranging materialistic and inauthentic norms of his (former) life. The freedom he achieves allows him to go wherever his needs and wants take him and to explore and be whomever he feels. However, traveling is a two-sided coin, also associated with a physical and psychological toll, which makes his journey not only exciting and eventful but also fatiguing and difficult when compared to the relative ease of his former life. Ultimately, the story ends dramatically, as he decides to cut bonds with society completely, attempting - unsuccessfully - to live off nature (Krakauer, 2018).

Although his story represents a perhaps drastic approach to the long-distance walk, it highlights topics that are relevant to understanding the meaning of long-distance walking more

The Wiley Blackwell Companion to Tourism, Second Edition. Edited by C. Michael Hall.

broadly. These topics include countering the demands of modern life and the appeal of nature, and it highlights how long-distance walking can be part of a search, not so much of hedonistic pleasure, but of deeper and more rewarding experiences. What transcends and unites Christopher McCandless extreme adventure with those of more common tourist long-distance walkers, is, that neither are discouraged by the *process*, but rather, find value in it, and what it may bring of challenging, formative, and unexpected experiences.

The aim of this chapter is to outline the state-of-the-art of research into long-distance walking, focusing on mental health, here broadly defined, relating to both the regaining of personal resources, e.g., countering mental stress, and to other, perhaps more long-term, goals of personal growth. To achieve the aim, we will sketch out the main bodies of research on the health outcomes of long-distance walking, both theoretically and empirically. To provide the necessary background, we will first, however, describe the historical and terminological roots of, and the different attempts at defining and differentiating between terms related to long-distance walking. Based on these accounts we will argue that long-distance walkers may be viewed as participating in the movement known as slow tourism, as this movement stands in opposition to and builds on the same elements of slowness and countering ills of modern life as research has identified as underlying motives behind long-distance walking.

Historical Roots and Terminological Changes

Although the story described above about Christopher McCandless is new, long-distance walking is in itself a very old activity (Edensor, 2000). While the psychological research is no more than 10–20 years old (Mau et al., 2021b), long-distance walking has been practiced across time and culture, for centuries or, arguably, for millenniums (Edensor, 2000). Terminologically, the long-distance walk has been (and still is) known under different labels. This includes the religiously or spiritually sounding *pilgrimage* (e.g., Kato & Progano, 2017) and the *hike*, which emphasises more of the outdoor aspects of walks (e.g., Nordbø & Prebensen, 2015)[1]. However, what differentiates these terms has over time become increasingly unclear, due to several significant sociological trends including changes in living and work conditions, and in belief systems.

The pilgrimage used to be, as stated earlier, a religious undertaking. In medieval times, leaving the security of one's settlement meant travelling along dangerous roads, potentially being misunderstood as, for example, a person with hostile intent (Edensor, 2000). Consequently, to make long-distance walking worthwhile, the motive must have appeared profound, and the very idea of long-distance walking has presumably relied on a different degree of belief when compared to today. The motives among past long-distance walkers also relied on different work and living conditions, already entailing some level of physical labour and being in nature. Aspects that are *today* categorised as 'recreational', including exercise and contact with nature, would have been part of normal daily life earlier. However, by the 18th and 19th centuries (in the Western world, i.e. Europe and North America), increasing urbanisation meant that walks in the woods, or *hiking*, became something one would do for pleasure (Warren, 2020; Nordbø & Prebensen, 2015); a shift, also made possible by safer travel options and changes in work conditions, including the regularised working week in which one had an apportioned amount of time for leisure (Edensor, 2000).

Nowadays, recreational aspects of long-distance walking are attributed to our increasingly sedentary lifestyle, disconnected from nature (Barton et al., 2009), but also the psychological demands posited by the fast-paced lifestyle that to some people is part of modern life.

Other related issues including alienation, or lack of authenticity, may also act as an underlying motive that causes people to embark on long-distance walks (Olsen, 2013; Edensor, 2000; Frey, 2004).

These societal changes have led to a change in what it means to go on a pilgrimage, and what differentiates a hike and a pilgrimage, is becoming increasingly unclear. A pilgrimage walk may nowadays still express a religious belief, in both the traditional sense, but today also in a more personalised syncretism of aspects from religious and non-religious practices (Kato & Progano, 2017), and/or a search for something authentic; a search for the 'really real' (Egan, 2011). Thus, we may travel as tourists along specific routes, for example, the Camino de Santiago, and be called pilgrims, but without being religious, or even spiritual, but rather on a journey of, for example, self-discovery (Kato & Progano, 2017). Moreover, the aspect of being outdoors, close to nature, may be a key component of a pilgrimage, along with a range of other experiences: 'Nakahechi, and other pilgrimage trails are visited by those who are not necessarily motivated by religious reasons, but rather, seek qualities such as positive elements of natural environment - serenity, beauty, cleanliness, a sense of spiritual and physical wellness, as well as a sense of achievement gained through completing a certain distance on foot' (Kato & Progano, 2017, p. 243).

In short, we have seen a blending of motives, and the motives that may be defining of a hike, e.g., a more intimate connection with nature, could as well be of a pilgrimage, and conversely, the pilgrimage may no longer entail religion[2]. Rather, it can be viewed as long walks through cultural landscapes (Egan, 2011), the same as hikes (Nordbø et al., 2014). To make matters more complicated, motives may change while walking; the motivation to *begin* may differ from the motivation to *keep going* (Egan, 2011), and most people travel due to many, and complex motivations, which are not mutually exclusive (Lowry & Lee, 2016).

With regards to, then, understanding what a long-distance walk is, the consequence of societal changes and the plurality and changeableness in motives have been terminological uncertainty. Over time, long-distance walks, regardless of their label, have been used as a means towards achieving a range of aims, from physical health to closeness to a deity, to self-exploration. Therefore, it is difficult, and perhaps not meaningful, to attempt a clear separation between, for example, hiking and pilgrimage walking. One may enjoy being in nature, have spiritual aspirations, and become more physically able at the same time. In this chapter, we apply long-distance walking, which may be used as the generic term, encompassing these other, related terms, i.e. the walks undertaken as part of pilgrimages and hikes (Mau et al., 2021b).

Approaches to Defining the Long-distance Walk in Research

As may perhaps be expected from the broadness of the term, the research we consequently refer to is diverse, covering different methods and methodologies (Mau et al., 2021b). What (paradoxically) unite the different studies in this field is uncertainty regarding exactly what long-distance walking is. Often studies do not clearly define long-distance walking, but rather rely on a method in which they encounter walkers at a point on a trail, on which they must have walked at least some distance to reach (e.g., Wolf & Wohlfart, 2014; den Breejen, 2007; Yang et al., 2018). This is viewed as a way of identifying the long-distance walkers, but it does not in itself explain what long-distance walking is. Other studies apply a similar logic by using a chain-sampling method, in which long-distance walkers refer to each other, thus helping the researcher establish contact (e.g., Coble et al., 2003; Mau et al., 2021a). The question of definition is to some extent left out, relying instead on the place of the walk (most often nature or rural areas), (Nordbø et al., 2014; Mau et al., 2021b) as the decisive parameter to finding long-distance walkers, or on peoples' self-understanding as long-distance walkers. These approaches have led to a large variety among studies, examining walks of between 10 to 100 miles or lasting from hours to days, or even weeks (Mau et al., 2021b).

The perhaps most obvious approach to scientifically defining long-distance walking is to outline the so-called long-distance that needs to be travelled. Building on Kay and Moxham (1996), Crust et al. (2011), states that long-distance walking is a walk of 20 miles or more, or which lasts several days. However, apart from the fact that this objective definition is not used stringently in the literature, it is not explained precisely why this should be the length of long-distance walks (perhaps explaining why this understanding has not taken root). Also, there is a subjective aspect to distance: what constitutes a long walk for a young person may diverge from what an older person thinks, making objectively measurable distances difficult to apply (see e.g., Mau et al., 2021a). Another branch of research focuses more on the motive behind long-distance walking or describes it in terms of what effect it has on mental states. Definitions coming from the literature on the pilgrimage tend to focus more on the underlying *meaning* of the walk, which may be religious aspirations (see e.g., Kato & Progano, 2017), but it can also more broadly be defined as a 'journey redolent with meaning' (Timothy & Olsen, 2006, p. 36).

In summation, understandings or definitions are usually not transparently outlined in research studies, and a common definition has yet to be accepted in the research field. However, according to the terminological and historical background outlined in the above sections, long-distance walking appears to be associated with 1) *walks*, which are 2) subjectively or objectively *long*, which 3) often take place in *nature*, and has 4) connotations relating to some form of *personal benefit,* including mental or spiritual health.

Slow Tourism and Long-distance Walking

In line with these four aspects, long-distance walking is used both as a mental health intervention, targeting symptoms of mental illness (see e.g., Sturm et al., 2012) and as a recreational, tourist activity (Egan, 2011). What motivates the long-distance walker, however, what benefits or experiences are hoped to come out of the walk, are not necessarily different depending on whether the walk is part of an intervention or a vacation. Stress relief and a more intimate connection with nature, for example, may underlie both interventions (e.g., Niedermeier et al., 2017a) and tourists' motivations (Lumsdon & McGrath, 2011). In particular, what may partly explain the increasing popularity of long-distance walking (Saunders et al., 2013), is that it can be viewed as part of the relatively new (Khan, 2015) but growing (Lumsdon & McGrath, 2011) movement known as *slow tourism* (Dickinson & Lumsdon, 2012).

The term slow tourism includes a collection of ideas, centering around the notion of *slow*. The emphasis in slow tourism, also labeled slow travelling (Lumsdon & McGrath, 2011), is placed on 'the journey and experience of slow travel itself rather than a destination and experience of place' (Caffyn, 2012, p. 78). In slow tourism, the *process* is emphasised (Timms & Conway, 2012), in contrast to more traditional forms of tourism characterised by, for example, standardised packaged tours and by visiting 'must-see' stops (Lowry & Lee, 2016; World Travel Market & Euromonitor, 2007), rendering the transportation, the process of *getting there*, as something to be overcome.

The achievement of personal, psychological, benefits is emphasised within slow tourism, through a slow process: the slow tourists, e.g., the long-distance walker, let the walk 'set the pace and context of development' (Kato & Progano, 2017, p. 243)[3]. The personal benefits of slow tourism, the overall goals to slow tourists, can be labeled *revitalisation*, including becoming refreshed or recharged, and *self-enrichment* including the discovering of self, understanding, and connecting with the surroundings, and expanding of one's perspective (Oh et al., 2014). These aspects illustrate that slow tourism may involve both recreation, in the form of restoring personal resources that have somehow been overused, and personal growth or development. No watertight definition of slow tourism exists (Caffyn, 2012; Oh et al., 2014), but 'tourists who opt for this type of tourism are pushing back against the fast pace of life and the ills (i.e., practices that contribute to unhealthy and unsustainable ways of life) that have befallen society in its race to modernity' (Lowry & Lee, 2016, p. 8).

Slow tourism is part of a movement (Khan, 2015), which stand upon a critique of modernity, especially the increasing pace of modern life: '[. . .] people's desire to "slow down" may be a consequential reflection of their mental resistance to the ever-accelerating pattern of lifestyle in search of opportunities to vent their stress, decelerate their pace, and ultimately restore and enrich the self' (Oh et al., 2014, p. 205–206). These topics reflect the analysis by the German sociologist Hartmut Rosa, who similarly views the challenges of modernity as consequences of society's continuously increasing acceleration (Lijster & Celikates, 2019). In parallel with the aspiration (or pressure) to do *more* in *less* time, follows a feeling of having lost control over, and contact with, our own life and the world that surrounds us. The result of this *societal* tendency is alienation, felt at the *individual* level (Lijster & Celikates, 2019). This issue reflects slow tourists' search for authentic experiences (Khan, 2015) as a response to alienated and alienating mass tourism (Conway & Timms, 2010). By slowing down, tourists may attain a more rewarding relationship with the world in which they travel (Kato & Progano, 2017), or with themselves (Oh et al., 2014), reflecting Rosa's suggested alternate term for the authentic: a *resonance* (Lijster & Celikates, 2019).

Although accounts of slow tourism also include dimensions which are objectively measurable, such as whether and how much one travels by plane or car (see e.g., Caffyn, 2012), the travel *experience*, the 'state of mind' (Lumsdon & McGrath, 2011, p. 266) is central. Perhaps it is the *felt* slowness, rather than the objectively measured slowness, that most aptly captures the meaning of slow tourism (Oh et al., 2014). Thus, fast and slow modes of travelling are not mutually exclusive as such; rather, they may co-exist, and be both complementary and competing parts of one journey, that is, however, *on the whole*, perceived of as slow (Oh et al., 2014). Walkers, who fly to pilgrimage trails, to experience the journeying by foot (Reader, 2007) can be seen as a reflection of this.

The focus of this chapter, the long-distance walk, can be seen as a form of slow tourism (e.g., Kato & Progano, 2017), in that:

- the aspect of movement is viewed as a quality;
- and therefore, the process is deliberately extended, by going slowly or over a long distance;
- both slow tourism and long-distance walking can be attempts at counterbalancing the challenges of modern life; and
- more broadly, be used as a way of experiencing personal growth or obtaining mental health benefits.

In the following, we will outline what these growth or health-oriented benefits may be more specifically according to research.

Research Perspectives on Mental Health Aspects of Long-distance Walking

The personal benefits that may come out of long-distance walking relate to a number of psychological theories[4]. In this section, we will outline three of the most significant strands of theory, grouped together by what influence long-distance walks are assumed to have on mental states, i.e. what long-distance walking 'does'. The first regards a state of *flow*, in which self and action seem to merge, while the walker is intensely involved in the walking; the second regards a state of *calmness*, in which the exposure to nature may de-stress and increase the individual's ability to focus; and the third regards a state of *contemplation*, in which aspects such as time provides the walker with the opportunity to reflect on deeper-lying topics, related to existence.

Flow: Overcoming the Challenges of Long-distance Walking

One aspect of the long-distance walk is the challenge that it is associated with. Overcoming such parts of the walk (covering the distance, enduring the weather, physical pain) may bring about the psychological state known as *flow* (Csikszentmihalyi, 1990). Along with other related and/or positive psychological concepts and theories (Rich, 2013), flow theory is among the most used frames of reference, when explaining the psychological benefits of long-distance walking (Mau et al., 2021b).

Flow is a 'state in which action follows upon action according to an internal logic which seems to need no conscious intervention on our part. We experience it as a unified flowing from one moment to the next, in which we feel in control of our actions, and in which there is little distinction between self and environment; between stimulus and response; or between past, present, and future' (Csikszentmihalyi, 1975, p. 43). The flow state is autotelic[5] and is characterised by concentration, a heightened sense of control, and a transformed sense of time (either speeding up or slowing down). As indicated in the quote, action and awareness may seem to merge (Wöran & Arnberger, 2012), and thoughts, feelings, and bodily performance may all seem to become part of the same, embodied, motion (Redick, 2016). This flow state overlaps conceptually with meditational practices (Swann et al., 2012), also examined in long-distance walking studies (Feliu-Soler et al., 2021).

The prerequisites for entering the flow state is that there is:

- a balance between the long-distance walker's skills and the challenges of the task;
- that the goals are unambiguous; and
- that the feedback, informing the walker on their progress, is ongoing (Csikszentmihalyi, 1996; Swann et al., 2012).

Flow has proven to be a suitable concept to apply the long-distance walk as the balance between skill and challenge can be adjusted to the capabilities of the individual walker; that the goal of walking can be viewed as evident (reaching the destination); and that the feedback, how the walk is progressing, is ongoing, for example, through bodily signals (Wöran & Arnberger, 2012; Redick, 2016; Crust et al., 2011). Other positive psychology concepts and theories studied as part of long-distance walking are the broaden-and-build theory (Crust et al., 2011) and the PERMA model (Saunders et al., 2013)[6], of which flow may be viewed as one aspect (Rich, 2013). Relatedly, self-efficacy, which has also been used as a lens to understanding the mental benefits of long-distance walking (e.g. Lesser et al., 2020), also emphasises the effects of overcoming the physical challenges associated with long-distance walking, and the subsequent feelings of control or competence the walker may achieve. In summation, flow, and concepts related to flow, are one group of common theoretical notions used to explain the advantages of long-distance walking, emphasising how overcoming the requirements through self-reliance may lead to an altered state of mind or self-perception.

Calmness: The De-stressing Effects of Walking in Nature

As noted, one of the characterising features of long-distance walking is the surroundings through which the walker moves. The second mental health benefit of long-distance walking focuses on the influence of exposure to nature (D'souza & Puttaraju, 2020; Niedermeier et al., 2019; Wöran & Arnberger, 2012), which regards stress recovery and the restoration of attentional capacities. Within this group of studies, mainly two theories are used: Attention Restoration Theory, by Kaplan (1995), and Stress Restoration Theory, by Ulrich et al. (1991).

According to Attention Restoration Theory, attention can, borrowing from a distinction by William James, be divided into two forms: *directed* attention and *involuntary* attention. Using directed attention is necessary both emotionally, cognitively, and behaviorally, but humans have a limited ability to use directed attention, and we may overuse or exhaust this attentional resource, resulting in fatigue (Kaplan, 1995; Kaplan & Kaplan, 1989). Nature provides a remedy for such mental fatigue, as it is not necessary to use directed attention in nature. Instead, one can linger on whatever appears attractive, and this paucity in demands for directed attention restores our capacity to exert it. Moreover, this also allows for reflections on topics, which have otherwise not been thought through (Basu et al., 2019). The sound of the birds can gather one's attention, but in a so-called 'soft' way, leaving room for other thoughts to emerge and be processed – in contrast to watching television, which is subject to 'hard' (more of an all-absorbing) fascination (Kaplan, 1995, p. 172).

According to Stress Reduction Theory, humans have an evolutionary preference for natural settings, which may reduce stress (Ulrich et al., 1991). This theory is focused on the visual aspects of nature such as depth, curvilinearity, and complexity, which have provided clues to prehistoric humans, that the setting was preferable to survival (Ulrich, 1983). It would, for example, have been more stressful, theoretically, for earlier humans, to walk, hunt, or be in places with no overview of the landscape, as dangerous animals or other risks could remain unseen. Underlying this evolutionary framework is Wilson's (2017) biophilia-hypothesis, according to which humans have a genetic preference for natural settings, as this is the setting which we have developed in, in a large part of human history. Applying Ulrich et al's (1991) theory to the field of long-distance walking, the visual landscape that the walker is surrounded by may offer a less stressful environment than an urban environment, which does not accommodate our evolutionary needs in the way that nature does (Niedermeier et al., 2019).

In differentiating between Attention Restoration theory and Stress Reduction theory, the focus in the latter is more explicitly on emotional processes, when compared to the first, in which emotional benefits come as a consequence for cognitive restoration. Both, however, are central in the psychological research on long-distance walking and point to the stress-reducing effects of the surroundings in which long-distance walkers move (Mau et al., 2021b; Berto, 2014).

Contemplation: Using the Walk as an Opportunity for In-depth Reflections

The third, and last, mental health benefit, regards reflection processes. A somewhat diverse group of studies examines how long-distance walking gives people the opportunity to reflect, look back, prioritise, or otherwise 'walk' through memories or impressions from their lives. In this perspective, it is emphasised how long-distance walkers are reflexive (Ólafsdóttir & Runnström, 2013). Existential considerations, which has historically (Edensor, 2000) and may (still) be characteristic of the contemplative processes during long-distance walks, include finding meaning (Fredrickson & Anderson, 1999) or taking stock of a lived life (Mau et al., 2021a), through reflective practices, which may be termed self-discovery (Eichberg et al., 2017), finding oneself (Reader, 2007) or self-examination (Jørgensen et al., 2020).

Different aspects of the walk may provide this opportunity for contemplation. As described earlier, long-distance walking entails walking a distance, which is either objectively or subjectively regarded as long. This takes time[7], and having time to spend in solitude offers the opportunity to address deeper-lying topics, perhaps otherwise (un)consciously left or driven out of awareness (Fredrickson & Anderson, 1999; Jørgensen et al., 2020). In addition, being unavailable to the usual demands and expectations of others, perhaps even being outside the range of one's telephone signal, may lead to a state of disengagement from common stresses and problems in life (Saunders et al., 2013). And lastly (though this list is not exhaustive, but only emphasises a few underlying factors) long-distance walking is a relatively simple activity when compared to the tasks of modern daily life (Svarstad, 2010), which may enable reflections that may otherwise have been brushed aside, for the sake of navigating through the complexities of daily life.

To theoretically describe these processes, one concept applied is *liminality* (Mau et al., 2021a), which has been updated by Victor and Edith Turner in the context of pilgrimages and leisure activities (Turner & Turner, 1978; Turner, 1974), but was originally part of Arnold van Gennep's (1909.1960) model of *rites de passage*. According to this perspective, long-distance walkers may enter a so-called liminal state (Mau et al., 2021a), characterised being 'betwixt-and-between' or 'neither-this-nor-that' (Turner, 1974, p. 55), in which the walker is allowed to freely contemplate life, without preoccupation with how potential realisations fit into already existing world views (Thomassen, 2009). By being freed from usual frames of reference, by being able to freely contemplate life choices or other topics that seem important, the walker may undergo personal transformation (Mau et al., 2021a). The notion of liminality is one way of representing the contemplative state among long-distance walkers examined in the research, including what potential lies in this opportunity for reflection.

Theoretical Co-existence: The Being, Thinking, and Doing of Long-distance Walking

These states of flow, calmness, and contemplation, along with their potential for personal growth, stress-reduction, and so on, represent three main bodies of research covering the mental benefits of long-distance walking. To an extent, these three states are distinct, emphasising *being in* nature (calmness) to *thinking about* life in the extended time that long-distance walking takes (contemplation) to *doing* the walk, overcoming the challenges one meets underway (flow). This is what we term the being, thinking, and doing of long-distance walking. Although they overlap to a degree, they emphasise different processes: contemplation *can* happen while in flow, but to an extent; thinking, especially in the third-person (e.g., reflecting on the reflections), may break the flow state (Redick, 2016). Being in nature is, according to the theories presented here, first and foremost about restoring attention, regaining the cognitive ability to focus, rather than using attention to think on or contemplate, deeper-lying topics. Flow is more about being engaged in the task at hand, and bodily overcoming straining aspects than being calm or consumed by contemplations.

What we will suggest, is that although these states are different, they are not necessarily mutually exclusive, but may rather co-exist and all be present in the same walk[8]. Thus, one may contemplate, reflect on deeper-lying topics in life, in moments of the walk. At other times, one may be occupied by a calmer state, a state of relaxation and perhaps enjoyment of the natural world. And after this, one may enter a state resembling flow, in which the walker feels at one with their actions and feels confident in managing and overcoming the physical challenges of long-distance walking. Thus, the walker may go from calmness to contemplation, to a flow state, or what we term the being, thinking, and doing states, like a pendulum swinging between the three corners of a triangle.

Viewing the long-distance walk as an activity that allows the co-existence of different mental states is, in our assessment, reasonable. A flow state is not necessarily something one is in, all the time (Crust et al., 2011). It may appear in moments, it may come and go, but thereby also be divided by states of, for example, calmness. Calmness is, however, likely neither a permanent state, as long-distance walking is, at least at times, tough: the weather may be bad, and become something that the walker needs to endure, rather than enjoy, or one may have blisters or sores after a long day of walking (Egan, 2011). Finally, contemplation may not be a non-stop, prolonged process either, but rather be accompanied by pauses; by moments of silence: of moments of *not* thinking, but merely of being or doing (Redick, 2016, 2021). Thus, these different mental states may all be present during one individual's long-distance walk, and it is perhaps together, as a dynamic moving back and forth from state to state, that the most precise picture of the overall experience is painted (as illustrated in Figure 16.1).

The State of the Art within Empirical Studies

This plurality of states has been examined using a plurality of research methods. Some studies on long-distance walking and mental health examine processes that are automatic; that regards reactions to walking, to the setting of the walk, etc., which does not necessitate reflecting about one's life; others examine contemplative processes, building on active awareness, on the individual thinking, or reflecting, on their life. Mirroring these foci, the research field

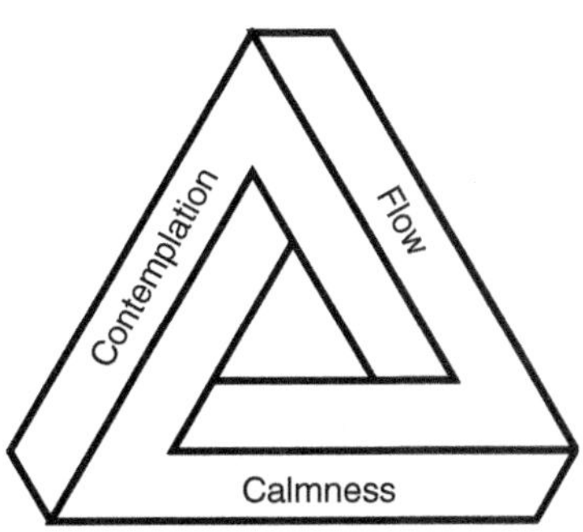

Figure 16.1. A Penrose triangle, modified to illustrate how the three states of minds among long-distance walkers may successively follow each other.

contains both a qualitative branch, examining subjective experiences, and a quantitative, using cross-sectional and interventional designs, examining a range of correlations.

Most notably within the quantitative paradigm is a relatively large group of longitudinal Austrian studies, measuring the effect of hiking on physiological stress responses (Niedermeier et al., 2019, 2017b) or affective states (depression, hopelessness) (Sturm et al., 2012; Niedermeier et al., 2017a). The walks examined in these studies take place in the mountainous, alpine regions of Austria, and draw on theories related to the beneficial aspects of exposure to *nature*. The Attention Restoration Theory and Stress Reduction Theory is used either directly (Wöran & Arnberger, 2012; Niedermeier et al., 2019) or indirectly, through the notion of 'green exercise' (Niedermeier et al., 2017a, 2017b), which combines these nature-oriented theories, with theories on the beneficial effects of exercise (Pretty, 2004). Another main group of studies is qualitative and originate from Northern Europe (including the UK). These examine experiences of walking, and how they relate to such notions as the psychological journey (Crust et al., 2011) or the transformative potential of long-distance walking (Saunders et al., 2013; Mau et al. 2021a). These studies focus particularly on contemplative processes and on flow or flow-like states (e.g., Jørgensen et al., 2020; Mau et al., 2021a). Aside from these two main groups of studies, two Belgian studies qualitatively examine how body awareness and identity may be affected by long-distance walks among people with multiple sclerosis (Calsius et al., 2015, 2019). The remaining research field is mostly comprised of a number of individual studies (Mau et al., 2021b). Overall, the empirical research on long-distance walking and mental health stems mainly from Western countries (Mau et al., 2021b) (with a few exceptions, e.g., Yang et al., 2018). The research field is also generally young, and growing, as can be seen from the above-mentioned studies, of which several have been published within the last five years.

This relative recentness of studies, and its roots in Western, particularly European, research milieus, may be explained by the perspective on long-distance walking and how it relates to *other* activities, which we have applied in this chapter. Long-distance walking was earlier, as described, more firmly integrated with religious practices (a practice which also exists today). However, in this chapter, we have not considered studies that examine the outcomes of, for example, pilgrimage walking, where the outcomes or influence on mental states, are not reported for the aspect of walking alone. Walks may however, be combined with visits to shrines or churches and/or with certain religiously accentuated practices (prayers, meditations), and findings reported on the *joint* influence of these activities. This choice of dissociating long-distance walking from a broader, religious journey, may reflect a bias of overlooking the still noteworthy role that religious or sacred practices other than walking play for some pilgrims (e.g., see Eade, 2020). Our focus on slow tourism, a concept originating from the Slow Food movement out of Italy (Lowry & Lee, 2016), may reflect the same tendency to focus on recent trends and concepts that may to some level have a Western turn. This naturally limits the scope of this chapter. However, this choice, to view long-distance walking in isolation and in the context of mental health, may merely reflect the bendability of the underlying meanings of long-distance walking; a reflection of the fact, that long-distance walking has been historically, and can nowadays, be used to obtain a host of purposes (Kaelber, 2006). Alternatively, long-distance walking could be examined through the lens of religious tourism (Olsen, 2013), and could then have been regarded as a subset of a collection of practices, and not as the sole or primary activity. However, long-distance walking as a means towards the improvement of mental health is a focus, likely to persist and grow in popularity.

Future Directions and Conclusions

We expect that long-distance walking will grow in popularity and receive increasing interest from research both inside and outside of European research milieus. The issues that long-distance walking may combat as described by Hartmut Rosa regarding acceleration and hectic lifestyles are not only restricted to the Western world (Lijster & Celikates, 2019), and may become subjects of research more broadly. We have already seen a steep increase in visitors on pilgrimage routes (Horváth & Szakolczai, 2017), and the chance for introspection, for testing one's capabilities, and for achieving a more intimate connection with nature may turn out to match the deficiencies of modern society also going forward. Overall, the topics of nature and of movement that long-distance walking tap into, and how these relate to mental health, are already receiving increasing attention; examining activities that combine them seems natural.

In conclusion, this chapter has outlined the history of long-distance walking along with the different approaches to defining this activity within research. We have established, that a common definition has yet to be agreed upon, which may explain why a number of studies rely on a more pragmatic method of, for example, encountering walkers on trails, thus not defining the walk as such, but relying on the place of the walk. Although an agreed-upon definition is still missing, there appear to be four characteristics of long-distance walking, being that it is a walk, which is either objectively or subjectively long, typically taking place in nature, and that has connotations to some form of personal benefit, for example, personal growth or stress relief. We have shown how these characteristics overlap with the emerging movement of slow tourism, of which the countering of society-induced ills are among the defining features. To outline what ills long-distance walking may combat, we outlined the three main mental states, which research applies to understand what long-distance walking does, vis-a-vis mental health. These are applied within a relatively large group of quantitative Austrian studies and a group of North European qualitative studies. These three states, however, may not be mutually exclusive, which is why we suggested that it may be the co-existence of these states, that best illustrates the comprehensive experiences of long-distance walkers.

The purposes of long-distance walking have changed and probably will change, in the coming years. What we use long-distance walking for – or against – reflects contemporary broader societal trends, and follows changes in belief systems, broadly defined, including notions of mental health. The individual walk, the journey undertaken by a tourist walking, for example, the Camino, may also contain a blend of motivations: the chance to reflect in an otherwise busy or distracted everyday life, a chance to relax and bond with nature, and a chance to be in a state of flow. Perhaps it is long-distance walking's very broadness, its facilitation of both being, thinking, and doing states of mind, which has encouraged its usage from earlier times to today, and what may underpin its resiliency or adaptability in facing the changes that society globally will undergo going forward.

Notes

1. More terms exists, including trekking or rambling, but for the sake of simplicity, and as these are used interchangeably with hiking (Nordbø et al., 2014), we in this brief historical exposition focus on the pilgrimage and on the hike.
2. The, to some seemingly oxymoronic, secular pilgrimage (Egan, 2011).

3. Considerations for the environment are also emphasised as key to the emergence of slow tourism (Kato & Progano, 2017; Dickinson & Lumsdon, 2010), but is not explained further in this chapter, as we focus on mental health aspects.
4. We focus here on benefits the individual walker may attain through long-distance walking. The *social* dynamics may also be altered during long-distance walking, with the benefit of, for example, making the sharing of difficult emotions more doable (Egan, 2011). However, the main body of research focus on individual benefits, and so this is what we will follow in this chapter.
5. Perceived of as valuable in itself, rather than because of some exotelic, means-end expectation (Coble et al., 2003).
6. An abbreviation of positive emotion, engagement (which overlaps with flow), relationships, meaning, and accomplishment (Rich, 2013).
7. And time may in fact be a more precise indicator of the distance in long-distance walking, rather than the length of the walk: one may cover a given length more or less fast, but regardless of speed, long-distance walking seems to take a long time (Mau et al., 2021b).
8. Redick (2016) portrays long-distance walking in a, to some degree, similar fashion.

References

Barton, J., Hine, R., & Pretty, J. (2009). The health benefits of walking in greenspaces of high natural and heritage value. *Journal of Integrative Environmental Sciences*, *6*(4), 261–278. https://doi.org/10.1080/19438150903378425

Basu, A., Duvall, J., & Kaplan, R. (2019). Attention restoration theory: Exploring the role of soft fascination and mental bandwidth. *Environment and Behavior*, *51*(9–10), 1055–1081. https://doi.org/10.1177/0013916518774400

Berto, R. (2014). The role of nature in coping with psycho-physiological stress: a literature review on restorativeness. *Behavioral Sciences*, *4*(4), 394–409. https://doi.org/10.3390/bs4040394

Caffyn, A. (2012). Advocating and implementing slow tourism. *Tourism Recreation Research*, *37*(1), 77–80. https://doi.org/10.1080/02508281.2012.11081690

Calsius, J., Courtois, I., Feys, P., Van Asch, P., De Bie, J., & D'hooghe, M. (2015). "How to conquer a mountain with multiple sclerosis". How a climbing expedition to Machu Picchu affects the way people with multiple sclerosis experience their body and identity: A phenomenological analysis. *Disability and Rehabilitation*, *37*(26), 2393–2399. https://doi.org/10.3109/09638288.2015.1027003

Calsius, J., Van Den Noortgate, M., Roncada, G., Van Asch, P., & D'hooghe, M. (2019). Wandering through the desert with multiple sclerosis: How outdoor life recalibrates body awareness and self-identity. *Journal of Interdisciplinary Sciences*, *3*(1), 37–55.

Coble, T. G., Selin, S. W., & Erickson, B. B. (2003). Hiking alone: Understanding fear, negotiation strategies and leisure experience. *Journal of Leisure Research*, *35*(1), 1–22. https://doi.org/10.18666/jlr-2003-v35-i1-608

Conway, D., & Timms, B. F. (2010). Re-branding alternative tourism in the Caribbean: The case for 'slow tourism'. *Tourism and Hospitality Research*, *10*(4), 329–344. https://doi.org/10.1057/thr.2010.12

Crust, L., Keegan, R., Piggott, D., & Swann, C. (2011). Walking the walk: A phenomenological study of long distance walking. *Journal of Applied Sport Psychology*, *23*(3), 243–262. https://doi.org/10.1080/10413200.2010.548848

Csikszentmihalyi, M. (1975). *Beyond boredom and anxiety*. Jossey Bass.

Csikszentmihalyi, M. (1990). *Flow: The psychology of optimal experience*. Harper & Row.

Csikszentmihalyi, M. (1996). *Creativity: Flow and the psychology of discovery and invention*. Harper Collins.

D'souza, T. B., & Puttaraju, S. (2020). Examining the influence of trekking in nature on adaptive response to stress in individuals. *The International Journal of Indian Psychology*, 8, 2349–3429.

den Breejen, L. (2007). The experiences of long distance walking: A case study of the West Highland Way in Scotland. *Tourism Management*, *28*(6), 1417–1427. https://doi.org/10.1016/j.tourman.2006.12.004

Dickinson, J., & Lumsdon, L. (2010). *Slow travel and tourism*. Routledge.

Dickinson, J., & Lumsdon, L. (2012). Slow travel. In A. Holden & D. A. Fennell (Eds), *The Routledge handbook of tourism and the environment* (pp. 393–403). Routledge.

Eade, J. (2020). The invention of sacred places and rituals: A comparative study of pilgrimage. *Religions*, *11*(12), 649. https://doi.org/10.3390/rel11120649

Edensor, T. (2000). Walking in the British countryside: reflexivity, embodied practices and ways to escape. *Body & Society*, *6*(3–4), 81–106. https://doi.org/10.1177/1357034X00006003005

Egan, K. (2011). "I want to feel the Camino in my legs". Trajectories of walking on the Camino de Santiago. In A. Fedele & R. L. Blanes (Eds.), *Encounters of body and soul in contemporary religious practices: Anthropological reflections* (pp. 3–22). Berghahn Books.

Eichberg, H., Kosiewicz, J., & Contiero, D. (2017). Pilgrimage: Intrinsic motivation and active behavior in the eldery. Physical culture and sport. *Studies and Research*, *75*(1), 35–42. https://doi.org/10.1515/pcssr-2017-0019

Feliu-Soler, A., Pérez-Aranda, A., Luciano, J. V., Demarzo, M., Mariño, M., Soler, J., Van Gordon, W., García-Campayo, J., & Montero-Marín, J. (2021). Psychometric properties of the 15-item five facet mindfulness questionnaire in a large sample of Spanish pilgrims. *Mindfulness*, *12*(4), 852–862. https://doi.org/10.1007/s12671-020-01549-6

Fredrickson, L. M., & Anderson, D. H. (1999). A qualitative exploration of the wilderness experience as a source of spiritual inspiration. *Journal of Environmental Psychology*, *19*(1), 21–39. https://doi.org/10.1006/jevp.1998.0110

Frey, N. L. (2004). Stories of the return: Pilgrimage and its aftermaths. In E. Badone & S. Roseman (Eds), *Intersecting journeys: The anthropology of pilgrimage and tourism* (pp. 89–109). University of Illinois Press.

Horváth, A., & Szakolczai, A. (2017). *Walking into the void: A historical sociology and political anthropology of walking*. Routledge.

Jørgensen, N. N., Eade, J., Ekeland, T. J., & Lorentzen, C. (2020). The processes, effects and therapeutics of pilgrimage walking the St. Olav way. *The International Journal of Religious Tourism and Pilgrimage*, *8*(1), 33–50. https://doi.org/10.21427/v0cc-7135

Kaelber, L. (2006). Paradigms of travel: From medieval pilgrimage to the postmodern virtual tour. In D. Timothy & D. Olsen (Eds), *Tourism, religion and spiritual journeys* (pp. 65–79). Routledge.

Kaplan, R., & Kaplan, S. (1989). *The experience of nature*. A psychological perspective. Cambridge University Press.

Kaplan, S. (1995). The restorative benefits of nature: Toward an integrative framework. *Journal of Environmental Psychology*, *15*(3), 169–182. https://doi.org/10.1016/0272-4944(95)90001-2

Kato, K., & Progano, R. N. (2017). Spiritual (walking) tourism as a foundation for sustainable destination development: Kumano-kodo pilgrimage, Wakayama, Japan. *Tourism Management Perspectives*, *24*, 243–251. https://doi.org/10.1016/j.tmp.2017.07.017

Kay, G., & Moxham, N. (1996). Paths for whom? Countryside access for recreational walking. *Leisure Studies*, *15*(3), 171–183. https://doi.org/10.1080/026143696375594

Khan, S. (2015). How slow is 'slow'? Dilemmas of Slow Tourism. *TEAM Journal of Hospitality and Tourism*, *12*(1), 39–49. https://www.researchgate.net/profile/Khan-Sonia/publication/327261306_How_slow_is_'slow'_Dilemmas_of_slow_tourism/links/5b895f3992851c1e123f49ec/How-slow-is-slow-Dilemmas-of-slow-tourism.pdf

Krakauer, J. (2018). *Into the wild*. Pan Macmillan.

Lesser, I. A., Prystupa, J., Belanger, L., Thomson, C., & Nienhuis, C. P. (2020). A mixed-methods evaluation of a group based trail walking program to reduce anxiety in cancer survivors. *Applied Cancer Research*, *40*(10). https://doi.org/10.1186/s41241-020-00094-x

Lijster, T., & Celikates, R. (2019). Beyond the echo chamber: An interview with Hartmut Rosa. *Krisis Journal for Contemporary Philosophy*, *1*.

Lowry, L. L., & Lee, M. (2016). CittaSlow, slow cities, slow food: Searching for a model for the development of slow tourism. *Travel and Tourism Research Association, 40*. https://scholarworks.umass.edu/ttra/2011/Visual/40/

Lumsdon, L. M., & McGrath, P. (2011). Developing a conceptual framework for slow travel: A grounded theory approach. *Journal of Sustainable Tourism, 19*(3), 265–279. https://doi.org/10.1080/09669582.2010.519438

Mau, M., Aaby, A., Klausen, S. H., & Roessler, K. K. (2021b). Are long-distance walks therapeutic? A systematic scoping review of the conceptualization of long-distance walking and its relation to mental health. *International Journal of Environmental Research and Public Health, 18*(15), 7741. https://doi.org/10.3390/ijerph18157741

Mau, M., Nielsen, D. S., Jakobsen, I. S., Klausen, S. H., & Roessler, K. K. (2021a). Mental movements: How long-distance walking influences reflection processes among middle-age and older adults. *Scandinavian Journal of Psychology, 62*(3), 365–373. https://doi.org/10.1111/sjop.12721

Niedermeier, M., Einwanger, J., Hartl, A., & Kopp, M. (2017a). Affective responses in mountain hiking - A randomized crossover trial focusing on differences between indoor and outdoor activity. *PLoS One, 12*(5), e0177719. https://doi.org/10.1371/journal.pone.0177719

Niedermeier, M., Grafetstätter, C., Hartl, A., & Kopp, M. (2017b). A randomized crossover trial on acute stress-related physiological responses to mountain hiking. *International Journal of Environmental Research and Public Health, 14*(8), 905. https://doi.org/10.3390/ijerph14080905

Niedermeier, M., Grafetstätter, C., Kopp, M., Huber, D., Mayr, M., Pichler, C., & Hartl, A. (2019). The role of anthropogenic elements in the environment for affective states and cortisol concentration in mountain hiking - a crossover trial. *International Journal of Environmental Research and Public Health, 16*(2), 290. https://doi.org/10.3390/ijerph16020290

Nordbø, I., & Prebensen, N. K. (2015). Hiking as mental and physical experience. *Advances in Hospitality and Leisure, 11*, 169–186. https://doi.org/10.1108/S1745-354220150000011010

Nordbø, I., Engilbertsson, H. O., & Vale, L. S. R. (2014). Market myopia in the development of hiking destinations: The case of Norwegian DMOs. *Journal of Hospitality Marketing & Management, 23*(4), 380–405. https://doi.org/10.1080/19368623.2013.827608

Oh, H., Assaf, A. G., & Baloglu, S. (2014). Motivations and goals of slow tourism. *Journal of Travel Research, 55*(2), 205–219. https://doi.org/10.1177/0047287514546228

Ólafsdóttir, R., & Runnström, M. C. (2013). Assessing hiking trails condition in two popular tourist destinations in the Icelandic highlands. *Journal of Outdoor Recreation and Tourism, 3*, 57–67. https://doi.org/10.1016/j.jort.2013.09.004

Olsen, D. H. (2013). A scalar comparison of motivations and expectations of experience within the religious tourism market. *The International Journal of Religious Tourism and Pilgrimage, 1*(1), 41–61. https://doi.org/10.21427/D7ZQ51

Pretty, J. (2004). How nature contributes to mental and physical health. *Spirituality and Health International, 5*(2), 68–78. https://doi.org/10.1002/shi.220

Reader, I. (2007). Pilgrimage growth in the modern world: Meanings and implications. *Religion, 37*(3), 210–229. https://doi.org/10.1016/j.religion.2007.06.009

Redick, K. (2016). Spiritual rambling: Long distance wilderness sojourning as meaning-making. *Journal of Ritual Studies, 30*(2), 41–51. https://www.jstor.org/stable/44735510

Redick, K. (2021). Kenotic walking, wilderness sojourning, and hospitality. *Journal for the Study of Religious Experience, 7*(2), 114–139. https://rerc-journal.tsd.ac.uk/index.php/religiousexp/article/view/93

Rich, G. J. (2013). Finding flow: The history and future of a positive psychology concept. In J. D. Sinnott (Ed.), *Positive psychology: Advances in understanding adult motivation* (pp. 43–60). Springer Science + Business Media.

Saunders, R. E., Laing, J., & Weiler, B. (2013). Personal transformation through long-distance walking. In S. Felip & P. Pearce (Eds), *Tourist experience and fulfilment: Insights from positive psychology* (pp. 127–146). Routledge.

Sturm, J., Plöderl, M., Fartacek, C., Kralovec, K., Neunhäuserer, D., Niederseer, D., Hitzl, W., Niebauer, J., Schiepek, G., & Fartacek, R. (2012). Physical exercise through mountain hiking in high-risk suicide patients. A randomized crossover trial. *Acta Psychiatrica Scandinavica*, *126*(6), 467–475. https://doi.org/10.1111/j.1600-0447.2012.01860.x

Svarstad, H. (2010). Why hiking? Rationality and reflexivity within three categories of meaning construction. *Journal of Leisure Research*, *42*(1), 91–110. https://doi.org/10.1080/00222216.2010.11950196

Swann, C., Keegan, R. J., Piggott, D., & Crust, L. (2012). A systematic review of the experience, occurrence, and controllability of flow states in elite sport. *Psychology of Sport and Exercise*, *13*(6), 807–819. https://doi.org/10.1016/j.psychsport.2012.05.006

Thomassen, B. (2009). The uses and meaning of liminality. *International Political Anthropology*, *2*(1), 5–28. http://www.politicalanthropology.org/

Timms, B. F., & Conway, D. (2012). Slow tourism at the Caribbean's geographical margins. *Tourism Geographies*, *14*(3), 396–418. https://doi.org/10.1080/14616688.2011.610112

Timothy, D., & Olsen, D. (Eds). (2006). *Tourism, religion and spiritual journeys*. Routledge.

Turner, V. (1974). *Liminal to liminoid in flow, play and ritual: An essay in comparative symbology*. Rice Institute Pamphlet - Rice University Studies, 60, 3.

Turner, V., & Turner, E. (1978). *Image and pilgrimage in Christian culture*. Columbia University Press.

Ulrich, R. S. (1983). Aesthetic and affective response to natural environment. In I. Altman & J. F. Wohlwill (Eds), *Behavior and the natural environment* (pp. 85–125). Springer.

Ulrich, R. S., Simons, R. F., Losito, B. D., Fiorito, E., Miles, M. A., & Zelson, M. (1991). Stress recovery during exposure to natural and urban environments. *Journal of Environmental Psychology*, *11*(3), 201–230. https://doi.org/10.1016/S0272-4944(05)80184-7

van Gennep, A. (1909/1960). *The rites of passage*. University of Chicago Press.

Warren, A. (2020). *Hiking as nature's therapy*. [Doctoral dissertation]. Southern New Hampshire University. http://hdl.handle.net/10474/3587

Wilson, E. O. (2017). Biophilia and the conservation ethic. In D. J. Penn & I. E. Mysterud (Eds), *Evolutionary perspectives on environmental problems* (pp. 250–258). Routledge.

Wolf, I. D., & Wohlfart, T. (2014). Walking, hiking and running in parks: A multidisciplinary assessment of health and well-being benefits. *Landscape and Urban Planning*, *130*, 89–103. https://doi.org/10.1016/j.landurbplan.2014.06.006

Wöran, B., & Arnberger, A. (2012). Exploring relationships between recreation specialization, restorative environments and mountain hikers' flow experience. *Leisure Sciences*, *34*(2), 95–114. https://doi.org/10.1080/01490400.2012.652502

World Travel Market & Euromonitor (2007). *WTM global trends report* 2007. https://www.saphirnews.com/attachment/68715/

Yang, P., Dai, S., Xu, H., & Ju, P. (2018). Perceived environmental, individual and social factors of long-distance collective walking in cities. *International Journal of Environmental Research and Public Health*, *15*(11), 2458. https://doi.org/10.3390/ijerph15112458

Chapter 17

Tourism and Development Contact, Consumption, and Building Together

Jamie Gillen

Tourism and development share uncannily parallel roles in shaping global relations in the modern era. As a transnational industry, big 'D' development came of age following Bretton Woods through a consolidation of World War II's 'winners' into an unmatched political-economic powerhouse that charged itself with economically supporting the recovery of blitzed continents through a global financial system designed primarily by the United Kingdom and United States (Watts, 1993; Escobar, 1995). Along the way the allies created international development organizations like the World Bank, International Monetary Fund, and United Nations that cemented the categories of First, Second, Third, and Fourth-World nations into the global psyche (Escobar, 1995). Despite being destructive, distasteful, and outmoded designations, both development and tourism industries hinge on stratifying the world according to these categories that remain in place today (perhaps in more whispered but no less impactful form). The extent to which the development and tourism industries reproduce or challenge 'developed/developing' and centre/periphery oppositions through their products and marketing is subject to historical concern (some classics include Britton, 1982; Ferguson, 1990; Lacoste, 1977; Pigg, 1992; Silvey & Rankin, 2011; Simon, 1997; Watts 1993).

On the tourism industry side, while the Grand Tour preceded the World Wars and subsequent development machinery by more than a century, it set the stage for how Europe came to 'know' and classify itself and the 'other' (Towner, 1985). Although not labeled as such in the seventeenth century, it is not difficult to imagine that forerunners to categories such as First/Second/Third/Fourth world had their genesis in the Grand Tour. After all, 'the Low Countries' (modern day Belgium, Luxembourg, and the Netherlands) and 'exotic' destinations in southern Europe such as Italy and Spain (warm, laid back, mysterious) were included in most Grand Tour circuits (Towner, 1985).

Importantly for the purposes of connecting development and tourism it was not only British tourists who embarked on the 'Grand Tour' but Americans as well (Towner, 1985, p. 301). The 'special relationship' between the United Kingdom and United States, usually understood to exist in coordinated diplomatic, development, foreign policy, trade, and cultural

The Wiley Blackwell Companion to Tourism, Second Edition. Edited by C. Michael Hall.

realms, was also thickened through interlinked leisure travels that served as one important pretext to these countries' domination over the twentieth century development and tourism industries. Indeed, the goal of modernizing places and peoples has been fueled through the process of raising distinctions between 'us' and 'them' that are foundational to both development and tourism (Said, 1978). Said differently, it is crucial to point the finger at both development and tourism and dial in to their ongoing interconnectedness when we think about how inequality and the destructive dimensions of difference-making have played out over the past few centuries.

The point of this chapter is to present the argument that it is in part due to the Global North's intertwined production of development and tourism that narratives of 'haves' and the 'have-nots' remain so fundamental to the identity and practices of the world's places and peoples (Britton, 1982, 1991)[1]. These narratives are imperative to justifying and operationalizing actions meant to elevate, transform, develop, and modernize the Global South. And this project is more than an initiative intended to share 'best practices' with interested parties: discourses entangling tourism and development are for the governments and tourism operatives of the Global North narratives that confirm and buttress their authority over the world. The discussion in this chapter is maybe overly critical and certainly not celebratory but it is one that the world's sustained unevenness compels us to consider.

I showcase three moments in the tourism/development life cycles over the past fifty years to underpin my argument. The first moment of the tourism-development nexus is the concept of *contact*, a globalized form of interaction between the Global North and Global South that emerged in the aftermath of direct colonialism. This era, in the 1970s and 1980s, ushered in new forms of domination anchored in the language and practice of contact, which I treat as a one-way, relativist view of the world generated from the Global North that deems it possible to 'correct' colonial inequities by relativizing the past and making amends through selective, value-laden tourism investments.

A second moment that dominated the 1990s and 2000s sphere of development-tourism influence is *consumption* (Chua, 2003; Hart, 2001; Jackson, 1999; Mansvelt, 2008; Miller, 1991). This period embraced consumerism as a more ethical and reciprocal arrangement between the Global North and the Global South. In this era the commodification of local practices, beliefs, rituals, and goods was meant to 'save' hidden or eroding cultures by offering them to new audiences in the Global North.

Third is what I call *building together* (Li, 2007; Brickell, 2012), a joint quest for authenticity among hosts and guests that characterizes the present tourism development moment (Rickly, 2022; Wang, 1999). In this example hosts and guests agree to commit energy to a meaningful and lasting relationship by incorporating a holistic mutual examination of their 'true' selves and venturing out together to actualize it. The 'building together' exercise is intended to move beyond monetary exchange and treat development as reciprocal and focused on wellbeing, though this goal is largely unmet due to the misunderstandings over the host-guest relationship and more importantly the financial privilege needed to undertake a tour of this nature. The brief conclusion reflects on the main arguments in the chapter.

Contact

The documentary film *Cannibal Tours* (O'Rourke, 1988), perhaps the sole classic in the tourism studies oeuvre, stands out for how it weaves together the historical exploitation of colonialism with the ongoing trauma of tourism-led efforts to shape development according to Western tastes (Silverman, 2012)[2]. It introduces tourism in the aftermath of direct colonization (MacCannell, 1990), in this case in Papua New Guinea, which gained independence in 1975. Very briefly the film follows North American and European tourists along the Sepik River as they interact with local communities who seem at once unsurprised by their reappearance and resigned to follow through on the unspoken development contract set by tourists themselves: to perform 'backwardness' characterized by the local (and disputed) cultural practice of cannibalism. The film also features Papuans' frustration and confusion at their largely failed efforts to sell handicrafts to foreign tourists.

The film can be interpreted in a number of ways (Silverman, 2013) but I want to centre this section of the chapter on what is for me the most haunting section of the film: the introduction of a German tourist, speaking to himself on an audio recording device (dictaphone) in a mixture of German and English, reflecting on the history of German colonialism of the area and the cannibalism that occurs/ed there. Beginning at the 3:44 mark of the film the tourist waxes nostalgically about the impact of German colonialism when he says in German, 'They were good times with the Germans—yes?' After short mumbling by an unnamed respondent the man continues, 'Yes, I have also heard this about Africa, that they (Germans) were very popular' but 'lost the colonies in the First World War' (O'Rourke, 1988). The exchange is the first evidence in the film that tourism contact in this era facilitates postcolonialism, by which I mean a reconfiguration of the unequal power relations between the Global North and Global South steeped in past colonial encounters. The tourist in this passage promotes German colonial power as a benevolent (even fun) ruler that could or should be missed by colonized peoples. In this tourist's narrative the people of the German colonies had as much reason to lament Germany's losses in World War I as German citizens because it weakened them and forced them into a subservient position to the Allied 'winners'.

Later the film focuses on the transactional nature of contact between tourists and Papuans, one in which both parties feel misunderstood since the unequal commercial core of the post-colonial relationship ensures the distance created during the colonial enterprise is extended through tourism. These economic conflicts extend from earlier micro-scale colonial relationships. For Papuans the fact that tourists are not readily purchasing their handicrafts is a violation of the unwritten contract surrounding the commodification of their authentic and traditional ways. Whereas in the colonial era locals were conditioned to 'perform primitiveness' to receive objectionable terms of trade, financial aid, military/arms support, or protection, in the postcolonial era locals were expected to perform a 'staged authenticity' of material heritage in exchange for tourism dollars (MacCannell, 1973; MacCannell, 1990, p. 15). Tourists therefore took it upon themselves to 'delay their (locals') entry into the modern world', a world the tourists believe local Papuans are unprepared for (MacCannell, 1990, p. 15). The resultant emotions expressed in the film are bemusement on the part of tourists and anger and loathing among locals.

When considered against the international development project I want to suggest that *Cannibal Tours*—insofar as tourists dominate discourses that depict themselves as modern and Papuans as primitive—is representative of the tensions inherent to the contact between the Global North and Global South driven by the UK-US development machinery of the late twentieth century. The undertones of superiority in *Cannibal Tours* resemble the rationale driving 'big D' development's mandate to restructure the world according to its own free market principles (Blaikie, 2000; Escobar, 2001; Glassman, 2011; Willis & Kumar, 2009). The catch-all phrase 'neoliberalism' has been applied to this kind of development contact though one concept does not adequately capture the extraordinary transformations that occurred during this time, with states in the Global North pushing through tourism as a key driver to generate massive profits on the backs of labor and the environment in the Global South (Britton, 1982). That tourism is considered 'fun', 'apolitical', and 'win-win' only quickened its implementation by sidestepping time-consuming and costly bureaucratic and political hurdles. Development initiatives, of course, are often driven by similar "win-win" narratives.

For those critical tourism scholars making broader connections about global inequality while watching *Cannibal Tours*, the pressure on Papuans and many others in the Global South to submit to a narrow and destructive form of financial 'instruction' on how to develop better and faster has a clear history in earlier colonial eras and must have been easily recognizable to the intended recipients as a form of neocolonialism.

Zeroing in on broader critiques of the punitive discourses of development in an era of contact shows how closely it hews to the takeaway messages of host-guest engagement in *Cannibal Tours*. Escobar (1995) argues similarly when thinking about the conditions of global inequality: it is as important to pay attention to development's discursive fastening around oppositions like developed/underdeveloped and First World/Third World as it is to the actual development projects created by the Global North. He describes the second half of the twentieth century this way: 'countries started to see themselves as underdeveloped, how 'to develop' became a fundamental problem for them, and how they embarked upon the task of 'un-developing' themselves by subjecting their societies to increasingly systematic, detailed, and comprehensive interventions' (Escobar, 1995, p. 6).

During this era the 'successes' of both tourism and development were only possible if the rich and poor were clear about where they stood in relation to one another. With respect to tourism, the tourism industry of the 1970s and 1980s promised financial windfall if both tourists and locals agree to play out these discursive conditions of inequality on stage (Britton, 1982). The problem in *Cannibal Tours* is that both sides performed their roles well enough but the tourists did not live up to their end of the resultant financial bargain. And just as countries throughout the Global South were subsumed in even poorer financial conditions when structural adjustment loans and development aid ran out so too did the Papuans suffer even greater financial harm when the initial exoticism and allure that drew international tourists to Papua New Guinea collapsed and tourism dollars dried up (Silverman, 2012).

Consumption

Contact—an unadornedly raw and prejudiced type of tourism development characterizing 1970s and 1980s encounters—gave way in the 1990s and 2000s to a less abrasive form of relationalities centred on consumption (Goss, 2006; Jackson, 2004; Mansvelt, 2008, 2009; Su, 2011, 2015; Watson & Kopachevsky, 1994; Winter, 2009; Young & Markham, 2019). To be sure the reproduction of stereotypes surrounding exoticism and primitiveness ensured the binary separating the Global North and Global South remained intact, though in this era they were wrapped in a more easily digestible and arguably politically correct package of consumerism and commercialization.

From the perspective of tourism-driven inequalities, in the consumption era residents and communities of the Global South went from being in a subservient postcolonial position hoping that some basic cultural production would be enough to attract meager tourism dollars (what I call the contact period) to a place where their customary creative practices became for some significant financial assets (Crang, 2015; Erb & Ong, 2017; Oakes, 2013, 2016; Richter, 1999; Smith, 2006; Su, 2015; Winter, 2008). This shift was primarily due to the intensification of 'time space compression' (Harvey, 1989). Examples include the rapidly decreasing costs of air travel; the rise of the internet and the corresponding exposure of 'isolated' communities to wealthy urbanites revolted by the fakery associated with modernity and thus drawn to more authentic 'premodern' cultures (resonant of MacCannell's, 1976, thesis); more temporal distance from the outright devastation caused by colonialism; a growing interest in preservation and heritage; plain curiosity; competition/rivalry among the global financial elite; and most importantly the rise of investment opportunities in mass tourism consumption as an easy and purportedly mutually beneficial development strategy among investors and local communities.

The dominant thinking in the 1990s and 2000s development sphere went that tourism investment in the Global South was a sure thing and future-proofed. The reasons are manifold and were seen to have positive outcomes for all citizens of the Global South, including those not linked to tourism. Investment in upgrading local infrastructure to fulfill overseas tourist needs required substantial financial outlays but the use of inexpensive local materials and human resources meant costs would be lower than if similar projects were undertaken in the Global North. The construction of airports, roads, water and sewage improvements, and protections of nature demanded a large local workforce, were to benefit communities, and created a tourism-friendly environment, particularly when welcoming international tourists and investment (Dwyer, 2020; Hirsh, 2016; Logan & Molotch, 1987; Shatkin, 1998, 2008; Su & Li, 2021).

Wealthy elites and tourism boosters, always on the lookout for frontier markets, argued that accomplishing these grand plans would be much easier in the Global South than the Global North. After all, the requisite environmental impact assessments, health and safety protocols, and government oversight measures encumbering development projects in the Global North could be easily sidestepped amidst the 'lawlessness' of the Global South (Hoang, 2015, 2022; but see Gupta, 2012). 'Upgrades' in the Global South's (tourism) infrastructure were driven by intensive, opportunistic relationships between development experts, international capital, local governments, and on-the-ground knowledge brokers (Mitchell, 2002; Whitington, 2018). The labor market was fully stocked with locals who could fill industry roles as gardeners, builders, maintenance workers, craftspeople, guides, cooks, drivers, and the like. They would require minimal training as these jobs are low skill, low pay, seasonal, and usually lacking any benefits beyond monetary compensation, itself usually in cash and off the books (Chan, 2018; Gibson, 2009; Gmelch, 2003; Terry, 2009; Veijola & Jokinen, 1994).

While very much like how business gets done in the Global North, the sense of skirting the 'rules' of conventional capitalist relations whet the appetites of an ever-broader band of entrepreneurs and developers, leading to explosive growth in tourism destinations around the world. It is a core facet of this growth that the tourism industry grapples with infrastructure disaster after disaster (Calgaro & Lloyd, 2008; Ritchie, 2008; Ziegler et al., 2014) and heightened host-guest conflicts (Craven, 2016). 'Local' or 'cultural' problems/weaknesses usually received blame for these tragedies well before anyone thought to assign culpability to the structural conditions of global capital and entrepreneurial governance.

This is the juncture where tourist contributions to consumption come in. If in the 1990s and 2000s cultural producers began understanding that to engage in a market-based world meant commodifying their heritage and selling it to guests then tourists were gradually learning that earlier colonial-inspired tourist behavior of the kind we saw in *Cannibal Tours* is ethically repugnant and may limit their consumption options to the 'artificial', 'fake', or (gasp!) mass tourism parts of the tourism experience. That said, leaving 'foreign' societies alone is not an option in any era for tourists from the Global North who repackage basic development tropes like 'I want to help poor people' (e.g., volunteer tourism) or 'I want to take part and invest in X unique/mystical/authentic culture' (e.g., Bali, India) to moralize their spending (Mostafanezhad, 2014; Sin et al., 2015; Sin, 2009). Along the way tourists acquired an understanding that the easiest way to celebrate the 'Other' without making any judgements about them is to consume their tourism offerings.

Instead, divisive judgements have been lodged in the opposite direction, intended to criticize the Global North and laud the Global South. Foreign tourists carried with them stories that local cultures are more pure, untouched, happier, less needy, more community oriented, more stable, more skillful, and more interesting than their counterparts in the Global North (Conran, 2011; Hall, 2007, 2008). Said differently, instead of making demands on local hosts to conform to their shifting and unpredictable demands tourists in the consumption era were now to self-flagellate by narrating the weaknesses of their worlds and seeking atonement by investing in the goods and services of the Global South. The still lingering mantra from that era—take only photos and leave only footprints—captures the feigned innocence and implicit anti-Western critique many wealthy tourists in the Global North constructed at the time.

Speaking more broadly to the progression of tourism development interactions over time, rendering themselves invisible in the host-guest relationship outside of the 'apolitical' act of consumption has done very little to overcome Global North/Global South, developed/developing, and wealthy/poor binaries that tourists may have wished to accomplish on their trips. This is because the underlying conditions of tourism development inequality are produced by capital and baked into even the most well-meaning of tourists: arbitrary and unpredictable expenditures on places and people that follow trends and fads rather than need or local interest in developing a tourism market; postcolonial path dependency (e.g., the British penchant to holiday in the commonwealth countries of the Caribbean); a tendency to 'bundle' time and investment to attain highest returns (e.g., lumping a number of different places and features into one homogenous whole, like Southeast Asian countries or Latin American countries), and many others. The failures of the tourist consumption era to create an equitable world has led to a more dialogic, relational approach favored today, one that utilizes a shared quest for authenticity based on more meaningful host-guest experiences. In this era development initiatives are orchestrated to conform to individual tourist interests so that they can find deeper meaning beyond monetary exchange and the commodification of difference.

Building Together

Authenticity has shaped critical tourism studies and marks host-guest encounters from the 2010s to present (Brown, 2013; Cohen & Cohen, 2012a, 2012b; Gillen, 2016; Knudsen et al., 2016; Mkono, 2020; Rickly, 2022; Wang, 1999; West & Carrier, 2004). Tourism today is associated with finding personal fulfillment—our authentic or 'true selves'—through the work of direct engagement between hosts and guests. In this arrangement development initiatives underpinning "building together" are much more than financial improvement, rather they constitute a mutually beneficial holistic learning journey designed to embolden and uplift both parties in a spirit of meaning-making and compromise (Notar, 2006). Non-profit organizations, corporate social responsibility packages, and 'do-gooder' tour groups advertise personalized, boutique visits that have clear mandate to forge close connections among hosts in the Global South and guests from the Global North. Thought to be a correction to earlier eras in which tourists dominated the conditions of a transactional, exploitative relationship, the quest for authenticity is for many tourists and hosts meant to be a two-way street: tourists and residents work together to get to know each other, understand each other's lives and desires, and exert some collaborative energy to make these dreams a reality (Kim & Jamal, 2007). The quest to find existential authenticity collaboratively requires significant emotional energy—awareness, empathy, and sacrifice among them—and has led to a novel version of tourism as a 'feel good industry' (Steiner & Resinger, 2006).

The development and tourism industries of recent years share a motivation to look inward as much as gazing at the 'other' by firstly asking questions about their respective roles (direct or not) in perpetuating cultural homogeneity, cultural 'freezing', economic inequality, and ecocide (Neef, 2021). Their response—resulting from critiques in the media, academia, some corners of the public sector, and interested citizen groups—has been to do development and tourism differently and equip themselves with more self-awareness, more reflection about explicit and implicit bias, and incorporate sensitivities to the agency, interests, and decisions of communities outside of their own when generating their plans. It seems tourism and development are now as much in the business of improving themselves as using their resources to address impoverishment in other places and societies (Sin et al., 2015).

The notion of 'building together' is a shorthand for the kind of excursions that dominate much of the contemporary tourism landscape (Scheyvens, 2011). The uneasy mix of undertaking tourism for self-interested reasons and seeking out tourism activities that support local needs and sensibilities results in some unfamiliar arrangements and outcomes (Thompson, 2022). From a critical perspective 'building together' requires a suspension of reality on the part of host and guest. Spending money for the tourist becomes a way to support mutual development and is thought to be a means to the greater end of mutual care and bridging 'us' versus 'them' binaries (Oakes 2013). The host must present themselves as involved in tourism for more than its financial benefits and tourism cannot be seen as a simply an occupation for guides, hosts, brokers, even drivers and cooking staff.

After all and thinking back to my earlier point made in the 'contact' section, many people in the Global South are attracted to tourism for its money-making potential and do not often enter into the tourism industry to become self-actualized. Some tourists flinch when their motivation to improve themselves and the world around them is reduced to monetary expenditures. Meanwhile local groups may be confused by tourists who want to establish a genuine relationship, visit their homes, involve themselves in local customs, and work alongside them rather than be introduced to the conventional tourism attractions, food, and performance. Hosts, most of whom are caring and want to do an exemplary job representing their business, family, community, and country run into problems when wealthy tourists from the Global North do not seem to understand that revenue rather than friendship lies at the heart of much of their interest in working in the tourism industry (Bianchi 2018).

Thus much of the boutique, existential tourism industry shepherding tourists from the Global North to the Global South is brokered by Western expatriates, diasporic communities who have returned to their homelands, and globe-trotting locals who have spent time overseas as a student or businessperson. Finding oneself on a tourism journey away from home is also expensive. 'Eco-friendly' accommodation is some of the most costly in the world and going on excursions like safaris, homestays, wellness retreats, and cycling/motorbike tours require a great deal of coordination and labor to implement. It should also be noted that MICE tourism (Meetings, Incentives, Conferences, and Exhibitions) has adopted wellness and reciprocity into their outings to far away destinations as well, normalizing and popularizing existential authenticity in big ticket tourism offerings.

An example that brings all of the facets of the 'building together' leisure moment together is Song Saa Private Island, a high-end, boutique tourist destination on the coast of Cambodia (https://www.songsaa-privateisland.com/, accessed 27 March 2024). Billed as an 'ethical luxury resort', its least expensive option (a one-bedroom jungle villa) starts at 890USD/night[3]. Featured activities for guests include snorkeling, yoga, kayaking, meditation, scuba diving, a nature treasure hunt, and Buddhist prayer. With respect to the tourism-development nexus, Song Saa also provides guests with an opportunity to give back to the local community by joining conservation efforts with members of its staff in the neighboring village of Prek Svay. This 'DIY' tactic to assist in development while on tour is one many resorts, hotels, and even package tours now feature for 'responsible' tourists from overseas (Liljeblad, 2020; Sin, 2010).

Due to its relative novelty the results of a 'building together' approach are incomplete and mixed. On the plus side reflexivity about the ways their industries shape the worlds they wish to support go a long way in improving the development and tourism industries' contributions to powerlessness (Hughes & Scheyvens, 2018; Nogues-Pedregal, 2019). Climate change and stewardship of the environment are taken much more seriously in tourism (and development) than they ever have before, though the carbon footprint brought on by long-haul air travel belies an increased environmental consciousness (Font et al., 2019; Higham & Font, 2020). And the irony about the rise of authenticity in tourism development is that its bedrock is capitalist exchange, with ever more remote or unique practices marked as 'exclusive' and commanding spectacular sums to participate in. These are sums that hardly ever 'trickle down' to rank and file tourism laborers. So despite what seems to be an honest interest in bridging differences through 'building together', the distinctions between the Global North and Global South are prolonged through tourism's most essential and longstanding mechanism, capital.

Conclusions

Epitomized through both the tourism and development industries, the late twentieth and early twenty-first century have seen refinement but certainly not separation from earlier colonial encounters between host and guest. The peoples and governments of the Global South experienced the force of Global North growth and development projects in wholesale if not universal ways and often defied them through counter-projects to globalization like the non-aligned movement, import-substitution initiatives, protective real estate and business operations maneuvers, restrictive visa conditions, and tightened border security. I would argue it is impossible to separate the tourism and development industries because their shared sensibilities have shaped relations between the Global North and Global South through time.

Outlined through the three eras above, I have difficulty seeing a borderless world of free movement of ideas, people, goods and services, and infrastructure facilitated through tourism and development. If the development industry has captured modernity's penchant for geographical categories, hierarchies, and oppositions like First World/Third World and developed/developing then tourism has implanted these divisions into the lived realities of host and guest over time. The glue of tourism projects—heritage, cultural performance, and tourism planning initiatives—are all contested outcomes of what, where, and who should be worthy of development's investments.

This chapter can be read as a skeptical take on the idea that tourism and development are tenets of globalization, a borderless world. This does not make the intersection of tourism and development any less interesting (quite the contrary!) or worthy of study. It is the unevenness of their entanglements that requires scholars of tourism (and perhaps of development?) to acknowledge that while a better, more just and more equitable world is certainly possible, the language and tools tourist and development practitioners in both the Global North and Global South have been using, even recently, to accomplish this ambitious goal have fallen short of their intentions.

Acknowledgement

Thank you to Sarah Turner for helping me clarify my ideas in the chapter. I also recognize inspiration from the late David Harrison's chapter on tourism and development in the first edition of this volume.

Notes

1. Other essential factors contributing to the divisions of the Global North and Global South not included here are public health, access to debt/credit, education, the purported importance of Gross National Product and Gross Domestic Product, and legions more.
2. The film is available here for viewing: https://www.youtube.com/watch?v=KUQ_8wl93HM, accessed 27 March 2024.
3. For comparison per capita Gross Domestic Product in Cambodia is 1, 590USD/year.

References

Bianchi, R. (2018). The political economy of tourism development: A critical review. *Annals of Tourism Research*, *70*, 88–102. https://doi.org/10.1016/j.annals.2017.08.005

Blaikie, P. (2000). Development, post-, anti-, and populist: a critical review. *Environment and Planning A*, *32*, 1033–1050. https://doi.org/10.1068/a3251

Brickell, K. (2012). Visual critiques of tourist development: Host-employed photography in Vietnam. *Tourism Geographies*, *14*(1), 98–116. https://doi.org/10.1080/14616688.2011.593187

Britton, S. G. (1982). The political economy of tourism in the third world. *Annals of Tourism Research*, *9*(3), 331–358. https://doi.org/10.1016/0160-7383(82)90018-4

Britton, S. (1991). Tourism, capital, and place: towards a critical geography of tourism. *Environment and Planning D: Society and Space*, *9*, 451–478. https://doi.org/10.1068/d090451

Brown, L. (2013). Tourism: A catalyst for existential authenticity. *Annals of Tourism Research*, *40*, 176–190. https://doi.org/10.1016/j.annals.2012.08.004

Calgaro, E., & Lloyd, K. (2008). Sun, sea, sand and tsunami: examining disaster vulnerability in the tourism community of Khao Lak, Thailand. *Singapore Journal of Tropical Geography*, *29*(3), 288–306. https://doi.org/10.1111/j.1467-9493.2008.00335.x

Chan, C. (2018). The politics of leisure and labor mobilities: discourses of tourism and transnational migration in Central Java, Indonesia. *Mobilities*, *13*(3), 325–336. https://doi.org/10.1080/17450101.2017.1356436

Chua, B. H. (2003). *Life is not complete without shopping: Consumption culture in Singapore*. NUS Press.

Cohen, E., & Cohen, S. A. (2012a). Authentication: Hot and cool. *Annals of Tourism Research*, *39*(3), 1295–1314. https://doi.org/10.1016/j.annals.2012.03.004

Cohen, E., & Cohen, S. A. (2012b). Current sociological theories and issues in tourism. *Annals of Tourism Research*, *39*(4), 2177–2202. https://doi.org/10.1016/j.annals.2012.07.009

Conran, M. (2011). They really love me! Intimacy in volunteer tourism. *Annals of Tourism Research*, *38*(4), 1454–1473. https://doi.org/10.1016/j.annals.2011.03.014

Crang, M. (2015). Travelling ethics: Valuing harmony, habitat and heritage while consuming people and places. *Geoforum*, *67*, 194–203. https://doi.org/10.1016/j.geoforum.2015.06.010

Craven, C. E. (2016)."Refusing to be toured: Work, tourism, and the productivity of "life" in the Colombian Amazon. *Antipode*, *48*(3), 544–562. https://doi.org/10.1111/anti.12208

Dwyer, M. (2020). "They will not automatically benefit": The politics of infrastructure development in Laos's Northern Economic Corridor. *Political Geography*, *78*, 1–12. https://doi.org/10.1016/j.polgeo.2019.102118

Erb, M., & Ong, C. E. (2017). Theming Asia: culture, nature, and heritage in a transforming environment. *Tourism Geographies*, *19*(2), 143–167. https://doi.org/10.1080/14616688.2016.1238501
Escobar, A. (1995). *Encountering development: The making and unmaking of the Third World*. Princeton University Press.
Escobar, A. (2001). Beyond the search for a paradigm? Post-development and beyond. *Development*, *43*(4), 11–14. https://doi.org/10.1057/palgrave.development.1110188
Ferguson, J. (1990). *The anti-politics machine: Development, depoliticization, and bureaucratic power in Lesotho*. Cambridge, Cambridge University Press.
Font, X., Higham, J., Miller, G., & Pourfakhimi, S. (2019). Research engagement, impact, and sustainable tourism. *Journal of Sustainable Tourism*, *27*(1), 1–11. https://doi.org/10.1080/09669582.2019.1560673
Gibson, C. (2009). Geographies of tourism: critical research on capitalism and local livelihoods. *Progress in Human Geography*, *33*(4): 527–534. https://doi.org/10.1177/0309132508099797
Gillen, J. (2016). Urbanizing existential authenticity: motorbike tourism in Ho Chi Minh city, Vietnam. *Tourist Studies*, 16(3), 258–275. https://doi.org/10.1177/1468797615618035
Glassman, J. (2011). Critical geography III: Critical development geography. *Progress in Human Geography*, *35*(5), 705–711. https://doi.org/10.1177/0309132510385615
Gmelch, G. (2003). *Behind the smile: The working lives of Caribbean tourism*. Indiana University Press.
Goss, J. (2006). Geographies of consumption: the work of consumption. *Progress in Human Geography*, *30*(2), 237–249. https://doi.org/10.1191/0309132506ph604pr
Gupta, A. (2012). *Red tape: Bureaucracy, structural violence, and poverty in India*. Duke University Press.
Hall, C. M. (Ed.) (2007). *Pro-poor tourism: Who benefits?* Channel View.
Hall, C. M. (2008). Pro-poor tourism: Do 'tourism exchanges benefit primarily the countries of the south'? *Current Issues in Tourism*, *10*(2 & 3): 111–118. https://doi.org/10.1080/13683500708668426
Hart, G. (2001). Development critiques in the 1990s: *culs de sac* and promising paths. *Progress in Human Geography*, *25*(4), 649–658. https://doi.org/10.1191/030913201682689002
Harvey, D. (1989). *The condition of postmodernity*. Basil Blackwell.
Higham, J., & Font, X (2020). Decarbonising academia: confronting our climate hypocrisy. *Journal of Sustainable Tourism*, *28*(1), 1–9. https://doi.org/10.1080/09669582.2019.1695132
Hirsh, M. (2016). *Airport urbanism: Infrastructure and mobility in Asia*. University of Minnesota Press.
Hoang, K. K. (2015). *Dealing in desire: Asian ascendancy, western decline, and the hidden currencies of global sex work*. University of California Press.
Hoang, K. K. (2022). *Spiderweb capitalism: How global elites exploit frontier markets*. Princeton University Press.
Hughes, E., & Scheyvens, R. (2018). Development alternatives in the Pacific: How tourism corporates can work more effectively with local communities. *Tourism Planning and Development*, *15*(5), 516–534. https://doi.org/10.1080/21568316.2018.1478881
Jackson, P. (1999). Commodity cultures: the traffic in things. *Transactions of the Institute of British Geographers*, *24*, 95–108. https://doi.org/10.1111/j.0020-2754.1999.00095.x
Jackson, P. (2004). Local consumption cultures in a globalizing world. *Transactions of the Institute of British Geographers*, *29*(2), 165–178. https://doi.org/10.1111/j.0020-2754.2004.00123.x
Kim, H., & Jamal, T. (2007). *Touristic quest for existential authenticity. Annals of Tourism Research*, *34*(1), 181–201. https://doi.org/10.1016/j.annals.2006.07.009
Knudsen, D. C., Rickly, J. M., & Vidon, E. S. (2016). The fantasy of authenticity: Touring with Lacan. *Annals of Tourism Research*, *58*, 33–45. https://doi.org/10.1016/j.annals.2016.02.003
Lacoste, Y. (1977). Self-critical reflections and critique of "a geography of underdevelopment". *Antipode*, *9*(3), 117–124. https://doi.org/10.1111/j.1467-8330.1977.tb00098.x
Li, T. M. (2007). *The will to improve: Governmentality, development, and the practice of politics*. Duke University Press.
Liljeblad, J. (2020). Tour guides and the transnational promotion of human rights: Agency, structure, and norm translators in responsible travel. *Tourist Studies*, *20*(3), 314–335. https://doi.org/10.1177/1468797620920991

Logan, J. R., & Molotch, H. L. (1987). The city as a growth machine. In S. Fainstein & S. Campbell (Eds.), *Urban fortunes: The political economy of place* (pp. 291–337). University of California Press.

MacCannell, D. (1973). Staged authenticity: Arrangements of social space in tourist settings. *American Journal of Sociology*, *79*(3), 589–603.

MacCannell, D. (1976). *The Tourist: A new theory of the leisure class*. University of California Press.

MacCannell, D. (1990). Cannibal tours. *Visual Anthropology Review*, *6*(2), 14–24.

Mansvelt, J. (2008). Geographies of consumption: citizenship, space and practice. *Progress in Human Geography*, *32*(1), 105–117. https://doi.org/10.1177/0309132507080623

Mansvelt, J. (2009). Geographies of consumption: the unman*age*able consumer? *Progress in Human Geography*, *33*(2), 264–274. https://doi.org/10.1177/0309132508094080

Miller, D. (1991). Material Culture and Mass Consumption. Blackwell.

Mitchell, T. (2002). *Rule of experts: Egypt, techno-politics, modernity*. University of California Press.

Mkono, M. (2020). Eco-hypocrisy and inauthenticity: Criticisms and confessions of the eco-conscious tourist/traveller. *Annals of Tourism Research*, *84*, 1–12. https://doi.org/10.1016/j.annals.2020.102967

Mostafanezhad, M. (2014). *Volunteer tourism: Popular humanitarianism in neoliberal times*. Routledge.

Neef, A. (2021). *Tourism, land grabs and displacement: The darker side of the feel-good industry*. Routledge.

Nogues-Pedregal, A.-M. (2019). Anthropological contributions to tourism studies. *Annals of Tourism Research*, *75*, 227–237. https://doi.org/10.1016/j.annals.2019.02.005

Notar, B. E. (2006). Authenticity anxiety and counterfeit confidence: Outsourcing souvenirs, changing money, and narrating value in reform-era China. *Modern China*, *32*(1), 64–98. https://doi.org/10.1177/0097700405282348

O'Rourke, D. (1988). Cannibal Tours. Papua New Guinea, Institute of Papua New Guinea Studios: 72 minutes.

Oakes, T. (2013). Heritage as improvement: Cultural display and contested governance in rural China. *Modern China*, *39*(4), 380–407. https://doi.org/10.1177/0097700412467011

Oakes, T. (2016). Villagizing the city: Turning rural ethnic heritage into urban modernity in southwest China. *International Journal of Heritage Studies*, *22*(10), 751–765. https://doi.org/10.1080/13527258.2016.1212387

Pigg, S. L. (1992). Inventing social categories through place: Social representation and development in Nepal. *Comparative Studies in Society and History*, *34*(3), 491–513. https://doi.org/10.1017/S0010417500017928

Ritchie, B. (2008). Tourism disaster planning and management: From response and recovery to reduction and readiness. *Current Issues in Tourism*, *11*(4): 315–348. https://doi.org/10.1080/13683500802140372

Richter, L. (1999). The politics of heritage tourism. In D. G. Pearce & R. Butler (Eds.), *Contemporary issues in tourism development* (pp. 108–126). Routledge.

Rickly, J. M., (2022). A review of authenticity research in tourism: Launching the Annals of Tourism Research Curated Collection on authenticity. *Annals of Tourism Research*, 92, 1–33. https://doi.org/10.1016/j.annals.2021.103349

Said, E. (1978). *Orientalism*. Routledge.

Scheyvens, R. (2011). The challenge of sustainable tourism development in the Maldives: Understanding the social and political dimensions of sustainability. *Asia Pacific Viewpoint*, *52*(2), 148–164. https://doi.org/10.1111/j.1467-8373.2011.01447.x

Shatkin, G. (1998). 'Fourth world' cities in the global economy: The case of Phnom Penh, Cambodia. *International Journal of Urban and Regional Research*, *22*(3), 378–393. https://doi.org/10.1111/1468-2427.00147

Shatkin, G. (2008). The city and the bottom line: urban megaprojects and the privatization of planning in Southeast Asia. *Environment and Planning A*, *40*(2), 383–401. https://doi.org/10.1068/a38439

Silverman, E. K. (2012). From *Cannibal Tours* to cargo cult: On the aftermath of tourism in the Sepik River, Papua New Guinea. *Tourist Studies*, *12*(2), 109–130. https://doi.org/10.1177/1468797612454511

Silverman, E. (2013). After *Cannibal Tours*: Cargoism and marginality in a post-touristic Sepik River society. *The Contemporary Pacific*, *25*(2), 221–257. https://doi.org/10.1353/cp.2013.0031

Silvey, R., & Rankin, K. (2011). Development geography: Critical development studies and political geographic imaginaries. *Progress in Human Geography*, *35*(5), 696–704.

Simon, D. (1997). Development reconsidered; New directions in development thinking. *Geografiska Annaler*, *79B*(4), 183–201. https://doi.org/10.1111/j.0435-3684.1997.00018.x

Sin, H. L. (2009). Volunteer tourism–"involve me and i will learn"? *Annals of Tourism Research*, *36*(3), 480–501. https://doi.org/10.1016/j.annals.2009.03.001

Sin, H. L. (2010). Who are we responsible to? Locals' tales of volunteer tourism. *Geoforum*, *41*, 983–992. https://doi.org/10.1016/j.geoforum.2010.08.007

Sin, H. L., Oakes, T., & Mostafanezhad, M. (2015). Traveling for a cause: Critical examinations of volunteer tourism and social justice. *Tourist Studies*, 15(2): 119–131. https://doi.org/10.1177/1468797614563380

Smith, L. (2006). *Uses of heritage*. Routledge.

Steiner, C. J., & Resinger, Y. (2006). Understanding existential authenticity. *Annals of Tourism Research*, *33*(2), 299–318. https://doi.org/10.1016/j.annals.2005.08.002

Su, X. (2011). Commodification and the selling of ethnic music to tourists. *Geoforum*, *42*(4), 496–505. https://doi.org/10.1016/j.geoforum.2011.03.006

Su, X. (2015). Urban entrepreneurialism and the commodification of heritage in China. *Urban Studies*, *52*(15): 2874–2889. https://doi.org/10.1177/0042098014528998

Su, X., & C. Li (2021). Bordering dynamics and the geopolitics of cross-border tourism between China and Myanmar. Political Geography, *86*, 1–10. https://doi.org/10.1016/j.polgeo.2021.102372

Terry, W. (2009). Working on the water: On legal space and seafarer protection in the cruise industry. *Economic Geography*, *85*(4), 463–482. https://doi.org/10.1111/j.1944-8287.2009.01045.x

Thompson, B. (2022). Ecotourism anywhere? The lure of ecotourism and the need to scrutinize the potential competitiveness of ecotourism developments. *Tourism Management*, 92, 1–13. https://doi.org/10.1016/j.tourman.2022.104568

Towner, J. (1985). The grand tour: A key phase in the history of tourism. *Annals of Tourism Research*, *12*(3): 297–333. https://doi.org/10.1016/0160-7383(85)90002-7

Veijola, S., & Jokinen, E. (1994). The body in tourism. *Theory, Culture & Society*, *11*, 125–151. https://doi.org/10.1177/026327694011003006

Wang, N. (1999). Rethinking authenticity in tourism experience. *Annals of Tourism Research*, *26*(2), 349–370. https://doi.org/10.1016/S0160-7383(98)00103-0

Watson, G. L., & Kopachevsky, J. P. (1994). Interpretations of tourism as commodity. *Annals of Tourism Research*, *21*(3), 643–660. https://doi.org/10.1016/0160-7383(94)90125-2

Watts, M. J. (1993). Development I: power, knowledge, discursive practice. *Progress in Human Geography*, *17*(2), 257–272. https://doi.org/10.1177/030913259301700210

West, P., & Carrier, J. G. (2004). Ecotourism and authenticity: Getting away from it all? *Current Anthropology*, *45*(4), 483–491.

Whitington, J. (2018). *Anthropogenic rivers: The production of uncertainty in Lao hydropower*. Cornell University Press.

Winter, T. (2009). The modernities of heritage and tourism: interpretations of an Asian future. *Journal of Heritage Tourism*, *4*(2), 105–115. https://doi.org/10.1080/17438730802366490

Winter, T. (2008). Post-conflict heritage and tourism in Cambodia: The burden of Angkor. *International Journal of Heritage Studies*, *14*(6), 524–539. https://doi.org/10.1080/13527250802503274

Willis, K., & Kumar, M. S. (2009). Development I. In R. Kitchin & N. Thrift (Eds.), *International encyclopedia of human geography* (pp. 111–116). Elsevier.

Young, M., & Markham, F. (2019). Tourism, capital, and the commodification of place. *Progress in Human Geography*, 44(2), 276–296. https://doi.org/10.1177/0309132519826679

Ziegler, A. D., Wasson, R. J., Bhardwaj, A., Sundriyal, Y. P., Sati, S. P., Juval, N., Nautiyal, V., Srivastava, P., Gillen, J., & Saklani, U. (2013). Pilgrims, progress, and the political economy of preparedness — the example of the 2013 Uttarakhand flood and Kedarnath disaster. *Hydrological Processes*, 28(24), 5985–5990. https://doi.org/10.1002/hyp.10349

Chapter 18

The Tourist Gaze 1.0, 2.0, 3.0 and 4.0[1]

Jonas Larsen

Introduction

Few concepts stemming from tourism research are more cited and employed than 'the tourist gaze' both within tourist studies and beyond. This impact is down to several partly interlinked factors. Firstly, it illustrates the importance of a good book title. John Urry's book *The Tourist Gaze* (1990, 2002; and 2011, with Jonas Larsen) is famous for the idea of the tourist gaze (even though it covers much ground). The concept and title nicely support each other and when we think of the tourist gaze, we automatically think of John Urry! Few other tourist scholars have had quite such a monopoly over a concept and a powerful brand. Secondly, Urry writes in a clear and fluid prose and since photographing, sightseeing, and sights are everywhere in the spaces of tourism, the notion makes sense intuitively. Thirdly, John Urry was much more than just a tourism scholar. Nor was he like the tourist who returned to the same place year after year. He was one of Europe's leading social scientists with an incredible eye for discovering new societal problems and research fields. Urry's sociological gaze was forever on the move; he was like an adventurous traveller who went to a new place before the hordes went there. But once he had reported from it, many would follow in his footsteps.

The publication of *The Tourist Gaze* in 1990 is an example of this. At that time tourism was not seen as a serious sociological matter and it was largely neglected despite its increasing social, cultural, and economic significance. There were some great tourism books at the time – most notably Dean MacCannell's *The Tourist: A New Theory of the Leisure Class* (1976) – but *The Tourist Gaze* must be read as an attempt to establish the sociology of tourism (in fact this was the book's title until the editor suggested otherwise). Since then, Urry has co-authored/edited several tourism books (Rojek & Urry, 1997; Bærenholdt et al., 2004; Sheller & Urry, 2004) and articles but he has mainly travelled elsewhere, writing about nature, mobilities, globalisation, complexity theory, cars, aeromobilities, mobile methods, places, climate change, and oil. Yet many of these debates have been incorporated into later versions of *The Tourist Gaze*. A new chapter was added in 2002 (Urry, 2002) while the 2011 or the

The Wiley Blackwell Companion to Tourism, Second Edition. Edited by C. Michael Hall.

3.0 version – where John Urry invited me on board as co-author – restructured and updated the previous versions and added three new chapters (Urry & Larsen, 2011). This came to be the last update of the Tourist Gaze as John Urry sadly passed away in 2016.

This chapter discusses the four moments or versions – which I call 1.0, 2.0, 3.0 and 4.0 – of the tourist gaze. The tourist gaze and its author(s) we may say were on the move. It gives an account of this double-sided movement. I pay particular attention to how these different moments of tourist gazes have been applied, or have influenced, tourism research but also how Urry (and Larsen) have incorporated new tourism ideas – such as embodiment and performance – and gazes identified by other tourist scholars in the conceptualisation and reconceptualisation of the tourist gaze. I also discuss much of the critique raised against it. In the final part – the *Tourist Gaze 4.0* – I speculate on subjects and discussions that John Urry and I might have included in a new edition.

1.0

One of the most influential thinkers helped the tourist gaze into being. Urry (1990) begins with writing that the tourist gaze is inspired by the French philosopher Michel Foucault's concept of the medic gaze. He cites Foucault (1976), who in the *Birth of the Clinic* argued that 'the clinic was probably the first attempt to order a science on the exercise and decisions of the gaze' (Foucault, 1976, p. 89 cited in Urry, 1990, p. 1). *The Tourist Gaze* is essentially about how tourism as a leisure activity and an industry is formed according to 'the exercise and decisions of the gaze' and employed by multiple societal agents and institutions. For Urry (1990), tourism is bound up with power relations to see and make things visible. Focusing on the gaze brings out how the organising sense in tourism is visual. And this mirrors the general privileging of the eye within the long history of Western societies, as discussed by Foucault (1976) and others. Tourism is a 'way of seeing' where business stages visual experiences and tourists consume them visually. This means that the notion of tourist gaze refers to the practice of gazing and how it is organised by powerful actors and its impact on destinations and locals.

Taking inspiration from Foucault (1976), Urry (1990) argues that gazes are organised and systematised across time and space. Therefore, 'the tourist gaze is a matter of socially and technologically patterned and learned "ways of seeing" ...' (Urry & Larsen, 2011, p. 2). Different kinds of gazes are authorised by various discourses. According to Urry (1990), these different discourses include '*health*, *group solidarity*, *pleasure and play*, *heritage and memory*, *education*, and *nation*' (Urry & Larsen, 2011, p. 19). While obviously not produced and sanctioned by a knowledge-generating institution and science (like the gaze of the medic), the varied gazes of tourists are discursively organised and circulated by many forms of media (e.g., visual art, books, television, the Internet, social media), professionals, and experts over time: romantic poets and painters, photographers, writers, bloggers, guides, local councils, experts in the heritage industry, travel agents, hotel owners, designers, tour operators, travel programs, Hollywood, the photography industry, tourism developers, architects, planners, tourism academics, and so on. This making of discourses and gazes is a form of institutional mediation by experts within which spectacle and surveillance overlap and power–knowledge relations are present (Dann, 1996; Hollingshead, 1999; Cheong & Miller, 2000; and McGregor, 2000 further develop this idea).

Urry (2002, p. 129) spoke of a 'vicious hermeneutic circle' to capture this discursive framing or choreographed nature of gazing. This circle suggests that tourists gaze upon places that they have already encountered as a text or image through a representational medium.

Gazing is about consuming and photographing signs or markers, so tourists are often framed before they begin framing the world with their vision and camera. This explains the many studies of how commercial images frame tourist gazes (e.g., Dann, 1996; Marckwick, 2001; Osborne, 2000; Waitt & Head, 2002). The tourist gaze is mediated by specific representations and technologies, and the guidebook reading and photograph-taking tourist is the personalisation of the tourist gaze. The coupling between photography and gazing is explored further by Taylor (1994) and Osborne (2000) (for more recent discussions, see Höckert, et al., 2018; Cilkin & Cizel, 2021).

If Foucault (1976) plays the disciplining part in *The Tourist Gaze*, then postmodernism, this intellectual hipster of the 1980–1990s, plays the playful one. Postmodernism involves dedifferentiation. It dissolves boundaries between high and low cultures, between scholarly or auratic art and popular pleasures, and between elite and mass forms of consumption. Postmodernism also undermines the distinction between representations and reality. We are said to live in an oversaturated image culture where representations are everywhere and often more exciting than reality. What we increasingly consume are signs or representations. Urry (1990) argues that the tourist gaze, even in the past, prefigures postmodernism 'because of its particular fusion of the visual, the aesthetic, the themed, the commercial and the popular' (Urry, 1990, p. 86). The gaze is not least postmodern because it implies that gazing tourists are absorbed by texts, images, and representational technologies when gazing upon the world. Urry (1990, p. 86) writes: 'Much of what is appreciated is not directly experienced reality but representations, particularly through the medium of photography. What people "gaze upon" are ideal representations of the view in question they internalise from various media'.

Finally, Urry (1990) writes that postmodernism, due to the many circulating images of distant places, undermines clear-cut distinctions between tourism and everyday life, and home and away. Postmodernism means people become everyday tourists, daily gazing upon the extraordinary places on TV screens: 'People are tourists most of the time, whether they are literally mobile or only experience simulated mobility through the incredible fluidity of multiple signs and electronic images' (Lash & Urry, 1994, p. 259). The notion of the tourist gaze captures both the consumption of places through mobile images and the actual travel to and embodied appreciation of specific places. The tourist gaze may be virtual and it may be corporeal, but it is always bound up with the pleasure of gazing upon places that are out of the ordinary. Indeed, tourists often travel to places that are made special through the media. Much subsequent work on movie-induced (Law et al., 2007) and postmodern tourism (Pretes, 1995) draws on Urry's (1990) understanding of mediatised gazing.

Tourists, we may say, travel in mediascapes as much as physical worlds and they largely appreciate them through the visual sense. This is the controversial and generalising part of the book that caused so much later debate (discussed below). In the article 'The tourist gaze 'revisited" published in 1992 (Urry, 1992), Urry clarifies that it would be absurd to argue that tourists only gaze and that the visual sense is the only sense employed by tourists. But he insists the visual sense is the master sense and that gazing is the emblematic and most widespread tourism practice. The popularity of the tourist gaze as a concept is in part down to the fact that Urry (1990, 1992) was bold enough and successful in arguing that tourism is largely about gazing. The tourist gaze belongs not just to the visual culture of tourism but also to its centre. Tourism is about organising places as *sights* and gazing is about 'consuming places' visually (Urry, 1995) as others have explored in relation to spectacular ecotourism (Ryan et al., 2000), polar bears (Ryan et al., 2000; Lemelin, 2006), Ground Zero (Lisle, 2004), and the 'hip-hop hood' (Xie et al., 2007).

2.0

The newness of the second version of *The Tourist Gaze* published in 2002 was the new chapter 'Globalising the gaze' (Urry, 2002, Chapter 8). It must be seen in relation to Urry's edited book *Touring Cultures* (with Chris Rojek; (Rojek & Urry, 1997)) and especially his new interest in mobilities and globalisation (see *Sociology Beyond Societies* (Urry, 2000) and *Global Complexity* (Urry, 2003)). In the latter, Urry (2003) formulates a sociological manifesto for our globalized age of mobile people, objects, images, information, and much more. In that new chapter, Urry (2002) writes that the world has become much more mobile and globalised, and this to an extent that he speaks of global tourism. He argues: 'there are. . . countless ways in which huge numbers of people and places get caught up within the swirling vortex of global tourism. There are not two separate entities, the 'global' and 'tourism' bearing some external connections with each other' (Urry, 2002, p. 144). Since the first edition, the Internet and mobile phones have become widespread and afforded what Urry (2002) terms 'virtual travel' and 'communicative travel'.

Urry (2002) notes how there are now many more screens now allowing virtual gazing. And there is more corporeal travel due to cheap flights, more airports, and more destinations. So Urry (2002) sums up the situation by saying that 'the amount of 'traffic'. . . has magnified over this last decade, there is no evidence that virtual and imaginative travel is replacing corporeal travel, but there are complex intersections between these different modes of travel that are increasingly de-differentiated from one another' (Urry, 2002, p. 141). The 'roaring nineties', we may say, was the decade of unprecedented levels of global mobility. The tourist gaze travelled abroad when especially package tours and budget airlines made air travel cheap and easy. In Europe, Berlin, Paris, Barcelona, and London became weekend playgrounds to what we might call the EasyJet generation. During that period, travel time and cost were reduced, and many places were within reach quickly and cheaply, so many people, especially in the affluent West, came to develop mobile leisure lifestyles and a cosmopolitan sensibility (see Larsen, 2014).

The 2.0 edition (Urry, 2002, pp. 142–143) further discusses how far more places are now part of global tourism. This includes very distant places and places thought to have little tourist appeal, such as places of war, disasters, and famine. This is part of a reflexive process whereby places and nations enter a new neoliberal global order where places compete to attract businesspeople and tourists (and cheap labour). Urry (2002) further notes that gazers increasingly emanate from many different countries. In particular, the new Asian middle classes visit places of the West. The tourist gaze is everywhere and socially much more widespread.

Then Urry (2002) responds to criticisms raised against the tourist gaze. This includes Veijola and Jokinen's (1994) assertion that the tourist gaze neglects the multisensuous tourist *body* and that there is a male basis to it (see also Pritchard & Morgan, 2000). Urry (2002) acknowledges this by highlighting that gazing involves corporeal movement and mobile bodies that encounter places multisensuously, often through movement (see Urry, 1999, on the multiple senses involved in tourism; Larsen, 2001, on the mobile travel glance). He discusses gazing in relation to embodiment and movement: 'the body senses as it *moves*. It is endowed with kinesthetics, the sixth sense that informs one what the body is doing in space through the sensations of movement registered in its joints, muscles, tendons and so on' (Urry, 2002, p. 152). The sense of touch is crucial here. The walker's feet on the mountain path and the climber's hands on a rock face are explored in *Bodies of Nature* (Macnaghten & Urry, 2000).

The 2.0 edition (see also Urry, 2007) also introduces the notions of 'corporeal proximity' and 'compulsion to proximity' to highlight that tourists travel long distances and pay much money to experience places *corporeally*. Virtual travel is a poor substitute for the sensuous richness of corporeal travel: 'To be there oneself is what is crucial in most tourism. . . . Co-presence then involves seeing or touching or hearing or smelling or tasting a particular place' (Urry, 2002, p. 154). For Urry (2003, pp. 164–165), 'much travel results from a powerful 'compulsion to proximity' that makes it 'obligatory, appropriate or desirable'.

3.0

Much happened between 2002 and 2011 when the third edition was published (Urry & Larsen, 2011). In the field of tourism studies there was a great deal of talk about replacing the tourist gaze with a new paradigm of performing (Perkins & Thorns, 2001). This paradigm suggests that the doings of tourism are physical or corporeal and not merely visual (see Edensor, 1998; Bærenholdt et al., 2004; Coleman & Crang, 2002; Pons et al., 2009; Haldrup & Larsen, 2010, among others). The tourist gaze was critiqued for reducing tourism to sight-*seeing*. Where are other senses, bodily experiences, and adventure, the critics asked?

Urry continued his work on developing the new mobilities paradigm, but now in a more critical and pessimistic tone that spelled out the many social, economic, and environmental problems that multiple mobilities generate, such as surveillance, overconsumption, and climate change. So, a new tourism paradigm challenged the tourist gaze as Urry became increasingly concerned about environmental problems to which the global tourist gaze contributes. These issues frame the 3.0 version that illuminates some of the embodied and *darker* sides of the tourist gaze (with inspiration from Hollingshead, 1999; Morgan & Pritchard, 2005; Elliott & Urry, 2010).

Urry and I argue that there are many similarities between the paradigms of gazing and performing (see Urry & Larsen, 2011, Chapter 8). Drawing in part on the performance paradigm, the 3.0 edition rethinks the tourist gaze by developing a relational understanding that recognises the integration of the senses in people's visual encounters with places (something Urry touched upon in the 2.0 version but did not develop much) (Urry & Larsen, 2011; see also Larsen & Urry, 2011). We demonstrate that gazing is embodied and involves the other senses as tourists may touch, stroke, walk on, or climb upon and even collect the object of the gaze while sights often require appropriate soundscapes and smellscapes. Gazers are not disembodied travelling eyes (see also Larsen & Urry, 2011) (for more recent discussions of embodiment and performance that departs from the tourist gaze, see Edensor (2020), Rickly (2021), and Everingham et al. (2021)).

The 3.0 version also examines the social relations of gazing (Urry & Larsen, 2011). Urry and Larsen (2011, pp. 201–205) state that tourists never just gaze upon places and things, but do so in the presence of other tourists and locals, as many tourist places are busy public places. They also show that much gazing involves negotiations between friends and family members about what to see, how to see it, and for how long (as Haldrup and Larsen (2003) show regarding family tourist photography). Whom we happen to gaze with can be as important to the quality of the experience as the object of the gaze (see also Larsen & Urry, 2011).

The 3.0 version also explores the *relational gazes* of gazers and gazes and how the eyes of *gazers* and *gazees briefly* meet when the tourist gaze is performed. Larsen and Urry (2011, p. 1120) explain how 'earlier moments of the tourist gaze stressed that tourists exercised much power over the places and locals become the 'mad one' behind bars, relentlessly gazed upon and photographed'. Indeed, scholars show that a pervasive tourist gaze can create feelings of being constantly watched and objectified (Maoz, 2006; Chhabra, 2010). Yet the 3.0 version also discusses how gazees are not necessarily passive and powerless but exercise power through their gaze (this is inspired by Maoz, 2006; see also Hollingshead, 1999; Cheong & Miller, 2000). The embodied 'gazer' is not invisible but gazed upon and evaluated through a so-called 'local gaze' and 'mutual gaze' (Wassler & Kirillova, 2019). Maoz (2006, p. 222) states that 'the local gaze is based on a more complex, two-sided picture, where both the tourist and local gazes exist, affecting and feeding each other, resulting in what is termed "the mutual gaze"' (for more recent studies of the mutual gaze, see Monterrubio (2019), Ranjan and Tuhina (2018), and Sroypetch (2016), for example). Locals return the gaze of tourists and consequently, tourists too can be objectified (although many are oblivious to this gaze). However, other studies reveal that many female tourists are aware that they are subject to a sexualised male gaze (Jordan & Aitchison, 2008). The uneasy awareness of being caught gazing is viewed in a different way with photography. Gillespie's (2006) notion of the reverse gaze (inspired by mutual gaze) highlights the discomfort tourists taking photographs feel when caught in the reverse gaze and positioned as voyeuristic gazer. Lastly, the 'intratourist gaze' (Holloway et al., 2011; Wang & Xu, 2016) highlights that tourists' influence and discipline each other's gaze (see also, Larsen & Svabo, 2014).

The 3.0 version ends dramatically. The final chapter, entitled '*Risk and futures*,' questions whether tourist travel under the regime of the tourist gaze may soon come to an end. Urry and Larsen (2011) consider the risks and damages of the tourist gaze to the environments being visited, of long-term climates, and of the supplies of oil that fuelled the tourist gaze over the mobile twentieth century. This also reflects that Urry was increasingly critical of the global tourist gaze, especially because of its disastrous impact upon the environment through resource extensive and polluting air travel, air-conditioned hotels, luxury consumption, and so on.

As the tourist gaze has gone global it has caused some significant new risks and institutional gazes. Terrorism is one such risk. Destinations for Western tourists and especially airports are the front line of the war on terror, where both terrorists and tourists are on the move. Airports have developed a new sophisticated panoptic surveillance gaze of cameras, face-recognition biometric cameras, iris recognition, and much more. So 'in order that one can enter paradise for a week, systems of personal security are morphing into a new Big Brother where gazing tourists are subject to omnipresent surveillance' (Urry & Larsen, 2011, p. 222).

The surveillance gaze is increasingly also in place in cities. In London, one is likely to be recorded over 300 times a day by CCTV cameras (Morgan & Pritchard, 2005). The flaneur was the forefather of the gazing tourist. Yet the flaneur's anonymity and liminality are now largely impossible: almost all movements are recorded, tracked, and databased: 'Tourists are now routinely captured by and subject to a powerful digital panoptic machine justified by the perceived risks of crime, violence, and terrorism' (Urry & Larsen, 2011, p. 223). The first versions of the tourist gaze highlighted the significance of tourism companies (like Thomas Cook) in affording safe journeys. Now so-called independent travellers always face the new surveillance gaze to secure a safe journey without terrorists! This is the new securitisation ordering of the tourist gaze.

Then Urry and Larsen (2011, pp. 228–231) turn to oil. What has oil to do with tourist gaze? For Urry and Larsen (2011), cheap and plentiful oil has been the foundation, the very resource that has allowed the tourist gaze to conquer the world. It has shipped tourist goods and tourists across nations and even continents. Whereas the previous editions have explored the social and discursive ordering of the tourist gaze, oil represents the resource ordering of it.

In his books on climate change and oil, Urry (2011, 2013) argues that modern societies and their industrial systems require a plentiful supply of oil. Cheap and plentiful oil has afforded cheap journeys and more people in the rich part of the world have become used to more frequent holidays abroad.

Whereas the 2.0 version speaks of the necessity of travel, in the 3.0 version, it is stressed how tourist mobility has become excessive, a consumer addiction (like much else in the consumer society), and how this binge mobility is incredibly destructive for the environment. Travel figures have soared and soared. There has been a move from low-carbon to high-carbon tourism as trips to neighbouring destinations are supplemented or replaced with far-away destinations (see Larsen, 2014). The tourist gaze 1.0 is largely about the rise and fall of British resorts and local tourists. The 3.0 version is about international tourists and faraway destinations. It ends with a polemic about Dubai, this neoliberal place of consumption excess and mega projects. Dubai's ambition was to be the number one luxury-consumer paradise in the most inhospitable of environments. Huge amounts of oil have been used to build the biggest and most spectacular islands, beaches, hotels, and attractions, to transport in and out many visitors and workers and provide cooled environments (including outdoor air conditioners and indoor skiing) in the desert. All that luxury was built with borrowed money and by thousands on thousands of poorly paid migrant labourers lured into a life of squalor and exploitation, hidden away from the tourist gaze (Urry & Larsen, 2011, pp. 237–240).

If 'business was as normal' then we would expect that Dubai and other tourist places to be physically travelled to for a long while. Yet, as Urry (2011, 2013) argued, this is not certain. Oil is running out and tourism travel is causing severe climate change (as also discussed by Scott et al., 2012). On top of that there was the financial crisis during 2008/2009 that resulted in huge debts and financial collapses. So, in the future, it seems probable that travel is likely to become increasingly expensive, which makes the long-term growth in travel and tourism less likely. Urry and Larsen (2011) saw signs that this bleak future had already arrived in Dubai and predicted that there would be *less* tourism in the future (Urry & Larsen, 2011). They end the book by asking:

> Is this history of the rise and fall of Dubai a forerunner of the history of the global present as in the next few decades the spreading of the tourist gaze comes to a shuddering halt or even a reversal beginning in an Arabian desert? Was the tourist gaze on a mass scale a feature of the twentieth century hubris that will gradually disappear once the oil begins to run down and sea levels rise further? The decline and fall of Dubai may thus be the start of a much more general decline in the significance of the tourist gaze. Will there still be a relatively widespread and common "tourist gaze" operating away in 2050? (Urry & Larsen, 2011, p. 240).

But the 3.0 edition also discusses some possible future scenarios of the gaze that are less harmful to the environment. One is 'local sustainability' and a localised gaze. It involves 'a dramatic global shift towards lifestyles and travel patterns more local and smaller in scale' (Urry & Larsen, 2011, p. 234): long-distance travel will be a source of low status, not unlike tobacco. People will realise that air travel damages the planet's health (Cohen et al., 2011). This scenario requires that people contest travel and the environmental ills that it generates, as well as articulating discourses that promote the *positive* outcomes of low-carbon holidays, not imagining that the distant is necessarily extraordinary and appealing. This ties into Dickinson and Lumsdon's (2010) idea of slow travel in tourism and many articles in the journal *Sustainable Tourism* and beyond. The exotic gaze needs to be dispensed with and flying needs to be heavily taxed. A marked *de*globalisation of the tourist gaze must take place so that many international tourism places and transport systems disappear. 'And when people do travel longer distances, this needs to be undertaken collectively and, where possible, by sustainable higher-speed trains' (Urry & Larsen, 2011, pp. 236–237).

Another – but related – scenario is that of digital gazing. One aspect of this scenario is that 'much travel would be replaced by virtual travel' (Urry, 2011, p. 151). So far, various types of screens have poorly substituted corporeal gazing. This scenario requires the innovation of communications that effectively stimulate the sensuous, emotional, and social richness of corporeal gazing and discourses that promote the joys of virtual sightseeing and meetings. Somehow, the Internet and software must substitute corporeal travel with virtual travel. The Internet, media and experts need to promote the pleasures of the nearby, strengthen localness and low-carbon travel (Urry & Larsen, 2011, pp. 233–237).

4.0

Much has happened to the world of tourism since 2011. However the notion of the tourist gaze is still highly cited, discussed, and used as a theoretical lens in tourist studies and beyond in relation to photography, sightseeing, marketing, destination development, tourist behaviour, social media, place images, experiences, performances, and much more. A young generation of scholars from universities in both the minority and majority world, combined with a broader focus on gazing outside the Western world and by non-Western tourists, add new perspectives and vibrancy to the discussions (see Lee, 2001; Shono et al., 2006; Chen et al., 2021; Li et al., 2021). This pushes the tourist gaze forward in all sorts of exciting directions that break with somewhat ethno-centric and male oriented accounts. The tourist gaze is very much alive (as demonstrated by comprehensive recent reviews (Lin & Fu, 2020; Samarathunga & Cheng, 2020; Samarathunga, 2022).

I now discuss a few issues that may have made the cut into the 4.0 edition. As discussed, the 3.0 edition questioned the long-term viability of global tourism. However, much of the tourism debate in the 2010s revolved around an *intensification* of tourism (and Dubai did not collapse either). For instance, according to many locals, some major European cities arguably became overrun by too many tourists and suffered from over-tourism (Milano et al., 2019). While the notion of over-tourism signifies the ever-increasing expansion of tourism, John Urry and I would probably have argued that as a discursive force, it contests and de-legitimises the tourist gaze with anti-tourism attitudes and an impervious 'local gaze' (Gössling et al., 2020a); in a time of over-tourism and tourism induced gentrification, tourists are less likely to be met with open arms or a submissive smile (Gant, 2016; Füller & Michel, 2014; Novy & Colomb, 2017). More broadly, as tourism spreads beyond typical tourism destinations, middle class urban inhabitants realise that they are not only tourists but also hosts; their neighbours rent out on Airbnb, restaurants are full of tourists and the neighbourhood may be promoted as an exciting new destination off the beaten track (as in Copenhagen where the DMO states that 'locals are the destination' in so-called 'localhoods', see Larsen (2019)).

The global tourist gaze and the right to roam the world are also de-legitimised by the public awareness that flying is exceptionally bad for the environment. 'Flight shame' (Coffey, 2019) is slowly catching up on the wider, especially younger, populations, so faraway tourists are increasingly met with a contesting local gaze or by fellow tourists that travel by train or bus. While flying is still the dominant regime, alternatives are gathering force. For instance, rail tourism has gained momentum, the mobile glance from the train window (Larsen, 2001) is back in fashion (Jensen & Larsen, 2021), with researchers, industries, and politicians talking about a rail renaissance (Wang et al., 2017). Making rail travel and other sustainable forms of travel competitive with flying is a major task for critical tourism researchers and organisations.

And then, as we all know, came the Covid-19 pandemic. It 'wiped out' the global tourist gaze with unprecedented, devastating speed and effect, with borders, airports, and cities closing to prevent the spread of this highly *mobile* virus; the war on terror gave way to a war against a virus. Foreign tourists were tested or banned from entering while locals were immobilised and isolated within their homes (Gössling et al., 2020b). The foreign tourist was no longer an anticipated guest, but an enemy, a potential threat. This undermined the accept of the tourist gaze. As Korstanje and George (2021, p. 88) state, 'Tourists, who were seen as ambassadors of the prosperous capitalism, were suspiciously labelled as potential mortal disease carriers. In Urry's terms, the tourist gaze became a wicked gaze'. The tourist gaze did, at least momentarily, collapse, but for reasons other than described in the 3.0 edition.

The Covid-19 lockdowns simultaneously revitalised two other discussions in the previous editions. First, the lockdowns emboldened 'virtual meetings' or videoconferencing conducted through software platforms like Zoom and Teams and nourished peoples' skills in meeting and socialising on screens. What we might call the 'Zoom gaze' flourished during the pandemic. While the 'Zoom gaze' is still a somewhat poor substitute for physical meetings in terms of sociality and body language, many have experienced that it can indeed be a good enough substitute for at least some physical meetings. Virtual meetings are attractive to employees and organisations since they do not require costly, time-consuming, and polluting travel (Standaert et al., 2022). Second, the Covid-19 restrictions on foreign travel also boosted localised forms of tourism. For instance, cottage owners fled the locked-down cities and prices for cottages skyrocketed in many places. Moreover, those vacationing in their own region or country discovered unknown places of interest and beauty in their backyard. When coupled with discourses of flight shaming and homeworking, one lasting effect of the COVID-19 lockdown may be less business travel and more localised forms of tourism (Becken & Hughey, 2022). This would be a welcome transition towards less carbon-intensive forms of travel.

However, powerful forces work to get tourism back on the wing again, securing a return to the old normal (Gössling & Schweiggart, 2022). Many tourists are longing for exotic beaches and warmer climates, and they are ready to fly, perhaps despite their nagging flight shame. They are encouraged by public institutions and organisations. For instance, while the City of Copenhagen aims to be carbon neutral by 2025 and has launched a tourism recovery plan where sustainability is a key term, the reintroduction of lost international flights and the return of international tourists also play a significant part in that plan (Wonderful Copenhagen, 2022). There are indeed many paradoxes involved and different interests are at stake when it comes to the future of global tourism.

Conclusion

The tourist gaze is on the move. This is in part because how the tourist gaze is legitimised and performed is contingent on larger social, cultural, economic, environmental and technical discourses and changes. It also changes because John Urry and the wider community of gaze scholars constantly move(d) on and push(ed) the concept forward, shedding light on new aspects of tourism. The 1.0 version explored the social and cultural – the discursive – ordering of the tourist gaze. The 2.0 version turned to its global ordering. The 3.0 version was concerned with the embodied and resource ordering of the tourist gaze. Urry's take on and interest in tourist gaze shifted over the years. It has been gazed upon through changing theoretical lenses (for instance, visual theory, postmodernism, and the mobilities paradigm) and new empirical realities such as media saturation, globalisation, the peak of oil, and climate change. We have seen how the 3.0 and 4.0 editions question the sustainability and very existence of a widespread and globalised tourist gaze. These editions suggest that future research regarding the tourist gaze must explore ways in which it can be localised and de-exoticised as well as more virtual. There is a need to tame the global tourist gaze and invent more sustainable ways of gazing at the world.

Note

1. This chapter is a revised and expanded version of the chapter *The Tourist Gaze 1.0, 2.0, 3.0* (Larsen 2014). The section 4.0 is specifically written for this version of the article. Both chapters draw on Urry and Larsen (2011) and Larsen and Urry (2011).

References

Bærenholdt, J. O., Haldrup, M., Larsen, J., & Urry, J. (2004). *Performing tourist places*. Ashgate.

Becken, S., & Hughey, K. F. (2022). Impacts of changes to business travel practices in response to the COVID-19 lockdown in New Zealand. *Journal of Sustainable Tourism*, *30*(1), 108–127. https://doi.org/10.1080/09669582.2021.1894160

Chen, G., Shi, H., Li, Z., & Huang, S. S. (2021). Examining tourist gaze in a highly controlled destination: A study of Chinese tourists to North Korea. *Journal of Hospitality and Tourism Management*, *49*, 287–295. https://doi.org/10.1016/j.jhtm.2021.09.022

Cheong, S.-M., & Miller, M. (2000). Power and tourism: a Foucauldian observation. *Annals of Tourism Research*, *27*(2), 371–390. https://doi.org/10.1016/S0160-7383(99)00065-1

Chhabra, D. (2010). How they see us: perceived effects of tourist gaze on the old order Amish. *Journal of Travel Research*, *49*(1), 93–105. https://doi.org/10.1177/0047287509336475

Cilkin, R., & Cizel, B. (2021). Tourist gazes through photographs. *Journal of Vacation Marketing*, *28*(2), 188–210. https://doi.org/10.1177/13567667211038955

Coffey, H. (2019). Flygskam: what is the flight-shaming environmental movement that's sweeping Europe? *The Independent*. https://www.independent.co.uk/travel/news-and-advice/flygskam-anti-flying-flight-shaming-sweden-greta-thornberg-environment-air-travel-train-brag-tagskryt-a8945196.html

Cohen, S., Higham, J., & Cavaliere, C. (2011). Binge flying: Behavioural addiction and climate change. *Annals of Tourism Research*, *38*(3), 1070–1089. https://doi.org/10.1016/j.annals.2011.01.013

Coleman, S., & Crang, M. (Eds). (2002). Tourism: *Between place and performance*. Berghahn Books.

Dann, G. (1996). *The language of tourism*, CABI Publishing.

Dickinson, J., & Lumsdon, L. (2010). *Slow travel and tourism*. Routledge.

Edensor, T. (1998). *Tourists at the Taj*. Routledge.

Edensor, T. (2020). Seeking sensuous mobilities. In H. Endo (Ed.), *Understanding tourism mobilities in Japan*. (pp. 13–26). Routledge.

Elliott, A., & Urry, J. (2010). *Mobile lives*. Routledge.

Everingham, P., Obrador, P., & Tucker, H. (2021). Trajectories of embodiment in *Tourist Studies*. *Tourist Studies*, *21*(1), 70–83. https://doi.org/10.1177/1468797621990300

Foucault, M. (1976). *The birth of the clinic*. Tavistock.

Füller, H., & Michel, B. (2014). 'Stop being a tourist!' New dynamics of urban tourism in Berlin-Kreuzberg. *International Journal of Urban and Regional Research*, *38*(4), 1304–1318. https://doi.org/10.1111/1468-2427.12124

Gant, A. C. (2016). Holiday rentals: The new gentrification battlefront. *Sociological Research Online*, *21*(3), 112–120. https://doi.org/10.5153/sro.4071

Gillespie, A. (2006). Tourist photography and the reverse gaze. *Ethos*, *34*(3), 343–366. https://doi.org/10.1525/eth.2006.34.3.343

Gössling, S., & Schweiggart, N. (2022). Two years of COVID-19 and tourism: what we learned, and what we should have learned. *Journal of Sustainable Tourism*, *30*(4), 915–931. https://doi.org/10.1080/09669582.2022.2029872

Gössling, S., McCabe, S., & Chen, N. C. (2020a). A socio-psychological conceptualisation of overtourism. *Annals of Tourism Research*, *84*, 102976. https://doi.org/10.1016/j.annals.2020.102976

Gössling, S., Scott, D., & Hall, C. M. (2020b). Pandemics, tourism and global change: a rapid assessment of COVID-19. *Journal of Sustainable Tourism*, *29*(1), 1–20. https://doi.org/10.1080/09669582.2020.1758708

Haldrup, M., & Larsen, J. (2003). The family gaze. *Tourist Studies*, *3*(1), 23–46.

Haldrup, M., & Larsen, J. (2010). *Tourism, performance and the everyday: Consuming the Orient*. Routledge.

Höckert, E., Lüthje, M., Ilola, H., & Stewart, E. (2018). Gazes and faces in tourist photography. *Annals of Tourism Research*, *73*, 131–140. https://doi.org/10.1016/j.annals.2018.09.007

Hollingshead, K. (1999). Surveillance of the worlds of tourism: Foucault and the eye-of-power. *Tourism Management*, *20*(1), 7–23. https://doi.org/10.1016/S0261-5177(98)00090-9

Holloway, D., Green, L., & Holloway, D. (2011). The intratourist gaze: grey nomads and 'other tourists'. *Tourist Studies*, *11*(3), 235–252. https://doi.org/10.1177/1468797611432043

Jensen, M. T., & Larsen, J. (2021). Rail tourism transitions: A sociological framework. *Annals of Tourism Research*, *87*, 102992. https://doi.org/10.1016/j.annals.2020.102992

Jordan, F., & Aitchison, C. (2008). Tourism and the sexualisation of the gaze: solo female tourists experiences of gendered power, surveillance and embodiment. *Leisure Studies*, *27(3)*, 329–349. https://doi.org/10.1080/02614360802125080

Larsen, J. (2001). Tourism mobilities and the travel glance: experiences of being on the move. *Scandinavian Journal of Hospitality and Tourism*, *1(2)*, 80–98. https://doi.org/10.1080/150222501317244010

Larsen, J. (2014). The tourist gaze 1.0, 2.0, and 3.0. In A. A. Lew, C. M. Hall, A. M. Williams (Eds), *The Wiley Blackwell companion to tourism*, (pp. 304–313). Wiley-Blackwell.

Larsen, J. (2019). Ordinary tourism and extraordinary everyday life: Re-thinking tourism and cities. In T. Frisch, C. Sommer, L. Stoltenberg & N. Stors (Eds), *Tourism and everyday life in the contemporary city* (pp. 24–41). Routledge.

Larsen, J., & Svabo, C. (2014). The tourist gaze and 'family treasure trails' in museums. *Tourist Studies*, *14*(2), 105–125. https://doi.org/10.1177/1468797614532178

Larsen, J., & Urry, J. (2011). Gazing and performing. *Environment and Planning D: Society and Space*, *29*(6), 1110–1125. https://doi.org/10.1068/d21410

Lash, S., & Urry, J. (1994). *Economies of signs and space*. Sage.

Law, L., Bunnell, T., & Ong, C.-E. (2007). The beach, the gaze and film tourism. *Tourist Studies*, *7*(2), 141–164. https://doi.org/10.1177/1468797607083499

Lee, Y.-S. (2001). Tourist gaze: Universal concept. *Tourism Culture & Communication*, *3*(2), 93–99. https://doi.org/10.3727/109830401108750724

Lemelin, R. H. (2006). The gawk, the glance, and the gaze: ocular consumption and polar bear tourism in Churchill, Manitoba, Canada. *Current Issues in Tourism*, *9*(6), 516–534. https://doi.org/10.2167/cit294.0

Li, M., Tucker, H., & Chen, G. (2021). Chineseness and behavioural complexity: rethinking Chinese tourist gaze studies. *Tourism Review*, *77*(3), 841–858. https://doi.org/10.1108/TR-02-2021-0088

Lin, B., & Fu, X. (2020). Gaze and tourist-host relationship–state of the art. *Tourism Review*, *76*(1), 138–149. https://doi.org/10.1108/TR-11-2019-0459

Lisle, D. (2004). Gazing at ground zero: tourism, voyeurism and spectacle. *Journal for Cultural Research*, *8*(1), 3–21. https://doi.org/10.1080/1479758042000797015

Korstanje, M. E., & George, B. (2021). *Mobility and globalization in the aftermath of COVID-19: Emerging new geographies in a locked world.* Palgrave Macmillan.
MacCannell, D. (1976). *The tourist: A new theory of the leisure class.* University of California Press.
Macnaghten, P., & Urry, J. (Eds). (2000). Bodies of nature: Introduction. *Body and Society*, *6*(3–4), 1–11. https://doi.org/10.1177/1357034X00006003001
Maoz, D. (2006). The mutual gaze. *Annals of Tourism Research*, *33*(1), 221–239. https://doi.org/10.1016/j.annals.2005.10.010
Marckwick, M. (2001). Postcards from Malta: Image, consumption, context. *Annals of Tourism Research*, *28*(2), 417–438. https://doi.org/10.1016/S0160-7383(00)00049-9
McGregor, A. (2000). Dynamic texts and tourist gaze – death, bones and buffalo. *Annals of Tourism Research*, *27*(1), 27–50. https://doi.org/10.1016/S0160-7383(99)00034-1
Milano, C., Novelli, M., & Cheer, J. M. (2019). Overtourism and tourismphobia: A journey through four decades of tourism development, planning and local concerns. *Tourism Planning & Development*, *16*(4), 353–357. https://www.tandfonline.com/doi/full/10.1080/21568316.2019.1599604
Monterrubio, C. (2019). Hosts and guests' social representations of nudism: A mutual gaze approach. *Annals of Tourism Research*, *75*, 18–28. https://doi.org/10.1016/j.annals.2018.12.011
Morgan, N., & Pritchard, A. (2005). Security and social 'sorting: Traversing the surveillance–tourism dialectic. *Tourist Studies*, *5*(2), 115–132. https://doi.org/10.1177/1468797605066923
Novy, J., & Colomb, C. (2017). Urban tourism and its discontents: An introduction. In J. Novy & C. Colomb (Eds), *Protest and resistance in the tourist city* (pp. 1–30). Routledge.
Osborne, P. (2000). *Travelling light. Photography, travel and visual culture.* Manchester University Press.
Perkins, H. C., & Thorns, D. C. (2001). Gazing or performing? Reflections on Urry's Tourist Gaze in the context of contemporary experience in the Antipodes. *International Sociology*, *16*(2), 185–204. https://doi.org/10.1177/0268580901016002004
Pons, P. O., Crang, M., & Travlou, P. (Eds). (2009). *Cultures of mass tourism: doing the Mediterranean in the age of banal mobilities.* Ashgate.
Pretes, M. (1995). Postmodern tourism: the Santa Claus industry. *Annals of Tourism Research*, *22*(1), 1–15. https://doi.org/10.1016/0160-7383(94)00026-O
Pritchard, A., & Morgan, N. (2000). Privileging the male gaze: gendered tourism landscapes. *Annals of Tourism Research*, *27*(4), 884–905. https://doi.org/10.1016/S0160-7383(99)00113-9
Ranjan, B., & Tuhina, G. (2018). Situating the tourist gaze: from appropriation to negotiation. *Current Issues in Tourism*, *21*(6), 599–615. https://doi.org/10.1080/13683500.2015.1110118
Rickly, J. M. (2021). Performing beyond the view: Embodiment and tourist experiences. In R. Sharpley (Ed.), *Routledge handbook of the tourist experience* (pp. 140–152). Routledge.
Rojek, C., & Urry, J. (Eds). (1997). *Touring cultures:* Transformations of travel and theory. Routledge.
Ryan, C., Hughes, K., & Chirgwin, S. (2000). The gaze, spectacle and ecotourism. *Annals of Tourism Research*, *27*(1), 148–163. https://doi.org/10.1016/S0160-7383(99)00061-4
Samarathunga, W. H. M. S. (2022). Gaze. In D. Buhalis (Ed.), *Encyclopedia of tourism management and marketing* (forthcoming). Edward Elgar Publishing.
Samarathunga, W. H. M. S., & Cheng, L. (2020). Tourist gaze and beyond: state of the art. *Tourism Review*, *76*(2), 344–357. https://doi.org/10.1108/TR-06-2020-0248
Scott, D., Hall, C. M., & Gössling, S. (2012). *Tourism and climate change: Impacts, adaptation & mitigation.* London.
Sheller, M., & Urry, J. (Eds). (2004). *Tourism mobilities: Places to play, places in play.* Routledge.
Shono, S., Fisher, D., & McIntosh, A. (2006). The changing gaze of Japanese tourists. *Tourism Review International*, *9*(3), 237–246. https://doi.org/10.3727/154427206776886241
Sroypetch, S. (2016). The mutual gaze: Host and guest perceptions of socio-cultural impacts of backpacker tourism: A case study of the Yasawa Islands, Fiji. *Journal of Marine and Island Cultures*, *5*(2), 133–144. https://doi.org/10.1016/j.imic.2016.09.004
Standaert, W., Muylle, S., & Basu, A. (2022). Business meetings in a postpandemic world: When and how to meet virtually. *Business Horizons*, *65*(3), 267–275. https://doi.org/10.1016/j.bushor.2021.02.047

Taylor, J. (1994). *A dream of England: Landscape, photography and the tourist's imagination*. Manchester University Press.

Urry, J. (1990). *The tourist gaze: Leisure and travel in contemporary societies*. Sage Publications.

Urry, J. (1992). The tourist gaze "revisited". *American Behavioral Scientist*, *36*(2), 172–186. https://doi.org/10.1177/0002764292036002005

Urry, J. (1995). *Consuming places*. Routledge.

Urry, J. (1999). Sensing leisure spaces. In D. Crouch (Ed.), *Leisure/tourism geographies: Practices and geographical knowledge* (pp. 34–45). Routledge.

Urry, J. (2000). *Sociology beyond societies: Mobilities for the twenty-first century*. Routledge.

Urry, J. (2002). *The tourist gaze*, (2nd ed.) Sage Publications.

Urry. J. (2003). *Global complexity*. Polity.

Urry, J. (2007). *Mobilities*. Polity.

Urry, J. (2011). *Climate change and society*. Polity Press.

Urry, J. (2013). *Societies beyond oil*: *Oil dregs and social futures*. Zed Books.

Urry, J., & Larsen, J. (2011). *The tourist Gaze 3.0* (3rd ed.) Sage.

Veijola, S., & Jokinen, E. (1994). The body in tourism. *Theory, Culture and Society*, *11*(3), 125–151. https://doi.org/10.1177/026327694011003006

Waitt, G., & Head, L. (2002). Postcards and frontier mythologies: sustaining views of the Kimberley as timeless. *Environment and Planning D: Society and Space*, *20*(3), 319–344. https://doi.org/10.1068/d269t

Wang, H., & Xu, S. (2016). Intratourist moral gaze and its significance for discipline: based on content analysis of online text. *Tourism Tribune*, *31*(5), 45–49.

Wang, K., Xia, W., Zhang, A., & Zhang, Q. (2017). *Effects of high-speed rail speed on airline demand and price: Theoretical analysis and empirical evidence from a quasi-natural experiment*. Sauder School of Business, University of British Columbia.

Wassler, P., & Kirillova, K. (2019). Hell is other people? An existential-phenomenological analysis of the local gaze in tourism. *Tourism Management*, *71*, 116–126. https://doi.org/10.1016/j.tourman.2018.10.005

Wonderful Copenhagen (2022). *Comeback Copenhagen* 2023. https://www.wonderfulcopenhagen.dk/wonderful-copenhagen/genopretningsplan/comeback-copenhagen-2023

Xie, P., Osumare, H., & Ibrahim, A. (2007). Gazing the hood: Hip-hop as tourism attraction. *Tourism Management*, *28*(2), 452–460. https://doi.org/10.1016/j.tourman.2006.03.009

Chapter 19

What's Left to Say? Late Modern Ruminations on Tourism and Critical Social Theory

Claudio Minca & Tim Oakes

Returning to texts that we have authored ten or more years earlier is always an odd and somewhat unsettling experience. This possibly has to do with the fact that texts do have a life of their own and, after some time, become strangely disentangled from us. It also has to do with the obvious fact that we change, as individuals and as academics. Those pages drafted in a different time of our personal and professional lives may look like a strange evocation of how we positioned ourselves compared to our respective interpretative communities in a specific conjunction of time and space. However, those texts remain there, unchanged, and produce an unintended inertia that holds the authors responsible for what they wrote and for what they thought while writing.

In the first edition of this fortunate and impactful *Companion* we were asked to write a chapter on tourism and the post/modern. We did so by critically engaging with the existing literature focused on different conceptualisations of modernity, and emphasising how this approach could reveal so much about the deep ties between tourism, travel theory and the modern, including its colonial and postcolonial implications. Ten years later, in the second edition of this volume, we revised some of our claims and tried to engage with what, in the meantime, had become a far more sophisticated and articulated field of tourism studies. We did this by acknowledging the field's extraordinary achievements (including that of taking tourism seriously as a field of critical inquiry):

> Tourism is indeed 'real' [. . .] as both a constitutive force in the social world, and a set of actual practices that challenge conventional modern categorizations or disembodied abstractions. This reading of tourism is thus focused on the actual practices of contemporary travel and requires a theoretical pluralism capable of going beyond orderly ideas of authenticity or modernity, and engaging in full with their transgressions, but also with the processes that give them the forms that we experience as tourists (Minca & Oakes 2014, p. 298).

The Wiley Blackwell Companion to Tourism, Second Edition. Edited by C. Michael Hall.

So, when we were recently asked to return to these two texts and 'update' them, after some hesitation, since we were not entirely sure that there was something left to say on tourism and post/modernity, we have begun to ask ourselves: where to start? Perhaps the best way to reconnect to those past conversations, we thought, is to recall the closing statement of 'Tourism After the Postmodern Turn':

> While our objective in this essay is not aimed at any kind of institutional dismantling of tourism studies, we nevertheless believe that the deghettoization of tourism might productively lead to the integration of a tourist analytic across the social sciences and humanities. It might, in other words, lead to a "tourist turn" of sorts (Minca & Oakes 2014, p. 301).

Have we experienced a 'tourist turn' of sorts in the past decade of social theory? Possibly not, and this should not come as a surprise since that closing sentence was a provocation rather than a statement. Having said that, with that closing sentence in mind, it may be useful to return in detail to those past review essays and critically reflect and problematise some of the key points that they advanced.

While we maintain that 'The tourism experience remains an important field through which to interrogate theory' (Minca & Oakes, 2014, p. 294), as we re-read in 2022 our initial essay entitled 'Tourism, modernity and postmodernity' (Oakes & Minca, 2004) we realise how far we all are now from discussions on the postmodern in relation to different interpretations of modernity and the modern that dominated key humanities and social science debates in the 1990s. That essay was very much an attempt to precisely connect those debates to tourism studies, and possibly to all formulations of travel theory associated with the study of modernity's tensions, ambivalence, and paradoxes. Today, no one ever mentions the postmodern. That powerful turn in social theory is a thing of the past, although its long-term impact certainly is not. Quite the contrary. The fast but intense 'postmodern season' in the social sciences and humanities has in many ways opened the door, or at least prepared the ground, for the development of several other 'turns' that will be at the core of academic debates and perspectives in the decades to come. In this very brief essay, then, we will sketch out some of these and offer our thoughts on how and why questions about late modernity and modern theory remain relevant in tourism studies today. Our focus lies with developments that emerged from critical interrogations with, as well as responses to, post/modernity in social theory during the 2010s in particular. These have included increasing attention to tourist practice (as opposed to more representational and symbolic dimensions of tourism), biopolitics, and a deepening interest in materiality and post-human approaches. We find an increasing convergence of the biopolitical and the technopolitical in tourism practice, and suggest that, arguably, critiques of the modernist rationality underlying the ongoing commodification, securitisation and even colonisation of bodies via tourism are now as important as ever.

Decentring the Human Subject in Tourist Practice

While ruminating on those pages that we penned in the early 2000s, we had the impression that perhaps too much emphasis was placed on the twists and turns of the interpretation of modern theory, and perhaps not enough space was dedicated to questions of gender, the body, and even the biopolitical in our reflections on tourism subjectivity, or, on tourism and the modern subject. In light of recent developments, we may also add that questions of intersectionality would have probably helped us to better understand the figure of the tourist and its imbrications with identity politics, the social and the cultural, to say the least. In that chapter, we also spent a bit of time analysing the link between modern travel and the colonial, particularly focusing on Timothy Mitchell's (1988) pathbreaking contribution to the conceptualisation of the colonial gaze and the tourist approach to the world 'as-an-exhibition'. We still find those passages useful for any exploration of the genealogies of the modern traveller; however, we could have possibly said more on the colonial in tourism, perhaps complicating even further the picture by incorporating contributions from debates in postcolonial studies. And finally, still in the spirit of a hopefully productive self-criticism, while that essay placed particular emphasis on the ways in which the tourist experience reflected several modern paradoxes (somewhat reverberating a book on the same topic that we published a few years earlier, see Minca & Oakes (2011)), at the same time, that emphasis reads today as implicitly problematic, as if the process of modern social dis/ordering at the origin of such tensions was something that was inherently and negatively linked to tourism and its histories.

While 'updating' that earlier intervention, 'Tourism After the Postmodern Turn' (Minca & Oakes, 2014) was written about 10 years later and was arguably a more original piece. It reads today much closer to the then contemporary debates, although remaining relatively shy in its engagement with some key developments in critical social theory that would become highly influential in subsequent years. In those pages, we insisted on the need 'to further problematize [the] very possibility of a modern subjectivity (even one rendered in all the nuanced and paradoxical terms that we insisted on a decade ago)' (Minca & Oakes, 2014, p. 299). We were also adamant about the fact that any attempt to identify or even conceptualise a 'tourist subject' should be dismissed, a position that we perhaps emphasise even more today for reasons that we will elaborate in the paragraphs below.

As we considered our previous work, though, we reflected as well on the importance of understanding the tourist experience as a key element in the production of individual and collective subjectivities, and this is why we proposed to acknowledge the importance in tourism, and in tourism studies, of questions of care and responsibility, and of the politicisation of the body – the tourist body and the body of the worker in tourism, also related to broader questions of 'hospitality' and highly gendered forms of care. Therefore, we somehow anticipated the growing role that would be played by biopolitics in the management of mobility, and in the spatial and temporal regimes characterising many tourist enclaves. We also suggested to avoid studying tourism as an isolated phenomenon, and to incorporate it in the broader discussions populating the emerging field of mobility studies (see, among many others, Sheller & Urry, 2006; Cresswell, 2006). What is more, we placed a rather strong emphasis on the need to study *tourist practice*, instead of simply limiting our focus to what tourists presumably 'think' or 'want'. Such emphasis on practice, we claimed, would indeed allow us to explore the banal and mundane elements that make the 'tourist experience', but also to hint at its emotional and affective dimensions, something that will be extensively analysed in the relevant literature in the years that followed.

In hindsight, however, our impulse to emphasise a 'practice turn' in tourism studies did not go nearly far enough. While pointing toward what was clearly a developing interest in affect and embodiment, we failed to appreciate the larger 'material turn' that was more fundamentally undermining the very coherence of the social as a conceptual foundation for theoretical work. Beyond the study of tourism, increasing interest in materiality (Ingold, 2007; Coole & Frost, 2010), material powers (Bennett & Joyce, 2010), and in the ways the more-than-human world conditions, constrains, and enables the social world (Whatmore, 2006; Bennett, 2010) was fuelled by a sense of frustration with the limits of the postmodern as itself a deconstruction of the modern. But it was also fuelled by a deepening awareness of, as Bowker (2018) succinctly put it, the relationship between the form of energy we consume and the social theory that we produce. Reflecting upon and extending this idea, the carbon dependence of tourist practice was perhaps too obvious to be of interest to those of us hoping to glean from that practice new insights into the social world. And this lack of interest was precisely what advocates of a 'material turn' sought to challenge. Tourism has thrived in the shadow of the Anthropocene; it has been perhaps the most exemplary form of a modern rationality that has sought to 'pillage past sunlight stored over many millions of years within several generations while knowing the very limitations of that resource' (Bowker, 2018, p. 204). Tourism may indeed be the ultimate statement of a modernity that indexes 'the toxic legacies of radically human-centred thinking and action' (Boyer, 2018, p. 226).

Covid-19 is merely one of these toxic legacies, and the disruptions that the coronavirus pandemic imposed on global tourism should only reinforce the need for academic approaches to tourism practice which radically *decentre* the human element of that practice. That a global pandemic would emerge to make this need painfully clear was of course something we did not see coming. Spawned by a zoonotic virus that has itself been called 'a paradigmatic example of an Anthropocene disease' (O'Callaghan-Gordo & Anto, 2020), the material power of Covid as a non-human agent in the constitution of the social is by now so obvious as to be almost banal. But clearly the role of the non-human, and of the powers of the material world in determining how the social is formed and, indeed, how it is theorised, is certainly a major development for a post-post/modern tourism practice.

The Convergence of Biopolitics and Technopolitics

There were many other developments that we also did not see coming. Some of these are extensions of, or at least conversations with, the material turn and its implications. This is certainly to be ascribed to the partiality of our analysis, but also to the fact that a number of recent global events have highly impacted the ways in which tourism has been managed and reconceptualised. More specifically, what we did not see coming, or at least what was not mentioned in our earlier essays, is the current debate around and the political reactions to overtourism (see Milano et al., 2019), a phenomenon that was only mitigated by the measures taken during the Covid-19 pandemic. However, this phenomenon is re-emerging along with all its problematic implications, as we write. Most obvious of these implications is the kind of ecosystem disturbance that spurs zoonosis, but in broader terms, what we conventionally call 'the environment' is too often understood to exist on the other side of some boundary separating humans and non-humans. And yet, the embodied materiality of the human, subject to viral transmissions and other transgressions of the material world, makes clear the fiction of such a boundary. 'More than human' theorisations of overtourism thus seem on the one hand obvious and yet, on the other, still sorely needed (Gibson, 2019).

Scrutinising the flimsy boundary between human tourist bodies and the non-human worlds through which they move points us toward a number of other developments worth considering at this juncture. Perhaps most prominent among these concerns is the digital revolution in tourism, and the associated algorithms and forms of ranking, rating, and metrics, together with the related spatial, social, and urban transformations (see, among others, Minca & Roelofsen, 2021). That we did not anticipate this in our earlier work is somehow surprising if one considers how the postmodern helped in problematising the 'metaphysics of representation' at the origin of modern European travel. But now, the digital mediation of tourism has been somehow taken to its extreme consequences by the role virtual spaces are playing today in determining tourist desires, practices and broader geographical imaginations. What also emerges, in this challenging exercise of connecting more recent developments back to our earlier conceptions and critiques of the modern, is the fact that the digital turn in tourism has been tightly associated with the development of tourism platforms, which in many ways have deepened and further transformed the working of the tourist colonial gaze.

If one considers, for example, the kind of tourist experience offered by a platform like Airbnb, they will note that the promise to penetrate 'the life of others' goes beyond the observation of smiling 'locals' in their real or deliberately performed daily practices. Rather, Airbnb claims that the tourist will become 'the local', will live like them, will even sleep in their bed and cook in their kitchen (see Roelofsen & Minca, 2018, among many others). This *vertical penetration of otherness* represents, to some extent, a deepening of the modern colonial gaze, a deepening that, again, could have been perhaps anticipated by our early discussion on the post/modern and the metaphysics of representation in tourism. Such platforms also serve as technologies through which precarious and informal labor is rationalised and made available for capital accumulation at scales previously unforeseen.

The platformisation of tourism, however, suggests not simply an extension into the metaphysics of representation, or a new level of labour exploitation at global scales, but can also be explored as another dimension of the material turn in tourism studies. Over the past decade, for example, 'smart tourism' has emerged as a rapidly expanding field of tourism practice. This has particularly been the case in China, where tourists are among the world's most digitally literate and where the state has embraced tourism as a leading mechanism in its efforts to blanket both urban and rural spaces with infrastructures of digital surveillance and algorithmic governance (Wang, 2020; Salimjan, 2021). Indeed, as Tong (2021) has argued, many of China's 'ancient' towns and tourist villages, some of which are still relatively remote, lie at the cutting edge of global smart city development. Tong (2021) notes, for example, that the ancient water town of Wuzhen has the highest density of WiFi transmitters of any place in the country. Indeed, Wuzhen is where the Chinese government has held its annual 'World Internet Conference' and where Chinese state policy pronouncements of data sovereignty and cyber-separation from the West have been officially promulgated.

In China, 'smart tourism' involves popular mobile phone apps that facilitate travel bookings and other forms of electronic payment, mediate the tourist experience in specific places, and connect tourists to local service providers and local state tourism and security administrations. In some places where overtourism has threatened to degrade the tourist experience, smart tourism has meant regulating tourist access to certain sites. The ubiquitous QR-code has been enrolled as the most basic feature of tourist practice. It has blended the material and digital spaces of tourism in fundamentally new ways.

On a more basic level, smart tourism turns tourists, and tourist spaces, into data. And data is a resource, for both capital accumulation and for state security. This is no different from the everyday ways our smart phones turn us into data for what Zuboff (2019) has called surveillance capitalism. Smart phones can be understood, then, as non-human prostheses that turn human bodies into revenue generating machines for digital capital. In China, smart phones turn tourists into data not just for revenue generation, but for state surveillance as well. Digital tourism infrastructures thus facilitate and enhance not just the commodification of tourism spaces, but the policing and politicising of those spaces as well (Minca & Roelofsen, 2021).

The political effects of smart tourism suggest that tourism studies should be attuned not just to the *biopolitical* effects of tourist practices, but also their *technopolitical* effects as well.

'Technopolitics' has been defined as 'the ways in which political actions are embedded within technical forms and, conversely, the ways in which the technical shapes political questions' (von Schnitzler, 2016, p. 10). One promising version of this approach has been explored in Rowen's (2022) analysis of the political effects of airline booking platforms which have been enrolled to advance China's territorial claims on Taiwan. In smart tourism, Tong (2021) has suggested a technopolitical framing in order to better understand how the state's objective for digital infrastructure development 'is to convert various relationships within tourism space, tourists, and tourism into data to facilitate the digital governance of tourism' (Tong, 2021, p. 9).

The technopolitics of smart tourism involve more than digital governance and state control, however. An additional dimension involves the ways digital platforms have been socialised as part of the tourist's everyday practice, and how the tourist's interaction with place-based service provision and affective labour is digitally mediated. Booking platforms like Airbnb and others, for example, have dramatically transformed the nature of service provision for tourists in local communities, often resulting in political contests over access to affordable housing (see Cocola-Gant & Gago, 2021; Roelofsen, 2022). Additionally, the ranking aspects of these platforms, as noted above, can wreak havoc on service providers who are compelled by intense market competition to accept increasingly narrow, or even nonexistent, profit margins in order to stay in business. Here, the platformisation of the relationship between 'tourist' and 'local' is one that increasingly involves more-than-human relationships. Indeed, when tourists become data sensors and evaluators in the service of capital and of the state, then we have a convergence of the technopolitical and the biopolitical.

On the Lingering Significance of the Modern

In our 2014 essay we presented the biopolitical as a key element to be considered in order to understand the regimes of spatio-temporal control of the tourist in some specific spaces. We, however, did not envisage the current triumph of biometric technologies and the associated forms of bordering, body scanning and profiling that have characterised, and will possibly continue to characterise, the traveller 'in time of pandemic', as well as, in what promises to be an unpredictable post-covid-19 longue durée. This may produce two consequences in the decades to come, both somehow related to the fact that some of the tensions of the post/modern may be taken to their extreme consequences. First, the merging of global technologies of digital and material control and profiling of travellers associated with the pandemic, and with those implemented after 9–11, may lead to the production of an idealised 'trusted traveler', a figure which may determine, in crucial ways, who has the right to travel and under which conditions. Already in China, for example, the pandemic has in many respects cleared the way for popular acceptance of biometric screening. If there was resistance among tourists to relinquishing their privacy for the convenience of a digitally mediated and more carefully controlled experience, Covid-19 has relegated such concerns to a distant memory. For most tourists in China now, a smart city is a safe city. The pandemic paved the way for the embrace of biopolitical technologies in ways the Chinese state could only previously dream of.

The second possible development has to do with the experimental nature of some covid-free tourist spaces, which, again, may take to a different level the long-standing tendency towards the realisation of enclavic purified total resorts. But this seems like the subject of a chapter that will have to be written in the coming years, in the hope that no further major terrorist attacks or future pandemics will accelerate the abovementioned processes of individual bordering, profiling and control.

Despite all of these new developments, that we did not see coming, we still agree with our past positionings on one point:

> To conclude this brief review, we would like to suggest that tourism be understood as a comprehensive analytic capable of reassembling our understandings of contemporary life, an analytic through which new insights into the "social," "cultural," and "political," along with their related conceptualizations, are gained. (Minca & Oakes, 2014, p. 301).

Here, we would suggest – albeit with some trepidation – that the convergence of biopolitics and technolopolitics explored briefly above can be viewed as something of a triumphal return of high modernist rationality. This is the same rationality that drives states and companies to manipulate the logics of algorithmic governance for both profit and securitisation. It is the same rationality that drives colonial appropriations of local communities or marginalised identity groups. And clearly we are seeing varied ways in which tourism is now enrolled in these modernist rationalising projects of territorial claims and control (Rowen, 2022), dispossession and surveillance (Salimjan, 2021), and suppression of ethnoreligious identity (Zhu, 2020). Perhaps this is not so much a new role for tourism as one that we previously might have expected to wither as the story of modernity diffused into one about something that came 'after' modernity. Yet trends over the past decade – the questioning and indeed reversal of 'globalisation', the rising power of authoritarian states, and the seemingly unchecked power of surveillance capitalism – have made it clear that this expectation was not simply premature but ill informed.

Accordingly, we may conclude this chapter with a few questions on how the most recent developments in tourism are still related to modern theory, and, more generally, to the multiple articulations of this long, messy, late modernity (whatever it is that this term evokes at this point). In other words, is reflecting on modernity and modern theory in relation to tourism still useful in understanding mobility in time of pandemic or even in time of war? (As we are authoring this piece we are confronted with the tragic consequences of Russia's invasion of Ukraine and all its unpredictable implications in terms of individual mobility). Clearly, we think so. On a more theoretical ground, is 'tourism as an analytic' still key to understanding the 'colonial present', and what can we learn about race and social exclusion by studying tourism? Again, we have tried to convey in the brief discussion above how thinking about and theorising the social through the lens of tourism remains highly productive.

But our somewhat morose take on the current state of tourism practice as wrapped up in the technopolitical and biopolitical logics of increasingly powerful state and market actors should, finally, produce another question regarding tourism as a *political* analytic. And this is perhaps the most important question of all: how will decolonial approaches and practices affect tourism, particularly tourism's highly racialised practices, as well as tourism studies, and the ways in which tourism valorises and exploits local cultures and local knowledges? How will these changes impact the questions of gender, intersectionality, power, and eurocentrism that we have discussed 10 and 20 years ago, and what remains to be said about tourism and modernity in such a transformed global context?

And finally, to recall one of our key claims in the 2014 piece, what did we 'learn from Asia' in the past decade or so? While this is difficult to assess, partially because Covid hit Asia first and it continues to have an enormous impact on the mobility of what used to be the most important origin/source of global tourism, at the same time, we maintain that this is possibly still a crucial question to envisage the future of tourism globally. For this reason, we have highlighted some recent developments in China's deployment of smart tourism and related technologies of biometric screening and data gathering. Indeed, we would argue that China has recently adopted the mantle of high modernist rationality more than any other country. However, China remains, as we write, under the relatively immobilising cloud of the state's 'zero-Covid' approach to managing the pandemic, and for this reason its influence as the world's preeminent source of out-bound tourists has declined considerably. Having said that, the tourist ecosystem within China remains highly developed and increasingly mediated digitally. In many ways China remains at the forefront of developing technologies governing tourist practice that will ultimately have a global impact. How the rest of the world responds to, appropriates, embraces, or rejects these technologies will remain a significant question to explore in the coming years.

References

Bennett, J. (2010). *Vibrant matter: A political ecology of things*. Duke University Press.

Bennett, T., & Joyce, P. (Eds). (2010). *Material powers: Cultural studies, history, and the material turn*. Routledge.

Bowker, G. (2018). Sustainable knowledge infrastructures. In N. Anand, A. Gupta & H. Appel (Eds), *The promise of infrastructure* (pp. 203–222). Duke University Press.

Boyer, D. (2018). Infrastructure, potential energy, revolution. In N. Anand, A. Gupta & H. Appel (Eds), *The promise of infrastructure* (pp. 223–243). Duke University Press.

Cocola-Gant, A., & Gago, A. (2021). Airbnb, buy-to-let investment and tourism-driven displacement: A case study in Lisbon. *Environment and Planning A: Economy and Space*, *53*(7), 1671–1688. https://doi.org/10.1177/0308518X19869012

Coole, D., & Frost, S. (Eds). (2010). *New materialisms: Ontology, agency, and politics*. Duke University Press.

Cresswell, T. (2006). *On the move*. Routledge.

Gibson, C. (2019). Critical tourism studies: new directions for volatile times. *Tourism Geographies*, *23*(4), 659–677. https://doi.org/10.1080/14616688.2019.1647453

Ingold, T. (2007). Materials against materiality. *Archaeological Dialogues*, *14*(1), 1–16. https://doi.org/10.1017/S1380203807002127

Milano, C., Novelli, M., & Cheer, J. M. (2019). Overtourism and degrowth: a social movements perspective. *Journal of Sustainable Tourism*, *27*(12), 1857–1875. https://doi.org/10.1080/09669582.2019.1650054

Minca, C., & Oakes, T. (2014). Tourism after the postmodern turn. In A. A. Lew, C. M. Hall & A. M. Williams (Eds), *Wiley Blackwell companion to tourism* (2nd ed., pp. 294–303). Wiley-Blackwell.

Minca, C., & Oakes, T. (Eds.). (2011). *Real tourism: Practice, care, and politics in contemporary travel culture*. Routledge.

Minca, C., & Roelofsen, M. (2021). Becoming Airbn*beings*. On datafication and the quantified self in tourism. *Tourism Geographies*, *23*(4), 743–764. https://doi.org/10.1080/14616688.2019.1686767

Mitchell, T. (1988). *Colonizing Egypt*. Cambridge University Press.

Oakes, T., & Minca, C. (2004). Tourism, modernity, and postmodernity. In A. A. Lew, C. M. Hall & A. M. Williams (Eds), *A Companion to tourism* (pp. 280–290). Blackwell.

O'Callaghan-Gordo, C., & Anto, J. (2020). COVID-19: The disease of the Anthropocene. *Environmental Research*, *187*, 109683. https://doi.org/10.1016/j.envres.2020.109683

Roelofsen, M. (2022). *Hospitality, home and life in the platform economies of tourism*. Palgrave Macmillan.

Roelofsen, M., & Minca, C. (2018). The superhost. Biopolitics, place and community in the Airbnb dream-world of global hospitality. *Geoforum*, *91*, 170–181. https://doi.org/10.1016/j.geoforum.2018.02.021

Rowen, I. (2022). Booking engines as battlefields: contesting technology, travel, and territory in Taiwan and China. *Geopolitics*. https://doi.org/10.1080/14650045.2022.2027914

Salimjan, G. (2021, September 1). Camp land: Eettler ecotourism and Kazakh dispossession in contemporary Xinjiang. *Lausan*. https://xinjiang.sppga.ubc.ca/critical-scholarship/academic-publications/camp-land-settler-ecotourism-and-kazakh-dispossession-in-contemporary-xinjiang-2/

Sheller, M., & Urry, J. (2006). The new mobilities paradigm. *Environment and Planning A: Economy and Space*, *38*(2), 207–226. https://doi.org/10.1068/a37268

Tong, Q. 童祁 (2021). 中国智慧古镇: 旅游和技术的文化政治 [China's smart ancient town: The cultural governance of tourism and technology]. [Unpublished doctoral dissertation]. Chinese University of Hong Kong.

von Schnitzler, A. (2016). *Democracy's infrastructure: Techno-politics and protest after apartheid*. Princeton University Press.

Wang, X. (2020). *Blockchain chicken farm and other stories of tech in China's countryside*. Farrar, Straus & Giroux.

Whatmore, S. (2006). Materialist returns: practising cultural geography in and for a more-than-human world. *Cultural Geographies*, *13*(4), 600–609. https://doi.org/10.1191/1474474006cgj377oa

Zhu, Y. (2020). Memory, homecoming and the politics of diaspora tourism in China. *Tourism Geographies*. https://doi.org/10.1080/14616688.2020.1844286

Zuboff, S. (2019). *The age of surveillance capitalism: The fight for a human future at the new frontier of power*. Hatchette.

Part 4 Producing Tourism and Tourism Spaces

Introduction
Producing Tourism and Tourism Spaces

C. Michael Hall

This section of the volume discusses the various ways in which tourism and tourism spaces are produced for consumption. Chapter 20 by Hultman et al. examine the processual understanding of a resource, what they describe as the relationship between resourcification and tourism. They observe how the four dimensions of resourcification: context, conditions, modes, and temporalities, contribute to the valuation process. However, as they note, there are different valuation regimes, and 'What tourism is, and how it becomes a resource, depends upon how we understand tourism per se, and which perspectives we apply to describe and evaluate its processes' (Hultman et al., Chapter 20).

Dwyer's chapter on transnational corporations and the globalisation of tourism (Chapter 21) presents a multidimensional interpretation of globalization, emphasizing that it is simultaneously an economic and social phenomenon, before focusing on the embedded and relational role of transnational corporations (TNCs) in global tourism development. The chapter identifies four main drivers of globalization as (1) economic and financial capital drivers seeking new opportunities to grow, (2) information and telecommunications and transport technology drivers overcoming the constraints of space, (3) social drivers exhibited in human mobilities that enable the flexible flow of labour and changes in values, attitudes and lifestyles, and (4) the political liberalization of trade and monetary policies (political drivers) making tourism an attractive development option for many places. However, as Dwyer observes, the costs (e.g., environmental impacts) and benefits (e.g., employment and incomes) of these processes are unevenly distributed.

One crucial element of globalisation is transport. However, issues of climate and environmental change and the desire for greater sustainability has led to increased interest in tourism's use of public transport. Azman et al. (Chapter 22) highlight both the enablers of tourist public transport as well as the constraints. The chapter details tourist use of public transport at tourism destinations, emphasising the various factors that influence tourists' travel-transport behaviour, in terms of personal characteristics (socio-demographic and psychographic

The Wiley Blackwell Companion to Tourism, Second Edition. Edited by C. Michael Hall.

factors), trip characteristics and public transport attributes. The chapter then discusses the relationship between tourism and public transport and public transport as a tourist experience and attraction.

A vital element in producing tourism is that of entrepreneurship and, increasingly associated with tourism and entrepreneurship, innovation. According to Aquino (Chapter 23), as a system, tourism is fuelled by networks of enterprises, directly and indirectly, interacting with each other to address market needs and opportunities. Tourism requires enterprises that can innovate in order to accommodate ever-changing consumer demands and environmental changes, such as those due to socio-technological advancements (Hall & Williams, 2020). Significantly, Aquino notes that tourism entrepreneurship research has expanded beyond its Anglo-American base to include a much wider coverage by geographical area, as well as a much broader understanding to also include social and non-profit dimensions. In addition, he argues that tourism enterprises have the potential to transform tourism into a more sustainable industry, especially by being innovative and initiating more environmentally focused and community-engaged CSR activities. Chapter 24 by Pikkemaat and Bichler provides a specific example of tourism innovation that may fulfil Aquino's hopes for more sustainable tourist enterprises when they discuss the role of Living Labs as a means of accelerating innovation in tourism.

Another crucial element in the production of tourist spaces is the attitude of destination residents towards tourism. The substantial body of literature in this area is reviewed in Chapter 25 by Hateftabar. As she notes, residents' attitudes significantly influence the satisfaction and loyalty of tourists to destinations, with unwelcoming or even hostile behaviour deterring tourists from returning or visiting in the first place (Ribeiro et al., 2017). In general, tourists are averse to visiting or revisiting places where they do not feel welcome. Such issues have become even more significant since COVID-19 and the reopening of destinations to tourists (Rey-Carmona et al., 2022). The chapter concludes by observing that the effects of tourism on communities appear to be localised, both in time and place. However, Hateftabar also urges caution, as she suggest that although the large number of case-based studies conducted in various geographical locations has added greatly to the collective level of knowledge on resident attitudes to tourism it has not led to the development a broader conceptual foundation for the understanding residents' perceptions of tourism nor their willingness to support its development (see also Chapter 14 on sense of place). As such, a comprehensive model of residents' perceptions and willingness to support tourism development is still waiting to be developed.

Chapter 26 by Chen et al. discusses a specific dimension of one of the world's largest outbound markets, the Chinese outbound market, and the vital role of shopping in the production of tourism experiences for this market. As Chen et al. suggest, tourism research is very much contextual and in investigating the production of tourism and tourism spaces it can matter substantially as to which country and/or culture tourists are from (Tiwari et al., 2018), given their role in coproducing tourism space. Perhaps nowhere is this more clearly seen than in the activity of shopping, given its centrality to both the tourist experience and the extraction of money from the visitor in destination economies. By building a conceptual foundation on Chinese outbound tourism and associated specific shopping behaviours, the chapter provides both an emic and an etic approach for researching tourist psychology and behaviours that are highly contextual and vary in home countries/places and destination countries/regions. For example, in the case of Chinese tourist shopping the concept of *daigou*.

The final chapter in the section discusses the importance of tourism workers in the production of tourism and tourism spaces. However, Ioannides (Chapter 27) gives his discussion of tourism work and workers a more specific focus by noting the surprising lack of the context of sustainable development. As Ioannides notes, this is surprising given the increased precarity of the tourism labour force, whereby workers' rights and security have gradually been eroded as part of the global neoliberal economy. As the chapter suggests, it appears that in spite of the hype surrounding the appearance of digital platforms as a means to expand individuals' work flexibility and which have often been enthusiastically embraced in tourism (Hall & Williams, 2020), these innovations have merely reinforced patterns of exploitation that have long surrounded the working conditions of many people (Ioannides et al., 2021). As such, this chapter brings the section to an end by returning to the issues associated with the tourism and globalisation relations noted by Dwyer, as well as the role of resourcification in tourism (Hultman et al., Chapter 20), whereby rather than being regarded as people, tourism workers instead become human resources. As such, as Ioannides et al. (Chapter 27, this volume) suggest, we have a narrow understanding of who tourism workers are, why they have chosen to work in particular destinations, and what their experiences of being in those destinations amount to.

References

Hall, C. M., & Williams, A. M. (2020). *Tourism and innovation* (2nd ed.). Routledge.

Ioannides, D., Gyimóthy, S., & James, L. (2021). From liminal labor to decent work: A human-centered perspective on sustainable tourism employment. *Sustainability, 13*(2), 851; https://doi.org/10.3390/su13020851

Rey-Carmona, F. J., Núñez-Tabales, J. M., Durán-Román, J. L., & Pulido-Fernández, J. I. (2022). Open the doors to tourism or remain cautious: residents' dilemma amidst a pandemic. *Current Issues in Tourism*. https://doi.org/10.1080/13683500.2022.2047162

Ribeiro, M. A., Pinto, P., Silva, J.vA., & Woosnam, K.vM. (2017). Residents' attitudes and the adoption of pro-tourism behaviours: The case of developing island countries. *Tourism Management, 61*, 523–537. https://doi.org/10.1016/j.tourman.2017.03.004

Tiwari, A. K., Dash, A. K., & Narayanan, B. G. (2018). Foreign tourist arrivals in India from major source countries: An empirical analysis. *Current Issues in Tourism, 21*(10), 1137–1156. https://doi.org/10.1080/13683500.2017.1296415

Chapter 20

Tourism Resourcification

Johan Hultman, Hervé Corvellec, & Malin Zillinger

Introduction

A conventional claim within tourism research, practice and policy is that tourism is a powerful engine for economic growth across all spatial scales, accounting for slightly more than 10 percent of the global number of jobs and gross domestic product (GDP) (World Travel and Tourism Council (WTTC), 2022). What does this tell us? First, that the tourism economy is heavily dependent on mobilising spatial, intangible, material and human resources. Second, that tourism is a crucial resource for economic development. Tourism engages and mobilises resources from every conceivable part of non-human and human environments. Simultaneously, tourism is itself a resource, for example, to support regional recovery from industrial decline, or to open people's eyes for cultural diversity.

There is a double resource-dynamic in play, which makes it difficult to separate tourism from other constructions of society-environment interactions. Such links between tourism and other parts of human lives have been deepened by globalisation as well as by digitalisation (Zillinger, 2021). In this chapter, we trace and discuss this double resource-dynamics in terms of resourcification: the social processes by which material and immaterial entities become resources (Hultman et al., 2021; Corvellec et al., 2021). We discuss the contexts, conditions, modes, and temporalities of resourcification. By doing so, we demonstrate how resourcification constitutes the core of tourism development.

A condition of this double dynamics is that tourism takes the consumer to the places where the resources are. In contradistinction to conventional export industries that take the resources to the consumer, tourism is the global circulation of consumers of place (Ferrario, 1979, p. 18):

> As in other industrial processes, tourism requires the transformation of raw material (the available natural/cultural resources) into a finished tourist product. The fact that such resources are often intangible and difficult to define does not detract from their real productive value. The final tourist product into which these resources are transformed must then be offered to national or international

The Wiley Blackwell Companion to Tourism, Second Edition. Edited by C. Michael Hall.

> consumer markets, through a concerted effort of sales organization and efficient servicing. Tourism's main singularity is that the final product is not shipped to its markets, but requires customers to come and consume it *in loco*.

This begs the question: What are these tourism resources that are attractive enough to power a circulation of consumers on all geographical scales? (Cole, 2015). Tourism literature answers by labelling and categorising resources. Employing regional analysis, Smith (1987) arrives at the four resource-pattern categories of 'urban tourism', 'outdoor recreation', 'cottaging/boating', and 'urban fringe tourism', a kind of geo-functional matrix. This study guided Backman et al. (1991), who distinguish between regions characterised by natural, historical, and cultural resources respectively. Following up on regional analysis, Spotts (1997, p. 5) states that 'A tourism resource was defined as anything that played a major role in attracting tourists to the study region, including natural resources, tourism events, outdoor recreation facilities, and tourist attractions'.

Focusing on nature-based tourism, Priskin (2001), followed by Alaeddinoglu and Can (2011), categorise tourism resources according to the four parameters of attraction (e.g., biodiversity and scenic diversity), accessibility, supporting infrastructure, and level of environmental degradation. Further, forging the bond between resources and (sustainable) development, Garrod et al. (2006) redefine 'the rural resource' in tourism as countryside capital in order to make resources such as biodiversity, historical features, landscapes and traditions, and ways of life visible and subject to economic and social investment. In a similar vein, cultural heritage has been understood as a tourism resource through management practices (Ashworth, 2000; Carter & Bramley, 2002; Ho & McKercher, 2004), or in terms of authenticity (Park et al., 2019). Recently, the quest for sustainable exploitations of natural tourism resources has led to further and increasingly multi-variable categorisations (Ruban, 2021). Of course, human resources have also been acknowledged as crucial for tourism development (Baum, 2007, 2015; Madera et al., 2017).

But, we argue, tourism resources are not simply being there, ripe for abstraction, assemblage, refinement, packaging, marketing, and consumption. As Hultman et al. (2021) detail in their resourcification manifesto, resources become. Therefore, rather than essentialising resources and classifying what they are, what matters for tourism research and practice is to provide theoretical means to show how things become tourism resources, and how tourism can become a resource. A resource for what? This chapter explains why the answer to this question is a double resource-dynamic, which is as limitless as the managerial contexts and policy purposes at hand. As we will show, a resourcification approach is contextual and flexible, yet – as it is processual – able to guide both tourism research and policy.

Part 1: Essentialist vs. Constructionist Understanding of Resources

Before we detail the double resource-dynamics of tourism development, this section presents key tenets of resource theories. We are ending this part with the resourcification theory, but start off in the Garden of Eden.

'The Garden of Eden story has shaped Western culture since earliest times and the American world since the 1600s' (Merchant, 2013, p. 1). In particular, the myth of the Garden of Eden has contributed to an understanding of Nature as a pool of pristine resources available for the benefits of Humankind. About the colonisation of the American continent:

> Using science, technology, and biblical imagery, [Euroamerican men] first changed the eastern forests and then the western deserts into cultivated gardens. Sanctioned by the Genesis origin story, they

> subdued the "wilderness," replenished the earth, and appropriated Indian homelands as free lands for settlement. (. . .) Timber, barrel staves, animal hides, herbal medicines, tobacco, sugar, and cotton were extracted from nature as part of the great project of "improving" the land (Merchant, 2013, p. 81).

This (anthropo-, andro-, and ethno-centric) view of an Eden-like nature as an abundance of resources open for exploitation still sub-tends the notion of natural resources. For the Organisation for Economic Co-operation and Development (OECD) (2005), 'Natural resources are natural assets (raw materials) occurring in nature that can be used for economic production or consumption', suggesting that natural resources exist for their exploitation. Typical of an anthropocentric view of nature, at an equal distance to ecocentric and theocentric views (Hoffman & Sandelands, 2005), are soils, water, forests, energy sources, minerals, and all forms of life seen as ways to serve human wants, needs, or desires. Moreover, natural resources include a vast range of ecosystem services that contribute to economic production and/or the maintenance of life, for example via carbon sequestration or the maintenance of biodiversity, or as sinks for waste or amenity for recreation (Bridge & Wyeth, 2020), that humans have not needed to work or pay for (so far, anyway, the Anthropocene changes this in a radical manner, spelling the end of cheap nature (Hamilton, 2016)). Water resources are representative of how geological, biological, and social dimensions are interwoven in the provision and use of a natural resources, with major impact of cultural representation and consumption patterns, and difficulties to manage them in a sustainable way (Wescoat, 2015).

Globally, the extraction of natural resources has fuelled economic growth (Behrens et al., 2007), with coal, oil and uranium as paradigmatic examples. The uneven distributions of the burdens and benefits of an unlimited natural resource exploitation has been severely criticised, though. Extractivism has been accused to impede the development of the places where an intense extraction of resources takes place (Burchardt & Dietz, 2014; Villalba-Eguiluz & Etxano, 2017). This is why calls have been made to develop principles for natural resource governance able to make the exploitation of these resources environmentally and socially sustainable (Lockwood et al., 2010). Such calls for a sustainable governance of resources refer, for example, back to German scientific forestry, allegedly the starting point for a sustainable exploitation of natural resources (Hölzl, 2010). Developed since the 18th century to protect the forest from an over-exploitation that would endanger its capacity to serve as a resource for future generations, scientific forestry builds on the intrinsic dynamics of forest eco-system to develop management schemes geared toward financial efficiency, leading both to a centralised control of nature, and an advocacy of the commonwealth of a nation. Looking forward, one could consider the Sustainable Development Goals (SDGs) a global call for a sustainable governance of natural resources.

Other resources that have received a considerable attention are humans. Considering that people are a key to wealth goes long back in history, as slavery should remind us. Predating Malthus's dystopic warning with a few centuries, Jean Bodin claimed back in 1576 that the only genuine form of wealth is people (Martin, 2014). However, it was not before the interwar period that corporations systematically developed human resource departments and adopted the corresponding techniques of recruitment, skill identification, motivation, competence development, work organisation, compensation and benefits, firing, and legal compliance to their employees (e.g., Armstrong & Taylor, 2020). An efficient human resource management (HRM) is considered indispensable today to organisational success. However, parallel to the emergence of a myriad of decent jobs characterised by predictable wage and social benefits (Acemoglu, 2001) that also make the most of the resources of humans, a myriad of bad jobs, sometimes called bullshit jobs (Graeber, 2019), have emerged in wake of an anytime, anywhere, and anyone organising of labour that hardly does justice to human resourcefulness.

Inside as well as outside the world of work, time is a particularly important resource for humans. For example, time has created cultural heritage. Some cultural heritage resources are tangible, for example archaeological, historical, architectural, and engineering properties, but also museums, libraries, shipwrecks, cultural landscapes, religious sites, and locations where plants and animals are utilised by indigenous peoples. Others are intangible, for example traditional lifeways, religious practices, and languages. Besides having a cultural value, cultural resources often have a particular value for tourism. They are protected by the 1970 Convention Concerning the Protection of World Cultural and Natural Heritage adopted by UNESCO (Green, 2015). But time is also important in the sense that the allocation of time between labour and leisure, between consumption and non-consumerist activities, and to activities such as transportation or participation to democratic life raise key questions about how the use of natural resources related to the quality of life and sustainability in contemporary societies (Rau, 2015).

Recently, resources have become a way to understand organisational behaviour. The resource-based view on the firm suggest analysing the success and failure of organisations out of their capacity to bundle heterogeneous resources (e.g., capital, people, technology, legal rights) into competitive ways of creating and retaining value over time (Barney, Wright, & Ketchen, 2001). One of these resource is the physical environment (Hart, 1995), and there are today two dominant business trends: 'business as usual' that considers that organisations have a right of access to natural resources, and 'greening of business' that aspire to a closer alignment of resource uses with planetary boundaries and biophysical cycles (Mathews, 2015). Among the latter, one of the latest trends in management is to consider waste as a resource for the circular economy (Corvellec, 2019).

Common to all these resources is the question of what resources are and what they can be used for. However, a key question to ask about resources is how they become resources. This question was originally posed by the resource economist Erich Zimmermann (1951 [1933]) who meant that resources are not, they become. His understanding of resource as something becoming led Austrian economists (see: Bradley, 2007) to investigate the institutional conditions that favoured the becoming of resources. But it is human geography that most explicitly has articulated that resources are not given but shaped by economic factors, political institutions, social attributes, spatial contingencies, and belief systems (Hayter & Patchell, 2015). Bridge (2009, p. 1220) means that:

> Whether a material is classed by society as a resource or not depends on the way it is related to other things, to knowledge, to the opportunity to realize value by exchange, and to other materials that can fulfil the same function. Thus, the fact that something – whales, tar sands, genetic diversity – is regarded as a resource (or not) tells us rather more about a society than it does about the substance itself.

Zimmerman (1951 [1933]) initiated a shift from seeing resources as belonging to the static conditions of 'being' to a non-essentialist understanding of resources as belonging to the dynamic category of 'becoming' (Corvellec et al., 2021). Empirical studies of how people use artefacts (e.g., De Laet & Mol, 2000; Orr, 1996) illustrate how people use their cognitive capacities and problem-solving competence to fluidly adapt resources to their needs, objectives, and conditions of practices. Drawing on the recent resourcification manifesto

(Hultman et al., 2021), it is this shift toward a processual understanding of resource that we intend to use in the remainder of this chapter, in order to analyse the making and unmaking of resources in tourism.

Four dimensions of resourcification processes are in focus here. First that these processes take place in contexts, for example, specific world views, spatial circumstances, or times. Second, that resourcification rests on specific conditions, for example of knowledge, technique, or available infrastructure. Third, that resourcification takes place along diverse modes that entail different relationships among, for example, people, money and materiality. Fourth, resourcification is dynamic and thus temporal. As an unstable processes, a resourcification can be followed by a de-resourcification. A change of context, a political decision, a technological change, or the overuse of a resource can put an end to any status as a resource. And even a re-resourcification can take place, in the same or another context.

The reader needs to keep in mind that resourcification processes take place in connection to other resourcification processes. As Hultman et al. (2021) explain, resourcification processes cannot be reduced to discrete instances of resources becoming or un-becoming. Resources become along chains of resourcification where something becomes a resource because it has been able to mobilise other things that have been turned into resources. Inversely, becoming a resource is but a step for a further resourcification.

Part 2: Four Approaches to Resourcification

Context

Nothing is a tourist resource per se, but anything can become one, as resourcification is a contextual realisation of a potential. Places, artefacts, social practices, or symbols not previously or currently understood as tourism assets can be invested with meaning and relevance. This is done by inscribing them in relevant contexts, for example, leisure activities, cultural heritage knowledge, or technological achievements. The European Union (EU) rural development programmes are an example of this (EU, 2023). Other examples: In response to funding opportunities, rural actors already tied together in social relations (e.g., a municipality losing its tax base due to industrial relocations) mobilise to retrofit or develop discourse and infrastructure for tourism purposes. A farmhouse or factory is turned into a gallery; a cluster of villages is narrated together through their connection to a famous author; an old and unusually large tree is redefined through storytelling as part of a cut path through local lore. These examples point to the double resourcification dynamics in tourism, where things become resourcified for tourism purposes at the same time as tourism actors use tourism as a vehicle and resource for chosen development paths.

The inventiveness of tourism resourcification is a rejection of sufficiency and finitude, and an embrace of development and growth. Tourism resourcifications has long been welcomed by people living in social, cultural, and economic peripheries who can draw benefits from contextual circumstances that transform their immediate environments into tourist attractions. However, tourism resourcification is also increasingly criticised. Gentrification of historical quarters, urban and rural overtourism, and near-shore cruise ship dumping of human waste are side effects of the inherent drive towards more and more intensive tourism resourcification processes. With the Anthropocene, the environmental limits of resourcification have become clear. Such limits are particularly inacceptable if they are not contributing to social justice, or to a fair distribution of wealth.

The *Slumdog Millionaire* tours by local guides in Dharavi, Mumbai is an example of the rejection of finitude that underlies tourism resourcification, because it extracts value in a place where this was seen as impossible before (Tours By Locals, 2023). It is also an example of how tourism resourcification is an extractive industry, driven by economic activities that remove resources from somewhere with little or no processing for the benefit of somewhere else. Extractive industries such as mining are typical of operations that rest on the assumption that natural and anthropogenic environments are abundant, cheap and ready-to-resourcify stocks of input, an assumption that they share with many tourism activities.

Tourism is fuelled by the displacement of consumers to localised natural and cultural resources, but also by the displacement of other categories of people, for example service and guest workers. Correspondingly, tourism resourcification is grounded on places, natural environments, people, traditions, and more being seen as reservoirs of resources, open for legitimate revenue generation. This includes expressions of abstraction, extraction, assemblage, refinement, management, commodification, and financialisation of natural and human entities. The rationale at work in tourism, assuming that anything can be considered a legitimate object for tourism resource extraction, is the same.

A final condition of tourism resourcification is that it is temporally, spatially and socially distributed. Temporally, tourism resourcification processes are embedded in the past at the same time as they are bets on future revenues. History is a typical resource for tourism, but tourism is also a typical resource for the future. Spatially, tourism resourcification has re-drawn maps of dwelling and mobilities, from up-market seaside resorts such as Monaco or Cannes, to massive ski tourism development in the Alps, whose success is often elevated by past glory. Socially, the distribution of tourism resourcification can create hierarchies among locals and tourists. Hierarchies are also manifest among service workers, among different tourist sites, and among time periods, some time-slices being more valued than others. The so-called brat circuit among Swedish upper-classers illustrates the intricacies of these three dimensions of distribution of resourcification. Participants are young and well-off, and they visit the prestigious seaside sites of Båstad during the annual tennis championship in mid-July, followed by a week each in medieval Visby and culturally perceived Tylösand, before heading south to St Tropez in France in August. The circuits' efforts of exclusive distinctiveness demonstrate the cross resourcification rationales of time, place and class status.

Inversely, failures to mobilise, situate, and maintain something in relevant time-space contexts explains failures of tourism development. Resourcification gives answers about (silenced) win-losses as much as (celebrated) win-wins.

Conditions

Resourcification depends on specific conditions that decide the becoming or non-becoming of a tourism resource. The first of these conditions pertains to the nature of tourism, since a resource for tourism depends on the definition thereof. Historically, the definition of tourism has shifted from an economic, via a geographical, to a holistic perspective. Tourism has been described as the social science of mobility, and for future definitions it may be important to include digital possibilities of mobility to the physical geographical approach that has been dominating so far. What can or cannot be a tourism resource depends on future tourism business models where the definition of a tourist includes both physical and digital mobility, and what it requires to be a tourist under such circumstances.

Likewise, what constitutes a resource for tourism depends on tourism knowledge. The knowledge that tourists have about a destination influences their behaviour before as well as during, and after their stay. Travel information is a constituting factor for the resources that are identified, giving a central resourcifying role to smartphones, guidebooks, and other information channels. And tourism as a field of knowledge is in constant evolution, with tourism research and academic tourism education building on theories and knowledge frames of their own, as witnessed by the more than 120 academic journals dedicated to tourism today.

Two other key conditions of resourcification are technology and infrastructure. For instance, ski tourism is conditioned by and a driving force for a dedicated technology cluster that encompasses the manufacturing of ski apparels and equipment, ski lifts, slope preparations, snow canons, and ski pedagogic. Other technology clusters encompass transport, housing, and health care infrastructures, in addition to support systems such as credit cards and insurance systems that are not specific to ski tourism, but that nonetheless offer essential conditions for its resourcification processes.

A further condition is legislation. Legal rules can promote or demote resourcification. Tourism resourcification has historically benefitted from legislative decisions at all geographical and governance levels, from the recognition of passports, to paid vacation, and detailed local planning rules. Supportive legislation can be questioned, though, for example that air transportation benefits from exemptions from environmental tax, or that not all countries recognise sex abuse committed abroad as domestic crime.

A supportive discourse of tourism resourcification is the development paths it generates. Examples are tourism's contributions to peace and love by enabling people to visit each other's everyday lives and become and/or remain emotionally attached, and its contribution to employment, wealth, and local development. Resourcification inserts tourism into a value regime that determines what is valuable or not. Valuation practices through which people decide what is worth visiting or not are moments of truth in tourism resourcification. These practices shift in personal as well as in societal time, and are determined by the dynamics of individual tastes and collective habitus. Valuation practices are also culturally influenced, with the Romantics placing a nostalgic lens over nature that is very different from the caring eye of conservationists or the defying approach of extreme sport practitioners.

Consequently, the resourcification of tourist destinations and attractions are in a constant flux, with unpredictable developments as a consequence. For example, the tourist destinations of Røros (Norway) and the Ruhr Area (Germany) are today attractive tourist destinations not because they produce copper or mine bituminous coal, but because they communicate the local history and heritage experiences of such soil resources in museums.

Modes

The dynamics of tourism resourcification are relational. Old social-environmental relations are broken up while new ones are forged. When demersal fish stocks collapse as a result of over-fishing, marine pollution and/or unchecked growth of predatory populations such as seal or cormorant, fish is de-resourcified through regulation that inhibits professional fishing efforts. Small-scale fishers might then see the potential in a resourcification of themselves and their assets by engaging in the relations and materiality of tourism, re-defining themselves as seal-spotting or up-market leisure fishing guides. Another example is the intimate historical relationships between the film industry and tourism resourcification. From Jacques Tati's gentle mockery of Monsieur Hulot and other guests at the holiday beach resort in the 1950s, to the glamour brought to Monaco by Grace Kelly when she married to become Princess Grace of Monaco, via innumerable movies that feature Paris as the City of Love, the film industry has taught the audience how to bind resources for tourism, and for tourism as a resource. The Taj Mahal is such a tourism resource that we all relate to it, regardless of whether we have seen it or not.

The nexus between tourism and society is captured by a resourcification approach, and this becomes even clearer when the performativity, dynamics, and hybridity of tourism resourcification are considered. Unfolding the modes through which resource become demonstrates the need to go beyond essentialising notions of tourism resources. Tourism resourcification is performative when every road sign directing motorists towards the 'scenic route' resourcifies the landscape as an attraction, but the sign has to be there for this attractiveness to hold true. The resourcification of the landscape leads to the resourcification of other human and non-human entities, such as farms into farm shops, farmers into artisans, infrastructure into narratives, and non-human life into breeds, varieties and local history. Today, these dynamics in the organisation between place and people are increasingly performed through digital technology. The algorithms of Airbnb blur the fundamental distinction between home and hotel, while the possibilities of peer-to-peer interactions make the hand of tourism management and accountability truly invisible. Tourism resourcification is hybrid: combinatory; multiple; and non-binary. A wilderness area or coral reef is not only just a state of nature when an ecotourism venture turns it into a commodity such as the last place on earth, a repository for lost knowledge. . . a place marketed for consumption by reasons explaining why it should not be consumed.

Market-orientation is a key mode of tourism resourcification. Public sector initiatives and financing are important, but the displacement and circulation of consumers takes place primarily through market-based framings. To be resourcified, something is abstracted from its environment (qualified as something different, unique or valuable), appropriated (through property or access rights), monetised (use-value is turned into exchange-value), commercialised (inserted into a tourism regime of value), and marketed. And digitalisation takes this market-orientation of resourcification a step further. By engaging with digital peer-to-peer platforms, consumers become co-producers of economic value for the firms operating them, while also consenting to be directed to similar or complementary objects of attraction where further consumption is facilitated through convenience, discounts, and the feeling of being noticed for their personal preferences. Corporate extractivism takes an added dimension when the resource that is mined actively participates in its extraction. This iterative, non-depleting extraction of economic value from resourcified tourists is yet another illustration of the double resourcification dynamics. While tourists resourcify places by mobilising digital skills, the ownership of – and knowledge how to apply – technology allow other actors to resourcify tourists by engaging them in multiple circuits of value co-creation.

When discussing modes of tourism resourcification in this way, issues of power and ethics follow. Where, when, why, by whom and for what purpose is power executed in tourism resourcification? These are important questions that let the political dimensions of resourcification surface. To stress the processual dimension of resourcification, we would further ask: What kind of control is needed to make liminality and potential amendable to resourcification? Whose interests are strengthened or weakened in tourism resourcification? How are privileges and injustices distributed? What conflicts do tourism resourcification generate, and which conflicts does it cease? What must be the ethical considerations? The point of asking such questions is the need to enquire about the moral and political consequences of tourism resourcification processes, beyond aprioristic praise or condemnation. Just like other communities, the tourism community has an urgent need to draw practical conclusions from its contribution to the threats of the Anthropocene, deriving from the unfettered resourcifications of the last two centuries.

Temporalities

Processes of tourism resourcification are inherently unstable. What has once been resourcified can be de-resourcified, and later re-resourcified, along paths that are difficult to foresee. It is therefore relevant to discuss the temporalities of tourism resourcification.

The origins of tourism resourcification processes are manifold. A mix of individual micro-initiatives and globalisation processes have enabled networks to arise, hence generating and intensifying mobility in and between places, with tourism being one form of mobility. Farmers may decide to offer tourism services on their farm holdings when crop yields are decreasing. Individual households may provide a room for tourists in their apartment, or entire houses, when the space is not needed by themselves. Tourism resourcification is spreading itself both in time and in space. While it does so, it is pushing back limits thought to be stable. They, too, are changing over time.

This pushing back of limits may involve reversing resourcification processes. De-resourcification can be caused by major political decisions such as closing borders during a pandemic, or a terrorist attack. De-resourcification can also be deliberate. In places marked by overtourism, inhabitants try to banish tourists from their neighbourhoods. Others condition access to their favourite place by film teams to a clause of geographical confidentiality. In some cases, de-resoursification appears as one of few ways to protect a resource from over-exploitation. It can be total, or selective, when a destination takes measures to deter an existing customer segment to attract a new one, for example by raising prices, or by conditioning access upon a sustainable behaviour.

The possibility of de-resoursification adds to the instability of tourism resourcification as it expresses a radical shift in developmental priorities. Some see it as a threat, others as a necessity to preserve the resource or even the planet. Symbolically strong events such as melting glaciers or jellyfish invasions on beaches, call into question the more and better that has implicitly sub-tended tourism resoursification, including the borders of legitimate mobility behaviour in the future.

This brings us to the dynamic impermanence of resourcification in tourism. The notion of what things can do, and what they are used for, changes consistently, and so does the value that is ascribed to them. The resort Prora, built by the Nazi regime 1936–1939 on the German island Rügen, was first used as a tourist destination, then disdained and abandoned, and then again re-resourcified as a heritage tourist destination to learn about German history. Sunbathing has been something that tourists have first carefully avoided when a fair skin was a certain class marker, then avidly searched for when sunlight was considered health-bringing, then again selectively avoided to reduce the risk of malign melanoma – an illustration of what is seen as a threat to another resource can be turned into a resource and again de-resourcified, in constant flux.

Conclusions

This book chapter is about the processual understanding of a resource. Tourism is engaging resources as it develops, and it is itself a resource for processes taking place outside the tourism realm. We have shown how the four dimensions of resourcification: context; conditions; modes; and temporalities, are contributing to the double resource-dynamics between tourism and other parts of society in many different ways. The characteristic of tourism resourcification is that it includes human mobility between places, including people's consumption in loco.

We have illustrated that anything can become a tourism resource – or cease to be. Tourism resourcification is the conceptual realisation of underlying capacities. A tourist destination, service or attraction is not becoming a resource in itself, but is always dependent on the time and space in which it is embedded. Such settings are defined by context, condition, mode, and temporality. Resourcification can be welcomed or not, depending on who is evaluating the process, in which time, and by which perspective. What is regarded as a resource says not only something about the resource itself, but also about the observer: understanding a forest as a place we can experience by a visit, instead of a place we lumber, says something about the society we live in.

What tourism is, and how it becomes a resource, depends upon how we understand tourism per se, and which perspectives we apply to describe and evaluate its processes. We can understand the resourcification of tourism in one way if we think about it in terms of economic development, or in another, if we think about it in terms of spatial relations, or in a holistic point of view. No matter which perspective we use, resourcification lies within selected valuation regimes. Such regimes are constantly shifting: they are always in flux, never still.

Being an inherently dynamic process, tourism resourcification is also shaping relations: to places you have been, and to places you have heard of. It is worth mentioning here that tourism is an enabler of emotional attachment, as in the mobility in order to visit friends and family. In a broader sense, tourism resourcification is also influencing structures we thought were stable, as to what is home and away, or what is work and leisure.

Tourism resourcification is not neutral. Processes will inevitably lead to power shifts, hence we need to ask questions about power relations and ethical influences resulting from shifting processes. Just as important, we need to answer the questions we ask, and draw consequences therefrom. That, by the way, is true for all resourcification processes, not only those related to tourism. We hope that our elaborations on these themes will guide future policy makers to adopt this kind of processual thinking, embracing a thought that builds on the constant becoming of tourism resources, instead of considering tourism as something that is, static and fixed. We also hope that by this chapter, we can lay a ground for future research on resourcification in the realm of tourism, including theory development. There is already research on processes, developments, and procedures of becoming in tourism research. Hopefully, the perspective of resourcification can strengthen this stream of research.

References

Alaeddinoglu F., & Can, A. S. (2011). Identification and classification of nature-based tourism resources: Western Lake Van basin, Turkey. *Procedia Social and Behavioural Sciences*, *19*, 198–207. https://doi.org/10.1016/j.sbspro.2011.05.124

Acemoglu, D. (2001). Good jobs versus bad jobs. *Journal of Labor Economics*, *19*(1), 1–21. https://doi.org/10.1086/209978

Armstrong, M., & Taylor, S. (2020). *Armstrong's handbook of human resource management practice*. KoganPage.

Ashworth, G. J. (2000). Heritage, tourism and places: A review. *Tourism Recreation Research*, *25*(1), 19–29. https://doi.org/10.1080/02508281.2000.11014897

Backman, S. J., Uysal, M., & Backman, K. (1991). Regional analysis of tourism resources. *Annals of Tourism Research*, *18*(2), 323–327. https://doi.org/10.1016/0160-7383(91)90016-5

Barney, J., Wright, M. & Ketchen, D. J. (2001). The resource-based view of the firm: Ten years after 1991. *Journal of Management*, 27(6), 625–641. https://doi.org/10.1177/014920630102700601

Baum, T. (2007). Human resources in tourism: Still waiting for change. *Progress in Tourism Management*, *28*(6), 1383–1399. https://doi.org/10.1016/j.tourman.2007.04.005

Baum, T. (2015). Human resources in tourism: Still waiting for change? – A 2015 reprise. *Tourism Management*, *50*, 204–212. https://doi.org/10.1016/j.tourman.2015.02.001

Behrens, A., Giljum, S., Kovanda, J., & Niza, S. (2007). The material basis of the global economy: Worldwide patterns of natural resource extraction and their implications for sustainable resource use policies. *Ecological Economics*, *64*(2), 444–453. https://doi.org/10.1016/j.ecolecon.2007.02.034

Bradley, R. L. (2007). Resourceship: An Austrian theory of mineral resources. *The Review of Austrian Economics*, 20(1), 63–90. https://doi.org/10.1007/s11138-006-0008-7

Bridge, G. (2009). Material worlds: Natural resources, resource geography and the material economy. *Geography Compass*, *3*(3), 1217–1244. https://doi.org/10.1111/j.1749-8198.2009.00233.x

Bridge, G., & Wyeth, R. (2020). Natural resources. In A. Kobayashi (Ed.), *International encyclopedia of human geography* (2nd ed., pp. 249–258). Elsevier.

Burchardt, H. J., & Dietz, K. (2014). (Neo-) extractivism–a new challenge for development theory from Latin America. *Third World Quarterly*, *35*(3), 468–486. https://doi.org/10.1080/01436597.2014.893488

Carter, R. W., & Bramley, R. (2002). Defining heritage values and significance for improved resource management: An application to Australian tourism. *International Journal of Heritage Studies*, *8*(3), 175–199. https://doi.org/10.1080/13527250220000/18895

Cole, S. (2015). Space tourism: Prospects, positioning, and planning. *Journal of Tourism Futures*, 1(2), 131–140. https://doi.org/10.1108/JTF-12-2014-0014

Corvellec, H. (2019). Waste as scats: For an organizational engagement with waste. *Organization*, 26(2), 217–235. http://doi.org/10.1177/1350508418808235

Corvellec, H., Hultman, J., Jerneck, A., Arvidsson, S., Ekroos, J., Wahlberg, N., & Luke, T. W. (2021). Resourcification: A non-essentialist theory of resources for sustainable development. *Sustainable Development*, *29*(6), 1249–1256. https://doi.org/10.1002/sd.2222

De Laet, M., & Mol, A. (2000). The Zimbabwe bush pump: Mechanics of a fluid technology. *Social Studies of Science*, 30(2), 225–263. https//doi.org/10.1177/030631200030002002

European Union (EU). (2023, October 2)*EU CAP Network*. https://eu-cap-network.ec.europa.eu/index_en

Ferrario, F. F. (1979). The evaluation of tourist resources: An applied methodology. Part 1. *Journal of Travel Research*, *17*(3), 18–22. https://doi.org/10.1177/004728757901700405

Garrod, B., Wornell, R., & Younell, R. (2006) Re-conceptualising rural resources as countryside capital: The case of rural tourism. *Journal of Rural Studies*, *22*(1), 117–128. https://doi.org/10.1016/j.jrurstud.2005.08.001

Graeber, D. (2019). *Bullshit jobs: A theory*. Penguin Books.

Green, T. J. (2015). Cultural resource management: Conservation of cultural heritage. In J. D. Wright (Ed.), *International encyclopedia of the social & behavioral sciences* (2nd ed., pp. 488–490). Elsevier.

Hamilton, C. (2016). The Anthropocene as rupture. *The Anthropocene Review*, 3(2), 93–106. https://doi.org/10.1177/2053019616634741

Hart, S. L. (1995). A natural-resource-based view of the firm. *Academy of Management Review*, *20*(4), 986–1014. https://doi.org/10.5465/amr.1995.9512280033

Hayter, R., & Patchell, J. (2015). Resource geography. In J. D. Wright (Ed.), *International encyclopedia of the social & behavioral sciences* (2nd ed., pp. 568–575). Elsevier.

Ho, P., & McKercher, B. (2004). Managing heritage resources as tourism products. *Asia Pacific Journal of Tourism Research*, *9*(3), 255–266. https://doi.org/10.1080/1094166042000290655

Hoffman, A. J., & Sandelands, L. E. (2005). Getting right with nature: Anthropocentrism, ecocentrism, and theocentrism. *Organization & Environment*, *18*(2), 141–162. https://doi.org/10.1177/1086026605276197

Hultman, J., Corvellec, H., Jerneck, A., Arvidsson, S., Ekroos, J., Gustafsson, C., Nilsson, F. L., & Wahlberg, N. (2021). A resourcification manifesto: Understanding the social process of resources becoming resources. *Research Policy*, *50*(9), 104297. https://doi.org/10.1016/j.respol.2021.104297

Hölzl, R. (2010). Historicizing sustainability: German scientific forestry in the eighteenth and nineteenth centuries. *Science as Culture*, 19(4), 431–460. https://doi.org/10.1080/09505431.2010.519866

Lockwood, M., Davidson, J., Curtis, A., Stratford, E., & Griffith, R. (2010). Governance principles for natural resource management. *Society & Natural Resources*, *23*(10), 986–1001. https://doi.org/10.1080/08941920802178214

Madera, J. M., Dawson, M., Guchait, P., & Belarmino, A. M. (2017). Strategic human resources management in hospitality and tourism: A review of current literature and suggestions for the future. *International Journal of Contemporary Hospitality Management*, *29*(1), 48–67. https://doi.org/10.1108/IJCHM-02-2016-0051

Martin, G. (2014). «Il n'est de richesse que d'hommes». *Idées économiques et sociales*, *177*(3), 1–1. https://doi.org/10.3917/idee.177.0001

Mathews, J. A. (2015). Greening of business. In J. D. Wright (Ed.), *International encyclopedia of the social & behavioral sciences* (2nd ed., pp. 392–396). Elsevier

Merchant, C. (2013). *Reinventing Eden. The fate of nature in Western Culture* (2nd ed.). Routledge.

Organisation for Economic Co-operation and Development (OECD). (2005). *Glossary of statistical terms. Natural Resources.* https://stats.oecd.org/glossary/detail.asp?ID=1740

Orr, J. E. (1996). *Talking about machines: An ethnography of a modern job.* ILR Press.

Park, E., Choi, B. K., & Lee, T. J. (2019). The role and dimensions of authenticity in heritage tourism. *Tourism Management*, *74*, 99–109. https://doi.org/10.1016/j.tourman.2019.03.001

Priskin, J. (2001). Assessment of natural resources for nature-based tourism: The case of the Central Coast region of Western Australia. *Tourism Management*, *22*(6), 637–648. https://doi.org/10.1016/S0261-5177(01)00039-5

Rau, H. (2015). Time use and resource consumption. In J. D. Wright (Ed.), *International encyclopedia of the social & behavioral sciences* (2nd ed., pp. 373–378). Elsevier.

Ruban, D. A. (2021). Natural resources of tourism: Towards sustainable exploitation on a regional scale. *Sustainability*, *13*(12), 6685. https://doi.org/10.3390/su13126685

Tours By Locals (2023). *Real Slumdog Millionaire Tour.* https://toursbylocals.com/slum-dog-millionaireprivatetour

Smith, S. (1987). Regional analysis of tourism resources. *Annals of Tourism Research*, *14*(2), 254–273. https://doi.org/10.1016/0160-7383(87)90088-0

Spotts, D. M. (1997). Regional analysis of tourism resources for marketing purposes. *Journal of Travel Research*, *35*(3), 3–15. https://doi.org/10.1177/004728759703500302

Villalba-Eguiluz, C. U., & Etxano, I. (2017). Buen Vivir vs Development (II): The Limits of (Neo-) Extractivism. *Ecological Economics*, 138, 1–11. https://doi.org/10.1016/j.ecolecon.2017.03.010

Wescoat, J. L. (2015). Water resources and sustainable water management. In J. D. Wright (Ed.), *International encyclopedia of the social & behavioral sciences* (2nd ed., pp. 437–442). Elsevier.

World Travel and Tourism Council (WTTC). (2022). *Economic impact reports.* https://wttc.org/Research/Economic-Impact

Zillinger, M. (2021). *Tourism revisited: The influence of digitalisation on tourism concepts.* (Report series ETOUR Working Paper). Mid Sweden University. https://www.miun.se/Forskning/forskningscentra/etour/nyheter/nyhetsarkiv/2021-9/malin-zillinger/

Zimmermann, E. W. 1951 [1933]. *World resources and industries: A functional appraisal of the availability of agricultural and industrial resources.* Harper & Brothers.

Chapter 21

Transnational Corporations and the Globalisation of Tourism

Larry Dwyer

Introduction

Globalisation involves the growing interconnectedness of the world, as reflected in the expanded flows of products, information, technology, capital, goods, services and people and political processes (Aspers & Kohl, 2015). It involves two interrelated elements: the opening of international boundaries to more rapid movements of commodities, services, money, people, and ideas, as well as changes in national and international institutions and policies that support or encourage such flows (Meethan, 2004). Economic globalisation, the standard concept, refers to the increased economic interdependence of countries worldwide resulting from increased volume and variety of goods and services transactions across borders, freer international capital flows and faster and wider diffusion of technology (Song et al., 2017). The growing interdependence between national economies has resulted in a trend towards global markets, global production, global competition and global communication (Mowforth & Mundt, 2015). Globalisation, however, includes, and goes beyond, simple internationalisation of markets. It is a multidimensional process, taking place simultaneously within the spheres of the economy, of politics, of technological developments – particularly media and communications technologies – of environmental change and of culture (Urata & Doan, 2022).

While globalisation has reduced economic, social and political barriers to the free flow of people, goods, capital, information, communication and lifestyles on an international scale, it favours the larger organisations that can take advantage of global economies of scale in order to expand operations into new markets to offer products at competitive prices to consumers. An important driver of globalisation has been the transnational corporation (TNC), with headquarters in one country and operating wholly or partially owned subsidiaries in one or more other countries (Narula & Pineli, 2019; Ushakov et al., 2020). TNCs alter the structure of international economic activity by increasing commerce, establishing global supply chains, and facilitating access to natural resources and labour markets. TNCs facilitate the globalisation process through their moves and counter moves to different markets, evidenced by the

The Wiley Blackwell Companion to Tourism, Second Edition. Edited by C. Michael Hall.

numerous cross border mergers and acquisitions, strategic alliances, franchising, licensing and other forms of partnerships. Operations in different countries necessitate substantial investments in terms of foreign direct investment (FDI), which invest in the host country (Urata & Doan, 2022).

Tourism, a service industry, (Bohn et al., 2018) is both a force for, as well as a product of, globalisation (Timothy, 2019). The growth of TNCs and the rise in the significance of global brands, has been central to the emergence of globalisation in the tourism industry and particularly FDI in the global industry (Song et al., 2017; Dwyer, 2022). Prominent TNC in the tourism industry include airlines, hotels, tour operators, cruise ships, travel intermediaries and so on. These corporations are characterised by high levels of vertical and horizontal integration and strong links with other organisations along the tourism supply chain (Dwyer & Čavlek, 2019; Ivanov & Ivanova, 2016).

There is ongoing debate as regards the effects of globalisation (Intriligator, 2005; Yan, 2019). TNCs are claimed to be inter alia, a catalyst for economic growth, the basis for disseminating technical knowledge and education, improving global communication and transport, creating new products and markets for firms and destinations, generating innovation and product quality, creating economies of scale enabling more efficient operations and lower prices, fostering democratic ideals and cultural exchange and understanding (Mowforth & Munt, 2015; Narula and Pineli, 2019).

Critics of globalisation, on the other hand, claim that it generates income inequalities, destroys local firms and industries, creates greater dependency of developing economies on developed ones, and results in environmental and cultural deterioration within developing destinations (Mowforth & Munt, 2015; Talani & Roccu, 2019). Given their market power, TNCs have the ability to disrupt traditional economies, impose monopolistic practices, minimise their tax liability through 'transfer pricing', and assert a political and economic agenda on a destination. Anti-competitive agreements and conduct can impose substantial costs particularly on developing economies eliminating many of the benefits of globalisation (Ravallion, 2018; Kyove et al., 2021). Some critics advance stronger claims that the effects of globalisation result in TNCs having undue political influence in the host country, displacement of indigenous peoples from lands, human rights abuses, unfair labour and wages, loss of community, commodification of cultures, and loss of biodiversity (Clark & Kwon, 2018).

Like most topics for debate, both sides contain truths, and generalisations are difficult (Yan, 2019). The effects of globalisation differ case by case and by context. This is particularly true for the two- way link between the tourism industry and globalisation.

Drivers of Globalisation

Four main types of forces that drive globalisation can be identified.

Economic Drivers

A rapid expansion of the private sector in many emerging market economies, along with deregulation and privatisation, spurs economic growth by generating competitive pressures to use resources more efficiently (Dwyer, 2015). Globalisation of international financial markets has created wider access to external funding sources. Increasing capital mobility acts as a stimulus to globalisation. Easier access to foreign capital and increased FDI lays down the foundation for import substitution and export promotion to generate competitive markets

(Song et al, 2017). When capital can move freely from country to country, firms can locate and invest abroad, and repatriate profits. FDI in any industry has the potential to contribute to growth and development, create jobs, build exports, provide additional sources of finance for commercial expansion, facilitate technology transfer and innovation, and further increase opportunities for global networking (Dwyer, 2022).

Broadly speaking, the economic paradigm of neoliberalism globalisation has driven industrial development including tourism for the past several decades (Dwyer, 2018; Wearing et al., 2019). The various motives for FDI in tourism can be explained in large part using a widely accepted 'eclectic' paradigm of international production (United Nations (UN), 1989; Dwyer, 2022). The eclectic paradigm asserts that the extent, pattern, and growth of value-adding activities undertaken by TNCs outside their home countries are dependent on the value of and interaction between three main variables - ownership, location and market internalisation. The interrelations of these variables, and the response to them by firms, vary according to industry, countries of origin, destination of investment, and firm specific characteristics. They also vary over time as changes in technology and the entrepreneurial and economic environment affect the competitive position of firms and the location of their value-adding activities (Dwyer, 2022). Recent years have witnessed substantial growth in global value chain activity, an important driver of productivity growth (Wearing et al., 2019; Findlay & Roelfsema, 2021).

Technology Drivers

Developments in information and communications technology (ICT) and transportation have accelerated the pace of globalisation, lowering operating costs making it more economically feasible for a firm to locate different phases of production in different countries. The increasing size of the computer-literate, information-oriented workforce in many developing countries, and efforts by global corporations to diversify their high-tech operations, fuel the spread of new technologies. Enhanced digitalisation of business operations has favoured the growth of TNC over small to medium enterprises (SME) (Sułkowski, 2020).

The internet has been a vital development within the tourism industry itself as consumers use it to gather information on destinations, products and services and make bookings online. The dramatic growth of social media platforms, such as Facebook, Twitter, Instagram and YouTube is increasingly being felt in the travel and tourism sector. Younger travellers retain their technology skills with them as they age, making online activity ever more important for travel and tourism along their lifecycle.

Travel and tourism companies and organisations increasingly determine windows of opportunities offered by ICT to integrate it into their sales and marketing activities. The internet makes it possible for marketing activity to be undertaken on a more level playing field whereby small businesses can connect directly to consumers and compete for market share on an even footing with larger firms (Dredge et al., 2019).

Social drivers

The pace of social change and transformation worldwide has risen substantially, with implications for social life and human culture. This social change is related to increasing connectivity among peoples and countries worldwide. The globalisation of social networks, whether through TNCs, higher education, or international civil society, can be expected to drive further growth of tourism and travel. Mass tourism, and associated over-tourism, is one of the most visible manifestations of globalisation (Timothy, 2019).

The migration and movement of people – either from rural to urban environments within countries, or migration between countries, is a prominent feature of the globalisation process. Immigration between countries offers better opportunities for people to gain access to jobs globally. The globalising economy and technological change inevitably require a more highly skilled labour force (European Commission (EU), 2019). Social infrastructure maximises opportunities for individuals and businesses to be innovative, learn and develop knowledge, skills and access knowledge services.

Important changes in values, attitudes and lifestyles are occurring which are directly affected by some of these demographic changes with implications for tourism management, marketing and new product development. New consumer values are also emerging, with a rise in ethical consumerism (Szmigin et al., 2009). A new and growing demographic of individuals is 'values aspirational', placing a higher value on healthy living, environmental and social justice, and ecological sustainability in the products and services they purchase, the companies in which they invest, the politicians and policies they support, the companies for which they work and, ultimately, the lifestyles they lead (Szmigin et al., 2009). TNCs are at the forefront of the new marketing effort to access the 'new consumer' (Dwyer, 2015).

Political Drivers

Political drivers of globalisation include the increasing liberalisation of trade and capital markets. There is ongoing political pressure for higher living standards in developing societies with a growing global middle class creating a cycle of rising aspirations. The continued deregulation and liberalisation of air transport and 'open skies' policy supports the growth of trade and tourism (Timothy, 2019). Globalisation results in easier access across borders, which for the tourism industry implies increased numbers of foreign tourists as well as increased global competition from international tourist destinations. Tourism is now being pursued as a serious development strategy for developing countries. Political realignments in the European Union and North American free-trade zone are encouraging greater travel within each region. Democracy has spread to countries previously under the control of communist, military led or otherwise authoritarian regimes giving political power to formerly disenfranchised citizens. The strong growth of global value chains (GVCs) in the past decades has dramatically challenged existing economic insights and policy implications related to globalisation. The organisation of production in international networks which stretch out across multiple borders has resulted in a growing interconnectedness between countries with important impacts on a broad range of national and international policies (Urata & Doan, 2022) The process of economic globalisation has been linked with the adoption of specific reforms aimed at improving economic growth. The reforms mostly revolve around issues such as trade liberalisation, attracting FDI, privatisation of state-owned enterprises, lifting of capital controls, and fiscal adjustment, aimed at reducing fiscal deficits and public debt levels (Roukanos, 2020).

The above discussion has highlighted only some of the relevant drivers of globalisation. Because of its complex and multifaceted structure, globalisation shapes many contemporary factors determining the development of tourism. Each of the drivers has tourism relevance and no single driver or trend will dominate. Each, to a greater and lesser degree, influences the other and will have varying impacts in tourism destination different regions, countries, industry sectors and tourist behaviour. The drivers are not necessarily mutually reinforcing; in some cases they will work at cross-purposes. Taken together they set the context in which the global tourism industry may be expected to develop in a context of continued globalisation.

Effects of Globalisation

Economic Development

Outcomes of globalisation are argued to include economic growth and economic performance, job creation, lower prices for consumers; greater competition; greater consumer choice of goods, bigger export markets for domestic suppliers; and economies of scale enabling more efficient production with lower prices (Roukanos, 2020) Globalisation and associated long-term economic trends influence tourism in a number of ways. Many claim that globalisation increases the economic prosperity of developing nations, as rising trade and investment flows generate increases in world GDP. Rising income has long been recognised as the most powerful generator of tourism flows (Song et al, 2017). Analysing 124 cross-country data sets for 1971–2010, Iamsiraroj (2016) found that FDI and economic growth positively affect each other. Through FDI, TNCs provide opportunities for developing countries to improve their balance of payments with critical financial infrastructure for economic and social development. Studies of the effects of tourism FDI on economic growth display mixed results (Sokhanvar, 2019; Roudi et al., 2019). In particular, tourism has helped to create the modern global transportation system through which remote destinations have become more accessible and has also contributed to the establishment of the 'moorings' which support travel mobility such as airports, hotels, and resorts (Cohen, 2012).

On the negative side, there are several reasons to be sceptical of globalisation's alleged economic benefits, particularly in the tourism context. Arguments have been advanced that globalisation leads to income inequality, destruction of local industry, economic dependency and tourism expenditure leakages that greatly reduce the economic impacts of tourism growth. We consider these issues in turn.

Income Inequality

The question of whether global inequality is rising or declining has been hotly debated. Over several decades, the standard view has been that the spread of industrialisation leads to the industrialisation of poorer regions of the world and the resulting economic gains lead, in turn, to an overall reduction in global income inequality (Roukanos, 2020). In contrast, a growing number of recent studies suggest that globalisation increases income inequality in both developing and developed destinations (Heimberger, 2019; Huh & Park, 2019). Studies consistently confirm rising *within-countries* inequality over the last three decades, a period defined by intensifying globalisation (Ravallion, 2018; Horner et al., 2018). In advanced countries, losers typically include low-skilled workers, or workers that are no longer able to use specialised skills and/or are not geographically mobile. (Rygh, 2019). Globalisation causes unemployment in industrialised countries as businesses outsource work to developing countries with lower labour costs. To the extent that jobs are lost because of the structural changes arising from globalisation, this increases inequalities in the international distribution of income (Narula & Van der Straaten, 2021). On the other hand, TNCs can potentially mitigate existing inequalities in the destinations in which they operate through philanthropic activities (Rygh, 2019). Economic inequalities result in a self-reinforcing vicious spiral of economic stagnation, recession, and limited economic development (Roukanos, 2020). Despite the substantial volume of research, however, the magnitude of the relationship between globalisation and income inequality remains unclear with factors such as technology, education, labour market institutions and specificities of the welfare state complicating analyses of the causal links (Heimberger, 2019).

Technological development and an increased 'skill premium' have played an important role in the rise in economic inequality over the past decades. Several empirical studies indicate that FDI is associated with higher income inequality in the host countries (Rygh, 2019). Rising income inequality creates profound changes in the workforce and society, leading to asymmetric access to knowledge and skills. Some tourism studies support the pessimistic view. Tax avoidance through practices such as transfer pricing, debt shifting, and the use of tax haven services and tax evasion clearly contribute to global inequality and are emerging as important corporate social responsibility (CSR) issues (Payne & Raiborn, 2018).

Does embracing globalisation allow a country's tourism industry to better contribute to its real per capita economic growth? A causal relationship between inbound tourism and economic growth may not necessarily lead to higher living standards in developing countries, as these are also dependent on the distribution of income and the quality of services such as health care and education. An investigation of 167 countries and territories for the years 2000–2010 to determine if globalisation has been linked with increases in per capita tourism's contribution to economic growth found no correlation between globalisation and the average per capita economic growth contribution of tourism over that period (Ivanov & Webster, 2013). Some studies (Li et al., 2018) suggest caution when generalising the effects of tourism growth on poverty within a destination. A destination's openness to the world in respect of economic, social and political aspects appears to influence how much per capita economic growth its tourism industry will generate. Ultimately government intervention will be essential to address inequalities (Rygh, 2019). Further analysis of this issue in tourism contexts would be helpful.

Dependency

Most developing economies are highly dependent on limited overseas markets and export products, making their economies vulnerable to any economic disruptions that occur in the developed countries that purchase their products. A significant criticism of globalisation highlights the increased risk to smaller economies associated with the interdependence of economies and financial markets. Two types of risks from overspecialisation relate to tourism (Dwyer & Čavlek, 2019). One type relates to a dependency on tourism in general as an export market. The global financial crisis has demonstrated the risks involved in tourism dependency arising from sudden unfavourable changes in demand from world markets (Sheldon & Dwyer, 2010). Another type of risk involves too much reliance on tourism from particular origin markets or too much reliance on a particular tourism product (e.g. conventions, gaming, cruise tourism). Given the discretionary nature of tourism expenditure, the industry is extremely sensitive to crises of every type (economic, environmental, political) (Hopkins, 2021). Changes adversely affecting these markets, making them less competitive and adaptable, can result in substantial costs to tourism destinations that overspecialise in them. Thus, some diversification of production and exports can be prudent even if it entails a temporary decrease in trade from particular tourism markets. Both of the above types of risk are compounded by the reality that TNCs firms are increasingly 'footloose', with ability to switch their investments between destinations at very short notice in search of the most favourable regulatory regimes, creating uncertainty for the host destination.

In many developing countries, local economic activities and resources are used less for the benefit and development of communities and increasingly for export and the enjoyment of others (i.e., tourists, consumers based in other destinations). This occurs in tourism when the domestic market is neglected. To avoid overdependence on the international tourism market and related problems, destinations can attempt to tap the potential of domestic tourism. In geographically large destinations in particular, domestic tourism can effectively absorb the excessive supply resulting from the slump and seasonality of inbound tourism. This also helps to redistribute the national income, thus reducing interregional gaps in the level of economic development and contributing to social equality. Typically, the domestic tourism segment has much lower entry barriers and broader and deeper linkages with local economy than the international sector dominated by TNCs, thus greatly facilitating local community participation. These special features make domestic tourism a promising countermeasure to the negative impacts of globalisation associated with TNCs (Dwyer, 2015).

Destruction of Local Industries

Consolidation processes are changing the nature of the global tourism industry. Globalisation increases the risk of failure for smaller companies that cannot compete globally, many of which are family owned. The crowding out effects of tourism growth are well recognised (Dwyer, 2022). The winners in this process are the powerful vertically integrated tourism businesses that are creating travel value chains that provide customers with distribution, tour operation, flights, accommodation, insurance, entertainment etc. from a single source. The losers are the many small and medium sized, independent companies lacking the resources to compete at global scale. In sectors where the existing range of facilities is limited and of low quality, crowding out effects may be substantial.

None of this implies that TNCs and SME cannot successfully work together on the same tourism market. SME can be highly flexible to respond to changes of demand and to increased innovation. Instead of competing with multinational enterprises (MNE) they should affiliate with themselves in a variety of ways or engage in niche marketing (Meethan, 2004). It should be noted also that crowding-out is not a problem per se. If the new players are simply displacing the least efficient competitors, there will be overall gains to the destination economy (Narula & Pineli, 2019).

Leakages

As noted above, tourist consumer habits are being forged by globalisation. TNCs are sometimes criticised for their allegedly high 'leakage effect' including repatriation of profits. Such leakages imply a loss of economic benefits outside of the community back to the companies and countries that have invested in travel infrastructure. A good example of this is the low economic impact to destinations generated by cruise ship visitation (Dwyer, 2015). The size of any additional leakages overseas due to TNC presence depends importantly on how foreign-owned facilities source their inputs (land, labour, capital) as compared to domestically-owned facilities (Dwyer, 2022). The weaker industrial bases of developing destinations implies that most commercially produced products demanded by tourists cannot be manufactured domestically, and are thus heavily dependent on imports. The traditional leakages issue has evolved to encompass off-shoring. With respect to tourism as a tool for development and economic growth, off-shoring is perhaps the major challenge. Off-shoring is the collective term to capture the movement, relocation and concealment of resources associated with global capitalism (Fuster et al., 2018). De-localised virtual environments enable information, money, resources and so on to move globally, making it hard for local jurisdictions to collect taxes, apply labour laws, environmental and consumer protection and other laws. Increasingly liberalised trade systems have sped up processes of off-shoring making it difficult for local destinations to keep the money spent by tourists within their localities (Fuster et al, 2018).

Ultimately leakages must be analysed in the context of the potential for tourism development to generate greater income and employment in the host destination (Dwyer, 2022). Policies that extend the domestic supply chain should be adopted where feasible (Song et al, 2017).

Employment Opportunities

A particular advantage claimed for globalisation is that it fosters job creation due to the emergence of new companies and new markets, where both skilled and unskilled labour is required. By providing employment, globalisation helps in increasing the standard of living of the people, and also alleviating poverty (Sen, 2019). Globalisation can be a catalyst to the creation of jobs that require higher skill sets with individuals throughout the world able to access information and additional educational opportunities through the internet. In this way, TNCs make a significant contribution to the development of human capital which remains critical for sustainable tourism development (Agarwal & Al Qouyatahi, 2018; Dwyer, 2023).

However, globalisation may cause unemployment in industrialised countries because firms move their factories to places where they can get cheaper workers. While TNCs can increase the productivity of labour by supplying foreign technology and employing better training methods, the jobs created may be low-skilled with the TNC employing expatriate workers for the more senior and skilled roles (Sarfraz et al., 2018). It has been argued that the inflow of imports that replaces domestic production leads to reduced employment (Urata & Doan, 2022). At the same time, TNCs may operate as local monopsonies of labour, pushing wages lower than the free market equilibrium.

Globalisation is causing people in developed countries to lose their livelihood because of outsourcing, while people in developing countries often face poor working conditions and poor compensation (Tridico, 2018). Given that globalisation enables workers to move more freely, some countries find it difficult to hold onto their best-skilled workers, who are attracted by higher wages elsewhere. This 'brain drain' particularly affects emerging tourism destinations. Studies show that the tourism-related informal sector contracts as the tourism-related formal sector expands, and in addition, displaced informal sector individuals are less likely to be absorbed into the formal economy, which is dominated by multinationals (Dwyer & Čavlek, 2019).

Tourism is a relatively labour-intensive industry, tending to employ relatively more females and young people than other industries (Robinson et al., 2019) and increased employment due to the effects of economic globalisation can enhance this. However, the quality of tourism jobs created is of particular concern. The jobs created in tourism typically are low-skilled and with poor career paths, and with employment of expatriate workers for the more senior and skilled roles (Robinson et al., 2019).

New Technologies

Supply-chain intermediaries represent the largest and fastest-growing component of the corporate revenue pool (Fritz, 2019). This has increased competition in virtually every industry. FDI can help connect host destination businesses to GVC. Active involvement in these value chains brings many benefits to domestic firms including reductions in production and transport costs, increased information technology capabilities, and the capacity to take advantage of niche markets (Kyove et al., 2021). With scale economies generating lower marginal costs and with new types of assets, such as data and algorithms, tech firms are able to bypass traditional value chains, reduce search costs and information asymmetries, and find new ways to source or deliver products and provide new services. In the process, they are transforming the business models of established firms (Dredge et al., 2019).

The digital disruption of value chains is most visible in consumer-facing industries such as tourism, where traditional intermediaries have come under pressure or have been crowded out by tech firms and their digital platforms. Technology-driven consolidation processes are changing the nature of the global tourism industry. TNCs in hospitality (Airbnb); transportation (Uber); payments (PayPal); and retail (Amazon) are likely to face a far more competitive environment, in part owing to the growth of digital trade, which has allowed new technology giants to enter different industries. The new technologies with sophisticated database management systems provide businesses with the tools to respond to individual preferences and stimulate tourism purchases. In contrast, local SME lack the necessary financial and technical resources to fully realise digital potentials (Dredge et al., 2019).

Technological spillovers are particularly important for developing countries, providing opportunities for them to narrow the productivity gap with the developed economies. TNCs assist in technology and knowledge transfer, involving also a 'demonstration effect' on local entrepreneurs (Dwyer, 2015). Since the ability to 'internalise' the spillovers generated by the presence of TNCs is unlikely to be uniformly distributed across domestic firms, policies to improve the absorptive capacity of domestic firms (ability to internalise knowledge created by others) are likely to increase the chances of positive FDI spillovers (Narula & Pineli, 2019). A barrier to the ability of developing countries to gain the full benefits from technology transfer is that they do not have a proper infrastructure in place to leverage technological benefits. This can be disadvantageous to the developing countries and can lead to economic disparity as indicated by the Travel and Tourism Competitiveness rankings (World Economic Forum, 2019).

Destination Marketing

Emerging destinations characterised by weak and under-resourced national tourism offices, often lack resources to mount sustained marketing campaigns in major potential long-haul markets in which awareness of their region as a tourism destination is low. The marketing arms of TNCs can help to overcome this (Dwyer, 2022). Multinational hotel companies in particular provide market connections including marketing ties through computer reservation systems with a variety of firms within a destination. These types of market connections related to global marketing and access to global distribution networks can generate greater flows of tourists to those countries that host FDI and TNCs involved in the tourist industry. TNCs may also be better able to monitor and control the quality of the services they provide to international tourists. Standards in hospitality management and the hotel sector are now judged by global comparisons, even at the more localised level of accommodation (Dwyer & Čavlek, 2019; Ivanov & Ivanova, 2016).

Social Effects

Since the cultures of those countries that have greater economic power are likely to dominate others, it is no surprise that global tourism threatens indigenous knowledge, social structures and relationships (Higgins-Desboilles & Whyte, 2015). Critics have identified the phenomenon of 'cultural globalisation', or 'McDonaldisation', the notion that there is now a global and common monoculture transmitted and reinforced by the internet, popular entertainment, transnational marketing of specific brands, and international tourism, that transcends local cultural traditions and lifestyles and shapes people's perceptions, aspirations, tastes, and daily activities (Al-Zo'by, 2019). In many developing countries, tourism's effects on indigenous peoples have been profound including eviction from their lands, economic dislocation, breakdown of traditional values and environmental degradation. Many critics point out that tourists' quest for 'authenticity' often leads to a 'commodification' and erosion of the local culture for the demand and enjoyment of the tourists and that ghettoism and alienation are often the product of large 'tourist enclaves', isolated from local lifestyles (Higgins-Desboilles & Whyte, 2015).

On a brighter note, populations are responding to the globalisation of economies, markets, systems and cultures by exploring their own identities (Zajda & Majhanovich, 2021). This tension between 'modernity' and 'identity' is leading to the increasing questioning by local communities of the form and scale of tourism development and appropriate type of destination marketing to underpin sustainable development. Increased community involvement in tourism planning and development can result in a growing dispersion of tourism away from mass tourism dominated by TNCs, into diverse niche markets. From this perspective, globalisation appears to pull in two ways, both towards the creation of TNCs by the extension of operations through take-overs, the formation of strategic alliances or franchising agreements, and, in terms of consumption, towards more localised or regional 'branding' and niche marketing with greater local involvement.

TNCs acquire legitimacy and a social license to operate when they are able to create local benefits for the communities in which they operate, offer socially valuable goods, engage with the local community and participate in CSR-based activities (Sinkovics & Archie-acheampong, 2020). The broader roles and responsibilities of TNCs that require going beyond traditional CSR are now being addressed in debates over the nature and implementation of social value creation and how social value creation can be better embedded within TNCs (Lind et al., 2018).

Environmental Protection

All destinations face intensified environmental and social problems as a result of population growth, economic development and rapid urbanisation. Globalisation is argued to foster the overuse and exploitation of natural resources to meet new higher demands in the production of goods (Yan, 2019) Increased consumption leads to an increase in the production of goods, which in turn puts stress on the environment via its carbon footprint. The natural environment and climate conditions are very important in determining the viability and attractiveness of a region as a tourist destination. Tourism transport, especially aviation, is a high emitter of greenhouse gases. Human, animal, and plant diseases can spread more quickly through globalisation. In terms of urban health risks, the steady increase in international travel is a driving force in the global emergence and spread of infectious diseases. The main environmental trends such as climate change, depletion of natural resources and loss of biodiversity (Dwyer, 2023), are all associated with globalisation. A TNC may set up operations in particular destinations because environmental laws are not as strict as they are elsewhere.

The fact that TNCs are footloose does not mean they are environmental vandals. Globalisation has brought about an understanding for the protection of the environment from which they benefit so much (Dwyer, 2015). Globalisation enhances the ability of numerous countries to work together to overcome natural disasters and global challenges, such as global warming and carbon emissions. There is evidence of a rising degree of environmental conscientiousness, as evinced, for example, through the engagement of respondents in organisations concerned with environmental issues. This rising environmental awareness leads to new patterns of tourist behaviour, as exemplified by the development of alternative tourism, in opposition to mass tourism (Dwyer, 2018).

The OECD *Guidelines for Multinational Enterprises* - recommendations to promote positive contributions by enterprises to economic, environmental, and social progress worldwide - are adhered to by a large proportion of TNCs involved in tourism (Reinert et al., 2022). The Guidelines, covering issues such as the environment, human rights, fairness in operating practices, organisational governance, labour practices, community participation and development, and consumer protection, are supported by a unique implementation mechanism of National Contact Points (NCPs), agencies established by adhering governments to promote and implement responsible business conduct. While TNCs are active in the process of legitimising CSR through formulating goals and targets in their subsidiaries (Wrana & Diez, 2018), the Guidelines are non-binding on signatories. However, governments have different tools at their disposal to promote responsible business conduct and to lead by example through economic diplomacy instruments, public procurement, state-owned enterprises, and trade and investment policies (Nieuwenkamp, 2018).

Power and Influence

Foreign ownership of tourist facilities renders the conduct of firms in the tourism industry subject to political and economic agenda developed outside the host nation. As a result of the leverage provided by economic liberalisation, TNCs are increasingly able to influence intergovernmental negotiations and rule-making, with continuing erosion of community engagement and control (Ruggie, 2018).

The high concentration of firms in some tourism sectors creates market power and the potential for abuse by large firms as evidenced by exclusionary agreements, price fixing, market sharing among dominant operators or boycott and refusal to deal with operators in developing countries (Dwyer, 2015). Anti-competitive agreements and conduct can impose substantial costs particularly on developing economies eliminating many of the benefits of globalisation (Timothy, 2018). A related concern is that developing countries can become over dependent on vertically integrated multinationals for the development of their tourism industry given the market connections enjoyed by international tour operators, hotel companies and airlines. The concern is that tourism receiving countries may be placed in a dependent, vulnerable situation because international tourism demand is determined by the actions of stakeholders based in other countries.

Conclusions

The discussion indicates the hazards of generalising too broadly regarding globalisation's impacts on tourism and the effects of TNC operations on economies, communities and natural environments. Globalisation has positive and negative effects in many areas of activity.

The latter may be reduced by developing sustainable forms of tourism through initiatives such as commitments from tourism companies to employ local people at fair wages and provide job training; enhancement of local economic linkages, establishment of environmental charters and industry codes of conduct, educating both visitors and businesses on traditional cultures and behaviour; and community participation in planning and decision-making processes. It is likely, however, that developing countries may need a minimum level of economic development in order to capture the growth-enhancing effects of FDI and TNC operations. More case studies are required to identify the effects in particular locations.

The tourism market is consistently heading towards further consolidation. Some critics (Cohen, 2012) argue that the growing incidence of crises - environmental, political and economic - encourages a counter trend to the globalisation process, with possible significant implications for the future of the global travel system. Large travel and tourism operators will continue to cater for a larger volume of tourist movements, accommodation, mainstream or mass tourism. At the other end of the scale, niche operators will further offer special products and services, experiences to individuals, and tourists demanding customisation. TNCs may be expected to increase their market share by using more and more newly specialised products which they will be able to deliver in close connection with SME specialised in niche marketing.

We agree with Cohen (2012) that the most significant emergent counter trend to globalisation with far reaching consequences for tourism is the threat of a profound structural crisis in the global economic system. However, it is impossible to determine the likelihood of a transformation of the dominant global political economic system or the direction in which such a transformation might take place. Even a disruptive event the size of the Covid pandemic is unlikely to result in the type of transformation required. Neither can we predict the future strength of the anti-consumerist movement and of a consequent 'de-consumerisation' of Western societies. In the meantime, a host of opportunities await tourism researchers to identify the effects of globalisation in different tourism contexts and to inform policy making to minimise adverse effects.

References

Agarwal, S., & Al Qouyatahi, K. M. S. (2018) HRM challenges in the age of Globalisation. *International Research Journal of Business Studies*, *10*(2), 89–98. https://doi.org/10.21632/irjbs.10.2.89-98

Al-Zo'by, M. (2019). Culture and the politics of sustainable development in the GCC: identity between heritage and globalisation. *Development in Practice*, *29*(5), 559–569. https://doi.org/10.1080/09614524.2019.1602110

Aspers, P., & Kohl, S. (2015). Economic theories of globalization. In B. Turner & R. Holton (Eds), *The Routledge international handbook of globalization studies* (pp. 61–79). Routledge.

Bohn, T., Brakman, S., & Dietzenbacher, E. (2018). The role of services in globalisation. *The World Economy*, *41*(10), 2732–2749. https://doi.org/10.1111/twec.12640

Clark, R., & Kwon, R. (2018). Taking stock of flow: Revisiting the link between FDI and human rights. *Sociology of Development*, *4*(4), 346–373. https://doi.org/10.1525/sod.2018.4.4.346

Cohen, E. (2012). Globalization, global crises and tourism. *Tourism Recreation Research*, *37*(2), 103–111. https://doi.org/10.1080/02508281.2012.11081695

Dredge, D., Phi, G. T. L., Mahadevan, R., Meehan, E., & Popescu, E. (2019). *Digitalisation in tourism: In-depth analysis of challenges and opportunities*. Executive Agency for Small and Medium-sized Enterprises (EASME), European Commission. https://ec.europa.eu/docsroom/documents/33163/attachments/1/translations/en/renditions/native

Dwyer, L. (2015). Globalization of tourism: Drivers and outcomes. *Tourism Recreation Research*, *40*(3), 326–339. https://doi.org/10.1080/02508281.2015.1075723

Dwyer, L. (2018). Saluting while the ship sinks: the necessity for tourism paradigm change. *Journal of Sustainable Tourism*, *26*(1), 29–48. https://doi.org/10.1080/09669582.2017.1308372

Dwyer, L. (2022). Foreign direct investment in tourism: Host destination opportunities and challenges. In H. C. Jonsson (Ed.), *Tourism and foreign direct investment: Issues, challenges and prospects* (pp. 11–30). Routledge.

Dwyer, L. (2023). Resident well-being and sustainable tourism development: the 'capitals Approach'. *Journal of Sustainable Tourism*, 31(9), 2119–2135. https://doi.org/10.1080/09669582.2021.1990304

Dwyer, L., & Čavlek, N. (2019). Economic globalisation and tourism. In D. Timothy (Ed.), *Handbook of globalisation and tourism* (pp. 12–26). Edward Elgar.

European Commission (EU). (2019). *Transnational business cooperation*. https://ec.europa.eu/growth/sectors/tourism/business-portal/internationalisation-tourism-businesses/transnational-business-cooperation_en

Findlay, C., & Roelfsema, H. (2021). Services trade in the ASEAN region and participation in GVCs. *Journal of the Asia Pacific Economy*. https://doi.org/10.1080/13547860.2021.1920658

Fritz, M. M. (2019). Sustainable supply chain management. In W. Leal Filho, A. M. Azul, L. Brandli, P. G. Özuyar, & T. Wall (Eds), *Responsible consumption and production . Encyclopedia of the UN sustainable development goals*. Springer. https://doi.org/10.1007/978-3-319-71062-4_21-1

Fuster, B., Lillo-Bañuls, A., & Martínez-Mora, C. (2018). Offshoring of services as a competitive strategy in the tourism industry. *Tourism Economics*, *24*(8), 963–979. https://doi.org/10.1177/1354816618785536

Heimberger, P. (2019). Does economic globalisation affect income inequality? A meta-analysis. Working paper 165, *The Vienna Institute for International Economic Studies*, wiiw.

Higgins-Desbiolles, F., & Whyte, K. P. (2015). Tourism and human rights. In A. Holden & D. A. Fennell (Eds), *The Routledge handbook of tourism and sustainability* (pp. 105–116). Routledge.

Hopkins, D. (2021). Crises and tourism mobilities. *Journal of Sustainable Tourism*, *29*(9), 1423–1435. https://doi.org/10.1080/09669582.2021.1905969

Horner, R., Schindler, S., Haberly, D., & Aoyama, Y. (2018). Globalisation, uneven development and the North–South 'big switch'. *Cambridge Journal of Regions, Economy and Society*, *11*(1), 17–33. https://doi.org/10.1093/cjres/rsx026

Huh, H. S., & Pak, C. Y. (2019). A new index of globalization: Measuring impacts of integration on economic growth and income inequality. *Asian Development Bank Economics Working Paper Series*, *587*. https://www.proquest.com/docview/2780427256?pq-origsite=gscholar&fromopenview=true&sourcetype=Working%20Papers

Iamsiraroj, S. (2016). The foreign direct investment–economic growth nexus. *International Review of Economics & Finance*, *42*, 116–133. https://doi.org/10.1016/j.iref.2015.10.044

Intriligator, M. (2005). Globalisation of the world economy: Potential benefits and costs and a net assessment. In P. Gangopadhyay & M. Chatterji (Eds), *Economics of globalisation* (pp. 85–94). Ashgate Publishing.

Ivanov, S., & Ivanova, M. (2016). Do hotel chains stimulate globalisation? *Tourism Management Perspectives*, *19*, 102–108. https://doi.org/10.1016/j.tmp.2016.05.004

Ivanov, S., & Webster, C. (2013). Tourism's impact on growth: The role of globalisation. *Annals of Tourism Research*, *41*, 231–236. https://doi.org/10.1016/j.annals.2012.12.008

Kyove, J., Streltsova, K., Odibo, U., & Cirella, G. T. (2021). Globalization impact on multinational enterprises. *World*, *2*(2), 216–230. https://doi.org/10.3390/world2020014

Li, K. X., Jin, M., & Shi, W. (2018). Tourism as an important impetus to promoting economic growth: A critical review. *Tourism Management Perspectives*, *26*, 135–142. https://doi.org/10.1016/j.tmp.2017.10.002

Lind, C. H., Kang, O., Ljung, A., & Forsgren, M. (2018). MNC involvement in social innovations: the issue of knowledge, networks and power. *Critical Perspectives on International Business*, *16*(1), 79–99. https://doi.org/10.1108/cpoib-09-2017-0060

Meethan, K. (2004). Transnational corporations, globalization, and tourism. In A. A. Lew, C. M. Hall & A. M. Williams (Eds), *Companion to tourism*, (pp. 110–121). Blackwell.

Mowforth, M., & Munt, I. (Eds).(2015). *Tourism and sustainability: Development, globalisation and new tourism in the third world* (4th ed.). Routledge.

Narula, R., & Pineli, A. (2019). Improving the developmental impact of multinational enterprises: policy and research challenges. *Journal of Industrial and Business Economics*, *46*(1), 1–24. https://doi.org/10.1007/s40812-018-0104-2

Narula, R., & Van der Straaten, K. (2021). A comment on the multifaceted relationship between multinational enterprises and within-country inequality. *Critical Perspectives on International Business*, *17*(1), 33–52. https://doi.org/10.1108/cpoib-10-2019-0080

Nieuwenkamp, R. (2018). Responsible FDI is no longer optional. *Columbia FDI Perspectives, No. 220*, Columbia University, Columbia Center on Sustainable Investment (CCSI). https://www.econstor.eu/bitstream/10419/254054/1/fdi-perspectives-no220.pdf

Payne, D. M., & Raiborn, C. A. (2018). Aggressive tax avoidance: A conundrum for stakeholders, governments, and morality. *Journal of Business Ethics*, *147*(3), 469–487. https://doi.org/10.1007/s10551-015-2978-5

Ravallion, M. (2018). Inequality and globalization: A review essay. *Journal of Economic Literature*, *56*(2), 620–42. https://doi.org/10.1257/jel.20171419

Reinert, K. A., Reinert, O. T., & Debebe, G. (2022). OECD guidelines for multinational enterprises. In K. A. Elliott (Ed.), *Handbook on globalisation and labour standards* (pp. 340–358). Edward Elgar.

Robinson, R. N., Martins, A., Solnet, D., & Baum, T. (2019). Sustaining precarity: Critically examining tourism and employment. *Journal of Sustainable Tourism*, *27*(7), 1008–1025. https://doi.org/10.1080/09669582.2018.1538230

Roudi, S., Arasli, H., & Akadiri, S. S. (2019). New insights into an old issue–examining the influence of tourism on economic growth: evidence from selected small island developing states. *Current Issues in Tourism*, *22*(11), 1280–1300. https://doi.org/10.1080/13683500.2018.1431207

Roukanos, S. (2020). Measuring economic development and the impact of economic globalisation. *Studies in Business & Economics*, *15*(3), 185–207. https://doi.org/10.2478/sbe-2020-0053

Ruggie, J. G. (2018). Multinationals as global institution: Power, authority and relative autonomy. *Regulation & Governance*, *12*(3), 317–333. https://doi.org/10.1111/rego.12154

Rygh, A. (2019). Multinational enterprises and economic inequality: A review and international business research agenda. *Critical Perspectives on International Business*, *17*(1), 72–102. https://doi.org/10.1108/cpoib-09-2019-0068

Sarfraz, I., Rajendran, D., Hewege, C., & Mohan, M. D. (2018). An exploration of global employability skills: a systematic research review. *International Journal of Work Organisation and Emotion*, *9*(1), 63–88. https://www.researchgate.net/profile/Isra-Sarfraz/publication/324579805_An_exploration_of_global_employability_skills_a_systematic_research_review/links/5b9f00de299bf13e6037c57e/An-exploration-of-global-employability-skills-a-systematic-research-review.pdf

Sen, K. (2019). What explains the job creating potential of industrialisation in the developing world? *The Journal of Development Studies*, *55*(7), 1565–1583. https://doi.org/10.1080/00220388.2017.1404033

Sheldon, P., & Dwyer, L. (2010). The global financial crisis and tourism: Perspectives of the academy. *Journal of Travel Research*, *49*(1), 3–4. https://doi.org/10.1177/0047287509353191

Sinkovics, N., & Archie-acheampong, J. (2020). The social value creation of MNEs–a literature review across multiple academic fields. *Critical Perspectives on International Business*, *16*(1), 7–46. https://doi.org/10.1108/cpoib-06-2017-0038

Sokhanvar, A. (2019). Does foreign direct investment accelerate tourism and economic growth within Europe? *Tourism Management Perspectives*, *29*, 86–96. https://doi.org/10.1016/j.tmp.2018.10.005

Song, H., Li, G., & Cao, Z. (2017). Tourism and economic globalization: An emerging research agenda. *Journal of Travel Research*, *57*(8), 999–1011. https://doi.org/10.1177/0047287517734943

Sułkowski, Ł. (2020). Covid-19 pandemic; recession, virtual revolution leading to de-globalization? *Journal of Intercultural Management*, *12*(1), 1–11. https://www.ceeol.com/search/article-detail?id=846210

Szmigin, I., Carrigan, M., & McEachern, M. G. (2009). The conscious consumer: Taking a flexible approach to ethical behaviour. *International Journal of Consumer Studies*, *33*(2), 224–231. https://doi.org/10.1111/j.1470-6431.2009.00750.x

Talani, L., & Roccu, R. (Eds).(2019). *The dark side of globalisation*. Springer.

Timothy, D. (Ed.). (2019). *Handbook of globalisation and tourism*. Edward Elgar.

Timothy, D. J. (2018). Globalization, supranationalism and tourism. In C. Cooper, S. Volo, W. C. Gartner, & N. Scott (Eds.), *The Sage handbook of tourism management* (pp. 569–581). Sage.

Tridico, P. (2018). The determinants of income inequality in OECD countries. *Cambridge Journal of Economics*, *42*(4), 1009–1042. https://doi.org/10.1093/cje/bex069

United Nations (UN). (1989). *Foreign direct investment and transnational corporations in services*. Centre on Transnational Corporations (UN).

Urata, S., & Doan, H. T. T. (Eds). (2022). *Globalisation and its economic consequences: Looking at APEC economies*. Taylor & Francis.

Ushakov, D., Fedorchenko, V., Fedorchenko, N., Rybachok, V., & Bazhenov, M. (2020). Brief geographical and historic overview of tourism transnationalization. *Geographical Journal of Tourism and Geosites*, *31*(3), 1180–1185. https://doi.org/10.30892/gtg.31333-556

Wearing, S., McDonald, M., Taylor, G., & Ronen, T. (2019). Neoliberalism and global tourism. In D. Timothy (Ed.), *Handbook of globalisation and tourism* (pp. 27–42). Edward Elgar.

World Economic Forum (2019). *The travel & tourism competitiveness report 2020*. WTTC.

Wrana, J., & Diez, J. R. (2018). Multinational enterprises or the quality of regional institutions–What drives the diffusion of global CSR certificates in a transition economy? Evidence from Vietnam. *Journal of Cleaner Production*, *186*, 168–179. https://doi.org/10.1016/j.jclepro.2018.03.113

Yan, J. (2019). Globalisation has double sides on developing countries: The benefits and the harm. *Journal of Business and Economic Development*, *4*(1), 11–14. https://doi.org/10.11648/j.jbed.20190401.12

Zajda, J. I., & Majhanovich, S. (2021). (Eds.) *Globalisation, cultural identity and nation-building: The changing paradigms*. Springer.

Chapter 22

Tourism and Public Transport

Inoormaziah Azman, Lim Xin Jean, C. Michael Hall, & Girish Prayag

Introduction: Tourism and Transport

Transport is essential for the tourism industry, as it facilitates mobility and movement of tourists from one place to another. A destination with better connections has a better chance to use tourist capacities and for overall area development (Kovačić & Milošević, 2016). If a destination has a poor transportation system that hinders tourists' movement, tourists will potentially seek other alternative destinations (Prideaux, 2000). However, transport and tourism development projects are often planned in isolation from each other, despite the crucial role of transport in tourism (Page & Ge, 2009; Yusoff et al., 2021). Investing in tourism infrastructure, marketing efforts and increasing air services may not be sufficient for tourism destination development if there is no efficient transport infrastructure provided (Khadaroo & Seetanah, 2008).

Although transport is vital for tourism, it also has some serious environmental impacts, including emissions, noise, and land use change (Le-Klähn et al., 2014b; Le-Klähn & Hall, 2015; Hall et al., 2017; Høyer, 2004, Miravet et al., 2021). Therefore, finding solutions to make travelling more environmentally friendly is essential (Kovačić & Milošević, 2016; Hall, 2022). A well planned public transport system is therefore essential for sustainable transport and mobility in many countries around the world (Le-Klähn et al., 2014b; Ban, 2016). From a sustainable tourism perspective, the use of public and active transport by tourists is encouraged to mitigate the negative environmental impacts (Le-Klähn & Hall, 2014). Given the need to develop sustainable transport for tourism (United Nations (UN), 2016), the use of public transport by tourists at destinations would appear to be essential.

Despite the important interrelationships between transport and tourism, research on public transport and tourism remains limited. The role of transport can be generally seen in two ways:

- An essential/utility of bringing the tourists to tourism destinations and enabling them to move between places at the destinations (transport for tourism); and/or
- transport that is used as part of the tourism attractions to enrich the tourists' experiences (transport as tourism).

The Wiley Blackwell Companion to Tourism, Second Edition. Edited by C. Michael Hall.

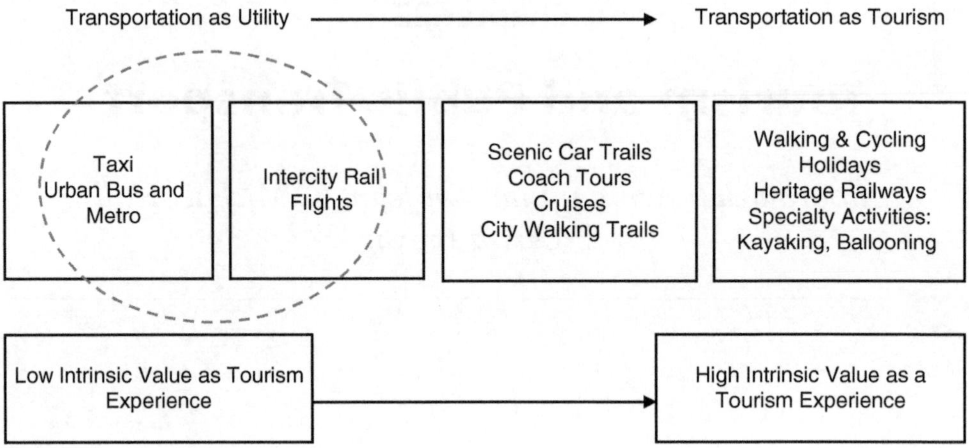

Figure 22.1. The tourism transportation continuum.

Public transport is usually considered as a basic transport service (Figure 22.1) (Lumsdon & Page, 2004). Public transport research has also largely focused on local travel, and there is a need for more knowledge about the specifics of regional public transport (Hansson et al., 2019). This circumstance raises the question, if public transport is or can become the tourists' choice for travelling long-haul, especially if more time-consuming or expensive than other transport modes.

In this chapter, the authors examine tourist use of public transport at tourism destinations, emphasising the various factors that influence tourists' travel-transport behaviour, in terms of personal characteristics (socio-demographic and psychographic factors), trip characteristics and public transport attributes. The chapter then discusses the relationship between tourism and public transport and public transport as tourist experience and attraction before concluding.

Tourist Use of Public Transport

From the context of demand, understanding tourist use of public transport is essential for transport and tourism planning. Public transport is usually perceived to have low intrinsic value as a tourism experience (Lumsdon & Page, 2004). Thus, customer orientation (from a demand perspective) is an important aspect to elaborate the transport supply for leisure purposes in tourism regions (Le-Klähn et al., 2014a; Gronau & Kagermeier, 2007). Demand related factors that influenced tourist use and non-use of public transport at tourism destinations are discussed and encompasses tourists' socio-demographic and psychographic factors, their trip characteristics and public transport attributes that influence travel/public transport behaviour.

In some studies, the tourist/visitor public transport users can be differentiated from the non-users based on certain distinctive socio-demographic factors (Table 22.1). The socio-demographic factors which were used in earlier studies in profiling research respondents were discovered to have significant associations with broader travel and transport behaviours in subsequent public transport/tourism studies (i.e., Masiero & Zoltan, 2013; Le-Klähn et al., 2014a; Bieland et al., 2017; Gross & Grimm, 2018).

Table 22.1. Tourists' socio-demographic characteristics and use of public transport.

Author(s)	*Year*	*Characteristics*								
		Car Ownership in a household	*Age*	*Gender*	*Income*	*Occupation*	*Tourist origin*	*Education level*	*Marital status*	*Driving license*
Downward & Lumsdon	2004	X	–	–	–	–	–	–	–	–
Lumsdon et al.	2006	O	X	O	X	X	–	–	–	–
Thompson & Schofield	2007	–	X	X	–	X	X	–	–	–
Su & Wall	2009	–	X	X	X	X	X	X	–	–
Chang & Lai	2009	–	X	–	–	–	X	X	X	–
Farag & Lyons	2012	–	O	O	O	X	–	O	–	–
Hergesell & Dickinger	2013	–	X	X	X	–	X	–	–	–
Huang et al.	2013	–	–	–	–	–	X	–	–	–
Masiero & Zoltan	2013	–	X	X	–	–	X	–	–	–
Parahoo et al.	2014	–	–	X	–	X	X	–	X	–
Le-Klähn et al.	2014a	X	O	X	–	–	X	X	–	O
Le-Klähn et al.	2014b	X	X	X	–	–	X	X	–	X
Le-Klähn et al.	2015	–	–	–	–	–	X	O	–	–
Gutiérrez & Miravet	2016	–	X	X	X	–	O	–	–	–
Yuksek et al.	2016	–	X	X	X	X	–	X	–	–
Bieland et al.	2017	–	X	–	–	–	–	–	X	–
Bajada & Titheridge	2017	–	X	X	–	–	–	–	–	–
Mulley et al.	2017	–	X	X	–	–	–	–	–	X
Nutsugbodo et al.	2018	–	O	O	O	O	O	O	X	–
Gross & Grimm	2018	O	O	X	O	O	O	X	O	–
Nwachukwu et al.	2019	X	X	X	–	–	X	X	X	–
Yeboah et al.	2019	X	X	X	–	O	–	–	–	X
Tang et al.	2020	O	O	O	–	–	–	–	–	–
Qi et al.	2020	X	–	–	–	X	X	–	–	X
Zamparini & Vergori	2021	–	X	X	X	–	X	–	–	–

(*Continued*)

Table 22.1. *(Continued)*

Author(s)	*Year*	*Characteristics*								
		Car Ownership in a household	*Age*	*Gender*	*Income*	*Occupation*	*Tourist origin*	*Education level*	*Marital status*	*Driving license*
Miravet et al.	2021	–	X	X	–	–	X	X	–	–
Downward & Lumsdon	2004	–	–	–	–	–	–	–	–	–
Lumsdon et al.	2006	–	–	–	–	–	–	–	–	–
Thompson & Schofield	2007	–	–	–	–	–	–	–	–	–
Su & Wall	2009	–	–	–	–	–	–	–	–	–
Chang & Lai	2009	–	–	–	–	–	–	–	–	–
Farag & Lyons	2012	–	–	–	–	O	O	–	–	–
Hergesell & Dickinger	2013	–	–	–	–	–	–	–	–	–
Huang et al.	2013	–	–	–	–	–	–	–	–	–
Masiero & Zoltan	2013	–	–	–	–	–	–	–	–	–
Parahoo et al.	2014	–	–	–	–	–	–	–	–	–
Le-Klähn et al.	2014a	X	–	–	–	–	–	–	–	–
Le-Klähn et al.	2014b	X	–	–	–	–	–	–	–	–
Le-Klähn et al.	2015	–	–	–	–	–	–	–	–	–
Gutiérrez & Miravet	2016	–	–	–	–	–	–	–	–	–
Yuksek et al.	2016	–	–	–	–	–	–	–	–	–
Bieland et al.	2017	–	X	–	–	–	–	–	–	–
Bajada & Titheridge	2017	–	–	–	–	–	–	–	–	–
Mulley et al.	2017	–	–	–	–	–	–	–	X	–
Nutsugbodo et al.	2018	–	–	–	–	–	–	–	–	–
Gross & Grimm	2018	–	O	X	O	–	–	–	–	–
Nwachukwu et al.	2019	–	–	–	–	–	–	–	–	–
Yeboah et al.	2019	–	–	–	–	–	–	X	–	–
Tang et al.	2020	–	–	–	–	–	–	–	–	–
Qi et al.	2020	–	–	–	–	–	–	–	–	–
Zamparini & Vergori	2021	–	–	–	–	–	–	–	–	X
Miravet et al.	2021	–	–	–	–	–	–	–	–	–

o = significant; x = not significant

Age and Health

Age is one of the most important demographic variables that affect tourists' public transport behaviour (Le-Klähn et al., 2014a; Gutiérrez & Miravet, 2016; Gross & Grimm, 2018). Age may also be related to other factors, such as the state of physical health and leisure time availability. Tourist public transport users tend to be of younger age (less than 30 years old), as they are more active (without health restrictions) in their movement. Younger tourists, who tend to be more tech-savvy, show a greater interest in digital information services and applications that allow them better access to public transport information (Gross & Grimm, 2018), something also identified in the active transport literature (Kim & Hall, 2022a, 2022b; Kim et al., 2023a). Meanwhile, older people tend to use public transport to avoid driving and road congestion (Tang et al., 2020) and buses rather than subways, as the stairways at subway stations are often a hindrance for the elderly and they prefer more comfortable public transport modes. Older tourists (especially those who are retired) also show interest in participating in day trips by public transport as they have more leisure time to do it and are found to be more likely to enjoy their trips for relaxation, sightseeing and the company of others while travelling on public transport (Lumsdon et al., 2006). Health restrictions are also an effect of age and are an issue in relation to tourist use of public transport (Le-Klähn et al., 2014a, 2014b), highlighting the importance of public transport infrastructure and facilities that are accessible for all (Hall et al., 2017).

Gender

Gender has high and moderate significant effects on transport mode choice and preferences, and pre-trip public transport information searches. Studies suggest that males tend to prefer driving cars or private vehicles and are less frequent users of public transport when they travel (Masiero & Zoltan, 2013; Tang et al., 2020), while females tend to be the public transport users and willing to search for public transport information prior to their trips (Lumsdon et al., 2006; Farag & Lyons, 2012). In Gutiérrez and Miravet's (2016) study, male tourists were found to be more likely to use private cars than female tourists when travelling. While at the destination, gender was not found to be significant in influencing tourists' transport mode choice. The different gender effects in the studies could potentially be related to differences in male and female travel patterns (Farag & Lyons, 2012, Nobis & Lenz, 2005).

Level of Education

There are positive relationships between education level and tourists/visitors use of public transport at tourism destinations, as younger and educated people are more likely to be well travelled and more familiar with the public transport system (Le-Klähn et al., 2015; Nutsugbodo et al., 2018). Additionally, more educated persons consult public transport information more often before making a leisure trip (Farag & Lyons, 2012).

Level of Income/Household Income

Public transport users are generally assumed to be middle- and low-income earners, and public transport is often viewed as the cheaper alternatives for tourists to move around at tourism destinations. However, tourists with high levels of income, who indicated a higher level of spending have been found to have a greater tendency to use public transport at tourism destinations (Gutiérrez & Miravet, 2016). Although those with higher incomes have been found to prefer more comfortable modes of transport like buses and taxis over cheaper alternatives (Nutsugbodo et al., 2018).

Occupation/Profession

Students are more likely to use public transport (Qi et al., 2020). Nutsugbodo et al. (2018) confirmed a significant relationship between occupation and tourists' mode of public transport, and it is evident that students preferred to use buses and minibuses (trotro) compared to formal and informal workers, who were more likely to take taxis. Employment status also has a significant effect on pre-travel information seeking behaviour, with unemployed persons being more likely to seek pre-travel information on public transport (Yeboah et al., 2019).

Tourist Origin

Tourist origin is an important factor that may affect the tourists' transport behaviour. There are two interpretations of tourist origin seemingly based on the literature. Tourist origin may indicate their familiarity with and frequency of use of public transport in their home or the distance that the tourists need to travel to arrive at the destinations. This may determine the initial mode of travel upon arriving at their holiday destination and tourist use of public transport during their stay (Gutiérrez & Miravet, 2016). Tourists who arrived by plane are more likely to use public transport to travel around the destinations, compared to those who arrived by private cars (Miravet et al., 2021). The choice of transport mode is distance related, and the automobile is the more preferred mode of transport in short-distance trips which may be less than 1,000km (Bieland et al., 2017).

Car Ownership and Valid Driving License

Car ownership can be considered as one of the important factors that influence an individuals' travel mode choice (He & Thøgersen, 2017), including use of public transport. People who have more favorable attitudes towards cars are more likely to travel by car, because they are most likely to own a car. People who do not own a car in their household are more familiar with public transport and more likely to use public transport at tourism destinations (Gross & Grimm, 2018) and are also more willing to use bikes, buses, or subways (Tang et al., 2020).

Marital Status and Household Size

Young singles and couples, and adult singles are most likely to use public transport at tourism destinations (Bieland et al., 2017; Gross & Grimm, 2018). For families, car travel is the most preferred way to travel (Tang et al., 2020). In addition, considering the baggage that they will need to bring for the trips, travel by private cars is deemed convenient for tourists who travel with family. Nerhagen (2003) pointed out that the individuals who attached great importance to flexibility on the actual trip were less likely to choose public transport.

Language

One of the greatest challenges of being independent international tourists is language barriers (Chang & Lai, 2009). A language barrier may make public transport appear less attractive to tourists/visitors as they are not familiar with the destinations and even the language (Le-Klähn & Hall, 2014). Public transport with a bilingual or multiple language information (e.g., combination of native language and English) environment is a basic requirement for international visitors to be comfortable to use public transport at destinations (Chang & Lai, 2009; Le-Klähn & Hall, 2014).

Internet/Information and Communications Technology (ICT)

Frequent internet users and people who have easy access to the internet (especially at home) are more likely to use public transport information for various trips. Those who never use public transport information tend to have a negative public transport attitude, infrequently or never travel by train, and tend to be less educated. Well published online or offline public transport information can be considered important to encourage public transport use among tourists and their wayfinding at destinations. Public transport related ICT use may be appealing to tourists from younger age groups as they show greater overall interest in digital information services (Gross & Grimm, 2018), and perhaps are more likely to use smartphones, tablets, or other electronic devices.

Tourists' socio-demographic characteristics affect their use and non-use of public transport but there are variations between studies. Nevertheless, there does appear to be differences between public transport users and non-users among different groups of domestic and international tourists. Independent socio-demographic variables are not the sole factors that influence visitors/tourists public transport tourism behaviour, but instead are associated with other independent variables such as the individuals' psychographic factors, their trip characteristics and public transport attributes.

Tourists' psychographic characteristics encompass their perceptions, motivations, preferences, and attitudes towards public transport. Unlike socio-demographic factors, psychographic variables tend to be multidimensional, and associations exist between dimensions and variables. A number of these factors help provide explanations for tourists' travel and transport behaviour, particularly in relation to their perceptions related to transport mode choice, which may be influenced by a range of internal and external factors. The psychographic variables that influence the tourist use of public transport at tourism destinations are presented in Tables 22.2 and 22.3.

Public Transport Use in Home Environment

Tourist use of public transport in their home environment indicates sustainable behaviour and familiarity with public transport, and this influences their behavior when they travel. There are studies which reported that frequent use of public transport in tourists' everyday lives, resulted in a higher likelihood for them to use public transport when they travel. However, there are also studies that show limited associations between individuals' everyday mobility and transport behaviour (Downward & Lumsdon, 2004; Lumsdon et al., 2006; Le-Klähn et al., 2014b; Nwachukwu et al., 2019; Mulley et al., 2017; Yeboah et al., 2019).

Table 22.2. Tourists' psychographic characteristics and their use of public transport.

Characteristics	*Author(s) and year*	*Remarks*
Public transport use in home environment	Downward & Lumsdon (2004) Lumsdon et al. (2006) Le-Klähn et al. (2014a, 2014b) Bieland et al. (2017) Nwachukwu et al. (2019) Mulley et al. (2017) Yeboah et al. (2019)	Related to how frequent respondents use public transport at home which may indicate sustainable travel behaviour and familiarity with public transport.
Travel motivations	Su & Wall (2009) Masiero & Zoltan (2013) Le-Klähn et al. (2015) Gutiérrez & Miravet (2016)	Travel motivation indicates tour activities that the tourists partake during their trips, which further implies their travel movements.
Travel experiences	Chang & Lai (2009)	Travel experiences may be related to tourists' familiarities with the destination and may influence tourist behaviour like transport mode choice.
Experiences of using public transport when travelling	Chang & Lai (2009) Gross & Grimm (2018)	This indicates tourists' familiarities with public transport provision when they travel.
Public transport satisfaction	Su & Wall (2009) Chang & Lai (2009) Le-Klähn et al. (2014b) Parahoo et al. (2014) Nwachukwu et al. (2019)	Level of satisfaction of using public transport at tourism destinations that were measured by using public transport attributes or service quality.
Public transport expectation	Bajada & Titheridge (2017)	Expectations derived from tourists' previous public transport experiences.
Value	Parahoo et al. (2014)	Public transport users' perceived value of public transport quality services. It refers to convenient means of transport, convenient to reach places and accessibility.
Travel attitude	Farag & Lyons (2012) Mulley et al. (2017)	Related to the travellers' attitude towards travel modes (bus, car, or train) and their willingness to use public transport or preference for cars.
Travel behaviour	Farag & Lyons (2012) Mulley et al. (2017)	Travel behaviour is related to how frequently the respondents make certain trips (business or leisure) and travel using specific transport modes.
Social surrounding	Farag & Lyons (2012)	It indicates the influence of other people towards the individual's use of public transport (i.e., receive a recommendation to use certain public transport information service or knowing many people who use public transport regularly).
Environmental attitude	Hergesell & Dickinger (2013)	This factor was tested by adopting the New Ecological Paradigm (NEP) scale (Dunlap et al., 2000).
Environmental behaviour	Hergesell & Dickinger (2013)	Self-assessment with regards to tourists' environmental friendliness based on 5-point rating scale.

Table 22.2. (*Continued*)

Characteristics	*Author(s) and year*	*Remarks*
Use environmentally friendly transport options at home and destinations	Zamparini & Vergori (2021)	The modal options included environmentally friendly transport modes (walking, cycling and public transport) and private motorised vehicles (campers, cars, and motorbikes).

Table 22.3. Tourists' reasons to use and not use public transport.

Author(s) & year	*Study context/area*	*Reasons to use public transport*	*Reasons to not use public transport*
Lumsdon et al. (2006)	Wayfarer ticket users in the Greater Manchester, UK.	To avoid stress *Drive free benefits Facilitate recreational activity Price *Convenience & flexibility *Integration through ticket Tourism/transport experience	–
Su & Wall (2009)	Passengers on-board the trains between Beijing & Lhasa, China.	Novelty Safety View Relaxation Altitude adaptation	–
Chang & Lai (2009)	Inbound tourists at Taipei MRT stations, Taiwan.	To commute To transfer Price Tourism/transport experience	–
Le-Klähn et al. (2014a)	Visitors at tourist sites in Munich, Germany.	*Drive free benefits Advantage of local public transport Car unavailability Traffic reduction Environmental concern	Inconvenience and restrictions Lack of information Disadvantage of local public transport Personal preference
Bieland et al. (2017)	Visitors at the Northern Hesse, City of Kassel, Germany.	Price Convenience & flexibility Attached to a mode of transport	Inconvenience and restrictions Luggage handling service Transit Transport offer
Nutsugbodo et al. (2018)	Inbound tourists leaving Ghana at the Kotoka International Airport.	Price Convenience & flexibility Safety Reliability Accessible Availability Waiting time Speed Comfort	–

* Significant factors

Travel Motivation

Travel motivation is associated with the tourist destination choice, and tourism products and activities they participate in when they travel. Tourists' activities influence their movement and need to (i) travel to, and (ii) travel around, the destination. Le-Klähn et al. (2015) discovered that the activity driven motivation (push factors) had significant positive effects on visitor use of public transport. The visitors who used public transport were the ones that travel to attend activities or events, have rest and relaxation, or shopping and stayed in the city areas. The visitors who travelled beyond the city areas for scenery and visited heritage and cultural sites, were less likely to use public transport due to locational factors.

Public Transport Value, Expectation, and Satisfaction

Satisfaction with public transport services is often associated with its service quality and these perceived dimensions are also related to tourists' personal and trip characteristics. For instance, perceived level of comfort, the importance of public transport frequencies, and reliability and prices among public transport commuters vary depending on the distance travelled and time spent on the trip (Hansson et al., 2019). Additionally, tourist's characteristics, for instance, the country of origin, may influence tourists' perceptions, expectations, and satisfaction with public transport. Familiarity with the places and public transport (networks, systems, and services) can also influence the individuals' movements and activities at tourism destinations. Arguably, tourists who have more travel experiences and had used public transport in other countries tend to use public transport when they travel elsewhere.

Travel Attitude and Behaviour

Tourism transport behaviour at destinations can be influenced by tourists' own attitudes and behaviour. In relation to public transport use, those who never use public transport information to compare travel mode tend to have negative public transport attitudes and behaviour (Farag & Lyons, 2012). People perceive public transport to have low leisure connotation, and lack of function factors compared to cars, and poor services (in terms of extra time to travel and lack of public transport information) (Gronau & Kagermeier, 2007).

Motivations to Use and Not to Use Public Transport

Numerous studies have attempted to explain the reasons or motivations for tourist use and non-use of public transport at tourism destinations in terms of tourist (benefit and movement related purposes) and public transport (attributes and quality) characteristics. Table 22.3 illustrates various reasons for tourists to use and not use public transport identified from previous research.

Tourist use and non-use of public transport may involve three situational factors:

- personal preferences;
- movement related purposes; and
- public transport attributes.

Drive free benefits and other personal factors and preferences, including avoiding stress, to relax and enjoy the scenery, and environmental concern influence tourists to use public transport. They also use public transport for movement purposes; to commute to tourist attractions, to transfer, and to avoid congestion, which is affected by public transport attributes such as price, integration of ticket, convenience, and flexibility. Many studies were in urban areas. However, this does raise questions about the capacity to transfer findings between different locational contexts.

Psychographic factors on tourists' travel and transport behaviour presents a more complex and dynamic explanation of tourist use and non-use of public transport at tourism destinations. The variables are multidimensional, which may provide different interpretations of behaviour and may be influenced by other situational and locational factors. The realisation of psychographic factors in relation to travel and transport behaviour is twofold:

- the intrinsic and extrinsic factors affect tourist use and non-use of public transport; and
- the interactions between psychographic variables offer another perspective in understanding tourists' travel and transport behaviour.

For instance, tourists' perceptions and expectations influence their travel and transport behaviour. Such factors could also be influenced by other psychographic variables such as value, motivations, attitude, experience, satisfaction, and behaviour. Hence, the associations between psychographic factors provide different interpretations for tourist travel and transport behaviour.

While Le-Klähn et al. (2015) contended that most demographic characteristics are not useful in predicting visitors' travel and transport behaviour, Farag and Lyons (2012) and Nutsugbodo et al. (2018) discovered a strong relationship between tourists' socio-demographic factors and their public transport behaviour (mode preference and information search). Nutsugbodo et al. (2018) argued that these factors significantly explain the variance in tourists' public transport mode choice if combined with other psychographic factors (perceived public transport quality). There are many studies that suggest that profiling tourists within their categories (excursionist, domestic or international tourists) could be useful in relation to their use of public transport in a tourism context (Tables 22.2 and 22.3). Thus, identification of target groups at tourism destinations to encourage public transport use is suggested by some authors (Gronau & Kagermeier, 2007) and it is important to assess the characteristics of the targeted visitor groups and customise services according to their needs (Chang & Lai, 2009).

In addition to tourist socio-demographic and psychographic factors, trip characteristics also influence tourist use and non-use of public transport when they travel, as illustrated in Table 22.4. The complex behaviours of tourists at destinations resulted from the interactions between:

- their past behaviours and decisions;
- the characteristics of the destination they are visiting; and
- types of touristic activities they partake (Moore et al., 2012).

While there is little direct examination of business traveller's use of public transport, visitors who travelled for personal purposes (holiday and visiting friends and family (VFR)) show greater usage of public transport when they travel to urban areas, and they generally visit attractions and participate in activities that occur in the city center.

Table 22.4. Trip characteristics and tourist use of public transport.

Author(s)	*Year*	*Characteristics*									
		Purpose of trip	*Travelling partners*	*Public transport availability*	*Access to car during trip*	*Length of stay*	*Repeat or first–time visitors*	*Transport mode for entire journey*	*Travel organisation*	*Trip expenses*	*Types of accommodation*
Downward & Lumsdon	2004	X	X	X	–	–	–	–	–	X	–
Lumsdon et al.	2006	O	–	–	X	–	–	–	–	–	–
Thompson & Schofield	2007	–	–	–	X	–	–	–	–	–	–
Su & Wall	2009	X	–	–	–	X	X	X	–	–	–
Chang & Lai	2009	X	X	–	–	–	–	–	O	–	–
Farag & Lyons	2012	X	–	–	–	–	–	–	–	–	–
Hergesell & Dickinger	2013	–	X	–	–	–	–	–	–	X	X
Masiero & Zoltan	2013	–	–	–	–	X	X	–	–	X	X
Le-Klähn et al.	2014a	O	X	–	X	O	X	–	–	–	–
Le-Klähn et al.	2014b	X	X	–	–	X	X	–	–	–	–
Le-Klähn et al.	2015	–	O	–	–	X	X	–	–	–	–
Gutiérrez & Miravet	2016	–	X	–	–	O	X	–	X	X	X
Bieland et al.	2017	–	X	–	–	O	–	–	–	–	–
Mulley et al.	2017	–	–	–	X	–	–	–	–	–	–
Bajada & Titheridge	2017	X	–	–	–	X	X	–	–	–	X
Nutsugbodo et al.	2018	–	–	–	–	X	–	–	–	–	–
Gross & Grimm	2018	O	O	–	–	X	–	–	O	X	–
Nwachukwu et al.	2019	X	–	–	–	X	X	–	–	–	–
Tang et al.	2020	–	O	–	–	O	–	–	–	–	–
Qi et al.	2020	–	O	–	–	X	–	–	–	–	–
Miravet et al.	2021	–	X	–	–	X	X	–	–	X	X
Zamparini & Vergori	2021	X	X	–	–	–	–	–	–	–	–
Downward & Lumsdon	2004	–	–	–	–	–	–	–	–	–	–
Lumsdon et al.	2006	–	–	–	–	–	–	–	–	–	–

Thompson & Schofield	2007	–	–	–	–	–	X	–	–	X	–
Su & Wall	2009	–	–	–	–	X	–	–	–	–	–
Chang & Lai	2009	–	–	–	–	–	–	–	–	–	–
Farag & Lyons	2012	–	–	–	–	–	–	–	–	–	–
Hergesell & Dickinger	2013	–	–	–	–	–	–	–	–	–	X
Masiero & Zoltan	2013	X	X	–	–	–	–	–	–	–	–
Le-Klähn et al.	2014a	–	–	–	–	–	–	–	–	–	–
Le-Klähn et al.	2014b	–	–	–	–	–	–	–	–	–	–
Le-Klähn et al.	2015	O	–	O	–	–	–	–	–	–	–
Gutiérrez & Miravet	2016	–	–	–	X	–	X	–	–	–	–
Bieland et al.	2017	–	–	–	X	–	–	–	–	–	–
Mulley et al.	2017	–	–	–	–	–	–	–	–	–	–
Bajada & Titheridge	2017	–	–	–	–	–	–	–	–	–	–
Nutsugbodo et al.	2018	–	–	–	–	–	–	–	–	–	–
Gross & Grimm	2018	–	–	X	–	–	O	O	–	–	–
Nwachukwu et al.	2019	–	–	–	–	–	–	–	–	–	–
Tang et al.	2020	–	–	–	–	O	–	–	–	–	–
Qi et al.	2020	–	–	–	–	–	–	–	–	–	–
Miravet et al.	2021	–	–	–	–	–	X	–	X	–	–
Zamparini & Vergori	2021	–	–	–	–	–	X	–	–	–	–

o: significant, x: not significant

Public Transport Availability

Public transport accessibility may vary at different types of tourism destinations, which may contribute to problems/challenges for tourists to use public transport, particularly when comparing rural and urban areas (Tang et al., 2020). According to Dickinson and Robbins (2008), tourists do not use public transport in rural areas because of availability issues. As Le-Klähn and Hall (2014) asserted, the higher quality public transport systems available in cities makes them more attractive to use. However, this may be as much because of its perceived availability and ease of use as opposed to its actual availability.

Travel Partners and Group Travel

Travel partner(s) or companion(s) influence tourist mobility. Tourists who travel with families and/or friends were more likely to use private vehicles (Le-Klähn et al., 2015; Tang et al., 2020), and those who travel alone (solo independent tourists) were likely to use public transport (Gross & Grimm, 2018). Visitors travelling with partner(s) are more likely to use public transport than single travellers (Le-Klähn et al., 2015). In addition, the size, or numbers of people in a group influence transport behaviour (see independent travellers or group travellers).

Independent tourists are different from group travellers in terms of their travel behaviour. Independent tourists may spend more time gathering information and had more opportunities to receive help from local people while using public transport (Chang & Lai, 2009). In the case of group travellers, the larger the groups of people travelling together the less likely they are to use public transport (Gross & Grimm, 2018). However, Qi et al. (2020) reported a contrasting finding; the more individuals in a group performing tour activities, the higher the probability for them to use public transport, with individuals being more likely to use public transport if they are part of an organised group holiday. This also reflects other factors that influence tourists to use public transport, such as public transport attributes (e.g., availability, conveniences), and tourists' characteristics (e.g., age, income).

Length of Stay

The duration of the stay influences the behaviours of travel and transport, in which longer stays allow tourists to have more time to explore the destination. Studies showed that tourists who stay longer at the destination have a higher probability to use public transport (Gutiérrez & Miravet, 2016; Qi et al., 2020) and a greater tendency to travel to other regions (Le-Klähn et al., 2015; Tang et al., 2020). However, the use of public transport is not always viable when tourists travel to other regions due to poor public transport connectivity between places, especially those located away from the urban area. Past studies also denote that there is a relationship between choice of transport mode by overnight visitors and same-day visitors. The same-day visitors used motorised private transport more frequently than overnight visitors, and they preferred fast and direct connections for their arrival and departure presumably because of the limited time they have at the tourism destinations.

Repeat or First Time Visitors, Travel Frequency and Familiarity to Destination

Questions regarding first time visitor/repeat visitors and the number of previous trips to a destination are often used interchangeably in studies that examine visitor familiarity with the destination they visit, including the public transport system and services available at the sites. This is a potentially significant variable as tourists' past travel experiences, may influence their travel and transport behaviour. According to Gutiérrez and Miravet (2016), repeating a stay increased the probability of tourists to use public transport, which may be due to the assumption that tourists have greater knowledge of the public transport network and other transport alternatives to move around. However, in contrast, Le-Klähn et al. (2014a, 2015) reported that visitors who are more familiar with Munich are less likely to use public transport. Masiero and Zoltan (2013) also found that repeat visitors are more likely to use private transport and visit more than one region. Tang et al. (2020) discovered that people prefer public transport or bikes when they go to where public transport infrastructure is highly developed; but the tendency weakens when people travel to where the bus connections are weak. Besides familiarity with the destination, it is notable that travel frequency reflects tourists being travel active (Hergesell & Dickinger, 2013), which presume tourists may be familiar with travel related decisions (i.e., choices of attractions and transport modes). Thus, repeat visit, travel frequency and familiarity with destinations are related to past travel and transport experiences.

Initial Mode of Transport, Access to Cars and Transport Mode for Entire Journey/During Trip

The mode of transport used to reach the destination has proven to be one of the most important factors in determining tourist use of public transport (Gutiérrez & Miravet, 2016; Bieland et al., 2017). Gutiérrez and Miravet (2016) reported that the initial transport mode travel to Costa Darauda by private cars or plane depended on the tourists' origin and influenced their movement once at the destination. Their study shows that tourists who arrived by private cars were the ones who least used public transport at the destination. Meanwhile the tourists who arrived by plane were the ones who used public transport most. Gross and Grimm (2018) also noted that the arrival transport means had a significant influence on the choice of transport at the holiday destination. Their study reported that tourists who travelled by train and bus/coach showed high usage of public transport at the destinations, in which the train users were more than two-times higher than bus/coach users. Similarly, Bieland et al. (2017) reported that overnight visitors who travel to Kassel by using public transport have a five times higher probability to use public transport in Kassel.

Travel Destinations and Destination Characteristics

The characteristics of travel destinations are important when considering tourists' travel and transport behaviour. Generally, public transport in the urban areas is more attractive than in the rural areas. However, Gross and Grimm (2018) discovered higher usage of public transport in the mountain areas and other destinations (inland lowlands and urban destinations) compares to coastal or island destinations, which is perhaps due to public transport availability and connectivity between places in these areas. According to Gutiérrez and Miravet (2016), tourists who travelled to 'sun, sea and sand' destinations (coastal areas or islands) probably have what they need without having to travel outside of their holiday accommodation and thus have less need to use public transport. Tang et al. (2020) noted that the use of public transport decreased when the travel destination is located outside the tourist area (which has better access to public transport).

Types of Accommodation

Masiero and Zoltan (2013) reported that staying in commercial accommodation has a positive impact on the likelihood of visiting more than one region and is related to movement patterns and is influenced by the purpose of trips. Gutiérrez and Miravet (2016) found that accommodation was one of the fundamental factors that influenced transport mode choice (in the case of travelling to and within destinations). Tourists who stayed at one- to three-star hotels were more likely to use private cars to reach Costa Darauda, compared to those who stayed at four- or five-star hotels and second residences, in which planes and public transport were the most probable modes of transport. Accordingly, the type of accommodation appears strongly related to the initial mode of transport used to reach the destination, which influences transport mode choice within the destination.

Trip Expense

Studies do not find any significant relationships between the cost of a trip and tourist use of public transport or transport behaviour (Masiero & Zoltan, 2013; Gross & Grimm, 2018; Gutiérrez & Miravet, 2016). However, studies show the possibility of direct or indirect relationships between trip expenses and tourist use of public transport at destinations. Downward and Lumsdon (2004) found that about 22.5 percent of tourists' total spending in the North York Moors National Park was transport related, with about 70 percent of that linked to car borne travel. The authors concluded that visitors who travelled by car were likely to spend more per group in the national park because they had the flexibility to travel around and were able to spend more time at different stops in the national park.

Travel Activity/Visits During Trip

Movement patterns and choice of transportation mode are interrelated (Masiero & Zoltan, 2013; Le-Klähn et al., 2015). The activity participation at tourism destinations is influenced by tourists' travel motivations. According to Masiero and Zoltan (2013), activity participation and motivation increase the understanding of tourist behaviour related to the spatial extent of the destination visited and to the transport mode choice. The locational factor (where the places of interest are located) and means of accessibility are likely to determine the tourists' choice of transport mode. The literature reveals substantial interdependency between tourists' travel related decisions with regards to travel and transport choice behaviour. For instance, the initial mode of transport use to reach the destination is related to the type of accommodation tourists stay in and influences the mode of transport they use to travel around the destination. For example, tourists who travel to destination by cars are more likely to travel around and visit more attractions during their stays. They have access to cars and thus focus on convenience and flexibility to travel to attractions or plan their trips.

Trip characteristics provide a meaningful understanding of tourist travel and transport behaviour in a broader sense and highlight the situational factors of tourists at the time they decide to use or not use public transport. It therefore means that these factors may either hinder or encourage the tourists to use public transport depending on the situation:

- the tourists' situation during the trip (trip related personal conditions); and
- the transport environment at the destination.

Trip characteristics are therefore another key factor in deciding whether tourists use public transport at tourism destinations.

Gronau and Kagermeier (2007) suggested that a demand-oriented approach leads to new understandings of traffic planning (public transport provision) by identifying and meeting tourist needs. Most studies on public transport attributes in relation to tourism are demand focused. These attributes are also multidimensional and suggest various interrelationships between attributes and other variables. Table 22.5 presents the various attributes of public transport identified in previous studies which investigated tourist transport behaviours from different perspectives or contexts, e.g., transport mode preferences and choices, motivation to

Table 22.5. Public transport attributes and tourist use of public transport.

Author(s)/year	*Study sites/context*	*Dimension(s)*	*Items*
Thompson & Schofield (2007)	Tourist sites in the Greater Manchester, UK	Ease of use	Easy to use
			Easy to find information
			Public transport problems or issues were dealt with effectively
			Easy to buy tickets
			Staff are helpful
		Efficiency and safety	A fast way to travel
			Arrive on time/punctuality
			Safe
			Feel safe travelling alone on public transport
		Good parking	Easy to park
			Safe to park
Chang & Lai (2009)	Urban area, Taipei MRT stations, Taiwan	Public transport source of information	Travel guide
			Relatives and friends
			Brochures
			Online
		Holistic attraction	Efficient
			Waiting time
			On time
			Safe and reliable
			Comfortable
			Price
			Cleanliness
		Service attraction	Multi-language interpretation
			Handicapped facility
			Pedestrianisation
			Diversified function of land use
			Convenient for shopping
			Friendly service/people
		Information attraction	Easy to understand
			Convenient transfer system
			Information is abundant
			Signposts are clear

(*Continued*)

Table 22.5. (*Continued*)

Author(s)/year	*Study sites/context*	*Dimension(s)*	*Items*
		Tourism image	Interior design Landscape Distinctive landmark
		Station location	Close to hotel Convenient to sight-seeing spot Attached to tourism spot
Huang et al. (2013)	Urban-rural, Pearl River Delta, China	Perceived utility function of public transport	Travel time* Number of transfers Fare* Comfort*
Le-Klähn et al. (2014b)	Top ten tourist sites in Munich, Germany	Travelling comfort	Space on vehicle Cleanliness of the vehicle Seat availability Comfort while waiting at bus stops or train stations Safety on-board
		Service quality	Punctuality Reliability Service frequency* Convenience of time schedule Network connection
		Accessibility	Accessibility of train stations or bus stops Accessibility of the vehicles
		Additional features	Ticket price* Ease of use* Staff service Information*
Parahoo et al. (2014)	Metro stations in Dubai	Reputation	Corporate image A smart way to travel Safety
		Dubai Metro quality	Customer service Physical appearance and infrastructure are visually appealing Safe
Le-Klähn et al. (2015)	Airport, train and bus stations, Munich, Germany	Ease of travel	Ordinal scale
		Transport decision-making time Importance of price* Importance of time	Binary scale
Yuksek et al. (2016)	Metropolitan Municipality in Eskisehir, Turkey	Ease of use	A fast way to travel Able to find information Easy to use public transport in Eskisehir Public transport is technological supported

Table 22.5. (*Continued*)

Author(s)/year	*Study sites/context*	*Dimension(s)*	*Items*
		Physical condition	Public transport vehicles are clean Crowded public transport Safe Temperature adequacy Comfort
		Infrastructure	Public transport is accessible Adequate number of bus/train stops Train/bus stops are conveniently located Train/bus stops shelters Accessibility to the public transport network in Eskisehir
		Staff	Helpful Respectful Knowledgeable and skilful
		Timing	Easy to buy ticket Punctual Service frequency
Bajada & Titheridge (2017)	Localities in Malta	Time	Punctuality issues and unreliability Amount of time spent waiting for the bus, travelling by the bus, and waiting on the bus (for people paying the driver)
		Information	Schedule about the bus service both online, on-board the fleet, and bus stops and interchanges
		Bus driver	Men and women driving the bus, their behaviour influence the passengers' experiences
		Fare	The fee paid for the journey, reflecting the fare structure
		Fleet	The vehicles used to provide the bus services
		Impacts on the environment	Pollution emitted from the fleet, particularly air and noise pollution

(*Continued*)

Table 22.5. (*Continued*)

Author(s)/year	*Study sites/context*	*Dimension(s)*	*Items*
Nutsugbodo et al. (2018)	Kotoka International Airport, Ghana	Affordability/ pricing Accessibility Availability Waiting time Safety/security Comfort/ convenience Reliability Speed limits	Frequency measurement
Nwachukwu et al. (2019)	Major tourist attractions in Lagos, Nigeria	Accessibility	Transport price/fare* High frequency of service* Short waiting time Availability Short travel time Route network coverage Short walking distance to bus stops or ferry stops
		Journey comfort	Enough leg-space in vehicles Vehicles are well maintained Safety of passengers on-board Seats are available Vehicles are clean inside* Comfort while waiting*
		Travelling security	Personal security during trip* Security while waiting* Security of luggage on-board
		Travelling information	Ease of finding information about routes, destinations and stops* Availability of information about service delays Availability of arrival information for vehicles
		Customer service	Willingness of staff to help* Attitude of driver/staff Availability of staff on-board

* Significant factors

use or hindering factors to use public transport, destination satisfaction, sustainable transport behaviour, and linkages between tourists' transport mode choice and their movement pattern.

Perceived public transport attributes are significant factors that encourage tourists to use public transport. However, the relative importance of these factors appears to differ depending on the situations of the tourist; the destinations they visit; the access to transport mode; and their own conditions during the trips. Accordingly, differences in perceptions lead to variations in conation, or behavioural intent, and it is important to identify the individuals' involvement, destination image, satisfaction, and perceptions of service quality (Cohen et al., 2013). The public transport attributes identified from past empirical studies indicate these dimensions, and thus provide different interpretations of public transport quality in various study settings.

Time

Time is an item found in several studies related to public transport services. Punctuality is commonly identified in most studies in which tourists/visitors were expecting the public transport services to be on-time according to arrival and departure schedules (Thompson & Schofield, 2007; Chang & Lai, 2009; Le-Klähn et al., 2014b; Yuksek et al., 2016; Bajada & Titheridge, 2017). Time is also associated with waiting-time (Chang & Lai, 2009; Bajada & Titheridge, 2017; Nutsugbodo et al., 2018; Nwachukwu et al., 2019), travel time/duration (Huang et al., 2013; Le-Klähn et al., 2015; Nwachukwu et al., 2019), service frequency (Le-Klähn et al., 2014b; Yuksek et al., 2016; Nwachukwu et al., 2019), and the ease of ticket purchase (Yuksek et al., 2016). While past studies revealed the importance and influences of time in tourist transport behaviour, Le-Klähn et al.'s (2015) study conducted in Munich, Germany found that time factors (in terms of transport decision-making time and travel time) did not influence the tourists' mode choice in their study. In contrast, Huang et al. (2013) found that travel time is one of the most important attributes that affect transport mode choice among recreationists in Pearl River Delta, China and tourists are willing to spend more money to reduce travel time when travelling by coach, subway, and train.

Comfort

Travel comfort affects tourist use of public transport at tourism destinations. While some past studies examined comfort as a single survey item, travelling or journey comfort may encompass various factors. Le-Klähn et al. (2014b) and Nwachukwu et al. (2019) revealed that tourists/visitors demonstrated a strong preference for travel comfort, in which cleanliness is one of the most important factors that contribute to their satisfaction with public transport. Additionally, comfort while waiting for public transport should also be considered (Nwachukwu et al., 2019). Huang et al. (2013) denoted that comfort is the most important factor for tourists who travel by coach among other public transport (light rail, subway, and train), considering the long time they spend on board, while lack of comfort in travelling by public transport prompted tourists to use private vehicles.

Source and Provision of Information

The ease of obtaining required public transport information appears to be a significant influence on tourists' transport behaviour (Hall et al., 2017). Tourists potentially get public transport information from travel guides, brochures, the internet, and/or friends and relatives (Chang & Lai, 2009). Studies of public transport information covers access to information (Thompson & Schofield, 2007; Le-Klähn et al., 2014b; Yuksek et al., 2016), and trip/travel related information (Bajada & Titheridge, 2017; Nwachukwu et al., 2019), while the ease in accessing the required information also influences the perceived reliability of the public transport service.

Safety and Security

Several studies indicate that safety and security affect tourist use of public transport (Hall et al., 2017). The sense of safety includes feeling safe when travelling alone on public transport (Thompson & Schofield, 2007), and onboard transport (Le-Klähn et al., 2014b; Nutsugbodo et al., 2018). The safety and security aspects of public transport account for both:

- public transport provider's staff, assets, and operations; and
- users' lives and belongings.

However, an area of future research is the need to examine gender-based perceptions of safety and security in public transport given that this has been identified as a potentially important factor in public transport use (Hall et al., 2017).

Price/Fare

Past studies suggest that price/fare influences tourist use of public transport in terms of modal choice (Huang et al., 2013) and satisfaction with public transport (Le-Klähn et al., 2014b, 2015). Accordingly, affordable public transport ticket prices may draw tourists to use public transport. Le-Klähn et al. (2014b) suggested that fare promotion and special ticket schemes, which have been proven positive to encourage local use of public transport, can also be implemented to encourage tourist use.

Ease of Use

The concept of ease of use consists of different factors that are related to public transport performance (Thompson & Schofield, 2007):

- public transport is easy to use;
- ablity to find public transport information;
- public transport problems or issue are dealt with effectively;
- ease of buying ticket for the journey; and
- public transport staff are helpful.

Le-Klähn et al. (2014b) found that ease of use of public transport is independent from most demographic and trip related characteristics. For instance, repeat visitors, frequent users of public transport, those who have a valid driving license, and those who recommend others to use public transport tend to find it easier to use public transport. Ease of use of public transport is also related to information and ticketing system improvement (information being widely available in English, and clear ticketing systems) (Yuksek et al., 2016).

While previous research was conducted through intercept surveys at various tourist sites/attractions (mainly in urban areas), the study of Ram et al. (2021) identified four similar public transport service quality attributes by using data mining methods:

i. Pricing and ticketing.
ii. Accessibility.
iii. Trip durations.
iv. Service availability, which seemingly conform to public transport attributes discovered in the past studies.

Moreover, while many studies examined tourists' travel and transport behaviour by using generic public transport attributes (i.e., public transport service quality and perceived utility function), it is noteworthy that Chang and Lai (2009) investigated inbound tourists' consideration to use public transport in Taipei, Taiwan from the perspective of public transport as a tourism attraction. It was found that Taipei Mass Rapid Transit (MRT) attributes appeal to different characteristics of tourists (based on their socio-demographics profile, types of tourists – independent or group travellers, and frequent users of MRT during their stay in Taiwan).

Past studies also demonstrated a range of findings with regards to public transport attributes that are important for different target groups and study settings. Time, price (or fare), comfort, safety, information, and accessibility (service frequency and ease of use) are mostly found to be significant for tourist use of, and satisfaction with public transport. However, for tourist users of public transport the prioritisation of these attributes depends on their travel situations. For instance, travel time, fare and comfort are important for those who travel longer distance with public transport (Huang et al., 2013). Likewise, travel fare and comfort are also important to tourists who travel in the urban areas, with service frequency, ease of use and public transport information also being important (Le-Klähn et al., 2014b; Nwachukwu et al., 2019).

Tourism and Public Transport Relationships

Previous studies seem to suggest that tourist trip characteristics and public transport attributes influence tourist behaviours for using public transport. Some key premises of studies related to public transport and tourism can be drawn based on past research. Firstly, tourist decisions to use public transport to or at tourism destinations can be explained by analysing their profile, their trip, and public transport characteristics. For example, tourists who use public transport at home will generally use public transport when they travel. Tourist preferences of using public transport may be related to their travel motivation. There are also direct and indirect associations for factors that influence tourist use of public transport when they travel (multi-layered effects) which may conform as situational or locational factors that affect behaviours (Cohen et al., 2013; Andrades Caldito et al., 2015).

Tourism public transport related studies are mostly undertaken in developed countries like the United Kingdom and Germany (i.e., Downward & Lumsdon, 2004; Lumsdon et al., 2006; Le-Klähn et al., 2014a, 2014b, 2015; Bieland et al., 2017; Gross & Grimm, 2018). Very little research has been carried out in the developing or less developed countries. Past studies have also mostly focused on public transport in:

- urban areas compared to rural areas;
- four-season countries (or conducted during visitor months); and
- tourist sites, the latter probably because of convenience sampling.

Much research is conducted on the targeted groups of public transport users (local users, domestic and international tourists), but very few studies make comparisons between these groups. There is also not much attention given to the public transport non-users group. Additionally, some research, such as Thompson and Schofield (2007), exclude domestic tourists on the basis that their perceived attributes of public transport performance could be like the local users. However, the extent to which this is true is unknown, and there is little comparison between local users (domestic use of public transport) and domestic tourists transport behaviour.

Many studies were conducted using quantitative approaches with very few qualitative studies being found in relation to public transport and tourism. To date, the work of Hall et al. (2017) is perhaps the most comprehensive in providing insights on public transport and tourism. It is also notable that the vast majority of studies are demand focused with little research on supply issues. Although Gronau's (2016) study partially examined public transport and tourism from a supply perspective, previous studies related to public transport and tourism mostly addressed concerns that emerged from tourists' transport mode choice and preferences to encourage tourist use of public transport for modal shift from private vehicles. Research is therefore responding to the need for sustainable transport and tourism that promotes protection of the environment and sustainable behavioural change among tourists and tourism providers. However, there are few studies that addressed questions regarding environmental concerns in relation to public transport use (Hergesell & Dickinger; 2013; Le-Klähn et al., 2014a; Zamparini & Vergori, 2021). In addition, tourist use of public transport may have economic spillover effects for regional development although there are few direct studies (Tian et al., 2021).

Tourism destinations with well-integrated and accessible public transport networks (and services) are arguably more attractive for tourists to visit and provides easy access to tourism attractions. Tourist use of public transport has the potential to promote sustainable transport and tourism behaviour, but it is not well implemented, especially in developing and less developed countries. Tourist use of public transport may:

- promote modal shift from private vehicles to public transport;
- encourage sustainable behaviour;
- has the potential to promote 'slow tourism' and disperse tourists to other areas; and
- contribute to sustainable regional development.

Studies of public transport attributes, customer satisfaction, modal choice and preferences are commonly done in a local setting and less so in regional public transport, which have different characteristics (longer distances, often lower service frequencies, fewer stops) (Hansson et al., 2019). Much of the literature focuses on the use of public transport in urban areas. Based on past research, the literature on tourist use of public transport is usually constituted around three dimensions:

- tourist profiles;
- trip characteristics; and
- public transport attributes.

These dimensions influence tourists' behaviour to use public transport when they travel and contribute to their satisfaction towards public transport service, to the overall destination performance, and to tourists' travel experiences.

The Promotion of Public Transport as a Tourist Experience and Attraction

Transport has the potential to enable culture contact and allow (or constrain) tourist-host interaction (Hall, 1999). The opportunity of taking local public transport contributes to the tourist experience, offering host-tourist interaction, appreciation of scenic views along the journey, and cultural observation. Very few studies examine the touristic value of public transport. Although some specific modes of public transport can be attractive for tourism, e.g., ferries and trams, most public transport appears to be perceived to have low intrinsic value to tourists and is usually framed in utilitarian terms (Hall et al., 2017). However, public tourism transport can possess an intrinsic value as a tourism attraction for its spatial movement and scenic opportunities. For example, McGrath et al. (2020) and Kizielewicz et al. (2017) identified ferry crossings as being attractive to some tourists as a travel experience, while Tarkowski and Puzdrakiewics (2021) identified the benefits of autonomous urban ferries by increasing public transport network connectivity, reducing travel distance, and creating modal shifts towards walking and cycling.

Some mass public transport like ferries, trains and buses are especially popular due to their unique infrastructure and architecture, historical and heritage values (Hall et al., 2017). They represent local culture to tourists and may possess high intrinsic values that tourists want to experience when they travel. For example, trains and rail networks across European countries, the ferry networks in Istanbul, Turkey, or Sydney, Australia, all appear to hold high intrinsic values for tourism. In addition, in some countries mass public transport is often supplemented with local paratransit which could be iconic for tourism promotion and makes the tourists' travel experience more interesting. Local paratransit such as three-wheelers found in Southeast Asia and South Asia countries (e.g., Tuktuk in Thailand, Bajaj in Indonesia, Kuliglig in Philippines), and taxis (black cabs in London, UK, and yellow cabs in New York, USA) show the potential of public transport for touristic activities (Hall et al., 2017).

Some public transport may also have heritage values (Peira et al., 2022). Lee and Chen (2016) suggest that provision of heritage railway attractions constitutes an essential and important component of railway tourism experience, with the smoothness and comfort of a rail journey playing a supporting role in enhancing railway tourism attractiveness. Modern high-speed rail may also hold intrinsic values to some tourists, particularly with respect to novelty. Kurihara and Wu (2016) found that the opening of high-speed rail had greatly influenced tourism demand but note that the impacts of high-speed rail for tourism growth decline in the second year and that improvement of destination attractiveness is necessary.

Summary and Conclusion

Tourists' socio-demographic and psychographic factors, their trip characteristics, and the public transport attributes at the destination, are all of importance for enabling a modal shift from private vehicle use to public and active transport in tourism. Public transport usage may potentially enhance tourists' satisfaction and travel experience, provide funding for public transport development and service improvement, and is better for the environment (Thompson & Schofield, 2007; Albalate & Bel, 2010; Hall et al., 2017; Ram et al., 2021). In addition, some studies have postulated new opportunities for public transport use in tourism, (e.g., Domènech & Gutiérrez, 2017; Lu et al., 2019; Wu et al., 2015; Xue et al., 2014) suggesting that the study of tourists' travelling patterns could improve services and create new business opportunities at destinations, while Korean research has focused on the potential use of smart apps and AI in promoting public transport use by domestic tourists (Kim et al., 2023b, c, d, e).

Despite public transport primarily being regarded as attracting passengers (tourists) for their utility value), the anticipation and experience (e.g., tourist-host interactions and encounters) of being in the public transport adds value to the tourist experience. A focus on the measures to improve tourists' experiences can be one of the effective ways to encourage tourist use of public transport at tourism destinations with the public transport experience contributing to tourists' overall travel experience. Consequently, memorable, and good travel experiences on public transport may prompt future travel behavior among tourists. Although tourists often do not have the intentions to use public transport prior to their arrivals at the destinations, the conditions of a public transport network, the integration between the various transport modes, and the quality of service may encourage public transport use.

While tourist users might find intrinsic value in travelling by public transport, tourist non-users might have prior experiences for choosing not to use it at the tourism destination. It is therefore important to discover the reasons behind tourist use and non-use of public transport, which may be indicative for future sustainable travel behaviour and public transport development. The roles and integrations of public transport and tourism as the potential for development appears implicit rather than explicit in the context of developing and less developed countries. Research on the multifaceted variables of tourists', their trip, and public transport characteristics, together with other factors in the local settings may present new perspectives on tourist use and non-use of public transport at tourism destinations.

References

Albalate, D., & Bel, G. (2010). Tourism and urban public transport: Holding demand pressure under supply constraints. *Tourism Management*, *31*(3), 425–433. https://doi.org/10.1016/j.tourman.2009.04.011

Andrades Caldito, L., Dimanche, F., & Ilkevich, S. (2015). Tourist behavior and trends. In F. Dimanche, & L. Andrades Caldito (Eds), *Tourism in Russia: A management handbook* (pp. 101–130). Emerald.

Bajada, T., & Titheridge, H. (2017). The attitudes of tourists towards a bus service: implications for policy from a Maltese case study. *Transportation Research Procedia*, *25*, 4110–4129. https://doi.org/10.1016/j.trpro.2017.05.342

Ban, K. (2016). Summary report on global sustainable transport conference [Paper presentation]. *Global Sustainable Transport Conference*, Ashgabat, Turkmenistan. https://sustainabledevelopment.un.org/Global-Sustainable-Transport-Conference-2016

Bieland, D., Sommer, C., & Witte, C. (2017). Uncommon leisure traffic – Analyses of travel behavior of visitors. *Transportation Research Procedia*, *25*, 3971–3984. https://doi.org/10.1016/j.trpro.2017.05.236

Chang, H. H., & Lai, T.-Y. (2009). The Taipei MRT (Mass Rapid Transit) tourism attraction analysis from the inbound tourists' perspectives. *Journal of Travel & Tourism Marketing*, *26*(5–6), 445–461. https://doi.org/10.1080/10548400903162964

Cohen, S. A., Prayag, G., & Moital, M. (2013). Consumer behavior in tourism: Concepts, influences, and opportunities. *Current Issues in Tourism*, *17*(10), 872–909. https://doi.org/10.1080/13683500.2013.850064

Dickinson, J. E., & Robbins, D. (2008). Representations of tourism transport problems in a rural destination. *Tourism Management*, *29*(6), 1110–1121. https://doi.org/10.1016/j.tourman.2008.02.003

Domènech, A., & Gutiérrez, A. (2017). A GIS-based evaluation of the effectiveness and spatial coverage of public transport networks in tourist destinations. *ISPRS International Journal of Geo-Information*, *6*(3), 83. https://doi.org/10.3390/ijgi6030083

Downward, P., & Lumsdon, L. (2004). Tourism transport and visitor spending: A study in the North York Moors National Park, UK. *Journal of Travel Research*, *42*(4), 415–420. https://doi.org/10.1177%2F0047287504263038

Dunlap, R. E., Van Liere, K. D., Mertig, A. G., & Jones, R. E. (2000). New trends in measuring environmental attitudes: measuring endorsement of the new ecological paradigm: a revised NEP scale. *Journal of Social Issues*, *56*(3), 425–442. https://doi.org/10.1111/0022-4537.00176

Farag, S., & Lyons, G. (2012). To use or not to use? An empirical study of pre-trip public transport information for business and leisure trips and comparison with car travel. *Transport Policy*, *20*, 82–92. https://doi.org/10.1016/j.tranpol.2011.03.007

Gronau, W. (2016). Encouraging behavioural change towards sustainable tourism: a German approach to free public transport for tourists. *Journal of Sustainable Tourism*, *25*(2), 265–275. https://doi.org/10.1080/09669582.2016.1198357

Gronau, W., & Kagermeier, A. (2007). Key factors for successful leisure and tourism public transport provision. *Journal of Transport Geography*, *15*(2), 127–135. https://doi.org/10.1016/j.jtrangeo.2006.12.008

Gross, S., & Grimm, B. (2018). Sustainable mode of transport choice at the destination – public transport at German destinations. *Tourism Review*, *73*(3), 401–420. https://doi.org/10.1108/TR-11-2017-0177

Gutiérrez, A., & Miravet, D. (2016). The determinants of tourist use of public transport at the destination. *Sustainability*, *8*(9), 908. https://doi.org/10.3390/su8090908

Hall, C. M. (2022). Sustainable tourism beyond BAU (Brundtlund as usual): Shifting from paradoxical to relational thinking? *Frontiers in Sustainable Tourism*, *1*, 927946. https://doi.org/10.3389/frsut.2022.927946

Hall, C. M., Le-Klähn, D-T., & Ram, Y. (2017). *Tourism, public transport and sustainable mobility*. Channel View Publications.

Hall, D. R. (1999). Conceptualising tourism transport: Inequality and externality issues. *Journal of Transport Geography*, *7*(3), 181–188. https://doi.org/10.1016/S0966-6923(99)00001-0

Hansson, J., Pettersson, F., Svensson, H., & Wretstrand, A. (2019). Preferences in regional public transport: a literature review. *European Transport Research Review*, *11*(38). https://doi.org/10.1186/s12544-019-0374-4

He, S. Y., & Thøgersen, J. (2017). The impact of attitudes and perceptions on travel mode choice and car ownership in a Chinese megacity: The case of Guangzhou. *Research in Transportation Economics*, *62*, 57–67. https://doi.org/10.1016/j.retrec.2017.03.004

Hergesell, A., & Dickinger, A. (2013). Environmentally friendly holiday transport mode choices among students: the role of price, time and convenience. *Journal of Sustainable Tourism*, *21*(4), 596–613. https://doi.org/10.1080/09669582.2013.785555

Høyer, K. G. (2004). From sustainable mobility to sustainable tourism. In F. D. Pineda, C. A. Brebbia & M. Mugica (Eds), *Sustainable tourism* (pp. 285–294). WIT Press.

Huang, J., Li, J., Lai, X., & Xu, H. (2013). Modal choice of recreational tourists under regional transportation integration: A case study of Pearl River Delta. *Applied Mechanics & Materials*, *253–255*, 287–292. https://doi.org/10.4028/www.scientific.net%2FAMM.253-255.287

Khadaroo, J., & Seetanah, B. (2008). Transport and economic performance: The case of Mauritius. *Journal of Transport Economics and Policy*, *42*(2), 255–267.

Kim, M. J., & Hall, C. M. (2022a). Is walking or riding your bike when a tourist different? Applying VAB theory to better understand active transport behavior. *Journal of Environmental Management*, *311*, 114868. https://doi.org/10.1016/j.jenvman.2022.114868

Kim, M. J., & Hall, C. M. (2022b). The influence of personal and public health and smart applications on biking behavior in South Korea. *Journal of Consumer Behavior*. https://doi.org/10.1002/cb.2076

Kim, M.J., Hall, C.M., & Kim, M. (2023a). What is significant for engagement in cycling and walking in South Korea? Applying value-belief-norm theory. *Travel Behaviour and Society*, 32, 100571, https://doi.org/10.1016/j.tbs.2023.02.008

Kim, M.J., Hall, C.M., & Chung, N. (2023b). The influence of AI and smart apps on tourist public transport use: Applying mixed methods. *Information Technology & Tourism*, https://doi.org/10.1007/s40558-023-00272-x

Kim, M.J., Hall, C.M., Chung, N., Kim, M., & Sohn, K. (2023c). Why do tourists use public transport in Korea? The roles of artificial intelligence knowledge, environmental, social, and governance, and sustainability. *Asia Pacific Journal of Tourism Research*, 8(5), 467–484, https://doi.org/10.1080/10941665.2023.2247099.

Kim, M.J., Hall, C.M., Chung, N.H., Kim, M., & Sohn, K. (2023d). What makes tourists use public transport? Value-belief-norm theory, environmental, social, and governance factors, and the Sustainable Development Goals. *Journal of Travel Research*, https://doi.org/10.1177/0047287523119170

Kim, M.J., Hall, C.M., Chung, N., Kim, M., & Sohn, K. (2023e). Does using public transport affect tourist subject well-being and behavior relevant to sustainability? VAB theory and AI benefits. *Current Issues in Tourism*, https://doi.org/10.1080/13683500.2023.2214721

Kizielewicz, J., Haahti, A., Luković, T., & Gračan, D. (2017). The segmentation of the demand for ferry travel – A case study of Stena Line. *Economic Research-Ekonomska Istraživanja*, *30*(1), 1003–1020. https://doi.org/10.1080/1331677X.2017.1314789

Kovačić, M., & Milošević, T. (2016). Interdependence of transport and tourism. *Journal of Maritime and Transportation Science*, *52*(1), 99–111. https://doi.org/10.18048/2016.52.06

Kurihara, T., & Wu, L. (2016). The impact of high-speed rail on tourism development: A case study of Japan. *The Open Transportation Journal*, *10*, 35–44. http://dx.doi.org/10.2174/1874447801610010035

Le-Klähn, D-T., & Hall, C. M. (2014). Tourist use of public transport at destinations – a review. *Current Issues in Tourism*, *18*(8), 785–803. https://doi.org/10.1080/13683500.2014.948812

Le-Klähn, D. T., & Hall, C. M. (2015). Tourist use of public transport at destinations–a review. *Current Issues in Tourism*, 18(8), 785–803. https://doi.org/10.1080/13683500.2014.948812

Le-Klähn, D-T., Gerike, R. & Hall, C. M. (2014a). Visitor users vs. non-users of public transport: The case of Munich, Germany. *Journal of Destination Marketing Management*, *3*(3), 152–161. https://doi.org/10.1016/j.jdmm.2013.12.005

Le-Klähn, D-T., Hall, C. M., & Gerike, R. (2014b). Analysis of visitor satisfaction with public transport in Munich. *Journal of Public Transportation*, *17*(3), 68–85. http://doi.org/10.5038/2375-0901.17.3.5

Le-Klähn, D-T., Roosen, J., Gerike, R. & Hall, C. M. (2015). Factors affecting tourists' public transport use and areas visited at destinations. *Tourism Geographies*, *17*(5), 738–757. https://doi.org/10.1080/14616688.2015.1084527

Lee, C. F., & Chen, K. Y. (2016). Exploring factors determining the attractiveness of railway tourism. *Journal of Travel & Tourism Marketing*, *34*(4), 461–474. https://doi.org/10.1080/10548408.2016.1182460

Lu, Y., Wu, H., Xin, L., Chen, P., & Zhang, J. (2019). TourSense: A Framework for tourist identification and analytics using transport data. *IEEE Transactions on Knowledge and Data Engineering*, *31*(2), 2407–2422. https://doi.org/10.1109/TKDE.2019.2894131

Lumsdon, L., & Page, S. J. (2004). Progress in transport and tourism research: Reformulating the transport-tourism interface and future research agendas. In L. M. Lumsdon & S. J. Page (Eds), *Tourism and transport. Issues and agenda for the new millennium* (pp. 1–27). Elsevier.

Lumsdon, L., Downward, P., & Rhoden, S. (2006). Transport for tourism: Can public transport encourage a modal shift in the day visitor market? *Journal of Sustainable Tourism*, *14*(2), 139–156. https://doi.org/10.1080/09669580608669049

Masiero, L., & Zoltan, J. (2013). Tourists intra-destination visits and transport mode: A bivariate probit model. *Annals of Tourism Research*, *43*, 529–546. https://doi.org/10.1016/j.annals.2013.05.014

McGrath, E., Harmer, N., & Yarwood, R. (2020). Ferries as travelling landscapes: Tourism and water mobilities. *International Journal of Culture, Tourism and Hospitality Research*, *14*(3), 321–334. http://dx.doi.org/10.1108/IJCTHR-10-2019-0184

Miravet, D., Domènech, A., & Gutiérrez, A. (2021). What prompts tourists to become public transportation users at their destination? The case of a Mediterranean city. *Travel Behavior & Society*, *24*, 10–21. https://doi.org/10.1016/j.tbs.2021.01.007

Moore, K., Smallman, C., Wilson, J., & Simmons, D. (2012). Dynamic in-destination decision-making: An adjustment model. *Tourism Management*, *33*(3), 635–645. https://doi.org/10.1016/j.tourman.2011.07.005

Mulley, C., Clifton, G. T., Balbontin, C., & Ma, L. (2017). Information for travelling: Awareness and usage of the various sources of information available to public transport users in NSW. *Transportation Research Part A: Policy & Practice*, *101*, 111–132. https://doi.org/10.1016/j.tra.2017.05.007

Nerhagen, L. (2003). Travel mode choice: effects of previous experience on choice behavior and valuation. *Tourism Economics*, *9*(1), 5–30. https://doi.org/10.5367%2F000000003101298240

Nobis, C., & Lenz, B. (2005). Gender differences in travel patterns: Role of employment status and household structure. *Research on Women's Issues in Transportation, Conference Proceedings*, 35(2), 114–123.

Nutsugbodo, R. Y., Amenumey, E. K., & Mensah, C. A. (2018). Public transport mode preferences of international tourists in Ghana: Implications for transport planning. *Travel Behavior & Society*, *11*, 1–8. https://doi.org/10.1016/j.tbs.2017.11.002

Nwachukwu, A. A., Gladys, N. I., & Chikezie, O. K. (2019). Tourists' satisfaction with public transport services in Lagos, Nigeria. *AUC Geographica*, *54*(1), 67–80. https://doi.org/10.14712/23361980.2019.7

Page, S., & Ge, Y. (2009). Transportation and tourism: a symbiotic relationship? In T. Jamal & M. Robinson (Eds.), *The SAGE handbook of tourism studies* (pp. 371–395). SAGE Publications.

Parahoo, S. K., Harvey, H. L., & Abdelrahim Radi, G. Y. (2014). Satisfaction of tourists with public transport: An empirical investigation in Dubai. *Journal of Travel & Tourism Marketing*, *31*(8), 1004–1017. https://doi.org/10.1080/10548408.2014.890158

Peira, G., Giudice, A. L., & Miraglia, S. (2022). Railway and tourism: A systematic literature review. *Tourism & Hospitality*, *3*(1), 69–79. https://doi.org/10.3390/tourhosp3010005

Prideaux, B. (2000). The role of the transport system in destination development. *Tourism Management*, *21*(1), 53–63. https://doi.org/10.1016/S0261-5177(99)00079-5

Qi, C., Zhu, Z. Guo, X., Lu, R., & Chen, J. (2020). Examining interrelationships between tourist travel mode and trip chain choices using the nested logit model. *Sustainability*, *12*(18), 7535. https://doi.org/10.3390/su12187535

Ram, Y., Gal-Tzur, A., & Rechavi, A. (2021). Identifying attributes of public transport services for urban tourists: A data-mining method. *Journal of Transport Geography*, *93*, 103069. https://doi.org/10.1016/j.jtrangeo.2021.103069

Su, M. M., & Wall, G. (2009). The Qinghai-Tibet railway and Tibetan tourism: Travelers' perspectives. *Tourism Management*, *30*(5), 650–657. https://doi.org/10.1016/j.tourman.2008.02.024

Tang, X., Wang, D., Sun, Y., Chen, M., & Waygood, E. O. D. (2020). Choice behavior of tourism destination and travel mode: A case study of local residents in Hangzhou, *China. Journal of Transport Geography*, *89*, 102895. https://doi.org/10.1016/j.jtrangeo.2020.102895

Tarkowski, M., & Puzdrakiewics, K. (2021). Connectivity benefits of small zero-emission autonomous ferries in urban mobility – Case of the coastal city of Gdańsk (Poland). *Sustainability*, *13*(23), 13183. https://doi.org/10.3390/su132313183

Thompson, K., & Schofield, P. (2007). An investigation of the relationship between public transport performance and destination satisfaction. *Journal of Transport Geography*, *15*(2), 136–144. https://doi.org/10.1016/j.jtrangeo.2006.11.004

Tian, M., Li, T., Ye, X., Zhao, H., & Meng, X. (2021). The impact of high-speed rail on service industry agglomeration in peripheral cities. *Transportation Research Part D: Transport & Environment*, *93*, 102745. https://doi.org/10.1016/j.trd.2021.102745

United Nations (UN) (2016). Mobilizing sustainable transport for development. *Secretary-General's High-Level Advisory Group on Sustainable Transport*. https://sustainabledevelopment.un.org/content/documents/2375Mobilizing%20Sustainable%20Transport.pdf

Wu, H., Tan, J. A., Ng, W. S., Xue, M., & Chen, W. (2015). FTT: A system for finding and tracking tourists in public transport services. *Proceedings of the ACM SIGMOD International Conference on Management of Data* (pp. 1093–1098). https://doi.org/10.1145/2723372.2735367

Xue, M., Wu, H., Chen, W., Ng, W. S., & Goh, G H. (2014). Identifying tourists from public transport commuters. *Proceedings of the ACM SIGKDD International Conference on Knowledge Discovery and Data Mining* (pp. 1779–1788). https://doi.org/10.1145/2623330.2623352

Yeboah, G., Cottrill, C. D., Nelson, J. D., Corsar, D., Markovic, M., & Edwards, P. (2019). Understanding factors influencing public transport passengers' pre-travel information seeking behavior. *Public Transport*, *11*(1), 135–158. https://doi.org/10.1007/s12469-019-00198-w

Yuksek, G., Akkoç, I. T., & Bayer, R. U. (2016). The effects of public transport performance on destination satisfaction. *African Journal of Hospitality, Tourism & Leisure*, *5*(4).

Yusoff, I, Ng, B.-K., & Azizan, S. A. (2021). Towards sustainable transport policy framework: A rail-based transit system in Klang Valley, Malaysia. *PLoS ONE*, *16*(3), e0248519. https://doi.org/10.1371/journal.pone.0248519

Zamparini, L., & Vergori, A. S. (2021). Sustainable mobility at tourist destinations: The relevance of habits and the role of policies. *Journal of Transport Geography*, *93*, 103088. https://doi.org/10.1016/j.jtrangeo.2021.103088

Chapter 23

Tourism Entrepreneurship

Richard S. Aquino

Introduction

Undoubtedly, entrepreneurship is an integral activity in many industries, including tourism. Entrepreneurship is a complex phenomenon that received multiple yet complementary conceptualisations, such as the classical views of entrepreneurship anchored on notions of the 'entrepreneur'. Schumpeter (1934) proposed entrepreneurship is founded on the entrepreneurs' ability to disturb existing economic equilibrium through innovation, or by introducing new goods or better quality of goods, ways of production, resources, organisations, and markets. Another view was provided by Kirzner (1973), which was based on the concept of 'entrepreneurial alertness', or the entrepreneurs' discovery of previously untapped opportunities and needs in the market. However, entrepreneurs and opportunities – whether created or discovered – are inseparable, especially in the creation of new ventures with the goal of wealth creation under risky and uncertain conditions (Dollinger, 2008; Gartner, 1985). Based on these views, *entrepreneurship* is conceptualised as 'an activity that involves the discovery, evaluation and exploitation of opportunities to introduce new goods and services, ways of organizing, markets, processes, and raw materials through organizing efforts that previously had not existed' (Shane, 2003, p. 4).

The modern tourism industry was developed through entrepreneurship, particularly through the efforts of entrepreneurial individuals and communities who create products and services to satisfy the needs of escapist and novelty-seeking tourists (Koh, 2002; Koh & Hatten, 2002). As a system, tourism is fuelled by networks of enterprises, directly and indirectly, interacting with each other to address market needs and opportunities. Tourism requires enterprises that can innovate in order to accommodate ever-changing consumer demands and environmental changes, such as those due to socio-technological advancements (Hall & Williams, 2020; Solvoll et al., 2015). Entrepreneurship also plays a crucial role in the delivery of desired tourism development outcomes (Altinay & Alrawadieh, 2022; Wanhill, 2000).

The Wiley Blackwell Companion to Tourism, Second Edition. Edited by C. Michael Hall.

Places that rely on tourism require businesses with sound entrepreneurial practices to help them achieve their development goals (Ateljevic & Page, 2009; Pollock, 2015).

Because of entrepreneurship's role and importance in tourism development, tourism entrepreneurship is probably one of the most studied topics in tourism and hospitality scholarship. This is evident in the significantly growing number of publications on tourism entrepreneurship in the past decade (see Fu et al. (2019) and Ochoa Jiménez et al. (2022) for recent publication trends). Earlier works focused on the impacts of tourism enterprises on destinations, especially the economic contributions of large transnational enterprises and small locally-owned businesses (Rodenburg, 1980), including analyses of the creation, characteristics, and role of small tourism enterprises in destination development (Shaw & Williams, 1987; Williams et al., 1989). Strong scholarly attention to understanding small tourism firms in various contexts (e.g., Dahles & Bras, 1999; Wanhill, 2000) has been given up to date, mainly because the tourism industry is mostly composed of micro-, small, and medium-sized enterprises (MSMEs). There is also established scholarly research on tourism entrepreneurs and the antecedents of their entrepreneurial behaviour (Fu et al., 2019; Koh & Hatten, 2002), including lifestyle, cultural, and gender factors in tourism entrepreneurship (Ateljevic & Doorne, 2000, 2003), and the contributions of entrepreneurship towards sustainable tourism (Lordkipanidze et al., 2005; Thomas et al., 2011). Also, gaining more attention in recent decades are knowledge management, technological innovation, family businesses, digital value creation (e.g., Ratten, 2019), and the adoption of social and environmental sustainability-oriented forms of entrepreneurship in tourism (Aquino et al., 2018).

In addition, the scholarship has moved from merely analysing entrepreneurship concepts as they apply in tourism, to embracing 'the unique characteristics of the tourism industry' in an 'attempt to gain knowledge of the phenomenon of entrepreneurship in this industry as something distinctive' (Solvoll et al., 2015, p. 127). The aim of this chapter is to examine the scholarship on tourism entrepreneurship according to the following interrelated themes: the tourism entrepreneur; entrepreneurial processes; and entrepreneurial forms and their implications on tourism development. In doing so, this chapter synthesises established and current trends in tourism entrepreneurship research and suggests some areas that need further scholarly attention.

Tourism Entrepreneurs

The tourism entrepreneur is the 'persona causa of tourism development' (Koh & Hatten, 2002, p. 22). The potential of places for tourism development and attracting tourists would not be realised without entrepreneurial individuals. As Hall and Williams (2020, p. 288) emphasised, tourism entrepreneurs are the 'heroes in the evolution of tourism'.

Tourism entrepreneurs are simply described as founders of tourism enterprises 'motivated by monetary and/or non-monetary reasons' (Koh & Hatten, 2002, p. 25), or having growth and/or profit orientations (Getz & Petersen, 2005). They come in many 'guises', such as copreneur, ethnic/indigenous, female/women, inventive, innovative, imitative, lifestyle, marginal, nascent, serial, social, and sustainable entrepreneurs (Koh & Hatten, 2002; Morrison, 2006). Early research on tourism entrepreneurship focused on understanding tourism entrepreneurs' motivations and characteristics (Williams et al., 1989). Apart from economic motives, tourism entrepreneurs could have growth-oriented motives which may include providing employment, risk-taking, creating value, scaling-up businesses, and capitalising on underutilised resources (Fu et al., 2019; McGehee & Kim, 2004).

'Lifestyle' is perhaps the most commonly identifiable factor when studying tourism entrepreneurs' motivations, as evident in seminal (e.g., Ateljevic & Doorne, 2000; Williams et al., 1989) and most recent studies of tourism entrepreneurship (Sun & Xu, 2017). Others argue that 'lifestyle' motives subordinate financial pursuits in tourism entrepreneurship (e.g., Morrison, 2006). Also known as 'lifestyle tourism entrepreneurs,' these individuals could be previous tourists, motivated to pursue a slow pace of life usually in rural areas, and seek more autonomy and balance over their lives, or embarking on tourism entrepreneurship in their host locations (Cunha et al., 2020; Sun & Xu, 2017). Significant scholarly attention has been given to understanding lifestyle tourism entrepreneurs and their non-monetary motives, due to the nature of the touristic enterprises they create, which reject 'the market-driven paradigm in favour of reciprocity and lifestyle' (Ateljevic & Doorne, 2000, p. 389) and advocate for sustainability-oriented tourism practices (Cunha et al., 2020).

Garnering scholarly attention recently are the pro-social and pro-environmental motives of tourism entrepreneurs for starting-up businesses, usually in studies of social and sustainable entrepreneurship in tourism (Aquino et al., 2018; Porter et al., 2018) with the pursuit of addressing neoliberal tourism entrepreneurship models. For example, known as having the dual mission of creating economic and social values, social entrepreneurs in tourism are found to be motivated by altruistic reasons – to make a difference in others, particularly stakeholders (e.g., usually socio-economically disadvantaged groups) involved in their activities (Mody et al., 2016). Nonetheless, motivations for tourism entrepreneurship vary according to individuals and the goals that they pursue. Tourism entrepreneurs' motivations are not static, and may evolve from non-monetary (e.g., lifestyle) reasons initially to more growth-oriented (e.g., scaling-up tourism business) motives (see Marchant & Mottiar, 2011).

Usually understood by demographic variables and industry experience (Fu et al., 2019), the characteristics of tourism entrepreneurs vary depending on the context and forms of tourism entrepreneurship. For example, lifestyle tourism entrepreneurs may come from educated backgrounds, are well-travelled, and in some cases, have previously owned businesses (Ateljevic & Doorne, 2000; Marchant & Mottiar, 2011). While entrepreneurs in the so-called, 'informal sector', are found to have low education levels and come from low socioeconomic status (Dahles & Prabawa, 2013). It is difficult to draw generalisable conclusions due to the subjectivity of motives and contexts where tourism entrepreneurs come from and operate.

Further enriching knowledge on tourism entrepreneurs are recent studies that examine these individuals' personality traits, entrepreneurial mindsets, and entrepreneurial success. Analysing entrepreneurs in the Italian tourism industry, Presenza et al. (2020, p. 2158) revealed:

> the tourism entrepreneur seems to be particularly attractive (narcissism), extroverted and friendly (extraversion), disciplined (agreeableness), self-assured, efficient and organized (conscientiousness), unconventional having many interests (openness to experience), and are proactive and vigorous (internal locus of control).

Amongst these traits, 'extraversion and openness to experience' (Presenza et al., 2020, p. 2,158) were shown to shape their abilities to identify market opportunities and establish tourism businesses and networks. Adding to these are Yachin's (2019) findings from a study of Swedish rural micro-tourism firms, wherein creativity, awareness, and other cognitive properties were depicted to induce the identification of market opportunities. From a survey in

Thailand, entrepreneurs' having a creative mindset and capabilities to innovate were found to influence business success (Yodchai et al., 2022).

Risk and uncertainty are integral to the entrepreneurship phenomenon. In negotiating different types of risk (monetary, functional, social, and psychological), tourism entrepreneurs were found to cultivate various coping mechanisms, namely: 'survival intelligence (eases monetary risks), system intelligence (reduces functional risk), emotional intelligence (improves social risks), and spiritual intelligence (decreases psychological risks)' (Power et al., 2019, p. 1101). Other responses employed by tourism entrepreneurs to navigate risks and uncertainties include obtaining knowledge and forming networks (Williams et al., 2020) – processes that are integral in tourism entrepreneurial phenomena.

Tourism Entrepreneurial Processes

If tourism entrepreneurs are defined as the creators of touristic enterprises, then tourism entrepreneurship can be simply conceptualised as the process of creating these ventures. However, regardless of the motives for venture creation, the entrepreneurial phenomenon often starts with entrepreneurs scanning the *environment* (the context) in which they are situated, making the 'destination environment' a key antecedent of tourism entrepreneurship (Fu et al., 2019). The destination environment and the larger tourism system consist of economic, social, cultural, political, and socio-technical factors (e.g., Hall, 2022), affecting the availability of market opportunities and the subsequent entrepreneurial processes and activities faced by tourism entrepreneurs (e.g., Aquino, 2022; Aquino et al., 2018). It should be noted, however, that the entrepreneurial process is not linear or static.

Borrowing and synthesising concepts from the general entrepreneurship literature, Koh (1996) illustrated that the tourism entrepreneurial process is composed of interconnected stages, namely cognitive orientation, opportunity search, opportunity assessment, consideration, opportunity pursuit, birth (of the enterprise), operation, and evaluation. Contextualising entrepreneurship through demonstrating small family tourism enterprise creation, Morrison (2006) suggested that the *entrepreneurial process* is formed by the interplay of the entrepreneurs' traits, culture, and entrepreneurship that is informed by economic and social purpose. These factors are then suggested to form a lens for evaluating the environment and the feasibility of the tourism industry to fulfil economic, social and lifestyle motives (Morrison, 2006). These earlier conceptualisations point to the role of 'opportunities' found in the destination environment, which is also a fundamental concept in the entrepreneurial phenomenon.

The dominant entrepreneurial logic that manifests in the tourism industry was the subject of recent research. As signalled at the outset, two schools of thought inform the understanding of the entrepreneurial process: the 'Schumpeterian' (innovation-based) and 'Kirznerian' (opportunity-based) views of entrepreneurship (Hall & Williams, 2020). But which of these entrepreneurial 'paradigms' underpin touristic venture creation and product innovation? Yachin (2019) found none of the tourism products offered by rural micro-tourism firms in Sweden followed a Schumpeterian logic; rather, most of the products were construed to be developed from Kirznerian thought. This could be partly explained by the structure of micro-tourism enterprises and their place in the tourism supply chain. Being micro-to-small in scale means most tourism enterprises are part of a larger system that possesses 'value gaps' or 'opportunities' waiting to be discovered and addressed through product offerings. Apart from being in such a position in the supply chain, micro-tourism entrepreneurs' formation of networks with other actors – such as other entrepreneurs and tourists – in the tourism system could facilitate the discovery of opportunities, which precedes product innovation in tourism (Yachin, 2019). Shedding light on the long-contested debate surrounding the 'entrepreneur–opportunity' nexus in tourism, this latter study has further illuminated the importance of network and knowledge in tourism entrepreneurship.

Networking and collaboration are essential for entrepreneurs in tourism since an aggregate of actors and clusters form the industry. Earlier studies on this topic demonstrated strong evidence of internal cooperation amongst tourism businesses, although these organisations are competing for customers (e.g., Tinsley & Lynch, 2007). More recent studies focused on tourism entrepreneurs' networking practices and the implications of network formation (e.g., exchange, communication, social, and political) on enterprise performance and destination development in different socio-cultural contexts (Hall & Williams, 2020). Such investigations are necessary since social capital utilisation is integral in network formation, especially of tourism MSMEs (e.g., Thomas et al., 2011). For example, in the Chinese context, the networking abilities of tourism entrepreneurial migrants were found to be built from *guanxi* networks, and in addition, 'effort was exerted to seek and maintain these networks in an attempt to gain resources and support at the start-up stage' (Zhou et al., 2017, p. 348). Investigating *veza*, an informal form of a personal network created on the basis of economic exchange in the Croatian context, Skokic et al. (2019) showed that such informal ties are helpful for acquiring resources and a competitive market position at the latter stage of business development. In the Ghanaian context, a survey of women tourism entrepreneurs showed that their entrepreneurial orientation positively shapes their abilities to create formal networks with government agencies (political ties), which in turn positively influences their resource acquisition capabilities (Ribeiro et al., 2021).

Apart from tangible resource constraints, tourism micro-enterprises also have limited skills for certain 'in-house' functions necessary to operate businesses. Examining tourism entrepreneurs in Sweden, Yachin (2021) found that entrepreneurs turn to networks within their immediate environment to find solutions for basic company requirements, including marketing and practical help. In a study of creative entrepreneurs in China (Zhou et al., 2020), the formation of personal networks through social capital utilisation was found to proliferate 'organic placemaking.' Conversely, it was depicted that informal entrepreneurial networks transform into formal ones as tighter competition emerges at later phases of tourism development (Zhou et al., 2020). Nonetheless, tourism researchers have long advocated for institutional policies that support the creation of formal entrepreneurial networks in tourism (e.g., Kimbu et al., 2019; Komppula, 2014).

Other key aspects of the tourism entrepreneurial process and value creation are the acquisition and utilisation of knowledge (Thomas et al., 2011). Much of past research examined the link between entrepreneurial knowledge acquisition and business performance (e.g., Dias et al., 2020). Entrepreneurial knowledge can be obtained through formal education (e.g., Alford & Jones, 2020; Williams et al., 1989). However, tourism entrepreneurial knowledge can also be acquired through previous experience working in tourism and its related industries. In a recent study of immigrant tourism entrepreneurs in three Eastern Canary Islands, Calero-Lemes and García-Almeida (2021) revealed that formal education and previous experience in working in tourism in immigrants' host countries are vital knowledge sources for establishing, operating, and managing tourism businesses. In addition, immigrant entrepreneurs who are well integrated and knowledgeable were found to obtain know-how through tourism experience in their host localities.

Apart from formal education and experience, knowledge for starting up and running tourism enterprises can be acquired from networks (Calero-Lemes & García-Almeida, 2021). Actors and firms within a tourism firm's local and extended network serve as sources of knowledge, such as tourism-related expertise, business know-how, and local knowledge (Calero-Lemes & García-Almeida, 2021; Yachin, 2021). Important for making tourism enterprises competitive and innovative is their ability to create networks and utilise and share knowledge within these networks (Hall & Williams, 2020).

Related to network embeddedness, a recent study of digital tourism entrepreneurship showed that digital marketing knowledge can be acquired through 'knowledge sharing' and 'learning-by-example' leading to the co-creation of opportunities for firms involved (Alford & Jones, 2020), signalling the implications of knowledge sharing on cooperation. However, this is not always the case. It was found that a high level of distrust and high competition amongst tourism MSMEs that dominate a destination are barriers to knowledge transfer and cooperation (Czernek, 2017). Nevertheless, 'socially-embedded' tourism entrepreneurs or those who have strong bonds in their social communities could gain benefits for cooperation within a destination, such as access to knowledge including tacit knowledge (knowledge that is acquired over time and experience) about business activities (Czernek-Marszałek, 2020).

Social and community-embeddedness have also been illustrated as important factors for acquiring local knowledge. Local knowledge is important in tourism entrepreneurship mainly because this phenomenon is place-based, and in many respects, community-based. In a recent study of tourism lifestyle entrepreneurs, Dias et al. (2020) depicted local knowledge as key in building sustainable business models. They suggested that the acquisition of local knowledge through informal ties within a place can be incorporated into organisational processes, tourism products, and communication strategies. In other words, having a community-centred strategy ensures the acquisition of local knowledge that can be capitalised on for tourism innovation (Dias et al., 2020).

Tourism Entrepreneurial Forms, Outcomes, and Implications

One way of classifying tourism enterprises is through their size. As pointed out earlier, the majority of tourism enterprises are MSMEs. However, tourism enterprises can also be understood by the type of *ownership* and *activities* including their *value propositions*. Entrepreneurial forms classified using such factors were previously outlined and labelled as 'entrepreneur guises' in the work of Morrison (2006). Signalled by the present scholarship, some of the dominant entrepreneurial forms in tourism are shown in Table 23.1.

While these entrepreneurial forms seem distinct due to their descriptors, note that some of these contain some conceptual and practical overlaps. For example, community-based enterprises, especially those founded by local communities themselves, are considered social enterprises (and perhaps, the most authentic form of social enterprises). Likewise, the community is at the centre of most social enterprises (e.g., Aquino et al., 2018), and thus, these can be understood as community-based enterprises, as well. Some indigenous enterprises can be considered community-based enterprises and in some cases, family businesses too, due to their embeddedness in families and communities. Recent evidence also shows that community and environment-centric missions are found in lifestyle entrepreneurship (e.g., Dias et al., 2020), and depict some form of social entrepreneurial behaviour. Nonetheless, entrepreneurial pursuits in tourism have intended, as well as unanticipated, outcomes (Table 23.1).

Table 23.1. Entrepreneurial forms in tourism and their key outcomes and implications.

Entrepreneurial form	*Description*	*Key outcomes and implications*
Digital entrepreneurship	Recognition and 'pursuit of opportunities based on the use of digital media and other information and communication technologies' (Davidson & Vaast, 2010, p. 2).	• Transforms tourism into a 'technology-enabled co-creation system' (Buhalis et al., 2019). • Introduces disruptions in service experiences, specifically 'extra-sensory experiences, hyper-personalized experiences and beyond-automation experiences' (Buhalis et al., 2019, p. 484). • The availability of digital platforms (e.g., Airbnb, Uber) democratises participation in tourism entrepreneurship.
Community-based entrepreneurship	Engagement in commercial activities to improve the life of either a geographical or non-geographical 'community'. Community-based enterprises belong (founded by a community) or are dedicated (externally founded) to a community (Parwez, 2017).	• Central mechanism of community-based tourism (CBT) development. • Usually manifested in the rural tourism context, with the aim to develop alternative or supplemental forms of livelihoods, alleviate poverty, and improve the quality of life and natural environment of local communities (Manyara & Jones, 2007; Stone & Stone, 2011). • Externally founded CBT enterprises may hinder community empowerment and participation, and the creation of positive outcomes if their missions are not well-embedded in the communities and local priorities (Manyara & Jones, 2007).
Family entrepreneurship	Ownership and/or management of venture by a family (Chua et al., 1999).	• Main goals or outcomes include improving family wellbeing, reputation, and legacy (Camilleri & Valeri, 2021; Chua et al., 1999). • Branding tourism enterprises as family-owned may increase competitive advantage (Getz & Carlsen, 2005). • In some cases (e.g., B&Bs, homestays, and guesthouses), interactions of visitors with the host families are part of the tourism experience (Doan et al., 2022).

(*Continued*)

Table 23.1. (*Continued*)

Entrepreneurial form	*Description*	*Key outcomes and implications*
Indigenous entrepreneurship	'Creation, management and development of new ventures by Indigenous peoples for the benefit of Indigenous peoples' (Hindle & Lansdowne, 2005, p. 132).	• Tourism becomes a dominant market activity for indigenous enterprises (Macpherson et al., 2021). • Indigenous products become cultural artefacts that could enhance tourists' cultural experiences (Macpherson et al., 2021). • Strong embeddedness in family, community, tribe (iwi/hapu), and the natural environment. • Can be a means for 'self-determined development' of indigenous peoples (Scheyvens et al., 2021). • Integration of indigenous values in the development, management, and operation of ventures enhances business sustainability and sustainability of outcomes (Ransfield & Reichenberger, 2021; Scheyvens et al., 2021).
Informal (sector) entrepreneurship	Recognition and pursuit of business opportunities outside of the formal legislated economy (Webb et al., 2009).	• Contributes to poverty alleviation as informal entrepreneurship involves individuals in economic activities (Slocum et al., 2011). • Because informal entrepreneurs operate in a non-legislated economy, they are often omitted in tourism policy and planning (Çakmak et al., 2018).
Lifestyle entrepreneurship	Formation of businesses in accordance or to create a chosen lifestyle (including hobbies), balancing income and need to sustain a particular way of life, and in some cases, totally rejecting profit orientation (Ateljevic & Doorne, 2000).	• Creation of niche products (Ateljevic & Doorne, 2000). • Integration of sustainability goals in lifestyle tourism businesses since the success of these enterprises is based on their environmental and social performance, as well as, keeping the desired lifestyle (Dias et al., 2020). • Formation of sustainable entrepreneurial ecosystems in destinations through forming networks of firms that promote community wellbeing and inclusive supply chains (Cunha et al., 2020).

Social entrepreneurship	A market-based approach to addressing social problems; entails the creation of enterprises having the dual purpose of generating economic and social value, and in many instances, positive environmental outcomes.	• Alternative approach to traditional capitalist (solely for-profit) tourism entrepreneurship and development (Aquino et al., 2018). • It is a means of solving local social problems (e.g., poverty, lack of livelihoods, etc.) and destination issues (Sheldon et al., 2017). • Can provide positive and equitable outcomes for host communities involved in social enterprise-led tourism development (Aquino et al., 2022). • Provides a model for commercial tourism enterprises to turn their business models into sustainable ones (de Lange & Dodds, 2017).
Women or female entrepreneurship	Establishment and operation of business ventures by women owner-managers.	• Self-employment through tourism entrepreneurship leads to women's economic empowerment (Vukovic et al., 2021). • Women tourism entrepreneurs develop self-esteem and gain status in the family and wider community (Tajeddini et al., 2017). • Institutional and social contexts enable or hinder the participation of women in tourism entrepreneurship (Kimbu & Ngoasong, 2016; Kimbu et al., 2019). • In some instances, women entrepreneurship in tourism may reinforce gender inequality (Trupp & Sunanta, 2017).

Small and medium tourism enterprises are understood to be community-embedded and produce valuable socio-economic impacts, such as local job creation (Wanhill, 2000). Furthermore, small tourism enterprises are found to behave responsibly through their corporate social responsibility programmes. Specifically, tourism enterprises that are community-embedded are found to support corporate social responsibility (CSR) initiatives that benefit the community and the environment (Wen et al., 2021). In their analysis, Fu et al. (2019) showed that tourism entrepreneurial activities have different impacts on the:

- firm, such as sales growth, increase in market share, and innovation capability; and
- destination, including economic development, destination competitiveness, and sustainable tourism and sustainable development.

A survey of the current tourism entrepreneurship literature by form reveals some of the key outcomes and implications of this phenomenon (Table 23.1). However, these impacts can also be analysed through a multi-level approach, which means that apart from the impacts on destinations and communities (macro-level) and firms (meso-level) as construed by Fu et al. (2019), tourism entrepreneurship has evident impacts on individuals (micro-level).

On the macro-level, certain forms of tourism entrepreneurship have implications on destination and community development. Gaining attention in recent years is the application of social entrepreneurship as an alternative approach to developing tourism while addressing local social needs and problems (Aquino et al., 2018; Sheldon et al., 2017). By challenging capitalist tourism development paradigms, social entrepreneurship through tourism was depicted to create transformative outcomes for host communities (Aquino et al., 2022). In a similar vein, community-based entrepreneurship in tourism is argued to improve community wellbeing through the provision of alternative or supplemental livelihood sources, poverty alleviation, and environmental conservation programmes (Stone & Stone, 2011). Community-involved tourism ventures, such as lifestyle enterprises, are analysed to increase the participation of local community members in tourism as the networks they build promote inclusive supply chains (Cunha et al., 2020). Such ventures are depicted to create sustainable enterprise ecosystems as they outsource material and services from their host communities (Aquino, 2022; Aquino et al., 2022; Cunha et al., 2020).

Some forms of tourism entrepreneurship are construed to contribute to making tourism a sustainable industry. Early scholarship referred to lifestyle entrepreneurship as having a strong alignment with sustainable tourism principles as they deliberately reject profit-oriented motives to maintain a 'lifestyle' while creating positive social and environmental impacts (Ateljevic & Doorne, 2000). It was found that lifestyle entrepreneurs explicitly incorporate 'sustainability goals' because their success is based on their social and environmental impacts (Dias et al., 2020). Moving away from mainly creating economic and social value for their beneficiaries, tourism social enterprises also integrate environmental sustainability in their missions (e.g., Aquino, 2022; Aquino et al., 2022), and by acting as 'sustainable businesses', social enterprises may put pressure on commercial tourism enterprises to become more sustainable (de Lange & Dodds, 2017).

Another form of entrepreneurship that has increasing importance due to its linkages to sustainability paradigms is indigenous entrepreneurship (e.g., Scheyvens et al., 2021). As evident in a study of Māori tourism enterprises in Aotearoa New Zealand, the integration of indigenous people's values in the development, management, and operation of tourism ventures enhances the sustainability of their businesses, as well as, the sustainability of their impacts on the economy, their iwi/hapu, and the environment (Ransfield & Reichenberger, 2021). For example, the Māori value of *tino rangatiratanga* (self-determination) and kotahitanga (relationships, unity, and alliances) inform social sustainability impacts in the wider community. As emphasised in this latter study, 'for their [Māori] business to be successful and contribute, there needed to be a concerted effort to include the wider community predominantly based upon kotahitanga' (Ransfield & Reichenberger, 2021, p. 55).

In terms of the meso-level impacts, numerous studies examined the influence of entrepreneurial activities, entrepreneurs' motivations and individual characteristics, and other antecedents (e.g., place attachment, business environment) on the firm's performance (e.g., Fu et al., 2019; Ochoa Jiménez et al., 2022). However, studies also reveal different organisational outcomes depending on the nature of tourism entrepreneurial form and activities pursued. Signalled earlier is the way lifestyle tourism entrepreneurs limit the growth of their ventures to sustain the lifestyles they envision (Ateljevic & Doorne, 2000; Cunha et al., 2020). Maintaining or improving a family's reputation and continuing its legacy through business are some of the implications of pursuing family businesses in tourism (Camilleri & Valeri, 2021). Some entrepreneurial forms are known to create specific types of product offerings, as well. For instance, the 'family' and 'community' are key aspects of the tourism experience created by family and community-based enterprises, respectively. As shown in a study of Vietnamese homestay businesses – ventures that are family-owned and integrated into CBT programmes (Doan et al., 2022) – interactions with the host families and communities are depicted to make tourists' experiences more authentic. Similar formation of niche products is offered by indigenous tourism enterprises in the forms of tours to ancestral lands and traditional crafts, enhancing tourists' cultural experiences (Macpherson et al., 2021). It should be noted, however, that not all indigenous-owned tourism enterprises offer products that have cultural or indigenous elements (see the portfolio of Ngāi Tahu Tourism in Aotearoa New Zealand at https://www.ngaitahutourism.co.nz/experiences/).

Tourism firms' product offerings are also shaped by the mission or central value proposition of the enterprise. This is prominent in the case of for-purpose tourism social enterprises, where the tourism products they create align with social impact missions (Aquino et al., 2018). Another entrepreneurial form that has direct implications on the way tourism and its products are created is digital entrepreneurship. Buhalis et al. (2019, p. 484) suggested digital technologies and platforms disrupt service experiences, specifically in terms of creating 'extra-sensory experiences, hyper-personalized experiences and beyond-automation experiences'. Conversely, these technologies and platforms give rise to digital business models in tourism. Analysing 80 mobile applications (apps), Ammirato et al. (2021) found digital business models in cultural tourism are designed to create value for tourists in terms of travel convenience, reducing risks while travelling, experience enhancement, sociability, and practicality, throughout the visitor experience lifecycle. Specifically, most of these apps are for 'sociability' or the sharing of travel images and stories of those who have lived them, intended at the travel 'dreaming' phase (24 apps); and 'practicality' or keeping reservations, quotations, and planned itineraries during the 'planning' stage (27 apps). Still, Ammirato et al. (2021) emphasised the underutilised virtual and augmented realities in creating immersive tourism experiences.

On the micro-level, tourism entrepreneurship can create profound impacts on individuals, such as tourism entrepreneurs and the people around them. After all, self-employment leading to financial independence is what primarily motivates people to establish commercial businesses. This form of economic empowerment was strongly evident in tourism entrepreneurship research in the informal economy. Comprising a large percentage of the visitor economy, especially in rural or developing economies, informal entrepreneurship provides direct economic benefits for individuals (Slocum et al., 2011). In turn, these financial outcomes may help alleviate poverty and create other intangible outcomes, such as increased self-esteem and lifestyle change (e.g., Aquino et al., 2022).

Women owner-managers of tourism enterprises are especially understood to experience social and political empowerment, especially in patriarchal societies (Seyfi et al., 2022). In a study of women tourism entrepreneurship in Bali, Indonesia (Tajeddini et al., 2017, p. 57), it was evident that 'women who own and manage small business often gain a higher degree of self-esteem and gain the respect of their family and the larger community'. In some contexts, empowerment may manifest in the ability to exercise 'self-determined development,' especially in the case of indigenous tourism enterprises that are strongly linked to families and communities (Ransfield & Reichenberger, 2021; Scheyvens et al., 2021). Finally, the availability of digital technologies to both tourism providers and tourists opened up opportunities for entrepreneurship. The introduction of sharing economy platforms in tourism, such as Airbnb, has made participation in commercial tourism activities more accessible. While these platforms may incur supplemental economic benefits, several sources of conflicts while hosting (e.g., blurring of private and social spaces) may negatively impact hosts' wellbeing (Ozanne & Prayag, 2022).

Concluding Thoughts

This chapter reviewed established areas and current trends in tourism entrepreneurship research, by synthesising existing knowledge according to notions of entrepreneurial individuals, processes, and forms. It discussed a diversity of topics pertaining to micro-level antecedents and macro- and societal influences (e.g., culture) of tourism entrepreneurship. In terms of context, earlier observers denoted that 'when reviewed through 2011 eyes and minds it is apparent that content and literature reviewed can be criticised as being strongly UK-centric, with some North American influence' (Carmichael & Morrison, 2011, p. 115–116). This is certainly not the case anymore given the increasing number of publications on tourism entrepreneurship in other continents, especially in Asia following the rise of Asian tourism scholarship. Research on forms of entrepreneurship designed to alleviate social and destination problems is also strongly situated in the context of low-income countries rather than in developed ones (Boluk & Aquino, 2022).

'Sustainability', specifically the contributions of tourism entrepreneurship in sustainable tourism, is still one of the consistently dominant foci in tourism entrepreneurship research. This was signalled by emerging streams of research on alternative and sustainable forms of entrepreneurship that challenge traditional and capitalist entrepreneurial approaches depicted as the key causes of negative externalities and wicked problems in the tourism system. Although, it should be noted that traditional (mainly commercial) tourism enterprises have the potential to transform tourism into a more sustainable industry, by transforming their processes and initiating more environmentally focused and community-engaged CSR activities. Tourism scholars need to further investigate how to harness entrepreneurial mindsets in introducing social and sustainable innovations in tourism. Similarly, more investigations are needed to examine tourism entrepreneurs' orientation, altruistic behaviours, and sustainable business practices.

The enormous impact of the COVID-19 pandemic on tourism ventures regardless of the size, ownership, and location, amplified the need to enrich the understanding of tourism entrepreneurial resilience, especially how organisations harness their networks and knowledge, and how they learn and adapt from crises (Tanner et al., 2022). While this is an important research undertaking, tourism entrepreneurship scholars investigating this entrepreneurial resilience are encouraged to conduct longitudinal studies to better understand such a construct in the tourism context. In contrast to explorations of resilient tourism enterprises, scholars can also turn their attention to investigating 'entrepreneurial failure' in tourism which has received very little scholarly attention. It is assumed that there is quite a number of enterprises that did not survive the pandemic. Although it is a negative experience and may impose some degree of psychological risk on participants, some scholars suggested that analysing entrepreneurial failures could enrich the understanding of resilience (e.g., Corner et al., 2017).

Entrepreneurship is one of the most studied areas in tourism studies, reflecting the important roles of entrepreneurs and enterprises in the tourism industry. This review of present scholarship indicates that topics of tourism entrepreneurship research evolve according to contemporary issues and technological advancements in tourism. This could still be the case in the coming years especially because tourism research is often real-world problem-oriented. However, echoing Solvoll et al.'s (2015) inferences, tourism scholars are in a position to contribute to the wider entrepreneurship scholarship and debates given the unique characteristics of the tourism phenomenon.

References

Alford, P., & Jones, R. (2020). The lone digital tourism entrepreneur: Knowledge acquisition and collaborative transfer. *Tourism Management*, *81*, 104139. https://doi.org/10.1016/j.tourman.2020.104139

Altinay, L., & Alrawadieh, Z. (2022). Entrepreneurship in hospitality and tourism. In D. Buhalis (Ed.), *Encyclopedia of tourism management and marketing* (in press). Edward Elgar Publishing.

Ammirato, S., Felicetti, A. M., Linzalone, R., & Carlucci, D. (2021). Digital business models in cultural tourism. *International Journal of Entrepreneurial Behavior & Research* (ahead-of-print). https://doi.org/10.1108/IJEBR-01-2021-0070

Aquino, R. S. (2022). Community change through tourism social entrepreneurship. *Annals of Tourism Research*, *95*, 103442. https://doi.org/10.1016/j.annals.2022.103442

Aquino, R. S., Lück, M., & Schänzel, H. A. (2018). A conceptual framework of tourism social entrepreneurship for sustainable community development. *Journal of Hospitality and Tourism Management*, *37*, 23–32. https://doi.org/10.1016/j.jhtm.2018.09.001

Aquino, R. S., Lück, M., & Schänzel, H. A. (2022). Mapping the outcomes of social entrepreneurship and tourism on host communities: A three-dimensional approach. *Journal of Sustainable Tourism*, *30*(8), 1799–1820. https://doi.org/10.1080/09669582.2021.1986515

Ateljevic, I., & Doorne, S. (2000). 'Staying within the fence': Lifestyle entrepreneurship in Tourism. *Journal of Sustainable Tourism*, *8*(5), 378–392. https://doi.org/10.1080/09669580008667374

Ateljevic, I., & Doorne, S. (2003). Unpacking the local: A cultural analysis of tourism entrepreneurship in Murter, Croatia. *Tourism Geographies*, *5*(2), 123–150. https://doi.org/10.1080/1461668032000068270

Ateljevic, J., & Page, S. J. (2009). *Tourism and entrepreneurship: International perspectives*. Taylor & Francis.

Boluk, K. A., & Aquino, R. S. (2022). Assessing the impacts of social entrepreneurs and social enterprises in tourism. In A. Stoffelen & D. Ioannides (Eds), *Handbook of tourism impacts: Social and environmental perspectives* (pp. 117–133). Edward Elgar Publishing.

Buhalis, D., Harwood, T., Bogicevic, V., Viglia, G., Beldona, S., & Hofacker, C. (2019). Technological disruptions in services: Lessons from tourism and hospitality. *Journal of Service Management*, *30*(4), 484–506. https://doi.org/10.1108/JOSM-12-2018-0398

Çakmak, E., Lie, R., & McCabe, S. (2018). Reframing informal tourism entrepreneurial practices: Capital and field relations structuring the informal tourism economy of Chiang Mai. *Annals of Tourism Research*, *72*, 37–47. https://doi.org/10.1016/j.annals.2018.06.003

Calero-Lemes, P., & García-Almeida, D. J. (2021). Immigrant entrepreneur knowledge in the tourism industry of island destinations. *Tourism Geographies*, *23*(3), 527–551. https://doi.org/10.1080/14616688.2020.1786152

Camilleri, M. A., & Valeri, M. (2021). Thriving family businesses in tourism and hospitality: A systematic review and a synthesis of the relevant literature. *Journal of Family Business Management*. https://doi.org/10.1108/JFBM-10-2021-0133

Carmichael, B. A., & Morrison, A. (2011). Tourism entrepreneurship research. *Tourism Planning & Development*, *8*(2), 115–119. https://doi.org/10.1080/21568316.2011.573910

Chua, J. H., Chrisman, J. J., & Sharma, P. (1999). Defining the family business by behavior. *Entrepreneurship Theory and Practice*, *23*(4), 19–39. https://doi.org/10.1177/104225879902300402

Corner, P. D., Singh, S., & Pavlovich, K. (2017). Entrepreneurial resilience and venture failure. *International Small Business Journal*, *35*(6), 687–708. https://doi.org/10.1177/0266242616685604

Cunha, C., Kastenholz, E., & Carneiro, M. J. (2020). Entrepreneurs in rural tourism: Do lifestyle motivations contribute to management practices that enhance sustainable entrepreneurial ecosystems? *Journal of Hospitality and Tourism Management*, *44*, 215–226. https://doi.org/10.1016/j.jhtm.2020.06.007

Czernek-Marszałek, K. (2020). Social embeddedness and its benefits for cooperation in a tourism destination. *Journal of Destination Marketing & Management*, *15*, 100401. https://doi.org/10.1016/j.jdmm.2019.100401

Czernek, K. (2017). Tourism features as determinants of knowledge transfer in the process of tourist cooperation. *Current Issues in Tourism*, *20*(2), 204–220. https://doi.org/10.1080/13683500.2014.944107

Dahles, H., & Bras, K. (1999). *Tourism and small entrepreneurs. Development, national policy, and entrepreneurial culture: Indonesian cases*. Cognizant Communication Corporation.

Dahles, H., & Prabawa, T. S. (2013). Entrepreneurship in the informal sector. The case of the pedicab drivers of Yogyakarta, Indonesia. *Journal of Small Business & Entrepreneurship*, *26*(3), 241–259. https://doi.org/10.1080/08276331.2013.803672

Davidson, E., & Vaast, E. 2010, (January 5–8). *Digital entrepreneurship and its sociomaterial enactment* [Paper presentation]. 43rd Hawaii International Conference on System Sciences.

de Lange, D., & Dodds, R. (2017). Increasing sustainable tourism through social entrepreneurship. *International Journal of Contemporary Hospitality Management*, *29*(7), 1977–2002. https://doi.org/10.1108/IJCHM-02-2016-0096

Dias, Á., Silva, G. M., Patuleia, M., & González-Rodríguez, M. R. (2020). Developing sustainable business models: local knowledge acquisition and tourism lifestyle entrepreneurship. *Journal of Sustainable Tourism*. https://doi.org/10.1080/09669582.2020.1835931

Doan, T., Aquino, R. S., & Qi, H. (2022). Homestay businesses' strategies for adapting to and recovering from the COVID-19 pandemic: A study in Vietnam. *Tourism and Hospitality Research*. https://doi.org/10.1177/14673584221103185

Dollinger, M. J. (2008). *Entrepreneurship: Strategies and resources* (4th ed.). Marsh Publications.

Fu, H., Okumus, F., Wu, K., & Köseoglu, M. A. (2019). The entrepreneurship research in hospitality and tourism. *International Journal of Hospitality Management*, *78*, 1–12. https://doi.org/10.1016/j.ijhm.2018.10.005

Gartner, W. B. (1985). A conceptual framework for describing the phenomenon of new venture creation. *The Academy of Management Review*, *10*(4), 696–706. https://doi.org/10.5465/amr.1985.4279094

Getz, D., & Carlsen, J. (2005). Family business in tourism: State of the Art. *Annals of Tourism Research*, *32*(1), 237–258. https://doi.org/10.1016/j.annals.2004.07.006

Getz, D., & Petersen, T. (2005). Growth and profit-oriented entrepreneurship among family business owners in the tourism and hospitality industry. *International Journal of Hospitality Management*, *24*(2), 219–242. https://doi.org/10.1016/j.ijhm.2004.06.007

Hall, C. M. (2022). Sustainable tourism beyond BAU (Brundtlund as usual): Shifting from paradoxical to relational thinking? *Frontiers in Sustainable Tourism*, *1*, 927946. https://doi.org/10.3389/frsut.2022.927946

Hall, C. M., & Williams, A. M. (2020). *Tourism and innovation* (2nd ed.). Routledge.

Hindle, K., & Lansdowne, M. (2005). Brave spirits on new paths: Toward a globally relevant paradigm of indigenous entrepreneurship research. *Journal of Small Business & Entrepreneurship*, *18*(2), 131–141. https://doi.org/10.1080/08276331.2005.10593335

Kimbu, A. N., & Ngoasong, M. Z. (2016). Women as vectors of social entrepreneurship. *Annals of Tourism Research*, *60*, 63–79. https://doi.org/10.1016/j.annals.2016.06.002

Kimbu, A. N., Ngoasong, M. Z., Adeola, O., & Afenyo-Agbe, E. (2019). Collaborative networks for sustainable human capital management in women's tourism entrepreneurship: The role of tourism policy. *Tourism Planning & Development*, *16*(2), 161–178. https://doi.org/10.1080/21568316.2018.1556329

Kirzner, I. M. (1973). *Competition and entrepreneurship*. University of Chicago Press.

Koh, K. (2002). Explaining a community touristscape. An entrepreneurism model. *International Journal of Hospitality & Tourism Administration*, *3*(2), 29–62. https://doi.org/10.1300/J149v03n02_03

Koh, K. Y. (1996). The tourism entrepreneurial process: A conceptualization and implications for research and development. *The Tourist Review*, *51*(4), 24–41. https://doi.org/10.1108/eb058234

Koh, K. Y., & Hatten, T. S. (2002). The tourism entrepreneur: The overlooked player in tourism development studies. *International Journal of Hospitality & Tourism Administration*, *3*(1), 21–48. https://doi.org/10.1300/J149v03n01_02

Komppula, R. (2014). The role of individual entrepreneurs in the development of competitiveness for a rural tourism destination – A case study. *Tourism Management*, *40*, 361–371. https://doi.org/10.1016/j.tourman.2013.07.007

Lordkipanidze, M., Brezet, H., & Backman, M. (2005). The entrepreneurship factor in sustainable tourism development. *Journal of Cleaner Production*, *13*(8), 787–798. https://doi.org/10.1016/j.jclepro.2004.02.043

Macpherson, W. G., Tretiakov, A., Mika, J. P., & Felzensztein, C. (2021). Indigenous entrepreneurship: Insights from Chile and New Zealand. *Journal of Business Research*, *127*, 77–84. https://doi.org/10.1016/j.jbusres.2021.01.013

Manyara, G., & Jones, E. (2007). Community-based tourism enterprises development in Kenya: An exploration of their potential as avenues of poverty reduction. *Journal of Sustainable Tourism*, *15*(6), 628–644. https://doi.org/10.2167/jost723.0

Marchant, B., & Mottiar, Z. (2011). Understanding lifestyle entrepreneurs and digging beneath the issue of profits: Profiling surf tourism lifestyle entrepreneurs in Ireland. *Tourism Planning & Development*, *8*(2), 171–183. https://doi.org/10.1080/21568316.2011.573917

McGehee, N. G., & Kim, K. (2004). Motivation for agri-tourism Entrepreneurship. *Journal of Travel Research*, *43*(2), 161–170. https://doi.org/10.1177/0047287504268245

Mody, M., Day, J., Sydnor, S., & Jaffe, W. (2016). Examining the motivations for social entrepreneurship using Max Weber's typology of rationality. *International Journal of Contemporary Hospitality Management*, *28*(6), 1094–1114. https://doi.org/10.1108/IJCHM-05-2014-0238

Morrison, A. (2006). A contextualisation of entrepreneurship. *International Journal of Entrepreneurial Behavior & Research*, *12*(4), 192–209. https://doi.org/10.1108/13552550610679159

Ochoa Jiménez, S., García García, A. R., Valdez del Río, S., & Jacobo Hernández, C. A. (2022). Entrepreneurship in tourism studies in the 21st Century: A bibliometric study of WoS and Scopus. *SAGE Open*, *12*(2). https://doi.org/10.1177/21582440221102438

Ozanne, L. K., & Prayag, G. (2022). Exploring Airbnb host wellbeing and host-guest conflicts in network hospitality. *Frontiers in Psychology*, *13*. https://doi.org/10.3389/fpsyg.2022.805761

Parwez, S. (2017). Community-based entrepreneurship: Evidences from a retail case study. *Journal of Innovation and Entrepreneurship*, 6, 14. https://doi.org/10.1186/s13731-017-0074-z

Pollock, A. (2015). *Social entrepreneurship in tourism: The conscious travel approach*. TIPSE. http://www.conscious.travel/wp-content/uploads/2018/03/Conscious-Tourism-TIPSE-2016-1.pdf

Porter, B. A., Orams, M. B., & Lück, M. (2018). Sustainable entrepreneurship tourism: An alternative development approach for remote coastal communities where awareness of tourism is low. *Tourism Planning & Development*, *15*(2), 149–165. https://doi.org/10.1080/21568316.2017.1312507

Power, S., Di Domenico, M., & Miller, G. (2019). Risk types and coping mechanisms for ethical tourism entrepreneurs: A new conceptual framework. *Journal of Travel Research*, *59*(6), 1091–1104. https://doi.org/10.1177/0047287519874126

Presenza, A., Abbate, T., Meleddu, M., & Sheehan, L. (2020). Start-up entrepreneurs' personality traits. An exploratory analysis of the Italian tourism industry. *Current Issues in Tourism*, *23*(17), 2146–2164. https://doi.org/10.1080/13683500.2019.1677572

Ransfield, A. K., & Reichenberger, I. (2021). Māori Indigenous values and tourism business sustainability. *AlterNative: An International Journal of Indigenous Peoples*, *17*(1), 49–60. https://doi.org/10.1177/1177180121994680

Ratten, V. (2019). Tourism entrepreneurship research: a perspective article. *Tourism Review*, *75*(1), 122–125. https://doi.org/10.1108/TR-06-2019-0204

Ribeiro, M. A., Adam, I., Kimbu, A. N., Afenyo-Agbe, E., Adeola, O., Figueroa-Domecq, C., & de Jong, A. (2021). Women entrepreneurship orientation, networks and firm performance in the tourism industry in resource-scarce contexts. *Tourism Management*, *86*, 104343. https://doi.org/10.1016/j.tourman.2021.104343

Rodenburg, E. E. (1980). The effects of scale in economic development: Tourism in Bali. *Annals of Tourism Research*, *7*(2), 177–196. https://doi.org/10.1016/0160-7383(80)90003-1

Scheyvens, R., Carr, A., Movono, A., Hughes, E., Higgins-Desbiolles, F., & Mika, J. P. (2021). Indigenous tourism and the sustainable development goals. *Annals of Tourism Research*, *90*, 103260. https://doi.org/10.1016/j.annals.2021.103260

Schumpeter, J. A. (1934). *The theory of economic development*. Oxford University Press.

Seyfi, S., Hall, C. M., & Vo-Thanh, T. (2022). The gendered effects of statecraft on women in tourism: Economic sanctions, women's disempowerment and sustainability? *Journal of Sustainable Tourism*, *30*(7), 1736–1753. https://doi.org/10.1080/09669582.2020.1850749

Shane, S. (2003). *A general theory of entrepreneurship: The individual–opportunity nexus*. Edward Elgar.

Shaw, G., & Williams, A. (1987). Firm formation and operating characteristics in the Cornish tourist industry — The case of Looe. *Tourism Management*, *8*(4), 344–348. https://doi.org/10.1016/0261-5177(87)90092-6

Sheldon, P. J., Pollock, A., & Daniele, R. (2017). Social entrepreneurship and tourism: Setting the stage. In P. J. Sheldon & R. Daniele (Eds), *Social entrepreneurship and tourism: Philosophy and practice* (pp. 1–18). Springer International Publishing.

Skokic, V., Lynch, P., & Morrison, A. (2019). Veza: An informal network of tourism entrepreneurs. *Annals of Tourism Research*, *77*, 26–37. https://doi.org/10.1016/j.annals.2019.05.004

Slocum, S. L., Backman, K. F., & Robinson, K. L. (2011). Tourism pathways to prosperity: Perspectives on the informal economy in Tanzania. *Tourism Analysis*, *16*(1), 43–55. https://doi.org/10.3727/108354211X12988225900045

Solvoll, S., Alsos, G. A., & Bulanova, O. (2015). Tourism entrepreneurship – Review and future directions. *Scandinavian Journal of Hospitality and Tourism*, *15*(1), 120–137. https://doi.org/10.1080/15022250.2015.1065592

Stone, L. S., & Stone, T. M. (2011). Community-based tourism enterprises: challenges and prospects for community participation; Khama Rhino Sanctuary Trust, Botswana. *Journal of Sustainable Tourism*, *19*(1), 97–114. https://doi.org/10.1080/09669582.2010.508527

Sun, X., & Xu, H. (2017). Lifestyle tourism entrepreneurs' mobility motivations: A case study on Dali and Lijiang, China. *Tourism Management Perspectives*, *24*, 64–71. https://doi.org/10.1016/j.tmp.2017.06.004

Tajeddini, K., Ratten, V., & Denisa, M. (2017). Female tourism entrepreneurs in Bali, Indonesia. *Journal of Hospitality and Tourism Management*, *31*, 52–58. https://doi.org/10.1016/j.jhtm.2016.10.004

Tanner, S., Prayag, G., & Coelho Kuntz, J. (2022). Psychological capital, social capital and organizational resilience: A Herringbone Model perspective. *International Journal of Disaster Risk Reduction*, *78*, 103149. https://doi.org/10.1016/j.ijdrr.2022.103149

Thomas, R., Shaw, G., & Page, S. J. (2011). Understanding small firms in tourism: A perspective on research trends and challenges. *Tourism Management*, *32*(5), 963–976. https://doi.org/10.1016/j.tourman.2011.02.003

Tinsley, R., & Lynch, P. A. (2007). Small business networking and tourism destination development: A comparative perspective. *The International Journal of Entrepreneurship and Innovation*, *8*(1), 15–27. https://doi.org/10.5367/000000007780007380

Trupp, A., & Sunanta, S. (2017). Gendered practices in urban ethnic tourism in Thailand. *Annals of Tourism Research*, *64*, 76–86. https://doi.org/10.1016/j.annals.2017.02.004

Vukovic, D. B., Petrovic, M., Maiti, M., & Vujko, A. (2021). Tourism development, entrepreneurship and women's empowerment – Focus on Serbian countryside. *Journal of Tourism Futures*. https://doi.org/10.1108/JTF-10-2020-0167

Wanhill, S. (2000). Small and medium tourism enterprises. *Annals of Tourism Research*, *27*(1), 132–147. https://doi.org/10.1016/S0160-7383(99)00072-9

Webb, J. W., Tihanyi, L., Ireland, R. D., & Sirmon, D. G. (2009). You say illegal, I say legitimate: Entrepreneurship in the informal economy. *Academy of Management Review*, *34*(3), 492–510. https://doi.org/10.5465/amr.2009.40632826

Wen, T., Zhang, Q., & Li, Y. (2021). Why small tourism enterprises behave responsibly: using job embeddedness and place attachment to predict corporate social responsibility activities. *Current Issues in Tourism*, *24*(10), 1435–1450. https://doi.org/10.1080/13683500.2020.1797648

Williams, A. M., Rodríguez Sánchez, I., & Škokić, V. (2020). Innovation, risk, and uncertainty: A study of tourism entrepreneurs. *Journal of Travel Research*, *60*(2), 293–311. https://doi.org/10.1177/0047287519896012

Williams, A. M., Shaw, G., & Greenwood, J. (1989). From tourist to tourism entrepreneur, from consumption to production: Evidence from Cornwall, England. *Environment and Planning A: Economy and Space*, *21*(12), 1639–1653. https://doi.org/10.1068/a211639

Yachin, J. M. (2019). The entrepreneur–opportunity nexus: Discovering the forces that promote product innovations in rural micro-tourism firms. *Scandinavian Journal of Hospitality and Tourism*, *19*(1), 47–65. https://doi.org/10.1080/15022250.2017.1383936

Yachin, J. M. (2021). Alters & functions: exploring the ego-networks of tourism micro-firms. *Tourism Recreation Research*, *46*(3), 319–332. https://doi.org/10.1080/02508281.2020.1808933

Yodchai, N., Ly, P. T. M., & Tran, L. T. T. (2022). How the creative mindset affects entrepreneurial success in the tourism sector: The mediating role of innovation capability. *International Journal of Contemporary Hospitality Management*, *34*(1), 279–298. https://doi.org/10.1108/IJCHM-06-2021-0695

Zhou, L., Chan, E., & Song, H. (2017). Social capital and entrepreneurial mobility in early-stage tourism development: A case from rural China. *Tourism Management*, *63*, 338–350. https://doi.org/10.1016/j.tourman.2017.06.027

Zhou, L., Cheng, X., Wall, G., & Zhang, D. (2020). Entrepreneurial networks in creative tourism placemaking: Dali village, Wuhan, China. *Tourism Geographies*. https://doi.org/10.1080/14616688.2020.1866056

Chapter 24

Accelerating Innovation in Tourism
Living Labs as a Way Forward

Bernhard F. Bichler & Birgit Pikkemaat

Introduction

Innovation has become a buzzword when discussing economic, ecological and social challenges. However, as we face climate change and other major crises, we need to accelerate innovations and explore new ways and spaces to achieve sustainable solutions. Currently, sustainable innovations are promoted on various levels from businesses to destinations and at the country levels.

Innovations in the field of hospitality and tourism can take multiple forms, extending from product to social innovations (Hjalager, 2010). Surprisingly, the understanding of what constitutes an innovation differs strongly from tourism research to fields of practice. While many studies exist that propose definitions of innovation (Marasco et al., 2018; Pikkemaat et al., 2019; Souto, 2015) some highlight drivers of innovation (Nordli, 2018; Pikkemaat et al., 2018) or report on successful innovation projects (Carlisle et al., 2013; Kofler et al., 2018). Little research and practical examples exist for tourism that uses and reports on the concept of living labs (Boes et al., 2016; Campos, 2022). This is surprising, given the rising interest in living labs as 'user-centred, open innovation ecosystems based on systematic user co-creation approach, integrating research and innovation processes in real-life communities and settings' (European Network of Living Labs (ENoLL), 2022).

In particular, tourism destinations appear well suited to act as 'user-centric innovation milieus' (Bergvall-Kåreborn et al., 2009, p. 1) as they often span across various municipalities or regions. This was also recently confirmed by Campos (2022) stating that 'living labs boost destination management organizations' capabilities by connecting tourism stakeholders in collaborative action'. Additionally, living labs promote an open approach and allow inclusion of various stakeholders such as locals, tourists and entrepreneurs. This also aligns with the fragmented composition of stakeholders providing the tourism experience. Additionally, at the destination level, the shift from destination marketing organisations to destinations management organisations (DMOs) which take over additional responsibilities and not

The Wiley Blackwell Companion to Tourism, Second Edition. Edited by C. Michael Hall.

only represent the interests of tourists but also the local population and businesses provides an interesting starting point for the living lab concept.

Following Beaudoin et al. (2022), the living lab approach has its roots in studies on information and communication technology but is nowadays used in many fields such as health, energy, mobility and education. In particular, living labs are promoted in agricultural studies (McPhee et al., 2021) and represent a declared policy objective of the European Commission. Despite the links between agriculture and tourism, few studies have taken a closer look at the potential of living labs for accelerating innovation in tourism and destinations. This is important since pivotal to other industries, innovation in tourism not only addresses technological progress but also includes changing market and customer needs (Breier et al., 2021; Thees et al., 2020). Therefore, in this chapter, we recall the foundations of innovation in tourism and focus on what potentially constitutes a living lab in tourism and hospitality and how lighthouses fit into the concept. Furthermore, we discuss if living labs are more than a lab and discuss potential challenges and opportunities in the light of current destination management literature.

Literature Review - Current State of Knowledge

What We Understand as Innovation

In this chapter, we will not re-define innovation, instead, we focus on classic and well-established definitions. For example, important definitions are provided by the OECD/Eurostat (2018, p. 20) defining innovation as 'a new or improved product or process (or a combination thereof) that differs significantly from the unit's previous products or processes and that has been made available to potential users (product) or brought into use by the unit (process)'. Although innovation is defined differently depending on the research tradition, it can have diverse forms, drivers and scales (Cooper, 2006; Hjalager, 2010).

Adams et al. (2006) noted newness, focus and attributes as possible ways to categorise innovations. Classifying innovations in terms of 'newness' relates to Schumpeter (1934), who used an entrepreneurial approach to characterise innovation as creating new knowledge or new amalgamations of knowledge, which are then converted into innovation at the firm level. Following the definition of the OECD/Eurostat (2018) 'focus' refers to product, process or technological innovations. The OECD/Eurostat (2018) uses a more strategic approach that emphasises the importance of market orientation, while the work of Schumpeter (1934) concentrated on the entrepreneur and the willingness to innovate.

In terms of 'attributes', the work of Adams et al. (2006) relates to the work of Rogers (1995), which underlines compatibility, observability, relative advantage, trialability and complexity. From a customer's point of view, Rogers (1995, p. 5) defines innovation as 'an idea, practice, or object that is perceived as new by an individual or another unit of adoption' and describes adoption as a social process in which some people adopt innovations earlier than others do. Since customer or adaptor groups can be characterised differently, different strategies to address and influence their adoption behaviour of innovation are needed. Rogers' (1995) 'diffusion theory' has also been used in many hospitality and tourism settings so far (Brooker et al., 2012; Fuchs et al., 2010; Smerecnik & Andersen, 2011).

Previous literature also distinguished innovations in terms of radical and incremental, while the latter address smaller modifications in products or processes (Damanpour, 1991). Product, process, organisational/managerial and market(ing) innovations represent the main innovation categories in hospitality and tourism (Hjalager, 2010; OECD/Eurostat, 2018). Product innovation relates to delivering new products and services to the marketplace. In contrast, process innovations refer to process improvements (e.g. improved production, new transport methods) (OECD/Eurostat, 2018). Organisational innovations include the introduction of a new organisational approach within the organisation or market innovations. These innovations are also related to the use of market-oriented skills to better respond to customers' needs (Nicolau & Santa-María, 2013). Two categories of innovation are specific to the hospitality and tourism field: distribution innovations, due to their pivotal role in distribution for the hospitality and tourism industry and, institutional innovations, which involve new comprehensive community structures or legal frameworks that enhance management in certain areas (Hjalager, 2010). These different innovation categories highlight that tourism, in contrast to other industries, is not primarily driven by technological progress that drives innovation, but by changes in markets, tourist behaviour and trends that demand innovation.

In rural areas, where small and medium-sized family businesses dominate the tourism supply side, characteristics of entrepreneurs and their willingness to cooperate as well as their social embeddedness are in many cases decisive for sustainable success (Bichler et al., 2022; Czernek, 2014). Moreover, many tourism entrepreneurs in rural areas still lack knowledge about market trends and innovation processes, amongst others, due to topographical restrictions and low mobility but also the perception of a never-ending tourism growth till COVID-19. Living labs could therefore provide an important starting point to collaboratively find new ways and solutions to promote innovation in community-oriented tourism destinations.

What We Understand as Living Labs

The interest in living labs is growing worldwide. The concept emerged in the early 2000s in information and communication technology and was adopted in 2006 by the European Commission. Since then, living labs have become a strong tool to increase the level of innovativeness of European countries. While first attempts of living labs focused merely on the introduction of new information and communication technologies, more recently they are extended to other fields as well, such as sustainable energy, health care, and safety (van Geenhuizen, 2018).

Various review papers (Schuurman et al., 2015; Westerlund & Leminen, 2011) addressed the development of living labs in the past. While early writings in tourism only mentioned living labs (Boes et al., 2016), recently Campos (2022) introduced the concept to tourism research on a broader basis. An additional search in Web of Science™ reveals only seven papers, published between 2016 and 2021, which address living labs and tourism in the title. Among them, only two have been published in tourism journals (see for example Jernsand, 2019; Lapointe et al., 2021).

Next to academic journals, living labs have also become a major topic in communities of practice. One of the largest initiatives on living labs is the European Network of Living Labs (ENoLL) (2022) with worldwide members and more than 450 participating living labs since its start in 2006. Little surprisingly, one of the most established definitions is that living labs 'are defined as user-centred, open innovation ecosystems based on systematic user co-creation approach, integrating research and innovation processes in real-life communities and settings' (ENoLL, 2022).

Beaudoin et al. (2022, p. 1) add that living labs provide 'a collaborative approach to solving complex environmental challenges' by bringing 'together actors through social interactions to achieve a particular goal' (Beaudoin et al., 2022, p. 2). Often living labs are understood

Table 24.1. Living lab definitions.

Author(s)	*Definitions*
Cigir (2019, p. 569)	'a user centred open-innovation research concept and environment that integrates concurrent research and innovation processes'.
Leminen et al. (2012, p. 7)	'physical regions or virtual realities in which stakeholders form public–private–people partnerships of firms, public agencies, universities, institutes, and users all collaborating for creation, prototyping, validating, and testing of new technologies, services, products, and systems in real-life contexts'.
Thees et al. (2020, p. 13)	'cooperative environments that follow a user-driven and cross-stakeholder approach in order to provide a shared space for experiments and tests of innovative products and services in a real-life environment'.
Westerlund and Leminen (2011, p. 23)	'a real-life test and experimentation environment where users and producers co-create innovations'.

as physical spaces but Hossain et al. (2019) also highlight the virtual component, which can be used to bring together stakeholders. Following Mastelic et al. (2015, p. 12) living labs are 'spaces for innovative and participative research, development and activity deployment' that are 'using multi-disciplinary methods and approaches and bringing people together in social contexts around a range of themes'(Mastelic et al., 2015, p. 12). More recently, Campos (2022) provided an overview of living lab definitions suited for the tourism context (Table 24.1).

Existing definitions are rather broad. In general, there might be overlaps with other concepts such as open innovation, action research or innovation ecosystems. Living labs differ from open innovation approaches since 'open innovation focuses on bipolar relations between firms, whereas living labs assume multiple stakeholders and consider the importance of the real-life environment' (Hossain et al., 2019, p. 985). However, especially in Europe, living labs are associated with open innovation (Spagnoli et al., 2019). Additionally, living labs promote 'iterative ways of learning-by-doing and are less focused on developing a common idea of the problem and solution' (Dekker et al., 2020). Previous research also stressed that in contrast to action research, users' opinions about products or services are more at the centre (Bartels & Wittmayer, 2014). Since many overlaps exist, the ENoLL (2022) promotes five components which are used to narrow down the concept of living labs:

- active user involvement
- real-life setting
- multi-stakeholder
- multi-method approach
- co-creation

Active User Involvement

A recent study by Thees et al. (2020) explores living labs as a tool to improve residents' participation and cross-stakeholder involvement. This is an interesting argument since user involvement has been identified as a central element of living labs (Hossain et al., 2019; Mulder et al., 2008). Unlike previous tourism literature, living lab literature discovered that the type of participation depends on the type of living lab, but in general, a living lab is designed to implement and test new products, services or innovations in real-life environments. Therefore, active involvement of tourism stakeholders appears of utmost importance (Sheehan & Ritchie, 2005).

Real-life Setting

Usually, living labs exist in real-life settings, but literature also offers cases which show a more lab-oriented case (Almirall et al., 2012). Discussing the real-life setting of living labs also raises questions about its scale. While Thees et al. (2020) situate living labs mostly at the 'geographical scale of cities or the sphere of single companies', other researchers have a more open definition and perceive living labs as 'a real-life test and experimentation environment where users and producers co-create innovations' (Westerlund & Leminen, 2012, p. 24). Many living labs are located at a specific scale and follow an 'in situ' approach, differentiating them from other innovation approaches (Edvardsson et al., 2012). From a European perspective, living labs may be situated at the regional level of *nomenclature of territorial units for statistics* 2 (NUTS 2) (Eurostat, 2022) with facilities at the level of destinations, cities or single companies.

Multi-stakeholder

To overcome complex environmental problems, as well as to achieve development outcomes, stakeholder involvement is essential. So-called multi-stakeholder platforms (MSPs) bring together a group of stakeholders, which are working in different sectors (Sartas et al., 2018). Multi-stakeholder initiatives are used in various contexts and by a variety of actors, which makes them a flexible concept. Soundararajan et al. (2019) also call multi-stakeholder initiatives a promising direction toward the evolution of the global economy (Soundararajan et al., 2019). In rural tourism, which is prototypeized by the community-model approach, a multi-stakeholder perspective is without alternative (Bichler, 2021; Bichler & Lösch, 2019). Strong leadership and networks of cooperative entrepreneurs are the basis of successful destination developments (Zehrer et al., 2014). Although some destinations already approach stakeholders, this move is just at the beginning and will grow in the future, also due to new ways of digital collaboration (Lalicic, 2018).

Multi-method Approach

Multi-method research, also known as mixed-method research, has risen in popularity over the last two decades. This aligns well with the multi-method approach in living labs. It usually contains both qualitative, as well as quantitative methods to gather a broad range of information about the researched topic. Multi-method research can involve either concurrent or sequential use of at least two, sometimes even more methods. By using a multi-method approach the strengths of more methods are combined while their respective weaknesses should be reduced (Stange et al., 2006). In the past, qualitative and quantitative researchers were often separated in their procedures and analysis, neglecting the benefits of each other. Although quantitative researchers still may doubt the results of qualitative researchers, e.g. concerning validity, reliability and representativeness (Aguinis & Solarino, 2019), nowadays leading scientific tourism journals call explicitly for multi-method studies, e.g. Annals of Tourism Research and Tourism Management (Stange et al., 2006).

Co-creation

Customers serve as an important source of innovation and value creation. Basic definitions of co-creation include the participation of end-users in the process of developing a product or a service (von Hippel, 1987). Following Lusch et al. (2007) co-creation refers to the joint value creation of businesses and customers. Another aspect of co-creation is empowerment, where participants take part in the value creation process and hence add value to existing solutions or exploit the benefits of new products and services. The participation in the creation of a core product through shared production of related offerings is perceived as co-production whereas customers advance to co-creators of value through their interaction with the company (Prahalad & Ramaswamy, 2004).

Since we have now extensively discussed the main building blocks of living labs, a remaining question is how to set up a living lab. In this context, previous research offers various promising guidelines. In general, a living lab consists of four phases (Spagnoli et al., 2019):

- contextualisation
- concretisation
- implementation
- feedback gathering

In the first phase, users and participants of the living lab are identified and the framework is described. This phase is also referred to as developing a working and evaluation plan, which addresses intended results, and budgets but also sets out criteria for the evaluation of the living lab's success (van Geenhuizen, 2018). In the second phase, a detailed observation of participants' everyday behaviour and perceptions of the test group is made. When it comes to the third phase, the implementation phase, users are involved in the co-creation process and finally, the prototype solution is introduced. Finally, in the evaluation phase, users are requested to provide their feedback to evaluate the change in attitudes and perceptions concerning products and services developed.

Discussion

Open Innovation

Many aspects of living labs are not entirely new to tourism research. In general, the call for open innovation in tourism aligns with the focus of living labs to open innovation processes to multiple stakeholders. In this context, Pikkemaat and Peters (2016) highlighted (Figure 24.1, external circle) that 'customers' and guests' opinions should be treated as valuable sources for improving products and services' and that 'DMOs can play a mediating role when external stimuli need to be transferred into the destination' (Pikkemaat & Peters, 2016, p. 166). Additionally, Pikkemaat and Peters (2016) also underlined DMOs' capability to support collaborative innovation networks by providing innovation project structure and a framework for innovation processes (Figure 24.1, inner circle).

Collaborative Innovation

Collaborative innovation can be defined as pursuing innovations across firms' boundaries through the sharing of ideas, knowledge, expertise and opportunities (Ketchen et al., 2007). It includes a broad spectrum of people and organisations (e.g. customers, suppliers, competitors,

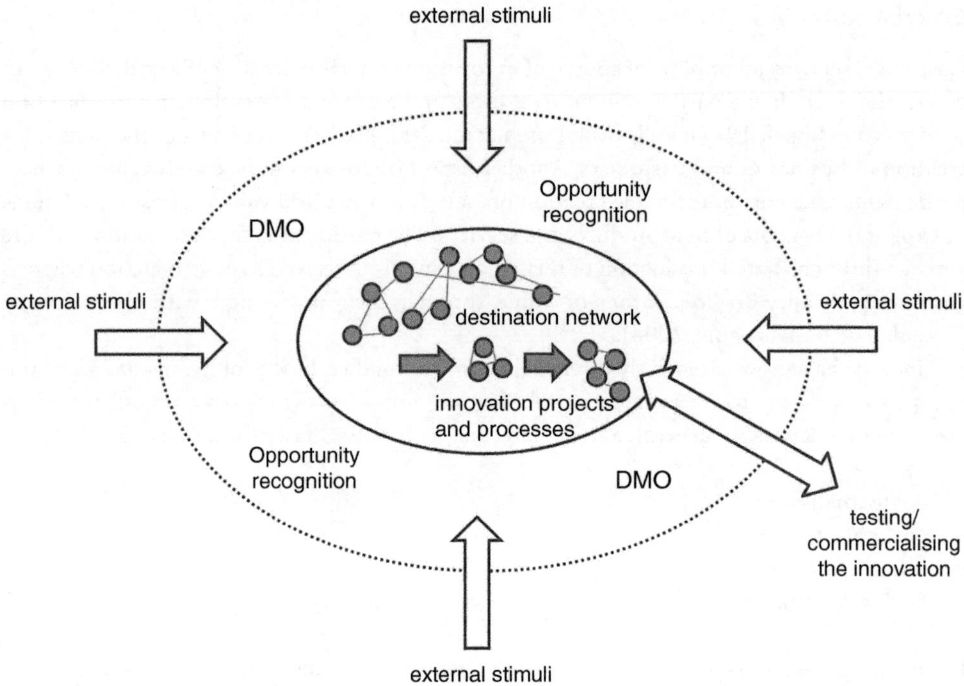

Figure 24.1. Open innovation model.

universities and research institutes) and stands for diverse collaboration types and concepts following an interactive, open innovation approach (Marasco et al., 2018). Collaborative networks help find innovative solutions to complex problems through the exchange of knowledge, competencies, and ideas besides the constructive management of knowledge exchange processes among stakeholders. They can be organised in public organisations or private businesses, whereas in tourism a public-private partnership context seems favourable for smart destination development (Xiang et al., 2013). Following Hjalager (2010) collaborative structures are a special typology of innovation in tourism (institutional innovations) because they enhance the business in specific fields of tourism. Marasco et al. (2018) present a review of research on collaborative innovation in tourism resulting in several categories: most papers deal with collaborative networks; followed by cooperative behaviour of innovating firms; co-creation and knowledge transfer for innovation; and innovation policies.

How Living Labs Could Look Like in Community-oriented Tourism Destinations

Previous literature provided numerous examples of living labs (cf. Boes et al., 2016). However, most of these existing living labs do not primarily tackle tourism issues. For example, Boes et al. (2016) reported on innovation districts in Barcelona, Amsterdam and Helsinki which are embedded in the smart city initiatives of these cities. One of the best practice examples for tourism is the SmartCulTour project (SmartCulTour, 2022) funded by the European Union (EU) under Horizon 2020. The project also includes several living labs, which range from metropolitan regions (e.g., Rotterdam Metropolitan Region), to provinces (e.g., Huesca) and cities (e.g. Vicenza). Again, this illustrates how flexible living labs can be adapted to various sizes but also raises questions about how living labs could look at the ground. Previous literature already highlighted that living labs are more established in urban settings and less used at the destination scale (Hossain et al., 2019; Thees et al., 2020). Regardless of the field of application, literature has already indicated a precise idea of what needs to be taken into account in a living lab (see Hossain et al. (2019), for a recent review):

- real-life environments
- stakeholders
- activities
- business models and networks
- methods, tools and approaches

Living Lab Performance Indicators in Tourism

Measuring the success of living labs provides important information for deciding upon the continuation of living labs. van Geenhuizen (2018) identified five key performance indicators for living labs. Distilling from several case studies, it emerged that a working and evaluation plan is the basis for performance. Additionally, the study identified boundary spanning (e.g., co-creation, joint set of values, broader networking) and performance factors indirectly related to boundary-spanning (e.g., real-life environment) (for a recent summary see van Geenhuizen (2018), p. 1291).

What Problems We Could Tackle With Living Labs in Tourism

Previous research has highlighted that community-oriented destinations would benefit from knowledge brokers and intermediaries to also establish connections beyond already established strong ties that can limit innovation (Bichler et al., 2022). Therefore, mediated activities such as living labs could help to foster innovations in emerging and established businesses (Feld, 2012; van Rijnsoever, 2020). When setting up living labs at the higher regional level (NUTS 2) exchange of ideas between various research and innovation facilities at the destination level could be ensured. This may also involve the exchange of ideas across borders and strengthen the collaboration between border regions (Makkonen et al., 2018). In this context, already existing platforms for collaboration should be used, e.g., the European region Tyrol, South Tyrol and Trentino (EUREGIO, 2022) could be used to establish a living lab focused on tackling joint issues and challenges in Alpine tourism (Weiermair et al., 2007).

Another aspect that living labs could support community-oriented destinations (Flagestad & Hope, 2001), is in handling so-called wicked problems (Buchanan, 1992). For example, participation is known to represent one of the most difficult challenges for tourism destinations (Bichler, 2021). In a well-designed living lab, residents would not only be consulted and requested for feedback but could take part in the co-creation of new tourism offers and services which also benefit residents (Joppe, 2018). Additionally, many community-oriented destinations are located in rural areas. In these areas, mobility is not only an issue for tourists who visit by public transport but also for residents and business operators. Living labs could act as regional innovation and transfer centres by raising awareness, sharing ideas, building networks and connecting stakeholders to solve jointly complex problems. Regional innovation systems focus on benefits from a common regional culture, trust, and social networks that facilitate knowledge exchange by using, e.g. open innovation approaches. Amongst others, one main challenge to installing living labs in community-oriented destinations is assumed to be the individualistic character and often missing collaboration between tourism stakeholders in destinations (Pikkemaat et al., 2018).

The Role That Lighthouses Can Play in Living Labs

Living labs are places for co-design and co-construction of innovations which go beyond the current state of understanding. This approach is accompanied by scientific methodologies and approaches and can therefore range from experimental and lab approaches to the implementation of innovation activities. Another concept which is currently much discussed in the European policy context is lighthouses. Lighthouses display existing solutions, training options or provide communication about specific topics. For example in Austria, lighthouses are an established measure of the state's funding policy and are therefore also financially supported (Rauch-Beran, 2021). One of the funded projects addresses systemic leadership in tourism and belongs to an innovative tourism destination in Tyrol, Austria. In 2017, the destination 'Wilder Kaiser' initiated a joint strategy process for sustainable tourism until 2024, including other projects targeting quality of life, sustainable transport but also regional consumption and emission reduction. These destinations are not yet recognised as living labs but their strategic co-creation approach and the multitude of activities qualify them as such.

The Role That Universities and Other Research Infrastructure Can Play in Living Labs

Over the last years, academic education in tourism is putting a stronger emphasis on practical application and hands-on knowledge transfer. In this context, universities and research institutions can become members of living labs. Thereby, researchers not only have the potential to develop new and courageous research ideas from the collaboration between industry and academia (Kock et al., 2020) but also accelerate knowledge transfer. In particular, van Geenhuizen (2013, p. 1126) highlighted the following benefits of living labs:

- reduce differences in values, needs, etc. between university and industry
- improve financial factors and reduce financial risks
- compensation for lack of 'critical mass' of regional economy and for lack of industrial specialization
- increase speed and efficiency of valorization

Proposal for a Living Lab Framework in Community-oriented Tourism

Before we present a framework model for living labs in community-oriented destinations we first give an overview of existing living lab initiatives in both tourism research and industry. Looking at Google Scholar results in more than three million results when searching for living labs in general, and more than 91,000 results when searching for living labs in tourism. In the 'Google Scholar result list' of publications on tourism living labs since 2021, it seems noticeable that on the one hand many are discussed in the area of sustainability, especially social sustainability, e.g. community participation (e.g. Lapointe et al., 2021). On the other hand, the focus is concentrated on technological innovations and smart approaches (e.g. Boes et al., 2016). There is research on both urban and rural living labs: e.g. while Alam and Porras

(2018) elaborate on the design of sustainable smart city services in a living lab environment, Zavratnik et al. (2019) contextualise living labs for smart rural developments.

A more analytical review on living labs identifies 40 relevant research papers and results in five clusters (Thees et al., 2020): by far the most research has been done in the area of characteristics and design of living labs (21 papers), whereby the subclusters *open innovation & user-driven* as well as *research & university* and, thirdly, *community & collaboration* were identified. Six papers can be assigned to the cluster *smart city*, five to *life and work*, and three each to the *urban transition* and *infrastructure & data cluster*. Regarding the fields of application of living labs only a few studies have concentrated on tourism destinations as the subject of analysis (Calzada, 2019; Guimont et al., 2017; Guimont & Lapointe, 2016).

Following Schuurman et al. (2015) three levels of analysis for living labs can be differentiated: the macro (living lab constellation); the meso (living lab innovation project); and the micro (living lab methodological research steps) level. This three-level living lab analysis model has been used by Guimont and Lapointe (2016) to analyse the case of a living lab in tourism. Their research shows the need to reflect on scales and levels of actions that emerged in living labs in strongly territorial industries, such as tourism and agriculture. They propose 'a descriptive grid that merges the levels of the innovation process with the scales of the territory associated with the living lab's areas of intervention' (Guimont & Lapointe, 2016, p. 24) which is presented in Table 24.2.

Another study carried out in tourism focuses on the analysis of a living lab that was implemented in the Basque Country (Spain) in 2017 to foster local entrepreneurship and overcome the growing mass tourism trend and its negative side effects affecting various stakeholders. Their results show that residents' priority is the impact on the village, even before reflecting on potential opportunities of an increasing tourism boom and therefore a multi-stakeholder approach is needed to overcome the concerns and claims of the different stakeholders in a tourism destination. The third tourism study about living labs uses an action research methodology for analysing a case study in Canada (Guimont et al., 2017). Their results show that the living lab was able to raise innovation capabilities of the involved tourism stakeholders

Table 24.2. Links between levels and characteristics of living labs.

	Innovation Process Level	*Territory Levels*	*Living Lab Characteristics*
Macro	Living lab constellation	Concertation/ governance	• Ecosystem approach • Real-world context • Technical infrastructure • Lifespan • Scale • Level of openness
Meso	Living lab project	Intervention/steering	• Technical infrastructure • Community • Level of openness
Micro	Living lab methodologies	Socialization/encounter	• Real-world context • User role • Level of openness • Evaluation, context research, co-creation

Source: Guimont et.al, 2016/Talent First Network/CC BY 3.0/Public domain.

but struggled with the following barriers: limited IT skills and resources; small size of businesses; lack of knowledge about change management culture; and rudimentary level of trust and collaboration. Despite these difficulties, the main goal of this living lab was achieved as a mobile app was developed for the region.

Another, more practical tourism living lab case, is recently provided by Slovenias' 'Tourism 4.0. Living Lab' which is promoted as an inspirational hands-on learning environment that encourages the cooperation, innovation and development of the entire tourism industry (for more information see Tourism 4.0 (2019) https://tourism4-0.org/ll/. Another living lab is SmartCulTour (2022) (Smart Cultural Tourism as a Driver of Sustainable Development of European Regions), which is a four-year project, funded by the European Union under Horizon 2020. The goal of this project with six living labs in different areas in Europe is to support regional development with cultural assets through sustainable cultural tourism (for more information see SmartCulTour (2022), http://www.smartcultour.eu/living-labs/).

Notably, research on living labs in tourism predominantly employs case study methods alongside action research methodologies (Thees et al., 2020). These studies are therefore very limited in terms of territory. In addition, no framework puts the various stakeholders in a destination context. The latter seems surprising as the fragmented nature of the tourism industry limits destinations' capabilities to innovate and living labs seem to be appropriate tools to boost innovation processes among tourism stakeholders (Guimont et al., 2017). Conceptual approaches that are used to explain living labs are missing in the tourism industry and have so far been little related to the three levels and the territorial components of tourist destinations.

Consequently, based on the literature review, we distilled the following prototype framework (Figure 24.2) which aims to support destination managers, tourism policy makers and researchers when setting up a living lab in tourism and tries to build on the work of Guimont and Lapointe (2016). Next to important tourism stakeholders (Sheehan & Ritchie, 2005),

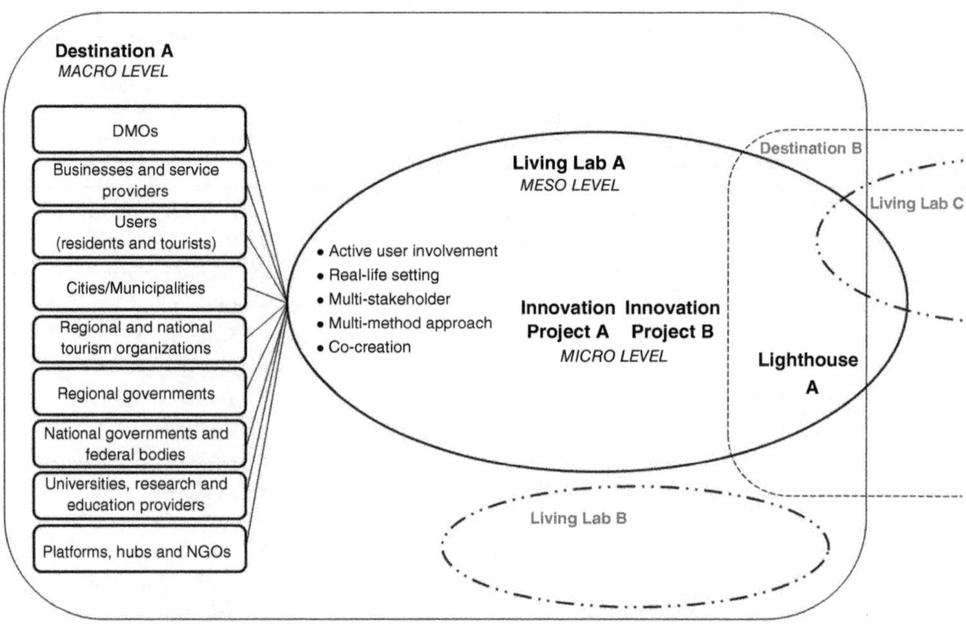

Figure 24.2. Prototype framework for living labs in community-oriented destinations.

Figure 24.2 also highlights the potential regional scale of tourism living labs which potentially span across various destinations and can also involve different destinations and administrative regions. The overall idea is to involve multiple stakeholders in a real-life setting to co-create, develop and test innovations and research in open communities and settings. Therefore, projects within a living lab can be of various sizes, various topics and involve a broad range of stakeholders. Finally, since tourism is not an isolated phenomenon, living labs may span across various destinations and co-exist next to other living labs in the same or other destinations as from a tourist experience point of view, tourism does not end with administrative borders, e.g. cycle routes often include cycle paths to a wide variety of destinations in different regions.

Conclusion: Theoretical and Methodological Reflections

In this chapter, we highlighted the potential value of living labs for rural and community-oriented tourism destinations. A major benefit of living labs is that the concept allows for flexibility. Living labs can be designed depending on the desired outcome. A major benefit is that living labs usually not only answer questions but also provide new discussions. Especially for tourism, living labs offer the possibility of transparency and involvement of various stakeholders – a topic that is still posing a challenge to tourism destinations. Based on our elaborations and previous literature, we developed a prototype framework model for living labs in community-oriented tourism destinations (Figure 24.2). Future research is necessary to collect empirical data on living labs in tourism and to trace the development and success or failure of living labs over time. A potential implication of this study is to create a platform at the national level to promote networking between living labs, provide information about existing living labs to users and stimulate knowledge exchange from the local to the national level. Now would be the time to focus on living labs in tourism as they will become an important topic in the future.

References

Adams, R., Tranfield, D. R., & Denyer, D. (2006). Innovation types: Configurations of attributes as a basis for innovation classification. *Advanced Institute of Management Research Paper*, 46. https://doi.org/10.2139/ssrn.1309554

Aguinis, H., & Solarino, A. M. (2019). Transparency and replicability in qualitative research: The case of interviews with elite informants. *Strategic Management Journal*, 6(3), 1291–1315. https://doi.org/10.1002/smj.3015

Alam, M., & Porras, J. (2018). Architecting and designing sustainable smart city services in a living lab environment. *Technologies*, 6(4), 99. https://doi.org/10.3390/technologies6040099

Almirall, E., Lee, M., & Wareham, J. (2012). Mapping living labs in the landscape of innovation methodologies. *Technology Innovation Management Review*, 2(9), 12–18. https://timreview.ca/article/603

Bartels, K. P., & Wittmayer, J. M. (2014). Symposium introduction: usable knowledge in practice. What action research has to offer to critical policy studies. *Critical Policy Studies*, 8(4), 397–406. https://doi.org/10.1080/19460171.2014.957335

Beaudoin, C., Joncoux, S., Jasmin, J.-F., Berberi, A., McPhee, C., Schillo, R. S., & Nguyen, V. M. (2022). A research agenda for evaluating living labs as an open innovation model for environmental and agricultural sustainability. *Environmental Challenges*, 7, 100505. https://doi.org/10.1016/j.envc.2022.100505

Bergvall-Kåreborn, B., Ihlström Eriksson, C., Ståhlbröst, A., & Svensson, J. (2009). *A milieu for innovation: defining living labs*. 2nd ISPIM Innovation Symposium. https://www.diva-portal.org/smash/get/diva2:1004774/FULLTEXT01.pdf

Bichler, B. F. (2021). Designing tourism governance: The role of local residents. *Journal of Destination Management and Marketing*, *19*, 100389. https://doi.org/10.1016/j.jdmm.2019.100389

Bichler, B. F., & Lösch, M. (2019). Collaborative governance in tourism: Empirical insights into a community-oriented destination. *Sustainability*, *11*(23), 6673. https://doi.org/10.3390/su11236673

Bichler, B. F., Kallmuenzer, A., Peters, M., Petry, T., & Clauss, T. (2022). Regional entrepreneurial ecosystems: Exploring the role of embeddedness for entrepreneurship in family firms. *Review of Managerial Science*, *16*, 15–44. https://doi.org/10.1007/s11846-020-00434-9

Boes, K., Buhalis, D., & Inversini, A. (2016). Smart tourism destinations: Ecosystems for tourism destination competitiveness. *International Journal of Tourism Cities*, 2(2), 108–124. https://doi.org/10.1108/IJTC-12-2015-0032

Breier, M., Kallmuenzer, A., Clauss, T., Gast, J., Kraus, S., & Tiberius, V. (2021). The role of business model innovation in the hospitality industry during the COVID-19 crisis. *International Journal of Hospitality Management*, *92*, 102723. https://doi.org/10.1016/j.ijhm.2020.102723

Brooker, E., Joppe, M., Davidson, M. C., & Marles, K. (2012). Innovation within the Australian outdoor hospitality parks industry. *International Journal of Contemporary Hospitality Management*, *24*(5), 682–700. https://doi.org/10.1108/09596111211237246

Buchanan, R. (1992). Wicked problems in design thinking. *Design Issues*, *8*(2), 5–21.

Calzada, I. (2019). Local entrepreneurship through a multistakeholders' tourism living lab in the post-violence/peripheral era in the Basque Country. *Regional Science Policy & Practice*, *11*(3), 451–466. https://doi.org/10.1111/rsp3.12130

Campos, A. C. (2022). Living labs in tourism. In D. Buhalis (Ed.), *Encyclopedia of tourism management and marketing*. Edward Elgar Publishing.

Carlisle, S., Kunc, M., Jones, E., & Tiffin, S. (2013). Supporting innovation for tourism development through multi-stakeholder approaches: Experiences from Africa. *Tourism Management*, *35*, 59–69. https://doi.org/10.1016/j.tourman.2012.05.010

Cigir, K. (2019). Creating a living lab model for tourism and hospitality businesses to stimulate CSR and sustainability innovations. *WIT Transactions on Ecology and the Environment*, *217*, 569–583. https://www.witpress.com/elibrary/wit-transactions-on-ecology-and-the-environment/217/36972

Cooper, C. (2006). Knowledge management and tourism. *Annals of Tourism Research*, *33*(1), 47–64. https://doi.org/10.1016/j.annals.2005.04.005

Czernek, K. (2014). The role of social embeddedness in tourist region cooperation. *European Journal of Tourism, Hospitality and Recreation*, *5*(2), 61–81.

Damanpour, F. (1991). Organizational innovation: A meta-analysis of effects of determinants and moderators. *Academy of Management Journal*, *34*(3), 555–590. https://www.jstor.org/stable/256406

Dekker, R., Franco Contreras, J., & Meijer, A. (2020). The living lab as a methodology for public administration research: A systematic literature review of its applications in the social sciences. *International Journal of Public Administration*, *43*(14), 1207–1217. https://doi.org/10.1080/01900692.2019.1668410

Edvardsson, B., Kristensson, P., Magnusson, P., & Sundström, E. (2012). Customer integration within service development—A review of methods and an analysis of insitu and exsitu contributions. *Technovation*, *32*(7–8), 419–429. https://doi.org/10.1016/j.technovation.2011.04.006

EUREGIO (2022). *Tyrol, South Tyrol, Trentino - 3 territories, 1 Euregio*. https://www.europaregion.info/en/

European Network of Living Labs (ENoLL). (2022). *What are living labs*. https://enoll.org/about-us/what-are-living-labs/

Eurostat (2022). *NUTS - Nomenclature of territorial units for statistics.Background*. https://ec.europa.eu/eurostat/web/nuts/background

Feld, B. (2012). *Startup communities: Building an entrepreneurial ecosystem in your city*. Wiley.

Flagestad, A., & Hope, C. (2001). Strategic success in winter sports destinations: A sustainable value creation perspective. *Tourism Management*, *22*, 445–461. https://doi.org/10.1016/S0261-5177(01)00010-3

Fuchs, M., Höpken, W., Föger, A., & Kunz, M. (2010). E-Business readiness, intensity, and impact: An Austrian destination management organization study. *Journal of Travel Research*, *49*(2), 165–178. https://doi.org/10.1177/0047287509336469

Guimont, D., & Lapointe, D. (2016). Empowering local tourism providers to innovate through a living lab process: Does scale matter? *Technology Innovation Management Review*, *6*(11). https://www.timreview.ca/article/1031

Guimont, D., Lapointe, D., & Sévigny, A. (2017). Changing collaborative practices in tourism - a living lab case. *Canadian Journal of Regional Science / Revue Canadienne Des Sciences Régionales*, *40*(1), 63–69.

Hjalager, A.-M. (2010). A review of innovation research in tourism. *Tourism Management*, *31*(1), 1–12. https://doi.org/10.1016/j.tourman.2009.08.012

Hossain, M., Leminen, S., & Westerlund, M. (2019). A systematic review of living lab literature. *Journal of Cleaner Production*, *213*, 976–988. https://doi.org/10.1016/j.jclepro.2018.12.257

Jernsand, E. M. (2019). Student living labs as innovation arenas for sustainable tourism. *Tourism Recreation Research*, *44*(3), 337–347. https://doi.org/10.1080/02508281.2019.1613299

Joppe, M. (2018). Tourism policy and governance: Quo vadis? *Tourism Management Perspectives*, *25*, 201–204. https://doi.org/10.1016/j.tmp.2017.11.011

Ketchen, D. J., Ireland, R. D., & Snow, C. C. (2007). Strategic entrepreneurship, collaborative innovation, and wealth creation. *Strategic Entrepreneurship Journal*, *1*(3–4), 371–385. https://doi.org/10.1002/sej.20

Kock, F., Assaf, A. G., & Tsionas, M. G. (2020). Developing courageous research ideas. *Journal of Travel Research*, *59*(6), 1140–1146. https://doi.org/10.1177/0047287519900807

Kofler, I., Marcher, A., Volgger, M., & Pechlaner, H. (2018). The special characteristics of tourism innovation networks: The case of the Regional Innovation System in South Tyrol. *Journal of Hospitality and Tourism Management*, *37*, 68–75. https://doi.org/10.1016/j.jhtm.2018.09.004

Lalicic, L. (2018). Open innovation platforms in tourism: How do stakeholders engage and reach consensus? *International Journal of Contemporary Hospitality Management*, *30*(6), 2517–2536. https://doi.org/10.1108/IJCHM-04-2016-0233

Lapointe, D., Guimont, D., Guillemard, A., & Benjamin, C. (2021). People, place, values: Living labs as social innovation processes for tourism communities. *Revista Anais Brasileiros De Estudios Turísticos*, *11*. https://doi.org/10.5281/zenodo.5771002

Leminen, S., Westerlund, M., & Nyström, A.-G. (2012). Living labs as open-innovation networks. *Technology Innovation Management Review*. https://timreview.ca/article/489

Lusch, R. F., Vargo, S. L., & O'Brien, M. (2007). Competing through service: Insights from service-dominant logic. *Journal of Retailing*, *83*(1), 5–18. https://doi.org/10.1016/j.jretai.2006.10.002

Makkonen, T., Williams, A. M., Weidenfeld, A., & Kaisto, V. (2018). Cross-border knowledge transfer and innovation in the European neighbourhood: Tourism cooperation at the Finnish-Russian border. *Tourism Management*, *68*, 140–151. https://doi.org/10.1016/j.tourman.2018.03.008

Marasco, A., Martino, M. de, Magnotti, F., & Morvillo, A. (2018). Collaborative innovation in tourism and hospitality: A systematic review of the literature. *International Journal of Contemporary Hospitality Management*, *30*(6), 2364–2395. https://doi.org/10.1108/IJCHM-01-2018-0043

Mastelic, J., Sahakian, M., & Bonazzi, R. (2015). How to keep a living lab alive? *Info*, *17*(4), 12–25. https://doi.org/10.1108/info-01-2015-0012

McPhee, C., Bancerz, M., Mambrini-Doudet, M., Chrétien, F., Huyghe, C., & Gracia-Garza, J. (2021). The defining characteristics of agroecosystem living labs. *Sustainability*, *13*(4), 1718. https://doi.org/10.3390/su13041718

Mulder, I., Velthausz, D., & Kriens, M. (2008). The living labs harmonization cube: Communicating living labs' essentials. *EJOV Executive - the Electronic Journal of Virtual Organizations and Networks*, *10*, 1–15.

Nicolau, J. L., & Santa-María, M. J. (2013). The effect of innovation on hotel market value. *International Journal of Hospitality Management*, *32*, 71–79. https://doi.org/10.1016/j.ijhm.2012.04.005

Nordli, A. J. (2018). Information use and working methods as drivers of innovation in tourism companies. *Scandinavian Journal of Hospitality and Tourism*, *18*(2), 199–213. https://doi.org/10.1080/15022250.2017.1343682

OECD/Eurostat. (2018). *Oslo manual 2018: Guidelines for collecting, reporting and using data on innovation* (4th ed.). OECD Publishing.

Pikkemaat, B., & Peters, M. (2016). Open innovation: A chance for the innovation management of tourism destinations? In R. Egger, I. Gula, & D. Walcher (Eds.), *Open tourism* (pp. 153–169). Springer.

Pikkemaat, B., Peters, M., & Bichler, B. F. (2019). Innovation research in tourism: Research streams and actions for the future. *Journal of Hospitality and Tourism Management*, *41*, 184–196. https://doi.org/10.1016/j.jhtm.2019.10.007

Pikkemaat, B., Peters, M., & Chan, C.-S. (2018). Needs, drivers and barriers of innovation: The case of an alpine community-model destination. *Tourism Management Perspectives*, *25*, 53–63. https://doi.org/10.1016/j.tmp.2017.11.004

Prahalad, C. K., & Ramaswamy, V. (2004). Co-creating unique value with customers. *Strategy & Leadership*, *32*(3), 4–9. https://doi.org/10.1108/10878570410699249

Rauch-Beran, S. (2021). Siegerprojekte als Leuchttürme der Nachhaltigkeit und Resilienz im Tourismus. *Österreichische Hotel- und Tourismusbank GmbH (ÖHT) TourismusBank*. https://www.oeht.at/siegerprojekte-als-leuchttuerme-der-nachhaltigkeit-und-resilienz-im-tourismus/

Rogers, E. M. (1995). *Diffusion of Innovations*. The Free Press.

Sartas, M., Schut, M., Hermans, F., van Asten, P., & Leeuwis, C. (2018). Effects of multi-stakeholder platforms on multi-stakeholder innovation networks: Implications for research for development interventions targeting innovations at scale. *PloS One*, *13*(6), e0197993. https://doi.org/10.1371/journal.pone.0197993

Schumpeter, J. A. (1934). *The theory of economic development*. Harvard University Press.

Schuurman, D., De Marez, L., & Ballon, P. (2015). *Living labs: A systematic literature review*. https://biblio.ugent.be/publication/7026155/file/7026171

Sheehan, L. R., & Ritchie, B. J. (2005). Destination stakeholders exploring identity and salience. *Annals of Tourism Research*, *32*(3), 711–734. https://doi.org/10.1016/j.annals.2004.10.013

SmartCulTour (2022). *Living labs*. http://www.smartcultour.eu/living-labs/

Smerecnik, K. R., & Andersen, P. A. (2011). The diffusion of environmental sustainability innovations in North American hotels and ski resorts. *Journal of Sustainable Tourism*, *19*(2), 171–196. https://doi.org/10.1080/09669582.2010.517316

Soundararajan, V., Brown, J. A., & Wicks, A. C. (2019). Can multi-stakeholder initiatives improve global supply chains? Improving deliberative capacity with a stakeholder orientation. *Business Ethics Quarterly*, *29*(3), 385–412. https://doi.org/10.1017/beq.2018.38

Souto, J. E. (2015). Business model innovation and business concept innovation as the context of incremental innovation and radical innovation. *Tourism Management*, *51*, 142–155. https://doi.org/10.1016/j.tourman.2015.05.017

Spagnoli, F., van der Graaf, S., & Brynskov, M. (2019). The paradigm shift of living labs in service co-creation for smart cities: SynchroniCity validation. In A. Lazazzara, R. C. Nacamulli, C. Rossignoli, & S. Za (Eds.), *Lecture Notes in information systems and organization. Organizing for digital innovation* (Vol. 27, pp. 135–147). Springer International Publishing.

Stange, K. C., Crabtree, B. F., & Miller, W. L. (2006). Publishing multimethod research. *Annals of Family Medicine*, *4*(4), 292–294. https://doi.org/10.1370/afm.615

Thees, H., Pechlaner, H., Olbrich, N., & Schuhbert, A. (2020). The living lab as a tool to promote residents' participation in destination governance. *Sustainability*, *12*(3), 1120. https://doi.org/10.3390/su12031120

Tourism 4.0 (2019). *Tourism 4.0 living lab*. https://tourism4-0.org/ll/

van Geenhuizen, M. (2013). From ivory tower to living lab: Accelerating the use of university knowledge. *Environment and Planning C: Government and Policy*, *31*(6), 1115–1132. https://doi.org/10.1068/c1175b

van Geenhuizen, M. (2018). A framework for the evaluation of living labs as boundary spanners in innovation. *Environment and Planning C: Politics and Space*, *36*(7), 1280–1298. https://doi.org/10.1177/2399654417753623

van Rijnsoever, F. J. (2020). Meeting, mating, and intermediating: How incubators can overcome weak network problems in entrepreneurial ecosystems. *Research Policy*, *49*(1), 103884. https://doi.org/10.1016/j.respol.2019.103884

von Hippel, E. (1987). Cooperation between rivals: Informal know-how trading. *Research Policy*, *16*(6), 291–302. https://doi.org/10.1016/0048-7333(87)90015-1

Weiermair, K., Peters, M., & Schuckert, M. (2007). Destination development and the tourist life-cycle: Implications for entrepreneurship in Alpine tourism. *Tourism Recreation Research*, *32*(1), 83–93. https://doi.org/10.1080/02508281.2007.11081526

Westerlund, M., & Leminen, S. (2011). Managing the challenges of becoming an open innovation company: Experiences from living labs. *Technology Innovation Management Review*, 19–25. https://timreview.ca/article/489

Xiang, Z., Tussyadiah, I., Buhalis, D., & Amaranggana, A. (Eds.) (2013). *Smart tourism destinations: Information and communication technologies in tourism 2014*. Springer International Publishing.

Zavratnik, V., Superina, A., & Stojmenova Duh, E. (2019). Living labs for rural areas: Contextualization of living lab frameworks, concepts and practices. *Sustainability*, *11*(14), 3797. https://doi.org/10.3390/su11143797

Zehrer, A., Raich, F., Siller, H., & Tschiderer, F. (2014). Leadership networks in destinations. *Tourism Review*, *69*(1), 59–73. https://doi.org/10.1108/TR-06-2013-0037

Chapter 25

Residents' Attitudes Toward Tourism

Fahimeh Hateftabar

Introduction

The tourism industry and its development rely on resources and particular infrastructure to be fulfilled. One of the most crucial features of a destination is its human inhabitants and their attitudes toward tourism. Several scholars assert that the success of the tourism industry is largely reliant on local residents' hospitality and pro-tourism behaviour (Gursoy et al., 2010; Liu et al., 2022; Ribeiro et al., 2017). Local residents' attitudes significantly influence the satisfaction and loyalty of tourists, meaning that unwelcoming and hostile behaviour deters tourists from returning or visiting in the first place (Ribeiro et al., 2017). In general, tourists are averse to visiting or revisiting places where they do not feel welcome. To ensure the success of local tourism, locals must be dedicated to its development, as local opposition typically results in low destination competitiveness and the loss of potential tourists (Gannon et al., 2021; Liu et al., 2022; Thyne et al., 2022).

'Attitude' is a psychological term used to denote an individual's positive or negative opinions (favourable or unfavourable) toward objects, places, or issues (Allport, 1937; Gannon et al., 2021). There have been many studies conducted in recent decades on residents' attitudes toward tourism based on the premise that attitudes are generally the causal factors underlying behaviours. This premise has resulted in residents' attitudes toward tourism and its development coming under extensive scrutiny by academics and practitioners (García et al., 2015), who now generally acknowledge its significance in determining local residents' support for tourism development and its sustainable development (Li et al., 2020).

This acknowledgment came over 30 years ago (García et al., 2015; Hateftabar & Chapuis, 2020), the main reason behind it being a belief that understanding resident perceptions and responses is a precondition for successful sustainable tourism (Hall & Lew, 2009; Šegota et al., 2017). The research derived from this acknowledgment has extended to destinations across many countries and territories. However, a significant portion of this research has

The Wiley Blackwell Companion to Tourism, Second Edition. Edited by C. Michael Hall.

been conducted in developed countries or within western cultures (Liu et al., 2022) such as North America (with a focus on rural tourism and resident perceptions near recreation areas) and, to a lesser extent, in Australia, New Zealand, and the United Kingdom (Nunkoo & Gursoy, 2012).

Evidence suggests that residents' perceptions of tourism's impact on their living environment is predictive of their attitudes toward tourism (Truong et al., 2014). However, there is a lack of agreement within the extant literature regarding the classification and measurement of residents' perceptions. Our literature review indicates that there are three main approaches exhibited by previous studies: the cost-benefit approach; the domain-related cost-benefit approach; and the non-forced approach. The most common approach has been the cost-benefit approach (Nunkoo & Ramkissoon, 2012). Scholars employing this approach classify tourism's impacts as 'blessings' or 'blights', arguing that there is a direct relationship between residents' perceived impacts of tourism and behaviours toward tourism's development (Gannon et al., 2021; Hall & Lew, 2009). This approach is straightforward and uncomplicated, but it overemphasises the impact of tourism on various dimensions of the lives of local residents (economic, sociocultural, environmental). Thus, it offers only partial comprehension of how perceived effects influence residents' pro-tourism behaviour, hindering its predictive strength (Gursoy et al., 2010; Nunkoo & Ramkissoon, 2012) and limiting insight into sustainable development and the marketing of new projects (Prayag et al., 2013).

Researchers employing the domain-related cost-benefit approach (Gursoy et al., 2010; Gursoy & Rutherford, 2004; Prayag et al., 2013) aim to obtain a deep understanding of the relationship between residents' perceptions of tourism's impacts and pro-tourism behaviours by considering two means of assessment: positive/negative and categorical (the triple-bottom-line impacts: economic, sociocultural, environmental) (Table 25.1). This approach, in line with the cost-benefit approach, hypothesises direct positive relationships between pro-tourism behaviour and perceived economic, sociocultural, and environmental benefits as well as direct negative relationships between pro-tourism behaviour and perceived economic, socio-cultural, and environmental costs. It is worth noting that scholars employing the costs-benefits approach and the domain-related costs-benefits approach rely on a priori impact categorisation. In both approaches, respondents are unable to express the degree to which they perceive impacts as positive or negative; they are confined to expressing their level of agreement with pre-coded answers (Andereck et al., 2005). Aiming to overcome this shortcoming, some scholars have employed the non-forced approach, through which residents are provided with neutral statements and asked to express their perceptions of directionality, or the degree to which they consider tourism to constitute a positive or negative influence on various aspects of their community (Andereck et al., 2005; Deccio & Baloglu, 2002; Jurowsk et al., 1997; Upchurch & Teivane, 2000; Stylidis et al., 2014). This approach facilitates a more nuanced classification of perceived impacts. Studies using this approach indicate that, as residents perceive tourism's impacts more positively, they are more likely to display support for tourism's development. In line with this dynamic, residents who perceive tourism as having unfavorable economic, socio-cultural, and environmental impacts on their community are less likely to support its development. In contrast to the two other approaches, which predetermine the nature of impacts (positive or negative), the non-forced approach suggests direct positive relationships between all domains of impact and support. In other words, impacts' perceived nature (extent of positive/negative perception) is reflected in the overall assessment of each impact (i.e., scores higher/lower than the scale's midpoint indicate more positive/negative assessments, respectively).

Table 25.1. Classification and measurement of tourism perceived impacts.

Approach	*Classification*	*Impacts on resident pro-tourism behaviour*	*Advantages and disadvantages*
Benefits-costs	• positive/negative impacts. • Agreement/disagreement to positively and negatively stated impacts.	• Positive impact (benefit) is directly and positively associated with resident support. • Negative impact (costs) is directly and negatively associated with resident support.	• Simple and uncomplicated. • Ignores the potential 'trade off' between different domains of impacts. • Predictive strength of the model is limited. • Impacts' directionality is a priori determined.
Domain related costs-benefits	• Economic, socio-cultural and environmental costs/benefits. • Agreement/disagreement to positively and negatively stated impacts.	• Economic, socio-cultural, and environmental benefits are directly and positively associated with resident support. • Economic, socio-cultural, and environmental costs are directly and negatively associated with resident support.	• Reflects the potential 'trade off' between different domains of impacts. • Higher predictive strength of the model. • Complex models. • Impacts' directionality is a priori determined.
Non-forced	• Economic, socio-cultural and environmental impacts, • Or 'opportunities' and 'concerns'. • The extent tourism is perceived to have negative or positive effect on neutrally stated economic, socio-cultural and environmental impact items.	• The more positively residents perceive the economic, socio-cultural and environmental impacts, they display greater support for tourism (and vice versa). • The more residents perceive tourism to improve the state of opportunities and concerns, they display greater support (and vice versa).	• Reflects the potential 'trade off' between different domains of impacts. • Less complex models compared to the domain related costs-benefits approach. • Potential for greater predictive strength compared to the costs-benefits approach. • Impacts' directionality is reflective of resident's evaluation.

Source: Adapted from Stylidis et al. (2014).

Factors Influencing Residents' Attitudes

In a review of factors that directly or indirectly influence residents' attitudes toward tourism, Sharpley (2014) indicates that there is significant flexibility in the considered independent variables pertaining to residents, explaining that we can classify these independent variables into two main categories: extrinsic and intrinsic factors.

a. Extrinsic factors

The literature on residents' attitudes suggests extrinsic factors such as the level of tourism development in the destination (Butler, 1980; Doxey, 1975; Gu & Ryan, 2008; Lepp, 2007; Nunkoo & Ramkissoon, 2010; Vargas-Sánchez et al., 2011), the density of tourists (Bestard & Nadal, 2007; Vargas-Sánchez et al., 2011), and tourists' type (Vargas-Sánchez et al., 2015; Weaver & Lawton, 2001; Wu & Chen, 2015) can be influential.

The 'irridex' of Doxey (1975) is regarded as one of the most valid and influential indexes and models for explaining the relationship between ***the level of tourism development*** and residents' responses (Mason & Cheyne, 2000). The author argues that the extent of this irritation is determined by how incompatible the residents and the tourists are in the community under scrutiny. The model fundamentally postulates that with the number of tourists rising and tourist destinations expanding and developing, the residents' attitude ranges from great excitement (euphoria) to indifference (apathy), then to annoyance or dislike, and on to outright irritation and hatred.

Doxey's (1975) irridex resonates with the well-known tourism area life-cycle (TALC) model of Butler (1980). The TALC model explains the relationship between ***the density of tourists*** and residents' responses. In view of the fact that tourist destinations undergo constant upheavals and changes over time, and based on the product cycle concept, according to which product sales proceed slowly at first, then increase rapidly, stabilise, and subsequently decline, Butler's (1980) model maintains that tourism destinations follow a similar pattern of evolution. By that virtue, the tourist destination will initially be visited by a small number of tourists, as there are too limited access and facilities to allow more visitors. Then, with the growth of awareness of the area and facilities, more visitors start pouring in. Then, it states that the area grows rapidly due to marketing efforts and dissemination of information, and provision of more and more facilities. It is followed by the number of tourists declining as the carrying capacity of the area reaches its limit.

Despite the value of such models, one of the most predominant limitations of these models is their unidirectional and stagnant nature, in that it is directed at representing the current status of an existing host community as a whole (Sharpley, 2014). Moreover, the model is based on the assumption of a certain degree of homogeneity in the community under study, disregarding the intrinsic qualities and factors associated with the individual members of the community. It has thus been acknowledged that these models are not adequate to explain in detail the exclusive intrinsic differences that exist among the residents within the same community (Zhang et al., 2006). Also, the models do not explain situations where visitor management plans or schemes are in place in order to minimise tourist insurgence or in destinations where the local community can participate effectively in directing tourism development (Sharpley, 2014).

b. Intrinsic Factors

Previous studies have concluded that intrinsic factors such as residents' socio-demographic profile (Almeida-García et al., 2016; Fredline & Faulkner, 2000; Mason & Cheyne, 2000; Nunkoo & Gursoy, 2012; Rasoolimanesh et al., 2015; Sinclair-Maragh, 2017), their community

Table 25.2. Extrinsic and intrinsic variables affecting resident perception.

Variables	*Examples of studies*
Extrinsic factors	
Level of tourism development in the destination	Allen et al. (1988); Gu & Ryan (2008); Kuvan & Akan (2005); Nunkoo & Ramkissoon (2010); Smith & Krannich (1998); Vargas-Sánchez et al. (2011); Vargas-Sánchez et al. (2015).
Tourists' density	Brida et al. (2011); Bestard & Nadal (2007); Vargas-Sáchez et al. (2011); Vargas-Sánchez et al. (2015).
Type of tourists	Vargas-Sánchez et al. (2015); Weaver & Lawton (2001); Wu & Chen (2015).
The state of local economy	Gursoy et al. (2002); Gursoy et al. (2010); Hateftabar & Chapuis (2020); Lepp (2007); Nunkoo & Ramkissoon (2010)
Intrinsic factors	
Economic/employment dependency on tourism	Andereck et al. (2005); Boley et al. (2014); Gu & Ryan (2008); Hateftabar & Chapuis (2020); Ko & Stewart (2002); McGehee & Andereck (2004); Vargas-Sánchez et al. (2011); Ward & Berno (2011).
Community attachment	Almeida-García et al. (2016); Andereck et al. (2005); Choi & Murray (2010); Gursoy et al. (2002); Jurowski et al. (1997); Mason & Cheyne (2000); Rasoolimanesh et al. (2015); Woosnam (2012).
Distance from tourism zone	Andereck et al. (2005); Gu & Ryan (2008); Khoshkam et al. (2016); Rasoolimanesh et al. (2019); Sharma & Dyer (2009).
Interaction with tourists	Andereck et al. (2005); Eusébio et al. (2018); Teye et al. (2002); Weaver & Lawton (2001)
Length of residency	Andriotis & Vaughan (2003); Dedeoğlu et al., (2021); Lee et al. (2020); Tatoglu et al. (2002); Tosun (2002).
Demographic: age, gender, education	Almeida-García et al. (2016); Andereck et al. (2005); Andriotis & Vaughan (2003); Fredline & Faulkner (2000); Rasoolimanesh et al. (2015); Mason & Cheyne (2000); McGehee & Andereck (2004); Tosun (2002).
Emotional solidarity residents experience with tourists	Erul et al. (2020); Joo et al. (2021); Phuc & Nguyen (2020); Woosnam (2012); Zaman et al. (2022)

attachment (Almeida-García et al., 2016; Gursoy et al., 2002; Gursoy & Rutherford, 2004; Rasoolimanesh et al., 2015), community concern (Gursoy et al., 2010; Gursoy et al., 2002; Gursoy & Rutherford, 2004), residents' empowerment (Boley et al., 2014; Strzelecka et al., 2016), distance from tourism zone (Andereck et al., 2005; Gu & Ryan, 2008; Khoshkam et al., 2016; Rasoolimanesh et al., 2019), interaction with tourists (Andereck et al., 2005; Eusébio et al., 2018), place attachment (Eusébio et al., 2018; Gu & Ryan, 2008; Wu & Chen, 2015) influence residents' attitudes toward tourism. Table 25.2 summarises the principal variables in resident perceptions under these two headings.

Theoretical Foundations for the Assessment of Residents' Attitudes

The extant literature on residents' attitudes toward and perceptions of tourism exhibits tentative signs of the development of a nascent theoretical foundation for the evaluation of the impacts of tourism on communities. Within the literature on residents' attitudes, social exchange theory (SET) has been extensively employed to understand the formation of residents' attitudes and their pro-tourism behaviour (Boley et al., 2014; Hateftabar & Chapuis, 2020; Sharpley, 2014). That is, if the residents consider the social, economic, and environmental costs of tourism to outweigh its benefits, they will develop negative attitudes toward tourism and stop supporting it. Although SET is still the predominant theoretical framework, some studies have criticised SET's ability to fully understand residents' attitudes and behaviours toward tourism (Andereck et al., 2005; Boley et al., 2014; Chen et al., 2021; Gannon et al., 2021, Ward & Berno, 2011; Woosnam et al., 2018). They assert that not only is this argument speculative and merely intuitive, but it also disregards the hidden processes suggested by SET that underlies why the exchange process may be regarded as unsuccessful. In fact, Pearce et al. (1996) maintain that SET's ability to explain residents' perceptions of tourism is limited by three factors:

- people are rational, reasonable, and rule-abiding processors of information;
- knowledge is often derived socially rather than solely by personal experience; and
- perceptions are developed within a wide sociocultural and historical context.

In other words, the linear, rational manner in which perceptions are formed according to SET is rarely accurate; additionally, it effectively ignores the sociocultural context within which SET operates - where the exchanges actually occur - neglecting the extrinsic influences on the exchange process. More specifically, SET is based on the assumption that both parties voluntarily and proactively engage in the exchange process. However, for many residents in tourist communities, interactions with tourists are often unintentional and involuntary, meaning they result in no serious or genuine interaction or exchange. In other words, for SET to acquire academic and practical relevance, it requires further testing that considers the complexity of the interrelationships proposed in the model (Zhang et al., 2006) as well as the complexity of human nature and behaviours, both at the personal and social levels. Thus, while SET aims to explain why local residents may develop positive or negative attitudes toward tourism and tourists, it is not a reliable or viable research framework. In fact, its inadequacies reflect some of the wider limitations of research into residents' attitudes toward tourism in general. To tackle SET's drawbacks, several scholars have integrated SET with other theories and frameworks (Gursoy et al., 2019; Látková & Vogt, 2012; Rasoolimanesh et al., 2015).

Bestard and Nadal (2007) integrated SET with the theory of reasoned action; Ward and Berno (2011) integrated SET with threat theory; Boley et al. (2014) and Strzelecka et al. (2016) tested the resident empowerment through tourism scale alongside the combined Weber's (1978) theory of formal and substantive rationality and SET theoretical perspective; Shakeela and Weaver (2018) integrated it with social representations theory; Eusébio et al. (2018) combined it with place attachment theory; and Gannon et al. (2021) combined Weber's (1978) theory of substantive and formal rationality and SET. These revised frameworks for SET were the building blocks for the maturation of resident attitudes studies. Moreover, several scholars endeavored to shed light on residents' attitudes by adopting other theories and frameworks, such as community attachment theory (Almeida-García et al., 2016; Harrill & Potts, 2003), growth machine theory (Harrill et al., 2011), emotional solidarity (Erul et al., 2020; Fredline & Faulkner, 2000; Joo et al., 2021; Li et al., 2020; Phuc & Nguyen, 2020; Woosnam, 2012; Zaman et al., 2022), tourist area life cycle (Diedrich & García-Buades, 2009; Lepp, 2007; Uysal et al., 2012), institutional theory (Earl & Hall, 2021; Sinclair-Maragh &

Gursoy, 2015), bottom-up spillover theory (Bimonte & Faralla, 2016), social representation theory (Andriotis & Vaughan, 2003; Thyne et al., 2022; Shakeela & Weaver, 2018), stakeholder theory (Stylidis et al., 2016), theory of planned behavior (Erul et al., 2020; Erul & Woosnam, 2022; Shen & Shen, 2020; Wu & Chen, 2015), theory of reasoned action (Bestard & Nadal, 2007; Lepp, 2007), and Weber's (1978) theory of substantial and formal rationality (Boley et al., 2014; Gannon et al., 2021; Hateftabar & Rasoolimanesh, 2023). Therefore, scholars asserted that "new theoretical perspectives by integrating the related theories to develop unified models of antecedents and consequences of residents' perceptions and attitudes are needed" (Rasoolimanesh & Seyfi, 2020, p. 54).

In Table 25.3 we present a summary of some recent studies mentioned above, which can be considered to be representative of resident attitudes analysis. The table provides a brief description of these papers in terms of the geographic area of analysis, the chosen framework, and the main empirical results.

Conclusion

Residents' positive perceptions of tourism and their support for its development have critical roles to play in community-based tourism and sustainable tourism development (Li et al., 2022). The number of publications in this area has steadily increased over time, and researchers have identified various variables in the formation of residents' perceptions and their pro- and anti-tourism behaviours. However, extant literature in English has been heterogeneous in geographic and thematic scope with a range of different findings (Table 25.3).

It is to be noted that the current literature does not consider the factors affecting perceptions as confounding variables of each other. For instance, research considers the duration of residency in the area, community attachment, employment in tourism-related jobs, and involvement in the decision-making process as independent variables affecting residents' perceptions of tourism, giving rise to conflicting and contradictory results. It is totally possible, however, that should the effects of these factors on one another be studied the discrepancies and conflicting results encountered may be reconciled. For instance, the individuals that have lived their entire lives in an area are likely to have greater and stronger bonds and attachment to the community and have closer views on various issues in decision-making; therefore, they are more likely to try to find employment in tourism so as to have greater involvement and control in various cases. It follows that the main effect on residents' perceptions of tourism arises not from their community attachment or employment in the tourism sector or involvement in decision-making but from the duration of their residence in the area. It is, therefore, essential that for a better understanding of residents' perceptions of tourism further analysis into how these variables impact one another come under scrutiny. Moreover, there could even be factors and variables other than tourism, such as the lifestyle and social interactions of residents that have not been studied that could influence their perceptions and their pro-tourism behaviours.

In addition, the effects of tourism on communities appear to be localised, both in time and place. In other words, not all communities present with the same attitudes towards tourism in all circumstances and with the same intensity. This discrepancy in the results of various studies can be attributed to the particular attributes of the various locations and communities studied (Sharpley, 2014). Although it has been shown that these discrepancies could also be the result of the various theories and models used in the studies, from the point of view of theoretical advancement, these anomalies, whether major or minor, have had a significant impact on the

Table 25.3. Studies on resident perception and willingness to support tourism development.

Reference	*Geographic area*	*Framework*	*Empirical results*
Chen (2001)	Virginia, USA	n/a	Economic and social costs/benefits, cultural enrichment, and environmental deterioration influence residents' pro-tourism behaviours.
Andriotis & Vaughan (2003)	Crete, Greece	Social exchange and social representations theories	Residents' education and employment in tourism are significant explanatory variables of their attitudes toward tourism development.
Gursoy & Rutherford (2004)	Washington and Idaho, USA	n/a	The level of community concern, ecocentric values, utilisation of tourism resource base, community attachment, sensitivity to the environment, the state of the local economy, social costs, and economic, cultural, and social benefits were found to be the major factors affecting residents' pro-tourism behaviours.
McGehee & Andereck (2004)	Arizona, USA	Social exchange theory	Personal characteristics (except age and having lived in the community as a child) did not predict attitudes toward tourism but personal benefit from tourism was found to be the major factor affecting residents' perceptions of tourism impact and their support of its development.
Andriotis (2005)	Crete, Greece	Social exchange theory	Tourism employment, non-reliant residents, and tourism business people have a higher inclination to perceive tourism impacts positively.
Andereck et al. (2005)	Arizona (state-wide), USA	Social exchange theory	Residents who feel tourism is important for economic development, and receive a higher personal benefit from it show a greater tendency to perceive positive impacts of tourism and support its development.
Bestard & Nadal (2007)	Balearic Islands, Spain	Social exchange theory and theory of reasoned action	The level of dependence of the family's incomes on tourism and some socioeconomic characteristics of resident's influence on their perception. Besides, in municipalities with a higher density of tourist accommodation residents are usually less prone to perceived negative impacts of tourism and have greater tolerance.
Lepp (2007)	Bigodi, Uganda	Theory of reasoned action	Residents have positive attitudes towards tourism and positive attitudes would lead to pro-tourism behaviour. Positive attitudes result from resident's belief that tourism creates community development, improves agricultural markets, generates income, and finally, that tourism brings random good fortune.

(*Continued*)

Table 25.3. (*Continued*)

Reference	*Geographic area*	*Framework*	*Empirical results*
Gursoy et al. (2010)	Sunshine Coast, Australia	n/a	Residents' level of community concern, community attachment, ecocentric values, use of the tourism resource base, state of the local economy, and the perceived impacts of tourism development are determinants of their pro-tourism behaviour.
Woosnam (2012)	Galveston, County, Texas, USA	Emotional solidarity	Each of the three emotional solidarity scale factors (welcoming nature, emotional closeness, and sympathetic understanding) significantly predicted residents' perceived impacts, their support toward tourism and contributions tourism makes to the community.
Zamani-Farahani & Musa (2012)	Masooleh and Sare'in, Iran	n/a	Islam is not against the development of tourism industry. 'Islamic Belief' and 'Islamic Practice' have negative relationships with the perceptions of socio-cultural impacts of tourism are not supported.
Boley et al. (2014)	Floyd, Botetourt, and Franklin County, Virginia	Social exchange theory and Weber's (1978) theory of substantive and formal rationality	Residents' perceived positive impacts, psychological empowerment and personal economic benefit having direct and positive effects on support for tourism, and perceived negative impacts has negative effect on support for tourism. In addition, psychological empowerment, social empowerment, and political empowerment have direct and positive effects on residents' perceived positive impacts of tourism, and negative effects on their perceived negative impacts.
Wu & Chen (2015)	Macau and Singapore	n/a	Macau residents tended to have higher scores regarding the consequences of gambling in their living area (more than Singaporeans
Strzelecka et al. (2016)	Choczewo, Poland	Social exchange theory and Weber's (1978) theory of substantive and formal rationality	Psychological empowerment was the best predictor of STD, however, political empowerment and economic benefits from tourism did not predict residents support of tourism development.
Sinclair-Maragh (2017)	Jamaica	Identity theory	Female residents support tourism development more than males. Also, those who are between the ages of 18 and 25 and are of black ethnicity are more supportive of tourism development.

Eusébio et al. (2018)	Boa Vista Island, Cape Verde, Africa	Social exchange theory and place attachment theory	Residents' attitudes are positively affected by place attachment host-tourist interaction, and perceived positive impacts of tourism and negatively affected by perceived negative impacts of tourism. Residents' perceived impacts have significant impacts on their support of its development.
Shakeela & Weaver (2018)	Maldives	Social exchange theory, social representations theory	In the more tourism-affiliated destination, social exchange and social representations influenced how tourism is rationalised as a 'managed evil'.
Rasoolimanesh et al. (2019)	Kashan and Tabriz, Iran	n/a	There is significant differences between the perceptions of tourism impacts, economic gain, environmental and cultural attitudes, and involvement between residents living within the vicinity of heritage tourism sites and those living further afield.
Erul et al. (2020)	Izmir, Turkey	Emotional solidarity and theory of planned behaviour	The three factors of emotional solidarity (welcoming nature, sympathetic understanding, and emotional closeness) uniquely explained 45% of the variance in attitudinal support for tourism. In turn, attitudinal support, along with perceived behavioural control and subjective norms, explained 42% of the variance in behavioral intentions to support tourism.
Gannon et al. (2021)	Kashan and Tabriz, Iran	Social exchange theory and Weber's (1978) theory of substantive and formal rationality	Community attachment, environmental attitude, and economic gain were found to be the major factors affecting residents' perceptions of tourism impacts. Moreover, residents' perceptions play a significant mediating role in shaping the relationships between community attachment, environmental attitudes, and economic gain on support for tourism development.
Hateftabar & Chapuis (2020)	Isfahan, Iran	Social exchange theory and place attachment theory	Residents who perceive a higher level of economic crisis are more likely to view the impacts of tourism positively and support its development.
Phuc & Nguyen (2020)	Ho Chi Minh, Vietnam	Emotional solidarity	Residents' perceptions of tourism impacts, value, collaboration and emotional solidarity with their community were found to be the major factors affecting their support of tourism development.

(*Continued*)

Table 25.3. (*Continued*)

Reference	*Geographic area*	*Framework*	*Empirical results*
Vinerean et al. (2021)	Sibiu, Romania	Social exchange theory	There is a negative relationship between the host community's perceptions of the coronavirus pandemic and their subsequent support for sustainable tourism development.
Erul & Woosnam (2022)	Izmir, Turkey	Emotional solidarity and theory of planned behavior	Emotional solidarity, attitudinal support for tourism, attitudinal contributions to community, perceived behavioural control, and subjective norms collectively are determinants of residents' behavioural intentions to support tourism. Residents' behavioural intentions has significant impact on their actual support of tourism.
García-Buades et al. (2022)	Majorca, Spain	n/a	Overtourism significantly reduced residents' perceptions of the destination as a good place to live, increased withdrawal, and strengthened attitudes towards reducing tourism in the high season.
Rey-Carmona et al. (2022)	Cordoba, Spain	Social exchange theory	Perceived risk of contagion was found to be a stronger predictor of residents' support for tourism development than perception of the economic crisis; an effect that increases notably among residents of the most popular tourist areas of the city.
Li et al. (2022)	Macao, China	Weber's (1978) theory of formal and substantive rationality	The study examines how resident perceptions of economic and noneconomic benefits of tourism and their support for gaming tourism development have changed as a result of the pandemic. Perceived economic benefits and trust in government were significant determinants before and during the pandemic; however, the effects of empowerment seem to be mixed, depending on how controversial gaming tourism was at the time of data collection.
Thyne et al. (2022)	Tokyo, Osaka, Fukuoka, Kyoto and Sapporo Japan	Social distance theory	More negative attitudes concerning tourism were found to have the strongest impact on less sympathetic understanding of tourists and vice versa, revealing that residents' attitudes toward tourists depend in part upon whether attitudes about tourism are positive or negative.

advancement and expansion of tourism studies, as it does not lend itself to any significant extrapolation of study findings. In other words, the increasing number of case-based studies conducted in various geographical locations and on various topics has added greatly to the collective level of knowledge but this has not created the development of a broader conceptual foundation for the understanding of residents' perceptions of tourism and their willingness to support its development. Thus, it follows that a multidimensional and comprehensive approach is needed so that it takes into account the many theories, models, and variables, for a more comprehensive account of residents' perceptions. This is very much required because of the conflicting, at times contradictory, results that have been reported from different studies, preventing research consensus from being achieved. The development of a comprehensive model of the residents' perceptions and willingness to support tourism development therefore awaits.

References

Allen, L. R., Long, P. T., Perdue, R. R., & Kieselbach, S. (1988). The impact of tourism development on residents' perceptions of community life. *Journal of Travel Research*, *27*(1), 16–21. https://doi.org/10.1177%2F004728758802700104

Allport, F. H. (1937). Toward a science of public opinion. *Public Opinion Quarterly*, *1*(1), 7–23. https://doi.org/10.1086/265034

Almeida-García, F., Peláez-Fernández, M. Á., Balbuena-Vazquez, A., & Cortés-Macias, R. (2016). Residents' perceptions of tourism development in Benalmádena (Spain). *Tourism Management*, *54*, 259–274. https://doi.org/10.1016/j.tourman.2015.11.007

Andereck, K., Valentine, K., Knopf, R., & Vogt, C. (2005). Residents' perceptions of community tourism impacts. *Annals of Tourism Research*, *32*(4), 1056–76. https://doi.org/10.1016/j.annals.2005.03.001

Andriotis, K. (2005). Community groups' perceptions of and preferences for tourism development: Evidence from Crete. *Journal of Hospitality & Tourism Research*, *29*(1), 67–90. https://doi.org/10.1177%2F1096348004268196

Andriotis, K., & Vaughan, R. D. (2003). Urban residents' attitudes toward tourism development: The case of Crete. *Journal of Travel Research*, *42*(2), 172–185. https://doi.org/10.1177%2F0047287503257488

Bestard, A. B., & Nadal, J. R. (2007). Modelling environmental attitudes toward tourism. *Tourism Management*, *28*(3), 688–695. https://doi.org/10.1016/j.tourman.2006.04.004

Bimonte, S., & Faralla, V. (2016). Does residents' perceived life satisfaction vary with tourist season? A two-step survey in a Mediterranean destination. *Tourism Management*, *55*, 199–208. https://doi.org/10.1016/j.tourman.2016.02.011

Boley, B. B., McGehee, N. G., Perdue, R. R., & Long, P. (2014). Empowerment and resident attitudes toward tourism: Strengthening the theoretical foundation through a Weberian lens. *Annals of Tourism Research*, *49*, 33–40. https://doi.org/10.1016/j.annals.2014.08.005

Brida, J. G., Osti, L., & Faccioli, M. (2011). Residents' perception and attitudes towards tourism impacts: A case study of the small rural community of Folgaria (Trentino–Italy). *Benchmarking: An International Journal*, *18*(3), 359–385. https://doi.org/10.1108/14635771111137769

Butler, R. W. (1980). The concept of a tourist area cycle of evolution: implications for management of resources. *Canadian Geographer/Le Géographe canadien*, *24*(1), 5–12.

Chen, J. S. (2001). Assessing and visualizing tourism impacts from urban residents' perspectives. *Journal of Hospitality & Tourism Research*, *25*(3), 235–250. https://doi.org/10.1177%2F109634800102500301

Chen, N., Hsu, C. H., & Li, X. (2021). Resident sentiment toward a dominant tourist market: scale development and validation. *Journal of Travel Research*, *60*(7), 1408–1425. https://doi.org/10.1177%2F0047287520947799

Choi, H. C., & Murray, I. (2010). Resident attitudes toward sustainable community tourism. *Journal of Sustainable Tourism*, *18*(4), 575–594. https://doi.org/10.1080/09669580903524852

Deccio, C., & Baloglu, S. (2002). Nonhost community resident reactions to the 2002 Winter Olympics: The spillover impacts. *Journal of Travel Research, 41*(1), 46–56. https://doi.org/10.1177%2F004728750204100106

Dedeoğlu, B. B., Karakuş, Y., Çalışkan, C., & Aydın, Ş. (2021). A complexity perspective for antecedents of support for tourism development. *Journal of Hospitality and Tourism Insights*. https://doi.org/10.1108/JHTI-02-2021-0048

Diedrich, A., & García-Buades, E. (2009). Local perceptions of tourism as indicators of destination decline. *Tourism Management, 30*(4), 512–521. https://doi.org/10.1016/j.tourman.2008.10.009

Doxey, G. V. (1975, September). A causation theory of visitor-resident irritants: Methodology and research inferences. *In Travel and Tourism Research Associations Sixth Annual Conference Proceedings*, 3, 195–198.

Earl, A., & Hall, C. M. (2021). *Institutional theory in tourism and hospitality*. Routledge.

Erul, E., Woosnam, K. M., & McIntosh, W. A. (2020). Considering emotional solidarity and the theory of planned behavior in explaining behavioral intentions to support tourism development. *Journal of Sustainable Tourism, 28*(8), 1158–1173. https://doi.org/10.1080/09669582.2020.1726935

Erul, E., & Woosnam, K. M. (2022). Explaining residents' behavioral support for tourism through two theoretical frameworks. *Journal of Travel Research, 61*(2), 362–377. https://doi.org/10.1177%2F0047287520987619

Eusébio, C., Vieira, A. L., & Lima, S. (2018). Place attachment, host-tourist interactions, and residents' attitudes towards tourism development: The case of Boa Vista Island in Cape Verde. *Journal of Sustainable Tourism, 26*(6), 890–909. https://doi.org/10.1080/09669582.2018.1425695

Fredline, E., & Faulkner, B. (2000). Host community reactions: A cluster analysis. *Annals of Tourism Research, 27*(3), 763–784. https://doi.org/10.1016/S0160-7383(99)00103-6

Gannon, M., Rasoolimanesh, S. M., Taheri, B. (2021). Assessing the mediating role of residents' perceptions toward tourism development. *Journal of Travel Research, 60*(1), 149–171. https://doi.org/10.1177%2F0047287519890926

García-Buades, M. E., García-Sastre, M. A., & Alemany-Hormaeche, M. (2022). Effects of overtourism, local government, and tourist behavior on residents' perceptions in Alcúdia (Majorca, Spain). *Journal of Outdoor Recreation and Tourism, 39*, 100499. https://doi.org/10.1016/j.jort.2022.100499

García, F. A., Vázquez, A. B., & Macías, R. C. (2015). Resident's attitudes towards the impacts of tourism. *Tourism Management Perspectives, 13*, 33–40. https://doi.org/10.1016/j.tmp.2014.11.002

Gu, H., & Ryan, C. (2008). Place attachment, identity and community impacts of tourism—the case of a Beijing hutong. *Tourism Management, 29*(4), 637–647. https://doi.org/10.1016/j.tourman.2007.06.006

Gursoy, D., Jurowski, C., & Uysal, M. (2002). Resident attitudes: A structural modeling approach. *Annals of tourism research, 29*(1), 79–105. https://doi.org/10.1016/S0160-7383(01)00028-7

Gursoy, D., & Rutherford, D. G. (2004). Host attitudes toward tourism: An improved structural model. *Annals of Tourism Research, 31*(3), 495–516. https://doi.org/10.1016/j.annals.2003.08.008

Gursoy, D., Chi, C. G., & Dyer, P. (2010). Locals' attitudes toward mass and alternative tourism: The case of Sunshine Coast, Australia. *Journal of Travel Research, 49*(3), 381–394. https://doi.org/10.1177%2F0047287509346853

Gursoy, D., Ouyang, Z., Nunkoo, R., & Wei, W. (2019). Residents' impact perceptions of and attitudes towards tourism development: A meta-analysis. *Journal of Hospitality Marketing & Management, 28*(3), 306–333. https://doi.org/10.1080/19368623.2018.1516589

Hall, C. M., & Lew, A. A. (2009). *Understanding and managing tourism impacts: An integrated approach*. Routledge.

Harrill, R., Uysal, M., Cardon, P. W., Vong, F., & Dioko, L. (2011). Resident attitudes towards gaming and tourism development in Macao: Growth machine theory as a context for identifying supporters and opponents. *International Journal of Tourism Research, 13*(1), 41–53. https://doi.org/10.1002/jtr.798

Harrill, R., & Potts, T. D. (2003). Tourism planning in historic districts: Attitudes toward tourism development in Charleston. *Journal of the American Planning Association, 69*(3), 233–244. https://doi.org/10.1080/01944360308978017

Hateftabar, F., & Chapuis, J. M. (2020). How resident perception of economic crisis influences their perception of tourism. *Journal of Hospitality and Tourism Management*, *43*, 157–168. https://doi.org/10.1016/j.jhtm.2020.02.009

Hateftabar, F., & Rasoolimanesh, S. M. (2023). Crisis-driven shifts in resident pro-tourism behaviour. *Current Issues in Tourism*, 1–24. https://doi.org/10.1080/13683500.2023.2277905

Joo, D., Xu, W., Lee, J., Lee, C. K., & Woosnam, K. M. (2021). Residents' perceived risk, emotional solidarity, and support for tourism amidst the COVID-19 pandemic. *Journal of Destination Marketing & Management*, *19*, 100553. https://doi.org/10.1016/j.jdmm.2021.100553

Jurowski, C., Uysal, M., & Williams, D. R. (1997). A theoretical analysis of host community resident reactions to tourism. *Journal of Travel Research*, *36*(2), 3–11. https://doi.org/10.1177%2F004728759703600202

Khoshkam, M., Marzuki, A., & Al-Mulali, U. (2016). Socio-demographic effects on Anzali wetland tourism development. *Tourism Management*, *54*, 96–106. https://doi.org/10.1016/j.tourman.2015.10.012

Ko, D. W., & Stewart. W. P. (2002). A structural equation model of residents' attitudes for tourism development. *Tourism Management*, *23*(5), 521–530. https://doi.org/10.1016/S0261-5177(02)00006-7

Kuvan, Y., & Akan, P. (2005). Residents' attitudes toward general and forest-related impacts of tourism: the case of Belek, Antalya. *Tourism Management*, *26*(5), 691–706. https://doi.org/10.1016/j.tourman.2004.02.019

Látková, P., & Vogt, C. A. (2012). Residents' attitudes toward existing and future tourism development in rural communities. *Journal of Travel Research*, *51*(1), 50–67. https://doi.org/10.1177%2F0047287510394193

Lee, D., Kim, S., Han, H., & Wong, A. K. F. (2020). Casino tourism development is blessing or curse? Assessment of casino tourism impacts and suggestions for sustainable casino tourism development. *Asia Pacific Journal of Tourism Research*, *25*(11), 1168–1184. https://doi.org/10.1080/10941665.2020.1837891

Lepp, A. (2007). Residents' attitudes towards tourism in Bigodi village, Uganda. *Tourism Management*, *28*(3), 876–885. https://doi.org/10.1016/j.tourman.2006.03.004

Li, X., Boley, B. B., & Yang, F. X. (2022). Resident empowerment and support for gaming tourism: Comparisons of resident attitudes pre-and amid-Covid-19 pandemic. *Journal of Hospitality & Tourism Research*. https://doi.org/10.1177%2F10963480221076474

Li, Z., Chen, H., Huang, S. S., Wanichwasin, P., & Cui, R. (2020). Resident perceptions of Chinese tourists in Thailand. *Tourism Review*, *76*(5), 1154–1163. https://doi.org/10.1108/TR-10-2019-0443

Liu, J., Lin, H., Hu, B., Zhou, Z., Agyeiwaah, E., & Xu, Y. (2022). Advancing the understanding of the resident pro-tourism behavior scale: An integration of item response theory and classical test theory. *Journal of Business Research*, *141*, 113–125. https://doi.org/10.1016/j.jbusres.2021.12.013

Mason, P., & Cheyne, J. (2000). Residents' attitudes to proposed tourism development. *Annals of Tourism Research*, *27*(2), 391–411. https://doi.org/10.1016/S0160-7383(99)00084-5

McGehee, N. G., & Andereck, K. L. (2004). Factors predicting rural residents' support of tourism. *Journal of Travel Research*, *43*(2), 131–40. https://doi.org/10.1177%2F0047287504268234

Nunkoo, R., & Gursoy, D. (2012). Residents' support for tourism: An identity perspective. *Annals of Tourism Research*, *39*(1), 243–268. https://doi.org/10.1016/j.annals.2011.05.006

Nunkoo, R., & Ramkissoon, H. (2010). Modeling community support for a proposed integrated resort project. *Journal of Sustainable Tourism*, *18*(2), 257–277. https://doi.org/10.1080/09669580903290991

Nunkoo, R., & Ramkissoon, H. (2012). Power, trust, social exchange and community support. *Annals of Tourism Research*, *39*(2), 997–1023. https://doi.org/10.1016/j.annals.2011.11.017

Pearce, P., Moscardo, G., & Ross, G. (1996). *Tourism community relationships*. Oxford: Pergamion Press.

Phuc, H. N., & Nguyen, H. M. (2020). The importance of collaboration and emotional solidarity in residents' support for sustainable urban tourism: case study Ho Chi Minh City. *Journal of Sustainable Tourism*. https://doi.org/10.1080/09669582.2020.1831520

Prayag, G., Hosany, S., Nunkoo, R., & Alders, T. (2013). London residents' support for the 2012 Olympic Games: The mediating effect of overall attitude. *Tourism Management*, *36*, 629–640. https://doi.org/10.1016/j.tourman.2012.08.003

Rasoolimanesh, S. M., & Seyfi, S. (2020). Residents' perceptions and attitudes towards tourism development: a perspective article. *Tourism Review*, *76*(1), 51–57. https://doi.org/10.1108/TR-11-2019-0461

Rasoolimanesh, S. M., Taheri, B., Gannon, M., Vafaei-Zadeh, A., & Hanifah, H. (2019). Does living in the vicinity of heritage tourism sites influence residents' perceptions and attitudes? *Journal of Sustainable Tourism*, *27*(9), 1295–1317. https://doi.org/10.1080/09669582.2019.1618863

Rasoolimanesh, S., Jaafar, M., Kock, N., & Ramayah, T. (2015). A revised framework of social exchange theory to investigate the factors influencing residents' perceptions. *Tourism Management Perspectives*, *16*, 335–45. https://doi.org/10.1016/j.tmp.2015.10.001

Rey-Carmona, F. J., Núñez-Tabales, J. M., Durán-Román, J. L., & Pulido-Fernández, J. I. (2022). Open the doors to tourism or remain cautious: residents' dilemma amidst a pandemic. *Current Issues in Tourism*. https://doi.org/10.1080/13683500.2022.2047162

Ribeiro, M. A., Pinto, P., Silva, J. A., & Woosnam, K.M. (2017). Residents' attitudes and the adoption of pro-tourism behaviours: The case of developing island countries. *Tourism Management*, *61*, 523–537. https://doi.org/10.1016/j.tourman.2017.03.004

Shakeela, A., & Weaver, D. (2018). "Managed evils" of hedonistic tourism in the Maldives: Islamic social representations and their mediation of local social exchange. *Annals of Tourism Research*, *71*, 13–24. https://doi.org/10.1016/j.annals.2018.04.003

Sharma, B., & Dyer, P. (2009). An investigation of differences in residents' perceptions on the Sunshine Coast: tourism impacts and demographic variables. *Tourism Geographies*, *11*(2), 187–213. https://doi.org/10.1080/14616680902827159

Sharpley, R. (2014). Host perceptions of tourism: A review of the research. *Tourism Management*, *42*, 37–49. https://doi.org/10.1016/j.tourman.2013.10.007

Shen, K., & Shen, H. (2020). Chinese traditional village residents' behavioural intention to support tourism: An extended model of the theory of planned behaviour. *Tourism Review*, *76* (2), 439–459. https://doi.org/10.1108/TR-11-2019-0451

Sinclair-Maragh, G. (2017). Demographic analysis of residents' support for tourism development in Jamaica. *Journal of Destination Marketing & Management*, 6(1), 5–12. https://doi.org/10.1016/j.jdmm.2016.03.005

Sinclair-Maragh, G., & Gursoy, D. (2015). Imperialism and tourism: The case of developing island countries. *Annals of Tourism Research*, *50*, 143–158. https://doi.org/10.1016/j.annals.2014.12.001

Smith, M. D., & Krannich, R. S. (1998). Tourism dependence and resident attitudes. *Annals of Tourism Research*, *25*(4), 783–802. https://doi.org/10.1016/S0160-7383(98)00040-1

Strzelecka, M., Boley, B. B., & Strzelecka, C. (2016). Empowerment and resident support for tourism in rural Central and Eastern Europe (CEE): the case of Pomerania, Poland. *Journal of Sustainable Tourism*, *25*(4), 554–572. https://doi.org/10.1080/09669582.2016.1224891

Stylidis, D., Biran, A., Sit, J., & Szivas, E. M. (2014). Residents' support for tourism development: The role of residents' place image and perceived tourism impacts. *Tourism Management*, 45, 260–274. https://doi.org/10.1016/j.tourman.2014.05.006

Stylidis, D., Sit, J., & Biran, A. (2016). An exploratory study of residents' perception of place image: The case of Kavala. *Journal of Travel Research*, *55*(5), 659–674. https://doi.org/10.1177%2F0047287514563163

Šegota, T., Mihalič, T., & Kuščer, K. (2017). The impact of residents' informedness and involvement on their perceptions of tourism impacts: The case of Bled. *Journal of Destination Marketing & Management*, *6(3)*, 196–206. https://doi.org/10.1016/j.jdmm.2016.03.007

Tatoglu, E., Erdal, F., Ozgur, H., & Azakli, S. (2002). Resident attitudes toward tourism impacts: The case of Kusadasi in Turkey. *International Journal of Hospitality & Tourism Administration*, *3*(3), 79–100. https://doi.org/10.1300/J149v03n03_07

Teye, V., Sirakaya, E., & Sönmez, S. F. (2002). Residents' attitudes toward tourism development. *Annals of Tourism Research*, *29*(3), 668–688. https://doi.org/10.1016/S0160-7383(01)00074-3

Thyne, M., Woosnam, K. M., Watkins, L., & Ribeiro, M. A. (2022). Social distance between residents and tourists explained by residents' attitudes concerning tourism. *Journal of Travel Research*, *61*(1), 150–169. https://doi.org/10.1177%2F0047287520971052

Tosun, C. (2002). Host perceptions of impacts: A comparative tourism study. *Annals of Tourism Research, 29*(1), 231–253. https://doi.org/10.1016/S0160-7383(01)00039-1

Truong, V. D., Hall, C. M., & Garry, T. (2014). Tourism and poverty alleviation: Perceptions and experiences of poor people in Sapa, Vietnam. *Journal of Sustainable Tourism, 22*(7), 1071–1089. https://doi.org/10.1080/09669582.2013.871019

Upchurch, R. S., & Teivane, U. (2000). Resident perceptions of tourism development in Riga, Latvia. *Tourism Management, 21*(5), 499–507. https://doi.org/10.1016/S0261-5177(99)00104-1

Uysal, M., Woo, E., & Singal, M. (2012). The tourist area life cycle (TALC) and its effect on the quality-of-life (QOL) of destination community. In M. Uysal, R. R. Perdue & M. J. Sirgy (Eds), *Handbook of tourism and quality-of-life research: Enhancing the lives of tourists and residents of host communities* (pp. 423–443). Dordrecht: Springer.

Vargas-Sánchez, A., do Valle, P. O., da Costa Mendes, J., & Silva, J. A. (2015). Residents' attitude and level of destination development: An international comparison. *Tourism Management, 48*, 199–210. https://doi.org/10.1016/j.tourman.2014.11.005

Vargas-Sánchez, A., Porras-Bueno, N., & Plaza-Mejía, M. (2011). Explaining residents' attitudes to tourism. Is a universal model possible? *Annals of Tourism Research, 38*(2), 460–480. https://doi.org/10.1016/j.annals.2010.10.004

Vinerean, S., Opreana, A., Tileagă, C., & Popşa, R. E. (2021). The impact of covid-19 pandemic on residents' support for sustainable tourism development. *Sustainability, 13*(22), 12541. https://doi.org/10.3390/su132212541

Ward, C., & Berno, T. (2011). Beyond social exchange theory: Attitudes toward tourists. *Annals of Tourism Research, 38*(4), 1556–1569. https://doi.org/10.1016/j.annals.2011.02.005

Weaver, D. B., & Lawton, L. J. (2001). Resident perceptions in the urban–rural fringe. *Annals of Tourism Research, 28*(2), 439–458. https://doi.org/10.1016/S0160-7383(00)00052-9

Weber, M. (1978). *Economy and society*, 2nd ed. University of California Press.

Woosnam, K. M., & Aleshinloye, K. D. (2018). Residents' emotional solidarity with tourists: Explaining perceived impacts of a cultural heritage festival. *Journal of Hospitality & Tourism Research, 42*(4), 587–605. https://doi.org/10.1177%2F1096348015584440

Woosnam, K. M. (2012). Using emotional solidarity to explain residents' attitudes about tourism and tourism development. *Journal of Travel Research, 51*(3), 315–327. https://doi.org/10.1177%2F0047287511410351

Wu, S. T., & Chen, Y. S. (2015). The social, economic, and environmental impacts of casino gambling on the residents of Macau and Singapore. *Tourism Management, 48*, 285–298. https://doi.org/10.1016/j.tourman.2014.11.013

Zaman, U., Aktan, M., Agrusa, J., & Khwaja, M. G. (2022). Linking regenerative travel and residents' support for tourism development in Kaua'i Island (Hawaii): Moderating-mediating effects of travel-shaming and foreign tourist attractiveness. *Journal of Travel Research*. https://doi.org/10.1177%2F00472875221098934

Zamani-Farahani, H., & Musa, G. (2012). The relationship between Islamic religiosity and residents' perceptions of socio-cultural impacts of tourism in Iran: Case studies of Sare'in and Masooleh. *Tourism Management 33*(4), 802–814. https://doi.org/10.1016/j.tourman.2011.09.003

Zhang, J., Inbakaran, R. J., & Jackson, M. S. (2006). Understanding community attitudes towards tourism and host—Guest interaction in the urban-rural border region. *Tourism Geographies, 8*(2), 182–204. https://doi.org/10.1080/

Chapter 26

Chinese Outbound Tourism and Shopping

Ning (Chris) Chen, Edward Commons, & Girish Prayag

Tourism research is very much contextual. When we study destination marketing or branding, it matters which destination it is; when we investigate tourist behaviours, it matters which country and/or culture they are from; when industry practitioners analyse tourism markets, they intend to comprehend how much the expenditure is to each region. To understand outbound tourism economy, it is thus crucial to investigate each scenario in terms of where the tourists are from and who they are. In the long history of tourism research, we have witnessed the evolution of several major tourist markets with significant impacts to global tourism business, their vicissitudes, from USA and the UK, to Japan, then to China (Tiwari et al., 2018).

Prior to the COVID pandemic, high-speed economic growth in China had led to 130 million outbound Chinese travellers, spending an estimated 115.29 billion $USD overseas in 2017 (ChinaTravelNews, 2018). The number of outbound Chinese travellers increased to 149.7 million in 2018, with a total of 120 billion $USD outbound tourist expenditure (Huang et al., 2021). These numbers cemented China as the world's leading outbound tourism market back then, which has unsurprisingly received notable attention from worldwide marketers (The Nielson Company, 2017; UNWTO, 2017; Wyman, 2017). This growth has also attracted a wide body of research literature, all attempting to explain this phenomenon and understand the evolving needs of this fast-paced transformative market (Arlt, 2013; Li et al., 2010). Because of this boom in outbound tourism, Chinese tourists are now travelling further abroad, experiencing unique diverse destinations, observing new cultures, and engaging in a variety of shopping activities. This has resulted in an interest in Chinese tourist shopping behaviour, which has been identified as one of the most important activities for their travel experiences (Huang & Hsu, 2005; Li et al., 2011a; Li et al., 2011b; Meng & Zhang, 2016). Studies on Chinese outbound tourism have encompassed a number of different research areas, including Chinese tourists' motivations, purchasing decisions, attitudes, perceptions, and overall experiences in different destinations.

Understanding any topic relevant to an outbound tourism economy such as China reflects an attempt in examining important tourism issues from a context perspective, and implies a

The Wiley Blackwell Companion to Tourism, Second Edition. Edited by C. Michael Hall.

cultural study approach, which can be taken via both an emic and an etic approach. The emic approach of cultural study focuses on an internal perspective from the native's perspective, while the etic approach follows an external perspective, often based on an analytical framework that is universally applicable to various cultures (Morris et al., 1999). Accordingly, this chapter reviews literature on Chinese outbound tourism, providing an overview of the relevant studies to this topic, as well as a framework to analyse other cultural and/or regional contexts relevant to tourism studies.

Chinese Outbound Tourism: A Brief History

Until the introduction of the 'open-door' policy in 1979 by the Chinese government, only Chinese government representatives and diplomats were permitted to leave the country (Zhang & Heung, 2002). The 'open-door' policy since 1979 has driven a shift from a Soviet style centrally controlled economy to a more free-market orientated system (Lim & Wang, 2008). Relevant movements enabled the Chinese government in 1983 to allow recreational travel to the neighbouring so-called Chinese satellites, Hong Kong and Macau (Keating et al., 2015). The purpose of the policy was to allow foreign investment into China and to initiate foreign trade capabilities. Because of these favourable government policies and subsequent economic growth, the Chinese people became increasingly able to travel abroad freely and explore life outside of China (Huang et al, 2015; Li et al., 2011; Li et al., 2011b).

Literature on Chinese outbound tourism has the consensus that it has previously undergone three stages of development (Zhang & Heung, 2002). The first stage began in 1983, when Macau and Hong Kong become the destinations for the Chinese to visit family and friends. The second stage began in 1990, when Asian neighbouring countries, Singapore, Thailand, and Malaysia became popular destinations to visit due to their significant Chinese populations. The third stage commenced in 1999, when non-Asian countries such as Australia and New Zealand were recognised as approved destinations in the Approved Destination System (ADS) government scheme in China. The ADS system is a bilateral tourism agreement between China and overseas destinations, allowing the Chinese to travel freely to those destinations (Lim & Wang, 2008). The introduction of the ADS system was the birth of long-haul Chinese outbound travel to Western countries. Until the COVID pandemic, the Chinese outbound tourism market may be described as being in the 3.5th stage, as travellers are becoming richer, better educated, and increasingly curious to experience the unknown (Arlt, 2013). Overall, China had experienced explosive growth in outbound tourism numbers in the past decade. However, since the COVID pandemic outbreak, the Chinese government has been following a 'zero-COVID' policy with strict border controls, putting any outbound tourism activity on hold, and thus fundamentally changing the status quo of Chinese outbound tourism as well as the general Chinese public's tourism preferences (Huang et al., 2021). How this will evolve will significantly affect the post COVID-19 international tourism development.

Chinese Outbound Tourism: The Etic View

The growth in Chinese outbound travel explains why it has received an overwhelming interest from academic researchers and practitioners. In accordance with the stages of development of Chinese outbound tourism, literature in this field has also evolved, changed and adapted with time, which can be classified into three stages (Huang et al., 2015). The first stage was from 1983 – 1992, when the literature had only just started to recognise the potential of China as a source of international tourists. The second stage was from 1993 – 2002, focusing on describing the potential of the market, with the late part of the decade focusing on the needs and motivations of the Chinese travellers. The third stage was from 2003 – 2019 prior to COVID, which transformed the literature into the behavioural and organisational issues that have been fundamentally important for international tourism providers to understand how to exploit the Chinese tourism opportunity. Table 26.1 illustrates the different research areas studied recently, specifically from a behavioural perspective.

These studies illustrated in Table 26.1 reflect an etic approach in understanding Chinese outbound tourists, and are thus limited in not sufficiently exploring the specific Chinese culture or psyche or attitude towards travel, venturing out to unknown parts of the world, whether there is a national sense of curiosity, spirit of adventure, or innate drive to explore.

In reviewing Chinese outbound tourism, Jin and Wang (2016) found one common theme throughout the literature: the use of Western paradigms in the Chinese setting. This academic neo-colonialism could result in lost opportunities for developing new theoretical perspectives

Table 26.1. A review of Chinese outbound tourist behaviours.

Research Focus	*Author(s)*
Motivation	Assiouras et al., 2015; Chang et al., 2010; Chen & Lin, 2012; Corigliano, 2011; Fu et al., 2015; Hsu et al., 2010; Hua & Yoo, 2011; Jiang et al., 2012; King & Gardiner, 2015; Lee et al., 2011; Li & Ryan, 2015; Li & Cai, 2012; Li et al., 2011a; Li et al., 2011b; Li et al., 2015; Park et al., 2015; Prayag et al., 2015a; Prayag et al., 2015b; Prideaux et al., 2012; Ruhanen et al., 2015; Song et al., 2016; Wong & Rosenbaum, 2012; Wu, 2015; Yang et al., 2011; Ye et al., 2011; Zeng et al., 2014; Zhang & Peng, 2014
Satisfaction	Chen et al., 2009; Chen et al., 2013; Lee et al., 2011; Li & Ryan, 2018; Lim & Bendle, 2012; Mao & Zhang, 2014; Pan et al., 2014; Park et al., 2015; Song & Hsu, 2013; Truong & King, 2009
Barriers/Constraints	Chen & Gassner, 2012; Lai et al., 2013; Sparks & Pan, 2009; Wu, 2015
Attitude	Agrusa et al., 2011; Chen & Gassner, 2012; Chow & Murphy, 2008; Huang & Hsu, 2009; Liu et al., 2011; Mohsin, 2008; Packer et al., 2014; Ruhanen et al., 2015, Sparks & Pan, 2009
Travel Preferences and Destination Image	Hsu et al., 2010; Lin et al., 2012; McCartney et al., 2009; Sparks & Pan, 2009; Stepchenkova & Li, 2012; Wang & Davidson, 2010
Destination Choice/ Experience	Cheng & Foley, 2018; Hsu et al., 2010; Li et al., 2011a; Li et al., 2011b; Li et al., 2017; Lin et al., 2015
Perception	Hsu & Song, 2012; Pan et al., 2014; Yang et al., 2011; Yu & Ko, 2012; Yun & Joppe, 2011
Information Sources	Choi et al., 2016; Kambele et al., 2015
Shopping	Lloyd et al., 2011; Meng & Zhang, 2016; Xu & McGehee, 2012; Zhu et al., 2016
Research Reviews	Huang et al., 2015; Jørgensen et al., 2017; Keating et al., 2015; Leung et al., 2014; Tse, 2015
Behavioural Studies	Agrusa et al., 2011; Chen & Lin, 2012; Chiang et al., 2012; Gu et al., 2018; Huang & Tian, 2013; Li & Cai, 2012; Li et al., 2011a; Li et al., 2011b; Wen et al., 2018; Wu, 2015; Zeng et al., 2014

based on Asian cultural traditions (Pearce, 2014), especially as research practice treats non-Western scenarios as a field of theory extension rather than a source of theory generation. Jørgensen et al. (2017) further found that current research does not reflect the evolving travel patterns of the Chinese outbound tourist and therefore, future research needs to be carried out in Asian destinations and equally outside Asia to reduce ambiguity, and to increase the reliability of the findings. Jørgensen et al. (2017) also call for an increase in qualitative studies and mixed-method studies, because 70 percent of all Chinese outbound tourism literature uses quantitative methods. An emic approach is essential in this case too, since the Chinese are unique and culturally distinct from other nationalities.

Chinese Outbound Tourism: The Emic View

One important aspect in investigating Chinese outbound tourism specifically from an emic view is to understand where they travel to and what they do. It is natural to assume that motives for travel, general experiences, and perceptions of the destinations vary in the actual tourism destinations, in studying Chinese outbound tourism. For instance, the Chinese behave differently in familiar Chinese markets such as Macau and Hong Kong (Huang & Hsu, 2005) in comparison to Western markets such as the USA (Agrusa et al., 2011), or Europe (Zhu et al., 2016). The unique cultural elements in Chinese behaviour have received a plethora of studies, and have been evolving along with the Chinese becoming more sophisticated and experienced travellers. Table 26.2 is an example of recent studies on Chinese outbound tourism in terms of various destinations.

Table 26.2. A review of Chinese outbound tourism in terms of destination.

Research Focus	*Author(s)*
Hong Kong	Chen et al., 2016; Choi et al., 2008; Correia et al., 2018; Huang & Hsu, 2005, 2009; Li et al., 2011a; Li et al., 2011b; Liu & McKercher, 2016; Liu et al., 2011; Liu et al., 2008; Qiu & Lam, 1999; Qu & Lam, 1997; Qu & Li, 1997; Shen et al., 2017; Tsang et al., 2014; Tse & Tse, 2015; Wang, 2004; Wong & Law, 2003; Ye et al., 2011; Zhang & Chow, 2004; Zhang & Heung, 2002; Zhang et al., 2009
Macau	Park et al., 2015; Wong & Rosenbaum, 2012; Zeng et al., 2014
Asia	Chan, 2006; Chen & Lin, 2012; Chen et al., 2009; Kim & Prideaux, 2005; Lee et al., 2011; Li & Ryan, 2018; Lim & Bendle, 2012; Lin et al., 2012; Lin & Lin, 2006; Pan et al., 2014; Song & Hsu, 2013; Song et al., 2016; Truong & King, 2009; Untong et al., 2015; Wang et al., 2016; Yu & Ko, 2012
Australia	Breakey et al., 2008; Chen et al., 2015; Chow & Murphy, 2008; Gu et al., 2018; Mao & Zhang, 2014; Pan & Laws, 2002; Pham et al., 2017; Ruhanen et al., 2015; Weiler & Yu, 2008; Wu, 2015; Zhang & Peng, 2014
New Zealand	Fountain et al., 2010; Howison et al., 2017; Ryan & Mo, 2002; Sun et al., 2014; Sun et al, 2015; Zhang & Shelton, 2015
North America	Agrusa et al., 2011; Cai et al., 2001; Hua & Yoo, 2011; Jang et al., 2003; Johanson, 2008; Lai et al., 2013; Li et al., 2011a; Li et al., 2011b; Lu, 2011; Meng & Zhang, 2016; Xu & McGehee, 2012
Europe	Assiouras et al., 2015; Corigliano, 2011; Huang & Tian, 2013; Lojo, 2016; Prayag et al., 2015; Prayag et al., 2015b; Skivalou & Filippidi, 2017; Wen et al., 2018; Yang et al., 2011; Zhu et al., 2016

Outbound Tourist Shopping

In understanding any outbound tourism economy, another crucial aspect is their expenditure in tourism activities. Outbound tourist shopping refers to 'an activity which tourists purchase goods to take home during their travel' (Jin et al., 2017, p. 120). It was not until the early 1990's that academics started to believe shopping abroad was an important motivation for travel. Shopping has now been described as a crucial part of tourists' overseas experience (Mehta et al., 2014; Wong & Wan, 2013). The idea of shopping in a tourism setting was first investigated by Jansen-Verbeke (1991) who proposed the concept of shopping tourism and claimed that 'leisure shopping as a tourist activity has always existed, but the shopping element has recently become an important instrument in the promotion of tourist places' (Jansen-Verbeke, 1991, p. 10).

Literature on shopping tourism has expanded into three key terms:

- shopping tourism;
- tourism shopping; and
- tourist shopping (Jin et al., 2017).

Timothy (2005) defined shopping tourism as a form of tourism with the distinct purpose of travelling to a destination for shopping alone. This entails that people who travel abroad for the sole purpose to shop or to a destination where shopping is recognised as an important tourist activity (Timothy, 2005). Jin et al. (2017) has suggested that the other two terms, tourism shopping and tourist shopping have yet to be consistently defined, as authors have used both terms interchangeably based on different research objectives and settings.

Besides being a primary and major tourism motive, shopping has also been classified as a major leisure and social activity for many tourists (Yüksel & Yüksel, 2007), with several studies suggesting shopping to be one of the major reasons for outbound travel (Huang & Hsu, 2005; Timothy, 2005; Meng & Zhang, 2016; Zhu et al., 2016). Furthermore, a trip without shopping opportunities may not provide the overall experience that people desire (LeHew & Wesley, 2007; Wu et al., 2014). Contrary to this are tourism studies that have found reason to believe that shopping is not as important as what other studies have claimed (Murphy et al., 2011). This implies that shopping is still valued and considered an important activity when travelling but not to the same extent previously identified in the earlier literature (Huang & Hsu, 2005; Jansen-Verbeke, 1991; Timothy, 2005; Timothy & Butler, 1995; Yüksel & Yüksel, 2007). As this research is focussed on outbound tourist shopping, it will be able to identify whether shopping is still considered an important tourist activity or whether it has lost value like the literature suggests. In exploring general trends and characteristics of outbound shopping, Park et al. (2010) find shopping participation and experience as a main reason for outbound travel, while Murphy et al. (2011) identify women in spending more and participating more frequently in shopping activities as opposed to males. Furthermore, Lyu and Noh (2017) found younger tourists are more willing to participate in shopping activities overseas and have a list of the shopping that they would like to do in a sequential manner.

Chinese Outbound Tourist Shopping

Overseas shopping has become a popular activity for outbound Chinese tourists (Song et al., 2016). Chinese outbound tourist shopping has followed a similar path to their outbound tourism, through the development and growth of outbound travel opportunities. There has been a steady growth and development of consumer culture in China, largely attributed to

Table 26.3. A review of Chinese outbound tourist shopping.

Country/Region	*Author(s)*
Taiwan, North Korea, Thailand	Hsieh & Chang, 2006; Li & Ryan, 2018; Lin & Lin, 2006
China	Choi et al., 2016
Europe	Zhu et al., 2016
Hong Kong	Choi et al., 2008; Correia et al., 2018; Heung & Cheng, 2000; Liu et al., 2008; Lo & Qu, 2015: Tsang et al., 2014; Tse & Tse, 2015; Yeung & Yee, 2012
Macau	Kong & Chang, 2016; Wong & Wan, 2013
USA	Meng & Zhang, 2016; Xu & McGehee, 2012

their growing interest in shopping overseas (Maguire & Hu, 2013). The Chinese interest in overseas shopping first began in the neighbouring regions, such as Hong Kong and Macau, with Hong Kong as an early shopping paradise in the eyes of many Chinese tourists (Huang & Hsu, 2005). This is attributed to the availability of luxury Western brands that were of great interest (Huang & Hsu, 2005).

With the opportunity for travel expanding to developed countries in the past decade, shopping has now become an integral activity for Chinese travellers (Zhu et al., 2016). Along with this expansion in tourism there has also been an increase in literature, investigating shopping behaviours in Western countries, such as the USA, France, Switzerland, and Germany (Meng & Zhang, 2016; Xu & McGehee, 2012; Zhu et al., 2016). Collectively, these studies have explored different aspects of the shopping experience of Chinese tourists, including their motivations, shopping attributes, what they purchased, and overall shopping experiences. Furthermore, these studies have been useful in providing real world point of view experiences from Chinese tourists to generate and conceptualise the understanding of their behaviour and preferences. Table 26.3 presents a summary of the studies based on the destination.

In understanding the motivations for shopping during international travels, Jansen-Verbeke (1995) divides the motivations into four distinct categories:

- strengthening social ties;
- taking advantage of the unique goods provided or bargain prices offered;
- purchasing goods and products that represent the identity of the destination; and
- taking advantage of a favourable exchange rate.

Yuan et al. (2013) find that people shop abroad for two main reasons: economic reasons; and personal or social reasons. Jin et al. (2017) group tourist motivations into two:

- instrumental motives as shopping for necessities associated with travel to meet social obligations; and
- expressive motives communicating the need to escape, relax, and change the pace of life through shopping.

Yüksel and Yüksel (2007) suggest that shopping is driven by either hedonic or utilitarian shopping values. Lastly, Sundström et al. (2011) find the possibility of getting a bargain and the chance to explore a wide variety of goods as motivators. Collectively, these studies have introduced why consumers shop overseas, building an etic foundation in the Chinese outbound shopping research. However, phenomena specifically derived from Chinese culture has complicated the case in fully understanding Chinese outbound tourist shopping behaviours. Specifically, amongst all shopping activities, daigou shopping (i.e. shopping on behalf of someone else) has become a powerful trend in Chinese outbound tourism (Zhang et al., 2021). This is because the Chinese culture perceive various values besides functional and utility values in the goods overseas.

In relation to destination for shopping, literature shows structural differences in Chinese tourist shopping preferences. Zhu et al. (2016) find that the Chinese viewed shopping as an essential activity in Europe for product quality and world known brands. Meng and Zhang (2016), and Xu and McGehee (2012) find shopping for world known brands with competitive pricing and high quality and trustworthy products are the main reason for shopping in the USA. In New Zealand, Chinese tourists prefer health and beauty products (Sun et al., 2014). Choi et al. (2008) summarise that luxury brands in Hong Kong are the main attraction for Chinese tourists. Collectively, all these studies suggest that quality, price and world known brands are the driving factors behind their avid shopping in these destinations, while country of origin remains a key variable in identifying distinctive shopping preferences.

Country of origin refers to 'any influence that the country of production, assembly, or country which designed the product has on consumer perception and behaviour' (Anastasiei & Chiosa, 2014, p. 219). This concept is one of the most widely researched topics in marketing literature exploring the impacts of perception and behaviour on product selection (Pharr, 2005). Before 1960, consumers only used price and brands to evaluate the quality of products (Kalicharan, 2014), but since then country of origin has caught the eye of many academics and practitioners who have all helped grow the literature to where it stands today. Literature on country of origin can be traced back to 1965, and has synthesised developments in several key areas, including consumer perceptions towards product quality (Hanzaee & Khosrozadeh, 2011), consumer buying decisions (Diamantopoulos et al., 2011), and brands (Agrawal & Kamakura, 1999). The origin of a product or brand influences the shopping decisions of Chinese tourists (Xu & McGehee, 2012; Zhu et al., 2016). Xu and McGehee (2012) found that Chinese tourists shopping in the USA felt dissatisfied after purchasing Nike products, a USA brand, because they learned after their purchase, that these products were manufactured in China or some other Asian country rather than the USA. Similar situations occur in Europe, with Zhu et al. (2016) finding tourists felt disappointment when they realised that the Liverpool football team mascot was made in China. These examples affirm the importance of originality of overseas products to Chinese tourists, as they take great pleasure from buying products from the places where they come from. In contrast, as mentioned in Xu and McGehee (2012), and Zhu et al. (2016), country of origin can be a contributing factor towards positive and negative shopping experiences.

Previous literature has uncovered some interesting features in the Chinese tourist demographic. Meng and Zhang (2016) suggest that free independent travellers and the growing Chinese middle class become the most prolific demographic of Chinese tourists. With a focus on brands, Choi et al. (2008) found that Shanghai tourists in Hong Kong were most concerned about brand image when making their shopping decisions in Hong Kong. The key omission in the academic literature is that no studies to date have specifically explored the impact of demographics on the shopping behaviour of Chinese tourists overseas.

In comparison, Western tourists in Europe are known to buy handcrafts at small local shops, while the Chinese are more attracted to shop in modern shopping malls, duty free shops, and boutique stores (Zhu et al., 2016). Chinese tourists have been identified as information gatherers as they are continually wandering the streets to different shops and comparing prices (Zhu et al., 2016). In Hong Kong, mainland Chinese tourists do not value service or product quality the most, as the local shoppers, but rather perceive risk and price as the most influential to their satisfaction and behavioural intentions (Lloyd et al., 2011). A further significant difference is when Chinese tourists make shopping decisions overseas, they are strongly influenced by the recommendations that they are given from friends and family (Beerli & Martin, 2004; Gao et al., 2017; Hsu et al., 2006; Sparks & Pan, 2009). Zhu et al. (2016) further discuss the impact of family and friends recommendations in the purchasing decisions made overseas.

Chinese are drawn to a wide range of different products (Xu & McGehee, 2012; Zhu et al., 2016). Zhu et al. (2016) found that the products that Chinese tourists purchase in Europe are in three categories:

- functional goods for everyday life, e.g., milk powder, garments, tobacco pipes, shoes, and kitchenware;
- adornment goods used for decoration, e.g., jewellery, watches, wine, cosmetics, and perfume; and
- social goods as gifts for family and friends back in China.

However, other studies highlight that fashion and apparel were the most popular products abroad (Meng & Zhang, 2016). In addition, half of the Chinese tourists purchase health and beauty products (Meng & Zhang, 2016).

A Unique Cultural Phenomenon: Daigou

Various studies have particularly focused on the distinctive Chinese outbound tourist shopping culture. For instance, Xu and McGehee (2012) found three main reasons for Chinese outbound tourists to shop in the USA:

- purchasing gifts for friends and family;
- taking advantage of unique products and price differences; and
- making good use of travel time.

Upchurch and Liu (2014) list four motives:

- to purchase goods for themselves, family, friends or other close acquaintances;
- to build or maintain networks by showing respect and concern for others;
- to gain social status; and
- to enhance foreign social and cultural experiences.

Furthermore, Tsang et al. (2014) found five motivations to shop in Hong Kong:

- physiological;
- safety and security;
- enhancement of relationship;
- self-esteem; and
- self-development and self-actualisation.

These studies have implied a list of motives (especially the ones relevant in the societal and cultural structure and networks) in the Chinese outbound tourist shopping culture. For instance, many foreign products available in China are more expensive than if purchased overseas due to taxes, and thus Chinese tourists enjoy taking advantage of the opportunity to purchase goods at lower prices when abroad (Xu & McGehee, 2012). The importance of quality of product is also confirmed in Chinese outbound shopping literature, which arguably has been the result of Western brands being of a superior quality to that of Chinese brands (Klein et al., 1998; Zhou & Hui, 2003). In addition, many of these motives summarised above are socially relevant, which can explain the increasing trend of daigou.

For Chinese tourists, purchasing souvenirs as gifts is a culturally embedded tradition (Gao et al., 2017; Mok & Defranco, 2000) and is considered an obligatory activity in Chinese society (Li & Ryan, 2018; Gao et al., 2017). In addition, Meng and Zhang (2016) found that 77 percent of the total expenditure of Chinese tourists shopping in the USA was spent on gifts for family and friends. They further summarise that this shopping behaviour is performed more based on an obligation rather than a personal interest. Also acknowledged throughout the literature is that Chinese gift purchase behaviour is different from Western consumers in the way that purchasing decisions are made (Guo et al., 2007; Timothy, 2005), because the Chinese will spend more time and effort making decisions on gifts to ensure that they are well received by the receivers (Gao et al., 2017). Lastly, in the Chinese culture, because not everyone in China can travel overseas, it is crucial that those who do bring back a special token to remember their travels to give to those that do not have the opportunity to travel. Xu and McGehee (2012), and Zhu et al. (2016) suggested that this was a reason why the Chinese tourists shopped in the USA, New Zealand and Europe. All the motivations outlined are consistent with Jansen-Verbeke's (1995) four categories of shopping motivations.

There is a lot of overlap with culture and the purchase of souvenirs or gifts for family and friends because culture is a driving factor behind ensuring that social relationships are maintained and further enhanced through bringing back gifts for family and friends (Gao et al., 2017; Jiang et al., 2012). Tsang et al. (2014) noted that the enhancement of relationships with family and friends was an important motivation to shop in Hong Kong. Because China is a culture driven society, one of the major drivers of this behaviour is Confucianism with 'Face' acknowledged as the key reason for this (Chow & Murphy, 2008; Mok & Defranco, 2000). 'Face' is one of the most important motivations for Chinese tourists to shop in Hong Kong (Tsang et al., 2014). As the participants in Tsang et al. (2014) highlighted, psychological personal developments including self-esteem, establishing a self-identity, and feeling more confident are through their shopping experience in Hong Kong. Furthermore, because of their cultural orientation, Chinese people place a great emphasis on authority, respectfulness to the elderly and the maintenance of interpersonal relationships. This has meant that personal and social factors influence their purchasing decisions abroad (Fugmann & Aceves, 2013). This literature on culture as a motivation for Chinese outbound shopping is in line with previous introductory academic articles (Ackerman & Tellis, 2001; Schütte & Ciarlante, 2016) that found cultural orientation plays an instrumental role in shaping the shopping motives of consumers.

In its core, daigou is largely in association with social connotations in Chinese culture. The actual goods are often seen as a symbolic gift for others (Boncinelli et al., 2019), as in Chinese culture, gift giving is a proxy to developing long term social networks, otherwise known as guanxi (关系 in Chinese characters) (Feng et al., 2011; Peng & Wang, 2021). Accordingly, outbound travels provide Chinese travellers a valuable opportunity to prepare symbolic gifts for various receivers in China (Jin et al., 2020). The implied values in these gifts are from the fact that these souvenirs may only be available in those destinations and are in line with the consumerism value of a high social class in those destinations (Lin, 2017; Zhang et al., 2021). Further, daigou has become a reciprocal activity, functioning in building and maintaining relational networks existing between individuals in China (Gao et al., 2017; Jin et al., 2020).

A Concluding Note

In spite of trade wars and the COVID-19 pandemic, China remains with great potential in international tourism markets. Therefore, understanding Chinese outbound tourism provides great practical implications for international tourism planners and operators, especially those who work closely with retailing industry for expanding shopping to this market. Theoretically speaking, tourist psychology and behaviours are highly contextual, vary in home countries/places and destination countries/regions. By building a conceptual foundation on Chinese outbound tourism, its history, and specific shopping behaviours, this chapter provides both an emic and an etic approach for studying a particular context in tourism research. It provides theoretical values in linking with various disciplines related to the topic, such as culture, economics, social and environmental psychology, and marketing, illustrating the interdisciplinary nature of tourism research. It encourages exploring various phenomena in Chinese outbound tourism such as daigou, in relation with Chinese culture in depth specifically, as well as smart technology and social media, which expedites the popularity of this shopping behaviour.

References

Ackerman, D., & Tellis, G. (2001). Can culture affect prices? A cross-cultural study of shopping and retail prices. *Journal of Retailing*, *77*(1), 57–82. https://doi.org/10.1016/S0022-4359(00)00046-4

Agrawal, J., & Kamakura, W. A. (1999). Country of origin: A competitive advantage? *International Journal of Research in Marketing*, *16*(4), 255–267. https://doi.org/10.1016/S0167-8116(99)00017-8

Agrusa, J., Kim, S. S., & Wang, K-C. (2011). Mainland Chinese tourists to Hawaii: Their characteristics and preferences. *Journal of Travel & Tourism Marketing*, *28*(3), 261–278. https://doi.org/10.1080/10548408.2011.562853

Anastasiei, B., & Chiosa, A. (2014). Influence of country of origin on foreign product evaluation. *Management & Marketing Journal*, *12*(2), 219–225.

Arlt, W. G. (2013). The second wave of Chinese outbound tourism. *Tourism Planning & Development*, *10*(2), 126–133. https://doi.org/10.1080/21568316.2013.800350

Assiouras, I., Skourtis, G., Koniordos, M., & Giannopoulos, A. A. (2015). Segmenting East Asian tourists to Greece by travel motivation. *Asia Pacific Journal of Tourism Research*, *20*(12), 1389–1410. https://doi.org/10.1080/10941665.2014.982140

Beerli, A., & Martin, J. D. (2004). Factors influencing destination image. *Annals of Tourism Research*, *31*(3), 657–681. https://doi.org/10.1016/j.annals.2004.01.010

Breakey, N., Ding, P., & Lee, T. (2008). Impact of Chinese outbound tourism to Australia: Reviewing the past; implications for the future. *Current Issues in Tourism*, *11*(6), 587–603. https://doi.org/10.1080/13683500802475935

Boncinelli, F., Dominici, A., Gerini, F., & Marone, E. (2019). Consumers wine preferences according to purchase occasion: Personal consumption and gift-giving. *Food Quality and Preference*, *71*, 270–278. https://doi.org/10.1016/j.foodqual.2018.07.013

Cai, L. A., Lehto, X. Y., & O'leary, J. (2001). Profiling the US-bound Chinese travelers by purpose of trip. *Journal of Hospitality & Leisure Marketing*, *7*(4), 3–16. https://doi.org/10.1300/J150v07n04_02

Chan, Y. W. (2006). Coming of age of the Chinese tourists: The emergence of non-Western tourism and host—guest interactions in Vietnam's border tourism. *Tourist Studies*, *6*(3), 187–213. https://doi.org/10.1177/1468797607076671

Chang, R. C., Kivela, J., & Mak, A. H. (2010). Food preferences of Chinese tourists. *Annals of Tourism Research*, *37*(4), 989–1011. https://doi.org/10.1016/j.annals.2010.03.007

Chen, C. C., & Lin, Y. H. (2012). Segmenting mainland Chinese tourists to Taiwan by destination familiarity: A factor-cluster approach. *International Journal of Tourism Research*, *14*(4), 339–352. https://doi.org/10.1002/jtr.864

Chen, C. M., Chen, S. H., & Lee, H. T. (2009). The influence of service performance and destination resources on consumer behaviour: A case study of mainland Chinese tourists to Kinmen. *International Journal of Tourism Research*, *11*(3), 269–282. https://doi.org/10.1002/jtr.687

Chen, N., Dwyer, L., & Firth, T. (2015). Factors influencing Chinese students' behaviour in promoting Australia as a destination for Chinese outbound travel. *Journal of Travel & Tourism Marketing*, *32*(4), 366–381. https://doi.org/10.1080/10548408.2014.897299

Chen, S. C., & Gassner, M. (2012). An investigation of the demographic, psychological, psychographic, and behavioral characteristics of Chinese senior leisure travelers. *Journal of China Tourism Research*, *8*(2), 123–145. https://doi.org/10.1080/19388160.2012.677340

Chen, Y., Mak, B., & Li, Z. (2013). Quality deterioration in package tours: The interplay of asymmetric information and reputation. *Tourism Management*, *38*, 43–54. https://doi.org/10.1016/j.tourman.2013.02.010

Chen, Y., Schuckert, M., Song, H., & Chon, K. (2016). Why can package tours hurt tourists? Evidence from China's tourism demand in Hong Kong. *Journal of Travel Research*, *55*(4), 427–439. https://doi.org/10.1177/0047287515612597

Cheng, M., & Foley, C. (2018). Understanding the distinctiveness of Chinese post-80s tourists through an exploration of their formative experiences. *Current Issues in Tourism*, *21*(11), 1312–1328. https://doi.org/10.1080/13683500.2017.1406462

Chiang, C. C., King, B. E., & Nguyen, T. H. (2012). Information searching and the travel behaviours of MICE travellers: a cross-cultural study. *International Journal of Tourism Research*, *14*(2), 103–115. https://doi.org/10.1002/jtr.833

ChinaTravelNews (2018). *130 million Chinese tourists spend $115 billion overseas in 2017*. https://www.chinatravelnews.com/article/120312

Choi, M. J., Heo, C. Y., & Law, R. (2016). Developing a typology of Chinese shopping tourists: An application of the Schwartz model of universal human values. *Journal of Travel & Tourism Marketing*, *33*(2), 141–161. https://doi.org/10.1080/10548408.2014.997961

Choi, T.-M., Liu, S.-C., Pang, K.-M., & Chow, P.-S. (2008). Shopping behaviors of individual tourists from the Chinese mainland to Hong Kong. *Tourism Management*, *29*(4), 811–820. https://doi.org/10.1016/j.tourman.2007.07.009

Chow, I., & Murphy, P. (2008). Travel activity preferences of Chinese outbound tourists for overseas destinations. *Journal of Hospitality & Leisure Marketing*, *16*(1–2), 61–80. https://doi.org/10.1080/10507050802096885

Corigliano, M. A. (2011). The outbound Chinese tourism to Italy: The new graduates' generation. *Journal of China Tourism Research*, *7*(4), 396–410. https://doi.org/10.1080/19388160.2011.627015

Correia, A., Kozak, M., & Kim, S. (2018). Luxury shopping orientations of mainland Chinese tourists in Hong Kong: Their shopping destination. *Tourism Economics*, *24*(1), 92–108. https://doi.org/10.1177/1354816617725453

Diamantopoulos, A., Schlegelmilch, B., & Palihawadana, D. (2011). The relationship between country-of-origin image and brand image as drivers of purchase intentions: a test of alternative perspectives. *International Marketing Review*, *28*(5), 508–524. https://doi.org/10.1108/02651331111167624

Feng, H., Chang, H., & Holt, R. (2011). Examining Chinese gift-giving behaviour from the politeness theory perspective. *Asian Journal of Communication*, *21*(3), 301–317. https://doi.org/10.1080/01292986.2011.559257

Fountain, J., Espiner, S., & Xie, X. (2010). A cultural framing of nature: Chinese tourists' motivations for, expectations of, and satisfaction with, their New Zealand tourist experience. *Tourism Review International*, *14*(2–3), 71–83. https://doi.org/10.3727/154427211X13044361606298

Fu, X., Cai, L., & Lehto, X. (2015). A Confucian analysis of Chinese tourists' motivations. *Journal of Travel & Tourism Marketing*, *32*(3), 180–198. https://doi.org/10.1080/10548408.2014.896233

Fugmann, R., & Aceves, B. (2013). Under control: Performing Chinese outbound tourism to Germany. *Tourism Planning & Development*, *10*(2), 159–168. https://doi.org/10.1080/21568316.2013.783734

Gao, H., Huang, S. S., & Brown, G. (2017). The influence of face on Chinese tourists' gift purchase behaviour: The moderating role of the gift giver–receiver relationship. *Tourism Management*, *62*, 97–106. https://doi.org/10.1016/j.tourman.2017.03.013

Gu, Q., Qiu Zhang, H., King, B., & Huang, S. (2018). Wine tourism involvement: a segmentation of Chinese tourists. *Journal of Travel & Tourism Marketing*, *35*(5), 633–648. https://doi.org/10.1080/10548408.2017.1401031

Guo, Y., Seongseop Kim, S., & Timothy, D. J. (2007). Development characteristics and implications of mainland Chinese outbound tourism. *Asia Pacific Journal of Tourism Research*, *12*(4), 313–332. https://doi.org/10.1080/10941660701760995

Hanzaee, K. H., & Khosrozadeh, S. (2011). The effect of the country-of-origin image, product knowledge and product involvement on information search and purchase intention. *Middle-East Journal of Scientific Research*, *8*(3), 625–636.

Heung, V. C., & Cheng, E. (2000). Assessing tourists' satisfaction with shopping in the Hong Kong special administrative region of China. *Journal of Travel Research*, *38*(4), 396–404. https://doi.org/10.1177/004728750003800408

Howison, S., Higgins-Desbiolles, F., & Sun, Z. (2017). Storytelling in tourism: Chinese visitors and Māori hosts in New Zealand. *Anatolia*, *28*(3), 327–337. https://doi.org/10.1080/13032917.2017.1318296

Hsieh, A-T., & Chang, J. (2006). Shopping and tourist night markets in Taiwan. *Tourism Management*, *27*(1), 138–145. https://doi.org/10.1016/j.tourman.2004.06.017

Hsu, C. H., Cai, L. A., & Li, M. (2010). Expectation, motivation, and attitude: A tourist behavioral model. *Journal of Travel Research*, *49*(3), 282–296. https://doi.org/10.1177/0047287509349266

Hsu, C. H., Kang, S. K., & Lam, T. (2006). Reference group influences among Chinese travelers. *Journal of Travel Research*, *44*(4), 474–484. https://doi.org/10.1177/0047287505282951

Hsu, C. H., & Song, H. (2012). Projected images of major Chinese outbound destinations. *Asia Pacific Journal of Tourism Research*, *17*(5), 577–593. https://doi.org/10.1080/10941665.2011.630674

Hua, Y., & Yoo, J. J.-E. (2011). Travel motivations of mainland Chinese travelers to the United States. *Journal of China Tourism Research*, *7*(4), 355–376. https://doi.org/10.1080/19388160.2011.627000

Huang, R., & Tian, X. (2013). An investigation of travel behavior of Chinese international students in the UK. *Journal of China Tourism Research*, *9*(3), 277–291. https://doi.org/10.1080/19388160.2013.812898

Huang, S., & Hsu, C. H. (2005). Mainland Chinese residents' perceptions and motivations of visiting Hong Kong: Evidence from focus group interviews. *Asia Pacific Journal of Tourism Research*, *10*(2), 191–205. https://doi.org/10.1080/10941660500135977

Huang, S., & Hsu, C. H. (2009). Effects of travel motivation, past experience, perceived constraint, and attitude on revisit intention. *Journal of Travel Research*, *48*(1), 29–44. https://doi.org/10.1177/0047287508328793

Huang, S., Keating, B. W., Kriz, A., & Heung, V. (2015). Chinese outbound tourism: an epilogue. *Journal of Travel & Tourism Marketing*, *32*(1–2), 153–159. https://doi.org/10.1080/10548408.2014.986018

Huang, S. S., Shao, Y., Zeng, Y., Liu, X., & Li, Z. (2021). Impacts of COVID-19 on Chinese nationals' tourism preferences. *Tourism Management Perspectives*, *40*, 100895. https://doi.org/10.1016/j.tmp.2021.100895

Jang, S., Yu, L., & Pearson, T. (2003). Chinese travellers to the United States: a comparison of business travel and visiting friends and relatives. *Tourism Geographies*, *5*(1), 87–108. https://doi.org/10.1080/1461668032000034097

Jansen-Verbeke, M. (1991). Leisure shopping: a magic concept for the tourism industry? *Tourism Management*, *12*(1), 9–14. https://doi.org/10.1016/0261-5177(91)90024-N

Jansen-Verbeke, M. (1995). The synergy between shopping and tourism: The Japanese experience. In W. F. Theobald (Ed.), *Global tourism: The next decade* (pp. 347–361). Butterworth-Heinemann Ltd.

Jiang, S., Scott, N., Ding, P., & Zou, T. T. (2012). Exploring Chinese outbound tourism motivation using means–end chains: A conceptual model. *Journal of China Tourism Research*, *8*(4), 359–372. https://doi.org/10.1080/19388160.2012.729425

Jin, H., Moscardo, G., & Murphy, L. (2017). Making sense of tourist shopping research: A critical review. *Tourism Management*, *62*, 120–134. https://doi.org/10.1016/j.tourman.2017.03.027

Jin, H., Moscardo, G., & Murphy, L. (2020). Exploring Chinese outbound tourist shopping: A social practice framework. *Journal of Travel Research*, *59*(1), 156–172. https://doi.org/10.1177/0047287519826303

Jin, X., & Wang, Y. (2016). Chinese outbound tourism research: A review. *Journal of Travel Research*, *55*(4), 440–453. https://doi.org/10.1177/0047287515608504

Johanson, M. M. (2008). The outbound mainland China market to the United States: Uncovering motivations for future travel to Hawaii. *Journal of Hospitality & Leisure Marketing*, *16*(1–2), 41–59. https://doi.org/10.1080/10507050802096836

Jørgensen, M. T., Law, R., & King, B. E. (2017). Understanding the past, anticipating the future–a critical assessment of China outbound tourism research. *Journal of Travel & Tourism Marketing*, *34*(7), 880–891. https://doi.org/10.1080/10548408.2016.1220889

Kalicharan, H. D. (2014). The effect and influence of country-of-origin on consumers' perception of product quality and purchasing intentions. *International Business & Economics Research Journal (IBER)*, *13*(5), 897–902. https://doi.org/10.19030/iber.v13i5.8760M

Kambele, Z., Li, G., & Zhou, Z. (2015). Travelers' information-seeking behaviors. *Journal of Travel & Tourism Marketing*, *32*(1–2), 141–152. https://doi.org/10.1080/10548408.2014.986017

Keating, B. W., Huang, S., Kriz, A., & Heung, V. (2015). A systematic review of the Chinese outbound tourism literature: 1983–2012. *Journal of Travel & Tourism Marketing*, *32*(1–2), 2–17. https://doi.org/10.1080/10548408.2014.986009

Kim, S. S., & Prideaux, B. (2005). Marketing implications arising from a comparative study of international pleasure tourist motivations and other travel-related characteristics of visitors to Korea. *Tourism Management*, *26*(3), 347–357. https://doi.org/10.1016/j.tourman.2003.09.022

King, B., & Gardiner, S. (2015). Chinese international students. An avant-garde of independent travellers? *International Journal of Tourism Research*, *17*(2), 130–139. https://doi.org/10.1002/jtr.1971

Klein, J. G., Ettenson, R., & Morris, M. D. (1998). The animosity model of foreign product purchase: An empirical test in the People's Republic of China. *Journal of Marketing*, *62*(1), 89–100. https://doi.org/10.1177/002224299806200108

Kong, W. H., & Chang, T.-Z. (2016). Souvenir shopping, tourist motivation, and travel experience. *Journal of Quality Assurance in Hospitality & Tourism*, *17*(2), 163–177. https://doi.org/10.1080/1528008X.2015.1115242

Lai, C., Li, X. R., & Harrill, R. (2013). Chinese outbound tourists' perceived constraints to visiting the United States. *Tourism Management*, *37*, 136–146. https://doi.org/10.1016/j.tourman.2013.01.014

Lee, S., Jeon, S., & Kim, D. (2011). The impact of tour quality and tourist satisfaction on tourist loyalty: The case of Chinese tourists in Korea. *Tourism Management*, *32*(5), 1115–1124. https://doi.org/10.1016/j.tourman.2010.09.016

LeHew, M. L., & Wesley, S. C. (2007). Tourist shoppers' satisfaction with regional shopping mall experiences. *International Journal of Culture, Tourism and Hospitality Research*, *1*(1), 82–96. https://doi.org/10.1108/17506180710729628

Leung, D., Li, G., Fong, L. H. N., Law, R., & Lo, A. (2014). Current state of China tourism research. *Current Issues in Tourism*, *17*(8), 679–704. https://doi.org/10.1080/13683500.2013.804497

Li, C., McCabe, S., & Chen, Y. (2017). Destination choice of Chinese long-haul outbound tourists and market segmentation. *Journal of China Tourism Research*, *13*(3), 298–315. https://doi.org/10.1080/19388160.2017.1398118

Li, F. S., & Ryan, C. (2018). Souvenir shopping experiences: A case study of Chinese tourists in North Korea. *Tourism Management*, *64*, 142–153. https://doi.org/10.1016/j.tourman.2017.08.006

Li, M., & Cai, L. A. (2012). The effects of personal values on travel motivation and behavioral intention. *Journal of Travel Research*, *51*(4), 473–487. https://doi.org/10.1177/0047287511418366

Li, M., Wen, T., & Leung, A. (2011a). An exploratory study of the travel motivation of Chinese female outbound tourists. *Journal of China Tourism Research*, 7(4), 411–424. https://doi.org/10.1080/19388160.2011.627020

Li, M., Zhang, H., Xiao, H., & Chen, Y. (2015). A grid-group analysis of tourism motivation. *International Journal of Tourism Research*, *17*(1), 35–44. https://doi.org/10.1002/jtr.1963

Li, X. R., Harrill, R., Uysal, M., Burnett, T., & Zhan, X. (2010). Estimating the size of the Chinese outbound travel market: A demand-side approach. *Tourism Management*, *31*(2), 250–259. https://doi.org/10.1016/j.tourman.2009.03.001

Li, X., Lai, C., Harrill, R., Kline, S., & Wang, L. (2011b). When east meets west: An exploratory study on Chinese outbound tourists' travel expectations. *Tourism Management*, *32*(4), 741–749. https://doi.org/10.1016/j.tourman.2010.06.009

Lim, C., & Wang, Y. (2008). China's post-1978 experience in outbound tourism. *Mathematics and Computers in Simulation*, *78*(2–3), 450–458. https://doi.org/10.1016/j.matcom.2008.01.033

Lim, C. C., & Bendle, L. J. (2012). Arts tourism in Seoul: tourist-orientated performing arts as a sustainable niche market. *Journal of Sustainable Tourism*, *20*(5), 667–682. https://doi.org/10.1080/09669582.2011.636817

Lin, Y-H., Chen, C-C., & Park, C. W. (2012). The salient and organic images of Taiwan as perceived by Mainland Chinese tourists. *Asia Pacific Journal of Tourism Research*, *17*(4), 381–393. https://doi.org/10.1080/10941665.2011.628329

Lin, L. (2017). Food souvenirs as gifts: Tourist perspectives and their motivational basis in Chinese culture. *Journal of Tourism and Cultural Change*, *15*(5), 439–454. https://doi.org/10.1080/14766825.2016.1170841

Lin, Y-H., & Lin, K. Q. (2006). Assessing mainland Chinese visitors' satisfaction with shopping in Taiwan. *Asia Pacific Journal of Tourism Research*, *11*(3), 247–268. https://doi.org/10.1080/10941660600753281

Lin, Z., He, G., & Vlachos, I. P. (2015). Britain in bloom? A study into Chinese tourists' experience. *EuroMed Journal of Business*, *10*(3), 297–310. https://doi.org/10.1108/EMJB-09-2014-0030

Liu, A., & McKercher, B. (2016). The impact of visa liberalization on tourist behaviors—The case of China outbound market visiting Hong Kong. *Journal of Travel Research*, *55*(5), 603–611. https://doi.org/10.1177/0047287514564599

Liu, S-C., Choi, T-M., Au, R., & Hui, C-L. (2011). Individual tourists from the Chinese mainland to Hong Kong: Implications for tourism marketing in fashion. *Tourism Economics*, *17*(6), 1287–1309. https://doi.org/10.5367/te.2011.0090

Liu, S-C., Choi, T-M., & Lee, W. T. (2008). Tourists' satisfaction levels and shopping preferences under the solo travel policy in Hong Kong. *Journal of Fashion Marketing and Management*, *12*(3), 351–364. https://doi.org/10.1108/13612020810889308

Lloyd, A. E., Yip, L. S., & Luk, S. T. (2011). An examination of the differences in retail service evaluation between domestic and tourist shoppers in Hong Kong. *Tourism Management*, *32*(3), 520–533. https://doi.org/10.1016/j.tourman.2010.04.004

Lojo, A. (2016). Chinese tourism in Spain: An analysis of the tourism product, attractions and itineraries offered by Chinese travel agencies. *Cuadernos de Turismo*, *37*, 243–268. https://doi.org/10.6018/turismo.37.256231

Lu, Z. (2011). The study of Chinese tourists' motivations to Canada. *Journal of China Tourism Research*, 7(4), 345–354. https://doi.org/10.1080/19388160.2011.626998

Lyu, S. O., & Noh, E. J. (2017). Shopping decisions of international tourists to Korea: The Heckman sample selection approach. *Journal of Destination Marketing & Management*, 6(4), 436–443. https://doi.org/10.1016/j.jdmm.2016.07.001

Maguire, J. S., & Hu, D. (2013). Not a simple coffee shop: local, global and glocal dimensions of the consumption of Starbucks in China. *Social Identities*, *19*(5), 670–684. https://doi.org/10.1080/13504630.2013.835509

Mao, I. Y., & Zhang, H. Q. (2014). Structural relationships among destination preference, satisfaction and loyalty in Chinese tourists to Australia. *International Journal of Tourism Research*, *16*(2), 201–208. https://doi.org/10.1002/jtr.1919

McCartney, G., Butler, R., & Bennett, M. (2009). Positive tourism image perceptions attract travellers—fact or fiction? The case of Beijing visitors to Macao. *Journal of Vacation Marketing*, *15*(2), 179–193. https://doi.org/10.1177/1356766708100823

Mehta, S., Jain, A., & Jawale, R. (2014). Impact of tourism on retail shopping in Dubai. *International Journal of Trade, Economics and Finance*, *5*(6), 530–535. https://doi.org/10.7763/IJTEF.2014.V5.428

Meng, F., & Zhang, P. (2016). Chinese outbound tourists' shopping behavior. In X. Li (Ed.), *Chinese outbound tourism 2.0* (pp. 267–282). Apple Academic Press.

Mohsin, A. (2008). Analysis of Chinese travellers' attitudes toward holidaying in New Zealand: The impact of socio-demographic variables. *Journal of Hospitality & Leisure Marketing*, *16*(1–2), 21–40. https://doi.org/10.1080/10507050802096794

Mok, C., & Defranco, A. L. (2000). Chinese cultural values: Their implications for travel and tourism marketing. *Journal of Travel & Tourism Marketing*, *8*(2), 99–114. https://doi.org/10.1300/J073v08n02_07

Morris, M. W., Leung, K., Ames, D., & Lickel, B. (1999). Views from inside and outside: Integrating emic and etic insights about culture and justice judgment. *Academy of Management Review*, *24*(4), 781–796. https://doi.org/10.5465/amr.1999.2553253

Murphy, L., Benckendorff, P., Moscardo, G., & Pearce, P. L. (2011). *Tourist shopping villages: Forms and functions*. Routledge.

Packer, J., Ballantyne, R., & Hughes, K. (2014). Chinese and Australian tourists' attitudes to nature, animals and environmental issues: Implications for the design of nature-based tourism experiences. *Tourism Management*, *44*, 101–107. https://doi.org/10.1016/j.tourman.2014.02.013

Pan, G. W., & Laws, E. (2002). Tourism marketing opportunities for Australia in China. *Journal of Vacation Marketing*, *8*(1), 39–48. https://doi.org/10.1177/135676670200800105

Pan, S., Ting, P-H., & Bau, D-Y. (2014). Perceptions of Taipei as revealed in travel blogs: A tale from two sides. *Asia Pacific Journal of Tourism Research*, *19*(6), 700–720. https://doi.org/10.1080/10941665.2013.779590

Park, K. S., Reisinger, Y., & Noh, E. H. (2010). Luxury shopping in tourism. *International Journal of Tourism Research*, *12*(2), 164–178. https://doi.org/10.1002/jtr.744

Park, S. H., Lee, C-K., & Miller, J. C. (2015). A comparative study of the motivations, activities, overall satisfaction, and post-trip behaviors of international tourists in Macau: Mainland Chinese, Hongkongese, Taiwanese, and Westerners. *Asia Pacific Journal of Tourism Research*, *20*(10), 1174–1193. https://doi.org/10.1080/10941665.2014.965184

Pearce, D. G. (2014). The internationalization of tourism research. *Journal of Travel Research*, *53*(3), 267–280. https://doi.org/10.1177/0047287513491336

Peng, W., & Wang, W. Y. (2021). Buying on Weixin/WeChat: Proposing a sociomaterial approach of platform studies. *Media, Culture & Society*, *43*(5), 945–956. https://doi.org/10.1177/0163443720968460

Pham, T. D., Nghiem, S., & Dwyer, L. (2017). The determinants of Chinese visitors to Australia: A dynamic demand analysis. *Tourism Management*, *63*, 268–276. https://doi.org/10.1016/j.tourman.2017.06.015

Pharr, J. (2005). Synthesizing country-of-origin research from the last decade: Is the concept still salient in an era of global brands? *Journal of Marketing Theory and Practice*, *13*(4), 34–45. https://doi.org/10.1080/10696679.2005.11658557

Prayag, G., Cohen, S. A., & Yan, H. (2015a). Potential Chinese travellers to Western Europe: segmenting motivations and service expectations. *Current Issues in Tourism*, *18*(8), 725–743. https://doi.org/10.1080/13683500.2013.868413

Prayag, G., Disegna, M., Cohen, S. A., & Yan, H. (2015b). Segmenting markets by bagged clustering: Young Chinese travelers to Western Europe. *Journal of Travel Research*, *54*(2), 234–250. https://doi.org/10.1177/0047287513514299

Prideaux, B., Cave, J., Thompson, M., & Sibtain, J. (2012). Recognising new market opportunities and selecting appropriate segments: Targeting Chinese outbound tourists. *Journal of Vacation Marketing, 18*(4), 287–299. https://doi.org/10.1177/1356766712461102

Qiu, H., & Lam, T. (1999). An analysis of Mainland Chinese visitors' motivations to visit Hong Kong. *Tourism Management, 20*(5), 587–594. https://doi.org/10.1016/S0261-5177(99)00028-X

Qu, H., & Lam, S. (1997). A travel demand model for Mainland Chinese tourists to Hong Kong. *Tourism Management, 18*(8), 593–597. https://doi.org/10.1016/S0261-5177(97)00084-8

Qu, H., & Li, I. (1997). The characteristics and satisfaction of mainland Chinese visitors to Hong Kong. *Journal of Travel Research, 35*(4), 37–41. https://doi.org/10.1177/004728759703500406

Ruhanen, L., Whitford, M., & McLennan, C.-l. (2015). Exploring Chinese visitor demand for Australia's indigenous tourism experiences. *Journal of Hospitality and Tourism Management, 24*, 25–34. https://doi.org/10.1016/j.jhtm.2015.07.002

Ryan, C., & Mo, X. (2002). Chinese visitors to New Zealand—Demographics and perceptions. *Journal of Vacation Marketing, 8*(1), 13–27. https://doi.org/10.1177/135676670200800103

Schütte, H., & Ciarlante, D. (2016). *Consumer behaviour in Asia*. Springer.

Shen, H., Li, X., Luo, J. M., & Chau, K. Y. (2017). One country, two strata: Implications of social and cultural conflicts of Chinese outbound tourism to Hong Kong. *Journal of Destination Marketing & Management* 6(3), 166–169. https://doi.org/10.1016/j.jdmm.2016.10.003

Skivalou, M., & Filippidi, E. (2017). Chinese tourism: Development and prospects for Greece. *Tourism and Hospitality Research, 17*(3), 325–335. https://doi.org/10.1177/1467358415610372

Song, H., & Hsu, C. H. (2013). The image of Taiwan as a travel destination: Perspectives from mainland China. *Journal of Travel & Tourism Marketing, 30*(3), 253–271. https://doi.org/10.1080/10548408.2013.774919

Song, H., Liu, K-N., & Huang, Y-T. (2016). Understanding mainland Chinese tourists' motivation and constraints of visiting Taiwan. *Journal of China Tourism Research, 12*(3–4), 394–413. https://doi.org/10.1080/19388160.2016.1262804

Sparks, B., & Pan, G. W. (2009). Chinese outbound tourists: Understanding their attitudes, constraints and use of information sources. *Tourism Management, 30*(4), 483–494. https://doi.org/10.1016/j.tourman.2008.10.014

Stepchenkova, S., & Li, X. (2012). Chinese outbound tourists' destination image of America: Part II. *Journal of Travel Research, 51*(6), 687–703. https://doi.org/10.1177/0047287512451137

Sun, M., Ryan, C., & Pan, S. (2014). Assessing tourists' perceptions and behaviour through photographic and blog analysis: The case of Chinese bloggers and New Zealand holidays. *Tourism Management Perspectives, 12*, 125–133. https://doi.org/10.1016/j.tmp.2014.09.007

Sun, M., Ryan, C., & Pan, S. (2015). Using Chinese travel blogs to examine perceived destination image: The case of New Zealand. *Journal of Travel Research, 54*(4), 543–555. https://doi.org/10.1177/0047287514522882

Sundström, M., Lundberg, C., & Giannakis, S. (2011). Tourist shopping motivation: go with the flow or follow the plan. *International Journal of Quality and Service Sciences, 3*(2), 211–224. https://doi.org/10.1108/17566691111146104

The Nielson Company (2017). *2017 outbound Chinese tourism and consumption trends*. No longer available.

Timothy, D. J. (2005). *Shopping tourism, retailing, and leisure*. Channel View Publications.

Timothy, D. J., & Butler, R. W. (1995). Cross-border shopping: A North American perspective. *Annals of Tourism Research, 22*(1), 16–34. https://doi.org/10.1016/0160-7383(94)00052-T

Tiwari, A. K., Dash, A. K., & Narayanan, B. G. (2018). Foreign tourist arrivals in India from major source countries: An empirical analysis. *Current Issues in Tourism, 21*(10), 1137–1156. https://doi.org/10.1080/13683500.2017.1296415

Truong, T. H., & King, B. (2009). An evaluation of satisfaction levels among Chinese tourists in Vietnam. *International Journal of Tourism Research, 11*(6), 521–535. https://doi.org/10.1002/jtr.726

Tsang, N. K. F., Lee, L. Y-S., & Liu, C. K. L. (2014). Understanding the shopping motivation of mainland Chinese tourists in Hong Kong. *Journal of China Tourism Research*, *10*(3), 323–346. https://doi.org/10.1080/19388160.2013.873329

Tse, T. S. (2015). A review of Chinese outbound tourism research and the way forward. *Journal of China Tourism Research*, *11*(1), 1–18. https://doi.org/10.1080/19388160.2014.984798

Tse, T. S., & Tse, Q. K. (2015). The legal aspects of "zero-fare" tour in shopping tourism: A case of Chinese visitors in Hong Kong. *Journal of China Tourism Research*, *11*(3), 297–314. https://doi.org/10.1080/19388160.2015.1086709

Untong, A., Ramos, V., Kaosa-Ard, M., & Rey-Maquieira, J. (2015). Tourism demand analysis of Chinese arrivals in Thailand. *Tourism Economics*, *21*(6), 1221–1234. https://doi.org/10.5367/te.2015.0520

UNWTO (2017). *Chinese tourists spent 12% more in travelling abroad in 2016*. No longer available.

Upchurch, R. S., & Liu, D. (2014). Does shopping matter? A research note discussing China's revised tariff policy and resultant outbound shopping behaviors. *Tourism Analysis*, *19*(2), 241–247. https://doi.org/10.3727/108354214X13963557455883

Wang, D. (2004). Tourist behaviour and repeat visitation to Hong Kong. *Tourism Geographies*, *6*(1), 99–118. https://doi.org/10.1080/1461668032000172235

Wang, L., Fong, D., & Law, R. (2016). Travel behaviors of mainland Chinese visitors to Macao. *Journal of Travel & Tourism Marketing*, *33*(6), 854–866. https://doi.org/10.1080/10548408.2015.1068266

Wang, Y., & Davidson, M. C. (2010). Chinese holiday makers' expenditure: Implications for marketing and management. *Journal of Hospitality Marketing & Management*, *19*(4), 373–396. https://doi.org/10.1080/19368621003667101

Weiler, B., & Yu, X. (2008). Case studies of the experiences of Chinese visitors to three tourist attractions in Victoria, Australia. *Annals of Leisure Research*, *11*(1–2), 225–241. https://doi.org/10.1080/11745398.2008.9686794

Wen, J., Meng, F., Ying, T. Y., Qi, H. X., & Lockyer, T. (2018). Drug tourism motivation of Chinese outbound tourists: Scale development and validation. *Tourism Management*, *64*, 233–244. https://doi.org/10.1016/j.tourman.2017.08.001

Wong, I. A., & Rosenbaum, M. S. (2012). Beyond hardcore gambling: Understanding why mainland Chinese visit casinos in Macau. *Journal of Hospitality & Tourism Research*, *36*(1), 32–51. https://doi.org/10.1177/1096348010380600

Wong, I. A., & Wan, Y. K. P. (2013). A systematic approach to scale development in tourist shopping satisfaction: Linking destination attributes and shopping experience. *Journal of Travel Research*, *52*(1), 29–41. https://doi.org/10.1177/0047287512457263

Wong, J., & Law, R. (2003). Difference in shopping satisfaction levels: A study of tourists in Hong Kong. *Tourism Management*, *24*(4), 401–410. https://doi.org/10.1016/S0261-5177(02)00114-0

Wu, M.-Y. (2015). Driving an unfamiliar vehicle in an unfamiliar country: Exploring Chinese recreational vehicle tourists' safety concerns and coping techniques in Australia. *Journal of Travel Research*, *54*(6), 801–813. https://doi.org/10.1177/0047287514532364

Wu, M.-Y., Wall, G., & Pearce, P. L. (2014). Shopping experiences: international tourists in Beijing's silk market. *Tourism Management*, *41*, 96–106. https://doi.org/10.1016/j.tourman.2013.09.010

Wyman, O. (2017). *Chinese travellers spend more but shop less*. http://www.oliverwyman.com/media-center/2017/jul/chinese-traveler-spend-more-but-shop-less.html

Xu, Y., & McGehee, N. G. (2012). Shopping behavior of Chinese tourists visiting the United States: Letting the shoppers do the talking. *Tourism Management*, *33*(2), 427–430. https://doi.org/10.1016/j.tourman.2011.05.003

Yang, X., Reeh, T., & Kreisel, W. (2011). Cross-cultural perspectives on promoting festival tourism-An examination of motives and perceptions of Chinese visitors attending the Oktoberfest in Munich (Germany). *Journal of China Tourism Research*, *7*(4), 377–395. https://doi.org/10.1080/19388160.2011.627009

Ye, B. H., Qiu, H. Z., & Yuen, P. P. (2011). Motivations and experiences of mainland Chinese medical tourists in Hong Kong. *Tourism Management*, *32*(5), 1125–1127. https://doi.org/10.1016/j.tourman.2010.09.018

Yeung, R. M. W., & Yee, W. M. S. (2012). A profile of the mainland Chinese cross-border shoppers: Cluster and discriminant analysis. *Tourism Management Perspectives*, *4*, 106–112. https://doi.org/10.1016/j.tmp.2012.06.001

Yu, J. Y., & Ko, T. G. (2012). A cross-cultural study of perceptions of medical tourism among Chinese, Japanese and Korean tourists in Korea. *Tourism Management*, *33*(1), 80–88. https://doi.org/10.1016/j.tourman.2011.02.002

Yuan, J. C., Fowler, D. K., Goh, B., & Lauderdale, K. M. (2013). Mexican cross-border shoppers' motivations to the USA. *International Journal of Culture, Tourism and Hospitality Research*, *7*(4), 394–410. https://doi.org/10.1108/IJCTHR-02-2013-0004

Yüksel, A., & Yüksel, F. (2007). Shopping risk perceptions: Effects on tourists' emotions, satisfaction and expressed loyalty intentions. *Tourism Management*, *28*(3), 703–713. https://doi.org/10.1016/j.tourman.2006.04.025

Yun, D., & Joppe, M. (2011). Chinese perceptions of seven long-haul holiday destinations: Focusing on activities, knowledge, and interest. *Journal of China Tourism Research*, *7*(4), 459–489. https://doi.org/10.1080/19388160.2011.627029

Zeng, Z., Prentice, C., & King, B. E. (2014). To gamble or not? Perceptions of Macau among mainland Chinese and Hong Kong visitors. *International Journal of Tourism Research*, *16*(2), 105–112. https://doi.org/10.1002/jtr.1902

Zhang, H. Q., & Chow, I. (2004). Application of importance-performance model in tour guides' performance: evidence from mainland Chinese outbound visitors in Hong Kong. *Tourism Management*, *25*(1), 81–91. https://doi.org/10.1016/S0261-5177(03)00064-5

Zhang, H. Q., & Heung, V. C. (2002). The emergence of the mainland Chinese outbound travel market and its implications for tourism marketing. *Journal of Vacation Marketing*, *8*(1), 7–12. https://doi.org/10.1177/135676670200800102

Zhang, H. Q., Heung, V. C., & Yan, Y. Q. (2009). Play or not to play—An analysis of the mechanism of the zero-commission Chinese outbound tours through a game theory approach. *Tourism Management*, *30*(3), 366–371. https://doi.org/10.1016/j.tourman.2008.08.006

Zhang, J. J., & Shelton, E. J. (2015). Ordering the disordered subject: A critique of Chinese outbound tourists as New Zealand seeks to become China ready. *Tourism Analysis*, *20*(3), 343–353. https://doi.org/10.3727/108354215X14362097173837

Zhang, X., Tham, A., Liu, Y., Spinks, W., & Wang, L. (2021). 'To market, to market': uncovering Daigou touristscapes within Chinese outbound tourism. *Journal of China Tourism Research*, *17*(4), 549–569. https://doi.org/10.1080/19388160.2021.1968554

Zhang, Y., & Peng, Y. (2014). Understanding travel motivations of Chinese tourists visiting Cairns, Australia. *Journal of Hospitality and Tourism Management*, *21*, 44–53. https://doi.org/10.1016/j.jhtm.2014.07.001

Zhou, L., & Hui, M. K. (2003). Symbolic value of foreign products in the People's Republic of China. *Journal of International Marketing*, *11*(2), 36–58. https://doi.org/10.1509/jimk.11.2.36.20163

Zhu, D., Xu, H., & Jiang, L. (2016). Behind buying: The Chinese gaze on European commodities. *Asia Pacific Journal of Tourism Research*, *21*(3), 293–311. https://doi.org/10.1080/10941665.2015.1048263

Chapter 27

Tourism Work and Workers in the Context of Sustainable Development

Dimitri Ioannides

Introduction

Sustainability has undoubtedly emerged over the decades to become one of the (if not the) most dominant research topics in tourism related scholarship (Butler, 1999; Krippendorf, 1987; Lu & Nepal, 2009; Nepal, 2022). Countless articles, books, reports and conference proceedings have explored, dissected and analysed terms such as 'sustainable tourism' and 'tourism in the context of sustainable development', while volumes of writings offer recommendations on how tourism can thrive in destinations and benefit the host community while simultaneously minimising adverse damage to their ecological and cultural resources (Ioannides, 2008).

The vast majority of writings on this topic have focused overwhelmingly on environmental aspects or the balance between economic growth objectives and preserving the natural and/or built environment (Chok & Macbeth, 2009; Neto, 2003; Saarinen, 2006). By comparison, the third integral part of the so-called sustainability triangle (Campbell, 1996), which deals with social issues and matters revolving around social justice and equity, has been addressed in a somewhat superficial manner. Most importantly, the topic of tourism work and workers as it relates to sustainable development has remained largely underexplored, an issue that Baum et al. (2016) characterise as surprising, especially when one seeks to tie tourism to the Sustainable Development Goals (SDGs) of the United Nations' (UN) 2030 Agenda for Sustainable Development (UN, 2015).

The previous *Wiley Blackwell Companion to Tourism* (Lew et al. 2014) included a chapter relating to matters of labour mobility and labour markets (Ladkin, 2014) but the connections between these issues and sustainable development were not fleshed out in that particular contribution. Instead, Ladkin (2014) focused mostly on the movement of tourism workers, including these individuals' migration experiences. Ladkin (2014) called for further research on how workers may, through employment in the sector, build on their skills. Ladkin (2014) also recommended the need to: better define what tourism jobs actually mean to improve our ability to conduct empirical studies on the topic; learn what tourism labour mobility means in

The Wiley Blackwell Companion to Tourism, Second Edition. Edited by C. Michael Hall.

terms of contemporary shifts in working life overall; study what the effects of international migration of tourism-related workers imply for 'integration, cohesion, and national identity' (Ladkin, 2014, p. 138).

In this particular chapter, my aim is to specifically highlight the role that tourist related work and workers play in the context of sustainable development. Among the matters I bring up, is the thorny issue of how the excessive levels of precarity that characterise many jobs in the tourism sector match the need to create so-called decent work as set out in the UN SDGs.

Labour Precarity: A Major Challenge to Sustainable Development

Labour precarity, whereby workers' rights and security have gradually been eroded, has become increasingly common in numerous sectors of the global neoliberal economy in recent years (Lambert & Herod, 2016; Booyens, 2022; Robinson et al., 2019). Specifically, this precarity translates into a growing army of people working on irregular schedules for an hourly wage without benefits such as paid or sick leave, overtime wages, a pension plan or other perks associated with regular salaried labour. Reduced power of unions associated with many employers' growing antipathy towards organised labour, have not helped matters (Shaw & Williams, 2004).

In countries, such as the United Kingdom, it is not unusual for workers to be employed on so-called zero-hour contracts, with essentially no set guarantees of work (Head & Lucas, 2004; Zampoukos & Ioannides, 2011). This means that, on certain occasions, workers must work long hours when demand levels are high while in other instances, they may remain idle for days or even weeks waiting for the next call to take up a job. These individuals have minimal protections and can easily be laid-off (Elliot, 2013; Lambert & Herod, 2016). The growing tendency to outsource certain peripheral functions to temporary agencies (e.g., housekeeping or linen laundering services for hotels or catering and baggage-handling services in the airline sector) as well as the recent proliferation of so-called platform-based employers such as UBER and Deliveroo are all trends that have accelerated this precarity. As Van Doorn and Badger (2020) point out, online food delivery companies such as Deliveroo rely excessively on migrant workers who are particularly vulnerable to exploitation and discrimination (see also, Van Doorn et al., 2020).

Zero-hour contracts and other forms of precarity have long been highlighted as problems that are specifically prevalent in the tourism and hospitality sector. Indeed, Robinson et al. (2019) describe how the very nature of the tourism and hospitality sector, which often tends to be characterised by seasonal activities and irregular demand patterns, lends itself to a numerically but also functionally flexible workforce (see also Booyens, 2022). It is common in many destinations for hotels to employ agency-based workers subject to variations in room occupancy levels (Head & Lucas, 2004). Further, since many tourism-related jobs are associated with minimal skill requirements (Zampoukos & Ioannides, 2011) this tends to lower the entry barriers into several job opportunities in the sector (Ladkin, 2014). This, in turn, means that employers take advantage of their workers by exploiting their weak bargaining power (Baum, 2007). Thus, while it may be relatively easy to find employment in a tourism related sector during periods of high demand, it is also easy to lose one's job when business is slow.

Further, in many tourism related sectors, employers tend to supress wages, a situation which fuels worker turnover. This is especially the case in situations where there is an ample supply of individuals (e.g., university students) who are willing to take on a part-time job to help cover their expenses (Zampoukos & Ioannides, 2011). Thus, rather than treat their staff as a long-term investment, whom they can train and build their skill set, eventually helping these individuals climb the employment ladder, employers have traditionally tended to view their workers as a cost and, therefore, as highly substitutable (Baum, 2007; Shaw & Williams, 2004). To make matters worse, tourism workers rarely receive the respect they deserve, a situation which in tandem with substandard work conditions, adversely influences a business' success levels (Robinson et al., 2019; Winchenbach et al., 2019).

While some observers have fleshed out the positive qualities associated with precarity, in that certain workers prefer performing contingent work due to the flexibility this might afford them (e.g., working from home during hours of their choosing instead of having to adhere to strict office guidelines or taking on a part-time position as a ski instructor in order to pursue one's hobby), this certainly does not hold true for everyone. On the one hand, as Lambert and Herod (2016, p. 16) maintain 'some might prefer part-time or temporary work because they have increasingly adopted a mindset that challenges... the "ideology of work", which for much of the history of capitalism has encouraged people to think that if they are not working they are somehow social failures'. Then there are persons, such as retirees, who willingly take on part-time or fixed term employment as a means to remain active, especially during their earlier retirement years. On the other hand, however, there has been a trend in recent years where capitalism has prompted employers 'to look to contingent workers as a mainstay of their labor forces and to force contingent work on people involuntarily' (Lambert & Herod, 2016, p. 17).

Florida (2014) has also extolled the positive aspects of contingent work in their discussions surrounding the creative class. On the one hand, there are those persons who can easily be classified as creatives (e.g., software designers, musicians, and artists) who are highly skilled and can command high fees for the work they perform on a contingent basis. On the other hand, this narrative does not account for the fact that these creatives often depend on the services of an army of unskilled or semi-skilled workers such as waiters, bartenders, food delivery personnel or platform-based drivers, all of whom are regularly the victims of the worst forms of precariousness (Baum, 2018; Zampoukos & Ioannides, 2011). Making matters worse, a significant number of such contingent workers are particularly vulnerable because of their characteristics, which reveal the 'intersectionality of migrants, women, ethnicity, race, youth, class and other marginal social groups' (Robinson et al., 2019, p. 1012; see also, McDowell et al., 2009).

It is worth adding at this juncture that the effects of the Covid-19 pandemic, which spread globally in early 2020 have exposed even further the vulnerability of millions of persons who worked in many tourism and hospitality related sectors. Zampoukos (2022) indicates that unlike many persons who were able to work from home during the worst period of the pandemic, workers in sectors related to tourism and hospitality, if they did not lose their jobs because of the devastating effects of the pandemic, were required due to the nature of their job to be physically present at their workplaces. Refusal to go to work because of the workers' fear of exposure to the virus resulted in them being laid off, often without the ability to collect unemployment benefits (Cohen & Hsu, 2020). Similarly, Booyens (2022) points out that, during the pandemic, Scottish hospitality workers were verbally and, sometimes, physically abused, mostly by customers. The reported cases of attacks by unruly passengers against flight attendants, especially in the United States, usually related to mask wearing mandates have reached unprecedented levels in recent times (Laris & Aratani, 2021).

The Invisibility of Tourism Workers in the Sustainability Debate

Considering the critique levelled against the nature of tourism work, not to mention the poor conditions many workers in the sector witness, it is surprising that the sustainability dialogue that has dominated tourism studies for more than three decades has only paid lip service to these problems (Baum et al., 2016; Ioannides et al., 2021; Robinson et al., 2019; Zampoukos, 2022). Largely, this lack of attendance relates to the ongoing tendency, especially by the tourism industry itself, but also policymakers, to adopt a narrow view of sustainability, whereby the predominant objective at most destinations is to achieve a sustainable growth rate of arrivals and receipts. Preserving the environmental quality of a destination is often thought of as a means of ensuring that the numbers of arrivals and receipts will not drop off (Ioannides, 2008). In other words, what we end up with is a treadmill approach to sustainable tourism development as defined by Kousis (2001).

Several observers have noted the academic neglect of the social side of the sustainability triangle compared to its economic and environmental components (Campbell, 1996). This problem is particularly evident in the case of tourism studies (Neto, 2003). Robinson et al. (2019) express surprise that while those who advocate for workers' rights normally place enhanced social equity at the forefront of their demands, tourism scholars have remained largely silent when it comes to the relationship of this issue specifically to sustainable development (see also Baum et al., 2016; Baum, 2018). Robinson et al. (2019) believe that this is particularly odd, when one considers that tourism employees themselves often serve 'roles as agents of sustainability practices' (Robinson et al., 2019., p. 1013) in that they implement the environmental agendas of the companies they work for (e.g., in terms of the reduced use of towels and bed sheets when following corporate social responsibility programmes of various hotel companies).

What lies behind this inattention towards labour related issues when we talk about tourism and sustainable development? In several of their writings, Baum (most recently in 2018), believes that the lack of clarity as to what actually constitutes a tourism job, a result of the tourism sector's fuzziness within the industrial classification systems (see also Ioannides & Debbage, 1998; Ladkin, 2014) complicates matters enormously, especially when trying to undertake empirical work. Also, Saarinen's (2006) well known resource and community based approach to viewing tourism sustainability makes it hard to identify where exactly tourism workers fit in since, in many instances, these individuals are highly mobile and only within a tourism destination for short periods of time. In other words, in many places, those who work in tourism related jobs are not local inhabitants (Salazar, 2020). This situation, whereby many workers are from elsewhere, complicates matters when focusing on the local conditions of a community (Ioannides et al., 2021).

In the case of Saarinen's (2006) resource based approach, the emphasis is mostly on damage limitation related to the growth of tourism in a particular community. The often used carrying capacity framework, which is applied in discussions as to how to limit tourism's negative externalities (Cocossis et al., 2001), might occasionally be used to examine ratios of migrant workers versus the local population but ignores who these workers are, why they do what they do, their employment situation and the social conditions they encounter in the host communities. Thus, the resource based perspective of sustainability (Saarinen, 2006) ends up divorcing the workers from the places where they perform their tasks by treating them as part of the global tourism production system (Ioannides et al., 2021; Robinson et al., 2014).

Meanwhile, the community based perspective of tourism sustainability as defined by Saarinen (2006) also sidesteps how tourism workers relate to the sustainable development of destinations. Since most authors who adopt this framework focus on which permanent community members gain or lose because of tourism, the workers who are often temporarily in such places are largely overlooked. When they are examined, this is done in general terms and somewhat superficially, whereby the focus is on their precarity (Robinson et al., 2019). One rare exception, which examines the relationship of tourism workers to sustainability in a local community is provided by James et al. (2020) in their comparative study of the effects of cruises in an Icelandic and a Greenlandic community. However, once again, this does not specifically examine the circumstances in which migrant workers find themselves in.

The preceding discussion highlights that since in numerous destinations, tourism workers are only temporary and their numbers fluctuate with the seasons, they become largely invisible in examinations that predominantly concentrate on how local communities (the permanent residents) exist and react to the global forces of tourism development. As Ioannides et al. (2021) have mentioned, these workers are 'rendered 'homeless' in different sustainability discussions'. The question then becomes, how can tourism scholarship embrace the thorny issues of tourism work and workers in meaningful discussions about the role of tourism in sustainable development?

Tourism Work and Sustainability: Moving the Discussion Further

Several organisations including the UN and the International Labor Organization (ILO) have emphasised the importance of promoting 'decent work' (Booyens, 2022). For example, the UN Agenda for Sustainable Development encompasses among its 17 Sustainable Development Goals (SDGs), Goal number 8, which is to 'promote sustained, inclusive and sustainable economic growth, full and productive employment and decent work for all' (UN, 2015, p. 14). Additional SDGs indirectly relate to the issue of work and workers by covering topics like poverty reduction and gender equality (de Jong & Figueroa-Domecq, 2022; Robinson et al., 2019; Winchenbach et al., 2019). Winchenbach et al. (2019) acknowledge that while it is not entirely clear what decent work means, a broad definition of the term should incorporate 'aspects of respect and self-respect, as well as safe, fair, productive and meaningful work in conditions of freedom' (Winchenbach et al., 2019, p. 1027). Further, the ILO (1999) sees decent work as incorporating the elimination of forced labour including child labour, the ability to work without being subject to discrimination and harassment and the right for workers to have access to collective bargaining (see also Scheyvens, 2018). It goes without saying that access to social security contributions by the employers, a guaranteed maximum number of work hours per week and benefits such as sick leave and paid annual leave should also be encompassed within the realm of decent work.

The problem is, however, that many aspects relating to tourism work and the conditions, which millions of tourism and hospitality workers around the globe face, hardly match the above definitions of what constitutes decent work. My previous discussion relating to labour precarity in the sector gives testimony to this problem. To add insult to injury, if we carefully dissect the UN SDG 8, which calls for decent work in a context of sustained economic growth and also examine the *Tourism and the Sustainable Development Goals* published by the UNWTO (2017) we immediately see the dominance of language that focuses on growth-oriented topics, like total number of jobs created and employment multipliers. Scant attention is actually revealed in these documents about the type and quality of work related to the sector. This critique is further fleshed out by Bianchi and de Man (2020) who, by focusing on the UNWTO's sustainable tourism indicators, reveal that these are all about numbers and growth rates.

The ongoing obsession of most international organisations, policymakers, destination management organisations (DMOs) and business lobbies to simplistically treat tourism in the context of sustainable development in terms of the sector's perpetual growth in various destinations does little to move us forward in the sustainability debate. As Ioannides et al. (2021) have pointed out, we should shift beyond the view of sustaining tourism businesses and destinations at all costs and, instead, begin to explore how tourism workers and the work they perform can relate to sustainable development (see Salazar, 2020). The rest of this chapter offers some thoughts as to how this may be achieved.

Fitting Tourism Work and Workers into the Sustainability Agenda

Winchenbach et al. (2019) note that the focus on decent work as promoted by the ILO and the UN ends up being largely ineffective during the neoliberal era that touts the dominance of the free market in an increasingly deregulated environment. How can we talk about decent work when the indications in numerous countries are that people are working in a context of eroding bargaining power to command living wages and other benefits? How precisely does the replacement of salaried employees by workers employed by temporary agencies lead to this goal?

To move us forward in this discussion, Winchenbach et al. (2019) propose shifting our attention by embracing the concept of dignity in academic discussions relating to tourism work and workers. As Booyens (2022, p. 199) mentions, 'the notion of dignity offers a nuanced view vis-à-vis social justice and human development as it manifests in the workplace'. Nevertheless, one of the problems that tourism and many other service-oriented jobs encounter, is that such work is often dismissed as labour that is not real since 'it does not contribute to the physical production of industrial goods' (Booyens, 2022, p. 199–200). Both from a neoclassical and a Marxist economic perspective, this work ends up been considered as unproductive or, as Zampoukos and Ioannides (2011) have highlighted, it is reproductive since it entails several functions traditionally associated with home keeping such as washing clothes, cooking, and serving food, and room cleaning. Booyens (2022) suggests that such a narrow view of tourism work is problematic. Taking a leaf out of Standing (2010), Booyens (2022) contends that the best way to look at tourism related jobs in a post-manufacturing society is to begin thinking of these as occupations. In Booyens' (2022) mind the term occupation 'evokes a sense of purpose', which 'in turn, leads to an enhancement of a sense of autonomy, agency and self-determination, which is central to dignity in the workplace' (Booyens, 2022, p. 201).

To embrace the concept of dignity in relation to tourism work and workers, Winchenbach et al. (2019, p. 1032) urge us to shift our attention respectively to the 'individual worker', the 'organizational context' and the 'wider socio-economic and policy context'. Each of these perspectives encompasses both obstacles that hinder the attainment of dignity but also opportunities that promote dignity. Thus, while treating workers as factors of production can be regarded as a challenge to the attainment of dignity (see Bianchi & de Man, 2020), a dignity promoting approach would be to embrace policies advocating a living wage. In order to achieve dignity in terms of these three perspectives then, it is important to understand how 'individual and institutional actors simultaneously generate, perform and adapt to the regulative, organizational and technological systems that frame dignified employment' (Ioannides et al. 2021). Thus, tourism work and workers can be brought into the centre stage of the sustainability dialogue when we begin to better understand how workers themselves can disrupt or adapt to the organisational and technological contexts in which they are embedded.

To do this, we could seek to better comprehend how various processes including international migration, the decline of union power and membership and other emergent effects of neoliberalism relate to tourism work. Doing so, would redirect us away from the tendency in much of the sustainability dialogue to focus on destinations as individual cases while overlooking factors that are external to the local community such as guest workers (see earlier discussion). A useful point of departure is, according to Underthun and Jordhus-Lier (2018), to embrace liminality as a concept, which allows us to see how persons working in the tourist sector negotiate their interaction with the aforementioned major trends (e.g., the decline of union power).

Underthun and Jordhus-Lier (2018) contend that the lens of liminality allows us to understand how, at least certain temporary and other precarious workers themselves often sabotage collective efforts to improve their working conditions precisely because they regard their tourism job as a short-term solution to raise funds for their travels, to further their studies or to cover other expenses (Ioannides et al., 2021). Meanwhile, others, who work in tourism jobs since they lack realistic alternatives (such as new immigrants with limited language or other skills or those whose presence in a new country is not strictly legal) might choose to distance themselves from any form of organised labour movement, because of fears of losing their job (or, worse yet, being deported). Also, as Underthun and Jordhus-Lier (2018, p. 25) point out, the current set-up of trade unions that represent tourism and hospitality workers is often seen as 'static and rigid' and, thus largely powerless in terms of catering to the needs of an extremely flexible labour force that is constantly on the move. The fact that even in countries such as Sweden or Norway, with their strong welfare state tradition, union membership in the tourism and hospitality sector remains quite low, reflects these realities.

We must also be careful so as not to always blame precarity as a one size fits all problem that consistently removes agency from the workers, rendering them powerless to cope with their vulnerability to poor working conditions and the lack of dignity in the workplace. As Rydzik and Anitha (2019) have shown, UK based female migrants who work in sectors such as tourism and hospitality, demonstrate surprising levels of resilience, which the authors define as 'a limited consciousness of the relations of oppression and are aimed at enabling workers to survive oppressive conditions' (Rydzik & Anitha, 2019, p. 884–885). In the face of the often appalling workplace conditions these women encounter they might break certain minor rules as a means of resistance. Alternatively, they choose to cope, temporarily at least, with discriminatory acts or even sheer harassment since their ambition is to move on to bigger and better things. On occasion, the reaction is to eventually quit and seek out new employment opportunities, a process that Rydzik and Anitha (2019, p. 885) describe as 'reworking'. In Rydzik and Anitha's (2019) study, as these immigrant women began to better comprehend their rights and gain confidence in their use of English, several became more vocal in fighting for better working conditions. Others, for example, Janta and Ladkin (2013) and Lugosi et al. (2016), have also revealed how Eastern European hospitality workers demonstrate their agency in the workplace by networking through Internet chat groups. Janta and Ladkin (2013) discuss the phenomenon of migrant workers themselves engaging in worker recruitment practices through social media and other online sites to attract others from their country to move to the UK and take up employment in the tourism and hospitality sector. Essentially, employers provide latitude to their migrant workers to seek to attract new recruits, thus empowering them to make decisions as to whom they wish to hire as colleagues. Moreover, migrant workers can use social media to post warnings about potentially problematic work conditions (e.g., employers who should be avoided at all costs).

These examples reveal that when investigating the role of tourism work and workers in the sustainability debate, it is important for us to better understand how these workers themselves identify ways to enhance their dignity, for example, through resilience or so-called reworking practices. Meanwhile, as Ioannides et al. (2021) have recently argued it is imperative that various labour organisations and unions can identify ways to better protect the expanding precarious workforce in the tourism and hospitality sector.

With this backdrop in mind, it is also vital to examine what role, if any, the rapid technological transformations of the past 15 years or so and, especially the appearance of the so-called platform-based economy, play in influencing the sustainability of tourism work. We are now aware that online platforms can enable peer-to-peer transactions in the overall economy and, particularly the tourism and hospitality sector (Christensen, 2022; Farmaki et al., 2022). These new business arrangements have, in several cases, led to novel employment structures. Airbnb, the homesharing platform, urges people to take the plunge and become self-employed entrepreneurs, by earning a living from renting out their personal space to short-term visitors. Similarly, ride-sharing platforms such as Uber and Lyft extol the virtues of self-employment that can be afforded by turning personal vehicles into money earning ventures. Unfortunately, however, despite all the positive rhetoric put forth by companies such as Airbnb and Uber, the reality is that these disruptive innovations have often resulted in expanding the vulnerability of the workforce as has been commented on by several observers (e.g., Van Doorn & Badger, 2020).

Moreover, although it might be true that platforms such as Airbnb allow individuals to emerge as self-employed micro-entrepreneurs, this certainly does not necessarily mean that they are entirely independent to do as they wish. As Roelofsen and Minca (2018) have explained, these individuals are constantly under surveillance through artificial intelligence to ensure that they have certain self-disciplining and self-regulating standards in place to maintain prescribed standards. The aim is to enable them as hosts of peer-to peer accommodation to receive positive reviews from their guests and avoid being blacklisted by the host online platforms they are using. This results in a situation where 'new, cunning forms of economic, physical and emotional vulnerabilities are created' (Ioannides et al., 2021).

On the one hand, then, it appears that in spite of the hype surrounding the appearance of digital platforms as a means to expand individuals' flexibility and the ability to be released from the nine to five work schedule, these innovations have merely reinforced patterns of exploitation that have long surrounded the working conditions of many people. Nevertheless, on the other hand, there are certain positive examples of situations where workers have shown resilience by seeking out ways to enhance their dignity in the workplace (Ioannides et al., 2021). For instance, depending of course on their personal circumstances (e.g., skill levels, family situation and motivation), some workers demonstrate considerable flexibility in moulding their work tasks to better suit their personal interests or hobbies. Those who seek to combine their own personal hobbies with their work (e.g., extreme adventure tour guides or digital nomads) often find it easier to adapt their employment situation in a manner that best fits their interests.

These instances of what can be termed 'job crafting' (Berg et al. 2007) as well as other proactive labour related practices, such as the formation of networks through social media to deal with common problems at the workplace, demonstrate that despite the growing vulnerability that threatens many facets of tourism related work, there are glimmers of hope that all is not lost in the pursuit of enhancing dignity in the workplace.

Concluding Remarks

The aim of this chapter was to flesh out some of the most pertinent issues that relate to the position of tourism work and workers in the sustainable development dialogue. Recognising that this topic has largely remained under the radar in most discussions relating to the interlinkages of tourism to sustainable development, the chapter sought to understand why this has been the case. Two fundamental problems exist. The first is that most stakeholders involved either directly or indirectly in the tourism sector (e.g., policymakers, DMOs, industry representatives) continue to advocate a sustainable tourism perspective, which essentially amounts to the sector's constant growth at any given destination. This ongoing neoliberal tendency of talking about sustainable tourism as opposed to tourism within a context of overall sustainable development is hardly conducive to improving our understanding of the social dimensions of sustainability, including matters relating to tourism work and workers. This is particularly surprising considering our widespread understanding that precarity is a main problem when it comes to the tourism labour market.

Second, the chapter has explained how Saarinen's (2006) resource and community based approaches to sustainable development, which focus specifically on various factors at the level of a destination, is somewhat limiting when it comes to better understanding labour related issues. This is because, in many instances, it is not unusual for highly mobile tourism workers to have limited ties to destinations in which they work on a temporary basis. This means that other than reducing these individuals to mere statistics, we have a narrow understanding of who they are, why they have chosen to work in a particular destination, and what their experiences of being there amount to.

To move the discussion forward, the chapter has suggested that by coupling dignity to tourism work as proposed by writers such as Winchenbach et al. (2019), and Booyens (2022), the sustainability debate relating to tourism can benefit considerably. The focus should shift towards matters such as improving the quality of work for people engaged in the tourism sector, building on their skill sets, and allowing them to move up the employment ladder rather than constantly berating the sector for catering to those who seek to perform sub-standard unproductive (or reproductive) labour. An important point of departure is to begin to reflect that despite the many problems tourism workers face because of issues such as the growing dominance of precarity in the labour force, these individuals are not entirely devoid of agency. Rather, even during this period of rapid change, which has been underlined by the uncertainties regarding new forms of doing business through online digital platforms, there are instances where tourism workers have been able to demonstrate resourceful and resilient behaviour through which they can hope to enhance dignity in the workplace.

The ambition as put forward in the UN (2020) *Policy Brief: COVID-19 and Transforming Tourism* (see Ioannides et al., 2021) is to shift our way of thinking so as tourism in the aftermath of the pandemic can better contribute to sustainable development while adopting a trajectory that is both resilient and inclusive. For this to happen, it is vital for those of us who engage in the sustainability dialogue to focus on social equity matters - including those surrounding tourism work and workers – far more actively.

Epilogue

It is fascinating how the travel sector in many regions, including much of Europe, has been seriously struggling because of the extreme labour shortages that have resulted in the aftermath of the COVID-19 pandemic. Essentially, the lack of demand for travel, especially during 2020 and 2021 led to an enormous reduction in the tourism related workforce. The growing demand during the summer of 2022, during a period when people have become tired of the pandemic and feel they are now in a position to risk travelling once more because of vaccinations and/or prior exposure to the virus, has led to chaotic scenes in many travel venues, especially major international airports. Serious staff shortages relating to airport security, baggage handling and check-in functions, among others, have resulted in huge lines, unprecedented delays and flight cancellations. Indeed, the situation has become so dire at London's Heathrow that the airport administration has begged airlines to stop booking any more holiday travel for the rest of the summer and has capped the number of travellers who depart the airport at a daily total of 100,000 (Kollewe, 2022).

The question emerging from these circumstances is this: given the current extreme labour shortages in the sector, could workers exploit the situation and significantly enhance their bargaining power in negotiations with their employers? Could they demand terms that are far more beneficial than they were before the outbreak of the pandemic? Many people have left the sector and enticing them back might not be so easy if the various employers in the sector do not fail to improve their working conditions. It remains to be seen how this situation eventually plays out and what it will eventually mean for enhancing the dignity associated with tourism work and workers.

References

Baum, T. (2007). Human resources in tourism: Still waiting for change. *Tourism Management*, *28*(6), 1383–1399. https://doi.org/10.1016/j.tourman.2007.04.005

Baum, T. (2018). Sustainable human resource management as a driver in tourism policy and planning: a serious sin of omission? *Journal of Sustainable Tourism*, *26*(6), 873–889. https://doi.org/10.1080/09669582.2017.1423318

Baum, T., Cheung, C., Kong, H., Kralj, A., Mooney, S., Nguyen, T. T. H., Ramachandran, S., Ruzik, M. D., & Siow, M. L. (2016). Sustainability and the tourism and hospitality workforce; A thematic analysis. *Sustainability*, *8*(8), 809. https://doi.org/10.3390/su8080809

Berg, J. M., Dutton, J. E., & Wrzesniewski, A. (2007). *What is job crafting and why does it matter?* Center for Positive Organizational Scholarship, Michigan Ross School of Business. https://www.researchgate.net/profile/Justin-Berg-4/publication/266094577_What_is_Job_Crafting_and_Why_Does_It_Matter/links/542a720d0cf27e39fa8e925c/What-is-Job-Crafting-and-Why-Does-It-Matter.pdf

Bianchi, R. V., & de Man, F. (2020). Tourism, inclusive growth and decent work: A political economy critique. *Journal of Sustainable Tourism*, *29*(2–3), 353–371. https://doi.org/10.1080/09669582.2020.1730862

Booyens, I. (2022). Tourism's development impacts: an appraisal of workplace issues, labour and human development. In A. Stoffelen & D. Ioannides (Eds), *Handbook of tourism impacts: Social and environmental perspectives* (pp. 197–211). Edward Elgar Publishing.

Butler, R. (1999). Sustainable tourism: A state-of-the-art review. *Tourism Geographies*, *1*(1), 7–25. https://doi.org/10.1080/14616689908721291

Campbell, S. (1996). Green cities. Growing cities. Just cities? Urban Planning and the Contradictions of Sustainable Development. *Journal of the American Planning Association*, *62*(3), 296–312. https://doi.org/10.1080/01944369608975696

Chok, S., & Macbeth, J. (2009, June 15–18). *Labour justice and sustainable tourism: The centrality of equity as a sustainability principle* [Paper presentation]. BEST Education Network's Think Tank IX. Newcastle upon Tyne, UK. https://researchrepository.murdoch.edu.au/id/eprint/9983/

Christensen, M. D. (2022). Peer-to-peer hospitality, gentrification and local entanglements. In A. Stoffelen & D. Ioannides (Eds), *Handbook of tourism impacts: Social and environmental perspectives* (pp. 88–102). Edward Elgar Publishing.

Cocossis, H., Mexa, A., Collovini, A., Parpairis, A., & Konstantoglou, M. (2001). *Defining, measuring and evaluating carrying capacity in European tourism destinations. B4-3040/2000/294577/MAR/D2. Final Report*. European Commission.

Cohen, P., & Hsu, T. (2020, May 14). 'Rolling shock' as job losses mount even with reopenings. *The New York Times*. https://www.nytimes.com/2020/05/14/business/economy/coronavirus-unemployment-claims.html

de Jong, A., & Figueroa-Domecq, C. (2022). Assessing the UNWTO's Global Report on Women in Tourism: Tourism's impact on gender equality. In A. Stoffelen & D. Ioannides (Eds), *Handbook of tourism impacts: Social and environmental perspectives* (pp. 151–165). Edward Elgar Publishing.

Elliot, L. (2013, August 4). Zero-hours contract workers – the new reserve army of labour? *The Guardian*. https://www.theguardian.com/uk-news/economics-blog/2013/aug/04/zero-hours-contract-workers-reserve-army-labour

Farmaki, A., Ioannides, D., & Kladou, S. (Eds). (2022). *Peer-to-peer accommodation and community resilience*: Implications for sustainable development. Wallingford.

Florida, R. (2014). *The rise of the creative class revisited*. Basic Books.

Head, J., & Lucas, R. (2004). Employee relations in the non-union hotel industry: A case of determined opportunism? *Personnel Review*, *33*(6), 693–710. https://doi.org/10.1108/00483480410561565

International Labour Organization (ILO). (1999). *Report of the Director General: Decent work*. Geneva: ILO, 87th Session. https://www.ilo.org/public/english/standards/relm/ilc/ilc87/rep-i.htm

Ioannides, D. (2008). Hypothesizing the shifting mosaic of attitudes through time: A dynamic framework for sustainable tourism development on a 'Mediterranean Isle'. In S. McCool & R. N. Molsey (Eds), *Tourism, recreation and sustainability: Linking culture and the environment(* 2nd ed., pp. 51–75). CABI.

Ioannides, D., & Debbage, K. G. (1998). *The economic geography of the tourist industry: A supply-side analysis*. Routledge.

Ioannides, D., Gyimóthy, S., & James, L. (2021). From liminal labor to decent work: A human-centered perspective on sustainable tourism employment. *Sustainability*, *13*(2), 851; https://doi.org/10.3390/su13020851

James, L., Olsen, L. S., & Karlsdóttir, A. (2020). Sustainability and cruise tourism in the Arctic: Stakeholder perspectives from Isafjödur, Iceland and Qaqortoq, Greenland. *Journal of Sustainable Tourism*, *28*(9), 1425–1441. https://doi.org/10.1080/09669582.2020.1745213

Janta, H., & Ladkin, A. (2013). In search of employment: Online technologies and Polish migrants. *New Technology, Work and Employment*, *28*(3), 241–253. https://doi.org/10.1111/ntwe.12018

Kollewe, J. (2022). Heathrow asks airlines to stop selling summer flights as it caps passengers. *The Guardian*. https://www.theguardian.com/uk-news/2022/jul/12/heathrow-airlines-stop-selling-summer-flights-airport-staff

Kousis, M. (2001). Tourism and the environment in Corsica, Sardinia, Sicily, and Crete. In D. Ioannides, Y. Apostolopoulos & S. Sonmez (Eds), *Mediterranean Islands and sustainable tourism development* (pp. 214–233). Cassell.

Krippendorf, J. (1987). *The holiday makers: Understanding the impact of leisure and travel*. Butterworth-Heinemann.

Ladkin, A. (2014). Labor mobility and labor market structures in tourism. In A. A. Lew, Hall C. M., & A. M. Williams (Eds), *The Wiley Blackwell companion to tourism* (pp. 132–142). Wiley Blackwell.

Lambert, R., & Herod, A. (2016). *Neoliberal capitalism and precarious work: Ethnographies of accommodation and resistance*. Elgar.

Laris, M., & Aratani, L. (2021). Unruly airplane passengers are straining the system for keeping peace in the sky. *The Washington Post.* https://www.washingtonpost.com/transportation/2021/07/18/airplane-fight-behavior-faa-arrests/

Lew, A. A., Hall, C. M., & Williams, A. M. (Eds). (2014). *The Wiley Blackwell companion to tourism.* Wiley Blackwell.

Lu, J., & Nepal, S. K. (2009). Sustainable tourism research: An analysis of papers published in the *Journal of Sustainable Tourism.* Journal of Sustainable Tourism, *17*(1), 5–16. https://doi.org/10.1080/09669580802582480

Lugosi, P., Janta, H., & Wilczek, B. (2016). Work(ing) dynamics of migrant networking among Poles employed in hospitality and food production. *The Sociological Review*, *64*(4), 894–911. https://doi.org/10.1111/1467-954X.12393

McDowell, L., Batnitzky, A., & Dyer, S. (2009). Precarious work and economic migration: Emerging immigrant divisions in Greater London's service sector. *International Journal of Urban and Regional Research*, *33*(1), 3–25 https://doi.org/10.1111/j.1468-2427.2009.00831.x

Nepal, S. K. (2022). Studies in tourism impacts: past trends and emerging trajectories. In A. Stoffelen & D. Ioannides (Eds), *Handbook of tourism impacts: Social and environmental perspectives* (pp. 20–34). Edward Elgar Publishing.

Neto, F. (2003). A new approach to sustainable tourism development: Moving beyond environmental protection. *Natural Resource Forum*, *27*(3), 212–222. https://doi.org/10.1111/1477-8947.00056

Robinson, R. N., Martin, A., Solnet, D., & Baum, T. (2019). Sustaining precarity: Critically examining tourism and employment. *Journal of Sustainable Tourism*, *27*(7), 1008–1025. https://doi.org/10.1080/09669582.2018.1538230

Robinson, R. N., Ritchie, B. W., Kralj, A., Solnet, D. J., Baum, T., & Ford, R. C. (2014). An Asia-Pacific core-periphery futures paradox: Divergent workers and tourist mobilities. *Journal of Travel Research*, *53*(6), 805–818. https://doi.org/10.1177/0047287513513164

Roelofsen, M., & Minca, C. (2018). The superhost. Biopolitics, home and community in the Airbnb dream-world of global hospitality. *Geoforum*, *91*, 170–181. https://doi.org/10.1016/j.geoforum.2018.02.021

Rydzik, A., & Anitha, S. (2019). Conceptualizing the agency of migrant women workers: Resilience, reworking and resistance. *Work, Employment Society*, *34*(5), 883–899. https://doi.org/10.1177/0950017019881939

Saarinen, J. (2006). Traditions of sustainability in tourism studies. *Annals of Tourism Research*, *33*(4), 1121–1140. https://doi.org/10.1016/j.annals.2006.06.007

Salazar, N. (2020). Labour migration and tourism mobilities: Time to bring sustainability into the debate. *Tourism Geographies*, *24*(1), 141–151. https://doi.org/10.1080/14616688.2020.1801827

Scheyvens, R. (2018). Ecotourism and the empowerment of local communities. *Tourism Management*, *20*(2), 245–249. https://doi.org/10.1016/S0261-5177(98)00069-7

Shaw, G., & Williams, A. M. (2004). *Tourism and tourism spaces.* SAGE Publications.

Standing, G. (2010). *Work After globalization: Building occupational citizenship.* Edward Elgar Publishing.

Underthun, A., & Jordhus-Lier, D. C. (2018). Liminality at work in Norwegian hotels. *Tourism Geographies*, *20*(1), 11–28. https://doi.org/10.1080/14616688.2017.1314546

United Nations (UN). (2015). *Transforming our world: The 2030 agenda for sustainable development. General Assembly Resolution A/RES/70/1.* UN. https://www.un.org/en/development/desa/population/migration/generalassembly/docs/globalcompact/A_RES_70_1_E.pdf

United Nations (UN). (2020). *Policy brief: COVID-19 and transforming tourism.* https://unsdg.un.org/sites/default/files/2020-08/sg_policy_brief_covid-19_tourism_august_2020.pdf

United Nations World Tourism Organization (UNWTO). (2017). *Tourism and sustainable development goals – Journey to 2030.* UNWTO. https://www.e-unwto.org/doi/book/10.18111/9789284419340

Van Doorn, N., & Badger, A. (2020). A platform capitalism's hidden abode: Producing data assets in the gig economy. *Antipode*, *52*(5), 1475–1495. https://doi.org/10.1111/anti.12641

Van Doorn, N., Ferrari, F., & Graham, M. (2020). Migration and migrant labour in the gig economy: An intervention. *SSRN*. http://dx.doi.org/10.2139/ssrn.3622589

Winchenbach, A., Hanna, P., & Miller, G. (2019). Rethinking decent work: The value of dignity in tourism employment. *Journal of Sustainable Tourism*, *27*(7), 1026–1043. https://doi.org/10.1080/09669582.2019.1566346

Zampoukos, K. (2022). Precarious jobs, precarious people in times of a pandemic: The impact on tourism workers and spaces of work. In A. Stoffelen & D. Ioannides (Eds), *Handbook of tourism impacts: Social and environmental perspectives* (pp. 183–195). Edward Elgar Publishing.

Zampoukos, K., & Ioannides, D. (2011). The tourism labour conundrum: Agenda for new research in the geography of hospitality workers. *Hospitality and Society*, *1*(1), 25–45. https://doi.org/10.1386/hosp.1.1.25_1

Part 5 Tourist Attractions
Place, Space and Forms

Introduction
Tourist Attractions: Place, Space and Forms

C. Michael Hall

Attractions are a central element in what draws tourists to destinations. There is a long-standing literature on attractions in tourism studies. In a seminal work, Leiper (1990) referred to a tourist attraction system which consists of an empirical connection between tourist, nucleus, and marker. The nucleus is the physical element of the attraction while the marker in the more intangible promoted and value-attached elements. Leiper (1990) built on the earlier work of Lew (1987) who developed a typology of attractions and MacCannell who defined an attraction in semiotic and systemic terms as a: 'tourist, a sight, and a marker' (1976, p.41).

Lew (1987) classified the elements of an attraction as ideographic, organisational, and cognitive. The ideographic element describes the uniqueness of a site and includes not just physical elements but also price, attitudes toward tourists, and accessibility. The organisational dimension described the spatial characteristics, capacity, and temporal aspects. The cognitive element describes the degree to which tourists are willing to experience different attractions (Lew, 1987). More recent research has also identified significant cultural (Richards, 2002) and temporal (Caccomo & Solonandrasana, 2002; Botti et al., 2008; Ćorluka et al., 2021) components, with the latter highlighting issues of seasonality. Richards (2002) concluded that tourists are 'pushed' towards attractions by their motivations with visitation strongly related to motivation, attraction markers, media use, and touristic characteristics. Nevertheless, considerable debate exists over the extent that specific attractions attract tourists (McKercher & Tolkach, 2020; Nicolau et al., 2023; Navarro-Ruiz & McKercher, 2020). For example, McKercher (2017) argues that the specificity of tourist need should be matched with the number of attraction sets and individual attractions within each set that can satisfy the need in order to understand the relative importance of specific attractions, leading McKercher (2017, p. 120) to conclude, 'Individual attractions will play an increasingly important role in destination choice as needs become more singular and specific. Alternately, if needs are broad, the number of attraction sets and attractions within each set becomes increasingly substitutable, making it difficult to ascribe visitation to any specific attraction'.

The Wiley Blackwell Companion to Tourism, Second Edition. Edited by C. Michael Hall.

Given continued expansion in tourist visitation, there is increasing interest in the sustainability of attractions. Adie (Chapter 28) addresses this issue in relation to tangible and intangible cultural heritage which, for many destinations, is often a key element of tourism marketing campaigns. However, as she notes, there has been a shift towards the blending of cultural heritage tourism and sustainable development, particularly with a focus on poverty alleviation. The chapter discusses the integration of these two principles and the importance of including local communities in planning and development processes and the acknowledgement and integration of local identity in relation to local heritage.

Chapter 29 by Nilsson notes how urban tourism has expanded rapidly in recent decades, account for approximately 40 percent of all international incoming tourists in Europe. The chapter notes that the boundaries of urban tourism are becoming blurred as many 'tourist' activities are undertaken by residents. High quality demands by local residents may serve to improve local competition and performance of tourist businesses and the cultural sector, with Nilsson arguing that the transformation of the food scene in Copenhagen is a good example of how local innovation and improved standards has resulted in the development of successful niche tourism. Nevertheless, he notes that urban destinations have adapted to a situation with low transport costs. Given that global fuel prices may increase in the future, the dependency on tourism may cause problems for the future economic sustainability of cities. In addition, Nilsson argues that urban tourism is increasingly dependent on actors outside local control, e.g., Airbnb, with potentially negative consequences for urban social sustainability (Hall et al., 2022).

Chapter 30 by Müller on second homes also notes that the sector is undergoing significant change. Müller notes that while large numbers of second homes can be found in resort areas, where they play an important role by contributing to the accommodation supply, the relationship between second-home locations and major tourism areas is not pre-determined. Furthermore, changes related to geopolitics, digitalization, and a changing work life, as well as a growing awareness of the consequences of climate and environmental change all pose challenges to the tradition understanding of second homes and tourism. The increasing dissolution of the need for physical presence at a workplace for many in the labour market opens for an intensified use of second homes. Such change challenges sedentarist ideas of 'permanent' residence and instead calls for greater research on the emotional, functional, and corporeal attachment to multiple places. Related to this, second homes may also have a greater investment function than in the past, particularly given the potential for their use as short-term rentals. Nevertheless, the second home will remain an important component of the attractiveness of many destinations.

Another longstanding research interest in the tourism literature with respect to destination attractions is that of hallmark or megaevents. Chapter 31 by Chen observes that mega-events have had, and will continue to have, significant impacts on the global and regional tourism environment, regardless of economic, social, and environmental circumstances. The chapter notes that international mega-events that have the capability of self-promotion and are associated with substantial political and commercial interests often take a profit driven approach, overlooking social equity but are nevertheless portrayed as a substantial contributor to the local economy. However, he cautions that these events, as with most contemporary tourism, are not as sustainable as they claim to be in their current form. However, smaller localised community-based events tend to be more sustainable, carefully incorporating local welfare over the long-term.

Food has long been regarded as core element of the attractiveness of destinations and as an attraction in its own right (Chen & Huang, 2019; Huang et al., 2022, 2023). Nevertheless, there have been calls for research in the area to be expanded with respect to its scope (Hall, 2020). Wang et al. (Chapter 32) provide a review of the food and tourism literature paying particular attention to different forms of food tourism, consumption behaviour, destination development, and the growing problem of food waste and destination sustainability.

Historically, it has been argued that popular gay travel destinations developed because they demonstrated certain characteristics, i.e. permissive or liberal attitudes towards queer identified travellers; friendly infrastructure (bars, businesses, restaurants, hotels, nightlife, entertainment, media); and create opportunities to socialise with similarly identified individuals thereby providing a sense of community. However, Visser (Chapter 33) argues that the connections that currently exist between place, space, sexuality and tourism demonstrate an evolution in both form and function. He notes that the identity markers have historically mostly been white gay men, yet in a contemporary setting few of the queer markers remain relevant as analytic categories and they are simply geographically too diverse given the diversity of LGBTQ or, perhaps more inclusively, the queer community overall. As Visser concludes, it is undeniably clear that the nature of queer place/space has changed and it is left to us to unravel the new articulations of queer sexuality, tourism and community in this new age in which understandings of queer socialisation and the various relationships to consolidated space is shifting dramatically.

Shopping and retail is often the major activity taken by tourists in many destinations. The final chapter in the section is that of Timothy (Chapter 34) in which he revisits trends in tourism, shopping, and retailing. Nevertheless, it is important to distinguish between tourist shopping, e.g., buying souvenirs, browsing in malls, buying supplies unintentionally left at home, purchasing gifts for friends and family members, procuring a bottle of wine following a wine tasting, buying picnic items from a supermarket, and any number of retail-related consumer activities, which is typically a secondary activity undertaken by travellers whose principal purpose of travel is something else; and shopping tourism which is travel motivated fundamentally by a desire to shop. In the latter, the destination is chosen because of its retail opportunities and advantages, including its reputation and image, emphasis on shopping tourism, unique retail environment, reasonable prices, or merchandise selection (Timothy, 2024). As Timothy concludes, despite the placelessness seemingly embedded in online shopping, there is still an element of place value in shopping and retail. As a result, shopping opportunities continue to function as an important attraction for many destinations and remain a vital element in marketing and promotion.

References

Botti, L., Peypoch, N., & Solonandrasana, B. (2008). Time and tourism attraction. *Tourism Management*, *29*(3), 594–596. https://doi.org/10.1016/j.tourman.2007.02.011

Caccomo, J.L., & Solonandrasana, B. (2002). Réflexions autour du concept d'attraction touristique: Analyse et taxonomie. *Revue de Recherche en Tourisme*, *3*, 68–71.

Chen, Q., & Huang, R. (2019). Understanding the role of local food in sustaining Chinese destinations. *Current Issues in Tourism*, *22*(5), 544–560. https://doi.org/10.1080/13683500.2018.1444020

Ćorluka, G., Vitezić, V., & Peronja, I. (2021). The temporal dimension of tourist attraction. *Tourism: An International Interdisciplinary Journal*, *69*(3), 443–453. https://doi.org/10.37741/t.69.3.9

Hall, C. M. (2020). Improving the recipe for culinary and food tourism? The need for a new menu. *Tourism Recreation Research*, *45*(2), 284–287. https://doi.org/10.1080/02508281.2019.1694243

Hall, C.M., Prayag, G., Safonov, A., Coles, T., Gössling, S., & Naderi Koupaei, S. (2022). Airbnb and the sharing economy. Current Issues in Tourism, *25*(19), 3057–3067, https://doi.org/10.1080/13683500.2022.2122418

Huang, Y., Hall, C. M., Bryden, A., Du, Y., Koupaei, S. N., Lyes, K., Marquez, V., Shaw, M., & Thomas, D. (2022). The characteristics of destination restaurants. *Journal of Hospitality and Tourism*, *20*, 1–30.

Huang, Y., Hall, C. M., & Chen, N. C. (2023). The characteristics of Michelin-starred restaurants in Mainland China, Hong Kong, Macau, and Taiwan. *International Journal of Gastronomy and Food Science*, *31*, 100673. https://doi.org/10.1016/j.ijgfs.2023.100673

Leiper, N. (1990). Tourist attraction systems. *Annals of Tourism Research*, *17*(3), 367–384. https://doi.org/10.1016/0160-7383(90)90004-B

Lew, A. (1987). A framework of tourist attraction research. *Annals of Tourism Research*, *14*, 533–575. https://doi.org/10.1016/0160-7383(87)90071-5

MacCannell, D. (1976). *The tourist: A new theory of the leisure class*. Macmillan.

McKercher, B. (2017). Do attractions attract tourists? A framework to assess the importance of attractions in driving demand. *International Journal of Tourism Research*, *19*(1), 120–125. https://doi.org/10.1002/jtr.2091

McKercher, B., & Tolkach, D. (2020). Influence of attractions on destination selection. *International Journal of Tourism Research*, *22*(6), 767–775. https://doi.org/10.1002/jtr.2371

Navarro-Ruiz, S., & McKercher, B. (2020). The usability of visitor attractions: State-of-the-art. *Tourism Review*, *75*(3), 497–509. https://doi.org/10.1108/TR-02-2019-0057

Nicolau, J. L., Casado-Díaz, A. B., & Navarro-Ruiz, S. (2023). Assessing the effects of interaction with attractions and types of visit on day trippers' satisfaction. *Current Issues in Tourism*, 1–17. https://doi.org/10.1080/13683500.2023.2204396

Richards, G. (2002). Tourism attraction systems: Exploring cultural behavior. *Annals of Tourism Research*, *29*(4), 1048–1064. https://doi.org/10.1016/S0160-7383(02)00026-9

Timothy, D.J. (Ed.) (2024). *Contemporary perspectives on shopping, retail and tourism*. Channel View Publications.

Chapter 28

Cultural Heritage Tourism
A Future for Sustainable Development?

Bailey Ashton Adie

Introduction

Cultural heritage is, by definition, the contemporary use, interpretation, and valuation of both tangible and intangible inheritances from the past. Tangible heritage is the more easily defined of the two heritage types and encompasses things such as buildings, industrial remnants, works of art, monuments, and smaller objects, including, for example, clothing, jewellery, books, and documents. In comparison, intangible heritage is less easy to describe as it includes elements which are constantly changing and evolving, such as language, customs, traditions, and beliefs. Regardless, both of these types of cultural heritage have been of keen interest for people across the world for centuries, and thus it is unsurprising that cultural heritage tourism is considered one of the 'oldest forms of travel' (Timothy & Boyd, 2006, p. 1). While the precursor to modern heritage tourism is pilgrimage, which has been undertaken for millennia, cultural heritage tourism in its current form can trace its origins to Europe's Grand Tour as well as a more democratized form of heritage visitation that emerged in the 19th century (Timothy, 2021). Today, cultural heritage tourism is one of the largest, if not the largest, niche tourism type in the world. According to the UNWTO (2018, p. 74), cultural tourism comprises approximately 40 percent of all global tourist trips and is one of the most rapidly growing niche forms of tourism (Du Cros & McKercher, 2014).

Unsurprisingly, then, cultural heritage is often understood to be a potential point of interest for tourists (Richards, 2007; Silberberg, 1995), and, for many destinations, cultural heritage attributes are often key elements of tourism marketing campaigns. However, there has also been a more recent shift towards the blending of cultural heritage tourism and sustainable development, particularly with a focus on poverty alleviation (Timothy & Nyaupane, 2009). To that end, this chapter discusses the integration of these two principles. In particular, the chapter seeks to discuss what the author has found to be key elements of sustainable cultural heritage tourism development, namely the inclusion of both local communities in planning and development processes and the acknowledgement and integration of local identity in

The Wiley Blackwell Companion to Tourism, Second Edition. Edited by C. Michael Hall.

relation to local heritage. There is also a brief presentation of the risks faced by sustainable heritage tourism development, both from external forces as well as from tourism itself. Finally, the chapter concludes with a discussion of the need for a shift from sustainable to regenerative tourism.

Cultural Heritage Tourism and Sustainable Development

The management of cultural heritage and tourism are often viewed as opposing forces with diverging goals, particularly as tourism itself poses a risk to the conservation and protection of cultural heritage assets. For example, the conflict between cultural heritage management and tourism development is clearly visible in Rodwell's (2002) study on industrial heritage in the United Kingdom. In their study, the tourism elements of the site management plans were overtly driven by the economic benefits of tourism to the detriment of the local community and site protection. Similarly, in Botswana, the Tsodilo Hills site has experienced uncontrolled tourism growth which has had an adverse impact on not only the site itself but also local infrastructure and resources (Mbaiwa, 2016). Regardless, in many cultural heritage contexts, tourism is often a key element of management policies, which may be due to economic need as well as the site's educational purview. Furthermore, cultural heritage tourism is often viewed by policy makers as a means by which to not only promote heritage but also reinvigorate local economies, as can be observed in the EU's heritage tourism policy (Lähdesmäki, 2016). Thus, regardless of the apparent oppositional roles that heritage and tourism have, contemporary policy debates in both arenas have placed a clear emphasis on the symbiotic nature of cultural heritage and tourism.

In particular, cultural heritage is viewed as an important way through which countries can integrate tourism into their wider sustainable development policies. This is visible in UNESCO's Sustainable Tourism Programme policy. Within this document, World Heritage sites are given guidance on how to manage tourism development in line with UNESCO World Heritage principles emphasizing protection and preservation (UNESCO, n.d.). Additionally, the current emphasis on the UN's Sustainable Development Goals (SDGs) adds further support to the establishment of sustainable development protocols in relation to cultural heritage tourism, even though heritage is not a key focus of the SDGs (Labadi et al., 2021). Therefore, sustainable development has been, and will continue to be, a key element of cultural heritage tourism, especially when considering the development of new tourism destinations. However, it should be noted that this is not without criticism, particularly as it relates to cultural heritage tourism. Sustainable development policies at the international level have been criticized for placing growth and industry ahead of heritage conservation as well as local populations (Hall, 2019; Saarinen & Rogerson, 2014). Thus, it is necessary to delve more deeply into several critical elements which play a role in sustainable heritage tourism development.

Community and Sustainable Cultural Heritage Tourism

As has been noted, community actors play an important role within the sustainable development of cultural heritage tourism. However, their relationship to their cultural heritage and tourism development can be quite complex. In some cases, they may feel, and potentially rightly so, that tourism poses a threat to their heritage (Litvin et al., 2020). These communities will be more likely to want to prioritize site conservation over tourism development (Vong, 2015). However, in some instances, the negative impacts of tourism can be felt so strongly that residents who live close to the heritage sites may start to view their heritage locations in a more negative light as was observed in Iran by Rasoolimanesh et al. (2019). They hypothesised that this would result in a lack of willingness to engage in protection or conservation activities as the heritage site and tourism had become inextricably interlinked and thus a combined source of unpleasantness. Issues surrounding local communities' complicated relationships with their heritage and tourism are particularly visible in the conversations surrounding overtourism, especially given the rise of the concept within tourist-historic cities. In Europe, for example, approximately one-third of a large-scale sample felt that while tourism was important for the local economy, it also posed a significant risk to local heritage sites (Adie & Falk, 2021). Given these results, they emphasised the importance of community inclusion in tourism planning processes in order to avoid the problems related to overtourism as residents were not inherently opposed to the development of heritage tourism.

In order to ensure that local populations voices are heard in the planning and development process, good governance is absolutely essential, particularly that which provides explicit agency to the local community. The necessity for inclusivity has been acknowledged at the policy level in the *Lusaka Declaration on Promoting Sustainable Tourism* (UNWTO, 2017) which stresses the need for the development of equitable partnerships. However, this is often difficult to implement in existing national governance structures wherein tourism and/or cultural heritage may have highly centralized systems. While this is not to imply that community inclusion within planning systems cannot occur with more hierarchical structures but instead that it will require a concerted effort from all partners. This is visible at Herulaneum in Italy where, although there has been a successfully developed public-private partnership, local government bodies have resisted increasing the involvement of the local community (Thompson, 2007). In fact, even when systems are designed to include local voices, these may be more performative. Human (2015, p. 173) noted this exact scenario in Turkey where 'the primary strategy is to produce evidence of participation, rather than to substantively incorporate local communities into the management process.' However, inclusive planning processes are possible, but they require a significant investment of time and energy in order to ensure that they continue to function for the betterment of all. A good example of this is the Nizamuddin Urban Renewal Initiative in New Delhi, India wherein the relevant projects were designed not only for the benefit of the local community but also in collaboration with the local community from the outset (Adie et al., 2022).

Additionally, although full community inclusion should be the primary goal when developing sustainable heritage tourism initiatives, it needs to be understood that communities are often heterogenous with varied viewpoints as to how best to develop and manage local heritage tourism. An example of this can be seen in the discussions around the development of a cross-village tourism initiative near Saadani National Park in Tanzania. While all villages involved expressed an interest in the development of village tourism, there was a general unwillingness among the different villages to collaborate, particularly as some villages were better placed to attract heritage tourists regardless of collaboration (Sirima & Mngulwi, 2020). Furthermore, local community members may have differing opinions regarding tourism development, as was noted by Neuts and Nijkamp (2012). This may be due to the fact that heritage tourism can be viewed differently dependent on the costs and benefits which are directly felt by the local community (Andereck et al., 2005; Chen & Chen, 2010; Janusz et al., 2017) with those who benefit more financially being generally more pro-tourism growth (Eckert et al., 2019). There may also be power dynamics between communities and non-governmental partnerships, as has been noted by Schéou and Southon (2013). They found that in a series of partnerships between French NGOs and West African communities, the income from heritage tourism was often used in a neo-colonial, paternalistic manner wherein the French partners chose how to allocate funding, regardless of community desires (Schéou & Southon, 2013). Thus, sustainable heritage tourism development planning processes need to be not only inclusive but also cognizant of existing power dynamics and various community viewpoints.

In certain contexts, particularly those wherein the goal is not merely sustainable tourism development but also poverty alleviation, there are additional essential elements of community inclusion that must be considered. Timothy and Nyaupane (2009, p. 64) noted that 'local residents often lack skills and investment abilities needed to establish tourism-related businesses and often end up with low-investment businesses and low-paying jobs.' Therefore, it is necessary to ensure that local populations are not only provided opportunities through which to actively take part in the local tourism economy but also to have access to training and skill building activities to support their involvement (Spenceley & Nelson, 2013). This, though, requires a sufficient amount of funding as well as consistent and long-term stakeholder engagement (Truong et al., 2016; Zapata et al., 2011). In cases where funding and time are not an issue, successful outcomes can be achieved, as was observed in the Nizamuddin Urban Renewal Initiative. This project not only provided direct tourism and heritage livelihoods, through the creation of guiding programmes and women's crafting groups, but it also provided wider skills training to the local community which would be applicable in relevant industries outside the localized tourism system, increasing overall local employment (Adie et al., 2022). However, without a clear succession plan following the withdrawal of project partners, communities may be unable to maintain the tourism businesses and services that were developed with the support of the initial development project, as occurred at the Kabaka Heritage Trail in Uganda (Ahebwa et al., 2016). This reinforces the need not only for continuous community inclusion, but also for community empowerment throughout the heritage tourism development process in order to avoid the pitfalls highlighted here.

Local Identity and Sustainable Cultural Heritage Tourism

A local community's identity and sense of connectedness to a cultural heritage site can have a significant impact on the overall conservation and protection of the site (Vong, 2015). In many ways, the site can be viewed as a physical manifestation or locus of this local identity construct with protection preserving both the site and the community's sense of self. This was observed by Lwoga (2017) in their survey of local residents at 3 different Tanzanian heritage locations. According to their findings, 'a local resident is likely to have the intention to conserve if she or he feels a sense of oneness with the built heritage, and has a spiritual, ancestral and historic bond with it' (Lwoga, 2017, p. 45). However, the inverse of this is also true in that individuals who have their site connections severed or who become distanced, either by design or simply the passage of time, from a local heritage element will generally exhibit little interest in the site. For example, in Zimbabwe at the Khami World Heritage Site, the traditional owners' ties to the site were severed through the colonial heritage management process which prioritized Western conservation techniques over local ones (Sinamai, 2019). At the Khami site, this is combined with a political disinterest in site promotion and protection, due in part to a low level of current visitation, which has rendered a wholly unsustainable site conservation ecosystem.

Within the tourism context, the construction, or destruction in certain circumstances, of connectedness to heritage is often an element of site narrative development. There are multiple reasons for the construction of a unified narrative which eliminates certain identity-defining elements, not least in that heritage is inherently political and thus is susceptible to manipulation. In the context of contested or difficult heritage, certain local identity-defining elements may be deliberately suppressed as a result of political decision-making. For example, in Melaka, part of the Historic Cities of the Straits of Malacca World Heritage Site, the official marketing material emphasizes the city's multi-cultural heritage, while, at the more local level, government officials continue to focus on specifically Malay and Islamic heritage which in turn minimizes the role of the local Chinese community in the city's history (King, 2016).

In some cases, the lack of inclusion of the community whose heritage is being promoted as a tourism attractor may be the cause of this heritage disconnect, as was observed at the Khami World Heritage Site (Sinamai, 2019). This may be due to a reliance on a focus on a site's ancient heritage as opposed to its modern use, as can be seen at Angkor Wat where the official narrative ignores the contemporary religious use of the temple complex (Winter, 2005). This lack of inclusion is also visible in the treatment of indigenous heritage in multiple contexts. For instance, in the case of Mexico, 'Mayan' heritage is a major element of Mexican tourism development, but the heritage presented is not only hyper commodified but also highly sanitized due to a long history of marginalization of Mayan ethnic groups (Evans, 2005). The narratives, in the Mexican case, have also created the illusion that the Maya are a long-dead civilization as opposed to a still living indigenous community (Ely, 2013). Thus, it is clear that tourism often plays a significant role in the decisions around how to build a heritage narrative, more specifically developing ones which are appealing to, predominantly, international visitors and fulfil their expectations around what the site is.

This heritage narrative construction that caters to international tourists' pre-conceptions is nowhere more visible than within post-colonial contexts. In certain destinations, post-colonial heritage narrative building may be entrenched in colonial interpretations of identity, where heritage managers may be engaging in what Feighery (2012) refers to as self-Orientalization. While this process may also be used to partly erase difficult histories, as was the case in Oman (Causevic & Neal, 2019), self-orientalization is predominantly undertaken to appeal to international tourism markets whose pre-conceptions of post-colonial states may be almost entirely derived from colonial-era travel or historical narratives. The Incredible India marketing campaign is an excellent example of this process whereby the imagery presented within the international marketing materials was visibly calling back to earlier colonial representations of India while simultaneously ignoring the quotidian existence of the local population (Kerrigan et al., 2012). In some contexts, this international tourist driven emphasis not only impacts narrative development but also what heritage is actually protected. For example, the Moroccan sites on the World Heritage List were noted as being taken almost entirely from colonial discourse on what defined Moroccan heritage (Adie, 2020). However, given the importance of local identity for the successful implementation of sustainable heritage tourism development, particularly in regard to conservation and preservation of the site, this then begs the question as to how this can occur within national, regional, or even local systems which are antithetical to the inclusion of local identity within site narratives?

Risks to the Future of Sustainable Cultural Heritage Tourism

As has been observed, local community involvement in tourism planning as well as a sense of local identity tied to a heritage element can play a significant role in the sustainable development of heritage tourism. In particular, both of these elements are vital to ensure the conservation and protection of cultural heritage assets. The previous sections have discussed the problematic impacts of non-local inclusion into heritage tourism development, but it is important to also discuss the risks with which cultural heritage tourism locations have to contend, including those posed by both tourism itself as well as external forces. It should be noted that, while the majority of this chapter has emphasised the impacts of sustainable development on physical cultural heritage, there are certain risks to which intangible cultural heritage attributes are significantly more susceptible. Most notably, intangible heritage, as an ever-changing element of living culture, faces the risk of commodification whereby the intangible heritage element is transformed into something *other*, potentially losing its authenticity (Dippon & Moskaliuk, 2020; Kim et al., 2019), a concept which is discussed in greater detail elsewhere in this book (See Chapter 13). Commodification can be as simple as changing beadwork patterns to better suit non-local tastes to as complex as changing the very social fabric of a historic city. This alteration in local atmosphere is what occurred in Hoi An, Vietnam. Whereas previously heritage centres were home to shops and amenities designed for the local communities, the development of heritage tourism led to a touristified area wherein all activity is effectively designed to generate income from tourists, including locals performing daily life for tourist photos (Caust & Vecco, 2017). Thus, in this instance, heritage tourism development led to the direct destruction of intangible heritage elements of a local site.

Tangible heritage also faces significant threats from the development of cultural heritage tourism, with one of the biggest worries for a site being protection and conservation. More specifically, tourism can pose a direct threat to these goals through uncontrolled visitor numbers, which can cause significant damage to the physical integrity of a site (Yang & Lin, 2014). This can include damage caused by not only an increase in foot traffic but also an indirect increase in pollution, litter, and even human waste (Hejazi, 2008; Timothy & Nyaupane, 2009). Another risk of uncontrolled visitation is the inability to properly protect a site from visitor misbehaviour. For example, in Petra, Jordan, visitors caused significant damage to an architectural element by hurling stones at it, which constitutes a deliberate act of vandalism (Vella et al., 2015). Additionally, this growth in visitor numbers may also provide a market for local looters, as was observed in Petra wherein artifacts stolen from the site were later marketed to visitors (Vella et al., 2015). These issues, even with the inclusion of the local community, will still pose a problem provided that visitor numbers are not strictly controlled so as to ensure that a heritage site does not exceed carrying capacity.

While the previous identified threats are a direct result of tourism development in the first place, perhaps the largest and most existential threat to sustainable heritage tourism as a whole is anthropogenic climate change. At the international level, UNESCO (2015) has already incorporated climate change-related policies into its World Heritage and sustainable development initiatives. These emphasise the importance not only of 'reduc[ing] the vulnerability of World Heritage properties and their settings' but also of 'promot[ing] the social and economic resilience of local and associated communities to disasters and climate change' (UNESCO, 2015, p. 5). This has also included the development of disaster risk management guidelines as well as a further push to moving towards sustainable development. This is, in many ways, a necessity given the impact that climate change is projected to have on World Heritage properties in the future, including threats from sea level rise, storm surge, desertification, and shifting biodiversity, which would have a particularly detrimental impact on cultural landscapes. In order to respond to climate change threats, the Neolithic Heart of Orkney World Heritage site undertook the testing of a climate vulnerability index to ascertain its suitability for cultural heritage sites. In so doing, the site was not only confirmed to be highly vulnerable, but the specific method also included risks to the local community, both in terms of economic loss from damage to local tourism but also from a potential loss of an integral element of local identity (Day et al., 2019). Thus, heritage climate change policy appears to reiterate the importance of the local community and local identity in the response to the threat of climate change, supporting Hall et al.'s (2016) statement that it is impossible to separate heritage tourism risks from the wider human context.

Is There a Future for Sustainable Cultural Heritage Tourism?

As has been observed throughout this chapter, the development of sustainable cultural heritage tourism has to balance the intersecting dynamics of community involvement, local identity, and risks, both internal and external. This has been shown to be a difficult process, and one which is not often successful due to a plethora of issues, including lack of sufficient funding, insufficient community inclusion, and exclusion of local heritage narratives, to name a few. This is doubly compounded by the fact that, although sustainable tourism theory originally had more of an ecological focus, its contemporary usage is considerably more industry driven (Saarinen, 2021a). More specifically, the contemporary policy use of the term sustainable has shifted from its original intent in an attempt to be inclusive of the capitalist focus which drives much of the tourism industry and includes a strong emphasis on growth-driven development. While the creation of truly sustainable systems would require a significant amount of time, capital, and policy development, the growth-driven narrative tends to prioritize short-term and easy to implement initiatives aimed at maximizing economic impact. In cultural heritage tourism, this may also be the result of a site's conservation needs which cannot be met by existing funding. Given the problematic nature of the modern use of the term sustainable combined with the aforementioned problems with sustainable cultural heritage tourism, it is perhaps necessary to move towards something beyond 'sustainable'.

Therefore, it is proposed that the future of cultural heritage tourism research lies within the paradigms of regenerative tourism and, by extension, degrowth. Degrowth, as an alternative economic model, places resource sufficiency and conservation at the forefront, over a pure profit-driven focus (Hall, 2009; Higgins-Desbiolles et al., 2019), and it is essential for the implementation of regenerative tourism development (Saarinen, 2021b). Many of the failures identified with the use of heritage tourism for sustainable development have been in the problematic integration of local community into the associated development processes while simultaneously respecting the local identity and narratives tied to these heritage elements. Regenerative tourism, by its very design, is community focused through its use of a collaborative inclusionary planning process (Healey, 2006). Furthermore, it emphasises the inclusion of a diversity of viewpoints in the understanding of what is regenerative tourism (Bellato et al., 2022), and, when extended to the heritage context, could be understood as the co-creation of localized heritage narratives. This then would not only counteract the issues around inclusion but also around the minimization or erasure of local heritage narratives and the subsequent impact on local identity. However, regenerative tourism goes beyond this to allow for a system wherein local values can be used as the basis and focal point for heritage tourism development that resonates with the local population's own value system (Becken & Kaur, 2022). Thus, the future of cultural heritage tourism *must* be regenerative tourism in order to uphold the principles of conservation, education, and community inclusion.

Conclusions

This chapter has focused on a discussion of the integration of sustainable development, cultural heritage, and tourism and the associated problems that can occur. This issue is of particular importance given the increasing usage of heritage tourism as a proposed panacea to a variety of socio-economic issues, particularly in the less developed world. However, as has been observed, the implementation of sustainable principles is not without its issues. The inclusion of community members within heritage tourism planning and development has been shown to be essential while also exceedingly difficult to achieve in a way that is respectful and empowering. Furthermore, ensuring that communities remained connected to their heritage and that their identity narratives are included within broader heritage ones also appears to be quite difficult to achieve. Given the identified risks and the need for both community inclusion and a strong local identity in order to combat them, it was then proposed that it may be necessary to shift away from the contemporary form of sustainable development, which emphasises a pro-growth mindset. Instead, there needs to be an emphasis on local places and people and a re-centring of these local elements within the context of cultural heritage tourism while shifting towards a model of sufficiency, namely regenerative cultural heritage tourism underpinned by a degrowth economic model. Without this paradigm shift, the current issues faced by sustainable cultural heritage tourism initiatives and the risks associated with this type of tourism will continue to worsen.

References

Adie, B. A. (2020). Touring "our" past: World Heritage tourism and post-colonialism in Morocco. In S. Seyfi & C. M. Hall (eds). *Cultural and Heritage Tourism in the Middle East and North Africa* [pp. 102–114]. Routledge.

Adie, B. A., Amore, A., & Hall, C. M. (2022). Just because it seems impossible, doesn't mean we shouldn't at least try: The need for longitudinal perspectives on tourism partnerships and the SDGs. *Journal of Sustainable Tourism*, 30(10), 2282–2297. https://doi.org/10.1080/09669582.2020.1860071

Adie, B. A., & Falk, M. (2021). Residents' perception of cultural heritage in terms of job creation and overtourism in Europe. *Tourism Economics*, 27(6), 1185–1201. https://doi.org/10.1177/1354816620943688

Ahebwa, W. M., Aporu, J. P., & Nyakaana, J. B. (2016). Bridging community livelihoods and cultural conservation through tourism: Case study of Kabaka heritage trail in Uganda. *Tourism and Hospitality Research*, 16(2), 103–115. https://doi.org/10.1177/1467358415589659

Andereck, K. L., Valentin, K. M., Knopf, R. C., & Vogt, C. A. (2005) Residents' perceptions of community tourism impacts. *Annals of Tourism Research*, 32(4) 1056–1076. https://doi.org/10.1016/j.annals.2005.03.001

Becken, S., & Kaur, J. (2022). Anchoring "tourism value" within a regenerative tourism paradigm – a government perspective. *Journal of Sustainable Tourism*, 30(1), 52–68. https://doi.org/10.1080/09669582.2021.1990305

Bellato, L., Frantzeskaki, N., & Nygaard, C. A. (2022). Regenerative tourism: a conceptual framework leveraging theory and practice. *Tourism Geographies*. https://doi.org/10.1080/14616688.2022.2044376

Causevic, S., & Neal, M. (2019). The exotic veil: Managing tourist perceptions of national history and statehood in Oman. *Tourism Management*, *71*, 504–517. https://doi.org/10.1016/j.tourman.2018.10.001

Caust, J., & Vecco, M. (2017). Is UNESCO World Heritage recognition a blessing or burden? Evidence from developing Asian countries. *Journal of Cultural Heritage*, 27, 1–9. https://doi.org/10.1016/j.culher.2017.02.004

Chen, C.-F., & Chen, P.-C. (2010). Resident attitudes toward heritage tourism development. *Tourism Geographies*, 12(4), 525–545. https://doi.org/10.1080/14616688.2010.516398

Day, J. C., Heron, S. F., Markham, A., Downes, J., Gibson, J., Hyslop, E., Jones, R. H., & Lyall, A. (2019). *Climate Risk Assessment for Heart of Neolithic Orkney World Heritage property: An application of the Climate Vulnerability Index*. Historic Environment Scotland.

Dippon, P., & Moskaliuk, J. (2020). Sharing intangible cultural heritage: disparities of distribution. *Journal of Heritage Tourism*, 15(4), 450–471. https://doi.org/10.1080/1743873X.2019.1682003

Du Cros, H., & McKercher, B. (2014). *Cultural Tourism* (2nd ed.). Routledge.

Eckert, C., Zacher, D., Pechlaner, H., Namberger, P., & Schmude, J. (2019). Strategies and measures directed towards overtourism: a perspective of European DMOs. *International Journal of Tourism Cities*, 5(4), 639–655. https://doi.org/10.1108/IJTC-12-2018-0102

Ely, P. A. (2013). Selling Mexico: Marketing and tourism values. *Tourism Management perspectives*, 8, 80–89. https://doi.org/10.1016/j.tmp.2013.07.003

Evans, G. (2005). Mundo Maya: From Cancun to City of Culture. World Heritage in Postcolonial Meso-America. In Harrison, D., & Hitchcock, M. (Eds.), *The politics of world heritage: Negotiating tourism and conservation* (pp. 35–49). Channel View Publications.

Feighery, W. G. (2012). Tourism and self-Orientalism in Oman: a critical discourse analysis. *Critical Discourse Studies*, 9(3), 269–284. https://doi.org/10.1080/17405904.2012.688210

Hall, C. M. (2009). Degrowing tourism: Décroissance, sustainable consumption and steady-state tourism. *Anatolia*, *20*(1), 46–61. https://doi.org/10.1080/13032917.2009.10518894

Hall, C. M. (2019). Constructing sustainable tourism development: The 2030 agenda and the managerial ecology of sustainable tourism. *Journal of Sustainable Tourism*, 27(7), 1044–1060. https://doi.org/10.1080/09669582.2018.1560456

Hall, C. M., Baird, T., James, M., & Ram, Y. (2016). Climate change and cultural heritage: Conservation and heritage tourism in the Anthropocene. *Journal of Heritage Tourism*, 11(1), 10–24. https://doi.org/10.1080/1743873X.2015.1082573

Healey, P. (2006). *Collaborative planning: Shaping places in fragmented societies*. (2nd ed.). Palgrave Macmillan.

Hejazi, M. (2008). The risks to cultural heritage in Western and Central Asia. *Journal of Asian Architecture and Building Engineering*, 7 (2), 239–245. https://doi.org/10.3130/jaabe.7.239.

Higgins-Desbiolles, F., Carnicelli, S., Krolikowski, C., Wijesignhe, G., & Boluk, K. (2019). Degrowing tourism: Rethinking tourism. *Journal of Sustainable Tourism*, 27(12), 1926–1944. https://doi.org/10.1080/09669582.2019.1601732

Human, H. (2015). Democratising world heritage: The policies and practices of community involvement in Turkey. *Journal of Social Archaeology*, 15(2), 160–183. https://doi.org/10.1177/1469605314566557

Janusz, K., Six, S., & Vanneste, D. (2017). Building tourism-resilient communities by incorporating residents' perceptions? A photo-elicitation study of tourism development in Bruges. *Journal of Tourism Futures*, 3(2), 127–143. https://doi.org/10.1108/JTF-04-2017-0011

Kerrigan, F., Shivanandan, J., & Hede, A.-M. (2012). Nation branding: A critical appraisal of Incredible India. *Journal of Macromarketing*, 32(3), 319–327. https://doi.org/10.1177/0276146712445788

Kim, S., Whitford, M., & Arcodia, C. (2019). Development of intangible cultural heritage as a sustainable tourism resource: the tangible cultural heritage practitioners' perspectives. *Journal of Heritage Tourism*, 14(5–6), 422–435. https://doi.org/10.1080/1743873X.2018.1561703

King, V. T. (2016). Melaka as a World Heritage site: Melaka as Malaysia? In V. T. King (ed.), *UNESCO in Southeast Asia: World Heritage Sites in Comparative Perspective* [pp. 140–168]. NIAS.

Labadi, S., Giliberto, F., Rosetti, I., Shetabi, L., & Yildirim, E. (2021). *Heritage and the Sustainable Development Goals: Policy Guidance for Heritage and Development Actors*. ICOMOS.

Lähdesmäki, T (2016). Politics of tangibility, intangibility, and place in the making of a European cultural heritage in EU heritage policy. *International Journal of Heritage Studies*, 22(10), 766–780. https://doi.org/10.1080/13527258.2016.1212386

Litvin, S. W., Smith, W. W., & McEwen, W. R. (2020). Not in my backyard: Personal politics and resident attitudes toward tourism. *Journal of Travel Research*, 59(4), 674–685. https://doi.org/10.1177/0047287519853039

Lwoga, N. B. (2017). Factors influencing local residents' intentions to conserve the built heritage in Tanzania. *Journal of Heritage Tourism*, 12(4), 394–409. https://doi.org/10.1080/1743873X.2016.1213846

Mbaiwa, J. E. (2016). The commodification of World Heritage Sites: The case study of Tsodilo Hills in Botswana. In H. Manwa, N. Moswete, & J. Saarinen (eds.). *Cultural Tourism in Southern Africa* [pp. 101–120]. Routledge.

Neuts, B., & Nijkamp, P. (2012). Tourist crowding perception and acceptability in cities: An applied modelling study on Bruges. *Annals of Tourism Research*, 39(4), 2133–2153. https://doi.org/10.1016/j.annals.2012.07.016

Rasoolimanesh, S. M., Taheri, B., Gannon, M., Vafaei-Zadeh, A., & Hanifah, H. (2019). Does living in the vicinity of heritage tourism sites influence residents' perceptions and attitudes? *Journal of Sustainable Tourism*, 27(9), 1295–1317. https://doi.org/10.1080/09669582.2019.1618863

Richards, G. (2007). *Cultural Tourism: Global and Local Perspectives*. The Haworth Hospitality Press.

Rodwell, D. (2002). The World Heritage Convention and the exemplary management of complex sites. *Journal of Architectural Conservation*, 8(3), pp. 40–60. https://doi.org/10.1080/13556207.2002.10785326

Saarinen, J. (2021a). Sustainable growth in tourism? Rethinking and resetting sustainable tourism for development. In C. M. Hall, L. Lundmark, & J. J. Zhang (eds.). *Degrowth and Tourism: New Perspectives on Tourism Entrepreneurship, Destinations and Policy* [pp. 135–151]. Routledge.

Saarinen, J. (2021b). Is being responsible sustainable in tourism? Connections and critical differences. *Sustainability*, 13(12), 6599. https://doi.org/10.3390/su13126599

Saarinen, J., & Rogerson, C. M. (2014). Tourism and the millennium development goals: Perspectives beyond 2015. *Tourism Geographies*, 16(1), 23–30. https://doi.org/10.1080/14616688.2013.851269

Schéou, B., & Southon, A. (2013). Ambiguities and difficulties in partnership relations: The case of 'fair tourism' in Western Africa. *Current Issues in Tourism*, 16(7–8), 753–772. https://doi.org/10.1080/13683500.2013.785486

Silberberg, T. (1995). Cultural tourism and business opportunities for museums and heritage sites. *Tourism Management*, 16(5), 361–365. https://doi.org/10.1016/0261-5177(95)00039-Q

Sirima, A., & Mngulwi, M. (2020). Community readiness to green tourism initiatives in less developed parks: Lessons from communities around Saadani National Park. In Walia, S. K. (ed.). *The Routledge Handbook of Community-Based Tourism Management: Concepts, Issues and Implications* [pp. 264–272]. Routledge.

Sinamai, A. (2019). *Memory and Cultural Landscape at the Khami World Heritage Site*, Zimbabwe: *An Uninherited Past*. Routledge.

Spenceley, A., & Nelson, F. (2013). Heritage tourism in Africa. In R. Staiff, R. Bushell, & S. Watson (eds.). *Heritage and Tourism: Place, Encounter, Engagement* [pp. 251–273]. Routledge.

Thompson, J. (2007). Engagement in public-private partnership in cultural heritage: The case of Herculaneum, Italy. In proceedings of *ICCROM: International Forum on Privatisation and Cultural Heritage*, Catania, Italy, 13 - 15 September.

Timothy, D. J. (2021). *Cultural Heritage and Tourism: An Introduction* (2nd ed.). Channel View Publications.

Timothy, D. J., & Boyd, S. W. (2006). Heritage Tourism in the 21st Century: Valued Traditions and New Perspectives. *Journal of Heritage Tourism*, 1 (1), pp. 1–16. https://doi.org/10.1080/17438730608668462

Timothy, D. J., & Nyaupane, G. P. (eds.), (2009). *Cultural Heritage and Tourism in the Developing World: A Regional Perspective*. Routledge.

Truong, V. D., Slabbert, E., & Nguyen, V. M. (2016). Poverty in tourist paradise? A review of pro-poor tourism in South and South-East Asia. In C. M. Hall & S. J. Page (eds.), *The Routledge Handbook of Tourism in Asia* [pp. 101–117]. Routledge.

UNESCO (n.d.). *UNESCO World Heritage and Sustainable Tourism Programme*. UNESCO. https://whc.unesco.org/uploads/activities/documents/activity-669-7.pdf

UNESCO (2015). *Policy Document for the Integration of a Sustainable Development Perspective into the Processes of the World Heritage Convention*. UNESCO. https://whc.unesco.org/en/sustainabledevelopment/

UNWTO (2017). Lusaka Declaration on Promoting Sustainable Tourism, a Tool for Inclusive Growth and Community Engagement in Africa. *UNWTO Declarations*, 26(3). https://doi.org/10.18111/unwtodeclarations.2017.26.03.

UNWTO (2018). *Tourism and Cultural Synergies*. UNWTO. https://doi.org/10.18111/9789284418978

Vella, C., Bocancea, E., Urban, T. M., Knodell, A. R., Tuttle, C. A., & Alcock, S. E. (2015). Looting and vandalism around a World Heritage Site: Documenting modern damage to archaeological heritage in Petra's hinterland. *Journal of Field Archaeology*, 40 (2), 221–235. https://doi.org/10.1179/0093469015Z.000000000119

Vong, L. T.-N. (2015). The mediating role of place identity in the relationship between residents' perceptions of heritage tourism and place attachment: The Macau youth experience. *Journal of Heritage Tourism*, 10(4), 344–356. https://doi.org/10.1080/1743873X.2015.1026908

Winter, T. (2005). Landscape, memory and heritage: New Year celebrations at Angkor, Cambodia. In D. Harrison & M. Hitchcock (eds.), *The Politics of World Heritage: Negotiating Tourism and Conservation* [pp. 50–65]. Channel View.

Yang, C.-H., & Lin, H.-Y. (2014). Revisiting the relationship between World Heritage sites and tourism. *Tourism Economics*, 20(1), 73–86. https://doi.org/10.5367/te.2013.03599

Zapata, M. J., Hall, C. M., Lindo, P., & Vanderschaeghen, M. (2011). Can community-based tourism contribute to development and poverty alleviation? Current Issues in Tourism, 14(8), 725–749. https://doi.org/10.1080/13683500.2011.559200

Chapter 29

Urban Tourism

Jan-Henrik Nilsson

Introduction

Cities have a large population size, a high density of people and activities and they are heterogeneous, which stimulates a large variety of experiences (Taylor, 2021). The size and multi-functionality of cities mean that they house a variety of high order central functions, urban functions whose influence reach out far beyond the city itself and create connections to other cities and places. The centrality of a place can be defined as 'the significance of a place in relation to its surrounding area' (Christaller, 1933, p. 27, author's translation), i.e., by the interaction created by its central functions. Interaction can be understood as interconnected flows of goods, capital, people and messages, which constitutes a city's particular space of flows (see Castells, 1996).

The central functions of major cities are important attractors of travel, since they tend to generate a large variety of national and international connections. A large part of that travel consists of meetings in private enterprise, the public sector and in political life where large concentrations of control functions are located in major cities. There are plenty of data pointing at the co-dependence of corporate and public central functions and international pattern of interaction. Major cities are connected to international traffic networks by air and land. Public transport, going through airports and railway stations, are necessary means for sustaining the central functions, including the tourism and meetings sectors (Taylor, 2004). As centres of national political life, capital cities tend to be particularly favoured, combining important domestic and international connections with concentrations of media, culture and communications.

The specific characteristics of cities are not only a matter of scale; cities represents a way of life, urbanity. Tourism is a constitutive part of urbanity in the contemporary world: 'Virtually no place is left without the presence of tourism imaginations, practices, tourists' (Stock, 2019, p. 46). The concentration of urban functions is shared between locals and visitors; most of them are not primarily intended for tourists. Both locals, business tourists and

The Wiley Blackwell Companion to Tourism, Second Edition. Edited by C. Michael Hall.

leisure tourists use cafés, restaurants, museums and other cultural amenities. These combinations of demand makes cities develop 'a high density of diverse cultural offerings in a relatively small area' (Bock, 2015, p. 20). The role of central functions is crucial for understanding the dynamics behind the rapid development of urban tourism.

Urban tourism has gone through a period of rapid growth during the latest decades (Bock, 2015; Zekan & Wöber, 2022). In Europe, urban tourism has grown faster than other forms of tourism, and the largest cities show particularly high growth (Eurostat, 2019). Heeley (2011) estimated that large cities account for around 40 percent of all international incoming tourists in Europe. Some selected data on accommodation may illustrate the development of urban tourism. Between 2002 and 2018, the number of hotels in Stockholm increased from 247 to 328, bed numbers increased by 77 percent. In Copenhagen, the number of guest nights grew from 3.5 million in 1992 to 8.5 million in 2017 (Nilsson, 2018a). Since these data only show registered guest nights the growth of urban tourism is likely to be underestimated, considering recent growth of Airbnbs and other unofficial accommodation. This means that tourism today is an important contributor to the local economy of most cities, as well as an important factor for understanding contemporary economic, social, cultural and environmental transformation in urban contexts.

Despite the rapid development of urban tourism, it was for a long time seen as a neglected field in tourism studies, which tended to focus on resorts and rural destinations (Aal & Koens, 2019; Maxim, 2016). Cities were generally viewed as the generating, not receiving, parts of the tourist system. This lack of interest was matched by a similar disinterest of tourism in urban studies (Ashworth & Page, 2011). One reason for this was the difficulty to define urban tourism, as urban attractions, transport, restaurants, and other facilities are used both by tourists and non-tourists. This has changed, there has been a considerable increase in research on urban tourism in recent years (see Pasquinelli, 2015; and van der Borg, 2022). This likely reflects both the growth of urban tourism, its economic significance, and its consequences for local environments and livelihoods. At the same time, tourism has become an important topic in urban studies and urban sociology, due to tourism's increasingly forceful impact on urban transformation. Contemporary issues such as urban gentrification and urban sustainability are closely connected to tourism growth, to 'overtourism' (Nilsson, 2020).

This introduction is followed by an account of the main supply side categories of urban tourism, based on a historical perspective on the development of tourist cities. This perspective is developed further in the second half of the chapter that focuses on contemporary developments in urban tourism, often termed *new urban tourism*. The main emphases are on the driving forces of new urban tourism, ontological changes of who the tourists are, cultural changes in urban tourism experiences, and urban spatial change as a result of tourism development. The chapter concludes by discussing some contemporary challenges for urban tourism.

Urban Heritage Tourism

The beginnings of modern tourism are generally associated with the Grand Tours. They commenced in the 17th century, but had their peak in the period 1760–1790, prior to the French Revolution. The travellers were known as 'tourists' and came from the upper classes of Northern Europe, mainly Great Britain. Grand Tours were in essence educational tours, where young men acquired knowledge of languages, arts and foreign cultures. The most popular destinations were Paris, Rome, Florence, and Naples (Zuelow, 2016). Paris was the cultural and political centre of 18th century Europe; the Italian cities hold important cultural heritage from the antiquity and the Italian renaissance in arts and architecture. Early tourism was thus mainly targeting urban destinations. It was in this context that people learned to become tourists; that the 'tourist gaze' developed (Urry & Larsen, 2011). This created a set of itineraries that tourists tended to follow, based on the experiences of previous travellers. In the 19th century, guidebooks played an important role in teaching tourists about culturally significant attractions, sometimes very ambitiously. For example, Baedeker (1878) spent 56 pages solely on the collections at the Louvre, Paris. This kind of cultural tourism still attracts millions of tourists every year to heritage cities all over the World, from Lübeck to Xian.

A large variety of central functions and opportunities for meetings, generating large flows of people, combined with concentrations of cultural amenities and historic heritage give many capital cities a unique position for attracting visitors. Many well-known tourist attractions hold political functions, for example the Forbidden City with the Tiananmen Square in Beijing and central Washington DC with the White House and Capitol Hill. These centres of power generally include famous monuments such as statues of prominent historical figures or monuments representing significant historical events. Close to the Houses of Parliament in London, you can find statues of prominent men like Winston Churchill and Mahatma Gandhi together with monuments commemorating Great Britain's many war efforts. Today, such monuments are often regarded as spatial representations of a contested past. In the late 2010s, there were debates in both USA and Great Britain concerning monuments of people involved in slave trade or representing slavery in the US South. This resulted in the removal of a number of statues, for example of Confederate generals. In similar ways, statues of Lenin and other communist icons were removed from central cities when East Central Europe became democratic around 1990. However, memories of dark pasts may still be interesting for visitors (Nilsson, 2007). For example, a communist monument park in the outskirts of Budapest has become a major tourist attraction.

Monuments are not the only attractions that have their roots in capital cities' roles as political centres. Obviously, capitals had a practical function as the place where political power was exercised: at court, in parliaments and in military establishments. Rulers also developed their symbolic power by building palaces and other impressive buildings to become 'architecture as theatre, design intended to impress' (Hall, 1988, p. 202). The area around Unter den Linden and Schlossplatz in central Berlin is a fine example of this. In the 17th to 19th centuries, Prussian kings built palaces, churches, art museums, monuments, and other prestigious buildings including the main building of the Humboldt University in the area. This marked the cultural centre of Prussia and later Germany. Interestingly this process has continued after Germany's re-unification in 1990. In 2020, the former Royal Palace was finally rebuilt, after war damage and demolition by the German Democratic Republic, to house the Humboldt Forum, a centre for culture and science (Figure 29.1). This exemplifies that investment in high culture was, and still is, a way of showing prestige. Ensembles like the heritage buildings in central Berlin attract visitors, strengthening the role of Berlin as a destination. There is thus a strong continuity in urban heritage tourism; architecture, palaces, museums, and other forms of cultural heritage have attracted visitors since the beginning of modern tourism. Urban heritage tourism tend to concentrate around iconic sights, contrary to the new forms of urban tourism discussed below.

Figure 29.1. Humboldt forum in Berlin (Photo: Jan-Henrik Nilsson).

Restructuring Urban Tourism

Globalisation and the structural crisis in the 1970s and 1980s had a huge impact on many cities. There was a considerable loss of workplaces in manufacturing and services directly related to manufacturing, in some cities populations declined. However, there were large differences between cities; manufacturing and port cities faced the worst problems whereas university cities and capitals were less affected by the crisis (Cheshire & Hay, 1989). There were thus a need for city governments to take action to turn the development around. Cities became more entrepreneurial and market oriented, which changed the role of local governments. This went hand in hand with an ideological change towards a neo-liberal view on governance. Cities increasingly encouraged investments in urban renewal, transport infrastructure and other projects, often with the aim to promote culture, leisure and entertainment. There was also a growing sense of competition between cities, not the least for attracting business meetings, conventions and major events (Spirou, 2011). As a consequence, place marketing has become vital for cities' efforts to increase incoming investment and visitor numbers, and to increase media visibility. In her study of post-1989 Berlin, Colomb (2012) shows an interesting example of how powerful such measures might be.

Major redevelopments in the physical structure of cities were a common part of urban strategies. As a result of de-industrialisation and containerisation there were often large areas of derelict land in cities, open for brownfield development. With harbours and shipyards leaving inner cities, waterfront developments were a common solution for upgrading urban environments. In some cases, such as in the London Docklands, the main purpose was to create space for expanding service industries and the financial sector (Hall, 1998). The effect on tourism was indirect, mainly promoting business tourism and meetings. In other cases, tourism was part of the projects from the start. Baltimore's harbour area is often mentioned as an early example of integration between urban development and leisure tourism (Spirou, 2011). Later examples of development districts include Hamburg's Speicherstadt with the new opera house, and the development around Albert Docks in Liverpool. In the Basque city of Bilbao, the cutting-edge architecture of the Guggenheim Museum has attracted large numbers of visitors to the former industrial city. Such developments are meant to act as flagship projects, and hubs of further growth. The impact on the hospitality industry is particularly important, as it is a major source of employment.

Increasing global accessibility by air has increased potential market areas for urban tourism. Thereby, some cities have developed growth strategies emphasising tourism and hospitality. Las Vegas was probably the first mega tourist city, taking advantage of Nevada's liberal legislation on gambling. Today, the city has also become a prominent destination for conventions and meetings (Spirou, 2011). The Emirate city of Dubai has grown into a major metropolis in the last couple of decades, focusing on tourism and hospitality. It houses the World's tallest building, numerous luxury hotels, artificial islands, beaches, gardens and resorts together with numerous facilities for shopping, culture and entertainment. Building on sizeable domestic financial resources, this strategy is a means to reduce the nation's dependency on the oil industry (Stephenson, 2014). To house major events has been another important urban growth strategy. Olympic Games and other global sports events require large investment, and potentially also attract large numbers of visitors. However, their long-term impact is often questioned. Athens, Greece, hosted the 2004 Olympics but had problems taking advantage of the investments (Ziakas & Boukas, 2012). Sports are not only about big global events. For instance, the European Champions League is played for a large part of the season, during which hundreds of thousands of followers travel to watch the matches. Interestingly from a tourist perspective, the successful European teams are increasingly concentrated in major cities in Western Europe. Cities like Paris, London, Madrid, Barcelona, Munich, Manchester, and Milan are homes for the largest clubs. It has become very difficult for clubs in minor cities to raise funding large enough to compete. It is hard to believe that the 1979 final between Nottingham Forest and Malmö FF will ever be repeated.

Malmö (population 350,000) in the south of Sweden is, however, an illustrative example of the kind of development discussed in this section. It went from being a successful industrial city in the 1970s to a city in deep crisis by the end of the 1980s. It had lost almost all its manufacturing industry due to international competition; unemployment had grown to above 20 percent. The need for new strategies was imminent. The tide slowly turned when the bridge to Copenhagen in Denmark was finished in the year 2000. In the same period, a major urban development project started in the former shipbuilding area, a sustainable flagship project housing Scandinavia's tallest building and a new university. The city also started to attract events and to invest in cultural amenities (Sydsvenskan, 2022). For example, the latest major development, Malmö Live (Figure 29.2) includes a concert hall for the city's symphony orchestra, a large hotel and a conference centre. It is located almost above the railway station; the travel time to Copenhagen airport is only 21 minutes. Thereby, it is very time efficient for business travellers. The vicinity to Copenhagen, and its international airport, has made it possible for the city to take advantage of peoples' ability to take shorter vacations, e.g., during weekends (cf. Gössling et al., 2018). The spirit of the city has become increasingly positive during the latest decades. It has not become a major tourist destination, but it has gone from attracting hardly any tourists to becoming a secondary tourist city.

Figure 29.2. Malmö Live (Photo: Jan-Henrik Nilsson).

Contemporary Dynamics of Tourist Mobility

In most parts of the world, conditions for travel and trade were liberalised in the 1980s and 1990s leading to an increasingly globalised economy. This was accompanied by a shift in economic policy advocating market-based solutions involving de-regulation of sectors like welfare, media, transport, and infrastructure (Dicken, 2015). One important effect of this was the de-regulation of international aviation, from the 1990's. It created opportunities for new business models to develop, most importantly low-cost aviation. The effects have been much lower prices on fares, a much denser network of connections than earlier, involving a large number of new destinations, and rapidly growing numbers of air passengers. The impact on destinations that previously had few international arrivals, such as in East-Central Europe and parts of the Mediterranean, has been substantial (Dobruszkes, 2013; Nilsson, 2018b). Both capital cities and larger provincial cities have seen large increases in visitor arrivals, for example Prague, Riga, Ljubljana, and Krakow.

Beginning when the Internet became public in 1993, digitalisation reached new levels of sophistication with substantial impact on tourism. In the beginning on-line travel agents developed, which allowed the public to book air tickets and accommodation through websites. People were able to compare travel opportunities and prices from home. Following the next generation of Internet technology, web 2.0, innovative business models emerged, building on the possibility for people to interact directly with one another through digital platforms (Dredge & Gyimóthy, 2017). The introduction of smart phones, from 2007, made digital travel information portable. If connected to the Internet, the tourist can reach immense quantities of information and most booking opportunities, increasing tourists' opportunities for immediate information searches and the possibility to change plans at short notice (Mieli, 2022).

Several categories of platforms are relevant for analysing the growth of urban tourism (cf. Gössling, 2017): On-line travel agents and booking sites (e.g., Expedia); Peer-to-peer marketplaces for the collaborative economy (e.g., Airbnb); information sharing platforms (e.g., Tripadvisor); and social media platforms (e.g., Facebook and Instagram). Peer-to-peer marketplaces, such as Airbnb, brought radical change to the tourist accommodation business by increasing the reach of potential hosts and guests (Guttentag, 2015). As business models, platforms like Airbnb have a series of advantages in mobilising assets, lower transaction costs through standardised mediation processes, enhanced methods to create trust and quality control using feedback mechanisms (Gyimóthy, 2017). Airbnb had a huge impact on urban tourism by radically transforming the accommodation sector. It is an important driver of overtourism (Nilsson, 2020).

The technological advances have had far-reaching consequences for the behaviour of tourists. Most of them are no longer dependent on travel agents or tour operators for planning and managing their travel. The almost ubiquitous availability of information and booking possibilities for individual tourists has made it possible for them to design their own itineraries, allowing for more flexible and individual forms of travel (Vainikka, 2013). The number of independent travellers has increased, particularly to cities with large supplies of travel opportunities and accommodation. Technological innovation has driven aggregate demand, particularly for urban tourism. It has also profoundly changed the organisation of the tourist industry, and the functions of tourist marketing channels. Figure 29.3 aims to show relevant drivers of growth in contemporary tourism, following a tentative timeline from the left to the right.

In the context of urban tourism, the degree of structural change resulting from the developments discussed so far in this chapter has been so important that many tourism scholars speak in terms of *new urban tourism* (Ba et al., 2022; Frisch et al., 2019). It goes 'beyond the narrow notion of a tourist city as an enclavic tourist space disconnected from the ordinary life of the city [and] requires a shift from 'staged authenticity' to the promotion of 'the ordinary as authentic" (Pasquinelli, 2015, p. 15). In relation to the tourist experience, we can see a decreasing interest in the forms of staged authenticity that are typically present in sites tailor made for tourists, as in heritage tourist cities and tourist business districts. The new urban tourists

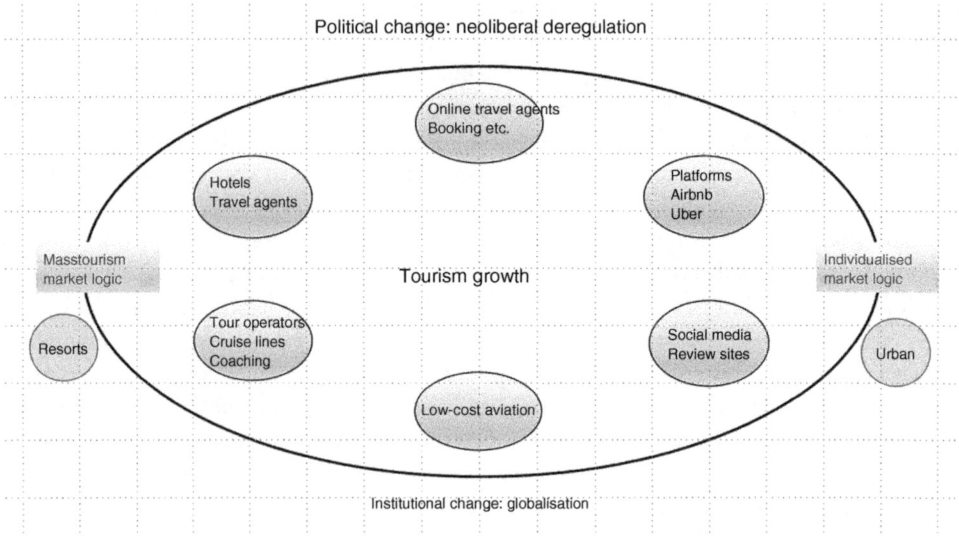

Figure 29.3. Structural factors leading to tourist growth.

tend to value authentic, less scripted, experiences in close relation to the city they visit (Füller & Michel, 2014). Still, the search for authenticity seems to be a vital part of tourism; but now it is directed in new directions, visitors tend to find new districts to visit 'off the beaten track' (Maitland & Newman, 2009b) where visitors perform tourism in new and untraditional ways. The notion of new urban tourism is highly relevant for how contemporary urban tourism is discussed. Based on the literature, it is possible to find three characteristics of new urban tourism that most researchers focus on:

- changes in the characterisation of urban tourists;
- changes in the cultural expressions of urban tourism; and
- urban spatial change.

These perspectives form the points of departure for the following part of this chapter.

The Changing Face of Urban Tourists

In the 20th century, urban tourism was generally viewed as a set of practices that were different from the lives of the resident population, taking place in 'tourist enclaves', designated for the needs of visitors (Judd, 2003). This tendency to differentiate between tourism and the everyday life of residents has been criticised in recent tourism literature (Larsen, 2008; Larsen, 2019). In a world where more and more people are mobile, sharing their lives between different places, tourism has become an ordinary practice rather than something extraordinary (Stock, 2019). This means that being a tourist is an integrated part of many peoples' lives. Tourists perform ordinary, often work related, tasks when they are away and they take part in the everyday life of the places where they stay. The borders between different kinds of visitors, between visitors and locals, and between different ways of 'doing tourism' is increasingly blurred. According to Stock (2019), we think of tourists as temporary inhabitants, and 'the city as a place inhabited by mobile and temporary inhabitants' (Stock, 2019, p. 44). Despite these tendencies, it is still meaningful to describe contemporary patterns of urban (tourism-) mobility and place consumption by discussing different categories of urban visitors depending on their length of stay and their relation to the city where they stay (see Hall, 2005; Novy, 2018). This clarifies that locals, temporary migrants and tourists use similar place-based amenities, thereby together shaping the tourist city.

When new urban tourism is discussed, we need to remember that there is a high degree of inertia in peoples' choice of travel. Not all urban tourists are 'new'; traditional categories are still important. Business travellers play an important part in urban tourism and for the development of tourist infrastructure. They constitute a substantial part of the demand for hotel accommodation and other up-market services, particularly during weekdays. There is also a large group of 'ordinary' leisure tourists taking part in packaged tours. They tend to concentrate in cities during weekends, when hotel prices are lower than during the working week. This group tend to go unnoticed, possibly because they are taken for granted. A particularly 'invisible' group of tourists are the ones coming in packaged tours by coach. Considering the practical reach of coach travel, they mostly come from neighbouring regions and countries.

Individual mass tourists are a much more visible category. The 'EasyJet-set' tend to use low-cost aviation and budget accommodation such as hostels or Airbnbs. These low-cost alternatives generate high demand for travel to cities that are well connected by air. When concentrations of tourists become too high, the daily lives of residents are negatively inflicted by difficulties moving around their neighbourhoods, higher price levels on food and drinks, and difficulties finding service and retail facilities serving local needs (e.g. Dumbrovská, 2017). There are also problems connected to inappropriate behaviour by tourists that result in disturbance and a sense of insecurity (Koens et al., 2018). It seems that some cities in East-central Europe, such as Krakow and Budapest, have particularly severe problems due to their popularity as nightlife destinations (Smith et al., 2019; Thurnell-Read, 2012). Some of the debate on over-tourism is a result of such problems.

Individual mobility within Europe has profited from the European citizens' right to work and travel freely within the European Union. This affects all sections of society, cosmopolitan elites and artists as well as manual workers in construction and services. The impact of mobile workers has been particularly important for the hospitality industry (Janta et al., 2011). Higher education is another area where the rate of internationalisation has increased significantly during the latest decades. Science driven mobility is simplified by international agreement on exchange and curricula, the Bologna process. Growing numbers of international students and researchers has a particularly large impact in small university cities, like Lund in Sweden. (Eskilsson & Nilsson, 2022). Globalisation of private enterprise has also increased the number of employees temporarily based outside their country of residence (Beaverstock et al., 2010).

To make things even more vague, the difference between work-oriented mobility and lifestyle migration is increasingly blurred. To take an example, an acquaintance of this author, living in Sweden, is very interested in dancing tango. He therefore moves to Buenos Aires in the winter months, to avoid the Swedish winter and to enjoy the Argentinian tango culture. Digitalisation makes it possible for him to retain his business as a consultant while away from 'home'. As a temporary inhabitant of Buenos Aires, he is a fine representative of what new urban tourism can include. The remote working trend makes many people less dependent on one specific place of residence; and possibly encourages people to acquire multiple (second) homes (Domínguez-Mujica et al., 2021). The growing importance of these forms of international mobility means that temporary migration has become recognised as an important factor for urban tourism and for urban place consumption (Novy, 2018; Stock, 2019). There is also evidence that temporary migration induces secondary mobility, in similar ways that local people are visited by friends and relatives. The difference is likely that highly mobile people retain a higher number of international connections and thereby induce more secondary mobility (Frändberg, 2014).

The blurring boundaries between different city users means that the role of permanent locals for tourism becomes more interesting. In many cities, permanent locals are an important group of visitors to tourist attractions and guests in restaurants and nightlife. Thereby, they act as 'tourists in their own city' (Richards, 2016; Richards, 2017). Their place consumption may in many respects function as a form of avant-garde. High quality demands by local residents may improve local competition and performance of tourist businesses and the cultural sector. The transformation of the food scene in Copenhagen is a fine example of how local innovation and improved standards has resulted in the development of successful niche tourism. This is in line with earlier research on competition in the manufacturing industry (Porter, 1990) which shows the importance of advanced domestic demand for the success in exports.

Cultural Change

Some aspects of urban tourism are relatively stable. Tourists travel, stay at hotels and other accommodation, eat at restaurants and visit attractions. However, the ways tourism is performed have changed. Some analysts have talked about post-tourism as a way of understanding the development away from organised mass tourism. '[T]he post-tourist represents the middle-class ambition to escape what tourism mobilities are generally associated with: the mass movement of people following standardized taste patterns' (Janson, 2018, p. 102). The interests of tourists seem to have become increasingly individualist and diverse. This tendency is a result of both the blurring of tourist categories discussed in the previous section and of the recognition that it is no longer reasonable to differentiate between tourists and non-tourists, and between tourist and non-tourist activities.

Larsen (2008, 2019) has frequently argued that tourism today is an integrated part of many peoples' everyday lives. In today's highly mobile society, tourism is something many people do often, for various reasons. There is thus a necessary connection between everyday life and tourist activities. For example, many people travel together with friends and family to enjoy 'quality time' together away from home. As social networks tend to be increasingly spread out, visiting friends and relatives in other places, and receiving them in one's own home, becomes a means to sustain social ties. In different ways, tourism is also a way to extend everyday practices into other places. It could be for leisure, as in taking part in cultural events, or for work related purposes, for instance when academics attend conferences (Larsen, 2008; Larsen, 2019). What does this perspective mean for how urban tourism is performed? Pasquinelli (2015) suggests that a new profile of urban tourist is emerging, 'that of the *urban traveller*, which is to be understood as flanking (and not replacing) the traditionally studied *urban tourist*' (Pasquinelli, 2015, p. 11). The urban traveller represents a more engaged relation to the tourist experience, to the host city and to the resident population. In these relations, new ways of performing urban tourism develop.

Frisch et al. (2019) propose three dimensions that characterise how new urban tourism is performed: 'the extra-ordinary mundane, encounters and contact zones, and urban co-production' (Frisch et al., 2019, p. 3). In searching for authentic experiences, the everyday, mundane, lives of others become interesting. A slightly different form of tourist gaze evolves, resembling that of the classical flaneur, focusing on the life in the streets rather than on iconic attractions. There is also a geographical aspect of this, since being interested in local peoples' whereabouts means that you need to move outside tourist areas and into residential parts of the city. This is where encounters with locals are possible. Places like local shops, bars, restaurants, markets, parks and squares, together with the street itself, become touristic objects. There is also a growing interest in taking part in tourism in other peoples' homes, for instance in couch surfing (see Germann Molz, 2013). In the efforts to be temporarily local, urban travellers tend to develop an engaged relation to the city. For example, Richards (2022) sees tendencies of a general shift in cultural tourism towards creative tourism, where tourists take direct part in tourist activities as co-creators. This is particularly evident in alternative tourism where new and innovative ways of approaching the local evolve (see Russo & Richards, 2016). Free guided tours, where locals show different aspects of their cities to outsiders is one example of such an interactive activity. There are many examples of tours that show 'less known' aspects of a cities' history and culture to visitors (Nilsson & Zillinger, 2020).

Spatial Change

When discussing the spatial impact of new urban tourism, we need to acknowledge that the structure of urban tourism does not change overnight. Spatial concentrations of heritage attractions, culture and entertainment together with hotels and restaurants creates an infrastructure that last over time, creating relative spatial stability. With this in mind, it is fair to say that some of the most interesting dynamics connected to urban tourism takes place outside traditional tourist precincts (cf. Judd & Fainstein, 1999). As pointed out by Füller and Michel (2014, p. 1306) 'it is a very specific kind of urbanity or urban experience that this new urban tourism focuses on. The sought-after imaginary of urbanity is often connected to former working-class and post-industrial inner-city neighbourhoods'. They normally have very few traditional tourist attractions, and underdeveloped tourist infrastructure. The importance of such districts for tourism was first recognised in connection to developments in East London, from Islington to Deptford (Maitland & Newman, 2004, Maitland & Newman, 2009a; Smith & Pappalepore, 2015). Similar developments can now be seen both in large cities like Berlin and Paris (Novy, 2018; Freytag & Bauder, 2018) and in less prominent cities such as Krakow (Matoga & Pawłowska, 2016). Today, there are examples of such new tourist areas in cities around the world, gaining increasing international attention (cf. Lonely Planet, 2020).

Based on the literature used above, some common characteristics of such areas can be found. They are areas of transformation, often with a mix of poor people, immigrant population and young people with alternative lifestyles. The atmosphere is described as bohemian, creative and edgy. On the street level, there are signs of change in the composition of functions and services when fashion shops, art galleries, bars and hip restaurants open up, catering for the creative class and their visitors. Since there are normally few hotels in these districts, they leave room for alternative forms of accommodation, such as hostels (Richards, 2016) and Airbnbs. It was in this context that new urban tourism started to be discussed. There is a particular change in the place specific qualities taking place off the beaten track. When the composition of the population slowly change and the number of temporary locals increase, the face of these districts transforms. They are longer as rough and edgy as they used to be. Instead, an increasingly comfortable and culturally hip form of authenticity develops. These *Hipstervilles* are also pleasing and welcoming to the curious visitors, since there are 'similar demands for an authentic urban experience amongst the new urban tourists and the urban middle classes' (Füller & Michel, 2014, p. 1306).

There is a dynamic relationship between gentrification and tourism in these areas. It is obvious that early gentrification creates breeding grounds for tourism development as the areas are upgraded, and become attractive to visitors. It is also obvious that tourism acts as a driver of gentrification. The term tourism gentrification was first coined in a study of transformation in the French quarters of New Orleans (Gotham, 2005). Tourism driven gentrification has recently gained increasing interest following the growing presence of Airbnb (Dolnicar, 2019). There is evidence suggesting that residents with a weak position in the housing market are pushed out by peer-to-peer accommodation. For example, Cócola Gant (2016) shows that there has been a significant change in the ethnic composition of inhabitants in the central district of Barcelona between 1998 and 2015. Half the population in the age groups 20–50 years are now non-Spanish. Wachsmuth and Weisler (2018) conclude that Airbnb works as a means for property owners in New York to extract a higher income from short-term rentals than from renting to permanent residents. They also show that residential areas 'with strong cultural cachet' (Wachsmuth & Weisler, 2018, p. 1155) were more likely to attract high numbers of Airbnb listings. Thus, Airbnb has become a means of spreading tourist accommodations into urban residential districts. This is not only the case in large cities like Berlin (Stors & Kagemeier, 2017) or Paris (Gravari-Barbas & Jacquot, 2017) but also in middle-sized cities like Utrecht (Ioannidis et al., 2018). The impact of Airbnb has become very visible, thereby highlighting the relation between tourism and gentrification (see: Gravari-Barbas & Guinand, 2017).

Many interrelated factors drive transformation in gentrifying tourist districts. Local actors buy and transform properties to run a Bed and Breakfast or similar business. Foreigners may buy a property and move permanently as lifestyle migrants, or use it as a second home and re-let it when they are not present themselves. Taken together, this increases mobility and the overall number of visitors in the districts. In some case, the number of visitors, and the impact they cause, reach a level where we begin to talk about overtourism. 'Overtourism describes destinations where hosts and guests, locals and visitors, feel that that there are too many visitors and that the quality of life in the area or the quality of the experience has deteriorated unacceptably' (Goodwin, 2017, p. 1). The concept generally describes localised problems, related to the carrying capacity of a destination (Nilsson, 2020). Urban problems associated with increasing tourism, for example in cities like Barcelona and Berlin, sometimes cause anti-tourism sentiments, which also influence local politics (Colomb & Novy, 2017; Novy & Colomb, 2019). It needs to be said that gentrification is not only about tourism. Structural transformation and forces related to housing politics are integral to the impact on local social sustainability. However, as tourists are a highly visible, and sometimes disturbing, ingredient in the local scene, they are likely to become the target of negative attention.

Concluding Discussion

This chapter gives a structural perspective on the development of urban tourism. With a history dating back to the Enlightenment, urban tourism has gone through several phases that relate to transport development, forms of organisation, visitor numbers, and the cultural interests of visitors. In this process, older forms of tourism seldom disappear, newer ones add to the former. Churches, museums and other heritage remain important, at the same time as alternative forms of tourism develop, and new cities and districts gain in popularity. The concept of new urban tourism points to the importance of viewing urban tourism in a historically based structural perspective.

However, the most significant change in recent decades is connected to the unprecedented growth of urban tourist travel; which has important consequences on several geographical scales, not only for the sustainability of cities and of urban tourism. The most important challenge concerns the ecological footprints caused by travel to cities. Aviation emits large quantities of greenhouse gases, with severe consequences for the global climate. Local traffic causes congestion and air pollution. These problems are to a large degree driven by new business models in low-cost aviation and by the breakthrough of platform businesses. Significant reductions on fares combined with lower transaction costs have increased demand levels.

Urban destinations have adapted to a situation with low transport costs. As there are signs that global fuel prices may increase in the future, the dependency on tourism may cause problems for the future economic sustainability of cities. It also seems to be the case that urban tourism is increasingly dependent on actors outside local control. Profits of global corporations like Airbnb are transferred to foreign shareholders and not re-invested locally. Potentially negative consequences for urban social sustainability need to be managed locally. There are also some interesting cultural developments taking place in tourist cities. The proliferation of alternative forms of tourist services tend to follow a highly internationalised formula. In the hip parts of cities, local services like micro-breweries, fair trade coffee shops, and vegan restaurants tend to look similar wherever you are. Local authentic tourism is thus a highly cosmopolitan cultural phenomenon, following role models from Brooklyn and other leading metropolitan districts in the Western World. These examples imply that the issues discussed in this section are challenging for urban tourist research, since the connections and dependencies between different geographical scales are significant and difficult to disentangle.

References

Aal, C., & Koens, K. (2019). The discourse on sustainable urban tourism: The need for discussing more than overtourism. *Sustainability*, *11*(15), 4228. https://doi.org/10.3390/su11154228

Ashworth, G., & Page, S. (2011). Urban tourism research. Recent progress and current paradox. *Tourism Management*, *32*(1), 1–15. https://doi.org/10.1016/j.tourman.2010.02.002

Ba, C., Frank, S., Müller, C., Raschke, A. L., Wellner, K., & Zecher, A. (Eds). (2022). *The power of new urban tourism. Spaces, representations and contestations*. Routledge.

Baedeker, K. (1878). *Paris und seine Umgebungen, nebst den Eisenbahn-Routen nach Paris*. Leipzig: Karl Baedeker.

Beaverstock, J. V., Derudder, B., Faulconbridge, J., & Witlox, F. (Eds). (2010). *International business travel in the global economy*. Ashgate.

Bock, K. (2015). The changing nature of city tourism and its possible implications for the future of cities. *European Journal of Futures Research*, 3(20). https://doi.org/10.1007/s40309-015-0078-5

Castells, M. (1996). *The information age, vol.1. The rise of the network society*. Blackwell.

Cheshire, P. C., & Hay, D. G. (1989). *Urban problems in Western Europe: An economic analysis*. Unwin Hyman.

Christaller, W. (1933). *Die Zentralen Orte in Süddeutschland*. Fischer.

Cócola Gant, A. (2016). Holiday rentals: The new gentrification battlefront. *Sociological Research Online*, *21*(3). 112–120. https://doi.org/10.5153/sro.4071

Colomb, C. (2012) *Staging the New Berlin. Place marketing and the politics of urban reinvention post-1989*. Routledge.

Colomb, C., & Novy, J. (Eds). (2017). *Protest and resistance in the tourist city*. Routledge.

Dicken, P. (2015). *Global shift. Mapping the changing contours of the world economy* (7th ed.). The Guildford Press.

Dobruszkes, F. (2013). The geography of European low-cost airline networks: a contemporary analysis. *Journal of Transport Geography*, *28*, 75–88. https://doi.org/10.1016/j.jtrangeo.2012.10.012

Dolnicar, S. (2019). A review of research into paid online peer-to-peer accommodation: Launching the *Annals of Tourism Research* Curated Collection on peer-to-peer accommodation. *Annals of Tourism Research*, *75*, 248–264. https://doi.org/10.1016/j.annals.2019.02.003

Domínguez-Mujica, J., McGarrigle, J., & Parreño-Catellano, J. M. (Eds). (2021). *International residential mobilities. From lifestyle migration to tourism gentrification*. Springer.

Dredge, D., & Gyimóthy, S. (Eds). (2017). *Collaborative economy and tourism: Perspectives, politics, policies and prospects*. Springer.

Dumbrovská, V. (2017). Urban tourism development in Prague: From tourist mecca to tourist ghetto. In N. Bellini & C. Pasquinelli (Eds), *Tourism in the city: Towards an integrative agenda on urban tourism* (pp. 275–284). Springer.

Eskilsson, L., & Nilsson, J. H. (2022). Science driven mobility as a form of new urban tourism: Insights from student and research internationalisation in Lund, Sweden. In C. Ba, S. Frank, C. Müller, A. L. Raschke, K. Wellner & A. Zecher (Eds), *The power of new urban tourism. Spaces, representations and contestations* (pp. 233–248). Routledge

Eurostat (2019). *Night spent at tourist accommodation establishments by NUTS 2 regions*. https://data.europa.eu/data/datasets/clfxgiiz02xfa2mtjwgjsq?locale=en

Frändberg, L. (2014). Temporary transnational youth migration and its mobility links. *Mobilities*, 9(1), 146–164. https://doi.org/10.1080/17450101.2013.769719

Freytag, T., & Bauder, M. (2018). Bottom-up touristification and urban transformations in Paris. *Tourism Geographies*, *20*(3), 443–460. https://doi.org/10.1080/14616688.2018.1454504

Frisch, T., Sommer, C., Stoltenberg, L, & Stors, N. (2019). *Tourism and everyday life in the contemporary city*. Routledge.

Füller, H., & Michel, B. (2014). 'Stop being a tourist!' New dynamics of urban tourism in Berlin-Kreuzberg. *International Journal of Urban and Regional Research*, *38*(4), 1304–1318. https://doi.org/10.1111/1468-2427.12124

Germann Molz, J. (2013). Social networking technologies and the moral economy of alternative tourism: The case of couchsurfing.org. *Annals of Tourism Research*, *43*, 210–230. https://doi.org/10.1016/j.annals.2013.08.001

Goodwin, H. (2017). The challenge of overtourism. *Responsible Tourism Partnership Working Paper* 4, October. http://www.millennium-destinations.com/uploads/4/1/9/7/41979675/rtpwp4overtourism012017.pdf

Gössling, S. (2017). Tourism, information technologies and sustainability: An exploratory review. *Journal of Sustainable Tourism*, *25*(7), 1024–1041. https://doi.org/10.1080/09669582.2015.1122017

Gössling, S., Scott, D., & Hall, C. M. (2018). Global trends in length of stay: implications for destination management and climate change. *Journal of Sustainable Tourism*, *26*(12), 2087–2101. https://doi.org/10.1080/09669582.2018.1529771

Gotham, K. F. (2005). Tourism gentrification: The case of New Orleans´ Vieux Carre (French Quarter). *Urban Studies*, *42*(7), 1099–1121. https://doi.org/10.1080/00420980500120881

Gravari-Barbas, M., & Guinand, S. (Eds). (2017). *Tourism and gentrification in contemporary metropolises. International perspectives*. Routledge.

Gravari-Barbas, M., & Jacquot, S. (2017). No conflict? Discourses and management of tourism-related tension in Paris. In C. Colomb & J. Novy. (Eds), *Protest and resistance in the tourist city* (pp. 45–65). Routledge.

Guttentag, D. (2015). Airbnb: disruptive innovation and the rise of an informal tourism accommodation sector. *Current Issues in Tourism*, *18*(12), 1192–1217. https://doi.org/10.1080/13683500.2013.827159

Gyimóthy, S. (2017). Business models of the collaborative economy. In D. Dredge & S. Gyimóthy (Eds), *Collaborative economy and tourism: Perspectives, politics, policies and prospects* (pp. 31–39). Springer.

Hall, C. M. (2005). *Tourism. Rethinking the social science of mobility*. Prentice Hall.

Hall, P. (1988). *Cities of tomorrow. An intellectual history of urban planning and design in the twentieth century*. Basil Blackwell.

Hall, P. (1998). *Cities in civilization. Culture, innovation, and urban order*. Weidenfeld & Nicolson.

Heeley, J. (2011). *Inside city tourism. A European perspective*. Channel View Publications.

Janson, A. (2018). Rethinking post-tourism in the age of social media. *Annals of Tourism Research*, *69*, 101–110. https://doi.org/10.1016/j.annals.2018.01.005

Janta, H., Ladkin, A., Brown, L., & Lugusi, P. (2011). Employment experiences of Polish migrant workers in the UK hospitality sector. *Tourism Management*, *32*(5), 1006–1019. https://doi.org/10.1016/j.tourman.2010.08.013

Judd, D. R. (2003). Visitors and the spatial ecology of the city. In L. M. Hoffman, S. S. Fainstein & D. R. Judd (Eds), *Cities and visitors. Regulating people, markets, and city space* (pp. 23–38). Blackwell.

Judd, D. R., & Fainstein, S. S. (Eds). (1999). *The tourist city*. Yale University Press.

Koens, K., Postma, A., & Papp, B. (2018). Is overtourism overused? Understanding the impact of tourism in a city context. *Sustainability*, *10*(12), 4384. https://doi.org/10.3390/su10124384

Larsen, J. (2008). De-exoticizing tourist travel. Everyday life and sociality on the move. *Leisure Studies*, *27*(1), 21–34. https://doi.org/10.1080/02614360701198030

Larsen, J. (2019). Ordinary tourism and extraordinary everyday life: Re-thinking tourism and cities. In T. Frisch, N. Stors, L. Stoltenberg & C. Sommer (Eds). *Tourism and everyday life in the city* (pp. 24–41). Routledge.

Lonely Planet (2020). *Secret city. The insider's guide to the world's coolest neighbourhoods*.

Maitland, R., & Newman, P. (2004). Developing metropolitan tourism on the fringe of central London. *International Journal of Tourism Resarch*, *6*(5), 339–348. https://doi.org/10.1002/jtr.496

Maitland, R., & Newman, P. (2009a). London: Tourism moving east? In R. Maitland & P. Newman (Eds). *World tourism cities. Developing tourism off the beaten track* (pp. 80–100). Routledge.

Maitland, R., & Newman, P. (Eds). (2009b). *World tourism cities*. Developing tourism off the beaten track. Routledge.

Matoga, Ł., & Pawłowska, A. (2016). Off-the-beaten-track tourism: a new trend in the tourism development in historical European cities. A case study of Krakow, Poland. *Current Issues in Tourism*, *21*(14), 1644–1669. https://doi.org/10.1080/13683500.2016.1212822

Maxim, C. (2016). Sustainable tourism implementation in urban areas: a case study of London. *Journal of Sustainable Tourism*, *24*(7), 971–989. https://doi.org/10.1080/09669582.2015.1115511

Mieli, M. (2022). *Smartphoned tourism. Tourist behaviour and experiences in a phygital world*. [Doctoral dissertion]. Lund University.

Nilsson, J. H. (2007). Taggtråd, betong och Spreewaldgurkor – Historiebruk kring den synliga och osynliga Berlinmuren. In R. Ek & J. Hultman (Eds), *Plats som produkt* (pp. 77–93). Studentlitteratur.

Nilsson, J. H. (2018a). Globalisering och urban turismutveckling – drivkrafter och konsekvenser. *Geografiska Notiser*, *76(3)*, 1–10.

Nilsson, J. H. (2018b). Mobility and regionalisation: Changing patterns of air traffic in the Baltic Sea Region in connection to European integration. *Geographia Polonica*, *91*(1), 77–93. https://doi.org/10.7163/GPol.0092

Nilsson, J. H. (2020). Conceptualizing and contextualizing overtourism: the dynamics of accelerating urban tourism. *International Journal of Tourist Cities*, *6*(4), 657–671. https://doi.org/10.1108/IJTC-08-2019-0117

Nilsson, J. H., & Zillinger, M. (2020). Free guided tours: storytelling as a means of glocalizing urban places. *Scandinavian Journal of Hospitality and Tourism*, *20*(3), 286–301. https://doi.org/10.1080/15022250.2020.1772866

Novy, J. (2018). 'Destination' Berlin revisited. From (new) tourism towards a pentagon of mobility and place consumption. *Tourism Geographies*, *20*(3), 418–442. https://doi.org/10.1080/14616688.2017.1357142

Novy, J., & Colomb, C. (2019). Urban tourism as a surce of contention and social mobilisations: A critical review. *Tourism Planning & Development*, *16*(4), 358–375. https://doi.org/10.1080/21568316.2019.1577293

Pasquinelli, C. (2015). Urban tourism(s): Is there a case for a paradigm shift? *GSSI Working Papers* #14. http://dx.doi.org/10.2139/ssrn.2584894

Porter, M. (1990). *The competitive advantage of nations*. Macmillan.

Richards, G. (2016). Hostels and the making of new urban spaces. In A. P. Russo & G. Richards (Eds), *Reinventing the local in tourism. Producing, consuming and negociating place* (pp. 171–184). Channel View Publications.

Richards, G. (2017) Tourists in their own city – considering the growth of a phenomenon. *Tourism Today*, *16*, 8–16.

Richards, G. (2022). Urban tourism as a special type of cultural tourism. In J. van der Borg (Ed.), *A research agenda for urban tourism* (pp. 31–50). Edward Elgar.

Russo, A. P., & Richards, G. (2016). *Reinventing the local in tourism. Producing, consuming and negociating place*. Channel View Publications.

Smith, A., & Pappalepore, I. (2015). Exploring attitudes to edgy urban destinations: the case of Deptford, London. *Journal of Tourism and Cultural Change*, *13*(2), 97–114. https://doi.org/10.1080/14766825.2014.896371

Smith, M., Pinke-Sziva, I., & Olt, G. (2019). Overtourism and resident resistance in Budapest. *Tourism Planning & Development*, *16*(4), 376–392. https://doi.org/10.1080/21568316.2019.1595705

Spirou, C. (2011). *Urban tourism and urban change. Cities in a global economy*. Routledge.

Stephenson, M. L. (2014). Tourism, development and 'destination Dubai': cultural dilemmas and future challenges. *Current Issues in Tourism*, *17*(8), 723–738. https://doi.org/10.1080/13683500.2012.754411

Stock, M. (2019). Inhabiting the city as tourist: Issues for urban and tourism theory. In T. Frisch, C. Sommer, L. Stoltenberg & N. Stors (Eds), *Tourism and everyday life in the contemporary city* (pp. 42–66). Routledge.

Stors, N., & Kagemeier, A. (2017). The sharing economy and its role in metropolitan tourism. In M. Gravari-Barbas & S. Guinand (Eds), *Tourism and gentrification in contemporary metropolises. International perspectives* (pp. 181–206). Routledge

Sydsvenskan (2022). Repaluu: Det är ett högt pris man får betala för din här typen av uppdrag. Sydsvenskan. Link no longer available.

Taylor, P. (2004). *World city network*. A global urban analysis. Routledge.

Taylor, P. (2021). *Advanced introduction to cities*. Edward Elgar.

Thurnell-Read, T. (2012). Tourism, place and space. British stag tourism in Poland. *Annals of Tourism Research*, *39*(2), 801–819. https://doi.org/10.1016/j.annals.2011.09.009

Urry, J., & Larsen, J. (2011). *The tourist gaze 3.0*. Sage.

Vainikka, V. (2013). Rethinking mass tourism. *Tourist Studies*, *13*(3), 268–286. https://doi.org/10.1177/1468797613498163

van der Borg, J. (Ed.) (2022). *A research agenda for urban tourism*. Edward Elgar.

Wachsmuth, D., & Weisler, A. (2018). Airbnb and the rent gap: Gentrification through the sharing economy. *Environment and Planning A: Economy and Space*, *50*(6), 1147–1170. https://doi.org/10.1177/0308518X18778038

Zekan, B., & Wöber, K. (2022). Urban tourism: major trends. In J. van der Borg (Ed.), *A research agenda for urban tourism* (pp. 19–30). Edward Elgar.

Ziakas, V., & Boukas, N. (2012). A neglected legacy. Examining the challenges and potential for sport tourism developmentin post-Olympic Athens. *International Journal of Event and Festival Management*, *3*(3), 292–316. https://doi.org/10.1108/17582951211262710

Zuelow, E. G. E. (2016). *A history of modern tourism*. Palgrave.

Chapter 30

Second-home Tourism in Troubled Times

Dieter K. Müller

Introduction

While the antecedents of second-home tourism can be traced back to the antique, as a modern phenomenon second-home tourism developed since the early decades of the 20th century (Hall & Müller, 2004; Hall & Müller, 2018). The scientific literature recognized second home tourism initially in the Nordic countries and North America and particularly the work of Wolfe (1951, 1952, 1962, 1965, 1970) on second homes in Canada established an early understanding of the spatial dimensions of the phenomenon. An increasing personal mobility owing to car ownership entailed a growth in second homes in the Western world. Coppock's *Second Homes: Curse or Blessing?* (1977) marked an early peak in the growing concern regarding the impact of second homes on rural communities and the environment. The volume is a classic review that highlighted many aspects of second home tourism that are still valid, but also expressed fears that never realized (Müller & Hoogendoorn, 2013).

Since the 2000s second home tourism research has revealed an increasingly comprehensive picture of a global phenomenon (Gallent et al., 2005; Hall & Müller, 2004; Hall & Müller, 2018; McIntyre et al., 2006; Paris, 2011; Roca, 2013). Even the previous editions of this Companion offered state of the art reports (Müller, 2004; Müller, 2014) and thus, this chapter focuses primarily the literature on the topic since the publication of the previous edition in 2014. Before the content of the research contributions is addressed, it can be noted that the development of scientific inquiry is developing geographically and indeed, today research on second homes can be seen from all parts of the world, also nuancing the phenomenon in relation to geographical settings and preconditions (Back & Marjavaara, 2017; Müller et al., 2004; Müller & Hall, 2018). Nevertheless, Nordic research on second homes continues to hold a dominant position (Müller, 2021a).

The Wiley Blackwell Companion to Tourism, Second Edition. Edited by C. Michael Hall.

Mapping and Understanding Second-home Tourism

A major academic undertaking after the rejuvenation of interest in second homes in the 1990s has been the establishment of baseline data (Hall & Müller, 2004; Hall & Müller, 2018). While many national overviews have been presented during the years, detailed and nuanced contributions acknowledging geographical differences in second-home landscapes are still added (Back & Marjavaara, 2017, Mowl et al., 2020). This can be seen as an indication of a maturing field of research that moves beyond crude generalizations of a complex phenomenon.

A core challenge in this context has been the geographical assessment of temporary population redistributions caused by second-home tourism beyond the static accounts provided by statistical bureaus. An overall lack of data capturing mobility limited any more exact account of the presence of population in different places. Recently, access to cell phone data has allowed for novel ways of assessing this (Willberg et al., 2021), but reporting based on such data remains the exception. Hence, more conventional methods estimating use patterns of second homes are the base for the analysis of scope and the patterns of second-home mobility. Several contributions continue to provide knowledge and conceptualizations of this, and repeatedly, second homes are addressed in the context of urban sprawl, seasonal suburbanization, and counterurbanization (Adamiak, 2016, Adamiak et al., 2017, Hiltunen & Rehunen, 2014). The emerging geographical patterns reflect migration dynamics, accessibility, and the location of amenities (Nouza et al., 2013). However, even planning affected the occurrence and shape of second-home landscapes (Soto & Clavé, 2017; Soto & Clavé, 2018).

In this context even the relationship of second-home tourism and other forms of tourism is addressed. While large numbers of second homes can be found in resort areas, where they play an important role by contributing to the accommodation supply (Flognfeldt & Tjørve, 2013), the relationship between second-home locations and major tourism areas is not pre-determined. Indeed, second-home areas may occur in distinct locations, for example, in littoral zones, while other tourism areas may lack second-home properties (Opačić & Koderman, 2019).

Why certain places are attractive for second home ownership remains thus unclear. Cultural representations have previously been identified as important reasons for national second-home imaginaries (Vepsäläinen & Pitkänen, 2010). This topic is revisited by Lagerqvist (2014) who focuses representations of second homes as an issue affecting their attractiveness. Accordingly, in Sweden second homes have not always held a strong symbolic position, partly because they initially were a privilege for the bourgeois class as well. However, a focus on the materiality of the second homes created an image that seems uncontroversial, though the related norms may indeed cause exclusion of certain groups. A similar though opposite transformation of representations is observed in New Zealand, where second homes increasingly are related to luxury and wealth (Walters & Carr, 2019). Independent of context a common denominator is that second-home owners attach themselves to place. Tuulentie and Kietäväinen (2020) argue that social contacts, media and nature experiences influence how such attachment takes place. Access to social meeting grounds affect thus the degree of integration into community. However, though second homes may be a blessing for many, they may also form a burden in relation to maintenance, inheritance, and shared use in larger families (Lundmark & Marjavaara, 2013; Pers et al., 2018).

Motivations and Mobilities

While motivations for owning a second home have been well documented already (Müller, 2014) – many of them can be found in the microcosm of family life and personal aspirations – the recent decade offered some new insights into why people purchase second homes. For many owners a second home represents a link to family origins and thus place attachment has been a factor explaining why people engage in second-home living. However, an Icelandic study indicates that it is the length of ownership rather than previous linkages that affects how people relate to places (Nouza et al., 2018).

Even a growing ecological awareness has contributed to the demand for rural second homes (Lipkina, 2013; Mamonova & Sutherland, 2015). In this context possibilities for gardening and a close-to-nature lifestyle are important assets and are expected to entail health benefits for their owners (Åkerlund et al., 2015b; Pitkänen et al., 2020b). Related to this notion is the so-called compensation hypothesis. Accordingly, second homes are a compensation for a lack of green spaces in urban living (Muñiz et al., 2013). While dense urban housing correlates with second-home use, ownership patterns seem to be far more complex and cannot be explained by dense urban living solely (Strandell & Hall, 2015).

Motivations for second-home tourism have been particularly scrutinized from an international perspective. Rogerson and Hoogendoorn (2014) and González-Herrera and colleagues (2021) address the topic of second homes in relation to VFR-tourism and the Global South. They reveal that second homes can be both, an expression of wealth and an international class commuting between the South and the North and a working class commuting between different homes in the South. The latter resembles also the notion that second-home tourism comprises wide spectra of society (Hoogendoorn & Visser, 2015). Furthermore, the studies indicate that second homes are related to both leisure and work and offer opportunities to adapt to a spatial division of these functions.

In relation to international residential tourism an imagined proximity – a home away from home – is identified as a major motivation as well (Bringas-Rábago et al., 2021). Åkerlund (2017) shows that second-home owners hereby attempt to combine the assets of different places, while Hannonen (2018) argues that second-home owners try to live like locals and thus de-differentiate themselves from other tourists.

However, even economic motivations related to taxation and property markets have been identified as important drivers of international second-home tourism (Montezuma & McGarrigle, 2019). This applies not least in a European context, but even elsewhere particular programs have been established in some countries to attract second-home owners with tax and other economic benefits (Åkerlund, 2017; Wong & Musa, 2014). Related to this, political stability has been highlighted among other motivations for second-home ownership abroad (Lipkina, 2013; Wong & Musa, 2014). Accordingly, second-home ownership can secure investment (Paris, 2018) or simply be a potential refuge during times of political turmoil. Crossing borders can imply both language and cultural barriers but at the same time open opportunity spaces not available in the home country (Hannonen et al., 2015). Sometimes this may also be related to a displaced diaspora where a return to the home country for political reasons is impossible but compensated by second-home ownership in culturally related destinations (Abbasian, 2018; Abbasian & Müller, 2019). Overall, it has been argued that push rather than pull factors create satisfaction with the decision for a second home abroad (Wong et al., 2017). Getting away from everyday life is thus a major factor for spending time and money for a property abroad.

Besides these individual motivations, structural preconditions, for example caused by national economic strategies or migration policies, influence second-home tourism and govern flows into certain destinations (Åkerlund et al., 2015a; Clausen & Garcia, 2018). However, even the role of property ownership for the accumulation of capital in various countries affect the likelihood for second-home ownership (Wind et al., 2020).

Second Home Impacts and Planning Issues

The impacts of second-home ownership have been a major focus of research into the phenomenon (Hall & Müller, 2004; Hall & Müller, 2018). Usually, distinctions are made between economic, social, and environmental impacts. However, not least social impacts of second-home tourism in rural destinations have yielded research attention. Furthermore, the governance and planning efforts for optimizing impact have been other topics for research (Hall, 2015). These systems are usually geared for managing stationary rather mobile populations and thus, new approaches are needed to cope with increasingly mobile societies.

As shown in previous research impacts of second-home tourism vary significantly between different places (Müller et al., 2004). Back (2020, 2021) reports how planners deal with second homes, and he notes that in some locations second homes are absent from the planners' agendas, while in other places second homes cause rapid development of the property market and new constructions of primary and secondary residences. In attractive locations second-homes construction consumes space and form a considerable challenge for planners (Gerber & Tanner, 2018). In such context densification of second home settlements may occur as solution but at the same time, this practice challenges other sustainability goals, e.g., biodiversity (Hjalager et al., 2022).

Several studies offer an interesting insight into the debate on the social impacts of second-home tourism on rural communities. They show that second-home owners do not differ significantly from their community neighbours when it comes to activities (Pitkänen et al., 2014; Fialová et al., 2018) and apply a similar 'not in my backyard' attitude towards change (Farstad & Rye, 2013). Hence, in relation to change, this means simply an increasing number of guardians of local interests. Still, there are dissonances and second home owners need to balance their multiple roles in the rural context (Farstad, 2018). Even differences in relation to environmental concern are thus more dependent on how people attach themselves to place, rather than on their urban or rural background or other socio-economic characteristics (Armstrong & Stedman, 2019; Jennings & Krannich, 2013). Blondy and colleagues (2018) argue however that second-home owners contribute in three ways to their destination community; they invest emotionally into place, engage socially, and contribute to the attractiveness of the area by becoming targets for VFR-tourism themselves. Even other studies concerning the potential loss of social values in rural communities indicate an increase of social capital through bridging strategies involving the temporary residents and adding to the available capacities of the rural community (Gallent, 2014; Robertsson & Marjavaara, 2015). And indeed, the presence of mobile populations has been important to many peripheral communities particularly, though developing places for people not registered requires awareness and political will (Pitkänen et al., 2017). Additionally, Larsson and Müller (2019) note that communities increasingly adapt to second-home tourists' demand, seizing them as asset rather than burden. Stiman (2020a) explains these place contingent reactions on second-home tourism by pointing at local overarching discourses of resource dependency that allocate particular roles to second-home tourism in the making of places.

However, it should be noted that these observations are geographically contingent. Cultural clash between permanent residents and second-home owners may occur though reasons are complex and cannot be simply found in urban-rural contrasts alone (Armstrong & Stedman, 2013). A case study from Russia reports for example 'super-gentrification' of rural areas featuring gated communities and a strict separation from the local (Mamonova & Sutherland, 2015), and even in China, the advent of an upper-class in rural areas challenges social and political local fabrics (Wu et al., 2015). Second-home owners' participation in local planning is thus not self-evident but contingent of their formal residential status, personal preferences, and informal positions in host communities (Huete & Mantecón, 2012; Rinne et al., 2015). Furthermore, particularly international second-home owners are not always welcomed as economic opportunity since they are seen as threat for property market and, regarding Russians in Finland, security as well (Honkanen et al., 2016).

Few studies have assessed the economic impacts of second-home tourism. An exception is from Croatia where it is shown that second-home tourism correlates positively with a dynamic local economy but also with educational levels and populations development (Miletić et al., 2018). Furthermore, it has been shown that economic impacts depend on the standards of the second homes and the business supplies in the destination communities (Velvin et al., 2013), as well as the second-home households' wealth and use patterns (Czarnecki, 2018). Hence, substantial impacts on local labour markets are not guaranteed since second home tourism tends to sustain existing rather than entailing new employment. Second-home owners are consumers of rural services and products as well. However, a great variation has been shown regarding their willingness to purchase, e.g., rural tourism products, where younger high-income households seem more inclined to spend additionally in the countryside (Tangeland et al., 2013). Still, a study from Spain counteracts stereotypes claiming that second-home destinations perceive less economic impacts than hotel-based destinations (Perles-Ribes et al., 2018).

Second-home demand has impacts on the property market as well. Back and colleagues (2022) investigate how second-homes prices vary in relation to other rural property and are in this context able to identify variegated development in different landscapes and locations. Long and Hoogendoorn (2013) highlight the complex relationship between environmental degradation and property markets. They show that the consequences of such environmental decline have percussions on the local economy but may also trigger a more environmentally sound behaviour among tourists, locals and government.

Even environmental impacts of second-home tourism have been increasingly recognized during the recent decade. It has been shown that domestic second-home tourism only partly replaces long-haul travel. Instead, it seems that second-home ownership for many is an integrated part of a hypermobile lifestyle (Adamiak et al., 2016). Although most second-home owners are aware of negative impacts related to their property ownership, they generally see them as less serious than non-owners and pose thus a challenge to ambitions for a more sustainable development (Hiltunen et al., 2016; Long & Hoogendoorn, 2014). Large scale development can also challenge biodiversity and eco-system as demonstrated for the Caspian Sea coast, where laissez-faire planning allowed for a rapid and uncontrolled growth (Alipour et al., 2017).

An emerging topic is related to second homes and their owners' responsiveness to environmental change. For example, it has been shown that second-home owners in a littoral region are less sensitive towards risks caused by sea-level rise and sometimes not ready to adapt, which creates challenges for local planning authorities (Hoogendoorn & Fitchet, 2018; Rey-Valette et al., 2015). Adie (2020) identified three potential strategies for second-home owners responding on environmental disaster, i.e., relocate, return, or return and adapt, and argues that decisions are dependent on place attachment (see also Adie and Hall, 2023).

Theoretical Approaches and Methodologies

Hall (2015) calls for a greater focus on policy and governance issues but acknowledges as well that these theoretical and approaches hitherto mainly grasp immobility rather than flows and mobility. Even Visser and Hoogendoorn (2015) argue along those lines and indeed ask for nuanced understandings of second home tourism for various social strata. This notion is echoed regarding geographical settings (Back & Marjavaara, 2017; Müller et al., 2004). Therefore, increasingly national reviews can be accessed (Treivish, 2014; Visser & Hoogendoorn, 2015).

A major concern for second-home research continues to be related to dichotomies of home and away, urban and rural, and core and periphery. It has been argued that these distinctions increasingly are dissolved because of changing practices and homes in multiple locations (Ellingsen & Hidle, 2013) but also that administrative practices uphold distinctions (Müller, 2021b). In fact, Ellingsen (2018) argues that second homes trigger a decentralisation of population that is in contrast to national discourses of urbanization. However, even the idea of second homes being leisure domiciles for urbanities is challenged, indicating that second-home mobility flows can be directed towards cities as well (Stiman, 2020b).

In this context, the relation between second-home ownership and migration has been revisited (Marjavaara & Lundholm, 2016; Stergiou et al., 2017). It has been shown that migration towards second homes is result of complex decision-making accounting for multifaceted factors, often contingent of place and time specific properties. While second-home owners seem often to be inclined to migrate to their rural properties, this is not motivated by amenities and natural beauty but rather by challenges in everyday life (Rye, 2015). Against this backdrop second homes are increasingly seen as nodes in multilocal lifestyles implying attachments to multiple places and utilizing their resources in various ways (Schier et al., 2015). Multiple home ownership is however also interpreted as a way of accumulating wealth. In this context second home do not only hold a consumption and use value but serve increasingly as investment, too (Kadi et al., 2020).

A significant contribution is also made by researchers from rural studies mainly interested in the psychological dimensions of place attachment (Armstrong & Stedman, 2013; Armstrong & Stedman, 2019; Jennings & Krannich, 2013) or sociological issues of integration in the countryside (Farstad & Rye, 2013). This research reveals that previous distinctions of primary and secondary homes are likely overrated and open up for a new understanding of rural communities as nodes in networks of flows particularly.

While second-home research early was guided by large datasets, trends pointed increasingly into more qualitative direction. Many of the reviewed studies are based on interviews, though surveys and database studies continue to form an important share of the literature. However, the literature mirrors even innovative methodological approaches for assessing the whereabouts of second-home owners. For example, Willberg and colleagues (2021) utilize geopositioning data from cell phones, while Sheludkov and Starikova (2021) employ nighttime satellite imagery to identify second-home hotspots. Utilizing these kinds of data sources will enable substantial progress in our understanding of the second-home phenomenon in the next decade.

(Post)Pandemic Second-home Tourism

The Covid-19 pandemic will certainly lead to a rejuvenated interest for second-home research. One interesting aspect of the pandemic is what has happened to second-home tourism during the spread of the disease. An international outlook reveals that impacts on second-home tourism varied as did national approaches to manage the pandemic (Pitkänen et al., 2020a).

Even during the pandemic perceptions of second-home tourism diverged, understanding it as threat to rural communities on one hand and as safe haven for their owners on the other hand. Evidence from Poland suggests however that usage patterns increased only slightly in cases where remote wok solutions could be applied (Czarnecki et al., 2021). In the Finnish case a greater presence of people in second-home destinations could be detected (Willberg et al., 2021). The experience of the pandemic and the related de-internationalization of tourism can by displacing demand at least temporarily exert pressure on the market for second homes (Seraphin & Dosquet, 2020; Zoğal et al., 2022). Hitherto far from all pandemic-induced changes have been analyzed. Not least international second-home tourism has been affected by the collapse and eventually by a post-pandemic reconfiguration of transportation networks. However, the strong emotional bonds of many second-home owners to their properties may indeed limit these impacts.

A major impact of the pandemic is the dissemination and adoption of digital technology enabling remote work solution. While the international literature seems to be mainly concerned with the consequences for work-life balance and productivity, there seems to be agreement that remote work arrangements will increase (e.g., Foss, 2020). This notion resembles trends of digital nomadism increasingly dissolving norms of sedentarism and long-term residency (Thompson, 2019; Hannonen, 2020). In this context second homes may develop into important nodes for remote work arrangements (Müller, 2021b, p. 97):

> The pandemic has manifested that online work from home offices can serve as a substitute for, or at least supplement, traditional work patterns for an increasing number of people. This insight opens up new opportunities to utilize second homes, and may indeed entail a step toward a truly multilocal and mobile society.

Departing from this notion it is important to consider the long-term consequences of the pandemic and the role of second homes in the context of multilocal lifestyles (Schier et al., 2015; Halseth et al., 2018). An increasing dissolution of the need for physical presence at a workplace for greater shares of the labor market opens for an intensified use of second homes. Such change challenges sedentarist ideas of 'permanent ' residence and instead it calls for understanding emotional, functional and corporeal attachment to multiple places. Such thinking is not new for second-home research (e.g., Kaltenborn, 1998; McIntyre et al., 2006), but the changes triggered by the pandemic highlight and potentially expand the phenomenon of multiple dwelling and locality.

Conclusion

While second-home tourism research by now can look back onto a century-long history, it is obvious that the topic still triggers vibrant and innovative research efforts all over the world. Indeed, recent changes related to geopolitical struggle, digitalization and a changing work life, as well as a growing awareness of the consequences of climate change all contribute to warrant fresh examinations of the second-home phenomenon. In this context access to new data sources grasping mobility in a hitherto unseen way enable more qualified empirical assessments and open up for new theorization of mobility, tourism and being at home and away, and thus call for a rejuvenation of the mobility turn in the social sciences.

References

Abbasian, S. (2018). Political crises and destination choice: An exploratory study of Swedish-Iranian second-home buyers. *Tourism, Culture & Communication*, *18*(3), 205–218. https://doi.org/10.3727/109830418X15319363084481

Abbasian, S., & Müller, D. K. (2019). Displaced diaspora second-home tourism: An explorative study of Swedish-Iranians and their second-home purchases in Turkey. *Tourism*, *67*(3), 239–252. https://hrcak.srce.hr/225739

Adamiak, C. (2016). Cottage sprawl: Spatial development of second homes in Bory Tucholskie, Poland. *Landscape and Urban Planning*, *147*, 96–106. https://doi.org/10.1016/j.landurbplan.2015.11.003

Adamiak, C., Hall, C. M., Hiltunen, M., & Pitkänen, K. (2016). Substitute or addition to hypermobile lifestyles? Second home mobility and Finnish CO_2 emissions. *Tourism Geographies*, *18*(2), 129–151. https://doi.org/10.1080/14616688.2016.1145250

Adamiak, C., Pitkänen, K., & Lehtonen, O. (2017). Seasonal residence and counterurbanization: the role of second homes in population redistribution in Finland. *GeoJournal*, *82*(5), 1035–1050. https://doi.org/10.1007/s10708-016-9727-x

Adie, B. A. (2020). Place attachment and post-disaster decision-making in a second home context: A conceptual framework. *Current Issues in Tourism*, *23*(10), 1205–1215. https://doi.org/10.1080/13683500.2019.1600475

Adie, B. A., & Hall, C. M. (Eds.). (2023). *Second homes and climate change*. Taylor & Francis.

Åkerlund, U. (2017). Strategic lifestyle management in later life: Swedish lifestyle movers in Malta seeking the 'best of both worlds'. *Population, Space and Place*, *23*(1), e1964. https://doi.org/10.1002/psp.1964

Åkerlund, U., Lipkina, O., & Hall, C. M. (2015a). Second home governance in the EU: in and out of Finland and Malta. *Journal of Policy Research in Tourism, Leisure and Events*, *7*(1), 77–97. https://doi.org/10.1080/19407963.2014.933229

Åkerlund, U., Pitkänen, K., Hiltunen, M., Overvåg, K., Müller, D., & Kahila, P. (2015b). Health, well-being and second homes: An outline of current research and policy challenges. *Matkailututkimus*, *11*(1), 43–54.

Alipour, H., Olya, H., Hassanzadeh, B., & Rezapouraghdam, H. (2017). Second home tourism impact and governance: Evidence from the Caspian Sea region of Iran. *Ocean & Coastal Management*, *136*, 165–176. https://doi.org/10.1016/j.ocecoaman.2016.12.006

Armstrong, A., & Stedman, R. C. (2013). Culture clash and second home ownership in the US Northern Forest. *Rural Sociology*, *78*(3), 318–345. https://doi.org/10.1111/ruso.12010

Armstrong, A., & Stedman, R. C. (2019). Understanding local environmental concern: the importance of place. *Rural Sociology*, *84*(1), 93–122. https://doi.org/10.1111/ruso.12215

Back, A. (2020). Temporary resident evil? Managing diverse impacts of second-home tourism. *Current Issues in Tourism*, *23*(11), 1328–1342. https://doi.org/10.1080/13683500.2019.1622656

Back, A. (2021). Endemic and diverse: Planning perspectives on second-home tourism's heterogeneous impact on Swedish housing markets. *Housing, Theory and Society*, 39(3), 317–334. https://doi.org/10.1080/14036096.2021.1944906

Back, A., & Marjavaara, R. (2017). Mapping an invisible population: the uneven geography of second-home tourism. *Tourism Geographies*, *19*(4), 595–611. https://doi.org/10.1080/14616688.2017.1331260

Back, A., Marjavaara, R., & Müller, D. K. (2022). The invisible hand of an invisible population: Dynamics and heterogeneity of second-home housing markets. *International Journal of Tourism Research*, https://doi.org/10.1002/jtr.2520

Blondy, C., Plumejaeud, C., Vacher, L., Vye, D., & Bontet, C. (2018). Do second home owners only play a secondary role in coastal territories?: A case study in Charente-Maritime (France). In C. M. Hall & D. K. Müller (Eds.), *The Routledge handbook of second home tourism and mobilities* (pp. 233–244). Routledge.

Bringas-Rábago, N. L., Sosa-Ferreira, A. P., & Osorio-García, M. (2021). International mobility: An approach to the imaginaries of residential tourism from the northern and south-eastern borders of Mexico. In J. Domínguez-Mujica, J. McGarrigle, & J. M. Parreño-Castellano (Eds.), *International residential mobilities: From lifestyle migrations to tourism gentrification* (pp. 87–104). Springer.

Clausen, H. B., & García, M. A. V. (2018). National Mexican tourism policy and North American second home owners in Mexico: Local tourism development and Mexican identity. In C. M. Hall & D. K. Müller (Eds.), *The Routledge handbook of second home tourism and mobilities* (pp. 64–74). Routledge.

Coppock, J. T. (Ed.) (1977). *Second homes: Curse or blessing?* Pergamon.

Czarnecki, A. (2018). *Going local? Linking and integrating second-home owners with the community's economy*. Peter Lang.

Czarnecki, A., Dacko, A., & Dacko, M. (2021). Changes in mobility patterns and the switching roles of second homes as a result of the first wave of COVID-19". *Journal of Sustainable Tourism*, https://doi.org/10.1080/09669582.2021.2006201

Ellingsen, W. (2017). Rural second homes: A narrative of de-centralisation. *Sociologia Ruralis*, *57*(2), 229–244. https://doi.org/10.1111/soru.12130

Ellingsen, W. G., & Hidle, K. (2013). Performing home in mobility: Second homes in Norway. *Tourism Geographies*, *15*(2), 250–267. https://doi.org/10.1080/14616688.2011.647330

Farstad, M. (2018). The moral dilemma of second home owners' position in the host community. In C. M. Hall & D. K. Müller (Eds.), *The Routledge handbook of second home tourism and mobilities* (pp. 258–266). Routledge.

Farstad, M., & Rye. J. F. (2013). Second home owners, locals and their perspectives on rural development. *Journal of Rural Studies*, *30*, 41–51. https://doi.org/10.1016/j.jrurstud.2012.11.007

Fialová, D., Vágner, J., & Kůsová, T. (2018). Second homes, their users and relations to the rural space and the resident communities in Czechia. In C. M. Hall & D. K. Müller (Eds.), *The Routledge handbook of second home tourism and mobilities* (pp. 222–232). Routledge.

Flognfeldt, T., & Tjørve, E. (2013). The shift from hotels and lodges to second-home villages in mountain-resort accommodation. *Scandinavian Journal of Hospitality and Tourism*, *13*(4), 332–352. https://doi.org/10.1080/15022250.2013.862440

Foss, N. J. (2021). The impact of the Covid-19 pandemic on firms' organizational designs. *Journal of Management Studies*, *58*(1), 270–274. https://doi.org/10.1111/joms.12643

Gallent, N. (2014). The social value of second homes in rural communities. *Housing, Theory and Society*, *31*(2), 174–191. https://doi.org/10.1080/14036096.2013.830986

Gallent, N., Mace, A., & Tewdwr-Jones, M. (2005). *Second homes: European perspectives and UK policies*. Ashgate.

Gerber, J.-D., & Tanner, M. B. (2018). The role of Alpine development regimes in the development of second homes: Preliminary lessons from Switzerland. *Land Use Policy*, *77*, 859–870. https://doi.org/10.1016/j.landusepol.2017.09.017

González-Herrera, M. R., Rodríguez-Rodríguez, M. Á., & Santana-Rivero, C. (2021). The contribution of international residential mobility to tourism Development: Cienfuegos City, Cuba. In J. Domínguez-Mujica, J. McGarrigle, & J. M. Parreño-Castellano (Eds.), *International residential mobilities: From lifestyle migrations to tourism gentrification* (pp. 215–239). Springer.

Hall, C. M. (2015). Second homes planning, policy and governance. *Journal of Policy Research in Tourism, Leisure and Events*, *7*(1), 1–14. https://doi.org/10.1080/19407963.2014.964251

Hall, C. M., & Müller, D. K. (Eds.) (2004). *Tourism, mobility and second homes: Between elite landscape and common ground*. Channel View Books.

Hall, C. M., & Müller, D. K. (Eds.) (2018). *The Routledge handbook of second home tourism and mobilities*. Routledge.

Halseth, G., Pitkänen, K., Adamiak, C., & Vepsäläinen, M. (2018). Rights to the rural: Comparison of political and property/land rights of second home owners in Canada, Finland and Poland. In C. M. Hall & D. K. Müller (Eds.), *The Routledge handbook of second home tourism and mobilities* (pp. 52–63). Routledge.

Hannonen, O. (2018). Second home owners as tourism trend-setters: A case of residential tourists in Gran Canaria. *Journal of Spatial and Organizational Dynamics*, *6*(4), 345–359.

Hannonen, O. (2020). In search of a digital nomad: Defining the phenomenon. *Information Technology & Tourism*, *22*(3), 335–353. https://doi.org/10.1007/s40558-020-00177-z

Hannonen, O., Tuulentie, S., & Pitkänen, K. (2015). Borders and second home tourism: Norwegian and Russian second home owners in Finnish border areas. *Journal of Borderlands Studies*, *30*(1), 53–67. https://doi.org/10.1080/08865655.2015.1012736

Hiltunen, M. J., & Rehunen, A. (2014). Second home mobility in Finland: Patterns, practices and relations of leisure oriented mobile lifestyle. *Fennia*, *192*(1), 1–22.

Hiltunen, M. J., Pitkänen, K., & Halseth, G. (2016). Environmental perceptions of second home tourism impacts in Finland. *Local Environment*, *21*(10), 1198–1214. https://doi.org/10.1080/13549839.2015.1079701

Hjalager, A.-M., Staunstrup, J. K., Sørensen, M. T., & Steffansen. R. N. (2022). The densification of second home areas—sustainable practice or speculative land use? *Land Use Policy*, *118*, 106143. https://doi.org/10.1016/j.landusepol.2022.106143

Honkanen, A., Pitkänen, K., & Hall, C. M. (2016). A local perspective on cross-border tourism: Russian second home ownership in Eastern Finland. *International Journal of Tourism Research*, *18*(2), 149–158. https://doi.org/10.1002/jtr.2041

Hoogendoorn, G., & Visser, G. (2015). Focusing on the 'blessing' and not the 'curse' of second homes: Notes from South Africa. *Area*, *47*(2), 179–184. https://doi.org/10.1111/area.12156

Hoogendoorn, G., & Fitchett, J. M. (2018). Perspectives on second homes, climate change and tourism in South Africa. *African Journal of Hospitality, Tourism and Leisure*, *7*(2), 1–18.

Huete, R., & Mantecón, A. (2012). Residential tourism or lifestyle migration: Social problems linked to the non-definition of the situation. In O. Moufakkir & P. M. Burns (Eds.), *Controversies in tourism* (pp. 160–173). CABI.

Jennings, B. M., & Krannich, R. S. (2013). A multidimensional exploration of the foundations of community attachment among seasonal and year-round residents. *Rural Sociology*, *78*(4), 498–527. https://doi.org/10.1111/ruso.12019

Kadi, J., Hochstenbach, C., & Lennartz, C. (2020). Multiple property ownership in times of late homeownership: A new conceptual vocabulary. *International Journal of Housing Policy*, *20*(1), 6–24. https://doi.org/10.1080/19491247.2019.1697514

Kaltenborn, B. P. (1998). The alternate home-motives of recreation home use. *Norsk Geografisk Tidsskrift*, *52*(3), 121–134. https://doi.org/10.1080/00291959808552393

Lagerqvist, M. (2014). The importance of an old rural cottage: Media representation and the construction of a national idyll in post-war Sweden. *Journal of Rural Studies*, *36*(1), 33–41. https://doi.org/10.1016/j.jrurstud.2014.06.006

Larsson, L., & Müller, D. K. (2019). Coping with second home tourism: responses and strategies of private and public service providers in western Sweden. *Current Issues in Tourism*, *22*(16), 1958–1974. https://doi.org/10.1080/13683500.2017.1411339

Lipkina, O. (2013). Motives for Russian second home ownership in Finland. *Scandinavian Journal of Hospitality and Tourism*, *13*(4), 299–316. https://doi.org/10.1080/15022250.2013.863039

Long, D. P., & Hoogendoorn, G. (2013). Second home owners' perceptions of a polluted environment: the case of Hartbeespoort. *South African Geographical Journal*, *95*(1), 91–104. https://doi.org/10.1080/03736245.2013.806112

Long, D., & Hoogendoorn, G. (2014). Second home owner perceptions of their environmental impacts: The case of Hartbeespoort. *Urban Forum*, *25*(4), 517–530. https://doi.org/10.1007/s12132-013-9208-y

Lundmark, L., & Marjavaara, R. (2013). Second home ownership: A blessing for all? *Scandinavian Journal of Hospitality and Tourism*, *13*(4), 281–298. https://doi.org/10.1080/15022250.2013.862439

Mamonova, N., & Sutherland, L. (2015). Rural gentrification in Russia: Renegotiating identity, alternative food production and social tensions in the countryside. *Journal of Rural Studies*, *42*, 154–165. https://doi.org/10.1016/j.jrurstud.2015.10.008

Marjavaara, R., & Lundholm, E. (2016). Does second-home ownership trigger migration in later life? *Population, Space and Place*, *22*(3), 228–240. https://doi.org/10.1002/psp.1880

McIntyre, N., Williams, D. R., & McHugh, K. E. (Eds.) (2006). *Multiple dwelling and tourism: Negotiating place, home and identity*. CABI.

Miletić, G.-M., Žmuk, B., & Mišetić, R. (2018). Second homes and local socio-economic development: The case of Croatia. *Journal of Housing and the Built Environment*, *33*(2), 301–318. https://doi.org/10.1007/s10901-017-9562-5

Montezuma, J., & McGarrigle, J. (2019). What motivates international homebuyers? Investor to lifestyle 'migrants' in a tourist city. *Tourism Geographies*, *21*(2), 214–234. https://doi.org/10.1080/14616688.2018.1470196

Mowl, G., Barke, M., & King, H. (2020). Exploring the heterogeneity of second homes and the 'residual' category. *Journal of Rural Studies*, *79*, 74–87. https://doi.org/10.1016/j.jrurstud.2020.08.016

Müller, D. K. (2004). Mobility, tourism and second homes. In A. A. Lew, C. M. Hall, & A. M. Williams (Eds.), *A companion to tourism* (pp. 387–398). Wiley-Blackwell.

Müller, D. K. (2014). Progress in second-home tourism research. In A. A. Lew, C. M. Hall, & A. M. Williams (Eds.), *The Wiley companion to tourism* (pp. 389–400). Wiley.

Müller, D. K. (2021a). 20 years of Nordic second-home tourism research: a review and future research agenda. *Scandinavian Journal of Hospitality and Tourism*, *21*(1), 91–101. https://doi.org/10.1080/15022250.2020.1823244

Müller, D. K. (2021b). Tourism and lifestyle-led mobilities. In J. Domínguez-Mujica, J. McGarrigle, & J. M. Parreño-Castellano (Eds.), *International residential mobilities: From lifestyle migrations to tourism gentrification* (pp. 339–351). Springer.

Müller, D. K., & Hall, C. M. (2018). Second home tourism: An introduction. In C. M. Hall & D. K. Müller (Eds.), *The Routledge handbook of second home tourism and mobilities* (pp. 3–14). Routledge.

Müller, D. K., Hall, C. M., & Keen, D. (2004). Second home tourism impact, planning and management. In C. M. Hall & D. K. Müller (Eds.), *Tourism, mobility and second homes: Between elite landscape and common ground* (pp. 15–32). Channel View.

Müller, D. K., & Hoogendoorn, G. (2013). Second homes: Curse or blessing? A review 36 years later. *Scandinavian Journal of Hospitality and Tourism*, *13*(4), 353–369. https://doi.org/10.1080/15022250.2013.860306

Muñiz, I., Calatayud, D., & Dobaño, R. (2013). The compensation hypothesis in Barcelona measured through the ecological footprint of mobility and housing. *Landscape and Urban Planning*, *113*, 113–119. https://doi.org/10.1016/j.landurbplan.2013.02.004

Nouza, M., Ólafsdóttir, R., & Müller, D. K. (2013). A new approach to spatial–temporal development of second homes: Case study from Iceland. *Scandinavian Journal of Hospitality and Tourism*, *13*(1), 20–37. https://doi.org/10.1080/15022250.2013.764512

Nouza, M., Ólafsdóttir, R., & Sæþórsdóttir, A. D. (2018). Motives and behaviour of second home owners in Iceland reflected by place attachment. *Current Issues in Tourism*, *21*(2), 225–242. https://doi.org/10.1080/13683500.2015.1072139

Opačić, V. T., & Koderman, M. (2019). Interrelations between spatial distribution of tourism and the second homes in Croatia and Slovenia. *Geografski Vestnik*, 91(1). 49–79.

Paris, C. (2011). *Affluence, mobility and second home ownership*. Routledge.

Paris, C. (2018). Australian holiday homes: Places of escape and sites of investment. In C. M. Hall & D. K. Müller (Eds.), *The Routledge handbook of second home tourism and mobilities* (pp. 152–166). Routledge.

Perles-Ribes, J. F., Ramón-Rodríguez, A. B., Sevilla-Jiménez, M., & Moreno-Izquierdo, L. (2018). Differences in the economic performance of hotel-based and residential tourist destinations measured by their retail activity. Evidence from Spain. *Current Issues in Tourism*, *21*(18), 2076–2107. https://doi.org/10.1080/13683500.2016.1240155

Pers, A. S., Lagerqvist, M., & Nordin, U. (2018). The family and the second home: On building sandcastles, sharing places and the passing of time. In C. M. Hall & D. K. Müller (Eds.), *The Routledge handbook of second home tourism and mobilities* (pp. 267–277). Routledge.

Pitkänen, K., Adamiak, C., & Halseth, G. (2014). Leisure activities and rural community change: Valuation and use of rural space among permanent residents and second home owners. *Sociologia Ruralis*, *54*(2), 143–166. https://doi.org/10.1111/soru.12023

Pitkänen, K., Sireni, M., Rannikko, P., Tuulentie, S., & Hiltunen, M. J. (2017). Temporary mobilities regenerating rural places: Case studies from northern and eastern Finland. *Journal of Rural and Community Development*, *12*(2/3), 93–113.

Pitkänen, K., Hannonen, O., Toso, S., Gallent, N., Hamiduddin, I., Halseth, G., Hall, C. M., Müller, D. K., Treivish, A., & Nefedova. T. (2020a). Second homes during Corona – Safe or unsafe haven and for whom? Reflections from researchers around the world. *Finnish Journal of Tourism Research*, *16*(2), 20–39.

Pitkänen, K., Lehtimäki, J., & Puhakka, R. (2020b). How do rural second homes affect human health and well-being? Review of potential impacts. *International Journal of Environmental Research and Public Health*, *17*(18), art. 6748. https://doi.org/10.3390/ijerph17186748

Rey-Valette, H., Rulleau, B., Hellequin, A., Meur-Férec, C., & Flanquart, H. (2015). Second-home owners and sea-level rise: The case of the Languedoc-Roussillon region (France). *Journal of Policy Research in Tourism, Leisure and Events*, 7(1), 32–47. https://doi.org/10.1080/19407963.2014.942734

Rinne, J., Paloniemi, R., Tuulentie, S., & Kietäiväinen, A. (2015). Participation of second-home users in local planning and decision-making–a study of three cottage-rich locations in Finland. *Journal of Policy Research in Tourism, Leisure and Events*, *7*(1), 98–114. https://doi.org/10.1080/19407963.2014.909818

Robertsson, L., & Marjavaara, R. (2015). The seasonal buzz: Knowledge transfer in a temporary setting. *Tourism Planning & Development*, *12*(3), 251–265. https://doi.org/10.1080/21568316.2014.947437

Roca, Z. (Ed.) (2013). *Second home tourism in Europe: Lifestyle issues and policy responses*. Ashgate.

Rogerson, C. M., & Hoogendoorn, G. (2014). VFR travel and second home tourism: The missing link? The case of South Africa. *Tourism Review International*, *18*(3), 167–178. https://doi.org/10.3727/154427214X14101901317156

Rye, J. F. (2015). Moving to the countryside? The case of second home users in Norway. *Norsk Geografisk Tidsskrift*, *69*(3), 166–177. https://doi.org/10.1080/00291951.2015.1040832

Schier, M., Hilti, N., Schad, H., Tippel, C., Dittrich-Wesbuer, A., & Monz, A. (2015). Residential multi-locality studies: The added value for research on families and second homes. *Tijdschrift voor economische en sociale geografie*, *106*(4), 439–452. https://doi.org/10.1111/tesg.12155

Seraphin, H., & Dosquet, F. (2020). Mountain tourism and second home tourism as post COVID-19 lockdown placebo? *Worldwide Hospitality and Tourism Themes*, *12*(4), 485–500. https://doi.org/10.1108/WHATT-05-2020-0027

Sheludkov, A., & Starikova, A. (2021). Nighttime-lights satellite imagery reveals hotspots of second home mobility in rural Russia (a case study of Yaroslavl Oblast). *Regional Science Policy & Practice*, https://doi.org/10.1111/rsp3.12441

Soto, M. T. R., & Clavé, S. A. (2017). Second homes and urban landscape patterns in Mediterranean coastal tourism destinations. *Land Use Policy*, *68*, 117–132. https://doi.org/10.1016/j.landusepol.2017.07.018

Soto, M. T. R., & Clavé, S. A. (2018). The role of second homes in a Mediterranean coastal mass tourism destination: An evolutionary perspective. In C. M. Hall & D. K. Müller (Eds.), *The Routledge handbook of second home tourism and mobilities* (pp. 27–38). Routledge.

Stergiou, D. P., Papatheodorou, A., & Tsartas, P. (2017). Second home conversion during the economic crisis: The case of Artemida, Greece. *Social & Cultural Geography*, *18*(8), 1129–1151. https://doi.org/10.1080/14649365.2016.1242151

Stiman, M. L. (2020a). Discourses of resource dependency: Second homeowners as "lifeblood" in vacationland. *Rural Sociology*, *85*(2), 468–494. https://doi.org/10.1111/ruso.12296

Stiman, M. (2020b). Second homes in the city and the country: A reappraisal of vacation homes in the twenty-first century. *International Journal of Housing Policy*, *20*(1), 53–74. https://doi.org/10.1080/19491247.2019.1627842

Strandell, A., & Hall, C. M. (2015). Impact of the residential environment on second home use in Finland–Testing the compensation hypothesis. *Landscape and Urban Planning*, *133*(1), 12–23. https://doi.org/10.1016/j.landurbplan.2014.09.011

Tangeland, T., Vennesland, B., & Nybakk, E. (2013). Second-home owners' intention to purchase nature-based tourism activity products – A Norwegian case study. *Tourism Management*, *36*, 364–376. https://doi.org/10.1016/j.tourman.2012.10.006

Thompson, B. Y. (2019). The digital nomad lifestyle: (Remote) work/leisure balance, privilege, and constructed community. *International Journal of the Sociology of Leisure*, 2(1), 27–42. https://doi.org/10.1007/s41978-018-00030-y

Treivish, A. I. (2014). 'Dacha studies' as the science on second homes in the West and in Russia. *Regional Research of Russia*, *4*(3), 179–188.

Tuulentie, S., & Kietäväinen, A. (2020). New rural community? Narratives from second home owners about everyday life in a tourist region in Finnish Lapland. *Sociologia Ruralis*, *60*(2), 357–374. https://doi.org/10.1111/soru.12288

Velvin, J., Kvikstad, T. M., Drag, E., & Krogh, E. (2013). The impact of second home tourism on local economic development in rural areas in Norway. *Tourism Economics*, *19*(3), 689–705. https://doi.org/10.5367/te.2013.0216

Vepsäläinen, M., & Pitkänen, K. (2010). Second home countryside: Representations of the rural in Finnish popular discourses. *Journal of Rural Studies*, 26(2), 194–204. https://doi.org/10.1016/j.jrurstud.2009.07.002

Visser, G., & Hoogendoorn, G. (2015). A decade of second home tourism research in South Africa: research prospects for the developing world? *South African Geographical Journal*, 97(2), 111–122. https://hdl.handle.net/10520/EJC171562

Walters, T., & Carr, N. (2019). Changing patterns of conspicuous consumption: Media representations of luxury in second homes. *Journal of Consumer Culture*, *19*(3), 295–315. https://doi.org/10.1177/1469540517717778

Willberg, E., Järv, O., Väisänen, T., & Toivonen, T. (2021). Escaping from cities during the covid-19 crisis: Using mobile phone data to trace mobility in Finland. *ISPRS International Journal of Geo-information*, *10*(2): art. 103.

Wind, B., Dewilde, C., & Doling, J. (2020). Secondary property ownership in Europe: Contributing to asset-based welfare strategies and the 'really big trade-off'. *International Journal of Housing Policy*, 20(1), 25–52. https://doi.org/10.1080/19491247.2019.1573961

Wolfe, R. I. (1951). Summer cottagers in Ontario. *Economic Geography*, 27(1), 10–32. https://doi.org/10.2307/141807

Wolfe, R. I. (1952). Wasaga Beach: The divorce from the geographic environment. *Canadian Geographer*, *1*(2), 57–66. https://doi.org/10.1111/j.1541-0064.1952.tb01711.x

Wolfe, R. I. (1962). The summer resorts of Ontario in the Nineteenth Century. *Ontario History*, *54*, 149–161.

Wolfe, R. I. (1965). About cottages and cottagers. *Landscape*, 15(1), 6–8.

Wolfe, R. I. (1970). Discussion of vacation homes, environmental preferences and spatial behavior. *Journal of Leisure Research*, 2(1), 85–87. https://doi.org/10.1080/00222216.1970.11969986

Wong, B. K. M., & Musa, G. (2014). Retirement motivation among 'Malaysia my second home' participants. *Tourism Management*, *40*, 141–154. https://doi.org/10.1016/j.tourman.2013.06.002

Wong, B. K. M., Musa, G., & Taha, A. Z. (2017). Malaysia my second home: The influence of push and pull motivations on satisfaction. *Tourism Management*, *61*, 394–410. https://doi.org/10.1016/j.tourman.2017.03.003

Wu, Y., Xu, H., & Lew, A. A. (2015). Consumption-led mobilized urbanism: Socio-spatial separation in the second-home city of Sanya. *Mobilities*, *10*(1), 136–154. https://doi.org/10.1080/17450101.2013.853952

Zoğal, V., Domènech, A., & Emekli, G. (2022). Stay at (which) home: Second homes during and after the COVID-19 pandemic. *Journal of Tourism Futures*, 8(1), 125–133. https://doi.org/10.1108/JTF-06-2020-0090

Chapter 31

Mega-events and Tourism

Ning (Chris) Chen

Mega-events have been widely recognised for their role in driving travelling and tourism activities in both event and tourism literature (Getz, 2008; Getz & Page, 2016). Hosting events, especially mega-events, at a tourism destination brings in various benefits, from enhancing destination image (Deng & Li, 2014), leveraging destination competitiveness (Dickson et al., 2017), to resulting in short-term and long-term economic, tourism, social, cultural, and/or environmental outcomes for the host region (Preuss, 2007). The process of convincing various stakeholders, including event owners and local community, prior to bringing any mega-event to a destination is long and complex. Accordingly, previous research on mega-events and tourism has been largely focused on the legacy (Dansero & Puttilli, 2010), the positive and negative impacts of the mega-events (Dwyer et al., 2000), the acceptance of local community (Gursoy & Kendall, 2006; Gursoy et al., 2017), the coordination with various stakeholders (Van Niekerk & Getz. 2019), and of course, the planning and management of these mega-events (Bramwell, 1997). These topics are rooted in disciplines of economics, management, marketing, civil management, and environmental psychology.

This chapter thus reviews works on linking mega-events to tourism in these disciplines, and portrays a paradigm allowing us to examine key relevant topics in this linkage, from mega-event management, mega-event tourism, mega-event experience, mega-event stakeholders, to mega-event legacy and impacts, and local stakeholders, residents, and volunteers. Discussion is based on reviewing management theories such as stakeholder theory, as well as social theories including social exchange theory, social alliance theory, social representations theory, interdependence theory, and resource theory. From this sense, mega-events are viewed as both a motivator driving travelling and a contributor to destination economy and development in contemporary tourism. The effects of mega-events on tourism are discussed in the economic, social, cultural, environmental, political, and psychological aspects.

The Wiley Blackwell Companion to Tourism, Second Edition. Edited by C. Michael Hall.

Mega-events, Tourism, and Event Experience: The Triangle

In a tourism context, planned events comprise a key element in both supply and demand sides of the market. At the tourist origin end, events are an important driver of travelling and tourism activities, for professional sport athletes (Weber et al., 2012), event participants (Akgunduz & Coşar, 2018), and spectators and visitors (Jiang & Chen, 2019). At the tourism destination end, events contribute to local tourism development such as building infrastructure and establishing international reputations, and feature in tourism and destination marketing plans (Getz & Page, 2016). Event tourism, as indicated by Getz (2014, p. 401), 'concentrates on the roles that events can play within tourism and other development initiatives'. It is a spatial-temporal phenomenon, in terms of the specification of time and location, which is often integrated in the tourism development of this location. Various fundamental aspects of event tourism have been elaborated upon in previous editions of this book (Getz, 2004; Getz, 2014).

The main categories of (planned) events, in the scope of event tourism in terms of their form, include: cultural celebration events (e.g. festivals and carnivals); political and state events (e.g. summits and political conventions); arts and entertainment events (e.g. concerts and ceremonies); business events (e.g. trade shows and meetings); educational and scientific events (e.g. seminars and conferences); sport and recreational events (e.g. Olympics and city marathons); and private events (e.g. weddings and parties) (Getz, 2008). Once the scope of these events reach a certain capacity and/or size in terms of visitor attractiveness, mediated reach, cost and transformative impact, they are referred to as mega-events and imply a different approach for study (Müller, 2015). Accordingly, mega-events can be defined as 'ambulatory occasions of a fixed duration that:

- attract a large number of visitors;
- have large mediated reach;
- come with large costs; and
- have large impacts on the built environment and the population' (Müller, 2015, p. 629).

Events function as driving social and economic exchange, symbolic value, and personal development (Getz, 2004), and naturally provide an opportunity to bring people together, meeting and interacting with each other, whether they are from the neighbourhood or overseas. Mega-events, specifically, contribute to the host regions in various impacts such as regional economic development and enhancing national pride (Lee & Taylor, 2005). The success of any mega-event, and its potential long-term impact, is largely determined in terms of the quality of the participant experience per se (Florek et al., 2008). From a consumer perspective, event experience is an important outcome of a mega-event, and thus should be considered as a key performance evaluator of the planning, management, and implementation of an event (Lee & Min, 2016). Event experience further influences participants' overall satisfaction which in addition affects their intention to return and revisit a host destination (Sorrentino et al., 2020).

Mega-event tourism sits in the intersection of two disciplinary fields, event and tourism, and thus is supported by theories and frameworks from both fields in studying its relevant topics. These theories are rooted in project management, stakeholder theory, social exchange theory, heritage management, destination marketing and management, regional development, and so on. Discussion on participants' experience in mega-events, on the other hand, can trace

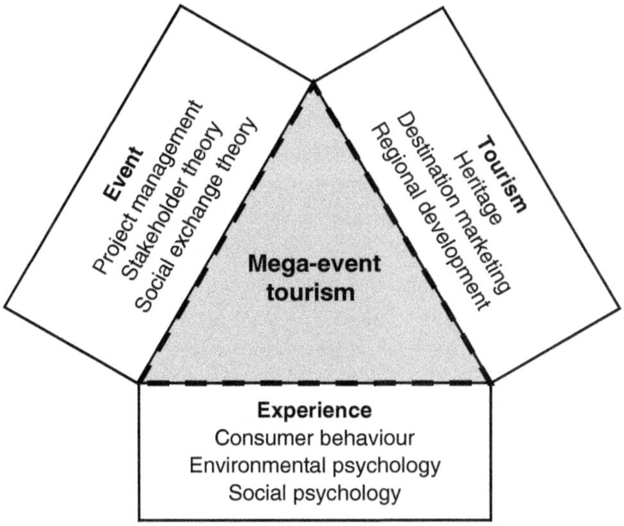

Figure 31.1. The interdisciplinary nature of mega-event tourism.

its roots in consumer psychology and behaviour, as well as generic environmental psychology and social psychology. The interdisciplinary nature of mega-event tourism is illustrated in Figure 31.1.

Project management principles fit in the context of mega-event tourism for mainly two reasons, both of which relates to the nature of mega-events. For one, a mega-event, no matter its size and scope, is temporary, and thus line with the definition of project, 'a temporary endeavour undertaken to create a unique product, service, or result' (Schwalbe, 2009, p. 6). Managing a mega-event, in its nature, is to organise a number of single-shot, time-bound, goal-driven activities (Grabher & Thiel, 2015). To achieve a vision, usually embedded in initiating a mega-event, mega-event organisers and managers need to overcome overall six main constraints, namely the triple constraints of scope, time, and cost, plus risk, quality, and resources (Schwalbe, 2009). These constraints are difficult to deal with due to their nature of uncertainty, difficulty in control, and affecting each other. For instance, no one can be 100% accurate in predicting potential risks, and event quality is in the end determined and defined by the customers. In addition, to increase the scope or shorten the time for preparation may result in significant increases in financial costs and human resources. In particular, mega-events, similar to projects, create an artificial countdown clock time, putting pressure on each of the tasks included for successfully delivering the mega-events.

On the other hand, mega-events are similar to complex projects for the difficulty in transferring knowledge between successive organising committees, as well as the difficulty in managing the power in its organisational body (Grabher & Thiel, 2015). Due to high demand in resources and scale, mega-event owners intentionally leave a large gap between hosting mega-events to allow sufficient time for planning and preparation. For instance, Summer Olympic Games are organised every four years, each of which lasts for around a month; the interval between two World Expos has differed through time but was fixed to five years in most recent instances. Accordingly, each mega-event creates a 'project ecology' with a complex multi-scalar relational space where individual, organisational, and institutional resources are accumulated and coordinated for performing this mega-event (Grabher & Thiel, 2015).

In summary, the focus of managing mega-events as projects is to professionalise and optimise actions coordinating inputs and outputs of knowledge. In spite of their temporary nature, mega-events still imply a continuity of past/present/future, as any project cannot be isolated from the society and its goal is to link and connect through a specific form, such as uniting people utilising sports in the case of the Olympic Games. Mega-events in almost every case specifically bear the responsibility in promoting the hosting place as a tourism destination.

Considering its project and relational nature, it is appropriate to apply stakeholder theory in managing and practising any mega-event, for coordinating among various participating individuals and organisations during hosting a mega-event (Laplume et al., 2008). These stakeholders do not only include the organising committee, event organisations, local community, sponsors and other commercial entities with interests, but also include organisations and institutions from the region that may bear the costs and/or expect benefits. These organisations and institutions such as destination management organisations (DMOs), local government, police, and hospitality sectors all have interests with intrinsic value, such as tourism development of the region. Figure 31.2 illustrates a typical stakeholder structure for hosting a mega-event. Besides economic benefits, mega-events can be levered to address different issues including regeneration, tourism, social integration, and quality of life, in terms of local needs and assets (Ziakas, 2015). Potential benefits derived from hosting mega-events for a tourism destination include enhanced place image as an attract for visit, promoted regional brands associated with the events, financial income from visitor expenditure and increase in employment, newly built infrastructure, and high-yield future tourists. (Getz, 2014).

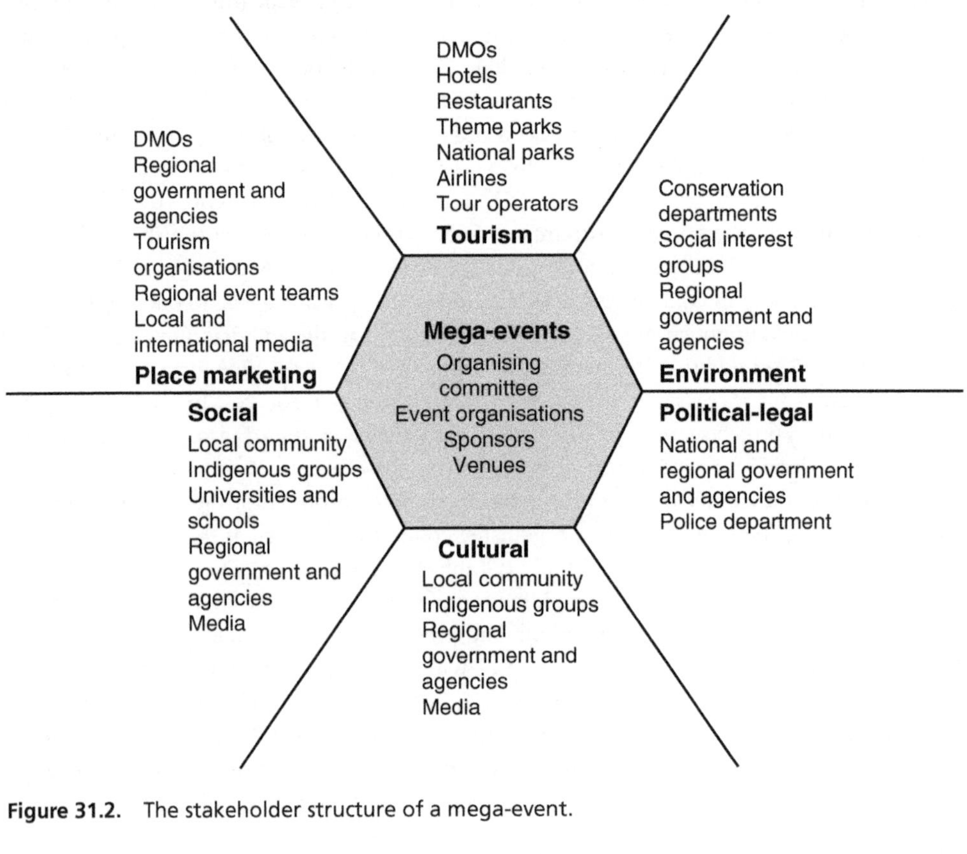

Figure 31.2. The stakeholder structure of a mega-event.

Stakeholder approach in investigating mega-events and tourism implies a multi-level social exchange process, which is coordinated and achieved via different stakeholder mechanisms such as asset-based community development (ABCD) framework (Misener & Schulenkorf, 2016). Through an established stakeholder network and structure, various stakeholders build processes of resource exchanges, and evaluate these social exchanges upon the calculation between costs and benefits (Cropanzano & Mitchell, 2005). This is in line with the principles of social exchange theory (SET), which represents a conceptual paradigm bridging key premises from social alliance theory, social representations theory, interdependence theory, and resource theory (Li et al., 2015).

The key benefits for regional tourism development from mega-events include leveraged destination image and increased interests of visiting from the intensive media exposure of the events, and relevant guerrilla marketing (Getz & Page, 2016). The long process of hosting a mega-event provides several key opportunities for tourism and destination marketing. The first marketing opportunity is toward event owners that are usually an internationally formed audience, such as International Olympic Committee (IOC) representatives, at the event bidding stage. In spite of a potentially small audience, the effect of marketing the potential host destination may draw global media attention, which naturally contribute to marketing the host region as a tourism destination. The second marketing opportunity lies in the pre-event period in promoting the mega-event to its target audience, from active participants (such as athletes for the Olympics) and audience (such as general visitors and spectators). The association of the host destination with the mega-event affects how potential visitors view and/or change their views of the destination, further driving them towards visiting the destination (Jiang & Chen, 2019). At the post-event stage, visitors have the potential to promote the destination as ambassadors, to a wider (even global) audience via a word-of-mouth approach (Chen & Dwyer, 2018; Nichols et al., 2017).

However, to encourage this voluntary promotion for the destination, previous research has validated the importance of creating unique and positive event experiences for visitors and participants (Florek et al., 2008; Sorrentino et al., 2020; Weber et al., 2012). Specifically, upon intensive interactions with the host destination and people within, visitors and participants build a unique sense of the place via mega-events, which forms the foundation of future revistation and interactions from an environmental psychological sense (Chen et al., 2021). This strong sense of place drives intimate relationship with the place e.g. attachment, and motivates storytelling in future occasions when the name of the place or the mega-events are mentioned. The experience creation is further co-created with locals, which adds meanings to the place (as illustrated in Figure 31.3), and in return affects how local residents perceive visitors and evaluate hosting the mega-events. Specific behaviours that are beneficial to the host destination may generate initiatives by the organising committee and/or event organisations, such as ambassador behaviours (Florek et al., 2008) and pro-environmental behaviours (Han et al., 2015). If coordinated well, mega-events may contribute to regional sustainability (Hall, 2012), as well as positive perceptions and attitude toward tourism, in addressing various issues such as over-tourism (Gössling et al., 2020).

It Is All About Legacy, Or Is It?

One key standpoint arguing for hosting mega-events, is to leave legacies for the host destination, especially for its urban (re)development and tourism development (Kassens-Noor et al., 2015). Legacy from mega-events can be defined in terms of three dimensions, whether

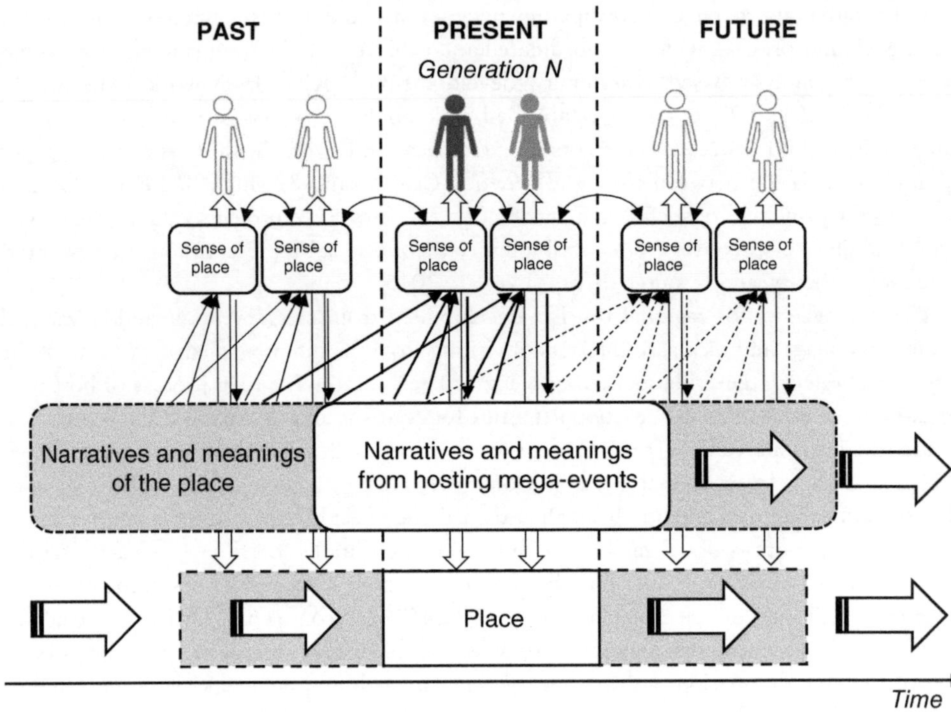

Figure 31.3. Sense of place evolves along with hosting mega-events.

it is planned or unplanned, whether it is positive or negative, and whether it is tangible or intangible. It refers to all structures created for and by a mega-event that remain in the host destination longer than the event itself, irrespective of the time of production and space (Preuss, 2015). Legacies of mega-events adopt many forms, such as economic impacts or infrastructural changes (Preuss, 2007, 2015), enhanced physical facilities and infrastructure, and increased social and cultural opportunities (Ritchie, 1984), information and education, public life, politics and cultures, sport, memories and history (Cashman & Hughes, 1999). These legacies are further distinguished in terms of pre-, during and post-events stages, and in terms of different impacts especially in the tourism context. Figure 31.4 summarises a typology of mega-event legacies in terms of impacts with examples.

Previous literature has summarised a number of typologies, which inevitably have overlapping categories and should be evaluated from an integrative sense. For instance, one typology defining a tourism legacy and another defining a sport legacy may consist of the same categories of infrastructure and image, and built infrastructure for a mega-event may contribute to both the physical impact and urban impact, as well as being a vehicle for social activities.

The legacy of mega-events can also be investigated in terms of its driving for sustainable development, focusing on the economic, socio-cultural, and environmental impacts, widely known as the triple bottom line. The United Nations (UN) in September 2015 launched its Sustainable Development Goals (SDGs) which covers a vast agenda of 17 goals and 169 targets (UN, 2015). Upon its launch, SDGs have been immediately adopted by various mega-event owners and organising committees, such as the IOC. In 2017, the IOC and the UN agreed on a direct partnership, which aims to help sport fulfil its role as an important enabler of sustainable development (IOC, 2017). However, sustainable development has long become

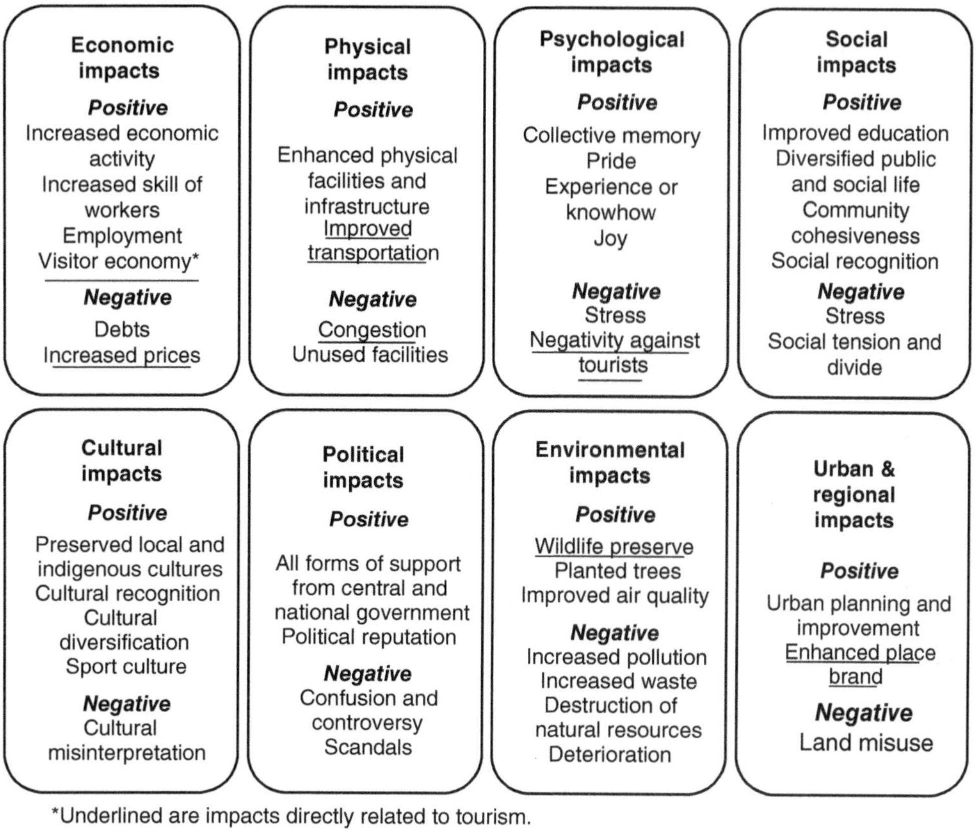

Figure 31.4. Legacies and impacts of mega-events and examples.

integrated into the goals of hosting mega-events. For example, the 2010 FIFA Football World Cup explicitly created a sustainable development program named 'Football for Hope' to achieve sustainable social development (Hall, 2012). In order to achieve overall sustainability in a tourism context, Hall (2012) suggests mega-event hosts seek long-term return on investment (ROI) of economic capital, focusing on contribution to economic growth as well as visitor numbers and expenditure. This pursuit of economic growth should be balanced with social benefits focusing on equity and social justice. Furthermore, a sustainable mega-event should develop a steady-state approach paying attention to the natural and ecological system. Overall, host destinations of mega-events should take a pragmatic approach to sustainability, optimising benefits within environmental parameters locally determined (Sharpley, 2009). Furthermore, each impact from hosting mega-events should be clarified for evaluating their potential to generate benefits to local stakeholders.

In spite of the mass focus on legacies and impacts in academia, mega-events start to lose their appeal as tools for urban development in practice, due to the 'white elephant' effect in the form of difficult-to-repurpose facilities, abandoned or vacant stadiums, and underperforming infrastructure (Lauermann, 2022). One notable issue is the often-low return on municipal investment in mega-events, due to either overestimated revenue or underestimated costs or cost overruns. This calls for more accurate assessment of the economic impacts, derived from hosting mega-events. Dwyer (2022) distinguishes three main approaches to the economic evaluation of events, namely standard economic impact analysis (EIA), computable general equilibrium modelling (CGE), and cost benefit analysis (CBA), and argues that EIA is fatally flawed. Dwyer and Forsyth (2019) suggest a 'hybrid' approach integrating both CGE and CBA for dealing with different aspects of the evaluation problem. In addition, for successfully bringing a mega-event to a destination, the bidding and proposing team frequently de-emphasises the potential negative impacts to the region and local community, leading to objections from local residents in hosting this mega-event. Therefore, the following question for careful evaluation is whether it is all about the legacy in hosting mega-events, and how to make sure a mega-event is hosted in a real sustainable sense for driving sustainable urban and tourism development.

An Emphasis On Local Stakeholders

Following the previous discussion on the assessment and evaluation of mega-events, research on mega-events and tourism should put an emphasis on local stakeholders, such as residents. Hosting mega-events such as the FIFA World Cup undoubtedly has short- and long-term effects on the host communities, from increasing residents' social capital and psychic income (Gibson et al., 2014), to their individual and social well-being (Yolal et al., 2016). The explicit benefits of hosting a mega-event have been widely acknowledged by most stakeholders in most cases. However, these benefits, burdens, and costs are not always equally distributed to and shared by different stakeholder groups.

Initiatives and pushes due to hosting mega-events put certain resident groups in more difficult situations, such as the pushed dislocation of impoverished residents in the city of Rio de Janeiro prior to the 2014 FIFA World Cup and 2016 Rio Olympics (Ivester, 2017). This, along with complaints against the high cost of hosting either mega-event, has resulted in a series of protests and demonstrations in at least ten Brazilian cities (Watts, 2014). Due to the large scale and tight time constraints of mega-events, organising committees have to make major adjustments proactively or passively to cope with changing circumstances, causing issues in the coordination of stakeholder interests. For example, the COVID pandemic has brought significant challenges to various industries since early 2020, especially those requiring international travel and large public assemblies such as tourism and events (Gössling et al., 2020; Hall et al., 2020). These challenges triggered mixed feelings among local residents in hosting the 2020 Tokyo Olympic Games, resulting in a backlash against this mega-event. Japanese Olympics sponsors were also frustrated by the 'very last minute' decision on postponement by organisers, and had to cancel or scale back promotional events and booths tied to the Olympic Games (Yamamitsu & Shiraki, 2021).

To address these challenges, the bidding teams and the organising committees of mega-events need to evaluate and assess the expected impacts of these events more carefully, and candidly communicate and negotiate with different stakeholder groups. In this process, they also need to realise the importance of the 'perceived' impacts, rather than the actual ones. In mega-event and tourism contexts, residents' perceived impacts influence their attitudes toward the events, and further cultivate their relevant behaviours, according to SET (Li et al., 2015). For example, previous studies have validated the relationship between residents' perceived impacts of mega-events and their support for these events based on SET (Gursoy et al., 2017; Kaplanidou et al., 2013). Residents who acknowledge the opportunities derived from mega-events to promote local cultures and traditions, to meet and engage people from different backgrounds, and to enhance their community pride are more inclined to support hosting such events (Gursoy et al., 2017). In principle, the more benefits and values residents expect from a mega-event, the more likely they support hosting this event. If residents perceive significant negative impacts to their community, such as in the case of Rio de Janeiro, they tend to not support hosting it.

In event literature, events are found in an environment stimulating positive psychological outcomes such as subjective well-being (Yolal et al., 2016) and sense of belonging (Jeon et al., 2014). This is due to event attendance improving and facilitating people's understanding of themselves via self-acceptance, of others via social coherence, and of the world via personal growth and mastery (Yolal et al., 2016). In particular, economic benefits derived from event hosting contribute to improving local standards of living, as well as quality of life and confidence among residents (Uysal et al., 2016). On the other hand, hosting mega-events often consumes large amounts of environmental and social resources at the local level, putting more pressure on local communities such as tax burdens, misuse and mismanagement of public resources, and expenditure of public funds (Nunkoo & Gursoy, 2016). Negatively perceived impacts of hosting mega-events therefore impinge on local quality of life.

Another direct impact of hosting mega-events on residents and other local stakeholders lies in volunteering. Event volunteers are defined as those who offer their time, knowledge, skills, labour, and experience at no wage cost to an event (Mair et al., 2021). Events, especially mega-events, rely heavily on the contribution by volunteers, who work episodically to fit the time restraints and limitation of mega-events. In continuing the discussion on event legacy, well-justified benefits from hosting mega-events encourage local residents in taking on volunteer roles, which further contributes in social inclusion legacy, in the form of social cohesion, civic pride and social capital (Nichols & Ralston, 2011).

Given the impacts and perceived impacts of hosting mega-events, and various sustainability issues, a significant dilemma exists. Popular mega-events that have the capability in self-promotion and with political and commercial interests often take a profit driven approach, overlooking social equity but being portrayed as big contributor to the local economy. These events, as with most contemporary tourism, are not as sustainable as they claim to be in their current form. However, smaller localised community-based events tend to be more sustainable, carefully incorporating local welfare over the long-term, but they are not as appealing as the mega-events. It is thus important to think whether hosting a mega-event in order to be competitive in a declining global economy should be the solution or rather a symptom of the problem.

Theoretical and Methodological Reflections

This chapter aims to provide insights on mega-event hosting, its implications to tourism, and more widely regional development, in terms of its context and interdisciplinary nature, its relevance to theories, and topics such as legacy, impact, and sustainability. The current paradigm allows us to examine the relationship between mega-events and tourism around key disciplines of economics, management, sociology, environmental psychology, and social psychology. Depending on the actual research question, relevant research methods include a case study approach in most works that investigate event legacy; economic evaluative approaches such as EIA, CGE, and CBA; anthropological and ethnographical approaches in understanding cultures and social communities; as well as psychology research methods investigating perceptions and behaviours. Mega-events have had, and will continue to have, significant impacts on the global and regional tourism environment, regardless of economic, social, and environmental circumstances. There are, of course, many other aspects of mega-event and tourism discussion that could have been considered. An all-around view at hosting mega-events and its implications for tourism will potentially open up new conversations, in fulfilling its economic, societal, cultural, political, geographical, and environmental role for future research.

References

Akgunduz, Y., & Coşar, Y. (2018). Motivations of event tourism participants and behavioural intentions. *Tourism and Hospitality Management*, *24*(2), 341–358. https://doi.org/10.20867/thm.24.2.4

Bramwell, B. (1997). Strategic planning before and after a mega-event. *Tourism Management*, *18*(3), 167–176. https://doi.org/10.1016/S0261-5177(96)00118-5

Cashman, R. I., & Hughes, A. (Eds). (1999). *Staging the Olympics: the event and its impact*. University of New South Wales Press.

Chen, N. C., Hall, C. M., & Prayag, G. (2021). *Sense of place and place attachment in tourism*. Routledge.

Chen, N., & Dwyer, L. (2018). Residents' place satisfaction and place attachment on destination brand-building behaviors: Conceptual and empirical differentiation. *Journal of Travel Research*, *57*(8), 1026–1041. https://doi.org/10.1177/0047287517729760

Cropanzano, R., & Mitchell, M. S. (2005). Social exchange theory: An interdisciplinary review. *Journal of Management*, *31*(6), 874–900. https://doi.org/10.1177/0149206305279602

Dansero, E., & Puttilli, M. (2010). Mega-events tourism legacies: the case of the Torino 2006 Winter Olympic Games–a territorialisation approach. *Leisure Studies*, *29*(3), 321–341. https://doi.org/10.1080/02614361003716966

Deng, Q., & Li, M. (2014). A model of event–destination image transfer. *Journal of Travel Research*, *53*(1), 69–82. https://doi.org/10.1177/0047287513491331

Dickson, T., Misener, L., & Darcy, S. (2017). Enhancing destination competitiveness through disability sport event legacies: Developing an interdisciplinary typology. *International Journal of Contemporary Hospitality Management*, *29*(3), 924–946. https://doi.org/10.1108/IJCHM-10-2015-0530

Dwyer, L. (2022). Economic impacts of events. In N. Wise & K. Maguire (Eds), *A research agenda for event impacts* (pp. 19–39). Edward Elgar Publishing.

Dwyer, L., & Forsyth, P. (2019). Evaluating special events: merging two essential approaches. *Event Management*, *23*(6), 897–911. https://doi.org/10.3727/152599519X15506259856417

Dwyer, L., Mellor, R., Mistilis, N., & Mules, T. (2000). A framework for assessing "tangible" and "intangible" impacts of events and conventions. *Event Management*, *6*(3), 175–189. https://doi.org/10.3727/096020197390257

Florek, M., Breitbarth, T., & Conejo, F. (2008). Mega event = Mega impact? Travelling fans' experience and perceptions of the 2006 FIFA World Cup host nation. *Journal of Sport & Tourism*, *13*(3), 199–219. https://doi.org/10.1080/14775080802310231

Getz, D. (2004). Geographic perspectives on event tourism. In A. A. Lew, C. M. Hall, & A. M. Williams (Eds), *A companion to tourism*, (pp. 410–422). Blackwell.

Getz, D. (2008). Event tourism: Definition, evolution, and research. *Tourism Management*, *29*(3), 403–428. https://doi.org/10.1016/j.tourman.2007.07.017

Getz, D. (2014). Timing tourism. In A. A. Lew, C. M. Hall, & A. M. Williams (Eds), *The Wiley Blackwell companion to tourism* (pp. 401–412). John Wiley & Sons.

Getz, D., & Page, S. J. (2016). Progress and prospects for event tourism research. *Tourism Management*, *52*, 593–631. https://doi.org/10.1016/j.tourman.2015.03.007

Gibson, H. J., Walker, M., Thapa, B., Kaplanidou, K., Geldenhuys, S., & Coetzee, W. (2014). Psychic income and social capital among host nation residents: A pre–post analysis of the 2010 FIFA World Cup in South Africa. *Tourism Management*, *44*, 113–122. https://doi.org/10.1016/j.tourman.2013.12.013

Gössling, S., McCabe, S., & Chen, N. C. (2020). A socio-psychological conceptualisation of overtourism. *Annals of Tourism Research*, *84*, 102976. https://doi.org/10.1016/j.annals.2020.102976

Grabher, G., & Thiel, J. (2015). Projects, people, professions: Trajectories of learning through a mega-event (the London 2012 case). *Geoforum*, *65*, 328–337. https://doi.org/10.1016/j.geoforum.2015.02.006

Gursoy, D., & Kendall, K. W. (2006). Hosting mega events: Modeling locals' support. *Annals of Tourism Research*, *33*(3), 603–623. https://doi.org/10.1016/j.annals.2006.01.005

Gursoy, D., Milito, M. C., & Nunkoo, R. (2017). Residents' support for a mega-event: The case of the 2014 FIFA World Cup, Natal, Brazil. *Journal of Destination Marketing & Management*, *6*(4), 344–352. https://doi.org/10.1016/j.jdmm.2017.09.003

Hall, C. M. (2012). Sustainable mega-events: Beyond the myth of balanced approaches to mega-event sustainability. *Event Management*, *16*(2), 119–131. https://doi.org/10.3727/152599512X13343565268294

Hall, C. M., Scott, D., & Gössling, S. (2020). Pandemics, transformations and tourism: Be careful what you wish for. *Tourism Geographies*, *22*(3), 577–598. https://doi.org/10.1080/14616688.2020.1759131

Han, J. H., Nelson, C. M., & Kim, C. (2015). Pro-environmental behavior in sport event tourism: Roles of event attendees and destinations. *Tourism Geographies*, *17*(5), 719–737. https://doi.org/10.1080/14616688.2015.1084037

International Olympic Committee (IOC) (2017, May 5). *IOC welcomes enhancement of close cooperation with the United Nations.* https://olympics.com/ioc/news/ioc-welcomes-enhancement-of-close-cooperation-with-the-united-nations

Ivester, S. (2017). Removal, resistance and the right to the Olympic city: The case of Vila Autodromo in Rio de Janeiro. *Journal of Urban Affairs*, *39*(7), 970–985. https://doi.org/10.1080/07352166.2017.1355665

Jeon, M., Shin, M., & Lee, S. J. (2014). An analysis of factors influencing individually and socially oriented happiness: The case of Seoul. *Journal of Policy Studies*, *29*(2), 43–69. https://hdl.handle.net/10371/93295

Jiang, Y., & Chen, N. C. (2019). Event attendance motives, host city evaluation, and behavioral intentions: An empirical study of Rio 2016. *International Journal of Contemporary Hospitality Management*, *31*(8), 3270–3286. https://doi.org/10.1108/IJCHM-06-2018-0501

Kaplanidou, K., Karadakis, K., Gibson, H., Thapa, B., Walker, M., Geldenhuys, S., & Coetzee, W. (2013). Quality of life, event impacts, and mega-event support among South African residents before and after the 2010 FIFA World Cup. *Journal of Travel Research*, *52*(5), 631–645. https://doi.org/10.1177/0047287513478501

Kassens-Noor, E., Wilson, M., Müller, S., Maharaj, B., & Huntoon, L. (2015). Towards a mega-event legacy framework. *Leisure Studies*, *34*(6), 665–671. https://doi.org/10.1080/02614367.2015.1035316

Laplume, A. O., Sonpar, K., & Litz, R. A. (2008). Stakeholder theory: Reviewing a theory that moves us. *Journal of Management*, *34*(6), 1152–1189. https://doi.org/10.1177/0149206308324322

Lauermann, J. (2022). The declining appeal of mega-events in entrepreneurial cities: From Los Angeles 1984 to Los Angeles 2028. *Environment and Planning C: Politics and Space*, 23996544211066101. https://doi.org/10.1177/23996544211066101

Lee, C. K., & Taylor, T. (2005). Critical reflections on the economic impact assessment of a mega-event: the case of 2002 FIFA World Cup. *Tourism Management*, *26*(4), 595–603. https://doi.org/10.1016/j.tourman.2004.03.002

Lee, J. S., & Min, C. K. (2016). Examining the quality antecedents and moderating effects of experiential value in a mega-event. *Journal of Travel & Tourism Marketing*, *33*(3), 326–347. https://doi.org/10.1080/10548408.2015.1051642

Li, X., Hsu, C. H., & Lawton, L. J. (2015). Understanding residents' perception changes toward a mega-event through a dual-theory lens. *Journal of Travel Research*, *54*(3), 396–410. https://doi.org/10.1177/0047287513517422

Mair, J., Chien, P. M., Kelly, S. J., & Derrington, S. (2021). Social impacts of mega-events: A systematic narrative review and research agenda. *Journal of Sustainable Tourism*. https://doi.org/10.1080/09669582.2020.1870989

Misener, L., & Schulenkorf, N. (2016). Rethinking the social value of sport events through an asset-based community development (ABCD) perspective. *Journal of Sport Management*, *30*(3), 329–340. https://doi.org/10.1123/jsm.2015-0203

Müller, M. (2015). What makes an event a mega-event? Definitions and sizes. *Leisure Studies*, *34*(6), 627–642. https://doi.org/10.1080/02614367.2014.993333

Nichols, G., & Ralston, R. (2011). Social inclusion through volunteering: The legacy potential of the 2012 Olympic Games. *Sociology*, *45*(5), 900–914. https://doi.org/10.1177/0038038511413413

Nichols, G., Ralston, R., & Holmes, K. (2017). The 2012 Olympic Ambassadors and sustainable tourism legacy. *Journal of Sustainable Tourism*, *25*(11), 1513–1528. https://doi.org/10.1080/09669582.2017.1291648

Nunkoo, R., & Gursoy, D. (2016). Rethinking the role of power and trust in tourism planning. *Journal of Hospitality Marketing & Management*, *25*(4), 512–522. https://doi.org/10.1080/19368623.2015.1019170

Preuss, H. (2007). The conceptualisation and measurement of mega sport event legacies. *Journal of Sport & Tourism*, *12*(3–4), 207–228. https://doi.org/10.1080/14775080701736957

Preuss, H. (2015). A framework for identifying the legacies of a mega sport event. *Leisure Studies*, *34*(6), 643–664. https://doi.org/10.1080/02614367.2014.994552

Ritchie, J. R. B. (1984). Assessing the impact of hallmark events: conceptual and research issues. *Journal of Travel Research*, *23*(1), 2–11. https://doi.org/10.1177/004728758402300101

Schwalbe, K. (2009). *Introduction to project management*. Course Technology Cengage Learning.

Sharpley, R. (2009). *Tourism development and the environment: Beyond sustainability?* Earthscan.

Sorrentino, A., Fu, X., Romano, R., Quintano, M., & Risitano, M. (2020). Measuring event experience and its behavioral consequences in the context of a sports mega-event. *Journal of Hospitality and Tourism Insights*, *3*(5), 589–605. https://doi.org/10.1108/JHTI-03-2020-0026

United Nations (UN). (2015). *Transforming our world: The 2030 Agenda for sustainable development*. UN.

Uysal, M., Sirgy, M. J., Woo, E., & Kim, H. L. (2016). Quality of life (QOL) and well-being research in tourism. *Tourism Management*, *53*, 244–261. https://doi.org/10.1016/j.tourman.2015.07.013

Van Niekerk, M., & Getz, D. (2019). *Event stakeholders*. Goodfellow Publishers.

Watts, J. (2014). *Anti-World Cup protests in Brazilian cities mark countdown to kick-off*. The Guardian. https://www.theguardian.com/football/2014/jun/12/anti-world-cup-protests-brazilian-cities-sao-paulo-rio-de-janeiro

Weber, K., Ali-Knight, J., & Nicol, L. (2012). Athletes' event experiences of the XIX Commonwealth Games in Delhi, India. *International Journal of Event and Festival Management*, *3*(1), 12–29. https://doi.org/10.1108/17582951211210915

Yamamitsu, E., & Shiraki, M. (2021). *Frustrated by delays, Tokyo 2020 sponsors cancel booths, parties*. Reuters. https://www.reuters.com/lifestyle/sports/exclusive-olympics-frustrated-by-delays-tokyo-2020-sponsors-cancel-booths-2021-07-08/#:~:text=Some%2060%20Japanese%20companies%20paid,involved%20in%20the%20Tokyo%20Games

Yolal, M., Gursoy, D., Uysal, M., Kim, H. L., & Karacaoğlu, S. (2016). Impacts of festivals and events on residents' well-being. *Annals of Tourism Research*, *61*, 1–18. https://doi.org/10.1016/j.annals.2016.07.008

Ziakas, V. (2015). For the benefit of all? Developing a critical perspective in mega-event leverage. *Leisure Studies*, *34*(6), 689–702. https://doi.org/10.1080/02614367.2014.986507

Chapter 32

Food and Tourism

Ling-en Wang, Guiqiang Qiao, Xiaoyu Zhang, Qi Zhang, & Junshuo Li

Introduction

As a basic element of human existence, food plays a particularly important role in the development of human society. Regardless of individuals' identity or origin, at work or as a tourist, food is a daily necessity. However, differing from other survival elements such as water or air, people have put forward more diverse requirements for food. People are often not satisfied with 'edible' food and are more looking forward to 'delicious' food or a different dining experience, especially when leaving home for travel purposes. In fact, for many places, food itself is the attraction that motivate people to visit. Thus, food and tourism are interlinked and have a natural connection.

The relationship between food and tourism has attracted the attention of many researchers from different areas (Cohen & Avieli, 2004; Bertella, 2011; Everett & Slocum, 2013; Hall, 2020). Based on a literature analysis, this chapter illustrates the topic of food and tourism from four perspectives, namely gourmet tourism, tourists' food consumption behaviour, food waste in hospitality, and food and destination development.

Gourmet Tourism

Definition

Gastronomy/gastronomic Tourism, Culinary Tourism, and Food Tourism

To express the connection between food and tourism, terms such as 'food tourism', 'cuisine tourism', 'food and wine tourism', 'culinary tourism', 'gastronomy/gastronomic tourism', 'tasting tourism', and 'gourmet tourism' have been coined (Horng & Tsai, 2012; Ellis et al., 2018). The noun 'cuisine' (a loan word from the French language) refers to a style of food preparation (e.g., 'French cuisine'), and the adjective 'culinary' (from the Latin word for 'kitchen': culina) can be used to modify nouns associated with the practice of cooking (e.g., 'culinary school')

The Wiley Blackwell Companion to Tourism, Second Edition. Edited by C. Michael Hall.

or its social context (e.g., 'culinary experience'). 'Gastronomy' (an anglicisation of the French word gastronomie, itself derived from the Greek portmanteau word gastronomia) can be defined as the science or art of eating and drinking well and has the connotation of a high-status form of gracious or elegant living (Horng & Tsai, 2010; Pavlidis & Markantonatou, 2020).

Culinary Tourism

The growing research field of culinary tourism can be seen as a form of cultural tourism, although there are many different definitions of what the term means, with it often being used interchangeably with food tourism and gastronomic tourism (Hjalager & Richards, 2002; Long, 2004; Ignatov & Smith, 2006; Horng & Tai, 2010). Culinary tourism is not only about consuming food on a trip, but also about the memorable experience of a tourism destination's unique 'food culture' (including cooking and dining) and 'local knowledge'. The phrase 'culinary tourism' was coined in 1998 and refers to international tourists who plan their travels partly or largely based on the desire to experience different and 'exotic' culinary specialties and traditions (Long, 2004). Long (2004, p. 21) describes culinary tourism as '. . .involving the intentional, exploratory participation in the food way of an "Other", participation including the consumption — or preparation and presentation for consumption — of a food item, cuisine, meal system, or eating style considered as belonging to a culinary system not one's own'. Lockwood (2008) expanded on Long's (2004) view, noting that food can be seen as 'a subject and medium, [both a] destination and [a] vehicle, for tourism' (Long, 2004, p. 20 cited in Lockwood, 2008, p. 362). Tourists travel to countries and cities renowned for their cuisines, restaurants, and chefs, such as France, Italy, and Spain. These tourists might be accurately thought of as pilgrims, in which the traveller holds an appreciation of and respect for a particular type of cuisine and is seeking it in its authentic form (Long, 2012). According to Ignatov and Smith (2006), culinary tourism may be defined as follows: 'trips during which the purchase or consumption of regional foods (including beverages), or the observation and study of food production (from agriculture to cooking schools), represent a significant motivation or activity' (Ignatov & Smith, 2006, p. 235).

Food Tourism

In contrast to culinary tourism, the concept of food tourism focuses more on the physical experiences of tourists. Hall and Sharples (2003) provides an excellent preliminary understanding of food tourism as 'visitation to primary and secondary food producers, food festivals, restaurants, and specific locations for which food tasting and/or experiencing the attributes of specialist food production regions are the primary motivation for travel' (Hall & Sharples, 2012, p. 10). Thus, when defining the term food tourism, a distinction needs to be made between types of tourists: the tourists who consider food as part of their travel experience; and the food motivated tourists who chooses relevant activities and destinations based on food interests (Hall, 2012).

The World Food Travel Association (WFTA) (2022) classifies the historical evolution of food tourism into three phases: in the first phase, from 2001 to 2011, food tourism was understood as the pursuit and enjoyment of unique and memorable food and drink experiences; in the second phase, from 2012 to 2018, food tourism was promoted and developed through social media and TV shows; and in the third phase, from 2018 to the present, food tourism includes different kinds of experiences, such as learning cookery and visiting farmlands.

Indeed, food tourism involves activites of visiting food producers, food festivals and various locations/space for food tasting and experiencing (Hall & Mitchell, 2001, p. 308). Tourists travel for a range of reasons; hence, for their travel experience, food tourism may be one of the components. The popular films *Eat Pray Love* (Murphy, 2010) and *The Trip to Italy* (Winterbottom, 2014) motivated many of their viewers to travel to Italy to experience food tourism. However, food and cuisine cannot be separated from culture and heritage, and Italy as a tourist destination is an evocative example of this. For these tourists, the cuisine of Italy is not only exotic and satisfying, but it is also emotive and experiential (Frost et al., 2016).

Through the proses of reviewing relevant literature, it has become apparent that food tourism has a close relationship with wine tourism (Hall et al., 2000). Some academic studies point out that wine tourism is a subset of food tourism. Hall (1996, p. 150) defined wine tourism as 'visits to vineyards, wineries, wine festivals, and wine shows, where wine tasting and experiencing the properties of wine regions are major motivators for visitors'. Additionally, Laing and Frost (2014) have identified an independent traveller known as the food explorer, who is a person with a keen interest in learning about the different cuisines and traditions of the world, and who is therefore motivated to travel to destinations with distinctive culinary and winemaking traditions and experience the local fare.

Gastronomy or Gastronomic Tourism

As noted above, 'gastronomy' can be defined as the science or art of eating and drinking well, with the gastronome being a connoisseur of fine food and wines. While the term gastronomic tourism or gastronomy tourism have been utilised far less than both 'culinary tourism' and 'food tourism' (Ellis et al. 2018), Gheorghe et al. (2014) pointed out that gastronomic tourism is a new trend within contemporary tourism. The United Nations World Tourism Organization (UNWTO, 2022) defines Gastronomy tourism as: 'a type of tourism activity, which is characterized by the visitors' experience linked with food and related products and activities while travelling'. Additionally, gastronomy tourism can also involve activities such as attending food festivals, visiting local food growers, and taking lessons in cooking local dishes (UNWTO, 2022). Dixit (2019) summarised gastronomic tourism as having its roots in agriculture, culture, and tourism. Kivela and Crotts (2006) stated that gastronomy tourism focusses on the relationship between culture and food, tourists are attracted and motivated to taste food; participate in preparing food processes; experience and experiment with food; search and discover food; as well as understand and write about food. This can add value to the tourist experience, which yields a high level of tourism satisfaction. Wine tourism is a good example, in Australia and New Zealand, attractive tourism packages have been promoted, offering the gourmet and wine weekend-escapes, which created demand for short or long weekend holidays. The implication for food-related industries is that gastronomy ought to be regarded as a type of new tourism (Kivela & Crotts, 2006). Furthermore, some Asian countries such as China, Japan, and Thailand have promoted gastronomy tours to attract international visitors who want to participate in gourmet-themed activities, such as shopping for produce at local markets, learning about cooking indigenous cuisine, dining out, and visiting local wineries (Kivela & Crotts, 2006).

Although three of the most frequently used terms describing food-related tourism activity are: 'food tourism', 'culinary tourism', and 'gastronomy tourism', the WFTA (2022) now regard these terms as synonymous. In defining food tourism stakeholders, the association views 20 sectors which can be grouped together into a 'food tourism industry cluster', including agricultural producers, food and beverage retailers, the hospitality industry (e.g., bars and restaurants), culinary schools, destination management organisations, tourism operators, tourist accommodation providers, trade groups, government departments, non-governmental organisations, academia, and the media (WFTA, 2022).

Many academics argue that the three terms: 'food tourism', 'culinary tourism', and 'gastronomy/gastronomic tourism', are used interchangeably in some instances (Horng & Tsai, 2012). Ellis et al. (2018, p. 252) argued that 'these terms do commonly appear to be used in slightly different contexts, and the meaning of each term represents different perspectives within the host-guest structure of tourism'. Culinary tourism and food tourism are conceptually related by the tourist's motivation for food-related activities, although they differ in their respective focus. For the 'culinary tourist', such activities provide an opportunity to learn about the local culture through participation in its culinary traditions, whereas for the 'food tourist', the motivation for activities is to experience sensory pleasures from physical interactions with the local fare (Ellis et al., 2018). Kim and Eves (2012) summarised that food tourism to tourists is about 'exciting experience, escape from routine, sensory appeal (sensory pleasure through taste, smell, touch, etc.), health concern, learning knowledge, authentic experience, togetherness (with families and friends), and prestige (to have high standing in the eyes of surrounding people)' (Kim & Eves, 2012, p. 1461). Culinary tourism and gastronomic/gastronomy tourism are conceptually related by the tourist's motivation for food-related cultural experiences (Wondirad et al., 2021; Horng & Tsai, 2012; Sánchez-Cañizares & López-Guzmán, 2012; Sormaz et al., 2016), however, the conceptual focus differ. For the culinary tourist, the focus is on the overall cultural experience (López-Guzmán & Sánchez-Cañizares, 2012), whereas for the gastronomic tourist, the focus is on the role food plays in the destination's culture (Chaney & Ryan, 2012; Ellis et al., 2018). From this discussion of the conceptual differences and relations between food tourism, culinary tourism, and gastronomic tourism, it is evident that what is important for tourists, is how food and culture are integrated to enrich their travel experience.

Tourists' Food Consumption Behavior

Eating, whether at home, or away from home, is arguably the most fundamental activity in a human being's life (Capaldi, 1996). As an away-from-home activity, tourism becomes a popular lifestyle in modern society; thus, eating or food consumption acts as an integral part of tourism (Liu et al., 2017a; Liu et al., 2017b). The importance of food consumption is justified by its one-third proportion of total travel expenditure (Kim et al., 2009), its essential function of shaping the overall travel experience (Mak et al., 2017) and its attractive role in motivating people to visit a specific destination (Tsai & Wang, 2017). Therefore, tourist food consumption behaviour brings the food and tourism together from the consumers' perspective.

Factors Affecting Tourists' Food Consumption

As previously cited by Hall and Mitchell (2001, p. 308), food tourism is the 'visitation to primary and secondary food producers, food festivals, restaurants, and specific locations for which food tasting and/or experiencing the attributes of specialist food production region are the primary motivating factor for travel'. The definition specifies the locations/settings and the activities of food consumption in tourism. Regardless of the consuming setting, eating behaviour is influenced by various motives (Schupp & Renner, 2011; Steptoe et al., 1995), such as physiological factors (hunger), emotional factors (pleasure), social factors (celebration), and external factors (smell and appearance), which are generally critical for food consumption.

As a special category, tourists' food consumption is influenced by similar motivational factors. In tourism and hospitality, Mak et al. (2012) identified five factors which might affect tourists' food consumption: 'cultural and religious influences, socio-demographic factors, food-related personality traits, exposure affect and past experience, and motivational factors' (Mak et al., 2012, p. 928). In their study, factors that are beyond motivations are identified and discussed, which generate significant impacts on tourists' eating behaviours. In respect to motivational factors, they further suggested to group them into five categories: symbolic (culture, authenticity, learning, status, etc.); obligatory (health, essential need); contrast (exotic experience, different food); extension (familiarity); and pleasure (sensory arousal, togetherness) (Mak et al. 2012). However, the levels of significance in the motivation mix are distinctive between the daily, and tourism food consumption. For example, people eat to satisfy physiological needs in both circumstances, yet it is the primary need for daily eating and arguably the minor need for the later.

It has repeatedly been confirmed that a good comprehension of tourist preferences or factors that affect tourists' food consumption is vital for food tourism (Hall & Mitchell, 2001; Kim et al., 2009; Liu et al., 2017a; Liu et al., 2017b). To better understand the phenomenon, Komariah et al. (2020) discovered three key factors which influence tourists' intention of traditional food consumption: personal attitude; subjective/social norms; and behavioural control, which was informed by the Theories of Planned Behaviour (TPB). The attitude of tourists is in line with the motivational factors that potentially encourage the food consumption intention; subjective norms link with the 'cultural and religious influences', that explains the influence of social assessment (e.g., important others and social media) on the consumption of food. While behavioural control is well supported by 'food-related personality traits' (food neophobia or food neophilic), which illustrates tourists perceived ease and difficulty to consume local food while travelling.

In addition to qualitative studies, empirical research on food consumption has been carried out in both general and specific (tourism) areas. Renner et al. (2012) developed an Eating Motivation Survey Scale, which empirically identified 15 factors that motivate people to have daily/general food consumption: Liking; Habits; Need & Hunger; Health; Convenience; Pleasure; Traditional Eating; Natural Concerns; Sociability; Price; Visual Appeal; Weight Control; Affect Regulation; Social Norms; and Social Image. It arguably covers the most factors in a scale, while the factors are not fully attributed within the category of motivation; factors like habits and price are just factors that affect people's food choice. Similar studies were conducted in food tourism, to construct and validate a scale for local food consumption (Kim & Eves, 2012), and to develop a multi-dimensional scale for measuring food tourist motivations (Kim et al., 2010). According to Kim and Eves (2012), cultural interest, interpersonal relations, exciting experience, sensory arousal, and health pursuit, are the essential dimensions motivating local food consumption at a tourism destination. When attending food events, Kim et al. (2010) found out that tourists were motivated by similar reasons, such as a food event as an escape, socialisation, innovative/exciting experience, and family togetherness. More recently, González et al. (2020) inherited the five factors suggested by Mak et al. (2012), and empirically tested how these factors influence tourists' food choices. Findings indicated that motivational factors and food-related personality traits were consistently significant predictors of local food consumption.

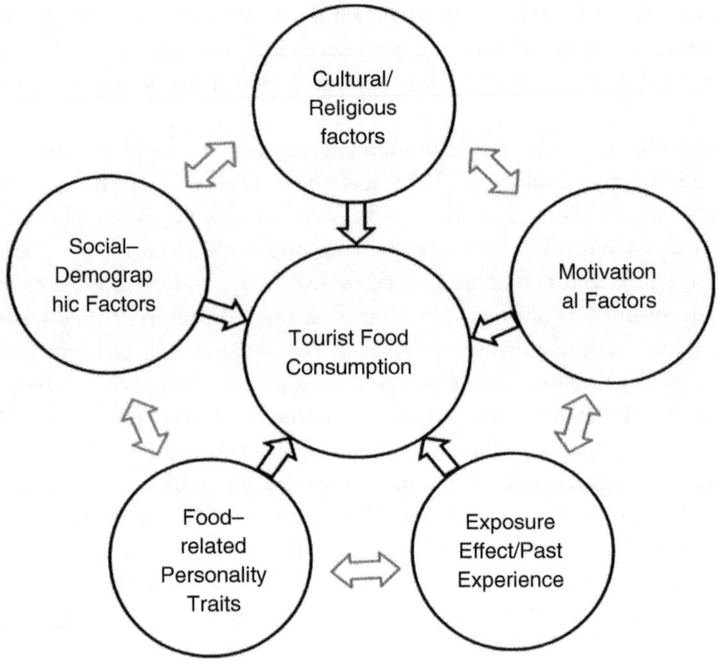

Figure 32.1. Factors affecting tourist food consumption and their relationship.

These studies have significantly improved the understanding of tourists' food consuming behaviour. The categorisation of and the relationship between the affecting factors of tourists' food consumption are illustrated in Figure 32.1 (Renner et al., 2012; Mak at al., 2012). The figure first displays the relationship between the factors and tourist food consumption; then the relationship between the five different factors, which are interconnected, and proposed to have considerable impact on tourists' food choices and consumption experience.

Tourists' Food Consumption Experience

Tourists spend time and money during their trips or at food events to experience food services, from local markets to high-end restaurants, street food stands to gourmet schools, homestay families to mass celebrations. Meanwhile, local food serves as an attraction to increase tourism sales and assist local economic development (Kumar & Smith, 2018). Food and related services consequently satisfy tourists' various needs and create food consumption experiences. In the interlinked relationship of the three parties, that is, tourists, gastronomy, and tourism, food plays multiple roles (Cheng & Huang, 2015): representing the local culture; acting as tourism attraction; being a tourist product; and consisting of a tourist experience. The abovementioned roles contribute to food consumption experience, which is either considered as a secondary or a primary activity within the tourism context (Smith & Costello, 2009).

Food consumption experience extends beyond the act of purchasing and eating food, but rather refers to a process wherein food is consumed in different phases (pre-trip search, onsite eating, post-trip memories) and various places (Björk & Kauppinen-Räisänen, 2016). During the process, sensory, emotional, social, and intellectual stimulation are recognised as the sources that food experiences originate from (Dubé & Le Bel, 2003). Thus, the food consumption experience is multi-dimensional in nature, as concluded by Björk and

Kauppinen-Räisänen (2016), there are five dimensions to explore: the food (factual and associative); the social (self and others); the external (physical setting and destination); the service (interior and exterior design); and time (routine or important moments).

Though the above dimensions have been partially confirmed in several studies (Mak et al., 2012; Walter et al., 2010), a study by Björk and Kauppinen-Räisänen (2016) indicated that 'food', 'place', and 'behaviour' are the key dimensions affecting tourists' gastronomic experiences. It is proposed that these three factors are central aspects of 'foodscape construct' highlighting the interactive relationship among food, place, and people.

Food attributes affect the consumption experience (Kim et al., 2009; Mak et al., 2012) by displaying the features of local cuisines and forming the ambiances of a destination. It is summarised that the attributes of local foods encompass various dimensions, for example food quality, food novelty, and restaurant quality. Specifically, attributes associated with food quality consist of multiple features, which includes, but are not limited to the following: attractiveness; sensory appeal; new tastes; freshness; spiciness; cleanness; food diversity; colour; health; hygiene; inexpensiveness; and nutrition (Ab Karim & Chi, 2010; Kim & Eves, 2012). Recognising the importance of these attributes for travellers' food experience (Rozin, 1996; Filimonau et al., 2018); food dimension should be considered when designing tourism products.

Place dimension refers to the servicescape that was initially defined by Bitner (1992, p. 58), as 'the man-made, physical surroundings as opposed to the natural or social environment'. In a food consumption context, the servicescape specifically covers the external place, for example, physical setting of a destination or restaurant, and the service place, for example, interior, and exterior ambience. This is termed as foodscape (Björk & Kauppinen-Räisänen, 2019), which usually focuses on the man-made and built consumption environment. However, on some occasions, the external environment is not organised or controlled by service providers, suggesting the process of tourists' food consumption are either created (Grönroos, 2008) or co-created (Vargo & Lusch, 2004) by the consumer and the service provider.

The people or behaviour dimension mainly refers to the food-promoted or food-mediated social interaction between the tourists and the local residents, or between the tourists and co-travellers. Stone et al. (2018) stressed that social interactions that occur during the process of food consumption are a significant part of tourists' recollection of the event. Furthermore, time dimension should be considered here, because people might travel and dine with different parties on different time-oriented occasions. Therefore, the behaviour dimension suggests designing a foodscape that promotes interpersonal communication and interactions.

However, technologies that enhance consumption experience and food quality were not well addressed within existing food tourism literature, which have contributed greatly to transform or improve tourists' food consumption experience. Yet this has been a traditional topic in food studies, for example, a conference that focuses on Food Technology and Processing has been organised for more than a quarter of a century. In this conference, food technology covers an extensive scope of aspects, such as food microbiology, food preservation, food packaging, food authentication, food consumption, and health mapping techniques. The mentioned technologies mainly enhance the core quality or attractiveness of food (Food Technology, 2022).

The importance of technology, from the customer experience perspective, have been confirmed by Panasonic (2022). The more favourable experience the customer has, the more repeated patrons occur; however, the key to experience is technology. Consumption data mining and analysis can achieve personalised food experience; provision of ecommerce can increase food convenience; and the application of Internet of things (IoT), Artificial intelligence (AI) and automation can predict food tourists' behaviour and enhance operational efficiencies. Therefore, technology, as an integral part of food consumption, requests more attention from academia in food tourism when addressing the food consumption experience.

The consumption of food by tourists generates various impacts beyond the consumption experience, which are mainly related to the destination, travel consumer, food, and consumption environment. Though prolonged impacts exist, most of the impacts occur at a tourism destination. One of the striking impacts of food consumption is food waste at destinations, which will be discussed in the following section (Filimonau et al., 2022; Wang et al. 2015; Wang et al., 2021).

Food Waste in Hospitality Industry

Food Waste and Impact

Food waste caused by tourism is a common problem within the economy, society, and ecological environment. Reducing food waste has become an important means to promote tourism to be more sustainable (Gössling et al., 2011; Gretzel et al., 2020). Bio Intelligence Services (2010) estimated that hospitality accounts for about 14 percent of the total food waste generated in the EU. The U.S. wastes 31–40 percent of its post-harvest food supply, of which a high percentage is happening at the consumer level (Neff et al., 2015). The problem of food waste not only occurs in developed countries, but equally persists in developing countries (Liwei et al., 2013; Thi et al., 2015). As far as the waste of family food is concerned, the social costs related to the waste of food in South Africa is estimated to be about 21.7 billion Rand (about US$2.7 billion), accounting for 0.82 percent of the total annual domestic production value of South Africa (Nahman et al., 2012). At a hotel in Bangkok, 1.3 tonnes of edible food were wasted in a week (Lephilibert, 2016). Food waste has produced many adverse effects (Hamerschlag, 2011). For example, a kilogram of food waste produces nearly two kilograms of CO_2 equivalent (CO_2e) emissions and leads to the depletion of 2.9 tonnes of natural resources (Bio Intelligence Services, 2010). The impact of food waste on global warming is almost the same as global road transportation emissions (FAO, 2015). Some foods buried directly in the soil produce methane, which is 25 times more harmful than carbon dioxide (United States Environmental Protection Agency (EPA), 2016). Consequently, reducing food waste is considered an effective way to reduce greenhouse gases (Garnett, 2011). According to the United Nations (UN) sustainable development goals, food waste should be reduced by 50 percent by 2030 (Gössling & Hall, 2022). Although this topic is very important, it has not been paid enough attention to in academic research (Filimonau & De Coteau, 2019).

Definition and Type

In view of tourist food waste, insufficient attention on the definition and classification of food waste has been identified. Two often used terms are food loss and food waste (Parfitt et al., 2010). The former refers to removing parts that cannot or do not need to be used, resulting in both the waste of food and the environment (Grandhi & Appaiah Singh, 2016). The latter refers to the waste of edible food due to the irresponsible behaviour of suppliers and consumers. Neff et al. (2015) put forward the term 'wasted food', emphasising that food is wasted by people, and turned into garbage. Furthermore, unusable materials in the production process (Wang et al., 2017), such as vegetable peelings, bones, and coffee dregs are eliminated (Silvennoinen et al., 2015), and the remaining avoidable food waste accounts for 73–79 percent of tourist destinations' food waste (Oliveira et al., 2016). Food waste occurs at all stages of the food supply chain, including food storage, transportation, processing, retailers, restaurants, and home kitchens (Lundqvist et al., 2008). This occurs due to reasons such as food which cannot be used owing to poor storage and preparation technology (Heikkila et al., 2016), hotel managers lacking management philosophy and corporate responsibility (McWilliams & Siegel, 2001), and 'plate garbage' in the consumption process (Lorenz et al., 2017). Therefore, according to the different stages of the supply chain (Legrand et al., 2013), food waste in tourism can be divided into pre-consumption (such as manufacturing and retail) and post-consumption (table) food waste.

Reasons

The cause of food waste is multi-faceted. From the perspective of destination development, encouraging tourism food consumption has an important impact on the social and economic development of destinations (Liu et al., 2017a; Liu et al., 2017b). Food consumption is a key travel experience (Mitchell & Hall, 2003), and many food destinations view food tourism experience as an element which enhances the authenticity of local tourism (Min et al., 2020). The rise of 'taste landscapes' increasingly explains the importance of food development in destination tourism. In the absence of constraints, the enthusiasm of destinations for tourism food consumption will lead to an increase in food waste (Everett, 2012).

In tourism-related industries, the hospitality industry is the one that suffers most from food waste. Hospitality food waste has been defined as unwanted and processed food, such as leftovers from guests' plates and cooking oil (Pirani & Arafat, 2016). According to the survey of the Waste and Resources Action Programme (WRAP) (2013), avoidable food waste can be divided into three stages: food deterioration (21 percent), preparation (45 percent), and food consumption (34 percent). At the consumption stage, the hotel's buffet offers a wide variety of options at no extra cost and leads to more food consumption (Kuo & Shih, 2016; Wansink & van Ittersum, 2013) and plate litter (Wansink & van Ittersum, 2013).

From the perspective of consumer behaviour, tourists' food consumption (or plate waste) is affected by economic development, cultural background, personal experience, psychological characteristics, and so on. With the growth of tourism and the increase in residential income, leading to higher consumption of food outside of the home (Wang et al., 2017), the 'face' culture of Chinese tourists will lead to more tourism food purchase and consumption (Juvan et al., 2018). Network live broadcast, eat broadcast, and other phenomena also lead to the rise of 'buy but do not eat' and similar acts (Qian et al., 2019). Additionally, education and social experience will also have an impact on the number of meals consumed, and studies have found that people with lower education levels are more likely to waste food (Li & Wang, 2020). Household structure also influences food waste, with surveys (Koivupuro et al., 2012) showing that single female households produce more food waste per person than single male households. Households in which women were the primary shoppers also generated more food waste than those in which men were the primary shoppers. At the same time, interpersonal behaviours such as parties easily causes higher levels of food waste (Zhang et al., 2018). Food attraction is the key element of cultural tourism experience, the psychological factors of food novelty seeking (Kim et al., 2009; Sthapit, 2017) and diversity seeking (Derinalp Çanakçı & Birdir (2020), will encourage tourists to constantly explore new food experiences. To solve these problems, we need to understand the scale and pattern of food waste within different consumer groups (Wang et al., 2017), for example, by determining the scale of food waste by weighing the weight and types of meals before and after consumption through practical investigation (Wang et al., 2018), and accurately quantify the problem of tourism food waste, providing a data basis for subsequent cause exploration.

Solutions

Waste management is to standardise a better utilisation of waste throughout the entire process, from food generation to final location (Ball & Taleb, 2011); therefore, management of food waste involves all stages of the food chain (Cheyne & Purdue, 1995), including food production mode, value composition, supply mode, marketing process, service process, and consumption mode. In the early stage of food production and preparation, legislation and the establishment of penalties are considered to be effective measures to reduce food waste in the catering industry (Gretzel et al., 2020). Improving the value of the dining environment is also seen as an effective method of reducing food waste. By promoting the consumption of a single resource to the comprehensive consumption of a wide range of environments and cultures, tourists' satisfaction with the authenticity of tourism can be enhanced and help tourists to convert simple taste sense to the pleasure brought by the environment and culture (Sims, 2009).

In the process of food production, on the one hand, we can change the cooking process and reduce food waste by investigating the waste of food by customers (WRAP, 2017a). On the other hand, it is important to enhance the hospitality industry's ability to predict and manage food waste through technological innovation. For example, companies such as IKEA use artificial intelligence bins and the IoT to identify, track, and reduce food waste (Parsons, 2019). Similarly, 'waste' smart meter application developed by Unilever for chefs, measure, monitor, and manage food waste in commercial kitchens (Gould, 2016). Alternatively, designing effective dining experiences based on surveys and forecasts (Gössling et al., 2011) and developing new menus and ordering systems, designing smaller portions for groups such as children and women (Duursma et al., 2016) are effective ways to reduce food waste in hospitality.

Changing consumer behaviour during the meal phase is critical to curbing food waste. Hotels can set up tasting tables in the cafeteria, as well as install screens aimed at changing consumer dining psychology (Juvan et al., 2018). As the constructors of the local food culture, the receptionists should help tourists construct reasonable food consumption concepts from the aspects of health, emotion, and prestige value, such as promoting the prevention of obesity (Wansink & van Ittersum, 2013) and similar innovations.

Employee-consumer interaction is also seen as an effective way to reduce food waste (Sirieix et al., 2017). Measures in place include the redistribution of unsold food to charities or vulnerable groups, thereby protecting the environment or fostering poverty. For example, some KFC restaurants in the UK will regularly donate meals to British charities every month (WRAP, 2017b). There are also businesses that recycle and compost food to produce environmental and economic value (Pirani & Arafat, 2016), such as Disney World, which compost food waste into soil additives or digest food waste, and generate electricity from captured methane gas (Waste360, 2018).

Food and Destination Development

Food and Destination Image

As an important part of tourism activities, food is the key to tourists' awareness of the destination. Some countries and regions have built their own unique tourism portrait with their food and unique cooking methods. For example, the religious food in the Chinese, Japanese, and Korean temples is a symbol of local prestige (WRAP, 2013). Similarly, France is famous for its excellent food and Thailand attracts tourists with high-quality restaurant services and suitable prices (Ab Karim & Chi, 2010). Whereas Italian pasta is not only a cultural symbol, but also considered a national idol (Ellis et al., 2018), while Houston is considered to be a port city full of fresh seafood (Nelson, 2015). Food is used to describe and explain the image of the city, giving these cities significance (Matwick & Matwick, 2018).

Visitors' preference for local cuisine will affect the composition of destination terrain (Guan & Jones, 2015). In addition to personal experience, the organic photos of social media and the induced photos of official websites can build a destination food image to tourists through the title, pictures, sounds, and other factors. Movies and documentaries are also important in connecting food images to destinations (Bolan & Williams, 2008), and which have made positive contributions to the change of destination terrain (Min et al., 2020).

Food and Place Attachment

Local attachment (Tuan, 1978) represents the connection between individuals and specific environments (Lee et al., 2012), emphasising location dependence and emotional participation. Tourists tend to remember the beautiful moments in life. By participating in local cuisine activities, it has attracted tourists' curiosity about local cuisine, understanding local culture and knowledge, and obtain a sense of identity and belonging (Clark & Chabrel, 2007). Deep and lasting positive memories further enhance their recognition or strong attachment to local attractions (Tsai, 2016).

Food constructs are increasingly becoming a mixed space for local attachment. Cooking sites have become a space for stage performances and tourism experience, through scenes, explanation performances, narrative and ceremony activities, the sense of gaming, and sense of space-time transcendence (Walter, 2017). Food festivals construct a temporary space of the tourism experience, affecting the attachment of tourists to the local and the recommendation intention to the destination (Culha, 2020). Religious cuisine constructs the real tourism space of mental leisure and physical experience during the pilgrimage of the Holy devotion (Ron & Timothy, 2013; WRAP, 2013). Food production sites are transformed into a tourism experience space to meet the needs of product production, tourism experience, leisure, and relaxation (Everett, 2012). Food production is gradually becoming a mixed space that meets consumer needs and production requirements, producing position dependence, and emotional expression.

Food and Destination Marketing

As a marketing tool, food can effectively attract and retain tourists, which should be taken into account when formulating brand strategy (Freire & Gertner, 2021). Many national tourism policy plans have proposed marketing strategies for culinary tourism (Okumus et al., 2007; Plummer et al., 2005). The Caribbean islands, for example, promote their destinations by using their authentic cuisine (Okumus et al., 2013). The Malaysian government created 'desire' and enhances 'sensory attraction' of tourists through tourism brochures, tourism guides, and web pages (Jalis et al., 2014). South Africa developed TOURPAT (a tourism and culinary atlas linked to a geospatial database) and PAT (a product potential and attractiveness tool) (Rand & Heath, 2006). Houston examined a place reputation management strategy, through the 'objective' description of the destination in media articles, people can better appreciate the local quality. Hong Kong and Singapore promote local brands by developing more diverse and attention-grabbing cuisines (Horng & Tsai, 2012).

In order to achieve better marketing, in every stage of food tourism, including communication (before leaving home), logistics (getting there), the experience(s) (while on-site), effective marketing activities must be carried out at different stages (Williams et al., 2014), this is closely related to the construction of tourism information websites (Boyne & Hall, 2004).

Food and Destination Development

Food has great potential in attracting tourists and improving the overall tourism experience. Developers regard local food as a means of differentiation and an important means to attract tourists and investment (Henderson, 2016). Steering from economic commodities to a means to convey social identity and cultural landscape in destinations (Stalmirska, 2021). Food is particularly important in tourism destination development and restaurants may even be attractions in their own right (Huang et al., 2022; Huang et al., 2023).

The role of food in destination development has attracted attention, including the use of food as a pleasant social means to build a destination tourism brand (Freire & Gertner, 2021) and promote the generation of positive and correct destination decision-making behaviour (Sohn et al., 2014). Food is also an important means to protect local culture, as the promotion process of food heritage will re-activate local memory and define a social structure through various developments, which helps the construction of social identity and enhance community pride (Bessiere, 2013). This also helps to promote cooperation and relations among various cooking enterprises in different cities and geographical regions, improve competitiveness at the local level, and create better living conditions for citizens (Sánchez-Cañizares & López-Guzmán, 2012). In terms of rural development and environmental protection, tourists' interest in agricultural products can promote the attention of local society to this type of agricultural production. Additionally, 'slow tourism' can promote the use of funds within the local area (Hall, 2006), developing rural tourism while protecting history, society, and the environment (Sidali et al., 2015). Proper culinary tourism is a form of protection for fragile and dangerous lands and a contemporary form of sustainable development (Hall, 2006).

Food and Destination Sustainability

Food is very important at multiple levels of sustainable tourism. Food can make tourists have the intention to recommend and revisit by creating unforgettable travel experiences (Chang et al., 2010), enhancing the national authenticity of the destination (Everett, 2012; Sims, 2009), transmitting local culture, and building a restaurant social platform (Mitchell & Hall, 2003). It can become a tool for external marketing (Karim et al., 2009) and help the sustainable development of local tourism (Boyne & Hall, 2004). At the same time, food tourism is also an important method of maintaining community relations and enhancing the availability of resources.

Major dimensions to measure the quality of culinary experience with local food, including the authenticity and uniqueness of food, the interaction between tourists and local people, and the physical environment of dining (Widjaja et al., 2020). Therefore, destinations should improve the quality of food and beverages by adding a sensory experience design, enriching menu variety, strengthening restaurant sanitation management, design feasible culinary routes, carry out appropriate crop cultivation, and implementation of environmental protection, so as to make food tourism an important means for sustainable development of the destination (Wan et al., 2018; Hall, 2020).

Conclusion

Food is an eternal topic, with the deepening of tourism research, research on food in the context of tourism has gradually become an important focus of attention for scholars internationally. Using literature analysis, this chapter comprehensively analysed the relationship between tourism and food from four perspectives. Food as tourist attraction was first introduced, which we called 'gourmet tourism'. We elaborated this part based on different definitions of gastronomy/gastronomic tourism, culinary tourism, and food tourism, as we believe different 'terms' reflect different research perspectives. Food is not a tourist attraction in all cases, however, in all cases tourists eat food. Secondly, we discussed tourists' food consumption behaviours, including their characteristics in comparison to other consumption occasions and influencing factors. Food waste was separated as an individual portion within this chapter, owing to its serious consequences and widespread concern. Lastly, we discussed the food with destination from the perspective of destination image, place attachment, destination marketing and sustainability, in which we intended to reveal the regional impact of food production and consumption in the context of tourism.

For future research, we believe more empirical research in different regions or diverse cultural backgrounds from different disciplines would be beneficial. In accordance with this, the basic theoretical framework for tourism and food research should be gradually established.

References

Ab Karim, S., & Chi, C. G. (2010). Culinary tourism as a destination attraction: An empirical examination of destinations' food image. *Journal of Hospitality Marketing & Management*, *19*(6), 531–555. https://doi.org/10.1080/19368623.2010.493064

Ball, S., & Taleb, M. A. (2011). Benchmarking waste disposal in the Egyptian hotel industry. *Tourism & Hospitality Research*, *11*(1), 1–18. https://doi.org/10.1057/thr.2010.16

Bertella, G. (2011). Knowledge in food tourism: The case of Lofoten and Maremma Toscana. *Current Issues in Tourism*, *14*(4), 355–371. https://doi.org/10.1080/13683500.2010.489638

Bessiere, J. (2013). 'Heritagisation', a challenge for tourism promotion and regional development: An example of food heritage. *Journal of Heritage Tourism*, *8*(4), 275–291. https://doi.org/10.1080/1743873X.2013.770861

Bio Intelligence Services. (2010). Preparatory study on food waste across EU 27. *European Commission*. https://ec.europa.eu/environment/eussd/pdf/bio_foodwaste_report.pdf

Bitner, M. (1992). Servicescapes. The impact of physical surroundings on customers and employees. *Journal of Marketing*, *56*(2), 57–71. https://doi.org/10.1177/002224299205600205

Björk, P., & Kauppinen-Räisänen, H. (2016). Exploring the multi-dimensionality of travellers' culinary-gastronomic experiences. *Current Issues in Tourism*, *19*(12), 1260–1280. https://doi.org/10.1080/13683500.2013.868412

Björk, P., & Kauppinen-Räisänen, H. (2019). Destination foodscape: a stage for travelers' food experience. *Tourism Management*, *71*, 466–475. https://doi.org/10.1016/j.tourman.2018.11.005

Bolan, P., & Williams, L. (2008). The role of image in service promotion: focusing on the influence of film on consumer choice within tourism. *International Journal of Consumer Studies*, *32*(4), 382–390. https://doi.org/10.1111/j.1470-6431.2008.00672.x

Boyne, S., & Hall, D. (2004). Place promotion through food and tourism: Rural branding and the role of websites. *Place Branding*, *1*(1), 80–92. https://doi.org/10.1057/palgrave.pb.5990007

Capaldi, E. D. (1996). *Why we eat what we eat: The psychology of eating*. American Psychological Association.

Chaney, S., & Ryan, C. (2012). Analyzing the evolution of Singapore's World Gourmet Summit: An example of gastronomic tourism. *International Journal of Hospitality Management*, *31*(2), 309–318. https://doi.org/10.1016/j.ijhm.2011.04.002

Chang, R., Kivela, J., & Mak, A. (2010). Food preferences of Chinese tourists. *Annals of Tourism Research*, *37*(4), 989–1011. https://doi.org/10.1016/j.annals.2010.03.007

Cheng, Q., & Huang, R. (2015). Understanding the importance of food tourism to Chongqing, China. *Journal of Vacation Marketing*, *22*(1), 42–54. https://doi.org/10.1177/1356766715589427

Cheyne, I., & Purdue, M. (1995). Fitting definition to purpose: The search for a satisfactory definition of waste. *Journal of Environmental Law*, *7*(2), 149–168. https://heinonline.org/HOL/LandingPage?handle=hein.journals/jenv7&div=16&id=&page=

Clark, G., & Chabrel, M. (2007). Measuring integrated rural tourism. *Tourism Geographies*, *9*(4), 371–386. https://doi.org/10.1080/14616680701647550

Cohen, E., & Avieli, N. (2004). Food in tourism, attraction and impediment. Annals of Tourism Research, 31(4), 755–778. https://doi.org/10.1016/j.annals.2004.02.003

Culha, O. (2020). The effect of food festival quality on place attachment and destination recommendation intention through festival experience and festival satisfaction: The case of the Didim International Olive Festival. *Journal of Convention & Event Tourism*, *21*(5), 387–416. https://doi.org/10.1080/15470148.2020.1775743

Derinalp Çanakçı, S., & Birdir, K. (2020). The relation among food involvement, food variety seeking and food neophobia: A study on foreign tourists visiting Turkey. *Current Issues in Tourism*, *23*(8), 917–928. https://doi.org/10.1080/13683500.2019.1602114

Dixit, S. K. (2019). Gastronomic tourism: A theoretical construct. In S. K. Dixit (Ed.), *The Routledge handbook of gastronomic tourism* (pp. 13–23). Abingdon: Routledge.

Dubé, L., & Le Bel, J. (2003). The content and structure of laypeople's concept of pleasure. *Cognition & Emotion*, *17*(2), 263–295. https://doi.org/10.1080/02699930302295

Duursma, G., Vrenegoor, F., & Kobus, S. (2016). Food waste reduction at Restaurant De Pleats: Small steps for mankind. *Research in Hospitality Management*, *6*(1), 95–100. https://doi.org/10.2989/RHM.2016.6.1.13.1301

Ellis, A., Park, E., Kim, S., & Yeoman, I. (2018). What is food tourism? *Tourism Management*, *68*, 250–263. https://doi.org/10.1016/j.tourman.2018.03.025

Everett, S. (2012). Production places or consumption spaces? The place-making agency of food tourism in Ireland and Scotland. *Tourism Geographies*, *14*(4), 535–554. https://doi.org/10.1080/14616688.2012.647321

Everett, S., & Slocum, S. L. (2013). Food and tourism: An effective partnership? A UK-based review. *Journal of Sustainable Tourism*, *21*(6), 789–809. https://doi.org/10.1080/09669582.2012.741601

Filimonau, V., & De Coteau, D. A. (2019). Food waste management in hospitality operations: A critical review. *Tourism Management*, *71*, 234–245. https://doi.org/10.1016/j.tourman.2018.10.009

Filimonau, V., Coşkun, A., Derqui, B., & Matute, J. (2022). Restaurant management and food waste reduction: factors affecting attitudes and intentions in restaurants of Spain. *International Journal of Contemporary Hospitality Management*, *34*(3), 1177–1203. https://doi.org/10.1108/IJCHM-07-2021-0899

Filimonau, V., Matute, J., Mika, M. & Faracik, R. (2018). National culture as a driver of pro-environmental attitudes and behavioural intentions in tourism. *Journal of Sustainable Tourism*, *26*(10), 1804–1825. https://doi.org/10.1080/09669582.2018.1511722

Food and Agriculture Organization (FAO). (2015). *Food wastage footprint & climate change.* https://www.fao.org/fileadmin/templates/nr/sustainability_pathways/docs/FWF_and_climate_change.pdf

Food Technology (2022). *26th international conference on food technology & processing.* https://foodtechnology.insightconferences.com/#section_Market%20Analysis

Freire, J. R., & Gertner, R. K. (2021). The relevance of food for the development of a destination brand. *Place Branding and Public Diplomacy, 17*, 193–204. https://doi.org/10.1057/s41254-020-00164-5

Frost, W., Laing, J., Best, G., Williams, K., Strickland, P., & Lade, C. (2016). *Gastronomy, tourism and the media.* Channel View Publications.

Garnett, T. (2011). Where are the best opportunities for reducing greenhouse gas emissions in the food system (including the food chain)? *Food Policy, 36*, S23–S32. https://doi.org/10.1016/j.foodpol.2010.10.010

Gheorghe, G., Tudorache, P., & Nistoreanu, P. (2014). Gastronomic tourism, a new trend for contemporary tourism. *Cactus Tourism Journal, 9*(1), 12–21.

González, A. F., Curtis, C., Washburn, I. J., & Shirsat, A. R. (2020). Factors in tourists' food decision processes: A US-based case study. *Journal of Tourism Analysis: Revista de Análisis Turístico, 27*(1), 2–19. https://doi.org/10.1108/JTA-01-2019-0002

Gössling, S., & Hall, C. M. (2022). *The sustainable chef: The environment in culinary arts, restaurants, and hospitality.* Routledge.

Gössling, S., Garrod, B., Aall, C., Hille, J., & Peeters, P. (2011). Food management in tourism: Reducing tourism's carbon 'foodprint'. *Tourism Management, 32*(3), 534–543. https://doi.org/10.1016/j.tourman.2010.04.006

Gould, H. (2016). Restaurants have a huge food waste problem; could an app help? *The Guardian.* https://www.theguardian.com/sustainable-business/2016/may/27/restaurants-huge-food-waste-problem-smart-meter-winnow-hugh-fearnley-whittingstall

Grandhi, B., & Appaiah Singh, J. (2016). What a waste! A study of food wastage behavior in Singapore. *Journal of Food Products Marketing, 22*(4), 471–485. https://doi.org/10.1080/10454446.2014.885863

Gretzel, U., Murphy, J., Pesonen, J., & Blanton, C. (2020). Food waste in tourist households: a perspective article. *Tourism Review, 75*(1), 235–238. https://doi.org/10.1108/TR-05-2019-0170

Grönroos, C. (2008). Service logic revisited: Who creates value? And who co-creates? *European Business Review, 20*(4), 298–314. https://doi.org/10.1108/09555340810886585

Guan, J., & Jones, D. L. (2015). The contribution of local cuisine to destination attractiveness: An analysis involving Chinese tourists' heterogeneous preferences. *Asia Pacific Journal of Tourism Research, 20*(4), 416–434. https://doi.org/10.1080/10941665.2014.889727

Hall, C. M. (1996, January). Wine tourism in New Zealand. [Paper presentation]. Proceedings of Tourism Down Under II: A Tourism Research Conference, University of Otago, Dunedin, New Zealand. https://www.researchgate.net/publication/302506378_Wine_tourism_in_New_Zealand

Hall, C. M. (2006). Introduction: Culinary tourism and regional development: From slow food to slow tourism? *Tourism Review International, 9*(4), 303–305.

Hall, C. M. (2012). Boosting food and tourism-related regional economic development. In *Food and the tourism experience: The OECD-Korea workshop, OECD studies on tourism* (pp. 49–62). OECD Publishing. http://dx.doi.org/10.1787/9789264171923-en

Hall, C. M. (2020). Improving the recipe for culinary and food tourism? The need for a new menu. *Tourism Recreation Research, 45*(2), 284–287. https://doi.org/10.1080/02508281.201.1694243

Hall, C. M., & Mitchell, R. (2001). Wine and food tourism. In N. Douglas & R. Derret (Eds), *Special interest tourism* (pp. 307–329). Wiley.

Hall, C. M., & Sharples, L. (2003). The consumption of experiences or the experience of consumption? An introduction to the tourism of taste. In C. M. Hall, L. Sharples, R. Mitchell, N. Macionis & B. Cambourne (Eds), *Food tourism around the world: Development, management and markets* (pp. 1–24). Routledge.

Hall, C. M., Longo, A. M., Mitchell, R., & Johnson, G. (2000). Wine tourism in New Zealand. In C. M. Hall, L. Sharples, B. Cambourne, N. Macionis, R. Mitchell & G. Johnson (Eds.). *Wine tourism around the world: Development, management and markets* (pp. 150–174). Butterworth-Heinemann.

Hall, C. M., Sharples, L., Cambourne, B., Macionis, N., Mitchell, R., & Johnson, G. (Eds). (2000). *Wine tourism around the world: Development, management and markets*. Butterworth-Heinemann.

Hamerschlag, K. (2011). *Meat eaters guide to climate change + health*. https://static.ewg.org/reports/2011/meateaters/pdf/report_ewg_meat_eaters_guide_to_health_and_climate_2011.pdf

Heikkila, L., Reinikainen, A., Katajajuuri, J. M., Silvennoinen, K., & Hartikainen, H. (2016). Elements affecting food waste in the food service sector. *Waste Management*, *56*, 446–453. https://doi.org/10.1016/j.wasman.2016.06.019

Henderson, J. C. (2016). Local and traditional or global and modern? Food and tourism in Singapore. *Journal of Gastronomy and Tourism*, *2*(1), 55–68. https://doi.org/10.3727/216929716X14546365943494

Hjalager, A., & Richards, G. (2002). *Tourism and gastronomy*. Routledge.

Horng, J., & Tsai, C. (2010). Government websites for promoting East Asian culinary tourism: A cross-national analysis. *Tourism Management*, *31*(1), 74–85. https://doi.org/10.1016/j.tourman.2009.01.009

Horng, J., & Tsai, C. (2012). Culinary tourism strategic development: an Asia-Pacific perspective. *International Journal of Tourism Research*, *14*(1), 40–55. https://doi.org/10.1002/jtr.834

Huang, Y., Hall, C. M., Bryden, A., Du, Y., Koupaei, S. N., Lyes, K., Marquez, V., Shaw, M., & Thomas, D. (2022). The characteristics of destination restaurants. *Journal of Hospitality and Tourism*, *20*, 1–30.

Huang, Y., Hall, C. M., & Chen, N. C. (2023). The characteristics of Michelin-starred restaurants in Mainland China, Hong Kong, Macau, and Taiwan. *International Journal of Gastronomy and Food Science*, *31*, 100673. https://doi.org/10.1016/j.ijgfs.2023.100673

Ignatov. E., & Smith, S. (2006). Segmenting Canadian culinary tourists. *Current Issues in Tourism*, *9*(3), 235–255. https://doi.org/10.2167/cit/229.0

Jalis, M. H., Che, D., & Markwell, K. (2014, May 19–21). Utilising local cuisine to market Malaysia as a tourist destination. *Procedia Social and Behavioral Sciences*, *144*, 102–110. https://doi.org/10.1016/j.sbspro.2014.07.278

Juvan, E., Grün, B., & Dolnicar, S. (2018). Biting off more than they can chew: Food waste at hotel breakfast buffets. *Journal of Travel Research*, *57*(2), 232–242. https://doi.org/10.1177/0047287516688321

Karim, M. A., Chua, B. L., & Salleh, H. (2009). Malaysia as a culinary tourism destination: International tourists' perspective. *Journal of Tourism, Hospitality & Culinary Arts*, *1*(3), 1–16.

Kim, Y. G., & Eves, A. (2012). Construction and validation of a scale to measure tourist motivation to consume local food. *Tourism Management*, *33*(6), 1458–1467. https://doi.org/10.1016/j.tourman.2012.01.015

Kim, Y. G., Eves, A., & Scarles, C. (2009). Building a model of local food consumption on trips and holidays: A grounded theory approach. *International Journal of Hospitality Management*, *28*(3), 423–431. https://doi.org/10.1016/j.ijhm.2008.11.005

Kim, Y. H., Goh, B. K., & Yuan, J. (2010). Development of a multi-dimensional scale for measuring food tourist motivations. *Journal of Quality Assurance in Hospitality & Tourism*, *11*(1), 56–71. https://doi.org/10.1080/15280080903520568

Kivela, J., & Crotts, J. C. (2006). Tourism and gastronomy: Gastronomy's influence on how tourists experience a destination. *Journal of Hospitality and Tourism Research*, *30*(3), 354–377. https://doi.org/10.1177/1096348006286797

Koivupuro, H.-K., Hartikainen, H., Silvennoinen, K., Katajajuuri, J.-M., Heikintalo, N., Reinikainen, A., & Jalkanen, L. (2012). Influence of socio-demographical, behavioural and attitudinal factors on the amount of avoidable food waste generated in Finnish households. *International Journal of Consumer Studies*, *36*(2), 183–191. https://doi.org/10.1111/j.1470-6431.2011.01080.x

Komariah, K., Razzaq, A. R. B. A., Nugraheni, M., Lastariwati, B., & Mahfud, F. (2020). The antecedent factor of tourists' intention to consume traditional food. *GeoJournal of Tourism and Geosites*, *32*(4), 1209–1215. 5. https://doi.org/10.30892/gtg.32403-559

Kumar, A., & Smith, S. (2018). Understanding local food consumers: Theory of planned behavior and segmentation approach. *Journal of Food Products Marketing*, *24*(2), 196–215. https://doi.org/10.1080/10454446.2017.1266553

Kuo, C. F., & Shih, Y. (2016). Gender differences in the effects of education and coercion on reducing buffet plate waste. *Journal of Foodservice Business Research, 19*(3), 223–235. https://doi.org/10.1080/15378020.2016.1175896

Laing, J., & Frost, W. (2014). *Explorer travellers and adventure tourism.* Channel View Publications.

Lee, J., Kyle, G., & Scott, D. (2012). The mediating effect of place attachment on the relationship between festival satisfaction and loyalty to the festival hosting destination. *Journal of Travel Research, 51*(6), 754–767. https://doi.org/10.1177/0047287512437859

Legrand, W., Sloan, P., & Chen, J. S. (2013). *Sustainability in the hospitality industry: Principles of sustainable operations* (2nd ed.). Routledge.

Lephilibert, B. (2016, January 15). Food waste the elephant in the room for the hospitality industry. *SustainableBrands.*https://sustainablebrands.com/read/waste-not/food-waste-the-elephant-in-the-room-for-the-hospitality-industry

Li, N., & Wang, J. Y. (2020). Food waste of Chinese cruise passengers. *Journal of Sustainable Tourism, 28*(11), 1825–1840. https://doi.org/10.1080/09669582.2020.1762621

Liu, B., Wang, T., & Chen, Z. (2017a). Review on tourist food consumption in foreign studies. *Human Geography, 32*(3), 20–26.

Liu, B., Wang, T., & Chen, Z. (2017b). A review of studies on foreign tourists' diet consumption. *Human Geography, 32*(3), 8

Liwei, G., Shengkui, C., Xiaochang, C., Dan, Z., Xiaojie, L., Qi, Q., & Yao, L. (2013). An overview of the resources and environmental issues from wasted food in urban catering across China. *Journal of Resources and Ecology, 4*(4), 337–343. https://doi.org/10.5814/j.issn.1674-764x.2013.04.006

Lockwood, Y. R. (2008). Culinary tourism [review]. *Journal of American Folklore, 121*(481), 362–363. https://doi.org/10.1353/jaf.0.0018

Long, L. M. (2012). Culinary tourism. In J. M. Pilcher (Ed.), *The Oxford handbook of food history* (pp. 389–408). Oxford University Press.

Long, L. M. (Ed.). (2004). *Culinary tourism.* The University Press of Kentucky.

López-Guzmán, T., & Sánchez-Cañizares, S. (2012). Culinary tourism in Córdoba (Spain). *British Food Journal, 114*(2), 168–179. https://doi.org/10.1108/00070701211202368

Lorenz, B. A. S., Hartmann, M., & Langen, N. (2017). What makes people leave their food? The interaction of personal and situational factors leading to plate leftovers in canteens. *Appetite, 116*, 45–56. https://doi.org/10.1016/j.appet.2017.04.014

Lundqvist, J., Fraiture, C. D., & Molden, D. (2008). *Saving water: From field to fork: Curbing losses and wastage in the food chain.* SIWI Policy Brief. https://hdl.handle.net/10535/5088

Mak, A. H., Lumbers, M., Eves, A., & Chang, R. C. (2017). The effects of food-related personality traits on tourist food consumption motivations. *Asia Pacific Journal of Tourism Research, 22*(1), 1–20. https://doi.org/10.1080/10941665.2016.1175488

Mak, A., Lumbers, M., Eves, A., & Chang, R. (2012). Factors influencing tourist food consumption. *International Journal of Hospitality Management, 31*(3), 928–936. https://doi.org/10.1016/j.ijhm.2011.10.012

Matwick, K., & Matwick, K. (2018). Culinary tourism in Central America: A cross-analysis of government tourism websites. *Journal of Culinary Science & Technology, 16*(3), 286–309, https://doi.org/10.1080/15428052.2017.1378601

McWilliams, A., & Siegel, D. (2001). Corporate social responsibility: A theory of the firm perspective. *Academy of Management Review, 26*(1), 117–127. https://doi.org/10.5465/amr.2001.4011987

Min, X., Sangkyun, K., & Stijn, R. (2020). From food to feet: Analysing *A Bite of China* as food-based destination image. *Tourist Studies, 20*(2), 145–165. https://doi.org/10.1177/1468797619888305

Mitchell, R., & Hall, C. M. (2003). Consuming tourists: food tourism consumer behaviour. In. C. M. Hall, L. Sharples, R. Mitchell, N. Macionis & B. Cambourne (Eds), *Food tourism around the world: Development, management and markets* (pp. 60–80). Routledge.

Murphy, R. (Director). (2010). *Eat pray love.* [Film]. Columbia Pictures.

Nahman, A., de Lange, W., Oelofse, S., & Godfrey, L. (2012). The costs of household food waste in South Africa. *Waste Management, 32*(11), 2147–2153. https://doi.org/10.1016/j.wasman.2012.04.012

Neff, R. A., Spiker, M. L., Truant, P. L., & Wiley, A. S. (2015). Wasted food: U.S. consumers' reported awareness, attitudes, and behaviors. *PLoS ONE, 10*(6): e0127881. https://doi.org/10.1371/journal.pone.0127881

Nelson, V. (2015). Place reputation: representing Houston, Texas as a creative destination through culinary culture. *Tourism Geographies, 17*(2), 192–207. https://doi.org/10.1080/14616688.2014.986751

Okumus, B., Okumus, F., & McKercher, B. (2007). Incorporating local and international cuisines in the marketing of tourism destinations: The cases of Hong Kong and Turkey. *Tourism Management, 28*(1), 253–261. https://doi.org/10.1016/j.tourman.2005.12.020

Okumus, F., Kock, G., Scantlebury, M. M. G., & Okumus, B. (2013). Using local cuisines when promoting small Caribbean Island destinations. *Journal of Travel & Tourism Marketing, 30*(4), 410–429. https://doi.org/10.1080/10548408.2013.784161

Oliveira, B., Moura, A. P. D., & Cunha, L. M. (2016). Reducing food waste in the food service sector as a way to promote public health and environmental sustainability. In B. Oliveria, A. P. de Moura & L. M. Cunha (Eds), *Climate change and health improving resilience and reducing risks* (pp. 117–132). Springer.

Panasonic (2022). *Food service technology*. https://na.panasonic.com/us/industries/food-service-technology

Parfitt, J., Barthel, M., & MacNaughton, S. (2010). Food waste within food supply chains: quantification and potential for change to 2050. Philosophical Transactions of the Royal Society B. *Biological Sciences, 365*(1554), 3065–3081. https://doi.org/10.1098/rstb.2010.0126

Parsons, J. (2019). Artificially intelligent bin aims to cut down on food waste. *Metro*. https://metro.co.uk/2019/03/22/artificially-intelligent-bin-aims-cut-food-waste-8979586/

Pavlidis, G., & Markantonatou, S. (2020). Gastronomic tourism in Greece and beyond: A thorough review. *International Journal of Gastronomy and Food Science, 21*, 100229. https://doi.org/10.1016/j.ijgfs.2020.100229

Pirani, S. I., & Arafat, H. A. (2016). Reduction of food waste generation in the hospitality industry. *Journal of Cleaner Production, 132*, 129–145. https://doi.org/10.1016/j.jclepro.2015.07.146

Plummer, R., Telfer, D., Hashimoto, A., & Summers, R. (2005). Beer tourism in Canada along the Waterloo–Wellington Ale Trail. *Tourism Management, 26*(3), 447–458. https://doi.org/10.1016/j.tourman.2003.12.002

Qian, J., Shen, H., Law, R., Chau, K. Y., & Wang, X. (2019). Examination of Chinese tourists' Unsustainable food consumption: Causes and solutions. *Sustainability, 11*(12), 3475. https://doi.org/10.3390/su11123475

Rand, G., & Heath, E. (2006). Towards a framework for food tourism as an element of destination marketing. *Current Issues in Tourism, 9*(3), 206–234.

Renner, B., Sproesser, G., Strohbach, S., & Schupp, H. T. (2012). Why we eat what we eat. The eating motivation survey (TEMS). Appetite, 59(1), 117–128. https://doi.org/10.1016/j.appet.2012.04.004

Rozin, P. (1996). The socio-cultural context of eating and food choice. In H. Meiselman & H. J. H. MacFie (Eds), *Food choice, acceptance and consumption* (pp. 83–104). Blackie.

Sánchez-Cañizares, S. M., & López-Guzmán, T. (2012). Gastronomy as a tourism resource: profile of the culinary tourist. *Current Issues in Tourism, 15*(3), 229–245. https://doi.org/10.1080/13683500.2011.589895

Schupp, H. T., & Renner, B. (2011). Food deprivation. A neuroscientific perspective. In V. Preedy, R. Watson & C. R. Martin (Eds), *International handbook of behavior, food and nutrition* (pp. 2239–2257). Springer.

Sidali, K. L., Kastenholz, E., & Bianchi, R. (2015). Food tourism, niche markets and products in rural tourism: Combining the intimacy model and the experience economy as a rural development strategy. *Journal of Sustainable Tourism* 23(8–9), 1179–1197. https://doi.org/10.1080/09669582.2013.836210

Silvennoinen, K., Heikkilae, L., Katajajuuri, J. M., & Reinikainen, A. (2015). Food waste volume and origin: Case studies in the Finnish food service sector. *Waste Management, 46*, 140–145. https://doi.org/10.1016/j.wasman.2015.09.010

Sims, R. (2009). Food, place and authenticity: local food and the sustainable tourism experience. *Journal of Sustainable Tourism, 17*(3), 321–336. https://doi.org/10.1080/09669580802359293

Sirieix, L., Lala, J., & Kocmanova, K. (2017). Understanding the antecedents of consumers' attitudes towards doggy bags in restaurants: Concern about food waste, culture, norms and emotions. *Journal of Retailing and Consumer Services*, *34*, 153–158. https://doi.org/10.1016/j.jretconser.2016.10.004

Smith, S., & Costello, C. (2009), Segmenting visitors to a culinary event: motivations, travel behavior, and expenditures, Journal of Hospitality Marketing & Management, 18 (1): 44–67.

Sohn, E., Yuan, J., & Jai, T. M. (2014). From a tea event to a host destination: Linking motivation, image, satisfaction and loyalty. *International Journal of Tourism Sciences*, *14*(3), 1–23. https://doi.org/10.1080/15980634.2014.11434697

Sormaz, U., Akmese, H., Gunes, E., & Aras, S. (2016). Gastronomy in tourism. *Procedia Economics and Finance*, *39*, 725–730. https://doi.org/10.1016/S2212-5671(16)30286-6

Stalmirska, A. M. (2021). Local food in tourism destination development: the supply-side perspectives. *Tourism Planning & Development*. https://doi.org/10.1080/21568316.2021.1928739

Steptoe, A., Pollard, T. M., & Wardle, J. (1995). Development of a measure of the motives underlying the selection of food. The food choice questionnaire. *Appetite*, *25*, 267–284. https://www.ucl.ac.uk/hbrc/diet/SteptoeandWardleFCQ.pdf

Sthapit, E. (2017). Exploring tourists' memorable food experiences: A study of visitors to Santa's official hometown. *Anatolia*, *28*(3), 404–421. https://doi.org/10.1080/13032917.2017.1328607

Stone, M. J., Soulard, J., Migacz, S., & Wolf, E. (2018). Elements of memorable food, drink, and culinary tourism experiences. *Journal of Travel Research*, *57*(8), 1121–1132. https://doi.org/10.1177/0047287517729758

Thi, N. B. D., Lin, C.-Y., Kumar, G. (2015). An overview of food waste management in developing countries: Current status and future perspective. *Journal of Environmental Management*, *157*, 220–229. https://doi.org/10.1016/j.jenvman.2015.04.022

Tsai, C. T. S., & Wang, Y. C. (2017). Experiential value in branding food tourism. *Journal of Destination Marketing and Management*, *6*(1), 56–65. https://doi.org/10.1016/j.jdmm.2016.02.003

Tsai, C.-T. (2016). Memorable tourist experiences and place attachment when consuming local food. *International Journal of Tourism Research*, *18*(6), 536–548. https://doi.org/10.1002/jtr.2070

Tuan, Y. F. (1978). Space and place: The perspective of experience. *Leonardo*, *68*(3).

United States Environmental Protection Agency (EPA). (2016). *Greenhouse gas overview*. https://www.epa.gov/enviro/greenhouse-gas-overview

Vargo, S. L., & Lusch, R. F. (2004). Evolving a services dominant logic. *Journal of Marketing*, *68*(1), 1–17. https://doi.org/10.1509/jmkg.68.1.1.24036

Walter, P. (2017). Culinary tourism as living history: staging, tourist performance and perceptions of authenticity in a Thai cooking school. *Journal of Heritage Tourism*, *12*(4), 365–379. https://doi.org/10.1080/1743873X.2016.1207651

Walter, U., Edvardsson, B., & Öström, Å. (2010). Drivers of customers' service experiences: A study in the restaurant industry. *Managing Service Quality: An International Journal*, *20*(3), 236–258. https://doi.org/10.1108/09604521011041961

Wan, H., Ismail, F., Som, A., & Rahman, N. (2018). Food choice determinants of island tourists: an implication towards sustainable island tourism. *International Journal of Engineering & Technology*, *7*(4), 163–166. https://eprints.unisza.edu.my/6088/

Wang, L. E., Cheng, S. K., Liu, G., Liu, X. J., Bai, J. F., Zhang, D., Gao, L.W., Cao, X. C., & Liu, Y. (2015). Study on theories and methods of Chinese food waste. *Journal of Natural Resources*, *30*(*5*): 715–724.

Wang, L. E., Filimonau, V., & Li, Y. (2021). Exploring the patterns of food waste generation by tourists in a popular destination. *Journal of Cleaner Production*, *279*, 123890. https://doi.org/10.1016/j.jclepro.2020.123890

Wang, L. E., Liu, G., Liu, X., Liu, Y., Gao, J., Zhou, B., Gao, S., & Cheng, S. (2017). The weight of unfinished plate: A survey based characterization of restaurant food waste in Chinese cities. *Waste Management*, *66*, 3–12. https://doi.org/10.1016/j.wasman.2017.04.007

Wang, L. G., Xue, L., Li, Y. Y., Liu, X. J., Cheng, S. K., & Liu, G. (2018). Horeca food waste and its ecological footprint in Lhasa, Tibet, China. *Resources Conservation and Recycling*, *136*, 1–8. https://doi.org/10.1016/j.resconrec.2018.04.001

Wansink, B., & van Ittersum, K. (2013). Portion size me: Plate-size induced consumption norms and win-win solutions for reducing food intake and waste. *Journal of Experimental Psychology: Applied*, *19*(4), 320–332. https://doi.org/10.1037/a0035053

Waste and Resources Action Programme (WRAP). (2013). WRAP - research on waste in the hospitality and Food Service Sector. *Sustainability Exchange*. https://www.sustainabilityexchange.ac.uk/research-on-waste-in-the-hospitality-and-food-serv

Waste and Resources Action Programme (WRAP). (2017a). Taking action on waste information sheets. Retrieved May 13, 2022, from https://wrap.org.uk/resources/case-study/hospitality-and-food-service-agreement-taking-action-waste-final-report#:~:text=The%20Hospitality%20and%20Food%20Service%20Agreement%20(HaFSA)%20was%20an%20ambitious,the%20sector%20in%20the%20UK

Waste and Resources Action Programme (WRAP). (2017b). *The hospitality and food service agreement: Taking action on waste (final report)*. https://wrap.org.uk/resources/case-study/hospitality-and-food-service-agreement-taking-action-waste-final-report

Waste360. (2018). *Local businesses benefit from Disney's biofuel facility*. https://www.waste360.com/food-waste/local-businesses-benefit-disney-s-biofuel-facility

Widjaja, D. C., Jokom, R., Kristanti, M., & Wijaya, S. (2020). Tourist behavioural intentions towards gastronomy destination: evidence from international tourists in Indonesia. *Anatolia-International Journal of Tourism and Hospitality Research*, *31*(3), 376–392. https://doi.org/10.1080/13032917.2020.1732433

Williams, H. A., Williams, R. L., & Omar, M. (2014). Gastro-tourism as destination branding in emerging markets. *International Journal of Leisure and Tourism Marketing*, *4*(1), 1–18. https://www.researchgate.net/file.PostFileLoader.html?id=54aa74bfd2fd6409728b45d0&assetKey=AS%3A273664501583881%401442258109366

Winterbottom, M. (Director). (2014). *The trip to Italy*. [Film]. British Broadcasting Corporation (BBC) & Revolution Films.

Wondirad, A., Kebete, Y., & Li, Y. (2021). Culinary tourism as a driver of regional economic development and socio-cultural revitalization-evidence from Amhara National Regional State, Ethiopia. *Journal of Destination Marketing and Management*, *19*, 100482. https://doi.org/10.1016/j.jdmm.2020.100482

World Food Travel Association (WFTA). (2022). *What is food tourism?* https://worldfoodtravel.org/what-is-food-tourism/

World Tourism Organization (UNWTO). (2022). *Gastronomy and wine tourism*. https://www.unwto.org/astronomy-wine-tourism

Zhang, P., Wang, L. E., Bai, J., Liu, X., & Shaopeng, F. U. (2018). Study on food waste behavior of catering consumers in tourist cities. *Resource Science*, *40*(6), 1186–1195.

Chapter 33

Where to With Evolving Gay and Lesbian Tourism Practices?

Gustav Visser

Introduction

There is an extensive history focused on gay tourism ever since it became a focus of geographers and tourism scholars in the 1990s (Hughes, 2006; Waitt & Markwell, 2006; Visser, 2014). The investigation of this 'type' of tourism has been challenging as the definition of gay tourism is complex (Vorobjovas-Pinta, 2021a; Vorobjovas-Pinta, 2021b). For the most part in exploring 'gay tourism', scholars have in the main actually focused on gay men and lesbians – the emphasis being that of white gay men. This chapter regretfully (owing to data shortage) follows in this tradition, although key themes are, at least in part, relevant to all other sexual orientations or gender identities under the broader LGBTQI+ (lesbian, gay, bisexual, transgender, queer, intersex) banner. Much of how broader 'gay' tourism practices is articulated in tourism spaces is, however, dependent of capturing particular tourism niches, usually via some form of the neoliberal market place, framed by the societal realities – often, if not dominated by gay male realities - diversely different geographical contexts.

Ten years ago, Gordon Waitt (2012) argued that queer perspectives on tourism geographies have been grappling with the gradual normalisation of particular forms of homosexuality within the neoliberal market place, and the implications for more radical imperatives of queer politics to blur and reveal the inconsistencies of sexual and gender categories. The whole movement towards queer tourism now having expanded within mainstream academic and in some cases policy debate to read LGBTQI+ is telling. Since then however, much of the former gay dynamics of the queer argument have changed. The core of the 'old' arguments and concerns remain prescient, although the focus is perhaps now more concerned with making visible minorities within the non-heterosexual minority. Nevertheless, the past three decades concerns focused on the various relationships between tourism, place, space and sexuality (Doan, 2015) remain. These debates have demonstrated a particular developmental trajectory in that they begin with the intersections between male gay sexual identities in

The Wiley Blackwell Companion to Tourism, Second Edition. Edited by C. Michael Hall.

relation to place and space and then slowly draw in other non-heteronormative identities and spaces over time (Gorman-Murray, 2007; Waitt & Markwell, 2006; Vorobjovas-Pinta, 2021a; Vorobjovas-Pinta, 2021b).

Currently, the LGBTQ or, perhaps more inclusively, the queer community[1] is diverse and it is thus problematic to make reference to it as a unified grouping or collective – for it is not. Historically, it has been argued that popular gay travel destinations existed because they demonstrate certain characteristics and served several key functions, i.e. usually they have permissive or liberal attitudes towards queer identified travellers; they provide friendly infrastructure (bars, businesses, restaurants, hotels, nightlife, entertainment, media, organisations, etc.); they create opportunities to socialise with similarly identified individuals and they provide a sense of community. Much of the current body of research explores the development of gay male tourism in various well-known cities with large queer resident communities. This chapter argues that the connections that currently exist between place, space, sexuality and tourism demonstrate an evolution in both form and function (Anderson & Knee, 2021; Brown, 2009; Llewellyn, 2021). These changes have mainly been viewed through the lens of leisure space and various forms of spatially consolidated 'queer community(ies)', although the identity markers have for the most part been white gay men. On the whole, few of the queer markers remain relevant as analytic categories, since they are simply geographically too diverse.

This investigation explores this claim relative to tourism linked to sexual identity. In so doing, this chapter first provides some brief insights which illustrate the rise of debates regarding the intersections of gay identification, place and space. The long-established nature of these debates are highlighted and some form of movement in discourse to the idea of homonormalisation in a changing heterosexual society (in some places) is discussed. The most significant contribution of this chapter is located in the final section whereby consideration is given to new directions of enquiry that might be fruitful in the foreseeable future in the investigation of the intersection(s) between tourism, place, space and sexuality – changing use of technology and aging queer populations is argued to present the greatest new frontiers in investigating the queer/tourism interface(s).

Sexuality, Place, Space and Tourism Clusters

The current linage of debates concerning queer sexualities, place, space and tourism began with reference to gay men in the early 1960s (while really very present-minded and certainly ignorant of other cultural contexts differently understood, see Luongo, 2007) that academically only found traction in the 1990s (Waitt & Markwell, 2006; Visser, 2014). In this style of literature it was generally suggested that gay men use space to adopt separate identities – identities often assumed away from home and work places, encouraging the consumption of leisure space (for example, Hughes, 1997, 1998, 2006). These investigations were set against the backdrop of mainly large Western cities (see Brown and Browne (2016) for an expansive exploration of this literature). It was argued that urban areas facilitated the assumption of a gay self in many ways, not least through the number and concentration of gay leisure places available. Pioneers in this line of argument (Binnie, 1997; Knopp, 1995) noted that such leisure-related freedoms have been widely embraced as a validation of gay identity, if only a reaction to a more general powerlessness. Nevertheless, it was argued that the use of space may have the effect of transforming those spaces so that it becomes coded as gay (note that little was really said about lesbian spaces).

Drawing on these observations, and subsequently expanded on by Hughes (2006) as well as Waitt and Markwell (2006), Visser (2003b, p. 171) noted that 'leisure tourism by gay men at the time may have been the ultimate manifestation of the use of space in order to separate identities'. This argument suggested that studies of gay leisure spaces interpret the clustering of such activity as part of a homosexual challenge to the heterosexual coding of public space, empowering those who are currently excluded and disenfranchised (see also Binnie and Valentine (2000)). At the time Pritchard et al. (1998, p. 274) suggested:

> These are places which enable not only the display of behaviour and affection but also access to a variety of gay services and facilities including shops, bars, housing and legal and medical services . . . Gay spaces, in essence, provide community and territory as well as a sense of order and power. They are sites of cultural resistance with enormous symbolic meaning, providing cultural and emotional support for a political movement comprised of an increasingly diverse and geographically scattered community.

Drawing on Jansen-Verbeke (1994), Visser (2003a, 2003b) noted that, historically, not only do gay leisure spaces tend to cluster, but so too does tourism activity. It was observed that tourism in a city is related to a combination of its primary and secondary tourism elements. It was suggested that the activity place is composed of attractions that characterise most cities, such as museums and historic sites, whereas the leisure setting is the physical and socio-cultural context within which the attractions are set: the overall spatial structure of the city and its ambience. The secondary element in Jansen-Verbeke's (1994) view includes shops, cafés, restaurants and bars, hotels and entertainment.

Primary elements that are proximate to each other, as clusters, may have a greater appeal to the tourist than non-proximate elements. Hughes (1998) argued that several tourist clusters, spatially separate and with distinct features, may exist in a city to give it polycentricity. The identification of clusters ultimately depends on tourist use. These elements are, however, usually consumed by both tourists and local residents. The over-lapping use of gay leisure cluster(s) and primary, as well as secondary, tourism elements may also cause a city, in whole or in part, to be transformed both materially and symbolically (Visser, 2003a, 2003b). These observations were made more than two decades ago and much has subsequently changed, if not entirely globally, then at least significantly so, in many developed Western countries and a range of countries in the Global South.

Current Debates on the Link Between Queer Sexualities, Place, Space and Tourism

During the ensuing decades, many developments occurred, particularly for gay men who travel. A starting point is to highlight that the empirical base from which conclusions of links/ or not between queer, place, space and tourism has notably expanded in spatial reach (see the collection of Vorobjovas-Pinta, 2021a). For example, the developing world has seen far greater exposure in the academic press, including diverse contexts such as the Argentina (Jeffrey & Sposato, 2021), Asia Pacific region (Nghiêm-Phú & Suter, 2021; Vorobjovas-Pinta, 2021b), China (Gao, 2021), Mexico (Monterrubio, 2021), some Muslim countries (Coşkun, 2021; Doan, 2021) and South Africa (Hattingh, 2021). In addition, the scale of analysis has over the past decade broadened even further beyond large metropolitan areas to smaller cities (Forstie, 2020) and rural areas (Lewis & Vorobjovas-Pinta, 2021), trends identified years ago (Waitt & Gorman-Murray, 2008). Internal to urban areas there has been debates concerning the reasons behind why gaybourhoods (Ghaziani, 2021) and gay villages retain relevance (Gorman-Murray, 2021) even in a supposed potential/anticipated post-gay Western world (Ghaziani, 2011). In some ways debates concerning consolidated gay space continues to be inspired by idea of 'the life and afterlife of gay neighbourhoods' with at least a tacit and sometimes more explicated desire for a 'renaissance and resurgence' of the gay neighbourhood (Bitterman & Hess, 2021).

Arguably, the zenith of queer space development in the form of physical infrastructure was reached by the end of the first decade of the new millennium expressed at many spatial scales of anaylsis. This era solidified and perpetuated some of the historical assumptions concerning the sexuality, physical and symbolic nexus consolidated in queer space narrative till recently inspiring collections such as Bitterman and Hess's (2021) *The life and afterlife of gay neighbourhoods: Renaissance and resurgence.* In recent years the number and concentration of gay infrastructure that supported gay leisure spaces and tourism started to decline across the global North, as well as, in those places in the global South where it did register as a physical phenomenon (Hattingh, 2021). Anderson and Knee (2021) argue that there are at least three intersecting factors that account for this trend of spatial retreat, which in the end interacts with ideas of gay tourism.

The first issue relates to the positioning of a gay (though not all members of the queer cohort) identity in an increasingly changing heterosexual leisure world. Visser (2007, 2008a, 2008b, 2010, 2013, 2016) hinted some time ago that these changes were already visibly underway in South Africa, for instance – arguing the case for the idea of homonormalisation of leisure spaces. In Western society, this idea was supported by scholars such as Ghaziani (2011, 2015), confirming that the LGBTQ community (in its broadest sense) has become more accepted and that 'queer enclaves' were not as necessary as in the past – resulting in the so-called demise of the gayborhood (Bitterman, 2021; Coffin, 2021; Doan, 2021; Gorman-Murray, 2021; Hess, 2021). Collectively, these investigators note the declining (perhaps mixed) importance of the fixed spatialisation of sexual orientation in the lives of queer community(ies) and a desire to assimilate and participate in the general public social spaces that once denied them (Doan & Higgins, 2011). These claims, I would suggest are true for only a handful of countries, regions and cities globally, although those locations have formed the backbone of gay tourism scholarship and theorisation. Within these claims are embedded suggestions of a 'post-gay' society - a construct that needs to be considered with caution. Some of these ideas are located in the role of technology in 'queer community' formation (Collins & Drinkwater, 2017; Llewellyn, 2021; Miles, 2021).

The second factor, as some have argued, is that the increased use of technology to form community(ies) provides a further explanation to the decline of queer spaces (see Anderson and Knee (2021)). For some time there have been conversations about the role of online platforms as a replacement for dedicated physical queer spaces (Miles, 2021). This, in particular, rings true for younger generations (Collins & Drinkwater, 2017). Smartphones have fundamentally re-imagined the physical and virtual space interface. Renninger (2019) explored these ideas in the provocatively entitled 'Grindr killed the gay bar, and other attempts to blame social technologies for urban development'. Anderson and Knee (2021) argue that the proliferation of location-based social applications in the past decade has allowed the queer population to become comfortable creating and maintaining social connection virtually. The important point here is that not only is queer space(s) for locals in a particular city, for example, negotiated by technological interfaces but the ability of queer consolidated spaces that, at least traditionally, form the bedrock of gay travel motivations to gay travel destinations. There is an argument to be made that the importance of these applications has surged against the backdrop of the current COVID-19 pandemic (Miles et al., 2021) and the various restrictions placed on the operations of many queer venues, many of which did not survive this pandemic.

Third, there are debates that have essentially argued that gay-led gentrification, not technology, has undermined the maintenance of queer places and spaces. 'Gay villages/spaces' were historically located in less-desirable areas where the investment of various forms of capital often led to urban renewal. Subsequently a significant literature has emerged of this spatialisation that has been turned on its head. Gay presence became an indicator of actual and/or potential gentrification (Ghaziani, 2021; Visser, 2016). This, was overlaid by a greater acceptance of queer individuals in mainstream society, correlates to the commercialisation of a once queer space and a resulting attraction of new heterosexual residents seeking 'in vogue' physical locations centred on consumptive practices. This phenomenon in many popular gay tourist destinations has led to the in-migration of heterosexual cohorts and, in some places, tourists. This has, however, resulted in the degaying of some formerly gay neighbourhoods, making them less friendly to queer individuals, as well as leading to businesses favouring a wider audience attraction (Orne, 2016; Visser, 2016). Anderson and Knee (2021, p. 120), nevertheless, observe that 'the changing nature of queer leisure spaces and . . . the decline of some, does not diminish the role that such spaces play in the lives of queer persons' (see also Mattson, 2019a; Mattson, 2019b). These concerns have focused on gay male spaces, but certainly are true for lesbian spaces too. Although historically there has not been a particularly strong spatialisation of lesbian leisure (as compared with gay men) to dedicated physical spaces, in recent years, looking towards the United States of America over the past decade, there has been an unmitigated decline in lesbian infrastructure in the form of bars and social venues (The Lesbian Bar Project, 2021). The potential deployment of such spaces as tourist destinations both gay and lesbian are diminished. The question now arises as to how the combination these remarks point to different ways of thinking about the future of gay space and tourism?

Where to with Evolving Gay and Lesbian Tourism Practices Going Forward?

There are many new avenues to consider about where current debates are in terms of 'gay tourism' and where they might head. In this section a few possibilities are proffered.

The first issue that will require greater clarification to make any sense of queer travel is to be more certain about definition. The brief review provided is certainly framed by a rather binary approach to queer tourism. It should not be so. The discussion observed issues concerning gay men and tourism/travel. The intellectual issue is that much of what now constitutes LGBQTIA+ discourse is the very notion of fluidity of sexual and gender identification which is not reflected in tourism scholarship. The idea of celebrating the appropriation of queer difference in many ways (Ooi, 2021) is certainly not without its challenges – in theory, policy and practice. It is, however, central to have some semblance of clarity if tourism studies generally and tourism geography in particular aims to offer any meaningful analyses. This then feeds into the following observation.

This observation is set against the opening idea in this investigation of Gordon Waitt (2012) that queer perspectives on tourism geographies have been grappling with the gradual normalisation of particular forms of homosexuality and/or non-heteronormative identities within neoliberal market places and the implications for more radical imperatives of queer politics to blur and reveal the inconsistencies of sexual and gender categories. This notion has changed in significant ways, and not necessarily positively, depending on the approach taken to address such questions. The key concern is the relationship between the increasingly/registered sexuality identity(ies) markers to specific places/spaces and their/its relationship to tourism. One thing that has been percolating to the surface in recent years is the decrease in physical queer infrastructure and actual tourist activities clustering in 'gay' tourist destinations. In contrast to claims made by Visser (2003a), among others, that historically, not only do gay leisure spaces tend to cluster, but so too do tourism activities, this tendency seems no longer to be the case.

Queer places/spaces have increasingly become momentary inscriptions of the queer body on sexually ambiguous public/heterosexual places, enabled through interactions on internet platforms that is now the backbone of queer socialisation. Dedicated queer (but mainly meaning gay) places and spaces are now fewer and not necessarily clustered. In this regard there is much to be considered. Motivators for queer leisure, what that might constitute, and travel, echo much of what has already been highlighted. Many of the observations to follow are not confined to the queer community(ies), but relate to broader leisure activities and tourist accommodation that have developed over at least the past decade.

In terms of entertainment built around nightlife in 'party locations' such as London, Paris, Amsterdam, New York, etc. there has been a general decline in club and bar culture, which holds true for both the queer and mainstream heterosexual communities (Anderson, 2007; Anderson, 2009; *International National Music Summit*, 2021). The spatially static club/party/bar, or gay/'gay-friendly' accommodations, restaurants, etc. are more fluid now and 'sexuality tribe identity' has for some time become increasingly rare to find in one consolidated physical space (The Lesbian Bar Project, 2021). In fact, many queer identified persons would find it somewhat problematic that adjectives are required. In addition, temporality in terms of physical engagement is not governed by the rhythm of the week/weekend of the past, nor climate seasons of summer, etc. Collectively, this is dissonant with former gay discourses concerning gay place, space and tourism.

Another issue that has entered the leisure space/sexual identity/tourism debate, though not remarked upon by Waitt (2012) a decade ago and which is not limited to non-heteronormative cohorts concerning those kinds of links, is that the imprint of neoliberal marketisation of leisure in what was once described as 'queer community space', has come to bare on queer space development and hence tourism potential. Property development with better returns has become a key factor in the lack of stationary queer spaces, which in turn halts the development of permanent clustering and potential tourism linked to other non-heteronormative identities. This is significantly tied to the role of mobile technologies.

The popular use of social media has changed the notion of 'clubbing or cruising bars'. From an investigatory point of view, the importance of alternative data points come to mind, shifting from more traditional qualitative and quantitative techniques in gaining information that concerns issues dealing with intersections between sexuality, places and space and tourism to the digital realm. This new data source of online information now ranges considerably from relatively public to extremely private, hidden in far-flung corners of the internet. This information is not universally accessible, queer or otherwise. Physical place and space, as a context in which 'identity learning' took place three or four decades ago, has now evolved into a very uneven platform of information. New knowledge about queer communities are emerging and alternative 'clubbing venues' have emerged, however these 'places and spaces' are not very visible. In addition, the ethics of these investigations need consideration as the manner in which information is obtained becomes inadvertently more covert. How does this play out in queer tourism as physical spaces?

Mobility of club and party places and spaces and the demise of megaclubs in favour of more bespoke, intermittent music and identity sharing experiences are becoming important analytic classes (Faber, 2019; Kimbrell, 2020; *The Economist*, 2020). They do not 'speak to' the former notions of consolidated place/space and the monetisation thereof for tourism – these spaces are now fluid and 'placeless'. The re-emergence of the 'private party' and transient places and spaces of an earlier age (or even still part in many places in the Muslim world, e.g., Luongo (2007)), seems relevant to current realities. Club and party scenes in tourism hotspots such as Ibiza, Amsterdam, Berlin or San Francisco, for example, are being re-thought, along with the notion of gay mass tourism. There are many reasons for this, cost being one and changes in Millennial's and Generation Z's relationship to the notion of leisure being another (*International National Music Summit*, 2021). This appears to be particularly pertinent to gay social/sexual interaction (Miles, 2021). Again, in the diminished number of bars and clubs available, a greater social acceptance of queer persons feeds the further declines of dedicated queer spaces (Doan, 2021) and tourism activity.

Related to 'historic' concerns surrounding dedicated gay accommodation or restaurants (gay spaces generally, see Miller, 2021), the physical and symbolic landscape has changed dramatically since, for example, Hughes (2006); Waitt (2012) and Waitt and Markwell (2006) wrote about queer (mainly gay men) perspectives on tourism geographies. Currently, in mainstream queer tourist destinations, 'gay-friendly' designations do not exist. Broadly-speaking, homonormalisation has made that type of vocabulary redundant and even offensive. In this regard, one has to return to the role of technology in the sexuality, tourism, space and place interface within the contemporary context.

Developments in mobile digital technologies are disrupting conventional understandings of space and place for smartphone users. One way in which location-based media are reconfiguring previously taken-for-granted spatial traditions is via GPS-enabled online dating and hook-up apps (Miles, 2021). For sexual minorities, these apps can reconfigure any street, park, bar or home into a queer space through a potential meeting between mutually attracted individuals, but what does this signify for already-existing queer spaces? Smartphone apps including Grindr, Tindr, and Blued synthesize online queer encounters with offline physical space in order to create a new hybrid terrain predicated on availability, connection and encounter (Miles, 2021). This terrain is able to sidestep established gay neighbourhoods entirely which, in turn, impacts older, physically rooted gay neighbourhoods and the role that these neighbourhoods have traditionally played in brokering social and sexual connection for sexual minorities. Few would deny that location-based apps have come to play a valuable role in multiplying opportunities for sexual minorities. However, the stratospheric rise of these technologies also provokes questions about their influence on embodied encounter, queer community, and a sense of place and tourism.

There is the issue of disentangling the queer discourse. This chapter has mainly followed debates on gay male sexuality, place, space and tourism intersections that historically has been the most remarked upon in the academic press - it purposefully does not address subsets of 'LBTQ+'. What should be underscored is that non-heteronormative and/or non-binary identities have become obsolete as a collective noun for spatial analysis vis-à-vis tourism (Bitterman, 2021). We are talking about minorities within minorities and their relationship to place, space, leisure and tourism, whether or not these investigations take place in large cities or smaller urban places (Browne, 2009; Forstie, 2020) it cannot be generalised, methodologically or epistemologically and probably should not be. Leading on from this observation is a critical need for far greater sensitivity regarding inter-generational changes in broadly queer leisure and various cohorts' relationship to place, space and tourism. What was in the 1960s and 1970s, regarding the search for shared space of 'the queer', and the major consolidating effects of AIDS in the 1980s, has changed. The 1990s and early 2000s were different in terms of what binds the non-heterosexual leisure seekers and tourism, a focus within the gay and lesbian cohorts as some form of a community. From industry sources, rather than the academic press, it has become apparent that different age cohorts within the broad denotation of 'queer' are fundamentally relating to place, space, leisure and sexuality in very different ways, and hence tourism as well. Binding links between these cohorts are fleeting at best, both in meaning and place.

The seemingly inevitable fractured nature of these intersections reflected through the prism of a broadly speaking empirical base of Northern and Southern understandings of the relationship between sexuality, place, space and tourism holds several major challenges (Visser, 2015). This harks back to the notion of a methodology and the manner in which scholars access information as aforementioned. Additionally, it relates to the epistemic issue of what would qualify as research questions and comparable data points across different locational realities and building theory(ies) (Coffin, 2021). How comparable are the realities of a seemingly 'online' dating space and perhaps resultant tourism between generally uniform Northern contexts and significantly more diverse technological possibilities in the full range of countries currently 'clumped' into the Global South categorisation? There is gay tourism, if not always visible in the Global South, regardless of whether this categorisation is not particularly useful in queer thinking about sexuality and tourism. The realities of Southern contexts such as South Africa, Brazil and Morocco are, for example, vastly different empirical contexts and will therefore demonstrate a variation of possible relationships to sexuality, place, space and tourism. Finally, it is necessary to beg the question, what does this mean to the ontology and what queer space and place is and its relationship to tourism aspire to? These questions aside, it is undeniably clear that the nature of queer place/space has changed and it is left to us to unravel the new articulations of queer sexuality, tourism and community in this new age.

Finally, the elephant in the room relates to that one thing that brings all of these actual and potential tourist practices and possibilities or challenges together. Current debates concerning gay leisure and tourism practices have not made a single reference to the fact that the general population of this grouping (excluding Africa) is aging across the globe and rapidly so. So do their relationship to queers spaces. The demise of gay space, for example, is noted in various media might be explained by a demographic shift and accompanying expressions of queerness rather than only a greater acceptance of queerness, technology and changing consumption patterns. No consideration is given to the fact that the median age of actual and potential actors in those important gay villages and neighbourhoods that are so often the focus of gay travel are increasingly changing because of age. For example, the median age in the European Union is 40 years, North America is 37 and Asia, 40. We are dealing with an aging gay population in the developed North and by many measures in the global South too. The academic community has built much of its engagement with the 'sexuality, tourism and space nexus' on an ever-lasting (even growing) grouping of essentially young people that do 'typical young gay things' that can be packaged in experiences often related to physical space. This is not an empirical demographic reality to build on. What if the older gay groupings are not adherent to such understandings of socialisation and various relationship to consolidated space? The biggest research question going forward in dealing with the evolution of gay and lesbian tourism practices is not only technology but the universal aging of the global population and what it is they actually engage as queer tourists.

Note

1. Following Anderson and Knee (2021), I use 'queer' as an umbrella term inclusive of all minoritised sex, gender and sexual identities that draws distinction from hegemonic/heteronormative society. The acronym 'LGBTQ' (lesbian, gay, bisexual, transgender, queer) is also used as a whole or in part throughout the chapter to specify segments of the community.

References

Anderson, A., & Knee, E. (2021). Queer isolation or queering isolation? Reflecting upon the ramifications of COVID-19 on the future of queer leisure spaces. *Leisure Sciences*, *43*(1–2), 118–124. https://doi.org/10.1080/01490400.2020.1773992

Anderson, C. (2007). *The long tail: How endless choice is creating unlimited demand*. Random House.

Anderson, T. (2009). Better to complicate, rather than homogenize, urban nightlife: A response to Grazian. *Sociological Forum*, *24*(4), 918–925. https://doi.org/10.1111/j.1573-7861.2009.01144.x

Binnie, J. (1997). Coming out of geography: Towards a queer epistemology? *Environment and Planning D: Space and Society*, *15*(2), 223–237. https://doi.org/10.1068/d150223

Binnie, J., & Valentine, G. (2000). Geographies of sexuality - a review of progress. *Progress in Human Geography*, *23*(2), 175–187. https://doi.org/10.1177/030913259902300202

Bitterman, A. (2021). Understanding generational gaps in LGBTQ+ communities: Perspectives about gay neighbourhoods among heteronormative and homonormative generational cohorts. In A. Bitterman & D. Hess (Eds), *The life and afterlife of gay neighbourhoods: Renaissance and resurgence* (pp. 307–338). Springer.

Bitterman, A. & Hess, D. (Eds). (2021). *The life and afterlife of gay neighbourhoods: Renaissance and resurgence*. Springer.

Brown, G. (2009). Thinking beyond homonormativity: Performative explorations of diverse gay economies. *Environment and Planning A: Economy and Space*, *41*(6), 1496–1510. https://doi.org/10.1068/a4162

Brown, G., & Browne, K. (Eds). (2016). *The Routledge research companion to geographies of sex and sexualities*. Routledge.

Coffin, J. (2021). Plateaus and afterglows: Theorizing the afterlives of gayborhood as post-places. In A. Bitterman & D. Hess (Eds), *The life and afterlife of gay neighbourhoods: Renaissance and resurgence* (pp. 371–389). Springer.

Collins, A., & Drinkwater, S. (2017). Fifty shades of gay: Social and technological change, urban deconcentration and niche enterprise. *Urban Studies*, *54*(3), 765–785. https://doi.org/10.1177/0042098015623722

Coşkun, G. (2021). The search for gay space, relationship dynamics and travel decisions among same-sex couples in a secular Muslim country. In O. Vorobjovas-Pinta (Ed.), *Gay tourism: New perspectives* (pp. 103–116). Channel View Publications.

Doan, P. (2015). *Planning and LGBTQ communities: The need for inclusive queer spaces*. Routledge.

Doan, P. (2021). After the life of LGBTQ urbanism and generation from Atlanta and Istanbul. In A. Bitterman & D. Hess (Eds), *The life and afterlife of gay neighbourhoods: Renaissance and resurgence* (pp. 261–285). Springer.

Doan, P., & Higgins, H. (2011). The demise of queer space? Resurgent gentrification and the assimilation of LGBT neighbourhoods. *Journal of Planning Education and Research*, *31*(1), 6–25. https://doi.org/10.1177/0739456X10391266

Faber, T. (2019). Is London's club scene really under threat? *Financial Times*. https://www.ft.com/content/9e45e870-f668-11e9-bbe1-4db3476c5ff0

Forstie, C. (2020). Theory making from the middle: Researching LGBTQ communities in small cities. *City and Community*, *19*(1), 153–168. https://doi.org/10.1111/cico.12446

Gao, X. (2021). (Re)discovering Chinese lesbian and gay travel. In O. Vorobjovas-Pinta (Ed.), *Gay tourism: New perspectives* (pp. 89–102). Channel View Publications.

Ghaziani, A. (2011). Post-gay collective identity construction. *Social Problems*, *58*(1), 99–125. https://doi.org/10.1525/sp.2011.58.1.99

Ghaziani, A. (2015). *There goes the gaybourhood?* Princeton University Press.

Ghaziani, A. (2021). Why gayborhoods matter: The street empirics of urban sexualities. In A. Bitterman & D. Hess (Eds), *The life and afterlife of gay neighbourhoods: Renaissance and resurgence* (pp. 87–113). Springer.

Gorman-Murray, A. (2007). Rethinking queer migration through the body. *Social & Cultural Geography*, *8*(1), 105–21. https://doi.org/10.1080/14649360701251858

Gorman-Murray, A. (2021). Recovering the gay village: A comparative historical geography of urban change and planning in Toronto and Sydney. In A. Bitterman & D. Hess (Eds), *The life and afterlife of gay neighbourhoods: Renaissance and resurgence* (pp. 239–260). Springer.

Hattingh, C. (2021). Understanding African LGBT traveller motivation: A non-Western perspective. In O. Vorobjovas-Pinta (Ed.), *Gay tourism: New perspectives* (pp. 67–88). Channel View Publications.

Hess, D. (2021). Who are the people in your gayborhood? Understanding population change and cultural shifts in LGBTQ+ neighorhoods. In A. Bitterman & D. Hess (Eds), *The life and afterlife of gay neighbourhoods: Renaissance and resurgence* (pp. 3–39). Springer.

Hughes, H. (1997). Holidays and homosexual identity. *Tourism Management*, *18*(1), 3–7. https://doi.org/10.1016/S0261-5177(96)00093-3

Hughes, H. (1998). Sexuality, tourism and space: The case of gay visitors to Amsterdam: In D. Tyler, M. Roberson & Y. Guerrier (Eds), *Managing tourism in cities: Policy, process and mobility* (pp. 163–178). Wiley.

Hughes, H. (2006). *Pink tourism: Holiday of gay men and lesbians*. CABI.

International National Music Summit (2021). *Where is club culture headed?* https://www.youtube.com/watch?v=sk5Yosev4vk

Jansen-Verbeke, M. (1994). The synergy between shopping and tourism. In W. Theobald (Ed.), *Global tourism: The next decade* (pp. 347–362). Butterman-Heidemann.

Jeffrey, H., & Sposato, M. (2021). Buenos Aires: Authentically straight-friendly. In O. Vorobjovas-Pinta (Ed.), *Gay tourism: New perspectives* (pp. 201–210). Channel View Publications.

Kimbrell, G. (2020). 4 predictions for nightlife after Covid-19. *Rolling Stone*. https://www.rollingstone.com/culture-council/articles/predictions-nightlife-after-covid-1091856/

Knopp, L. (1995). Sexuality and urban space: A framework for analysis. In D. Bell & G. Valentine (Eds), *Mapping desire* (pp. 149–161). Routledge.

Lewis, C., & Vorobjovas-Pinta, O. (2021). Out of the closet and into the streets: Reflection on the considerations in housing rural pride events. In O. Vorobjovas-Pinta (Ed.), *Gay tourism: New perspectives* (pp. 211–223). Channel View Publications.

Llewellyn, A. (2021). "A space where queer is normalized": The online world and fanfictions as heterotopias for WLW. *Journal of Homosexuality*. https://doi.org/10.1080/00918369.2021.1940012

Luongo, M. (2007). *Gay travels in the Muslim World*. Harrington Park Press.

Mattson, G. (2019a). Small city gay bars, big city urbanism. *City and Community*, *19*(1), 76–97. https://doi.org/10.1111/cico.12443

Mattson, G. (2019b). Are gay bars closing? Using business listings to infer rates of gay bar closure in the United States, 1977-2019. *Socius: Sociological Research for a Dynamic World*, *5*(1–2). https://doi.org/10.1177/2378023119894832

Miles, S. (2021). Let's (not) go outside: Grindr, hybrid space and digital queer neighourhoods. In A. Bitterman & D. Hess (Eds), *The life and afterlife of gay neighbourhoods: Renaissance and Resurgence* (pp. 203–220). Springer.

Miles, S., Coffin, J., Ghaziani, A., Hess, D. B., & Bitterman, A. (2021). After/lives: Insights from the COVID-19 pandemic for gay neighbourhoods. In A. Bitterman & D. Hess (Eds), *The life and afterlife of gay neighbourhoods: Renaissance and resurgence* (pp. 393–418). Springer.

Miller, S. (2021). Commemorating historically significant gay places across the United States. In A. Bitterman & D. Hess (Eds), *The life and afterlife of gay neighbourhoods: Renaissance and resurgence* (pp. 339–370). Springer.

Monterrubio, C. (2021). The significance of gay tourism spaces for local gay men: Empirical evidence from Mexico. In O. Vorobjovas-Pinta (Ed.), *Gay tourism: New perspectives* (pp. 34–51). Channel View Publications.

Nghiêm-Phú, B., & Suter, R. (2021). Sensual and sexual experiences come first: An analysis of gay patrons' comments on gay saunas, spas and cruising clubs in the Asia Pacific regions. In O. Vorobjovas-Pinta (Ed.), *Gay tourism: New perspectives* (pp. 134–152). Channel View Publications.

Ooi, C-S. (2021). Gay tourism: A celebration and appropriation pf queer difference. In O. Vorobjovas-Pinta (Ed.), *Gay tourism: New perspectives* (pp. 15- 21). Channel View Publications.

Orne, J. (2016). *Boystown: Sex and community in Chicago*. University of Chicago Press.

Pritchard, A., Morgan, N., Sedgely, D., & Jenkins, A. (1998). Reaching out to the gay tourist: opportunities and threat in an emerging market segment. *Tourism Management*, *19*(3), 273–282. https://doi.org/10.1016/S0261-5177(98)80016-2

Renninger, B. (2019) Grindr killed the gay bar, and other attempts to blame social technologies for urban development: A democratic approach to popular technologies and queer sociality. *Journal of Homosexuality*, 66(12), 1736–1755. https://doi.org/10.1080/00918369.2018.1514205

The Economist (2020). Berliners fear "club death". But nightlife in the world's clubbing capital is changing, not dying. https://www.economist.com/europe/2020/01/30/berliners-fear-club-death

The Lesbian Bar Project (2021). *The Lesbian bar project*. https://www.lesbianbarproject.com/

Visser, G. (2003a). Gay men, leisure space and South African cities: The case of Cape Town. *Geoforum*, *34*(1), 123–137. https://doi.org/10.1016/S0016-7185(02)00079-9

Visser, G. (2003b). Gay men, tourism and urban space: Reflections on Africa's 'gay capital'. *Tourism Geographies*, *5*(2), 168–189. https://doi.org/10.1080/1461668032000068261

Visser, G. (2007). Homonormalising (white) heterosexual middle-class leisure spaces: The case of white gay men in Bloemfontein, South Africa. *International Journal of Diversity in Organisations, Communities and Nations*, 7(1), 217–228.

Visser, G. (2008a). The homonormalisation of white heterosexual leisure spaces in Bloemfontein, South Africa. *Geoforum*, *39*(3), 1347–1361. https://doi.org/10.1016/j.geoforum.2007.11.004

Visser, G. (2008b). Exploratory notes on the geography of black gay leisure spaces in Bloemfontein, South Africa. *Urban Forum*, *19*(4), 413–423. https://doi.org/10.1007/s12132-008-9039-4

Visser, G. (2010). Leisurely lesbians in a small city in South Africa. *Urban Forum*, *21*(2), 171–185. https://doi.org/10.1007/s12132-010-9084-7

Visser, G. (2013). Challenging the gay ghetto in South Africa: Time to move on? *Geoforum*, *49*, 268–274. https://doi.org/10.1016/j.geoforum.2012.12.013

Visser, G. (2014). Gay and lesbian tourism practices. In A. A. Lew, C. M. Hall & A. M. Williams (Eds), *The Wiley Blackwell companion to tourism*, (pp. 435–443). Wiley Blackwell.

Visser, G. (2015). Thinking beyond exclusionary gay male spatial frames in the developing world. In P. Doan (Ed.), *Planning and the LGBTQ community: The need for inclusive queer spaces* (pp. 81–93). London.

Visser, G. (2016). Urban leisure and tourism-led redevelopment frontiers in central Cape Town since the 1990s. *Tourism: An International Interdisciplinary Journal*, 64(4), 397–408. https://hrcak.srce.hr/170570

Vorobjovas-Pinta, O. (2021a). Gay tourism: New perspectives. In O. Vorobjovas-Pinta (Ed.), *Gay tourism: New perspectives* (pp. 1–12). Channel View Publications.

Vorobjovas-Pinta, O. (Ed.). (2021b). *Gay tourism: New perspectives.* Channel View Publications.

Waitt, G. (2012). Queer perspectives on tourism geographies. In J. Wilson (Ed.), *The Routledge Handbook of tourism geographies* (pp. 82–89). Routledge.

Waitt, G., & Gorman-Murray, A. (2008). Camp in the country: Re-negotiating sexuality and gender through a rural lesbian and gay festival. *Journal of Tourism and Cultural Change*, 6(3), 185–207. https://doi.org/10.1080/14766820802647616

Waitt, G., & Markwell, K. (2006). *Gay tourism: Culture and context.* Haworth.

Chapter 34

Tourism, Shopping and Connotations of Place

Dallen J. Timothy

Introduction

Shopping is a fundamental tourist pursuit and one of the most essential activities people engage in while travelling. Tourist shopping is not the same as shopping tourism. Tourist shopping manifests in many ways and entails buying souvenirs, browsing in malls, buying supplies unintentionally left at home (e.g., sunglasses, backpacks, bathing suits, or vitamin supplements), purchasing gifts for friends and family members, procuring a bottle of wine following a wine tasting, buying picnic items from a supermarket, and any number of retail-related consumer activities. Tourist shopping is, therefore, typically a secondary activity undertaken by travelers whose principal purpose of travel is something else: conducting business, taking a beach vacation, participating in a sports event, or participating in a cultural heritage tour (Frost & Frost, 2024; Timothy, 2014). Shopping tourism, on the other hand, denotes travel that is motivated fundamentally by a need or desire to shop (Timothy, 2005; Timothy, 2024), and the destination is chosen because of its retail opportunities and advantages, including its reputation and image, emphasis on shopping tourism, unique retail environment, reasonable prices, or merchandise selection.

Globalization and neoliberal trade and economics have opened the doors wide to global shopping experiences (Timothy, 2019). No longer are people relegated to shopping only in their home regions, as technological globalization has enabled people to shop anywhere in the world through online retailing. In addition, for the most part, people are more physically mobile than they have ever been and, equipped with better knowledge of places and retail opportunities and greater levels of affluence, more people have shown a growing interest in travelling for the purpose of shopping.

Like all other forms of tourism, there is a strong element of place, placeness, or placelessness associated with shopping tourism, for it is the connection between place and retail that creates attractive shopping landscapes and destinations that draw millions of consumers each

The Wiley Blackwell Companion to Tourism, Second Edition. Edited by C. Michael Hall.

year. Acknowledging the vital importance of tourist shopping and shopping tourism, this chapter focuses on the second of these two phenomena — shopping tourism — and examines it from a place perspective. First, the chapter describes the appeal of shopping as a travel motivator. It then examines several contexts of shopping tourism and their unique characteristics in the broader phenomenon of tourism and how place is an important part of their appeal. These contexts include cross-border shopping, luxury shopping, shopping festivals, and second home-buying as a form of shopping tourism. Attention is then turned to several current trends in shopping tourism that emphasize place, place-making and sense of place.

Retail Opportunities: The Appeal of Shopping Tourism

Modes of shopping as a motive for travel have existed since ancient times. Some of the goals of early colonialism and the European 'discovery' of new worlds was to acquire valuable merchandise, such as spices, silk, perfumes, and gold. Such valued commodities motivated many a journey to buy, trade, and even conquer and colonize. Even before early modern colonialism, many ancient traders bought or traded for merchandise not available in their home regions. Many accounts tell of ancient 'shoppers' in the old-time empires and along transcontinental trade routes (Casson, 1985; Charlesworth, 2016; Holleran, 2012; McLaughlin, 2010). Even buying souvenirs for self and gifts for others was a common retail activity in ancient sojourns (Popkin, 2022; Rabotić, 2014).

Today, many destinations realize the value of shopping as a primary or secondary attraction in the global tourism marketplace and are positioning themselves to become retail destinations, with other place-based characteristics providing a more scenic and leisure backdrop to the retailscapes they are working to develop (Liberato et al., 2020; Muro-Rodríguez, 2020). Among their efforts are attracting large retailers, encouraging small-scale entrepreneurs and developing superlative shopping venues (e.g., world's largest, tallest, most expensive), often in conjunction with resort development or other tourism foci. These actions are only a few of the current manifestations of 'placeness' in shopping tourism. The following sections detail several trends and patterns that have significant implications for place-meaning, place-making and shopping.

Cross-border Shopping

All international outshopping is by definition a form of cross-border shopping, but this section focuses on the retail phenomenon that physically occurs in transborder regions. Cross-border shopping entails people travelling to nearby areas abroad to purchase items for their own use or for others. This notion has recently been extended to include petty trade, wherein people from one country cross a border to buy merchandise (e.g., clothing, toys, car parts, tools, household goods, and food), which they then take home to sell for a profit (Michalkó et al., 2014; Michalkó & Timothy, 2001; Szytniewski et al., 2020). Although this does not resemble 'tourism' in the traditional sense, it is certainly part of the tourism system as people cross international borders, spend time abroad (sometimes even overnight), have impacts on the other side, undertake various activities, eat, and stimulate local economies (Timothy & Teye, 2005). The importance of cross-border petty trade notwithstanding, the focus of this section is the traditional manifestations of cross-border shopping.

When consumers are free to cross a permeable national boundary, possess knowledge about the retail opportunities on the other side, have some level of familiarity with the neighboring region, and can afford to take advantage of price differentials abroad, shopping tourism tends to thrive in borderlands (Spierings & van der Velde, 2013; Studzińska et al., 2018). This is one of the clearest manifestations of shopping tourism. Even if the journey and products of their attention are utilitarian in nature, the actions of cross-border shoppers are a form of tourism. Consumers who live near an international boundary may cross frequently for such things as dog food, grocery items, car repairs, gasoline and clothing, and their daily action space may in fact extend beyond the border into foreign territory (Ramsey et al., 2019). Those who live further from a border checkpoint typically cross less often, buy bigger-ticket items when they do cross (e.g., furniture, electronics, and appliances), and their journeys tend to have a greater leisure component, such as sightseeing, dining out, or visiting museums and national parks (Timothy, 1999; Timothy & Butler, 1995).

Cross-border shopping occurs in all parts of the world, wherever the opportunity and motive exist. Research shows that the main motivation for shopping across a border is lower prices (Makkonen, 2016; Makkonen, 2023; Timothy & Butler, 1995; Tömöri, 2010). This is not to suggest that all cross-border shopping is utilitarian in nature, as noted above, as shopping abroad can also involve many other activities, and shopping itself is often considered a pleasurable activity. Lower prices are typically a result of economy of scale in a neighboring country, where a larger population means goods can be produced or imported in greater volumes and therefore have reduced costs. Prices are often lower because of lesser taxes and favorable exchange rates. Some countries have no sales or VAT taxes on food or clothing items, whereas neighboring states might have high taxes on the same consumer goods. A second major draw is a greater merchandise selection with products that are unavailable at home. This too may be a result of economy of scale. In small economies, there tend to be fewer products to choose from. Third, cross-border shoppers frequently cite high-quality service as another reason for shopping abroad. Finally, the existence of other attractions and activities besides shopping are additional reasons, in conjunction with shopping, why people choose to shop in a neighboring country, and the experience is sometimes enhanced when it takes place in interesting environments such as historic cities or unique natural areas (Ramsey et al., 2019; Timothy & Butler, 1995).

Even in the European Union, cross-border shopping occurs between member states despite the single market's efforts to equalize prices and taxes as much as possible. Germans regularly shop in Poland for cigarettes, meat, and various other items. Norwegians (non-EU) shop in Sweden (EU) for candy and soft drinks, which have a high sugar tax in Norway. Norwegians also shop in Sweden for groceries, tobacco, and alcohol. Finns shop in Sweden for many things, most notably chewing tobacco pouches known as snus, which are not sold in Finland. French consumers shop in Spain for vegetables, fruit, wine, prepared foods, and ordinary household merchandise. Belgian drivers frequently cross into Luxembourg to fill up with petrol at one of Luxembourg's ubiquitous border petrol stations, where gas can be 20–30 cents cheaper per litre. The degree of openness of a border and cost differences will determine how frequently people extend their retail spaces across a national frontier. The price differentials between Eastern Europe and Western Europe are still considerable, even though most countries of the former Eastern Bloc have joined the EU. Thus, there is still a strong competitive advantage for retailers located inside the borderlands of Eastern Europe (Bar-Kołelis & Wendt, 2018; Tömöri, 2010).

Likewise, 'free-trade zones' have been established throughout the world in border areas to cater to the needs of neighbors. These are typically located at land borders, international seaports, and a few international airports. These zones are usually granted either extraterritorial status or are otherwise customs-free areas where goods can be stored, transshipped or exported without normal customs enforcement or duties charged. Some of these are particularly geared towards shopping and retail tourism, including the Corozal Free Zone in Belize. This duty-free shopping multiplex is located in the 700-meter stretch between the Mexican border and Belize's border checkpoint. Given its location outside Belize's customs area but inside Belizean territory, the area offers foreign consumers (i.e., Mexican shoppers), opportunities to buy many imported items at tax-free rates.

The Corozal Free Zone is an example of the growing diversity of duty-free shopping opportunities that are typically located at border crossings or other international ports of entry and exit. The theory behind duty-free shopping is that if a product will not be consumed in the country where it is sold, the buyers should not have to pay import taxes. That is why most duty-free retail opportunities are positioned where consumers leave a country, such as near departure gates at airports, in cruise ports, and even international railway stations. Duty-free is a common feature of border retail landscapes and draws many consumers across national borders to save money on high-taxed items such as chocolate, perfume, tobacco, and alcohol (Hussin et al., 2012; Timothy, 2005).

Luxury Shopping

Certain social classes have always sought luxury retail opportunities. Today, the notion of luxury shopping has grown considerably with the vast accumulation of wealth in certain societies. Many destinations (e.g., Paris, Dubai, New York, Singapore, Hong Kong, and London) have made considerable efforts to brand themselves as luxury shopping destinations. Luxury shopping refers to consumers seeking indulgent or superior-quality products that are unique, exclusive and prestigious (Hung et al., 2021). The most common products associated with this trend include jewelry and watches, footwear, cosmetics, designer handbags and luggage, fragrances, vehicles, high-end apparel, and fashion accessories. Although luxury shopping takes place in consumers' home countries, the majority of it occurs abroad, and many trips are taken specifically for this purpose, not only to acquire exclusive items but also to add personal prestige, bragging rights, and interpersonal influence (Wiedmann et al., 2007).

Certain market segments have arisen in recent decades with increased wealth and prosperity. The fossil fuel-based wealth of the Arabian Gulf states (e.g., Kuwait, Qatar, Bahrain, Saudi Arabia, Oman, and the UAE) has spurred a significant outshopping trend from those countries, whose citizens regularly shop at high-end malls in their own countries but favor travelling abroad, especially to Europe and other Middle Eastern countries for luxury shopping. Citizens of the Gulf countries are known to be especially brand conscious and spend billions of dollars every year on luxury products. Because of its critical mass of high-end shopping malls, exclusive retail landscapes, extensive brand-name merchandise, and Muslim-friendly entertainment opportunities, Dubai is the preferred shopping destination for wealthy Saudis, Kuwaitis and Qataris (Alhosani & Zaidan, 2014).

With the opening of individual economic freedom during the past two decades, the affluent Chinese market has grown exponentially through business ventures and private investments. This has enabled Chinese citizens to travel abroad in large numbers, with deepening desires to buy luxury merchandise and experience high-end destinations (Yao, 2023; Zhang & Zhao, 2019). Chinese high-end shopping has been the focus of much research in recent years, with studies revealing that most Chinese luxury shopping takes place outside of China and that Chinese consumers seek materialism, pragmatism, and social consciousness in their retail activities abroad (Bu et al., 2017; Zhang & Zhao, 2019).

The luxury retail market is growing in many parts of the world, and traditional shopping destinations are competing to become the world's top shopping destinations. Even less-traditional shopping regions (e.g., Cambodia and the Philippines) are scrambling to get a foothold in the sector and establish themselves as alternative luxury retail destinations.

Shopping Festivals

A new fad in shopping is the 'online shopping festival'—large-scale promotional events held by e-commerce retailers and sales platforms on specified occasions to increase mass sales. These online shopping events are becoming increasingly popular in China but are also debuting in other regions for their ability to raise product awareness and stimulate mass sales (Kim et al., 2023; Yulianto et al., 2021). These cyber occasions tend to remove the place element from the shopping experience, are overwhelmingly utilitarian in nature, and they do not resemble the sorts of in situ events held in popular shopping tourism destinations, which are the focus of this section.

Festivals are a deep-rooted human activity designed to celebrate important causes in a region. Tens of thousands of festivals are held every year to commemorate successful harvests, defeating an enemy in war, important cultural events, spectacular happenings from history, religious and spiritual occurrences, and many more celebratory causes. In the past two decades, certain destinations that are closely associated with shopping have initiated shopping festivals as a way of increasing brand awareness, bringing retailers and community stakeholders together, and providing a stimulus for increased visitation (Peter & Anandkumar, 2014; Peter & Anandkumar, 2024).

Shopping festivals not only bring consumers together en masse and are a primary tourist attraction, their purpose is to enhance a shopping destination's image abroad. Most shopping events are serious collaborative endeavors with networks of retailers, producers, catering and lodging establishments, airlines, local transportation, and municipal services that provide a holistic shopping experience. They typically last from a week to several weeks and involve international chain stores, small boutique shops, and local marketplaces. They are known for offering deep discounts and sales, prize giveaways, entertainment, food and drink, musical events, and competitions.

The best-known and most researched global shopping event is the Dubai Shopping Festival, which takes place annually during the coolest time of the year (December-January) and offers a vast array of activities beyond retail only (Mehta et al., 2014; Peter & Anandkumar, 2013; Peter & Anandkumar, 2016; Zaidan, 2016; Zaidan, 2019). According to the event's website, 'Residents and visitors are invited to enjoy 38 days of unmissable daily shopping deals, raffle draws, immersive competitions, exciting concerts, memorable events, bustling outdoor markets, exclusive pop-up events, and exceptional experiences across Dubai. Expect big-name performances and concerts, while shoppers will discover incredible savings and huge grand wins. Make unforgettable memories with vibrant New Years Eve celebrations and fireworks across the city. Get ready to enjoy top thrills with a series of indoor and outdoor experiences, all set to be bigger, better and even more spectacular than ever before' (Visit Dubai, 2023). Other popular shopping festivals (recent and ongoing) include, the Andorra Shopping Festival, the Thailand Grand Sale, the Istanbul Shopping Festival, the Hong Kong Shopping Festival, the Cannes Shopping Festival, and the Great Singapore Sale. Many other localities have begun to brand themselves as key retail destinations by hosting shopping festivals at different scales and for different markets (Timothy, 2018; Timothy, 2021).

The primary motive of people attending these sorts of mega-events is, unsurprisingly, to shop in a mass retail environment where all senses are stimulated and where discounts, especially on famous brand-name items, are plentiful. Other motives include relaxation and enjoying local culture and food (Peter & Anandkumar, 2016), and the festival characteristics most appreciated by attendees include cleanliness, discounts, available parking, and merchandise variety, as well as a critical mass of retailers in a given area and accessibility (Mehta et al., 2014).

The Real Estate Market: Shopping for Second Homes Abroad

Given the extremes of commerce and retail in the 21st century, it is not too unrealistic to suggest that purchasing real estate abroad can also be considered a type of shopping tourism, perhaps a subcategory of luxury retail and of course a crossover with second-home tourism. People purchasing vacation homes abroad is a longtime phenomenon, especially in Europe's Nordic countries (Hall & Müller, 2004; Hiltunen et al., 2013; Müller, 2021) and in the Mediterranean Basin (Soto & Clavé, 2017), although it is an increasingly salient socioeconomic activity in other parts of the world as well, such as Americans buying recreational homes in Canada and Mexico (Timothy, 2004).

Some countries have restrictions on property ownership by non-citizens, which has curtailed second-home buying, but through various globalization processes, such as supranationalism, especially in the European Union, owning a property abroad is becoming increasingly popular and allowable. This has been especially trendy in Eastern Europe where, since the fall of communism in the early 1990s and the subsequent land privatization, and after many countries in the former Eastern Bloc have joined the European Union, Western Europeans have bought up large numbers of rural second homes (Illés & Michalkó, 2008; Jagodic, 2012; Opačić & Koderman, 2018).

Hungary's Lake Balaton has been a magnet of foreign home ownership for the past three decades, particularly among German and Austrian 'home shoppers', owing to the area's relatively lower prices coupled with its scenic landscape and easy access (Lőrincz et al., 2021; Sulyok & Lőrincz, 2016). Finland has recently been the target destination for Russian second home buyers who prefer to invest their money in a more secure marketplace, which is located nearby and gives them access to the European Union, at least on a part-time basis (Hannonen, 2023; Hannonen et al., 2015; Honkanen et al., 2016).

Other Trends in Shopping Tourism

Whether a primary motivation for travel or a secondary activity while travelling, shopping is a natural fit with other types of tourism. For example, shopping frequently accompanies heritage tourism, as experiencing cultural heritage can get people in the mood to buy tangible memories of their cultural experiences (Su, 2010; Timothy, 2005). The success of museum shops and souvenir vendors at cultural attractions attests to this phenomenon where retail opportunities tend to enhance the visitor experience (McIntyre, 2010). Likewise, the proliferation of religious souvenirs in pilgrimage destinations, and food and drink souvenirs in historic city centers, wineries/distilleries, and rural agritourism regions is another manifestation of the close relationships between heritage, shopping and retail. In heritage tourism settings, such as archaeological sites, historic monuments, and museums, retailing is becoming an increasingly important part of the experience, largely because it can enhance the visitor experience, but also because it provides a justification and funding source for archaeological work and heritage management.

Online souvenir shopping is a growing trend. Although not necessarily a purely 'touristic' activity, the online market for place-based souvenirs is growing (Abendroth, 2011; Seaman & Schroeder, 2022). Many consumers see this as an opportunity to buy the memento they might have missed while in the destination or to acquire more gifts to give away to friends and family after a trip (He, 2022). It also provides an opportunity for tourists in the destination to have something shipped home rather than having to carry it with them. Although the online platform diminishes the element of place (Punzon, 2021), the souvenirs purchased may in fact help enliven a sense of place for people who have been to the destination commemorated by the item, or it might serve even as a surrogate visit for people who might not otherwise have an opportunity to go (He, 2022).

On the subject of souvenirs, the nature of these 'material culture' elements of tourism (Hitchcock & Teague, 2019) has seen considerable change in recent years, with a greater emphasis on perceived authenticity and place embeddedness (He, 2022), as consumers are increasingly dissatisfied with generic mementos made in faraway factories that scarcely even memorialize the place or event where they are sold. There appears to be a move towards tourists buying more place-entrenched souvenirs that reflect subtler, albeit more meaningful, aspects of their journey, including food and drink souvenirs, utilitarian items, and higher quality products (Collins-Kreiner & Zins, 2011; Swanson & Timothy, 2012). To increase place-embeddedness, many tourist attractions now offer on-demand souvenirs, including 3D printed souvenirs, which enable visitors to design their own mementos based on what they perceived to be the most important or most memorable part of the destination. Likewise, culture and creative products (CC products) are a new genre of souvenir that is gaining particular popularity in China and elsewhere in Asia. CC products are souvenirs that are somewhat mass produced but also depict genuine themes that are germane to the place where they are sold, such as famous museum pieces (He, 2022). CC souvenirs, too, are seen to have the appeal of both mass production and a close connection to place.

Branding and Place-making

A final consideration of how retail can appeal to tourists as a strong motivation for travel, or even a secondary motivation, is the related notions of branding and place-making. Place branding can happen in various ways, but the most common are organic imaging, intentional/purposive branding, or a combination of the two. Organic place imaging occurs as a natural process through time as a destination grows and develops. Word spreads and its popularity swells with the outcome that a place becomes well-established as a tourist destination, often following a life cycle of discovery, development and possible decline (Butler, 1980). A place brand or image can also be developed intentionally through various promotional media and extensive advertising. In the planning and policy arena, communities or regions can carry out intentional image-creating programs, such as marketing campaigns, hosting special events, developing a strong online presence, courting travel writers and other media representatives, and collaborating with industry stakeholders and service providers (e.g., tour operators and airlines) to create travel packages that will expose the most positive features of the destination to potential visitors.

As noted previously, certain destinations have become synonymous with shopping. This is particularly the case with high-end retail venues such as Dubai, Paris, Rome and Singapore.

Many of these localities have become prime shopping destinations through a natural touristification process that encompasses, initially at least, high quality retail, plentiful retail space, convenient access, and otherwise attractive tourist environments. Singapore and Paris are examples of this type, as are most of the European microstates (e.g., Monaco, Andorra, San Marino, and Liechtenstein), which have taken advantage of their unique border situations, sovereignty over taxation and import regulations, and open borders with their larger neighbors (with the exception of Andorra, which maintains border checkpoints with France and Spain). Andorra is one of Europe's hottest shopping destinations, whose retail appeal began as an organic process triggered primarily by the country's low taxes. Even today, Andorra has Europe's lowest VAT rate (4.5%), and compared to its neighbors, France (20%) and Spain (21%), shopping in the microstate becomes a lucrative day-trip, weekend getaway, or longer holiday activity (Timothy, 2021). The microstate is particularly popular for French and Spanish consumers who cross the border for tobacco, alcohol, sweets, perfumes and food items. This natural growth has been enhanced in recent years with purposive marketing, including a popular shopping festival every year, intensive marketing emphasis on retail, and enticing luxury brands to establish stores in the co-principality.

As already discussed, borders are also natural areas for shopping owing to their role as dividers of economic, social, and political systems. This enables many other types of borderland tourism besides shopping (e.g., prostitution, alcohol and drug consumption, medical tourism, and gambling), but retail opportunities continue to permeate the sense of place of many borderlands. On the US-Mexico frontier, the shopping emphases have changed over time. From the early 1900s with the United States' prohibition against the production, sale, and consumption of alcohol (1920–1933), Mexican border towns became destinations of deviant behavior, with most shopping focusing on alcohol and tobacco. Through the 1900s, and as alcohol was no longer prohibited in the US, the Mexican border towns evolved into 'curioscapes' or landscapes of 'Mexicanness', where American and Canadian (or other) tourists could get a minuscule taste of Mexico without having to go too far abroad. During this period from the 1950s until the 2000s, the retailscapes of Mexico's northern border were dominated by curios/souvenir shops, craft and artisan markets, bars, and pharmacies. However, with the rise in medical costs in the United States and expensive insurance premiums, as well as growing security concerns on the Mexican side of the border, the ludic nature of the border retailscape diminished and was replaced by medical services (e.g., dentists, chiropractors, optometrists, and medical doctors) and ubiquitous pharmacies (Cuevas Contreras & Zizaldra Hernandez, 2023; Oberle & Arreola, 2004; Timothy, 2020). Each of these three phases of the retail environment has left a mark on Americans' views of Mexican border towns. The current image is one of a dangerous place to be avoided unless one needs cheaper healthcare services and inexpensive pharmaceuticals (Cruz-Milán, 2022). Although the US-Mexico border has a long history of retail tourism and an established image around shopping activities, this image has changed significantly in recent decades.

Another trend in place-making, although not new, is now receiving additional research attention for its shopping component. Small towns on the margins of society and economy that sprang up because of industrialization (e.g., factories and processing plants) and resource extraction (e.g., mining and forestry) boomed during their heyday, but with deindustrialization processes and the rise of the service and information economies in the mid-twentieth century, many have been abandoned or teetered on the verge of abandonment. Tourism is often a tool selected to save dying communities or resurrect them from ghost towns to tourist villages into what Getz (1993) calls 'tourist shopping villages' (TSVs). These are nearly abandoned settlements that reinvent themselves into tourist destinations, and their appeal is a combination of shopping and other elements of the community's past (Correia & Kozak, 2019; Moscardo et al., 2020; Murphy et al., 2011; Timothy, 2005). TSVs are typically located near access highways and within reasonable proximity of population centers that support day-trips or short weekend getaways in B&Bs, 'haunted hotels', or other heritage-related lodging. Shops in TSVs tend to focus on antiques, wine and spirits, ice cream and candy, second-hand clothes, cafes and restaurants. TSVs have become popular for the complementarity between their heritage ambience and the leisure shopping and dining opportunities they provide. TSVs frequently host events and festivals, heritage museums, antique car shows, and competitions.

Like TSVs, many historic city centers are home to plethoric shopping opportunities that complement their heritage appeal. A unique perspective on heritage and retail is the refocus in many urban destinations on traditional marketplaces (fish markets, farmers markets, spice markets, craft markets, or general periodic markets) as both heritage sites and shopping venues (Noaime et al., 2022). Traditional markets have long been part of the heritage milieu of places, the focus of urban life, and have attracted the attention of tourists for their sensual displays of color, fragrance, and flavor, as well as their sociocultural functions (Hall, 2016). European towns and villages are well-known for their weekend markets, which attract local consumers and tourists alike, providing glimpses into ordinary life, agricultural heritage, and different cultural practices (Rodwell, 2008).

The markets (souqs) of the Middle East and North Africa were traditional places of trading, buying, and socializing. While their functions have diminished through processes of modernization, urbanization and touristification, their traditional role as heritage spaces (living and built) is now being renewed to attract tourists and provide space for residents' renewed social and economic exchanges. Despite the rapid mall development of the region, several countries in the Arabian Gulf area have realized the importance of keeping the markets alive for locals and for tourists. The market of Muscat, Oman, has recently been reoriented into a tourist attraction (Fatma, 2023; Gutberlet, 2024). The old market in Doha, is one of Qatar's most important heritage sites that reflects a sense of yesteryear in a city that is undergoing rapid modernization and mega-development. Likewise, Dubai, once famous for its gold souqs, now has separate areas in mega-malls that are designated as 'traditional' gold markets, and some of the city's new mega-malls and shopping centers are patterned architecturally after the aesthetics of the old marketplaces. Retail as part of the heritagescape of cities (old markets, souqs, city squares) can create a heritage-retail ambience that becomes an important part of the urban milieu and appealing feature of the destination (Correia & Kozak, 2016; Gutberlet, 2024; Zaidan, 2019, 2019; Zandieh & Seifpour, 2020).

Conclusion

Sense of place and place-making lie at the core of tourism. It is the uniqueness of place, through intentional image-making and branding, or organic and unintentional processes, that destinations are made and become attractive to would-be visitors. This is no less true for shopping tourism than it is for any other type of tourism. In fact, shopping tourism is entirely reliant upon placeness and context for its success, including differential merchandising, taxes and prices, and service quality—all of which are place-specific.

Retail opportunities without the mitigating effects of image, place-making, or sense of place, including heritage and cultural contexts, would be extremely homogenous and unattractive indeed for all but a few devout consumers who might never venture beyond the shelves, corridors and escalators of stores and shopping centers. Even in that cases, however, the competitive advantage of any retail opportunity is bound by place. The interaction between place and retail is foundational to the very existence of shopping tourism, let alone its considerable success in borderlands, luxury destinations, shopping festivals, and real estate buying abroad.

Despite the placelessness seemingly embedded in online shopping, there is still an element of place value. This is especially true in the trend of buying souvenirs online, where even mass-produced mementos and souvenirs from places a consumer has never even visited, can exude a sense of place. Place continues to hold a significant spot in online shopping, even in terms of grocery items, as freshness and localness are still important factors that affect people's food purchase decisions. Online shopping, destination branding efforts through shopping, and the resurrection of dying rural communities through shopping tourism, and the repurposing of historic markets into shopping-heritage attractions are all important components of the evolving retail landscape of shopping tourism.

References

Abendroth, L. J. (2011). The souvenir purchase decision: Effects of online availability. *International Journal of Culture, Tourism and Hospitality Research*, *5*(2), 173–183. https://doi.org/10.1108/17506181111139582

Alhosani, A., & Zaidan, E. (2014). Shopping tourism and destination development: Dubai as a case study. *Arab World Geographer*, *17*(1), 66–81. https://meridian.allenpress.com/awg/article-abstract/17/1/66/59963/Shopping-Tourism-and-Destination-Development-Dubai

Bar-Kołelis, D., & Wendt, J. A. (2018). Comparison of cross-border shopping tourism activities at the Polish and Romanian external borders of the European Union. *Geographia Polonica*, *91*(1), 113–125. https://doi.org/10.7163/GPol.0103

Bu, L., Durand-Servoingt, B., Kim, A., & Yamakawa, N. (2017). *Chinese luxury consumers: The 1 Trillion Renminbi opportunity*. McKinsey & Company.

Butler, R. W. (1980). The concept of a tourist area cycle of evolution: Implications for management of resources. *Canadian Geographer*, *24*(1), 5–12. https://doi.org/10.1111/j.1541-0064.1980.tb00970.x

Casson, L. (1985) *Travel in the Ancient World*. Johns Hopkins University Press.

Charlesworth, M. P. (2016). *Trade-routes and Commerce of the Roman Empire*. Cambridge University Press.

Collins-Kreiner, N., & Zins, Y. (2011). Tourists and souvenirs: Changes through time, space and meaning. *Journal of Heritage Tourism*, *6*(1), 17–27. https://doi.org/10.1080/1743873X.2010.515312

Correia, A., & Kozak, M. (2016). Tourists' shopping experiences at street markets: Cross-country research. *Tourism Management*, *56*, 85–95. https://doi.org/10.1016/j.tourman.2016.03.026

Correia, A., & Kozak, M. (2019). Browsing and shopping. In P. L. Pearce (Ed.) *Tourist Behaviour: The Essential Companion* (pp. 175–198). Edward Elgar.

Cruz-Milán, O. (2022). Hotels' marketing mix responses at insecurity-stricken destinations: A study in the US–Mexico border. *International Hospitality Review*, *36*(1), 131–155. https://doi.org/10.1108/IHR-09-2020-0054

Cuevas Contreras, T., & Zizaldra Hernandez, I. (2023). Borders and healthcare: Medical mobility, globalization and borderland tourism. In D. J. Timothy, & A. Gelbman (Eds.), *Routledge handbook of borders and tourism* (pp. 281–295). Routledge.

Fatma, A. H. (2023). Social sustainability and globalization; and the role of Souq Mutrah in Muscat-Oman. *Environmental Science & Sustainable Development*, *8*(1), 10–24. https://doi.org/10.21625/essd.v8i1.961

Frost, J., & Frost, W. (2024). The ancillary role of shopping in other types of tourism. In D. J. Timothy (Ed.) *Contemporary Perspectives on Shopping, Retail and Tourism*. Channel View Publications

Getz, D. (1993). Tourist shopping villages: Development and planning strategies. *Tourism Management*, *14*(1), 15–26.

Gutberlet, M. (2024). *Overtourism and cruise tourism in emerging destinations on the Arabian Peninsula*. Routledge.

He, L. (2022). Place Representation through Souvenir Production and Consumption. Unpublished doctoral dissertation, School of Community Resources and Development, Arizona State University.

Hall, C. M. (2016). Heirloom products in heritage places: Farmers' markets, local food and food diversity. In D. J. Timothy (Ed.), *Heritage Cuisines: Traditions, Identities and Tourism* (pp. 88–103). Routledge.

Hall, C. M., & Müller, D. K. (Eds) (2004). *Tourism Mobility and Second Homes: Between Elite Landscape and Common Ground*. Channel View Publications.

Hannonen, O. (2023). Transboundary second-home tourism. In D. J. Timothy, & A. Gelbman (Eds.), *Routledge handbook of borders and tourism* (pp. 310–323). Routledge.

Hannonen, O., Tuulentie, S., & Pitkänen, K. (2015). Borders and second home tourism: Norwegian and Russian second home owners in Finnish border areas. *Journal of Borderlands Studies*, *30*(1), 53–67.

Hiltunen, M. J., Pitkänen, K., Vepsäläinen, M., & Hall, C. M. (2013). Second home tourism in Finland: Current trends and eco-social impacts. In A. Roca (Ed.), *Second home tourism in Europe: Lifestyle issues and policy responses* (pp. 165–198). Ashgate.

Hitchcock, M., & Teague, K. (2019). *Souvenirs: The material culture of tourism*. Routledge.

Holleran, C. (2012). *Shopping in ancient Rome: The retail trade in the Late Republic and the Principate*. Oxford University Press.

Honkanen, A., Pitkänen, K., & Hall, C. M. (2016). A local perspective on cross-border tourism: Russian second home ownership in eastern Finland. *International Journal of Tourism Research*, *18*(2), 149–158. https://doi.org/10.1002/jtr.2041

Hung, K., Lianping, R., & Hanqin, Q. (2021). Luxury shopping abroad: What do Chinese tourists look for? *Tourism Management*, *82*, 104182. https://doi.org/10.1016/j.tourman.2020.104182

Hussin, F., Abdullah, N., Maamor, S., & Abdullah, H. (2012). Border economy: Issues and problems faced by traders in the Rantau Panjang Duty-free Zone. *Journal of Sociological Research*, *3*(2), 46–56.

Illés, S., & Michalkó, G. (2008). Relationships between international tourism and migration in Hungary: Tourism flows and foreign property ownership. *Tourism Geographies*, *10*(1), 98–118. https://doi.org/10.1080/14616680701825271

Jagodic, D. (2012). Living (beyond) the border: European integration processes and cross-border residential mobility in the Italian-Slovenian border area. In D. J. Andersen, M. Klatt, & M. Sandberg (Eds.), *The border multiple. The practicing of borders between public policy and everyday life in re-scaling Europe* (pp. 201–218). Ashgate.

Kim, W. B., Xie, J., & Choo, H. J. (2023). Role of perceived benefits of online shopping festival in Vietnam: Differences between millennials and generation Z. *Journal of Retailing and Consumer Services*, *75*, 103530. https://doi.org/10.1016/j.jretconser.2023.103530

Liberato, D., Liberato, P., & Silva, M. (2020). Shopping tourism: Comparative analysis of the cities of Oporto and Lisbon as shopping destinations. In V. Katsoni, & T. Spyriadis (Eds.), *Cultural and tourism innovation in the digital era* (pp. 365–379). Springer.

Lőrincz, K., Grósz, A. S., & Csapó, J. (2021). Evaluation of quality of life and living space in the Balaton destination (Hungary) based on the perceptions of the local residents and the owners of second homes. *European Countryside*, *13*(1), 108–129. https://doi.org/10.2478/euco-2021-0007

Makkonen, T. (2016). Cross-border shopping and tourism destination marketing: The case of Southern Jutland, Denmark. *Scandinavian Journal of Hospitality and Tourism*, *16*(1), 36–50. https://doi.org/10.1080/15022250.2016.1244506

Makkonen, T. (2023). Outshopping abroad: Cross-border shopping tourism and the competitive advantage of borders. In D. J. Timothy, & A. Gelbman (Eds.), *Routledge handbook of borders and tourism* (pp. 269–280). Routledge.

McIntyre, C. (2010). Designing museum and gallery shops as integral, co-creative retail spaces within the overall visitor experience. *Museum Management and Curatorship*, *25*(2), 181–198. https://doi.org/10.1080/09647771003737299

McLaughlin, R. (2010). *Rome and the distant east: Trade routes to the ancient lands of Arabia, India and China*. Continuum.

Mehta, S., Jain, A., & Jawale, R. (2014). Impact of tourism on retail shopping in Dubai. *International Journal of Trade, Economics and Finance*, *5*(6), 530–535.

Michalkó, G., Rátz, T., Hinek, M., & Tömöri, M. (2014). Shopping tourism in Hungary during the period of the economic crisis. *Tourism Economics*, *20*(6), 1319–1336. https://doi.org/10.5367/te.2014.0387

Michalkó, G., & Timothy, D. J. (2001). Cross-border shopping in Hungary: Causes and effects. *Visions in Leisure and Business*, 20(1), 4–22. https://scholarworks.bgsu.edu/visions/vol20/iss1/2/

Moscardo, G., Murphy, L., Hughes, K., & Benckendorff, P. (2020). Shopping on the edge: Identifying factors contributing to tourist retail development in heritage villages. In I. Yeoman, & U. McMahon-Beattie (Eds.) *The future past of tourism: Historical perspectives and future evolutions*, Channel View Publications.

Müller, D. K. (2021). 20 years of Nordic second-home tourism research: A review and future research agenda. *Scandinavian Journal of Hospitality and Tourism*, *21*(1), 91–101. https://doi.org/10.1080/15022250.2020.1823244

Muro-Rodríguez, A. I., Pérez-Jiménez, I. R., & Sánchez-Araque, J. A. (2020). Impact of shopping tourism for the retail trade as a strategy for the local development of cities. *Frontiers in Psychology*, *11*, 67. https://doi.org/10.3389/fpsyg.2020.00067

Murphy, L., Benckendorff, P., Moscardo, G., & Pearce, P. L. (2011) *Tourist shopping villages: Forms and functions*. Routledge.

Noaime, E., Alnaim, M. M., Bay, M. A., Albaqawy, G. A., Abdelhafez, M. H. H., & Elkhayat, K. (2022). The rehabilitation of the historic Barzan traditional market and its impact on cultural tourism in Hail City. *Land*, *11*(11), 2058. https://doi.org/10.3390/land11112058

Oberle, A. P., & Arreola, D. D. (2004). Mexican medical border towns: A case study of Algodones, Baja California. *Journal of Borderlands Studies*, *19*(2), 27–44. https://doi.org/10.1080/08865655.2004.9695625

Opačić, V. T., & Koderman, M. (2018). From socialist Yugoslavia to the European Union: Second home development in Croatia and Slovenia. In C. M. Hall, & D. K. Müller (Eds) *The Routledge handbook of second home tourism and mobilities* (pp. 167–178). Routledge.

Peter, S., & Anandkumar, V. (2014). Dubai Shopping Festival: tourists' nationality and travel motives. *International Journal of Event and Festival Management*, *5*(2), 116–131. https://doi.org/10.1108/IJEFM-08-2013-0022

Peter, S., & Anandkumar, V. (2016). Travel motivation-based typology of tourists who visit a shopping festival: An empirical study on the Dubai shopping festival. *Journal of Vacation Marketing*, *22*(2), 142–153. https://doi.org/10.1177/1356766715607587

Peter, S., & Anandkumar, V. (2024). Shopping festivals. In D. J. Timothy (Ed.) *Contemporary perspectives on shopping, retail and tourism*. Channel View Publications.

Popkin, M. (2022). *Souvenirs and the experience of empire in ancient Rome*. Cambridge University Press.

Punzon, J. G. (2021). Augmented reality in shopping tourism: Boosting tourism development through innovation in Barcelona. *European Journal of Tourism, Hospitality and Recreation*, *11*(1), 1–10. https://doi.org/10.2478/ejthr-2021-0001

Rabotić, B. (2014). Special-purpose travel in ancient times: 'Tourism' before tourism? *Turističko Poslovanje*, (14), 5–17. https://scindeks.ceon.rs/Article.aspx?artid=0354-30991414005R

Ramsey, D., Thimm, T., & Hehn, L. (2019). Cross-border shopping tourism: A Switzerland-Germany case study. *European Journal of Tourism, Hospitality and Recreation*, 9(1), 3–17. https://doi.org/10.2478/ejthr-2019-0002

Rodwell, D. (2008). *Conservation and sustainability in historic cities*. Wiley.

Seaman, A. N., & Schroeder, L. H. (2022). Exploring the relationship between travel and online food souvenirs. *Anatolia*, *33*(2), 210–221. https://doi.org/10.1080/13032917.2022.2040913

Soto, M. T. R., & Clavé, S. A. (2017). Second homes and urban landscape patterns in Mediterranean coastal tourism destinations. *Land Use Policy*, *68*, 117–132. https://doi.org/10.1016/j.landusepol.2017.07.018

Spierings, B., & van der Velde, M. (2013). Cross-border differences and unfamiliarity: Shopping mobility in the Dutch-German Rhine-Waal Euroregion. *European Planning Studies*, *21*(1), 5–23. https://doi.org/10.1080/09654313.2012.716236

Studzińska, D., Sivkoz, A., & Domaniewski, S. (2018). Russian cross-border shopping tourists in the Finnish and Polish borderlands. *Norsk Geografisk Tidsskrift-Norwegian Journal of Geography*, *72*(2), 115–126. https://doi.org/10.1080/00291951.2018.1451365

Su, X. (2010). The imagination of place and tourism consumption: A case study of Lijiang Ancient Town, China. *Tourism Geographies*, *12*(3), 412–434. https://doi.org/10.1080/14616688.2010.494688

Sulyok, M. J., & Lőrincz, K. (2016). Building the future based on the past—Tourism around the Lake Balaton. *Pannon Management Review*, 6(1–2), 9–18.

Swanson, K. K., & Timothy, D. J. (2012). Souvenirs: Icons of meaning, commercialization, and commoditization. *Tourism Management*, 33(3), 489–499. https://doi.org/10.1016/j.tourman.2011.10.007

Szytniewski, B. B., Spierings, B., & van der Velde, M. (2020). Stretching the border: Shopping, petty trade and everyday life experiences in the Polish–Ukrainian borderland. *International Journal of Urban and Regional Research*, *44*(3), 469–483. https://doi.org/10.1111/1468-2427.12857

Timothy, D. J. (1999). Cross-border shopping: Tourism in the Canada-United States borderlands. *Visions in Leisure and Business*, 17(4), 4–18. https://scholarworks.bgsu.edu/visions/vol17/iss4/2/

Timothy, D. J. (2004). Recreational second homes in the United States: Development issues and contemporary patterns. In C. M. Hall, & D. Müller (Eds.), *Tourism, mobility and second homes: Between elite landscape and common ground* (pp. 133–148). Channel View Publications.

Timothy, D. J. (2005). *Shopping tourism, retailing and leisure*. Channel View Publications.

Timothy, D. J. (2014). Trends in tourism, shopping, and retailing. In A. A. Lew, C. M. Hall, & A. M. Williams (Eds.), *The Wiley Blackwell companion to tourism* (pp. 378–388). Wiley Blackwell.

Timothy, D. J. (2018). Shopping tourism. In S. Agarwal, G. Busby, & R. Huang (Eds.), *Special interest tourism: Concepts, contexts and cases* (pp. 134–144). CABI.

Timothy, D. J. (2019). Supranationalism and tourism: Free trade, customs unions, and single markets in an era of geopolitical change. In D. J. Timothy (Ed.), *Handbook of globalisation and tourism* (pp. 100–113). Edward Elgar.

Timothy, D. J. (2020). Borderscapes and tourismscapes: The place of postcards in Mexican border town tourism. *Geographia Polonica*, 93(4), 553–568. https://doi.org/10.7163/GPol.0188

Timothy, D. J. (2021). *Tourism in European microstates and dependencies: Geopolitics, scale and resource limitations*. CABI.

Timothy, D. J. (Ed.) (2024). *Contemporary perspectives on shopping, retail and tourism*. Channel View Publications.

Timothy, D. J., & Butler, R. W. (1995). Cross-border shopping: A North American perspective. *Annals of Tourism Research*, 22(1), 16–34. https://doi.org/10.1016/0160-7383(94)00052-T

Timothy, D. J., & Teye, V. B. (2005). Informal sector business travelers in the developing world: A borderlands perspective. *Journal of Tourism Studies*, 16(1), 82–92.

Tömöri, M. (2010). Investigating shopping tourism along the borders of Hungary: A theoretical perspective. *GeoJournal of Tourism and Geosites*, 6(2), 202–210.

Visit Dubai (2023). Dubai Shopping Festival. https://www.visitdubai.com/en/whats-on/dsf

Wiedmann, H., Hennings, N., & Siebels, A. (2007). Measuring consumers' luxury value perception: A cross-cultural framework. *Academy of Marketing Science Review*, 7, 1–21.

Yao, A. (2023). Uncovering heterogeneous prestige effect in luxury consumption: Insights from the Chinese luxury market. *Journal of Business Research*, *168*, 114235. https://doi.org/10.1016/j.jbusres.2023.114235

Yulianto, Y., Sisko, A., & Hendriana, E. (2021). The stimulus of impulse buying behavior on E-commerce shopping festival: A moderated-mediated analysis. *Journal of Business and Management Review*, *2*(10), 692–714. https://doi.org/10.47153/jbmr210.2152021

Zaidan, E. (2016). Tourism shopping and new urban entertainment: A case study of Dubai. *Journal of Vacation Marketing*, *22*(1), 29–41. https://doi.org/10.1177/135676671558942

Zaidan, E. (2019). Shopping, tourism and hyper-development in the Middle East and North Africa. In D. J. Timothy (Ed.) *Routledge handbook on tourism in the Middle East and North Africa* (pp. 365–377). Routledge.

Zandieh, M., & Seifpour, Z. (2020). Preserving traditional marketplaces as places of intangible heritage for tourism. *Journal of Heritage Tourism*, *15*(1), 111–121. https://doi.org/10.1080/1743873X.2019.1604714

Zhang, L., & Zhao, H. (2019). Personal value vs. luxury value: What are Chinese luxury consumers shopping for when buying luxury fashion goods? *Journal of Retailing and Consumer Services*, *51*, 62–71. https://doi.org/10.1016/j.jretconser.2019.05.027

Part 6 Tourism and Global Change
Impacts and Response

Introduction
Tourism and Global Change: Impacts and Response

C. Michael Hall

Tourism's interaction with the environment and issues of environmental change have long been a focus of research in tourism studies. Historically, such research has tended to be focussed on the destination level, but over time there has been a shift in focus to examine all stages of the tourist trip and, especially in the case of emissions, the greatest impacts are usually in travelling to and from the destination (Eijgelaar & Peeters, Chapter 35; Scott, Chapter 37).

However, interest in the relationships between tourism and the environment have also broadened out to include greater awareness of the way in which concept of nature and environment are socially constructed; the role of institutional structures in delimiting development pathways (Coles, Chapter 41), and the consequent effects on the drawdown of natural capital; as well as the very construct of an impact – a term which has also long been a focal point of tourism research.

Constructing Impacts

As in other areas of the study of environmental impacts, the classical empiricist approach has been extremely significant in teaching and research on tourism with respect to the notion of the impacts of tourism. Yet, as Castree (2002) observed, the framework it provides is often taken for granted without consideration of the 'assumptions that organise and, importantly, circumscribe the field of analysis' (Castree, 2002, pp. 116–117). In such a situation 'the metaphor of human impacts has come to frame our thinking and circumscribe debate about what constitutes explanation' (Head, 2008, p. 374). Derived from the material realist ontology of classical empiricism, the metaphor has several features, which can be identified in most accounts of tourism impacts. First, and most apparently in the discourse of impacts, is the emphasis on the moment(s) of 'collision; between two separate entities, i.e. the 'impact' between tourism and the environment. This has favoured explanations and methods that

The Wiley Blackwell Companion to Tourism, Second Edition. Edited by C. Michael Hall.

depend on correlation in time and space, and typically in relatively narrow temporal periods and constrained spaces, to the detriment of the search for mechanisms of connection and causation rather than simple correlation.

Second, the emphasis on the moment(s) of impact assumes a stable natural, social, or economic baseline, and an experimental method in which only one variable is changed. However, this approach is inappropriate for understanding complex and dynamic socio-environmental systems in which there are multiple sources of change and a host of different feedback mechanisms. While feedbacks occur in the absence of human influences, they are also altered by human activity. Such feedback is a key element in understanding human-influenced environments in ways that extends the traditional approach of considering humans as unidirectional drivers of change, especially with respect to the issue of resilience (Amore, Chapter 46). Nevertheless, enormous gaps and uncertainties in knowledge remain with respect to understanding impact-feedback loops within environmental systems with significant human alterations as well as socio-technical systems, where the impacted systems in turn affect people, including the tourist market.

Third, and perhaps most influentially, is the way the terms 'tourism impact(s)' or 'tourist impact(s)' ontologically positions tourism and tourists as 'outside' the system under analysis, as outside of nature, protected areas, wilderness (or whatever it is that is being impacted). However, this position is rather ironic given that research on global climate and environmental change demonstrates just how deeply entangled tourism is in environmental systems. Furthermore, it also fails to appreciate the extent to which tourism and environment relations are socially constructed, especially in relation to the world views of indigenous peoples whose traditional lands are often most at threat from environmental change. Yet despite such criticisms the metaphor remains in widespread use as it is integral to the dominant discourse on tourism and environment (Rasoolimanesh et al., 2023).

Finally, as Head (2008) suggests, one of the implications of dichotomous explanations that place a significant explanatory divide between humans and nature is that it requires the conflation of bundles of variable processes under headings such as 'human', 'climate', 'environment' and 'nature'. Furthermore, although such dichotomous explanations are characterised by their veneer of simplicity and elegance, 'the view that causality is simple' actually takes 'many more assumptions than the view that it is complex' (Head, 2008, p. 374).

The classical realist empiricism that dominates much research and certainly the policy discourse on the effects and response of tourism to change is strongly driven by materialist and mechanistic orientations that artificially divide facts and values. Ontologically and epistemologically, there is relatively little critique of the assumptions that underlie many studies of tourism impacts as well as research on the effects of environmental change on tourism, although research on the impacts of climate change has started to question such assumptions (Scott, Chapter 37). For example, the tourist response to climate change, especially changes in temperatures, is likely much more complex than suggested by econometric modelling given the role of numerous feedback mechanisms as well as shifting motivations and the condition of competing attractions and destinations (see Scott, Chapter 37). Yet ontological differences between competing understandings of tourism-environment relationships raise fundamental questions about how future trends and the extent to which desired futures can actually be understood; the ethical relationships between humans and the environment given the influence of ontology on values toward nature; and criticisms of natural science (Fennell, Chapter 39; Huijbens & Gren, Chapter 35).

Constructing Responses to Climate and Environmental Change

Even though there have been calls for greater social science information to be brought into the climate and environmental change assessment, process this has primarily been assessed in terms of neo-classical economic contributions, which have themselves been greatly influenced by the ontology of natural science, and which frame human behaviour as well as interventions within the context of individuals as rational utility-maximisers. Neo-classical economic accounts of climate change and the justification for action to minimise the effects of climate change are also dominated by formal modelling. As Demeritt and Rothman (1999, p. 404) noted with respect to the Intergovernmental Panel on Climate Change (IPCC) process, such a utilitarian view of value has been subjected to a number of different philosophical and moral critiques, including with respect to:

- anthropocentric assumptions and failure to recognise intrinsic value;
- reductionist and decisionist approach and failure to recognise holism;
- confusion of values and preferences; and
- the equality of exchange relations.

Such issues are therefore fundamental to thinking about the relationship between tourism and the environment as well as associated issues of sustainable tourism development and adaptation and mitigation of change (Scott, Chapter 37).

As several of the chapters in this section suggest, the choice of valuation procedure of tourism and the environment, including the cultural aspects of the environment, like the definition of value itself as used by governments and organisations such as the United Nations World Tourism Organization (WTO), World Travel and Tourism Council (WTTC) and the World Economic Forum (WEF), is part of a mediated and embodied social construction of knowledge and is ultimately personal and political rather than objective and rational (Hall, 2019). However, to acknowledge otherwise would potentially be to cast doubt on the objectivity and strength of their forecasts. Nevertheless, such issues are extremely important as they influence individual and organisational behaviour (Coles, Chapter 41; Tuan et al., Chapter 42), including with respect to the application of the precautionary principle in policy and decision-making (Fennell, Chapter 39).

The various chapter contributions to this section highlight the way in which tourism researchers are increasingly seeking to integrate social and natural science approaches in understanding the tourism and environment relationships. The contributions to this section highlight that a destination is more than just a geographic or promotional space and is instead a holistic system in which nature and culture co-evolve. Therefore, as an interdisciplinary or even post-disciplinary field of study, tourism studies is seeking to understand and improve the sustainability of destinations as well as contribute to sustainability at broader geographic scales, including the global. To move forward in the Anthropocene, tourism studies need to develop a more complete understanding of the social construction of nature from the perspectives of different audiences and reconnect culture with nature and environment more effectively in assessments, strategies and policy making (Saarinen, 2006, 2015; Saarinen and Lundén, Chapter 38; Huijbens & Gren, Chapter 36). To achieve this goal, tourism studies has much to learn from engagement with a range of social, natural and humanistic sciences. More pluralistic and integrated approaches to tourism are needed if destinations and the entire tourism system are not merely to be 'studied' but also become more resilient and sustainable.

References

Castree, N. (2002.) False antitheses? Marxism, nature and actor networks. *Antipode*, *34*, 111–146. https://doi.org/10.1111/1467-8330.00228

Demeritt, D. & Rothman, D. (1999). Figuring the costs of climate change: An assessment and critique. *Environment and Planning A*, *31*, 389–408. https://doi.org/10.1068/a310389

Hall, C. M. (2019). Constructing sustainable tourism development: The 2030 agenda and the managerial ecology of sustainable tourism. *Journal of Sustainable Tourism*, 27(7), 1044–1060. https://doi.org/10.1080/09669582.2018.1560456

Head, L. (2008). Is the concept of human impacts past its use-by date? *The Holocene*, *18*, 373–377. https://doi.org/10.1177/0959683607087927

Rasoolimanesh, S. M., Ramakrishna, S., Hall, C. M., Esfandiar, K., & Seyfi, S. (2023). A systematic scoping review of sustainable tourism indicators in relation to the sustainable development goals. *Journal of Sustainable Tourism*, *31*(7), 1497–1517. https://doi.org/10.1080/09669582.2020.1775621

Saarinen, J. (2006). Traditions of sustainability in tourism studies. *Annals of Tourism Research*, *33*(4), 1121–1140. https://doi.org/10.1016/j.annals.2006.06.007

Saarinen, J. (2015). Conflicting limits to growth in sustainable tourism. Current Issues in Tourism, 18(10), 903–907. https://doi.org/10.1080/13683500.2014.972344

Chapter 35

Tourism and the Anthropocene

Edward H. Huijbens and Martin G. Gren

Introduction

The Anthropocene was a suggested term for a new geological time unit distinguished by the fact that the collective of humans (i.e., the Anthropos) have become a geological force affecting the functioning of the Earth System. Since the Anthropocene was launched as a geological term in 2000, it has grown in its range and reach well beyond the confines of geology and the earth sciences. During the last decade it has made its way into the humanities and the social sciences where the concept has been both praised and criticized in different ways and for various reasons. It is now also part of the conceptual vocabulary in tourism in a range of ways this entry will explore.

At the outset, however, it needs to be stated clearly that the Anthropocene introduces a 'geo-bio-physical' ontology and a planetary dimension that challenge the Modern ontological understanding of tourism and tourists as predominantly social phenomena. Herein Modern is capitalised, drawing inspiration from Bruno Latour in signifying cultures and institutions rather than individuals. The notions of progress, emancipation, growth, and development often follow in the same paragraph of the Modern, denoting a particular historical cultural logic tied to the predominant understanding of the term's emergence in Europe in the 'long sixteenth century' (1450–1640 CE), as defined by Immanuel Wallerstein and picked up by Jason Moore (2015), and gained momentum through the birth of the Enlightenment during the 17th and the 18th centuries.

In this chapter we provide an overview of the concept of the Anthropocene and an examination of how it has been understood and promoted in the tourism literature by us. We also offer some critical observations on the Anthropocene in tourism scholarship and research, which leads us to suggest some potential future directions to accompany tourism in this field. One key issue to consider when it comes to the Anthropocene, is that it now has to be placed in the context of the planetary climate and ecological emergency.

The Wiley Blackwell Companion to Tourism, Second Edition. Edited by C. Michael Hall.

The Anthropocene

The Anthropocene, as a geological term, was originally launched by Paul Crutzen in 2000 and presented in a paper thus entitled (Crutzen & Stoermer, 2000). Whereas it was previously assumed that humans could not rival the 'great forces of nature', the Anthropocene recognizes instead that the *Anthropos* (the collective of humans, humanity, the human species) has become a geological force that is impacting the *Earth System*, and to an extent that will be potentially legible in the Earth's stratigraphic archive. In other words, as a geological term the Anthropocene was intended to signify a shift in the Earth's geological history.

In geology such shifts, from one geological time-period to another, are defined by so called 'stratigraphic markers', i.e., whereby the occurrence of a notable change in the qualities, properties or content of rock layers (strata) under our feet stretching back billions of years, can be detected. The smallest time unit of these periods is the '-cene'. For example, as the last period of glaciation ended some 12,000 years ago, when vast polar icecaps dominating Earth System dynamics started to recede, geology uses the periodization of a shift from the Pleisto-cene to the Holo-cene. The question now is if another more recent shift has occurred, this time caused by the human impact and dominance over geo-bio-physical processes also at planetary scale. If so, then this will qualify as a new geological time unit for which the Anthropo-cene is the proposed term although voted down in February 2024 by geologists of the International Commission on Stratigraphy (ICS) of the International Union of Geological Sciences (IUGS).

Regardless, there are in fact several possible stratigraphic markers of such a shift, but one of them is the 'golden spike' in radioactive isotopes following surface and oceanic nuclear testing in the wake of World War II. These isotopes will be observable for millions of year, as for instance the half-time of Uranium-235 used in atomic and hydrogen bombs is 703.8 million years. This marker enables some geologists to claim that the Anthropocene began at a particular date, namely in when the Trinity test of the Manhattan Project took place on 16 July 1945. The first detonation of a nuclear weapon (see Zalasiewicz et al., 2019). However, this geological dating also overlaps with other human historical processes and social developments that are not reducible to geology per se. The year 1945 also marks the beginning of a fundamental societal transformation whereby an ever more interconnected world shifted into production and consumption overdrive after World War II, powered by an enormously productive fossil-fuel post-colonial Western capitalist apparatus honed through years of war mobilisation. From the perspective of the humanities and the social sciences, this implies that an understanding of humans as geological beings, which is essential to the Anthropocene, must be understood as intrinsically entangled with the non-geological agency that derives from humans as social and cultural beings. In other words, the Anthropos of the Anthropocene denotes a being with a 'geo-social' planetary agency, however geographically and socially differentiated the agency of that being is.

The socio-economic developments since the 1950's have generally been understood as being driven by human processes of modernization, eventually leading the world to globalization. In a more natural science perspective, this rapid and far-reaching historical transformation has been labelled 'The Great Acceleration' (Steffen et al., 2015a), and its proponents claim that it needs to be seen in tandem with the onset of the Anthropocene. The Great Acceleration is based on the fact that there is an empirical correlation between 'socio-economic trends' and 'Earth System trends'. One example is the exponential growth by socio-economic measures in international tourism since the 1950s, which corresponds to effects on the Earth System, such as measured increase in fossil-fuel emissions or loss of biodiversity.

When socio-economic and Earth System trends are considered together the Anthropocene cannot simply be understood as referring only to a natural geological shift. This would imply that the Earth System has moved naturally on its own, or is in the process of moving, beyond the geological conditions of the Holocene (i.e., the 11,000 to 12,000 years since the end of the last glacial epoch). The Anthropocene *also* marks a shift whose signature is irreversible *human* impact ('anthropogenic forcing') on the Earth System, and on the geo-bio-physical processes that are essential for sustaining earthly human life.

The most well-known impact is the historically and geologically unprecedented increase in atmospheric CO_2, which is hazardous for the climate and ecological conditions that make the Earth habitable for humans. Although the dangers have often been voiced in terms of climate (such as 'climate change' and 'global warming/heating'), it is important to emphasize that it also precipitates the ecological, for example rampant biodiversity loss and even an emerging sixth mass extinction. The Anthropocene then also signals that humans (humanity, the human species) are on a trajectory towards an unprecedented and perilous planetary terrain. In the language of 'The Planetary Boundaries Approach', the collective of humans has already transgressed several boundaries that are crucially important for preserving the Earth System in a state that provides 'a safe operating space for humanity' (Rockström et al., 2009a; Rockström et al., 2009b; Steffen et al., 2015b). In this planetary perspective, the overall goal of 'climate action' is to try to preserve the Earth System, as much as that might be possible, in a Holocene-like state.

That the Anthropocene signals a dangerous territory for humans has more recently been further amplified by both science and the world community (i.e., United Nations), both proclaiming that we are now facing a planetary 'climate and ecological emergency' or even 'breakdown'. The greatest fear among climate and earth system scientists is that the Earth System may be set for an irrevocable trajectory towards a tipping point which will 'flip' it into another, and much hotter, system state with devastating consequences for earthly life conditions.

Of crucial importance to us, is the ways in which all of the above in effect introduces a *new being* into the fold and repertoire of human concerns; 'the Earth' as a mutating Earth System. The agency of this new Earth of the Anthropocene has namely become an active foregrounded matter of both fact and concern in scientific and academic discourse, as well as in our daily undertakings. For example, we should no longer care only for nature or the environment, but also make consumption choices that are 'good for the planet', as indeed there 'is no planet B'. The Earth thus conceptualized as a changing, or mutating, Earth System also presents Modern thought in general with a novel challenge, and calls for in-depth conceptual interrogations of all things, tourism included. As explained by Zalasiewicz et al. (2019, p. 36):

> Throughout recent history, *an underlying stable condition of the Earth System has been taken as a given* [our emphasis]. This is the premise upon which our legal and political structures have been created over the past several centuries. /…/ there has been an implicit assumption that current conditions form an objective and unchanging reality that has surrounded us since time immemorial.

In Modern thought-and-action the Earth has been foremost conceptualized as the Earth's surface, which was assumed to be a relatively stable backdrop that one therefore did not have to consider or take into account. It follows that many of the central concepts in Modern thought, like 'nature', 'culture', 'the environment', and 'society', have in fact had this stable Earth's surface as a taken-for-granted reference plane upon which the complexities of each and how they relate play out. For instance, in the Modern understanding of tourism, one travels to other places and visits tourism spaces on the Earth's surface. In this understanding, tourism may affect nature, society and the socio-spatial organization of tourism on the Earth's surface, but there is no conceptual room for incorporating a mutating Earth System. In the same vein, Modern understanding did not have to consider humans as beings that collectively have a geological agency that affects the functioning of the Earth System. The Modern assumption has been that humans can alter nature and the environment on Earth's surface, but not the Earth System. However, all that changes in the Anthropocene where the agencies of the Earth System and of humans are now intertwined, and where the compound effects of their agencies take us to the planetary scale and point of reference. As a dynamic mutating relational entity, the Earth System is thereby at one with our own undertakings affecting its functioning.

Faced with this Anthropocene conundrum, our current, and still very much Modern, mode of organizing the economy, society, our cultural and political domains, is wholly untenable. Business as usual is simply not an option when our present and future agency is partly in the hands of an Earth System that no longer provides a stable backdrop for our undertakings, if it ever did. As the timescales of geological and human history thereby coalesce into a common 'geo-story', one may argue that the ´becoming-Earth-and-humans´ in and of the Anthropocene needs to be further incorporated into the ways in which tourism is thought, framed, understood, and practiced. We will return to this at the end of this chapter.

Anthropocene Themes in the Tourism Literature

Around 2010 the authors of this chapter began thinking about how to make sense of the Earth as an emerging reciprocal relational force imbricated with human activities and society writ large. This reflected a long-standing fascination for us as geographers. After all geography literally means 'earth-writing', and the discipline is by default concerned with how the 'geo' is written and made sense of. However, this time our particular objective was to place the Earth in a tourism context. The resulting article (Gren & Huijbens, 2012) argued for 'Earth led priorities and perspectives on tourism', foregrounding the Earth as the primary concern when it comes to making sense of and conceptualizing tourism. In retrospect, it may well be the case that we too much took for granted that particular Earth which Modern geography has had as a primary reference plane, i.e., the Earth's surface. This was reflected also in our own contribution to 'Actor-network theory and tourism' (Huijbens & Gren, 2012).

These self-critical reflections notwithstanding, from these initial provocations to earthly tourism theorising, it was a relatively short step to the growing interest and broader debates around the Anthropocene within the humanities and social sciences. As these, at the time, were yet to be translated into a tourism theory context, we set out to explore and outline scientific, political, and ethical challenges of tourism and the Anthropocene. This included how these could be conceived in terms of tourism as a geophysical force censoriously interrelated with the capacity of the Earth to sustain the human species (Gren & Huijbens, 2014). These thoughts developed further into an edited volume on the topic, *Tourism and the Anthropocene*, wherein tourism was conceptualized as a ´geo-force'. We also stated that the Anthropocene in tourism eventually has to find its truth-value in terms of its usefulness in negotiating and navigating tourism futures in practice (Gren & Huijbens, 2016, p. 193). In our concluding section we outlined and explored three potential tourism futures in and of the Anthropocene inspired by contributions to the volume; non-carbon tourism, stay-home tourism, and destination stewardship of earthly priorities.

The challenge of the Anthropocene was later addressed in tourism in an edited volume by Mostafanezhad and Norum (2019), but from a slightly different angle. The focus in *Anthropocene Ecologies: Entanglements of Tourism, Nature and Imagination* was on the political-ecological drivers of the Anthropocene, conceptualized as a material and symbolic force increasingly becoming entangled with tourism. This edited volume is premised on understanding the Anthropocene as a fundamental social imaginary in contemporary global society and is grounded in ethnographic and qualitative research that tell a range of stories around tourism and leisure activities, and thereby how tourism enrols political ecological concerns with our leisure activities and aspirations. Another example of incorporating the Anthropocene into tourism is Moore's *Destination Anthropocene: science and tourism in Bahamas* (Moore, 2019). The book documents the emergence of new travel imaginaries forged at the intersection of the natural sciences and the tourism industry in a Caribbean archipelago and explores an experimental form of tourism developed in the name of sustainability. This particular vein of Anthropocene thinking thereby taps into the emerging 'abyssal thought', which is a distinct analytical approach putting coloniality at the heart of the Modern problematic, and turning to the Caribbean in particular as a gateway space of 'abyssal geographies', inviting a deconstruction and unmaking of the Modern world (Chandler & Pugh, 2022).

Our own explorations into the Anthropocene have been followed up in recent work by one of us. *Developing Earthly Attachments in the Anthropocene* (Huijbens, 2021a) explores and compares cases from the Netherlands and Iceland, with the aim of promoting a vitalist Earth-oriented progressive outlook woven from a critical engagement with post-humanistic new materialism ontology and debates about the Anthropocene. Further exploring life affirming theories and processes Huijbens (2022) has moved into the domains of 'conviviality', reckoning with the urban as a site of diversity, multiplicity under threat from commercialisation, not least through tourism, and at the same time exploring Earthly tourism and travel's contribution to a planetary genre de vie (Huijbens, 2021b). Given that the imperatives and the urgency of the planetary climate and ecological emergency are becoming ever more recognised, we have jointly recently explored a new kind of Earth politics (Huijbens & Gren, 2022), which arguably also needs to be understood in terms of tourism.

Anthropocene Themes in the Broader Tourism Literature and Research

For the purposes of mapping how the Anthropocene has manifested in broader tourism literature and research, we analysed articles selected from 15 top journals in the field. The Social Sciences Citation Index (SSCI) from 2021 was used, and the journals were identified by their impact factor. The search word 'Anthropocene' was used on each journal's homepage, yielding a total of 75 articles ranging from 2014 to date. We however found that not all are full research articles and these we did not include. More than half of search hits were found in the *Journal of Sustainable Tourism*, and a notable absence of the Anthropocene can be identified in the more marketing and management oriented tourism journals. In Table 35.1, the journals queried are listed alphabetically.

Note that this is not a systematic literature review. For the present purpose, we decided that a brief overall view of how the Anthropocene appears in the sampled 75 articles

Table 35.1. The tourism journals and articles explored for Anthropocene content.

Journal	*#Articles*	*Period*
Annals of Tourism Research	13	2016–2022
Current Issues in Tourism	3	2020–2022
Information Technology and Tourism	0	
International Journal of Contemporary Hospitality Management	0	
International Journal of Hospitality Management	0	
Journal of Destination Marketing and Management	0	
Journal of Hospitality and Tourism Management	2	2015–2022
Journal of Hospitality Marketing and Management	4	2017–2022
Journal of Sustainable Tourism	0	
Journal of Travel and Tourism Marketing	39	2014–2022
Journal of Travel Research	0	
Tourism Geographies	0	
Tourism Management	13	2016–2021
Tourism Management Perspectives	1	2022
Tourism Review	0	
	0	
Total	75	

would suffice. This also means that we have abstained from citing and referencing particular articles. On this basis, and by our own scholarly assessments, we have identified the following five tentative themes of how the Anthropocene is used in the reviewed articles:

- the Anthropocene as a framing concept;
- the Anthropocene as a container for re-packaging;
- the Anthropocene as a booster for posthumanism;
- the Anthropocene as naturalization;
- the Anthropocene as overruled by tourism.

(a) The Anthropocene as a Framing Concept

In the articles reviewed, the Anthropocene very often features only as a short-hand for framing a bigger and/or contemporary contexts for the article in question. For instance, the Anthropocene was often times only evoked at the outset in order to convey a sense of what humanity faces in terms of climate change and a variety of environmental challenges. For example, humanity is by now an environmentally transformative force of global reach and geological duration. Whether it is the sixth mass extinction, climate change or the generalised 'loss of nature', several articles thus start out by framing their topic as lodged within the Anthropocene on those terms. We surmise that these general references to the Anthropocene are often rather superficial, and do not really engage with or interrogate the substance, or the underlying problematics invested in the concept. At the same time, this framing of the Anthropocene often provides added justifications for pleas to transformations of tourism as practiced, governed or monitored, but still could arguably do better in engaging with the conceptual implications of evoking the concept.

(b) The Anthropocene as a Container for Re-packaging

This second theme is intimately related to the first and follows from the point of justifications. Adopting the Anthropocene as a framing concept often goes hand in hand with using it as a label for re-packaging one's own traditional concerns and areas of research. This often coincides with the use of already established terms and conceptualizations that are not problematized anew in the light of the Anthropocene. Examples include the usage of 'nature', 'environmental change', 'sustainability', 'global', as if these were to have a similar meaning in the Anthropocene. In this theme we find rather traditional investigations into of how complex relations with nature at particular tourist destinations are unravelled, and engagements with politics, ecology and tourism, with rather loose references to some kind of a new Anthropocene state of affairs. Here we found a re-packaging of tourism's traditional concerns with sustainable development and sustainable tourism, and yet where, for example, the global is not further problematized in relation to the planetary reach and duration of human undertakings.

(c) The Anthropocene as a Booster for Posthumanism

As framing in some instances justifies calls for transformation, the seething uncertainty invested in the Anthropocene, and the fact that all taken as given has been put into question, a strong emerging strand has to do with articles employing the Anthropocene to negotiate tourism's relationships with the more-than/other-than non-human, and, unsurprisingly, particularly in relation to specific tourism destinations. Through engagements with posthumanism, including the critique of human exceptionalism, the Anthropocene in this theme also serves to problematize the growth imperative invested in tourism, supporting claims and explorations of new forms of governance and the transformation of tourism. For example, details of the role of, e.g. civic activism, justice regimes, conscious travel, indigenous perspectives, buen vivir and degrowth agendas can be found under this theme. Also herein explorations into novel methods of accounting for tourism and destinations can be identified, framed under the imperative that we need to make space for more-than-human-centred aspirations and intentions in order to remedy our current state of affairs. However, most engagements with posthumanism in tourism research privileges the local and lack a consideration of the braided agencies of humans and the Earth System at planetary scale.

(d) The Anthropocene as Naturalization

Under this theme we find more critical engagements that implicitly or explicitly draw on the now well-established social science criticism of the Anthropocene as a form of naturalization. Something dubious is going on when humanity is presented as an undifferentiated 'Anthropos', as humanity or human species, placed on an equally undifferentiated planet. The critical theory tradition is not strong in tourism (Bianchi, 2009), but there is tourism research that have engaged critically with the Anthropocene by interrogating its relation with the dominant capitalist logic of tourism. References to the 'Capitalocene' or even the 'Cuthulucene', have been presented as alternatives to the Anthropocene (Moore, 2015). Topics herein deal with how 'the end of nature' becomes yet another commodity, fuelling the growth and heat engine of capital manifesting in the particularities of the valuing of wilderness or maintaining imaginary constructs of the pristine for tourism purposes. Post catastrophic tourism futures, be it through becoming one with the Earth, coping with COVID-19, or leaving it altogether for space feature as topics herein.

(e) The Anthropocene as Overruled by Tourism

Lastly, we speculate from reviewing the 75 articles in the journals how deep running and genuine the attempts to rethink and transform tourism under the concept of the Anthropocene really run. After all, tourism research is wedged between an industry and a scholarly focus, and while transformations are called for, and deemed necessary under the terms of the Anthropocene, an uneasy feeling remains that whatever it morphs into will remain tourism business-as-usual with people seeking out experiences they can consume with impunity. To what extent we can truly develop and employ down to earth-led priorities and a planetary perspective, which to our reckoning is the clarion call of the Anthropocene, as long as tourism is regarded as a specific sector and a particular industry that needs to be maintained in a still all too Modern and capitalist logic? If this maintenance is taken as a given premise, then the conceptual and epistemological terrain ploughed open by the Anthropocene will simply be overruled by tourism.

Conclusion: Tourism and the Anthropocene

In this chapter we have drawn on our own work and our own understanding of the Anthropocene, adding insights from a segment of the wider tourism research literature. We accept the critique that this contribution has perchance been too largely filtered through our own conceptual wranglings, but we hope this provides further opportunity to use and deploy the Anthropocene in tourism research. In the entry we identified five themes in the literature beyond our own concerns. In terms of the uses of the Anthropocene as a framing concept and the Anthropocene as a container for re-packaging; we remain wary and critical. In these particular evocations of the term in the research literature we do not see the conceptual and empirical observations grappling with the potential implications of earth-led priorities and perspectives to the extent possible to augment the message and weight of the research. However in the third theme, wherein the Anthropocene is a booster for posthumanism; we see more seething potential of novel explorations. Bringing more-than/non-humans into the fold of tourism scholarship allows for conceptual and empirical reorientations we see as the necessary starting points for earth-led priorities and perspectives. In the last two themes we identify, seeing the Anthropocene as naturalization and overruled by tourism; we go back to being critical on the one hand and seeing promise on the other. Anthropocene as naturalization we see as necessary in furthering and adding depth to a critical angle of tourism studies. Interrogating the Anthropocene with the dominant capitalist logic of tourism is imperative to challenge tourism as business as usual. However a fundamental stumbling block is how the Anthropocene is overruled by imagining tourism as a sector that somehow needs to be understood and negotiated according to the Modern lexicon.

All in all, our exploration of the concept of the Anthropocene in the tourism research literature suggests that it has been employed in rather fragmented and mostly superficial ways. It is probably where we see the Anthropocene used as a booster for posthumanism that the potential for earth-led priorities and perspectives emerge through engagement with stones, mites, sled-dogs and things in wait of being scaled up to the planetary through the meshwork of life. Also wherein the Anthropocene allows for an engagement with the imperatives of growth and legacies of colonialism, exploitation and commodification we see critical momentum gained in tourism research by the term, whereby nuanced place based insights are scalable to the planetary with an eye on all the transformations now required.

Future Directions in the Planetary Climate and Ecological Emergency

Based on the above critique, and our current state of thinking on the Anthropocene, we suggest some potential future directions for tourism and tourism research and scholarship. However, first we want to emphasize one common premise for all and any future directions. It is crucially important to note that the Anthropocene can no longer be understood only as a suggested term for a new geological time-unit, nor as simply a broad signifier for how the Anthropos (humans, humanity, the human species) has become recognized as a geological force that impacts the functioning and the state of the Earth System. It is now evident that any serious and substantial engagement with the Anthropocene and tourism herein after also has to consider that we now face a dire climate and ecological situation with wide open doors to a precarious planetary future.

Given this, the most important of all is that both tourism and the Anthropocene must be placed at our current juncture of the planetary climatic and ecological emergency (or even 'breakdown'). In addition, when evoking tourism and the Anthropocene in this time of planetary emergency, calls for urgent action inherently belong. Furthermore, the urgent calls for climate and ecological action, from both science and the world community (i.e., United Nations), translate to the need for immediate transformations of societies that historically unprecedented in scope and time. In the words of the UN secretary-general António Guterres (United Nations, 2021, p. 3):

> We are at an inflection point in history. In our biggest shared test since the Second World War humanity faces a stark and urgent choice: a breakdown or a breakthrough. /.../ The choice is ours to make; but we will not have this chance again.

Breakthrough or breakdown, the stakes could not be much higher at an inflection point in history filled with dire planetary scenarios. At the time of writing (2024) the average global surface temperature is now close to 1.5°C higher, relative to the pre-industrial baseline. It certainly seems that we are locked-in to exceeding 1.5°C global warming, which could trigger multiple tipping points in the planetary climate system (Armstrong McKay et al., 2022). The estimate is that currently the world is heading toward at best 2°C warmer, but more likely somewhere around 3°C warmer surface temperature compared to this baseline. That would mean almost a degree per century, and such rapid heating is only to be seen from estimates of the great environmental upheavals at the end of the last ice-age. Even the lower aspirational goal of 1.5°C in the Paris Agreement leads us into an uncharted highly dangerous Anthropocene territory. How this will play out, on a densely populated planet with rigorously set and heavily militarised borders, is anyone's guess. When it comes to tourism the dystopian vision of Urry and Larsen (2011), presented over a decade ago, seems ever more apt:

> Systems of secure long-range mobility and tourism would disappear except for the super-rich who will congregate in 'policed' enclaves or camps. As in the medieval epoch, long-distance travel would be risky and probably not undertaken unless armed. Mass tourism would disappear. The rich would travel mainly in the air in armed helicopters or light aircraft. Each warlord-dominated region would potentially be at war with their neighbours, especially for control of water, oil and gas (p. 234).

It goes without saying that the tourism sector is highly vulnerable to 'climate change', which in the context of the Anthropocene is rather to be understood as 'Earth System change'.

It is also a fact that the tourism sector itself is one of the drivers behind global heating by contributing directly and indirectly to the emission of greenhouse gases. A general observation suggests that the tourism sector is currently unprepared for the enormous challenges we face in the planetary climate and ecological emergency, and maybe it cannot be otherwise? Yet the sector persists, and tourism has even declared climate emergency[1] as well as embraced climate action. This is also manifested in policy documents that increasingly pay at least lip service to the 'planet' or the 'planetary'. The signatories of 'the Glasgow Declaration on Climate Action in Tourism'[2] have committed to accelerate climate action to cut global tourism emissions by at least a half over the next decade, and reach Net-Zero emissions as soon as possible before 2050. Another initiative from the sector is the 'One Planet Sustainable Tourism Programme'[3], which aims to accelerate sustainable consumption and production (SCP) in tourism policies and practices to address the challenges of pollution, biodiversity loss and climate change. However, waving the now almost ubiquitous magic wand of 'Net-Zero', or flogging the dead horse of sustainability, will not be enough. If such ambitions are seriously to be reached and transform into geo-bio-physical effectiveness, then it will entail very drastic immediate changes in the modes of travel and transportation, which would demand a truly radical reorientation of what we expect of tourism and what we think travel means.

However well-intended the tourism sector initiatives are, we are nevertheless left with the impression that they are still very much trapped in the Modern sustainable development agenda. This is a variant of reformistic modernization based on the premise that tourism can proceed by gradually combining social, economic and environmental development in ways that step by step pave the road toward a better and more sustainable tourism future. This particular Modern conceptualization of the future is in stark contrast to the dire planetary climate and ecological future of the Anthropocene that is now moving closer toward us, and at a speed that corresponds to the climate and ecological actions we do **not** undertake now. This reformistic ideology that underpins sustainable tourism development is incompatible with the planetary climate and ecological emergency, and its demands for immediate unprecedented action. As for future directions in the context of the Anthropocene, all this suggests is that there is an urgent need to rethink and revalue the ontological and conceptual nature and practical purpose of Modern tourism research in the time of planetary climate and ecological emergency.

A pertinent question emerges. Should, or must, we stop thinking about tourism as predominantly a social phenomenon, and as a sector in need to adapt or develop resilience in the face of these challenges? If tourism and travel are understood as part of everyday life and in everything we do, does not the need to go to far-flung places in order to tick of some cultural or nature experience become both redundant and ridiculous. In other words, to travel to other places on the Earth's surface for social leisurely purposes might well be a luxury pursuit that soon enough cannot be legitimated anymore because of planetary climate and ecological reasons and the expectations and experiences sought need to be rethought? Dissolving the sectoral thinking may then be a way forward to loosen the strictures of Modern aspirations and stay true to addressing tourism's geo-bio-physical nature in the Anthropocene.

Scaling this rethinking, in the Anthropocene, the Earth's surface is no longer the only reference plane for mapping tourism spaces and places and tourist destinations. In the Anthropocene, there is an urgent need to further develop conceptualizations and investigations of tourism as a geo-bio-physical phenomenon. Engagements with posthumanism has already opened up this direction by embracing more-than/non-human agencies as constitutive for tourism, but the planetary is still to be fully recognized. In particular, it is necessary critically explore in detail similarities and differences in relation to the still dominant 'Global' as well as to problematize the 'planetary'. The planetary introduces a new scale by which all tourism things that happen locally will have to be translated. How to translate these particular *Planetary Longings* (Pratt, 2022) thus become a fundamentally important challenge for future research directions.

In this chapter we have placed the ´Earth System´ as central to the understanding of the Anthropocene, but it should be noted that there are other conceptual alternatives that deserve to be investigated and developed, such as: The Anthropocene Geobiosphere, GAIA, and the Critical Zone. An essential characteristic of them all is that they provide different ways to place and investigate tourism's geo-bio-physical agency. This is, again, crucial because the ontological and epistemological default position in tourism and tourism theory is still very much to conceive tourism predominantly as a social phenomenon. This does not mean, however, that future directions will not have to pay attention to the social of tourism and tourists. For example, well-founded criticisms of conceptualizations of both the Anthropos and the Earth as an Earth System, showing how they tend to naturalize or neglect the social, will continue to be necessary and important. Put differently, one may claim that the geo-bio-physicality of tourism, as well as its planetary dimension, is intrinsically saturated by the social in the Anthropocene.

Finally, in the context of the planetary emergency that now poses an existential threat to humanity it is imperative to develop research that actively contributes to accelerating climate and ecological action in tourism, whether that might be mitigation, adaptation or other strategies. Most importantly, this necessitates a transgression of the habit of placing climate and the ecological as separate topics among others in tourism. In the Anthropocene they require an urgent fundamental revision of all other topics.

Notes

1. https://www.tourismdeclares.com/
2. https://www.oneplanetnetwork.org/programmes/sustainable-tourism/glasgow-declaration
3. https://www.oneplanetnetwork.org/programmes/sustainable-tourism

References

Armstrong McKay, D I., Staal, A., Abrams, J. F., Winkelmann, R., Sakschewski, B., Loriani, S., Fetzer, I., Cornell, S. E., Rockström, J. & Lenton, T. M. (2022). Exceeding 1.5°C global warming could trigger multiple climate tipping points. *Science*, *377*, eabn7950. https://doi.org/10.1126/science.abn7950

Bianchi, R. V. (2009). The 'critical turn' in Tourism Studies: A radical critique. *Tourism Geographies*, *11*(4), 484–504. https://doi.org/10.1080/14616680903262653

Chandler, D. & Pugh, J. (2022). Interstitial and abyssal geographies. *Political Geography*, *98*, https://doi.org/10.1016/j.polgeo.2022.102672

Crutzen, P. J. & Stoermer, E. F. (2000). The "Anthropocene". *Global Change Newsletter*, 41, 17.

Gren, M., & Huijbens, E. (2012). Tourism theory and the Earth. *Annals of Tourism Research*, *39*(1), 155–170.

Gren, M., & Huijbens, E. (Eds.) (2016). *Tourism and the Anthropocene*. Routledge.

Gren, M., & Huijbens, E. (2014). Tourism in the Anthropocene. *Scandinavian Journal of Hospitality and Tourism*, *14*(1), 6–22. https://doi.org/10.1080/15022250.2014.886100

Huijbens, E. H. (2021a). *Developing earthly attachments in the Anthropocene.* Routledge.

Huijbens, E. H. (2021b). Earthly tourism and travel's contribution to a planetary genre de vie. *Tourist Studies, 21*(1), 108–118. https://doi.org/10.1177/1468797620986345

Huijbens, E. (2022). Convivial encounters in the city. On welcoming the other. *Current Issues in Tourism,* https://doi.org/10.1080/13683500.2022.2027890

Huijbens, E. H. & Gren, M. G. (2012). Tourism, ANT and the Earth. In van der Duim, R., Ren, C., & Jóhannesson, G. T (Eds.), *Actor-network theory and tourism: ordering, materiality and multiplicity* (pp. 146–163). Routledge

Huijbens, E. H., & Gren, M. G. (2022). They say "our house is on fire - on the climate emergency and (new) Earth politics. In E. T. Harper & D. Specht (Eds.), *Imagining apocalyptic politics for the Anthropocene* (pp. 15–33). Routledge.

Moore, A. (2019). *Destination Anthropocene: science and tourism in the Bahamas.* University of California Press.

Moore, J. (2015). *Capitalism in the web of life. Ecology and the accumulation of capital.* Verso.

Mostafanezhad, M., & Norum, R. (Eds.) (2019). *Anthropocene ecologies: entanglements of tourism, nature and imagination.* Routledge.

Pratt, M. L. (2022) *Planetary longings.* Duke University Press.

Rockström, J., Steffen, W., Noone, K., Persson, Å., et.al. (2009a). Planetary boundaries: exploring the safe operating space for humanity. *Ecology and Society, 14*(2), 32.

Rockström, J., Steffen, W., Noone, K., Persson, Å., et.al. (2009b). A safe operating space for humanity. *Nature, 461,* 472–475 https://doi.org/10.1038/461472a

Steffen, W., Broadgate, W., Deutsch, L., Gaffney, O., & Ludwig, C. (2015a). The trajectory of the Anthropocene: The great acceleration. *The Anthropocene Review,* 2(1), 1–18. https://doi.org/10.1177/2053019614564785

Steffen, W., Richardson, K., Rockström, J., Cornell, S. E. et.al. (2015b). Planetary boundaries: Guiding human development on a changing planet. *Science, 347*(736), 1259855 https://doi.org/10.1126/science.1259855

United Nations (2021). *Our common agenda – Report of the Secretary-General.* United Nations.

Urry, J., & Larsen, J. (2011). *The tourist gaze 3.0.* Sage.

Zalasiewicz, J., Waters, C. N., Williams, M., & Summerhayes, C. P. (2019). *The Anthropocene as a geological time unit.* Cambridge University Press.

Chapter 36

The Global Footprint of Tourism

Eke Eijgelaar & Paul Peeters

Introduction

With the Industrial Revolution and, more profoundly, the globalisation process of the last decades, both tourism and environmental issues have become increasingly global in occurrence. Globalisation processes amplify or replace many regional or national forces of change (e.g. Lambin et al., 2001). Human activities have become the main cause of global environmental change (GEC), hence our era has been designated the 'Anthropocene' (Crutzen, 2002). The Earth System could be approaching a planetary threshold resulting in a rapid warming pathway (Steffen et al., 2015). A whole range of high-profile reports has emphasised that human pressures are driving this system to its limit, pointing to the multiple crises of climate change, species extinction, land degradation and water scarcity (FAO, 2022; Intergovernmental Science-Policy Platform on Biodiversity and Ecosystem Services (IPBES), 2018, 2019; IPCC, 2021, 2022b; UNEP, 2019). To determine the trajectories of where the Earth System is heading, it has become essential to take all anthropogenic impacts into account, which in turn has implications for the decisions we make as a society, about the future we want (Bai et al., 2016; Steffen et al., 2015).

Globalisation could be referred to as the 'worldwide interconnectedness of places and people' (Lambin et al., 2001, p. 266), a description that also applies to present-day tourism. In view of tourism's global importance, worldwide growth, and its tendency to create interdependencies between countries (Mowforth & Munt, 2015), it is appropriate to devote a chapter to the global environmental impacts of tourism. As participation in tourism increases in and between countries all over the globe, many of its environmental impacts are becoming scaled up to a global level. Minimising these guarantees a more stable system, which is also essential for tourism (Gössling & Peeters, 2015).

The Wiley Blackwell Companion to Tourism, Second Edition. Edited by C. Michael Hall.

Rockström et al. (2009) have identified nine key biophysical Earth-system processes that could lead to unacceptable GEC if 'safe' thresholds are crossed. Five of these thresholds, or planetary boundaries, have already been exceeded (Persson et al., 2022; Steffen et al., 2015). Symbolic of this development is the recent surpassing of all global living biomass by the total anthropogenic mass (Elhacham et al., 2020). The system processes to which tourism can be linked are climate change, rate of biodiversity loss, global freshwater use, and change in land use (Gössling, 2002; Gössling & Peeters, 2015). All four (climate, biodiversity, water, and land) are vital resources for tourism. Climate change, land-system change, and biodiversity loss are three key system processes where the planetary boundaries have already been crossed. The boundary for global freshwater use has not been crossed in large parts of the world, but is exceeded in some regions (Steffen et al., 2015), several of which are much visited tourism destinations. Hence, these four systems will be further discussed. It is important to note that many impacts and changes, specifically concerning biodiversity, freshwater, and land use, occur at the local to regional level, but can have 'pervasive effects on how the Earth-system functions, and it interacts with several other planetary boundaries' (Rockström et al., 2009, p. 474).

Tourism's contribution to climate change is addressed a little more extensively, as it dominates tourism's environmental costs (Peeters et al., 2007), climate change itself is seen as one of the major future drivers of change in tourism (Becken, 2013), and because climate change poses existential challenges for the other Systems (Bates et al., 2008; IPCC, 2021; IPCC, 2022b; Parry et al., 2008). It is also the Earth System with the most established national and international policy and legislation, following the Paris Climate Change Agreement of 2015. Table 36.1 shows that the contribution of tourism to global climate change, particularly when expressed in terms of radiative forcing – that is, when including all impacts on global warming – is considerable, and a further strong increase is expected (Lenzen et al., 2018; Scott et al., 2010). Tourism's impacts on land use, water and biodiversity are arguably as present, with similar high growth forecasts (Gössling & Peeters, 2015), but impacts are more of a regional nature.

Table 36.1. Human and tourism alteration of major earth systems.

	Degree of human alteration	*Tourism global share of resource use*	*Future tourism use (estimated)*
Land	75% of land surface used or affected	<0.5% of biologically productive area (62,000 km^2)	Strong increase surface used (2010–2050: +189%)
Freshwater	<50% of accessible surface fresh water used	3.6% of global freshwater consumption	Strong increase fresh water used (2010–2050: +92%)
Carbon dioxide	Concentration +47% since Industrial Revolution	5–8% of CO_2 emissions, 5.2–12.5% of radiative forcing	Strong increase emissions (2010–2050: +164%)
Biodiversity	Living Planet Index, –68%* ~25% of species are threatened with extinction	~15% of Red List species (also) threatened by tourism	Strong increase (following from tourism growth and other Systems' impacts)

*1970–2016.

Sources: Vitousek et al. (1997); Scott et al. (2010); Ellis et al. (2010); IPBES (2019); IPCC, (2021); Gössling & Peeters (2015); Lenzen et al. (2018); Lusseau & Mancini (2019); UNESCO (2022); WWF (2020).

Biodiversity

Biodiversity, 'the diversity of genes, populations, species, communities, and ecosystems' (Mace et al., 2005, p. 79), provides a wide range of ecosystem services, such as food crops, freshwater, and clean air, but also for instance the stability of natural systems. The tourism element, for instance an aesthetic landscape or charismatic species, is just one of these services (a 'cultural' one), albeit a vital one for tourism (UNWTO, 2010). Naturally, tourism needs all the other biodiversity dependent natural resources like food and freshwater too. Thus, the alarming rate of species extinction and biodiversity loss during the Anthropocene, 100 to 1,000 times more than natural (Mace et al., 2005; Rockström et al., 2009), is of concern for the tourism industry.

Biodiversity has faced a steady decline over the last four decades, as demonstrated by a range of state indicators such as the Living Planet Index (mean population trends of vertebrates) or the Red List Index (extinction risk for mammals, birds, amphibians, and corals) (see IPBES, 2019). Some one million species may face extinction: on average 25% of species in different animal and plant groups. Changes in land use have been the most important driver in biodiversity loss, followed by direct exploitation. Climate change is seen as a direct driver that is increasingly exacerbating the impact of other drivers (IPBES, 2019). Biodiversity pressures have led the United Nations (UN) to construct a Paris-like agreement, the Post-2020 Global Biodiversity Framework (GBF), which should be finalised in 2022. An official first draft is available (Convention on Biological Diversity (CBD), 2021). The GBF intends to install targets for protecting marine and land habitats, reducing pesticides and plastic waste, et cetera. Its ultimate vision of living in harmony with nature by 2050 is very relevant for tourism. Meanwhile, the (2011–2020) Aichi Global Biodiversity Targets have not been fully met (CBD, 2020).

Tourism contributes to all five principal pressures of biodiversity loss: habitat change, overexploitation, pollution, invasive alien species, and climate change (Hall, 2010). Its growth in recent decades has enormous impacts on nature overall (IPBES, 2019). Human mobility, including tourism transport, is a considerable factor, as it leads to species exchange on a large and global scale, disrupting and homogenising biota (Gössling, 2002; IPBES, 2019), and produces large quantities of greenhouse gases and air pollutants, whereas the supporting infrastructure destroys or fragments habitats. Habitat loss is also where tourism urbanisation comes in, with coastal and mountain areas as specific areas of concern (UNWTO, 2010). Closely related to the mobility impacts is the exchange and dispersion of diseases (IPBES, 2019). Tourism can also benefit biodiversity (conservation) through its 'help' in creating protected areas and conserving charismatic species (Balmford et al., 2015), but this is still more on an isolated, local level. On a global scale, tourism is not likely to be a net contributor to biodiversity conservation (Hall, 2010). Furthermore, the value of tourism per hectare is low compared to other – generally more damaging – economic activities (Verweij et al., 2009).

Proposed mitigation strategies mostly aim at the destination level to prevent further habitat loss, for example by applying land use planning and development controls (UNWTO, 2010). Mitigation of mobility related impacts are more likely to be addressed in a climate policy context.

Water

Various consumption needs and pollution, as well as population growth and climate change, have increased pressures on freshwater, to the extent that water issues are no longer only discussed on a local or national level, but also on a global one. Mekonnen and Hoekstra (2016) find that two-thirds of the world population is now exposed to severe water scarcity at least one month per year, while half a billion people face this situation year-round. Agriculture is the key factor in freshwater consumption, accounting for 69 percent of global water withdrawal, followed by industry and energy (19 percent) and municipalities (12 percent) (UNESCO, 2021).

Water is an essential ingredient for tourism. Tourists not only use it for daily hygiene, but also when engaging in a wide range of activities. Besides these, water is an important element of tourism landscapes and is used in building and maintaining tourism infrastructure, and for producing fuel and food (Gössling, 2015). As such, water (security) has become a factor in the location of economic services, including tourism (WWAP, 2012). Direct water use for tourism varies greatly, between 80 and 2,400 litres per tourist per day, but generally exceeds water use at home. Use tends to increase with the size and luxury of the accommodation. Taking all water-use categories into account (direct and indirect), 4,600–12,000 litres per tourist per day is more realistic, with the bulk being used through food (Gössling, 2015) (see Table 36.2). On a global level, the 138 km^3 consumed by tourism annually (Gössling & Peeters, 2015) would point to approximately 3.6 percent of all global freshwater withdrawal (cf. UNESCO, 2022). However, there are several countries where tourism is the main factor in water consumption and water security (see also Table 36.5, below). Furthermore, tourist consumption of water is often concentrated in the dry season, when water availability is limited in many tourist regions. In these regions, further growth of tourism will possibly lead to competition with other users, which can be exacerbated by a decrease in freshwater availability through depletion, pollution, and climatic changes (IPCC, 2022b; UNESCO, 2021). If biofuels are used as a way to reduce aviation's emissions, this will increasingly contribute to tourism's water footprint in future.

To adapt to future water situations and mitigate its use, the tourism industry needs to engage itself more in water (conservation) management, ranging from measuring and charging water consumption, with better indicators (Gössling, 2015), to management measures to reduce use of, reuse, and recycle water, investment in new technologies, and educating tourists and staff, among others. Most of these measures will even be economically sound, but, as with other environmental resource-related measures, strong policies will be needed for their full implementation (Gössling et al., 2012).

Table 36.2. Water-use categories and estimated use per tourist per guest night.

Water use category	*Litres per tourist per guest night*
Direct	
Accommodation	84–2,425
Activities	10–875
Indirect	
Fossil fuels for transport	5–2,500
Energy use at hotel	0.3–200
Biofuels	2,500
Food	4,500–8.000
Total per tourist/guest night	4,600–12. 000

Source: Gössling (2015, p. 234).

Land Use and Change

Changes to the Earth's surface are made to support a population with goods and services. This 'represents the most substantial human alteration of the Earth system', as it has a profound impact on the structure and functioning of ecosystems and 'interacts strongly with most other components of global environmental change' (Vitousek et al., 1997, p. 494). The anthropogenic degradation of the Earth's land surface has progressed to a level that is now negatively impacting the well-being of at least 3.2 billion people, besides pushing biodiversity loss and contributing to mass species extinction (IPBES, 2018). For the past decade, some 23 percent of human-induced greenhouse gasses are related to changes in land use and land cover, through agriculture, forestry and other activities. More than 75 percent of the Earth's ice-free land surface is used or in some way affected by human-induced processes (Ellis et al., 2010; IPCC, 2019). Again, agriculture is the main factor, using 28 percent of the global land surface (FAO, 2022).

Gössling and Peeters (2015) estimated that the area used by tourism for accommodation, traffic infrastructure, and activities was around 62,000 km^2 in 2010, which would represent a very small fraction of the Earth's terrestrial surface. However, this only concerns the directly built area, the so-called tourism urbanisation (Table 36.3). The ecological footprint of tourism land use may be much higher, especially in coastal areas (like the Mediterranean), due to pressures on water resources, biodiversity, waste issues, and so on (Hall, 2006). Despite pledges to apply land use planning and development controls for limiting or reducing biodiversity loss at destinations (UNWTO, 2010), tourism is a considerable factor in high-value land acquisition, particularly at coastal areas. Since 2000, around 27,000 km^2 or 4 percent of the global total of land deals have been for tourism purposes, mainly in Africa (Anseeuw et al., 2012). Indirectly, demand for aviation biofuels may become a huge land use factor (see next section of this chapter). Biofuels have a yield of between about 1 to 5 metric tons of fuel per hectare per year (Peeters, 2017). With the exception of microalgae, that may in the future be boosted to 15 metric tons. The annual aviation fuel consumption in 2019 was 95 billion gallon (Statista, 2021), which would translate to some 21 percent of global forest area (based on data from FAO & UNEP, 2020).

On the other hand, the area covered by protected recreational areas has also increased, a development that is illustrative for more intensive land transitions, stemming from changes in demography and economy (Foley et al., 2005). The global surface area of protected and conserved areas has increased by 42 percent over the last decade, to 22.5 million km^2 (16.6 percent) of land and inland water ecosystems, and 28.1 million km^2 (7.7 percent) of coastal waters and the ocean (UNEP-WCMC & IUCN, 2021). Nature-based tourism may have been an (economic) factor in this development, with the world's terrestrial protected areas possibly receiving 8 billion (day-) visits per year (Balmford et al., 2015). However, the role of tourism in

Table 36.3. Land use in tourism.

Land use	*Average estimate (km^2)*
Accommodation	4,264
Traffic infrastructure	38,882
Golf courses	18,680
Global total area for tourism	61,826

Source: Adapted from Gössling and Peeters (2015, p. 646).

nature protection has been found to vary a great deal and to depend on a range of other institutional and democratic factors (Mossaz et al., 2015). Furthermore, the recent decade's substantial area growth can probably more be attributed to political pressure in achieving related Sustainable Development Goals and Aichi Biodiversity Targets. Also, protected areas with recreation as the main management function are often of low biological diversity and their conversion will not have interfered with the acquisition of those (frequently high-biodiversity) coastal areas for tourism development (see discussion in Hall, 2010).

Apart from some spatial planning to decrease impacts on landscape, the response of the tourism industry on land use issues will most likely be indirect, coming from the angles of freshwater, biodiversity, waste (water), pollution, and climate change. Particularly the development of biofuels will be a crucial one in respect to tourism land use.

Climate Change

The strong growth of anthropogenic greenhouse gas emissions has led to significant changes in the global atmospheric composition, causing changes in our climate that occur much faster than when induced by natural processes (IPCC, 2021). These climatic changes drive other Earth-system processes, which are already affecting biodiversity and water security, among others. Climate-driven temperature rise, sea-level rise, more extreme weather events, and changes in precipitation have serious impacts on humanity, and without strong mitigation efforts these impacts will become more severe in the coming decades (IPCC, 2022b), or might cross tipping points and lead to irreversible climate changes (Lenton et al., 2019). 3.5 billion people live in hotspots of high vulnerability to climate change (IPCC, 2022b). In the 2015 Paris Agreement on Climate Change, the global community has therefore pledged to pursue emission pathways that are 'consistent with holding the increase in the global average temperature to well below 2 °C above pre-industrial levels and pursuing efforts to limit the temperature increase to 1.5 °C above pre-industrial levels' (UNFCCC, 2015, p. 2). Most tourism is at risk through direct impacts on tourism destinations, infrastructure, and resources, and changes in health and safety. Furthermore, the small share of high-carbon tourism (cruises, long-haul trips by air) is also at risk through impacts of climate policy and changing consumer perceptions (Becken, 2013), while most of short-haul and all public transport-based tourism potentially could benefit from climate policies.

Tourism, with its dependency on transport, is also a driver of climate change. The carbon dioxide emissions of global tourism contributed to around 5 percent of anthropogenic carbon dioxide emissions in 2005 (Scott et al., 2010). Three-quarters of these emissions are caused by tourism transport (see Table 36.4). Lenzen et al. (2018), using a wider scope, estimated tourism's share of global emissions at 8 percent in 2013. Growth in tourism emissions has been driven by the rapid increase in tourist numbers, the continuous increase in travel distance, and the rising share of air transport (Scott et al., 2010). The latter development is particularly critical when measuring the overall impact of tourism on climate change; that is, in terms of radiative forcing. Several air-transport related processes occurring in high altitude, some of the exact impacts of which are still scientifically uncertain, could raise tourism's contribution to radiative forcing to 12.5 percent (Scott et al., 2010).

Tourism emissions are certain to increase further. For the 2005–2100 period, Peeters (2017) forecasts the total trip number to increase by 340 percent, the number of guest-nights by 170 percent, distance travelled by 880 percent, total tourism revenues by 430 percent, and CO_2 emissions by 370 percent in his reference scenario for global tourism development and

Table 36.4. Distribution of emissions from tourism by subsector.

Subsector	*2015*		*2100*	
	CO_2 (Mt)	*%*	*CO_2 (Mt)*	*%*
Air transport	835	53%	3889	72%
Car transport	304	19%	589	11%
Other transport	49	3%	35	1%
Accommodation	333	21%	494	9%
Activities	61	4%	401	7%
Total	1582	100%	5408	100%
Total world	31 850			
Tourism contribution	5%			

Source: Tourism data based on the model published by Peeters (2017) and global emissions scaled from UNWTO-UNEP-WMO (2008) using data from Minx et al. (2021).

emissions. This scenario sees the sector using 28 percent of the total global CO_2 budget between 2015 and 2100. Air transport share of tourism's CO_2 emissions is 55 percent in 2015 and may grow to 76 percent in 2100 (see Table 36.4).

Tourism will need to drastically reduce its energy use. This is easier for accommodation than for transport, even more because energy-saving measures are often very economical (see e.g. EY Parthenon et al., 2021). In global modelled pathways that limit warming to 2°C, and assuming immediate action, net global GHG emissions are projected to fall from 2019 levels by 27 percent by 2030 and 63 percent by 2050 (IPCC, 2022c). The remaining 37 percent in 2050 is assumed to be captured by massive carbon removal from the atmosphere. This removal idea, however, might have prohibitive consequences for other planetary boundaries regarding fresh water and biosphere integrity, which would demand for a further reduction of anthropogenic emissions (Heck et al., 2018). When the pathway starts with further overshoot, or a 1.5°C limit is targeted, reductions will also need to be much higher. It is highly unlikely that tourism will fit into such a scenario without there being a reduction in the growth of air transport. Mitigation efforts will need to go beyond the current approach of trusting technological and operational changes to provide sufficient reduction (Peeters, 2017).

A mix of three strategies may be needed for aviation to achieve zero emissions by 2050: zero-emission aircraft, sustainable aviation fuels (SAF), and reduced volumes (van der Sman et al., 2021). For zero-emission aircraft, the principal technologies are battery-electric, hydrogen-fuel cell-electric and hydrogen-jets. Regardless of the technology, the time to 2050 is too short for developing new aircraft types for each segment (from regional to widebody long-haul) and replacing the global fleet of 26,000 aircraft. Both involve two to three decades. The SAF route is the most promising pathway to zero emissions. SAF comes in three forms: biofuels, fuels from waste and synthetic e-fuels. Land use, biodiversity, and food production constrain the total volume of biofuels (de Jong et al., 2018). Though potentially fuel from waste from landfills can be very effective in reducing emissions from landfills (Huq et al., 2021), the resource is still based on the same low-efficiency energy uptake of plants (all the suitable waste is bio-based) and thus will diminish in a zero emissions future. A relatively new option form synthetic e-fuels, which use the power-to-liquids process to produce fuels from Hydrogen, CO_2 and renewable energy (Schmidt et al., 2018). Potentially, this system can fully close the carbon-cycle. The primary constraint here is the energy required, as illustrated by Drünert et al. (2020). Whatever the technological and fuel pathways, unlimited aviation growth is implausible to be compatible with zero-emissions in 2050.

It is quite obvious that sacrifices, in terms of changes in travel behaviour (shorter distances, low-carbon transport modes, and henceforth reduced growth or less air travel), need to be made, with negative impacts for some countries and benefits for others as a result (Peeters & Eijgelaar, 2014; Peeters et al., 2019). For adversely affected (developing) countries some compensation, through the pricing of aviation emissions, seems to be a realistic option (Keen et al., 2012). Doing nothing will very likely cause worse climate impacts to some countries, which in turn will negatively affect tourism flows and economies in an even stronger and less predictable way. Meanwhile, it seems quite certain that several related developments, such as increasing cost of fossil fuels, increasing government control over emissions, and increasing public awareness of climate change and of the impacts of tourism on climate change, are going to have an impact on this discussion and tourism in the near future (Draper et al., 2009).

Geographical Scope

Tourism is a large player in several regions that have high biodiversity levels, and which are facing severe environmental threats (see Table 36.5). Lusseau and Mancini (2019) particularly identify the Caribbean, Southeast Asia and the Alps, and coastal marine areas in general, as seeing worrisome tourism and recreation conservation threats. In coastal areas, both tourism and the human population concentrate. Population density is twice the global average across all biodiversity hotspots (Cunningham & Beazley, 2018). The Mediterranean is a classic example, facing cumulative environmental threats and large pressures from tourism (Drius et al., 2019). Brazil, China, and the Indian subcontinent have been added in the face of severe environmental threats combined with (pre-pandemic) forecasts of strong tourism growth (UNWTO, 2011).

Methodological Progress

Internationally, progress in measuring impacts on the environment has been made, as countries and industries are working towards the 2030 Agenda for Sustainable Development (the Sustainable Development Goals (SDGs)) and the Paris Agreement under the United Nations Framework Convention on Climate Change (UN, 2015; UNFCCC, 2015). The focus of decision-makers, also in tourism, has broadened from primarily on climate change and calculating carbon footprints, to other systems (e.g., the Aichi Biodiversity Targets). While the SDGs are useful for goal-setting, their success depends on a range of institutional factors, not the least that of formalising commitment (Biermann et al., 2017). The SDGs also do not distinguish specific economic sectors. The tourism sector will have to adapt to a range of international and national policies that address specific sustainability themes, but that are not necessarily tailored to the situation of specific economic sectors. Sustainability themes that are more grounded in daily practice and policymaking may thus be more useful.

Calculating the footprints of tourism is essentially a straightforward process of multiplying emission or resource-use factors with volumes. Gössling and Peeters (2015) have applied this for energy, emissions, food, water, and land use on a global scale. Measuring these impacts is, however, still not practiced in the industry itself. Indicators are still being developed. In this respect the UN World Tourism Organization (UNWTO) initiative Towards a Statistical

Table 36.5. Important tourism regions facing GEC threats.

Region	*Tourism importance (% GDP)*	*Regional biodiversity hotspot*	*Biodiversity threat from tourism*	*Water security threat*	*Tourism >5% of domestic water use*	*Overall climate vulnerability in 2050*
Caribbean	High	Caribbean	Very high	High	Barbados	High
Mediterranean	High	Mediterranean Basin	High	Low-High	Eight countries	Moderate-High
Southeast Asia	Medium-High	Sundaland, Wallacea, Philippines & Indo-Burma	Very high	Low-High	Thailand, Indonesia	High
New Zealand & SW Australia	Medium-High	New Zealand & SW Australia	High	Low-High	no	Moderate
East Africa	High	Eastern Arc Mountains and Coastal Forests of Tanzania and Kenya	High	High	unknown	Extreme
West Coast USA	Medium	California Floristic Province	High	Low-High	no	High
Coastal zone Brazil	Medium	Atlantic Forest Region	High	Low-High	no	Moderate
Indian subcontinent	Low	Western Ghats and Sri Lanka	High	High	India	High
China	Medium	Mountains of South-Central China	High	High	no	High

Sources: Gössling et al. (2012), IPCC (2022a), Lusseau and Mancini (2019), Mekonnen and Hoekstra (2016).

Framework for Measuring the Sustainability of Tourism (SF-MST) has recently identified four core accounts for measurement: flows of water, energy, GHG emissions and solid waste (UNWTO, 2018). A useful method for emissions calculation in tourism is given by Peeters et al. (2007). Tourism water indicators have been proposed by Gössling (2015). The most difficult impact to assess is tourism's impact on biodiversity. There are direct impacts because of habitat destruction and distribution of invasive species by tourism transport and by trade in souvenirs of biological origin. Indirect impacts come from water use, climate change, noise, disturbing wildlife, and local air-quality issues.

Another difficult issue is to assess whether a certain impact is within the limits of sustainability. Scale is a problem, because it is possible to tell the sustainable level of greenhouse gas emissions at a global level, but this is impossible for local emissions or emissions from a specific sector. To avoid irreversible climate change, it is clear zero-emissions should be the goal. This helps to prioritise solutions that have the potential of zero-emissions over those that only reduce emissions by a certain amount. For example, priority should be given to developing radically new zero-emission aircraft over a new generation of 10–20 percent more efficient conventional aircraft. The latter is less desirable because it causes another three to five decades lock-in of kerosene- and high-energy SAF-dependency. For water and land use, the sustainability is more locally determined and easier to determine at scales of tourism where decision-makers have full responsibility. But the 'safe' levels of water and land use are much less objectively determined than for climate change. Biodiversity is again more complex as it has elements at all scales. For instance, a local extinction of a species can be evaluated as still being sustainable when the species is still thriving globally, but a global extinction risk can generally only be avoided by local action in several places. So here responsibilities become spread between different stakeholders, who are not necessarily economically or socially connected.

Theoretical and Methodological Reflections

Over the last decades, many environmental issues have gained enormous importance and are now discussed on a global scale. Biodiversity loss, climate change and land use change are the most critical ones, but many others, including freshwater availability, are almost as urgent (Steffen et al., 2015). Above all, the four processes are strongly related. Despite the global focus, local or regional issues are always very important (see e.g. Turner et al., 2007). Tourism's role in environmental change should also be seen in this light. Apart from the direct global impact of tourism mobility on climate change and biodiversity, tourism impacts are predominantly of a local or regional nature, thereby contributing to GEC (see Table 36.6). The scope of tourism

Table 36.6. Indication of level of tourism impacts on GEC (on a per-trip basis for typical consumption patterns for instance the various transport modes).

Component	*Biodiversity*	*Climate change*	*Freshwater*	*Land use*
Accommodation	Strong	Strong	Severe	Severe
Cruises	Strong	Severe	Strong	Low
Air transport	Severe	Severe	Moderate	Strong
Car transport	Strong	Moderate	Moderate	Severe
Maritime transport	Moderate	Moderate	Moderate	Low
Rail, coach, and public transport	Moderate	Low	Moderate	Moderate
Activities	Low-moderate	Strong	Moderate	Moderate

impacts from different components varies per system process. Air, car, and maritime transport are the main factors in relation to climate change and biodiversity loss, while accommodations and the infrastructure to support tourism mobility are the largest factors in land use change. The impacts of air and maritime transport on biodiversity are rated more severe than from car transport due to their global 'reach'. Freshwater is mainly affected by accommodations, and indirectly also through the food that is consumed in them. Fuel and food production are issues on their own, with high impacts on all processes (not shown in table).

Mitigation of impacts will be crucial, not only to reduce the risk of overstepping planetary boundaries of vital system processes, but also because it is essential for the sustainable development if not the survival of tourism. The mitigation challenge seems largest in relation to tourism energy use and emissions. Seriously reducing these in the face of the forecasted increases in tourism numbers and mobility will require radical, large-scale systemic changes in the way we travel (Becken, 2019). Careful strategies need to be implemented to align mitigation efforts with other industry goals, such as supporting poverty alleviation. In some cases, this will require letting go of one goal to make a more significant contribution to the other. In regard to climate change mitigation and poverty alleviation there are several realistic constructions available that could benefit both goals (Peeters & Eijgelaar, 2014). In terms of adaptation, the challenge for tourism will be the greatest in many coastal areas, where biodiversity loss, freshwater shortages, and several impacts from climate change may interfere with the very existence of tourism.

In many cases, mitigation in tourism is a relatively straightforward process of reducing distances, energy use, water use, and so on. This does not need be preceded by knowledge of exact impacts; it simply requires common sense. However, knowledge about exact impacts of tourism on GEC is necessary to establish policy and compare sectors, for example. In some areas, particularly biodiversity, accurate establishment of tourism impacts is largely absent. Methodological issues that hamper this establishment range from lacking data on waste/emission streams and uncertainties about possible effects of such streams on the environment, to a lack of sufficient and reliable data of tourism flows, among others. The chapter has shown that the level of detail about tourism impacts varies considerably per environmental system and tourism component.

Many of the GECs caused by tourism can be linked to growth, both in numbers and in luxury, speed, and distance/scale. Vitousek et al. (1997, p. 498) concluded that we 'live on a human- dominated planet – and the momentum of human population growth, together with the imperative for further economic development in most of the world, ensures that our dominance will increase'. Tourism's growth imperative, if held on to, will ensure it will remain a considerable factor in human-induced GEC. Becken (2019) identifies five other barriers to rapid and substantial progress of climate action – which may hold truth also for other GEC issues: 'the institutionalisation of interests, the nature of policy making, the inadequacy of incremental improvements, the focus on technological efficiency instead of (behavioural) conservation, and the global distribution of tourism' (Becken, 2019, p. 419). Despite the clear urgency of climate change, where emission reductions of more than 15 percent per year are needed to reach the 55 percent reduction goals by 2030 and zero emissions by 2050, the tourism sector is still increasing its emissions and avoiding any serious action. One of the biggest problems in tourism seems to be the sector's inability to prioritise current low-impact forms of tourism (short-haul, domestic, rail-based) over ever increasing long-haul air-based international travel, while the only secure way to significantly reduce emissions and a range of other global environmental issues is by such a shift.

References

Anseeuw, W., Alden Wily, L., Cotula, L., & Taylor, M. (2012). *Land rights and the rush for land: Findings of the global commercial pressures on land research project*. https://grassrootsjusticenetwork.org/resources/land-rights-and-the-rush-for-land-findings-of-the-global-commercial-pressures-on-land-research-project/#:~:text=This%20report%20synthesises%20the%20findings,civil%20society%20organisations%2C%20academics%2C%20and

Bai, X., van der Leeuw, S., O'Brien, K., Berkhout, F., Biermann, F., Brondizio, E. S., Cudennec, C., Dearing, J., Duraiappah, A., Glaser, M., Revkin, A., Steffen, W., & Syvitski, J. (2016). Plausible and desirable futures in the Anthropocene: A new research agenda. *Global Environmental Change*, *39*, 351–362. https://doi.org/10.1016/j.gloenvcha.2015.09.017

Balmford, A., Green, J. M. H., Anderson, M., Beresford, J., Huang, C., Naidoo, R., Walpole, M., & Manica, A. (2015). Walk on the wild side: Estimating the global magnitude of visits to protected areas. *PLOS Biology*, *13*(2), e1002074. https://doi.org/10.1371/journal.pbio.1002074

Bates, B. C., Kundzewicz, Z. W., Wu, S., & Palutikof, J. P. (Eds). (2008). *Climate change and water. Technical paper of the Intergovernmental Panel on Climate Change*. IPCC Secretariat.

Becken, S. (2013). Shapers and shifters for the future of travel and tourism. In J. Leigh, C. Webster & S. Ivanov (Eds.), *Future tourism: political, social and economic changes* (pp. 80–91). Routledge.

Becken, S. (2019). Decarbonising tourism: mission impossible? *Tourism Recreation Research*, *44*(4), 419–433. https://doi.org/10.1080/02508281.2019.1598042

Biermann, F., Kanie, N., & Kim, R. E. (2017). Global governance by goal-setting: the novel approach of the UN Sustainable Development Goals. *Current Opinion in Environmental Sustainability*, *26–27*, 26–31. https://doi.org/10.1016/j.cosust.2017.01.010

Convention on Biological Diversity (CBD). (2020). *Global biodiversity outlook 5*. Secretariat of the CBD.

Convention on Biological Diversity (CBD). (2021). *First draft of the post-2020 global biodiversity framework*. UN CBD.

Crutzen, P. J. (2002). Geology of mankind. *Nature*, *415*(6867), 23. https://doi.org/10.1038/415023a

Cunningham, C., & Beazley, K. F. (2018). Changes in human population density and protected areas in terrestrial global biodiversity hotspots, 1995–2015. *Land*, *7*(4), 136. https://doi.org/10.3390/land7040136

de Jong, S., Stralen, J., Londo, M., Hoefnagels, R., Faaij, A., & Junginger, M. (2018). Renewable jet fuel supply scenarios in the European Union in 2021–2030 in the context of proposed biofuel policy and competing biomass demand. *GCB Bioenergy*, *10(9)*, 661–682. https://doi.org/10.1111/gcbb.12525

Draper, S., Goodman, J., Hardyment, R., & Murray, V. (2009). *Tourism 2023: Four scenarios, a vision and a strategy for UK outbound travel and tourism*. Forum for the Future.

Drius, M., Bongiorni, L., Depellegrin, D., Menegon, S., Pugnetti, A., & Stifter, S. (2019). Tackling challenges for Mediterranean sustainable coastal tourism: An ecosystem service perspective. *Science of The Total Environment*, *652*, 1302–1317. https://doi.org/10.1016/j.scitotenv.2018.10.121

Drünert, S., Neuling, U., Zitscher, T., & Kaltschmitt, M. (2020). Power-to-Liquid fuels for aviation – Processes, resources and supply potential under German conditions. *Applied Energy*, *277*, 115578. https://doi.org/10.1016/j.apenergy.2020.115578

Elhacham, E., Ben-Uri, L., Grozovski, J., Bar-On, Y. M., & Milo, R. (2020). Global human-made mass exceeds all living biomass. *Nature*, *588*, 442–444. https://doi.org/10.1038/s41586-020-3010-5

Ellis, E. C., Klein Goldewijk, K., Siebert, S., Lightman, D., & Ramankutty, N. (2010). Anthropogenic transformation of the biomes, 1700 to 2000. *Global Ecology and Biogeography*, *19*(5), 589–606. https://doi.org/10.1111/j.1466-8238.2010.00540.x

EY Parthenon, Booking.com, & OC&C. (2021). *Global accommodation sector. The road to net zero emissions*. https://www.oneplanetnetwork.org/sites/default/files/from-crm/Global%2520accommodation%2520sector%2520-%2520The%2520road%2520to%2520net%2520zero%2520emissions_Full%2520report_digital_compressed.pdf

FAO & UNEP. (2020). *The state of the world's forests 2020. Forests, biodiversity and people*. https://doi.org/10.4060/ca8642en

FAO. (2022). *The state of the world's land and water resources for food and agriculture: Systems at breaking point. Main report*. https://doi.org/10.4060/cb9910en

Foley, J. A., DeFries, R., Asner, G. P., Barford, C., Bonan, G., Carpenter, S. R., Chapin, F. S., Coe, M. T., Daily, G. C., & Gibbs, H. K. (2005). Global consequences of land use. *Science*, *309*(5734), 570–574. https://doi.org/10.1126/science.1111772

Gössling, S. (2002). Global environmental consequences of tourism. *Global Environmental Change*, *12*(4), 283–302. https://doi.org/10.1016/S0959-3780(02)00044-4

Gössling, S. (2015). New performance indicators for water management in tourism. *Tourism Management*, *46*, 233–244. https://doi.org/10.1016/j.tourman.2014.06.018

Gössling, S., & Peeters, P. (2015). Assessing tourism's global environmental impact 1900–2050. *Journal of Sustainable Tourism*, *23*(5), 639–659. https://doi.org/10.1080/09669582.2015.1008500

Gössling, S., Peeters, P., Hall, C. M., Ceron, J.-P., Dubois, G., Lehmann, L. V., & Scott, D. (2012). Tourism and water use: Supply, demand, and security. An international review. *Tourism Management*, *33*(1), 1–15. https://doi.org/10.1016/j.tourman.2011.03.015

Hall, C. M. (2006). Tourism urbanisation and global environmental change. In S. Gössling & C. M. Hall (Eds.), *Tourism & global environmental change: ecological, social, economic and political interrelationships* (pp. 142–155). Routledge.

Hall, C. M. (2010). Tourism and biodiversity: more significant than climate change? *Journal of Heritage Tourism*, *5*(4), 253–266. https://doi.org/10.1080/1743873x.2010.517843

Heck, V., Gerten, D., Lucht, W., & Popp, A. (2018). Biomass-based negative emissions difficult to reconcile with planetary boundaries. *Nature Climate Change*, *8*(2), 151–155. https://doi.org/10.1038/s41558-017-0064-y

Huq, N. A., Hafenstine, G. R., Huo, X., Nguyen, H., Tifft, S. M., Conklin, D. R., Stück, D., Stunkel, J., Yang, Z., Heyne, J. S., Wiatrowski, M. R., Zhang, Y., Tao, L., Zhu, J., McEnally, C. S., Christensen, E. D., Hays, C., Van Allsburg, K. M., Unocic, K. A., Meyer, H. M., Abdullah, Z., & Vardon, D. R. (2021). Toward net-zero sustainable aviation fuel with wet waste–derived volatile fatty acids. *Proceedings of the National Academy of Sciences*, *118*(13), e2023008118. https://doi.org/10.1073/pnas.2023008118

Intergovernmental Science-Policy Platform on Biodiversity and Ecosystem Services (IPBES). (2018). *The IPBES assessment report on land degradation and restoration*. IPBES secretariat. https://doi.org/10.5281/zenodo.3237392

Intergovernmental Science-Policy Platform on Biodiversity and Ecosystem Services (IPBES). (2019). *Global assessment report on biodiversity and ecosystem services of the Intergovernmental Science-Policy Platform on Biodiversity and Ecosystem Services*. IPBES secretariat. https://doi.org/10.5281/zenodo.3831673

IPCC. (2019). Summary for policymakers. In Climate change and land: an IPCC special report on climate change, desertification, land degradation, sustainable land management, food security, and greenhouse gas fluxes in terrestrial ecosystems.

IPCC. (2021). Summary for policymakers. In Climate change 2021: The physical science basis. Contribution of Working Group I to the Sixth Assessment Report of the Intergovernmental Panel on Climate Change. Cambridge University Press.

IPCC. (2022a). *Climate change 2022: Impacts, adaptation and vulnerability. Working Group II contribution to the Sixth Assessment Report of the Intergovernmental Panel on Climate Change*. Cambridge University Press.

IPCC. (2022b). Summary for policymakers. In Climate change 2022: Impacts, adaptation and vulnerability. Working Group II Contribution to the Sixth Assessment Report of the Intergovernmental Panel on Climate Change. Cambridge University Press.

IPCC. (2022c). Summary for policymakers. In Climate change 2022: Mitigation of climate change. Contribution of Working Group III to the Sixth Assessment Report of the Intergovernmental Panel on Climate Change. Cambridge University Press.

Keen, M., Perry, I., & Strand, J. (2012). *Market-based instruments for international aviation and shipping as a source of climate finance*. http://www.imf.org/external/np/g20/pdf/110411a.pdf

Lambin, E. F., Turner, B. L., Geist, H. J., Agbola, S. B., Angelsen, A., Bruce, J. W., Coomes, O. T., Dirzo, R., Fischer, G., Folke, C., George, P. S., Homewood, K., Imbernon, J., Leemans, R., Li, X., Moran, E. F., Mortimore, M., Ramakrishnan, P. S., Richards, J. F., Skånes, H., Steffen, W., Stone, G. D., Svedin, U., Veldkamp, T. A., Vogel, C., & Xu, J. (2001). The causes of land-use and land-cover change: moving beyond the myths. *Global Environmental Change*, *11*(4), 261–269. https://doi.org/http://dx.doi.org/10.1016/S0959-3780(01)00007-3

Lenton, T. M., Rockström, J., Gaffney, O., Rahmstorf, S., Richardson, K., Steffen, W., & Schellnhuber, H. J. (2019). Climate tipping points—too risky to bet against. *Nature*, *575*, 592–595. https://doi.org/10.1038/d41586-019-03595-0

Lenzen, M., Sun, Y.-Y., Faturay, F., Ting, Y.-P., Geschke, A., & Malik, A. (2018). The carbon footprint of global tourism. *Nature Climate Change*, *8*, 522–528. https://doi.org/10.1038/s41558-018-0141-x

Lusseau, D., & Mancini, F. (2019). A global assessment of tourism and recreation conservation threats to prioritise interventions. *Current Biology*. https://doi.org/10.2139/ssrn.3279418

Mace, G., Masundire, H., & Baillie, J. (2005). Biodiversity. In R. Hassan, R. Scholes & N. Ash (Eds), *Ecosystems and human well-being: Current state and trends, Volume 1* (pp. 77–122). Island Press.

Mekonnen, M. M., & Hoekstra, A. Y. (2016). Four billion people facing severe water scarcity. *Science Advances*, *2*(2), e1500323. https://doi.org/10.1126/sciadv.1500323

Minx, J. C., Lamb, W. F., Andrew, R. M., Canadell, J. G., Crippa, M., Döbbeling, N., Forster, P. M., Guizzardi, D., Olivier, J., Peters, G. P., Pongratz, J., Reisinger, A., Rigby, M., Saunois, M., Smith, S. J., Solazzo, E., & Tian, H (2021). A comprehensive and synthetic dataset for global, regional, and national greenhouse gas emissions by sector 1970–2018 with an extension to 2019. *Earth System Science Data*, *13*(11), 5213–5252. https://doi.org/10.5194/essd-13-5213-2021

Mossaz, A., Buckley, R. C., & Castley, J. G. (2015). Ecotourism contributions to conservation of African big cats. *Journal for Nature Conservation*, *28*, 112–118. https://doi.org/http://dx.doi.org/10.1016/j.jnc.2015.09.009

Mowforth, M., & Munt, I. (2015). *Tourism and sustainability : development, globalisation and new tourism in the Third World*. Routledge.

Parry, M., Palutikof, J., Hanson, C., & Lowe, J. (2008). Squaring up to reality. *Nature Climate Change*, *1*, 68–71. https://doi.org/10.1038/climate.2008.50

Peeters, P. (2017). *Tourism's impact on climate change and its mitigation challenges. How can tourism become 'climatically sustainable'?* [Unpublished doctoral dissertation]. TU Delft.

Peeters, P. M., Szimba, E., & Duijnisveld, M. (2007). Major environmental impacts of European tourist transport. *Journal of Transport Geography*, *15*, 83–93. https://doi.org/10.1016/j.jtrangeo.2006.12.007

Peeters, P., & Eijgelaar, E. (2014). Tourism's climate mitigation dilemma: flying between rich and poor countries. *Tourism Management*, *40*, 15–26. https://doi.org/10.1016/j.tourman.2013.05.001

Peeters, P., Higham, J., Cohen, S., Eijgelaar, E., & Gössling, S. (2019). Desirable tourism transport futures. *Journal of Sustainable Tourism*, *27*(2), 173–188. https://doi.org/10.1080/09669582.2018.1477785

Persson, L., Carney Almroth, B. M., Collins, C. D., Cornell, S., de Wit, C. A., Diamond, M. L., Fantke, P., Hassellöv, M., MacLeod, M., Ryberg, M. W., Søgaard Jørgensen, P., Villarrubia-Gómez, P., Wang, Z., & Hauschild, M. Z. (2022). Outside the safe operating space of the planetary boundary for novel entities. *Environmental Science & Technology*, *56*(3), 1510–1521. https://doi.org/10.1021/acs.est.1c04158

Rockström, J., Steffen, W., Noone, K., Persson, Å., Chapin, F. S., Lambin, E. F., Lenton, T. M., Scheffer, M., Folke, C., Schellnhuber, H. J., Nykvist, B., de Wit, C. A., Hughes, T., van der Leeuw, S., Rodhe, H., Sörlin, S., Snyder, P. K., Costanza, R., Svedin, U., … Foley, J. A. (2009). A safe operating space for humanity. *Nature*, *461*(7263), 472–475. https://doi.org/10.1038/461472a

Schmidt, P., Batteiger, V., Roth, A., Weindorf, W., & Raksha, T. (2018). Power-to-liquids as renewable fuel option for aviation: A review. *Chemie Ingenieur Technik*, *90*(1–2), 127–140. https://doi.org/10.1002/cite.201700129

Scott, D., Peeters, P., & Gössling, S. (2010). Can tourism deliver its "aspirational" greenhouse gas emission reduction targets? *Journal of Sustainable Tourism*, *18*(3), 393 - 408. https://doi.org/10.1080/09669581003653542

Statista. (2021). *Total fuel consumption of commercial airlines worldwide between 2005 and 2022.* https://www.statista.com/statistics/655057/fuel-consumption-of-airlines-worldwide/

Steffen, W., Richardson, K., Rockström, J., Cornell, S. E., Fetzer, I., Bennett, E. M., Biggs, R., Carpenter, S. R., de Vries, W., de Wit, C. A., Folke, C., Gerten, D., Heinke, J., Mace, G. M., Persson, L. M., Ramanathan, V., Reyers, B., & Sörlin, S. (2015). Planetary boundaries: Guiding human development on a changing planet. *Science, 347*(6223). https://doi.org/10.1126/science.1259855

Turner, B. L., Lambin, E. F., & Reenberg, A. (2007). The emergence of land change science for global environmental change and sustainability. *Proceedings of the National Academy of Sciences, 104*(52), 20666–20671. https://doi.org/10.1073/pnas.0704119104

UN. (2015). *Transforming our world: The 2030 agenda for sustainable development.* https://sdgs.un.org/2030agenda

UNEP-WCMC, & IUCN. (2021). *Protected planet report 2020.* Updated: May 2021. https://www.unep.org/resources/protected-planet-report-2020

UNEP. (2019). *Global environment outlook – GEO-6: Healthy planet, healthy people.* https://wedocs.unep.org/handle/20.500.11822/27539

UNESCO. (2021). *The United Nations World Water Development Report 2021: Valuing Water.* https://www.unwater.org/un-world-water-development-report-2021-valuing-water/

UNESCO. (2022). *The United Nations world water development report 2022. Groundwater. Making the invisiblevisible.*https://www.undp.org/publications/united-nations-world-water-development-report-2022-groundwater-making-invisible

UNFCCC. (2015). *Adoption of the Paris Agreement. Proposal by the president.* https://unfccc.int/documents/9064

UNWTO-UNEP-WMO. (2008). *Climate change and tourism: Responding to global challenges.* https://www.e-unwto.org/doi/book/10.18111/9789284412341

UNWTO. (2010). *Tourism and biodiversity – Achieving common goals towards sustainability.* https://www.e-unwto.org/doi/book/10.18111/9789284413713

UNWTO. (2011). *Tourism towards 2030 / Global overview - Advance edition presented at UNWTO 19th General Assembly.* http://www.e-unwto.org/content/w45127/?p=88c7aa22a06c4c4bbebc2e1dea594b57&pi=0

UNWTO. (2018). *Linking the TSA and the SEEA: a technical note.* https://seea.un.org/content/linking-tsa-and-seea-technical-note

van der Sman, E., Peerlings, B., Kos, J., Lieshout, R., & Boonekamp, T. (2021). *Destination 2050 – A route to net zero European aviation.* Tech. Rep. NLR-CR-2020-510, Netherlands Aerospace Centre NLR. https://www.destination2050.eu/wp-content/uploads/2021/02/Destination2050_Report.pdf

Verweij, P., Schouten, M., van Beukering, P., Triana, J., van der Leeuw, K., & Hess, S. (2009). *Keeping the Amazon forests standing: a matter of values.* WWF-Netherlands.

Vitousek, P. M., Mooney, H. A., Lubchenco, J., & Melillo, J. M. (1997). Human domination of Earth's ecosystems. *Science*, 277(5325), 494–499. https://doi.org/10.1126/science.277.5325.494

WWAP. (2012). *The United Nations world water development report 4: Managing water under uncertainty and risk.* https://unesdoc.unesco.org/ark:/48223/pf0000217175.locale=en

WWF. (2020). *Living planet report 2020. Bending the curve of biodiversity loss.* https://icriforum.org/living-planet-report-2020-bending-the-curve-of-biodiversity-loss/

Chapter 37

Climate Change Implications for Tourism

Daniel Scott

Introduction

Global climate change represents a grand challenge for society. No nation or economic sector are immune to the impacts of changes in climate or the low carbon transition, including tourism where sector investment, planning, operations, and demand are increasingly being influenced.

Accumulating evidence of the ongoing anthropogenic disruption of the global climate system has been relentless over the first two decades of the 21st century. Atmospheric concentrations of carbon-dioxide continue to rise at rates unprecedented in the geologic record, exceeding 420 ppm in February 2022 (National Oceanic and Atmospheric Administration, 2022), a level that last occurred 3.5 million years ago. Land and sea surface temperatures continue the multi-decade warming trend, with the frequency, intensity, and duration of extreme heat, heavy precipitation, and drought events increasing in most continental regions of the world (IPCC, 2021). If you find the recurrent headlines of climatic disasters disconcerting, as many do, we are but in the early stages of climate disruption. The Intergovernmental Panel on Climate Change (IPCC) Sixth Assessment Report (IPCC, 2022a) issued its starkest warning ever that current trajectories of climate change would fundamentally reshape ecosystems on Earth as well as the development prospects of billions of people worldwide.

Global leaders of government, business, civil society, and science emphasise the imperative to respond to climate change. The United Nations (UN) Secretary-General Antonio Guterres (UN, 2021) declared a 'code red for humanity . . . (with) . . . a clear moral and economic imperative to protect the lives and livelihoods of those on the front lines of the climate crisis'. Social momentum for strong climate response has also intensified, with over 2000 national and sub-national jurisdictions that represent more than 1 billion people in 35 countries declaring a climate emergency, with calls for urgent and society-wide climate action (Climate Emergency Declaration, 2021). Hundreds of tourism organisations similarly declared a

The Wiley Blackwell Companion to Tourism, Second Edition. Edited by C. Michael Hall.

climate emergency in 2020 (Tourism Declares, 2022). The landmark Paris Climate Agreement (UN, 2015) represents the commitment of the international community (signed by 195 countries) to decarbonise the global economy in order to avoid the consequences of dangerous climate change by limiting global warming to 'well below 2°C' (UN, 2015) and seek to stabilise the climate system at +1.5°C above pre-industrial levels. The tourism sector has voiced its strong support for the Paris Agreement and its science-based emission targets, most recently in the 2021 *Glasgow Declaration: A Commitment to a Decade of Tourism Climate Action* (One Planet, 2021).

While the transition to decarbonise the global economy is unmistakably underway in some parts of the world, the current state of emission reduction progress and ambition remains appreciably out of pace with the transformations required by science-based targets (United Nations Environment Programme (UNEP), 2021). Recent estimates from the World Meteorological Organization (WMO) (2021) warn that the Paris Agreement 1.5°C threshold could be exceeded, at first temporarily, as early as 2024. The IPCC (2022b) estimated that without a strengthening of policies beyond those implemented in 2020, future emissions would lead to global warming of approximately 3.2°C by the end of the century. It is difficult to overstate the consequences of a high emissions future that would transform coastlines and ecosystems, displace millions, and reduce growth in the economy in many nations (IPCC, 2022a). Such outcomes are incompatible with progress on sustainable development.

Regardless of whether society achieves the Paris Agreement +2°C target through unprecedented decarbonisation that would represent one of the most consequential socio-economic transformations in human history or the more extensive impacts of a post +3°C climate disrupted world are realised; climate change will transform global tourism over the next three decades and beyond. It is self-evident that any phenomenon that would adversely affect economic growth in many areas of the world, greatly increase regional water and food insecurity, harm the health and displace more than a billion people, massively alter many ecosystems and increase extinctions, increase transportation costs, threaten cultural heritage, and increase security risks, will have far-reaching implications for tourism development (Scott et al., 2012a; Scott et al., 2012b; Scott, 2021). The imperative of understanding how climate change will transform global tourism over the next three decades and beyond is a foremost challenge facing the tourism academy.

A comprehensive examination of the two-way relationship between climate change and tourism is beyond the scope of a single chapter. The contributions of tourism to global climate change are examined in Chapter 36 by Eijgelaar and Peeters. This chapter restricts its focus to the impacts of climate change (climate risk) and the low carbon transition (carbon risk) on tourism. It provides a conceptual overview of the four main pathways by which climate change will impact the tourism system. It then explores the comparative vulnerability of major tourism regions and adaptive capacity of tourism stakeholders. The chapter concludes with consideration of how well prepared the tourism sector is for climate change and the need to accelerate tourism responses to the climate crisis.

The New Realties of Global Climate Change for Tourism

The interface between climate change and tourism is multi-faceted and highly complex. There are two overarching challenges that will transform tourism over the coming decades. The first is the transition to a net-zero emission economy (where anthropogenic carbon emissions are balanced by sequestration). This includes the difficult process of decarbonising tourism but also the diverse implications of the societal transition to a net-zero economy of mid-century, which may include substantial carbon prices, national or sectoral emission caps and penalties, or social change (e.g., 'flight shame'). The second overarching challenge is adapting to multiple physical climate risks in an increasingly climate disrupted world. Tourism is highly sensitive to the impacts of climate change with a comprehensive review of how ten major climate change impacts could affect 89 areas of human health, food, water, infrastructure, economy, and security, finding that tourism was one of only five areas impacted by all ten (Mora et al., 2018). Adapting to the negative and positive consequences of environmental and socio-economic impacts of climate change is highly place-based and market context specific (Scott & Becken, 2010; Scott et al., 2022).

Figure 37.1 sets out four broad pathways (arrows 1 to 4) by which climate change will affect the future prospects of tourism around the world (Scott et al., 2012a, b). It illustrates that all major components of the global tourism system will be differentially impacted by the regional manifestations of climate change and the capacity of the tourism sector to respond. The integrated effects of climate change will simultaneously interact with other major macro-scale drivers of tourism development to generate negative and positive impacts that will vary by market segment and geographic region, altering sustainability and competitiveness at the business, destination, and country scale. As a result, impacts in one part of the tourism system will have repercussions for competitors, so that a negative impact in one part of the tourism system may have a knock-on negative effect or constitute an opportunity elsewhere. This section illustrates some of the major risks and opportunities associated with each of these impact pathways.

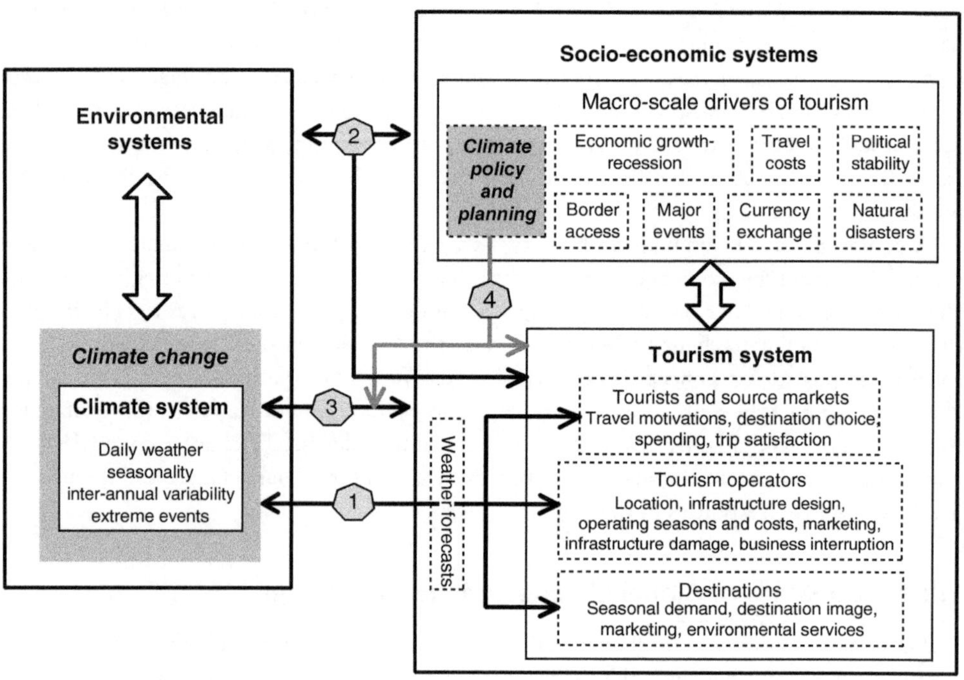

Figure 37.1. The complex interactions between climate change and the tourism sector.
Source: After Scott et al. (2012a, 2012b).

(1) Direct Impacts of a Changed Climate

For the tourism industry and tourists alike, climate represents both a resource to be exploited and an important constraint. As a consequence, the implications of climatic change (both changes in mean conditions and extremes) are anticipated to be far-reaching for some tourism segments and destinations, altering the length and quality of climate-dependent tourism seasons, operating costs (e.g., air conditioning, wildfire suppression, snowmaking, insurance, emergency evacuations), location decisions and infrastructure design, destination attractiveness, real estate value, causing infrastructure damage and business/supply-chain interruptions, and affecting tourist demand and destination choice.

It is anticipated that the distribution of climate resources for tourism will be altered over the 21st century (Amelung et al., 2007; Ma et al., 2021; Yu et al., 2022), with conditions ideal for most tourism activity expanding at higher latitudes. Climatic conditions for tourism were simultaneously expected to degrade in subtropical and tropical destinations, particularly in the peak summer tourism season. As a result, tourists from temperate regions that currently dominate international travel (e.g., Northern Europe, Japan, USA, China) would adapt their travel patterns to take advantage of new climatic opportunities closer to home, resulting in a gradual shift in international tourism demand to higher latitude countries (Bigano et al., 2006).

Changes in the length and quality of climate-dependent tourism seasons could have considerable implications for competitive relationships between destinations. For example, the risks posed by climate change to the ski tourism industry have received considerable attention (Steiger et al., 2019). In contrast, the season lengths for summer tourism opportunities are projected to increase substantially in higher latitude regions but have been studied far less (Scott et al., 2012a; Scott et al., 2012b; Scott & Gössling, 2022a). The economic impact of shifts in tourism seasonality will differ at the destination level, depending on local circumstances and adaptation strategies.

Recent record heat waves and summer drought and wildfires in Europe, North America, India, China, Australia, and Brazil illustrate that the impacts of changes in climate are not some distant future challenges. Projected changes in the number and/or intensity of extreme weather events under climate change (IPCC, 2021) also have important implications for tourism infrastructure damage, emergency preparedness requirements, higher operating expenses (e.g., insurance, backup water and power systems, and evacuations), and business interruptions. However, the effects of altered patterns of extreme events on tourism operations and demand patterns remain poorly understood (Scott & Gössling, 2022a).

(2) Indirect Climate-induced Environmental Change

Many environmental assets that define destination image and are critical attractions for tourists are sensitive to climate variability and extreme events. Climate also influences environmental conditions that can deter tourists, including infectious disease and wildfires, as well as alter operating costs, real estate values, supply-chains, and the capacity of tourism firms to do business sustainably (e.g., energy use). The wide-range of projected climate-induced environmental changes will have profound effects on tourism at the local and regional scale. For example, changes in water availability, biodiversity loss, reduced landscape aesthetic (e.g., glacial retreat, fire or disease impacted forest landscape, and algae blooms), altered agricultural production (e.g., wine tourism), increased natural hazards, coastal erosion and inundation, damage to infrastructure from flooding and wildfires, and the increasing incidence of vector-borne diseases will diminish the quality of the tourism product with implications for visitation and local economies. In contrast to the varied impacts of a changed climate on tourism (both risks and opportunities), the indirect effects of climate-induced environmental change are anticipated to be predominantly negative and substantially greater under the higher emission/warmer scenarios (Scott et al., 2012a, b, Scott & Gössling, 2022a).

(3) Indirect Climate-induced Socioeconomic Change

Climate change is anticipated to influence important socio-economic changes that have implications for the long-term growth and geographic distribution of tourism, such as decreased economic growth, increased poverty, increased political instability and security risks, and changing public attitudes toward travel (e.g., concerns over emission intensive holidays). Economic development is a central determinant of tourism. Tourism depends on economic prosperity and climate change is anticipated to reduce future economic growth. For example, the Organisation for Economic Co-operation and Development (OECD) (2015) estimated the annual negative impact of 2.5°C warming on the global economy would range between 0.7 and 2.5 percent of GDP to 2060 and become more impactful beyond mid-century. By comparison, a 1.4 percent reduction in annual global GDP in the United Nations World Tourism Organization (UNWTO) (2011) slower than expected economic growth scenario resulted in a large reduction (-22 percent) in international tourism arrivals in 2030 (1.4 billion instead of 1.8 billion).

Climate change is also considered a national and international security risk that will intensify under greater warming scenarios (World Bank, 2013, Mach et al., 2019), with several major international organisations recognising the connections between climate change, social disruptions, and violence (including the UN Security Council, North Atlantic Treaty Organization, the African Union). Climate change can undermine human security by reducing access to, and the quality of, natural resources that are important to sustain livelihoods (World Bank, 2021a) as well as displace large numbers of people (World Bank, 2021b), which under certain circumstances increase civil unrest and conflict. Potential climate-conflict hotspots have been identified in Central and South America, the Sahel region of Africa, as well as Central and Southeast Asia (Kohli et al., 2018).

Tourism makes a major contribution to the economies of several countries in these regions, and more are planning on tourism growth as an economic development strategy. Contemporary experience demonstrates very clearly that political instability and civil unrest lead to a decline in international tourism (Hall et al., 2004; Causevic & Lynch, 2013). While the negative repercussions for tourism demand to countries that become climate change security hotspots is obvious, no study has yet to examine the economic consequences for these nations.

(4) Impacts of Climate Change Mitigation Policy

The tourism sector is a major contributor to global greenhouse gas emissions and the process of decarbonisation will transform the sector in the coming decades. The scale of tourism emissions was first estimated at approximately 5 percent of global emissions of CO_2 in 2005 (Scott et al., 2008). A more recent estimate increased the contribution to approximately 8 percent in 2013 (Lenzen et al., 2018). Limited monitoring capabilities of sectoral emissions (a long existing capacity gap identified by Scott et al. (2008)) and disruption of the COVID-19 pandemic (Gössling et al., 2020) make projections difficult, but available evidence indicates that tourism emissions are expected to grow rapidly (e.g., 130 percent from 2005 through 2035 (Scott et al., 2008); 169 percent from 2010 through 2050 (Gössling & Peeters, 2015).

The tourism sector recognises this trend is incompatible with the Paris Climate Agreement and that future tourism growth must be decoupled from emissions growth. The World Travel and Tourism Council (WTTC) (2010) announced the first emission reduction ambition for the sector at -50 percent by 2035 (from 2005 levels). This ambition was later endorsed by the UNWTO. In 2021 the WTTC and the signatories of the 2021 *Glasgow Declaration: A Commitment to a Decade of Tourism Climate Action* (One Planet, 2021) pledged to cut tourism emissions in half by 2030 and achieve net-zero emissions before 2050. Overcoming the six systemic decarbonisation challenges identified by Becken (2019) will not be easy, particularly as the sector recovers from the disruption of the COVID-19 pandemic through the 2020s.

While many potential emission reduction strategies have been identified for the tourism sector, to date no credible plan has been presented by which emissions could be cut in half by 2030 (Scott & Gössling, 2022b). Without a detailed roadmap of how the tourism sector could pursue its net-zero commitment, the financial costs of pursuing deep decarbonisation remain uncertain. One analysis of the cost to achieve a -50 percent target (by 2050, not 2030) through abatement and some strategic offsetting (in aviation), represented less than 0.1 percent of the estimated global tourism economy in 2020 (approximately USD 600 million annually) and grew to 3.6 percent (approximately USD 351 billion annually) in 2050 as emission reductions became more difficult (Scott et al., 2015a). Achieving a 70 percent emission reduction target more closely aligned with IPCC targets requires greater investment of approximately 4.1 percent of the anticipated total global tourism economy near to 2050. While substantial, the cost of decarbonising tourism when distributed equally among all tourists (international and domestic) was roughly equivalent to many typical travel fees or approximately US$11 per trip. How the costs of sectoral decarbonisation are distributed among global travellers would have important implications for regional tourism development and destinations; important research that remains to be completed.

The transition to a post-carbon economy will be technologically, economically, politically, and socially complex. The manifold ways the pursuit of net-zero global emissions will impact tourism is not well recognised within the tourism sector and research on the transition risk remains virtually absent in the tourism literature (Scott & Gössling, 2022a; Scott & Gössling, 2022b). Examining the place of tourism in deep decarbonisation plans that have been developed to achieve the Paris Agreement +2°C policy target is revealing. A review of Paris Agreement compatible plans found that none explicitly mention tourism, but all assumed that demand management will play an important role in achieving required emission reductions from aviation (Scott & Gössling, 2015). A review of more recent plans, including the first net-zero emissions by 2050 pathway developed by the influential International Energy Agency (Scott & Gössling, 2022b), similarly found that tourism is not mentioned, and that demand management remains a central strategy to address aviation emissions. The post-carbon transition risk will influence tourism patterns unevenly around the world and the tourism sector and international development agencies that promote tourism have not adequately considered the long-term implications for international travel patterns and destination competitiveness (Gössling et al., 2013; Scott & Gössling, 2022b).

The Geography of Tourism Vulnerability to Climate Change

The four major types of climate change impacts identified in Figure 37.1 will unfold concurrently, creating a complex mosaic of 'winners and losers' at the business, destination, and country scales. Few studies consider the wide range of potential impacts and their interactions at the destination scale or consider regional or market segment interactions (Scott & Gössling, 2022a). As Mora et al. (2018) warn, the study of individual climate change impacts in isolation remains common in many sectors (not just tourism), which results in incomplete and potentially misleading understanding of the consequences of climate change.

Understanding regional patterns of differential climate change risk is important for insights into where longer-term tourism development is most likely to be hindered. A global analysis of the differential climate change risk for tourism across 181 countries, found the highest risk in Africa, the Middle East, South Asia, and Small Island Developing States (SIDS) in the Caribbean as well as Indian and Pacific Oceans (Scott et al., 2019). Importantly, when climate change risk was compared with the relative economic contribution (percent GDP) the countries with the most risk to their economies were mostly SIDS but included some major destinations such as Thailand and Mexico. Higher climate change risk also aligned strongly with regions where tourism growth was projected to be the strongest in the coming decades, including Sub-Saharan Africa and South Asia. The climate justice implications are inescapable. Disturbingly, the substantial regional imbalances in the information base on climate change and tourism reveal geographic gaps in precisely the regions identified as most at risk and more economically reliant on tourism (Scott et al., 2015b; Scott & Gössling, 2022a).

Although tourism regions will be differentially affected by climate change, in the globalised and highly interconnected world of tourism, the consequences of impacts in highly vulnerable regions will transverse political boundaries to influence travel patterns, value chains, and tourism investment worldwide (Scott et al., 2019). A negative impact in one part of the tourism marketplace is likely to constitute an opportunity elsewhere as climate change alters destination competitiveness (Scott, 2021).

Differential climate change vulnerability at the destination and enterprise scale is also very visible within most major tourism markets. One tourism market that has received considerable attention because of its very direct influence from changes in climate has been the ski industry (Steiger et al., 2019). Several studies have used common impact indicators to compare the implications of climate change for ski tourism in multiple destination communities within regional ski tourism marketplaces in Austria, Canada, China, France, Norway, Sweden, and the United States. Figure 37.2 illustrates Dawson and Scott's (2013) projected evolution of the New England ski tourism market under increased warming. It reveals the changing geographic distribution of operating ski areas over time, as it contracts and consolidates into the climatically advantaged higher elevation destinations with greater snowmaking potential. Assuming ski tourism demand stays relatively stable (as evidenced in record warm winters), the destination communities where ski areas are projected to remain operational could be in a position to take advantage of a changed business environment and gain market share from their lost competitors. Although these ski areas and associated communities are likely to benefit from increased tourism revenue, they will still need to adapt and prepare for the possibility of increased development pressures (e.g., water use for snowmaking, real estate development, slope expansion, congestion), crowding, and infrastructure deficiencies. In contrast, the communities that lose ski operations will need to prepare for lost winter tourism revenues and related jobs and could also see a drop in real estate value with secondary effects for

Figure 37.2. Projected contraction of the New England (USA) ski market under climate change.
Source: Adapted from Dawson and Scott (2013).

community services (Scott et al., 2012a, 2012b). These more vulnerable ski areas will need to determine if they should invest heavily in adaptations that will aid in the continuation of the ski industry at least in the short to medium term or if the investment should focus on other forms of tourism and economic diversification. These same questions about the future of tourism in an era of climate change will play out in thousands of destination communities around the world.

Adapting to Climate Change

Just as tourism has a critical decarbonisation challenge, so too it has a massive challenge adapting to the diverse impacts of a changing climate (Scott & Becken, 2010; Scott & Gössling, 2022a). Despite differential impacts of climate change, the reality is that all tourism enterprises and destinations will need to adapt to climate change, whether to manage risks or to capitalise on new market opportunities. Positively, the dynamic nature of the tourism sector and its ability to cope with a range of shocks including terrorism attacks, major natural disasters, financial crisis, and the COVID-19 pandemic, suggests the sector possesses relatively high adaptive capacity overall (Hall, 2010; Hall et al., 2020).

Over a decade ago, Scott and Becken (2010) concluded that research for and of climate change adaptation in tourism was underdeveloped relative to many other sectors. Recent reviews by Loehr (2020) and Scott and Gössling (2022a) reconfirm that assessment and argue that a greater focus on the solution space for climate risks is needed to inform feasible, effective, and just adaptation in the tourism sector. Scott and Gössling (2022a, p. 8) conclude that, 'There remains highly inadequate research on the limits to adaptation (physical, economic, social) under future climate scenarios, the scalability of adaptations, associated co-benefits, maladaptation, and adaptive capacity (including enablers and constraints).' This is not just a conceptual problem, because as climate risk disclosure requirements expand (a requirement endorsed by the G7 nations and subsequently the International Financial Reporting Standards Foundation (IFRS) in 2021), the lack of information on adaptation will be detrimental to the financial interests of tourism businesses.

Adaptation in the tourism sector can be pursued by tourists, tourism operators, tourism dependent communities (destinations), national and international tourism organisations (government or industry), government agencies involved in tourism, tourism focused non-governmental organisations, international development organisations, and the financial sector

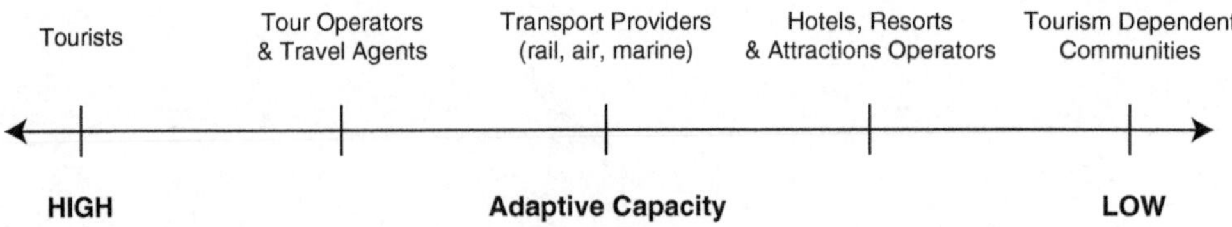

Figure 37.3. Relative ability of tourism stakeholders to adapt to climate change. Adapted from Scott and Jones (2006).

(investors and insurers). The ability to adapt to climate change varies substantially across the wide range of tourism stakeholders (Figure 37.3).

Tourists have the greatest capacity to adapt to the impacts of climate change, with freedom to avoid any destination negatively impacted by climate change. It is the response of tourists to the complexity of destination impacts that will reshape consumer demand patterns and play a pivotal role in the eventual impacts of climate change on the tourism sector worldwide. The perceptions of climate, environmental or security conditions will influence destination image and will be central to the decision-making of tourists (Gössling et al., 2012). Unfortunately, the impacts and sectoral responses to many early onset climate change impacts (e.g., heatwaves, major hurricanes, wildfires, landscape, and ecosystem changes) that serve as natural experiments to understand future impacts that are anticipated to occur more frequently and become more severe are not being well studied. In particular, the responses of tourists to these natural experiments offered by climate anomalies continue to be major missed learning opportunities that are essential to improve projections of changes in tourism demand patterns (Scott, 2021).

Most climate change adaptation happens at the local scale and is highly place-context specific (Berrang-Ford et al., 2021). Assessments of the state of climate change adaptation by tourism operators and local tourism officials have consistently found low, but improving, levels of climate change awareness and relatively low perceived climate change risk (Kaján & Saarinen, 2013; Loehr, 2020; Scott & Gössling, 2022a). Destination communities possess the least adaptive capacity of major tourism stakeholders (Figure 37.3), because their visitor-based economy can be greatly affected by the adaptation of other tourism stakeholders (e.g., the decision of tour operators to direct tourists to other destinations less affected by climate change), adaptation by other sectors (e.g., the decision of insurance companies to declare areas as uninsurable because of high risk), the immobility of high-capital infrastructure, and the limited capacity of some destinations to rebrand themselves with other tourist attractions. A changing climate will lead to a changing business environment and changing community relationships in many destinations, new realities that the tourism sector has yet to consider with any depth.

There is very little evidence of long-term strategic planning in anticipation of future changes in climate (Loehr, 2020; Scott & Gössling, 2022a). Scott et al. (2012) summarised several key barriers to climate change adaptation by tourism businesses and local tourism officials. There is not a long tradition of communication climate scientists and tourism businesses, and this poses a challenge for how to communicate weather and climate change risk. Uncertainty over the magnitude and timing of future climate change translates into low confidence in the current understanding of how climate change would impact tourism and the perception that it is too early for large investments in adaptation. Limited technical expertise, human resource and financial capacity are identified as critical barriers to climate change adaptation in the tourism sector.

Some tourism stakeholders have also voiced concerns that early adapters could be penalised by changes in adaptation legislation or the emergence of financial assistance for adaptation in the future. All studies have found that tourism stakeholders believe that leadership on climate change must come from government (Loehr, 2020; Gössling & Scott, 2018).

Government legislation (both existing and forthcoming) will strongly influence enterprise and community-level adaptation decisions (Berrang-Ford et al., 2021). The influence of adaptation strategies by government (particularly transportation, water resources, energy, land use planning, building codes, coastal and protected areas management) will have very important implications for tourism, but these policy and planning interactions have not been adequately explored (Scott, 2021).

Adaptation will pose particular challenges in developing nations, which have lower adaptive capacity overall (Berrang-Ford et al., 2021) yet will have to be undertaken to a sufficiently high quality to still attract international tourists. The UNWTO, UNEP and WTTC have all emphasised the important role of the tourism sector to develop strategies to assist local communities to understand climate change vulnerabilities and increase their adaptive capacity, but a strategy by which to do this has yet to be developed.

Conclusion – Is Tourism Prepared for Climate Change?

Accumulating evidence indicates that climate change, particularly high emission scenarios, will alter the competitiveness of tourism destinations and has the potential to transform major market segments (Scott et al., 2012; Scott & Gössling, 2022a). While the low carbon transition will be disruptive for some tourism businesses and destinations, a +3°C or warmer world will be far more disruptive for tourism. Global tourism leaders interviewed by Gössling and Scott (2018) were unanimous that a high emissions future represents a cataclysm for both society and tourism and must be avoided. Importantly, the significance of climate change to tourism is not solely in some distant future. Climate change is already affecting investment and operating decisions and consequently, adaptation to the integrated effects of climate change will be fundamental to the development of sustainable tourism systems in the decades ahead. Is the tourism sector building capacity and investing in the transition to climate resilient development, which the IPCC (2022a) defines as 'a process of implementing greenhouse gas mitigation and adaptation solutions to support sustainable development for all'?

The 2020s are widely considered by the UN and policy and scientific communities as a decisive decade for climate action. A decade ago, KPMG's (2008) assessment of the climate change regulatory, physical, and reputational risks posed to 18 major economic sectors found tourism to be one of six sectors most at risk and yet least prepared. Available evidence suggests the level of preparedness within the sector is improving but unfortunately remains low (Becken et al., 2020; Gössling & Scott, 2018; Conefrey & Hanrahan, 2022).

Climate change engagement of international tourism organisations followed the expanding scientific literature and broader advances in climate policy and social momentum for climate action (Scott & Gössling, 2022c). The first multilateral exploration of the complex relationships between climate and tourism occurred in 2003 and resulted in the *Djerba Declaration on Tourism and Climate Change*. The follow-up *Davos Declaration on Climate Change and Tourism* in 2007 was organised by UNWTO together with UNEP and WMO and included a watershed science report that estimated the scope of tourism emissions for the first time, provided an overview of the regional impacts and key vulnerabilities, and explored implications for tourism demand. The Declaration highlighted a sector pledge to develop a roadmap to become 'climate-neutral' and concluded that climate change '. . .must be considered the greatest challenge to the sustainability of tourism in the 21st century' (Scott et al. 2008, p. 38). A decade and a half later, the response has not matched this rhetoric (Scott & Gössling, 2022c). The signatories of the 2021 *Glasgow Declaration: A Commitment to a Decade of Tourism Climate Action* (One Planet, 2021) reaffirmed the salience of climate action and pledged to cut tourism emissions in half by 2030 and achieve net-zero emissions before 2050 (despite no discernable effect on the trajectory of tourism emissions or progress on developing a roadmap to become climate-neutral). The WTTC (2015, p. 5) emphasised that, 'The next 20 years will be characterized by our sector fully integrating climate change and related issues into business strategy, supporting the global transition to a low carbon economy, strengthening resilience at a local level against climate risks. . .' . Nonetheless, interviews with global tourism leaders demonstrated that many had not internalised the magnitude of carbon and climate risks to the sector or how the political imperatives of the moment have changed (Gössling & Scott, 2018).

Disconnects also persist between tourism and climate policy at national scales. While the tourism industry expects government to take the lead on climate change, an assessment of tourism sector policy in 18 countries found, 'The inescapable conclusion is that current [national tourism] policy [to adapt to climate change], with few exceptions, is inadequate to the scale of the challenge, both on mitigation and on adaptation' (OECD and UNEP, 2011, p. 4). A decade later, Becken et al.'s (2020) review of tourism and climate change policy documents from 61 countries reiterated this conclusion, finding little coherence and importantly, the consideration of climate change within tourism policy was shown to be weaker. Tourism is identified as a high priority in only 16 of the country Nationally Determined Contributions (NDC) submitted to the Paris Climate Agreement, and a priority in an additional 35 (Scott & Gössling, 2018). That many countries where tourism is most at risk to climate change and where tourism is a major component of the economy do not identify the tourism sector as a priority within their NDC submissions is disconcerting. Destination tourism plans have similarly been found to lack climate change risk recognition and adaptive response strategies (Freeman-Prince, 2021; Jarratt & Davies, 2020).

Loehr and Becken (2021) found important differences in climate change knowledge across academic, practitioner, and political domains, noting that availability, access, and transferability of knowledge were key barriers that must be addressed to enhance the impact of academic knowledge outputs on climate action. Interviews with global tourism leaders also emphasised the need for targeted decision relevant information for climate adaptation and emission reduction strategies at the destination scale, particularly in developing countries with high economic dependency on tourism (Gössling & Scott, 2018).

The information requirements to inform effective policy changes and investments to advance the transition to climate resilient tourism development are enormous and compels a priority research agenda for tourism scholars. Addressing the many key information gaps identified in this chapter and elsewhere (see Scott and Gössling (2022a) for recent review) and regional information asymmetry are prerequisites for evidence-based collective sectoral climate action and must be a priority for the tourism research community throughout the 2020s. Positively, there has been tremendous recent growth in climate change and tourism research in the 2010s. A bibliometric analysis by Scott and Gössling (2022a) found 49 percent of publications occurred between 2016–2020, and after the Paris Agreement was signed. While this rapid expansion at first appears impressive, it is important to put it in context, as these publications represent only 3.5 percent of papers published in the top five tourism journals between 2011 and 2020. We, as a scholarly community, continue to dedicate a meager proportion of our capacity to understand and respond to what is potentially the greatest influence on our sector in the next 30 years (and beyond). Going forward, our scale of research activity has to become commensurate to the grand challenge of the climate crisis.

By declaring a climate emergency and pledging strong support of the Paris Climate Agreement climate goals, the tourism sector is answerable for a commensurate action agenda. Tourism cannot afford to remain one of the global sectors least prepared for climate change if it is to have a meaningful place in the post-carbon, climate resilient economy of the future.

References

Amelung, B., Nicholls, S., & Viner, D. (2007). Implications of global climate change for tourism flows and seasonality. *Journal of Travel Research*, *45*(3), 285–296. https://doi.org/10.1177/0047287506295937

Becken, S. (2019). Decarbonising tourism: mission impossible? *Tourism Recreation Research*, *44*(4), 419–433. https://doi.org/10.1080/02508281.2019.1598042

Becken, S., Whittlesea, E., Loehr, J., & Scott, D. (2020). Tourism and climate change: Evaluating the extent of policy integration. *Journal of Sustainable Tourism*, *28*(10), 1603–1624. https://doi.org/10.1080/09669582.2020.1745217

Berrang-Ford, L., Siders, A. R., Lesnikowski, A., Fischer, A. P., Callaghan, M. W., Haddaway, N. R., Mach, K. J., Araos, M., Shah, M. A. R., Wannewitz, M., & Doshi, D. (2021). A systematic global stocktake of evidence on human adaptation to climate change. *Nature Climate Change*, *11*, 989–1000. https://doi.org/10.1038/s41558-021-01170-y

Bigano, A., Hamilton, J., & Tol, R (2006). The impact of climate on holiday destination choice. *Climatic Change*, *76*, 389–406. https://doi.org/10.1007/s10584-005-9015-0

Causevic S., & Lynch P. (2013). Political (in)stability and its influence on tourism development. *Tourism Management*, *34*, 145–157. https://doi.org/10.1016/j.tourman.2012.04.006

Climate Emergency Declaration (2021). *Climate emergency declarations in 2033 jurisdictions and local governments cover 1 billion citizens.* https://climateemergencydeclaration.org/climate-emergency-declarations-cover-15-million-citizens/

Conefrey, A., & Hanrahan, D. J. (2022). Comparative analysis of national tourism decarbonisation plans. *European Journal of Tourism Research*, *31*, 3105. https://doi.org/10.54055/ejtr.v31i.1979

Dawson, J., & Scott, D. (2013). Managing for climate change in the alpine ski sector. *Tourism Management*, *35*, 244–254. https://doi.org/10.1016/j.tourman.2012.07.009

Freeman-Prince, N. (2021). *An evaluation of national tourism plans in the Caribbean-SIDS region in the context of climate change* [Doctoral dissertation, University of Waterloo]. UWSpace. http://hdl.handle.net/10012/17552

Gössling, S., & Peeters, P. (2015). Assessing tourism's global environmental impact 1900–2050. *Journal of Sustainable Tourism*, *23*(5), 639–659. https://doi.org/10.1080/09669582.2015.1008500

Gössling, S., & Scott, D. (2018). The decarbonisation impasse: Global tourism leaders' views on climate change mitigation. *Journal of Sustainable Tourism*, *26*(12), 2071–2086. https://doi.org/10.1080/09669582.2018.1529770

Gössling, S., Scott, D., & Hall, C. M. (2013). Challenges of tourism in a low-carbon economy. *WIRES Climate Change*, *4*(6), 525–538. https://doi.org/10.1002/wcc.243

Gössling, S., Scott, D., & Hall, C. M. (2020). Pandemics, tourism and global change: a rapid assessment of COVID-19. *Journal of Sustainable Tourism*, *29*(1), 1–20. https://doi.org/10.1080/09669582.2020.1758708

Gössling, S., Scott, D., Hall, C. M., Ceron, J. P., & Dubois, G. (2012). Consumer behaviour and demand response of tourists to climate change. *Annals of Tourism Research*, *39*(1), 36–58. https://doi.org/10.1016/j.annals.2011.11.002

Hall, C. M. (2010). Crisis events in tourism: Subjects of crisis in tourism. *Current Issues in Tourism*, *13*(5), 401–417. https://doi.org/10.1080/13683500.2010.491900

Hall, C. M., Gössling, S., & Scott, D. (2020). Pandemics, transformations and tourism: Be careful what you wish for. *Tourism Geographies*, *22*(3), 577–598. https://doi.org/10.1080/14616688.2020.1759131

Hall, C. M., Timothy, D., & Duval, D. (2004). Security and tourism: Towards a new understanding? *Journal of Travel and Tourism Marketing*, *15*(2–3), 1–18. https://doi.org/10.1300/J073v15n02_01

International Financial Reporting Standards Foundation (IFRS). (2021). *IFRS foundation announces international sustainability standards board, consolidation with CDSB and VRF, and publication of prototype disclosure requirements.* https://www.ifrs.org/news-and-events/news/2021/11/ifrs-foundation-announces-issb-consolidation-with-cdsb-vrf-publication-of-prototypes/

IPCC (2021). Summary for policymakers. In *Climate change 2021: The physical science basis. Contribution of Working Group I to the Sixth assessment report of the Intergovernmental Panel on Climate Change.* https://www.ipcc.ch/report/ar6/wg1/#FullReport

IPCC (2022a). Summary for policymakers. In *Climate change 2021: Impacts, adaptation and vulnerability. Contribution of Working Group 2 to the Sixth assessment report of the Intergovernmental Panel on Climate Change.* https://www.ipcc.ch/report/ar6/wg2/

IPCC (2022b). Summary for policymakers. In *Climate change 2021: Mitigation of climate change. Contribution of Working Group 3 to the Sixth assessment report of the Intergovernmental Panel on Climate Change.* https://www.ipcc.ch/report/ar6/wg3/downloads/report/IPCC_AR6_WGIII_SPM.pdf

Jarratt, D., & Davies, N. J. (2020). Planning for climate change impacts: Coastal tourism destination resilience policies. *Tourism Planning & Development*, *17*(4), 423–440. https://doi.org/10.1080/21568316.2019.1667861

Kaján, E., & Saarinen, J. (2013). Tourism, climate change and adaptation: A review. *Current Issues in Tourism*, *16*(2), 167–195. https://doi.org/10.1080/13683500.2013.774323

Kohli, A., Steinemann, M., Denisov, N., & Droz, S. (2018). *Fragility and conflict.* Climate Change & Environment Nexus Brief. Swiss Agency for Development and Cooperation: Bern, Switzerland.

KPMG. (2008). *Climate changes your business.* http://www.kpmg.nl/sustainability.

Lenzen, M., Sun, Y. Y., Faturay, F., Ting, Y. P., Geschke, A., & Malik, A. (2018). The carbon footprint of global tourism. *Nature Climate Change*, *8*(6), 522–528. https://doi.org/10.1038/s41558-018-0141-x

Loehr, J. (2020). The Vanuatu Tourism Adaptation System: a holistic approach to reducing climate risk, *Journal of Sustainable Tourism*, *28*(4), 515–534. https://doi.org/10.1080/09669582.2019.1683185

Loehr, J., & Becken, S. (2021). The tourism climate change knowledge system. *Annals of Tourism Research*, *86*, 103073. https://doi.org/10.1016/j.annals.2020.103073

Ma, S., Craig, C., Scott, D., & Feng, S. (2021). Global climate resources for camping and nature-based tourism. *Tourism and Hospitality*, *2*(4), 365–379. https://doi.org/10.3390/tourhosp2040024

Mach, K. J., Kraan, C. M., Adger, W. N., Buhaug, H., Burke, M., Fearon, J. D., Field, C. B., Hendrix, C. S., Maystadt, J. F., O'Loughlin, J., & Roessler, P. (2019). Climate as a risk factor for armed conflict. *Nature*, *571*, 193–197. https://doi.org/10.1038/s41586-019-1300-6

Mora, C., Spirandelli, D., Franklin, E. C., Lynham, J., Kantar, M. B., Miles, W., Smith, C. Z., Freel, K., Moy, J., Louis, L. V., & Barba, E. W. (2018). Broad threat to humanity from cumulative climate hazards intensified by greenhouse gas emissions. *Nature Climate Change*, *8*(12), 1062–1071. https://doi.org/10.1038/s41558-018-0315-6

National Oceanic and Atmospheric Administration (2022). *Trends in atmospheric carbon dioxide.* https://gml.noaa.gov/ccgg/trends/weekly.html

One Planet (2021). *Glasgow declaration.* https://www.oneplanetnetwork.org/sustainable-tourism/glasgow-declaration

Organisation for Economic Co-operation and Development (OECD) and United Nations Environment Programme (UNEP). (2011). *Climate change and tourism policy in OECD countries.* OECD.

Organisation for Economic Co-operation and Development (OECD). (2015). *The opportunities and challenges of greener growth: Getting the whole policy package right.* https://www.oecd.org/greengrowth/opportunities-challenges-greener-growth.htm

Scott, D. (2021). Sustainable tourism and the grand challenge of climate change. *Sustainability, 13*(4), 1966. https://doi.org/10.3390/su13041966

Scott, D., Amelung, B., Ceron, J. P., Dubois, G., Gössling, S., Peeters, P., & Simpson, M. C. (2008). *Climate change and tourism: Responding to global challenges*. World Tourism Organization (UNWTO), United Nations Environment Programme (UNEP) and World Meteorological Organization.

Scott, D., & Becken, S. (2010). Adapting to climate change and climate policy: Progress, problems and potentials. *Journal of Sustainable Tourism, 18*(3), 283–295. https://doi.org/10.1080/09669581003668540

Scott, D., & Gössling, S. (2015). What could the next 40 years hold for global tourism? *Tourism Recreation Review, 40*(3), 269–285. https://doi.org/10.1080/02508281.2015.1075739

Scott, D., & Gössling, S. (2018). *Tourism and climate change mitigation: Embracing the Paris Agreement – Pathways to decarbonization*. European Travel Commission.

Scott, D., & Gössling, S. (2022a). A review of research into tourism and climate change - Launching the Annals of Tourism Research curated collection on tourism and climate change. *Annals of Tourism Research, 95*, 103409. doi.org/10.1016/j.annals.2022.103409

Scott, D., & Gössling, S. (2022b). Destination net-zero: What does the international energy agency roadmap mean for tourism? *Journal of Sustainable Tourism, 30*(1), 14–41. https://doi.org/10.1080/09669582.2021.1962890

Scott, D., & Gössling, S. (2022c). From Djerba to Glasgow: Have declarations on tourism and climate change brought us any closer to meaningful climate action? *Journal of Sustainable Tourism, 30*(1), 199–222. https://doi.org/10.1080/09669582.2021.2009488

Scott, D., Gössling, S., & Hall, C. M. (2012a). International tourism and climate change. *WIRES Climate Change, 3*(3), 213–232. https://doi.org/10.1002/wcc.165

Scott, D., Gössling, S., Hall, C. M., & Peeters, P. (2015a). Can tourism be part of the decarbonized global economy? The costs and risks of alternate carbon reduction policy pathways. *Journal of Sustainable Tourism, 24*(1). 52–72. https://doi.org/10.1080/09669582.2015.1107080

Scott, D., Hall, C. M., & Gössling, S. (2012b). *Tourism and climate change: Impacts, adaptation and mitigation*. Routledge.

Scott, D., Hall, C. M., & Gössling, S. (2015b). A review of the IPCC 5th Assessment and implications for tourism sector climate resilience and decarbonization. *Journal of Sustainable Touris, 24*(1), 8–30. https://doi.org/10.1080/09669582.2015.1062021

Scott, D., Hall, C. M., & Gössling, S. (2019). Global tourism vulnerability to climate change. *Annals of Tourism Research, 77*, 49–61. https://doi.org/10.1016/j.annals.2019.05.007

Scott, D., & Jones, B. (2006). *Climate change and seasonality in Canadian outdoor recreation and tourism - executive summary*. Report prepared for the Government of Canada Climate Change Action Fund. University of Waterloo.

Scott, D., Knowles, N., & Steiger, R. (2022). Is snowmaking climate change maladaptation? A multi-criteria Analysis. *Journal of Sustainable Tourism*. https://doi.org/10.1080/09669582.2022.2137729

Steiger, R., Scott, D., Abegg, B., Pons, M., & Aall, C. (2019). A critical review of climate change risk for ski tourism. *Current Issues in Tourism*, 22(11), 1343–1379. https://doi.org/10.1080/13683500.2017.1410110

Tourism Declares (2022). *Tourism declares a climate emergency*. https://www.tourismdeclares.com/

United Nations (UN). (2015). *United Nations treaty collection chapter XXVII environment, 7.d Paris Agreement, Paris, 12 December 2015* (C.N.92.201). https://unfccc.int/sites/default/files/english_paris_agreement.pdf

United Nations (UN). (2021). *IPCC report: 'code red' for human driven global heating, warns UN Chief*. https://news.un.org/en/story/2021/08/1097362

United Nations Environment Programme (UNEP). (2021). Emissions gap report 2021: The heat is on. https://wedocs.unep.org/bitstream/handle/20.500.11822/37350/AddEGR21.pdf

United Nations World Tourism Organization (UNWTO). (2011). *Tourism towards 2030: Global overview*. UNWTO General Assembly, 19th session, Gyeongju, Republic of Korea.

World Bank (2013). *Turn down the heat: Climate extremes, regional impacts and the case for resilience*.

World Bank (2021a). When poverty meets climate change: A critical challenge that demands cross-cutting solutions. https://blogs.worldbank.org/climatechange/when-poverty-meets-climate-change-critical-challenge-demands-cross-cutting-solutions

World Bank (2021b). Groundswell part 2: Acting on internal climate migration. https://openknowledge.worldbank.org/handle/10986/36248

World Meteorological Organization (WMO). (2021). New climate predictions increase likelihood of temporarily reaching 1.5 °C in next 5 years. https://wmo.int/news/media-centre/new-climate-predictions-increase-likelihood-of-temporarily-reaching-15-degc-next-5-years

World Travel and Tourism Council (WTTC). (2010). *Climate change – a joint approach to addressing the challenge*. WTTC.

World Travel and Tourism Council (WTTC). (2015). *Travel & tourism 2015: Connecting global climate action.* https://www.hospitalitynet.org/file/152005907.pdf

Yu, D., Matthews, L., Scott, D., Li, S., & Guo, Z. Y. (2022) Climate suitability for tourism in China in an era of climate change: A multiscale analysis using a tourist defined holiday climate index. *Current Issues in Tourism*, *25*(14), 2269–2284. https://doi.org/10.1080/13683500.2021.1956442

Chapter 38

Tourism and Tourists in Nature Governance of National Parks and Wilderness in an Era of Global Change

Jarkko Saarinen & Aapo Lundén

Introduction

Natural environments are key resources and attraction elements for tourism development in many regions. People are aware of places such as Grand Canyon, Yellowstone, Serengeti, Halong Bay, and Great Barrier Reef - even if not visited, those iconic sites that are deeply integrated to the global tourism industry and its economic and cultural circulation. Although nature and wilderness without the elements of civilisation were historically environments to be feared and avoided for many Westerners (Nash, 1982; Shields, 1990; Short, 1991), an early recreational interest in natural settings emerged gradually in the 18th century by the upper classes to get away from crowded urban environments (Neumann, 1998; Löfgren, 1999). Later, the experience of the natural environment in tourism and recreation began to refer or represent a past connection between people and the land and serve as a valued link that has disappeared from contemporary urban and suburban life (Nash, 1982; Oelschlaeger, 1991).

Natural and wild environments are 'universally regarded as a source of pleasure' (Wang, 2000, p. 80) and are actively promoted as sites for tourists' consumption and associated activities (Job et al., 2017; Slocum, 2017; Saarinen, 2022). This use for the growth needs of the industry has created a variety of positive but also negative externalities for natural environments. The touristic value of nature has been used for the conservation initiatives of many iconic and other natural environments. Historically, the designation of Yellowstone National Park (USA) in 1872 represents one of the very early examples of symbiotic i.e., positive relationship between nature conservation and tourism (Runte, 1987; Eagles & McCool, 2002). The designation process was influenced by increasing human impacts on nature in the American frontier lands and, thus, aimed to reserve and withdraw the last remaining wild areas in the Yellowstone region 'from settlement, occupancy, or sale' (Nash, 1982, p. 108), with an explicit emphasis on public use in tourism and recreation (Hall, 1992; Saarinen, 2019).

The Wiley Blackwell Companion to Tourism, Second Edition. Edited by C. Michael Hall.

Since that, as Hall (2006, p. 213) has further argued, tourism has provided 'an economic justification to establish conservation areas, such as national parks'. In this respect the historical growth of visitation numbers to Yellowstone National Park, for example, demonstrates the importance and attraction value of nature in tourism. Furthermore, it also indicates the motivation for the industry to support specific nature conservation efforts: nature creates new and increasing opportunities for economic development in tourism. Indeed, in the case of Yellowstone, annual visitor numbers have grown from 19,575 in 1910 to 4,860,242 in 2021 (National Park Service 2022a). From the late 18th century onwards, the accelerating disappearance of wild places resulted in an increasing global scale need to protect remaining natural areas (Hannah et al., 1994; Sanderson et al., 2002; Watson et al., 2018), in which process the tourism industry has often played locally an important role (Hall, 1992; Buckley, 2011; Whitelaw et al., 2014; Leung et al., 2018).

In his seminal work, Budowski (1976) indicated that the relationship between tourism and nature can be characterised by the above kind of symbiotic co-existence. However, he also described the relation as a potentially disruptive and exploitative conflict situation (Fennell & Weaver, 2005). Indeed, the increasing demand and massive scale of tourism growth in natural areas have also brought conflicts and major challenges in the relationship between tourism and natural environments. While natural environments have emerged as a key resource for the tourism industry, nature and especially national parks and wilderness areas cannot absorb very high tourist numbers and intensive tourism practices without negative consequences for the environment and the experiences nature-based tourists are searching for (Saarinen, 2013; Karhu et al., 2020). This has led to attempts to regulate tourist numbers and the creation of various management procedures to limit the growth of tourism, but despite these attempts, the growth of tourism supported by its' economic and political value has caused inevitable negative impacts on natural environments in many places. According to Van der Duim and Caalders (2002, p. 746) the impacts of tourism have been a major focal point in the policy discussions focusing on tourism and nature, and the impacts are often perceived as conflicting with nature conservation (Sæþórsdóttir, 2004).

In order to reduce the negative and optimise the positive impacts of a growing industry, new modes and forms of tourism utilising nature have been developed in recent decades. These new forms aiming to balance the negative and positive impacts of tourism are referring to the idea of sustainable development in tourism (Hall, 2019; Saarinen, 2022) and terms such as ecotourism, green tourism, or responsible and smart tourism. However, at the same time there have emerged other forms of nature-based tourism, such as adventure, wildlife, and safari tourism, which may not always have explicit ideas or practices aiming towards sustainability. This chapter takes an overview of the significance and role of natural environments for tourism and tourists. The special focus is on tourism and tourists in national parks and wilderness contexts. In addition, the chapter discusses the changing modes and forms of tourism in nature and existing and emerging issues of conflicts and their management in nature-based tourism.

The Role of Nature: Tourism in National Parks and Wilderness Areas

Demand for the Wild

Nature-based tourism is one of the central sub-sectors of the tourism industry today (Winter et al., 2020). Historically, increased demand for tourists to experience nature has been based on the changing positive attitudes towards the environment in general. Thus, the development of better accessibility to remote natural areas has integrated natural areas more closely to global tourism markets and the main tourist-generating regions. Ideally, nature-based tourism relies on the attractiveness of relatively underdeveloped areas characterised by natural attractions and environments and the associated activities that can be pursued there (Newsome et al., 2013). According to Goodwin (1996, p. 287), nature-based tourism is a type of tourism that 'encompasses all forms of tourism – mass tourism, adventure tourism, low-impact tourism, ecotourism – which use natural resources in wild or underdeveloped form'. Occasionally it is linked to the idea of alternative tourism or sustainable tourism (Whelan, 1991; Burton, 1998). However, the element of sustainability is not strictly a condition for nature-based tourism utilising the natural environment as its resource base, and for some specific nature-based tourism activities, such as snowmobiling or hunting tourism, it could be a limiting factor. Furthermore, the level of 'naturalness' may vary greatly in between different nature-based tourism products for different customers (Sæþórsdóttir & Saarinen, 2016).

It is estimated that nature-based tourism is one of the most rapidly growing sectors of tourism (Wearing et al., 2010). While this may be the case, it is rather challenging to verify, as national or international tourism statistics do not separate nature-based tourism from other forms or environments of tourism. In addition, the definitions of nature-based tourism, such as the one given above, are difficult to operationalise in a statistical sense: 'wild or underdeveloped' are not objective criteria but vary between places and cultures. Like nature (Beck, 1992), ideas of wild and wilderness are culturally constructed (Nash, 1982; Saarinen, 2013). In addition, many definitions of nature-based tourism also involve motivational aspects (Goodwin, 1996), which are important for experience-based definitions but, again, very difficult to transfer into statistics aiming to demonstrate the intensity and growth of nature-based tourism on a national or international scale.

The growth estimates of nature-based tourism are often based on the development of use levels (visitations) of certain kinds of nature-based tourism attractions. In this respect, national parks and wilderness areas have formed typical points of reference for such estimations (Cessford & Muhar, 2003; Fredman & Margaryan, 2021). National parks are protected natural or near-natural areas set aside to protect large-scale ecological processes, along with the complement of species and ecosystems characteristic to the area, which also provide a foundation for environmentally and culturally compatible spiritual, scientific, educational, recreational and visitor opportunities (International Union for Conservation of Nature (IUCN), 1978). Recreational and visitor opportunities refer to the tourism potential which has been noted as significant in many places. In Finland, for example, total visitor numbers to national parks more than doubled in the 1990s and again in 2000s reaching almost a million visitations annually (Saarinen, 2005; Puhakka, 2008). By 2021 the visitation number of Finnish national parks reached over four million visits that created an economic impact of some 310 million Euros and 2,452 jobs in the areas that surround national parks based on visitors' spending (Metsähallitus, 2022). These economic impacts have over tripled based on spending estimates and over doubled based on local employment impacts in the decade (see Huhtala et al., 2010). Similarly, national park visitation numbers have experienced a substantial growth in Norway and Sweden, for example (see Fredman & Margaryan, 2021).

However, tourists' interest is not evenly distributed between different areas: some national parks are receiving a much higher than average level of growth, while others may have more stable or even declining visitor numbers. In addition, different regions and countries are facing varying levels of growth. While Nordic and Southern African countries, for example, have reported increasing visitor numbers since the 1990s (see Sæþórsdóttir, 2004; Saarinen, 2005;

Saayman & Saayman, 2006; Fredman & Tyrväinen, 2010; Swedish Environmental Protection Agency (SEPA), 2019), many of the North American national parks and wilderness areas experienced much more stable or even decreasing interest (Table 38.1). In many cases in the United States, the growth of nature-based tourism in national parks occurred long before the 1990s and it stabilised in the 1990s and 2000s till the mid-2010s when many US national parks experienced a new growth in visitor numbers. Therefore, the estimated and promoted global growth of nature-based tourism should not be generalised automatically on a global basis. In addition to the geographical dimension, the touristic use of national parks and wilderness areas is also often highly seasonal, with intensive peak times and slow periods.

In recent years, the COVID-19 pandemic has also changed the demand for nature-based tourism. According to Fredman and Margaryan (2021) national parks in the Nordic countries have experienced a strong growth in visitation in 2020. Furthermore, they state that mobile tracking data from Norway indicated a 291% increase in outdoor recreational activity near the city of Oslo during the lockdown. For them, these kinds of drastic growth trends reflect not only the limited possibilities for international travel due to COVID-19 restrictions, but also 'a renewed interest towards domestic nature-based tourism and outdoor recreation' (Fredman & Margaryan, 2021, p. 14) calling for a new research agenda.

Indeed, it seems that nature-based tourism is quite resilient towards external changes and compared to urban areas nature perhaps represented a safer and, thus, an attractive environment in the early phase of the pandemic. However, compared to the Nordic context, COVID-19 had an opposite impact in the visitation levels of national parks in the US, in general (Table 38.1). While there has been a new major increase in the visitation levels of many large national parks (e.g., Glacier, Grand Canyon, Grand Teton, Rocky Mountain and Yellowstone National Parks) since the mid-2010s, most of the parks experienced a major decline in visitations in 2020. Still, the drop seems to be short-term as most of the US national parks returned to a growth path in the following year (National Park Service, 2022a). Nevertheless, these kinds of differences in responses towards external shocks and pressures call for a nuanced, policy-informed and context specific analysis in the future demand and trends of nature-based tourism.

Experiencing the Wild

Why do people go to national parks and other natural environments? Are the reasons still related to sublime, enlightening experiences and nostalgic escapism from urban environments? Based on the COVID-19 impacts on the visitation levels in the Nordic countries, for example, there seem to be some level of escapism involved from the urban crowds. However, based on international empirical studies, the spectrum of tourists' motivations to visit nature is much broader (Uysal et al., 1994; Holden & Sparrowhawk, 2002). The motivations are largely based on experiences and activities that natural settings can provide in general. Kaplan and Kaplan (1989), for example, have reported that experiencing nature and its different aspects are of key interest to people who visit natural environments. It is difficult to disagree with such an obvious conclusion. Thus, elements such as experiencing peace and quiet and natural beauty, wildlife viewing, physical exercise and spiritual encounters are among the main motivations for nature-based tourism.

In addition, studies have indicated many social reasons (e.g. shared time and experiences with family or friends) and heritage and identity issues (e.g. visiting nationally or personally important sites) to be important motivations for nature tourists. Wearing et al. (2010, p. 80)

Table 38.1. Development of visitor numbers in 1950–2020 and 2021 in selected National Parks in the United States.

National Park	*1950*	*1960*	*1970*	*1980*	*1990*	*2000*	*2010*	*2020*	*2021*
Acadia	485,220	1,638,200	2,776,300	2,779,666	2,339,591	2,469,238	2,504,208	2,669,034	4,069,098
Arches	16,257	71,600	178,500	290,519	620,719	786,429	1,014,405	1,238,083	1,806,865
Badlands	447,654	878,600	1,303,100	952,652	1,326,475	1,105,824	977,778	916,932	1,224,226
Bryce Canyon	212,976	272,000	345,900	571,541	862,659	1,099,275	1,285,492	1,464,655	2,104,600
Capitol Reef	3,066	102,500	225,900	342,788	562,477	612,656	662,661	981,038	1,405,353
Carlsbad Caverns	467,283	537,000	712,700	672,963	747,016	469,303	428,524	183,835	349,244
Crater Lake	310,796	397,700	535,000	455,143	384,941	426,883	448,319	670,500	647,751
Death Valley	189,045	355,900	580,500	618,140	690,965	1,179,094	984,775	820,023	1,146,551
Glacier	482,298	724,500	1,241,600	1,474,578	1,986,737	1,728,693	2,200,048	1,698,864	3,081,656
Grand Canyon	665,039	1,187,700	2,258,200	2,304,973	3,776,685	4,460,228	4,388,386	2,897,098	4,532,677
Grand Teton	189,286	1,429,900	3,352,500	2,555,627	1,588,253	2,590,624	2,669,374	3,289,638	3,885,230
Great Smoky Mountains	1,843,620	4,528,600	6,778,500	8,440,953	8,151,769	10,175,812	9,463,538	12,095,720	14,161,548
Hawaii Volcanoes	362,245	709,100	822,300	1,692,338	1,096,816	1,514,636	1,304,667	589,775	1,262,747
Hot Springs	294,955	719,100	2,092,400	1,160,588	1,123,175	1,338,156	1,311,807	1,348,215	2,162,884
Mammoth Cave	254,187	519,100	1,726,500	1,495,787	1,924,538	1,749,268	497,225	290,392	515,774
Mesa Verde	88,184	225,700	527,200	539,257	611,375	452,287	559,712	287,477	548,477
Mount Rainier	573,685	1,538,700	1,925,100	1,268,256	1,327,101	1,344,833	1,191,754	1,160,754	1,670,063
Olympic	404,125	1,160,400	2,283,100	2,032,418	2,794,903	3,327,722	2,844,563	2,499,177	2,718,925
Petrified Forest	352,889	911,500	1,151,400	683,121	844,592	605,192	664,725	384,483	590,334
Rocky Mountain	1,275,160	1,532,500	2,357,900	2,641,937	2,647,323	3,185,392	2,955,821	3,305,199	4,434,848
Saguaro	36,220	141,000	351,800	611,317	702,328	765,195	717,614	762,226	1,079,786
Sequoia	379,218	610,800	875,500	862,397	1,063,538	838,947	1,002,979	796,086	1,059,548
Shenandoah	1,279,387	1,780,100	2,411,500	1,699,228	1,771,780	1,419,579	1,253,386	1,666,265	1,592,312
Yellowstone	1,110,524	1,443,300	2,297,300	2,000,269	2,823,572	2,838,233	3,640,185	3,806,306	4,860,242
Yosemite	820,953	1,150,400	2,277,200	2,490,282	3,124,939	3,400,903	3,901,408	2,268,313	3,287,595
Zion	323,402	575,800	903,600	1,123,846	2,102,400	2,432,348	2,665,972	3,591,254	5,039,835

Source: Adapted from National Park Service, 2022a.

indicate that natural areas can 'represent, and are living embodiments of, the past, contributing to a sense of continuity and identity'. van der Merwe and Saayman's (2008) study of the travel motives of tourists visiting the Kruger National Park in South Africa, for example, identified six motives: nature; activities; attractions; nostalgia; novelty; and escape. Thus, nature-based tourism experience by definition focuses on nature, but it may go beyond to involve some other touristic experiences and dimensions. van der Merwe and Saayman (2008) also concluded that while there were commonalities with other similar studies conducted elsewhere internationally, 'different attractions and destinations feed different motives to travel. Thus, although nature or natural environments can be seen as a single attraction category for tourism, they can involve cultural dimensions, for example, among other elements' (van der Merwe & Saayman, 2008, p. 158). This further highlights the importance of contextuality in understanding motivations and demand in nature-based tourism.

In addition to national parks, one specific and culturally loaded environment for nature-based tourism is the wilderness. The International Union for Conservation of Nature (IUCN) has defined wilderness as protected areas that are usually large, unmodified, or slightly modified areas, retaining their natural character and influence, without permanent or significant human habitation, which are protected and managed so as to preserve their natural condition (Category Ib) (IUCN, 1978; see also IUCN/WCMC, 1994). This kind of definition, emphasising untouched and uncivilised characteristics of wilderness environments, are linked to the first wilderness legislation passed in the United States in 1964 (*The Wilderness Act*), which states that wilderness areas are places that provide primitive and unconfined recreational opportunities and experiences and a sense of remoteness, peace, challenge and escapism, and the areas should offer 'outstanding opportunities for solitude' (The Wilderness Act, 1964, p. 1).

Wilderness tourism is considered an independent form of tourism, consisting of activities such as backpacking and hiking (Saarinen, 2022). However, since the 1990s the growth of wilderness tourism has also been based on a rise in organised forms of nature-based and adventure tourism (Cloke & Perkins, 1998; Buckley, 2006; Sæþórsdóttir, 2010). Obviously, as nature-based tourist activities have changed and visitor numbers have grown, it has become increasingly challenging to provide 'primitive and unconfined recreational opportunities' for visitors by the wilderness area managers (Hendee et al., 1990; Higham, 1998). This is noted to cause conflicts among wilderness users and negative changes to wilderness experiences (see Vaske et al., 1986; Buckley, 1999; Manning, 1999; Hall & Boyd, 2005; Sæþórsdóttir & Hall, 2020).

Nature-based Tourism Industry

Old and New Modes of Nature-based Tourism

In tourism, any given environment, natural or urban, is not a static but constantly changing resource with various combinations of social, political and economic relationships that are specific in space and time. Thus, not even the attractions of nature are static and 'out there' waiting to be discovered and admired: in tourism they are our and others' constructions, dependent not only on individual perspectives but also changing societal contexts (Saarinen, 2005), i.e. what is seen and valued by nature-based tourists are often cultural projections created by tourism and modernisation in general (Pedersen & Viken, 1996). In tourism, nature and its specific elements have become products with certain qualities attached to particular places. For example, MacCannell (1992) describes Yosemite National Park (USA) as a human product, a piece of nature that is marked off, interpreted and museumised for the purposes of visitors and society.

In addition to being a source and force of change to nature, nature-based tourism has itself transformed. Current high visitor numbers in national parks and many wilderness areas are based on individual but also increasingly packaged, or organised tourism (Buckley, 1999). This represents a somewhat opposite trend to discussions about so called new tourism (Poon, 1993), which is said to indicate a shift or a step away from mass tourism towards individual and alternative travel, in general (Urry, 1990). While there has been a 'massification' of nature-based tourism in many destinations, which has resulted crowding and displacement of visitor segments searching for solitude and primitive nature experiences (Manning, 1999; Moyle & Croy, 2007; Fefer et al., 2021), the new tourism has also influenced how nature is produced and packaged for customers. In this respect the organised nature-based tourism can involve increasing possibilities of flexibility and hybridity. For example, international tour operators have created a product that links the Arctic Circle, Rovaniemi (Finland), and its winter activities and wilderness experiences with the tropical seaside resort of Phuket, Thailand, in the same 12-day package holiday (Hall et al., 2009).

The new tourism is supposed to be more innovative post-Fordist style of tourist production (Urry, 1990). Some of these innovations are strongly based on the new kind of uses of nature in tourism and the growth of the so-called 'experience economy' (Pine & Gilmore, 1999). The experience economy aims to interlink space, time and human experiences together in systems in which individuals seek to perform a wide and new variety of activities (Saarinen, 2004) involving excitement, exercise and speed (see Ryan et al., 2000; Taylor et al., 2013; Fletcher, 2014). One of the core forms of these new modes of nature-based tourism with high emphasis on activity levels and experiences is related to adventure tourism (see Gyimothy & Mykletun, 2004; Buckley, 2006). For some, adventure tourism is seen as an active recreational practice which requires deep involvement: 'being, doing and seeing' (Cloke & Perkins, 1998, p. 189) rather than 'touristic staying'. Thus, it resonates strongly with the mental characteristics of nature-based tourism, and according to Swarbrooke et al. (2003) adventure tourism is situated at the cutting edge of shifts in tourism production and consumption. This underlines the key role of nature-based tourism in the over-all tourism industry.

Evolving Conflicts and Governance Needs in Local-global Nexus

The establishment and management of national parks and wilderness areas has created better possibilities for the public to visit them. While this has provided new possibilities for nature-based tourism activities, the growing levels of use have also started to pose a threat to ecological bases of protected areas and their wilderness characteristics. Although tourism has historically benefited the establishment of national parks and wilderness areas (Butler & Boyd, 2000; Frost & Hall, 2010), the current situation is that increasing and transforming modes of nature-based tourism are targeting areas and places that exist primarily for nature conservation and protection purposes. In addition to potential challenges between new and increasing touristic pressures and biodiversity conservation, the growth of tourism can conflict with nature experiences focusing on solitude and an unconfined type of recreation (Higham, 1998; Saarinen, 2013). The increasing number of visitors require more infrastructure, services and environmental modifications, for example, which may be regarded negatively by other user groups that prefer opportunities for solitude and the experience of unregulated and wild settings. In some cases, this may lead to a displacement process, i.e. a replacement of former users and user types by new modes of nature-based tourists and tourism products (see Anderson & Brown, 1984; Vaske et al., 1986).

In this respect, the increasing numbers of nature-based tourism will also bring new kinds of tourist groups with new sets of motivations, activities, and user norms (see Patterson & Hammitt, 1990; Roggenbuck et al., 1993; Saarinen, 2022). This increased attractiveness and numbers of on-site visitors, along with diminishing wild or natural areas, creates substantial challenges for the governance and management needs of these areas (see Butler, 1996; Butler, 2010; Lindberg et al., 1997). In addition to tourism, various other localised human activities are also seen to be increasingly 'filtering in' to protected areas (see Duffy, 2015; Hall, 1992).

Recently, however, research needs and efforts beyond a single on-site or destination scale analysis have emerged in a conceptual level with an emphasis on 'a more global scale' focus in the research of tourism-environment nexus (Hall, 2022; see also Bramwell & Lane, 2008, Saarinen, 2018). There are many dimensions in global tourism-environment relations, but on one hand, it is about how the ambivalent relationship between tourism and nature has turned the adaptive capacity of the tourism system towards global environmental change, especially climate change. The management of tourism and protected areas is likely to become more complex due to rapid environmental change. For example, in 2022, Yellowstone National Park was closed to visitors because of record rainfalls, flooding, and mudslides in the beginning of the peak summer season, also closing tourism dependent hotels and restaurants in the park entrance areas (Traub et al., 2022). Moreover, due to climate change and changing ecological processes, protected area governance requires new tools and new thinking: in the era of the Anthropocene, both direct and indirect human activities have become the dominant forces on the environment (Rockström & Klum, 2012; Moore, 2015). Indeed, due to the global change processes, distant wilderness areas are increasingly perceived to be threatened, disappearing and, thus, unable to support the integrity of their 'wild' ecosystems (Saarinen, 2019).

Therefore, questions as to how tourism 'contributes' to that environmental change and what kind of policy structures and elements of control and change (adjustments) are required are emerging at local and global scales. One closely related indication of this is the discussion on new environmental ethics for tourism (Macbeth, 2005; Fennell, 2006; Jamal, 2019) with less tourism-centric views in development and planning thinking and practices. However, as noted by Buckley (2012) empirical examinations and management processes on tourism impacts and environmental relations in a global scale are rare. One reason for this is probably the challenge in collecting, scaling and analysing such system level phenomenon operating in and between local and global at the same time. In addition, the conceptual understanding of tourism-environment relations or the implications of 'truly' sustainable development thinking in the global tourism system are still highly contested.

Conclusions: Theoretical and Methodological Reflections on Growing Complexities of Nature-based Tourism

The link between tourism and nature is almost as old as the history of nature conservation (Frost & Hall, 2010), and in the past decades the tourism industry has become a very significant user and powerful stakeholder in the use of nature conservation areas. In this respect, nature-based tourism can provide positive contributions to environment and nature conservation, but it can also create negative impacts. Indeed, the relationship between tourism and nature has always been characterised by a certain level of contradictions and ambivalence, both supporting symbiotic and conflicting connections at the same time. Initially, the conflicting element were largely managed based on the idea of carrying capacity, which was abandoned in the 1980s and turned towards the idea of sustainable tourism in the 1990s (Butler, 1996; Butler, 2010).

While sustainability is a holistic and future-oriented perspective on the use of ecological, sociocultural, and economic resources, in tourism it has mostly focused on local-scale issues and management challenges. What appears to be clear in the local scale uses of nature in tourism is that in order for there to be areas that provide primitive and unconfined nature-based tourism opportunities and experiences in the future, there is a requirement for critical discussions and evaluations about the production modes of nature-based tourism. Key governance issues are how much and what kind of tourism taking place in natural environments we allow and support in future as some places are already ecologically endangered and socially in a process of becoming 'non-nature'.

In addition, the development policies, structures, and management models that aim to safeguard natural environments in tourism uses in a local-global nexus need to be re-evaluated and clarified. In other words, there is a need to shift or expand the focus of sustainability and environmental concerns in tourism from destinations to a global tourism system scale but also further develop planning processes, tools, and indicators for site specific, local scale management purposes in nature-based tourism development. All this calls for new theoretical and methodological perspectives, or a revival of previous approaches, in tourism studies which obviously are easier to plea than produce, especially when considering currently hegemonic modes of tourism-centric thinking in development policies and plans in many places and parts of the world.

To secure the natural resource, the impacts of rapid ecological changes in protected areas requires these changes more central integration to protected area management. For example, in the US, the National Park Service's new strategy 'Resist-Accept-Direct' (Schuurman et al., 2020) is an attempt to respond to the recent impacts of climate change in iconic national parks such as Yellowstone, Yosemite and Joshua Tree in the USA. The report provides new management guidelines for park managers as a response to the impacts of climate change and irreversible ecological change in American national parks to 'consider strategic, forward-looking actions, rather than structure management goals based on past conditions', in a reality where 'the past is no longer the guide it used to be' (National Park Service, 2022b). These ecological changes are forcing park managers to a new kind of governance 'triage' over resource allocation to decide what and where resources should be invested under rapid ecological change. The implications of such management efforts are still largely unknown for nature-based tourism; however, they indicate radical changes in the management of some of the most iconic national parks in the USA.

On a wider sense, nature-based tourism and tourism in more generable needs to be integrated into multiscale sustainability issues (Hall, 2022) and provide more actionable policies (Joppe, 2018). This entails aligning the use of natural areas in tourism with sustainability goals that are based on maintaining resilience of natural environments through governance instead of maintaining the status quo (see Agrawal et al., 2022) of the tourism industry. In addition to changing the management practices of protected areas, integration of tourism into broader sustainability goals is likely to require the re-evaluation of the tourism industry's scale and weight in justifying protected areas beyond their use as a touristic resource. Therefore, keeping the ideas of the limits to growth, carrying capacity and regulation on the table is crucial when considering the future relationship between tourism and protected areas.

References

Agrawal, A., Brandhorst, S., Jain, M., Liao, C., Pradhan, N., & Solomon, D. (2022). From environmental governance to governance for sustainability. *One Earth*, *5*(6), 615 621. https://doi.org/10.1016/j.oneear.2022.05.014

Anderson, D. H., & Brown, P. J. (1984). The displacement process in recreation. *Journal of Leisure Research*, *16*(1), 61–73. https://doi.org/10.1080/00222216.1984.11969573

Beck, U. (1992). *Risk society, towards a new modernity*. Sage.

Bramwell, B., & Lane, B. (2008). Priorities in sustainable tourism research. *Journal of Sustainable Tourism* *16*(1), 1–4. https://doi.org/10.2167/09669580803489612

Buckley, R. (1999, May 23–27). Wilderness in Australia. In S. F. McCool, D. N. Cole, W. T. Borrie & J. O'Loughlin (Eds), *Wilderness science in a time of change conference - Volume 2. Proceedings RMRS-P-15-VOL-2* (pp. 190–193). U.S. Department of Agriculture, Forest Service, Rocky Mountain Research Station.

Buckley, R. (2006). *Adventure tourism*. Oxfordshire: CABI.

Buckley, R. (2011). Tourism and Environment. *Annual Review of Environment and Resources*. *36*, 397–416. https://doi.org/10.1146/annurev-environ-041210-132637

Buckley, R. (2012). Sustainable tourism: research and reality. *Annals of Tourism Research* 39: 528–546. doi: 10.1016/j.annals.2012.02.003

Budowski, G. (1976). Tourism and environmental conservation: Conflict, coexistence, or symbiosis? *Environmental Conservation*, *3*(1), 27–31. https://doi.org/10.1017/S0376892900017707

Burton, R. (1998). Maintaining the quality of ecotourism: Ecotour operators' responses to tourism growth. *Journal of Sustainable Tourism*, *6*(2), 117–142. https://doi.org/10.1080/09669589808667306

Butler, R. (1996). The concept of carrying capacity for tourism destinations: Dead or merely buried? *Progress in Tourism and Hospitality Research*, *2*(3–4), 283–293. https://doi.org/10.1002/pth.6070020309

Butler, R. (2010). Carrying capacity in tourism. In D. Pearce & R. Butler (Eds), *Tourism research: A 20-20 vision* (pp. 53–64). Goodfellow Publishers.

Butler, R., & Boyd, S. (eds). (2000). *Tourism and national parks: Issues and implications*. Wiley & Sons.

Cessford, G., & Muhar, A. (2003). Monitoring options for visitor numbers in national parks and natural areas. *Journal for Nature Conservation*, *11*(4), 240–250. https://doi.org/10.1078/1617-1381-00055

Cloke, P., & Perkins, H. (1998). Cracking the canyon with the awesome foursome. *Environment and Planning D: Society and Space*, *16*, 185–218. https://doi.org/doi:10.1068/d160185

Duffy, R. (2015). Nature-based tourism and neoliberalism: concealing contradictions. *Tourism Geographies*, *17*(4), 529–543. https://doi.org/10.1080/14616688.2015.1053972

Eagles, P., & McCool, S. (2002). *Tourism in national parks and protected areas: planning and management*. CABI.

Fefer, J. P., Hallo, J. C., Collins, R. H., Baldwin, E. D., & Brownlee, M. T. (2021). From displaced to misplaced: Exploring the experience of visitors who were 'crowded out' of their recreation destination. *Leisure Sciences*. https://doi.org/10.1080/01490400.2021.1898497

Fennell, D. (2006). *Tourism ethics*. Channel View.

Fennell, D., & Weaver, D. (2005). The ecotourism concept and tourism-conservation symbiosis. *Journal of Sustainable Tourism*, *13*(4), 373–390. https://doi.org/10.1080/09669580508668563

Fletcher, R. (2014). *Romancing the wild: Cultural dimensions of ecotourism*. Duke University Press.

Fredman, P., & Margaryan, L. (2021). 20 years of Nordic nature-based tourism research: a review and future research agenda. *Scandinavian Journal of Hospitality and Tourism*, *21*(1), 14–25. https://doi.org/10.1080/15022250.2020.1823247

Fredman, P., & Tyrväinen, L. (2010). Frontiers in nature-based tourism. *Scandinavian Journal of Hospitality and Tourism*, *10*(3), 177–189. https://doi.org/10.1080/15022250.2010.502365

Frost, W., & Hall, C. M. (Eds). (2010). *Tourism and national parks*. Routledge.

Goodwin, H. (1996). In pursuit of ecotourism. *Biodiversity and Conservation*, *5*, 277–291. https://doi.org/10.1007/BF00051774

Gyimothy, S., & Mykletun, R. (2004). Play in adventure tourism: The case of arctic trekking. *Annals of Tourism Research*, *31*(4), 855–878. https://doi.org/10.1016/j.annals.2004.03.005

Hall, C. M., & Boyd, S. (Eds). (2005). *Nature-based tourism in peripheral areas: Development or disaster?* Channel View.

Hall, C. M. (1992). *Wasteland to world heritage: Preserving Australia's wilderness*. Melbourne University Press.

Hall, C. M. (2006). Tourism, biodiversity and global environmental change. In S. Gössling & C. M. Hall (Eds), *Tourism and global environmental change*, (pp. 211–226). Routledge.

Hall, C. M. (2019). Constructing sustainable tourism development: The 2030 agenda and the managerial ecology of sustainable tourism. *Journal of Sustainable Tourism*, *27*(7), 1044–1060. https://doi.org/10.1080/09669582.2018.1560456

Hall, C. M. (2022). Sustainable tourism beyond BAU (Brundtlund as usual): Shifting from paradoxical to relational thinking. *Frontiers in Sustainable Tourism*, *1*, 927–946. https://doi.org/10.3389/frsut.2022.927946

Hall, C. M., Müller, D., & Saarinen, J. (2009). *Nordic tourism: Cases and issues*. Channel View.

Hannah, L., Lohse, D., Hutchinson, C., Carr, J. L., & Lankerani, A. (1994). A preliminary inventory of human disturbance of world ecosystems. *Ambio*, *23*(4–5), 246–250. https://www.jstor.org/stable/4314213

Hendee, J., Stankey, G., & Lucas, B. (1990). *Wilderness management*. Fulcrum.

Higham, J. (1998). Sustaining the physical and social dimensions of wilderness tourism: the perceptual approach to wilderness management in New Zealand. *Journal of Sustainable Tourism*, *6*(1), 26–51. https://doi.org/10.1080/09669589808667300

Holden, A., & Sparrowhawk, J. (2002). Understanding the motivations of ecotourists: The case of trekkers in Annapurna, Nepal. *International Journal of Tourism Research*, *4*(6), 435–446. https://doi.org/10.1002/jtr.402

Huhtala, M., Kajala, L., & Vatanen, E. (2010). Local economic impacts of national park visitors' spending: The development process of an estimation method. *Metlan työraportteja* 149. http://urn.fi/URN:ISBN:978-951-40-2224-1

International Union for Conservation of Nature (IUCN). (1978). *Categories, objectives and criteria for protected areas*. Commission in National Parks and Protected Areas, Committee on Criteria and Nomenclature.

International Union for Conservation of Nature (IUCN)/World Conservation Monitoring Centre (WCMC). (1994). *Guidelines for protected area management categories*.

Jamal, T. (2019). *Justice and ethics in tourism*. Routledge.

Job, H., Becken, S., & Lane, B. (2017). Protected areas in a neoliberal world and the role of tourism in supporting conservation and sustainable development: an assessment of strategic planning, zoning, impact monitoring, and tourism management at natural World Heritage Sites. *Journal of Sustainable Tourism*, *25*(12), 1697–1718. https://doi.org/10.1080/09669582.2017.1377432

Joppe, M. (2018). Tourism policy and governance: Quo vadis? *Tourism Management Perspectives*, *25*, 201–204. https://doi.org/10.1016/j.tmp.2017.11.011

Kaplan, R., & Kaplan, S. (1989). *The experience of nature: A psychological perspective*. Cambridge University Press.

Karhu, J., Lähteenmäki, M., Ilmolahti, O., & Osipov, A. (2020). From threat to opportunity: sustainability and tourism in Koli National Park. *Tourism Geographies*. https://doi.org/10.1080/14616688.2020.1812112

Leung, Y. F., Spenceley, A., Hvenegaard, G., Buckley, R., & Groves, C. (2018). *Tourism and visitor management in protected areas: Guidelines for sustainability* (Vol. 27). IUCN.

Lindberg, K., McCool, S., & Stankey, G. (1997). Rethinking carrying capacity. *Annals of Tourism Research*, *24*(2), 461–465.

Löfgren, O. (1999). *On holiday: A history of vacationing*. University of California Press.

Macbeth, J. (2005). Towards an ethics platform for tourism. *Annals of Tourism Research*, *32*(4), 962–984. https://doi.org/10.1016/j.annals.2004.11.005

MacCannell, D. (1992). *Empty meeting ground: the tourist papers*. Routledge.

Manning, R. E. (1999). Crowding and carrying capacity in outdoor recreation: From normative standards to standards of quality. In E. L. Jackson & T. L. Burton (Eds), *Leisure studies: Prospects for the twenty-first century* (pp. 329). Venture Publishing.

Metsähallitus (2022). *Kansallispuistojen, valtion retkeilyalueiden ja eräiden muiden suojelu- ja virkisty-salueiden kävijöiden rahankäytön paikallistaloudelliset vaikutukset vuonna 2021*. (National parks, public hiking areas, and few other protected and recreational areas visitors economic impact on local economy in 2021). https://www.metsa.fi/wp-content/uploads/2022/03/Ptvaikutukset_2021.pdf

Moore, A. (2015). Tourism in the Anthropocene Park? New analytic possibilities. *International Journal of Tourism Anthropology*, *4*(2), 186–200. https://www.inderscienceonline.com/doi/abs/10.1504/IJTA.2015.070067

Moyle, B., & Croy, G. (2007). Crowding and visitor satisfaction during the off-season: Port Campbell National Park. *Annals of Leisure Research*, *10*(3–4), 518–531. https://doi.org/10.1080/11745398.2007.9686779

Nash, R. (1982). *Wilderness and the American mind*. Yale University Press.

National Park Service (2022a). *National Park Service visitor use statistics*. https://irma.nps.gov/STATS/Reports/Park/

National Park Service (2022b). *Resist-Accept-Direct (RAD). A Framework for the 21st-century natural resource manager*. www.nps.gov/subjects/climatechange/radframework.htm

Neumann, R. (1998). *Imposing wilderness: Struggles over livelihood and nature preservation in Africa*. University of California Press.

Newsome, D., Moore, S. A., & Dowling, R. K. (2013). *Natural area tourism. Ecology, impacts and management*. Channel View.

Oelschlaeger, M. (1991). *The idea of wilderness: from prehistory to the age of ecology*. Yale University Press.

Patterson, M., & Hammitt, W. (1990). Backcountry encounter norms, actual reported encounters, and their relationship to wilderness solitude. *Journal of Leisure Research*, *22*(3), 259–275. https://doi.org/10.1080/00222216.1990.11969829

Pedersen, K., & Viken, A. (1996). From Sami nomadism to global tourism. In M. F. Price (Ed.), *People and tourism in fragile environments* (pp. 69–88). John Wiley & Sons.

Pine, J., & Gilmore, J. (1999). *The experience economy: work is theatre and every business a stage: goods & services are no longer enough*. Harvard Business School Press.

Poon, A. (1993). *Tourism, technology and competitive strategies*. CABI.

Puhakka, R. (2008). Increasing role of tourism in Finnish national parks. *Fennia*, *186*(1), 47–58. https://fennia.journal.fi/article/view/3711

Rockström, J., & Klum, M. (2012). *The human quest – Prospering within planetary boundaries*. Bokförlaget Langenskiöld.

Roggenbuck, J., Williams, D., & Watson, A. (1993). Defining acceptable conditions in wilderness. *Environmental Management*, *17*, 187–197. https://doi.org/10.1007/BF02394689

Runte, A. (1987). *National parks: The American experience*. University of Nebraska Press.

Ryan, C., Hughes, K., & Chirgwin, S. (2000). The gaze, spectacle and ecotourism. *Annals of Tourism Research*, *27*(1), 148–163. https://doi.org/10.1016/S0160-7383(99)00061-4

Saarinen, J. (2004). Tourism and touristic representations of nature. In A. Lew, C. M. Hall & A. Williams (Eds), *Geography: A companion to tourism* (pp. 438–449). Blackwell.

Saarinen, J. (2005). Tourism in northern wildernesses: Nature-based tourism development in Northern Finland. In C. M. Hall & S. Boyd (Eds), *Nature-based tourism in peripheral areas: Development or disaster?* (pp. 36–49). Channel View.

Saarinen, J. (2013). 'Tourism into the wild': The limits of tourism in wilderness. In A. Holden. & D. Fennell (Eds), *The Routledge handbook of tourism and the environment* (pp. 145–154). Routledge.

Saarinen, J. (2018). Beyond growth thinking: The need to revisit sustainable development in tourism. *Tourism Geographies*, *20*(2), 337–340. https://doi.org/10.1080/14616688.2018.1434817

Saarinen, J. (2019). What are wilderness areas for? Tourism and political ecologies of wilderness uses and management in the Anthropocene. *Journal of Sustainable Tourism*, *27*(4), 472–487. https://doi.org/10.1080/09669582.2018.1456543

Saarinen, J. (2022). Wilderness tourism: Nature-based tourist experiences in wild places. In R. Sharpley (Ed.), *Handbook of tourist experiences* (pp. 521–532). London.

Saayman, M., & Saayman, A. (2006). Estimating the economic contribution of visitor spending in the Kruger National Park to the regional economy. *Journal of Sustainable Tourism*, *14*(1), 67–81. https://doi.org/10.1080/09669580608668592

Sæþórsdóttir, A. D. (2004). Adapting to change: Maintaining a wilderness experience in a popular tourist destination. *Tourism Today*, *4*, 52–65. https://www.cothm.ac.cy/_files/ugd/79301e_88714efe4d0d45eba23d5b130c62cb9f.pdf#page=53

Sæþórsdóttir, A. D. (2010). Planning nature tourism in Iceland based on tourist attitudes. *Tourism Geographies*, *12*(1), 25–52. https://doi.org/10.1080/14616680903493639

Sæþórsdóttir, A. D., & Hall, C. M. (2020). Visitor satisfaction in wilderness in times of overtourism: A longitudinal study. *Journal of Sustainable Tourism*, *29*(1), 123–141. https://doi.org/10.1080/09669582.2020.1817050

Sæþórsdóttir, A. D., & Saarinen, J. (2016). Changing ideas about natural resources: Tourists' perspectives on the wilderness and power production in Iceland. *Scandinavian Journal of Hospitality and Tourism*, *16*(4), 404–421. https://doi.org/10.1080/15022250.2015.1108866

Sanderson, E. W., Jaiteh, M., Levy, M. A., Redford, K. H., Wannebo, A. V., & Woolmer, G. (2002). The human footprint and the last of the wild: The human footprint is a global map of human influence on the land surface, which suggests that human beings are stewards of nature, whether we like it or not. *BioScience*, *52*(10), 891–904. https://doi.org/10.1641/0006-3568(2002)052[0891:THFATL]2.0.CO;2

Schuurman, G. W., Hawkins Hoffman, C., Cole, D. N., Lawrence, D. J., Morton, J. M., Magness, D. R., Cravens, A. E., Covington, S., O'Malley, R., & Fisichelli, N. A. (2020). *Resist–Accept–Direct (RAD): A framework for the 21st-century natural resource manager*. National Park Service. Natural resource report no. NPS/NRSS/CCRP/NRR—2020/2213.

Shields, R. (1990). The 'system of pleasure': Liminality and Carnevalesque at Brighton. *Theory, Culture & Society*, *7*(1), 39–72. https://doi.org/10.1177/026327690007001002

Short, J. (1991). *Imagined country: Society, culture and environment*. Routledge.

Slocum, S. L. (2017). Operationalising both sustainability and neo-liberalism in protected areas: implications from the USA's National Park Service's evolving experiences and challenges. *Journal of Sustainable Tourism*, *25*(12), 1848–1864. https://doi.org/10.1080/09669582.2016.1260574

Swarbrooke, J., Beard, C., Leckie, S., & Pomfret, G. (2003). *Adventure tourism: The new frontier*. Butterworth-Heinemann.

Swedish Environmental Protection Agency (SEPA). (2019). *National parks in Sweden*. Link no longer available.

Taylor, S., Varley, P., & Johnston, T. (Eds). (2013). *Adventure tourism: Meaning, experience and learning*. Routledge.

The Wilderness Act 1964. Pl (ss. 88–577).

Traub, A., Albeck-Ripka, L., & Hauser, C. (2022). Yellowstone Park, staggered by rain and floods, will stay closed for days. *The New York Times*. https://www.nytimes.com/2022/06/14/us/yellowstone-rain-flooding.html

Urry, J. (1990). *The tourist gaze: Leisure and travel in contemporary societies*. Sage.

Uysal, M., McDonald, G., & Martin, B. (1994). Australian visitors to US national parks and natural areas. *International Journal of Contemporary Hospitality Management*, *6*(3), 18–24.

Van der Duim, R., & Caalders, J. (2002). Biodiversity and tourism: impacts and interventions. *Annals of Tourism Research*, *29*(3), 743–761. https://doi.org/10.1016/S0160-7383(01)00087-1

van der Merwe, P., & Saayman, M. (2008). Travel motivations of tourists visiting Kruger National Park. *Koedoe*, *50*(1), 154–159. https://hdl.handle.net/10520/EJC139689

Vaske, J., Shelby, B., Graefe, A., & Heberlein, A. (1986). Backcountry encounter norms: Theory, method and empirical evidence. *Journal of Leisure Research*, *18*(3), 137–153. https://doi.org/10.1080/00222216.1986.11969653

Wang, N. (2000). *Tourism and modernity*. Pergamon.

Watson, J. E., Venter, O., Lee, J., Jones, K. R., Robinson, J. G., Possingham, H. P., & Allan, J. R. (2018). Protect the last of the wild. *Nature*, *563*, 27–30. https://doi.org/10.1038/d41586-018-07183-6

Wearing, S., Stevenson D., & Young, T. (2010). *Tourist cultures*. Sage.

Whelan, T. (Ed.). (1991). *Nature tourism: managing for the environment*. Island Press.

Whitelaw, P. A., King, B. E., & Tolkach, D. (2014). Protected areas, conservation and tourism financing the sustainable dream. *Journal of Sustainable Tourism*, *22*(4), 584–603. https://doi.org/10.1080/09669582.2013.873445

Winter, P. L., Selin, S., Cerveny, L., & Bricker, K. (2020). Outdoor recreation, nature-based tourism, and sustainability. *Sustainability*, *12*(1), 81. https://doi.org/10.3390/su12010081

Chapter 39

Tourism and the Precautionary Principle in Theory and Practice

David A. Fennell

Introduction

Over the past 70 years the tourism industry has been marked by change, innovation, and development. Increased foreign revenue, higher levels of income and employment, as well as greater public sector revenues, have been attractive forces catalysing governments to develop new destinations (Archer, 1996). Unfortunately, such growth has often been conceived as short-term financial gain, without due regard for long-term environmental or socio-cultural implications. Hence, the paradox of tourism has been revealed: economic gain at the expense of the natural world and local identity and traditional cultures (Convention on Biological Diversity (CBD), United Nations Environment Programme (UNEP), 2001). In recognising these problems, there has been a sustained call for better planning and management within the tourism industry, at all levels (Inskeep, 1991).

One planning instrument which has received considerable attention is the precautionary principle; a concept which has provided guidance on debates regarding health and safety, as well as environmental and resource management issues. 'Precaution' is often applied in circumstances where chemicals have potentially toxic or bio-accumulative effects, and where usage could lead to serious physical harm on humans or the environment. It has thus become an increasingly powerful mechanism for environmental groups to amass political and public support. Intuitively, precaution appeals to our sense of controlling risks and detrimental outcomes, designed to address scientific uncertainty in areas where failure to act may lead to future harm or disaster (Kaiser, 1997). Just like sustainable development, precaution puts the onus on the present population to address current actions which may incite potential risks and detrimental consequences for future generations. However, although examined in detail in the aforementioned fields, it has still received little attention in the realm of tourism. With this in mind, the aim of this chapter is to:

- provide a brief review of literature on the precautionary principle;
- explore the fundamental concepts which underlie the precautionary principle; and
- discuss its applicability to tourism.

The Wiley Blackwell Companion to Tourism, Second Edition. Edited by C. Michael Hall.

The Precautionary Principle

O'Riordan and Cameron (1994, p. 12) define the precautionary principle as 'a culturally framed concept that takes its cue from changing social conceptions about the appropriate roles of science, economics, ethics, politics and the law in pro-active environmental protection and management'. In this regard, precaution has been extended to include six basic concepts:

- preventative anticipation;
- safeguarding ecological space;
- restraint adopted is not unduly costly;
- duty of care, or onus of proof on those who propose change;
- promotion of the cause of intrinsic natural rights; and
- paying for past ecological debt.

More concisely, precaution is grounded in the need for a 'premium on a cautious and conservative approach to human interventions in environmental sectors that are:

- usually short on scientific understanding; and
- usually susceptible to significant injury, especially irreversible injury' (Myers, 1993, p. 74).

VanderZwaag (1994, p. 7) writes that there are a number of core elements associated with the precautionary principle, including:

- a willingness to take action (or no action) in advance of formal scientific proof;
- cost effectiveness of action, that is, some consideration of proportionality of costs;
- providing ecological margins of error;
- intrinsic value of non-human entities;
- a shift in the onus of proof to those who propose change;
- concern with future generations; and
- paying for ecological debts through strict/absolute liability regimes.

The precautionary principle was conceived in Germany (*Vorsorgeprinzip,* meaning precautionary principle) during the 1970s for the purpose of exercising foresight in matters of environmental policy and resource protection (see Boehmer-Christiansen, 1994). It was introduced internationally in 1984 at the First International Conference on Protection of the North Sea (Tickner & Raffensberger, 1998). Since then, the principle has been extended into national and international environmental policy by more than 40 countries, and is now affirmed in many international treaties and laws (e.g., 1990 Bergen Declaration; 1992 Rio Declaration; 1992 Maastricht Treaty on European Union; The Convention on International Trade in Endangered Species of Wild Flora and Fauna (Freestone & Hey, 1996; Rogers et al., 1997; Dickson, 1999; Ellis, 2000; Tapper, 2001)). In 1992, the precautionary principle was incorporated in the Earth Summit in Rio de Janeiro. Principle 15 states that:

> In order to protect the environment, the precautionary approach shall be widely applied by States according to their capabilities. Where there are threats of serious or irreversible damage, lack of full scientific certainty shall not be used as a reason for postponing cost-effective measures to prevent environmental degradation (Van Dyke, 1996, p. 10).

Responding to this document, VanderZwaag (1999) feels that phrases such as 'according to their capabilities' are wide open to interpretation. He also questions how threats are to be determined; the role of scientific assessments; who will make such determinations; and to what extent should economic costs be weighed against environmental benefits. There is also the question of how principles are refined into practice, in much the same way decision-makers have grappled with how to operationalise the principles of sustainable development.

Precaution appears to have more support in Europe, where it forms the basis of environmental law and policy in several European states (e.g., United Kingdom, Denmark, and Sweden). In North America, Canada has a long-standing history of implementing the precautionary approach in science-based programmes of health, safety and natural resources protection (Government of Canada, 2001a). For example, discussions on the value of precaution have been taken up in regard to oceans policy. Canada's Oceans Act now requires the Minister of Fisheries and Oceans to develop an oceans management strategy based on:

- sustainable development;
- integrated management; and
- the precautionary approach (as cited in VanderZwaag, 1999).

Although the Canadian government does not consider the precautionary principle to be a rule of customary international law, there may be sufficient state practice to allow a good argument for the principle's induction into international law (Government of Canada, 2001b). In the United States, Tickner and Raffensberger (1998) note that the precautionary principle is a relatively new concept, although the general principle of precaution underpins much legislation.

Tourism and Precaution

Tourism has consistently been shown to have an impact on air and water quality, erode soils, create noise pollution, expand the built environment, increase transport networks, disrupt species behaviour in any number of ways, and dislocate human communities – socially, politically, and economically. There are myriad examples where tourism has been instituted in an ad hoc fashion, and with little regard to appropriate socio-ecological planning. A classic example of poor development is Cancun, Mexico, where ineffective sewage management has polluted beaches, natural habitat has been reduced, economic benefits are unevenly distributed, changes in lifestyles and traditions occur, and increased competition for resources exists (see Daltabuit & Pi-Sunyer, 1990).

The numerous ecological impacts stemming from tourism are likely to increase in number and intensity as domestic and international travel increases. The World Tourism Organization's (WTO) Tourism 2020 Vision forecast predicts that travellers of the 21st century will go farther and farther, and that by 2020 one out of every three trips will be a long-haul journey to another region of the world (WTO, 2001). Such future tourist activities are forecasted to be most intense in 'unspoilt' natural areas and remoter places, taking people to the most ecologically fragile parts of the earth (Holden, 2000; Martin, 2000).

Tourism development, especially when it occurs in regions unaccustomed to the industry (e.g., Antarctica), holds many uncertainties and unknown impacts. Many of these areas are particularly sensitive to change - physically distinctive areas that are extremely vulnerable to increased human impact and environmental change due to tourism. deFur and Kaszuba

(2002) consider the precautionary principle to be an invaluable tool when policy makers are forced to make decisions with little or no experience or history to draw from. Thus, the precautionary principle may be most applicable in these areas of new tourism development, situations in which it may be especially difficult to predict and prove the full range of consequences. Adhering to the sustainability philosophy, the precautionary principle could be proactive in assessing impacts accruing from tourism while acknowledging the uncertainty inherent in these systems, instead of trying to modify systems to cope with resulting impacts as they occur.

Hall (1995) and Faulkner (1998) argue that conventional Newtonian/Cartesian approaches to tourism research are more in step with studying relatively stable systems, resulting in an inadequate understanding of the dynamics of change and chaotic phases of tourism development. In contrast to these traditional reductionist approaches to research on tourism destination development, Russell and Faulkner (1999) suggest that chaos and complexity theories provide a sound alternative perspective in recognising that systems are innately complex (i.e., non-linear), unstable, and dynamic/life-like. Tourism is an integrated system in which many elements are linked (e.g., environment protection, economic viability), thus changes in one element affect other elements (Swarbrooke, 1999). In line with these theories, the precautionary principle also acknowledges the possibility of change, instability, and uncertainty in systems.

There is still a dearth of literature on tourism and the precautionary principle. Many texts mention the concept, but most do so only in passing or as it relates to some broader concept like sustainability. Tribe et al. (2000), mention precautionary principle as a guiding principle to effective policy development, along with conservation of resources, improvement of environmental quality, preventing environmental damage, the polluter pays principle, and incentive-based policies. A book by Kirstges (1995) also mentions precaution as one of several principles that should be followed by tour operators. Kirstges (1995) observes that there should be an environmental audit for all tourism developments that cause negative impacts. (See also Jennings (2003), in the context of sport and adventure tourism; Throsby (2009), in the context of cultural heritage; Butler (2011), in the context of island tourism; and Soleimanpour (2012), in reference to environmental law and nature-based tourism.)

Two areas in tourism research that have focused on the precautionary principle include cetaceans and Antarctica/Arctic. In the former case, scholars have advocated precautionary principle for the better management of whale and dolphin tourism. In the absence of solid data on the real effects of boat noise and other related disturbances on cetaceans, a precautionary approach is suggested (see Garrod & Fennell 2004; Lusseau et al., 2006; Martinez & Orams, 2011). In regard to the latter, scholars argue that better management techniques, including the precautionary principle, are needed to regulate tourism activities in Arctic and Antarctic regions (see Scott, 2001; Bastmeijer & Roura, 2004; Stewart & Draper, 2006).

One of the most noteworthy applications of the precautionary principle and tourism is by Gössling (2001) in his work on sustainable water use in Zanzibar. He observes that tourism development has placed a significant level of pressure on the water resources of Zanzibar, including the lowering of the groundwater table, deteriorating water quality, and saltwater intrusion. Gössling (2001) recommends a precautionary approach, where water consumption would be reduced to 200 litres of water per bed space as compared to daily average consumption levels of 2,000 to 3,000 litres. Hall (2011) has used the precautionary principle in detail with reference to the management of tourism-related biological invasions, and Holden (1999) argues that the precautionary principle has a role to play in policies designed to limit the effects of downhill skiing at Cairngorm.

Precaution also appears to be frequently cited in tourism policy documents of various organisations. For example, The Wilderness Society (1999) of Australia, in their Tourism and Natural Areas Policy document, makes reference to precaution, under Policy 2:

> 2. Provision of visitor access to natural areas must not compromise or infringe on the environmental qualities of the area, or the normal and desired routine of local communities. It will be determined largely by the visitor carrying capacity of an area or the limits of acceptable change. Where difficulties are encountered in determining visitor carrying capacity, the precautionary principle should apply.

The World Wildlife Fund (WWF) (2002), in their tourism principles and aims, also makes reference to the precautionary principle, as follows:

> WWF will promote in particular the precautionary principle; the polluter pays principle; economic instruments; minimum standards; and environmentally sound technologies, especially in sustainable means of travel to reduce fuel consumption and pollution emissions.
>
> - *work with the tourism industry, governments and others to support the development of national and regional sustainable tourism policies.*

The WWF (2001) believes that action must be taken to reduce and, where possible, eliminate negative impacts on natural resources and processes. These actions include limiting tourism-related pollution so as not to exceed ecological carrying capacity (i.e. the robustness of habitats and their ability to replenish extracted resources), including waste assimilation processes. Tourism-related pollution and exploitation must therefore be carefully controlled and regulated, and the precautionary approach should be considered a fundamental principle in tourism development. The British Columbia Wilderness Tourism Association (2001), in their Draft Code of Practices for BC's Wilderness Tourism Operators, makes the following reference to the precautionary principle, as the last of 22 statements: 'Follow the ideal of the precautionary principle: When in doubt - Don't!' Finally, the Convention on Biological Diversity (CBD UNEP, 2001) has established International Guidelines for Sustainable Tourism, for the purpose of assisting stakeholders at all levels in sustainable management. The report indicates that the act of decision-making 'should be a transparent and accountable process to approve or refuse a proposal, and it should always apply the precautionary principle' (CBD UNEP, 2001, p. 21).

Given that none of these environmentally based organisations, above, fully articulate the conceptual basis of the precautionary principle suggests that there is either a tacit understanding of precaution, or perhaps a lack of understanding of how it might be infused in tourism decision-making. One could argue that principles such as low impact, sustainability, local control, and responsibility are implicit applications of the precautionary principle. In reality, however, what is needed is a more explicit understanding of how it applies in a tourism context. As such, the potential for integrating precaution into tourism planning has yet to be examined.

However, in the context of urban planning and development, Counsell (1999) found that proponents of a weak sustainable development see the precautionary principle as an obstacle and threat to urban planning and development, while those who maintain a strong sustainable development perspective view it as necessary. Howie (2003) argues that the precautionary principle has been criticised because it is thought to counter the spirit of entrepreneurialism, under the belief that risk is often good for business. In a study of town planning in five regions of England and Wales, Counsell (1999) found that respondents (various community stakeholders involved in planning) were divided over the use of the precautionary principle, with many suggesting that it has been misapplied, with others considering that it has no place in planning. The author attributes this to the notion that engaging with socio-economic themes and principles related to sustainable development pose many problems in a transition towards sustainable development. Counsell (1999) concludes by observing that, in a planning context, the precautionary principle has a role to play in safeguarding areas that are inviolable, for ensuring that the overall quality of the environment is maintained.

The regard for assets and integrity holds true for communities in general. For example, Rogers et al. (1997) identify a number of cases where local people have been willing to spend time and money defending their natural environments from damage from large developments, because they recognise that the benefits they receive far out weigh the costs of defence. Such attempts provide further credence to the notion that communities need not sit idly waiting for others to decide their fate. This means that for tourism to be truly representative of a broad number of stakeholder groups, the natural environment included, many hard questions need to be addressed in regard to appropriate tourism planning, development, and management, and how precaution may guide better decision-making. Some of these questions include:

- What in the community will be sacrificed for tourism development?
- What are the anticipated direct and indirect social, economic, and ecological impacts?
- Who inside and outside the community has been consulted, over what period of time?
- Who will be compensated for loss and how? Is the possibility for loss built into the proposal?
- What is the political and industry receptivity to the precautionary principle?
- How can precaution be built into these existing structures or visa versa?
- How can scientific data be more accessible to the public?
- Who has the knowledge to effectively plan with the interests of the community in mind?

Many of these questions can be addressed through the precautionary decision-making framework initially proposed at the Wingspread Conference attended by scientists, academics, policy makers, and environmental advocates (adapted here from Tickner & Raffensberger, 1998), and adapted in work specific to tourism by Fennell and Ebert (2004). Proposed steps for incorporating the precautionary principle for better tourism industry decision-making include:

1. *Define the general duty to take precautionary action.* This involves the adoption of a corporate or industry wide duty to take precautionary action in the face of scientific uncertainty where there is a threat to human health or the environment. The concept of human health could be expanded to include an assessment of how tourism developments have impacts on the ecology and customs of local communities.
2. *Set aggressive goals/vision for achieving sustainability (backcasting).* This step involves the establishment of clear and measurable goals from which to drive innovative best practices within the tourism industry.
3. *Assume responsibility for demonstrating the safety of products and processes.* Tourism industry stakeholders involved in the planning, development and management of the tourism industry must demonstrate the safety of their operations before engaging in such activities.

4. *Create criteria for decision-making under uncertainty.* Indicators of sustainability and other such tools will need to be employed for the purpose of determining how to assess, and what type of evidence to weigh in assessing, impacts.
5. *Use tools for implementing precautionary, preventative approaches.* There are numerous tools for carrying out precautionary policies related to the provision of services (e.g., environmental management systems).
6. *Use the 'polluter pays' principle.* Offending parties must pay the costs of the damage they cause. One mechanism is assurance bonding. Companies are required to pay a premium before undertaking a project, which is based on the worst potential damage that might occur from development. If no damage occurs, the bond is returned to the developer.
7. *Develop a scheme to systematically evaluate alternative activities, technologies, chemicals, etc.* In order to prevent an impact while creating another, developers must be careful that the substitutes they may use are not more harmful than the original product. This becomes especially salient in a tourism context where tourism developments have life cycles of many years.
8. *Assume a duty to monitor, understand, investigate, inform, and act.* Tourism businesses have an obligation to investigate and understand their potential impacts, continuously. This calls for more science to understand how developments impact people, sites, communities, and regions. Companies should be responsible for periodic assessments and audits of their initiatives, over the long-term.
9. *Employ participative corporate decision-making.* Just as many perspectives enrich decision-making within a firm, tourism industry development decisions must be open to those who are often affected by the initiative. This means involvement by the development firm, governments, community members, and so on.

The extent to which these nine factors are taken into account in is still an open question. Wildlife tourism appears to be the one sector where the precautionary principle has an established foothold, but generally in theory only. Examples include manta ray tourism in Ningaloo Marine Park (Venables et al., 2016), and the ethics of food provisioning of whale sharks by Ziegler et al. (2019). Ziegler et al. (2019) argue that the burden to prove that provisioning does not negatively impact sharks should be on the shoulders of proponents of this practice. In support of the framework discussed above, Tickle and von Essen (2020) argue that the precautionary principle is an essential tool to navigate the responsibilities that tourists and operators have in protecting communities form unethical hunting practices. The setting of hunting standards and practices is consistent with the precautionary principle in preventing several 'sins' and upholding hunter ethics and responsibility. The precautionary principle is one of seven principles for an ecocentric approach to wildlife tourism (Burns, et al., 2011). This principle (Principle 4), states that: 'If a wildlife tourism action has a suspected risk of causing harm to animals or their habitat, in the absence of scientific consensus that the action is harmful, then the burden of proof that it is not harmful falls on those proposing the action' (Burns et al., 2011, p. 185). New research on akrasia is providing inroads to understanding why tourists and service providers have a weakness of will in choosing to participate in these activities even though they know that such participation is morally wrong (Fennell, 2015).

The precautionary principle has also been referenced in discussions on the impacts of covid-19. Martin (2021) argues that the decision to leave open or close ski resorts in many different regions around the world became less a function of the precautionary principle and more about political decisions that elevated individual and organisational interests. Political and economic manoeuvring, therefore, has considerable influence on decisions that have broad social and ecological consequences.

Conclusion

There is general consensus pointing to the fact that precaution is a tool that is here to stay (VanderZwaag, 1999). The breadth of discussion on precaution and its inclusion as a principle in many international conventions, serves notice that it holds potential in standing up to the many uncertainties which exist in human-environment relationships. However, like sustainability, there appears to be a void between what industry, environmentalists, and governments want.

It is understandable that the precautionary principle has led to a backlash from industry, because it accentuates the process of pulling back the reins on unfettered growth (Howie, 2003). So, while no caution is dangerous, too much caution may be equally counterproductive. Opponents suggest that there is a fear that if taken too far in the other direction, science will have no role to play in qualifying the usefulness of certain products (see Cohen, 2001). Scientific proof has thus become a burden and a barrier in the protection of the environment and people. At the same time however, there must be the realisation that market forces, left on their own, cannot alone run economies. Slowly, and with little respect for time (time to ask the right questions, to follow process, and to consider impacts over longer periods), there has been a steady erosion of a market culture, which is no longer respectful of the rule of law. As such, corporate actions have significant effects on quality of life, particularly on Least Developed Countries (LDCs) (O'Riordan & Cameron, 1994).

Tourism has been criticised for failing to adopt practices aimed at achieving sustainability (McCool & Moisey, 2001; Swarbrooke, 1999), even though there are several different mechanisms at hand to meet these challenges, including regulations, codes of conduct, action plans, and so on. In relation to these other mechanisms, the precautionary principle is underrepresented in the literature and in practice. Precaution holds promise as a planning tool that actualises the imperative of sustainability, effectively managing tourism in a more proactive, future-focused manner and acknowledging the uncertainty inherent in tourism-related development and activities. Since tourism development is continuously stretching into less populated and more pristine environments, science is often unable to provide data or causal links connecting action to harm in these new, unique areas. Thus, the precautionary principle can be employed as a decision-making tool within tourism development for the purpose of safeguarding natural environments and securing human wellbeing. The proposed framework discussed earlier for incorporating the precautionary principle into better decision-making for the tourism industry illustrates the relevance and potential applicability of the principle.

One of the main drawbacks of the precautionary approach is that it does not directly offer explanations about how it should be applied. There are questions about: Who will implement? How will they implement? What level of expertise is required to implement? This final question is especially important. After all, one need only look at the legacy of tourism to question process and end. What passes for 'highly trained' and 'expert' has too often reduced complex systems or entities into individual pieces for the purpose of making decisions, without any effort to ask essential questions about the whole. In the absence of definitive expertise, politicians must still make decisions. They cannot sit idle and thus leave themselves open to criticism about not being active enough in policymaking, spending, and research and development. There comes a time for action, even though there is danger about the implications of such action. To politicians, it may be better to suffer criticism at a later date – when they may not be in power – than to suffer it at time when they are attempting to consolidate authority and popularity.

Another constraint to the implementation of precaution is that a precise definition is far from clear, even though it has acquired the standing of a political/moral norm (Kaiser, 1997). Thus, increased attention should be paid to what the precautionary principle actually means to tourism, and in particular, how it can be operationalised. Furthermore, the use of the precautionary principle for regulatory purposes is highly controversial. Some stakeholder's express trepidation in the misuse or abuse of the precautionary principle, as evidenced by the extensive recent debate on the concept. Certain groups feel that it could be applied to perceived risks for which there is no firm scientific foundation. Many European industries view the precautionary principle decreasingly as an acceptable risk management approach, and increasingly as a tool for the more radical environment and health advocates (Harremoës et al., 2001).

The foregoing suggests that we may never be able to fully understand and control the impacts that tourism developments have on social and ecological systems. There will be cases where the best way forward, or perhaps the only way forward, is by adopting a precautionary position. However, the adoption of such will no doubt be contentious in view of the interests of a broad array of different stakeholders – what they stand to gain and lose in balancing individual and organisational interests.

References

Archer, B. (1996). Sustainable tourism - Do economists really care? *Progress in Tourism and Hospitality Research*, *2*(3–4), 217–222. https://doi.org/10.1002/pth.6070020303

Bastmeijer, K., & Roura, R. (2004). Regulating Antarctic tourism and the precautionary principle. *The American Journal of International Law*, *98*(4), 763–781. https://doi.org/10.2307/3216699

Boehmer-Christiansen, S. (1994). The precautionary principle in Germany - enabling government. In T. O'Riordan & J. Cameron (Eds), *Interpreting the precautionary principle* (pp. 31–60). Earthscan.

British Columbia Wilderness Tourism Association (2001). *Draft code of practices for BC's wilderness tourism operators.* No link available.

Burns, G. L., MacBeth, J., & Moore, S. (2011). Should dingoes die? Principles for engaging ecocentric ethics in wildlife tourism management. *Journal of Ecotourism*, *10*(3), 179–196. https://doi.org/10.1080/14724049.2011.617450

Butler, R. W. (2011). Sustainable tourism in high-latitude islands: Shetland Islands. In J. Carlssen & R. W. Butler (Eds), *Island tourism: Sustainable perspectives* (pp. 140–154). CABI.

Cohen, B. R. (2001). The safety Nazis. *American Spectator*, *34*(6), 16.

Convention on Biological Diversity (CBD), United Nations Environment Programme (UNEP). (2001). *Biological diversity and tourism: International guidelines for sustainable tourism.* No link available.

Counsell, D. (1999). Sustainable development and structure plans in England and Wales: operationalizing the themes and principles. *Journal of Environmental Planning and Management*, *42*(1), 45–61. https://doi.org/10.1080/09640569911299

Daltabuit, M., & Pi-Sunyer, O. (1990). Tourism development in Quintana Roo, Mexico. *Cultural Survival Quarterly*, *14*(1), 9–13.

deFur, P., & Kaszuba, M. (2002). Implementing the precautionary principle. *The Science of the Total Environment*, *288*(1–2), 155–165. https://doi.org/10.1016/S0048-9697(01)01107-X

Dickson, B. (1999). The precautionary principle in CITES: a critical assessment. *Natural Resources Journal*, *39*, 211–228.

Ellis, J. (2000). The precautionary principle: from paradigm to rule of law. *International Law*, 2, 127–129.

Faulkner, B. (1998). Introduction. In E. Laws, B. Faulkner & G. Moscardo (Eds), *Embracing and managing change in tourism: International case studies* (pp. 1–10). Routledge.

Fennell, D. A., & Ebert, K. (2004). Tourism and the precautionary principle. *Journal of Sustainable Tourism*, *12*(6), 461–479. https://doi.org/10.1080/09669580408667249

Fennell, D. A. (2015). Akrasia and tourism: Why we sometimes act against our better judgment? *Tourism Recreation Research*, *40*(1), 95–106. https://doi.org/10.1080/02508281.2015.1006419

Freestone, D., & Hey, E. (1996). Origins and development of the precautionary principle. In D. Freestone & E. Hey (Eds), *The precautionary principle and environmental law: The challenge of implementation* (pp. 3–12). Kluwar Law International.

Garrod, B., & Fennell, D. A. (2004). An analysis of whalewatching codes of conduct. *Annals of Tourism Research*, *31*(2), 334–352. https://doi.org/10.1016/j.annals.2003.12.003

Gössling, S. (2001). The consequences of tourism for sustainable water use on a tropical island: Zanzibar, Tanzania. *Journal of Environmental Management*, *61*(2), 179–191. https://doi.org/10.1006/jema.2000.0403

Government of Canada (2001a). *A Canadian perspective on the precautionary approach/principle proposed guiding principles.* No link available.

Government of Canada (2001b). *A Canadian perspective on the precautionary principle/approach.* No link available.

Hall, C. M. (1995). In search of common ground: reflections on sustainability and complexity and process in the tourism system – a discussion between C. Michael Hall and Richard W. Butler. *Journal of Sustainable Tourism*, *3*(2), 99–105.

Hall, C. M. (2011). Biosecurity, tourism and mobility: institutional arrangements for managing tourism-related biological invasions. *Journal of Policy Research in Tourism, Leisure & Events*, *3*(3), 256–280. https://doi.org/10.1080/19407963.2011.576868

Harremoës, P., Gee, D., MacGarvin, M., Stirling, A., Keys, J., Wynne, B., & Vaz, S. G. (2001). *Late lessons from early warnings: the precautionary principle 1896–2000.* European Environment Agency, Office for Official Publications of the European Communities.

Holden, A. (1999). High impact tourism: a suitable component of sustainable policy? The case of downhill skiing development at Cairngorm, Scotland. *Journal of Sustainable Tourism*, *7*(2), 97–107. https://doi.org/10.1080/09669589908667329

Holden, A. (2000). *Environment and tourism.* Routledge.

Howie, F. (2003). *Managing the tourist destination.* Thomson.

Inskeep, E. (1991). *Tourism planning: An integrated and sustainable development approach.* Wiley and Sons.

Jennings, G. (2003). Marine tourism. In S. Hudson (Ed.), *Sport and adventure tourism* (pp. 125–164). Haworth Hospitality Press.

Kaiser, M. (1997). "The precautionary principle and its implications for science" - Introduction. *Foundations of Science*, 2, 201–205. https://doi.org/10.1023/A:1009641701933

Kirstges, T. (1995). *Sanfter Tourismus*, 2nd ed. Oldenbourg Verlag.

Lusseau, D., Slooten, L., & Currey, R. (2006). Unsustainable dolphin-watching tourism in Fiordland, New Zealand. *Tourism in Marine Environments*, *3*(2), 173–178. https://doi.org/10.3727/154427306779435184

Martin, A. (2000). Making tourism sustainable. *UNESCO Sources*, *120*, 10–12.

Martin, B. (2021). The responsibility of winter sports resorts in the global spread of COVID-19. *RSPP Workshop: Pandemics, location and mobility*, Ponta Delgada, Portugal.

Martinez, E., & Orams, M. (2011). Kia angi puku to hoe I te wai: Ocean noise and tourism. *Tourism in Marine Environments*, 7(3–4), 191–202. https://doi.org/10.3727/154427311X13195453162895

McCool, S. F., & Moisey, R. N. (Eds). (2001). *Tourism, recreation and sustainability: Linking culture and the environment.* CABI Publishing.

Myers, N. (1993). Biodiversity and the precautionary principle. *Ambio*, *22*(2–3), 74–79. https://www.jstor.org/stable/4314050

O'Riordan, T., & Cameron, J. (1994). The history and contemporary significance of the precautionary principle. In T. O'Riordan & J. Cameron (Eds), *Interpreting the precautionary principle* (pp. 12–30). Earthscan.

Rogers, M. F., Sinden, J. A., & De Lacy, T. (1997). The precautionary principle for environmental management: a defensive-expenditure application. *Journal of Environmental Management, 51*(4), 343–360. https://doi.org/10.1006/jema.1997.0154

Russell, R., & Faulkner, B. (1999). Movers and shakers: chaos makers in tourism development. *Tourism Management, 20*(4), 411–423. https://doi.org/10.1016/S0261-5177(99)00014-X

Scott, S. V. (2001). How cautious is precautious? Antarctic tourism and the precautionary principle. *International and Comparative Law Quarterly, 50*(4), 963–971. https://doi.org/10.1093/iclq/50.4.963

Soleimanpour, H. (2012). Legal implications for nature-based tourism. In J. A. Seba (Ed.), *Ecotourism and sustainable tourism: New perspectives and studies* (pp. 16–29). Apple Academic Press.

Stewart, E. J., & Draper, D. (2006). Sustainable cruise tourism in Arctic Canada: an integrated coastal management approach. *Tourism in Marine Environments, 3*(2), 77–88. https://doi.org/10.3727/154427306779435210

Swarbrooke, J. (1999). *Sustainable tourism management.* CABI Publishing.

Tapper, R. (2001). Tourism and socio-economic development: UK tour operators' business approaches in the context of the new international agenda. *International Journal of Tourism Research, 3*(5), 351–366. https://doi.org/10.1002/jtr.348

The Wilderness Society (1999). *Tourism in natural areas policy.* No link available.

Throsby, D. (2016). Tourism, heritage and cultural sustainability: Three 'golden rules'. In L. F. Girard, & P. Nijkamp (Eds.), *Cultural tourism and sustainable local development* (pp. 31–48). Routledge.

Tickle, L, & von Essen, E. (2020). The seven sins of hunting tourism. *Annals of Tourism Research, 84*, 102996. doi.org/10.1016/j.annals.2020.102996

Tickner, J., & Raffensberger, C. (1998). The precautionary principle: a framework for sustainable business decision-making. *Environmental Policy, 5*(4), 75–82. https://doi.org/10.1016/S1066-7938(00)80085-8

Tribe, J., Font, X., Griffiths, N., Vickery, R., & Yale, K. (2000). *Environmental management for rural tourism and recreation.* Cassell.

Van Dyke, J. M. (1996). The Rio principles and our responsibilities of ocean stewardship. *Ocean & Coastal Management, 31*(1), 1–23. https://doi.org/10.1016/0964-5691(95)00071-2

VanderZwaag, D. (1994). *CEPA and the precautionary principle/approach.* Minister of Supply and Services.

VanderZwaag, D. (1999). The precautionary principle in environmental law and policy: elusive rhetoric and first embraces. *Journal of Environmental Law and Practice, 8(3)*, 355–375.

Venables, S., McGregor, F., Brain, L., & van Keulen, M. (2016). Manta ray tourism management, precautionary strategies for a growing industry: a case study from the Ningaloo Marine Park, Western Australia. *Pacific Conservation Biology, 22*(4), 295–300.

World Tourism Organization (WTO). (2001) *Tourism 2020 vision - Global forecast and profiles of market segments.* WTO.

World Wildlife Fund (WWF). (2001). *Position statement on tourism.* No link available.

World Wildlife Fund (WWF). (2002) *Tourism.* No link available.

Ziegler, J., Silberg, J. N., Araujo, G., Labaja, J., Ponzo, A., Rollins, R., & Dearden, P. (2019). Applying the precautionary principle when feeding an endangered species for marine tourism. *Tourism Management, 72*, 155–158. https://doi.org/10.1016/j.tourman.2018.11.021

Chapter 40

Evolution of Concepts From Tourism Carrying Capacity to Overtourism

Anna Dóra Sæþórsdóttir & Margrét Wendt

Introduction

Increased numbers of tourists have created challenges for numerous destinations. Examples of common negative impacts from tourists and the tourism industry include littering and the pollution of air, water and marine systems. In natural areas, trampling, soil erosion, landscape destruction as well as wildlife and ecosystem disturbance are common; in urban settings, impacts such as a decline in the availability of residential housing and inflation are often mentioned. The negative impacts of tourism can affect residents and tourists alike. On the one hand, tourism can diminish the quality of life for local residents. On the other hand, it can decrease the quality of the tourist experience. Crowding, degraded destinations, an unsatisfied host community and displeased customers are regarded as manifestations of this situation and create significant management challenges.

The negative impacts of tourism pressure have long been a central issue within tourism and outdoor recreation research. Carrying capacity was one of the first concepts to address the problems arising from tourism overuse, and has played a key role as a research paradigm in the management of tourist destinations and the complex issues of high visitation. The concept of tourism carrying capacity is not based on a scientific theory, but it is rooted in the core idea of Malthusianism, which suggests that there is a finite limit to growth. In the context of tourist destinations, it means that there is a limit to tourism growth before the destinations' attractiveness will begin to decline due to excessive tourism development (Bramwell & Lane, 1993; Butler, 2019). The overutilisation of tourist destinations is closely related to Hardin's (1968) influential article *The Tragedy of the Commons* (Butler, 1991; Manning, 2007), in which it is stated that public resources will ultimately become exploited beyond their carrying capacity in the absence of careful management. The cause of this overexploitation is each individual's desire to maximise their own benefits while not taking responsibility for the overall depletion of the resource. Nevertheless, the subsequent consequences of the depletion of the resource is shared among all users ('the tragedy of the commons'). Many tourist destinations such as

The Wiley Blackwell Companion to Tourism, Second Edition. Edited by C. Michael Hall.

protected areas, green spaces and streetscapes are common resources and therefore in danger of 'overconsumption' by tourists and the tourism industry if their limits (carrying capacity) are left unmanaged and thus exceeded. Consequently, it is important to identify the carrying capacity of tourist destinations, and based on that, aim for their sustainable use.

This chapter provides an overview of the concept of tourism carrying capacity and its relation to the more recent concepts 'sustainable tourism' and 'overtourism'. Following the increased interest in sustainability in the latter half of the twentieth century, sustainable tourism replaced carrying capacity as the most commonly used concept to describe tourism pressure (Manning, 2007). More lately, overtourism has gained popularity as a concept to describe the negative impacts of excessive tourism development on destinations, visitors and residents (Milano, Cheer, et al., 2019). While the concepts of carrying capacity, sustainable tourism and overtourism may differ with regards to their terminology, they are based on the same core idea that unmanaged tourism growth results in negative impacts and eventually the decline of tourist destinations due to overconsumption. Using each concept's terminology, it means that if tourism grows beyond its limit, then the carrying capacity of the destination is exceeded, tourism becomes unsustainable, and the destination reaches a state of overtourism. This chapter introduces the history of these concepts, as well as their definitions, applications, and flaws.

Carrying Capacity in Outdoor Recreation and Tourism

The term carrying capacity originally derives from the field of ecology and wildlife management where biologists calculate how many individuals of a particular species can be supported in a defined ecosystem without its exploitation (Hendee et al., 1990; McCool & Lime, 2001; Seidl & Tisdell, 1999). This concept was applied for the first time to the field of outdoor recreation by the biologist Lowell Sumner, a warden at the US national park service in the 1930s. At that time, wilderness had been admired by relatively few, but visitation to wilderness areas and national parks had started to increase (Nash, 2014). In light of increasing public appreciation of wilderness, Sumner (1936) argued that national parks have a threshold with regards to the number of visitors they can tolerate and that access to such areas should thus be limited in order to conserve their natural qualities. The issue did not get much attention to begin with, as tourist numbers were low due to the economic recession in the early 1930s, and later on, due to the Second World War and its aftermath. In the 1960s when western economies started to recover from the war, recreational travel took off again, growing at a higher pace than ever (Butler, 1997). At the same time, concerns over environmental issues increased, manifested for instance through the publication of *Silent Spring* by Carson (1962) and Meadows et al.'s (1972) *The Limits to Growth*. With mass tourism growing at significant speed, researchers and environmentalists started to look critically at tourism development as a potential cause for concern (Butler, 1997).

National parks and wilderness are particularly prone to the impacts of excessive visitation as they are intended to simultaneously protect pristine nature and promote recreation so that visitors can experience solitude and unspoilt nature. Thus, the first research conducted on the concept of carrying capacity within the context of recreation paid special attention to areas with such qualities (Butler, 1997). Research by Wagar (1964) and Lucas (1964) argued that there is not only a physical, but a social layer to the concept of carrying capacity when applied in the context of recreation. Arguably, increased visitation does not only impact the environmental qualities of a destination, but also the experience of those who come to enjoy the area (Manning et al., 2002; O'Reilly, 1986). In that context, Lucas (1964, p. 5) claimed 'that the capacity of a recreational area is its ability to provide satisfaction'. Furthermore, Wagar (1964, p. 3) defined recreational carrying capacity as 'the level of recreational use an area can withstand while providing a sustained quality of recreation'. Wagar (1964) further claimed that social carrying capacity would differ based on visitors' motivations and expectations for their visit. In the following decade, considerable research was conducted on the subject, primarily in the United States (Butler, 1997).

After having at first mostly been studied in the field of outdoor recreation, the concept of carrying capacity was later also applied in the context of tourism (Butler, 1997; Lindberg et al., 1997). In 1981, the World Tourism Organization (UNWTO) (1981) was one of the first to do so with the publication of the report *Saturation of tourist destinations*. There, the concept of carrying capacity was referred to as 'the maximum number of tourists that a space can absorb without a lowering of the quality of the visitor's experience and without serious consequences for its ecology and its socio-economic structures' (UNWTO, 1981, p. 5). Later, the UNWTO defined carrying capacity as 'the level of visitor use an area can accommodate with high levels of satisfaction for visitors and few impacts on resources' (UNWTO & United Nations Environment Programme (UNEP), 1992, p. 18). In 1982, Mathieson and Wall wrote a fundamental book about the impacts of tourism, classifying them as environmental, economic, and social. They addressed the questions regarding user limits at tourist destinations and defined carrying capacity as 'the maximum number of people who can use a site without an unacceptable alteration in the physical environment and without an unacceptable decline in the quality of experience gained by visitors' (Mathieson & Wall, 1982, p. 21). As research on the concept evolved, more dimensions of carrying capacity were defined and it was recognised that tourist destinations have various forms of carrying capacities, such as physical, environmental, economic, social, perceptual and infrastructural (O'Reilly, 1986; Swarbrooke, 1999). Thus, there are numerous definitions of tourism carrying capacity, none of which is universally accepted. Furthermore, the concept is multidimensional and calls for a multidimensional approach.

Magic Numbers or Pure Fiction?

In the beginning, the concept of carrying capacity was used with the aim of identifying the maximum acceptable number of visitors to a particular area. In the 1980s, the search for the 'magic number' was mostly abandoned (Butler, 1997) as the limits were said to be unclear and highly subjective and thus the concept was not helpful for practical implementation (Lindberg et al., 1997; McCool et al., 2007). Moreover, carrying capacities could also be increased, for instance by building infrastructure to raise a destination's tolerance (Butler, 1997; Saarinen, 2006). So, rather than asking how many visitors a destination can sustain, some scholars argued that visitors inevitably impact the areas they visit. Thus, a more appropriate question would be to consider how much change is acceptable, what the desired conditions are and who the targeted visitor group should be (Lindberg et al., 1997; McCool & Lime, 2001; Stankey & McCool, 1984). Among the most influential publications in this context is Stankey et al.'s (1985) report on *The Limits of Acceptable Change* and Clark and Stankey's (1979) *Recreation Opportunity Spectrum*, which presented a framework for identifying permissible conditions within the target areas. Focus was put on addressing the needs and preferences of visitors and determining the desired conditions of tourist destinations (Butler, 1997; Graefe et al., 1984). This was, for example, reflected in the introduction of zoning systems whereby some areas were designated as high service areas for visitors while others were preserved by limiting access and development, ensuring low visitation.

According to Butler (2020), this fundamental shift in the research on carrying capacity, from identifying maximum acceptable numbers of visitors to approaches relying on norms and visitors preferences, was a step backwards. The dilemma of the normative approach is that visitors who are sensitive towards crowding will seek out other less crowded destinations and be replaced by new users with a higher tolerance to crowding. Therefore, the approach searching for the 'limits of acceptable change' would be based on the opinion of the new users, who would obviously have a higher tolerance towards crowding than the earlier visitors. Thus, approaches based on norms and visitor expectations will ultimately result in increasing levels of use and the decline of destinations (Butler, 1997; Butler, 2020).

Even though carrying capacity can be a useful concept in wildlife management, its normative application in the context of human behaviour can become very complex and fail to achieve the desired goal. As a normative concept, it is affected by the aims people set, values based on human preferences, governmental settings and management policies which again are based on values and beliefs. Thus, one destination can have a range of different possible capacities, all depending on the chosen aims nominated for the area, making the usefulness of the carrying capacity concept for management purposes highly debatable (McCool et al., 2007).

In addition to the normative application of the concept, the application of carrying capacity in the context of tourism is also challenging due to the vague nature of regulations and management plans. Such regulations and plans often lack the accuracy needed to set a carrying capacity. Moreover, the lack of scientific foundations weakens the concept's usefulness. In light of this, many (e.g. Lindberg et al., 1997; Manning et al., 1999; McCool et al., 2007; Seidl & Tisdell, 1999) have pointed out that the complexity of tourism and the imprecision of the carrying capacity concept limits its use and some (e.g. Lindberg et al., 1997; McCool et al., 2007) even go so far as to say that the concept should be avoided.

Sustainable Tourism

In the 1990s, the concept of sustainable tourism gained recognition and became prominent in the discussion on tourism's impacts on the environment, economy and society (Saarinen, 2006; Swarbrooke, 1999). The concept is based on the ideology of sustainable development, which according to the The World Commission on Environment and Development (1987) report *Our Common Future,* is defined as a development which 'meets the needs of the present without compromising the ability of future generations to meet their own needs'. By applying the concept of sustainable development to tourism, the UNWTO (1997) defined sustainable tourism as follows:

> Sustainable tourism development meets the needs of present tourists and host regions while protecting and enhancing opportunities for the future. It is envisaged as leading to management of all resources in such a way that economic, social and aesthetic needs can be fulfilled while maintaining cultural integrity, essential ecological processes, biological diversity and life support systems.

As such, sustainable tourism is about the fair distribution of benefits and costs of tourism and that decision-making concerning tourism development should include all stakeholders (Swarbrooke, 1999). However, the UNWTO's (1997) definition is one of many, as there exists no universally applied definition of the term (Butler, 1999; Sharpley, 2000; Swarbrooke, 1999).

Sustainable tourism has evolved into one of the most used concepts within tourism research (Bramwell & Lane, 1993). Yet, research has provided little evidence of how sustainable tourism plays out in practice and how the concept can be utilised to achieve the principles for which it stands (Butler, 1999). In other words, while many strategies for sustainable tourism have been developed, only a few have been implemented (Liu, 2003; Swarbrooke, 1999). According to Butler (1999), the sustainability concept is based on the fundamental idea of using resources without exploiting them, thus the implementation of sustainable tourism requires setting limits. To put it another way, Butler (2020) argues that the measurement of the carrying capacity of a tourist destination is the basis for sustainable tourism. Thus, the abandonment of the concept of carrying capacity has made it more difficult to achieve sustainable tourism (Butler, 2020). Moreover, there seems to be a considerable lack of integration of tourism with other sectors when it comes to tourism policy making, which further prevents sustainability from becoming a leader in tourism.

Despite its criticism, the concept of sustainable tourism has also proved valuable in that it has advanced the discussion of tourism impacts and growth. It underlines the multiplicity and diversity of tourism stakeholders, relating both to the economic, social and environmental components of sustainable tourism (Swarbrooke, 1999). Furthermore, it provides a basis for a dialogue among these diverse stakeholders on the impacts of their actions (Bramwell et al., 2017; Saarinen, 2006; Saarinen, 2014), reducing 'tensions and friction' (Bramwell & Lane, 1993, p. 2) between tourists, local residents, the environment and the tourism industry. It is also useful to shed light on the tourism-centric views which often dominate the discourse of tourism development. The concept of sustainable tourism assumes an equity between its three elements. In practice, however, economic aspects have received more attention compared to social and environmental ones (Saarinen, 2014). Furthermore, while sustainable tourism may be an ideology and an unachievable goal, it still provides an aim and a vision for what tourism stakeholders should be working towards. Even if sustainable tourism is not achieved, one can argue that efforts aimed at making tourism more sustainable will most likely result in improvements rather than additional problems (Bramwell & Lane, 1993).

Overtourism

Overtourism is the most recent concept used to describe the negative impacts of tourist visitation on destinations. The term 'overtourism' arose from media discourses without a theoretical foundation (Koens et al., 2018; Sæþórsdóttir et al., 2020) but quickly gained recognition among tourism researchers. It refers to a situation of excessive tourist visitation which results in negative impacts on the physical environment, the quality of experiences for tourists and the quality of life of residents (Capocchi et al., 2020; Goodwin, 2019; Koens et al., 2018). As with tourism carrying capacity and sustainable tourism, there is no one widely accepted definition of the term. Goodwin (2019), for instance, sees overtourism as the opposite of responsible tourism, which Goodwin (2019, p. 110) in turn defines as creating 'better places to live in and better places to visit'. According to Butler (2019), the concept places particular attention on resident attitudes and their perception of tourism impacts. In that context, overtourism has, for example, been defined as 'the excessive growth of visitors leading to overcrowding in areas where residents suffer the consequences of temporary and seasonal tourism peaks, which have caused permanent changes to their lifestyles, denied access to amenities and damaged their general well-being' (Milano, Cheer, et al., 2019, p. 1). However, there is a lack of a universally accepted definition and, consequently, the concept of overtourism has been criticised for being fuzzy and unclear (Capocchi et al., 2019; Koens et al., 2018). Still, it has become a buzz word both within academia and the international media to describe the problems resulting from tourism pressure (Milano, Cheer, et al., 2019).

Overtourism is not the same as mass tourism. What these two forms of tourism have in common is that they cause pressure due to high tourist numbers. However, unlike mass tourism, overtourism's manifestation is not limited to the number of tourists, as overtourism can also be experienced in destinations with few tourists in absolute terms. It can be felt in the form of negative tourist behaviour, unequal tourist distribution, problematic encounters etc., which exceed the tolerance of residents and tourists (Butler, 2019). In other words, a relatively small increase in tourist visitation can lead to overtourism, as overtourism 'is about perceived tourism encounters, environmental changes and infringements on people's lives' (Koens et al., 2018, p. 9).

Whereas overtourism can affect natural, rural and urban areas, the concept has primarily been used in the context of popular European cities (Butler & Dodds, 2022; Dodds & Butler, 2019b; Milano, Cheer, et al., 2019). Following the economic crisis of 2008, many cities made efforts to attract a high number of tourists with the aim of driving economic growth through tourism development (Koens et al., 2018). Since cities are generally characterised by having good infrastructure, it is often assumed that they are well equipped to cope with high visitation. However, as tourist numbers grew rapidly, many cities such as Barcelona and Venice struggled to cope with the impacts of increased visitation and ultimately became a posterchild for overtourism. Residents in overcrowded cities began to voice their concerns about tourism development and formed anti-tourism campaigns (Milano, Novelli, et al., 2019a, 2019b). In that context, the emergence of the term overtourism has been valuable, as it provides locals with 'a label that enables people, whether visitors or locals, to voice their concerns about too much tourism' (Capocchi et al., 2019, p. 7). In addition, the media has shown a particular interest in portraying the dissatisfaction of residents, which has further contributed to the popularity of the term overtourism.

Many have argued that overtourism is not a new phenomenon but simply a new term for a familiar research topic (e.g. Butler & Dodds, 2022; Wall, 2019). However, it can also be argued that overtourism provides a valuable contribution to the field of tourism research, as it places a different focus in its conceptualisation of tourism impacts, thereby reflecting current issues of tourism. Modern tourism has undergone major changes since the first conceptualisation of tourism impacts. Thus, while both carrying capacity and overtourism are concepts used to describe the problems created by tourism pressure, the circumstances during which these concepts emerged were vastly different. The initial studies on carrying capacity were first and foremost applied in national parks, wilderness and other areas under nature protection in the United States, focusing on recreation and domestic tourism. At that time, most visitors were rather homogenous in terms of expectations and needs, but today's tourists are an increasingly diverse group who engage in various activities (Butler, 2020). Moreover, international tourism has grown significantly, which has led to the creation of new destinations in areas which previously had not been visited by tourists (Dodds & Butler, 2019a). While there were 25 million international tourist arrivals in 1950, this number had increased to almost 1,500 million in 2019 (UNWTO, 2020b). Among the reasons for the increase in tourist numbers is that the global population has grown, at the same time as overall prosperity for many has increased and travel costs have diminished, which has made it possible for more people to travel (Dodds & Butler, 2019a). Recently, travel has become even more affordable due to the emergence of low-cost airlines and alternative forms of accommodation such as AirbnB (Oklevik et al., 2019; Hall et al., 2022). The media have traditionally played a key role in shaping tourism demand and with more and more people gaining access to the internet, social media's power to influence tourists' destination choice has only increased (Dodds & Butler, 2019a). The increase in information available online and in tourism supply means that 'the modern traveller has access to more information on destinations than ever before and can now get there more seamlessly at lower cost' (Milano, Cheer, et al., 2019, p. 228). Finally, it is also evident that the impacts of tourism, whether positive or negative, are not just a topic within tourism research anymore. They have become a focal point of public discussion, as reflected in the vast media coverage on overtourism and anti-tourism protests in overcrowded places.

Considering these changes that have occurred within the society and in tourism, it can be argued that the term overtourism emerged from the need to better reflect the complex issues of modern tourism. It has proven to be particularly useful in raising awareness about tourists' and residents' attitudes and preferences with regard to tourism, thereby shedding light – again - on the social and psychological carrying capacities (Butler, 2019; Mihalic, 2020). Overtourism and carrying capacity are tightly linked (Benner, 2020), but that does not mean that the concept of overtourism is redundant. At the very least, overtourism is bringing attention back to carrying capacity and applying it in new realities (Mihalic, 2020).

Current and Future Challenges

Whereas the concepts used to describe tourism pressure have changed, the underlying problem remains current and critical within tourism research, namely that tourism can grow beyond its limit, thereby exceeding the carrying capacity of destinations, making tourism unsustainable and putting destinations in a state of overtourism. What these three concepts have in common is that they point to the need to manage tourism and control its growth so that tourism development creates more benefits than costs. There are several ways in which tourism stakeholders can manage tourism and/or respond to the pressure that high visitation creates.

One approach is to change the type of tourist that visits a destination. Usually, the aim of this approach is to attract visitors who are educated, responsible and bring economic benefits to the destinations, for example, through longer stays and higher spending (Butler, 1991). In other words, the approach evolves around attracting fewer but 'better' visitors. Often, pricing strategies are used to appeal to different types of tourists (Butler & Dodds, 2022). However, high pricing denies the majority of tourists access to the destination, namely those who cannot afford the costs, and as such this strategy is seen by many as ethically unacceptable (Swarbrooke, 1999). Ultimately, it is also important to bear in mind that all types of tourists can cause negative impacts on destinations.

Another approach is to change the distribution of visitors. In that context, marketing is a helpful tool in that it can be used to steer attention away from overcrowded sites, while putting other less frequented sites in the spotlight (Oklevik et al., 2019). An example of this is removing crowded sites from guide books, maps and other promotional material (Butler & Dodds, 2022) or encouraging tourists to visit during off-season (Swarbrooke, 1999). However, as Swarbrooke (1999) points out, shifting tourism marketing to less visited destinations can result in an overall increase in demand which unavoidably leads to overconsumption of the destination. Contrary to their aim, the marketing attempts thus maintain high visitation in areas already under pressure, while also endangering new areas.

Instead of attempting to control what type of tourists come to a destination or where they go, the physical environment of destinations can be modified, thus increasing the destinations' tolerance to withstand impacts from crowding (Butler, 1991; Wall, 2019). Commonly, this includes infrastructure development such as the construction of bridges and roads or the provision of more accommodation. The problem with this approach is that it favours continuous development and tourism growth and may thus increase tourism pressure rather than relieving it. In addition, if only the physical carrying capacity is increased, it will most likely lead to the exceedance of other types of capacities such as the social and perceptual (O'Reilly, 1986).

Finally, it is possible to limit the number of tourists in an area (Butler, 1991; Butler & Dodds, 2022). This strategy can be compared to ticket sales in sports or for cultural events, where there is only a fixed number of tickets given the properties of the event (Butler & Dodds, 2022; Swarbrooke, 1999). Butler and Dodds (2022) argue that it is feasible to implement such strategies within the context of tourism, but they must be 'well-communicated, be non-exclusive and transparent and fairly enforced' (Butler & Dodds, 2022, p. 42). Whereas reducing the number of tourists would be the most effective way to reduce tourism pressure compared to these approaches, this management practice is particularly unpopular among tourism stakeholders and is rarely pursued (Butler, 2020). Attempts to limit tourist visitation have mostly been applied to single tourist sites rather than for whole destinations (Oklevik et al., 2019). There are multiple reasons for this.

1. One reason is that the tourism industry consists of various suppliers, all of whom aim for economies of scale. Consequently, these suppliers (e.g. ships, planes, hotels, resorts) are constantly increasing in size and seek to keep themselves fully booked rather than limiting their number of customers (Wall, 2020).
2. Closely related to the previous point is the fact that, historically, the primary objective of destinations has been to increase visitations in the hopes of generating high economic returns as well as jobs (Milano, Cheer, et al., 2019; Oklevik et al., 2019). Tourism remains first and foremost an economic activity and consequently, the creation of economic benefits for both private and public stakeholders continues to drive tourism development. Limiting tourist numbers is generally assumed to restrict the industry's economic returns and is thus often considered undesirable (Butler, 1997; Oklevik et al., 2019).
3. Due to the tourism industry's economic focus, many of the issues created by overtourism cannot be solved without participation from governments and policy makers. However, the government is often in a weak position to implement rules and regulations which would hinder uncontrolled tourism growth. The tourism industry consists largely of privately owned businesses. In line with neoliberal ideas, it is often argued that the tourism industry thrives best without external control from the government, thereby allowing the sector to develop within the principles of capitalism and competition. Consequently, policies and regulations issued by governments are often met with opposition (Butler, 1997; Butler, 2020).
4. As many tourism destinations are common resources there is often a lack of control and responsibility which can lead to overconsumption (Hardin, 1968), which in turn necessitates the question of who should be in charge of taking the required actions to limit tourism. Also, those wanting to damper tourism growth are not always in the position to do so. For instance, local authorities may lack the power to manage tourism effectively within their region, particularly if access to the destination is controlled by other actors, such as low-cost airlines wanting to increase their number of passengers and flights (Butler & Dodds, 2022; Dodds & Butler, 2019a).

5. The right to travel is firmly rooted in modern cultures. The UNWTO (2020a) has, for instance, declared that tourism is a right which should be available equally to everyone. Dodds and Butler (2019b, p. 525), however, argue that 'the idea that unlimited and unrestricted travel should be a right and thus available to all who can pay is neither acceptable any longer nor compatible with environmental problems, changing attitudes to energy use, increasing concerns over quality of life and the preservation of cultural heritage'.

To sum up, there are multiple reasons why overtourism and unsustainable tourism practices, which exceed the destinations' carrying capacity, continue to thrive. At its core, the problem centres around the imbalance of power among the different tourism stakeholders, most of whom have competing interests (Dodds & Butler, 2019a). Solving the issues created by tourism pressure would thus require establishing a fair and balanced distribution of power among all stakeholders. The industry's focus on growth-led neoliberalism would need to be abandoned in favour of a perspective which strikes a balance between economic, social and environmental issues.

Unexpectedly, tourism came to a standstill in 2020 due to the emergence of the Covid-19 pandemic. Suddenly, overcrowded destinations were empty and able to take a break from overtourism. Many scholars have argued that the pandemic provided the chance to reconsider current tourism practices and establish management strategies which redirect tourism development towards sustainability (e.g., Abbas et al., 2021; Jones & Comfort, 2020). However, the problems associated with high tourism pressure have not been solved and therefore it seems that many destinations will once again become overcrowded and suffer from excessive visitation, as travel restrictions have been lifted (Butler & Dodds, 2022). What the pandemic has suggested is that as long as tourism stakeholders and societies remain guided by neoliberal and growth-oriented principles, the only way that tourism growth will be limited is through the occurrence of events beyond human control.

Conclusion

Historically, there have been very few restrictions on tourism development and, as a consequence, too many tourists have visited the same destinations. In many cases, this has resulted in overcrowded destinations with damages to the environment and dissatisfied host populations. Tourism researchers have long pointed to the need for managing tourism growth, as reflected in the popularity of the concepts carrying capacity, sustainable tourism and overtourism. Clearly, the tourism phenomenon consists of too many complex interactions for a simple numerical solution, but complexity cannot be a reason for not taking action. Despite the fact that research on tourism carrying capacity has not been able to provide a magic number, it does throw light on the need for science and politics to set the foundation for policy makers to ensure that tourism growth is within its sustainable limits. It must be in everybody's interests to reap the benefits of tourism, while minimising the negative impacts. Uncontrolled tourism growth will inevitably result in the depletion of the resources on which tourism relies and it is in the interest of all stakeholders to prevent this fate. Therefore, it is vital that decision-makers and tourism managers withdraw from business as usual and instead employ a perspective that benefits not only the industry itself but the society as a whole. Since human nature does not hinder all individuals to limit their travel desire for the greater good, then a system needs to be put in place that will limit it for them. Still, it will take political courage to respond as nobody wants to stop a good party.

References

Abbas, J., Mubeen, R., Iorember, P. T., Raza, S., & Mamirkulova, G. (2021). Exploring the impact of COVID-19 on tourism: transformational potential and implications for a sustainable recovery of the travel and leisure industry. *Current Research in Behavioral Sciences*, *2*, 100033. https://doi.org/10.1016/j.crbeha.2021.100033

Benner, M. (2020). The decline of tourist destinations: An evolutionary perspective on overtourism. *Sustainability*, *12*(9), 3653. https://doi.org/10.3390/su12093653

Bramwell, B., Higham, J., Lane, B., & Miller, G. (2017). Twenty-five years of sustainable tourism and the *Journal of Sustainable Tourism*: looking back and moving forward. *Journal of Sustainable Tourism*, *25*(1), 1–9. https://doi.org/10.1080/09669582.2017.1251689

Bramwell, B., & Lane, B. (1993). Sustainable tourism: An evolving global approach. *Journal of Sustainable Tourism*, *1*(1), 1–5. https://doi.org/10.1080/09669589309450696

Butler, R. W. (1991). Tourism, environment, and sustainable development. *Environmental Conservation*, *18*(3), 201–209. https://doi.org/10.1017/S0376892900022104

Butler, R. W. (1997). The concept of carrying capacity for tourism destinations: Dead or merely buried? *Progress in Tourism and Hospitality Research*, 2(3–4), 283–293. https://doi.org/10.1002/pth.6070020309

Butler, R. W. (1999). Sustainable tourism: A state-of-the-art review. *Tourism Geographies*, *1*(1), 7–25. https://doi.org/10.1080/14616689908721291

Butler, R. W. (2019). Overtourism and the tourism area life cycle. In R. Dodds & R. W. Butler (Eds), *Overtourism: Issues, realities and solutions* (pp. 76–89). De Gruyter.

Butler, R. W. (2020). Tourism carrying capacity research: a perspective article. *Tourism Review*, *75*(1), 207–211. https://doi.org/10.1108/TR-05-2019-0194

Butler, R. W., & Dodds, R. (2022). Overcoming overtourism: a review of failure. *Tourism Review*, *77*(1) 35–53. https://doi.org/10.1108/TR-04-2021-0215

Capocchi, A., Vallone, C., Amaduzzi, A., & Pierotti, M. (2020). Is 'overtourism' a new issue in tourism development or just a new term for an already known phenomenon? *Current Issues in Tourism*, *23*(18), 2235–2239. https://doi.org/10.1080/13683500.2019.1638353

Capocchi, A., Vallone, C., Pierotti, M., & Amaduzzi, A. (2019). Overtourism: A literature review to assess implications and future perspectives. *Sustainability*, *11*(12), 3303. https://doi.org/10.3390/su11123303

Carson, R. (1962). *Silent spring*. Houghton Mifflin.

Clark, R. N., & Stankey, G. H. (1979). *The recreation opportunity spectrum: A framework for planning, management and research*. U.S. Department of Agriculture, Forest Service, Pacific Northwest Forest and Range Experiment Station.

Dodds, R., & Butler, R. W. (2019a). The enablers of overtourism. In R. Dodds & R. W. Butler (Eds), *Overtourism: Issues, realities and solutions* (pp. 6–26). De Gruyter.

Dodds, R., & Butler, R. W. (2019b). The phenomena of overtourism: a review. *International Journal of Tourism Cities*, *5*(4), 519–528. https://doi.org/10.1108/IJTC-06-2019-0090

Goodwin, H. (2019). Overtourism: Causes, symptoms and treatment. *Tourism Wissen - quarterly*. https://responsibletourismpartnership.org/wp-content/uploads/2019/06/TWG16-Goodwin.pdf

Graefe, A. R., Vaske, J. J., & Kuss, F. R. (1984). Social carrying capacity. *Leisure Sciences*, 6(4), 395–431.

Hall, C. M., Prayag, G., Safonov, A., Coles, T., Gössling, S., & Naderi Koupaei, S. (2022). Airbnb and the sharing economy. *Current Issues in Tourism*, *25*(19), 3057–3067. https://doi.org/10.1080/13683500.2022.2122418

Hardin, G. (1968). The tragedy of the Commons: The population problem has no technical solution; it requires a fundamental extension in morality. *Science*, *162*(3859), 1243–1248. https://doi.org/10.1126/science.162.3859.1243

Hendee, J. C., Stankey, G. H., & Lucas, R. C. (1990). *Wilderness management* (2nd ed.). North American Press.

Jones, P., & Comfort, D. (2020). The COVID-19 crisis, tourism and sustainable development. *Athens Journal of Tourism*, 7(2), 75–86. https://www.athensjournals.gr/tourism/2020-7-2-1-Jones.pdf

Koens, K., Postma, A., & Papp, B. (2018). Is overtourism overused? Understanding the impact of tourism in a city context. *Sustainability*, *10*(12), 4384. https://doi.org/10.3390/su10124384

Lindberg, K., McCool, S., & Stankey, G. (1997). Rethinking carrying capacity. *Annals of Tourism Research*, *24*(2), 461–465.

Liu, Z. (2003). Sustainable tourism development: A critique. *Journal of Sustainable Tourism*, *11*(6), 459–475. https://doi.org/10.1080/09669580308667216

Lucas, R. C. (1964). *The recreational capacity of the Quetico-Superior area.* U. S. Department of Agriculture, Forest Service.

Manning, R., Wang, B., Valliere, W., Lawson, S., & Newman, P. (2002). Research to estimate and manage carrying capacity of a tourist attraction: A study of Alcatraz Island. *Journal of Sustainable Tourism*, *10*(5), 388–404. https://doi.org/10.1080/09669580208667175

Manning, R. E. (2007). *Parks and carrying capacity. Commons without tragedy*. Island Press.

Manning, R. E., Valliere, W. A., & Wang, B. (1999). Crowding norms: Alternative measurement approaches. *Leisure Sciences*, *21*(2), 219–229. https://doi.org/10.1080/014904099273174

Mathieson, A., & Wall, G. (1982). *Tourism: Economic, physical, and social impacts*. Longman.

McCool, S. F., Clark, R., & Stankey, G. H. (2007). *An assessment of frameworks useful for public land recreation planning*. US Department of Agriculture, Forest Service, Pacific Northwest Research Station.

McCool, S. F., & Lime, D. W. (2001). Tourism carrying capacity: Tempting fantasy or useful reality? *Journal of Sustainable Tourism*, *9*(5), 372–388. https://doi.org/10.1080/09669580108667409

Meadows, D. H., Meadows, D. L., Randers, J., & Behrens, W. W. (1972). *The limits to growth*. Universe Books.

Mihalic, T. (2020). Conceptualising overtourism: A sustainability approach. *Annals of Tourism Research*, *84*, 103025. https://doi.org/10.1016/j.annals.2020.103025

Milano, C., Cheer, J. M., & Novelli, M. (2019). *Overtourism: Excesses, discontents and measures in travel and tourism*. CABI.

Milano, C., Novelli, M., & Cheer, J. M. (2019a). Overtourism and degrowth: a social movements perspective. *Journal of Sustainable Tourism*, *27*(12), 1857–1875. https://doi.org/10.1080/09669582.2019.1650054

Milano, C., Novelli, M., & Cheer, J. M. (2019b). Overtourism and tourismphobia: A journey through four decades of tourism development, planning and local concerns. *Tourism Planning & Development*, *16*(4), 353–357. https://doi.org/10.1080/09669582.2019.1650054

Nash, R. (2014). *Wilderness and the American mind* (5th ed.). Yale University Press.

O'Reilly, A. M. (1986). Tourism carrying capacity: concept and issues. *Tourism Management*, *7*(4), 254–258. https://doi.org/10.1016/0261-5177(86)90035-X

Oklevik, O., Gössling, S., Hall, C. M., Steen Jacobsen, J. K., Grøtte, I. P., & McCabe, S. (2019). Overtourism, optimisation, and destination performance indicators: A case study of activities in Fjord Norway. *Journal of Sustainable Tourism*, *27*(12), 1804–1824. https://doi.org/10.1080/09669582.2018.1533020

Saarinen, J. (2006). Traditions of sustainability in tourism studies. *Annals of Tourism Research*, *33*(4), 1121–1140. https://doi.org/10.1016/j.annals.2006.06.007

Saarinen, J. (2014). Critical sustainability: Setting the limits to growth and responsibility in tourism. *Sustainability*, *6*(1), 1–17. https://doi.org/10.3390/su6010001

Seidl, I., & Tisdell, C. A. (1999). Carrying capacity reconsidered: from Malthus' population theory to cultural carrying capacity. *Ecological Economics*, *31*(3), 395–408. https://doi.org/10.1016/S0921-8009(99)00063-4

Sharpley, R. (2000). Tourism and sustainable development: Exploring the theoretical divide. *Journal of Sustainable Tourism*, *8*(1), 1–19. https://doi.org/10.1080/09669580008667346

Stankey, G., Cole, D. N., Lucas, R. C., Peterson, M. E., & Frissell, S. S. (1985). *The limits of acceptable change (LAC) system for wilderness planning*. Intermountain Forest and Range Experiment Station, USDA Forest Service, General Technical Report INT-176.

Stankey, G. H., & McCool, S. F. (1984). Carrying capacity in recreational settings: Evolution, appraisal, and application. *Leisure Sciences*, 6(4), 453–473. https://doi.org/10.1080/01490408409513048

Sumner, L. (1936). *Special report on a wildlife study of the high sierra in Sequoia and Yosemite National Parks and adjacent territory*. US Department of the Interior, National Park Service.

Swarbrooke, J. (1999). *Sustainable tourism management*. CABI Publishing.

Sæþórsdóttir, A. D., Hall, C. M., & Wendt, M. (2020). From boiling to frozen? The rise and fall of international tourism to Iceland in the era of overtourism. *Environments*, 7(8), 59. https://doi.org/10.3390/environments7080059

The World Commission on Environment and Development. (1987). *Our common future*. Oxford University Press.

Wagar, J. A. (1964). The carrying capacity of wild lands for recreation. *Forest Science*, *10*(2), 1–24. https://doi.org/10.1093/forestscience/10.s2.a0001

Wall, G. (2019). Perspectives on the environment and overtourism. In R. Dodds & R. W. Butler (Eds), *Overtourism: Issues, realities and solutions* (pp. 27–45). De Gruyter.

Wall, G. (2020). From carrying capacity to overtourism: a perspective article. *Tourism Review*, *75*(1), 212–215. https://doi.org/10.1108/TR-08-2019-0356.

World Tourism Organization (UNWTO). (1981). *Saturation of tourist destinations*. https://www.e-unwto.org/doi/pdf/10.18111/unwtogad.1981.1.un406362r557g40k

World Tourism Organization (UNWTO). (1997). *Agenda 21 for the travel and tourism industry*. https://www.e-unwto.org/doi/book/10.18111/9789284403714

World Tourism Organization (UNWTO). (2020a). *Framework convention on tourism ethics*. https://www.e-unwto.org/doi/pdf/10.18111/9789284421671

World Tourism Organization (UNWTO). (2020b). *World tourism barometer*, *18*(7). https://www.e-unwto.org/doi/epdf/10.18111/wtobarometereng.2020.18.1.7

World Tourism Organization (UNWTO) and United Nations Environment Programme (UNEP). (1992). *Guidelines: Development of national parks and protected areas for tourism*.

Chapter 41

Tourism and CSR – Time to Shift the Dial

Tim Coles

Introduction

2020 was an auspicious year. For many, it will be remembered as the year of Coronavirus, COVID-19 and the shutdown of global society and economy. Scholars of corporate social responsibility (CSR) widely read and with a keen sense for history will realise it also marked the 50th anniversary of the appearance of Milton Friedman's (1970) famous remarks in *The New York Times* magazine. Famously adamant about the role of the corporation, he concluded that 'there is one and only one social responsibility of business – to use its resources and engage in activities designed to increase its profits so long as it stays within the rules of the game, which is to say, engages in open and free competition without deception or fraud' (Friedman, 1970, p. 126).

Coming from a future Nobel Prize-winning economist (in 1976), Friedman's (1970) remarks were influential, expounding his earlier (1962) views in mass circulation form, and their appearance reflected the Nixon administration's view at the time that 'the nation's economic welfare was best served by basic reliance on a fundamentally liberal, market-organised economic system (i.e. private enterprise)' (McCracken, 1996, p. 166). Since then, they have become an easy target for scholars of CSR, sometimes even being misrepresented by critics (Elrick & Thies, 2018). Yet half a century later and the Coronavirus pandemic has demonstrated, perhaps more emphatically than ever before, that the responsibilities of a firm do extend far beyond owners and investors.

Elusive to define and fuzzy in nature, CSR has become one of the major subject areas and mainstreamed in contemporary business and management, both in practice (Stobierski 2021), and scholarship (Crane et al., 2008; Hopkins, 2016). Often reduced, somewhat unhelpfully in shorthand to 'sustainable business' (Stobierski, 2021), CSR is a voluntary form of self-regulation that reflects the commitment of a business to wider societal goals. According to the World Business Council for Sustainable Development (WBCSD) in one of the earliest and enduring definitions, it refers to the 'continuing commitment by business to behave ethically

The Wiley Blackwell Companion to Tourism, Second Edition. Edited by C. Michael Hall.

and contribute to economic development while improving the quality of life, of the workforce, and their families, as well as the local community and society at large' (Holme & Watts, 2000, p. 10). Although 'the firm' and the private sector dominated early discourse (and still does in some cases), current thinking acknowledges the relevance and application of this approach beyond business to organisations across the public and voluntary sectors. Definition of CSR has been heavily contested and has resulted in an 'alphabet soup' of acronyms (cf. Coles et al., 2015) that grapple with how best to encapsulate organisational commitment to sustainable development. Recently, the 'definition debate' has been revisited by the emergence, alongside CSR, of Environment, Social and Governance (ESG) in corporate discourse. In essence, the difference is in emphasis, such that 'while CSR holds businesses accountable for their social commitments in a qualitative manner, ESG helps measure or quantify such social efforts' (Hung, 2021). Put another way, 'CSR aims to make a business accountable, [whereas] ESG criteria make its efforts measurable' (Alva, 2021).

Not surprisingly then, major reviews demonstrate that CSR has been the focus for hundreds, if not thousands of studies somehow connected to a topic defined by relatively wide parameters and interests (Lindgreen & Swaen, 2010; Aguinis & Glavas, 2012; Wickert, 2021; Fatima & Elbanna, 2022). This chapter critically examines a distinctive sub-set of them: research on the CSR of tourism businesses and organisations. The chapter does not become embroiled in debates about what does or does not constitute CSR and its precise form. Relatively wide interpretations of both 'CSR' and 'tourism' are used here to consider perspectives on CSR from businesses and organisations (not just 'firms') across the entire tourism-leisure-hospitality-events nexus. Thus, this chapter charts, and critically appraises, how the subject area touching on these terms of reference has developed since 2010 as a key moment in the trajectory of attention (Madanaguli et al., 2022), and because work to this point was the subject of the first major review published on the topic (Coles et al., 2013). While there has been continued growth in interest and a maturing of the body of knowledge on tourism CSR, familiar themes continue to be pursued and recent experiences suggest that the time may now be right to 'shift the dial'.

Tourism Research on CSR

Sketching Out the Body of Knowledge

As Madanaguli et al. (2022, p. 447) have observed, 'CSR in tourism is a promising area of study that is growing rapidly'. As Figure 41.1 reveals, their succinctness may underplay just how rapid the recent rate of growth in this topic has been. Basic searches of two bibliographical databases -Scopus and Web of Science- reveal that over 450 publications have appeared since 2003 including the terms 'corporate', 'social', 'responsibility', and 'tourism' variously in their titles, abstracts and keywords.

Of course, keyword searches are not without their limitations. Notwithstanding, as an indication of the size of corpus of knowledge on this topic, Web of Science records 451 studies defined by these parameters while Scopus reports 546 studies. In fact, estimates in this range may under-represent the overall magnitude of interest and engagement with the subject. Not all studies will be captured by these databases nor by the particular search parameters, and work may exist beyond English-language contributions. Moreover, some studies may employ different terms, such as corporate social environmental responsibility (CSER) (e.g. Lin et al., 2018), while others may selectively focus on certain aspects of CSR, such as 'community

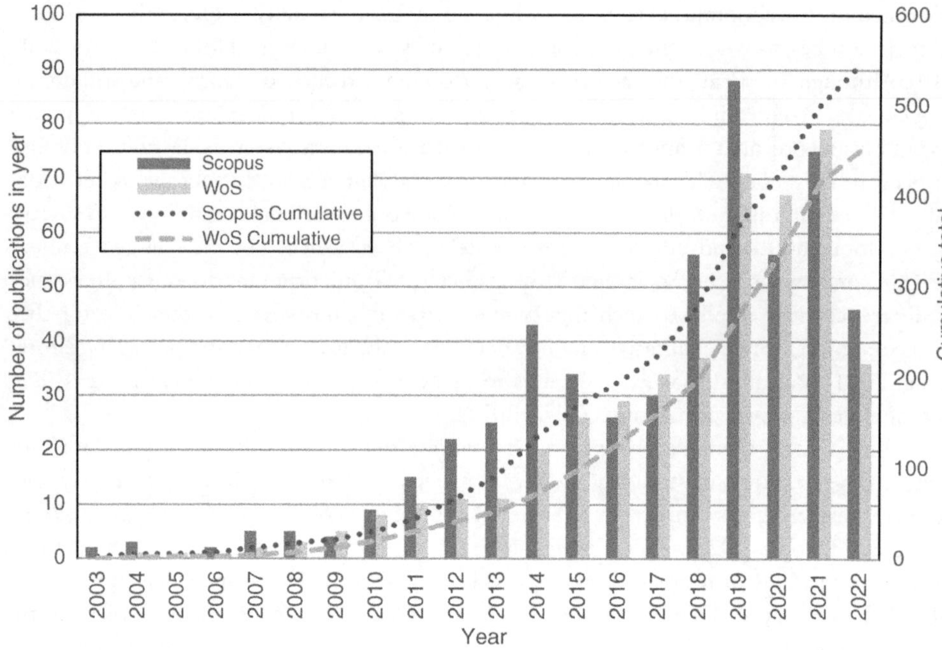

Figure 41.1. Publication trends for studies of 'Tourism CSR'.
Source: Author, data from Scopus and Web of Science searches (30/06/22).

responsibility' or 'environmental responsibility' that a search may not capture. Hence, a major challenge in surveying this intellectual terrain is delimiting it (Coles et al., 2013). Not only are there studies that start from a position of CSR and are firmly rooted in thinking on and around the topic (and perhaps influenced or inspired contributions in mainstream business and management studies). There are also studies that more loosely engage with the term such that CSR is a secondary concern or interest or the CSR implications of a study ostensibly on another topic are raised. Emblematic of this, Acampora et al. (2022) reported identifying over 600 articles dealing in some way with 'green hotels' and what they term 'sustainable practices in hotels'. Yet, as they lamented, the 'social and economic aspects are almost absent, and when they are present, they are investigated jointly with environmental aspects, under the umbrella of CSR' (Acampora et al., 2022, p. 512).

Of the two categories, the latter are numerous and present an invidious challenge to scholars interested in conducting large-scale, systematic reviews. Be this as it may, Figure 41.1 suggests a clear growth trend over time, with interest accelerating markedly in the later 2010s. From Scopus data alone, two-and-a-half times more publications appeared in 2021 than in the whole of the period from 2003 to 2010. Between 2010 and 2020, growth in cumulative interest was between eighteen- (Scopus) and twenty-one-fold (Web of Science). It is possible to speculate the reasons for such growth. Early contributions drew inspiration from thought-leading contributions on CSR in other sectors (e.g., Michael Porter, Arch Carroll) published in a wider setting of business and management studies where debate had already matured (Coles et al., 2013). Subsequent growth after 2010 may also have been catalysed by a 'responsibility turn' in scholarship on, and related to, (sustainable) tourism (cf. Mowforth et al., 2008), its management (Frey & George, 2010) and that of hospitality (cf. Hawkins & Bohndanowicz,

2012). It may also be a legacy of some of pioneering advocacy work by industry bodies and trade associations (Coles et al., 2013). Interestingly though, it does not appear that particular global challenges as potential foci for responsibility have been strong drivers for the growth of CSR research in tourism. Adding the terms 'climate change', 'sustainable development goal' and 'modern slavery' to those noted above returned no studies in Scopus. Similarly, the distinction ESG appears not to have penetrated into tourism research, specifically on CSR. In the case of the former, Scopus identifies 15 studies covering 'tourism' and 'ESG' and two for the latter ('corporate', 'social', 'responsibility', 'tourism', 'ESG'): Uyar et al.'s (2020) study linking CSR and firm performance with board characteristics, and Kim and Lee's (2020) examination of the impact of material and immaterial sustainability on firm performance.

The State of the State-of-the-art

In addition to very many empirical studies seeking to inform conceptual-cum-theoretical advances on tourism CSR, as befits a growing body of knowledge, frequent reviews have appeared appraising the state-of-the-art (Coles et al., 2011; Coles et al., 2013; Coles et al., 2015; Farrington et al., 2017; Font & Lynes, 2018; Serra-Cantallops et al., 2018; Guzzo et al., 2020; Qian et al., 2021; Madanaguli et al., 2022; Kaur et al., 2022; Wong et al., 2022). Empirical studies have proliferated at such a rate that, as times passes, the challenge of making sense of the body of knowledge becomes ever more testing.

Among the numerous reviews, there is broad consensus of the subject area developing apace but still with more to achieve. Each survey has its own particular terms of reference on what is, after all, a broad subject area and this leads to apparent differences in foci as well as in study classification, progress to date, and the priorities for future research. Reviewing CSR research on the hotel industry from 2006 to 2015, Serra-Cantallops et al. (2018) argued that of three streams of research, there has been a similar amount of work on 'practices' and 'impact' compared to reporting, with more attention required to consumer reactions to CSR and the link between CSR and corporate financial performance. Farrington et al. (2017), by contrast, observed what they regarded as 'a general tendency towards measurement of financial and other internal benefit impacts' (Farrington et al., 2017, p. 30). For them, CSR research in hospitality had adopted a largely uncritical approach to the 'instrumental emphasis on assessing processes, perceptions, and private profitability from the general management literature, without engaging on a contextually-specific and/or theoretical level' (Farrington et al., 2017, p. 30). Font and Lynes (2018) placed a special emphasis on stakeholders, noting the relative privileging of management, employees, shareholders, and consumers at the expense of communities, suppliers, NGOs, and government. In contrast, Guzzo et al. (2020) questioned this view, noting that hospitality researchers had only recently started to focus on understanding how CSR relates to employee behaviours and attitudes.

The production of knowledge has also come under the spotlight. Wong et al. (2021) employed bibliometric mapping to investigate trends among 172 tourism and hospitality articles from 2006 to 2019. Theoretically, they noted the importance of Social Identity Theory and, methodologically, the prevalence of structural equation modelling (cf. Qian et al., 2021). Thematically, from 2006 to 2012 they identified four clusters of work broadly on: motives, demands and effects of CSR implementation; evolution of CSR; financial performance; and integration of stakeholder management. From 2013 and 2019, they note a subtle shift in orientation, with a stronger emphasis on customer reactions to CSR instead of interest in motives, demands and effects. A predominant use of quantitative methods and approaches has been regularly commented upon (Coles et al., 2011; Coles et al., 2013; Farrington et al., 2017; Wong et al., 2021; Qian et al., 2021; Madanaguli et al., 2022). Moreover, there is some evidence that this will remain the case for some time to come. Madanaguli et al. (2022) identified what they described as three groups of studies and 'two typologies' [sic] of work on CSR and sustainability in the tourism sector. In addition to 'conceptual and theoretical studies' drawing inspiration from widely across the social sciences (see also Wong et al., 2021, Qian et al., 2021), they noted quantitative and qualitative empirical studies. Among the two types, which are mainly associated with quantitative analysis, the first addresses the direct effects of CSR on 'employees, customers and business performance' while the second group 'highlights the moderated and mediated relationships of variables, including customer trust, identification, trust and value with CSR' (Madanaguli et al., 2022, p. 447). Finally in one of the few historiographical reviews exploring the origins and genealogy of tourism CSR research Coles et al. (2013), noted the emergent nature of interest in, and research on, tourism CSR during the 2000s, as well as the flow of ideas and thinking from business and management studies into tourism studies. Much of the essential 'thought leadership' in terms of theory, concepts, analytical approaches and methods, and — most importantly — ways of knowing CSR (i.e. epistemology) originated outside tourism studies. Taking this as their cue, Lindgreen and Swaen (2010) use a fivefold categorisation developed in a major review of CSR in management studies to examine progress in understanding tourism CSR. Tourism studies of CSR had made notable advances in the three areas of CSR Implementation, the 'narrow' business case (i.e., covering firm performance), and on stakeholder engagement, especially at the destination level. In contrast, CSR measurement and communication had been largely overlooked (See also Coles et al., 2015). Other distinctive features included sub-sectoral emphases on airlines and aviation, accommodation providers and casino operations, as well as a predominance of interest in organisations headquartered in North America, Europe and the United Kingdom.

Back to the Future: Viewing Recent Progress From a 2010 Lens

As noted above, there is not the space here to review such a large and growing corpus of empirical work exhaustively. Nor is there scope to conduct a more thorough and extensive meta-analysis, comparing findings to produce a composite representation of how research on this subject has evolved. Suffice to note that, despite the routine identification of lacunae to be filled or possible trajectories for the future research agenda, in very few (if any) cases have the respective authors presented subsequent retrospectives to document progress over time.

In order to take the longest view of how the corpus has evolved, returning to the categorisation used by Coles et al. (2013) is instructive as a lens through which to inspect more recent progress. Specifically, its redeployment for critical analysis reveals three main trends that appear to characterise subsequent knowledge production about tourism CSR. The first is incremental additions of knowledge to paint 'fuller pictures' on key themes. Studies have continued to appear on the two most popular areas of the 'narrow' business case around 'firm performance' (Garay & Font, 2012; Fu et al., 2014; Theodoulidis et al., 2017; Marino-Romero et al., 2018; Franco et al., 2020) and 'CSR implementation' (de Grosbois, 2012, 2016; Iyer & Jarvis, 2019) and its effects on stakeholders, above all employees (Su & Swanson, 2019; Wang et al., 2020; Farmaki et al., 2022). The identification of stakeholders and their perceptions of responsibility continue to fascinate in a complex and changing world (Coles et al., 2013; Font & Lynes, 2018; Farmaki, 2019; Su et al., 2020). Building on early, pre-2010 progress (See Coles et al., 2013; Coles et al., 2105; Wong et al., 2022), further mining of sub-sector data has painted more even elaborate, vibrant and sometimes contrasting pictures of CSR behaviours and outcomes in particular among hotel (de Grosbois, 2012; Marino-Romero et al., 2018; Wang et al., 2020) and restaurant (Youn et al., 2016; Liu & Tse, 2021) businesses.

Second and closely connected, there is evidence of several major gaps being addressed. For instance, in 2010 there had been very little work on CSR communications relative to the other thematic areas. This comparative situation is still the case but a decade later at least the theme is more visible. The importance of social media posts about CSR has attracted research, especially on restaurant brands (cf. Kim & Stepchenkova, 2021; Sung et al., 2022). One analysis involving twitter revealed that customers perceived 'that CSR activities related to the local community and society were as important as the activities related to the environment' (Sung et al., 2022, p. 1). This should come as little surprise given the scope of CSR, nor should the apparent influence of firm-level social media engagement with CSR on customers' behavioural responses. Studies of this type resonate these days because carefully developed reputations and curated social images can be tarnished in no time. In 2010, two of the main sub-sectoral lacunae appeared to be CSR as related to cruise lines (cf. de Grosbois, 2016) and tour operators (van de Mosselar et al., 2012), both of which have now attracted attention.

It may seem contrived to differentiate between 'filling gaps' and consolidating initial gains, but a common result of both has been a widening of geographical coverage of the topic. In part this is reflected in Table 41.1 which points to the increasing internationalisation of research and authorship teams on tourism CSR. Up to and including 2010, there was an early dominance of scholars from the English-speaking world, especially the UK and North America (see also Coles et al., 2014; Farrington et al., 2017) which has remained from 2011 to 2022. During this period there has been strong growth in interest in both Asia (from China, South Korea, Malaysia and Hong Kong) and Europe (from Spain, and Italy). Emblematic of this

Table 41.1. Top-10 documents by countries or territories of authors for studies of 'Tourism CSR' up to and after 2010.

Country	*Status: pre-2011*	*Number of authors*	*Status: 2011–2022*	*Number of authors*
United Kingdom	1	9	3	60
United States	2	4	1	67
Australia	3	3	5	43
Canada	4	2	19	11
South Africa	5	2	12	16
Spain	6	2	2	61
Austria	7	1	17	12
Croatia	8	1	35	5
Denmark	9	1	52	2
Egypt	10	1	27	7
China	–	–	4	49
Italy	–	–	6	24
Malaysia	13	1	7	20
South Korea	–	–	8	20
Hong Kong	–	–	9	19
Turkey	–	–	10	18

Source: Author, data from Scopus searches (30/06/22).

gradual geographical diffusion, 16 studies have appeared exploring tourism CSR in Africa; nine on the Caribbean, with a special emphasis on the industry; and eight on the Middle East. More generally, Pena-Miranda et al. (2021) have argued that there has been comparatively little attention to strategic CSR practices in the hospitality sector in developing countries. From empirical insights of eight hotels in Columbia, they set out to develop a locally-informed CSR management model for the hotel industry. Work of this nature is invaluable in reminding us that CSR is a construct of the developed world, particularly from English-language discourse around corporate governance. The unchallenged assumption in much tourism CSR research is that this is the most appropriate approach for other parts of the world. Assumptions of this nature are at best lazy and inadvertent, at worse yet another case of post- or even neo-colonialism.

The final trend is the addition of further layers of (apparent) sophistication to investigation, including the application of more intricate data collection and analysis techniques. For example, Horng et al. (2018) employed what they termed a 'fuzzy Delphi method' involving the DEMATEL method (Decision-making, trial and evaluation laboratory) and the related Analytical Network Process (ANP). Uyar et al. (2021a, b) used one of an increasingly powerful array of text-mining tools to analyse the content of sustainability reports on the Global Reporting Initiative (GRI) Sustainability Disclosure Database. Taking advantage of performance gains, they examine 478 reports from the hospitality and tourism industry from 44 countries extending from 2018 back to 1999. Some reviews have synthesised current knowledge into new conceptual frameworks (Kaur et al., 2022) in order to drive further (structural equation) modelling of the relationship between activity and outcome (especially performance) and finessing of analysis through moderating and mediating variables (Kim et al., 2018; Uyar et al., 2021a; Uyar et al., 2021b; Meng-Chan et al., 2021; Kaur et al., 2022; Madanaguli et al., 2022). The need for integrative approaches of this nature also reflects the challenges of studying CSR empirically. Wong et al. (2022, p. 270) observe an 'overreliance on third-party data, incongruent CSR measurements, and unclear mechanisms between CSR and its outcome' as major problems affecting research on tourism CSR. Earlier contributions pointed to its being easier to study the CSR *of* an organisation rather than *in* it (Coles et al., 2014). For an ethos that should cut across activities and organisational divisions, access often precludes researchers embedded in-situ. Instead, many studies depend on aggregated, organisational-level views that have a propensity both to under-report the nature and quantum of activity as well as glossing-over the intricacies of implementing CSR in complex organisations.

Casting a critical eye on the literature in this way suggests very strongly that tourism research CSR has *not* moved on its intellectual agenda that far in the past decade. For example, notwithstanding the appliance of advanced technology to conduct a Delphi study, Horng et al. (2018) still concluded that one dimension — the environment — is perceived as the most influential. Despite their use of powerful analytical software and the latest visualisation techniques, content analysis of CSR statements is much the same topic and generating similar-looking research questions and conclusions, much as it ever has (Holcomb et al., 2007; de Grosbois, 2012; de Grosbois, 2016). In echoes of earlier work, Uyar et al. (2021a, p. 2614) concluded that 'stakeholders, investors, customers, and employees are the primary target audiences of SRs' and that 'generic environmental themes are sufficiently addressed. . ..', further tellingly adding that this latest result is '....in congruence with the findings of some earlier studies, which proved the dominance of environmental aspects over the other dimensions of sustainability'. In fact, it seems quite evident that a great many studies published in the last decade are fighting similar if not the same intellectual 'battles' as before, covering the same sorts of issues as the 2010s as the 2000s. Most prominent among them still appears to be the continuing fetish for examining the nature and strength of the relationship between firm performance and CSR behaviour (Franco et al., 2020; Moneva et al., 2020; Camilleri, 2021; Kaur et al., 2022); in other words, uncertainty appears to persist, the (possibly illusory) goal appears to be to overcome this (once and for all), and the challenge remains to justify why it is necessary for tourism business or organisations to act responsibly. Research has steadily appeared on the wider business case for CSR (Rhou & Singal, 2020; Camilleri, 2021) and whether it is possible to do well by doing good, or do good by doing well (Garay & Font, 2012; Fu et al., 2014; Marino-Romero et al., 2018). To a degree, revisiting this relationship makes sense, for instance when the purpose of acting in a socially-responsible manner has been challenged during the pandemic (Shin et al., 2021). By and large though, we find ourselves in the 2020s not the 1960s and 1970s; the case for CSR as a broadly 'good thing' for an enterprise to be engaged in, was made long ago (Lee, 2008, cited in Coles et al., 2015); and these days there is almost a social, even cultural expectation in many markets that businesses, organisations and those operating them will act in a responsible manner.

Of course, an obvious counter to this view is that work in the 2000s was foundational and further research in a similar vein during the 2010s was necessary to consolidate the body of knowledge. Furthermore, economic conditions and operating environments were different in the 2010s to the later 2000s when there was a major global financial crisis followed by an economic downturn (Coles, 2013). A keener counter-argument may be that this observation is the result of a self-fulfilling analytical prophecy because of the particularity of the original categorisation and its anchoring in management studies around 2010. However, even this argument does not hold. The frontiers of management studies of CSR have advanced in this time and tourism studies have not always kept pace. As one prominent commentator notes of research appearing in a leading journal (Wickert, 2021), this has shifted from a more business-centric to society-centric focus from 2006 to 2021. Moreover, turning the tables somewhat, Volgger and Haung (2019) have recognised the importance of scoping *irresponsible* behaviour as a means of widening perspectives of tourism CSR. Their study is one of four in total, published between 2016 and 2022. Within Scopus, 'corporate' 'social' 'irresponsibility' is the subject of 260 studies in total and 176 over the past six years.

The Future of Responsibility Scholarship on Tourism

Arguably then, relatively conservative progress has been made in understanding tourism CSR over the past decade, notwithstanding the quantum of attention. The question is, where next for studies of tourism CSR?

Pandemic and Post-pandemic Studies of Tourism CSR

The more obvious and somewhat predictable place to start is the Coronavirus pandemic, the unprecedented disruption this created through the virtual cessation of travel and tourism, and the need for tourism businesses and organisations to take on different roles, including public health which rarely featured in the discourse before. At a time of major and very rapid reductions in their revenues and set against a Friedman-esque backdrop, it is perhaps all the more remarkable that the tourism sector exhibited high levels of social responsibility and was able to contribute collectively to adapting to the challenge of Coronavirus and COVID-19. Whether this was in the form of arranging furlough schemes for employees or acting on behalf of the state to provide quarantine facilities (Choi & Choi, 2021), developing new safety procedures and innovations to combat the spread of the virus or monitoring the health status of customers through 'vaccine passports', certifications and other technologies, the 'grey literature' and popular media are replete with stories of how the tourism sector, writ large, contributed to overcoming this major public health disaster. There is also evidence of organisations laying-off or redeploying staff because slack or spare capacity is lacking (Coles et al., 2022); organisations unwilling to enter into job retention schemes; employees leaving the sector to find safer, less precarious employment elsewhere (Coles et al., 2022); and the closure of enterprises unable to sustain themselves without additional support.

The understandings and practices of social responsibilities by tourism sector organisations have not gone unnoticed in rapid publications, many of which are centred on enhancing future resilience among organisations in the sector (Razak, 2020; Colmekcioglu et al., 2022; Marco-Lajara et al., 2022; Shin et al., 2021). Lin et al. (2022) note the promptness and responsive of casinos to saving lives and uniting communities in what they term 'just in time' CSR as a throwback to traditional views of CSR that view it as a longer-term strategic and cultural concern for organisations (see also Lin et al., 2021). The effects of COVID-19 on employees and the workforce as well as their efforts to act responsibly, in particular to protect customers, clients and communities as major stakeholder groups have been a major coalescence point for attention (Choi & Choi, 2021; Lin et al., 2021; Mao et al., 2021). For Kim et al. (2022), innovations in handling customers, both in the form of communications and contactless interaction, 'foster consumer confidence and response efficacy, leading to positive consumers' attitudes and booking intentions' (Kim et al., 2022, p. 1).

Amid this burgeoning endeavour, Lee (2022) has suggested a potential research agenda, with topics including, inter alia: 'emerging CSR initiatives throughout the pandemic, a comparison between pre- and post-pandemic financial implications of CSR, a moderating role of CSR strategy, various performance measures including risk measures, and culture and industry as boundary conditions' (Lee, 2022, p. 863). Of itself, the Coronavirus pandemic is unprecedented and there is a clear need to address the sorts of issues that Lee (2022) identifies. However, if a longer view is taken the proposed agenda is quite conservative and largely replicates the sorts of interests characterising the body of knowledge under the 'old normal'. As noted above, previously considerable attention had been devoted to implementation, firm performance, and relative performance in and among sub-sectors, in particular because of their different strategies and approaches. Although these three topics are more overtly reminiscent of research prior to the pandemic, both culture as a boundary (i.e. cultural understandings of the firm and of responsibility etc.; e.g. Kang et al., 2016) and the need for more tailored, sub-sector-specific analysis (i.e. deeper dives into practices, activities and outcomes; e.g. Youn et al., 2016) also represent extensions of the existing academic gaze.

Tourism CSR and 'Grand Challenges': Another Way Forward?

At present the opportunities afforded by the pandemic for the study of tourism CSR have not been fully realised: whether they will be is an altogether different question. As is the extent to which critical scholarship will emerge in this area. Currently, the discourse focuses in a very literal manner on what was done to act responsibly, rather than what was not done or where the responsibility of firms was, for instance, suspended or called into question (for instance, as being potentially irresponsible).

The extra-ordinary conditions during the pandemic offered considerable untapped insights, almost in the form of natural experiments. Often one of the criticisms of CSR is that it is a feature of business administration for stable, predictable, even 'normal' operating conditions. In a crisis or catastrophe, higher level concerns, like acting responsibly and in a more principled manner, may not always seem to make sense in terms of firm performance (Lee, 2022), possibly having a deleterious effect (Shin et al., 2022). In the post-pandemic period, there will also be valid research questions about how tourism organisations will perceive their responsibilities (the same as and reverting to the 'old normal', or a new approach for new times?) and how they will embed them in their operations and cultures. However, to follow well-trodden paths yet again is to miss the chance to inject a new dynamic into tourism studies of CSR. Once more the focus would be on CSR almost as a business attribute, an input which ultimately results in outcomes, mainly in the form of firm performance. If the lived experiences of Coronavirus-COVID 19 demonstrate anything, it is the importance of considering questions of 'responsibility for what?' and 'what may be achieved together?' in response to the worst international public health crisis in nearly a century.

Put another way, it appears time to shift the dial away from 'how' or 'in what ways' is an organisation responsible, towards a greater understanding of the totality of contribution that the sector can make collectively towards addressing grand challenges; in brief, a shift from the business-centric to the society-centric seen elsewhere in CSR studies (Wickert, 2021). Major global issues including but not restricted to climate change, poverty alleviation, resource depletion, greater equality, diversity and inclusion, the eradication of modern slavery, and so on, will not be tackled in any other way. As Friedman (1970) reflected in one of the lesser-known passages of his famous work, 'the doctrine of "social responsibility" . . .. does not differ in philosophy from the most explicitly collectivist doctrine. It differs only by professing to believe that collectivist ends can be attained without collectivist means' (Friedman, 1970, cited in Crane et al., 2008, p. 32). In other words, it is fallacious to argue that collective goals can and will be achieved independently and disparately through individual firms acting separately.

Quite so. Yet the persistence of more granular approaches, especially those dealing with smaller or select groups of businesses, as the primary focal point for research on tourism CSR, is at odds with a mode of thinking that extols mobilisation to, and organisation around, shared interests and common imperatives. Reporting on an organisation (or organisations) acting responsibly only takes us so far; the far more urgent issue is what contribution(s) does the sector make to the greatest, often very urgent issues facing humanity? In a sense then, this echoes the ethos behind ESG and measuring how business outcomes contribute to more sustainable development. It is also similar to, and may be regarded by some as an extension of, the strand of work looking at CSR impacts. However, it is different in one key respect. Instead of trying to specify individual impacts that may accrue from responsibility — for instance, from particular 'programmes' or 'strategies' or organisational 'cultures'— future critical scholarship on tourism CSR requires more mature discussion about collective responsibility per se and who is able to contribute what, who is not and why. This may require revisiting the relationship between the state and CSR (Wickert, 2021). Not every organisation can or will contribute to each shared imperative, and to the same degree. As contemporary approaches to neo-liberal governance in the developed world reveal, government alone (i.e. the state) cannot tackle grand challenges alone 'on behalf of its citizens' but rather citizens (i.e. as employees, managers, owners, investors etc.) have to engage with, and contribute towards, solutions. Thus, a more pragmatic assessment of what will have to be done economy- or society-wide to tackle issues, is both required and helpful.

Conclusion – Time to Shift the Dial

Although not intended to be a retrospective on 'Friedmann's view at 50', there is a sense that, as the subject area of CSR in travel and tourism advances, it is essential to review its achievements periodically and cast a critical eye over them. Just as Friedman passed on the 'intellectual baton', as it were, to a new generation of thinkers many years ago, the principles and practices of CSR have featured in the research agenda and the curriculum of tourism studies for nearly a quarter of a century. Today ideas that appeared fresh, relevant even radically different in the 2000s or early 2010s are being conveyed to students, employees and other stakeholders by a wide array of academics, practitioners and instructors themselves trained at that time. The challenge then is to build on the past for a better, qualitatively different future.

Tourism studies of CSR are at a critical moment. There is ample evidence of a growing, even maturing body of knowledge that has consolidated our understanding of this major aspect of business and management. No doubt the corpus of work will continue to grow if left to its own devices. Many of the thematic, geographical or sectoral gaps evident just a decade ago have been, or are in the process of being, gradually filled by piecemeal additions and the inexorable march of empirical studies. An implicit subtext remains the extent to which businesses and organisations in travel and tourism conform to mainstream thinking developed from a range of other sectors. The risk is that tourism studies of CSR will become ever more rooted in the past and not keep pace with evolving times and new thinking. Herein lies an apparent paradox: there is a responsibility on those variously researching, learning and teaching about the administration and governance of travel and tourism organisations to ensure that their knowledge remains current and fit for purpose.

The time is right to shift the dial. The body of knowledge can — and may well — continue to grow in a conservative, incremental fashion with some familiar looking themes revisited, both in studies of the pandemic as well as the post-pandemic period. The question is what will we really learn from revisiting familiar and comfortable themes in this way? Is there really a need for studies of CSR that are fundamentally rooted in the past? The world moves on and there is a case for greater consideration to the 'collective responsibility' of tourism organisations. In this respect, the pandemic may yet turn out to be a transformative moment, a time when we move from describing 'what', 'by whom' and 'why' of tourism CSR to a far greater focus on the magnitude of collective endeavour to tackle significant (global) challenges, the effects the combined efforts have, and the contribution of the sector relative to others. An 'accountability turn' of this nature may help us overcome the persistence of approaches that are, in Friedman's (1970, cited in Crane et al., 2008, p. 31) once more prophetic words, 'short-sighted and muddle-headed in matters that are outside . . .. businesses but affect the survival of businesses in general'.

References

Acampora, A., Lucchetti, M., Merli, R., & Ali, F. (2022). The theoretical development and research methodology in green hotels research: a systematic literature review. *Journal of Hospitality and Tourism Management*, *51*, 512–528. https://doi.org/10.1016/j.jhtm.2022.05.007

Aguinis, H., & Glavas, A. (2012). What we know and don't know about corporate social responsibility: A review and research agenda. *Journal of Management*, *38*(4), 932–968. https://doi.org/10.1177/0149206311436079

Alva (2020, October 27). *What's the difference between CSR and ESG?* https://www.alva-group.com/blog/whats-the-difference-between-csr-and-esg/

Camilleri, M. (2021). Strategic attributions of corporate social responsibility and environmental management: the business case for doing well by doing good! *Sustainable Development*, *30*(3), 409–422. https://doi.org/10.1002/sd.2256

Choi, M., & Choi, Y. (2021). Employee perceptions of hotel CSR activities during the COVID-19 pandemic. *International Journal of Contemporary Hospitality Management*, *33*(10), 3355–3378. https://doi.org/10.1108/IJCHM-03-2021-0289

Coles, T. E. (2013). Much ado about nothing? Tourism and the financial crisis. In G. Visser, & S. Ferreira (Eds), *Tourism and crisis* (pp. 35–52). Routledge.

Coles, T. E., Fenclova, E. A., & Dinan, C. R. (2013). Tourism and corporate social responsibility: a critical review and research agenda. *Tourism Management Perspectives*, 6, 122–141. https://doi.org/10.1016/j.tmp.2013.02.001

Coles, T. E., Fenclova, E., & Dinan, C. (2014). Corporate social responsibility reporting among European low-fares airlines: challenges for the examination and development of sustainable mobilities. *Journal of Sustainable Tourism*, *22*(1), 69–88. https://doi.org/10.1080/09669582.2013.790391

Coles, T. E., Fenclova, E., & Dinan, C. R. (2011). Responsibilities, recession and the tourism sector: perspectives on CSR among low-fares airlines during the economic downturn in the UK. *Current Issues in Tourism*, *14*(6), 519–536. https://doi.org/10.1080/13683500.2010.544719

Coles, T. E., Fenclova, E., & Dinan, C. R. (2015). Tourism and corporate social responsibility. In C. M. Hall, S. Gössling & D. Scott (Eds), *The Routledge handbook of tourism and sustainability* (pp. 270–279). Routledge.

Coles, T. E., Garcia, G., O'Malley, E., & Turner, C. (2022). Experiencing event management during the Coronavirus pandemic: a public sector perspective. *Frontiers in Sport and Active Living (Sport, Leisure and Tourism)*, *3*, 814146. https://doi.org/10.3389/fspor.2021.814146

Colmekcioglu, N., Dineva, D., & Lu, X. (2022). "Building back better": the impact of the COVID-19 pandemic on the resilience of the tourism and hospitality industries. *International Journal of Contemporary Hospitality Management*. https://doi.org/10.1108/IJCHM-12-2021-1509

Crane, A., Matten, D., & Spence, L. (Eds). (2008). *Corporate social responsibility: Readings and cases in a global context.* Routledge.

de Grosbois, D. (2012). Corporate social responsibility reporting by the global hotel industry: commitment, initiatives and performance. *International Journal of Hospitality Management*, *31*(1), 896–905. https://doi.org/10.1016/j.ijhm.2011.10.008

de Grosbois, D. (2016). Corporate social responsibility reporting in the cruise tourism industry: a performance evaluation using a new institutional theory based model. *Journal of Sustainable Tourism*, *24*(2), 245–269. https://doi.org/10.1080/09669582.2015.1076827

Elrick, J., & Thies, C. (2018). The social responsibility of business. Milton Friedman reconsidered. *Journal of Markets and Morality*, *21*(2), 297–307. https://www.marketsandmorality.com/index.php/mandm/article/view/1356

Farmaki, A. (2019). Corporate social responsibility in hotels: a stakeholder approach. *International Journal of Contemporary Hospitality Management*, *31*(6), 2297–2320. https://doi.org/10.1108/IJCHM-03-2018-0199

Farmaki, A., Pappas, N., Kvasova, O., & Stergious, D. (2022). Hotel CSR and job satisfaction: a chaordic perspective. *Tourism Management*, *91*, 104526. https://doi.org/10.1016/j.tourman.2022.104526

Farrington, T., Curran, R., Gori, K., O'Gorman, K., & Queenan, C. (2017). Corporate social responsibility: reviewed, rated, revised. *International Journal of Contemporary Hospitality Management*, *29*(1), 30–47. https://doi.org/10.1108/IJCHM-05-2015-0236

Fatima, T., & Elbanna, S. (2022). Corporate social responsibility (CSR) implementation: A review and a research agenda towards an integrative framework. *Journal of Business Ethics*. https://doi.org/10.1007/s10551-022-05047-8

Font, X., & Lynes, J. (2018). Corporate social responsibility in tourism and hospitality. *Journal of Sustainable Tourism*, *26*(7), 1027–1042. https://doi.org/10.1080/09669582.2018.1488856

Franco, S., Caroli, M., Cappa, F., & del Chiappa, G. (2020). Are you good enough? CSR, quality management and corporate financial performance in the hospitality industry. *International Journal of Hospitality Management*, *88*, 102395. https://doi.org/10.1016/j.ijhm.2019.102395

Frey, N., & George, R. (2010). Responsible tourism management: the missing link between business owners' attitudes and behaviour in the Cape Town tourism industry. *Tourism Management*, *31*(5), 621–628. https://doi.org/10.1016/j.tourman.2009.06.017

Friedman, M. (1962). *Capitalism and freedom*. University of Chicago Press.

Friedman, M. (1970). The social responsibility of business is to increase its profits. *The New York Times Magazine*, (pp. 32–33 and 122–126).

Fu, H., Ye, B., & Law, R. (2014). You do well and I do well? The behavioural consequences of corporate social responsibility. *International Journal of Hospitality Management*, *40*, 62–70. https://doi.org/10.1016/j.ijhm.2014.03.004

Garay, L., & Font, X. (2012). Doing good to do well? Corporate social responsibility reasons, practices and impacts in small and medium accommodation enterprises. *International Journal of Hospitality Management*, *31*(2), 329–337. https://doi.org/10.1016/j.ijhm.2011.04.013

Guzzo, R., Abbott, J., & Madera, J. (2020). A micro-level view of CSR: a hospitality management systematic literature review. *Cornell Hospitality Quarterly*, *61*(3), 332–352. https://doi.org/10.1177/1938965519892907

Hawkins, R., & Bohdanowicz, P. (2012). *Responsible hospitality. Theory and practice*. Goodfellow.

Holcomb, J., Upchurch, R., & Okumus, F. (2007). Corporate social responsibility: what are the top hotel companies reporting? *International Journal of Contemporary Hospitality Management*, *19*(6), 461–475. https://doi.org/10.1108/09596110710775129

Holme, R., & Watts, P. (2000). *Corporate social responsibility: Making good business sense*. World Business Council for Sustainable Development (WBCSD).

Hopkins, M. (2016). *CSR and sustainability. From the margins to the mainstream: A textbook*. Taylor and Francis.

Horng, J-S., Hsu, H., & Tsai, C-Y. (2018). An assessment model of corporate social responsibility practice in the tourism industry. *Journal of Sustainable Tourism*, *26*(7), 1085–1104. https://doi.org/10.1080/09669582.2017.1388384

Hung, C. (2021, September 23). Three reasons why CSR and ESG matter to businesses. *Forbes*. https://www.forbes.com/sites/forbesbusinesscouncil/2021/09/23/three-reasons-why-csr-and-esg-matter-to-businesses/?sh=470e119739b9

Iyer, G., & Jarvis, L. (2019). CSR adoption in the multinational hospitality context: a review of representative research and avenues for future research. *International Journal of Contemporary Hospitality Management*, *31*(6), 2376–2393. https://doi.org/10.1108/IJCHM-06-2018-0451

Kang, K., Lee, S., & Yoo, C. (2016). The effect of national culture on corporate social responsibility in the hospitality industry. *International Journal of Contemporary Hospitality Management*, *28*(8), 1728–1758. https://doi.org/10.1108/IJCHM-08-2014-0415

Kaur, P., Talwar, S., Madanaguli, A., Srivastava, S., & Dhir, A. (2022). Corporate social responsibility(CSR) and hospitality sector: Charting new frontiers for restaurant businesses. *Journal of Business Research*, *144*, 1234–1248. https://doi.org/10.1016/j.jbusres.2022.01.067

Kim, B., & Lee, S. (2020). The impact of material and immaterial sustainability on firm performance: the moderating role of franchising strategy. *Tourism Management*, *77*, 103999. https://doi.org/10.1016/j.tourman.2019.103999

Kim, B., Lee, S., & Kang, K. (2018). The moderating role of CEO narcissism on the relationship between uncertainty avoidance and CSR. *Tourism Management*, *67*, 203–213. https://doi.org/10.1016/j.tourman.2018.01.018

Kim, H., Li, J., & So, K. (2022). Enhancing consumer confidence and response efficacy in tourism: typology and effectiveness of the hotel industry's responses to COVID-19. *Journal of Travel Research*. 10.1177/00472875221095211

Kim, M., & Stepchenkova, S. (2021). Do consumers care about CSR activities of their favourite restaurant brands? Evidence from engagement on social networks. *Journal of Hospitality Marketing and Management*, *30*(3), 305–325. https://doi.org/10.1080/19368623.2020.1817223

Lee, S. (2022). Corporate social responsibility and COVID-19: Research implications. *Tourism Economics*, *28*(4), 863–869. https://doi.org/10.1177/1354816620978136

Lin, L-P., Yu, C-Y., & Chang, F-C. (2018). Determinants of CSER practices for reducing greenhouse gas emissions: from the perspectives of administrative managers in tour operators. *Tourism Management*, *64*, 1–12. https://doi.org/10.1016/j.tourman.2017.07.013

Lin, Z., Wong, I., & Lin, S. (2022). Just-in-time CSR for urgent needs during the COVID-19 outbreak: The case of casinos. *Journal of China Tourism Research*. https://doi.org/10.1080/19388160.2022.2085221

Lin, Z., Wong, I., Lin, S., & Yang, Y. (2021). Delivering warmth and expediting love: just-in-time CSR in the midst of COVID-19 mega disruption. *International Journal of Contemporary Hospitality Management*, *33*(12), 4334–4354. https://doi.org/10.1108/IJCHM-05-2021-0678

Lindgreen, A., & Swaen, V. (2010). Corporate social responsibility. *International Journal of Management Reviews*, *12*(1), 1–7. https://doi.org/10.1111/j.1468-2370.2009.00277.x

Liu, P., & Tse, E. (2021). The impact of perceived corporate social responsibility on dining intention in US restaurants: focusing on customers' health. *Journal of Culinary Science and Technology*. https://doi.org/10.1080/15428052.2021.2016533

Madanaguli, A., Srivastava, S., Ferraris, A., & Dhir, A. (2022). Corporate social responsibility and sustainability in the tourism sector: a systematic literature review and future outlook. *Sustainable Development*, *30*(3), 447–461. https://doi.org/10.1002/sd.2258

Mao, Y., He, J., Morrison, A., & Andres Coca-Stefaniek, J. (2021). Effects of tourism CSR on employee psychological capital in the COVID-19 crisis: from the perspective of conservation of resource theory. *Current Issues in Tourism*, *24*(19), 2716–2734. https://doi.org/10.1080/13683500.2020.1770706

Marco-Lajara, B., Garcia, M., Ruiz-Fernandez, L. Poveda-Pareja, E., & Sanchez-Garcia, E. (2022). Rural hotel resilience during COVID-19: the crucial role of CSR. *Current Issues in Tourism*, *25*(7), 1121–1135. https://doi.org/10.1080/13683500.2021.2005551

Marino-Romero, J., Campon-Cerro, A., Hernandez-Mogollon, J., & Folgado-Fernandez, J. (2018). The relationship between revenue and environmental responsibility: a causal study using reputation in the hotel industry. In A. Leal-Millan, M. Peris-Ortiz & A. Leal-Rodríguez (Eds), *Sustainability in innovation and entrepreneurship. Innovation, technology, and knowledge management* (pp. 119–136). Springer.

McCracken, P. (1996). Economic policy in the Nixon Years. *Presidential Studies Quarterly*, *26*(1), 165–177. https://www.jstor.org/stable/27551556

Meng-Chan Lau, V., Ren, L., & Yang, F. (2021). CSR and casino hotel branding: the joint moderation of CSR misfit and corporate awareness. *Journal of Hospitality and Tourism Management*, *49*, 375–384. https://doi.org/10.1016/j.jhtm.2021.10.011

Moneva, J., Bonilla-Priego, M., & Ortas, E. (2020). Corporate social responsibility and organisational performance in the tourism sector. *Journal of Sustainable Tourism*, *28*(6), 853–872. https://doi.org/10.1080/09669582.2019.1707838

Mowforth, M., Charlton, C., & Munt, I. (2008). *Tourism and responsibility. Perspectives from Latin America and the Caribbean*. Routledge.

Pena-Miranda, D., Guevara, A., Fraiz, J., & Camilleri, M. (2021). A corporate social responsibility model for a competitive and resilient hospitality industry. *Sustainable Development*, *1*, 14. https://doi.org/10.1002/sd.2259

Qian, J., Law, R., Wei, J., & Li, X. (2021). Progress of hotel corporate social responsibility research in terms of theoretical, methodological and thematic development. *Journal of Hospitality Marketing and Management*, *30*(6), 717–737. https://doi.org/10.1080/19368623.2021.1883497

Razak, N. (2020). Health and tourism: implications of COVID-19 pandemic to the Malaysian travel and hospitality sector. *International Journal of Supply Chain Management*, *9*(4), 663–670.

Rhou, Y., & Singal, M. (2020). A review of the business case for CSR in the hospitality industry. *International Journal of Hospitality Management*, *84*, 102330. https://doi.org/10.1016/j.ijhm.2019.102330

Serra-Cantallops, A., Pena-Miranda, D., Ramon-Cardona, J., & Martorell-Cunill, O. (2018). Progress in research on CSR and the hotel industry (2006–2015). *Cornell Hospitality Quarterly*, *59*(1), 15–38. https://doi.org/10.1177/1938965517719267

Shin, H., Sharma, A., Nicolau, J., & Kang, J. (2021). The impact of hotel CSR for strategic philanthropy on booking behaviour and hotel performance during the COVID-19 pandemic. *Tourism Management*, *85*, 104322. https://doi.org/10.1016/j.tourman.2021.104322

Stobierski, T. (2021, April 8). Types of corporate social responsibility to be aware of. *Harvard Business School Online*. https://online.hbs.edu/blog/post/types-of-corporate-social-responsibility

Su, L., & Swanson, S. (2019). Perceived corporate social responsibility's impact on the well-being and supportive green behaviours of hotel employees: the mediating role of the employee-corporate relationship. *Tourism Management*, 72, 437–450. https://doi.org/10.1016/j.tourman.2019.01.009

Su, L., Swanson, S., & He, X. (2020). A scale to measure residents perceptions of destination social responsibility. *Journal of Sustainable Tourism*, *28*(6), 873–897. https://doi.org/10.1080/09669582.2019.1708372

Sung, K., Tao, C-W., & Slevitch, L. (2022). Do strategy and content matter? Restaurant firms' corporate social responsibility communication on Twitter: A social network theory perspective. *Tourism and Hospitality Research*. https://doi.org/10.1177/14673584221103177

Theodoulidis, B., Diaz, D., Crotto, F., & Rancati, E. (2017). Exploring corporate social responsibility and financial performance through stakeholder theory in the tourism industries. *Tourism Management*, *62*, 173–188. https://doi.org/10.1016/j.tourman.2017.03.018

Uyar, A., Kilic, M., Koseoglu, M., Kuzey, C., & Karaman, A. (2020). The link among board characteristics, corporate social responsibility performance, and financial performance: Evidence from the hospitality and tourism industry. *Tourism Management Perspectives*, *35*, 100714. https://doi.org/10.1016/j.tmp.2020.100714

Uyar, A., Koseoglu, M., & Kilic, M. (2021a). Thematic structure of sustainability reports of the hospitality and tourism sector: a periodical, regional, and format-based analysis. *Current Issues in Tourism*, *24*(18), 2602–2627. https://doi.org/10.1080/13683500.2020.1847050

Uyar, A., Kuzey, C., Kilic, M., & Karaman, A. (2021b). Does firms' CSR engagement support tourism sector development? Moderation effect of CSR committee and CEO duality. *Tourism Economics*. https://doi.org/10.1177/13548166211024502

Van de Mosselar, F., van de Duim, R., & van Wijk, J. (2012). Corporate social responsibility in the tour operating industry: the case of Dutch outbound tour operators. In D. Leslie (Ed.), *Tourism enterprises and the sustainability agenda across Europe*. Ashgate.

Volgger, M., & Huang, S. (2019). Scoping irresponsible behaviour in hospitality and tourism: Widening the perspective of CSR. *International Journal of Contemporary Hospitality Management*, *31*(6), 2526–2543. https://doi.org/10.1108/IJCHM-05-2018-0367

Wang, C., Hu, R., & Zhang, T. (2020). Corporate social responsibility in international hotel chains and its effects on local employees: Scale development and empirical testing in China. *International Journal of Hospitality Management*, *90*, 102598. https://doi.org/10.1016/j.ijhm.2020.102598

Wickert, C. (2021). Corporate social responsibility research in the *Journal of Management Studies*: a shift from a business-centric to a society-centric focus. *Journal of Management Studies*, *58*(8), E1–E17. https://doi.org/10.1111/joms.12775

Wong, A., Kim, S., & Lee, S. (2022). The evolution, progress, and the future of corporate social responsibility: comprehensive review of hospitality and tourism articles. *International Journal of Hospitality and Tourism Administration*, *23*(1), 1–33. https://doi.org/10.1080/15256480.2019.1692753

Wong, A., Koseoglu, M., & Kim, S. (2021). The intellectual structure of corporate social responsibility research in tourism and hospitality: A citation/co-citation analysis. *Journal of Hospitality and Tourism Management*, *49*, 270–284. https://doi.org/10.1016/j.jhtm.2021.09.015

Youn, H., Song, S., & Lee, S. (2016). Does restaurant type matter for investment in corporate social responsibility? *International Journal of Hospitality Management*, *58*, 24–33. https://doi.org/10.1016/j.ijhm.2016.07.004

Chapter 42

Green Human Resource Management in the Tourism and Hospitality Industry
Outlook and Further Research

Tran Hoang Tuan, Nhat Tan Pham, & Tan Vo-Thanh

Introduction

Due to harmful environmental consequences caused by human activities, the environment has become one of the top concerns in fields such as politics, law, and business (Howard-Grenville et al., 2014; Pham et al., 2022; Pham, Vo-Thanh, Tučková, et al., 2020), and is requiring an expanded task of environmental management proactively in organisations (Vo-Thanh et al., 2020). This growing environmental concern can easily be observed via the legal regulations and environmental standard guidelines from the government, and requirements from stakeholders, such as business partners and consumers (Paulet et al., 2021). Consequently, organisations need to integrate green and sustainable actions into their activities, helping improve their ecological footprint and economic objectives simultaneously (Ahmed et al., 2021; Tanova & Bayighomog, 2022; Vo-Thanh et al., 2020).

Changes to achieve green in the management strategy of the organisation have drawn attention of scholars from various areas, for instance from human resource management (HRM), organisational culture, marketing, and finance (Obeidat et al., 2020; Pham, Chiappetta Jabbour, et al., 2020; Pham et al., 2018; Pham et al., 2019; Shafaei et al., 2020). The utilisation of green HRM (GHRM) has attracted numerous scholars value to tackle environmental issues in a human resource context (Pham et al., 2019; Pham, Vo-Thanh, Tučková, et al., 2020; Ren et al., 2018; Renwick et al., 2013). The proposed theories and empirical findings show the importance of accomplishing ecologically sustainable development by the way of HRM practices (Aust et al., 2020; Pham et al., 2022). The target of GHRM is to upgrade employees' willingness and contribution to the environmental performance in their work (Pham et al., 2022). Accordingly, GHRM is described as a component of a sustainable HRM approach that considers environmental sustainability in the operations of organisations (Aust et al., 2020; Pham et al., 2022; Tanova & Bayighomog, 2022).

Recently, there has growth in interest in GHRM application in organisations in a range of different industries. Nonetheless, only a few theoretical and empirical works have been

The Wiley Blackwell Companion to Tourism, Second Edition. Edited by C. Michael Hall.

connected with GHRM in the service sector (Tanova & Bayighomog, 2022), particularly in the tourism and hospitality industry, despite the great harm provoked by this industry to the environment (Pham, Chiappetta Jabbour, et al., 2020; Pham, Vo-Thanh, Tučková, et al., 2020).

The tourism and hospitality sector is a significant driver of worldwide economic development, and it therefore affects other economic sectors (Siyambalapitiya et al., 2018). It creates tremendous benefits yet consumes much energy as well as emanates a lot of waste to the environment (Hall, 2009; Shafaei et al., 2020; Vo-Thanh et al., 2020). Meanwhile, preserving the landscape and natural environment is critical in guaranteeing the customers' requirements and satisfaction, and sustainable development of the tourism and hospitality industry (Tanova & Bayighomog, 2022). Moreover, clients are increasingly likely to decline tourism products and services that harm the environment (Paulet et al., 2021). Therefore, implementing GHRM is promising for tourism and hospitality organisations to truly change perceptions, attitudes, and behaviours of stakeholders, whether they are customers, business and social partners, governments, or employees (Pham et al., 2022; Shafaei et al., 2020; Tanova & Bayighomog, 2022). Given the significant role of GHRM in the tourism and hospitality industry, this chapter discusses the GHRM practices, their antecedents and consequences, and the barriers to implementing these practices within tourism and hospitality organisations, as well as potential research directions to enrich the current GHRM literature.

GHRM Practices

For tourism and hospitality organisations, laying out environmental management goals is central to achieving both environmental and economic performances, as the two have been found to be interrelated (Yusoff et al., 2020). The survival of a tourism and hospitality organisation requires it to be friendly to the ecosystem, thereby acquiring a competitive advantage by being sustainable (Pham et al., 2022). While implementing GHRM, employees play a crucial role. Accordingly, organisations need to make them be green through developing their green abilities, motivating them to engage in green behaviours, and providing them with opportunities to share green initiatives (Pham et al., 2022; Renwick et al., 2013). Employees should have a 'commitment to the issues of environmental sustainability' (Masri & Jaaron, 2017, p. 474). This implies that they need to clearly understand the organisation's expectations and requirements in terms of environmental protection, helping them adopt appropriate behaviours that are vital to the environment (Pham, Chiappetta Jabbour, et al., 2020; Vo-Thanh et al., 2020). However, from the organisation side, how to engage employees in green behaviours and encourage them to contribute to green initiatives is challenging (Yusoff et al., 2020), though there are some existing practices, such as avoiding wasting water and energy, recycling waste, reducing carbon dioxide emissions, and so on. Scholars stated that GHRM practices could play a significant role in influencing employees' environmental awareness, thus contributing to decreasing the costs caused by the impact of organisations' activities on the environment and enhancing success in organisations' environmental management exercises (Russo & Fouts, 1997; Siyambalapitiya et al., 2018).

Based on the extant literature, scholars have suggested five main GHRM practices: green recruitment; green training; green evaluation; green feedback handling; and green compensation.

Green Recruitment

Organisations carrying out GHRM should attract and recruit collaborators who tend to be supportive and committed to partaking in environmental protection (Renwick et al., 2013). In other words, they need to build a sustainable and environmentally-oriented workforce.

These collaborators are expected to adapt to the green organisational culture and promote it (Wehrmeyer, 1996). Bauer et al. (2012) argue that environmental performance development requires the organisation to hire employees more sensitive towards the environment and, consequently, more willing to engage in green activities. Green employee recruitment provides the organisation with several benefits. If organisations succeed in recruiting employees who have sensitivity towards and knowledge of environmental protection, they will gain economically, because it allows them to reduce costs and time spent training new employees to fit their green culture and engage in green behaviours. Moreover, green-oriented employees can assist organisations in expanding the employer brand (e.g., positive public image, differentiation), establishing better relationships with local community and authorities and business and social partners, and enhancing customer satisfaction (Pham et al., 2022).

Additionally, potential environment-oriented collaborators are likely to look for and prioritise working for reputable and green-oriented organisations. Indeed, they identify with green organisations' values and convictions and think that these green organisations can bring them solidarity, social sustainability, and work stability (Amrutha & Geetha, 2020). Therefore, through the implementation of GHRM, organisations have a green reputation and good environmental performance, which will help them attract the right environmentally conscious employees to work for them.

Green Training

To meet their green objectives, organisations require employees to be prepared to procure the critical green competencies. Green training programs are 'fundamental to any successful activity of environmental management' (Chiappetta Jabbour, 2013, p. 145). Empirically, Pham, Vo-Thanh, Shahbaz, et al. (2020) indicate that green training is an effective solution to improve employee green performance. When the organisation deals with green training programmes, focus should be put on developing green skills and competencies and making employees be aware of the importance of their green behaviours (Guerci et al., 2015). Thanks to green training programmes, employees are able to identify environmental problems and, as a consequence, to take appropriate actions to improve environmental performance (Chiappetta Jabbour, 2013). Green training has been shown to be the most adopted practice among GHRM practices (Guerci et al., 2015).

Employees well trained in environmental issues are willing to help organisations save resources, for example, energy and water and lessen waste, as well as to recommend innovative ideas to them (Zoogah, 2011). Since these employees actively participate in the organisation's environmental management activities, they may be recognised by the organisation, which is the foundation for them to foster their career (Vo-Thanh et al., 2021). Furthermore, effective green training programmes contribute to enhancing employees' motivation at work and their commitment to the organisation (Amrutha & Geetha, 2020).

Green Evaluation

During the working process, in addition to allotting tasks required to, and supporting and training employees to play out their job well, tourism and hospitality organisations also need to periodically carry out green employee evaluation. The green employee evaluation framework is often developed by organisations to assess execution standards as per the organisation's interest in both environmental management and economic management (Pham et al., 2022). Tourism and hospitality organisations applying GHRM ought to set up an evaluation framework to check the environmental tasks of employees in the working process. In this regard, they need to establish environmental performance standards for every employee (Alfred & Adam, 2009). Green employee evaluation therefore allows tourism and hospitality organisations to strike a balance between economic and environmental objectives (Amrutha & Geetha, 2020).

Moreover, since employees are evaluated on their green performance, on the one hand, the organisation can easily identify and rectify harmful outcomes for the environment (Barakat et al., 2016). On the other hand, the results of the green performance appraisal allow the organisation to detect the difficulties faced by every employee, thereby providing timely and appropriate improvements. Furthermore, based on the results of the green employee appraisal, the organisation is able to recognise its employees' performance timely, thereby making them satisfied, happy, motivated, creative and engaged (Barakat et al., 2016).

Green Feedback Handling

Practitioners in the tourism and hospitality industry have acknowledged the importance of applying GHRM in their activities; yet this is challenging and priorities given to various GHRM practices are not the same according to the organisation's characteristics and objectives (Pham et al., 2022). Thus, determining appropriate GHRM practices to implement is essential. Furthermore, organisations need to establish standards to promptly handle customers' complaints due to employees' flaws in dealing with the environment issue during the working process.

In addition, organisations should define a process for taking care of stakeholders' grievances for violation of guidelines or regulations that can harm the environment (Siyambalapitiya et al., 2018). This is critical for tourism and hospitality organisation's because they need to enhance themselves and show their listening, focus, and response to the increasingly requested necessities of numerous tourists and other stakeholders in environmental protection.

Finally, clients' feedback should be amassed in the organisation's environmental reports. Through reports, the organisation can manage environmental issues and appreciate the compliance and commitment of employees promptly (del Brío et al., 2007). Employees can also gain from their flaws and colleagues' environmental protection initiatives, thereby enhancing the organisation's environmental performance (Siyambalapitiya et al., 2018).

Green Compensation

The compensation received from the organisation due to the good performance regarding the assigned green work tasks are among the primary motivators to propel employees to work effectively, carrying advantages to the organisation and the environment (Govindarajulu & Daily, 2004). If employees do not complete or comply with the green work designed by the organisation, they will forfeit their right to the green compensation. The compensation can be both financial (e.g., bonus, cash) and non-financial (e.g., special gifts, time off, public recognition) (Renwick et al., 2013); yet they should effectively serve to support and encourage employees' environmental activities and initiatives (Masri & Jaaron, 2017). Previous empirical studies have shown the strong impact of green compensation on motivation of employees working in organisations that constantly seek to value environmental improvement initiatives and to engage in green behaviours (Forman & Jørgensen, 2001). The scholarship also points out that employees of these organisations have a higher commitment to the organisation's environmental management activities (Forman & Jørgensen, 2001).

Antecedents of GHRM

The current literature shows that there are two types of factors that influence the successful GHRM implementation to assist organisations with accomplishing sustainable development: organisation level factors (Tanova & Bayighomog, 2022); and employee level factors (Chams & García-Blandón, 2019).

Organization Level Factors

Research on the impact of organisation level factors on the successful implementation of GHRM in tourism and hospitality organisations has suggested three main factors, including leadership, top management support, and organisational green culture.

First, regarding leadership, scholars have focused on specific characteristics of leaders as being environment-oriented, ethical, servant, empowering and transformational (Ahmad et al., 2021; DuBois & Dubois, 2012; Luu, 2019; Moin et al., 2021; Yusliza et al., 2017). According to Luu (2019), different styles of leadership affect greening the workplace. Albeit these types of leaders have different characteristics, they all have attributes of integrity, fairness, concern, sympathy, and commitment to achieve the gathering and motivation objectives in order to engage employees in green behaviours (Moin et al., 2021). Facing these types of leaders, employees are willing to adopt a reciprocity response to them by becoming green, thus complying with GHRM practices implemented by organisations. On the other hand, environment-oriented, ethical, servant, and empowering leaders are often pioneers within organisations in environmental protection. In other words, they are often initiators and partisans of the implementation of GHRM within organisations.

Second, top management support also plays a critical role in effectively achieving environmental management goals of organisations (Daily & Huang, 2001), specifically GHRM. Top management assesses the seriousness of environmental issues to determine appropriate environmental management policy. Furthermore, the more top management supports implementing the environmental management strategy, the more effectively the organisation promotes the implementation of the proposed strategy. In particular, for industries that have a strong link with the environment like tourism and hospitality, the support of top management is critical. Since the environment is regarded as one of the main raw materials of any successful tourism and hospitality firm, top management should be interested in strategies to tackle environmental problems in order to develop sustainably, diminish waste, and make positive the organisation's image (Pham et al., 2022). At the same time, the support of top management also motivates employees to go along or take green initiatives (Zoogah, 2011). In addition, the study conducted by Obeidat et al. (2020) in the oil and gas industry has also found a positive relationship between top management support and GHRM.

Finally, under pressure from customers, business and social partners, legal regulations and requirements, and to comply with industry and international standards, tourism and hospitality organisations are forced to be active and reactive in protecting the environment. To that end, they seek to turn towards green strategy and values by creating a green organisational culture (Obeidat et al., 2020). Organisational culture embodies the 'values, beliefs, and behaviors of organizational employees' (Shafaei et al., 2020, p. 1043). Research has indicated that organisational culture is among the main predictors of GHRM (DuBois & Dubois, 2012). According to Ren et al. (2018), organisational conditions are key motivators for practicing green activities such as GHRM. If a firm values pro-environmental activities, and seeks ways to minimise the negative consequences of its activities on the environment, it is setting up a green culture that promotes, among others, GHRM and its associated practices (Shafaei et al., 2020). Green organisational culture, through creating an organisation in which green activities are highly valued, promotes green recruitment, training, evaluation, feedback handling, and compensation, which are the main practices of GHRM (Guerci et al., 2015; Pham et al., 2022). Therefore, a green organisational culture assists the organisation in obtaining a favourable environment, facilitating the implementation of GHRM. Empirically, recent research shows that green organisational culture is positively associated with GHRM (Shafaei et al., 2020).

Employee Level Factors

The primary purpose of GHRM is to mobilise HRM practices to shape employees' green behaviours so that they can recognise and care about the organisation's environmental protection objectives and activities, which allows the organisation to accomplish sustainable development both environmentally and financially (Shafaei et al., 2020). Employees play a vital role in helping organisations effectively carry out GHRM, which has been highlighted by Wehrmeyer (1996, p. 41): 'if a company is to adopt an environment-aware approach to its activities, the employees are the key to its success or failure'. Prior studies have shown that a number of personal characteristics of employees affect their behaviours towards the environment in the workplace. Those characteristics refer to the Big Five personality traits (Ma et al., 2021) and sociodemographic factors (Klein et al., 2012; Meyer, 2016).

The five personality traits include conscientiousness, agreeableness, openness to experience, extraversion, and neuroticism (Costa & McCrae, 1992). First, employees who are conscientious of work often comply with the requirements of the organisation, and define clear goals and endeavor to accomplish them (Ma et al., 2021). More specifically, employees with this personality type are portrayed as determined at work and particularly persistent, and consistently endeavor to work with high performance (Gudmunson & Danes, 2011). Thus, conscientious employees are willing to make great efforts to guarantee the successful implementation of environmental protection regulations and standards by the organisation (Hirsh, 2010). Accordingly, conscientious employees are more motivated to conform to the GHRM practices compared to those who are not. Second, employees with the agreeableness personality trait tend to support colleagues in their assigned tasks (Bernerth et al., 2007). Agreeable employees generally tend to collaborate, coordinate with others to ensure the performance of their own work and the common work of the organisation such as environmental management. Third, employees who are open to experience show that they are receptive to new and creative ideas and procedures (Bozionelos et al., 2014). Open-minded employees are intellectually curious. They actively seek and listen to suggestions about work performance from others, thereby acquiring new solutions and working initiatives, and improving their capacity to carry out required environment management practices of the organisation (Ma et al., 2021). Fourth, extroverted employees are motivated and ambitious at work (Milfont & Sibley, 2012). They actively look for opportunities from work in order to attain the greatest benefits. Organisational environmental management practices such as GHRM practices can, thus, constitute both challenges and opportunities that extroverted employees may want to reach. Finally, neurotic employees often have sensations of tension and are easily concerned about issues (McCrae & Costa, 1989). Therefore, neurotic employees are exceptionally worried about the consequences of environmental pollution and the adverse impact of their work on the environment (Hirsh, 2010). Other studies confirm the findings of Hirsh (2010) by showing a positive connection between neurotic employees and their propensity to protect the environment (Milfont & Sibley, 2012). Earlier studies have shown that the Big Five personality traits are the main predictors of environmental engagement, though the influence level of these five personality traits on environment engagement is different (Milfont & Sibley, 2012). Thus, organisations and researchers should pay more attention to employees' personality traits to advance the application of GHRM practices.

Moreover, sociodemographic factors such as age, gender, ethnicity, and level of education have been shown to relate to environmentally friendly behaviours or the effectiveness and adoption of environmental management programmes in organisations (Meyer, 2016). These sociodemographic factors have also been featured in the analysis of environmental concern in the workplace as 'the most powerful analyses of the social bases of environmental concern will likely be those which consider both its demographic and cognitive determinants.' (Liere & Dunlap, 1980, p. 194). In organisations, empirical studies have revealed that female employees are more interested in supporting and implementing environmental management (De Silva & Pownall, 2014). Besides, employees with higher education have a higher consciousness of the significance of environmental protection and the harms of their work in the tourism and hospitality industry to the environment. Highly educated employees more straightforwardly seize the process and how to carry out GHRM practices without training. Consequently, the level of education has a positive relationship with employees' concern for the environment and their pro-environmental behaviours that support organisational environmental management, such as GHRM (Meyer, 2016). Thus, tourism and hospitality organisations should prioritise the selection of employees with suitable characteristics to promote the application of GHRM.

Consequences of GHRM

Based on the current literature, the consequences of GHRM can also be split into two levels: the organisation level (O'Donohue & Torugsa, 2016); and the employee level (Pinzone et al., 2019).

Organization Level Consequences

Today, the organisation's HRM exercises must also aim to assist the organisation with accomplishing competitiveness and sustainable development through environmental management to accomplish more effective environmental outcomes (O'Donohue & Torugsa, 2016). The environmental performance of a firm is characterised as 'a firm's effectiveness in meeting and exceeding society's expectations and concerns for the natural environment' (Judge & Douglas, 1998, p. 245). Based on the organisation's environmental performance, we can appreciate its interest for and commitment to environmental protection (Roscoe et al., 2019). GHRM has been seen as a strategic management tool to enhance both corporate environmental and economic performances because the latter are interrelated (Yusoff et al., 2020). For example, the environmental performance by environmental management of a hotel can help it reduce waste, purchases of non-renewable chemicals, materials, and components, and overall costs (Kim et al., 2019). First, in the tourism and hospitality industry, in a general way, GHRM has been found to be directly associated with the corporate's environmental performance (Ahmed et al., 2021; Kim et al., 2019; Pham et al., 2022; Pham, Vo-Thanh, Tučková, et al., 2020; Shafaei et al., 2020; Yusoff et al., 2020). Second, GHRM can exert an indirect effect on environmental performance through employee organisational commitment (Kim et al., 2019), employee eco-friendly behaviour (Kim et al., 2019; Pham et al., 2022), employee organisational citizenship behaviour for the environment (Pham, Vo-Thanh, Tučková, et al., 2020), employee green culture (Ahmed et al., 2021), and employee environmental responsibility (Ahmed et al., 2021). At the same time, GHRM practices stimulate environmental improvement initiatives that reduce carbon footprint and increase work efficiency with lower work expenses (Ahmed et al., 2021; Tanova & Bayighomog, 2022). Therefore, the effective implementation of GHRM assists the organisation with having better environmental and financial performances.

Employee Level Consequences

First, GHRM has been shown to positively affect employee green commitment (Ansari et al., 2021; Pham et al., 2022). Employee green commitment is described as 'all about an individual's core internal motivation towards ameliorating the environment' (Ansari et al., 2021, p. 231). Accordingly, employees with a green commitment are willing to engage in and improve the environment while working for the organisation. These employees are ready to share the organisation's green objectives and values that are embedded into the working process. Besides, the organisation that effectively implements GHRM practices can assist them in improving economically and gaining reputation (e.g., via evaluation and compensation practices), which makes them feel proud of and increases their commitment to the organisation and its environmental strategy. At the same time, GHRM through green training provides employees with essential green knowledge, competencies, and skills, helping them be aware of the importance of environmental issues, better prepare for their green job, and become more attached to the organisation's green objectives and values (Roscoe et al., 2019).

Second, GHRM can lead to employees' job satisfaction (Moin et al., 2021). Indeed, the supplies-values fit theory (Edwards, 1996) argues that the congruence of the organisation's values with employees' ones results in employees' attachment and positive attitudes towards the organisation. Thus, GHRM practices implemented by the organisation, if positively judged by employees, perfectly fit with their own values, which means that they are attached to and satisfied with the organisation and their job. Moreover, a good implementation of GHRM practices also means that the organisation not only cares about the environment, but also its employees' values and well-being (Shafaei et al., 2020), which may contribute to employees' job satisfaction. Furthermore, for employees, the environmental task is an additional task compared with their main job. By participating in the environmental strategy through their job, they contribute to protecting the surrounding lives, which makes them feel that their job is meaningful and, as a consequence, be satisfied with it (Shafaei et al., 2020). Prior research supports the positive effect of GHRM on job satisfaction (Moin et al., 2021).

Finally, while on the one hand, the environment is one of the main raw materials of tourism and hospitality organisations. On the other hand, the tourism and hospitality industry is one of the industries considerably raising the burden on the environment. Hence, there is an urgent need for tourism and hospitality firms to pay attention to environmental issues by performing sustainable business practices such as the adoption of GHRM practices. GHRM practices aim to enhance employees' pro-environmental behaviours. Pro-environmental behaviours are described as the eagerness and willingness of employees to perform their job well, yet pursue

Figure 42.1. Antecedents and consequences of GHRM practices from the current literature.

environmental stewardship (Kim et al., 2019). Pro-environmental behaviours of employees working in tourism and hospitality organisations can refer to energy saving, equipment recycling, waste reduction, and so on (Kim et al., 2019). GHRM seeks to provide employees with green knowledge, competencies, and skills, motivate their green behaviours, and give to them opportunities to share their green initiatives, thus improving their attitudes towards the environment (Pham et al., 2022; Renwick et al., 2013). In organisations with good GHRM implementation, employees are rewarded for their effective pro-environmental behaviours (Ansari et al., 2021). Previous research clearly demonstrates that GHRM has a significant impact on employees' pro-environmental behaviours (Kim et al., 2019).

Figure 42.1 summarises antecedents and consequences of GHRM practices from the current literature.

Challenges

Despite the acknowledgement of GHRM as an effective strategic management tool to accompany tourism and hospitality organisations in their environmental strategy, its implementation on a large scale faces numerous challenges.

First, most tourism and hospitality organisations are faced with financial difficulty, especially in the COVID-19 context. Although GHRM has been found to effectively assist tourism and hospitality organisations in sustainably developing, the initial cost of implementing GHRM is a real challenge for many of them, particularly for tourism and hospitality organisations in developing economies (Yusliza et al., 2017). Investing in environmentally friendly equipment, training employees, and updating the working process are often very costly (Visvanathan & Kumar, 1999). In particular, approaching new technologies and techniques remains a significant hindrance for most tourism and hospitality organisations in developing countries (Foxon & Pearson, 2008).

Second, both employees and management of most tourism and hospitality organisations are, thus far, not completely ready for a successful implementation of GHRM. To date, it is still hard to have all employees believe in and adhere to GHRM's values and strategic importance (Amrutha & Geetha, 2020). Additionally, a recent empirical study conducted by Amran et al. (2017) has clearly indicated that hotel HR managers in Malaysia have gained a lot of progress about embracing green practices; however, the majority of them are still doubtful about these green practices. This study has also revealed that only 6 percent of respondents urge their employees to identify, suggest, and spread green practices. According to the authors, these findings may be due to a low sense of environmental awareness among the respondents, limited resources on green aspects, and low support from the local authorities. Therefore, raising awareness of the importance to protect the environment for sustainable development of the tourism and hospitality industry among both managers and employees is one of the main missions of administrations and social partners. Moreover, legal environmental regulations may need to be reinforced.

Third, tourism and hospitality organisations should also overcome pressures from various stakeholders in the implementation of GHRM (Guerci & Carollo, 2016). Indeed, the vision and interest about implementing GHRM vary from one stakeholder to another. For example, some green practices are beneficial for customers but not for business partners and vice versa. Finally, the implementation of GHRM is influenced by lawful regulations, organisational rules, and trade unions (Amrutha & Geetha, 2020).

Conclusions and Further Research

Environmental issues are becoming more urgent for the whole society than ever. Yet, production and business activities are among the most harmful sources of the environment. In particular, due to its rapid development, the tourism and hospitality industry is one of the industries mostly raising the burden on the environment (Vo-Thanh et al., 2020). Therefore, there is a critical need for tourism and hospitality organisations to react to environmental issues by rightly adopting sustainable business practices such as GHRM, providing a win-win relationship for them and their stakeholders (Pham et al., 2022; Siyambalapitiya et al., 2018; Yusoff et al., 2020). However, some challenges must be overcome in order to make the implementation of GHRM on a large scale.

From an academic perspective, research about GHRM has constantly been growing. In the tourism and hospitality literature, it is in its infancy phase (Tanova & Bayighomog, 2022). GHRM is described as a way for implementing GHRM practices according to the organisation's green goals. Undertaking GHRM should assist organisations in accomplishing its environmentally friendly goals, managing the environment effectively, bringing the best green products and services to clients, and decreasing their carbon footprint to get sustainable development. This chapter aims, on the one hand, to provide an outlook on GHRM by reviewing its practices, antecedents, and consequences, as well as challenges for implementing it within tourism and hospitality organisations. On the other hand, it sheds light on some main GHRM topics that need to be explored to enrich the current literature.

First, the majority of studies have focused on the positive consequences of GHRM at both the organisation and employee levels, such as corporate environmental and financial performances, employee green commitment, employee job satisfaction, and employee pro-environmental behaviours. Empirically, little is known about barriers tourism and hospitality organisations may encounter in implementing GHRM and other employee outcomes like employee turnover intention, employee green creativity, and so on.

Second, while researchers have concentrated their focus on the antecedents or consequences of GHRM as highlighted in the present chapter, the mediating and moderating mechanisms of GHRM need further research (Moin et al., 2021; O'Donohue & Torugsa, 2016).

Third, for any business organisations, the perceptions of main stakeholders about their business practices are extremely important. Since a significant number of studies have already been conducted from an employee perspective, more research on the perceptions about GHRM of two key stakeholders, i.e., business partners and customers, is needed. A recent qualitative study from both employer and employee points of view(Pham et al., 2022) showed that GHRM could provide the organisation with not only internal benefits but also external ones, including better relationships with the local community and authorities, partners, customer satisfaction, and competitive advantage (e.g., positive public image, differentiation). It is, therefore, suggested that more empirical studies are carried out in order to investigate the influence of GHRM on customers and business partners.

Fourth, research about GHRM was found much more in European economies (Amrutha & Geetha, 2020) and less in developing ones (Kim et al., 2019; Pham et al., 2022). In addition, research on corporate social responsibility (CSR) and trade union in the European context discovered that trade unions have a significant impact on CSR activities and eco-friendly initiatives of firms (Preuss et al., 2006). Indeed, in those countries, strict rules and regulations offer trade unions a real power, which influences management practices and employees of the firm. On the contrary, in countries where rules and regulations are lenient and the trade unions lack legitimacy to form a CSR agenda, their influence on management practices appears less (Preuss et al., 2006). Therefore, from an institutional logic, it would be desirable to have more studies examining government regulations, political conditions, and trade unions involvement as core influencers of shaping employee pro-environmental behaviours and implementing GHRM practices for achieving the environmental sustainability objective of a firm.

Finally, recent research done in developing countries highlighted that the influence of GHRM practices on employees' organisational citizenship behaviours for the environment and green performance is stronger when hotels are managed by Western corporations (Pham, Chiappetta Jabbour, et al., 2020). Therefore, cross-cultural research about GHRM based on the national cultural viewpoint would appear to be a relevant contribution to the current literature.

References

Ahmad, I., Ullah, K., & Khan, A. (2021). The impact of green HRM on green creativity: Mediating role of pro-environmental behaviors and moderating role of ethical leadership style. *The International Journal of Human Resource Management*. https://doi.org/10.1080/09585192.2021.1931938

Ahmed, U., Umrani, W. A., Yousaf, A., Siddiqui, M. A., & Pahi, M. H. (2021). Developing faithful stewardship for environment through green HRM. *International Journal of Contemporary Hospitality Management*, *33*(10), 3115–3133. https://doi.org/10.1108/IJCHM-09-2020-1066

Alfred, A. M., & Adam, R. F. (2009). Green management matters regardless. *Academy of Management Perspectives*, *23*(3), 17–26. https://doi.org/10.5465/amp.2009.43479261

Amran, N. A., Saad, N., & Aripin, N. (2017). Green practices in Northern region hotels of Malaysia. *Journal of Management and Marketing Review*, *2*(3), 61–66.

Amrutha, V. N., & Geetha, S. N. (2020). A systematic review on green human resource management: Implications for social sustainability. *Journal of Cleaner Production*, *247*, 119131. https://doi.org/10.1016/j.jclepro.2019.119131

Ansari, N. Y., Farrukh, M., & Raza, A. (2021). Green human resource management and employees pro-environmental behaviours: Examining the underlying mechanism. *Corporate Social Responsibility and Environmental Management*, *28*(1), 229–238. https://doi.org/10.1002/csr.2044

Aust, I., Matthews, B., & Muller-Camen, M. (2020). Common good HRM: A paradigm shift in sustainable HRM? *Human Resource Management Review*, *30*(3), 100705. https://doi.org/10.1016/j.hrmr.2019.100705

Barakat, S. R., Isabella, G., Boaventura, J. M. G., & Mazzon, J. A. (2016). The influence of corporate social responsibility on employee satisfaction. *Management Decision*, *54*(9), 2325–2339. https://doi.org/10.1108/MD-05-2016-0308

Bauer, T. N., Erdogan, B., & Taylor, S. (2012). Creating and maintaining environmentally sustainable organizations: Recruitment and on-boarding. In S. E. Jackson, D. S. Ones & S. Dilchert (Eds.), *Managing human resources for environmental sustainability* (pp. 222–240). Jossey-Bass.

Bernerth, J. B., Armenakis, A. A., Feild, H. S., Giles, W. F., & Jack Walker, H. (2007). Is personality associated with perceptions of LMX? An empirical study. *Leadership & Organization Development Journal, 28*(7), 613–631. https://doi.org/10.1108/01437730710823879

Bozionelos, N., Bozionelos, G., Polychroniou, P., & Kostopoulos, K. (2014). Mentoring receipt and personality: Evidence for non-linear relationships. *Journal of Business Research, 67*(2), 171–181. https://doi.org/10.1016/j.jbusres.2012.10.007

Chams, N., & García-Blandón, J. (2019). On the importance of sustainable human resource management for the adoption of sustainable development goals. *Resources, Conservation and Recycling, 141*, 109–122. https://doi.org/10.1016/j.resconrec.2018.10.006

Chiappetta Jabbour, C. J. (2013). Environmental training in organisations: From a literature review to a framework for future research. *Resources, Conservation and Recycling, 74*, 144–155. https://doi.org/10.1016/j.resconrec.2012.12.017

Costa, P. T., & McCrae, R. R. (1992). Four ways five factors are basic. *Personality and Individual Differences, 13*(6), 653–665. https://doi.org/10.1016/0191-8869(92)90236-I

Daily, B. F., & Huang, S. (2001). Achieving sustainability through attention to human resource factors in environmental management. *International Journal of Operations & Production Management, 21*(12), 1539–1552. https://doi.org/10.1108/01443570110410892

De Silva, D. G., & Pownall, R. A. J. (2014). Going green: Does it depend on education, gender or income? *Applied Economics, 46*(5), 573–586. https://doi.org/10.1080/00036846.2013.857003

del Brío, J. Á., Fernández, E., & Junquera, B. (2007). Management and employee involvement in achieving an environmental action-based competitive advantage: An empirical study. *The International Journal of Human Resource Management, 18*(4), 491–522. https://doi.org/10.1080/09585190601178687

DuBois, C. L. Z., & Dubois, D. A. (2012). Strategic HRM as social design for environmental sustainability in organization. *Human Resource Management, 51*(6), 799–826. https://doi.org/10.1002/hrm.21504

Edwards, J. R. (1996). An examination of competing versions of the person-environment fit approach to stress. *Academy of Management Journal, 39*(2), 292–339. https://www.jstor.org/stable/256782

Forman, M., & Jørgensen, M. S. (2001). The social shaping of the participation of employees in environmental work within enterprises—Experiences from a Danish context. *Technology Analysis & Strategic Management, 13*(1), 71–90. https://doi.org/10.1080/09537320120040455

Foxon, T., & Pearson, P. (2008). Overcoming barriers to innovation and diffusion of cleaner technologies: Some features of a sustainable innovation policy regime. *Journal of Cleaner Production, 16*(1), S148–S161. https://doi.org/10.1016/j.jclepro.2007.10.011

Govindarajulu, N., & Daily, B. F. (2004). Motivating employees for environmental improvement. *Industrial Management & Data Systems, 104*(4), 364–372. https://doi.org/10.1108/02635570410530775

Gudmunson, C. G., & Danes, S. M. (2011). Family financial socialization: Theory and critical review. *Journal of Family and Economic Issues, 32*(4), 644–667. https://doi.org/10.1007/s10834-011-9275-y

Guerci, M., & Carollo, L. (2016). A paradox view on green human resource management: Insights from the Italian context. *The International Journal of Human Resource Management, 27*(2), 212–238. https://doi.org/10.1080/09585192.2015.1033641

Guerci, M., Radaelli, G., Siletti, E., Cirella, S., & Rami Shani, A. B. (2015). The impact of human resource management practices and corporate sustainability on organizational ethical climates: An employee perspective. *Journal of Business Ethics, 126*(2), 325–342. https://doi.org/10.1007/s10551-013-1946-1

Hall, C. M. (2009). Degrowing tourism: Décroissance, sustainable consumption and steady-state tourism. *Anatolia, 20*(1), 4–61. https://doi.org/10.1080/13032917.2009.10518894

Hirsh, J. B. (2010). Personality and environmental concern. *Journal of Environmental Psychology, 30*(2), 245–248. https://doi.org/10.1016/j.jenvp.2010.01.004

Howard-Grenville, J., Buckle, S. J., Hoskins, B. J., & George, G. (2014). Climate change and management. *Academy of Management Journal, 57*(3), 615–623. https://doi.org/10.5465/amj.2014.4003

Judge, W. Q., & Douglas, T. J. (1998). Performance implications of incorporating natural environmental issues into the strategic planning process: An empirical assessment. *Journal of Management Studies*, *35*(2), 241–262. https://doi.org/10.1111/1467-6486.00092

Kim, Y. J., Kim, W. G., Choi, H.-M., & Phetvaroon, K. (2019). The effect of green human resource management on hotel employees' eco-friendly behavior and environmental performance. *International Journal of Hospitality Management*, *76*, 83–93. https://doi.org/10.1016/j.ijhm.2018.04.007

Klein, R. M., D'Mello, S., Wiernik, B. M., & Erdogan, B. (2012). Demographic characteristics and employee sustainability. In S. E. Jackson, D. S. Ones & S. Dilchert (Eds.), *Managing human resources for environmental sustainability* (pp. 117–154). Jossey-Bass.

Liere, K. D. V., & Dunlap, R. E. (1980). The social bases of environmental concern: A review of hypotheses, explanations and empirical evidence. *Public Opinion Quarterly*, *44*(2), 181–197. https://doi.org/10.1086/268583

Luu, T. T. (2019). Green human resource practices and organizational citizenship behavior for the environment: The roles of collective green crafting and environmentally specific servant leadership. *Journal of Sustainable Tourism*, *27*(8), 1167–1196. https://doi.org/10.1080/09669582.2019.1601731

Ma, Y., Chen, S.-C., & Ruangkanjanases, A. (2021). Understanding the antecedents and consequences of green human capital. *SAGE Open*, *11*(1). https://doi.org/10.1177/2158244020988867

Masri, H. A., & Jaaron, A. A. M. (2017). Assessing green human resources management practices in Palestinian manufacturing context: An empirical study. *Journal of Cleaner Production*, *143*, 474–489. https://doi.org/10.1016/j.jclepro.2016.12.087

McCrae, R. R., & Costa, P. T. (1989). Reinterpreting the Myers-Briggs type indicator from the perspective of the five-factor model of personality. *Journal of Personality*, *57*(1), 17–40. https://doi.org/10.1111/j.1467-6494.1989.tb00759.x

Meyer, A. (2016). Heterogeneity in the preferences and pro-environmental behavior of college students: The effects of years on campus, demographics, and external factors. *Journal of Cleaner Production*, *112*, 3451–3463. https://doi.org/10.1016/j.jclepro.2015.10.133

Milfont, T. L., & Sibley, C. G. (2012). The big five personality traits and environmental engagement: Associations at the individual and societal level. *Journal of Environmental Psychology*, *32*(2), 187–195. https://doi.org/10.1016/j.jenvp.2011.12.006

Moin, M. F., Omar, M. K., Wei, F., Rasheed, M. I., & Hameed, Z. (2021). Green HRM and psychological safety: How transformational leadership drives follower's job satisfaction. *Current Issues in Tourism*, *24*(16), 2269–2277. https://doi.org/10.1080/13683500.2020.1829569

Obeidat, S. M., Al Bakri, A. A., & Elbanna, S. (2020). Leveraging "green" human resource practices to enable environmental and organizational performance: Evidence from the Qatari oil and gas industry. *Journal of Business Ethics*, *164*(2), 371–388. https://doi.org/10.1007/s10551-018-4075-z

O'Donohue, W., & Torugsa, N. (2016). The moderating effect of 'Green' HRM on the association between proactive environmental management and financial performance in small firms. *The International Journal of Human Resource Management*, *27*(2), 239–261. https://doi.org/10.1080/09585192.2015.1063078

Paulet, R., Holland, P., & Morgan, D. (2021). A meta-review of 10 years of green human resource management: Is Green HRM headed towards a roadblock or a revitalisation? *Asia Pacific Journal of Human Resources*, *59*(2), 159–183. https://doi.org/10.1111/1744-7941.12285

Pham, N. T., Chiappetta Jabbour, C. J., Vo-Thanh, T., Huynh, T. L. D., & Santos, C. (2020). Greening hotels: Does motivating hotel employees promote in-role green performance? The role of culture. *Journal of Sustainable Tourism*. https://doi.org/10.1080/09669582.2020.1863972

Pham, N. T., Phan, Q. P. T., Tučková, Z., Vo, N., & Nguyen, L. H. L. (2018). Enhancing the organizational citizenship behavior for the environment: The roles of green training and organizational culture. *Management & Marketing*, *13*(4), 1174–1189. https://doi.org/10.2478/mmcks-2018-0030

Pham, N. T., Tučková, Z., & Chiappetta Jabbour, C. J. (2019). Greening the hospitality industry: How do green human resource management practices influence organizational citizenship behavior in hotels? A mixed-methods study. *Tourism Management*, *72*, 386–399. https://doi.org/10.1016/j.tourman.2018.12.008

Pham, N. T., Vo-Thanh, T., Shahbaz, M., Huynh, T. D. L., & Usman, M. (2020). Managing environmental challenges: Training as a solution to improve employee green performance. *Journal of Environmental Management*, *269*, 110781. https://doi.org/10.1016/j.jenvman.2020.110781

Pham, N. T., Vo-Thanh, T., & Tučková, Z. (2022). Green human resource management in hotels in developing countries: A practices- and benefits-related conceptual framework. In N. Nguyen, H. V. Nguyen, C. D'Souza & C. Strong (Eds.), *Environmental sustainability in emerging markets: Consumer, organisation and policy perspectives* (pp. 151–176). Springer Nature Singapore.

Pham, N. T., Vo-Thanh, T., Tučková, Z., & Vo, T. N. T. (2020). The role of green human resource management in driving hotel's environmental performance: Interaction and mediation analysis. *International Journal of Hospitality Management*, *88*, 102392. https://doi.org/10.1016/j.ijhm.2019.102392

Pinzone, M., Guerci, M., Lettieri, E., & Huisingh, D. (2019). Effects of 'green' training on pro-environmental behaviors and job satisfaction: Evidence from the Italian healthcare sector. *Journal of Cleaner Production*, *226*, 221–232. https://doi.org/10.1016/j.jclepro.2019.04.048

Preuss, L., Haunschild, A., & Matten, D. (2006). Trade unions and CSR: A European research agenda. *Journal of Public Affairs*, *6*(3–4), 256–268. https://doi.org/10.1002/pa.236

Ren, S., Tang, G., & Jackson, S. E. (2018). Green human resource management research in emergence: A review and future directions. *Asia Pacific Journal of Management*, *35*(3), 769–803. https://doi.org/10.1007/s10490-017-9532-1

Renwick, D. W. S., Redman, T., & Maguire, S. (2013). Green human resource management: A review and research agenda. *International Journal of Management Reviews*, *15*(1), 1–14. https://doi.org/10.1111/j.1468-2370.2011.00328.x

Roscoe, S., Subramanian, N., Jabbour, C. J. C., & Chong, T. (2019). Green human resource management and the enablers of green organisational culture: Enhancing a firm's environmental performance for sustainable development. *Business Strategy and the Environment*, *28*(5), 737–749. https://doi.org/10.1002/bse.2277

Russo, M. V., & Fouts, P. A. (1997). A resource-based perspective on corporate environmental performance and profitability. *Academy of Management Journal*, *40*(3), 534–559. https://doi.org/10.2307/257052

Shafaei, A., Nejati, M., & Mohd Yusoff, Y. (2020). Green human resource management: A two-study investigation of antecedents and outcomes. *International Journal of Manpower*, *41*(7), 1041–1060. https://doi.org/10.1108/IJM-08-2019-0406

Siyambalapitiya, J., Zhang, X., & Liu, X. (2018). Green human resource management: A proposed model in the context of Sri Lanka's tourism industry. *Journal of Cleaner Production*, *201*, 542–555. https://doi.org/10.1016/j.jclepro.2018.07.305

Tanova, C., & Bayighomog, S. W. (2022). Green human resource management in service industries: The construct, antecedents, consequences, and outlook. *The Service Industries Journal*, *42*(5–6), 412–452. https://doi.org/10.1080/02642069.2022.2045279

Visvanathan, C., & Kumar, S. (1999). Issues for better implementation of cleaner production in Asian small and medium industries. *Journal of Cleaner Production*, *7*(2), 127–134. https://doi.org/10.1016/S0959-6526(98)00050-X

Vo-Thanh, T., Seraphin, H., Okumus, F., & Koseoglu, M. A. (2020). Organizational ambidexterity in tourism research: A systematic review. *Tourism Analysis*, *25*(1), 137–152. https://doi.org/10.3727/108354220X15758301241701

Vo-Thanh, T., Vu, T.-V., Nguyen, N. P., Nguyen, D. V., Zaman, M., & Chi, H. (2021). How does hotel employees' satisfaction with the organization's COVID-19 responses affect job insecurity and job performance? *Journal of Sustainable Tourism*, *29*(6), 907–925. https://doi.org/10.1080/09669582.2020.1850750

Wehrmeyer, W. (1996). Green policies can help to bear fruit. *People Management*, *2*(4), 38–42.

Yusliza, M.-Y., Othman, N. Z., & Chiappetta Jabbour, C. J. (2017). Deciphering the implementation of green human resource management in an emerging economy. *Journal of Management Development*, *36*(10), 1230–1246. https://doi.org/10.1108/JMD-01-2017-0027

Yusoff, Y. M., Nejati, M., Kee, D. M. H., & Amran, A. (2020). Linking green human resource management practices to environmental performance in hotel industry. *Global Business Review*, *21*(3), 663–680. https://doi.org/10.1177/0972150918779294

Zoogah, D. B. (2011). The dynamics of green HRM behaviors: A cognitive social information processing approach. *German Journal of Human Resource Management*, *25*(2), 117–139. https://doi.org/10.1177/239700221102500204

Part 7 Tourism Policies, Planning and Governance

Introduction
Tourism Policies, Planning and Governance

C. Michael Hall

The role of government in tourism and the influence of public policy on tourism development have long been of interest to scholars. Although the political dimensions of tourism have not been a major focal point of research, there has nevertheless been sustained study of tourism planning and policies from various perspectives and approaches (Hall, 1994; Hall & Jenkins, 2004; Jenkins et al., 2014; Pforr, 2021). Since the 1990s there has also been a gradual shift in approach in the tourism policy literature from the notion of government to that of governance, a situation which reflects broader political movements in the political economy of Western society and changes in political philosophy with respect to the role of government and the nature of government intervention. However, the change in the discourse of the role of the state in tourism has not necessarily been reflected in a deeper understanding of political theory and public policy analysis and their connectedness to the study of tourism (Amore, Chapter 43).

One of the curiosities of research on policy, planning and governance is that in many languages there is no substantial difference between the terms used to describe politics and policy. However, in the Anglo-American public administration tradition there has long been an avowed division between politics and policy and politics and public administration, i.e. that politicians engage in politics and that the public service implements policies in an apolitical rational fashion. Yet all planning and policy is political in the sense that there is a contest between interests with respect to power and values and the determination of certain output and outcomes in the policy-making process. Although much of the tourism research literature does not deal with the politics of tourism directly, there is a significant history of research in the area (see Amore, Chapter 43). Nevertheless, the field is not as well served as it could be.

Attention to governance issues in tourism mirrors the broader growth of interest in, and development of, the concept in the policy sciences. Governance is the act of governing and is an increasingly significant issue in the tourism public policy and planning literature. It has assumed importance as researchers have sought to understand how the state can best act to

The Wiley Blackwell Companion to Tourism, Second Edition. Edited by C. Michael Hall.

mediate contemporary tourism related social, economic, political and environmental policy problems at a time when the role of the state has itself changed given the dominance of neo-liberal policy discourse in many developed countries (Bramwell & Lane, 2009; Spyriadis et al., 2011).

In developed countries the focus on economic development and destination competitiveness has often been approached via the lens of innovation (Weidenfield & Hall, Chapter 44). Government is recognised as having an important role in tourism innovation via its public policy settings, especially in relation to regional an industry innovation systems. Nevertheless, as Weidenfeld and Hall point out in Chapter 44, much more research is required on tourism innovation systems especially with respect to the relative roles of place embedded qualities versus wider networks and relationships. Such questions remain extremely important in relation to questions of where and how government should invest in tourism innovation and the overall nature of state intervention in the sector in the pursuit of policy goals.

The role of the state remains as important as ever in tourism, perhaps even more so given responses to COVID-19, economic and financial crises, and natural disasters. Considerable work has been undertaken on the nature and role of state intervention in tourism, including in specific areas such as innovation and poverty reduction. However, compared to other areas of sectoral public policy the field remains relatively undertheorised overall, or at least poorly connected to some of the extant public policy and political science literatures (Amore, Chapter 46). Nevertheless, the political capacity for individual mobility across borders, as well as the capacities of the state with respect to tourism remain important research opportunities. As some of the various contributions in this section suggest (Grønvold, Chapter 47), the foundations for future research in the public policy area are much firmer than what they were just a few years ago, although the character of research over the forthcoming decade will likely as much depend on the interests of individual researchers, as the political activities of the state.

References

Bramwell, B., & Lane, B. (1999). Editorial: collaborations and partnerships for sustainable tourism. *Journal of Sustainable Tourism*, 7(3–4), 179–181.

Hall, C.M. (1994). *Tourism and politics*. John Wiley.

Hall, C M., & Jenkins, J. M. (2004). Tourism and public policy. In A. Lew, C. M. Hall & A. M. Williams (Eds.), *A companion to tourism* (1st ed.), (pp. 525–540). Blackwell Publishing.

Jenkins, J. M., Hall, C. M., & Mkono, M. (2014). Tourism and public policy: contemporary debates and future directions. In A. Lew, C. M. Hall & A. M. Williams (Eds.), *A companion to tourism* (2nd ed.) (pp. 542–555). Wiley.

Pforr, C. (2021). Tourism and public policy. In A. Kellow, T. Porter & K. Ronit (Eds.), *Handbook of business and public policy* (pp. 378–391). Edward Elgar Publishing.

Spyriadis, T., Buhalis, D., & Fyall, A. (2011). Dynamics of destination governance: governance and metagovernance in the composite industrial environment of destinations. In E. Laws, H. Richins, J. Argusa & N. Scott (Eds.), *Tourist destination governance: Practice, theory, and issues* (pp. 187–202). CABI.

Chapter 43

Tourism, Public Policy and Governance

Alberto Amore

Introduction

The purpose of this chapter is to provide a timely overview of the most recent advancements in the study of policies, politics and governance in tourism over the last decade. It does so by illustrating prominent works and themes on this valuable yet often taken for granted topic in tourism research. Arguably, this is the ideal time for an appraisal of works focusing on aspects such as the role of government, overt and vested interests, lobbying and the role of stakeholders along the multi-scalar and cross-scalar governance of tourism. On the one hand, domestic and international politics have had undeniable repercussions on tourism mobilities across countries. Examples include Trump's travel bans (Executive Orders 13769 and 13780) restricting travel to the United States from countries including Iran, Iraq and Syria (Seyfi & Hall, 2020); the repercussion of the 2016 Brexit Referendum for the UK and the EU (Dutta et al., 2021, Hall, 2020); the response by governments in both developed and developing countries in the aftermath of mass shootings and terrorist attacks (Agarwal et al., 2021; Avraham, 2021); and, more recently, policy responses to the COVID-19 global pandemic (Khalid et al., 2021). On the other hand, we assisted to an increased pressure from the public opinion and the civic society at large on the negative impacts of tourism at the local, regional, national, and international levels. These eventually culminated with socio-political phenomena like overtourism, tourismphobia, the impact of Airbnb, recovery and resilience and the nexus between travel emissions and tourism, all of which have been researched from a tourism public policy and governance perspectives (Amore et al., 2022; Efthymiou and Papatheodorou, 2020, Hall et al., 2022; Milano et al., 2019).

Along with the trends and themes mentioned above, research on tourism, public policy and governance walks alongside research on tourism and sustainable development. The importance of policies and government action – at different scales – in the pursuit of the United Nations Sustainable Development Goals (UNSDGs) builds from the extensive research on tourism and sustainable development from the 1990s and the early 2000s (e.g., Bramwell, 1995;

The Wiley Blackwell Companion to Tourism, Second Edition. Edited by C. Michael Hall.

Sharpley, 2000). For example, the importance of networking and partnerships building across different stakeholders is still at the centre of debate nowadays (Scheyvens and Cheer, 2022) just as it was at the turn of the century (Bramwell & Lane, 1999). The emerging literature provides both aspirational appraisals on what policies in tourism should feature in order to pursue sound sustainability (Hall, 2019; Sharpley, 2020) as well as critical reflections on how politics and vested interests hinder and manipulate sustainable tourism development discourses (Farmaki et al., 2015; Glyptou et al., 2022; Muangasame & McKercher, 2015). Hall (2019) uses the term *Brundtland-as-usual* to show how the intended vagueness around the concept of sustainable development allowed the promotion of pro-growth policies to enhance economic development in tourism instead of ensuring environmental protection. This is similarly addressed in Amore (forthcoming) in relation to the observable policy failures and the need for a metagovernance shift in the implementation of the UNSDGs (Meuleman, 2020).

This chapter is structured as follows: the first section provides an overview of the current 'state of the art' in tourism, public policy, and governance literature with a selection of relevant themes and foci. The second section illustrates theoretical advancements that have occurred in this subfield over the last decade and critically discusses the hindrances and issues with conducting research in public policy, governance, and tourism within a politicized context. The concluding section provides a summary of the key points of the chapter while also highlighting future directions for research. Additionally, a final remark is provided on the positionality of researchers in tourism, public policy and governance and the need to overcome policy-pleasing rhetoric and fully embrace political action.

Trends and Themes in Research

Research on the political dimension of tourism is well established. Building from political theory and policy studies, we assisted to a proliferation of studies from different regions of the world providing extensive evidence of decision-making and policies implemented at different scales. Notwithstanding the variety of contexts and destinations featured in the literature, scholars widely agree on the fallacies of policy action and inaction when it comes to public policy and governance in tourism. Echoing the early studies at the turn of the century (e.g., Hall, 1999), research in governance, policymaking and tourism has further highlighted the limits and the paradoxes within the policy environments in which government action takes place and shapes tourism at the local (e.g., Amore & Hall, 2021), regional (e.g., Rogerson, 2015) and national level (e.g., Dredge & Jenkins, 2012). Overall, government decisions and actions with regards to tourism have been reiterating the dogmas of neoliberal-driven tourism development. For example, findings from Thailand and Madagascar show how the neoliberal discourse is key in promoting allegedly green nature-based tourism policies at the expenses of environmental conservation and wildlife protection (Duffy, 2015). Similarly, Torkington et al. (2020) review of national tourism policy documents in Croatia, Iceland, Ireland, Malta, Portugal, Romania, and the United Kingdom illustrates how the neoliberal discourse frame tourism and growth as unconditional dogmas.

More recently, we assisted to a further exacerbation of market-obeying policies impacting tourism policy and governance. This goes hand in hand with a wider policy shift in which governments further withdraw from providing public goods while encouraging private and not-for-profit organizations to supply services that ultimately expose societies to increasing global market competition (Bosman, 2007). Such policy shift – also referred to as hyperneoliberalism (Booth, 2006) – was first noticed in urban tourism contexts. Evidence from the

Gold Coast, Australia, shows how destination governance in the area overshadows local agencies and stakeholders in an increasingly de-territorialized and globalized policy environment that prioritizes the instances of city-users, temporary residents, and tourists alike (Bosman & Dredge, 2011; Dredge & Jamal, 2013). Similarly, findings from Christchurch, New Zealand, show how the 2008 Global Financial Crisis and the 2010–2011 Canterbury earthquakes have framed the disaster recovery of the city as a destination (Amore & Hall, 2017a). As the authors observe, 'hyperneoliberalism dominates the socio-technical system and frames the policy agenda that affect directly and indirectly impact tourism destinations and DMOs and sets destinations on particular development trajectories' (Amore & Hall, 2017a, p. 12). The nexus between hyperneoliberalism, governance and tourism policy has been similarly discussed in relation to the pursuit of sound good governance principles of equality, justice, and democracy among marginalized Mayan communities in Quintana Roo, Mexico (Jamal & Camargo, 2014).

Over the last decade, research on public policy and governance in the field has been covering different perspectives and roles of government in tourism. The latter have changed over time in response to the dynamic and complex tourism system as well as the broader environment (Pforr, 2021), but encompass most of the roles identified by Hall (2008). Research on the co-ordinating role of government in tourism convenes that the public sector can be a key player in the promotion of tourism, as in the case of local authorities in Tuscany (Gori et al., 2021). Evidence from Namibia shows how government enable and support structures and processes helping local communities to benefit from the tourism sector (Kavita & Saarinen, 2016). However, the co-ordinating role of the state is far from being effective. This is particularly evident with poorly cohesive climate policies in relation to tourism and the ability to co-ordinate between national and regional level (Brendehaug et al., 2017; Santos-Lacueva & Velasco Gonzalez, 2018). Development programmes encompassing tourism as one of the key outcomes have been key in fostering economic growth across cities in Brazil (Silva et al., 2023). Similarly, planning for leisure and tourism development has proved to be successful in the establishment of Singapore as an internationally established Meetings, Incentives, Conferences and Exhibitions (MICE) destination (Mena-Navarro et al., 2022). Spatial planning and tourism represent and important feature in the pursuit of sustainable development in island destinations, as in the case of the Azores, Portugal (Castanho et al., 2020).

The fragmented nature of tourism and the governance indicates that legislation and regulations — at all scales — are 'rarely exclusively devoted to tourism per se' (Hall, 2008, p. 14). An emerging body of literature has looked at the importance of regulation – and lack thereof – in the rise of Airbnb in cities across Europe and North America (Aguilera et al., 2021; Amore et al., 2022; Guttentag, 2015). Research has also focused on the factors that inhibit or facilitate policy implementation and the promulgation of laws that indirectly and directly impact tourism at the local level, as in the cases of rural tourism development in East Java, Indonesia (Aly et al., 2019), the establishment of enterprise zones in the Philippines (Dela Santa, 2018) and the regulation of accommodation standards in the Canary Islands, Spain (Cruz et al., 2018). Research on the entrepreneurial role of the state and tourism has mainly focused on allocation and dispute of land for anchor projects and urban redevelopment following natural hazards. Studies on the enactment of emergency legislation along with the establishment of recovery authorities include evidence from New Orleans and New York, United States (Gotham & Greenberg, 2014), Christchurch, New Zealand (Amore & Hall, 2016a) and Istanbul, Turkey (Saxena, 2014). The logics of the *tabula rasa* and *recovery as opportunity* permeate the policy discourse of government, with the recovery agencies acting as a Quasi

Non-Governmental Organization (QuaNGO) to clear land and resell it to private investors at very convenient real estate prices (Amore, 2019; Saxena, 2014).

Literature focusing on stimulation policies with relevance to tourism highlight the contingency nature of state action. Policies aimed at stimulating tourism positively influence regional resilience in rural areas. Evidence from Greece, in particular, shows how stimulation policies can be decisive during periods of economic crisis (Psycharis et al., 2022). Following the COVID19 pandemic, studies have focused on the study of stimulation policies for tourism. Evidence from Portugal stresses on the importance of tax incentives in areas where tourism represents an important source of income and regional development (Dinis & Pereira, 2022). During crises, public authorities responsible for tourism promotion launch demarketing campaigns to divert tourist mobility to other areas of the country (e.g., New Zealand following the Canterbury earthquakes) (Orchiston & Higham, 2016) or to exhort tourists not to travel (e.g., COVID19 lockdown) (Hudson, 2020; Hall & Wood, 2021). Looking at initiatives towards tourism promotion, research has steadily acknowledged the importance of the local community in destination image perception and building. Santos et al. (2017), study on the Guimarães European Capital of Culture (2012) stress on the importance of including the perspective of both residents and tourists in the evaluation of event-driven promotion strategies for the city. A comprehensive review of the literature on this topic acknowledges the role local government plays 'in acting as a host to a city's tourists, by taking initiatives to promote their welcome and striving to provide hospitality services and policies that enhance the city's appeal' (Del Gesso et al., 2022, p. 3870). It is argued that residents and tourists can potentially converge to co-create and re-brand cities within a smart and sustainable urban development (Amore et al., 2022; Pasquinelli & Trunfio, 2020).

Studies on the role of government in supporting social tourism policies are comparatively limited, but there have been advancements also in this segment of literature. Research encompasses social policies for senior residents (e.g., Spain) (Gonzáles et al., 2021), and people with moderate and severe disabilities (e.g., Germany) (Pagán, 2015). Research also stresses on the potential of social tourism within the frames of sustainability and community development to mitigate escalating socioeconomic problems (e.g., Greece) (Kakoudakis & McCabe, 2018). Research on the effect of government sanctions in tourism is gaining momentum. Seyfi and Hall's (2020) exploratory study concludes that sanctions can directly impact tourism and cause a domino effect in the economy of countries and businesses (i.e., disruption of financial investments and supply chains, inflation, currency exchange, trade balance). Xu et al. (2022) show how public policy is key in mitigating the impact of political risk in tourism. Their study suggests that political risk has a greater impact underdeveloped regions in the Asia-Pacific and Africa that much depend on international tourism revenues. Governments can intervene as public interest protectors in key tourism services. This was the case of the aviation industry in the peak of the COVID19 pandemic, with governments prioritising air transport connectivity and jobs in the sector (Abate et al., 2020). Similarly, government enabled business supporting schemes that proved to be vital for hospitality businesses in England (Ntounis et al., 2022).

Notwithstanding the rising number of episodes of political unrest and turmoil around the globe, research on governance, policy, and government action in relation to tourism is limited. Campiranon et al. (2011) shed light on the political turmoil during the anti-Thaksin protests in Thailand that eventually culminated with the Thai government decision to close the Bangkok Suvarnabhumi International Airport. Insights from the 1996–2009 Maoist Insurgency in Nepal illustrate the vulnerability of Nepalese tourism to chronic political instability in the country (Nepal, 2010). A great extent of tourism research on large-scale political changes like the Arab

Spring in the Middle East and North Africa have been predominantly researched from either tourism arrivals (Groizard et al., 2021) or receipts (Morakabati, 2020). Overall, reflections on the influence of macro-economic, political, and societal changes for tourism in these countries are overlooked in the field.

A relatively prolific body of research has emerged over the last decade to address tourism governance and public policy in relation to natural hazards, destination vulnerability and financial crises. Looking at natural hazards, studies on public policy and governance in tourism underpin the disaster capitalism complex (Klein, 2007). As Gotham (2012, p. 640) observe in post-Katrina rebuilding in New Orleans, emergency and recovery authorities 'created new political opportunities for corporations to leverage public resources for private gain'. Both in the cases of New Orleans and New York City, tourism was regarded as a flywheel to rebrand and redevelop the city (Gotham and Greenberg, 2014). Similarly, tourism and sport events have been central in the Canterbury Earthquake Recovery Authority (CERA) agenda following the earthquakes of 2010 and 2011 in Christchurch (Amore & Hall, 2017b; Hall & Amore, 2019). Research in vulnerable destinations has also looked at governance as a process in which human agency adapts in response to government action and structure. Evidence from the Adamant Coast, Thailand, shows how stakeholders responded and adapted to the impacts of the 2004 Boxing Day Tsunami and the relevance of tourism in the process of vulnerability reduction and resilience building (Calgaro et al., 2014; Larsen et al., 2011). Similarly, studies on financial crises and tourism from a public policy and governance perspectives look at local destination governance networks dynamics and evolution in times of public sector austerity (e.g., Giovanelli et al., 2015; Farmaki et al., 2015; Mesquita Nunes, 2014).

More recently, shifts in national and international politics have drawn attention on the actions of governments and their implications for tourism. The consequences of Brexit for tourism have been analysed in relation to wider political economy issues like the United Kingdom-Ireland border (Hall, 2020), freedom of movement of British tourists in the European Union (Anastasiadou, 2020) and of European workforce in tourism and hospitality (Amore, 2023). Research on economic and political sanctions has gained relevance in tourism studies following the exacerbation of international relations and global geopolitical frictions. One the one hand, Seyfi and Hall (2019, 2020) provide a taxonomy of political sanctions that can directly and indirectly impact tourism and hospitality (e.g., exports of goods, freezing of financial assets, air travel bans). Moreover, it is suggested that political sanctions in tourism can have wider societal impacts as well as foster a negative perception of the affected destinations. Overall, it is agreed that resolutions sanctioning countries have profound negative impacts on the tourism sector in the sanctioned country, particularly when sanctions are promulgated by key geopolitical powers and main tourism generating countries. For example, Khodadadi (2018, p. 29) notes how political frictions between Washington and Tehran 'could seriously jeopardise the progress made and the future of the now-flourishing tourism industry'. Similarly, Yu et al. (2020, p. 12) provide evidence on how boycotts in response to political tensions between China and neighbouring countries and territories (i.e., Hong Kong, Japan, South Korea, and Taiwan) 'can result in more enduring damage on long-term tourist arrivals'.

A final theme of research worth addressing is the one of overtourism. While it should be noted that 'coining the term 'overtourism' [means] simply resetting the clock on well-established debates' (Dredge, 2017), it is undeniable that policies and regulations as part of government action are key in the management and development of destinations. Overtourism has been mainly discussed as the consequence of poor destination planning and a lack of regulations and policies to manage tourism flows and uncontrolled touristificaiton of destinations. Established urban destinations like Amsterdam, Barcelona, Budapest, Prague, and Venice pay nowadays the result of lasseiz-faire and pro-growth policies of local governments and relevant tourism stakeholders (Goodwin, 2021; Pinkster & Boterman, 2017; Pixová & Sládek, 2017; Smith et al., 2019; Visentin & Bertocchi, 2019). Notwithstanding wide media coverage and interest from governments at different scales, 'there are few practical recommendations coming out of academic research . . . on how to curtail and responsibly manage tourism growth as part of a broader agenda of job creation' (Joppe, 2018, p. 203). As the malaise and antagonism of residents towards tourists increases (Milano et al., 2019), there is a broad agreement among scholars on the inadequacy of adaptive policies to address overtourism. In the view of Koens et al. (2019) and Pasquinelli and Trunfio (2020), long-terms solutions and 'smart' city-wide governance are required to address overtourism in cities. However, such shift in tourism governance should not be assimilated with digitalization and ICT. As Hernández et al. (2019, p. 29) put it:

> The potentiality of the technological solutions and their supposed efficiency cannot constitute a pretext to avoid measures of a greater scope such as limiting the growth of supply and tourism demand, and even de-growth. Smart solutions are compatible with these types of policies and can become a fundamental support.

Theoretical Advancements

Theories are 'a product of their time' (Allmendinger, 2009, p. 7) and the outcome of the reflective understanding of society within which scholars interact as social actors (Johnston, 1979). Theories in social sciences and political sciences are relative and 'not inherently predictive' (Hall, 1984, p. 12). Theoretical advancements in the field of tourism, public policy, and governance are no exception to the rule and subject to revalidation following major institutional and societal changes. A good extent of the research on tourism and public policy between the late 1990s and the early 2000s built from the shift from government to governance in public policy (Rhodes, 1996) (Hall & Jenkins, 2004). Similarly, studies on public policy, governance, and planning in tourism throughout the first decade of the 2000s underpinned epistemological and theoretical advancements in planning theory and policy (Allmendinger, 2009; Healey, 2006; Dredge & Jenkins, 2011; Jenkins et al., 2014). Overall, the political science tradition in the study of public policy and governance enables researchers to focus 'on the different forms of action and control in tourist destinations and puts the role of power – and its allocation among the actors – under scrutiny' (Pechlaner & Volgger, 2013, p. 7).

Theoretical advancements in the field of public policy and governance in tourism can be broadly grouped in six complementary streams of research. The first stream encompasses studies on destination governance and resilience (see Chapter 46, this volume). Research predominantly focuses on the perspectives of diverse tourism-relevant stakeholders and communities following non-linear natural hazards, adaptation and mitigation to climate change, and socio-political crises. Lew (2014) acknowledges the political and social embeddedness framing resilience and policy action in tourism. According to Luthe and Wyss (2014, p. 162) research on the political dimension of resilience in tourism should address the following questions:

- How do assessments based on structural and on functional tourism networks complement resilience understanding, including functional positions of stakeholders, different forms of ties (economic, social, social-ecological), and different scales of governance?

- On what scale of governance should processes of planning and managing resilience take place, and how do such processes interrelate between different governance scales?
- How can tourism systems best adapt to both slow change processes (such as climate change) as well as prompt change impacts (such as an economic crisis) requiring different forms of innovation, adaptation, and structural transformation?

Amore, Hall, and Prayag (2018) envisage socio-political and institutional dimensions in their conceptualisation of destination resilience. In their view, scale and community-driven collaborative planning are key to resilience building. However, in order to understand resilience and destination resilience, it is necessary to adopt 'a different approach in the management, policy making, and environmental awareness thus far conducted in tourism policy and planning practices' (Hall et al., 2017, p. 132).

A second stream of research builds from Jessop's (2008) strategic-relational approach on the roles and activities of the state and their implications for tourism (Bramwell, 2011). The strategic-relational approach conceives the political sphere associated to governance to be strongly tied to both social and economic spheres (Bramwell, 2011). To this end, Bramwell (2011) identifies a range of perspectives associated with strategic-relational approach in relation to research of public policy and governance in tourism and sustainability. In his view, the latter should 'adopt a broad horizon that considers the state within its wide relations with society, and for it to explore the multiple and reciprocal relations between tourism, the state, economy and society' (Bramwell, 2011: 474). The strategic-relational approach conceives a policy environment in which the protection of cultural and natural resources is traded off for economy growth and production of wealth through tourism (Bramwell, 2011). Such policy environment can be best assessed through longitudinal research of 'micro-scale agency, macro-scale structures and the dialectical dimensions between them' (Bramwell, 2011, p. 474).

A third stream of research is that of lobbying and public policy affairs in tourism The latter allows to critically examine the inherently fragmented nature of tourism policy making and how this undermines the pursuit of a holistic and collaborative dialogue between stakeholders. Lobbying and pro-growth economic agendas 'are usually limits to the extent to which policies in the public sphere will influence commercial businesses and business decisions' (Bramwell, 2011, p. 461). This was the case of Cyprus, where collaborations between the hospitality sector and government influenced the implementation of profit-maximising development plans despite previous commitments towards sustainability (Farmaki et al., 2015). Similarly, private interests prevailed over environmental protection in the development of a luxury ecolodge in Kangaroo Island, Australia (Higgins-Desbiolles, 2011). According to Glyptou et al. (2022), finally, the influence of lobbying groups on national governments exacerbates the fallibility of UNSDGs indicators and their application from both top-down hierarchy and network governance perspectives. These are just three of the many examples where the influence and lobby of private tourism stakeholders compromised the pursuit of sound sustainable development (Amore, 2021).

The fourth stream of research, instead, stems from Hall's (2011a) tourism governance framework. The latter distinguishes between types of actors (public and private) and steering modes (hierarchical and non-hierarchical) to identify four main typologies of governance. The first typology – hierarchies – encompasses top-down decision-making commonly associated to public policy action, promulgation of legislation and regulations and a neat distinction between public and private sector. Conversely, the second typology – markets – conceives exchange and negotiation as the key to efficient governance. Governance by markets has permeated the public sector through various reforms since the 1980s (Olesen, 2014). Yet, efficiency through market governance has proved to fail, 'especially with respect to the equity of policy outcomes and the achievement of more sustainable forms of tourism' (Hall, 2011a, p. 447). The third typology of governance – networks – 'has received considerable attention in tourism policy and planning because of the ways in which they may facilitate coordination of public and private interests and resources' (Hall, 2011a, p. 447). However, network openness and accessibility to all relevant stakeholders are often questioned among tourism scholars. For example, Amore et al. (2017) show how members of the civic society and key local advocacy groups in Christchurch, New Zealand, were kept at arm's length from key decisions on cultural and tourism redevelopment of the city following the 2010–2011 Canterbury Earthquakes. The fourth and final typology – communities – conceives public participation, decision making and deliberation central to governance. As Hall (2011a, p. 448) observes, 'community participation and even control over planning and decision-making remain an important issue in tourism planning and policymaking'. At the same time, it is acknowledged that this type of governance can be challenging to be implemented in practice.

A fifth, more recent advancement is that of metagovernance and tourism. The latter was first mentioned in the second edition of this Companion (Jenkins et al., 2014, p. 550) as 'an important future area of inquiry for tourism scholars'. Metagovernance – also referred to as 'governance of governance' (Jessop, 2011, p. 106) – conceives the state as 'just one participant among others in a pluralist guidance system, contributing its own resources to the negotiation process' (Jenkins et al., 2014, p. 550). The theoretical shift towards metagovernance enables scholars to have greater understanding of the dynamic policy environment in which decision making and haphazardous coalescences among tourism relevant stakeholders take place (Amore & Hall, 2016b; Jenkins et al., 2014). According to Amore and Hall (2016b, p. 118):

> a focus on metagovernance ... seeks to make governance more transparent and raises questions as to why some strategies are employed and not others and who wins and who loses in decision making, especially with respect to the values and interests in what is an overtly political context.

Central to the notion of metagovernance is the fallibility of policy systems. From an etymological perspective, in fact, metagovernance refers to what is after governance (from the Greek μετά, 'beyond', 'upon' or 'after'). From a conceptual standpoint Jessop (2011) and Sørensen (2006) argue that metagovernance is the response to the perceived failure of governance modes. Similarly, Amore and Hall (2016b, p. 112) convene that metagovernance is both the process and the outcome of state withdrawal from governance, 'whereby political authorities at national and other levels are regarded as encouraging the self-organisation of partnerships, networks and governance regimes'. Building from Hall (2011a) and Meuleman (2008), metagovernance in tourism conceives the withdrawal of the state as a new articulation of state power into what Scharpf (1994, p. 41) describes as a 'shadow of hierarchical authority'. The latter echoes Meuleman (2008, p. 68) definition of metagovernance as 'a means by which to produce some degree of coordinated governance, by designing and managing sound combinations of hierarchical, market and network governance'.

Amore and Hall (2016b) develop a metagovernance framework encompassing eight tourism metagovernance archetypes in addition to the typologies first introduced by Hall (2011a). Two of such metagovernance archetypes – New Public Management and Public-Private Partnerships – combine hierarchy, market and network governance and have been extensively research in tourism and public policy (e.g., Chaperon, 2017; Paddison & Walmsley, 2018).

Participatory and deliberative archetypes of metagovernance refer to interactive and problem-solving forms of policy making respectively and combine elements from each of the governance typologies (Amore & Hall, 2016b; Meuleman, 2008). Nodal metagovernance, instead, 'is a structured form of participatory governance in which state and nonstate actors co-operate to provide public services' (Meuleman, 2008, p. 36), while reflexive and adaptive archetypes of metagovernance reflect the dynamicity of policy making in response to influences within the governance environment and adaptive management building from experience and capacity to 'adapt to a range of anticipated and unanticipated conditions' (Meuleman, 2008, p. 37). Finally, community as metagovernance archetype foresees citizen-driven forms of associations, decision making and provision of services. In presenting the framework, the authors suggest that the hierarchical authority of the state still permeates concessions, compromises and decision making over a range of different spatial scales (Amore and Hall, 2016b). In their view, therefore, collaboration and coordination among stakeholders at the local/destination level are politically and socially embedded within hierarchies of governance.

Research on metagovernance and tourism has focused on different aspects of destination management and planning. Wan and Bramwell (2015) apply metagovernance theory to examine governance modes, steering mechanisms, and interactions among stakeholders in tourism policy making in Hong Kong. Giovanell et al. (2015) and Bregoli (2013) use metagovernance to respectively examine local tourism networks in Sardinia, Italy, and destination branding coordination initiatives in Edinburgh, United Kingdom. Lamers et al. (2014) adopts a metagovernance approach to understand steering modes in policy systems around conservation and tourism partnerships in Kenya, while Adie and Amore (2020) use metagovernance theory to analyse hierarchical policy failures within the national Italian heritage governance structure in the Prehistoric Pile Dwellings around the Alps transnational World Heritage site. Spyriadis et al. (2011) analyse the policy environment around Destination Management Organizations (DMOs) from a metagovernance perspective and conclude that DMOs are key 'in coordinating and integrating the development and implementation of policies and strategies at intersectoral (across sectors) and intergovernmental (across the levels of government) levels' (Spyriadis et al., 2011, p. 198). More recently, research has embraced metagovernance thinking to evaluate the potential towards sustainable tourism reconfiguration in response to both short-term (COVID-19 pandemic) and long-term (climate change) crises (Becken and Loehr, 2022). At the same time, metagovernance theory has been used to reflect on UNSDGs indicator selection, sustainable tourism governance and policy making (Rasoolimanesh et al., 2023).

A sixth and final theoretical advancement calls for a rethinking of public policy and governance in tourism from a political ecology perspective. This is a much-needed shift if we consider how important natural resources and land use are in sound sustainable tourism policies. Yet, as Nepal et al. (2016, p. 1) observe, 'very few tourism scholars have specifically used political ecology as a lens to examine tourism-centric or tourism- related natural resource management issues'. Themes such as ownership and control of natural resources are central to political ecology research. Scholars researching public policy and governance in tourism are encouraged to shed light on the many episodes of economic and political injustice at the expenses of indigenous communities, particularly in the so-called developing countries (Douglas, 2014; Scheyvens & Russell, 2012). The latter are forcibly kept at arm's length (Nepal et al., 2016) or manipulated through mechanisms of non-decision making on key issues like conservation and restoration of the environment (Hall, 2003). Greenwashing and the rhetoric of local development are common in development projects where a restricted elite benefits from the economic gains of tourism while undermining the real prospects of the local community (Dahal et al., 2014).

Future research in the field, therefore, has to recognize the politically charged context within which environmental policies impacting tourism take place and how these can potentially undermine sustainability in both the so-called developed world (e.g., United Kingdom) (Hall, 2020) and the Global South (e.g., Nicaragua) (Hunt, 2016). At the same time, it is crucial to prioritize ecological protection, rectify the inequalities caused by neoliberal-infused tourism policy and put 'tourism in the service to the public and to be accountable to the public' (Higgins-Desbiolles, 2020, p. 618). As Jamal and Camargo (2014, p. 26) put it:

> We lack theoretical and practical tools to tackle important ethical and justice-related issues related to destination development and marketing, particularly intangible aspects such as human–environmental (ecocultural) relationships, cultural commodification, and inequitable distribution of tourism opportunities (and costs) among disadvantaged groups .... We call for an ecocultural, participatory, and integrated framework of justice and care to guide sustainable tourism.

Conclusions

This chapter provided an overview of the main themes and trends in tourism public policy and governance research over the last decade and presented six theoretical advancements in the field. Albeit less prolific than other fields in tourism studies, research on the politics and decision-making processes in tourism represents a widely established and a theoretically cohesive body of literature. These are, understandably, interesting times. Recent triggering political events, natural hazards and climate change clearly show the inherent vulnerability of tourism. More research on the way policies and metagovernace for tourism adapt and evolve in time and space is needed. Nothing is static and policy systems, as societal constructs, are prone to fail and change. In such extremely fluid policy context, research on governance and public policy in tourism can foster a much-needed third-order change in sustainability and policy making. As Hall (2011b, p. 665) puts it:

> it is the growing awareness of the contradictions in, and policy failure of, liberal environmentalism that may also offer an opportunity for third-order change. The role of policy failure and learning with respect to sustainable tourism policy and governance issues in relation to first- and second-order change is important and clearly warrants further study. However, any understanding of the potential for third-order change needs to be grounded in research of the interrelationships between power, values, norms and interests and how they influence the selection of policy instruments, indicators and settings within broader frames of governance and change.

The readers of this chapter might wonder: how such third-order change can be achieved when policy actors are unwilling to acknowledge policy failure? Undeniably, this is a hefty task for tourism scholars to accomplish. Research in public policy occurs within a depoliticized environment of consensus, rationalization, and power (Cryle & Hillier, 2005; Flyvbjerg, 1998) and tourism is no exception. Nonetheless, it should be the job – and the duty – of scholars to critically conduct policy analysis and question the legitimacy and rationale of policy actions. The way forms of citizenry engagement and inclusive participation in tourism policy and governance are evolving as we speak (Erdmenger, 2022) reflect the obsolete modes of destination governance. At the same time, research should sideline with the power-less and those seeking to have their voices heard, in the pursuit of genuinely democratic, equal, and diverse governance arenas.

References

Abate, M., Christidis, P., & Purwanto, A. J. (2020). Government support to airlines in the aftermath of the COVID-19 pandemic. *Journal of Air Transport Management*, *89*, 101931. https://doi.org/10.1016/j.jairtraman.2020.101931

Adie, B. A., & Amore, A. (2020). Transnational World Heritage, (meta) governance and implications for tourism: an Italian case. *Annals of Tourism Research*, *80*, 102844. https://doi.org/10.1016/j.annals.2019.102844

Agarwal, S., Page, S. J., & Mawby, R. (2021). Tourist security, terrorism risk management and tourist safety. *Annals of Tourism Research*, *89*, 103207. https://doi.org/10.1016/j.annals.2021.103207

Aguilera, T., Artioli, F., & Colomb, C. (2021). Explaining the diversity of policy responses to platform-mediated short-term rentals in European cities: a comparison of Barcelona, Paris and Milan. *Environment and Planning A: Economy and Space*, *53*(7), 1689–1712. https://doi.org/10.1177/0308518X19862286

Allmendinger, P. (2009). *Planning Theory* (2nd ed.). Palgrave Macmillan.

Aly, M. N., Yuliawan, R., Noviyanti, U. D. E., Firdaus, A. A., & Prasetyo, A. (2019.) Public policy and rural tourism development in East Java Province, Indonesia. *African Journal of Hospitality, Tourism and Leisure* (Special Issue): 1–8.

Amore, A. (forthcoming). Metagovernance and tourism: advocating for a much-needed reframing in the study of tourism politics. In J. Saarinen & C. M. Hall (Eds). *Handbook of Tourism Governance*, Edward Elgar.

Amore, A. (2023). The importance of policies and planning in the tourism job market: A reflection on actions and inactions in Italy and the United Kingdom. In P. Xie (ed). *Handbook of tourism planning*, (pp. 182–194), Edward Elgar.

Amore, A. (2019). *Tourism and urban regeneration: Processes compressed in time and space*. Routledge.

Amore, A. (2021). Sustainable tourism, lobbying and public affairs. In P. Harris, A. Bitonti, C. S. Fleisher & A. S. Binderkrantz (Eds.), *The Palgrave encyclopedia of interest groups, lobbying and public affairs*, Palgrave Macmillan, https://doi.org/10.1007/978-3-030-13895-0_216-1.

Amore, A., Arvanitis, P., d'Angella, F., & De Carlo, M. (2022). Tourism destination metagovernance and smart governance in Milan, Italy. In D. Buhalis, B. Taheri and R. Rahimi (Eds.) *Smart Cities and Tourism: Co-creating experiences, challenges and opportunities*, (pp. 142–158). Goodfellow Publishers.

Amore, A., de Bernardi, C., & Arvanitis, P. (2022). The impacts of Airbnb in Athens, Lisbon and Milan: a rent gap theory perspective. *Current Issues in Tourism*, *25*(20), 3329–3342. https://doi.org/10.1080/13683500.2020.1742674

Amore, A., & Hall, C. M. (2016a). 'Regeneration is the focus now': anchor projects and delivering a new CBD for Christchurch. In C. M. Hall, S. Malinen, R. Vosslamber, & R. Wordsworth (Eds.), *Business and post disaster management: Business, organisational and consumer resilience and the Christchurch Earthquakes* (pp. 181–199). Routledge.

Amore, A., & Hall, C. M. (2016b). From governance to meta-governance in tourism? Re-incorporating politics, interests, and values in the analysis of tourism governance. *Tourism Recreation Research*, *41*(2), 109–122. https://doi.org/10.1080/02508281.2016.1151162

Amore, A., & Hall, C. M. (2017a). National and urban public policy in tourism. Towards the emergence of a hyperneoliberal script? *International Journal of Tourism Policy*, *7*(1), 4–22. https://doi.org/10.1504/IJTP.2017.082761

Amore, A., & Hall, C. M. (2017b). Sports and event-led regeneration strategies in post-earthquake Christchurch. In M. Wise & J. Harris (Eds.), *Sport, events, tourism and regeneration* (pp. 100–117). Routledge.

Amore, A., & Hall, C. M. (2021). Shock of the new: the rhetoric of global urban tourism in the rebuild of Christchurch, New Zealand. In S. Frank, C. Jürgens, C. Müller, A. L. Raschke & K. Wellner (eds), *The power of new urban tourism: Markets, representations and contestations* (pp. 175–188). Routledge.

Amore, A., Hall, C. M., & Jenkins, J. M. (2017). They never said 'Come here and let's talk about it': exclusion and non-decision-making in the rebuild of Christchurch, New Zealand. *Local Economy*, 32(7), 617–639. https://doi.org/10.1177/0269094217734326

Amore, A., Prayag, G., & Hall, C. M. (2018). Conceptualizing destination resilience from a multilevel perspective. *Tourism Review International*, *22*(3–4), 235–250. https://doi.org/10.3727/154427218X15369305779010

Anastasiadou, C. (2020). Tourism and the EU: retrospect. In Hall, D. K. (ed.), *Brexit and tourism: Process, impacts and non-policy* (pp. 28–40). Routledge.

Avraham, E. (2021). Combating tourism crisis following terror attacks: image repair strategies for European destinations since 2014. *Current Issues in Tourism*, *24*(8), 1079–1092. https://doi.org/10.1080/13683500.2020.1764510

Becken, S., & Loehr, J. (2022). Tourism governance and enabling drivers for intensifying climate action. *Journal of Sustainable Tourism*, 1–19. https://doi.org/10.1080/09669582.2022.2032099

Booth, W. J. (2006). *Communities of memory on witness, identity, and justice*. Cornell University Press.

Bosman, C. (2007). Building community places – Machizukuri – neoliberalism, suburbanization and 'Americanization'. *International Planning Studies*, *12*(4), 309–325. https://doi.org/10.1080/13563470701745447

Bosman, C., & Dredge, D. (2011). *Histories of placemaking in the Gold Coast City: The neoliberal norm, the State story, and the community narrative*. Griffith University.

Bramwell, B. (1995). Interventions and policy instruments for sustainable tourism. In W. F. Theobald (Ed.), *Global tourism* (3rd ed.), (pp. 406–425). Elsevier.

Bramwell, B. (2011). Governance, the state and sustainable tourism: a political economy approach. *Journal of Sustainable Tourism*, *19*(4–5), 459–477. https://doi.org/10.1080/09669582.2011.576765

Bramwell, B., & Lane, B. (1999). Editorial: collaborations and partnerships for sustainable tourism. *Journal of Sustainable Tourism*, *7*(3–4), 179–181.

Bregoli, I. (2013). Effects of DMO coordination on destination brand identity: a mixed-method study on the city of Edinburgh. *Journal of Travel Research*, *52*(2), 212–224. https://doi.org/10.1177/0047287512461566

Brendehaug, E., Aall, C., & Dodds, R. (2017). Environmental policy integration as a strategy for sustainable tourism planning: issues in implementation. *Journal of Sustainable Tourism*, *25*(9), 1257–1274. https://doi.org/10.1080/09669582.2016.1259319

Calgaro, E., Lloyd, K., & Dominey-Howes, D. (2014). From vulnerability to transformation: a framework for assessing the vulnerability and resilience of tourism destinations. *Journal of Sustainable Tourism*, *22*(3), 341–360. https://doi.org/10.1080/09669582.2013.826229

Campiranon, K., Laws, E., & Scott, N. (2011). Responding to crises in Thailand: a governance analysis. In E. Laws, H. Richins, J. Argusa & N. Scott (Eds.), *Tourist destination governance: Practice, theory, and issues* (pp. 91–102). CABI.

Castanho, R. A., Couto, G., Pimentel, P., Carvalho, C. B., & Sousa, Á. (2020). Territorial management and governance, regional public policies and their relationship with tourism. A case study of the Azores Autonomous Region. *Sustainability*, 12(15), Art. 6059. https://doi.org/10.3390/su12156059

Chaperon, S. (2017). Tourism industry responses to public-private partnership arrangements for destination management organisations in small island economies: a case study of Jersey, Channel Islands. *International Journal of Tourism Policy*, *7*(1), 23–41. https://doi.org/10.1504/IJTP.2017.082767

Cruz, M. S., Cruz, J. I. G., Greifemberg, C. A., & Zaragoza, M. P. P. (2018). Strategies to improve the quality and competitiveness of coastal tourism areas: applying tourism standards. *Journal of Tourism Analysis*, *25*(1), 68–90.

Cryle, D., & Hillier, J. (Eds) (2005). *Consent and consensus. Politics, media and governance in Twentieth Century Australia*. API Network, Australia Research Institute.

Dahal, S., Nepal, S. K., & Schuett, M. (2014). Marginalized communities and local conservation institutions: indicators of participation in Nepal's Annapurna Conservation Area. *Environmental Management*, *53*(1), 219–230. https://doi.org/10.1007/s00267-013-0204-8

Dela Santa, E. (2018). Fiscal incentives for tourism development in the Philippines: a case study from policy networks and advocacy coalition framework. *Tourism Planning & Development*, *15*(6), 615–632. https://doi.org/10.1080/21568316.2017.1360935

Del Gesso, C., Lodhi, R. N., & Cobanoglu, C. (2022). Local government's role in promoting city hospitality: a meta-ethnography with a public management perspective. *International Journal of Contemporary Hospitality Management*, *34*(10), 3855–3880. https://doi.org/10.1108/IJCHM-09-2021-1092

Dinis, A. A., & Pereira, L. (2022). Tax incentives for tourism: the case of Portuguese municipalities. In V. Costa and C. Costa (Eds.), *Sustainability and competitiveness in the hospitality industry* (pp. 95–118). IGI Global.

Douglas, J. A. (2014). What's political ecology got to do with tourism? *Tourism Geographies*, *16*(1), 8–13. https://doi.org/10.1080/14616688.2013.864324

Dredge, D. (2017). *Overtourism: Old wine in new bottles?* www.linkedin.com/pulse/overtourism-old-wine-new-bottles-dianne-dredge.

Dredge, D., & Jenkins, J. M. (2011). New spaces of tourism planning and policy. In D. Dredge and J. M. Jenkins (Eds.), *Stories of practice: Tourism policy and planning* (pp. 1–12). Ashgate.

Dredge, D., & Jenkins, J. M. (2012). Australian national tourism policy: influences of reflexive and political modernisation. *Tourism Planning & Development*, *9*(3), 231–251. https://doi.org/10.1080/21568316.2012.678379

Dredge, D., & Jamal, T. (2013). Mobilities on the Gold Coast, Australia: implications for destination governance and sustainable tourism. *Journal of Sustainable Tourism*, *21*(4), 557–579. https://doi.org/10.1080/09669582.2013.776064

Dutta, A., Mishra, T., Uddin, G. S., & Yang, Y. (2021). Brexit uncertainty and volatility persistence in tourism demand. *Current Issues in Tourism*, *24*(16), 2225–2232. https://doi.org/10.1080/13683500.2020.1822300

Duffy, R. (2015). Nature-based tourism and neoliberalism: concealing contradictions. *Tourism Geographies*, *17*(4), 529–543. https://doi.org/10.1080/14616688.2015.1053972

Efthymiou, M., & Papatheodorou, A. (2020). Environmental policies in European aviation: a stakeholder management perspective. In T. Walker, A. Bergantino, N. Munch & L. Loiacono (Eds.), *Sustainable aviation* (pp. 101–125). Palgrave Macmillan.

Erdmenger, E. C. (2022). The end of participatory destination governance as we thought to know it. *Tourism Geographies*, 1–23. https://doi.org/10.1080/14616688.2022.2086904

Farmaki, A., Altinay, L., Botterill, D., & Hilke, S. (2015). Politics and sustainable tourism: the case of Cyprus. *Tourism Management*, *47*, 178–190. https://doi.org/10.1016/j.tourman.2014.09.019

Flyvbjerg, B. (1998). *Rationality and power: Democracy in practice*. Chicago University Press.

Giovanelli, L., Rotondo, F., & Fadda, N. (2015). The evolution of governance networks in a time of crisis. Evidence from the Italian tourism sector. *Journal of Public Administration and Policy Research*, 7(4), 76–88. https://doi.org/10.5897/JPAPR2015.0317

Glyptou, K., Amore, A., &Adie, B. A. (2022). From aspirations to applications: the SDGs and the role of indicators in the measurement of sustainable tourism. In A. Farmaki, L. Altinay & X. Font (Eds), *Planning and managing sustainability in tourism: Empirical studies, best-practice cases and theoretical insights* (pp. 13–25). Springer.

Goodwin, H. (2021). City destinations, overtourism and governance. *International Journal of Tourism Cities*, 7(4), 916–921. https://doi.org/10.1108/IJTC-02-2021-0024

González, E. A., Domínguez Vila, T., & Sánchez, N. L. (2021). Social tourism for seniors in Spain: an example to be followed? In J Lima & C. Eusebio (Eds.), *Social tourism: Global challenges and approaches* (pp. 72–82). CABI.

Gori, E., Fissi, S., & Romolini, A. (2021). A collaborative approach in tourism planning: the case of Tuscany region. *European Journal of Tourism Research*, *29*, 2907–2907. https://doi.org/10.54055/ejtr.v29i.2426

Gotham, K. F. (2012). Disaster, inc.: privatization and post-Katrina rebuilding in New Orleans. *Perspectives on Politics*, 10(3), 633–646. https://doi.org/10.1017/S153759271200165X

Gotham, K. F., & Greenberg, M. (2014). *Crisis cities: Disaster and redevelopment in New York and New Orleans*. Oxford University Press.

Groizard, J. L., Ismael, M., & Santana-Gallego, M. (2022). Political upheavals, tourism flight, and spillovers: the case of the Arab Spring. *Journal of Travel Research*, *61*(4), 921–939. https://doi.org/10.1177/00472875211002652

Guttentag, D. (2015). Airbnb: disruptive innovation and the rise of an informal tourism accommodation sector. *Current Issues in Tourism*, *18*(12), 1192–1217. https://doi.org/10.1080/13683500.2013.827159

Hall, C. M. (1984). *Politics of Policy. The development of hydro-electric policy in South-West Tasmania* (Unpublished Master's Thesis). University of Waterloo.

Hall, C. M. (1999). Rethinking collaboration and partnership: a public policy perspective. *Journal of Sustainable Tourism*, *7*(3–4), 274–289. https://doi.org/10.1080/09669589908667340

Hall, C. M. (2003). Politics and place: an analysis of power in tourism communities. In S. Singh, D. J. Timothy & R. K. Dowling (Eds.), *Tourism in destination communities* (pp. 99–113). CABI.

Hall, C. M. (2008). *Tourism planning: Policies, processes, and relationships* (2nd ed). Prentice Hall.

Hall, C. M. (2011a). A typology of governance and its implications for tourism policy analysis. *Journal of Sustainable Tourism*, *19*(4–5), 437–457. https://doi.org/10.1080/09669582.2011.570346

Hall, C. M. (2011b). Policy learning and policy failure in sustainable tourism governance: from first-and second-order to third-order change? *Journal of Sustainable Tourism*, *19*(4–5), 649–671. https://doi.org/10.1080/09669582.2011.555555

Hall, C. M. (2019). Constructing sustainable tourism development: the 2030 agenda and the managerial ecology of sustainable tourism. *Journal of Sustainable Tourism*, *27*(7), 1044–1060. https://doi.org/10.1080/09669582.2018.1560456

Hall, C. M. (2020). The environmental implications of Brexit for tourism. In D. K. Hall (Ed.), *Brexit and tourism: Process, impacts and non-policy* (pp. 121–127). Channel View Publications.

Hall, C. M., & Amore, A. (2019). The 2015 Cricket World Cup in Christchurch: using an event for post-disaster reimagine and regeneration. *Journal of Place Management and Development*, *13*(1), 4–17. https://doi.org/10.1108/JPMD-04-2019-0029

Hall, C M., & Jenkins, J. M. (2004). Tourism and public policy. In A. Lew, C. M. Hall and A. M. Williams (Eds.), *A companion to tourism* (1st ed), (pp. 525–540). Blackwell Publishing.

Hall, C. M., Prayag, G., & Amore, A. (2017). *Tourism and resilience: Individual, organisational and destination perspectives*. Channel View Publications.

Hall, C. M., & Wood, K. J. (2021). Demarketing tourism for sustainability: Degrowing tourism or moving the deckchairs on the titanic? *Sustainability*, *13*(3), 1585. https://doi.org/10.3390/su13031585

Hall, C. M., Prayag, G., Safenov, A., Coles, T., Gössling, S., & Naderi Koupaei, S. (2022). AirBnB and the sharing economy. *Current Issues in Tourism*, *25*(19), 3057–3067. https://doi.org/10.1080/13683500.2022.2122418

Hall, D. K. (Ed.) (2020). *Brexit and tourism: Process, impacts and non-Policy*. Channel View Publications.

Healey, P. (2006). *Collaborative planning: Shaping places in fragmented societies*(2nd ed). Palgrave Macmillan.

Hernández, M. G., Baidal, J. I., & de Miguel, S. M. (2019). Overtourism in urban destinations: the myth of smart solutions. *Boletín de la Asociación de Geógrafos Españoles*, *83*(2830), 1–38. https://www.bage.age-geografia.es/ojs/index.php/bage/article/view/2830

Higgins-Desbiolles, F. (2011). Death by a thousand cuts: governance and environmental trade-offs in ecotourism development at Kangaroo Island, South Australia. *Journal of Sustainable Tourism*, *19*(4–5), 553–570. https://doi.org/10.1080/09669582.2011.560942

Higgins-Desbiolles, F. (2020). Socialising tourism for social and ecological justice after COVID-19. *Tourism Geographies*, *22*(3), 610–623. https://doi.org/10.1080/14616688.2020.1757748

Hudson, S. (2020). *COVID-19 and Travel: Impacts, responses and outcomes*. Goodfellow Publishers.

Hunt, C. (2016). A political ecology of tourism in the shadow of an inter-oceanic canal in Nicaragua: displacing poverty of displacing social and environmental welfare? In S. K. Nepal & J. Saarinen (Eds.), *Political ecology and tourism* (pp. 163–178). Routledge.

Jamal, T., & Camargo, B. A. (2014). Sustainable tourism, justice, and an ethic of care: toward the Just Destination. *Journal of Sustainable Tourism*, *22*(1), 11–30. https://doi.org/10.1080/09669582.2013.786084

Jamal, T., & Camargo, B. A. (2018). Tourism governance and policy: whither justice? *Tourism Management Perspectives*, *25*, 205–208. https://doi.org/10.1016/j.tmp.2017.11.009

Jenkins, J. M., Hall, C. M., & Mkono, M. (2014). Tourism and public policy: contemporary debates and future directions. In A. Lew, C. M. Hall & A. M. Williams (Eds.), *A companion to tourism* (2nd ed.) (pp. 542–555). Wiley.

Jessop, B. (2008). *State power: A strategic-relational approach*. Polity Press.

Jessop, B. (2011). Metagovernance. In M. Bevir (Ed.), *The Sage handbook of governance* (pp. 106–123). Sage.

Johnston, R. J. (1979). *Geography and geographers: Anglo-American human geography Since 1945*. Edward Arnold.

Joppe, M. (2018). Tourism policy and governance: quo vadis? *Tourism Management Perspectives, 25*, 201–204. https://doi.org/10.1016/j.tmp.2017.11.011

Kakoudakis, K. I., & McCabe, S. (2018). Social tourism as a modest, yet sustainable, development strategy: policy recommendations for Greece. *Journal of Policy Research in Tourism, Leisure and Events, 10*(3), 189–203. https://doi.org/10.1080/19407963.2018.1443938

Kavita, E., & Saarinen, J. (2016). Tourism and rural community development in Namibia: policy issues review. *Fennia-International Journal of Geography, 194*(1), 79–88. https://fennia.journal.fi/article/view/46331

Klein, N. (2007). *The shock doctrine: The rise of disaster capitalism*. Macmillan Publishers.

Khalid, U., Okafor, L. E., & Burzynska, K. (2021). Does the size of the tourism sector influence the economic policy response to the COVID-19 pandemic? *Current Issues in Tourism, 24*(19), 2801–2820. https://doi.org/10.1080/13683500.2021.1874311

Khodadadi, M. (2018). Donald Trump, US foreign policy and potential impacts on Iran's tourism industry: post-nuclear deal. *Tourism Management Perspectives, 26*, 28–30. https://doi.org/10.1016/j.tmp.2018.01.007

Koens, K., Postma, A., & Papp, B. (2019). Management strategies for overtourism: from adaptation to system change. In H. Pechlaner, E. Innerhofer & G. Ershbamer (eds), *Overtourism: Tourism management and solutions* (pp. 149–160). Routledge.

Lamers, M., Van der Duim, R., van Wijk, J., Nthiga, R., & Visseren-Hamakers, I. J. (2014). Governing conservation tourism partnerships in Kenya. *Annals of Tourism Research, 48*, 250–265. https://doi.org/10.1016/j.annals.2014.07.004

Larsen, R. K., Calgaro, E., & Thomalla, F. (2011). Governing resilience building in Thailand's tourism-dependent coastal communities: conceptualising stakeholder agency in social–ecological systems. *Global Environmental Change*, 21(2), 481–491. https://doi.org/10.1016/j.gloenvcha.2010.12.009

Lew, A. A. (2014). Scale, change and resilience in community tourism planning. *Tourism Geographies*, 16(1), 14–22. https://doi.org/10.1080/14616688.2013.864325

Luthe, T., & Wyss, R. (2014). Assessing and planning resilience in tourism. *Tourism Management, 44*, 161–163. https://doi.org/10.1016/j.tourman.2014.03.011

Mena-Navarro, A., Almeida-García, F., & Cortés-Macías, R. (2022). The role of the MICE sector in Singapore's tourism policy. A historical perspective. *Journal of Policy Research in Tourism, Leisure and Events*: 1–21. https://doi.org/10.1080/19407963.2022.2043880

Mesquita Nunes, A. (2014). Did the context of economic crisis affect the image of Portugal as a tourist destination? The government's perspective. *Worldwide Hospitality and Tourism Themes*, 6(5), 452–457. https://doi.org/10.1108/WHATT-09-2014-0035

Meuleman, L. (2008). *Public management and the metagovernance of hierarchies, networks and markets: The feasibility of designing and managing governance style combinations*. Springer

Meuleman, L. (2020). *Metagovernance for sustainability: A framework for implementing the Sustainable Development Goals*. Routledge.

Milano, C., Novelli, M., & Cheer, J. M. (2019). Overtourism and tourismphobia: a journey through four decades of tourism development, planning and local concerns. *Tourism Planning & Development, 16*(4), 353–357. https://www.tandfonline.com/doi/full/10.1080/21568316.2019.1599604

Morakabati, Y. (2020). A question of confidence. Is tourism as vulnerable to civil unrest as we think? A comparative analysis of the impact of Arab Spring on total reserves and tourism receipts. *International Journal of Tourism Research*, 22(2), 252–265. https://doi.org/10.1002/jtr.2333

Muangasame, K., & McKercher, B. (2015). The challenge of implementing sustainable tourism policy: a 360-degree assessment of Thailand's "7 Greens sustainable tourism policy". *Journal of Sustainable Tourism*, *23*(4), 497–516. https://doi.org/10.1080/09669582.2014.978789

Nepal, S. K. (2010). Tourism and political change in Nepal. In R. Butler & W. Suntikul (Eds.), *Tourism and political change* (1st ed., pp. 147–159). Goodfellow Publishers.

Nepal, S. K., Saarinen, J., & McLean-Puron, E. 2016. Introduction: political ecology and tourism – concepts and constructs. In S. K. Nepal & J. Saarinen (Eds.), *Political ecology and tourism* (pp. 1–15). Routledge.

Ntounis, N., Parker, C., Skinner, H., Steadman, C., & Warnaby, G. (2022). Tourism and hospitality industry resilience during the Covid-19 pandemic: evidence from England, *Current Issues in Tourism*, *25*(1), 46–59. https://doi.org/10.1080/13683500.2021.1883556

Olesen, K. (2014). The neoliberalisation of strategic spatial planning. *Planning Theory*, *13*(3), 288–303. https://doi.org/10.1177/1473095213499340

Orchiston, C., & Higham, J.E.S. (2016). Knowledge management and tourism recovery (de) marketing: the Christchurch earthquakes 2010–2011. *Current Issues in Tourism*, *19*(1), 64–84. https://doi.org/10.1080/13683500.2014.990424

Paddison, B., & Walmsley, A. (2018). New Public Management in tourism: a case study of York. *Journal of Sustainable Tourism*, *26*(6), 910–926. https://doi.org/10.1080/09669582.2018.1425696

Pagán, R. (2015). The contribution of holiday trips to life satisfaction: the case of people with disabilities. *Current Issues in Tourism*, *18*(6), 524–538. https://doi.org/10.1080/13683500.2013.860086

Pasquinelli, C., & Trunfio, M. (2020). Reframing urban overtourism through the Smart-City lens. *Cities*, *102*, 102729. https://doi.org/10.1016/j.cities.2020.102729

Pechlaner, H., & Volgger, M. (2013). Towards a comprehensive view of tourism governance: relationships between the corporate governance of tourism service firms and territorial governance. *International Journal of Globalisation and Small Business*, *5*(1–2), 3–19. https://doi.org/10.1504/IJGSB.2013.050484

Pforr, C. (2021). Tourism and public policy. In A. Kellow, T. Porter & K. Ronit (Eds.), *Handbook of business and public policy* (pp. 378–391). Edward Elgar Publishing.

Pinkster, F. M., & Boterman, W. R. (2017). When the spell is broken: gentrification, urban tourism, and privileged discontent in the Amsterdam canal district. *Cultural Geographies*, *24*(3), 457–472. https://doi.org/10.1177/1474474017706176

Pixová, M., & Sládek, J. (2017). Touristification and awakening civil society in post-socialist Prague. In C. Colomb & J. Novy (Eds.), *Protest and resistance in the tourist city* (pp. 73–89). Routledge.

Psycharis, Y., Panori, A., & Athanasopoulos, D. (2022). Public investment and regional resilience: empirical evidence from the Greek regions. *Tijdschrift voor Economische en Sociale Geografie*, *113*(1): 57–79. https://doi.org/10.1111/tesg.12499

Rasoolimanesh, S. M., Ramakrishna, S., Hall, C. M., Esfandiar, K., & Seyfi, S. (2023). A systematic scoping review of sustainable tourism indicators in relation to the sustainable development goals. *Journal of Sustainable Tourism*, 31(7), 1497–1517. https://doi.org/10.1080/09669582.2020.1775621

Rhodes, R. A. W. (1996). The new governance: governing without government. *Political Studies*, *44*(4), 652–667. https://doi.org/10.1111/j.1467-9248.1996.tb01747.x

Rogerson, C. M. (2015). Tourism and regional development: the case of South Africa's distressed areas. *Development Southern Africa*, *32*(3), 277–291. https://doi.org/10.1080/0376835X.2015.1010713

Santos, J. D. F., Vareiro, L., Remoaldo, P., & Cadima Ribeiro, J. (2017). Cultural mega-events and the enhancement of a city's image: differences between engaged participants and attendees. *Journal of Policy Research in Tourism, Leisure and Events*, 9(2), 129–151. https://doi.org/10.1080/19407963.2016.1157598

Santos-Lacueva, R., & Velasco Gonzalez, M. (2018). Policy coherence between tourism and climate policies: the case of Spain and the Autonomous Community of Catalonia. *Journal of Sustainable Tourism*, *26*(10), 1708–1727. https://doi.org/10.1080/09669582.2018.1503672

Saxena, G. (2014). Cross-sector regeneration partnership strategies and tourism. *Tourism Planning & Development*, *11*(1), 86–105. https://doi.org/10.1080/21568316.2013.839473

Scharpf, F. W. (1994). Games real actors could play: Positive and negative coordination in embedded negotiations. *Journal of Theoretical Politics*, 6(1), 27–53. https://doi.org/10.1177/0951692894006001002
Scheyvens, R., & Cheer, J. M. (2022). Tourism, the SDGs and partnerships. *Journal of Sustainable Tourism*, *30*(10), 2271–2281. https://doi.org/10.1080/09669582.2021.1982953
Scheyvens, R., & Russell, M. (2012). Tourism, land tenure and poverty alleviation in Fiji. *Tourism Geographies*, *14*(1), 1–25. https://doi.org/10.1080/14616688.2011.593188
Seyfi, S., & Hall, C. M. (2019). *Tourism, sanctions, and boycotts*. Routledge.
Seyfi, S., & Hall, C. M. (2020). Sanctions and tourism: effects, complexities and research. *Tourism Geographies*, 22(4–5), 749–767. https://doi.org/10.1080/14616688.2019.1663911
Sharpley, R. (2000). Tourism and sustainable development: exploring the theoretical divide. *Journal of Sustainable Tourism*, 8(1), 1–19. https://doi.org/10.1080/09669580008667346
Sharpley, R. (2020). Tourism, sustainable development and the theoretical divide: 20 years on. *Journal of Sustainable Tourism*, *28*(11), 1932–1946. https://doi.org/10.1080/09669582.2020.1779732
Silva, T. C., da Silva Neto, P. V., & Tabak, B. M. (2023). Tourism and the economy: evidence from Brazil. *Current Issues in Tourism*, *26*(6), 851–862. https://doi.org/10.1080/13683500.2022.2048804
Smith, M. K., Pinke, I., & Olt, G. (2019). Overtourism and resident resistance in Budapest. *Tourism Planning & Development*, *16*(4), 376–392. https://doi.org/10.1080/21568316.2019.1595705
Sørensen, E. (2006). Metagovernance: the changing role of politicians in processes of democratic governance. *The American Review of Public Administration*, *36*(1), 98–114. https://doi.org/10.1177/0275074005282584
Spyriadis, T., Buhalis, D., & Fyall, A. (2011). Dynamics of destination governance: governance and metagovernance in the composite industrial environment of destinations. In E. Laws, H. Richins, J. Argusa & N. Scott (Eds.), *Tourist destination governance: Practice, theory, and issues* (pp. 187–202). CABI.
Torkington, K., Stanford, D., & Guiver, J. (2020). Discourse (s) of growth and sustainability in national tourism policy documents. *Journal of Sustainable Tourism*, 28(7), 1041–1062. https://doi.org/10.1080/09669582.2020.1720695
Visentin, F. & Bertocchi, D. (2019). Venice: an analysis of tourism excesses in an overtourism icon. In C. Milano, J. M. Cheer & M. Novelli (Eds.), *Overtourism: Excesses, discontents and measures in travel and tourism* (pp. 18–38). CABI.
Wan, Y. K. P., & Bramwell, B. (2015). Political economy and the emergence of a hybrid mode of governance of tourism planning. *Tourism Management*, *50*, 316–327. https://doi.org/10.1016/j.tourman.2015.03.010
Xu, Q., Zhong, M., Cheng, H., & Li, X. (2022). Does public policy alleviate the impact of political risks on international tourism? *Journal of Policy Research in Tourism, Leisure and Events*, https://doi.org/10.1080/19407963.2022.2069115
Yu, Q., McManus, R., Yen, D. A., & Li, X. (2020). Tourism boycotts and animosity: a study of seven events. *Annals of Tourism Research*, 80, 102792. https://doi.org/10.1016/j.annals.2019.102792

Chapter 44

Tourism and Innovation
A Cross Sectoral and Interregional Perspective

Adi Weidenfeld and C. Michael Hall

Introduction

Knowledge management is pivotal in innovation using mechanisms to facilitate knowledge exchange and learning among competitive tourism organisations and destinations. Nevertheless, both the tourism industry and its scholars have been slow to embrace the field leaving it relatively underdeveloped (Cooper, 2015; Cooper, 2018; Raisi et al, 2020). Knowledge management, as a sequential process of capturing, developing, sharing, and leveraging from organisational knowledge, particularly from knowledge transfer and knowledge-based innovation (Cooper, 2015; Cooper, 2018), includes spillovers, flows, and transfers. The latter two are more intentional and therefore may be effectively managed through peer-to-peer exchanges, knowledge sharing and team learning. Knowledge exchange, learning and innovation can take place in-house (within organisation) at the micro level and at the macro, i.e., sectoral and territorial levels, where knowledge transferred around networks as a supply-side collaborative response to tourism demand and changing business environments (Cooper, 2018). This chapter focuses on the macro level by taking an evolutionary economic geographical approach.

With the growing use of social media and online technologies and platforms, the role of space and sectoral specialisation in shaping innovation processes is changing (Hall et al., 2022). The increasing role of technology in knowledge exchange and learning may mean that geographic proximity will be less important in mediating this process. However, this is not necessarily the case because the challenge is in developing a trusting, learning, and sharing culture at destination regions (Cooper, 2018) and between tourism actors, where local conditions may be pivotal for addressing current societal challenges, competing for customers and visitors with other sectors and remaining resilient to major crises through innovation (Cooper, 2015; Safonov et al., 2023).

Apart from the territorial dimension, which is likely to remain significant, there is a need to place importance on cross sectoral relations and networks. This importance lies in the established premise of tourism organisations' reluctance (or incapability) to invest in knowl-

The Wiley Blackwell Companion to Tourism, Second Edition. Edited by C. Michael Hall.

edge management because of their lack of understanding of its importance and/or available resources (Safonov et al., 2023). This explains why the role of networks and systems remains important.

This chapter explores the current research agenda for addressing the abovementioned challenges in tourism and begins with a discussion on the knowledge and system dimensions and their relevance to the tourism industry. It is followed by revisiting the innovation system approach and its sectoral and regional tourism perspectives as well as taking a critical approach towards being addressed in the tourism literature. At the end, an approach for future studies in tourism economic geography and management studies is suggested.

Tourism Knowledge Networks

The network perspective that emphasises the role of relations and structures in determining the behaviour of actors (Stuck et al., 2016) has been increasingly used in studies on the collaborative dynamics of innovation and knowledge exchange (Marasco et al, 2018; Raisi et al, 2020). In tourism, networks are antecedents and a necessary condition of tourism innovation (Brandão et al, 2018). *Knowledge networks* are conceptualised as 'pipelines' for individuals and groups of actors (Bathelt et al., 2004) for exchanging geographically dispersed information and knowledge (Owen-Smith & Powell, 2004). They derive from informal ties between individuals, or they can be formally defined as contracts or strategic alliances (Owen-Smith & Powell, 2004; Wanzenböck et al., 2014). Networks can consist of sets of at least three organizations linked by an innovation project, whereby firms benefit from complementarities, achieving economies of scale, sharing the costs and risks associated with developing an innovation, and ultimately making use of a competitive advantage (Favre-Bonte et al., 2019). Therefore, scrutinizing network characteristics becomes central to understand the dynamics and complexity of inter-organizational cooperation about the area of innovation at a regional level (Kofler et al., 2018), including destinations.

Unlike social networks in general, inter-organizational networks as collaborative mechanisms between enterprises require the consideration of organizational structures. Such cooperation, including horizontal and vertical networks, may be more complex and multi-faceted than coordinating relationships between individuals (Kofler et al, 2018). Thus, networks create and offer unique value, access to resources, skills and experience, timely access to external knowledge, improved trust, cooperation, social cohesion and rapid response to market opportunities, providing crucial conditions to innovate (Hall & Williams, 2020). This is particularly germane for tourism SMEs by providing them with external economies and shared resources such as Research and development (R&D), overcoming risks and costs, alignment of management processes, identifying and analysing market opportunities, development of human resources' competencies for innovating, and the ability to overcome competitors and the access to diversified sources of knowledge (Brandão et al., 2018; Hall & Williams, 2020). If it is well-organized, it can also be a valuable strategy to access resources and save costs (Kofler et al., 2018).

Regional networks are important mechanisms of growth for both individual businesses and for regions as a whole. Territories have an essential role, and innovation must be seen from the destination point of view and materialized in networks, characterized by the diversity and geographical coverage of its members. Destinations with greater participation of innovative tourism companies and great innovators require greater involvement in the creation and sharing of new knowledge (Brandão et al., 2019). However, tourism is fragmented in nature and involves several complementary activities that bundle together in an integrated experience at destination level (Hall, 2005). Therefore, when networks are diversified in locations and products, they are in a better position to achieve a higher innovation performance.

Networks allow access and use of valuable knowledge about markets and trends, funding, and opportunity of jointly creating new products and services or developing marketing strategies and creating organisational structures that improve destinations' overall functioning and competitiveness (Brandão et al., 2018; Brandão et al., 2019). These are enabled by the advantages that networks bring to the innovation process, as they can create and offer unique value, access to resources, skills, and experience, timely access to external knowledge, improved trust, cooperation, social cohesion and rapid response to market opportunities, providing crucial conditions to innovate. They are considered as a fundamental leverage for tourism SMEs by compensating the lack of internal R&D and helping to overcome the risks and the high investment associated with the innovation process (Raisi et al., 2020). The very systemic qualities of networks, which constitute their systemic behaviour are explored.

Tourism and the Systemic Features of Sectoral and Regional Innovations

Innovations in tourism constitutes minor and major adaptations of products and services, rarely involving entirely new products and/or new markets but rather differentiation, product line extension via brand policies, or changes in the cost (price)/quality ratio of the product (Hjalager, 2010; Brackenbury, 2006). The Innovation System (IS) approach explains the symbiosis between the application of new or combined knowledge and its accumulation and transformation into economically significant products and processes (Hall & Williams, 2020). It explains innovative capabilities at the geographical and sectoral levels and considers the determinants of innovation such economic, political, institutional and historical and the fact that systems differ in their knowledge base, R&D expenditure, organisations and institutions. It also assumes nonlinearity of the process learning observed in different activities performed by the various actors and interdependency among a wide range of actors (Gutiérrez1 et al., 2021).

Systemic Qualities and the Network Perspective

Innovation is a socially and institutionally embedded process, endowed with a systemic nature (Brandão et al., 2019). Every IS needs to have both complementary formal and informal institutions. These include formal ones such as bridging organisations and their regulatory regime, that define the 'rules of the game' for all agents in an explicit top-down manner, and informal institutions that usually emerge bottom-up from the interactions between actors, such as social and corporate 'praxis', ways of doing things and norms (Borrás, 2004; Earl & Hall, 2021). Informal institutions thus complement formal institutions by fine-tuning agents' behaviour and expectations – to perform a kind of matching function of conventions and norms established between groups of individual agents, but which cannot ultimately depart from the rules set by the formal institutions (Coriat & Weinstein, 2002). The network perspective also emphasises the role of relations and structures in determining the behaviour of actors (Stuck et al., 2016). Every network of actors which is always evolutionary in nature derive from informal ties between individuals, or they can be formally defined as contracts or strategic alliances (Owen-Smith and Powell, 2004; Sun, 2016; Wanzenböck et al., 2014) or can have both. From an evolutionary perspective, both formal and informal institutes that underlie systems can stimulate the emergence of one another (Weidenfeld et al., 2021).

In tourism, formal institutions exist at all levels, at the regional level. destination management organisations can act as repositories of knowledge and identifiers of new opportunities due to their position as intermediaries as they often have contacts with external actors (as part of broader associations) and simultaneously close to local firms. In particular, institutions in destination regions are crucial in creating the conditions for networks to develop and succeed (Brandão et al., 2018). They can therefore be considered as formal institutions which determine the rules followed by tourism actors and provide a platform for networking activities in which informal rules evolve. However, institutions are insufficient to determine whether networks can be defined or to what extent they 'behave' as systems and actually in most cases they operate as both. Sufficient systemic qualities including coherence, unified function, and boundedness determine the extent to which they operate as such (Weidenfeld et al., 2021) and their relevance to tourism is further explored below.

Coherence

This criterion refers to the extent to which a system's elements are consistent in terms of shared belief and values behind attitudes and social interaction that characterise the ways innovation processes takes place (Borrás, 2004). Tourism organisations create the necessary conditions such as establishing relationships with other organisations for the development of product and service innovation. Networks provide mechanisms whereby these relationships are established and engender the development of new products and services (Brandão et al., 2018; Brandão et al., 2019). Coherence implies the presence of feedback loops, common developmental trajectories and complementary competencies between agents (Roper et al., 2006). In this context, Related Variety (RV) between tourism sub-sectors and between them and non-tourism sectors is pivotal in determining their knowledge exchange linkages and drive diversification of their services and products (Weidenfeld, 2018). It refers to the sufficient cognitive difference between existing regional economic activities, which engenders novel re-combinations of different but complementary knowledge and therefore underlies regional diversification (Boschma & Frenken, 2011; Frenken, et al., 2007).

RV in tourism enables a composition of related technologies or economic activities with high cognitive proximity, which are neither too similar nor too different between businesses. It is an important yet understudied determinant of clustering of tourism firms for stimulating a chain of related activities, which are essential for the formation of tourism destination clusters and is positively correlated with the concentration of tourism and non-tourism firms (Lazzeretti et al., 2016). Thus, relatedness between sectors is positively related to coherence and given the multi-sectoral nature of tourism as interrelated sub-sectors, it can be a useful indicator of coherence in tourism ISs. Low coherence systems are characterized by loosely connected or isolated elements that are unlikely to produce a meaningful collective output (Rakas & Hain, 2019).

Another element is structure. Networks' structure in terms of density and strength of ties i.e. dense versus sparse and strong versus weak ties vary in benefitting innovation performance. Cohesive or dense networks occur at high connectivity leading to higher levels of trust and norms of reciprocity, which, in turn, facilitates both knowledge dissemination and the operation of supporting governance mechanisms. Here there is a paradox resulting in the need to balance cohesion resulting in over-embeddedness of the network. Given that people with different backgrounds and perspectives or working in different industries will exchange information, learning from each other, and enhancing the potential for new combinations of knowledge into innovative products and services, very cohesive networks with strong and dense ties may bring detrimental effects to innovation if it is not open enough to engaging with extra actors. However, if it consists of weak internal connections, it prevents strong ties which are necessary to the transfer of tacit knowledge. Therefore, balanced cohesion, i.e., the combination of both strong and weak ties within the same network is desirable for generating coherence resulting in the most fruitful scenario for innovation (Brandão et al., 2018).

Unified Function

In ISs a function includes identified themes, specific priorities and/or innovative activities of a specific sector and/or addressing common challenges or threats, which require innovative solutions (Edquist, 2004a; Edquist, 2004b; Edquist, 2006) and involves identifiable aims to which all elements of the system contribute (Rakas & Hain, 2019). In tourism knowledge networks, high centralisation fosters the coordination in innovation processes in terms of sharing the same principles and objectives especially in an industry with a high diversity of actors and businesses (Brandão et al., 2018). Similar to coherence, it might impede access to diverse and new sources of knowledge because centralised networks often depend on a small number of organizations and their loss or lack of functionality can affect the performance of the whole network (Raisi et al, 2020), particularly without the participation of extra regional sources. Unified function is likely to characterise actors working within the same sub-sectors and/or when subsectors are thematically related in destination regions. For example, the World Association of Zoos and Aquariums acts as a global network with considerable unified function stemming from well stated shared priorities among its members and partners. It guides, encourages and supports the zoos, aquariums and like-minded organisations of the world in animal care and welfare, environmental education and global conservation (https://www.waza.org/about-waza/).

Boundedness

ISs can have meaningful sectoral, technological, cognitive, and/or regional boundaries that generate and distribute innovations and are supported by specific institutions. This criterion is more commonly used in the Regional Innovation System (RIS) context when regions are characterised by homogenous specific criteria with pre-existing internal cohesion and cultural embeddedness (Doloreux & Parto, 2005), that allow a sensible delineation between the system itself and the 'rest of the world' (Edquist, 2006). The boundary between the system and its environment can be specified in many ways, largely depending on what is the purpose of the systems analysis (Edquist, 2004a). Administrative boundaries and/or functionality of i.e. frequency or intensity of economic interaction (Andersson & Karlsson, 2004) are often used. Functionality in terms of knowledge related interactions may also include localised mobility of skilled workers as knowledge carriers and a minimum proportion of innovation related collaborations among actors taking place with regional partners within certain regional boundaries (Edquist, 2006). In both ways, a sufficient scope for interaction and coherence to enhance collective external economies within a certain set of boundaries can delineate a RIS's boundaries (Asheim et al., 2011). In tourism, embeddedness may refer to specific local culture and natural resources, which contribute to knowledge of developing new niche tourism markets, e.g., health tourism, ecotourism, gastronomy (European Commission, 2015) and exists when there are linkages between consumers, producers, communities and institutions. Thus, boundedness may be based on such embeddedness and the geographic boundaries of the system. The systemic focus in tourism destinations tends to be both industrial (or sectoral), which is typical of more global and national systems and geographic at the destination regional level.

Tourism as a Sectoral Innovation System

The sectoral system framework (Malerba, 2002) has been used in the study of the determinants and performance of innovation in a wide range of sectors in the last two decades and its building blocks include actors, sectoral environments (knowledge, technologies, demand and markets, institutions and policies, and unexpected shock events) as well as interactions and networks (Gutiérrez et al., 2021; Li et al., 2021). A sector is defined as a 'set of activities that are unified by some linked product groups for a given or emerging demand and which share some common knowledge' (Malerba, 2005, p. 385). In each sector firms have some commonalities with others in the same sector but are also heterogeneous (Malerba, 2004). The term 'tourism sector' (albeit often defined as an industry) refers to the grouping of enterprises into specific tourism services, such as accommodation and tours (Hall, 2005; Raisi et al., 2020). When 'heterogeneous firms facing similar technologies, searching around similar knowledge bases, undertaking similar production activities, and embedded in the same institutional setting, share some common behavioral traits and develop a similar range of learning patterns, behavior, and organizational form' (Malerba, 2004, p. 387), they constitute a Sectoral Innovation System (SIS).

Tourism innovation systems are defined as 'the parts and aspects of the economic structure and institutional set-up affecting learning and innovation in tourism firms' (Sundbo et al., 2007, p. 93). Given that SISs' boundaries are marked largely by the interaction of the actors that interplay to achieve innovation in a sector (Li et al., 2021), it is plausible to assume that Tourism SISs (TSISs) exist in every geographic scale (i.e., local to global). The TSIS approach, regardless of its scale, has to consider the following aspects: the network of stakeholder relationships (networks), businesses' strategic decision making (competitiveness), the participation of organizational actors (Brandão et al., 2019) and the sustainable development of these systems (innovation and research). It is assumed the major international tourism institutions such as the International Air Transport Association and others constitute a TSIS at the global scale. However, the extent to which such network mechanisms have informal institutes is questionable and refers also to their systemic nature. This raises the question of whether tourism regional innovation really exists or perhaps only knowledge networks. The most obvious scale in tourism for identifying any form of networks or ISs is at the destination region level, which will be explored below in the context of the National Innovation System (NIS) and RIS.

The main three dimensions of the SIS are knowledge and technological domain; actors and networks; and institutions. In these cases, sectoral specificities or sectoral evolution play an essential role in explaining the behaviours and performance of different players within certain kinds of sectoral environments. Based on Gutiérrez et al, (2021) SISs of different industries and sectors can differ between sectors in the following:

i. The *knowledge base and learning processes* that support the innovative activities, which differ across sectors and sub sectors. In tourism like other service industries, symbolic knowledge base including the aesthetic attributes of regional products, designs, images, and the appeal of various cultural artifacts and narratives as a way of adding value to products (Asheim, 2007) is particularly relevant (Weidenfeld & Hall, 2014).

Other knowledge bases such as synthetic and analytic are relevant in the context of other industries and sectors which engage in the tourism industries.

ii. The *links and interactions* among actors. The diversity of actors and the interactions (market and non-market relations) vary across sectors including non-industry actors such as universities and government agencies when carrying out innovative activities. On rare occasions, firms perform their innovative activities in isolation and then their relations with other agents become relevant. In tourism, the importance of knowledge networks has already been in discussed in this chapter.

iii. *Formal and informal Institutions* include laws, regulations, habits and routines that shape the interactions among actors, both enhancing and hampering innovation (Earl & Hall, 2021; Gutiérrez et al., 2021). These may include any formal rules and regulations at the global scale such as those determined by the UN World Tourism Organisation (UNWTO) and other regulatory organisations at the national and regional levels, and informal institutions such as the social norms among the professional associations of destination actors. For example, according to Weidenfeld et al. (2010) attractions imitate and adopt ideas from those that are distant enough to refrain from direct competition for the same.

The Technological 'Turn'

The new wave of social-technological changes since the new Millennium has been largely reshaping sectoral systems along the above three dimensions and an additional market dimension (Li et al., 2021). More recently, the emergence of so-called 'fourth industrial revolution' technologies, such as artificial intelligence, cloud computing, additive manufacturing, advanced robotics, Internet of Things, smartphones, and autonomous vehicles is becoming a game changer in both the operationalisation and participation of knowledge exchange and actors' engagement respectively. As a result, the blurring of the boundaries of existing sectors, has brought both opportunities and challenges for sectoral systems research. In the current economic landscape, the development of many related technologies, such as artificial intelligence, cloud computing, additive manufacturing, advanced robotics, Internet of Things, smartphones, and autonomous vehicles, have been blurring the boundaries of existing sectors stimulating collaboration between previously unrelated sectoral systems (Li et al., 2021). Information and communication technologies create RVs between tourism and other sectors by facilitating collaborations between stakeholders and allowing for customer engagement, that, in turn, provides vital knowledge for developing new products and particularly new niche sectors and increases coherence. Such RVs take place in ISs at the national and regional scales and may involve the introduction of new products initially marketed and tested among tourists and then adjusted and modified to local markets contributing to market diversification at the national and regional levels (Weidenfeld, 2018). For example, similar technologies and marketing approaches used by tourism are tested and modified for their use in non-tourism sectors that are unrelated to tourism such as plastic and textiles (Bellini et al., 2017).

Tourism and Regional Innovation Systems

The term 'tourism region' refers to a specific geographical area where tourism enterprises are clustered (Raisi et al, 2020) and networks are important mechanisms of growth for all regional stakeholders. Tourism networks are characterized by the diversity and geographical coverage of its members (Brandão et al, 2019) and may be defined or evolve into systems.

The RIS concept (Cooke et al., 1998; Cooke, 2004) remains most widespread and consists of three significant elements: 'interactive learning' as a dialogic and recursive process producing knowledge and innovation; 'milieu' as a territorial context characterised by specific sets of values and norms; and 'embeddedness' as a relational perspective in socio-structural and territorial terms (Kofler et al., 2018). It highlights the regional dimension of innovation processes and emphasises how innovative and economic competitive advantages of regions relate to geographical proximity between actors. Organizations within a RIS belong either to the knowledge application and exploitation (i.e., firms and customers) or to the knowledge generation and diffusion subsystem (i.e., support organisations, public administration and educational organizations). Actors from the former subsystem are the major drivers of commercial innovation activities in an RIS and thus of vital importance (Stuck et al., 2016).

The relationship between tourism and RISs remains underexplored and tourism may be described as:

- another industry 'actor' in the RIS with a similar role or impact to that of others; and
- dominating the system and its components

The inter-sectoral nature and linkages with other industries, such as with agriculture and other service sectors, e.g., transport and retail, is based on growing evidence of inter-sectoral learning activities with other industries and depends on the way tourism actors and institutions are spatially and institutionally interconnected to non-tourism companies. This will be described in the next section. However, so far there has not been any compelling evidence of destinations identified as a tourism lead sectoral-regional innovation system wherein tourism acts as the main engine facilitating and/or engendering innovation through RV with most of the other regional sectors, that result in innovation and diversification of many of the other regional industries (Weidenfeld & Hall, 2014).

There is considerable overarching indirect socio-economic impact on the entire system and its actors, particularly in rural and peripheral regions where it is often a staple that generates enabling conditions for innovation activities. The economic contribution of tourism on RIS goes beyond innovation and depends, first of all, on whether it is considered as a regional staple (or one of the main ones) or the extent to which regional economies concentrated on tourism, which is often the case in rural and peripheral regions (Weidenfeld & Hall, 2014). Additionally, tourism's special service and experience related features are linked to specific localities and influenced by interactions with customers that make it distinct from other industries. Their direct and indirect contribution to RISs includes:

- provision of both soft (skills, knowledge, trust) and hard (transport, communication, finance) infrastructures;
- improving and sustaining quality of life and sense of place thereby creating an attractive place for human capital including enhancing place image and visibility, connectivity and high levels of amenity, which are pivotal in creating an attractive place for human capital and maintaining linkages with external actors;
- maintaining linkages of other industries and organisations with external actors;
- international tourists, suppliers or other producers are knowledge sources, allowing access to deeper knowledge flows and innovations; and
- tourism is in itself a source of innovation and creativity (Hall & Williams, 2020).

However, tourism is a predominantly small business dominant low-tech service industry and given the tendency to measure regional innovativeness in technological and/or scientific measures, tourism's regional contributions are often underestimated even when the tourism industry is innovative (Kofler et al., 2018). The destination regional level is the most applicable for the tourism industries which rely heavily on the unique and specific types of regional resources (cultural, natural, environmental, technological). The above contribution of tourism and its relationships emphasise the role of linkages across sectors within and beyond tourism. After exploring the regional (or territorial) and sectoral approach each, a combined regional-sectoral approach is suggested.

Cross Sectoral and Regional Approach

People with different backgrounds and perspectives or working in different industries exchange information and learn from each other whereby enhancing the potential for new combinations of knowledge into innovative products and services takes place (Brandão et al., 2018). The fact that tourism firms tend to collaborate and compete with the same tourism firms might explain why tourism entrepreneurs are often not overly keen to collaborate closely with other similar tourism enterprises for innovation purposes (Brandão et al., 2018; Brandão et al., 2019), which is evident in the recent tourism literature (Bellini et al., 2017; Weidenfeld, 2018; Biagi et al., 2021). Kofler et al. (2018), for example, argue that most hospitality and tourism enterprises that are strongly embedded in their regional context collaborate with other sectors for innovation processes in which new ideas and innovations are easily and widely disseminated and imitated (Kofler et al., 2018). This is particularly germane at the destination regional level where new tourism sectors, activities and products emerge from knowledge exchange among existing tourism sectors as well as with non-tourism sectors (Weidenfeld, 2018). Competition and the motivation to explore newness encourage them to do so with other organisations from other sectors that may diversify their innovation capability. This results in regional diversification that is enabled by RV and is positively related to compatibility, complementarities, and the extent of labour exchange. This cross sectoral diversification also exists across regions by developing RV between sectors in different regions defined as diversification across regions (Weidenfeld, 2018). This is particularly common in cross border tourism contexts, such as on the border between Norway and Sweden where sustainable solutions to increase accessibility for foreign travellers were designed and offered for mountain tourism (Jönsson, 2013).

Apart from intra- and inter-regional cross sectoral collaboration leading to diversification tourism is also a platform for diversifying non-tourism sectors at the regional level. New tourism markets provide new demands for new products and services that are consumed differently thereby diversifying the economy (Weidenfeld, 2018). This underlies regional diversification in destinations such as Catalunya, Andalusia and Canarias in Spain, Tuscany, Lazio, Emilia-Romagna, Tirol in Italy, Algarve in Portugal and Rhone-Alpes, Provence-Alpes and Cote d'Azur in France (Biagi et al., 2021).

Nevertheless, inter-sectoral networks might have their drawbacks. Their structure might bring detrimental effects to innovation for preventing strong ties, that are necessary to transfer tacit knowledge or too cohesive network where insular approach creates lock-in resulting in less creativity and innovation. Thus, as already mentioned an adequate combination of both strong and weak ties within the same network are considered as most fruitful scenario for innovation (Brandão et al, 2018).

This interregional approach in tourism innovation remains largely ignored. Tourism networks gain from regional cohesiveness and inter-sectoral links for innovation implementation. However, to achieve a higher level of inventiveness, tourism networks should increase the degree of nonregional links with tourism actors outside of the regional context (Kofler et al., 2018). This is because regional over-embeddedness, i.e., when networks are too closed, lock-in may dominate and prevent its actors from searching mew ideas from external actors. External knowledge from extra-regional relationships (national and international) plays a major role in learning and innovation, while networks based on local and regional actors can lead to the underdevelopment of RISs (Brandão et al., 2018).

Conclusions

This chapter explores the challenges that tourism faces in advancing innovation within its industrial boundary, i.e., among its sub-sectors and beyond. Whereas the Innovation System (IS) approach has been used as a theoretical framework in tourism, its applicability to destination regional scale is questionable as tourism tends to locate in rural and peripheral areas, which are often less advanced with low levels of innovation and knowledge exchange among their stakeholders and therefore the existence of innovation networks let alone knowledge networks is also questioned. However, at the global level, the IS approach, that is more sectoral or industrial is more relevant rather than questionable. This chapter suggests that the network system perspective, which has been dominating the RIS literature (Stuck et al., 2016) is applicable to tourism innovation, which requires networks and systems of cooperation because companies are mutually dependent within (Raisi et al., 2020) but also outside the industry. This is germane to both regional networks and sectoral (non-territorially-based networks that can have members regardless of their locations on the globe). Further, networks are essential to the creation of systems and most destination regions are assumed to have some territorial network mechanisms even if not cohesive.

The contribution of tourism destination region and sectoral networks does not only depend on the strength of their formal and informal institutions and their role in the regional system but also on their impact on each of the underlying criteria: coherence, unified function, and boundedness of the RIS or SIS. The nature and extent of such contribution remains to be examined and compared in future studies. For example, RISs consist of multi-sectoral tourism networks- some might be more regionally embedded and some would operate as at the interregional level or even global one in terms of knowledge exchange and other innovation related activities. Some of the networks maintain stronger ties among their players such as coherence and unified function with or without territorially based embeddedness and boundedness (i.e., sectoral networks) and therefore may also show high levels of systemic quality and be considered as systems. Systems like networks may be territorially based, i.e., RISs are more likely to be multi-sectoral. Others may be more sector focused and more interregional with low levels of boundedness and regional embeddedness with weaker intraregional ties and stronger interregional (and/or international) ones. Similarly studying the role of tourism networks and comparing it to other industrial networks will contribute to our understanding of the relative role of tourism in innovation systems.

The ever-changing role of technology may be an agent of change that needs to be explored in general and in possibly technological systems too, which is beyond the scope of this chapter. This chapter throws light on the pivotal role of the network-system perspective that may provide a better conceptual framework and empirical evidence to study the role of tourism in different types of innovation systems. Tourism companies and knowledge producers (universities and R&D organizations) should strengthen their ties and seek international partners to provide new and diversified knowledge (Brandão et al., 2019) through diversifying their networks for enhancing innovation (Raisi et al., 2020).

RISs in destinations can range from having tourism as its staple (or one of them) or tourism is one of many other industries. In the first instance, tourism is unlikely to create a viable 'tourism regional innovation system' but rather tourism actors are likely to participate interregional and multi-sectoral networks, that are essential to drive innovation processes. Moreover, given that SMEs constitute most of the tourism industry, their limited resources are likely to reduce their ability to create sustainable interregional knowledge linkages and this is likely to be the action of very few businesses or other tourism organisations. The nature and extent to which can establish interregional networks needs to be addressed. In practice, highly intersectoral learning has been the main approach for studying innovation in destination regions rather than the notion of unique RISs, dominated by tourism firms (Weidenfeld & Hall, 2014). Therefore, when studying innovation related activities at a cross sectoral-regional approach should be adopted.

This critical approach suggests a multi-scalar perspective that combines sectoral perspective that is relevant to various geographic scales for various stakeholders, within and outside the tourism industry. Given that perceiving networks and systems from the tourism lenses alone is feasible for both academic and practitioners alike, this chapter call for adopting this approach in the tourism literature. In other words, so far the most popular approach is to examine tourism innovation from a sectoral or regional (territorial) approach separately, with the latter struggling to relate to 'the real world'. This chapter suggests a combined approach in which a sectoral focus is related to its combined multi-scalar territorial dimension i.e., intraregional, inter-regional, global scales, and sectoral scales while acknowledging that innovations are also associated with non-territorial tourism innovation systems at the global scale.

Acknowledgement

This work was supported by the National Science Centre Poland (Nardowe Centrum Nauki) [grant numbers UMO-2019/35/B/HS5/04010].

References

Andersson, M., & Karlsson, C. (2004). *Regional innovation systems in small & medium-sized regions a critical review & assessment*. Royal Institute of Technology, CESIS - Centre of Excellence for Science and Innovation Studies, Working Paper Series in Economics and Institutions of Innovation.

Asheim, B. (2007). Differentiated knowledge bases and varieties of regional innovation systems. *Innovation*, *20*(3), 223–241. https://doi.org/10.1080/13511610701722846

Asheim, B. T., Smith, H. L., & Oughton, C. (2011). Regional innovation systems: Theory, empirics and policy. *Regional Studies*, *45*(7), 875–891. https://doi.org/10.1080/00343404.2011.596701

Bathelt, H., & Turi, P., (2011). Local, global and virtual buzz: the importance of face-to-face contact in economic interaction and possibilities to go beyond. *Geoforum*, *42*(5), 520–529. https://doi.org/10.1016/j.geoforum.2011.04.007

Bellini, N., Grillo, F., Lazzeri, G., & Pasquinelli, C. (2017). Tourism and regional economic resilience from a policy perspective: lessons from smart specialization strategies in Europe. *European Planning Studies*, *25*(1), 140–153. https://doi.org/10.1080/09654313.2016.1273323

Biagi, B., Brandano, M. G., & Ortega-Argiles, R. (2021). Smart specialisation and tourism: Understanding the priority choices in EU regions. *Socio-Economic Planning Sciences*, *74*, 100883. https://doi.org/10.1016/j.seps.2020.100883

Borrás, S. (2004). System of innovation theory and the European Union. *Science and Public Policy*, *31*(6), 425–433. https://doi.org/10.3152/147154304781779787

Boschma, R., & Frenken, K. (2011). Technological relatedness, related variety and economic geography. In P. Cooke, B. Asheim, R. Boschma, R. Martin, D. Schwartz, & F. Tödtling (Eds.), *Handbook of regional innovation and growth* (pp. 187–197). Edward Elgar Publishing.

Brackenbury, M. (2006). Has innovation become a routine practice that enables companies to stay ahead of the comprtition in the travel industry? In *Innovation and Tourism Policy* (pp. 71–83). OECD Publishing.

Brandão, F., Breda, Z., & Costa, C. (2019). Innovation and internationalization as development strategies for coastal tourism destinations: The role of organizational networks. *Journal of Hospitality and Tourism Management*, *41*, 219–230. https://doi.org/10.1016/j.jhtm.2019.10.004

Brandão, F., Costa, C., & Buhalis, D. (2018). Tourism innovation networks: a regional approach. *European Journal of Tourism Research*, *18*, 33–56.

Cooke, P. (2004). Introduction: Regional innovation systems - An evolutionary approach. In Cooke, P., Heidenreich, M., and Braczyk, H.-J. (Eds.), *Regional innovation systems - the role of governances in a globalized world* (pp. 1–18). Routledge.

Cooke, P., Uranga, M. G., & Etxebarria, G. (1998). Regional systems of innovation: an evolutionary perspective. *Environment and Planning A*, *30*, 1563–1584. https://doi.org/10.1068/a301563

Cooper, C. (2015). Managing tourism knowledge. *Tourism Recreation Research*, *40*(1), 107–119. https://doi.org/10.1080/02508281.2015.1006418

Cooper, C. (2018). Managing tourism knowledge: a review. *Tourism Review*, *73*(4), 507–520. https://doi.org/10.1108/TR-06-2017-0104

Coriat, B., & Weinstein, O. (2002). Organizations, firms and institutions in the generation of innovation. *Research Policy*, *31*(2), 273–290. https://doi.org/10.1016/S0048-7333(01)00141-X

Doloreux, D., & Parto, S. (2005). Regional innovation systems: Current discourse and unresolved issues. *Technology in Society*, *27*(2), 133–153. https://doi.org/10.1016/j.techsoc.2005.01.002

Earl, A., & Hall, C. M. (2021). *Institutional theory in tourism and hospitality*. Routledge.

Edquist, C. (2004a). Systems of innovation – a critical review of the state of the art. In Fagerberg, J., Mowery, D. C., Nelson, R. R. (Eds.), *The Oxford handbook of innovation*. Oxford University Press.

Edquist, C. (2004b). Reflections on the systems of innovation approach. *Science and Public Policy*, *31*(6), 485–489.

Edquist, C. (2006). Systems of innovation: Perspectives and challenges. In J. Fagerberg & D. C. Mowery (Eds.), *The Oxford Handbook of Innovation* (pp. 181–208). Oxford University Press.

European Commission (2015). *Perspectives for Research and Innovation Strategies for Smart Specialisation (RIS3) in the wider context of the Europe 2020 Growth Strategy*. Directorate General for Research and Innovation, Directorate B, Innovation Union and European Research Area, Unit B5.

Favre-Bonte, V., Gardet, E., & Thevenard-Puthod, C. (2019). The influence of territory on innovation network design in mountain tourism resorts. *European Planning Studies*, *27*(5), 1035–1057. https://doi.org/10.1080/09654313.2019.1588856

Gutiérrez, C., Roa P, V., & Smith, J. (2021). The Chilean sectoral innovation systems: An approach from the national innovation survey. *Journal of Technology Management & Innovation*, *16*(1), 3–13. http://dx.doi.org/10.4067/S0718-27242021000100003

Hall, C. M. (2005). *Tourism. Rethinking the social sciences of mobility*. Pearson.

Hall, C. M., Prayag, G., Safonov, A., Coles, T., Gössling, S., & Naderi Koupaei, S. (2022). Airbnb and the sharing economy. *Current Issues in Tourism*, *25*(19), 3057–3067. https://doi.org/10.1080/13683500.2022.2122418

Hall, C. M., & Williams, A. M. (2020). *Tourism and Innovation*. Routledge.

Hjalager, A. (2002). Repairing innovation defectiveness in tourism. *Tourism Management*, *23*(5), 465–474. https://doi.org/10.1016/S0261-5177(02)00013-4

Jönsson, M. (2013). Boosting innovation in a cross-border mountain area to develop tourism. *Nord RegionNews*,*5*,https://archive.nordregio.se/en/Metameny/Nordregio-News/2013/Cross-Border-Innovation-Policy/Case/index.html

Kofler, I., Marcher, A., Volgger, M., & Pechlaner, H. (2018). The special characteristics of tourism innovation networks: The case of the Regional Innovation System in South Tyrol. *Journal of Hospitality and Tourism Management*, *37*, 68–75. https://doi.org/10.1016/j.jhtm.2018.09.004

Lazzeretti, L., Capone, F., & Innocenti, N. (2016). The role of related variety and commons on heritage tourist destinations. In *Commons/Comune* (pp. 467–473). Società di Studi Geografici

Li, D., Liang, Z., Tell, F., & Xue, L. (2021). Sectoral systems of innovation in the era of the fourth industrial revolution: an introduction to the special section. *Industrial and Corporate Change*, *30*(1), 123–135. https://doi.org/10.1093/icc/dtaa064

Malerba, F. (2002). Sectoral systems of innovation and production. *Research Policy*, *31*(2), 247–264. https://doi.org/10.1016/S0048-7333(01)00139-1

Malerba, F. (2004). Sectoral systems of innovation: basic concepts. In F. Malerba (Ed.), *Sectoral systems of innovation. Concepts, issues and analyses of six major sectors in Europe* (pp. 9–41). Cambridge University Press.

Malerba, F. (2005). Sectoral systems: How and why innovation differs across sectors. In Mowery D. C. & Nelson R. R. (Eds.), *The Oxford handbook of innovation* (pp. 380–406). Oxford University Press.

Marasco, A., De Martino, M., Magnotti, F., & Morvillo, A. (2018). Collaborative innovation in tourism and hospitality: a systematic review of the literature. *International Journal of Contemporary Hospitality Management*, *30*(6), 2364–2395. https://doi.org/10.1108/IJCHM-01-2018-0043

Owen-Smith, J., & Powell, W. W. (2004). Knowledge networks as channels and conduits: The effects of spillovers in the Boston biotechnology community. *Organization Science*, *15*(1), 5–21. https://doi.org/10.1287/orsc.1030.0054

Raisi, H., Baggio, R., Barratt-Pugh, L., & Willson, G. (2020). A network perspective of knowledge transfer in tourism. *Annals of Tourism Research*, *80*, 102817. https://doi.org/10.1016/j.annals.2019.102817

Rakas, M., & Hain, D. S. (2019). The state of innovation system research: What happens beneath the surface? *Research Policy*, *48*(9), Art. 103787. https://doi.org/10.1016/j.respol.2019.04.011

Roper, S., Scott, L. J., Clifton, N., & Executive, S. (2006). *The Scottish innovation system: Actors, roles and actions*. Edinburgh: *Scottish Executive*. Department of Enterprise, Transport and Lifelong Learning.

Safonov, A., Hall, C. M., & Prayag, G. (2023). Non-collaborative behaviour of accommodation businesses in the associational tourism economy. *Journal of Hospitality and Tourism Management*, *54*, 98–107. https://doi.org/10.1016/j.jhtm.2022.12.007

Stuck, J., Broekel, T., & Revilla Diez, J. (2016). Network structures in regional innovation systems. *European Planning Studies*, *24*(3), 423–442. https://doi.org/10.1080/09654313.2015.1074984

Sundbo, J., Orfila-Sintes, F., & Sørensen, F. (2007). The innovative behaviour of tourism firms—Comparative studies of Denmark and Spain. *Research Policy*, *36*, 88–106. https://doi.org/10.1016/j.respol.2006.08.004

Wanzenboeck, I., Scherngell, T., & Brenner, T. (2014). Embeddedness of regions in European knowledge networks: a comparative analysis of inter-regional R&D collaborations, co-patents and co-publications. *The Annals of Regional Science*, *53*, 337–368. https://doi.org/10.1007/s00168-013-0588-7

Weidenfeld, A. (2018). Tourism diversification and its implications for smart specialisation. *Sustainability*, *10*(2), 319. https://doi.org/10.3390/su10020319

Weidenfeld, A., Butler, R. W., & Williams, A. M. (2010). Clustering and compatibility between tourism attractions. *International Journal of Tourism Research*, *12*(1), 1–16. https://doi.org/10.1002/jtr.732

Weidenfeld, A., & Hall, C. M. (2014). Tourism in the development of regional and sectoral innovation systems. In A. A. Lew, C. M. Hall, & A. M. Williams (Eds.), *The Wiley Blackwell Companion to Tourism* (pp. 578–588). Wiley Blackwell.

Weidenfeld, A., Makkonen, T., & Clifton, N. (2021). From interregional knowledge networks to systems. *Technological Forecasting and Social Change*, *171*, Art. 120904. https://doi.org/10.1016/j.techfore.2021.120904

Chapter 45

Reviewing Empowerment in Tourism
Conceptualisation to Future Research Directions

Marianna Strzelecka & B. Bynum Boley

What is Empowerment?

Empowerment can generally be thought of as individuals' or groups' capacity to determine their own affairs. Notions of empowerment originate from the community psychology literature. In their seminal paper on empowerment, Rappaport (1987) conceived of the construct as systematically securing authority over one's life and engaging in community life through democratic participation. Zimmerman (1995, p. 583) later added that empowerment involves 'gaining control, access to resources, and a critical understanding of one's sociopolitical context'.

Regarding the meaning of power in the community empowerment discourse, Kelly (1992) noted that the term 'empowerment' refers to 'power to', and it is achieved by increasing one's ability to resist and even change 'power over' (Rowlands, 1995). The 'power to/power over' perspective focuses simply on access to political structures and formal communication channels. In this conventional view, 'empowerment' is about bringing people who have been traditionally outside the decision-making process into the discussion.

Scheyvens and van der Watt (2021) stress it is more beneficial for sustainable development to also consider the perspectives of 'power with' and 'power within'. Citing Rowlands (1997, p. 13), Scheyvens and van der Watt (2021, p. 2) insist empowerment should be viewed as 'an enabling, liberating, productive force, which can include power 'to' (a generative form of power which can include resistance); power 'with' (the power of collective action); and power 'from within' (a sense of self-belief which leads to confidence to seek changes which could improve one's life)'.

Power is central to tourism research (Hall, 1994; Reed, 1997; Beritelli & Laesser, 2011, Cheong & Miller, 2000, Nunkoo & Ramkissoon, 2011), stemming from the Foucauldian perspective that power is omnipresent and manifests itself within all relationships between tourism stakeholders (Boley et al., 2014). For over two decades, tourism researchers have investigated when destination residents are empowered through tourism and what that means within the context of tourism development. Consequentially, resident empowerment has been

The Wiley Blackwell Companion to Tourism, Second Edition. Edited by C. Michael Hall.

increasingly recognised as an integral component of sustainable tourism (Murphy, 1985; Scheyvens, 1999; Scheyvnes & van der Watt, 2021; Sofield, 2003; Boley & McGehee, 2014; Aghazamani & Hunt, 2017; Strzelecka et al., 2017a; Strzelecka et al., 2017b; Joo et al., 2020).

In the above perspective, successful tourism development is more likely to be achieved when residents' perspective is recognised and represented in the development process (Nunkoo & Ramkissoon, 2011). Choi and Murray (2010, p. 589) even write, that 'If the government fails to empower residents, the success of tourism development and sustainability cannot be guaranteed'. Thus, since sustainability in tourism starts with residents' needs, concerns, and welfare, tourism must strive to empower residents before other benefits can occur (Scheyvens, 1999).

In their seminal paper on empowerment through tourism, Scheyvens (1999) built on the psychology and development literature to recognise empowerment as a multidimensional construct with psychological, social, political, and economic facets. Following Scheyvens (1999), other scholars, including Boley et al. (2014), Strzelecka et al. (2017a, b), and Joo et al. (2020), argued that conceptualising individual empowerment as a multidimensional concept can provide a more comprehensive approach to destination development. Such conception of empowerment from tourism extends the focus beyond stakeholder inclusion in tourism decision-making but the many other aspects of life that determine the community's residents' access to decision-making, such as their economic status (i.e., economic empowerment), their pride, and self-esteem (i.e., psychological empowerment), and the lasting strength of the community's social bonds (social empowerment) (Scheyvens, 1999; Boley & McGehee, 2014; Strzelecka et al., 2017b).

While Rappaport (1985, p. 17) jokes that defining empowerment is troublesome and that 'empowerment is a little bit like obscenity; you have trouble defining it, but you know it when you see it,' the tourism literature has attempted to define empowerment primarily within the sustainable tourism development discourse. One definition that we find valuable is Aghazamani and Hunt's (2017, p. 343) definition of empowerment which defines empowerment as a 'multidimensional, context-dependent, and dynamic process that provides humans, individually or collectively, with greater agency, freedom, and capacity to improve their quality of life as a function of engagement with the phenomenon of tourism.' Aghazamani and Hunt (2017) primarily focus on empowerment as a process. In contrast, Scheyvens (2020, p. 115) provides a more outcome focused approach which outlines empowerment as 'the activation of the confidence and capabilities of previously disadvantaged or disenfranchised individuals or groups so that they can exert greater control over their lives, challenge unequal power relations, mobilize resources to meet their needs, and work to achieve social justice'.

The debate around empowerment as a process vs. empowerment as the outcome has initially emerged in reference to political empowerment (the transfer of power to those who do not have it). The empowerment process aims to undo negative social constructions so that the people see themselves as having the ability to influence their surroundings. This can be seen in Fawcett et al. (1995), referring to empowerment as a process in which stakeholders gain influence over events and outcomes that are important to them. Through empowerment, people become aware of their interests and access intangible decision-making. In contrast, Cole (2006, p. 3) views empowerment as the 'capacity ... to determine their own affairs'. Empowerment emerges because of individual changes, interpersonal changes, or changes in social structures that can impact the individual (Kieffer, 1984; Simmons & Parsons, 1983; Wallerstein & Bernstein, 1994).

A more comprehensive conception of empowerment sees it as a process through which people seek to gain control over their own lives. Rappaport's (1987, p. 122) well-supported conception of empowerment describes empowerment as 'a process, a mechanism by which people, organizations, and communities gain mastery over their affairs'. Thus, empowerment is both, a process of acquiring control and the capability to act and an outcome of the empowering processes - the actual ability of individual stakeholders and communities to act (Staples, 1990). Similarly, Zimmerman (1995, p. 583) notes that '(a) personal empowerment manifests itself in different perceptions, skills, and behaviors across people; (b) different beliefs, competencies, and actions may be required to master various settings; and (c) empowerment may fluctuate over time'. Zimmerman (1995) integrated two elements: a) perception of socio-political control; and b) a critical understanding of individual and political systems with a proactive approach to solving personal and social problems.

Tourism initiatives in a community development context shape empowerment's intrapersonal, interactional, and behavioural components (Zimmerman, 1995). The intrapersonal component refers to perceived control in a specific domain and motivation to exercise that control. Perceived competence and self-efficacy, a belief of having influence in a particular domain, usually relate to exact domains of life. Such belief may lead to behavioural outcomes. The three components of empowerment remain in constant interaction and thereby become mutual causes and consequences (Kroeker, 1995). Individuals may develop skills to pursue desired goals due to their engagement in tourism decision-making (Zimmerman, 1995). They learn to interact effectively (e.g., leadership skills, problem-solving) (Kieffer, 1984). The final behavioural component is the action directly aimed at desired tourism outcomes.

Strategies that empower residents engage them in learning about opportunities to influence elements of their environment (Zimmerman et al., 1992; Christens, 2012). Actions aiming to increase resident empowerment can employ participation in local tourism development projects or partnership building (Miller & Campbell, 2006). Tourism practices have high potential to help people become in charge of the factors affecting their lives (e.g., income, political access, identity, community cohesion) (Cole, 2006; Rappaport, 1987; Scheyvens, 1999), but the quality of involvement is critical (Zimmerman, 1995).

Dimensions of Empowerment in Tourism

Regardless of definitional preferences and if one considers empowerment a process or outcome, Scheyvens and van der Watt (2021, p. 2) summarise the many vital points associated with empowerment as 'multidimensional, context-specific, both process and outcome-related ... redistribution of power to marginalized groups'. In their work on empowerment, Scheyvens (1999) incorporated the perspectives of Friedman (1992) to describe multidimensional empowerment in tourism encapsulating the dimensions of psychological, social, political, and economic empowerment.

Psychological Empowerment

Psychological empowerment is one of tourism's most important non-economic benefits (Stronza & Gordillo, 2008; Strzelecka et al., 2017a; Strzelecka et al., 2017b) and stems from residents of tourist destinations having increased confidence and pride from tourists desiring their unique culture and experiencing their community's natural beauty (Scheyvens, 1999). Psychological empowerment is an essential element of a thriving tourism destination because residents need to be on board with how the destination is promoted to tourists and experienced by tourists to ensure that their community's natural and cultural resources are not being commodified in a manner which leads to embarrassment or shame (e.g., Scheyven, 1999; Boley & McGehee, 2014). Psychological empowerment can occur when tourism initiatives promote resident self-esteem, pride in cultural traditions, and a feeling of being able to assume new roles in their communities (Ramos & Prideaux, 2014), as well as when tourists recognise the value of areas surrounding the host community they visit (Strzelecka et al., 2017a; Strzelecka et al., 2017b).

Social Empowerment

Social empowerment refers to a situation in which a community's sense of cohesion and integrity is strengthened by tourism. In other words, it happens when tourism can contribute to community development by supporting the community social fabric (Scheyvens, 1999). Tourism can catalyse social change by bringing rural residents together to realise community projects, for instance, promoting environmental activism or community art festivals (McHenry, 2011; Joo et al., 2020). Tourism can also transform traditional roles within the community (Stronza & Gordillo, 2008).

Strzelecka and Wicks (2010) demonstrated the potential of tourism mechanisms to foster community cohesion in emerging destinations. The authors adopted interactional field theory (Wilkinson, 1991) to discuss how tourism promotes interaction across different community interest groups and engages them in joint tourism-related activities. However, Scheyvens (2003) concludes that only socially empowered individuals and groups work together to build a local tourism sector that benefits them and will be supported by the local community. Arguably, social empowerment from tourism links back to the idea of individual empowerment.

Political Empowerment and Tourism Governance

Residents who become politically empowered exhibit political efficacy. Politically empowered residents are motivated to employ social and political resources to their advantage (Zimmerman & Rappaport, 1988, Zimmerman & Zahniser, 1991; Cole, 2006; Strzelecka et al., 2017a; Strzelecka et al., 2017b). Knowledge and information raise residents' awareness of available opportunities and thus help empower individuals to be tourism decision-makers (Cole, 2006, Tosun & Timothy, 2003). In reference to the political environment within a destination, Sofield (2003) noted that political empowerment occurs more often in democratic cultures that facilitate knowledge transfer and promote local leaders. Both Cole (2006) and Timothy (2007), recognised that empowerment is not mere inclusion in the participation process but is at the top rung of Arnstein's (1969) participation ladder, where people have the agency to steer and control the tourism development process. Indeed, Beaumont and Dredge (2010) argue that political empowerment is indispensable to effective local tourism governance.

On the one hand, political empowerment 'ensures community support and acceptance of tourism development projects' (Cole, 2006, p. 630). Without political empowerment, locals 'have the inconvenience of tourism without economic advantages' (Sofield, 2003, p. 634). On the other hand, tourism can motivate residents to participate in local decision-making. Because of tourism decision-making, residents and other local tourism stakeholders can change their participatory habits by acquiring familiarity and comfort with new, more democratic ways of destination development (Strzelecka & Wicks, 2010). However, Scheyvens (1999) warns us that empowering diverse interest groups politically means their voices and concerns should guide the development of any tourism project from the feasibility stage to its implementation.

The political domain of resident empowerment is probably the most discussed in the literature in reference to political action. Such empowerment is linked to individual, interpersonal, and social structural changes (Kieffer, 1984; Simmons & Parsons, 1983). The essence of this form of empowerment in tourism development is that individuals or groups enhance their competency for tourism governance (Strzelecka & Wicks, 2015; Strzelecka et al., 2017a). Politically empowered individuals or groups do not hesitate to solve significant problems they face (Perkins & Zimmerman, 1995; Rappaport, 1981; Rappaport, 1985). Political empowerment unfolds through divergent methods (Zimmerman, 1995), whereas empowering processes may yield diverse outcomes depending on the context in which empowerment is pursued (Rappaport, 1987). The core idea of this form of empowerment in tourism is to shift power to people and communities so that they can engage in the governance of tourism destinations.

Economic

Scheyvens' (1999) final dimension of empowerment from tourism - economic empowerment, is about formal and informal business opportunities. It concerns equity in the spread of economic benefits across the community (Scheyvens, 1999). The economic gains experienced by economically empowered residents usually provide a regular and reliable income. Some concerns may arise due to perceived inequity in the spread of economic benefits. Scheyvens (1999) noted that economic empowerment could also concern the local community's access to productive resources in tourism development areas. For example, protected areas established and used for tourism development may reduce access to hunting and agricultural lands in emerging protected area tourism regions. In terms of the equitable distribution of benefits from tourism, these situations are of concern. This is a crucial aspect of economic empowerment or disempowerment from tourism, given that local people are more likely to support protected areas if this assists with their own community development (Sindiga, 1995).

In this regard, tourism literature has broadly addressed the resident perception of economically benefiting from tourism (e.g., Perdue et al., 1990; Sharpley, 2014; Boley et al., 2014; Strzelecka et al., 2017a). However, little attention has been paid to understanding further what economic empowerment from tourism really means and how we can assess it at the individual and local community levels. Finally, research looking into the implications of resident empowerment for tourism destinations needs to consider economic empowerment in tandem with political empowerment.

Measuring Empowerment from Tourism

As with any construct, after it is theorised, scholars try to operationalise it. Building on Scheyvens' (1999) conceptual model of resident empowerment, Boley and McGehee (2014) developed the Resident Empowerment from Tourism Scale (RETS) to measure residents' perceived empowerment through tourism. Comprised of 12 items across three dimensions (i.e., psychological, social, and political empowerment), the RETS is intended for assessing resident empowerment through tourism and has been tested for its validity and reliability in multiple tourism contexts both within (Boley et al., 2014) and outside (Boley et al., 2017a; Boley et al., 2015; Strzelecka et al., 2017a; Strzelecka et al., 2017b) the USA. However, the RETS does not include a measure of economic empowerment and thus, economic empowerment is not considered or measured when the scale is used. Nevertheless, to date, the RETS has been frequently utilised to understand how residents perceive tourism development in their community (Boley et al., 2014; Strzelecka et al., 2017a; Strzelecka et al., 2017b; Joo et al., 2020; Aleshinloye et al., 2022).

Resident Empowerment and Resident Attitudes

Research on resident empowerment is primarily located within the discourse on resident attitudes toward tourism, linking empowerment with support for tourism (i.e., Boley et al., 2014; Strzelecka et al., 2017a; Strzelecka et al., 2017b). Residents' support for tourism is a core characteristic of locally sustainable tourism and has been previously used as a proxy for resident attitudes towards tourism (Cole, 2006; Nunkoo et al., 2013; Strzelecka et al., 2017a; Strzelecka et al., 2017b). Given the important implications of resident empowerment and support for tourism to tourism's sustainability, many scholars have studied the mechanisms governing resident empowerment in different tourism development contexts (Boley & McGehee, 2014; Joo et al., 2020; Mendoza-Ramos & Prideaux, 2018; Strzelecka et al., 2017a; Strzelecka et al., 2017b).

Psychological empowerment from tourism is linked to more positive attitudes toward visitors experiencing local culture or nature (e.g., Boley et al., 2014; Strzelecka et al., 2017a). This support for tourism matters, especially in remote rural destinations where interaction with locals is the core of the tourist experience (Cole, 2006; Strzelecka et al., 2017a). The general premise behind social empowerment is that increased cohesion and collaboration associated with social empowerment will have a significant impact upon residents' overall support for tourism. While the link between social empowerment and resident support for tourism has been extensively studied (Boley et al., 2014; Maruyama et al., 2016; Strzelecka et al., 2017a; Strzelecka et al., 2017b; Yeager et al., 2020), the findings have not been consistent. For example, Boley et al. (2014) have not found a link between these two constructs. In contrast, Strzelecka et al. (2017a) found the positive effect of social empowerment on support for tourism even though the effect size of the relationship was slightly masked by psychological empowerment. Similarly, political empowerment from tourism has been frequently hypothesised to have a positive relationship with support for tourism, but a positive and significant relationship has yet to materialise (Boley et al., 2014; Maruyama et al., 2016; Strzelecka et al., 2017a; Yeager et al., 2020). Research shows that residents who feel politically empowered in tourism tend to oppose tourism development, rather than support it (Strzelecka et al., 2017a; Strzelecka et al., 2017b; Strzelecka et al., 2021). This could mean that inclusive and participatory tourism decision-making enables opposition to tourism to make its voice visible.

Social Exchange Theory

The predominant theory used to explain resident support for tourism in conjunction with resident empowerment is Social Exchange Theory (SET), which posits that residents conduct an internal cost-benefit analysis of tourism's impacts to determine their support or opposition to tourism (Ap, 1992). Perdue et al. (1990) were the first to suggest SET to explain why residents support or resist tourism development. Their research was followed by numerous studies applying the concept of empowerment from tourism within the SET (i.e., Boley et al., 2014; Boley et al., 2015; Strzelecka et al., 2017a; Strzelecka et al., 2017b).

SET is argued to help researchers understand resident attitudes because it 'can account for both the positive and negative impacts of tourism as perceived by the host population' (Ap, 1992, p. 685). Power is an integral part of SET together with the theoretical understanding of how residents perceive the costs and benefits of tourism development and ultimately make the decisions to support or oppose tourism (Ap, 1992; Nunkoo & Ramkissoon, 2011). While SET is the most common theory used to explain resident attitudes in general, the literature highlights its many shortcomings (McGehee & Andereck, 2004; Woosnam et al., 2009; Boley et al., 2014; Strzelecka et al., 2017a; Strzelecka et al., 2017b). Other theories have been used in combination with SET to help explain attitudes toward tourism.

Weber's Theory of Formal and Substantive Rationality

Weber's (1968) theory of formal and substantive rationality (WFSR) holds promise in bridging the economic and non-economic factors influencing resident attitudes toward tourism (Andereck et al., 2005). Boley et al. (2014) first proposed WFSR to understand the effect of empowerment from tourism on residents' attitudes. Others quickly followed to study resident perceptions of tourism's economic and non-economic benefits through the lens of WFSR (Boley & McGehee, 2014; Strzelecka et al., 2017a; Strzelecka et al., 2017b; Yeager et al., 2020; Strzelecka et al., 2021).

The WFSR was developed as a means to explain a person's motives behind engaging in different forms of economic activity (Kalberg, 1980). Rather than limiting rationality to Western societies and their focus on capitalism and subsequently labeling non-Western cultures as irrational, Weber acknowledged that rationality is inherent in all and is comprised of means-end rational action (formal rationality) as well as value-rational action (substantive rationality). Thus, WFSR highlights that people's decision-making processes are more complicated than motivation for potential economic rewards (e.g., formal rationality) and that people are also driven by values, morals, and social norms (Kalberg, 1980; McGehee, 2007). WFSR illumines the context in which to assess residents' attitudes towards tourism as influenced by both non-economic factors such as non-economic empowering benefits from tourism as well as tourism's extrinsic economic rewards (Strzelecka et al., 2021).

The rationalisation process occurs in different domains of life and along a continuum between formal and substantive rationality. *Formal rationality* focuses on maximising economic gain through market laws (Kalberg, 1980). It is mainly motivated by the provision of economic needs (McGehee & Meares, 1998). In the case of *substantive rationality*, individuals consider a range of possible values and morals within their decisions on whether to engage in different economic transactions. According to Kalberg (1980), Weber's *substantive rationality* is about an individual's inherent capacity for value-rational action in certain areas of life, leaving all other areas, such as economic needs, untouched. While the *formal* component emphasises the importance of the relationship between resident perceptions of economically benefiting from tourism and their support for tourism development, the *substantive* component of Weber's theory provides credence for including non-economic benefits such as resident perceptions of how tourism empowers or disempowers them. Through the substantive rationality piece of Weber's theory, scholars argue that perceptions of empowerment from tourism are likely to influence their attitudes toward tourism.

Antecedents and Outcomes of Empowerment From Tourism

Place Attachment

Research has demonstrated that the affective connection to places and the meanings of places can influence residents' perceptions about tourism development and their empowerment through tourism (Strzelecka et al., 2021). Residents' emotional connections to and meanings of places cannot be overemphasised. In tourism, the components of place attachment that have been examined in relation to resident empowerment from tourism include place identity and place dependence (Boley et al., 2021).

Place identity is about place meanings that are then reinforced through local practices and rituals (Cresswell, 1996). People self-define through meaningful places (e.g., Greider & Garkovich, 1994) and develop a place identity (e.g., Williams et al., 1992). Such an identity arises from particular values, attitudes, and beliefs about the physical world and direct experiences of this environment (Proshansky et al., 1983). Finally, 'place identity is developed by thinking and talking about places through a process of distancing, which allows for reflection and appreciation of places' (Tuan, 1980, cited in Proshansky et al., 1983, p. 61). A place can also become a resource for satisfying goals, creating, in turn, a relationship of dependence (Williams et al., 1992). Individuals value places for their functional attributes in supporting individual goals (Schreyer et al., 1981; Stokols & Shumaker, 1981; Williams & Roggenbuck, 1989). It is 'embodied in the area's physical characteristics and may increase when the place is close enough to allow for frequent visitation' (Williams & Vaske, 2003, p. 831).

People's perceptions of 'place identity' and 'place dependence' have been shown to influence residents' perceptions of social and environmental conditions in natural settings (Kyle et al., 2004). On the other hand, changes occurring because of tourism and available activities may be reflected in the strength of place dependency or place identity (Gu & Ryan, 2008). Strzelecka et al. (2017a, b) further proposed that place bonds influence residents' perceptions of psychological empowerment. Namely, the more residents identify with their local community and depend on the local environment for their lifestyle, the greater the potential for them to benefit from tourism, which forges pride and self-esteem. Similarly, Strzelecka et al. (2017a, b) found the more one's identity is tied up in a specific locale, and the more one is dependent on a particular location, the more likely a person is socially empowered through tourism. Lastly, in reference to political empowerment, Manzo and Perkins (2006) noted that residents mobilise when events such as tourism development jeopardise their relationship with local places. In particular, those with a strong attachment to their communities respond strongly to place disruptions threatening their identity and lifestyle. Yet, Strzelecka's et al's (2017a, b) research shows that the relationship may only exist between the place dependence dimension of residents' attachment and political empowerment from tourism. In contrast, Lee et al. (2019) found that residents' empowerment predicted their attachment to using Airbnb as their hosting platform. This poses a question about the reciprocal relationship between empowerment and place attachment in the tourism development context warranting future studies.

Nature Connection

Informed by their research on the relationship between place bonds and perceived empowerment form tourism, Strzelecka et al. (2021) examined the link between nature connection and empowerment from rural nature tourism. To support their argumentation, they applied the Nature's Contributions to People framework which offers a robust perspective for understanding how the residents' subjective sense of their relationship with the natural world reinforces local challenges of developing tourism in rural natural areas (Chan et al., 2018; Ramos & Prideaux, 2014). The sense of connection with the natural world emerges from the value of experiencing nature and the emotional bond between the individual and nature (Beery & Wolf-Watz, 2014;

Schultz et al., 2004). The strength of one's connection to nature rests on individual geographical, historical knowledge of, and familiarity with the surrounding environment (Gustafson, 2001) or provision of other direct cultural ecosystem services (CES) (Wheeler et al., 2015).

Tourism can affect how people connect to everyday landscapes (e.g., Cole, 2006; Robinson & Picard, 2006). When tourism enhances natural attributes of the rural landscape that rural residents value, stronger connection to nature will likely correspond to a more positive view of such tourism, including personal and community benefits (Jurowski et al., 1997; Gursoy & Rutherford, 2004) or even tourism economic contributions to nature conservation.

Strzelecka et al. (2021) proposed that the more residents bond with nature and depend on its environment for their lifestyle, the more likely it is that they will feel empowered through tourism that cares for nature. In other words, when tourism aligns with residents' expectations, a robust nature connection may boost the sense of psychological, social, political, and economic empowerment. In contrast, stronger nature bonding may result in the opposite when tourism disrupts a community's relationship to its natural surroundings (Strzelecka et al., 2021). Specifically, disruptions in nature bonding are likely to lead to more negative perceptions of tourism within the rural community, such as growing awareness that tourism lacks positive empowering contributions and thus the possible rejection of tourism as a viable development strategy.

Quality of Life

Empowerment from tourism has also been linked to residents' quality of life (QoL). QoL is 'one's satisfaction with life and feelings of contentment or fulfillment with one's experience in the world' (Andereck & Nyaupane, 2011, p. 248). The premise of this discourse is that the more empowered residents feel in a community, the greater their QoL will be. For instance, Woo et al. (2018) argue that residents' perception of outcomes of tourism development involves economic, social, cultural, political, and environmental domains, which in turn influences overall judgment about the living quality within the community. Aleshinloye et al. (2022) employed the RETS to investigate, among other things, the influence of residents' empowerment on the perceived quality of life. Interestingly, the impact of political empowerment on QoL was negative, while psychological and social dimensions of empowerment influences were positive.

Future Research

Additional Dimensions of Empowerment

In the above-presented research, scholars employed the three-dimensional model of empowerment from tourism, including psychological, social, and political dimensions. Boley and McGeehe (2014) offered a clear operationalisation of individual empowerment and a reliable instrument to measure it. As the discourse focused on understanding empowerment from tourism by applying quantitative approaches is still evolving, new dimensions of empowerment are being put forth. For example, Scheyvens and van der Watt (2021) expand the existing conceptualisation of the tourism empowerment nexus by adding cultural and environmental dimensions of empowerment. The scholars considered cultural empowerment evident when residents value and respect multiple, diverse local heritage and feel they can express their cultural heritage in all life domains and have the power to determine how their cultural heritage is represented. The environmental dimension of empowerment from tourism focused on resident awareness of and connectedness with natural resources and pro-environmental behaviour, as well as environmental advocacy (Scheyvens & van der Watt, 2021). However, to date, no studies have operationalised cultural and environmental empowerment. Therefore, future research endeavors should also develop reliable scales for both dimensions.

Relevance of Empowerment from the Resident Perspective

A pertinent question that future studies on empowerment need to address is the relevance of empowerment for local communities worldwide, especially within local cultures that have not yet embraced a model of local democracy with empowerment as one of the essential virtues. While empowerment is hailed as the ultimate condition for local democratic communities to thrive according to the academic literature, some communities may feel empowerment is not a relevant aspect of who they are and how they envision local tourism development. Despite the literature's persistence on the importance of empowerment, communities may have more pressing matters than empowerment from tourism or may be just indifferent or occupied with other concerns of life. As such, the idea of empowerment from tourism may be irrelevant to their resident' sense of tourism sustainability or perceptions of QoL. This may explain the prevalent finding of no relationship between political empowerment and support for tourism (Boley et al., 2014). We propose that future research considers the diversity of developmental values and what QoL means in local terms before bluntly promoting empowerment. One such technique that holds promise is *Important Performance Analysis* (Boley et al., 2017b). Residents could be asked about how important various types of empowerment are to them and then asked how well the destination does at empowering them. This would provide four quadrants ranging from keep up the good work, concentrate here, potential overkill or low priority depending on the levels of importance and performance. If certain empowerment dimensions fell in the low priority or potential overkill quadrants indicating little importance to residents, other community initiatives that matter to residents could be undertaken. Such differences may exist between regions with different economic situations, political regimes, and cultural heritages. Similarly, we may expect other ideas on developing sustainable tourism and the importance of empowerment from tourism in urban vs. rural destinations.

Collective Empowerment

Conceptualised as a multilevel construct, empowerment is an interdependent phenomenon functional at the individual, organisational, and community levels (Zimmerman, 1995; Rodrigues et al., 2018). Scholars have commonly assumed a positive relationship between individual empowerment and collective empowerment (e.g., political action, community agency) (Joo et al., 2020). Furthermore, it is believed to connect individual characteristics with acts directed toward social change (Rappaport, 1981), linking individual wellbeing with a larger environment (Perkins & Zimmerman, 1995). In line with this view, individual empowerment from tourism is likely to be closely interlinked with group empowerment.

Strzelecka and Wicks (2010) noted that individual empowerment in tourism happens through developing social relationships leading to greater access to resources needed by individual actors. Arguably social capital is the means to achieving political empowerment from tourism, implying the overlap between social and political domains. Strzelecka and Wicks (2010) refer to Zimmerman (1995) to highlight that as a community, stakeholders work together 'in an organized fashion to improve their collective lives and linkages among community organizations and agencies that help maintain that quality of life' (Zimmerman, 1995, p. 582). Community empowerment is then a process in which community members gain influence over conditions that matter to them (Fawcett et al., 1995). Cole (2006) also observed that in community empowerment, residents become agents of change capable of solving problems they face. In tourism, stakeholders manifest individual empowerment through community action and the pursuit of individual tourism interests in a community field.

The study by Joo et al. (2020) followed Rocha's (1997) model that conceptualised empowerment as starting at the individual level and eventually welling up into the community level, with political action being the outcome of community empowerment. Results indicated that only social and political empowerment were likely to generate political action. Moreover, the tourism knowledge appears another factor that impacts community empowerment with social and political empowerment mediating this relationship.

Prior research points out that the social and political aspects of individual empowerment from tourism matter for a collective sense of empowerment in tourism development (Strzelecka & Wicks, 2010; Joo et al., 2020). However, we need more discussion and empirical evidence on how individual empowerment from tourism translates into community empowerment. This idea provokes future research to engage with:

- the mechanisms of individual empowerment;
- operationalise the notion of community empowerment from tourism; and
- exploring the interconnectedness between individual and collective forms of empowerment in tourism.

Conclusion

The chapter opens with the discussion of empowerment construct and its definitions. We first highlight different ways of conceptualising individual empowerment, either as a process or an outcome. It then presents a multidimensional definition of empowerment in tourism originally proposed by Scheyvens' (1999) that consists of psychological, social, political, and economic dimensions.

The following section provides an overview of core studies within the empowerment discourse, including Boley and McGehee's (2014) study that operationalised the psychological, social, and political dimensions and proposed quantitative scale to measure them. By reviewing significant publications operationalising empowerment through RETS, we highlight how those individual perceptions of being empowered from tourism matter for resident support of tourism and perceived QoL. We also discuss how residents' bonds with the community they live in and local nature affect how they feel about tourism and how empowered or disempowered they can be.

The final part of this chapter points towards three emerging directions for future research. First, the current empowerment model is limited in terms of how empowerment was initially conceptualised (four dimensions) and operationalised (3 dimensions), demanding that future studies expand the model to incorporate additional dimensions of empowerment such as environmental and cultural. The second issue that has been overlooked is the relevance of empowerment to local contexts and communities. Researchers have assumed that residents desire psychological, social, and political empowerment from tourism development. However, studies consistently showed that political empowerment does not affect residents' support for tourism development in their communities or those who feel tourism empowers them to oppose tourism (Maruyama et al., 2016; Strzelecka et al., 2017a; Strzelecka et al., 2017b; Yeager et al., 2020). The ambivalent results signal that we must pay more attention to local values and seek to understand how vital empowerment from tourism is for residents in different economic and political contexts. Third, few studies have been focused on mechanisms fostering collective empowerment in tourism development and the conditions needed for individual empowerment to materialise as collective forms of empowerment in tourism destinations. Given the three potential areas for future research, we would like to highlight that consistency in how tourism literature conceptualises empowerment and precision of definition whether it is the empowerment from tourism, empowerment in tourism or even empowerment for tourism can facilitate empowerment discourse in tourism to address its past limitations effectively.

References

Aghazamani, Y., & Hunt, C. A. (2017). Empowerment in tourism: A review of peer-reviewed literature. *Tourism Review International*, *21*(4), 333–346. https://doi.org/10.3727/154427217X15094520591321

Aleshinloye, K. D., Woosnam, K. M., Tasci, A. D. A., & Ramkissoon, H. (2022). Antecedents and outcomes of resident empowerment through tourism. *Journal of Travel Research*, *61*(3), 656–673. https://doi.org/10.1177/0047287521990437

Andereck, K. L., & Nyaupane, G. P. (2011). Exploring the nature of tourism and quality of life perceptions among residents. *Journal of Travel Research*, *50*(3), 248–260. https://doi.org/10.1177/0047287510362918

Andereck, K. L., Valentine, K. M., Knopf, R. C., & Vogt, C. A. (2005). Residents' perceptions of community tourism impacts. *Annals of Tourism Research*, *32*(4), 1056–1076. https://doi.org/10.1016/j.annals.2005.03.001

Ap, J. (1992). Residents' perceptions on tourism impacts. *Annals of Tourism Research*, *19*(4), 665–690. https://doi.org/10.1016/0160-7383(92)90060-3

Arnstein, S. R. (1969). A ladder of citizen participation. *Journal of the American Institute of Planners*, *35*(4), 216–224. https://doi.org/10.1080/01944366908977225

Beaumont, N., & Dredge, D. (2010). Local tourism governance: a comparison of three network approaches. *Journal of Sustainable Tourism*, *18*(1), 7–28. https://doi.org/10.1080/09669580903215139

Beery, T. H., & Wolf-Watz, D. (2014). Nature to place: Rethinking the environmental connectedness perspective. *Journal of Environmental Psychology*, *40*, 198–205. https://doi.org/10.1016/j.jenvp.2014.06.006

Beritelli, P., & Laesser, C. (2011). Power dimensions and influence reputation in tourist destinations: Empirical evidence from a network of actors and stakeholders. *Tourism Management*, *32*(6), 1299–1309. https://doi.org/10.1016/j.tourman.2010.12.010

Boley, B. B., & McGehee, N. G. (2014) Measuring empowerment: Developing and validating the resident empowerment through tourism scale (RETS). *Tourism Management*, *45*, 85–94. https://doi.org/10.1016/j.tourman.2014.04.003

Boley, B. B., Ayscue, E., Maruyama, N., & Woosnam, K. M. (2017a). Gender and empowerment: assessing discrepancies using the resident empowerment through tourism scale. *Journal of Sustainable Tourism*, *25*(1), 113–129. https://doi.org/10.1080/09669582.2016.1177065

Boley, B. B., Maruyama, N., & Woosnam, K. M. (2015). Measuring empowerment in an eastern context: Findings from Japan. *Tourism Management*, *50*, 112–122. https://doi.org/10.1016/j.tourman.2015.01.011

Boley, B. B., McGehee, N. G., & Hammett, A. T. (2017b). Importance-performance analysis (IPA) of sustainable tourism initiatives: The resident perspective. *Tourism Management*, *58*, 66–77. https://doi.org/10.1016/j.tourman.2016.10.002

Boley, B. B., McGehee, N. G., Perdue, R. R., & Long, P. (2014). Empowerment and resident attitudes toward tourism: Strengthening the theoretical foundation through a Weberian lens. *Annals of Tourism Research*, *49*, 33–50. https://doi.org/10.1016/j.annals.2014.08.005

Boley, B. B., Strzelecka, M., Yeager, E. P., Ribeiro, M. A., Aleshinloye, K. D., Woosnam, K. M., & Mimbs, B. P. (2021). Measuring place attachment with the abbreviated place attachment scale (APAS). *Journal of Environmental Psychology*, *74*, 101577. https://doi.org/10.1016/j.jenvp.2021.101577

Chan, K. M. A., Gould, R. K., & Pascual, U. (2018). Editorial overview: Relational values: What are they, and what's the fuss about? *Current Opinion in Environmental Sustainability*, *35*, A1–A7. https://doi.org/10.1016/j.cosust.2018.11.003

Cheong, S. M., & Miller, M. L. (2000). Power and tourism: A Foucauldian observation. *Annals of Tourism Research*, *27*(2), 371–390. https://doi.org/10.1016/S0160-7383(99)00065-1

Choi, H. C. & Murray, I. (2010). Resident attitudes toward sustainable community tourism. *Journal of Sustainable Tourism*, *18*(4), 575–594. https://doi.org/10.1080/09669580903524852

Christens, B. D. (2012). Toward relational empowerment. *American Journal of Community Psychology*, *50*, 114–128. https://doi.org/10.1007/s10464-011-9483-5

Cole, S. (2006). Information and empowerment: The keys to achieving sustainable tourism. *Journal of Sustainable Tourism*, *14(6)*, 629–644. https://doi.org/10.2167/jost607.0

Cresswell, T. (1996). *In place/out of place: Geography, ideology, and transgression*. University of Minnesota Press.

Fawcett, S. B., Paine-Andrews, A., Francisco, V. T., Schultz, J. A., Richter, K. P., Lewis, R. K., Williams, E. L., Harris, K. J., Berkley, J. Y., Fisher, J. L., & Lopez, C. M. (1995). Using empowerment theory in collaborative partnerships for community health and development. *American Journal of Community Psychology*, *23*(5), 677–697. https://doi.org/10.1007/BF02506987

Friedmann, J. (1992). *Empowerment: The politics of alternative development*. Blackwell.

Greider, T., & Garkovich, L. (1994). Landscapes: The social construction of nature and the environment. *Rural Sociology*, *59*(1), 1–24. https://doi.org/10.1111/j.1549-0831.1994.tb00519.x

Gu, H., & Ryan, C. (2008). Place attachment, identity and community impacts of tourism—the case of a Beijing hutong. *Tourism Management*, *29*(4), 637–647. https://doi.org/10.1016/j.tourman.2007.06.006

Gursoy, D., & Rutherford, D. G. (2004). Host attitudes toward tourism: An improved structural model. *Annals of Tourism Research*, *31*(3), 495–516. https://doi.org/10.1016/j.annals.2003.08.008

Gustafson, P. E. R. (2001). Meanings of place: Everyday experience and theoretical conceptualizations. *Journal of Environmental Psychology*, *21*(1), 5–16. https://doi.org/10.1006/jevp.2000.0185

Hall, C. M. (1994). *Tourism and politics: Policy, power and place*. John Wiley & Sons.

Joo, D., Woosnam, K. L. M., Strzelecka, M., & Boley, B. B. (2020). Knowledge, empowerment, and action: testing the empowerment theory in a tourism context. *Journal of Sustainable Tourism*, *28*(1). 69–85. https://doi.org/10.1080/09669582.2019.1675673

Jurowski, C., Uysal, M., & Williams, D. R. (1997). A theoretical analysis of host community resident reactions to tourism. *Journal of Travel Research*, *36*(2), 3–11. https://doi.org/10.1177/004728759703600202

Kalberg, S. (1980). Max Weber's types of rationality: Cornerstones for the analysis of rationalization processes in history. *American Journal of Sociology*, *85*(5), 1145–1179. https://doi.org/10.1086/227128

Kelly, J. (1992). Does job re-design theory explain job re-design outcomes? *Human Relations* *45*(8), 753–774. https://doi.org/10.1177/001872679204500801

Kieffer, C. H. (1984). Citizen empowerment: A developmental perspective. *Prevention in Human Services*, *3*(2–3), 9–36. https://doi.org/10.1300/J293v03n02_03

Kroeker, C. J. (1995). Individual, organizational, and societal empowerment: A study of the processes in a Nicaraguan agricultural cooperative. *American Journal of Community Psychology*, *23*(5), 749–764. https://doi.org/10.1007/BF02506990

Kyle, G., Graefe, A., Manning, R., & Bacon, J. (2004). An examination of the relationships between leisure activity involvement and place attachment among hikers along the Appalachian Trail. *Journal of Leisure Research*, *35*(3), 249–273. https://doi.org/10.1080/00222216.2003.11949993

Lee, H., Yang, S. B., & Koo, C. (2019). Exploring the effect of Airbnb hosts' attachment and psychological ownership in the sharing economy. *Tourism Management*, *70*, 284–294. https://doi.org/10.1016/j.tourman.2018.08.017

Manzo, L. C., & Perkins, D. D. (2006). Finding common ground: The importance of place attachment to community participation and planning. *Journal of Planning Literature*, *20*(40), 335–350. https://doi.org/10.1177/0885412205286160

Maruyama, N. U., Woosnam, K. M., & Boley, B. B. (2016). Comparing levels of resident empowerment among two culturally diverse resident populations in Oizumi, Gunma, Japan. *Journal of Sustainable Tourism*, *24*(10), 1442–1460. https://doi.org/10.1080/09669582.2015.1122015

McGehee, N. G. (2007). An agritourism systems model: A Weberian perspective. *Journal of Sustainable Tourism*, *15*(2), 111–124. https://doi.org/10.2167/jost634.0

McGehee, N. G., & Andereck, K. L. (2004). Factors predicting rural residents' support of tourism. *Journal of Travel Research*, *43*(2), 131–140. https://doi.org/10.1177/0047287504268234

McGehee, N. G., & Meares, A. C. (1998). A case study of three tourism-related craft marketing cooperatives in Appalachia: Contributions to community. *Journal of Sustainable Tourism*, *6*(1), 4–25. https://doi.org/10.1080/09669589808667299

McHenry, J. A. (2011). Rural empowerment through the arts: The role of the arts in civic and social participation in the Mid West region of Western Australia. *Journal of Rural Studies*, *27*(3), 245–253. https://doi.org/10.1016/j.jrurstud.2011.03.001

Mendoza-Ramos, A., & Prideaux, B. (2018). Assessing ecotourism in an Indigenous community: using, testing and proving the wheel of empowerment framework as a measurement tool. *Journal of Sustainable Tourism*, *26*(2), 277–291. https://doi.org/10.1080/09669582.2017.1347176

Miller, R. L., & Campbell, R. (2006). Taking stock of empowerment evaluation: An empirical review. *American Journal of Evaluation*, *27*(3), 296–319. https://doi.org/10.1177/109821400602700303

Murphy, P. E. (1985). *Tourism: A community approach*. Methuen.

Nunkoo, R., & Ramkissoon, H. (2011). Developing a community support model for tourism. *Annals of Tourism Research*, *38*(3), 964–988. https://doi.org/10.1016/j.annals.2011.01.017

Nunkoo, R., Smith, S. L. J., & Ramkissoon, H. (2013). Residents' attitudes to tourism: a longitudinal study of 140 articles from 1984 to 2010. *Journal of Sustainable Tourism*, *21*(1), 5–25. https://doi.org/10.1080/09669582.2012.673621

Perdue, R. R., Long, P., & Allen, L. (1990). Resident support for tourism development. *Annals of Tourism Research*, *17*(4), 586–599. https://doi.org/10.1016/0160-7383(90)90029-Q

Perkins, D. D., & Zimmerman, M. A. (1995). Empowerment theory, research, and application. *American Journal of Community Psychology*, *23*, 569–579. https://doi.org/10.1007/BF02506982

Proshansky, H. M., Fabian, A. K., & Kaminoff, R. (1983). Place-identity: Physical world socialization of the self. *Journal of Environmental Psychology*, *3*(1), 57–83. https://doi.org/10.1016/S0272-4944(83)80021-8

Ramos, M. A., & Prideaux, B. (2014). Indigenous ecotourism in the Mayan rainforest of Palenque: Empowerment issues in sustainable development. *Journal of Sustainable Tourism*, *22*(3), 461–479. https://doi.org/10.1080/09669582.2013.828730

Rappaport, J. (1981). In praise of paradox: A social policy of empowerment over prevention. *American Journal of Community Psychology*, *9*(1), 1–25. https://doi.org/10.1007/BF00896357

Rappaport, J. (1985). The power of empowerment language. *Social Policy*, *16*(2), 15–21.

Rappaport, J. (1987). Terms of empowerment/exemplars of prevention: Toward a theory for community psychology. *American Journal of Community Psychology*, *15*, 121–148. https://doi.org/10.1007/BF00919275

Reed, M. G. (1997). Power relations and community-based tourism planning. *Annals of Tourism Research*, *24*(3), 566–591. https://doi.org/10.1016/S0160-7383(97)00023-6

Robinson, M., & Picard, D. (2006). *Tourism, culture and sustainable development*. UNESCO. https://unesdoc.unesco.org/ark:/48223/pf0000147578

Rocha, E. M. (1997). A ladder of empowerment. *Journal of Planning Education and Research*, *17*(1), 31–44. https://doi.org/10.1177/0739456X9701700104

Rodrigues, M., Menezes, I., & Ferreira, P. D. (2018). Validating the formative nature of psychological empowerment construct: Testing cognitive, emotional, behavioral, and relational empowerment components. *Journal of Community Psychology*, *46*(1), 58–78. https://doi.org/10.1002/jcop.21916

Rowlands, J. (1995). Empowerment examined. *Development in Practice*, *5*(2), 101–107. https://doi.org/10.1080/0961452951000157074

Rowlands, J. (1997). *Questioning empowerment: Working with women in Honduras*. Oxfam.

Scheyvens, R. (1999). Ecotourism and the empowerment of local communities. *Tourism Management*, *20*(2), 245–249. https://doi.org/10.1016/S0261-5177(98)00069-7

Scheyvens, R. (2003). Local involvement in managing tourism. In S. Singh, D. J. Timothy & R. Dowling (Eds), *Tourism in destination communities* (pp. 229–252). CABI Publishing.

Scheyvens, R. (2020). Empowerment. In A. Kobayashi (Ed.), *International encyclopedia of human geography* (pp. 115–122). Elsevier.

Scheyvens, R., & van der Watt, H. (2021). Tourism, empowerment and sustainable development: A new framework for analysis. *Sustainability*, *13(22)*, 12606. https://doi.org/10.3390/su132212606

Schreyer, R., Jacob, G., & White, R. (1981). Environmental meaning as a determinant of spatial behavior in recreation. *Proceedings of the Applied Geography Conferences*, Volume 4, State University of New York (pp 294–300).

Schultz, P. W., Shriver, C., Tabanico, J. J., & Khazian, A. M. (2004). Implicit connections with nature. *Journal of Environmental Psychology*, *24*(1), 31–42. https://doi.org/10.1016/S0272-4944(03)00022-7

Sharpley, R. (2014). Host perceptions of tourism: A review of the research. *Tourism Management*, *42*, 37–49. https://doi.org/10.1016/j.tourman.2013.10.007

Simmons, C. H., & Parsons, R. J. (1983). Developing internality and perceived competence: The empowerment of adolescent girls. *Adolescence*, *18*(72), 917–922. https://www.proquest.com/openview/b2a67754b609e76e2bba78c26ce3d38a/1?pq-origsite=gscholar&cbl=1819054

Sindiga, I. (1995). Wildlife based tourism in Kenya: Land use conflicts and government compensation policies over protected areas. *Journal of Tourism Studies*, *6*(2), 45–55. https://search.informit.org/doi/abs/10.3316/ielapa.961010438

Sofield, T. H. B. (2003). *Empowerment for sustainable tourism development*. Pergamon.

Staples, L. H. (1990). Powerful ideas about empowerment. *Administration in Social Work*, *14*(2), 29–42. https://doi.org/10.1300/J147v14n02_03

Stokols, D., & Shumaker, S. A. (1981). People and places: A transactional view of settings. In J. Harvey (Ed.), Cognition, social behavior, and the environment (pp. 441–488). Erlbaum.

Stronza, A., & Gordillo, J. (2008). Community views of ecotourism. *Annals of Tourism Research*, *35*(2), 448–468. https://doi.org/10.1016/j.annals.2008.01.002

Strzelecka M., Prince, S., & Boley, B. B. (2021). Resident connection to nature and attitudes towards tourism: findings from three different rural nature tourism destinations in Poland. *Journal of Sustainable Tourism*. https://doi.org/10.1080/09669582.2021.1995399

Strzelecka, M., & Wicks, B. E. (2010). Engaging residents in planning for sustainable rural-nature tourism in post-communist Poland. *Community Development*, *41*(3), 370–384. https://doi.org/10.1080/15575330903359614

Strzelecka, M., & Wicks, B. E. (2015). Community participation and empowerment in rural post-communist societies: Lessons from the leader approach in Pomerania, Poland. *Tourism Planning & Development*, *12*(4), 381–397. https://doi.org/10.1080/21568316.2015.1013564

Strzelecka, M., Boley B. B., & Strzelecka, C. (2017a). Empowerment and resident support for tourism in rural Central and Eastern Europe (CEE): the case of Pomerania, Poland. *Journal of Sustainable Tourism*, *25*(4), 554–572. https://doi.org/10.1080/09669582.2016.1224891

Strzelecka, M., Boley, B. B., & Woosnam, K. M. (2017b). Place attachment and empowerment: Do residents need to be attached to be empowered? *Annals of Tourism Research*, *66*, 61–73. https://doi.org/10.1016/j.annals.2017.06.002

Timothy, D. J. (2007). Empowerment and stakeholder participation in tourism destination communities. In A. Church & T. Coles (Eds), *Tourism, power and space* (pp. 199–216). Routledge.

Tosun, C., & Timothy, D. J. (2003). Arguments for community participation in the tourism development process. *Journal of Tourism Studies*, *14*(2), 2–15. https://search.informit.org/doi/abs/10.3316/ielapa.200402849

Tuan, Y. F. (1980). Rootedness versus sense of place. *Landscape*, *24*(1), 3–8.

Wallerstein, N., & Bernstein, E. (1994). Introduction to community empowerment, participatory education, and health. *Health Education Quarterly*, *21*(2), 141–148. https://doi.org/10.1177/109019819402100202

Weber, M. (1968). *Economy and society*. G. Roth & C. Wittich. (Eds). Bedminister Press. Trans from Weber, M. (1956). *Wirtschaft und Gesellschaft* (4th ed.). J. Winckelmann (Ed.). J. C. B. Mohr.

Wheeler, B. W., Lovell, R., Higgins, S. L., White, M. P., Alcock, I., Osborne, N. J., Husk, K., Sabel, C. E., & Depledge, M. H. (2015). Beyond greenspace: An ecological study of population general health and indicators of natural environment type and quality. *International Journal of Health Geographics*, *14*. https://doi.org/10.1186/s12942-015-0009-5

Wilkinson, K. P. (1991). *Community in rural America*. Social Ecology Press.

Williams, D. R., & Roggenbuck, J.W. (1989). Measuring place attachment: Some preliminary results. In L. H. McAvoy & D. Howard (Eds), *Abstracts of the 1989 leisure research symposium, National Recreation and Park Association*, Arlington, VA (p. 32).

Williams, D. R., Patterson, M. E., Roggenbuck, J. W., & Watson, A. E. (1992). Beyond the commodity metaphor: Examining emotional and symbolic attachment to place. *Leisure Sciences*, *14*(1), 29–46. https://doi.org/10.1080/01490409209513155

Williams, R. D., & Vaske, J. J. (2003). The measurement of place attachment: Validity and generalizability of a psychometric approach. *Forest Science*, *49*(6), 830–840. https://academic.oup.com/forestscience/article/49/6/830/4617493

Woo, E., Uysal, M., & Sirgy, M. J. (2018). Tourism impact and stakeholders' quality of life. *Journal of Hospitality & Tourism Research*, *42*(2), 260–286. https://doi.org/10.1177/1096348016654971

Woosnam, K. M., Norman, W. C., & Ying, T. (2009). Exploring the theoretical framework of emotional solidarity between residents and tourists. *Journal of Travel Research*, *48*(2), 245–258. https://doi.org/10.1177/0047287509332334

Yeager, E. P., Boley, B. B., Woosnam, K. M., & Green, G. T. (2020). Modeling residents' attitudes toward short-term vacation rentals. *Journal of Travel Research*, *59*(6), 955–974. https://doi.org/10.1177/0047287519870255

Zimmerman, B. J., Bandura, A., & Martinez-Pons, M. (1992). Self-motivation for academic attainment: The role of self-efficacy beliefs and personal goal setting. *American Educational Research Journal*, *29*(3), 663–676. https://doi.org/10.3102/00028312029003663

Zimmerman, M. A. (1995). Psychological empowerment: Issues and illustrations. *American Journal of Community Psychology*, *23*(5), 581–599. https://doi.org/10.1007/BF02506983

Zimmerman, M. A., & Rappaport, J. (1988). Citizen participation, perceived control, and psychological empowerment. *American Journal of Community Psychology*, *16*(5), 725–750. https://doi.org/10.1007/BF00930023

Zimmerman, M. A., & Zahniser, J. K. (1991). Refinements of sphere-specific measures of perceived control: Development of a sociopolitical control scale. *Journal of Community Psychology*, *19*(2), 189–204. https://doi.org/10.1002/1520-6629(199104)19:2<189::AID-JCOP2290190210>3.0.CO;2-6

Chapter 46

Tourism and Resilience

Alberto Amore

Introduction

This chapter aims at providing a timely account of research focusing on tourism and resilience. Studies on the different forms of resilience in relation to tourism are very recent. In the first ever review of works on this topic, Hall (2016) showed how the concept of resilience in the field of tourism was very marginal until the early 2010s. However, as of September 2022, there are around 1,500 contributions featuring resilience and tourism. In the prior five years, there were around 1,000 works on tourism and resilience addressing conjunctural global events like climate change and COVID19, as well as tourism specific issues like overtourism and destination vulnerability. Undeniably, part of the success in such proliferation of works is due to the relatively flexible approach with the conceptualization of resilience among tourism scholars (Hall et al., 2018).

Research on resilience in tourism represents the way forward in the study of the mitigation and adaptation of destinations to triggers like climate change, environmental vulnerability, risk reduction, crisis management and post-disaster recovery. The UWNTO conceives resilience as one of the pillars for the achievement of the United Nations Sustainable Development Goals (UNSDGs) in tourism. Particularly in the so-called developing countries, building a resilient tourism sector enables destinations to reduce the likelihood of risks, enhance mitigation during times of crisis and ultimately foster recovery (UNWTO, 2022). Similarly, the EU conceives resilience in tourism as part of the wider ecological transition and post-COVID19 recovery agenda so that the industry is "better prepared for the challenges and opportunities of the green and digital transitions" (EU, 2022). Tourism and resilience are also included in a myriad of national and local tourism development strategies. On the one hand the Tourism Industry Aotearoa (TIA) puts business resilience as one of the key pillars in the achievement of long-term sustainable economic development for the tourism sector in the country (TIA, 2022). On the other hand, the Local Government Association (LGA) conceives resilience an essential

The Wiley Blackwell Companion to Tourism, Second Edition. Edited by C. Michael Hall.

trait in the planning and design of leisure, retail and tourism recovery in High Streets across the United Kingdom (LGA, 2022).

Notwithstanding the rise and popularity of resilience thinking in tourism, there are two key challenges that need to be addressed. From a research perspective, there are intrinsic contradictions and a weak framing in tourism studies around the notion of resilience. This can be mainly attributed to two factors. First, the persistency of what Sayer (1999) first referred to as disciplinary parochialism, a limitation that has been highlighted and addressed specifically in tourism research (Coles et al., 2006; Coles et al., 2016). Second, and more importantly, the abuse of the term resilience in tourism literature. This is particularly evident in the proliferation of research outputs either discussing or marginally addressing resilience in relation to post-COVID19 recovery. Nearly a quarter of the 1,500 contributions address COVID19 in relation to resilience. To a certain extent, the use of resilience as '*new wine* within disaster research' (Andresen, 2019, p. 170, emphasis in the original) can potentially lead to yet another empty concept as it is in the case of sustainability (Hall, 2011; Hall, 2022). At the same time, the positivistic notion of resilience as a trait that tourism employees and businesses show in times of crisis downplays ecological determinism. As Hayward (2013, p. 37) puts it, 'the idea of resilience as a personal quality must be treated with caution'.

This chapter is structured as follows. The first section provides an overview of the literature and the different insights that have become part of research in tourism and resilience. The next section illustrates the theories and concepts relevant to the advancement of research in tourism and resilience. The concluding section provides a summary of the chapter and a reflection on the way forward for research. A panarchy approach (Gunderson & Holling, 2002) is introduced develop 'a better understanding of the tourism system' (Hall, et al, 2018, p. 147) and, potentially, contribute towards a genuine rethinking of tourism as a phenomenon.

Overview of the Literature

The very first works acknowledging resilience in relation to tourism addressed biodiversity and sustainable development (Lovejoy, 1994) or the adaptation of destinations to international tourism market fluctuations and internal political crises (e.g., Sri Lanka) (O'Hare & Barrett, 1994). These studies, however, were published in natural sciences or economic geography, with very little visibility in tourism studies. As Hall (2016, 2018) notes, research on resilience among tourism scholars has only started to gain momentum over the last ten years. It has grown into a rather prolific and heterogeneous body of literature on environmental, political, psychological, and organizational implications. Overall, we can broadly distinguish between macro-level appraisal from an ecological perspective (e.g., Malinauskaite et al., 2022) and meso-level understanding from destination, business and community perspectives (e.g., Musavengane & Kloppers, 2020). Research on the resilience of individuals features a range of relevant studies on organizations (e.g., Fang et al., 2020) and travel behaviour (e.g., Hajibaba et al., 2015). In addition, there is a fourth strand of works that assimilates resilience as a conjunctural outcome in the wake of disasters and crises in the travel and tourism industry. However, it should be noted that this body of literature is mostly the result of crises and disasters in global travel and tourism. For example, more than half of the work published between 2020 and 2022 address resilience and tourism in the wake of the COVID19 pandemic. Similarly, the focus of most of research on tourism and resilience tends to be on management and business studies. As Hall (2018) noted, around half of tourism studies conceive resilience in relation to the economic dimensions of travel and tourism.

Research on tourism and resilience at the macro-level is centred on ecological and socio-ecological dimensions in relation to climate change and vulnerability. These studies predominantly address how ecological stressors affect tourism services and businesses. Examples include the impact of environmental pollution on leisure and hospitality sectors across the state of Florida, United States (Yang et al., 2022) and the correlation between sea-surface temperature and cetacean sightseeing in Skjálfandi Bay, Iceland (Malinauskaite et al., 2022). Other studies look at ecological resilience and tourism in coastal areas adjacent to hotel resorts, as in the cases of the barrier reefs in the Rivera Maya, Mexico (Secore International, 2017) and in the Caribbean (Blanco-Pimentel et al., 2022). Research also addresses ecological resilience and tourism with respect to restoration plans for marine protected areas in Southern Italy (Mirauda et al., 2022). Understandably, climate change is among the main ecological challenges with implications for resilience and tourism. For example, a study from the Lofoten Islands, Norway, sheds light on the use of climate scenarios to foster adaptive capacity of nature-based tourism activities (Antonsen et al., 2022). Overall, it is argued that climate change and climate-related vulnerabilities can severely impact tourism (Moreno & Becken, 2009; Espiner & Becken, 2014). It should however be noted that, according to Dogru et al. (2019), the tourism sector is likely to be more resilient than other sectors to the consequences of climate hazards and climate change.

A relevant body of literature in the field addresses resilience from a socio-ecological perspective. The latter builds from Adger et al.'s (2005, p. 1036) notion of resilience as 'the capacity to absorb recurrent disturbances such as hurricanes or floods so as to retain essential structures, processes, and feedbacks' and reduce the vulnerability to future hazards. Much of the focus of research on socio-ecological resilience and tourism is concerned with regime shifts thresholds. For instance, Calgaro (2010) study from coastal destinations in the Andaman coast of Thailand shows different regime shifts in the governance of Khao Lak, Patong and Phi Phi Don following the 2004 Boxing Day Tsunami. Evidence from Greenland, Iceland, and Norway, instead, sheds light on how governance and decision making adapt to changes in the ecosystem services (Cook et al., 2022). Insights on socio-ecological resilience can help understand how tourism can spur adaptive capacity to issues like water scarcity, as in the cases of San Andres Ixtlahuaca, Mexico (Reyes-Santiago et al., 2022) and Dunhuang City, China (Li et al., 2022). Similarly, a recent study on fishing communities in Japan following the 2011 Great Eastern Japan Earthquake and Tsunami sheds light on how tourism-oriented recovery has the potential to enhance socio-ecological resilience in the affected regions (Bui & Saito, 2022). Adaptability and preparedness can be possible when socio-ecological resilience thinking permeates the perceptions and opinions of communities and individuals whose livelihoods depend on the fragile equilibrium between tourism and the environment as shown in the case of coastal communities in the Dominica (Weis et al., 2021).

Ecological, social, economic and political shifts can trigger changes in government and organisations. As Amore et al. (2018, p. 243) observe, 'both informal and formal social networks contribute to resilience building'. Destination adaptation was observed in the case of Thailand following the 2004 Boxing Day Tsunami (Larsen et al., 2011), with informal and formal modes of governance shaping the recovery of destinations around the Andaman coast. Looking at shifts and changes at the government level, the work of Hulke et al. (2022) illustrates the role of national and local authorities in strengthening agriculture-tourism linkages and enhancing regional resilience. Resource use is a key issue around which governments and organizations are likely to focus their attention and foster resilience through effective policies, planning and management (Choi et al., 2021; Ricart & Rico-Amoros, 2022). Tourism businesses and management organisations can play a decisive role in fostering resilience.

Evidence from Maine, United States, suggests that tourism supply chain can directly enhance resilience to climate change (Horne et al., 2022). Similarly, Destination Management Organisations (DMOs) are regarded as an asset in the identification of vulnerabilities (Alvarez et al., 2022) and in steering destination resilience in response to adverse ecological and economic changes in the long run (Luthe & Wyss, 2016).

Genuine Community-Based Tourism (CBT) initiatives much depend on sufficiently resilient socio-ecosystems, as observed in Agua Blanca and in the Galapagos Islands, Ecuador (Ruiz-Ballesteros, 2011; Ruiz-Ballesteros & del Campo Tejedor, 2020). The role of the community at large is considered to a driver in disaster risk reduction and resilience building. This can be observed, for example, with the Tsou community in Taiwan following Typhoon Morakot in 2009 (Bayrak, 2022). In the long-term, community resilience can support collaborative land management and CBT, as in the case of KwaZulu Natal, South Africa (Musavengane & Kloppers, 2020). Community resilience enablers can either be socio-cultural or genuine bottom-up initiatives from its members. On the one hand, local arts festivals can contribute to resilience and sustainable development, as observed in the case of the Shiosai Art Festival in Mitarai, Japan (Qu & Cheer, 2021). On the other hand, individual and community support towards environmental conservation can foster community resilience building, as highlighted in a recent study by Holland et al. (2022) in the Maasai Mara National Reserve, Kenya.

The ability of systems to adapt in times of adversity and uncertainty walks alongside the capacity of groups of individuals to change and adapt. From a multi-level perspective, these groups are also referred to as niches (Geels, 2005) encompassing firms, organizations and businesses. Focusing on tourism, Amore et al. (2018) suggest that tourism organisations can enhance resilience in multiple ways and have the potential to transform the prevailing structures of the regime and landscape of destinations. Evidence on organisation and business resilience in tourism is well covered in the literature. Most of the focus is on Small and Medium Enterprises (SMEs) in tourism, with evidence from Australia (Nguyen et al., 2022), England (Ntounis et al., 2022), New Zealand (Tanner et al., 2022) and St Helen Island (Williams et al., 2020). The literature also features studies on organisational resilience among specific subsectors. Usher et al. (2019) analyse perceived vulnerability, strategic planning, preparedness and recovery planning across accommodation, foodservice and retail businesses in Virginia Beach, United States. Conversely, Elshaer and Saad (2022) explore the dynamics of business resilience of restaurants and travel agencies in Egypt following the COVID-19 pandemic. A study on hotels in Wellington and Hawke's Bay, New Zealand explores hazard preparedness and planned resilience and recommends integrated training for managers and employees (Brown et al., 2021). Finally, a study by Melián-Alzola et al. (2020) ascertains the organisational resilience determinants for hotels in the context of Canary Islands, Spain.

The size of the tourism sector, the context where it operates and the degree of both internal and external vulnerabilities have an impact on organisational resilience. A recent study on economic policy response to COVID-19 by Okafor et al. (2022) argues there is a direct correlation between organisational response in the sector and aggressive mitigation measures put in place for businesses. The study also states that tourism businesses operating in countries with high standards of risk preparedness and policies are less likely to require direct assistance from government. Within the literature on organisational resilience and tourism, research on roles and decisions of entrepreneurs in times of uncertainty and perceived risk encompasses studies on psychological traits and resilient entrepreneurial behaviour among bazaar stallholders in Cairo, Egypt (Haddoud et al., 2022), research on formal and informal enterprises and resilience in vulnerable marine environments in Thailand and Australia (Biggs et al., 2012; Biggs et al., 2015) and, more recently, entrepreneurial resilience and COVID-19 crisis in the hospitality sector in Tirol, Austria (Schwaiger et al., 2021). The way organisations and entrepreneurs plan ahead of crises and adapt in times of difficulty has direct implications for employee resilience. For example, research on tourism businesses in China shows how the adoption of Corporate Social Responsibility (CSR) principles can enhance employee resilience and psychological capital during the COVID-19 crisis (Mao et al., 2021).

An additional stream of research builds on studies looking at individual resilience. A great extent of these studies addresses employees' resilience to adverse working contexts. For example, a study by Koo et al. (2022) on the tourism sector in the United Arab Emirates (UAE) at the time of COVID-19 shows the importance of effective crisis communication in enhancing employee resilience and the adverse impacts of poor leadership in times of crisis. Conversely, Prayag and Dassanayake (2023) shed light on the influence of individual creative self-efficacy and resilience in enhancing planned resilience of tourism businesses in Sri Lanka during the pandemic. Sudden triggering events like COVID-19 can arguably have negative impacts on the mental health of employees and employers, yet there can be exceptions, as in the case of owners and operators in leisure and tourism in Victoria, Australia (Lindsay-Smith et al., 2022). Further studies on individual resilience from the supply side perspective look at the relationship between employee and organizational resilience. Findings from tourism businessowners and managers in Kaikoura, New Zealand, following the 2016 earthquakes show how problem-focused and emotion-focused coping mechanisms enabled individuals to activate psychological capital and, in turn, foster organizational resilience (Fang et al., 2020). Similarly, evidence from tourism businesses in Christchurch, New Zealand, in the aftermath of the 2010–2011 Canterbury Earthquakes show significant and positive relationships between psychological and employee resilience (Prayag et al., 2020).

A further stream of literature on individual resilience in tourism studies looks at tourists, temporary residents and the local community. Looking at tourists, Hajibaba et al. (2015, p. 48) identify 'crisis-resistant tourists as those that tend to absorb risks instead of engaging in risk avoidance strategies' and are prone to travel to places during or shortly after a crisis. A recent study by Veréb et al. (2022) states that individual features like cosmopolitan mindset, objectivity and open-mindedness are key in enhancing personal resilience amid the COVID-19 pandemic. The latter represents a timely triggering event to understand the behaviour of individuals from the lens of tourism and resilience. To this end, the works by Kim et al. (2022a, 2022b) show how non-pharmaceutical interventions (e.g., social distancing, wearing a mask) positively influence prosocial biosecurity behaviours and, in turn, resilience. Focusing on Peer-to-Peer (P2P) users, the insights from Wales, United Kingdom and Dalarna, Sweden show how Airbnb users can potentially provide a greater informative base and support for different actors towards sustainability, rural landscape preservation and resilience to short-and-long-term triggers (Adie et al., 2022). These findings underpin previous research on second home communities in Fire Island, United States, on place attachment as trigger for individual resilience (Adie, 2020; Adie, 2024). Place attachment and resilience are similarly addressed from the perspective of communities, with evidence from the Sichuan Province, China (Guo et al., 2018). Sheppard and Williams (2015), instead, address the role of the individual instances of residents in relation to destination resilience in Whistler, Canada in the aftermath of the Global Financial Crisis and the hosting of the 2010 Winter Olympics in Vancouver. The inclusion of the local community is key in enhancing dialogue and foster a vision for long-term resilience, as suggested in a recent study in Kemi, Finland (Partanen, 2022).

Theories and Concepts

Within the applied social sciences, we can broadly distinguish between concepts of resilience rooted in ecology and engineering. Arguably, the work of Holling (1973) is considered as the cornerstone of research on resilience as 'the persistence of systems and of their ability to absorb change and disturbance and still maintain the same relationships between populations or state variables' (Holling, 1973, p. 14). This was later built on by Walker et al. (2004) in one of the most cited socio-ecological definitions of resilience as 'the capacity of a system to absorb disturbance and reorganise while undergoing change so as to retain essentially the same function, structure, identity and feedbacks' (ibid: 6) that existed before the disturbance event. As Hall (2018, p. 39) observes, most of the studies on resilience and tourism 'have not clearly defined how their research relates to the different framings of resilience and the important assumptions made with respect to the nature of the system, equilibrium, and change'. This can be mainly attributed to two reasons. On the one hand theories on resilience and tourism are far from definitive as this novel body of literature is still in its infancy. On the other hand, the overreliance of tourism studies on business-related literature (Tribe, 2004) means that there is need for a theoretical scoping in other disciplines. This underpins the call for a postdisciplinary appraisal of resilience and tourism (Hall et al., 2018). As Tribe and Liburd (2016, p. 49) suggest, postdisciplinarity can be understood as:

> an effort to escape the possible over-disciplining of disciplines and the negative effects of their rigid boundaries and theoretical natures and instead allowing the problem to define a more eclectic and practical approach to the solution.

According to Lew (2014) very early studies on tourism and resilience were rooted on chaos theory (e.g., Faulkner, 2001; Ritchie, 2004) and complex adaptive systems (e.g., Farrell &Twining-Ward, 2004), but their application in research was limited. Over the last years, however, we assisted to the establishment of concepts and theories on resilience and tourism that closely align to the engineering resilience paradigm (Davoudi, 2012; Holling, 1973; McManus et al., 2007). Engineering resilience 'focuses on efficiencies in returning to a steady state that existed prior to a disturbance or change through readiness, response and recovery planning' (Lew, 2014: 15). Studies rooted on engineering resilience seek to ascertain the elasticity and speed of systems to bounce back to an arbitrarily chosen previous equilibrium. The term resilience is either used as a polysemic concept or a metaphor (Norris et al., 2008; Sharifi & Yamagata, 2016) to illustrate the processes of response and adaptation following a triggering event. Ultimately, the engineering paradigm of resilience conceives predictability and controllability inherent features of resilient systems (Hall et al., 2018).

A second body of research on tourism and resilience underpins the ecological resilience paradigm (Davoudi, 2012; Holling, 2010). Unlike engineering resilience, ecological resilience is about change and adaptation into 'a new or different equilibrium' (Lew, 2014, p. 15) than the one existing before the triggering event. From an ecological resilience perspective, there is no possibility of bouncing back to a previous equilibrium (Hall, 2018). This is due to the complexity of different systems and states to return to a given original stage at the same time. The emphasis in ecological resilience thinking is, therefore, 'on all the constituents of resilience and their intrinsic abilities to maintain, resume, or adaptively change in face of external disturbances' (Amore et al., 2018, p. 237). Within the ecological resilience paradigm, the socio-ecological approach is regarded as a much suitable theory to understand community development and practices in land use planning (Amore et al., 2018; Davoudi, 2012; Lew, 2014). As Hall (2018, p. 39) notes:

> From a socio-ecological resilience perspective, resilience practice is very much concerned with the thresholds at which one regime shifts to another. In such a shift the system behaves in a different way and has a different identity. ... Such shifts are often difficult to identify until after they have occurred because the variables in the system do not demonstrate a simple linear response.

Research rooted on socio-ecological resilience predominantly focuses on the roles and the perspectives of tourism stakeholders and communities following non-linear natural hazards (e.g., Adie, 2020) and the modes of adaptation and mitigation to climate change, and socio-political crises (e.g., Weis et al., 2021). Hopkins and Becken (2014, p. 490) observe 'how resilient different actors and cultures are to changes', while Lew (2014) acknowledges the political and social embeddedness framing resilience and policy action in tourism. This is best illustrated in the study of the case of Calgaro et al. (2014) which frames vulnerability, adaptivity and resilience building as dynamic socio-ecological constructs in space and time. Their findings from Khao Lak, Patong and Phi Phi Don, Thailand, show different vulnerability and resilience patterns in each of the destinations following the following the 2004 Boxing Day Tsunami.

The existence of two fundamentally different and contrasting paradigms (Holling, 2010) means that 'there is no single accepted definition of resilience' (Hall et al., 2018, p. 33). This is evident in tourism studies, where there is a 'lack of consistency in use of the term or in seeking to explain how it is being used in particular studies and publications' (Hall, 2018, p. 36). Hall et al. (2018) further point that, in many cases, studies on resilience and tourism do not provide a definition in the first place. A likely 'third way' is however possible based on the concept of resilience as panarchy encompassing ecological, social, organisational, and individual dimensions altogether. As Hall (2018, p. 34) observes, 'bridging concepts can be extremely useful in providing a conceptual umbrella under which different stakeholder viewpoints can be brought together'. Hall et al. (2018, p. 147) were the first to advocate for a much-needed shift in resilience thinking 'by integrating different ways of understanding different system structures from individuals to organisations and destinations'. The latter encompass 'a complex set of interactions among various stakeholders and systems' (Amore et al., 2018, p. 237) that can only be better appraised with a multi-scalar and multidisciplinary approach.

An example of resilience as panarchy can be applied in the conceptualisation of destination resilience. According to Amore et al. (2018, p. 237) the notion of destination resilience 'implies a complex set of interactions among various stakeholders and systems' ranging from tourism-relevant industries, communities and non-tourism sectors. Such a heterogeneous and seemingly diverse groups and networks can ultimately change the scale and structure of destinations. At the same time, destination resilience acknowledges change caused by an external trigger cascading down to the individual elements of the destination. Understanding destinations from the lenses of resilience as panarchy is crucial because 'change processes and their interrelations have become more complex in a globalized, accelerated world, placing tourism under pressure to respond and adapt to various factors' (Luthe & Wyss, 2014, p. 161). Resilience as panarchy conceives change as cross-scalar and how this affects what is happening at the landscape, regime, niche and actor levels (Amore et al., 2018; Hall et al., 2018).

Notwithstanding much of research on resilience and tourism are at the regional or community levels, there is 'surprisingly limited overt research on destinations' (Hall et al., 2018,

p. 56). One plausible reason is the empirically limited application of panarchy thinking in tourism (Hall, 2022). To this end, Amore et al (2018, p. 238) elaborate a six-point list on the quintessential features of resilient destinations. In their view, a destination is resilient when stakeholders:

- are aware of the vulnerabilities and the likely impact of potential hazards;
- embark on redevelopment paths that advantage the local community at large;
- engage in networked and collaborative forms of destination planning;
- reframe the metagovernance of destinations;
- operate predominantly at regional and local scale; and
- reflexively and willingly learn from previous crises and reduce destination vulnerability for the future.

In addition, destination resilience requires 'a different approach in the management, policy making, and environmental awareness thus far conducted in tourism policy and planning practices' (Hall et al., 2018, p. 132). For example, cross-scale interactions can both be bottom-up (e.g., group of individuals proactively contributing to a regime change) or top-down (e.g., natural hazards or triggering events that directly impact communities and organisations) (Hall, 2018).

One of the first studies on destination resilience (Luthe & Wyss, 2014; Luthe & Wyss, 2016) put policy and governance at the centre of the evaluation of steering mode and innovative capacity of regional Destination Management Organisations (DMOs) in the Uri and Grisons cantons of Switzerland. In their view, resilience thinking enables a functional appraisal of tourism networks and can help understand the different governance forms and ties embedded within the destination. In particular, the socio-ecological dimension of resilience is acknowledged alongside economic, social and political dimensions at different scales of governance. A further feature of this work is the acknowledgement of the different forms of adaptation and change between slow change processes (e.g., climate change) and prompt change (e.g., economic crisis). This is further elaborated in Lew (2014), with the identification of four generalized types of tourism contexts featuring the Scale Change and Resilience (SCR) model:

- *entrepreneur tourism slow change* occurs when actors with business interests in tourism explore opportunities to adapt in response to likely decline and demise of services (e.g., coral reef excursions and climate change in the Great Barrier Reef, Australia) (Biggs et al., 2015);
- *community tourism slow change* is when residents in a tourism-appealing region acknowledge the importance of natural conservation and environmental protection in the long-term (e.g., community-driven vision towards sustainability in the Dolomites, Italy) (Amore & Adie, 2020; Amore & Adie, 2021);
- *entrepreneur tourism sudden shock* is common in circumstances of sharp decline of visitors and often results in swift adaptation and staff training to foster business recovery (e.g., restaurant operativity following the 2010–2011 Canterbury Earthquakes in Christchurch, New Zealand) (Hall et al., 2018); and
- *community tourism sudden shock* is when residents coalesce in extemporary initiatives of mutual support in the aftermath major disasters and contribute towards destination recovery (e.g., Vietnamese community in post-Katrina New Orleans, United States) (Li et al., 2008).

Table 46.1. Fundamental questions related to destination resilience.

Who?	Who determines what is desirable in the destination system? Whose resilience is prioritised? Who is excluded and included in the destination system? (Who are the winners and losers, especially between sectors, and the balances between destination community and visitor interests?)
What?	What perturbations should the destination system be resilient to? What networks, interests and sectors are included in the destination system? Is the focus on resilience generic or specific?
When?	What is the relative focus on rapid-onset disturbances and slow-onset changes? Is the focus on short or long-term resilience? Is the focus on the resilience of present or future generations?
Where?	What are the spatial and jurisdictional boundaries of the destination system? Is the resilience of some areas prioritised over others? Does building resilience in some areas affect resilience elsewhere?
Why?	What is the goal of building destination resilience? What are the underlying motivations for building destination resilience? What is the relative balance between process and outcome?

The increasing interest in defining resilience in tourism can be somehow compared to the prolific contributions about tourism and sustainability. According to Hall et al. (2018, p. 149), 'resilience and sustainability are related concepts but they are not the same'. On the one hand, resilience does not contemplate intergenerational equity and does not explicit a desired state of use and conservation of natural resources. On the other hand, both resilience and sustainability advocate for resource conservation and sufficiency in the long run and stress on the vulnerability of ecosystems to human activities. A further aspect relating resilience and sustainability is that both concepts are inevitably contested. Similarly to what observed with sustainable tourism and sustainable development (Hall, 2011; Hall, 2022; Saarinen, 2006; Saarinen, 2015), the concepts of resilience and tourism resilience can be politically and socially defined in a context of vested interests, neoliberal ideology and steering modes of governance (Amore & Hall, 2017; Andersen, 2019; Hayward, 2013). In their appraisal of resilience and spatial planning, Porter and Davoudi (2012, p. 331) question issues like 'resilience from what, to what, and who gets to decide?' to critically challenge the conceptualisation of resilience in policymaking. This is further applied in resilience and tourism, with Hall et al. (2018) identifying a range of fundamental questions to critically appraise destination resilience in practice (Table 46.1).

Conclusions

Research on tourism and resilience has the potential to genuinely advance the field of tourism as a whole and to foster what Tribe and Liburd (2016) refer to as problem-centred knowledge in opposition to mainstream tourism business literature (Tribe, 2004). These aspirations, however, can be undermined when the term resilience is used more as a catchword than an actual concept. Arguably, there are many studies – and scholars – that do not provide a clear definition or understanding as to what exactly resilience is. This chapter sought to provide a clear picture in the nebulous and growing body of literature around resilience and tourism. It did so by illustrating the state of the art of the literature, the different perspectives research to date and the quintessential theoretical and conceptual underpinnings from key works.

Future research on resilience and tourism is expected to overcome positivist and individualistic conceptualisations of resilience that much align with the neoliberal agenda (Andresen, 2019; Chandler, 2013; Evans & Reid, 2014; Tierney, 2015) and 'the unsafe assumption that growth is normal' (Hall, 2018, p. 43). The challenge nowadays is to frame a counter-narrative of resilience that distances itself to the rhetoric of reigniting international tourism and enable a speedy recovery in an alleged post-pandemic new normality (e.g., Buhalis, 2020). Rather than reigniting tourism, it is paramount to critically rethink contemporary travel. As Bianchi (2020, p. 80) notes, this counter-narrative provides a more cautious appraisal of the current COVID19 pandemic as a 'critical juncture and a moment to challenge the current growth trajectory of tourism and re-align it with social and ecological limits'. Arguably, research on tourism and resilience can be the beacon for a genuine alternative approach in the management and planning of destinations as long as it acknowledges the centrality of collaborative and communicative modes of metagovernance across diverse and locally embedded stakeholders (Amore & Adie, 2020; Amore & Adie, 2021).

The chapter provided an overview of the theories and concepts in resilience and tourism research. Overall, it is agreed that the literature in tourism and resilience is far from cohesive. As Hall (2022) further argues, one of the greatest challenges that tourism research – at large – faces are the highly fragmentary appraisal between the different elements of tourism and non-tourism elements. To this end, the notion of resilience as panarchy comes into force to help develop 'a better understanding of the tourism system' (Hall et al., 2018, p. 147) and contribute towards genuine rethinking of tourism as a phenomenon. Arguably, connecting research at the microscale to the meso-and-macro scale and *vice versa* enables a more articulated understanding of tourism and non-tourism elements within the context of sustainable destination and tourism (Hall, 2022; Hall et al., 2023).

Moving forward, the task of future research on tourism and resilience is a challenging one. It is about advocating for longitudinal studies in a field that overwhelmingly consists of one-time studies that provide a rather blurred appraisal of the complexities of destinations and unit of analysis. At the same time, it is about embracing postdisciplinary thinking in the creation of new knowledge in tourism studies. Far from being a radical shift, this is a well overdue change in the study of tourism within the frame of sustainability and finally advance the field as we know it.

References

Adger, W. N., Hughes, T. P., Folke, C., Carpenter, S. R., & Rockstrom, J. (2005). Social-ecological resilience to coastal disasters. *Science*, *309*(5737), 1036–1039. https://doi.org/10.1126/science.1112122

Adie, B. A. 2020. Place attachment and post-disaster decision-making in a second home context: A conceptual framework. *Current Issues in Tourism*, *23*(10), 1205–1215. https://doi.org/10.1080/13683500.2019.1600475

Adie, B. A. (2024). Should I stay or should I go? Hurricane Sandy and second home tourism on Fire Island, New York. In C. M. Hall & G. Prayag (eds). *Tourism, cyclones, hurricanes and flooding*, Channel View Publications.

Adie, B. A., de Bernardi, C., & Amore, A. (2022). Reframing rurality: The Impact of Airbnb on second-home communities in Wales and Sweden. In A. Farmaki, D. Ioannides & S. Kaldou (Eds.), *Peer-to-peer accommodation and community resilience: Implications for Sustainable Development* (pp. 81–92). CABI.

Alvarez, S., Bahja, F., & Fyall, A. (2022). A framework to identify destination vulnerability to hazards. *Tourism Management*, *90*, 104469. https://doi.org/10.1016/j.tourman.2021.104469

Amore, A., & Adie, B. A. (2020). Global importance, local problems: Degrowth in Italian World Heritage destinations. In C. M. Hall, L. Lundmark, & J. J. Zhang (Eds), *Degrowth and tourism: New perspectives on tourism entrepreneurship* (pp. 85–99). Routledge.

Amore, A., & Adie, B. A. (2021). (Re) igniting tourism in cities after COVID: The same old risks of the "new normal". In V. Pecorelli (Ed.), *From Overtourism to Undertourism: Any Sustainable Scenarios in the Post pandemic Time?* (pp. 13–31). Unicopli.

Amore, A., & Hall, C. M. (2017). National and urban public policy in tourism. Towards the emergence of a hyperneoliberal script? *International Journal of Tourism Policy*, 7(1): 4–22. https://doi.org/10.1504/IJTP.2017.082761

Amore, A., Prayag, G., & Hall. C. M. (2018). Conceptualizing destination resilience from a multilevel perspective. *Tourism Review International*, 22(3–4): 235–250. https://doi.org/10.3727/154427218X15369305779010

Andresen, S. A. (2019). R.I.P. Resilience. *Fennia 197(*1), 168–170.

Antonsen, K. M., Dale, B., & Mayer, S. (2022). "It's in Our DNA": Climate change and perceived resilience and adaptive capacity in nature-based tourism in Lofoten, Norway. *Weather, Climate, and Society*, *14*(1), 177–190. https://doi.org/10.1175/WCAS-D-21-0026.1

Bayrak, M. M. (2022). Does Indigenous tourism contribute to Indigenous resilience to disasters? A case study on Taiwan's highlands. *Progress in Disaster Science*, *14*, Art. 100220. https://doi.org/10.1016/j.pdisas.2022.100220

Bianchi, R. V. (2020). COVID-19 and the potential for a radical transformation of tourism? *ATLAS Tourism and Leisure Review*, 2, 80–86.

Biggs, D., Hall, C. M., & Stoeckl, N. (2012). The resilience of formal and informal tourism enterprises to disasters: reef tourism in Phuket, Thailand. *Journal of Sustainable Tourism*, *20*(5), 645–665. https://doi.org/10.1080/09669582.2011.630080

Biggs, D., Hicks, C.C., Cinner, J.E., & Hall, C.M. (2015). Marine tourism in the face of global change: The resilience of enterprises to crises in Thailand and Australia. *Ocean & Coastal Management*, *105*, 65–74. https://doi.org/10.1016/j.ocecoaman.2014.12.019

Blanco-Pimentel, M., Evensen, N. R., Cortés-Useche, C., Calle-Triviño, J., Barshis, D. J., Galvan, V., Harms, E., & Morikawa, M. K. (2022). All-inclusive coral reef restoration: How the tourism sector can boost restoration efforts in the caribbean. *Frontiers in Marine Science*, 9. https://doi.org/10.3389/fmars.2022.931302

Brown, N. A., Feldmann-Jensen, S., Rovins, J. E., Orchiston, C., & Johnston, D. (2021). Exploring disaster resilience within the hotel sector: A case study of Wellington and Hawke's Bay New Zealand. *International Journal of Disaster Risk Reduction*, *55*, Art. 102080. https://doi.org/10.1016/j.ijdrr.2021.102080

Buhalis, D. (2020). *Reigniting Mediterranean tourism: Welcome and introduction and smart practice.* Availableathttps://microsites.bournemouth.ac.uk/etourism-lab/event/1mediterranean-and-leisure-tourism-in-the-post-covid-19-era/

Bui, H. T., & Saito, H. (2022). Resource convergence for post disaster recovery. *Annals of Tourism Research*, *93*, 103375. https://doi.org/10.1016/j.annals.2022.103375

Calgaro, E. (2010). *Building resilient tourism destination futures in a world of uncertainty: Assessing destination vulnerability in Khao Lak, Patong and Phi Phi Don, Thailand, to the 2004 tsunami* (PhD thesis). Macquarie University, Sydney.

Calgaro, E., Lloyd, K., & Dominey-Howes, D. (2014). From vulnerability to transformation: A framework for assessing the vulnerability and resilience of tourism destinations. *Journal of Sustainable Tourism*, *22*(3), 341–360. https://doi.org/10.1080/09669582.2013.826229

Chandler, D. (2013). Resilience and the autotelic subject: Toward a critique of the societalization of security. *International Political Sociology*, 7(2), 210–226. https://doi.org/10.1111/ips.12018

Choi, Y. E., Oh, C. O., & Chon, J. (2021). Applying the resilience principles for sustainable ecotourism development: A case study of the Nakdong Estuary, South Korea. *Tourism Management*, *83*, 104237. https://doi.org/10.1093/cjres/rsac001

Coles, T., Hall, C. M., & Duval, D. T. (2006). Tourism and post-disciplinary enquiry. *Current Issues in Tourism*, 9(4–5), 293–319. https://doi.org/10.2167/cit327.0

Coles, T., Hall, C. M., & Duval, D. T. (2016). Tourism and postdisciplinarity: Back to the future? *Tourism Analysis*, *21*(4), 373–387. https://doi.org/10.3727/108354216X14679788636113

Cook, D., Malinauskaite, L., Davíðsdóttir, B., & Ögmundardóttir, H. (2022). Capital assets underpinning economic well-being–The example of whale ecosystem services in Arctic coastal communities. *Ecosystem Services*, *55*, 101432. https://doi.org/10.1016/j.ecoser.2022.101432

Davoudi, S. (2012). Resilience: A bridging concept or a dead end? *Planning Theory and Practice*, *13*(2), 299–333. https://doi.org/10.1080/14649357.2012.677124

Dogru, T., Marchio, E. A., Bulut, U., & Suess, C. (2019). Climate change: Vulnerability and resilience of tourism and the entire economy. *Tourism Management*, *72*, 292–305. https://doi.org/10.1016/j.tourman.2018.12.010

Elshaer, I. A., & Saad, S. K. (2022). Entrepreneurial resilience and business continuity in the tourism and hospitality industry: the role of adaptive performance and institutional orientation. *Tourism Review*, *77*(5), 1365–1384. https://doi.org/10.1108/TR-04-2021-0171

Espiner, S., & Becken, S. (2014). Tourist towns on the edge: Conceptualising vulnerability and resilience in a protected area tourism system. *Journal of Sustainable Tourism*, *22*(4), 646–665. https://doi.org/10.1080/09669582.2013.855222

European Union (EU) (2022). *Recovery and resilience facility*. https://single-market-economy.ec.europa.eu/sectors/tourism/funding-guide/recovery-and-resilience-facility_en

Evans, B. & Reid, J. (2014). *Resilient life: The art of living dangerously*. John Wiley & Sons.

Fang, S., Prayag, G., Ozanne, L. K., & de Vries, H. (2020). Psychological capital, coping mechanisms and organizational resilience: Insights from the 2016 Kaikoura earthquake, New Zealand. *Tourism Management Perspectives*, *34*, 100637.

Farrell, B. H., & Twining-Ward, L. (2004). Reconceptualizing tourism. *Annals of Tourism Research*, *31*(2), 274–295. https://doi.org/10.1016/j.annals.2003.12.002

Faulkner, B. (2001). Towards a framework for tourism disaster management. *Tourism Management*, 22(2), 135–147.

Geels, F. W. (2005). Processes and patterns in transitions and system innovations: Refining the co-evolutionary multilevel perspective. *Technological Forecasting and Social Change*, *72*(6), 681–696. https://doi.org/10.1016/j.techfore.2004.08.014

Gunderson, L. H., & Holling, C. S. (Eds.) (2002). *Panarchy: Understanding transformations in human and natural systems*. Island Press.

Guo, Y., Zhang, J., Zhang, Y., & Zheng, C. (2018). Catalyst or barrier? The influence of place attachment on perceived community resilience in tourism destinations. *Sustainability*, *10*(7), 2347. https://doi.org/10.3390/su10072347

Haddoud, M. Y., Onjewu, A. K. E., Al-Azab, M. R., & Elbaz, A. M. (2022). The psychological drivers of entrepreneurial resilience in the tourism sector. *Journal of Business Research*, *141*, 702–712. https://doi.org/10.1016/j.jbusres.2021.11.069

Hajibaba, H., Gretzel, U., Leisch, F., & Dolnicar, S. (2015). Crisis-resistant tourists. *Annals of Tourism Research*, *53*, 46–60. https://doi.org/10.1016/j.annals.2015.04.001

Hall, C. M. (2011). Policy learning and policy failure in sustainable tourism governance: from first-and second-order to third-order change? *Journal of Sustainable Tourism*, 19(4–5), 649–671. https://doi.org/10.1080/09669582.2011.555555

Hall, C M. (2016). *Tourism and resilience*. Paper presented at Linneaus University, Kalmar, Sweden, 4 November.

Hall, C. M. (2018). Resilience theory and tourism. In J. Saarinen and A. M. Gill (eds), *Resilient destinations and tourism: Governance strategies in the transition towards sustainability in tourism* (pp. 34–47). Routledge.

Hall, C. M. (2022). Sustainable tourism beyond BAU (Brundtland As Usual): Shifting from paradoxical to relational thinking? *Frontiers in Sustainable Tourism 1*, Art. 927946. https://doi.org/10.3389/frsut.2022.927946

Hall, C. M., Prayag, G., & Amore, A. (2018). *Tourism and Resilience: Individual*, Organisational and Destination Perspectives. Channel View Publications.

Hall, C. M., Safonov, A., & Naderi Koupaei, S. (2023). Resilience in hospitality and tourism: issues, synthesis and agenda. *International Journal of Contemporary Hospitality Management*, *35*(1), 347–368. https://doi.org/10.1108/IJCHM-11-2021-1428

Hayward, B. M. (2013). Rethinking resilience: Reflections on the earthquakes in Christchurch, New Zealand, 2010 and 2011. *Ecology and Society*, *18(4*), 37–42. https://www.jstor.org/stable/26269450

Holland, K. K., Larson, L. R., Powell, R. B., Holland, W. H., Allen, L., Nabaala, M., Tome, S., Seno, S., & Nampushi, J. (2022). Impacts of tourism on support for conservation, local livelihoods, and community resilience around Maasai Mara National Reserve, Keya. *Journal of Sustainable Tourism*, *30*(11), 2526–2548. https://doi.org/10.1080/09669582.2021.1932927

Holling, C. S. (1973). Resilience and stability of ecological systems. *Annual Review of Ecological Systems*, *4*, 1–23. https://doi.org/10.1080/09669582.2021.1932927

Holling, C. S. (2010). Engineering resilience versus ecological resilience. In L. H. Gunderson, C. R. Allen, & C. S. Holling (eds), *Foundations of ecological resilience*, pp. 51–66). Island Press.

Hopkins, D., & Becken, S. (2014). Socio-cultural resilience and tourism. In A. A. Lew, C. M. Hall, & A. M. Williams (Eds), *The Wiley Blackwell companion to tourism* (pp. 490–499). Wiley-Blackwell.

Horne, L., De Urioste-Stone, S., Rahimzadeh Bajgiran, P., & Seekamp, E. (2022). Understanding tourism suppliers' resilience to climate change in a rural destination in Maine. *Tourism Planning & Development*: 1–22. https://doi.org/10.1080/21568316.2022.2083222

Hulke, C., Kalvelage, L., Kairu, J., Revilla Diez, J., & Rutina, L. (2022). Navigating through the storm: conservancies as local institutions for regional resilience in Zambezi, Namibia. *Cambridge Journal of Regions, Economy and Society*, *15*(2), 305–322. https://doi.org/10.1093/cjres/rsac001

Kim, M. J., Bonn, M., & Hall, C. M. (2022a). What influences COVID-19 biosecurity behaviour for tourism? *Current Issues in Tourism*, *25*(1), 21–27. https://doi.org/10.1080/13683500.2021.1883558

Kim, M. J., Bonn, M., & Hall, C. M. (2022b). Traveler biosecurity behavior during the COVID-19 pandemic: Effects of intervention, resilience, and sustainable development goals. *Journal of Travel Research*, *61*(7): 1599–1618. https://doi.org/10.1177/00472875211034582

Koo, I., Anjam, M., & Zaman, U. (2022). Hell is empty, and all the devils are here: Nexus between toxic leadership, crisis communication, and resilience in COVID-19 tourism. *Sustainability*, *14*(17), 10825. https://doi.org/10.3390/su141710825

Larsen, R. K., Calgaro, E., & Thomalla, F. (2011). Governing resilience building in Thailand's tourism-dependent coastal communities: Conceptualising stakeholder agency in social–ecological systems. *Global Environmental Change*, *21*(2), 481–491. https://doi.org/10.1016/j.gloenvcha.2010.12.009

Lew, A. A. (2014). Scale, change and resilience in community tourism planning. *Tourism Geographies*, *16*(1), 14–22. https://doi.org/10.1080/14616688.2013.864325

Li, G., Jiao, H., & Wang, Q. (2022). Interannual variation and influencing factors of social-ecological system resilience of cultural tourism destination in arid area: A case of Dunhuang City, Gansu Province. *Arid Land Geography*, *45*(3), 935–945. http://alg.xjegi.com/EN/10.12118/j.issn.1000-6060.2021.354

Li, W., Airriess, C. A., Leon, K., Chen, A. C., & Keith, V. (2008). *Vietnamese Americans in New Orleans East: From Vietnamese village to Asian quarter?* Arizona State University.

Lindsay-Smith, G., Pyke, J., Gamage, A., Nguyen, V., & de Lacy, T. (2022). Tourism operator mental health and its relationship with SME organisational resilience during disasters. *Tourism Management Perspectives*, 42, 100961. https://doi.org/10.1016/j.tmp.2022.100961

Local Government Association (LGA) (2022). *Creating resilient and revitalised high streets in the 'new normal'*. https://www.local.gov.uk/publications/creating-resilient-and-revitalised-high-streets-new-normal

Lovejoy, T.E. (1994). The quantification of biodiversity: An esoteric quest or a vital component of sustainable development? *Philosophical Transactions of the Royal Society of London. Series B: Biological Sciences*, *345*(1311), 81–87.

Luthe, T., & Wyss, R. (2014). Assessing and planning resilience in tourism. *Tourism Management*, *44*, 161–163. https://doi.org/10.1016/j.tourman.2014.03.011

Luthe, T., & Wyss, R. (2016). Resilience to climate change in a cross-scale tourism governance context: A combined quantitative-qualitative network analysis. *Ecology and Society 21*(1), 27–44. https://www.jstor.org/stable/26270333

Malinauskaite, L., Cook, D., Davíðsdóttir, B., Karami, M. P., Koenigk, T., Kruschke, T., Ögmundardóttir, H., & Rasmussen, M. (2022). Connecting the dots: An interdisciplinary perspective on climate change effects on whales and whale watching in Skjálfandi Bay, Iceland. *Ocean & Coastal Management, 226*, 106274. https://doi.org/10.1016/j.ocecoaman.2022.106274

Mao, Y., He, J., Morrison, A. M., & Coca-Stefaniak, A. (2021). Effects of tourism CSR on employee psychological capital in the COVID-19 crisis: From the perspective of conservation of resources theory. *Current Issues in Tourism, 24*(19), 2716–2734. https://doi.org/10.1080/13683500.2020.1770706

McManus, S., Seville, E., Brunsdon, D., & Vargo, J. (2007). *Resilience management: A framework for assessing and improving the resilience of organizations.* (Resilient Organizations Research Report, 2007/01). University of Canterbury.

Melián-Alzola, L., Fernández-Monroy, M., & Hidalgo-Peñate, M. (2020). Hotels in contexts of uncertainty: Measuring organisational resilience. *Tourism Management Perspectives, 36*, 100747. https://doi.org/10.1016/j.tmp.2020.100747

Mirauda, D., De Donato, R., & Santandrea, G. (2022). Proposed improvement of coastal habitat resilience: The case study of Pantano forest of Policoro in southern Italy. *Frontiers in Marine Science*, 9, Art. 1528. https://doi.org/10.3389/fmars.2022.891251

Moreno, A., & Becken, S. (2009.) A climate change vulnerability assessment methodology for coastal tourism. *Journal of Sustainable Tourism, 17*(4), 473–488. https://doi.org/10.1080/09669580802651681

Musavengane, R., & Kloppers, R. (2020). Social capital: An investment towards community resilience in the collaborative natural resources management of community-based tourism schemes. *Tourism Management Perspectives, 34*, 100654. https://doi.org/10.1016/j.tmp.2020.100654

Nguyen, V. K., Pyke, J., Gamage, A., De Lacy, T., & Lindsay-Smith, G. (2022). Factors influencing business recovery from compound disasters: Evidence from Australian micro and small tourism businesses. *Journal of Hospitality and Tourism Management, 53*, 1–9. https://doi.org/10.1016/j.jhtm.2022.08.006

Norris, F., Stevens, S., Pfefferbaum, B., Wyche, K., & Pfefferbaum, R. (2008). Community resilience as a metaphor, theory, set of capacities, and strategy for disaster readiness. *American Journal of Community Psychology, 41*: 127–150. https://doi.org/10.1007/s10464-007-9156-6

Ntounis, N., Parker, C., Skinner, H., Steadman, C., & Warnaby, G. (2022). Tourism and hospitality industry resilience during the Covid-19 pandemic: Evidence from England. *Current Issues in Tourism, 25*(1): 46–59. https://doi.org/10.1080/13683500.2021.1883556

O'Hare, G., & Barrett, H. (1994). Effects of market fluctuations on the Sri Lankan tourist industry: Resilience and change, 1981–1991. *Tijdschrift voor Economische en Sociale Geografie, 85*(1), 39–52. https://doi.org/10.1111/j.1467-9663.1994.tb00672.x

Okafor, L. E., Khalid, U., & Burzynska, K. (2022). Does the level of a country's resilience moderate the link between the tourism industry and the economic policy response to the COVID-19 pandemic? *Current Issues in Tourism, 25*(2): 303–318. https://doi.org/10.1080/13683500.2021.1956441

Partanen, M. (2022). Social innovations for resilience—Local tourism actor perspectives in Kemi, Finland. *Tourism Planning & Development, 19*(2), 143–163. https://doi.org/10.1080/21568316.2021.2001037

Porter, L., & Davoudi, S. (2012). The politics of resilience for planning: A cautionary note. *Planning Theory & Practice, 13*(2): 329–333. https://doi.org/10.1080/14649357.2012.677124

Prayag, G., & Dassanayake, D. C. (2023). Tourism employee resilience, organizational resilience and financial performance: the role of creative self-efficacy. *Journal of Sustainable Tourism, 31*(10), 2312–2336. https://doi.org/10.1080/09669582.2022.2108040

Prayag, G., Spector, S., Orchiston, C., & Chowdhury, M. (2020). Psychological resilience, organizational resilience and life satisfaction in tourism firms: Insights from the Canterbury earthquakes. *Current Issues in Tourism, 23*(10), 1216–1233. https://doi.org/10.1080/13683500.2019.1607832

Qu, M., & Cheer, J. M. (2021). Community art festivals and sustainable rural revitalisation. *Journal of Sustainable Tourism 29*(11–12), 1756–1775. https://doi.org/10.1080/09669582.2020.1856858

Reyes-Santiago, M. D. R., Méndez-García, E., & Sánchez-Medina, P. S. (2022). A mixed methods study on community-based tourism as an adaptive response to water crisis in San Andrés Ixtlahuaca, Oaxaca, Mexico. *Sustainability*, *14*(10), Art. 5933. https://doi.org/10.3390/su14105933

Ricart, S., & Rico-Amorós, A. M. (2022). To be, to do, to share: The triple-loop of water governance to improve urban water resilience—Testing the Benidorm experience, Spain. *Land*, *11*(1), Art. 121. https://doi.org/10.3390/land11010121

Ritchie, B. W. (2004). Chaos, crises and disasters: A strategic approach to crisis management in the tourism industry. *Tourism Management*, *25*(6): 669–683. https://doi.org/10.1016/j.tourman.2003.09.004

Ruiz-Ballesteros, E. (2011). Social-ecological resilience and community-based tourism: An approach from Agua Blanca, Ecuador. *Tourism Management*, *32*(3): 655–666. https://doi.org/10.1016/j.tourman.2010.05.021

Ruiz-Ballesteros, E., & del Campo Tejedor, A. (2020). Community-based tourism as a factor in socio-ecological resilience. Economic diversification and community participation in Floreana (Galapagos). *Sustainability*, *12*(11), Art. 4724. https://doi.org/10.3390/su12114724

Schwaiger, K., Zehrer, A., & Braun, B. (2021). Organizational resilience in hospitality family businesses during the COVID-19 pandemic: A qualitative approach. *Tourism Review*, *77*(1), 163–176. https://doi.org/10.1108/TR-01-2021-0035

Saarinen, J. (2006). Traditions of sustainability in tourism studies. *Annals of Tourism Research*, *33*, 1121–1140. https://doi.org/10.1016/j.annals.2006.06.007

Saarinen, J. (2015). Conflicting limits to growth in sustainable tourism. *Current Issues in Tourism*, *18*, 903–907. https://doi.org/10.1080/13683500.2014.972344

Sayer, A. (1999). *Long live postdisciplinary studies! Sociology and the curse of disciplinary parochialism/imperialism*. Department of Sociology, Lancaster University.

Secore International (2017). *How to regrow corals along the Riviera Maya—insights from Project Mexico*. http://www.secore.org/site/de/newsroom/article/howto-regrow-corals-along-the-riviera-maya-insights-fromproject-mexico.151.html

Sharifi, A., & Yamagata, Y. (2016). Principles and criteria for assessing urban energy resilience: A literature review. *Renewable and Sustainable Energy Reviews*, 60, 1654–1677. https://doi.org/10.1016/j.rser.2016.03.028

Sheppard, V. A., & Williams, P. W. (2015). Systems-based and internal factors enhancing resort community resilience. Paper presented at *2015 TTRA International Conference: Advancing Tourism Research Globally*. Portland, OR, 15–17 June 2015.

Tanner, S., Prayag, G., & Coelho Kuntz, J. (2022). Psychological capital, social capital and organizational resilience: A Herringbone Model perspective. *International Journal of Disaster Risk Reduction*, *78*, 103149. https://doi.org/10.1016/j.ijdrr.2022.103149

Tribe, J. (2004). Knowing about tourism. In J. Phillimore & L. Goodson (Eds.), *Qualitative research in tourism: Ontologies, epistemologies and methodologies* (pp. 46–62). Routledge.

Tribe, J., & Liburd, J. J. (2016). The tourism knowledge system. *Annals of Tourism Research*, *57*, 44–61. https://doi.org/10.1016/j.annals.2015.11.011

Tierney, K. (2015). Resilience and the neoliberal project: discourses, critiques, practices – and Katrina. *American Behavioral Scientist*, *59*(10), 1327–1342. https://doi.org/10.1177/0002764215591187

Tourism Industry Aotearoa (TIA) (2022). *Achieving economic sustainable tourism*. https://www.sustainabletourism.nz/home/economic/

United Nations World Tourism Organisation (UNWTO) (2022). *Key priority: UNWTO agenda for Africa*. https://www.unwto.org/africa/fostering-resilience

Usher, L. E., Yusuf, J. E. W., & Covi, M. (2019). Assessing tourism business resilience in Virginia Beach. *International Journal of Tourism Cities*, 6(2), 397–414. https://doi.org/10.1108/IJTC-02-2019-0019

Veréb, V., Nobre, H., & Farhangmehr, M. (2022). Cosmopolitan tourists: the most resilient travellers in the face of COVID-19. *Service Business*, *16*(3), 503–527. https://doi.org/10.1007/s11628-022-00482-z

Walker, B., Holling, C. S., Carpenter, S. R., & Kinzig, A. (2004). Resilience, adaptability and transformability in social–ecological systems. *Ecology and Society*, 9(2). https://www.jstor.org/stable/26267673

Weis, K., Chambers, C., & Holladay, P. J. (2021). Social-ecological resilience and community-based tourism in the commonwealth of Dominica. *Tourism Geographies*, *23*(3), 458–478. https://doi.org/10.1080/14616688.2021.1878267

Williams, C., You, J. J., & Joshua, K. (2020). Small-business resilience in a remote tourist destination: exploring close relationship capabilities on the island of St Helena. *Journal of Sustainable Tourism*, *28*(7), 937–955. https://doi.org/10.1080/09669582.2020.1712409

Yang, E., Kim, J., & Hwang, C. S. (2022). The spatial moderating effect of environmental pollution on the relationship between tourism and community resilience. *Tourism Management*, *93*: 104554. https://doi.org/10.1016/j.tourman.2022.104554

Chapter 47

The Influence of Place-narratives on Destination Development

Dikte Grønvold

Introduction

Places make the frame for the stories we tell, they give us spaces and become deeply embedded in our identities, beliefs, and the sense-making of the world (Herman et al., 2012). As we make place, places make us (Binder, 2008; Cheng et al., 2021) and, in the context of destination development, the questions of how to market a place for tourism purposes, how to tell its stories or how to change them become present. Questions, which should encourage us to start listening to the stories of the destination already told, how these are communicated and when (Andersen & Grønvold, 2022). Yet, in tourism marketing contexts (Acikgoz & Tasci, 2020; Wong et al., 2016) the narratives about place often become one way directed towards a potential visitor or settler, the purpose of which is to inform about place and leave little space for the listener to interact and actively connect with the place's story (Rappaport, 1995; Chronis, 2012).

Therefore, this chapter will explore the concept of 'place-narratives' from a wide perspective based on a case-study conducted on the rural island Lolland in Denmark (Andersen & Grønvold, 2022). The aim of this chapter is to gain insights on the complexities of place-narratives in destination development that aims at increasing tourism and settlement. Before we get there, the following chapter begins by defining 'place', 'sense of place' and 'place-narratives' in tourism, urban planning and social theory. That will introduce us to three thematised definitions of 'place-narratives' which will work as points of reference for the introduction of Lolland and its application of 'place-narratives' in tourism and settlement development (Andersen & Grønvold, 2022). In concluding, the study will illustrate the complexities of 'place-narratives' and demonstrate in what ways the concept needs to be understood in its multilayers when managed in destination development and marketing.

The Wiley Blackwell Companion to Tourism, Second Edition. Edited by C. Michael Hall.

Place

In order to unfold the concept of 'place-narrative', it is first necessary to understand its components; 'place' and 'narratives'. In the context of destination management organisations (DMOs) and tourism policymakers, 'place' is often managed and marketed on presumptions of what the destination is and how it should be presented (Lichrou et al., 2008; Morgan et al., 2011; Wong et al., 2016). As these presumptions of a place have concrete impacts on the planning, infrastructure and development of a specific site, rural area, or region (Cheng et al., 2021), the theoretical key definitions of 'place' can help to provide a language and an awareness of the more intangible elements constituting 'place' (Lichrou et al., 2008).

The 'place' as a topic of study has its origin in the disciplines of geography, landscape architecture, urban planning or in literature (Relph, 1976; Liu & Freestone, 2016). The study of place as a cognitive and relational matter can also be found in the phenomenological school of thought, in which practices and lived experiences are understood in the context of places and personal interactions (Merleau-Ponty, 1962; Seamon, 2020). Drawing parallels to the phenomenological definition of place, Binder (2008) describes it in tourism as a dynamic, interactive relationship between visitor and place: 'Just as people are shaped by experiences and memories, so is place. Layered onto and absorbed into the physical fabric of place are intangibles, emotional and impressionistic elements generated by human interaction over time.' (Binder, 2008, p. 2). This implies that the inner more subjective mechanism and interactions with place over time form the actual place as it forms its people (Binder, 2008).

In the context of destination management, this perception of 'place' as an interactive relationship may seem slightly absent in how places are actually marketed for tourism purposes (Simonsen, 2016; Larsen & Urry, 2016; Dupre, 2019). To Lichrou et al. (2010), 'place' in destination marketing is often being looked and managed as a 'product'. This implies that 'place' is the host of a variety of consumption practices and activities, which make up the 'tourism experience' and hence there is less emphasis on the cocreation of the place (Binder, 2008; Acikgoz & Tasci, 2020). To Dupre (2019) the definition of 'place' is where spatial and social characteristics intersect in placemaking processes that are characterised by being in constant fluctuation. In this definition, one may notice that 'place' is not unified or stable but makes up the chaotic meeting point between the tangible and the intangible, the spatial and the social.

This cognitive internal understanding of 'place' is further defined by Smith (2015) as 'sense of place', who confirms that 'place' in a tourism context is a patchwork of different qualities, which include both *material qualities* such as environment, landscape and architecture and *immaterial qualities* such as heritage, history, social structures and relations. To Bærenholdt and Granås (2016) a place needs to be looked upon from the perspective of mobility as it consists of the materiality of the arena as well as the different forms of mobility practiced at that arena. Based on these definitions, 'place' is never a constant site in a tourism destination development (Simonsen, 2016; Larsen & Urry, 2016), but is always in the making by different kinds of for example environmental changes, infrastructural planning projects or by our ever-changing opinions of that place (Amsden et al., 2010; Dupre, 2019; Cheng et al., 2021).

Sense of Place

The relationship between the immaterial and the material in a place are dialectical processes that consist of symbols, meanings, and history (Smith, 2015; Seamon 2020). Now, behind this form of understanding lies a sensitivity to the cognitive inner processes, we all make when being in a place (Amsden et al., 2010). In the literary collection of *Letters from the headwaters* (Abeyta, 2014), the narrator Aaron Abeyta describes in a letter to his brother George: 'essentially what I'm saying is that a place defines us because the soul and the home are synonymous' (Abeyta, 2014, p. 4). This is not to be understood as if home is a static specific site, but more that it is an internal process formed in the meeting with place (Abeyta, 2014). So instead understanding 'sense of place' as only being influenced by factors such as our beliefs, our history or language (Amsden, et al., 2010; Smith, 2015). Moreover, Abeyta (2014) also put emphasis on how we understand our beliefs, our history and language are determined by place. Based on these definitions, one can say that 'sense of place' is more than merely 'the atmosphere of that place' but is a multidimensional cocreation of social constructions, cultural discourses, historical ties and everyday experiences in a physical setting, all of which make up the relation between subject and place (Abeyta, 2014; Campelo et al., 2014; d'Orey et al., 2019).

Narratives

A place can be understood not only in the physical sense but also through narratives and the stories we tell about where we come from and where we are going (Andersen & Grønvold, 2022). Narratives of places occur in our conversations (Herman et al., 2012), in the media, on social platforms and frequently work as a consumer-to-consumer communication in tourism (Cheng et al., 2021; Wong et al., 2016) and through storytelling in, for example, tourism marketing (Campelo et al., 2014; Stokowski, 2002). To Smith (2015), narratives are an effective medium to convey meaning and values, which inherently assign people and places certain attributes. Although it is important to note that stories are inherently selective and the choices made of what information, values and attributes are communicated can engender lasting connotations for a place (Smith, 2015; Cheng et al., 2021). In other words, narratives can be very broadly understood but have very impactful consequences and thus, in order to define more specifically how narratives work in tourism and destination development, this chapter will map out three thematic definitions of the concept *place-narratives*.

The Planned Place-narrative

In destination marketing, the focus for the DMO should be not only creating campaigns but also on managing a destination in terms of its brand, reason-to-go and its attractiveness for tourism or settlement purposes (Wong et al., 2016; Campelo et al., 2014; Chronis, 2012). This form of administrative and structured approach to a place's narrative forms the first of the three definitions that is: 'the planned place-narrative'. According to Smith (2015) two dominating destination marketing types of destination narratives exist: the event-based and the attraction-based. Both in practice resemble product promotion (Lichrou et al., 2017), which Smith (2015) critiques as these types may be informative but do not encourage for an immersive experience or interaction with place (Smith, 2015; Cheng, et al., 2021). To Campelo et al. (2014), the management of place-narratives should therefore always be guided by the theory of place in order to 'benefit from and contribute to the place's natural, cultural, social, and economic wealth' (Campelo et al., 2014, p. 154). Another approach to the planned narrative is to put emphasis on the role of the messenger of the brand-marketing (Chronis, 2012). This means to not only focus on communicating to the messenger but also on examining who the messenger talks to and how (Acikgoz & Tasci, 2020; Chronis, 2012). In planned place-narratives, the emphasis is on the management of a destination's brand, which is administrated and planed from a more top-down level in the advertisement and product-placement marketing campaigns (Morgan et al., 2011; Chronis, 2012). Despite being a focus on positive consumer-to-consumer marketing or storytellers of a destination (Wong et al., 2016; Chronis, 2012; Acikgoz & Tasci, 2020), the planned place-narrative has become the role of the DMOs to steward by nurturing, communicating and servicing the contacts and partnerships between the various tourism stakeholders (Chronis, 2012; Morgan et al., 2011).

The Collective Place-narrative

While the planned place-narrative may approach 'place' as a product with the consumer being a potential visitor or settler, the collective definition of place-narratives does not necessarily have one specific messenger in mind. This definition of place-narrative focuses more on the collective and social creation of place and place-narratives. For example, to Stokowski (2002) places can be categorised as 'social places' which are sites understood, created and reproduced in the interaction between people, in social behaviors and sustained in collective memory. Stokowski (2002) furthermore draws the parallel between narratives of place with the concept 'sense of place'. To Stokowski (2002) 'sense of place' is a socially constructed process and this construction is being sustained through narratives not only on an individual level but on a collective level. In other words, the stories *we* tell. This means that place-narratives introduce, give meaning, and reinforce habitus and cultural practices at places collectively, which help to form a 'sense of place'. This is further described by Stokowski (2002, p. 372): 'The derived symbols of place are formalized through use into coherent language structures and appear to people as narratives, myths, fables, and the like. One of the most common of these forms is the place narrative'. Following this definition, place-narratives are effective to confirm, validate and refresh memories, myths and traditions socially (Campelo et al., 2014; Stokowski, 2002). To Rappaport (1995) story creation, as well as listening to stories, help people to support their own personal stories and their collective narratives. In this way, it can be an empowering activity. Nonetheless, without any collective support to a place-narrative or branding of destination, there is a risk of the inhabitants not recognising the narrative, which is why Morgan et al. (2011) call attention to the point that tourists, residents, investors, students – everyone – are managers of a place reputation as place-narratives are seen as socially common-place. Therefore, narration comes from the commonplace, still as the case study will later illustrate, the commonplace is not necessarily an equal place in which every story is being listened to (Andersen & Grønvold, 2022).

The Individual Place-narrative

The last thematic definition of place-narratives is the 'individual place-narrative'. This definition is, as the name suggests, a subjective and internal process. The creation of the place-narratives occurs in the meeting between the individual and place, also known as 'sense of place' (Amsden et al., 2010). d'Orey et al. (2019) define the subjective 'sense of place' more directly from a tourism perspective as the 'psychological perceptions emerging from the expectations and experiences of the individuals in their tourism leisure activity' (d'Orey et al., 2019, p. 2). Thereby, 'sense of place' is not only formed at the destination, but is already formed in the mind of the individual or in the personal expectations to a place (d'Orey et al., 2019). This implies that the meeting between place and person does not only happen at the destination, in fact, there is a meeting already in the mind between imagined place and the visitor (Binder, 2008). Thus, in the individual place-narrative the stories happen in the memory of place and hence the creation of sense of place is an ongoing process that forms with time (Binder, 2008; Stokowski, 2002). The manifestation of these cognitive processes are personal narratives, which in tourism can be the sense-making and the valuation of experiences (Chronis, 2012; Cheng, et al., 2021). For this reason, Chronis (2012) names visitors 'story-builders' as people interact and form narratives about their experiences. In the context of the individual place-narrative, however, the concept of 'story-builders' can be wider understood to also apply to the person, who has never physically been at the destination but knows its story (d'Orey et al., 2019; Andersen & Grønvold, 2022).

Introduction to the Case-study

What will follow is a short introduction of the case, its results, and a wider discussion of the influence of place-narratives on Lolland's development for increasing tourism and settlement. Lolland is one of the geographically biggest municipalities in Denmark with approximately 43,000 residents (Lolland Kommune, 2016b) and hosts tourist attractions such as the holiday resort Lalandia with 520,000 annual visitors in 2018, closely followed by Knuthenborg Safaripark (Kvistgaard, 2019). By its many square meters of coastline, Lolland makes ideally an attractive beach house and second house-owner's destination nationally and internationally (Lolland Kommune, 2016b). At first glance with the amount of tourist offers and thereby job opportunities, one would not assume that Lolland has issues with attracting settlers, however, the number of residents has declined since 2007 with an average loss of 847 people per year (Lolland Kommune, n.d.). This number has stagnated since 2009, but the population forecast is still predicted to decline. Emigration to larger cities, decreasing birth rates and a high population of older people are reasons for the negative curve (Lolland Kommune, n.d).

In combination with the decline of residents on Lolland, it has been, and still is today, a custom for vulnerable people or families on social welfare subsidy to move to Lolland. From 2014–2019 Lolland municipality received 250 new recipients partly due to other municipalities in Denmark paying the recipients a financial bonus and help finding housing on Lolland as an incentive for the recipients to move (Arnholm, 2021). This has put Lolland and the nearby island Falster in an unwanted financial situation that culminated in 2015 with the documentary on the national TV2 channel *På Røven I Nakskov* (Screwed in Nakskov, no longer online available) portraying families having social, economic and/or health problems in the city of Nakskov on Lolland (Larsen, 2016). The documentary led to a harsh critique towards the TV station due to the stigmatisation of Lolland (Larsen & Urry, 2016). As a direct result of the documentary, *LollandFalster Lovestorm* became a response bottom-up movement that is a privately driven and initiated movement by residents on the islands of Lolland, Falster and Fejø (Lolland-Falster Lovestorm, 2020). *LollandFalster Lovestorm* primarily focuses on communicating the positive stories about Lolland on their Facebook-page, via networks and events (Lolland-Falster Lovestorm, 2020). Nevertheless, the negative connotation from the documentary was implemented in the minds of many people in Denmark and is kept up by negative representations in the media (Andersen & Grønvold, 2022; Larsen, 2016; Kaspersen 2020). These challenges created the foundation for a persistent negative place-narrative of Lolland to remain.

Currently, Lolland stands before a big infrastructural change with the EU-founded project of the water-tunnel, Femernbelt, which will connect the northern German island, Fehmarn, with

Rødbyhavn in Denmark (Stokholm, 2020). The tunnel will include a highway and railway, which will decrease the commute time between Scandinavia and the rest of Europe and is meant to provide jobs for 6,000 laborers to Lolland as a direct or indirect result of the large construction and implementation process (Stokholm, 2020; FemernBelt, 2021). Both Lolland's regional Visit Organization, Lolland Municipality, tourism businesses and stakeholders on the island see the infrastructural development as the momentum to turn the financial and settlement development into a positive curve by for example increased tourism revenue (Lolland Kommune, 2020; Lolland Kommune 2016a). As a result, it is believed that the coming general destination development of Lolland will help to change its negative place-narrative once and for all (Andersen & Grønvold, 2022). The following case demonstrates how place-narratives come in many other shapes and forms than what we in tourism otherwise manage, plan and market for (Figure 47.1).

Figure 47.1. Map of the Femernbelt Tunnel in the larger Trans-European Transport Network from South to North Europe.
Source: TEN-T Corridor, Femern A/S, 2022.

Methodological Framework

Before going deeper into the case study, it is necessary to elaborate on which data the conclusions for this chapter are drawn from. The study is based on a vast examination of Lolland and its nearby smaller island Fejø from a wide variety of methodological approaches to understand it as a tourism and settlement destination (Andersen & Grønvold, 2022).

There are multiple examples of development strategies to change settlement or tourism in rural areas of Denmark (Suenson et al., 2005), however, most of these are based on quantitative methods, such as surveys and statistics supported by structured interviews. Yet, there are not many examples of primary data, in which the practitioner or the researcher involved themselves as temporary settlers or tourists in a destination to gain a deeper understanding of the significance of that place.

For this reason, this study explores a wide qualitative and self-engaging approach to contribute new perspectives on destination development. Based on an iterative research-design, the primary data collected are grounded in six semi-structured interviews and excessive ethnographic fieldwork, in which the researchers temporarily moved to Lolland and Fejø in order to try out the new settlement- and tourism initiatives and strategy in practice. The fieldwork consisted of qualitative methods such as multi-sensory ethnography (Pink, 2009; Pink, 2012) and the element of play (Hall, 2010). The programme during the fieldwork as temporary settlers contained a mixture of tourism activities and settlement formalities and hence the researchers were oscillating between tourism activities to settlement practicalities. This was consciously decided on to explore the grey zone of the role between being a new settler and tourist, as well as participating in the activities set out by the destination development strategies by Lolland municipality and tourism stakeholders (Andersen & Grønvold, 2022). Most of these strategies were targeting both potential settlers and tourists with the same kind of activity offers, and were meant to attract the same demographic group of young families from middle to high income classes (Andersen & Grønvold, 2022; Lolland Kommune, 2020). The grey zone contributed to a rich dataset ranging from camping at an outdoor 'glamping' site, participating in open house visits, visits at local museums to attending a welcoming event for new settlers on Lolland, joining community association activities, participating in the local gym, being invited to people's homes and more. In other words, the study explored the destination development initiatives for Lolland firsthand to get as close to the experience of settling or visiting the place as possible (Andersen & Grønvold, 2022) (Figure 47.2).

The ethnographic methods became beneficial for the researchers to engage fully in the field as temporary settlers on Lolland. Alongside the self-engaging research, informal and unstructured interviews were conducted as well as creative conversation exercises (Westwood, 2007; Pole & Hillyard, 2016). The data collected were supported by reflexive practices, such as field-notes and daily journal entries and photos. The data was collected thematically coded based on a thematic network analysis (Attride-Stirling, 2001). The following lists six key findings from the analysis categorised as influential factors to why the negative place-narrative still sustains on Lolland (Andersen & Grønvold, 2022). This is meant to illustrate and question how we may understand and manage place-narratives as a concept in destination development and marketing today.

1. Centrality is Relative

The first key-finding will explore the concept of centrality in relation to Lolland's negative place-narrative. Now, to develop a destination by increasing its infrastructure is to Larsen and Urry (2016) based on the idea that transport and communication constantly need to be reduced in time. In urban planning this is manifested in large infrastructure projects, in which

Figure 47.2. Overview of the planned tourism and settlement areas and the location of the Femernbelt Tunnel on Lolland.
Source: Lolland Municipality, 2024.

the aim is to overcome physical distances with the message that the projects contribute with financial growth to a society (Larsen & Urry, 2016). This form of argumentation is similarly communicated in Lolland municipality's action plan and strategy for the Femern tunnel, in which the reduced transport time to Germany is emphasised to attract industries to locate on Lolland (Lolland Kommune, 2021). A representative from the tourism and settlement development in Lolland municipality argued that with the coming of the Femern tunnel Lolland will have a 'better strategic location in relation to transport entrepreneurs, as there are many entrepreneurs now recognizing the benefits of being located closer to Germany due to the tunnel connection' (Andersen & Grønvold, 2022, pp. 60).

To the initiator of the movement Lolland-Falster Lovestorm, however, Lolland is already placed perfectly geographically: 'If this had been anywhere else in the world, then Lolland would have been a Liebhaver place. It is only because of the negative reputation of Lolland that makes it seem like it is a bad place' (Andersen & Grønvold, 2022, p. 85). To this participant centrality does not mean to have a fast connection to the nearest capital. Instead, the destination development process on Lolland should be less about changing its infrastructure and transport systems and more about working with the negative narrative, which the participant confirms is not an easy process (Andersen & Grønvold, 2022). Furthermore, the aspects which made the participant perceive Lolland as 'central' were aspects such as fresh air, less traffic and less noise, elements which are in fact assets associated with being away from urban areas (Pastak & Kährik, 2021; Andersen & Grønvold, 2022). To draw on the researchers experiences as temporary settlers, Andersen & Grønvold corresponded with the uniqueness of Lolland as being peaceful, not crowded and having less urbanised areas close by (Andersen & Grønvold, 2022). In other words, what seemed attractive was exactly the alternative to the urban life's high pace and intensity of traffic (Andersen & Grønvold, 2022).

2. The Mental Distance

The second key finding is what can be described as the 'mental distance' (Andersen & Grønvold, 2022), also referred to as the perceptual distance (Hall, 2005). Reducing transport time to other larger cities is necessary, according to the chief consultant in the municipality of Lolland, in order to combat people's belief that Lolland is 'very far away' from Copenhagen or other larger cities in Denmark (Andersen & Grønvold, 2022). A reason for the general support of the Femern tunnel rests upon the belief that reduction of transit time between Lolland and the larger cities such as Copenhagen and Hamburg will increase tourism and settlement (Andersen & Grønvold, 2022; Lolland Kommune, 2016a; Lolland Kommune 2020). Travelling from the North part of Lolland, Nakskov, to Copenhagen with public transport, Andersen & Grønvold personally experienced a very easy ride, which took two hours in total (Andersen & Grønvold, 2022). So, the question changed from focusing on the reduced transport time with the Fermen tunnel to: how come other places situated two hours from Copenhagen are not perceived as just as 'far away?'

One of the key-findings of the study shows that the perception of Lolland as being 'far away' rests not only upon geographical measurements and collective transport in the practical sense but is a result of the negative place-narrative (Andersen & Grønvold, 2022). Cheng et al. (2021) support how a place-narrative can shape a place, regardless of the correctness of the narrative: 'Accordingly, language, culture, collectively experienced history, and their derived narratives and stories of a place shape and sustain the cultural and social values associated with place. However, an important point to raise here is that narratives and stories do not have to derive from facts, but can be fictitious' (Cheng et al, 2021, p. 26). As the dominating

Figure 47.3. Overview of Nakskov bay on Lolland. Credits: Rune Johansen 2020.
Source: Visit Lolland-Falster, 2024.

place-narrative is negative (Larsen & Urry, 2016), it might foster an alienation between people with no relation to Lolland and the semi- and permanent residents on Lolland (Andersen & Grønvold, 2022). To put it more precisely, the negative place-narrative created a 'mental distance', which is a cultural barrier for potential tourists and settlers to visit Lolland.

For this reason, this case study argues that the more a place becomes associated with negative values and less aspiring cultural identities, the greater the 'mental distance' becomes (Andersen & Grønvold, 2022). Arguably, the negative place-narrative conveys a *negative sense of place* on Lolland for the potential settlers and tourists that are familiar with its negative place-narrative. In this way, the 'mental-distance' rests upon a collective place-narrative and is sustained in memory and representation (Stokowski, 2002). As a result, the place-narrative of a place is not only about telling other stories about new development projects and information about Lolland as a tourism destination, but about identifying *what keeps the negative place-narrative alive*. Otherwise, the potential visitors and settlers believe that they already know Lolland as a place. Consequently, potential settlers or tourists familiar with the negative place-narrative do not see a reason-to-go as it is 'very far away' from them (Figure 47.3).

3. The Expat Storybuilders

One key finding that seems to keep the negative narrative alive was identified when Andersen & Grønvold participated at a Lolland-Falster Expat event in Copenhagen. The event was a gathering for people coming from Lolland or the nearby island Falster but were currently living in Copenhagen. Some participants identified themselves as 'expat Lollikker' and many argued that the documentary *På Røven i Nakskov* (Screwed in Nakskov) (2015) accurately corresponds to their perception of Lolland. Now, it is difficult to change a negative place-narrative, if the storyteller of that narrative has no personal desire to change it or that it becomes a part of the person's place-identity (Cheng et al., 2021). This study argues that the negative narrative of Lolland is, amongst other factors, being retold by people with a genealogical connection to Lolland (Cheng et al., 2021). The group of people moving away from a destination are shown to be one component of a place-narrative (Chronis, 2012), which is difficult for planners of tourism and destination development to change directly. To take it one step further, the Lolland-Falster Expat event in Copenhagen also illustrates how there exists a sense of belonging to a community outside Lolland, due to their common cultural background which re-unites them in Copenhagen. Paradoxically, the shared experience of coming from a place that is being associated with negative stories may build up another form of shared identity and makes it more difficult to combat the negative place-narrative (Andersen & Grønvold, 2022; Wilson, 2012; Andersson & Clausen, 2014.) Nevertheless, if the negative place-narrative keeps on being retold elsewhere of Lolland, the 'mental distance' is being reproduced also.

4. Place-narratives in Cultural Practices and Language

The fourth key finding is of a more invisible kind. The negative place-narrative seems to have internalised it into what this study defines as the 'humble attitude' (Andersen & Grønvold, 2022). The 'humble attitude' is a cultural attitude of 'humbleness,' which implicitly or explicitly states Lolland not to be the first choice as a tourism or settlement destination. The 'humble attitude' was expressed and confirmed by the interviewees as well as Andersen & Grønvold experienced it first-hand on several occasions such as at the welcoming event for new settlers (Andersen & Grønvold, 2022). For example, the tagline of one of the new marketing campaigns for Lolland indicates how the attitude can be found on multiple institutional levels: 'We know your search did not begin here [vacation on Lolland] but it was here you ended up' (Andersen & Grønvold, 2022, p. 75). This also marks the first definition of place-narratives as planned but in which the 'humble attitude' is still internalised despite the professionalisation and management of Lolland as a tourism brand (Wong et al., 2016).

The 'humble attitude' is not only a cultural habitual way of describing Lolland, but this study argues that it can be understood as the practice of a place-narrative, because it is a story told, retold, responded to, and recreated on various policy, administrative, local and marketing levels (Andersen & Grønvold, 2022). Thus, in destination development and marketing, one must look at cultural assets and habitual practices of a destination not only from the perspective of consumption for tourism purposes but also as messengers, in which place-narratives are being produced and consumed. Lichrou et al. (2008, p. 32) describe it as: 'culture cannot be viewed simply as a part of the place-product sum or as an asset/resource of the destination product, but rather as a dynamic context within which destinations are produced and consumed'. Following this perspective, culture is something that is constantly in creation, but as production and consumption can become habitual, so can a certain cultural attitude or identity of a place such as the 'humble attitude' has become on Lolland (Andersen & Grønvold, 2022).

5. The Internalised Prejudice

Another example of how place-narratives become internalised cultural attitudes is the case study of the Greek island Santorini by Lichrou et. al. (2010). In this study, the issue of the 'Greek mentality' is discussed as problematic for the destination development on the island, as it tells the story of a fundamental lack of trust towards the authorities and ultimately being associated with rule-bending practices (Lichrou et. al., 2010). The study discusses how the mentality depicts a negative brand of Greece abroad, well as how the mentality becomes internalised by the islanders on Santorini: 'In the case of Santorini, participants' stories provide associations of the 'Greek mentality' with the rural past of the island, a small agricultural community, which rapidly developed into a tourism destination' (Lichrou et al., 2010, p. 142). Thereby, the mentality from which the residents on the island wish to move away from is still internalised in their stories about Santorini as a place (Lichrou et. al., 2010).

The study of Santorini is meant to illustrate the tendency of internalising narratives and images of places into cultural costumes such as stories about where one comes from (Lichrou et al., 2010). Simultaneously, it also shows that like with culture in destination development, place-narratives cannot be viewed as merely products or assets, which must be communicated as much as possible to become manifest. Instead, of place-narratives being dynamic and occuring on various levels at a place (Andersen & Grønvold, 2022; Chronis, 2012). This resembles the second definition of place-narratives, namely the collective place-narrative. In this way, the negative place-narrative is adapted to ways of talking about personal heritage, in traditions and myths collective told and collectively understood (Binder, 2008; Stokowski, 2002). For this reason, one needs to remain sensitive to how place-narratives are communicated not only in marketing campaigns but also on a subtler level (Andersen & Grønvold, 2022). This is particularly important in the case of negative-place narratives to avoid the risk of reaffirming it (Lichrou, et al., 2010). In the example of Lolland, its cultural 'humble attitude' might be an additional reason for why the negative narrative is still not defeated within the municipality and community's collective place-narrative (Kerstetter & Bricker, 2009; Dupre, 2019) on Lolland (Andersen & Grønvold, 2022).

6. Infrastructure as not the Only Way Home

The reduced transit-time between Germany and Lolland with the Femern tunnel will result in a bigger incentive for Lollikkere to move back to 'home' after their studies according to participants from the destination and tourism development organisation on Lolland (Andersen & Grønvold, 2022). This will imply a change in the negative place-narrative of Lolland from being a place one moves away from to being a place one comes back to (Andersen & Grønvold, 2022). Interestingly, the infrastructural development was not mentioned by the participating resettlers of this study as a reason to move back to Lolland. Instead, what seemed difficult in the process of moving back was the lack of meeting places on Lolland for resettlers who were self-employed shown in the following statement by a young Lollik moving back after his studies: 'There is a lack of meeting places on Lolland, and those meeting places in cafes and other places are to meet people you already know'. Thus, for the resettler it was less about the lack of a fast connection to the nearest capital, and more about finding a place to meet other people, who shared the same interests and, in this case, be a part of a professional startup community on Lolland (Andersen & Grønvold, 2022). As Stokowski (2002) puts it, 'sense of place' refers to an individual's possibility of developing feelings of attachment, which are based on the combination of use, attentiveness and emotion to that site or place. Based on this definition, the study argues that the resettler calls for a 'sense of place' on Lolland, which speaks to the more intangible aspects of a destination development process (Andersen & Grønvold, 2022). In contrast to the representative for the tourism and settlement development on Lolland who defines a destination development process from a more tangible and public administrative point of view by focusing on infrastructure (Andersen & Grønvold, 2022). In this way, the tangible developments are managed and marketed as the 'attraction' in the brand marketing of Lolland instead of focusing on how the potential settler or tourist can engage in Lolland as a community place (Chronis, 2012).

Conclusion

This study introduced the case of Lolland in Denmark, which has faced a negative place-narrative on a national scale (Larsen & Urry, 2016) and now stands in front of infrastructural changes that have resulted in various destination development and urban planning projects. Common for all the projects are that they share the goal of attracting more settlers and tourists and once and for all change the negative place-narrative (Andersen & Grønvold, 2022). This study set out to explore this communality and examine Lolland as a tourism and settlement destination. As a result, the complexities of place-narratives were demonstrated and the need to understand its multiple layers in a destination management context highlighted as the steps to move forward in destination development.

Furthermore, this chapter illuminated the importance of identifying and learning a destination's place-narratives because these manifest on an individual, collective and planned level. In the example of Lolland, it was concluded that there is a focus on the tangible changes in the destination development processes, which are meant to help to change Lolland's negative place-narrative. Yet, if the negative narrative still manifests in the 'humble attitude' or is reproduced by negative 'storybuilders' (Chronis, 2012), it may not lead to an overall change as potential visitors and settlers are more familiar with its narrative than with the physical place (Andersen & Grønvold, 2022). This is because expectations to place form reasons-to-go or not to go (Wong, et al., 2016; Andersen & Grønvold, 2022). This chapter therefore calls for more research on identifying how place narratives are at work simultaneously on a planned, collective, and individual level. How these interfere with one another are case specific as seen by the example of the Greek Island Santorini, in which the 'Greek mentality' had become internalised cultural habitual ways of talking about Santorini as a place (Lichrou et. al, 2010). Thus, it is not only bottom-up or top-down communication and marketing of a place-narrative in the context of destination development. Rather, it is about creating the best fundament for positive place-narratives to be formed on all three levels, which is a process influenced by top administrative policy making, social cultural behaviors and the personal encounter with place (Andersen & Grønvold, 2022).

The results of this study can be applied as a useful grid to identify and critically question what the definition of place-narratives is and why in a destination development and marketing contexts as well as in the literature. Lastly, this study questions whether tangible changes in destination development, such as infrastructural projects, are the relevant components to market when wanting to increase tourism revenue, job creation and attracting settlement. Instead, this study calls for greater sensitivity towards the story-builders and the receiving end of the destination marketing campaigns, which do not only include the potential tourist or settler, but also the investor, the temporary settler, the current resident and the worker (Andersen & Grønvold, 2022). In conclusion, the intangible aspects of destination development were not prioritised in Lolland municipality's urban and tourism planning strategies and approaches to change the negative narrative (Andersen & Grønvold, 2022). Lastly, the analysis led to the conclusion that the theories of place and place-narratives should be applied in tourism marketing to stay aware of the intangible aspects, such as sense of place, to change Lolland's negative narrative once and for all and ensure a sustainable development which is attractive for tourists, settlers and residents (Andersen & Grønvold, 2022).

References

Abeyta, A. A. (2014). *Letters from the headwaters* (ed. J. Hausdoerffer). University Press of Colorado.

Acikgoz, F., & Tasci, A. D. A. (2020). A model of brand cocreation, brand immersion, their antecedents and consequences in café brand context. In A. Kavoura, E. Kefallonitis & P. Theodoridis (Eds), *Strategic innovative marketing and tourism: 8th ICSIMAT, Northern Aegean, Greece, 2019* (pp. 111–119). Springer.

Amsden, B. L., Stedman, R. C., & Kruger, L. E. (2010). The creation and maintenance of sense of place in a tourism-dependent community. *Leisure Sciences*, *33*(1), 32–51. https://doi.org/10.1080/01490400.2011.533105

Andersen, M., & Grønvold, D. (2022). *The influence of narratives on destination development. A case study of Lolland and Fejø as tourism and settlement destinations* [Master's thesis]. Aalborg Universitet.

Andersson, V. Gyimóthy, S., & Clausen, B. H. (2014) *Global mobilities and tourism development: a community perspective*. Aalborg Universitetsforlag.

Arnholm, D. (2021). Ny analyse: Lolland-Falster får de dyre tilflyttere. *Folketidende.dk*. Lokal nyt. https://folketidende.dk/Lokal-nyt/Ny-analyse-Lolland-Falster-faar-de-dyre-tilflyttere/artikel/594235

Attride-Stirling, J. (2001). Thematic networks: an analytic tool for qualitative research. *Qualitative Research*, *1*(3), 385–405. https://doi.org/10.1177/146879410100100307

Bærenholdt, J. O., & Granås, B. (2016). Places and mobilities beyond the periphery. In J. O. Bærenholdt & B. Grenås (Eds), *Mobility and place. Enacting Northern European peripheries* (pp. 17–26). Routledge.

Binder, R. (2008, September 29-October 4). *Sense of place in sustainable tourism development* [Paper presentation]. In 16th ICOMOS General Assembly and International Symposium: 'Finding the spirit of place – between the tangible and the intangible', Quebec, Canada.

Campelo, A., Aitken, R., Thyne, M., & Gnoth, J. (2014). Sense of place: The importance for destination branding. *Journal of Travel Research*, *53*(2), 154–166. https://doi.org/10.1177/0047287513496474

Cheng, N. C., Hall, C. M. & Prayag, G. (2021). *Sense of place and place attachment in tourism*. Routledge.

Chronis, A. (2012). Tourists as story-builders: Narrative construction at a heritage museum. *Journal of Travel & Tourism Marketing*, *29*(5), 444–459. https://doi.org/10.1080/10548408.2012.691395

d'Orey, F., Cardoso, A., & Abreu, R. (2019). "Tourist' sense of place", an assessment of the sense of place in tourism studies: The case of Portugal. *Academy of Strategic Management Journal*, *18*(1), 1–14. http://hdl.handle.net/11328/2722

Dupre, K. (2019). Trends and gaps in place-making in the context of urban development and tourism: 25 years of literature review. *Journal of Place Management and Development*, *12*(1), 102–120. https://doi.org/10.1108/JPMD-07-2017-0072

FemernBelt (2021) *Femern-forbindelsen er en vækstmotor*. Fonden Femern Belt Development. https://www.femern.info/da/news/femern-forbindelsen-er-en-vaekstmotor

Hall, C. M. (2005). *Tourism: Rethinking the social science of mobility*. Pearson.

Hall, C. M. (Ed.). (2010). *Fieldwork in tourism: Methods, issues and reflections*. Routledge.

Herman, D., Phelan, J., Rabinowitz, P. J., Richardson, B., & Warhol, R. R. (2012). *Narrative theory: Core concepts and critical debates*. Ohio State University Press.

Kaspersen, J. S. (2020). Ung iværksætter: Jeg flyttede hjem til Maribo for at hjælpe og forandre et stykke af udkantsdanmark. Jeg var nok naiv. Politiken. https://politiken.dk/debat/art7882200/Jeg-flyttede-hjem-til-Maribo-for-at-forandre-et-stykke-af-udkantsdanmark.-Jeg-var-nok-naiv

Kerstetter, D. & Bricker, K (2009). *Exploring Fijian's sense of place after exposure to tourism development*, Journal of Sustainable Tourism, 17:6, 691–708, DOI: 10.1080/09669580902999196

Kvistgaard, L. (2019). Vidensgrundlag for udviklingsplan for sjælland og øerne. *Kortlægning og analyser af potentialer i turismeudbuddet. Delrapport 1*. Wilke/KVEA – Kvistgaard Evaluering og Analyse. https://www.kystognaturturisme.com/sites/kystognaturturisme.com/files/2020-02/delrapport-1_kortlaegning-og-analyser-af-potentialer-i-turismeudbuddet_20191126.pdf

Larsen, N. (2016) *Dem på røven bliver endnu en gang taberne*. Information. 18. august. 2016. Sektion 1. Side 4.

Larsen, J., & Urry, J. (2016). Networking in mobile societies. In J. O. Bærenholdt & B. Grenås (Eds), *Mobility and place. Enacting Northern European peripheries* (pp 89–100). Routledge.

Lichrou, M., O'Malley, L., & Patterson, M. (2008). Place-product or place narrative (s)? Perspectives in the marketing of tourism destinations. *Journal of Strategic Marketing*, *16*(1), 27–39. https://doi.org/10.1080/09652540701794429

Lichrou, M., O'malley, L., & Patterson, M. (2010). Narratives of a tourism destination: Local particularities and their implications for place marketing and branding. *Place Branding and Public Diplomacy*, *6*, 134–144. https://doi.org/10.1057/pb.2010.10

Lichrou, M., Patterson, M., O'Malley, L., & O'Leary, K. (2017). Place branding and place narratives. In A. Campelo (Ed.), *Handbook on place branding and marketing* (pp. 160–177). Elgar.

Liu, E., & Freestone, R. (2016). Revisiting place and placelessness. In R. Freestone & E. Liu (Eds), *Place and placelessness revisited*, (pp. 1–19). Routledge.

Lolland Kommune (2016a). Turismestrategi for Lolland Kommune, 2016–2036. *Byrådet.* Link no longer available.

Lolland Kommune (2016b). Det nye Lolland. Plan- og udviklingsstrategi 2016–2030. *Byrådet Lolland kommune.* Link no longer available.

Lolland Kommune (2020). *Status på bosætning- og rekrutteringsindsats samt boligudvikling.* Juni. https://www.lolland.dk/document/8109eb0a-4b1c-4d9a-ae47-5f6bd5b55953

Lolland Kommune (2021) *Handleplan for implementering af Femernstrategien; perioden januar 2021-januar 2022.* Link no longer available.

Lolland Kommune (n.d.). *Befolkningsprognose 2021–2030.* https://www.lolland.dk/kommunen/oekonomi-og-fakta/befolkningsprognose

Lolland-Falster Lovestorm (2020). *Om Lolland-Falster Lovestorm.* https://www.lovestorm.dk/om-lolland-falster-lovestorm/

Merleau-Ponty, M. (1962). *Phenomenology of perception.* (C. Smith, Trans). Routledge.

Morgan, N., Pritchard, A., & Pride, R. (2011). Tourism places, brands, and reputation management. In N. Morgan, A. Pritchard, & R. Pride (Eds), *Destination brands: Managing place reputation* (3rd ed., pp. 3–19). Butterworth-Heinemann.

Pastak, I., & Kährik, A. (2021). Symbolic displacement revisited: Place-making narratives in gentrifying neighbourhoods of Tallinn. *International Journal of Urban and Regional Research*, *45*(5), 814–834. https://doi.org/10.1111/1468-2427.13054

Pink, S. (2009). *Doing sensory ethnography* (3rd ed.). SAGE Publications Ltd.

Pink, S. (2012). *Situating everyday life: Practices and places.* Sage Publications.

Pole, C., & Hillyard, S. (2016). *Doing fieldwork.* SAGE Publications Ltd.

Rappaport, J. (1995). Empowerment meets narrative: Listening to stories and creating settings. *American Journal of Community Psychology*, *23*(5), 795–807. https://doi.org/10.1007/BF02506992

Relph, E. (1976). *Place and placelessness.* Pion.

Seamon, D. (2020). Place attachment and phenomenology: The dynamic complexity of place. In L. C. Manzo & P. Devine-Wright (Eds), *Place attachment. Advances in theory, methods and applications* (2nd ed., pp. 29–44). Routledge.

Simonsen, K. (2016) Place as Encounters: Practice, Conjunction and Co-existence. In: Bærenholdt, J. O. & Grenås, B. (2016) *Mobility and place. Enacting Northern European Peripheries.* Routledge. Page 13–23.

Smith, S. (2015). A sense of place: place, culture and tourism. *Tourism Recreation Research*, *40*(2), 220–233. https://doi.org/10.1080/02508281.2015.1049814

Stokholm, R. (2020) *Nu kan Femern-byggeriet for alvor starte – få overblikket over Danmarks største anlægsprojekt.* TV2-Øst. Link no longer available.

Stokowski, P. A. (2002) *Languages of place and discourses of power: Constructing new senses of place.* Journal of leisure research 34.4: 368–382.

Suenson, V., Andersen, H. S., & Ærø, T. (2005). Bosætning i yderområder. *SBI Statens Byggeforskningsinstitut*, DK-2970 Hørsholm. 1–98. https://www.livogland.dk/files/dokumenter/publikationer/bosaetning_i_yderomraader_1.pdf

Westwood, S. (2007). What lies beneath? Using creative, projective and participatory techniques in qualitative tourism inquiry. In I. Ateljevic, A. Pritchard & N. Morgan (Eds), *The critical turn in tourism studies: Innovative research methodologies* (pp. 293–316) Elsevier.

Wilson, G. A. (2012) *Community resilience, globalization, and transitional pathways of decision-making.* Geoforum Volume 43, Issue 6, P: 1218–1231

Wong, J. Y., Lee, S. J., & Lee, W. H. (2016). 'Does it really affect me?' Tourism destination narratives, destination image, and the intention to visit: examining the moderating effect of narrative transportation. *International Journal of Tourism Research*, *18*(5), 458–468. https://doi.org/10.1002/jtr.2063

Part 8 Conclusions

Chapter 48

Conclusions
Challenges and Futures for Tourism

C. Michael Hall

Introduction

In concluding the previous edition of the companion Williams et al. (2014, p. 625) suggested, 'this book has sought to celebrate the richness and diversity of tourism studies, the meeting points and sometimes clashes between approaches, and new perspectives confronting traditional, sometimes dogmatic, approaches and beliefs.' Undoubtedly, the present volume also reflects such a perspective. As comprehensive and far reaching as this volume is, it presents only a beginning to understanding the diversity of tourism studies. Tourism studies, like its subject matter, is such a complex and evolving field that no single volume could hope to do it justice, even if focusing on the hegemonic English language literature alone. Nevertheless, the chapters in the current volume show tremendous insights into many of the leading topics, themes and issues of contemporary tourism scholarship.

Tourism is consistently affected by a change. Whether it be economic, social, and cultural, or whether it be technological and environmental, which are the focus of much contemporary tourism scholarship. Furthermore, change occurs at different scales and is uneven over space and time. Therefore, while all change is experienced locally, global overviews such as that provided in this book can only provide a more general appreciation of some of the factors that lead to change and their associated consequences. Indeed, the COVID-19 pandemic was a useful example of how something which occurred at a global scale was responded to differently by different countries and even within countries, as well as responded to in different ways by tourism academics, with a number framing it as an opportunity to transform tourism, although the direction of transformation was more problematic (Hall et al., 2020; Lončarić et al., 2022; Milano & Koens, 2022; Wijesinghe, 2022).

This concluding chapter aims to briefly outline some of the main factors influencing the trajectory of tourism scholarship as well as some of the major challenges it faces as it continues to progress. As such it engages with a number of the insights found in the chapters in this volume and those of earlier editions of the companion.

The Wiley Blackwell Companion to Tourism, Second Edition. Edited by C. Michael Hall.

The Institutional Environment of Tourism

No field of study is an island from the environment in which it is situated. This may be particularly important for scholarship in sectorally oriented subjects such as tourism. Of course, and has been made clear in several chapters in this volume as well as elsewhere, tourism is not just a business realm and is integral to understanding much of contemporary society, as well as social science perspectives on consumption and mobility. Nevertheless, the institutional dynamics for tourism, whether in terms of university organisation or government research assessment mechanisms, clearly push the business agenda. Arguably, this situation is also reflected in the emphasis given to rankings and impact in publishing in the now over 180 English language journals in tourism and hospitality (Cooper & Hall, 2024). This emphasis on tourism researchers focusing on publishing in tourism journals, and in particular in supposed 'top tourism journals' as defined in terms of impact factors or other metrics, was already well established a decade ago (Benckendorff & Zehrer, 2013; Xiao et al., 2013), but continues to gain momentum despite calls for greater open access to academic research and the damaging effects of rankings and the misuse and abuse of journal metrics by institutions, publishers and journal editors (Cooper & Hall, 2024).

Technologies

Another major factor in tourism and tourism research is the growth of the influence of new technologies. One rubric for such a trend is that of so-called smart tourism. Evolving from its origins in the concept of smart cities, the notion of smart tourism seeks to describe the integration of information technology with the tourism experience (Hunter et al., 2015; Borges-Tiago et al., 2022). Li et al. (2017) defines smart tourism in temporal terms as the pervasive tour information service that visitors receive during the touring process, with smart tourism being applied to an information service support system and associated technologies. However, more recently Liu et al. (2023) have redefined it as 'a tourism experience that involves technology, innovation, sustainability, accessibility, and is tourist-centered'. Indeed, they point out that although the notion of 'smart' has become ubiquitous (Mehraliyev et al., 2020) and that 'smart tourism' is approached from different perspectives, including suppliers (O'Connor, 2023), tourism destinations (Shafiee et al., 2019), and sustainability (Ivars-Baidal et al., 2023; Shafiee et al., 2019) the topic has often lacked critical assessment.

As important technological element of smart tourism is artificial intelligence along with its application in areas such as such as robotics, augmented reality (AR), virtual reality (VR). One area in which these developments have seen substantial interest is that of the metaverse. The metaverse is a shared virtual world based on virtual world technologies that allows and improves co-creation of experiences, communication, and interaction with more users in the digital world (Koo et al., 2023; Suh, 2023, Kim et al., 2023). The Metaverse is a network of digitally mediated spaces that immerse users in shared, real-time experiences' and is characterised by the convergence of five key elements—it is digitally mediated, spatial, immersive, shared, and operates in real-time (Hadi et al., 2023). The metaverse will undoubtedly bring enormous changes and development in hospitality and tourism since the metaverse can change tourists' experiences in the different stages of travel. However, whether the Metaverse acts as a substitute for conventional tourism activities or whether it enhances them is an issue that is still to be resolved.

Augmented reality (AR) is a technological experience that enhances the features of an attraction through information technology. It can be conceived as a real-world environment that is dynamically overlaid with coherent location- or context-sensitive virtual information and integrates both virtual and realistic scenes, thus retaining the authenticity of the scene (Zhu et al., 2023). In contrast, virtual reality provides consumers with a virtual travel experience without requiring their physical presence to substitute or compliment actual visitor experiences and, in the case of heritage sites for example, to encourage interest in conservation and reach stakeholder without them necessarily visiting the site physically. It has been argued that VR, through its virtual tours and experiences, can be used to improve booking processes and travel planning (Lee et al., 2020). However, the impact on tourist behaviours is not yet known. Augmented reality integrates virtual information with an actual environment (Kounavis et al., 2012), allowing users to easily adapt and generate real-time information about something they are interested in, e.g. an attraction (Park & Stangl, 2020; Yu et al., 2023). Shaping the behaviour and intention of a significant share of modern-day travellers, VR and AR are increasingly becoming mainstream consumer usage trends (Yung & Khoo-Lattimore, 2019).

COVID-19 had a substantial impact on industry and customer use of robotics and AI as contactless travel and service developed (Gaur et al., 2021; Kim et al., 2021), what Bae and Chang (2021) described as 'untact tourism'. AI and robotics are undoubtedly going to become of more importance to tourism (Li et al., 2021; Goel et al., 2022; Gursoy et al., 2023; Ivanov & Soliman, 2023), even if many consumers do not want them, as they hold the promise of significant cost savings for tourism businesses. However, they will undoubtedly have significant implications for the tourist experiences and behaviours, not all of which will be positive. As Hennessy (2023) observed, 'The closer ChatGPT comes to the "unparalleled precision" it promises, the further it will get from the guts of what makes travel so good. The best times are the messiest, the wildest, the most spontaneous and unexpected. And no, "this bar closed eight years ago", is not the unexpected I mean'

Some authors, such as Buhalis et al. (2023), claim that the metaverse is a disruptive technology. Similar claims were made about precursors of the Metaverse such as Second Life (Flavin, 2021) however many of proposed impacts of Second Life on business and education did not eventuate. Instead, disruption may be more akin to the impact of Airbnb or Uber, whereby it is the commercial model which becomes disruptive affecting preceding modes of consumption and production because of market dominance and capacity for regulatory change (Hall & Williams, 2020). Transformation is therefore as much a political process as it is a technological one.

Global Environmental Change

Another major influence on the future trajectory of tourism and tourism studies is global environmental change. This not only includes climate change, but also biodiversity loss at both ecosystem and species scales, as well as changes to specific types of environments such as coral reefs or boreal forests. The Tourism Panel on Climate Change (TPCC) (2023) report provided before COP28 highlighted that approximately 10 percent of global greenhouse gas (GHG) emissions are from tourism. The TPCC (2023) show that tourism emissions increased annually over 2010–2023, and tourism is not on track to achieve its sectoral interim target of reducing emissions by 50 percent by 2030. Tourist transport emissions increased by 65 percent between 1995 and 2019 with aviation being a key driver of this growth, with it contributing 26 percent of GHG emissions of all tourist trips (domestic and international), but 75 percent of tourist transport emissions.

Although tourism is a significant contributor to climate change it is also already being affected by global heating with, for example, many traditional ski resorts in lower altitudes now inoperable in winter. (Steiger et al., 2019). Some adaptation is possible, but there are limits to adaptation to climate change, depending on markets and location (Scott et al., 2019). As a result, new tourism landscapes are emerging, often marked by substantial economic uncertainty (Varnajot & Saarinen, 2022). Similarly, the loss of charismatic species for some destinations may also pose problems for ecotourism businesses and shifts in the seasonality of products may also create issues for destinations to attract tourist markets (Tervo-Kankare et al., 2013; Hall, 2014).

Economic and Political Uncertainty

The above concerns are all the more challenging as tourism now exists in a world of greater economic and political uncertainty than ever. However, as the COVID-19 experience demonstrated, this has meant that tourism has become recognised as being more important than ever as an economic force, especially with respect to employment generation. This has potentially therefore made tourism research more relevant to policy makers and other stakeholders than ever, given that tourism is seen as a mechanism to provide employment for those affected by economic restructuring. However, such issues of 'relevance' have long been an issue in tourism and will undoubtedly continue to be so in the future.

The previous edition of the companion identified the environment, risk and uncertainty, globalization, and mobility rights, as the four themes that are illustrative of some of the major challenges that confront, or should confront, tourism research (Williams et al., 2014). Undoubtedly, the environment remains a significant theme a decade on. In some ways, issues of globalization have become normalised in tourism studies and have become a 'given' about how much of the tourism world works. And, although mobility rights remain a contentious topic, especially given the importance of networks in tourism and the closure or restrictions of borders, numerous issues continue to exist in tourism if you are of the wrong passport, religion or gender. In addition, concerns over risk and uncertainty have become incorporated in the substantial growth in research on resilience (Amore, Chapter 46). These and the issues identified above will help determine the tourism research agenda over the next decade.

As noted by Williams et al. (2014), this framing of the key issues and key players in tourism is profoundly mediated by power relationships, built firmly, but not only, on social cleavages of class, income, gender, ethnicity, and spatiality. Importantly, this are also locked into a different set of power structures in academic institutions, the shape of which influences who does the researching and publishing as well as also who defines what constitutes the research agenda and the metrics used to assess what is produced (Cooper & Hall, 2024). An outsider reviewing the research undertaken and published in the tourism field would most likely form the impression of a vast array of largely incremental research being undertaken by researchers unevenly distributed across the globe, and little connected to each other. To some extent this is driven by the relationships between tourism researchers and those who seek answers, or at least insights, into the issues that they face. Nevertheless, it raises some difficult questions as to whether 'to avoid the conclusions that often we have been unable to see the wood for the trees: we have focused so much on what exists in terms of small parts of the tourism system that we have failed to ask the questions of why things exist in the form they do' (Williams et al., 2014, p. 631).

The chapters in this volume demonstrate the richness of tourism research. Nevertheless, massive challenges remain with respect to the continued relevance of such research especially with respect to the capacity to respond to globally significant issues of the impacts of technology, in which tourism as an industry is a major stakeholder and motivator. As with climate change, tourism operates on both sides of the ledger, causing as well as being affected. This latest edition of the companion reflects both continuity and change in tourism studies. Tourism continues to flourish despites COVID-19 and a range of other factors affecting tourism, including the growth of right-wing political conservatism and extremism and often xenophobic reactions to non-locals. Many of these political issues are not given the research coverage they deserve with, as Amore and Mosedale note in their respective chapters of this volume, such research and associated skill sets required for policy analysis remaining limited. Nevertheless, in a formal sense the tourism academy appears strong. Yet, overall, tourism studies remain rooted in a business studies setting with only limited links to the broader social sciences especially given the success of its own core journals. In the short run such relationships probably set the field a fairly firm base. However, over the long term it will be interesting to witness if some of the richness of the interplay between tourism and wider social sciences will be lost. Or, whether there will be a transformation in tourism studies that sees the field become much more engaged in questions that improve the human condition rather than the condition of capital.

References

Bae, S. Y., & Chang, P. J. (2021). The effect of coronavirus disease-19 (COVID-19) risk perception on behavioural intention towards 'untact' tourism in South Korea during the first wave of the pandemic (March 2020). *Current Issues in Tourism*, *24*(7), 1017–1035. https://doi.org/10.1080/13683500.2020.1798895

Benckendorff, P., & Zehrer, A. (2013). A network analysis of tourism research. *Annals of Tourism Research*, *43*, 121–149. https://doi.org/10.1016/j.annals.2013.04.005

Borges-Tiago, T., Veríssimo, J.M.C., & Tiago, F. (2022). Smart tourism: a scientometric review (2008–2020). *European Journal of Tourism Research*, 30, 1–18. https://doi.org/10.54055/ejtr.v30i.2593

Buhalis, D., Leung, D., & Lin, M. (2023). Metaverse as a disruptive technology revolutionising tourism management and marketing. *Tourism Management*, *97*, 104724. https://doi.org/10.1016/j.tourman.2023.104724

Cooper, C., & Hall, C.M. (Eds.) (2024). *How to publish in the best tourism journals*. Edward Elgar.

Flavin, M. (2021). Disruptive innovation, the episteme and technology-enhanced learning in higher education. *Prometheus*, *37*(2), 155–169. https://www.jstor.org/stable/10.13169/prometheus.37.2.0155

Gaur, L., Afaq, A., Singh, G., & Dwivedi, Y. K. (2021). Role of artificial intelligence and robotics to foster the touchless travel during a pandemic: a review and research agenda. *International Journal of Contemporary Hospitality Management*, *33*(11), 4079–4098. https://doi.org/10.1108/IJCHM-11-2020-1246

Goel, P., Kaushik, N., Sivathanu, B., Pillai, R., & Vikas, J. (2022). Consumers' adoption of artificial intelligence and robotics in hospitality and tourism sector: literature review and future research agenda. *Tourism Review*, *77*(4), 1081–1096. https://doi.org/10.1108/TR-03-2021-0138

Gursoy, D., Li, Y., & Song, H. (2023). ChatGPT and the hospitality and tourism industry: an overview of current trends and future research directions. *Journal of Hospitality Marketing & Management*, 32(5), 579–592. https://doi.org/10.1080/19368623.2023.2211993

Hadi, R., Melumad, S., & Park, E. S. (2023). The Metaverse: A new digital frontier for consumer behavior. *Journal of Consumer Psychology*, https://doi.org/10.1002/jcpy.1356.

Hall, C. M. (2014). Will climate change kill Santa Claus? Climate change and high-latitude Christmas place branding. *Scandinavian Journal of Hospitality and Tourism*, *14*(1), 23–40. https://doi.org/10.1080/15022250.2014.886101

Hall, C. M., Scott, D., & Gössling, S. (2020). Pandemics, transformations and tourism: be careful what you wish for. *Tourism Geographies*, *22*(3), 577–598. https://doi.org/10.1080/14616688.2020.1759131

Hall, C. M., & Williams, A. M. (2020). *Tourism and innovation* (2nd ed.). Routledge.

Hennessy, K. (2023). A two-hour walking tour with ChatGPT: 'I'd not suggest that to my worst enemy'. *The Guardian*, 11 November. https://www.theguardian.com/travel/2023/nov/12/chatgpt-travel-planning-intinerary-walking-ai

Hunter, W.C., Chung, N., Gretzel, U., & Koo, C. (2015). Constructivist research in smart tourism. *Asia Pacific Journal of Information Systems*, *25*(1), 105–120. http://dx.doi.org/10.14329/apjis.2015.25.1.105

Ivanov, S., & Soliman, M. (2023). Game of algorithms: ChatGPT implications for the future of tourism education and research. *Journal of Tourism Futures*, *9*(2), 214–221. https://doi.org/10.1108/JTF-02-2023–0038

Ivars-Baidal, J., Vera-Rebollo, J.F., Perles-Ribes, J., Femenia-Serra, F., & Celdrán-Bernabeu, M.A. (2023). Sustainable tourism indicators: what's new within the smart city/destination approach? *Journal of Sustainable Tourism*, *31*(7), 1556–1582. https://doi.org/10.1080/09669582.2021.1876075

Kim, J., Kim, J., Kim, S., and Hailu, T. (2023). Effects of AI ChatGPT on travelers' travel decision-making. *Tourism Review*. https://doi.org/10.1108/TR-07-2023-0489

Kim, S., Kim, J., Badu-Baiden, F., Giroux, M., & Choi, Y. (2021). Preference for robot service or human service in hotels? Impacts of the COVID-19 pandemic. *International Journal of Hospitality Management*, *93*(February), 102795. https://doi.org/10.1016/j.ijhm.2020.102795

Koo, C., Kwon, J., Chung, N., & Kim, J. (2023). Metaverse tourism: conceptual framework and research propositions. *Current Issues in Tourism*, 26(20), 3268–3274. https://doi.org/10.1080/13683500.2022.2122781

Kounavis, C. D., Kasimati, A. E., & Zamani, E. D. (2012). Enhancing the tourism experience through mobile augmented reality: Challenges and prospects. *International Journal of Engineering Business Management*, *4*, 10. https://doi.org/10.5772/51

Lončarić, D., Popović, P., & Kapeš, J. (2022). Impact of the COVID-19 pandemic on tourism: a systematic literature review. *Tourism: An International Interdisciplinary Journal*, *70*(3), 512–526. https://doi.org/10.37741/t.70.3.12

Lee, M., Lee, S. A., Jeong, M., & Oh, H. (2020). Quality of virtual reality and its impacts on behavioral intention. *International Journal of Hospitality Management*, *90*, 102595. https://doi.org/10.1016/j.ijhm.2020.102595

Li, M., Yin, D., Qiu, H., & Bai, B. (2021). A systematic review of AI technology-based service encounters: Implications for hospitality and tourism operations. *International Journal of Hospitality Management*, *95*, 102930. https://doi.org/10.1016/j.ijhm.2021.102930

Li, Y., Hu, C., Huang, C., & Duan, L. (2017). The concept of smart tourism in the context of tourism information services. *Tourism Management*, *58*, 293–300. https://doi.org/10.1016/j.tourman.2016.03.014

Liu, J., Hall, C.M., Zhu, C., & Cheng, V. T. P. (2023). Redefining the concept of smart tourism in tourism and hospitality. *Anatolia*. https://doi.org/10.1080/13032917.2023.2282712

Mehraliyev, F., Chan, I. C. C., Choi, Y., Koseoglu, M. A., & Law, R. (2020). A state-of-the-art review of smart tourism research. *Journal of Travel & Tourism Marketing*, *37*(1), 78–91. https://doi.org/10.1080/10548408.2020.1712309

Milano, C., & Koens, K. (2022). The paradox of tourism extremes. Excesses and restraints in times of COVID-19. *Current Issues in Tourism*, *25*(2), 219–231. https://doi.org/10.1080/13683500.2021.1908967

O'Connor, P. (2023). Small- and medium-sized tourism enterprises and smart tourism: Tourism Agenda 2030 perspective article. *Tourism Review*, *78*(2), 339–343. https://doi.org/10.1108/TR-09-2022-0431

Park, S., & Stangl, B. (2020). Augmented reality experiences and sensation seeking. *Tourism Management*, *77*, 104023. https://doi.org/10.1016/j.tourman.2019.104023

Scott, D., Steiger, R., Rutty, M., Pons, M., & Johnson, P. (2019). The differential futures of ski tourism in Ontario (Canada) under climate change: The limits of snowmaking adaptation. *Current Issues in Tourism*, *22*(11), 1327–1342. https://doi.org/10.1080/13683500.2017.1401984

Shafiee, S., Rajabzadeh Ghatari, A., Hasanzadeh, A., & Jahanyan, S. (2021). Smart tourism destinations: A systematic review. *Tourism Review*, *76*(3), 505–528. https://doi.org/10.1108/TR-06-2019-0235

Steiger, R., Scott, D., Abegg, B., Pons, M., & Aall, C. (2019). A critical review of climate change risk for ski tourism. *Current Issues in Tourism*, *22*(11), 1343–1379. https://doi.org/10.1080/13683500.2017.1410110

Suh, A. (2023). How users cognitively appraise and emotionally experience the metaverse: Focusing on social virtual reality. *Information Technology & People*, https://doi.org/10.1108/ITP-06-2022-0461

Tervo-Kankare, K., Hall, C. M., & Saarinen, J. (2013). Christmas tourists' perceptions to climate change in Rovaniemi, Finland. *Tourism Geographies*, *15*(2), 292–317. https://doi.org/10.1080/14616688.2012.726265

Tourism Panel on Climate Change (TPCC) (2023). *Tourism and climate change stocktake 2023: Executive report and key findings*. TPCC.

Varnajot, A., & Saarinen, J. (2022). Emerging post-Arctic tourism in the age of Anthropocene: case Finnish Lapland. *Scandinavian Journal of Hospitality and Tourism*, *22*(4–5), 357–371. https://doi.org/10.1080/15022250.2022.2134204

Wijesinghe, S. N. (2022). Neoliberalism, Covid-19 and hope for transformation in tourism: the case of Malaysia. *Current Issues in Tourism*, *25*(7), 1106–1120. https://doi.org/10.1080/13683500.2021.2012431

Williams, A. M., Hall, C. M., & Lew, A. (2014). Theoretical and methodological challenges for tourism: Just out of sight or just within our reach? In A. Lew, C.N. Hall, & A.M. Williams (Eds.), *The Wiley Blackwell companion to tourism* (1st ed.) (pp. 625–634). John Wiley & Sons.

Xiao, H., Jafari, J., Cloke, P., & Tribe, J. (2013). Annals: 40–40 vision. *Annals of Tourism Research*, *40*, 352–385. https://doi.org/10.1016/j.annals.2012.10.003

Yu, J., Kim, S., Hailu, T., Park, J., & Han, H. (2023). The effects of virtual reality (VR) and augmented reality (AR) on senior tourists' experiential quality, perceived advantages, perceived enjoyment, and reuse intention. *Current Issues in Tourism*. https://doi.org/10.1080/13683500.2023.2165483

Yung, R., & Khoo-Lattimore, C. (2019). New realities: a systematic literature review on virtual reality and augmented reality in tourism research. *Current Issues in Tourism*, *22*(17), 2056–2081. https://doi.org/10.1080/13683500.2017.1417359

Zhu, C., Hall, C. M., Fong, L. H. N., Liu, C. Y. N., & Naderi Koupaei, S. (2023). Does a good story prompt visit intention? Evidence from the augmented reality experience at a heritage site in China. *Journal of Heritage Tourism*, *18*(4),427–444. https://doi.org/10.1080/1743873X.2023.2188451

Index

The Wiley Blackwell Companion to Tourism, Second Edition. Edited by C. Michael Hall.